Frontiers in Complex Dynamics

Princeton Mathematical Series

EDITORS: PHILLIP A. GRIFFITHS, JOHN N. MATHER, AND ELIAS M. STEIN

Frontiers in Complex Dynamics

In Celebration of John Milnor's 80th Birthday

Edited by
Araceli Bonifant,
Mikhail Lyubich, and
Scott Sutherland

PRINCETON UNIVERSITY PRESS
PRINCETON AND OXFORD

Published by Princeton University Press
41 William Street, Princeton, New Jersey 08540

In the United Kingdom: Princeton University Press
6 Oxford Street, Woodstock, Oxfordshire, OX20 1TW

ISBN 978-0-691-15929-4
Library of Congress Control Number: 2013944768

British Library Cataloging-in-Publication Data is available

This book has been composed in LATEX

The publisher would like to acknowledge the editors of this volume for providing the camera-ready copy from which this book was printed.

Printed on acid-free paper. ∞

press.princeton.edu

Printed in the United States of America

10 9 8 7 6 5 4 3 2 1

Contents

Preface

In February of 2011, over 130 mathematicians gathered at the beautiful Banff Centre in the Canadian Rockies for a week of discussing holomorphic dynamics in one and several variables and other topics related to the work of John Milnor.

John Milnor is undoubtedly one of the most significant mathematicians of the second half of the twentieth century. He has made fundamental discoveries in many areas of modern mathematics, including topology, geometry, K-theory, and dynamical systems. Since in recent years his main interest has been in complex dynamics, it was only fitting that the conference had this as a primary focus.

The conference in Banff was a great success. In addition to all of the wonderful mathematics, the beautiful setting and friendly atmosphere at the Banff Centre inspired us all, both professionally and personally. All but one of the talks were videotaped and can be viewed or downloaded from the conference website, at http://www.math.sunysb.edu/jackfest (unfortunately, there was a camera malfunction at the start of Arnaud Chéritat's presentation, so only his slides are available).

This collection is an outgrowth of that conference, which was also organized by the editors of this book. Almost all of the authors whose papers appear here attended the conference. Both this collection and the conference were designed to honor John Milnor. But this volume is not merely a record of that conference; rather, it extends and complements that event. For example, some of the speakers gave primarily expository lectures but chose to contribute research papers to this volume; others went the other route. There is very little skiing in the book, and you'll have to bring your own food. But, it should last longer. We hope this volume will be valuable to any mathematician working in complex dynamics or related fields, whether or not they attended the Banff conference.

This volume is organized in five main parts: *I. One Complex Variable, II. One Real Variable, III. Several Complex Variables, IV. Laminations and Foliations*, and *V. Geometry and Algebra*. The first part is further subdivided into the areas of *arithmetic dynamics, polynomial dynamics, rational dynamics*, and *Thurston theory*, and the third part first covers *local dynamics in several complex variables* and then turns to *global dynamics*. In addition, there is a section containing color versions of those images for which color is essential; such images have references within the body of the main text, where a greyscale version appears for the reader's convenience.

The photograph of Jack Milnor at Lake Louise (Plate 1) was taken by Thomas Milnor. The images in Figure 1.2 on page 75 (which appears in modified form as Plate 4) are reprinted from John Milnor's article "Remarks on Iterated Cubic

Maps" in *Experimental Mathematics* 1, no. 1 (1992) by permission of Taylor & Francis (http://www.tandfonline.com). The image in Plate 15 (which also appears in modified form as Figure 3.11 on page 151) was produced by Hiroyuki Inou. The images in Plate 23 and Figure 1.1 on page 465 were produced by Vincent Pit. The group photo of conference participants (Plates 29 and 30) was taken by Photographic Services, The Banff Centre. The conference poster on page C-24 used elements from photographs taken by Marco Martens and by Tom Arban Photography (http://www.tomarban.com). All images are used by permission.

The editors are grateful to editorial staff at Princeton University Press for their help in the production of this volume. We are also indebted to the Banff Centre and the Banff International Research Station (BIRS) for hosting the conference, as well as to the National Science Foundation, the Simons Foundation, BIRS, IBM, the University of Rhode Island, and Stony Brook University for providing financial and other support to make the conference possible (and consequently also this volume). And finally, we are especially indebted to Jack Milnor for decades of inspirational mathematics, friendship, and personal support. We look forward to many more birthday celebrations!

Introduction

Holomorphic dynamics is one of the earliest branches of dynamical systems which is not part of classical mechanics. As a prominent field in its own right, it was founded in the classical work of Fatou and Julia (see [Fa1, Fa2] and [J]) early in the 20th century. For some mysterious reason, it was then almost completely forgotten for 60 years. The situation changed radically in the early 1980s when the field was revived and became one of the most active and exciting branches of mathematics. John Milnor was a key figure in this revival, and his fascination with holomorphic dynamics helped to make it so prominent. Milnor's book *Dynamics in One Complex Variable* [M8], his volumes of collected papers [M10, M11], and the surveys [L1, L5] are exemplary introductions into the richness and variety of Milnor's work in dynamics.

Holomorphic dynamics, in the sense we will use the term here, studies iterates of holomorphic maps on complex manifolds. Classically, it focused on the dynamics of rational maps of the Riemann sphere $\widehat{\mathbb{C}}$. For such a map f, the Riemann sphere is decomposed into two invariant subsets, the *Fatou set* $\mathcal{F}(f)$, where the dynamics is quite tame, and the *Julia set* $\mathcal{J}(f)$, which often has a quite complicated fractal structure and supports chaotic dynamics.

Even in the case of quadratic polynomials $Q_c \colon z \mapsto z^2 + c$, the dynamical picture is extremely intricate and may depend on the parameter c in an explosive way. The corresponding bifurcation diagram in the parameter plane is called the *Mandelbrot set*; its first computer images appeared in the late 1970s, sparking an intense interest in the field [BrMa, Man].

The field of holomorphic dynamics is rich in interactions with many branches of mathematics, such as complex analysis, geometry, topology, number theory, algebraic geometry, combinatorics, and measure theory. The present book is a clear example of such interplay.

⸺⸰⸺⸰⸺◆⸺⸰⸺⸰⸺

The papers **"Arithmetic of Unicritical Polynomial Maps"** and **"Les racines de composantes hyperboliques de M sont des quarts d'entiers algébriques,"** which open this volume,[1] exemplify the interaction of holomorphic dynamics with number theory. In these papers, John Milnor and Thierry Bousch study number-theoretic properties of the family of polynomials $p_c(z) = z^n + c$, whose bifurcation diagram is known as the *Multibrot set*.

In the celebrated Orsay Notes [DH1], Douady and Hubbard undertook a remarkable combinatorial investigation of the Mandelbrot set and the corresponding bifurcations of the Julia sets. In particular, they realized (using important contributions from Thurston's work [T]) that these fractal sets admit an explicit topological model as long as they are locally connected (see [D]). This led to the most famous conjecture in the field, on the local connectivity of the Mandelbrot set, typically

---

[1]Both these papers were originally written circa 1996 but never published. Milnor's paper is a follow-up to Bousch's note, but it was significantly revised by the author for this volume.

abbreviated as *MLC*. The MLC conjecture is still currently open, but it has led to many important advances, some of which are reflected in this volume.

In his thesis [La], Lavaurs proved the non-local-connectivity of the cubic connectedness locus, highlighting the fact that the degree two case is special in this respect. In attempt to better understand this phenomenon, Milnor came across a curious new object that he called the *tricorn*: the connectedness locus of antiholomorphic quadratic maps $q_c(z) = \bar{z}^2 + c$. In the paper **"Multicorns are not path connected,"** John Hubbard and Dierk Schleicher take a close look at the connectedness locus of its higher degree generalization, defined by $p_c(z) = \bar{z}^n + c$.

The paper by Alexandre Dezotti and Pascale Roesch, **"On (non-)local connectivity of some Julia sets,"** surveys the problem of local connectivity of Julia sets. It collects a variety of results and conjectures on the subject, both "positive" and "negative" (as Julia sets sometimes fail to be locally connected). In particular, in this paper the reader can learn about the work of Yoccoz [H, M7], Kahn and Lyubich [KL], and Kozlovski, Shen, and van Strien [KSvS]; the latter gives a positive answer in the case of "non-renormalizable" polynomials of any degree.

Related to connectivity, an important question that has interested both complex and algebraic dynamicists is that of the irreducibility of the closure of X_n, the set of points $(c, z) \in \mathbb{C}^2$ for which z is periodic under $Q_c(z) = z^2 + c$ with minimal period n. These curves are known as *dynatomic curves*. The irreducibility of such curves was proved by Morton [Mo] using algebraic methods, by Bousch [Bou] using algebraic and analytic (dynamical) methods, and by Lau and Schleicher [LS], using only dynamical methods. In the paper **"The quadratic dynatomic curves are smooth and irreducible,"** Xavier Buff and Tan Lei present a new proof of this result based on the *transversality theory* developed by Adam Epstein [E].

Similarly, in the case of the family of cubic polynomial maps with one marked critical point, parametrized by the equation $F(z) = z^3 - 3a^2z + (2a^3 + v)$, one can study the *period p-curves* $\mathcal{S}_p$ for $p \geq 1$. These curves are the collection of parameter pairs $(a, v) \in \mathbb{C}^2$ for which the marked critical point a has period exactly p; Milnor proved that $\mathcal{S}_p$ is smooth and affine for all $p > 0$ and irreducible for $p \leq 3$ [M9]. The computation of the Euler characteristic for any $p > 0$ and the irreducibility for $p = 4$ were proved by Bonifant, Kiwi and Milnor [BKM]. The computation of the Euler characteristic requires a deep study of the unbounded hyperbolic components of $\mathcal{S}_p$, known as *escape regions*. Important information about the limiting behavior of the periodic critical orbit as the parameter tends to infinity within an escape region is encoded in an associated *leading monomial vector*, which uniquely determines the escape region, as Jan Kiwi shows in **"Leading monomials of escape regions."**

As we have alluded to previously, a locally connected Julia set admits a precise topological model, due to Thurston, by means of a *geodesic lamination* in the unit disk. This model can be efficiently described in terms of the *Hubbard tree*, which is the "core" that encodes the rest of the dynamics. In particular, it captures all the cut-points of the Julia set, which generate the lamination in question. This circle of ideas is described and is carried further to a more general topological setting in the paper by Alexander Blokh, Lex Oversteegen, Ross Ptacek and Vladlen Timorin **"Dynamical cores of topological polynomials."**

The realm of general *rational dynamics* on the Riemann sphere is much less explored than that of polynomial dynamics. There is, however, a beautiful bridge connecting these two fields called *mating*: a surgery introduced by Douady and

Hubbard in the 1980s, in which the filled Julia sets of two polynomials of the same degree are dynamically related via external rays. In many cases this process produces a rational map. It is a difficult problem to decide when this surgery works and which rational maps can be obtained in this way. A recent breakthrough in this direction was achieved by Daniel Meyer, who proved that in the case when f is postcritically finite and the Julia set of f is the whole Riemann sphere, every sufficiently high iterate of the map can be realized as a mating [Me1, Me2]. In the paper **"Unmating of rational maps, sufficient criteria and examples,"** Meyer gives an overview of the current state of the art in this area of research, illustrating it with many examples. He also gives a sufficient condition for realizing rational maps as the mating of two polynomials.

Another way of producing rational maps is by "singular" perturbations of complex polynomials. In the paper **"Limiting behavior of Julia sets of singularly perturbed rational maps,"** Robert Devaney surveys dynamical properties of the families $f_{c,\lambda}(z) = z^n + c + \lambda/z^d$ for $n \geq 2$, $d \geq 1$, with c corresponding to the center of a hyperbolic component of the Multibrot set. These rational maps produce a variety of interesting Julia sets, including *Sierpinski carpets* and *Sierpinski gaskets*, as well as laminations by Jordan curves. In the current article, the author describes a curious "implosion" of the Julia sets as a polynomial $p_c = z^n + c$ is perturbed to a rational map $f_{c,\lambda}$.

There is a remarkable *phase-parameter relation* between the dynamical and parameter planes in holomorphic families of rational maps. It first appeared in the early 1980s in the context of quadratic dynamics in the Orsay notes [DH1] and has become a very fruitful philosophy ever since. In the paper **"Perturbations of weakly expanding critical orbits,"** Genadi Levin establishes a precise form of this relation for rational maps with one critical point satisfying the *summability condition* (certain expansion rate assumption along the critical orbit). This result brings to a natural general form many previously known special cases studied over the years by many people, including the author.

One of the most profound achievements in holomorphic dynamics in the early 1980s was *Thurston's topological characterization of rational maps*, which gives a combinatorial criterion for a postcritically finite branched covering of the sphere to be realizable (in a certain homotopical sense) as a rational map (see [DH2]). A wealth of new powerful ideas from hyperbolic geometry and Teichmüller theory were introduced to the field in this work. The *Thurston Rigidity Theorem*, which gives uniqueness of the realization, although only a small part of the theory, already is a major insight, with many important consequences for the field (some of which are mentioned later).

Attempts to generalize Thurston's characterization to the transcendental case faces many difficulties. However, in the exponential family $z \mapsto e^{\lambda z}$, they were overcome by Hubbard, Schleicher and Shishikura [HSS]. In the paper **"A framework towards understanding the characterization of holomorphic dynamics,"** Yunping Jiang surveys these and further results, which, in particular, extend the theory to a certain class of postcritically infinite maps. His paper includes an appendix by the author, Tao Chen, and Linda Keen that proposes applications of the ideas developed on the survey to the characterization problem for certain families of quasi-entire and quasi-meromorphic functions.

The field of *real one-dimensional dynamics* emerged from obscurity in the mid-1970s, largely due to the seminal work by Milnor and Thurston [MT], where they laid down foundations of the combinatorial theory of one-dimensional dynamics, called *kneading theory*. To any piecewise monotone interval map f, the authors associated a topological invariant (determined by the ordering of the critical orbits on the line) called the *kneading invariant*, which essentially classifies the maps in question. Another important invariant, the *topological entropy* $h(f)$ (which measures "the complexity" of a dynamical system) can be read off from the kneading invariant. One of the conjectures posed in the preprint version of [MT] was that in the real quadratic family $f_a : x \mapsto ax(1-x)$, $a \in (0,4]$, the topological entropy depends monotonically on a. This conjecture was proved in the final version using methods of holomorphic dynamics (the Thurston Rigidity Theorem alluded to earlier). This was the first occasion that demonstrated how fruitful complex methods could be in real dynamics. Much more was to come: see, e.g., [L4], a recent survey on this subject.

Later on, Milnor posed the general *monotonicity conjecture* [M6] (compare [DGMTr]) asserting that *in the family of real polynomials of any degree, isentropes are connected* (where an *isentrope* is the set of parameters with the same entropy). This conjecture was proved in the cubic case by Milnor and Tresser [MTr], and in the general case by Bruin and van Strien [BvS]. In the survey **"Milnor's conjecture on monotonicity of topological entropy: Results and questions,"** Sebastian van Strien discusses the history of this conjecture, gives an outline of the proof in the general case, and describes the state of the art in the subject. The proof makes use of an important result by Kozlovski, Shen, and van Strien [KSvS] on the density of hyperbolicity in the space of real polynomial maps, which is a far-reaching generalization of the Thurston Rigidity Theorem. (In the quadratic case, density of hyperbolicity had been proved in [L3, GrSw].) The article concludes with a list of open problems.

The paper **"Entropy in dimension one"** is one of the last papers written by William Thurston and occupies a special place in this volume. Sadly, Bill Thurston passed away in 2012 before finishing this work. In this paper, Thurston studies the topological entropy h of postcritically finite one-dimensional maps and, in particular, the relations between dynamics and arithmetics of e^h, presenting some amazing constructions for maps with given entropy and characterizing what values of entropy can occur for postcritically finite maps. In particular, he proves: *h is the topological entropy of a postcritically finite interval map if and only if $h = \log \lambda$, where $\lambda \geq 1$ is a weak Perron number, i.e., it is an algebraic integer, and $\lambda \geq |\lambda^\sigma|$ for every Galois conjugate $\lambda^\sigma \in \mathbf{C}$.*

The editors received significant help from many people in preparing this paper for publication, to whom we are very grateful. Among them are M. Bestvina, M. Handel, W. Jung, S. Koch, D. Lind, C. McMullen, L. Mosher, and Tan Lei. We are especially grateful to John Milnor, who carefully studied Bill's manuscript, adding a number of notes which clarify many of the points mentioned in the paper.

In the mid-1980s, Milnor wrote a short conceptual article *"On the concept of attractor"* [M5] that made a substantial impact on the field of real one-dimensional dynamics. In this paper Milnor proposed a general notion of *measure-theoretic attractor*, illustrated it with the *Feigenbaum attractor*, and formulated a problem of existence of *wild attractors* in dimension one. Such an attractor would be a Cantor set that attracts almost all orbits of some topologically transitive periodic interval. It turns out that the answer depends on the degree: in the quadratic case, there are no wild attractors [L2], while they can exist for higher degree unicritical maps $x \mapsto x^d + c$ [BKNvS]. This work made use of the idea of *random walk*, which describes transitions between various dynamical scales. In the paper **"Metric stability for random walks (with applications in renormalization theory),"** by Carlos Moreira and Daniel Smania, this idea was carried further to prove a surprising *rigidity* result: the conjugacy between two unimodal maps of the same degree with Feigenbaum or wild attractors is *absolutely continuous*.

One of the central results of classical *local* one-dimensional holomorphic dynamics is the *Leau-Fatou Flower Theorem* describing the local dynamics near a parabolic point (see [M8]). A natural problem is to develop a similar theory in higher dimensions. In the late 1990s important results in this direction were obtained by M. Hakim [Ha1, Ha2], but unfortunately, they are only partially published. In the paper **"On Écalle-Hakim's theorems in holomorphic dynamics,"** Marco Arizzi and Jasmin Raissy give a detailed technical account of these advances. This paper is followed by a note, **"Index theorems for meromorphic self-maps of the projective space,"** by Marco Abate, in which local techniques are used to prove three index theorems for global meromorphic maps of projective space.

The field of *global multivariable holomorphic dynamics* took shape in the early 1990s. It was pioneered by the work of Friedland and Milnor [FrM], Hubbard and Oberste-Vorth [HOV], Bedford and Smillie [BS1], [BS2], [BS3], Bedford, Lyubich, and Smillie [BLS1], [BLS2], Fornæss and Sibony [FS1], [FS2], [FS3], followed by many others. It focused on the dynamics of complex Hénon automorphisms,

$$f \colon \mathbb{C}^2 \to \mathbb{C}^2, \quad (x, y) \mapsto (x^2 + c + by, \, x)$$

and holomorphic endomorphism of projective spaces. It revealed a beautiful interplay between ideas coming from dynamics (such as *entropy, Lyapunov exponents, Pesin boxes,* ...), algebraic geometry (such as *dynamical degree, Bezout's theorem, blowups,* ...), and complex analysis (such as *pluripotential theory of currents*). In time, this theory was extended to a more general setting on complex manifolds.

The paper of Friedland and Milnor [FrM] made clear that in $\mathbb{C}^2$, the only interesting polynomial automorphisms are those that are conjugate to a Hénon automorphism or to compositions of Hénon automorphisms. In more generality, automorphisms of compact complex surfaces X with non-trivial dynamics have been classified by Cantat [C] and Gizatulin [Gi]. In [C], Cantat shows that if X is not Kähler, then the automorphisms of X have no interesting dynamics (in particular, X has topological entropy zero). He classifies the surfaces having *automorphisms with positive entropy* as *complex tori, K3 surfaces* (i.e., simply connected Kähler surfaces with a nonvanishing holomorphic 2-form), *Enriques surfaces* (i.e., quotients of K3 surfaces by a fixed-point free involution) or *non-minimal rational surfaces*. In the survey **"Dynamics of automorphisms of compact complex surfaces,"** Serge Cantat describes these and further developments in the study of the dynamics of automorphisms with positive entropy. In particular, he carries to this general setting the interplay between ideas from ergodic and pluripotential theories.

This interplay turned out to be important even for one-dimensional holomorphic dynamics since the parameter spaces of polynomials of degree > 2 (as well as the spaces of rational maps of any degree) are higher dimensional. In fact, this is one of the reasons why the bifurcations in the quadratic family $Q_c(z) = z^2 + c$ (described by the Mandelbrot set) are much better understood than the bifurcations in the space of cubic polynomials. An efficient approach to the problem appeared in the work of DeMarco who constructed a *bifurcation current* in any holomorphic family of one-dimensional rational maps [De]. (In the quadratic case, this current is just the harmonic measure on the boundary of the Mandelbrot set.) This construction was then generalized to higher-dimensional endomorphisms by Bassanelli and Berteloot [BaBe]. In the survey **"Bifurcation currents and equidistribution in parameter space,"** Romain Dujardin presents an overview of this circle of ideas and then describes his recent work with Bertrand Deroin, where the bifurcation current is constructed for holomorphic families of groups of Möbius transformations. This adds a new interesting line to *Sullivan's dictionary* between rational maps and Kleinian groups.

···+··+··+··❚··+··+··+···

A *(singular) holomorphic lamination* by holomorphic curves on a complex manifold is called *hyperbolic* if every leaf of the foliation is a *hyperbolic Riemann surface*. Each leaf of such a lamination is endowed with the canonical hyperbolic metric, but there is no a priori knowledge of how this metric depends on the transverse parameter. Starting with the work of Verjovsky [V], this important issue has been extensively studied by many authors, see [Gh], [Ca], [CaGM], [Glu], [Lin] and [FS4]. In the series of two papers **"Entropy for hyperbolic Riemann surface laminations I and II,"** Tien-Cuong Dinh, Viet-Anh Nguyên, and Nessim Sibony estimate the modulus of transverse continuity for these metrics. Then they apply these results to prove finiteness of the geometric entropy for singular foliations with linearizable singularities.

A notion of geometric entropy for regular Riemannian foliations was introduced by Ghys, Langevin, and Walczak [GhLnW]: roughly speaking, it measure the exponential rate at which nearby leaves diverge. They proved finiteness of the entropy and showed that if the entropy vanishes then the foliations admit a transverse measure. In the preceeding articles, Dinh, Nguyên, and Sibony modify the definition to make it suitable for singular laminations, and prove the finiteness result for this difficult setting.

Any real closed k-current on a compact manifold M^n, considered as a linear functional on the space of differential forms, naturally gives rise to a homology class in $H_{n-k}(M, \mathbb{R})$. When we have an oriented lamination $\mathcal{S}$ in M (also called a *solenoid*) endowed with a transverse measure μ, we obtain a *geometric current* by integrating forms along the leaves of $\mathcal{S}$ and then averaging over μ. The homology classes obtained this way are invariants of the solenoid. For flows, they appeared in the work of Schwartzman [Sc] under the name of *asymptotic cycles*. A general notion is due to Ruelle and Sullivan [RS]. (Note that these geometric ideas played an important motivational role in the study of currents in the higher-dimensional holomorphic dynamics discussed earlier.)

In a series of papers [MuPM1, MuPM2], Vincente Muñoz and Ricardo Pérez-Marco undertook a systematic study of this interplay between dynamics and algebraic topology. To this end, they refine a notion of a geometric current: impose a suitable transverse regularity and allow the solenoids to be *immersed*. Their paper, **"Intersection theory for ergodic solenoids,"** in this volume develops the geometric intersection theory for these objects.

A broad range of applications of the idea of asymptotic cycle to topology and physics (including Donaldson, Jones, and Seiberg-Witten invariants for foliations) is discussed in the paper **"Invariants of four-manifolds with flows via cohomological field theory,"** by Hugo García-Compeán, Roberto Santos-Silva and Alberto Verjovsky.

--·--+--·--◆--·--+--·--

The final part of this volume is devoted to two areas of *geometry* and *algebra* strongly influenced by earlier work of John Milnor. At the early stage they were not directly related to holomorphic dynamics, but some deep connections were discovered more recently.

One of Milnor 's impacts on contemporary mathematics lay in his careful treatment of various topological, combinatorial, and geometric *structures*, discovering striking distinctions and relations between them. His discovery of exotic spheres and his counterexample to Hauptvermutung are celebrated examples of this kind.

Less known to general mathematical public is Milnor's role in developing the theory of flat bundles and affine manifolds. This is the theme of William Goldman's article **"Two papers which changed my life: Milnor's seminal work on flat manifolds and bundles."** Specifically, these are *"On the existence of a connection with curvature zero"* [M1], which deeply influenced the theory of characteristic classes of flat bundles, and *"On fundamental groups of complete affinely flat manifolds"* [M4], which clarified the theory of affine manifolds, setting the stage for its future flourishing. Goldman's article describes the history of the *Milnor-Wood inequality* and the *Auslander Conjecture* and then proceeds to more recent developments, including a description of *Margulis space-times*, a startling example of an affine 3-manifold with free fundamental group.

Our volume concludes with Rostislav Grigorchuk's survey **"Milnor's problem on the growth of groups and its consequences."** The notion of *group growth* first appeared in 1955 in a paper of A. S. Schwarz [Sch], but it remained virtually unnoticed for over a decade. The situation changed after Milnor's papers from 1968 [M2, M3], which sparked significant interest in this area. Particularly influential were two problems raised in these papers: the characterization of groups of polynomial growth (Milnor conjectured that they must be virtually nilpotent) and the question of the existence of groups of intermediate growth (in between polynomial and exponential).

Both problems were solved in the early 1980s: Gromov proved Milnor's Conjecture on groups of polynomial growth [Gro] while Grigorchuk constructed examples of groups of intermediate growth [G1, G2]. These breakthroughs led to exciting advances in the past 30 years that changed the face of *Combinatorial Group Theory*. Grigorchuk's survey presents a broad picture of the area, and suggests a number of further interesting problems and directions of research.

Let us mention one relation between this area and holomorphic dynamics that was discovered recently. To a rational endomorphism f of the Riemann sphere, one can associate an *iterated monodromy group* that describes the covering properties of all the iterates of f. One way to define this group is as the holonomy of the affine lamination constructed in [LMin], a more algebraic approach was proposed by Nekrashevych in [N]. These groups exhibit many very interesting algebraic, combinatorial, and geometric properties (see [BGN]); in particular, it turned out that some of them have intermediate growth [BuPe].

--+--+--◆--+--+--

We hope this volume gives a glimpse into one beautiful field of research that was strongly influenced by Milnor's work. Read and enjoy!

1 Bibliography

[BGN]   L. Bartholdi, R. Grigorchuk, and V. Nekrashevych. *From fractal groups to fractal sets*. In: "Fractals in Graz 2001. Analysis — Dynamics — Geometry — Stochastics," Edited by P. Grabner and W. Woess, pp. 25–118. Birkhäuser Verlag, Basel, Boston, Berlin, 2003.

[BaBe]   G. Bassanelli and F. Berteloot. *Bifurcation currents in holomorphic dynamics on* $\mathbb{P}^k$. J. Reine Angew. Math. **608** 201–235 (2007).

[BLS1]   E. Bedford, M. Lyubich, and J. Smillie. *Polynomial diffeomorphisms of* $\mathbb{C}^2$. *IV. The measure of maximal entropy and laminar currents*. Invent. Math. **112** 77–125 (1993).

[BLS2]   E. Bedford, M. Lyubich, and J. Smillie. *Distribution of periodic points of polynomial diffeomorphisms of* $\mathbb{C}^2$. Invent. Math. **114** 277–288 (1993).

[BS1]   E. Bedford and J. Smillie. *Polynomial diffeomorphisms of* $\mathbb{C}^2$: *currents, equilibrium measure and hyperbolicity*. Invent. Math. **103** 69–99 (1991).

[BS2]   E. Bedford and J. Smillie. *Polynomial diffeomorphisms of* $\mathbb{C}^2$. *II. Stable manifolds and recurrence*. J. Amer. Math. Soc. **4** 657–679 (1991).

[BS3]   E. Bedford and J. Smillie. *Polynomial diffeomorphisms of* $\mathbb{C}^2$. *III. Ergodicity, exponents and entropy of the equilibrium measure*. Math. Ann. **294** 395–420 (1992).

[BKM]   A. Bonifant, J. Kiwi, and J. Milnor *Cubic polynomial maps with periodic critical orbit II: Escape regions*. Conform. Geom. Dyn. **14** 68–112 (2010).

[Bou]   T. Bousch. "Sur quelques problémes de dynamique holomorphe." Ph.D. thesis, Université de Paris-Sud, Orsay 1992.

[BrMa]   R. Brooks and J. P. Matelski, *The dynamics of 2-generator subgroups of* PSL(2, $\mathbb{C}$), In: "Riemann Surfaces and Related Topics: Proceedings of the 1978 Stony Brook Conference," Edited by I. Kra and B. Maskit, Ann. Math. Stud. **97**, pp. 65–71. Princeton Univ. Press, Princeton, NJ, 1981.

[BKNvS]   H. Bruin, G. Keller, T. Nowicki, and S. van Strien. *Wild Cantor attractors exist*. Ann. of Math. **143** 97–130 (1996).

[BvS]   H. Bruin and S. van Strien. *Monotonicity of entropy for real multimodal maps*. arXiv:0905.3377

[BuPe]   K.-U. Bux and R. Pérez. *On the growth of iterated monodromy groups*. In: "Topological and asymptotic aspects of group theory," pp. 61–76, Edited by R. Grigorchuk, M. Mihalik, M. Sapir, and Z. Šuniḱ. Contemp. Math. **394** Amer. Math. Soc. Providence, RI 2006.

[Ca]   A. Candel. *Uniformization of surface laminations*. Ann. Sci. École Norm. Sup. **26** 498–516 (1993).

[CaGM] A. Candel and X. Gómez-Mont. *Uniformization of the leaves of a rational vector field.* Ann. Inst. Fourier (Grenoble) **45** 1123–1133 (1995).

[C] S. Cantat. *Dynamique des automorphismes des surfaces projectives complexes.* C. R. Acad. Sci. Paris Sér. I Math., **328** 901–906 (1999).

[DGMTr] S. P. Dawson, R. Galeeva, J. Milnor, and C. Tresser. *A monotonicity conjecture for real cubic maps.* In: "Real and Complex dynamical systems," pp. 165–183. Edited by B. Branner and P. Hjorth. Kluwer Academic, 1995. Also available in: "Collected Papers of John Milnor VI," pp. 157–175 Amer. Math. Soc. 2012.

[De] L. DeMarco. *Dynamics of rational maps: a current on the bifurcation locus.* Math. Res. Lett. **8** 57–66 (2001).

[D] A. Douady. *Description of compact sets in* ℂ. In: "Topological Methods in Modern Mathematics, A Symposium in Honor of John Milnor's 60th Birthday," pp. 429–465 Edited by L. R. Goldberg and A. V. Phillips, Publish or Perish, 1993.

[DH1] A. Douady and J. H. Hubbard. "Étude dynamique des polynômes complexes. Parties I et II." Publications Mathématiques d'Orsay, 1984/1985.

[DH2] A. Douady and J. H. Hubbard, *A proof of Thurston's topological characterization of rational functions.* Acta. Math. **171** 263–297 (1993).

[E] A. Epstein. *Transversality Principles in Holomorphic Dynamics.* Preprint.

[Fa1] P. Fatou. *Sur les substitutions rationnelles.* Comptes Rendus de l'Académie des Sciences de Paris, **164** 806–808 and **165** 992–995 (1917).

[Fa2] P. Fatou, *Sur les équations fonctionnelles,* Bull. Soc. Math. France **47** 161–271 (1919).

[FS1] J. E. Fornæss and N. Sibony. *Complex Hénon mappings in* ℂ² *and Fatou-Bieberbach domains.* Duke Math. J. **65** 345–380 (1992).

[FS2] J. E. Fornæss and N. Sibony. *Complex dynamics in higher dimension. I.* In: "Complex analytic methods in dynamical systems (Rio de Janeiro, 1992)." 201–231, Edited by C. Camacho, Astérisque **222**, Soc. Math. France 1994.

[FS3] J. E. Fornæss and N. Sibony. *Complex dynamics in higher dimension. II.* In: "Modern methods in complex analysis (Princeton, NJ, 1992)," Edited by T. Bloom, D. Catlin, J. P. D'Angelo, and Y.-T. Siu. Ann. of Math. Stud. **137**, pp. 135–182, Princeton Univ. Press, Princeton, NJ, 1995.

[FS4] J. E. Fornæss and N. Sibony. *Riemann surface laminations with singularities.* J. Geom. Anal. **18** 400–442 (2008).

[FrM] S. Friedland and J. Milnor. *Dynamical properties of plane polynomial automorphisms.* Erg. Th. and Dyn. Syst., **9** 67–99 (1989).

[Gh] É. Ghys. *Gauss-Bonnet theorem for 2-dimensional foliations.* J. Funct. Anal. **77** 51–59 (1988).

[GhLnW] É. Ghys, R. Langevin, and P. Walczak. *Entropie géométrique des feuilletages.* Acta Math. **160** (1988) 105–142.

[Gi]   M. H. Gizatulin. *Rational G-surfaces.* Izv. Akad. Nauk SSSR Ser. Mat. **44** 110–144 (1980).

[Glu]   A. A. Glutsyuk. *Hyperbolicity of the leaves of a generic one-dimensional holo-morphic foliation on a nonsingular projective algebraic variety.* Tr. Mat. Inst. Steklova **213** Differ. Uravn. s Veshchestv. i Kompleks. Vrem., 90–111 (1997); translation in Proc. Steklov Inst. Math. 1996, **213** 83–103 (1996).

[G1]   R. I. Grigorchuk. *On the Milnor problem of group growth.* Dokl. Akad. Nauk SSSR **271** 30–33 (1983).

[G2]   R. I. Grigorchuk. *Degrees of growth of finitely generated groups and the theory of invariant means.* Izv. Akad. Nauk SSSR Ser. Mat. **48** 939–985 (1984).

[GrSw]   J. Graczyk and G. Swiatek. *Generic hyperbolicity in the logistic family.* Ann. of Math. **146** 1–52 (1997).

[Gro]   M. Gromov. *Groups of polynomial growth and expanding maps.* Publica-tions mathematiques I.H.E.S., **53** 53–78 (1981).

[Ha1]   M. Hakim. *Analytic transformations of $(\mathbb{C}^p, 0)$ tangent to the identity.* Duke Math. J. **92** 403–428 (1998).

[Ha2]   M. Hakim. *Stable pieces of manifolds in transformations tangent to the iden-tity.* Preprint 1998.

[H]   J. H. Hubbard. *Local connectivity of Julia sets and bifurcation loci: three the-orems of J.-C. Yoccoz.* In: "Topological Methods in Modern Mathematics, A Symposium in Honor of John Milnor's 60th Birthday," pp. 467–511. Edited by L. .R. Goldberg and A. V. Phillips, Publish or Perish, 1993.

[HOV]   J. H. Hubbard and R. W. Oberste-Vorth. *Hénon mappings in the complex domain. I: The global topology of dynamical space.* Inst. Hautes Études Sci. Publ. Math. **79** 5–46 (1994).

[HSS]   J. H. Hubbard, D. Schleicher, and M. Shishikura. *Exponential Thurston maps and limits of quadratic differentials.* J. Amer. Math. Soc. **22** 77–177 (2009).

[J]   G. Julia. *Mémoire sur l'iteration des fonctions rationnelles.* Journal de Mathématiques Pures et Appliquées **8** 47–245 (1918).

[KL]   J. Kahn and M. Lyubich. *The quasi-additivity law in conformal geometry.* Ann. of Math. **169** 561–593 (2009).

[KSvS]   O. Kozlovski, W. Shen, and S. van Strien. *Rigidity for real polynomials.* Ann. of Math. **165** 749–841 (2007).

[La]   P. Lavaurs. *Une description combinatoire de l'involution define par M sur les rationnels a denominateur impair.* C. R. Acad. Sci. Paris **303** 143–146 (1986).

[LS]   E. Lau and D. Schleicher. *Internal addresses in the Mandelbrot set and irre-ducibility of polynomials.* IMS preprint series **94/19** Stony Brook Univer-sity 1994. http://www.math.sunysb.edu/preprints/ims94-19.pdf. arXiv:math/9411238.

[Lin]   A. Lins Neto. *Uniformization and the Poincaré metric on the leaves of a foliation by curves.* Bol. Soc. Brasil. Mat. (N.S.) **31**, 351–366 (2000).

[L1]   M. Lyubich, *Back to the origin: Milnor's program in dynamics.* In: "Topological Methods in Modern Mathematics, A Symposium in Honor of John Milnor's 60th Birthday," pp. 85–92 Edited by L. .R. Goldberg and A. V. Phillips, Publish or Perish, 1993.

[L2]   M. Lyubich. *Combinatorics, geometry and attractors of quasi-quadratic maps.* Annals of Math., **140** 347–404 (1994).

[L3]   M. Lyubich. *Dynamics of quadratic polynomials. I, II.* Acta Math. **178** 185–247 and 247–297 (1997).

[L4]   M. Lyubich. *Forty years of unimodal dynamics: on the occasion of Artur Avila winning the Brin prize.* J. of Modern Dynamics., **6** 183–203 (2012).

[L5]   M. Lyubich. *John Milnor's work in dynamics.* In: "The Abel Prize", Edited by H. Holden and R. Piene, Springer-Verlag. To appear.

[LMin]   M. Lyubich and Y. Minsky. *Laminations in holomorphic dynamics.* J. Differ. Geom. **47** 17–94 (1997).

[Man]   B. Mandelbrot. *Fractal aspects of the iteration of $z \mapsto \lambda z(1 - z)$ for complex λ, z.* Annals N.Y. Acad. Sci. **357** 249–259 (1980).

[Me1]   D. Meyer. *Expanding Thurston maps as quotients.* arXiv:0910.2003

[Me2]   D. Meyer. *Invariant Peano curves of expanding Thurston maps.* Acta Math. **210** 95–171 (2013).

[M1]   J. Milnor. *On the existence of a connection with curvature zero.* Comment. Math. Helv. **32** 215–223 (1958). Also available in: "Collected Papers of John Milnor I," pp. 37–47 Amer. Math. Soc. 1994.

[M2]   J. Milnor. *A note on curvature and fundamental group.* J. Differential Geometry **2** 1–7 (1968). Also available in: "Collected Papers of John Milnor I," pp. 53–61 Amer. Math. Soc. 1994.

[M3]   J. Milnor. *Growth of finitely generated solvable groups.* J. Differential Geometry **2** 447–449 (1968). Also available in: "Collected Papers of John Milnor V," pp. 155–157 Amer. Math. Soc. 2011.

[M4]   J. Milnor. *On fundamental groups of complete affinely flat manifolds.* Advances in Math. **25** 178–187 (1977). Also available in: "Collected Papers of John Milnor I," pp. 119–130 Amer. Math. Soc. 1994.

[M5]   J. Milnor. *On the concept of attractor.* Comm. Math. Phys. **99** 177–195 (1985). *Correction and remarks* Comm. Math. Phys. **102** 517–519 (1985). Also available in: "Collected Papers of John Milnor VI," pp. 21–41 and 43–45 Amer. Math. Soc. 2012.

[M6]   J. Milnor. *Remarks on iterated cubic maps.* Experimental Math. **1** 5–24 (1992). Revised version in: "Collected Papers of John Milnor VI," pp. 337–369 Amer. Math. Soc. 2012.

[M7]   J. Milnor. *Local Connectivity of Julia Sets: Expository Lectures*, In: "The Mandelbrot set, Theme and Variations," pp. 67–199. Edited by Tan Lei, LMS Lecture Note Series **274** , Cambridge U. Press 2000. Revised version in: "Collected Papers of John Milnor VII," Amer. Math. Soc. In Press.

[M8]   J. Milnor. "Dynamics in One Complex Variable." Third Edition. Annals of Math. Studies, **160**. Princeton University Press, 2006.

[M9]   J. Milnor. *Cubic polynomial maps with periodic critical orbit I.* In: "Complex dynamics: Friends and Families," 333–411. Edited by D. Schleicher. A.K. Peters, 2009.

[M10]   J. Milnor. "Collected papers of John Milnor VI. Dynamical Systems (1953–2000)." Edited by A. Bonifant, Amer. Math. Soc. 2012.

[M11]   J. Milnor. "Collected papers of John Milnor VII. Dynamical Systems (1984–2012)." Edited by A. Bonifant, Amer. Math. Soc. In Press.

[MT]   J. Milnor and W. Thurston. *On iterated maps of the interval.* In: "Dynamical systems (College Park, MD, 1986–87)," pp. 465–563, Edited by J. C. Alexander, Lecture Notes in Math., **1342** Springer, Berlin, 1988. Revised version in: "Collected Papers of John Milnor VI," pp. 87–155 Amer. Math. Soc. 2012.

[MTr]   J. Milnor and Ch. Tresser. *On entropy and monotonicity for real cubic maps. With an appendix by Adrien Douady and Pierrette Sentenac.* Comm. Math. Phys. **209** 123–178 (2000). Also available in: "Collected Papers of John Milnor VI," pp. 177–232 Amer. Math. Soc. 2012.

[Mo]   P. Morton. *On certain algebraic curves related to polynomial maps.* Compositio Math. **103** 319–350 (1996).

[MuPM1]   V. Muñoz and R. Pérez-Marco. *Ergodic solenoids and generalized currents.* Rev. Mat. Complut. **24** 493–525 (2011).

[MuPM2]   V. Muñoz and R. Pérez-Marco. *Ergodic solenoidal homology: realization theorem.* Comm. Math. Phys. **302** 737–753 (2011).

[N]   V. Nekrashevych. "Self-similar groups." Mathematical Surveys and Monographs **117**. Amer. Math. Soc., Providence, RI, 2005.

[RS]   D. Ruelle and D. Sullivan. *Currents, flows and diffeomorphisms.* Topology **14** 319–327 (1975).

[Sc]   S. Schwartzman. *Asymptotic cycles.* Ann. of Math. **66** 270–284 (1957).

[Sch]   A. S. Schwarz. *A volume invariant of covering.* Dokl. Akad. Nauj SSSR (N.S.), **105** 32–34 (1955).

[T]   W. Thurston. *On the geometry and dynamics of iterated rational maps.* In: "Complex Dynamics: Friends and Families," pp. 3–110. Edited by D. Schleicher. A. K. Peters 2009.

[V]   A. Verjovsky. *A uniformization theorem for holomorphic foliations.* In: "The Lefschetz centennial conference, Part III (Mexico City, 1984)," pp. 233–253, Contemp. Math., **58** Amer. Math. Soc., Providence, RI, 1987.

Part I

One Complex Variable

Arithmetic of Unicritical Polynomial Maps

John Milnor

ABSTRACT. This note will study complex polynomials with only one critical point, relating arithmetic properties of the coefficients to those of periodic orbits and their multipliers and external rays.

1   INTRODUCTION

A complex polynomial maps of degree $n \geq 2$ with only one critical point can always be put in the standard normal form

$$f_c(z) \;=\; z^n + c \tag{1.1}$$

by an affine change of coordinates. The connectedness locus, consisting of all c for which the Julia set of f_c is connected, is sometimes known as the "multibrot set." (Compare [S].) It is not difficult to check that the power $\widehat{c} = c^{n-1}$ is a complete invariant for the holomorphic conjugacy class of f_c.

In §2 we will use the alternate normal form

$$w \;\mapsto\; g_b(w) \;=\; \frac{w^n + b}{n} \,, \tag{1.2}$$

with derivative $g_b'(w) = w^{n-1}$, and use the conjugacy invariant $\widehat{b} = b^{n-1}$. These normal forms (1.1) and (1.2) are related by the change of variable formula

$$w \;=\; n^{1/(n-1)} z \qquad \text{with} \qquad \widehat{b} \;=\; n^n \widehat{c},$$

and hence $b = n^{n/(n-1)} c$. (In particular, in the degree two case, $b = \widehat{b}$ is equal to $4c = 4\widehat{c}$.)

If A is any ring contained in the complex numbers $\mathbb{C}$, it will be convenient to use the non-standard notation $\overline{A}$ for the ***integral closure***, the ring consisting of all complex numbers which satisfy a monic polynomial equation with coefficients in A. (See, for example, [AM].)

Section 2 consists of statements about periodic orbits, which are proved in §3. The last section discusses the critically finite case. (For background in 1-dimensional complex dynamics see [M1].)

2   PERIODIC ORBITS

The following statement generalizes Bousch [Bo]. Here it is essential to work with the parameter b, rather than the classical parameter $c = b/n^{n/(n-1)}$.

I want to thank Thierry Bousch for his help with this manuscript and the NSF for its support under grant DMSO757856.

THEOREM 2.1. *If w is a periodic point for the map g_b, with multiplier μ, then the rings $\mathbb{Z}[\mu]$, $\mathbb{Z}[w]$, $\mathbb{Z}[b]$, and $\mathbb{Z}[\widehat{b}]$ all have the same integral closure.*

Following are some immediate consequences.

COROLLARY 2.2. *If any one of the four numbers μ, w, b, $\widehat{b}$ belongs to the ring $\overline{\mathbb{Z}}$ consisting of all algebraic integers, then all four of these numbers are algebraic integers. As an example, if the map g_b is **parabolic**, that is, if the multiplier of some periodic orbit is a root of unity, then the parameters b and $\widehat{b}$ are algebraic integers; hence every periodic point w is an algebraic integer, and the multiplier μ of every periodic orbit is an algebraic integer.*

(For a sharper version of this statement, see Remark 3.2.)

REMARK 2.3. It would be interesting to understand more generally which rational maps have the property that all multipliers are algebraic integers. The family of Lattès maps provides one well known collection of non-polynomial examples.

More generally, if $f \colon \mathbb{C} \to \mathbb{C}$ is any polynomial map with only one critical point, then we have the following.

COROLLARY 2.4. *If μ and μ' are the multipliers of two periodic orbits for f, then the integral closure $\overline{\mathbb{Z}[\mu]}$ is equal to $\overline{\mathbb{Z}[\mu']}$. It follows that this integral closure depends only on the holomorphic conjugacy class of f and not on the particular choice of periodic orbit.*

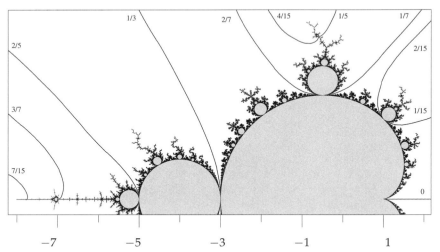

FIGURE 2.5: Parameter plane for quadratic maps, showing the external rays in the upper half-plane which have period at most four under doubling, and showing integer values of $b = 4c$.

Now suppose that the parameter value b is the landing point of an external ray to the connectedness locus in the b-parameter plane, with angle $p/q \in \mathbb{Q}/\mathbb{Z}$ which is periodic under multiplication by n. (See Figure 2.5 for the degree two case.) Then the associated map g_b has a parabolic orbit (compare [ES], as well as

There is a curious relationship between the denominator q of this angle and the parameter value b or $\widehat{b}$. Here are some examples in the quadratic case $n = 2$, as illustrated in Figure 2.5. For the landing points of the $1/3$, $2/5$, and $3/7$ rays we find that

$$b(1/3) = -3, \qquad b(2/5) = -5, \qquad b(3/7) = -7.$$

At first glance, this relationship between angles and landing points seems to disappear for the landing point of the $1/7$ and $2/7$ rays, with $b = (-1 + 3i\sqrt{3})/2$. However, this number satisfies the irreducible monic equation

$$b^2 + b + 7 = 0,$$

so the denominator $7 = 2^3 - 1$ again appears in the description of the landing point.

Here is a more general statement, working in the b parameter plane for polynomials of degree n. If b has degree d over the rational numbers $\mathbb{Q}$, define $\mathrm{Norm}(b) \in \mathbb{Q}$ to be the product of the d algebraic conjugates of b over $\mathbb{Q}$. Up to sign, this is just the constant term in the irreducible monic polynomial satisfied by b. If b belongs to the ring $\overline{\mathbb{Z}}$ of algebraic integers, note that $\mathrm{Norm}(b) \in \mathbb{Z}$.

THEOREM 2.6. *Consider an external ray of angle p/q in the b parameter plane with landing point $b = b(p/q)$. If p/q is periodic under multiplication by $n \pmod 1$ with period r, so that q divides $n^r - 1$, and if b has degree d over $\mathbb{Q}$, then it follows that the integer $\mathrm{Norm}(b)$ is a divisor of $(n^r - 1)^d$. Similarly, if $\widehat{b}$ has degree $\widehat{d}$ over $\mathbb{Q}$, then $\mathrm{Norm}(\widehat{b})$ divides $(n^r - 1)^{(n-1)\widehat{d}}$.*

Here are two examples of degree $n = 2$ and one example with arbitrary degree:

- For the landing point of the $1/7$ ray with ray period $r = 3$, we have

$$\mathrm{Norm}(b) = 7 \quad | \quad (2^3 - 1)^2 = 49.$$

- For the $1/5$ ray, the ray period is $r = 4$, and the irreducible equation is $b^3 + 9b^2 + 27b + 135 = 0$ of degree $d = 3$, with

$$|\mathrm{Norm}(b)| = 135 = 3^3 \cdot 5 \quad | \quad (2^4 - 1)^3 = 3^3 \cdot 5^3.$$

- For an arbitrary degree $n \geq 2$, let b be a fixed point of multiplier μ. Then the equations $g_b(w) = w$ and $g_b'(w) = \mu$ imply respectively that $b = w(n - \mu)$, and $w^{n-1} = \mu$, so that $\widehat{b} = b^{n-1} = \mu(n - \mu)^{n-1}$. If the multiplier is $\mu = 1$, so that $r = 1$ and $\widehat{d} = 1$, it follows that $\widehat{b} = (n - 1)^{n-1}$ is precisely equal to $(n^r - 1)^{(n-1)\widehat{d}}$. If the multiplier is $\mu = -1$, with ray period $r = 2$ and with $\widehat{d} = 1$, then we get $\widehat{b} = -(n + 1)^{n-1}$, which divides $(n^2 - 1)^{n-1}$.

3   PROOFS

The proofs of the statements of §2 will depend on some basic properties of the integral closure. Let u and v be complex numbers. Then clearly $\overline{\mathbf{Z}[u]} \subset \overline{\mathbf{Z}[v]}$ if and only if $u \in \overline{\mathbf{Z}[v]}$. Note also that $\overline{\mathbf{Z}[u^k]} = \overline{\mathbf{Z}[u]}$ for any integer $k > 0$. The following statement will be needed.

LEMMA 3.1. *Let u and v be complex numbers. If the integral closure $\overline{\mathbf{Z}[u]}$ is equal to $\overline{\mathbf{Z}[v]}$, then it is also equal to $\overline{\mathbf{Z}[uv]}$.*

PROOF. The product uv certainly belongs to the ring $\overline{\mathbf{Z}[u]} = \overline{\mathbf{Z}[v]}$; hence $\overline{\mathbf{Z}[uv]} \subset \overline{\mathbf{Z}[u]}$. Conversely, since u is an element of $\overline{\mathbf{Z}[v]}$, it satisfies an equation of the form

$$u^k = \sum_{i=0}^{k-1} \sum_{j=0}^{\ell} n_{i,j} u^i v^j \qquad \text{with} \qquad n_{i,j} \in \mathbf{Z} \,.$$

Multiplying both sides of this equation by u^ℓ, the result can be written as

$$u^{\ell+k} = \sum_{i=0}^{k-1} \sum_{j=0}^{\ell} n_{i,j} u^{\ell+i-j} (uv)^j \,,$$

which proves that $u \in \overline{\mathbf{Z}[uv]}$; hence $\overline{\mathbf{Z}[u]} \subset \overline{\mathbf{Z}[uv]}$.   □

PROOF OF THEOREM 2.1. We can write the k-fold iterate $g_b^{\circ k}(w)$ as a polynomial with integer coefficients divided by a common denominator as follows. Set

$$g_b^{\circ k}(w) = \frac{P_k(b, w)}{N_k}$$

where $N_1 = n$ and $N_{k+1} = n N_k^n = n^{1+n+n^2+\dots+n^k}$. Then $P_1(b, w) = w^n + b$, and a straightforward induction shows that

$$P_{k+1}(b, w) = P_k(b, w)^n + N_k^n b \,.$$

It follows easily that $P_k(b, w)$ is a polynomial in two variables with integer coefficients and that $P_k(b, w)$ is monic of degree n^k when considered as a polynomial in w with coefficients in $\mathbf{Z}[b]$, or monic of degree n^{k-1} when considered as a polynomial in b with coefficients in $\mathbf{Z}[w]$.

Now suppose that w is a periodic point for g_b with period h. Then

$$g_b^{\circ h}(w) - w = 0 \quad \text{or, in other words,} \qquad P_h(b, w) - N_h w = 0 \,.$$

This last polynomial equation is also monic in w and monic in b, so it follows that $w \in \overline{\mathbf{Z}[b]}$ and that $b \in \overline{\mathbf{Z}[w]}$. Thus the two rings $\mathbf{Z}[b]$ and $\mathbf{Z}[w]$ have the same integral closure. It follows that the ring $\mathbf{Z}[\widehat{b}]$ has this same integral closure.

Now let

$$w = w_0 \mapsto w_1 \mapsto \cdots \mapsto w_h = w_0$$

be any period h orbit for g_b. We know from the preceding argument that the rings $\mathbb{Z}[w_j]$ all have the same integral closure as $\mathbb{Z}[b]$. Setting $W = w_1 w_2 \cdots w_h$, it follows inductively from Lemma 3.1 that the ring $\mathbb{Z}[W]$ has the same integral closure. Since the multiplier of this periodic orbit can be written as $\mu = W^{n-1}$, it follows that $\mathbb{Z}[\mu]$ also has the same integral closure. this completes the proof of Theorem 2.1 and its corollaries. $\qquad\square$

REMARK 3.2. Here is a supplementary statement. By definition, an element $w \in \overline{\mathbb{Z}}$ is relatively **prime** to n if the ideal $w\overline{\mathbb{Z}} + n\overline{\mathbb{Z}}$ is equal to $\overline{\mathbb{Z}}$ or, in other words, if w maps to a unit in the quotient ring $\overline{\mathbb{Z}}/n\overline{\mathbb{Z}}$. Now suppose that $w \in \overline{\mathbb{Z}}$ is periodic with multiplier μ under the map g_b. If any one of the four numbers $w, \mu, b, \hat{b} \in \overline{\mathbb{Z}}$ is prime to n, then it follows that all four of these numbers are prime to n. As an example, if g_b has a parabolic orbit, then all of these numbers are prime to n, but if g_b is critically periodic, then none of them is prime to n. To prove this statement, consider an orbit $\{w_j\}$ of period h. Then

$$n w_{j+1} = w_j^n + b \,;$$

hence, $w_j^n \equiv -b \pmod{n\overline{\mathbb{Z}}}$ for all w_j. Taking the product over the orbit elements, this yields $W^n \equiv (-b)^n$, and hence $\mu^n \equiv (-b)^{(n-1)h} \pmod{n\overline{\mathbb{Z}}}$. The conclusion follows easily.

PROOF OF THEOREM 2.6. Suppose again that $\{w_1, \ldots, w_h\}$ is an orbit of period h for g_b, with multiplier $\mu = W^{n-1}$, where $W = w_1 w_2 \cdots w_h$. Then we have the congruence

$$n w_{j+1} = w_j^n + b \equiv w_j^n \pmod{b\overline{\mathbb{Z}}} \,.$$

In the situation of Corollary 2.2, where b and the w_j belong to the ring $\overline{\mathbb{Z}}$ of algebraic integers, we can take the product over j to obtain

$$n^h W \equiv W^n \pmod{b\overline{\mathbb{Z}}} \,.$$

If $\mu = W^n$ is a unit in the ring $\overline{\mathbb{Z}}$, then W is also a unit, and we can divide this congruence by W, yielding

$$n^h \equiv W^{n-1} = \mu \pmod{b\overline{\mathbb{Z}}} \,. \tag{3.1}$$

Now suppose that μ is a primitive mth root of unity. Then, raising this congruence to the mth power, we obtain

$$n^{hm} \equiv 1 \pmod{b\overline{\mathbb{Z}}} \,.$$

Here the product hm is precisely the smallest integer r such that the iterate $g_b^{\circ r}$ maps each w_j to itself and has derivative $+1$ at each w_j. If an external ray of angle p/q in the w-plane lands on w_j, then r is precisely equal to the ray period, that is the period of p/q under multiplication by n. (See, for example, [M2].) Using the usual Douady-Hubbard correspondence between parameter plane and dynamic plane, at least one of these p/q is also the angle of an external ray in the parameter plane which lands on b. (Compare [LS].) *Thus we see that the ratio* $(n^r - 1)/b$ *is an algebraic integer.*

Now taking the product over the d distinct embeddings of the field $Q(b)$ into $\mathbb{C}$, we see that the rational number

$$(n^r - 1)^d / \text{Norm}(b)$$

belongs to $\overline{\mathbb{Z}}$ and hence, belongs to the ring $\overline{\mathbb{Z}} \cap Q = \mathbb{Z}$. In other words, the integer $\text{Norm}(b)$ is a divisor of $(n^r - 1)^d$, as asserted. A completely analogous argument proves the corresponding statement for $\text{Norm}(\hat{b})$. $\square$

4   POSTCRITICALLY FINITE MAPS

The situation for parameter values corresponding to postcritically finite maps is rather different. In this case, it is more convenient to work with the classical normal form of Equation (1.1), with invariant $\hat{c} = c^{n-1} = \hat{b}/n^n$. First consider the critically periodic case.

LEMMA 4.1. *If the critical point* 0 *is periodic under* f_c, *then the following two equivalent properties are satisfied:*
* *The critical value* c *is either an algebraic unit*[1] *or zero.*
* *Every periodic point of* f_c *is either an algebraic unit or zero.*

As examples in the quadratic case $n = 2$, if $c = -1$, then the critical point has period 2, while if $c^3 + 2c^2 + c + 1 = 0$, it has period 3, with $\text{Norm}(c) = -1$ in both cases.

PROOF OF LEMMA 4.1. If $c = 0$, then all non-zero periodic points are roots of unity, so it suffices to consider the case $c \neq 0$. If c is periodic of period p with $c \neq 0$, then the polynomial equation $f_c^{\circ(p-1)}(c)/c = 0$ is monic with constant term $+1$, hence c is an algebraic unit. In particular, $\text{Norm}(c) = \pm 1$. Any period h point of f_c satisfies a polynomial equation $f_c^{\circ h}(z) - z$ with coefficients in $\mathbb{Z}[c]$ which is monic with constant term c. Taking the product over all Galois conjugates[2] of c, we obtain a monic equation with integer coefficients and with constant term $\text{Norm}(c) = \pm 1$. This proves that any periodic point is an algebraic unit whenever c is an algebraic unit. Conversely, the fixed point equation $z^n - z + c = 0$ implies that the product of the fixed points of f_c is $\pm c$. Hence if every fixed point is a unit in $\overline{\mathbb{Z}}$ it follows that c is a unit. $\square$

REMARK 4.2. If z is periodic under f_c, then a similar argument shows that $z \in \overline{\mathbb{Z}}$ if and only if the parameter c (or $\hat{c}$) belongs to $\overline{\mathbb{Z}}$. In this case, the multiplier μ of the orbit belongs to the ideal $n^h \overline{\mathbb{Z}}$, where h is the period.

More generally, consider the critically preperiodic case.

THEOREM 4.3. *For* $c \neq 0$, *if the orbit of the critical point is eventually periodic, then* c *and* $\hat{c} = c^{n-1}$ *are algebraic integers, with* $\text{Norm}(c)$ *and* $\text{Norm}(\hat{c})$ *dividing* n.

A number of critically preperiodic examples are shown in Figure 4.4 and described further in Tables 4.5 and 4.6. Note that there is no evident arithmetic relation between the external angles and the landing point c in these cases.

---

[1]In other words, both c and $1/c$ are algebraic integers.
[2]Here the Galois conjugation acts only on the coefficients, not on the indeterminate z.

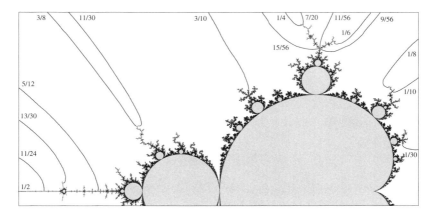

FIGURE 4.4: Some eventually periodic rays for the Mandelbrot set.

p/q	transient time	eventual period	degree(c)	|Norm(c)|
$\frac{1}{2}$	$t=1$	$h=1$	$d=1$	2
$\frac{1}{6}$	$t=1$	$h=2$	$d=2$	1
$\frac{1}{4}, \frac{5}{12}$	$t=2$	$h=1$	$d=3$	2
$\frac{1}{8}, \frac{9}{56}, \frac{11}{56}, \frac{15}{56}, \frac{3}{8}, \frac{11}{24}$	$t=3$	$h=1$	$d=7$	2
$\frac{1}{30}, \frac{1}{10}, \frac{7}{30}, \frac{3}{10}, \frac{11}{30}, \frac{13}{30}$	$t=1$	$h=4$	$d=12$	1

TABLE 4.5: Description of the corresponding landing points c. (Compare Table 4.6. Here the **transient time** is defined to be the smallest t such that $f_c^{ot}(c)$ is periodic.) Note that Norm(c) is always a divisor of $n=2$.

$$c + 2 = 0$$
$$c^2 + 1 = 0$$
$$c^3 + 2c^2 + 2c + 2 = 0$$
$$c^7 + 4c^6 + 6c^5 + 6c^4 + 6c^3 + 4c^2 + 2c + 2 = 0$$
$$c^{12} + 6c^{11} + 15c^{10} + 22c^9 + 23c^8 + 18c^7 + 11c^6 + 8c^5 + 6c^4 + 2c^3 + 1 = 0$$

TABLE 4.6: Corresponding irreducible equations.

There can be many different postcritically finite parameters which satisfy the same irreducible equation over $\mathbb{Q}$. This is related to the fact that the Galois group of $\overline{\mathbb{Q}}$ over $\mathbb{Q}$ may act in a highly non-trivial way on these points. (Compare [P] as well as Remark 4.8.)

PROOF OF THEOREM 4.3. Since $c \neq 0$, we can use the modified normal form $\xi \mapsto F_{\widehat{c}}(\xi)$, where

$$F_{\widehat{c}}(\xi) = \frac{f_c(c\,\xi)}{c} = \frac{(c\,\xi)^n + c}{c} = \widehat{c}\,\xi^n + 1\,.$$

This has critical orbit of the form

$$0 \;\mapsto\; 1 \;\mapsto\; \widehat{c}+1 \;\mapsto\; \widehat{c}\,(\widehat{c}+1)^n+1 \;\mapsto\; \cdots\,.$$

The kth point of this critical orbit can be expressed as a polynomial function $P_k(\widehat{c})$, with $P_1 = 1$ and

$$P_{k+1}(\widehat{c}) = \widehat{c}\,P_k(\widehat{c})^n + 1\,.$$

Evidently each $P_k(\widehat{c})$ is a monic polynomial with constant term $P_k(0) = +1$. Therefore, if $F_{\widehat{c}}$ has periodic critical point, then it follows that $\widehat{c}$ is a unit in the ring of algebraic integers, with $\mathrm{Norm}(\widehat{c}) = \pm 1$.

Now suppose that the orbit of zero is eventually periodic but not periodic. The transient time $t \geq 1$, and the eventual period $h \geq 1$ are defined as the smallest positive integers such that $F_c^{\circ t}(1)$ is periodic of period h. It follows that the two orbit points $P_t(\widehat{c})$ and $P_{t+p}(\widehat{c})$ are distinct and yet have the same image under the nth power map. In other words, the ratio

$$x = \frac{P_{t+p}(\widehat{c})}{P_t(\widehat{c})} \tag{4.1}$$

must be an nth root of unity, not equal to $+1$. Hence it must satisfy the equation

$$1 + x + x^2 + \ldots + x^{n-1} = 0\,.$$

Clearing denominators, we see that

$$\sum_{i+j=n-1} P_{t+p}(\widehat{c})^i P_t(\widehat{c})^j = 0\,.$$

It is not difficult to check that this is a monic polynomial equation in $\widehat{c}$ with constant term n. Therefore, $\widehat{c}$ is an algebraic integer, and $\mathrm{Norm}(\widehat{c})$ divides n. A similar argument shows that $\mathrm{Norm}(c)$ divides n. $\qquad\qquad\square$

REMARK 4.7. Let $\{z_j\}$ be a periodic orbit of period $h > 1$ so that

$$z_{j+1} = z_j^n + c\,,$$

where j ranges over $\mathbb{Z}/h\mathbb{Z}$. Using the polynomial expression

$$\phi(x, y) = \frac{x^n - y^n}{x - y} = x^{n-1} + x^{n-2}y + \ldots + xy^{n-2} + y^{n-1}\,,$$

note the identity

$$\frac{z_{j+2} - z_{j+1}}{z_{j+1} - z_j} = \frac{z_{j+1}^n - z_j^n}{z_{j+1} - z_j} = \phi(z_j, z_{j+1}) \,.$$

Taking the product over all j modulo h, it follows that $\prod_{j \bmod h} \phi(z_j, z_{j+1}) = 1$. (Compare [Be].) In particular, if $c \in \mathbb{Z}$ so that the z_j also belong to $\mathbb{Z}$, then it follows that each expression $\phi(z_j, z_{j+1})$ is a unit in the ring $\mathbb{Z}$. (For other "dynamical units", see [MS].)

REMARK 4.8 (Classical Problems). To conclude this discussion, we mention two well-known unsolved problems.

> If the maps f_{c_1} and f_{c_2} have parabolic orbits with the same period and the same ray period, does it follow that the corresponding invariants $\widehat{c}_1$ and $\widehat{c}_2$ satisfy the same irreducible equation over $\mathbb{Q}$?

In other words, does it follow that $\widehat{c}_1$ and $\widehat{c}_2$ are conjugate under some Galois automorphism of the field $\overline{\mathbb{Q}}$ over $\mathbb{Q}$?

> Similarly, if two maps f_{c_1} and f_{c_2} have critical orbits which are periodic with the same period, does it follow that $\widehat{c}_1$ and $\widehat{c}_2$ are Galois conjugate?

There is a similar question for the eventually periodic case, but the situation is more complicated. There is an extra invariant if the degree n is not prime, since the ratio of Equation (4.1) must be a primitive τth root of unity for some divisor τ of n, with $1 < \tau \leq n$.

> If two such parameter values have the same transient time t, the same eventual period h, and the same integer $1 < \tau \mid n$, does it follow that the corresponding invariants $\widehat{c}$ are Galois conjugate?

5   Bibliography

[AM] M. Atiyah and I. Macdonald, "Introduction to Commutative Algebra," Addison-Wesley 1969.

[Be] R. Benedetto, *An elementary product identity in polynomial dynamics*, Amer. Math. Monthly **108** (2001) 860–864.

[Bo] T. Bousch, *Les racines des composantes hyperboliques de M sont des quarts d'entiers algébriques* (manuscript, 1996). In this volume, 25–26.

[DH] A. Douady and J. H. Hubbard, "Étude dynamique des polynômes complexes I & II," Publ. Math. Orsay (1984–85).

[ES] D. Eberlein and D. Schleicher, *Rational parameter rays of multibrot sets*, in preparation.

[LS]   E. Lau and D. Schleicher, *Internal addresses in the Mandelbrot set and irreducibility of polynomials*, arXiv:math/9411238v1. Stony Brook IMS Preprint 1994/19, http://www.math.sunysb.edu/preprints/ims94-19.pdf.

[M1]   J. Milnor, "Dynamics in One Complex Variable," Princeton U. Press 2006.

[M2]   J. Milnor, *Periodic Orbits, Externals Rays and the Mandelbrot Set: An Expository Account*, In "Geometrie Complexe et Systemes Dynamiques," ed. M. Flexor, P. Sentenac, J. C. Yoccoz, Astérisque **261** (2000) 277–333.

[MS]   P. Morton and J. Silverman, *Periodic points, multiplicities, and dynamical units*, J. Reine Angew. Math. **461** (1995) 81–122.

[P]   K. Pilgrim, *Dessins d'enfants and Hubbard Trees*, Ann. Sci. École Norm. Sup. **(4) 33** (2000) 671–693.

[S]   D. Schleicher, *On fibers and local connectivity of Mandelbrot and Multibrot sets*, in "Fractal Geometry and Applications: a jubilee of Benoît Mandelbrot. Part 1," Proc. Sympos. Pure Math., **72**, Part 1, Amer. Math. Soc. (2004) 477–517.

Les racines des composantes hyperboliques de M sont des quarts d'entiers algébriques

Thierry Bousch

ABSTRACT. We show that for complex numbers c and z such that z is a periodic point of $z \mapsto z^2 + c$, the multiplier of the orbit of z is an algebraic integer if and only if $4c$ is an algebraic integer. In particular, if c is the root point of a hyperbolic component of the Mandelbrot set, then $4c$ is an algebraic integer.

THÉORÉME. *Soit c un nombre complexe, et z un point périodique du polynôme $z \mapsto z^2 + c$. Le multiplicateur ρ de l'orbite de z est un entier algébrique si et seulement si $4c$ est un entier algébrique.*

COROLLAIRE. *Si c est une racine de composante hyperbolique de l'ensemble de Mandelbrot, alors $4c$ est un entier algébrique.*

Notations. On pose $C = 4c$, et $f(z) = z^2 + c$. Considérons l'équation $f^n(z) = z$ comme équation en z sur le corps $K = \mathbb{C}(c)$; cette équation admet $N = 2^n$ solutions *distinctes* dans la clôture algébrique de K, que nous noterons $z_1, \ldots, z_N$. (Ceci correspond au fait que pour c nombre complexe générique, l'application quadratique $z \mapsto z^2 + c$ possède exactement N points n-périodiques dans $\mathbb{C}$.) Si on pose $Z_i = 2z_i$, le multiplicateur ρ_i de l'orbite de z_i s'écrit

$$\rho_i = 2z_i \times 2f(z_i) \times \cdots \times 2f^{n-1}(z_i)$$
$$= Z_{i_0} Z_{i_1} \cdots Z_{i_{n-1}}$$

où $z_{i_k} = f^k(z_i)$. Une des implications du Théorème est facile à démontrer; nous l'énoncerons sous la forme suivante:

LEMME 1. *Les Z_i sont des entiers algébriques sur $\mathbb{Z}[C]$, et par conséquent, les ρ_i aussi. En particulier, si on spécialise C en un nombre complexe qui est un entier algébrique sur $\mathbb{Z}$, alors les valeurs correspondantes des Z_i et ρ_i seront aussi des nombres complexes entiers algébriques sur $\mathbb{Z}$.*

PREUVE DU LEMME 1. Considérons une orbite n-périodique $(z_0, z_1, \ldots, z_{n-1})$ de $z \mapsto z^2 + c$ dans la clôture algébrique de $K = \mathbb{C}(c)$. Les z_i vérifient

$$\begin{cases} z_0^2 + c = z_1 \\ z_1^2 + c = z_2 \\ \quad \vdots \\ z_{n-1}^2 + c = z_0 \end{cases}$$

ce qui équivaut à

$$
\begin{cases}
Z_0^2 + C = 2Z_1 \\
(2Z_1)^2 + 4C = 8Z_2 \\
\quad\vdots \\
(2^{(2^k-1)}Z_k)^2 + 2^{(2^{k+1}-2)}C = 2^{(2^{k+1}-1)}Z_{k+1} \\
\quad\vdots \\
(2^{(2^{n-1}-1)}Z_{n-1})^2 + 2^{(2^n-2)}C = 2^{(2^n-1)}Z_0.
\end{cases}
$$

En éliminant $Z_1, \ldots, Z_{n-1}$ dans le système ci-dessus, on voit que Z_0 vérifie

$$
(\cdots((Z_0^2 + C)^2 + 4C)^2 + \ldots)^2 + 2^{2^n-2}C = 2^{2^n-1}Z_0.
$$

Cette équation est monique en Z_0 et à coefficients dans $\mathbb{Z}[C]$, donc Z_0 est un entier algébrique sur l'anneau $\mathbb{Z}[C]$. Evidemment, il en est de même pour tous les autres Z_i. Ceci démontre le Lemme 1. $\qquad\square$

La réciproque du Théorème sera donnée par le Lemme 2.

LEMME 2. *Le paramètre C est un entier algébrique sur $\mathbb{Z}[\rho]$. En particulier, si on spécialise ρ en un nombre complexe qui est un entier algébrique sur $\mathbb{Z}$, alors les valeurs correspondantes de C seront aussi des nombres complexes entiers algébriques sur $\mathbb{Z}$.*

PREUVE DU LEMME 2. Nous avons vu que les les multiplicateurs $\rho_1, \ldots, \rho_N$ sont des entiers algébriques sur $\mathbb{Z}[C]$. Ils vérifient donc une équation de la forme

$$
\prod_{i=1}^N (\rho - \rho_i) = \rho^N + \lambda_1(C)\rho^{N-1} + \ldots + \lambda_N(C) = 0,
$$

où les λ_i sont des polynômes à coefficients entiers. Pour prouver le Lemme 2, il suffit de voir que le coefficient dominant de C dans cette équation est égal à ± 1.

LEMME 3. *Le polynôme λ_N est de degré $Nn/2$ et son coefficient dominant est égal à ± 1. Tous les λ_k sont de degré au plus $kn/2$.*

En effet, quand C est un nombre complexe de module très grand, les (nombres complexes) Z_i sont approximativement égaux à $\pm\sqrt{-C}$. Les ρ_i sont donc de l'ordre de $\pm(-C)^{n/2}$. Le polynôme $\lambda_k(C)$ étant donné par

$$
\lambda_k(C) = (-1)^k \sum_{i_1 < \ldots < i_k} \rho_{i_1} \cdots \rho_{i_k}
$$

on a, pour tout $k \leq N$, la majoration

$$
\lambda_k(C) = O(|C|^{kn/2}) \qquad \text{quand } |C| \to \infty,
$$

qui montre que le degré de λ_k est $\leq kn/2$. On a également l'estimation

$$
\lambda_N(C) = (-1)^N \rho_1 \cdots \rho_N \sim \pm(-C)^{Nn/2} \qquad \text{quand } |C| \to \infty,
$$

qui montre que λ_N est de degré $Nn/2$ et de coefficient dominant ± 1. Ceci prouve le Lemme 3, et termine la démonstration du Lemme 2 et du Théorème. $\qquad\square$

Dynamical cores of topological polynomials

Alexander Blokh, Lex Oversteegen, Ross Ptacek,
and Vladlen Timorin

ABSTRACT. We define the (dynamical) core of a topological polynomial (and the associated lamination). This notion extends that of the core of a unimodal interval map. Two explicit descriptions of the core are given: one related to periodic objects and one related to critical objects.

1 INTRODUCTION AND THE MAIN RESULT

1.1 Motivation

Complex dynamics studies, among other topics, the limiting behavior of points under iteration of complex polynomials. It is reasonable to restrict our attention to the Julia set, as elsewhere the limiting behavior is easy to describe. Since in many cases polynomial Julia sets are one-dimensional continua, one can consider the problem as a far-reaching generalization of the similar problem for simple one-dimensional spaces such as an interval.

A popular one-dimensional family is that of *unimodal interval maps*, i.e., interval maps with unique turning point. Often such maps f are considered on $[0,1]$ and normalized by assuming that the turning point in question is a local maximum and that $f(0) = f(1) = 0$. It is easy to see that the only case when such a map f can exhibit non-trivial dynamics is when $f^2(c) < c < f(c)$. Moreover, any point of $[0,1]$ will either eventually map to $[f^2(c), f(c)]$ or it will converge to a fixed point of f. The interval $[f^2(c), f(c)]$ is often called the *core* of f; we prefer to call it the *dynamical core* of f. In the quadratic polynomial case, when 0 is a repelling fixed point, *all* points of $[0,1]$ except 0 and 1 are eventually mapped to the core of f. A similar notion can be introduced for self-mappings of *graphs* (i.e., one-dimensional branched manifolds).

In these cases the dynamical core is a "small" invariant subcontinuum which captures the limit sets of all but finitely many points of the space; clearly, subsets smaller than the entire space, which nevertheless contain limit sets of all but finitely many points, are of interest in dynamics. However, one can also think of the dynamical core as a small invariant subcontinuum which contains the limit sets of all *cutpoints* of the space. In this form the notion of the dynamical core can be extended to polynomials with locally connected Julia sets. Still, one should justify one's interest in the dynamics of cutpoints of connected Julia sets J because, except in special cases, the set of cutpoints is not a "big" subspace of J (e.g., in the

The first named author was partially supported by NSF grant DMS–0901038.
The second named author was partially supported by NSF grant DMS-0906316.
The fourth named author was partially supported by the Deligne fellowship, the Simons-IUM fellowship, RFBR grants 10-01-00739-a, 11-01-00654-a, MESRF grant MK-2790.2011.1, RFBR/CNRS project 10-01-93115-NCNIL-a, and AG Laboratory NRU-HSE, MESRF grant ag. 11 11.G34.31.0023.

locally connected case the set of cutpoints is of zero harmonic measure and of first category in J).

In our view, one reason for studying the set of cutpoints of J, despite its small size, is that the set of cutpoints carries the bulk of the structural information about J. This enables us in many cases to construct a topological model for the Julia set. Indeed, suppose that J is locally connected and neither an arc nor a Jordan curve. Then it follows from a result of Hausdorff [H37], that the set of endpoints of J is always homeomorphic to the set of all irrational numbers (this can also be seen directly by a straightforward argument). Loosely speaking, Julia sets differ in as much as their sets of cutpoints differ. Thus, two locally connected Julia sets can be distinguished by the fact that there does not exist a plane homeomorphism which maps cutpoints of one of them onto cutpoints of the other one. In fact, laminations (defined later in the text) are based upon cutpoints. This shows the importance of the dynamics of cutpoints and provides a justification for our interest in the dynamical core of a complex polynomial.

1.2   Preliminary version of main results

Topological polynomials are topological dynamical systems that generalize complex polynomials with locally connected Julia sets restricted to their Julia sets and considered up to topological conjugacy. Note that every complex polynomial f of degree d with locally connected Julia set J gives rise to an equivalence relation $\approx$ on the unit circle $S^1 = \{z \in \mathbb{C} \,|\, |z| = 1\}$ such that two points in S^1 are equivalent if and only if the corresponding external rays land at the same point of J. Such an equivalence relation $\approx$ is forward invariant under the map $\sigma_d : S^1 \to S^1$, $\sigma_d(z) = z^d$, in the sense that $\sigma_d(z) \approx \sigma_d(w)$ whenever $z \approx w$.

The topological dynamics of f on the Julia set can be recovered from the equivalence relation $\approx$ as follows: we consider the quotient space $J_\approx = S^1/\approx$ and the map $f_\approx : J_\approx \to J_\approx$ induced by σ_d. Then the map $f_\approx : J_\approx \to J_\approx$ is topologically conjugate to the map $f|_J : J \to J$. We define a σ_d-*invariant lamination* $\sim$ as an equivalence relation on S^1 subject to certain assumptions similar to those satisfied by $\approx$ (see Section 2 for a more complete description). The set $J_\sim = S^1/\sim$ is called the *topological Julia set*. Then the map $f_\sim : J_\sim \to J_\sim$ is defined as the map induced on $J_\sim$ by σ_d and is called a *topological polynomial*. There is a natural embedding of $J_\sim$ into the plane and a natural extension of $f_\sim$ as a branched self-covering of the plane. We will write $f_\sim$ for both the topological polynomial and its extension to the plane. The components of the complement of $J_\sim$ in the plane are called *Fatou components* (of $f_\sim$).

Define an *atom* of a topological polynomial $f_\sim$ as either a singleton in $J_\sim$ or the boundary of some bounded Fatou component. A *cut-atom* is by definition an atom whose removal disconnects the topological Julia set. In particular, a point $a \in J_\sim$ is a *cutpoint* if $\{a\}$ is a cut-atom.

An atom A of $J_\sim$ is said to be a *persistent cut-atom* if all its iterated $f_\sim$-images are cut-atoms. A periodic atom A of minimal period q is said to be *rotational* if either A is a cutpoint, and $f_\sim^q$ gives rise to a non-trivial permutation of the germs of complementary components of A in $J_\sim$, or A is the boundary of some Fatou component such that $f_\sim^q : A \to A$ is of degree one and different from the identity.

A continuum $C \subset J_\sim$ is said to be *complete* if, for every bounded Fatou component U of $f_\sim$, the intersection $\mathrm{Bd}(U) \cap C$ is either empty, or a singleton, or the entire boundary $\mathrm{Bd}(U)$. Let $\mathrm{IC}_{f_\sim}(A)$ (or $\mathrm{IC}(A)$ if $\sim$ is fixed) be the smallest complete *invariant continuum in* $J_\sim$ *containing a set* $A \subset J_\sim$; we call $\mathrm{IC}(A)$ the *dynamical span of* A. Recall that the *ω-limit set* $\omega(Z)$ of a set (e.g., a singleton) $Z \subset J_\sim$ is defined as

$$\omega(Z) = \bigcap_{n=1}^{\infty} \overline{\bigcup_{i=n}^{\infty} f_\sim^i(Z)}.$$

DEFINITION 1.1 (Dynamical core). *The (dynamical) core of* $f_\sim$, *denoted by* $\mathrm{COR}(f_\sim)$ *or just* COR, *is the dynamical span of the union of the ω-limit sets of all persistent cut-atoms. The union of all periodic cut-atoms of* $f_\sim$ *is denoted by* $\mathrm{PC}(f_\sim) = \mathrm{PC}$ *and is called the* periodic core *of* $f_\sim$. *Finally, the union of all periodic rotational atoms of* $f_\sim$ *is denoted by* $\mathrm{PC}_{rot}(f_\sim) = \mathrm{PC}_{rot}$ *and is called the* periodic rotational core *of* $f_\sim$.

One of the aims of our paper is to illustrate the analogy between the dynamics of topological polynomials on their *cutpoints* and *cut-atoms* and interval dynamics. For example, it is known that for interval maps, periodic points and critical points play a significant, if not decisive, role. The main purpose of this paper is to establish similar facts for topological polynomials. To give a flavor of the main results, we give next a non-technical version of one of them.

THEOREM 1.2. *The dynamical core of* $f_\sim$ *coincides with* $\mathrm{IC}(\mathrm{PC}(f_\sim))$. *If* $J_\sim$ *is a dendritethen* $\mathrm{COR} = \mathrm{IC}(\mathrm{PC}_{rot}(f_\sim))$.

In Section 2, we state a full version of Theorem 1.2 in which we deal with several types of the dynamical core. Observe that the so-called *growing trees* [BL02a] are related to the notion of the dynamical core.

Theorem 1.2 is related to the corresponding results for maps of the interval: if g is a piecewise-monotone interval map, then the closure of the union of the limit sets of all its points coincides with the closure of the set of its periodic points (see, e.g., [B95], where this is deduced from similar results which hold for all continuous interval maps, and references therein). Theorem 1.2 shows the importance of the periodic cores of $f_\sim$. We also introduce the notion of a *critical atom* and prove in Theorem 3.15 that the dynamical cores of a topological polynomial equal the dynamical spans of critical atoms of the restriction of $f_\sim$ onto these cores.

Acknowledgements. During the work on this project, the third author (R. P.) was a graduate student at University of Alabama at Birmingham, and the fourth author visited Max Planck Institute for Mathematics (MPIM), Bonn. The latter is very grateful to MPIM for inspiring working conditions. All the authors would like to thank the referee for useful and thoughtful suggestions.

2 PRELIMINARIES

Let $\mathbb{D}$ be the open unit disk and $\widehat{\mathbb{C}}$ be the complex sphere. For a compactum $X \subset \mathbb{C}$, let $U^\infty(X)$ be the unbounded component of $\mathbb{C} \setminus X$. The topological hull of X equals $\mathrm{Th}(X) = \mathbb{C} \setminus U^\infty(X)$. Often we use $U^\infty(X)$ for $\widehat{\mathbb{C}} \setminus \mathrm{Th}(X)$, including the point at infinity. If X is a continuum, then $\mathrm{Th}(X)$ is a *non-separating* continuum, and there exists a Riemann map $\Psi_X \colon \widehat{\mathbb{C}} \setminus \overline{\mathbb{D}} \to U^\infty(X)$; we always normalize it so that $\Psi_X(\infty) = \infty$ and $\Psi'_X(z)$ tends to a positive real limit as $z \to \infty$.

Consider a polynomial P of degree $d \geq 2$ with Julia set J_P and filled-in Julia set $K_P = \mathrm{Th}(J_P)$. Extend $z^d \colon \mathbb{C} \to \mathbb{C}$ to a map θ_d on $\widehat{\mathbb{C}}$. If J_P is connected then

$\Psi_{K_P} = \Psi\colon \mathbb{C}\setminus \overline{\mathbb{D}} \to U^\infty(K_P)$ is such that $\Psi\circ\theta_d = P\circ\Psi$ on the complement of the closed unit disk [DH85a, M00]. If J_P is locally connected, then Ψ extends to a continuous function $\overline{\Psi}\colon \widehat{\mathbb{C}}\setminus \mathbb{D} \to \widehat{\mathbb{C}}\setminus K_P$, and $\overline{\Psi}\circ\theta_d = P\circ\overline{\Psi}$ on the complement of the open unit disk; thus, we obtain a continuous surjection $\overline{\Psi}\colon \mathrm{Bd}(\mathbb{D}) \to J_P$ (the *Carathéodory loop*). Identify $\mathsf{S}^1 = \mathrm{Bd}(\mathbb{D})$ with $\mathbb{R}/\mathbb{Z}$.

2.1   Laminations

Let J_P be locally connected, and set $\psi = \overline{\Psi}|_{\mathsf{S}^1}$. Define an equivalence relation $\sim_P$ on S^1 by $x \sim_P y$ if and only if $\psi(x) = \psi(y)$, and call it the (σ_d-invariant) *lamination of P*. Equivalence classes of $\sim_P$ are pairwise *unlinked*: their Euclidean convex hulls are disjoint. The topological Julia set $\mathsf{S}^1/\sim_P = J_{\sim_P}$ is homeomorphic to J_P, and the topological polynomial $f_{\sim_P}\colon J_{\sim_P} \to J_{\sim_P}$ is topologically conjugate to $P|_{J_P}$. One can extend the conjugacy between $P|_{J_P}$ and $f_{\sim_P}\colon J_{\sim_P} \to J_{\sim_P}$ to a conjugacy on the entire plane.

An equivalence relation $\sim$ on the unit circle, with similar properties as $\sim_P$, can be introduced abstractly without any reference to the Julia set of a complex polynomial. In what follows we often call classes of equivalence of $\sim$ simply $\sim$-classes.

DEFINITION 2.1 (Laminations). *An equivalence relation $\sim$ on the unit circle S^1 is called a* lamination *if it has the following properties:*

(E1) *the graph of $\sim$ is a closed subset in $\mathsf{S}^1 \times \mathsf{S}^1$;*

(E2) *if $t_1 \sim t_2 \in \mathsf{S}^1$ and $t_3 \sim t_4 \in \mathsf{S}^1$, but $t_2 \nsim t_3$, then the open straight line segments in $\mathbb{C}$ with endpoints t_1, t_2 and t_3, t_4 are disjoint;*

(E3) *each equivalence class of $\sim$ is totally disconnected.*

DEFINITION 2.2 (Laminations and dynamics). *A lamination $\sim$ is called (σ_d-)invariant if*

(D1) *$\sim$ is forward invariant: for a $\sim$-class g, the set $\sigma_d(g)$ is a $\sim$-class too;*

(D2) *for any $\sim$-class g, the map $\sigma_d\colon g \to \sigma_d(g)$ extends to S^1 as an orientation preserving covering map such that g is the full preimage of $\sigma_d(g)$ under this covering map;*

(D3) *all $\sim$-classes are finite.*

Part (D2) of Definition 2.2 has an equivalent version. A *(positively oriented) hole* (a, b) *of a compactum* $Q \subset \mathsf{S}^1$ is a component of $\mathsf{S}^1\setminus Q$ such that movement from a to b inside (a, b) is in the positive direction. Then (D2) is equivalent to the fact that for a $\sim$-class g, either $\sigma_d(g)$ is a point or for each positively oriented hole (a, b) of g the positively oriented arc $(\sigma_d(a), \sigma_d(b))$ is a positively oriented hole of $\sigma_d(g)$.

For a σ_d-invariant lamination $\sim$ we consider the *topological Julia set* $\mathsf{S}^1/\sim = J_\sim$ and the *topological polynomial* $f_\sim\colon J_\sim \to J_\sim$ induced by σ_d. The quotient map $p_\sim\colon \mathsf{S}^1 \to J_\sim$ extends to the plane with the only non-trivial fibers being the convex hulls of $\sim$-classes. Using Moore's Theorem one can extend $f_\sim$ to a branched-covering map $f_\sim\colon \mathbb{C} \to \mathbb{C}$ of the same degree. The complement of the unbounded component of $\mathbb{C}\setminus J_\sim$ is called the *filled-in topological Julia set* and is denoted by $K_\sim$. If the lamination $\sim$ is fixed, we may omit $\sim$ from the notation.

A particular case is when $J_\sim$ is a *dendrite* (a locally connected continuum containing no simple closed curve) and so $\widehat{\mathbb{C}}\setminus J_\sim$ is a simply connected neighborhood of infinity. It is easy to see that if a lamination $\sim$ has no domains (i.e., if convex hulls of all $\sim$-classes partition the entire unit disk), then the quotient space $\mathbb{S}^1/\sim$ is a dendrite.

For points $a, b \in \mathbb{S}^1$, let $\overline{ab}$ be the (perhaps degenerate) *chord* with endpoints a and b. For $A \subset \mathbb{S}^1$ let $\mathrm{Ch}(A)$ be the *convex hull* of A in $\mathbb{C}$.

DEFINITION 2.3 (Leaves and gaps of a lamination). *If A is a $\sim$-class, we call an edge $\overline{ab}$ of $\mathrm{Bd}(\mathrm{Ch}(A))$ a* leaf *(if $a = b$, we call the leaf $\overline{aa} = \{a\}$ degenerate, cf. [T85]). All points of $\mathbb{S}^1$ are also called* leaves. *Normally, leaves are denoted as above, or by a letter with a bar above it ($\overline{b}, \overline{q}$, etc), or by ℓ. The family of all leaves of $\sim$, denoted by $\mathcal{L}_\sim$, is called the* geometric lamination (geo-lamination) *generated by $\sim$. Denote the union of all leaves of $\mathcal{L}_\sim$ by $\mathcal{L}_\sim^+$. The closure of a non-empty component of $\mathbb{D} \setminus \mathcal{L}_\sim^+$ is called a* gap *of $\sim$. If G is a gap, we talk about* edges *of G; if G is a gap or leaf, we call the set $G' = \mathbb{S}^1 \cap G$ the* basis *of G.*

Extend σ_d (keeping the notation) linearly over all *individual chords* in $\overline{\mathbb{D}}$, in particularly over leaves of $\mathcal{L}_\sim$. Note that, even though the extended σ_d is not well defined on the entire disk, it is well defined on $\mathcal{L}_\sim^+$ (as well as on every individual chord in the disk).

A gap or leaf U is said to be *(pre)periodic* if $\sigma_d^{m+k}(U') = \sigma_d^m(U')$ for some $m \geq 0, k > 0$. If m can be chosen to be 0, then U is called *periodic*. If U is (pre)periodic but not periodic then it is called *preperiodic*.

Infinite gaps of a σ_d-invariant lamination $\sim$ are called *Fatou gaps*. Let G be a Fatou gap; by [K02] G is (pre)periodic under σ_d. If a Fatou gap G is periodic, then by [BL02a] its basis G' is a Cantor set and the quotient map $\psi_G \colon \mathrm{Bd}(G) \to \mathbb{S}^1$, collapsing all edges of G to points, is such that ψ_G-preimages of points are points or single leaves.

DEFINITION 2.4 (Siegel gaps and gaps of degree greater than 1). *Suppose that G is a periodic Fatou gap of minimal period n. By [BL02a] ψ_G semiconjugates $\sigma_d^n|_{\mathrm{Bd}(G)}$ to a map $\hat{\sigma}_G = \hat{\sigma} \colon \mathbb{S}^1 \to \mathbb{S}^1$ so that either (1) $\hat{\sigma} = \sigma_k \colon \mathbb{S}^1 \to \mathbb{S}^1, k \geq 2$ or (2) $\hat{\sigma}$ is an irrational rotation. In case (1) call G a* gap of degree k. *In case (2) G is called a* Siegel gap. *A (pre)periodic gap eventually mapped to a periodic gap of degree k (Siegel) is also said to be of degree k (Siegel). Domains (bounded components of the complement) of $J_\sim$ are said to be of degree k (Siegel) if the corresponding gaps of $\mathcal{L}_\sim$ are such.*

Various types of gaps and domains described in Definition 2.4 correspond to various types of atoms of $J_\sim$; as with gaps and domains, we keep the same terminology while replacing the word "gap" or "domain" by the word "atom." Thus, the boundary of a (periodic) Siegel domain is called a *(periodic) Siegel atom*, the boundary of a (periodic) Fatou domain of degree $k > 1$ is called a *(periodic) Fatou atom of degree k*, etc.

2.2 Complete statement of main results

Let $f_\sim \colon J_\sim \to J_\sim$ be a topological polynomial. Recall that an atom is either a point of $J_\sim$ or the boundary of a bounded component of $\mathbb{C} \setminus J_\sim$.

A *persistent cut-atom of degree 1* is either a non-(pre)periodic persistent cut-atom, or a (pre)periodic cut-atom of degree 1; (pre)periodic atoms of degree 1 are either (pre)periodic points or boundaries of Siegel gaps (recall, that all Siegel gaps are (pre)periodic). A *persistent cut-atom of degree $k > 1$* is a Fatou atom of degree k. A recurring theme in our paper is the fact that in some cases the dynamical span of a certain set A and the dynamical span of the subset $B \subset A$ consisting of all periodic elements of A with some extra-properties (e.g., being a periodic cut-atom, a periodic cut-atom of degree 1, etc.) coincide. We call a periodic atom A of period n and degree 1 *rotational* if $\sigma_d^n|_{p_\sim^{-1}(A)}$ has **non-zero rotation number**. The following definition extends Definition 1.1.

DEFINITION 2.5 (Dynamical cores). *The* (dynamical) core *of* $f_\sim$, *denoted* $\mathrm{COR}(f_\sim)$, *or simply* COR, *is the dynamical span of the union of the ω-limit sets of all persistent cut-atoms. The union of all periodic cut-atoms of $f_\sim$ is denoted by* $\mathrm{PC}(f_\sim) = \mathrm{PC}$ *and is called the* periodic core *of* $f_\sim$.

The (dynamical) core *of degree 1* $\mathrm{COR}_1(f_\sim)$ *of* $f_\sim$ *is the dynamical span of the ω-limit sets of all persistent cut-atoms of degree 1. The union of all periodic cut-atoms of* $f_\sim$ *of degree 1 is denoted by* $\mathrm{PC}_1(f_\sim) = \mathrm{PC}_1$ *and is called the* periodic core of degree 1 *of* $f_\sim$.

The (dynamical) rotational core *$\mathrm{COR}_{rot}(f_\sim)$ of* $f_\sim$ *is the dynamical span of the ω-limit sets of all wandering persistent cutpoints and all periodic rotational atoms. The union of all periodic rotational atoms of* $f_\sim$ *is denoted by* $\mathrm{PC}_{rot}(f_\sim) = \mathrm{PC}_{rot}$ *and is called the* periodic rotational core *of* $f_\sim$.

Clearly, $\mathrm{COR}_{rot} \subset \mathrm{COR}_1 \subset \mathrm{COR}$ and $\mathrm{PC}_{rot} \subset \mathrm{PC}_1 \subset \mathrm{PC}$. Observe that in the case of dendrites, the notions become simpler and some results can be strengthened. First, it follows that in the dendritic case we can talk about cutpoints only. Second, $\mathrm{COR} = \mathrm{COR}_1$ and $\mathrm{PC} = \mathrm{PC}_1$. A priori, then, COR_{rot} could be strictly smaller than COR; however Theorem 1.2 shows that $\mathrm{COR}_{rot} = \mathrm{COR}$. The following theorem applies to all topological polynomials and extends Theorem 1.2.

THEOREM 2.6. *The dynamical core of* $f_\sim$ *coincides with* $\mathrm{IC}(\mathrm{PC}(f_\sim))$. *The dynamical core of degree 1 of* $f_\sim$ *coincides with* $\mathrm{IC}(\mathrm{PC}_1(f_\sim))$. *The rotational dynamical core coincides with* $\mathrm{IC}(\mathrm{PC}_{rot}(f_\sim))$. *If* $J_\sim$ *is a dendrite, then* $\mathrm{COR}(f_\sim) = \mathrm{IC}(\mathrm{PC}_{rot}(f_\sim))$.

There are two other main results in Section 3. As the dynamics on Fatou gaps is simple, it is natural to consider the dynamics of gaps/leaves which never map to Fatou gaps, or the dynamics of gaps/leaves which never map to "maximal concatenations" of Fatou gaps, which we call *super-gaps* (these notions are made precise in Section 3). We prove that the dynamical span of limit sets of all persistent cut-atoms which never map to the $p_\sim$-images of super-gaps coincides with the dynamical span of all periodic rotational cut-atoms located outside the $p_\sim$-images of super-gaps (recall that $p_\sim$ is the quotient map generated by $\sim$). In fact, the "dendritic" part of Theorem 2.6 follows from that result.

A result similar to Theorem 2.6, using critical points and atoms instead of periodic ones, is proven in Theorem 3.15. Namely, an atom A is *critical* if either A is a critical point of $f_\sim$, or $f_\sim|_A$ is not one-to-one. In Theorem 3.15 we prove, in particular, that various cores of a topological polynomial $f_\sim$ coincide with the dynamical spans of critical atoms of the restriction of $f_\sim$ onto these cores.

If $J_\sim$ is a dendrite, then critical atoms are critical points and Theorem 2.6 is closely related to the interval, even unimodal case. Thus, Theorem 2.6 can be viewed as a generalization of the corresponding results for maps of the interval.

2.3   Geometric laminations

The connection between laminations, understood as equivalence relations, and the original approach of Thurston's [T85] can be explained once we introduce a few key notions. Assume that a σ_d-invariant lamination $\sim$ and its associated geometric lamination $\mathcal{L}_\sim$ are given.

Thurston's idea was to study similar collections of chords in $\mathbb{D}$ abstractly, i.e., without assuming that they are generated by an equivalence relation on the circle with specific properties.

DEFINITION 2.7 (Geometric laminations, cf. [T85]). *A* geometric prelamination *$\mathcal{L}$ is a set of (possibly degenerate) chords in $\overline{\mathbb{D}}$ such that any two distinct chords from $\mathcal{L}$ meet at most in a common endpoint; $\mathcal{L}$ is called a* geometric lamination *(geo-lamination) if all points of S^1 are elements of $\mathcal{L}$, and $\bigcup \mathcal{L}$ is closed. Elements of $\mathcal{L}$ are called* leaves *of $\mathcal{L}$ (leaves may be degenerate). The union of all leaves of $\mathcal{L}$ is denoted by $\mathcal{L}^+$.*

DEFINITION 2.8 (Gaps of geo-laminations). *Suppose that $\mathcal{L}$ is a geo-lamination. The closure of a non-empty component of $\mathbb{D} \setminus \mathcal{L}^+$ is called a* gap *of $\mathcal{L}$. Thus, given a geo-lamination $\mathcal{L}$ we obtain a cover of $\overline{\mathbb{D}}$ by gaps of $\mathcal{L}$ and (perhaps, degenerate) leaves of $\mathcal{L}$ which do not lie on the boundary of a gap of $\mathcal{L}$ (equivalently, are not isolated in $\mathbb{D}$ from either side). Elements of this cover are called* $\mathcal{L}$-sets. *Observe that the intersection of two different $\mathcal{L}$-sets is at most a leaf.*

In the case when $\mathcal{L} = \mathcal{L}_\sim$ is generated by an invariant lamination $\sim$, gaps might be of two kinds: finite gaps, which are convex hulls of $\sim$-classes, and infinite gaps, which are closures of *domains* (of $\sim$), i.e., components of $\mathbb{D} \setminus \mathcal{L}_\sim^+$ which are not interiors of convex hulls of $\sim$-classes.

DEFINITION 2.9 (Invariant geo-laminations, cf. [T85]). *A geometric lamination $\mathcal{L}$ is said to be an* invariant *geo-lamination of degree d if the following conditions are satisfied:*

1. (Leaf invariance) *For each leaf $\ell \in \mathcal{L}$, the set $\sigma_d(\ell)$ is a (perhaps degenerate) leaf in $\mathcal{L}$. For every non-degenerate leaf $\ell \in \mathcal{L}$, there are d pairwise disjoint leaves $\ell_1, \ldots, \ell_d$ in $\mathcal{L}$ such that for each i, $\sigma_d(\ell_i) = \ell$.*

2. (Gap invariance) *For a gap G of $\mathcal{L}$, the set $H = \mathrm{Ch}(\sigma_d(G'))$ is a (possibly degenerate) leaf, or a gap of $\mathcal{L}$, in which case $\sigma_d|_{\mathrm{Bd}(G)}: \mathrm{Bd}(G) \to \mathrm{Bd}(H)$ is a positively oriented composition of a monotone map and a covering map (a* monotone *map is a map such that the full preimage of any connected set is connected).*

Note that some invariant geo-laminations are not generated by equivalence relations. We will use a special extension $\sigma_{d,\mathcal{L}}^* = \sigma_d^*$ of σ_d to the closed unit disk associated with $\mathcal{L}$. On S^1 and all leaves of $\mathcal{L}$, we set $\sigma_d^* = \sigma_d$ (in the paragraph right after Definition 2.3, σ_d was extended over all chords in $\overline{\mathbb{D}}$, including leaves of $\mathcal{L}$). Otherwise, define σ_d^* on the interiors of gaps using a standard barycentric construction (see [T85]). Sometimes we lighten the notation and use σ_d instead of σ_d^*. We will mostly use the map σ_d^* in the case $\mathcal{L} = \mathcal{L}_\sim$ for some invariant lamination $\sim$.

DEFINITION 2.10 (Critical leaves and gaps). *A leaf of a lamination $\sim$ is called* critical *if its endpoints have the same image. A $\mathcal{L}_\sim$-set G is said to be* critical *if $\sigma_d|_{G'}$ is k-to-1 for some $k > 1$. For example, a periodic Siegel gap is a non-critical $\sim$-set, on whose basis the first return map is not one-to-one because there must be critical leaves in the boundaries of gaps from its orbit. We define* precritical *and* (pre)critical *objects similarly to how (pre)periodic and preperiodic objects are defined in the second paragraph after Definition 2.3: i.e., a precritical object maps to a critical object but is not critical itself. On the other hand, a (pre)critical object is either precritical, or critical.*

We need more notation. Let $a, b \in \mathbb{S}^1$. By $[a, b], (a, b)$, etc., we mean the appropriate *positively oriented* circle arcs from a to b, and by $|I|$ the length of an arc I in $\mathbb{S}^1$ normalized so that the length of $\mathbb{S}^1$ is 1.

2.4 Stand-alone gaps and their basic properties

DEFINITION 2.11 (Return time and related notions). *Let $f: X \to X$ be a self-mapping of a set X. For a set $G \subset X$, define the return time (to G) of $x \in G$ as the least positive integer n_x such that $f^{n_x}(x) \in G$ or infinity if there is no such integer. Let $n = \min_{y \in G} n_y$, define $D_G = \{x: n_x = n\}$, and call the map $f^n: D_G \to G$ the return map (of G).*

For example, if G is the boundary of a periodic Fatou domain of period n of a topological polynomial $f_\sim$ whose images are all pairwise disjoint until $f^n_\sim(G) = G$, then $D_G = G$ and the corresponding return map on $D_G = G$ is the same as $f^n_\sim$.

We have already introduced the notion of a gap of a lamination or of a geo-lamination. Next we will describe a closed convex set in $\overline{\mathbb{D}}$ which has all the properties of a gap of a geo-lamination but for which no corresponding lamination is specified.

DEFINITION 2.12 (Stand-alone gaps). *If $A \subset \mathbb{S}^1$ is a closed set such that all the sets $\mathrm{Ch}(\sigma^i(A))$ are pairwise disjoint, then A is called wandering. If there exists $n \geq 1$ such that all the sets $\mathrm{Ch}(\sigma^i_d(A)), i = 0, \ldots, n-1$ have pairwise disjoint interiors while $\sigma^n_d(A) = A$, then A is called periodic of period n. If there exists $m > 0$ such that all $\mathrm{Ch}(\sigma^i(A)), 0 \leq i \leq m + n - 1$ have pairwise disjoint interiors and $\sigma^m_d(A)$ is periodic of period n, then we call A preperiodic. Moreover, suppose that $|A| \geq 3$, A is wandering, periodic or preperiodic, and for every $i \geq 0$ and every hole (a, b) of $\sigma^i_d(A)$ either $\sigma_d(a) = \sigma_d(b)$ or the positively oriented arc $(\sigma_d(a), \sigma_d(b))$ is a hole of $\sigma^{i+1}_d(A)$. Then we call A (and $\mathrm{Ch}(A)$) a σ_d-stand-alone gap.*

Recall that in Definition 2.4 we defined Fatou gaps G, of various degrees as well as Siegel gaps. Given a periodic Fatou gap G, we also introduced the monotone map ψ_G, which semiconjugates $\sigma_d|_{\mathrm{Bd}(G)}$ and the appropriate model map $\hat{\sigma}_G: \mathbb{S}^1 \to \mathbb{S}^1$. This construction can be also done for stand-alone Fatou gaps G.

Indeed, consider the basis $G \cap \mathbb{S}^1 = G'$ of G (see Definition 2.3) as a subset of $\mathrm{Bd}(G)$. It is well known that G' coincides with the union $A \cup B$ of two well-defined sets, where A is a Cantor subset of G' or an empty set and B is a countable set. In the case when $A = \varnothing$, the map ψ_G simply collapses $\mathrm{Bd}(G)$ to a point. However, if $A \neq \varnothing$, one can define a semiconjugacy $\psi_G: \mathrm{Bd}(G) \to \mathbb{S}^1$ which collapses all holes of G' to points. As in Definition 2.4, the map ψ_G semiconjugates $\sigma_d|_{\mathrm{Bd}(G)}$ to a circle map, which is either an irrational rotation or the map $\sigma_k, k \geq 2$. Depending on the type of this map we can introduce for periodic infinite stand-alone gaps terminology similar to Definition 2.4. In particular, if $\sigma_d|_{\mathrm{Bd}(G)}$ is semiconjugate to $\sigma_k, k \geq 2$ we say that G is a *stand-alone Fatou gap of degree k*.

DEFINITION 2.13 (Rotational sets). *If G is a periodic stand-alone gap such that G' is finite and contains no fixed points of the first return map, then G is said to be a finite (periodic) rotational set. Finite rotational sets and Siegel gaps G are called (periodic) rotational sets. If such G is invariant, we call it an invariant rotational set.*

The maps σ_k serve as models of return maps of periodic gaps of degree $k \geq 2$. For rotational sets, models of return maps are non-trivial rotations.

DEFINITION 2.14 (Rotation number). *A number τ is said to be the* rotation number *of a periodic set G if for every $x \in G'$ the circular order of points in the orbit of x under the return map of G' is the same as the order of points $0, \mathrm{Rot}_\varnothing(0), \ldots$, where $\mathrm{Rot}_\varnothing : \mathbb{S}^1 \to \mathbb{S}^1$ is the rigid rotation by the angle τ.*

It is easy to see that to each rotational set G, one can associate its well-defined rotation number $\tau_G = \tau$ (in the case of a finite rotational set, the property that endpoints of holes are mapped to endpoints of holes implies that the circular order on G' remains unchanged under σ). Since G' contains no points which are fixed under its return map by our assumption, $\tau \neq 0$. Given a topological polynomial $f_\sim$ and an $f_\sim$-periodic point x of minimal period n, we can associate to x the rotation number $\rho(x)$ of σ_d^n restricted to the $\sim$-class $p_\sim^{-1}(x)$, corresponding to x (recall, that $p_\sim : \mathbb{S}^1 \to J_\sim = \mathbb{S}^1/\sim$ is the quotient map associated to $\sim$); then if $\rho(x) \neq 0$ the set $p_\sim^{-1}(x)$ is rotational, but if $\rho(x) = 0$ then the set $p_\sim^{-1}(x)$ is not rotational.

The following result allows one to find fixed stand-alone gaps or points of specific types in some parts of the disk; for the proof see [BFMOT10]. It is similar in spirit to a fixed-point result by Goldberg and Milnor [GM93].

THEOREM 2.15. *Let $\sim$ be a σ_d-invariant lamination. Consider the topological polynomial $f_\sim$ extended over $\mathbb{C}$. Suppose that $e_1, \ldots, e_m \in J_\sim$ are m points, and $X \subset K_\sim$ is a component of $K_\sim \setminus \{e_1, \ldots, e_m\}$ such that for each i we have $e_i \in \overline{X}$ and either $f_\sim(e_i) = e_i$ and the rotation number at e_i is zero, or $f_\sim(e_i)$ belongs to the component of $K_\sim \setminus \{e_i\}$ which contains X. Then at least one of the following claims holds:*

1. *X contains an invariant domain of degree $k > 1$;*

2. *X contains an invariant Siegel domain;*

3. *$X \cap J_\sim$ contains a fixed point with non-zero rotation number.*

For dendritic topological Julia sets $J_\sim$ the claim is easier as cases (1) and (2) are impossible. Thus, in the dendritic case Theorem 2.15 implies that there exists a rotational fixed point in $X \cap J_\sim$.

Let U be the convex hull of a closed subset of $\mathbb{S}^1$. For every edge ℓ of U, let $H_U(\ell)$ denote the hole of U that shares both endpoints with ℓ (if U is fixed, we may drop the subscript in the preceding notation). Notice that in the case when U is a chord, there are two ways to specify the hole. The hole $H_U(\ell)$ is called the *hole of U behind (at) ℓ.* In this situation we define $|\ell|_U$ as $|H_U(\ell)|$.

LEMMA 2.16. *Suppose that $\ell = \overline{xy}$ is a non-invariant leaf such that there exists a component Q of the complement of its orbit whose closure contains $\sigma_d^n(\ell)$ for all $n \geq 0$. Then the leaf ℓ is either (pre)critical or (pre)periodic.*

PROOF. Suppose that ℓ is neither (pre)periodic nor (pre)critical. Then it follows that there are infinitely many leaves in the orbit of ℓ such that both complementary arcs of the set of their endpoints are of length greater than $1/2d$. Since the corresponding holes of Q are pairwise disjoint, we get a contradiction. $\qquad\square$

3 DYNAMICAL CORE

In this section we fix $\sim$, which is a σ_d-invariant lamination, study the dynamical properties of the topological polynomial $f_\sim : J_\sim \to J_\sim$, and discuss its dynamical core $\mathrm{COR}(f_\sim)$. For brevity, we write f, J, p, COR for $f_\sim$, $J_\sim$, $p_\sim : \mathbb{S}^1 \to J$, $\mathrm{COR}(f_\sim)$, respectively, throughout the rest of the paper. Note that, by definition, every topological Julia set J in this section is locally connected.

3.1 Super-gaps

Let ℓ be a leaf of $\mathcal{L}_\sim$. We equip $\mathcal{L}_\sim$ with the topology induced by the Hausdorff metric. Then $\mathcal{L}_\sim$ is a compact and metric space. Suppose that a leaf ℓ has a neighborhood (in $\mathcal{L}_\sim$), which contains at most countably many leaves of $\mathcal{L}_\sim$. Call such leaves *countably isolated* and denote the family of all such leaves $\mathrm{CI}_\sim = \mathrm{CI}$. Clearly, CI is open in $\mathcal{L}_\sim$. Moreover, CI is countable. To see this, note that CI, being a subset of $\mathcal{L}_\sim$, is second countable and, hence, Lindelöf. Hence there exists a countable cover of CI, all of whose elements are countable, and CI is countable as desired. In terms of dynamics, CI is backward invariant and almost forward invariant (it is forward invariant except for critical leaves in CI because their images are points which are never countably isolated).

It can be shown that if we remove all leaves of CI from $\mathcal{L}_\sim$ (this is in the spirit of *cleaning* of geometric laminations [T85]), the remaining leaves (if any) form an invariant geometric lamination $\mathcal{L}_\sim^c$ ("c" coming from "countable cleaning"). One way to see this is to use an alternative definition given in [BMOV11]. A geo-lamination (initially not necessarily invariant in the sense of Definition 2.9) is called a *sibling d-invariant lamination* or just *sibling lamination* if (a) for each $\ell \in \mathcal{L}$, either $\sigma_d(\ell) \in \mathcal{L}$ or $\sigma_d(\ell)$ is a point in $\mathbb{S}^1$, (b) for each $\ell \in \mathcal{L}$ there exists a leaf $\ell' \in \mathcal{L}$ with $\sigma_d(\ell') = \ell$, and (c) for each $\ell \in \mathcal{L}$ with non-degenerate image $\sigma_d(\ell)$ there exist d disjoint leaves $\ell_1, \ldots, \ell_d$ in $\mathcal{L}$ with $\ell = \ell_1$ and $\sigma_d(\ell_i) = \sigma_d(\ell)$ for all i. By [BMOV11, Theorem 3.2], sibling-invariant laminations are invariant in the sense of Definition 2.9. Now, observe that $\mathcal{L}_\sim$ is sibling invariant. Since CI is open and contains the full grand orbit of any leaf in it which never collapses to a point and the full backward orbit of any critical leaf in it, $\mathcal{L}_\sim^c$ is also sibling invariant. Thus, by [BMOV11, Theorem 3.2] $\mathcal{L}_\sim^c$ is invariant in the sense of Definition 2.9. Infinite gaps of $\mathcal{L}_\sim^c$ are called *super-gaps of* $\sim$. Note that all finite gaps of $\mathcal{L}_\sim^c$ are also finite gaps of $\mathcal{L}_\sim$.

Let $\mathcal{L}_\sim^0 = \mathcal{L}_\sim$ and define $\mathcal{L}_\sim^k$ inductively by removing all isolated leaves from $\mathcal{L}_\sim^{k-1}$.

LEMMA 3.1. *There exists n such that $\mathcal{L}_\sim^n = \mathcal{L}_\sim^c$; moreover, $\mathcal{L}_\sim^c$ contains no isolated leaves.*

PROOF. We first show that there exists n such that $\mathcal{L}_\sim^{n+1} = \mathcal{L}_\sim^n$ (i.e., $\mathcal{L}_\sim^n$ contains no isolated leaves). We note that increasing i may only decrease the number of infinite periodic gaps and decrease the number of finite critical objects (gaps or leaves) in $\mathcal{L}_\sim^i$. Therefore, we may choose m so that $\mathcal{L}_\sim^m$ has a minimal number of infinite periodic gaps and finite critical objects. If $\mathcal{L}_\sim^m$ has no isolated leaves, then we may choose $n = m$. Otherwise, we will show that we may choose $n = m + 1$.

Suppose that $\mathcal{L}_\sim^m$ has an isolated leaf ℓ. Then ℓ is a common edge of two gaps U and V. Since finite gaps of $\mathcal{L}_\sim$ (and, hence, of all $\mathcal{L}_\sim^i$) are disjoint, we may assume that U is infinite. Moreover, it must be that V is finite. Indeed, suppose that V is infinite and consider two cases. First assume that there is a minimal j such that $\sigma_d^j(U) = \sigma_d^j(V)$. Then $\sigma_d^{j-1}(\ell)$ is critical and isolated in $\mathcal{L}_\sim^m$. Hence $\mathcal{L}_\sim^{m+1}$ has fewer finite critical objects than $\mathcal{L}_\sim^m$, a contradiction with the choice of m. On the other

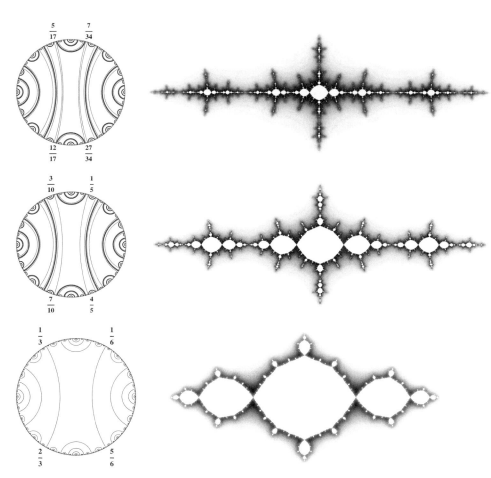

FIGURE 3.2: An example of 2-step cleaning for quadratic laminations. Above: the quadratic lamination $\mathcal{L}^0$ with a quadratic gap between the leaves $\overline{\frac{5}{17}\,\frac{12}{17}}$ and $\overline{\frac{7}{34}\,\frac{27}{34}}$, and the corresponding Julia set (basilica tuned with basilica tuned with basilica). In the middle: the first cleaning $\mathcal{L}^1$ of $\mathcal{L}^0$ — the lamination with a quadratic gap between the leaves $\overline{\frac{1}{5}\,\frac{4}{5}}$ and $\overline{\frac{3}{10}\,\frac{7}{10}}$, and the corresponding Julia set (basilica tuned with basilica). Below: the second cleaning $\mathcal{L}^2$ of $\mathcal{L}^0$ — the lamination with a quadratic gap between the leaves $\overline{\frac{1}{3}\,\frac{2}{3}}$ and $\overline{\frac{1}{6}\,\frac{5}{6}}$, and the corresponding Julia set (basilica). Note that the next cleaning $\mathcal{L}^3$ is empty and coincides with $\mathcal{L}^c$. Thus, the unique supergap of $\mathcal{L}^0$ coincides with the entire unit disk.

hand, if U and V never have the same image, then in $\mathcal{L}_{\sim}^{m+1}$ their periodic images will be joined. Then $\mathcal{L}_{\sim}^{m+1}$ would have a periodic gap containing both such images which contradicts the choice of m with the minimal number of infinite periodic gaps. Thus, V is finite.

Furthermore, the other edges of V are not isolated in any lamination $\mathcal{L}_{\sim}^{t}, t \geq m$. For if some other edge $\bar{q}$ of V were isolated in $\mathcal{L}_{\sim}^{t}$, then it would be an edge of an infinite gap H, contained in the same gap of $\mathcal{L}_{\sim}^{t+1}$ as U. If $\sigma_d^r(H) = \sigma_d^r(U)$ for some r, then V is (pre)critical and the critical image of V is absent from $\mathcal{L}_{\sim}^{t+1}$, a contradiction with the choice of m. Otherwise, periodic images of H and U will be contained in a bigger gap of $\mathcal{L}_{\sim}^{t}$, decreasing the number of periodic infinite gaps and again contradicting the choice of m. By definition this implies that all edges of V except for ℓ stay in all laminations $\mathcal{L}_{\sim}^{t}, t \geq m$ and that $\mathcal{L}_{\sim}^{m+1}$ has no isolated leaves.

Choose n so that $\mathcal{L}_{\sim}^{n}$ contains no isolated leaves. It is well-known that a compact metric space without isolated points is locally uncountable. Hence $\mathcal{L}_{\sim}^{n} = \mathcal{L}_{\sim}^{c}$, as desired. $\qquad \square$

LEMMA 3.3. *If $\sim$ is a lamination, then the following holds.*

1. *Every leaf of $\mathcal{L}_{\sim}$ inside a super-gap G of $\sim$ is (pre)periodic or (pre)critical; every edge of a super-gap is (pre)periodic.*

2. *Every edge of any gap H of $\mathcal{L}_{\sim}^{c}$ is not isolated in $\mathcal{L}_{\sim}^{c}$ from outside of H; all gaps of $\mathcal{L}_{\sim}^{c}$ are pairwise disjoint. Moreover, gaps of $\mathcal{L}_{\sim}^{c}$ are disjoint from leaves which are not their edges.*

3. *There are no infinite concatenations of leaves in $\mathcal{L}_{\sim}^{c}$. Moreover, the geo-lamination $\mathcal{L}_{\sim}^{c}$ gives rise to a lamination $\sim^{c}$ such that the only difference between $\mathcal{L}_{\sim}^{c}$ and $\mathcal{L}_{\sim^{c}}$ is as follows: it is possible that one edge of certain finite gaps of $\sim^{c}$ is a leaf passing inside an infinite gap of $\mathcal{L}_{\sim}^{c}$.*

4. *Any periodic Siegel gap is a proper subset of its super-gap.*

PROOF. (1) An isolated leaf ℓ of any lamination $\mathcal{L}_{\sim}^{k}$ is either (pre)periodic or (pre)-critical. Indeed, since two finite gaps of $\mathcal{L}_{\sim}^{k}$ are not adjacent, ℓ is an edge of an infinite gap V. Then by Lemma 2.16, ℓ is (pre)critical or (pre)periodic. Since in the process of constructing $\mathcal{L}_{\sim}^{c}$ we remove isolated leaves of laminations $\mathcal{L}_{\sim}^{k}$, we conclude that all leaves inside a super-gap of $\sim$ (i.e., a gap of $\mathcal{L}_{\sim}^{c}$) are either (pre)periodic or (pre)critical.

Let G be a gap of $\sim$. Since by Lemma 3.1 $\mathcal{L}_{\sim}^{c}$ contains no isolated leaves, every edge of G is a limit of leaves from outside of G. Hence the only gaps of $\mathcal{L}_{\sim}^{c}$ which can contain a critical leaf in their boundaries are those which collapse to a single point (if a gap H of $\mathcal{L}_{\sim}^{c}$ has a critical edge $\ell = \lim \ell_i$, then leaves $\sigma_d(\ell_i)$ separate the point $\sigma_d(\ell)$ from the rest of the circle, which implies that $\sigma_d(H) = \sigma_d(\ell)$ is a point). Thus, super-gaps have no critical edges, and by Lemma 2.16 all their edges are (pre)periodic.

(2) Every edge of any gap H of $\mathcal{L}_{\sim}^{c}$ is not isolated in $\mathcal{L}_{\sim}^{c}$ from outside of H by Lemma 3.1. Moreover, no point $a \in S^1$ can be an endpoint of more than two leaves of $\mathcal{L}_{\sim}$ because $\mathcal{L}_{\sim}$ is a geometric lamination generated by an equivalence relation. Hence, no point $a \in S^1$ can be an endpoint of more than two leaves of $\mathcal{L}_{\sim}^{c}$ either. This implies that if G is a gap, then it is disjoint from all leaves which are not

its edges. Indeed, suppose otherwise. Then there exists a vertex a of G and a leaf $\ell_1 = \overline{ax}$ which is not an edge of G. Then there is an edge ℓ_2 of G which emanates from a and by the preceding there is a gap H with edges ℓ_1 and ℓ_2 (H is squeezed in-between G and ℓ_1). This implies that ℓ_2 is isolated, a contradiction. Thus, two gaps cannot have edges which "touch" at an endpoint. The beginning of the paragraph implies that two gaps cannot have a common edge either. We conclude that all gaps of $\mathcal{L}_\sim^c$ are pairwise disjoint and that all gaps of $\mathcal{L}_\sim^c$ are disjoint from all leaves which are not their edges.

(3) There are no infinite concatenations of leaves in $\mathcal{L}_\sim^c$ because there are no such concatenations in $\mathcal{L}_\sim$. Now it is easy to see that the geometric lamination $\mathcal{L}_\sim^c$ gives rise to a lamination (equivalence relation), which we denote $\sim^c$. Two points $a, b \in \mathbb{S}^1$ are said to be $\sim^c$-*equivalent* if there exists a finite concatenation of leaves of $\mathcal{L}_\sim^c$ connecting a and b. Since all leaves of $\mathcal{L}_\sim^c$ are non-isolated, it follows that the geo-lamination $\mathcal{L}_\sim^c$ and the geo-lamination $\mathcal{L}_{\sim^c}$ associated to $\sim^c$ can differ only as claimed.

(4) Clearly a Siegel gap U is contained in a super-gap. To see that it does not coincide with a super-gap, observe that there exists a non-negative integer k such that $\sigma_d^k(U)$ has a critical edge. Then by (1), $\sigma_d^k(U)$ (and hence U itself) is *properly* contained in a super-gap. $\qquad\square$

PROPOSITION 3.4. *If X is a persistent cut-atom of J of degree 1 such that $p^{-1}(X)$ is a subset of some super-gap of $\sim$, then either X is the boundary of a Siegel domain, or X is a (pre)periodic point which eventually maps to a periodic cutpoint. In any case, X eventually maps to* PC_1.

PROOF. We may assume that $X = x$ is a persistent cutpoint. Then the $\sim$-class $p^{-1}(x)$ is non-trivial. If the boundary of this $\sim$-class consists of (pre)critical leaves only, then the entire class gets eventually collapsed, which is a contradiction with $f^n(x)$ being cut-atoms for all $n \geq 0$. Therefore, there is a leaf ℓ on the boundary of $p^{-1}(x)$ that is not (pre)critical. Then this leaf is (pre)periodic by Lemma 3.3; hence $p^{-1}(x)$ is also (pre)periodic and eventually maps to a periodic gap or leaf. $\qquad\square$

3.2 Proof of Theorem 2.6

Lemma 3.5 studies intersections between atoms and complete invariant continua.

LEMMA 3.5. *Let $X \subset J$ be an invariant complete continuum and A be an atom intersecting X but not contained in X. Then $A \cap X = \{x\}$ is a singleton, A is the boundary of a Fatou domain, and one of the following holds: (1) for some k we have*

$$f^i(A) \cap X = \{f^i(x)\},\ i < k$$

and $f^k(A) \subset X$, or (2) $f^i(A) \cap X = \{f^i(x)\}$ for all i and there exists the smallest n such that $f^n(A)$ is the boundary of a periodic Fatou domain of degree $r > 1$ with $f^n(x)$ being a point of $f^n(A)$ fixed under the return map.

PROOF. Since X is complete, we may assume that $A \cap X = \{x\}$ is a singleton and A is the boundary of a Fatou domain such that $f^i(A) \cap X = \{f^i(x)\}$ for all i. Choose the smallest n such that $f^n(A)$ is periodic. If $f^n(x)$ is not fixed by the return map of $f^n(A)$, then another point from the orbit of x belongs to $f^n(A) \cap X$, a contradiction. $\qquad\square$

Lemma 3.6 rules out certain dynamical behavior of points.

LEMMA 3.6. *Suppose that $x \in J$ is a non-(pre)critical cutpoint. Then there exists $n \geq 0$ such that at least two components of $J \setminus \{f^n(x)\}$ contain forward images of $f^n(x)$.*

PROOF. In the proof we will work with an equivalent statement. It can be given as follows: there exists no non-(pre)critical non-degenerate $\sim$-class X such that for every n, *all* the sets $f^{n+k}(X), k > 0$ are contained in the same hole of $f^n(X)$.

By way of contradiction suppose that such $\sim$-class X exists. Let us show that then the iterated images of X cannot converge (along a subsequence of iterations) to a critical leaf ℓ. Indeed, suppose that $\sigma_d^{n_k}(X) \to \ell$ so that $\sigma_d^{n_k+1}(X)$ separates $\sigma_d^{n_k}(X)$ from ℓ. Then $\sigma_d^{n_k+1}(X) \to \sigma_d(\ell)$, where $\sigma_d(\ell)$ is a point of S^1. Since

$$\sigma_d^{n_{k+1}+1}(X)$$

separates $\sigma_d(\ell)$ from $\sigma_d^{n_k+1}(X)$ for a sufficiently large k, it follows by the assumption that the entire orbit of $\sigma^{n_k}(X)$ must be contained in a small component of $\overline{\mathbb{D}} \setminus \sigma^{n_k}(X)$, containing $\sigma_d(\ell)$. As this can be repeated for all sufficiently large k, we see that the limit set of X has to coincide with the point $\sigma_d(\ell)$, a contradiction. Hence X contains no critical leaves in its limit set.

Note that the assumptions of the lemma imply that X is wandering. By [C04] if X is not a leaf, then it contains a critical leaf in its limit set. This implies that X must be a leaf. For each image $\sigma_d^n(X)$ let Q_n be the component of $\overline{\mathbb{D}} \setminus \sigma_d^n(X)$ containing the rest of the orbit of X. Let $W_n = \bigcap_{i=0}^n \overline{Q_i}$. Then W_n is a set whose boundary consists of finitely many leaves — images of X alternating with finitely many circle arcs. On the next step, the image $\sigma_d^{n+1}(X)$ of X is contained in W_n and becomes a leaf on the boundary of $W_{n+1} \subset W_n$.

Consider the set $W = \bigcap W_n$. For an edge ℓ of W, let $H_W(\ell)$ be the hole of W behind ℓ. When saying that a certain leaf is contained in $H_W(\ell)$, we mean that its endpoints are contained in $H_W(\ell)$, or, equivalently, that the leaf is contained in the convex hull of $H_W(\ell)$. If W is a point or a leaf, then the assumptions on the dynamics of X made in the lemma imply that X converges to W but never maps to W. Clearly, this is impossible. Thus, we may assume that W is a non-degenerate convex subset of $\overline{\mathbb{D}}$ whose boundary consists of leaves and possibly circle arcs. The leaves in $\mathrm{Bd}(W)$ can be of two types: limits of sequences of images of X (if ℓ is a leaf like that, then images of X which converge to ℓ must be contained in $H_W(\ell)$), and images of X. It follows that the limit leaves from this collection form the entire limit set of X; moreover, by the preceding there are no critical leaves among them.

Let us show that this leads to a contradiction. First assume that among boundary leaves of W there is a limit leaf $\bar{q} = \overline{xy}$ of the orbit of X (here $(x, y) = H_W(\bar{q})$ is the hole of W behind $\bar{q}$). Let us show that $\bar{q}$ is (pre)periodic. Indeed, since $\bar{q}$ is approached from the outside of W by images of X and since all images of X are disjoint from W, it follows that $(\sigma_d(x), \sigma_d(y))$ is the hole of W behind $\sigma_d(\bar{q})$. Then by Lemma 2.16 and because there are no critical leaves on the boundary of W (by the first paragraph of the proof) we see that $\bar{q}$ is (pre)periodic. Let ℓ be an image of $\bar{q}$ which is periodic. Since ℓ is a repelling leaf, we see that images of X approaching ℓ from within $H_W(\ell)$ are repelled farther away from ℓ inside $H_W(\ell)$. Clearly, this contradicts the properties of X.

Now assume that there are no boundary leaves of W which are limits of images of $X = \overline{uv}$. Then all boundary leaves of W are images of X. Let us show that then there exists N such that for any $i \geq N$ we have that if the hole $H_W(\ell)$ of W behind $\ell = \sigma_d^i(X)$ is (s, t), then $H_W(\sigma_d(\ell)) = (\sigma_d(s), \sigma_d(t))$. Indeed, first we show that if $H_W(\ell) = (s, t)$ while $H_W(\sigma_d(\ell)) = (\sigma_d(t), \sigma_d(s))$, then (s, t) contains a σ_d-fixed point. To see that, observe that in that case σ_d-image of $[s, t]$ contains $[s, t]$ and the images of s, t do not belong to (s, t). This implies that there exists a σ_d-fixed point in $[s, t]$. Since there are finitely many σ_d-fixed points, it is easy to see that the desired number N exists. Now we can apply Lemma 2.16 which implies that X is either (pre)periodic or (pre)critical, a contradiction. $\square$

Now we study dynamics of super-gaps and the map as a whole. Our standing assumption from here through Theorem 3.9 is that $\sim$ is a lamination and $\mathcal{L}_\sim$ is such that $\mathcal{L}_\sim^c$ is not empty (equivalently, $\mathbb{S}^1$ is not a super-gap). Denote the union of all periodic super-gaps by SG. Then there are finitely many super-gaps in SG, none of which coincides with $\mathbb{S}^1$, and, by Lemma 3.3, they are disjoint. Choose $N_\sim = N$ as the minimal number such that all periodic super-gaps and their periodic edges are σ_d^N-fixed. By Lemma 3.3 each super-gap has at least one σ_d^N-fixed edge and all its edges eventually map to σ_d^N-fixed edges.

Consider a component A of $J \setminus p(SG)$. There are several σ_d^N-fixed super-gaps bordering $p^{-1}(A)$, and each such super-gap has a unique well-defined edge contained in $p^{-1}(A)$. If all these edges are σ_d^N-fixed, we call A *settled*. By Theorem 2.15, a settled component A contains an element of $\text{PC}_{rot} \setminus p(SG)$ denoted by y_A. In this way, we associate elements of $\text{PC}_{rot} \setminus p(SG)$ to all settled components of $J \setminus p(SG)$.

LEMMA 3.7. *If ℓ is not a σ_d^N-fixed edge of a σ_d^N-fixed super-gap H, then the component of $J \setminus p(\ell)$ which contains $p(H)$ contains a settled component of $J \setminus p(SG)$. In particular, settled components exist.*

PROOF. Set $\ell_0 = \ell$. Choose a σ_d^N-fixed edge ℓ_0' of H and a component B_0 of $J \setminus p(SG)$ such that ℓ_0' is contained in the closure of $p^{-1}(B_0)$. If B_0 is settled, we are done. Otherwise find a super-gap H_1 with an edge ℓ_1 such that ℓ_1 is not σ_d^N-fixed and borders $p^{-1}(B_0)$, then proceed with ℓ_1 as before with ℓ_0.

In the end we will find a settled component of $J \setminus p(SG)$ in the component of $J \setminus p(\ell)$ containing $p(H)$. Indeed, on each step we find a new σ_d^N-fixed super-gap different from the preceding one. Since there are finitely many σ_d^N-fixed super-gaps, we either stop at some point or form a cycle. The latter is clearly impossible. Thus, there exists a non-empty collection of settled components A. $\square$

Given any subcontinuum $X \subset J$ and $x \in X$, a component of $X \setminus \{x\}$ is called an X-*leg of* x. An X-leg of x is called *essential* if x eventually maps into this leg. An X-leg is said to be *critical* if it contains at least one critical atom; otherwise a leg is called *non-critical*.

Recall that by Definition 2.13 we call a periodic atom *rotational* if it is of degree 1 and its rotation number is not zero. Then $\text{PC}_{rot} \setminus p(SG)$ is the set of all periodic rotational atoms x which are not contained in $p(SG)$ (any such x is a point by Lemma 3.3). Finally, define COR_s as the dynamical span of the limit sets of all persistent cut-atoms x (equivalently, cutpoints) which never map into $p(SG)$. Observe that if J is a dendrite, then $\text{COR}_s = \text{COR}$.

LEMMA 3.8. *Every x-essential J-leg of every point $x \in J$ contains a point of*

$$\text{PC}_{rot} \setminus p(SG).$$

PROOF. Let L be an x-essential J-leg of x. Denote by A the component of $L \setminus p(SG)$ which contains x in its closure. If L contains no p-images of σ_d^N-fixed super-gaps (which implies that $L = A$), then the claim follows from Theorem 2.15 applied to $\overline{A}$. Otherwise there are finitely many super-gaps $U_1, \ldots, U_t$ such that $p(U_i)$ borders A. For each i let us take a point $x_i' \in \mathrm{Bd}(p(U_i))$ that separates x from the rest of $p(U_i)$. If all the points x_i' are g-fixed, then we are done by Theorem 2.15 applied to $\overline{A}$. If there exists i such that the point x_i' is not g-fixed, then, by Lemma 3.7, x_i' separates the point x from some settled component B, which in turn contains an element $y_B \in \mathrm{PC}_{rot} \setminus p(SG)$. Thus, in any case *every x-essential leg of $x \in J$ contains a point of* $\mathrm{PC}_{rot} \setminus p(SG)$ and the lemma is proven.   □

Observe that by Lemma 3.8, for a non-(pre)periodic persistent cutpoint, x there exists n such that $f^n(x)$ separates two points of $\mathrm{PC}_{rot} \setminus p(SG)$. This follows from Lemma 3.6, since some iterated g-image of x has at least two x-essential J-legs.

THEOREM 3.9. *If x is a persistent cutpoint that is never mapped to $p(SG)$, then there is $n \geq 0$ such that $f^n(x)$ separates two points of $\mathrm{PC}_{rot} \setminus p(SG)$ and is a cutpoint of $\mathrm{IC}(\mathrm{PC}_{rot} \setminus p(SG))$ so that $\mathrm{COR}_s = \mathrm{IC}(\mathrm{PC}_{rot} \setminus p(SG))$.*

Moreover, there exist infinitely many persistent periodic rotational cutpoints outside $p(SG)$, $\mathrm{COR}_s \subset \mathrm{COR}_{rot}$, and any periodic cutpoint outside $p(SG)$ separates two points of $\mathrm{PC}_{rot} \setminus p(SG)$ and is a cutpoint of COR_s, of COR_1 and of COR.

PROOF. By Lemma 3.8 and the remark after that lemma we need to consider only the case of a g-periodic cutpoint y of J outside $p(SG)$ (a priori it may happen that y above is an endpoint of COR_s). We want to show that y separates two points of $\mathrm{PC}_{rot} \setminus p(SG)$. Indeed, a certain power $(\sigma_d^N)^k$ of σ_d^N fixes $p^{-1}(y)$ and has rotation number zero on $p^{-1}(y)$. Choose an edge ℓ of $p^{-1}(y)$ and consider the component B of $J \setminus \{y\}$ such that $\overline{p^{-1}(B)}$ contains ℓ.

Then $(\sigma_d^N)^k$ fixes ℓ while leaves and gaps in $\overline{p^{-1}(B)}$ close to ℓ are repelled away from ℓ inside $\overline{p^{-1}(B)}$ by $(\sigma_d^N)^k$. Hence their p-images are repelled away from y inside B by the map g^k. By Lemma 3.8, there is an element (a point)

$$t_B \in \mathrm{PC}_{rot} \setminus p(SG) \in B.$$

As this applies to all edges of $p^{-1}(y)$, we see that y separates two points of

$$\mathrm{PC}_{rot} \setminus p(SG).$$

As $\mathrm{PC}_{rot} \setminus p(SG) \subset \mathrm{COR}_s \subset \mathrm{COR}$, this proves that any periodic cutpoint outside $p(SG)$ separates two points of $\mathrm{PC}_{rot} \setminus p(SG)$ and is a cutpoint of COR_s (and, hence, of COR_1 and of COR). This also proves that for a (pre)periodic persistent cutpoint x, there exists n such that $f^n(x)$ separates two points of $\mathrm{PC}_{rot} \setminus p(SG)$; by the preceding, it suffices to take r such that $g^r(x) = f^{Nr}(x)$ is periodic and set $n = Nr$.

Let us prove that there are infinitely many points of PC_{rot} in any settled component A. Indeed, choose a g-periodic point $y \in A$ as before such that a certain power $(\sigma_d^N)^k$ of σ_d^N which fixes $p^{-1}(y)$ has rotation number zero on $p^{-1}(y)$. Choose an edge ℓ of $p^{-1}(y)$ and consider the component B of $A \setminus \{y\}$ such that $\overline{p^{-1}(B)}$ contains ℓ. Then leaves and gaps in $\overline{p^{-1}(B)}$, which are close to ℓ, are repelled away from ℓ inside $\overline{p^{-1}(B)}$ by $(\sigma_d^N)^k$. Hence their p-images are repelled away from y inside B by g^k. By Theorem 2.15, this implies that there exists a g^k-fixed point $z \in B$

with non-zero rotation number. Replacing A by a component of $A \setminus z$, we can repeat the same argument. If we do it infinitely many times, we will prove that there are infinitely many points of PC_{rot} in any settled component. □

Corollary 3.10 follows immediately from Theorem 3.9. Observe that if J is a dendrite, then $SG = p(SG) = \varnothing$, and hence $COR = COR_1 = COR_s$.

COROLLARY 3.10. *If J is a dendrite, then*

$$COR = COR_1 = COR_s = COR_{rot} = IC(PC_{rot}).$$

Furthermore, for any persistent cutpoint x there is $n \geq 0$ such that $f^n(x)$ separates two points from PC_{rot} (thus, at some point x maps to a cutpoint of COR). Moreover, any periodic cutpoint separates two points of PC_{rot} and therefore is itself a cutpoint of COR.

PROOF. Left to the reader. □

Corollary 3.10 implies the last, dendritic part of Theorem 2.6. The rest of Theorem 2.6 is proven next.

PROOF OF THEOREM 2.6. By definition,

$$IC(PC) \subset COR, \quad IC(PC_1) \subset COR_1 \quad \text{and} \quad IC(PC_{rot}) \subset COR_{rot}.$$

To prove the opposite inclusions, observe that by definition in each of these three cases, it suffices to consider a persistent cut-atom X which is not (pre)periodic. By Proposition 3.4 and because all Fatou gaps are eventually periodic, this implies that X never maps to $p(SG)$. Hence, in this case, by Theorem 3.9, there exists n such that $f^n(X)$ separates two points of $PC_{rot} \setminus p(SG)$ and therefore is contained in

$$IC(PC_{rot} \setminus p(SG)) \subset IC(PC_{rot}) \subset IC(PC_1) \subset IC(PC),$$

which proves all three inclusions of the theorem. □

3.3 Critical Atoms

Theorems 1.2 and 3.9 give explicit formulas for various versions of the dynamical core of a topological polynomial f in terms of various sets of periodic cut-atoms. These sets of periodic cut-atoms are most likely infinite. It may also be useful to relate the sets COR or COR_s to a finite set of critical atoms.

In the next lemma, we study one-to-one maps on complete continua. To do so we need a few definitions. The set of all critical points of f is denoted by $Cr_f = Cr$. The p-preimage of Cr is denoted by $Cr_\sim$. We also denote the ω-limit set of Cr by $\omega(Cr)$ and its p-preimage by $\omega(Cr_\sim)$. A *critical atom* is the p-image of a critical gap or a critical leaf of $\mathcal{L}_\sim$. Thus, the family of critical atoms includes all critical points of f and boundaries of all bounded components of $\mathbb{C} \setminus J$ on which f is of degree greater than 1, while boundaries of Siegel domains are not critical atoms. An atom of J is said to be *precritical* if it eventually maps to a critical atom.

LEMMA 3.11. *Suppose that $X \subset J$ is a complete continuum containing no critical atoms. Then $f|_X$ is one-to-one.*

PROOF. Otherwise, choose points $x, y \in X$ with $f(x) = f(y)$, and connect x and y with an arc $I \subset X$. If there are no Fatou domains with boundaries in X, then I is unique. Otherwise for each Fatou domain U with $\mathrm{Bd}(U) \subset X$, separating x from y in X, there are two points $i_U, t_U \in \mathrm{Bd}(U)$, each of which separates x from y in X, such that I must contain one of the two subarcs of $\mathrm{Bd}(U)$ with endpoints i_U, t_U. Note that $f(I)$ is not a dendrite since otherwise there must exist a critical point of $f|_I$. Hence we can choose a minimal subarc $I' \subset I$ so that $f(I')$ is a closed Jordan curve (then $f|_{I'}$ is one-to-one except for the endpoints x', y' of I' mapped into the same point).

It follows that $f(I')$ is the boundary of a Fatou domain U (otherwise there are points of J "shielded" from infinity by points of $f(I')$ which is impossible). The set $f^{-1}(U)$ is a finite union of Fatou domains, and I' is contained in the boundary of $f^{-1}(U)$. Therefore, there are two possible cases. Suppose that I' is a finite concatenation of at least two arcs, each arc lying on the boundary of some component of $f^{-1}(U)$. Then, as I' passes from one boundary to another, it must pass through a critical point, a contradiction. Suppose now that I' is contained in the boundary of a single component V of $f^{-1}(U)$ which implies that $\mathrm{Bd}(V) \subset X$ is a critical atom, a contradiction. $\square$

Clearly, (pre)critical atoms are dense. In fact, they are also dense in a stronger sense. To explain this, we need the following definition. For a topological space X, a set $A \subset X$ is called *continuum-dense* (or *con-dense*) in X if $A \cap Z \neq \varnothing$ for each non-degenerate continuum $Z \subset X$ (being non-degenerate means containing more than one point). The notion was introduced in [BOT06] in a different context.

Let us introduce a relative version of the notion of a critical atom. Namely, let $X \subset J$ be a complete continuum. A *point* $a \in X$ is *critical with respect to X* if in a neighborhood U of a in X the map $f|_U$ is not one-to-one. An *atom* $A \subset X$ is *critical with respect to X* if it is either a critical point with respect to X or a Fatou atom $A \subset X$ of degree greater than 1.

LEMMA 3.12. *Let $I \subset X$ be two non-degenerate continua in J such that X is complete and invariant. Suppose that X is not a Siegel atom. Then $f^n(I)$ contains a critical atom of $f|_X$ for some n. Thus, (pre)critical atoms are con-dense in J (every continuum in J intersects a (pre)critical atom).*

PROOF. If I contains more than one point of the boundary $\mathrm{Bd}(U)$ of a Fatou domain U, then it contains an arc $K \subset \mathrm{Bd}(U)$. As all Fatou domains are (pre)periodic, K maps eventually to a subarc K' of the boundary $\mathrm{Bd}(V)$ of a periodic Fatou domain V. If V is of degree greater than 1, then eventually K' covers $\mathrm{Bd}(V)$; since in this case $\mathrm{Bd}(V)$ is a critical atom of $f|_X$, we are done. If V is Siegel, then every point of $\mathrm{Bd}(V)$ is eventually covered by K'. Since we assume that X is not a Siegel atom, it follows that a critical point of $f|_X$ belongs to $\mathrm{Bd}(V)$, and again we are done.

By the preceding paragraph, from now on we may assume that non-empty intersections of I with boundaries of Fatou domains are single points. By [BL02a], I is not wandering; we may assume that $I \cap f(I) \neq \varnothing$. Set $L = \bigcup_{k=0}^{\infty} f^k(I)$. Then L is connected, and $f(L) \subset L$. If $\overline{L}$ contains a Jordan curve Q, then Q is the boundary of an invariant Fatou domain. If $L \cap Q = \varnothing$, then L is in a single component of $J \setminus Q$, hence $\overline{L} \cap Q$ is at most one point, a contradiction. Choose the smallest k with $f^k(I) \cap Q \neq \varnothing$; by the assumption $f^k(I) \cap Q = \{q\}$ is a singleton. Since the orbit of I cannot be contained in the union of components of $J \setminus \{q\}$ disjoint from Q, there exists m with $f^m(I) \cap Q$ not being a singleton, a contradiction.

Hence $\overline{L}$ is an invariant dendrite, and all cutpoints of $\overline{L}$ belong to images of I. Suppose that no cutpoint of $\overline{L}$ is critical. Then $f|_{\overline{L}}$ is a homeomorphism (if two points of $\overline{L}$ map to one point, there must exist a critical point in the open arc connecting them). However, it is proven in [BFMOT10] that if $D \subset J$ is an invariant dendrite, then it contains infinitely many periodic cutpoints. Hence we can choose two points $x, y \in \overline{L}$ and a number r such that $f^r(x) = x, f^r(y) = y$. Then the arc $I' \subset \overline{L}$ connecting x and y is invariant under the map f^r, which is one-to-one on I'. Clearly, this is impossible inside J (it is easy to see that a self-homeomorphism of an interval with fixed endpoints and finitely many fixed points overall must have a fixed point attracting from at least one side which is impossible in J). $\qquad \square$

We need new notation. Let $\mathrm{CrA}(X)$ be the set of critical atoms of $f|_X$; set $\mathrm{CrA} = \mathrm{CrA}(J)$. Similar to Definition 2.5, denote by $\mathrm{PC}(X)$ the union of all periodic cut-atoms of X and by $\mathrm{PC}_1(X)$ the set of all periodic cut-atoms of X of degree 1.

LEMMA 3.13. *Let $X \subset J$ be an invariant complete continuum which is not a Siegel atom. Then the following facts hold:*

1. *For every cutpoint x of X, there exists an integer $r \geq 0$ such that $f^r(x)$ either (a) belongs to a set from $\mathrm{CrA}(X)$, or (b) separates two sets of $\mathrm{CrA}(X)$, or (c) separates a set of $\mathrm{CrA}(X)$ from its image. In any case $f^r(x) \in \mathrm{IC}(\mathrm{CrA}(X))$, and in cases (b) and (c) $f^r(x)$ is a cutpoint of $\mathrm{IC}(\mathrm{CrA}(X))$. In particular, the dynamical span of all cut-atoms of X is contained in $\mathrm{IC}(\mathrm{CrA}(X))$.*

2. *$\mathrm{PC}(X) \subset \mathrm{IC}(\mathrm{CrA}(X))$. In particular, if $X = \mathrm{IC}(\mathrm{PC}(X))$, then $X = \mathrm{IC}(\mathrm{CrA}(X))$.*

PROOF. (1) Suppose that x does not eventually map to $\mathrm{CrA}(X)$. Then all points in the forward orbit of x are cutpoints of X, (in particular, there are at least two X-legs at any such point).

If $f^r(x)$ has more than one critical X-leg for some $r \geq 0$, then $f^r(x)$ separates two sets of $\mathrm{CrA}(X)$. Assume that $f^k(x)$ has one critical X-leg for every $k \geq 0$. By Lemma 3.11, each non-critical X-leg L of $f^k(x)$ maps in a one-to-one fashion to some X-leg M of $f^{k+1}(x)$. There is a connected neighborhood U_k of $f^k(x)$ in X so that $f|_{U_k}$ is one-to-one. We may assume that U_k contains all of the non-critical legs at $f^k(x)$. Hence there exists a bijection φ_k between components of $U_k \setminus f^k(x)$ and components of $f(U_k) \setminus f^{k+1}(x)$ showing how small pieces (*germs*) of components of $X \setminus \{f^k(x)\}$, containing $f^k(x)$ in their closures, map to small pieces (*germs*) of components of $X \setminus \{f^{k+1}(x)\}$, containing $f^{k+1}(x)$ in their closures.

By Lemma 3.12 choose $r > 0$ so that a non-critical X-leg of $f^{r-1}(x)$ maps to the critical X-leg of $f^r(x)$. Then the bijection φ_{r-1} sends the germ of the critical X-leg A of $f^{r-1}(x)$ to a non-critical X-leg B of $f^r(x)$. Let $C \subset A$ be the connected component of $X \setminus (\mathrm{CrA}(X) \cup \{f^{r-1}(x)\})$ with $f^{r-1}(x) \in \overline{C}$; let R be the union of $\overline{C}$ and all the sets from $\mathrm{CrA}(X)$ non-disjoint from $\overline{C}$. Then $f(R) \subset \overline{B}$ while all the critical atoms are contained in the critical leg $D \neq B$ of $f^r(x)$. It follows that $f^r(x)$ separates these critical atoms from their images. By definition of $\mathrm{IC}(\mathrm{CrA}(X))$ this implies that either $f^r(x)$ belongs to a set from $\mathrm{CrA}(X)$, or $f^r(x)$ is a cutpoint of $\mathrm{IC}(\mathrm{CrA}(X))$. In either case $f^r(x) \in \mathrm{IC}(\mathrm{CrA}(X))$. The rest of (1) easily follows.

(2) If x is a periodic cutpoint of X, then, choosing r as before, we see that $f^r(x) \in \mathrm{IC}(\mathrm{CrA}(X))$, which implies that $x \in \mathrm{IC}(\mathrm{CrA}(X))$ (because x is an iterated image of $f^r(x)$ and $\mathrm{IC}(\mathrm{CrA}(X))$ is invariant). Thus, all periodic cutpoints of

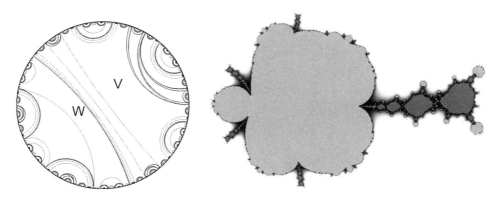

FIGURE 3.14: Example of a lamination and the corresponding Julia set, for which $COR_1 \neq IC(CrA(COR_1))$. The picture corresponds to a degree 6 polynomial. The invariant quadratic gap V corresponds to the largest dark grey region on the right. The invariant gap W corresponds to the large light grey "cauliflower" on the left. This is an invariant parabolic domain that contains a critical point on its boundary.

X belong to $IC(CrA(X))$. Now, take a periodic Fatou atom Y. If Y is of degree greater than 1, then it has an image $f^k(Y)$ which is a critical atom of $f|_X$. Thus, $Y \subset IC(CrA(X))$. Otherwise for some k the set $f^k(Y)$ is a periodic Siegel atom with critical points on its boundary. Since X is not a Siegel atom itself, $f^n(Y)$ contains a critical point of $f|_X$. Hence the entire Y is contained in the limit set of this critical point and again $Y \subset IC(CrA(X))$. Hence each periodic Fatou atom in X is contained in $IC(CrA(X))$. Thus, $PC(X) \subset IC(CrA(X))$, as desired.   □

We can now relate various dynamical cores to the critical atoms contained in these cores. First first let us consider the following heuristic example. Suppose that the lamination $\sim$ of sufficiently high degree has an invariant Fatou gap V of degree 2 and, disjoint from it, a super-gap U of degree 3. The super-gap U is subdivided ("tuned") by an invariant quadratic gap $W \subset U$ with a critical leaf on its boundary (or a finite critical gap sharing an edge with its boundary as in Figure 3.14) so that W concatenated with its appropriate pullbacks fills up U from within.

Also assume that the strip between U and V is enclosed by two circle arcs and two edges, a fixed edge ℓ_u of U and a prefixed edge ℓ_v of V (that is, $\sigma_d(\ell_v)$ is a fixed edge of V). Moreover, suppose that U and V have only two periodic edges, namely, ℓ_u and $\sigma_d(\ell_v)$, so that all other edges of U and V are preimages of ℓ_u and $\sigma_d(\ell_v)$. All other periodic gaps and leaves of $\sim$ are located in the component A of $\mathbb{D} \setminus \sigma_d(\ell_v)$ which does not contain U and V. In Figure 3.14 an example of this construction is shown for $d = 6$.

It follows from Theorem 1.2 that in this case COR_1 includes a continuum

$$K \subset p(A \cup V)$$

united with a connector-continuum L connecting K and $p(\ell_u)$. Moreover,

$$p(Bd(V)) \subset K.$$

Basically, all the points of L except for $p(\ell_u)$ are "sucked into" K while being repelled away from $p(\ell_u)$. Clearly, in this case, even though $p(\ell_u) \in COR_1$, $p(\ell_u)$

still does not belong to the set $IC(CrA(COR_1))$ because $p(Bd(W))$, while being a critical atom, is not contained in COR_1. This shows that some points of COR_1 may be located outside $IC(CrA(COR_1))$ and also that there might exist non-degenerate critical atoms intersecting COR_1 over a point (and hence not contained in COR_1). This example shows that the last claim of Lemma 3.13 cannot be established for $X = COR_1$.

Also, let us consider the case when COR is a Siegel atom Z. Then by definition there are no critical points or atoms of $f|_{COR}$, so in this case $CrA(COR)$ is empty. However this is the only exception.

THEOREM 3.15. *Suppose COR is not just a Siegel atom. Then* $COR = IC(CrA(COR))$, $COR_{rot} = IC(CrA(COR_{rot}))$, *and* $COR_s = IC(CrA(COR_s))$.

PROOF. By Theorem 1.2 we have $COR = IC(PC)$. Let us show that, in fact, COR is the dynamical span of its periodic cutpoints and its periodic Fatou atoms. It suffices to show that any periodic cutpoint of J either belongs to a Fatou atom or is a cutpoint of COR. Indeed, suppose that x is a periodic cutpoint of J which does not belong to a Fatou atom. Then by Theorem 2.15 applied to different components of $J \setminus \{x\}$ we see that x separates two periodic elements of PC. Hence x is a cutpoint of COR as desired. By Lemma 3.13 we have $COR = IC(CrA(COR))$. The remaining claims can be proven similarly. $\square$

There is a bit more universal way of stating a similar result. Namely, instead of considering critical atoms of $f|_{COR}$ we can consider critical atoms of f *contained in* COR. Then the appropriately modified claim of Theorem 3.15 holds without exception. Indeed, it holds trivially in the case when COR is an invariant Siegel atom. Otherwise it follows from Theorem 3.15 and the fact that the family of critical atoms of $f|_{COR}$ is a subset of the family of all critical atoms of f contained in COR. We prefer the statement of Theorem 3.15 to a more universal one because it allows us not to include "unnecessary" critical points of f which happen to be endpoints of COR; clearly, the results of Theorem 3.15 hold without such critical points. Notice, that the explanations given in this paragraph equally relate to COR_1 and COR_{rot}.

In the dendritic case the following corollary holds.

COROLLARY 3.16. *If J is a dendrite, then the following holds:*

$$COR = IC(PC_{rot}) = IC(CrA(COR)).$$

4 Bibliography

[B95]   A. Blokh, *The Spectral Decomposition for One-Dimensional Maps*, Dynamics Reported **4** (1995), 1–59.

[BFMOT10] A. Blokh, R. Fokkink, J. Mayer, L. Oversteegen, E. Tymchatyn, *Fixed point theorems in plane continua with applications*, Memoirs of the AMS **224**, no. 1053, (2013), 1–97. arXiv:1004.0214

[BL02a]   A. Blokh, G. Levin, *Growing trees, laminations and the dynamics on the Julia set*, Ergod. Th. and Dynam. Sys. **22** (2002), 63–97.

[BMOV11]   A. Blokh, D. Mimbs, L. Oversteegen, K. Valkenburg, *Laminations in the language of leaves*, Trans. Amer. Math. Soc. **365** (2013), 5367–5391. arXiv:1106.0273

[BOT06]   A. Blokh, L. Oversteegen, E. Tymchatyn, *On almost one-to-one maps*, Trans. Amer. Math. Soc. **358** (2006), 5003–5014.

[C04]   D. Childers, *Wandering polygons and recurrent critical leaves*, Ergod. Th. and Dynam. Sys. **27** (2007), no. 1, 87–107.

[DH85a]   A. Douady, J. H. Hubbard, *Étude dynamique des polynômes complexes* I, II, Publications Mathématiques d'Orsay **84-02** (1984), **85-04** (1985).

[GM93]   L. Goldberg and J. Milnor, *Fixed points of polynomial maps. II. Fixed point portraits*, Ann. Sci. École Norm. Sup. (4) **26** (1993), no. 1, 51–98.

[H37]   F. Hausdorff, *Die schlichten stetigen Bilder des Nullraums*, Fund. Math. **29** (1937), 151–158.

[K02]   J. Kiwi, *Wandering orbit portraits*, Trans. Amer. Math. Soc. **254** (2002), 1473–1485.

[M00]   J. Milnor, *Dynamics in One Complex Variable*, Annals of Mathematical Studies **160**, Princeton (2006).

[T85]   W. Thurston. *The combinatorics of iterated rational maps* (1985), published in: "Complex dynamics: Families and Friends," ed. D. Schleicher and A. K. Peters (2008), 1–108.

The quadratic dynatomic curves are smooth and irreducible

Xavier Buff and Tan Lei

ABSTRACT. We re-prove here both the smoothness and the irreducibility of the quadratic dynatomic curves $\{(c, z) \in \mathbb{C}^2 \mid z \text{ is } n\text{-periodic for } z^2 + c\}$.

The smoothness is due to Douady-Hubbard. Our proof here is based on elementary calculations on the pushforwards of specific quadratic differentials, following Thurston and Epstein. This approach is a computational illustration of the power of the far more general transversality theory of Epstein.

The irreducibility is due to Bousch and Lau-Schleicher, but with a different method. Our approach is inspired by the proof of Lau-Schleicher. We use elementary combinatorial properties of the kneading sequences instead of internal addresses.

1 INTRODUCTION

For $c \in \mathbb{C}$, let $f_c \colon \mathbb{C} \to \mathbb{C}$ be the quadratic polynomial

$$f_c(z) := z^2 + c.$$

A point $z \in \mathbb{C}$ is periodic for f_c if $f_c^{\circ n}(z) = z$ for some integer $n \geq 1$; it is of period n if $f_c^{\circ k}(z) \neq z$ for $0 < k < n$. For $n \geq 1$, let $X_n \subset \mathbb{C}^2$, be the *dynatomic curve* defined by

$$X_n := \{(c, z) \in \mathbb{C}^2 \mid z \text{ is periodic of period } n \text{ for } f_c\}.$$

The objective of this note is to give new proofs of the following known results.

THEOREM 1.1 (Douady-Hubbard). *For every $n \geq 1$, the closure of X_n in $\mathbb{C}^2$ is a smooth affine curve.*

THEOREM 1.2 (Bousch and Lau-Schleicher). *For every $n \geq 1$ the closure of X_n in $\mathbb{C}^2$ is irreducible.*

Theorem 1.1 has been proved by Douady-Hubbard in [DH] using parabolic implosion techniques. Milnor [Mi4], Section 5 reproved the result with a different approach. Milnor [Mi2] reformulated a proof of Tsujii in the language of quadratic differentials, showing that the topological entropy of the real quadratic polynomial $x \mapsto x^2 + c$ varies monotonically with respect to the parameter c. Our approach here to prove Theorem 1.1 is similar. We use elementary calculations on quadratic differentials and Thurston's contraction principle. Our calculation is a computational illustration of the far deeper and more conceptual transversality theory of Epstein [E].

The first author is partially supported by the Institut Universitaire de France.

Theorem 1.2 has been proved by Bousch [B] using a combination of algebraic and dynamical arguments, and independently by Lau and Schleicher [LS, Sc2] using dynamical arguments only. Morton [Mo] proved a generalized version of this result, using more algebraic methods. Our approach here follows essentially [LS, Sc2], except we replace their argument on internal addresses (Lemma 4.5 in [Sc2]) by a purely combinatorial argument on kneading sequences (see Lemma 4.4); also we make use of a result of Petersen-Ryd ([PR]) instead of Douady-Hubbard's parabolic implosion theory.

Interested readers may consult Bousch [B] or Milnor [Mi4] for other results on the curves X_n. Manes [Ma] proves that the irreducibility of the dynatomic curves may fail in some other families of rational maps, such as the family of odd quadratic rational maps.

The somewhat similar curve consisting of cubic polynomial maps with periodic critical orbit is smooth of known Euler characteristic, due to the works of Milnor and Bonifant-Kiwi-Milnor [Mi3, BKM]. But the irreducibility question remains open.

In Section 2 we prove that the topological closure $\overline{X}_n \subset \mathbb{C}^2$ is an affine curve by introducing *dynatomic polynomials* defining the curve, and in Section 3 we prove the smoothness, while in Section 4 we prove the irreducibility. Sections 3 and 4 can be read independently.

Acknowledgements. We wish to express our thanks to Adam Epstein and John Milnor for helpful discussions, to William Thurston for his encouragement, and to Schleicher, Silverman, and the anonymous referee for their useful comments.

2   DYNATOMIC POLYNOMIALS

In this section, we define the dynatomic polynomials $Q_n \in \mathbb{Z}[c,z]$ (see [Mi1] and [Si]) and show that
$$\overline{X}_n = \{(c,z) \in \mathbb{C}^2 \mid Q_n(c,z) = 0\}.$$
For $n \geq 1$, let $P_n \in \mathbb{Z}[c,z]$ be the polynomial defined by
$$P_n(c,z) := f_c^{\circ n}(z) - z.$$
The dynatomic polynomials Q_n will be defined so that
$$P_n = \prod_{k \mid n} Q_k.$$

EXAMPLE 2.1. For $n = 1$ and $n = 2$, we have

$$P_1(c,z) = z^2 - z + c, \qquad P_2(c,z) = z^4 + 2cz^2 - z + c^2 + c,$$
$$Q_1(c,z) = z^2 - z + c, \qquad Q_2(c,z) = z^2 + z + c + 1$$
$$P_1(c,z) = Q_1(c,z), \qquad P_2(c,z) = Q_1(c,z) \cdot Q_2(c,z).$$

Further examples may be found in [Si, Table 4.1].

With an abuse of notation, we will identify polynomials in $\mathbb{Z}[c,z]$ and polynomials in $\mathbf{Z}[z]$, with $\mathbf{Z} = \mathbb{Z}[c]$. In particular, we shall write $R(c,z)$ when $R \in \mathbf{Z}[z]$. Note that $P_n \in \mathbf{Z}[z]$ is a monic polynomial (its leading coefficient is 1) of degree 2^n.

PROPOSITION 2.2. *There exists a unique sequence of monic polynomials* $(Q_n \in \mathbf{Z}[z])_{n \geq 1}$
such that for all $n \geq 1$, *we have* $P_n = \prod_{k|n} Q_k$.

PROOF. The proof goes by induction on n. For $n = 1$, it is necessary and sufficient
to define
$$Q_1(c, z) := P_1(c, z) = z^2 - z + c.$$
Note that $Q_1 \in \mathbf{Z}[z]$ is indeed monic. Assume now that $n > 1$ and that the polynomials Q_k are defined for $1 \leq k < n$. Set
$$A := \prod_{k|n, k<n} Q_k.$$
Since the polynomials $Q_k \in \mathbf{Z}[z]$ are monic, the polynomial $A \in \mathbf{Z}[z]$ is also monic.
So, we may perform a Euclidean division to find a monic quotient $Q \in \mathbf{Z}[z]$ and
a remainder $R \in \mathbf{Z}[z]$ with degree$(R) <$ degree(A), such that $P_n = QA + R$. We
need to show that $R = 0$, which enables us to set $Q_n := Q$.

Let $\Delta \in \mathbb{Z}[c]$ be the discriminant of A. We claim that $\Delta(0) \neq 0$. Indeed, for each
$k < n$, the polynomial $Q_k(0, z) \in \mathbf{Z}[z]$ divides $P_k(0, z) = z^{2^k} - z$ whose roots are
simple. So, the roots of $Q_k(0, z)$ are simple. In addition, a root z_0 of $Q_k(0, z)$ is a
periodic point of f_0 whose period m divides k. We have $m = k$ since otherwise,
$$Q_k(0, z) \cdot P_m(0, z) = Q_k(0, z) \cdot \prod_{j|m} Q_j(0, z)$$
would have a double root at z_0 and would at the same time divide
$$P_k(0, z) = \prod_{j|k} Q_j(0, z),$$
whose roots are simple. So, if $1 \leq k_1 < k_2 < n$, then $Q_{k_1}(0, z)$ and $Q_{k_2}(0, z)$ do
not have common roots. This shows that the roots of $A(0, z)$ are simple, whence
$\Delta(0) \neq 0$.

Fix $c_0 \in \mathbb{C}$ such that $\Delta(c_0) \neq 0$ (since Δ does not identically vanish, this holds
for every c_0 outside a finite set). Then, the roots of $A(c_0, z) \in \mathbf{Z}[z]$ are simple.
Such a root z_0 is a periodic point of f_{c_0}, with period dividing n, whence a root
of $P_n(c_0, z)$. As a consequence, $A(c_0, z)$ divides $P_n(c_0, z)$ in $\mathbb{C}[z]$. It follows that
$R(c_0, z) = 0$ for all $z \in \mathbb{C}$. Since this is true for every c_0 outside a finite set, we have
that $R = 0$ as required.   □

REMARK 1. The proof we gave shows that the dynatomic polynomials Q_n have
no repeated factors (otherwise $Q_n(0, z) \in \mathbf{Z}[z]$ would have a double root) and
moreover, if $k_1 \neq k_2$, then Q_{k_1} and Q_{k_2} do not have common factors (otherwise
$Q_{k_1}(0, z) \in \mathbf{Z}[z]$ and $Q_{k_2}(0, z) \in \mathbf{Z}[z]$ would have a common root). Those facts
will be used later.

REMARK 2. The degree of Q_k is at most that of P_k, that is 2^k. It follows that the
degree of $A := \prod_{k|n, k<n} Q_k$ is at most $2^n - 2$, and so the degree of $Q_n = P_n/A$ is at
least 2. In particular, for $n \geq 1$, the set X_n is non-empty.

We extensively used the properties of roots of $Q_n(0,z) \in \mathbb{Z}[z]$. We will now study the properties of the roots of $Q_n(c_0,z) \in \mathbb{C}[z]$ for an arbitrary parameter $c_0 \in \mathbb{C}$.

PROPOSITION 2.3. *Let $n \geq 1$ be a positive integer and $c_0 \in \mathbb{C}$ be an arbitrary parameter. Then, $z_0 \in \mathbb{C}$ is a root of $Q_n(c_0,z) \in \mathbb{C}[z]$ if and only if one of the following three exclusive conditions is satisfied:*

1. *z_0 is periodic for f_{c_0}, the period is n and the multiplier is not 1; in that case $Q_n(c_0,z)$ has a simple root at z_0, or*

2. *z_0 is periodic for f_{c_0}, the period is n and the multiplier is equal to 1; in that case $Q_n(c_0,z)$ has a double root at z_0, or*

3. *z_0 is periodic for f_{c_0}, the period $m < n$ is a proper divisor of n and the multiplier of z_0 as a fixed point of $f_{c_0}^{\circ m}$ is a primitive (n/m)th root of unity; in that case $Q_n(c_0,z)$ has a root of order n/m at z_0.*

PROOF. If $Q_n(c_0,z_0) = 0$, then $P_n(c_0,z_0) = 0$ and so z_0 is periodic for f_{c_0} and the period m divides n. Conversely, if z_0 is periodic of period m for f_{c_0}, then $P_k(c_0,z_0) = 0$ if and only if k is a multiple of m. In particular, if k is not a multiple of m, then $Q_k(c_0,z_0) \neq 0$. Since

$$0 = P_m(c_0,z_0) = \prod_{k|m} Q_k(c_0,z_0),$$

we deduce that $Q_m(c_0,z_0) = 0$.

Case 1. If the multiplier ρ of z_0 as a fixed point of $f_{c_0}^{\circ m}$ is not a root of unity, then $P_n(c_0,z)$ has a simple root at z_0 whenever n is a multiple of m. In that case, $Q_m(c_0,z)$ is a factor of $P_n(c_0,z)$, so no other factor of P_n can vanish at z_0. As a consequence, $Q_n(c_0,z_0)$ vanishes if and only if $n = m$. In addition, $Q_m(c_0,z) \in \mathbb{C}[z]$ has a simple root at z_0.

Next, if the multiplier ρ of z_0 as a fixed point of $f_{c_0}^{\circ m}$ is a primitive sth root of unity, then the multiplier of z_0 as a fixed point of $f_{c_0}^{\circ mk}$ is ρ^k. It is equal to 1 if and only if k is a multiple of s. In that case, z_0 is a multiple root of $P_{mk}(c_0,z)$ of order $s + 1$. Indeed, f_{c_0} has only one cycle of attracting petals since this cycle must attract the unique critical point of f_{c_0}.

Case 2. If $s = 1$, then $P_n(c_0,z)$ has a double root at z_0 whenever n is a multiple of m. As before, $Q_n(c_0,z_0)$ vanishes if and only if $n = m$, but this time, $Q_m(c_0,z) \in \mathbb{C}[z]$ has a double root at z_0.

Case 3. If $s \geq 2$, then $P_n(c_0,z)$ has a simple root at z_0 whenever n is a multiple of m but not a multiple of ms, and a multiple root at z_0 of order $s + 1$ whenever n is a multiple of ms. So, $Q_n(c_0,z_0)$ vanishes if and only if $n = m$ or $n = ms$; the polynomial $Q_m(c_0,z) \in \mathbb{C}[z]$ has a simple root at z_0 and the polynomial $Q_{ms}(c_0,z) \in \mathbb{C}[z]$ has a root of order s at z_0. $\square$

PROPOSITION 2.4. *For all $n \geq 1$, we have*

$$\overline{X}_n = \{(c,z) \in \mathbb{C}^2 \mid Q_n(c,z) = 0\}.$$

If $(c,z) \in \overline{X}_n - X_n$, then z is periodic for f_c, its period m is a proper divisor of n, and the multiplier of z as a fixed point of $f_c^{\circ m}$ is a primitive (n/m)th root of unity.

PROOF. Let $Y_n \in \mathbb{C}^2$ be the affine curve defined by Q_n:

$$Y_n := \left\{ (c,z) \in \mathbb{C}^2 \mid Q_n(c,z) = 0 \right\}.$$

According to the previous proposition,

- if (c,z) belongs to X_n, then $Q_n(c,z) = 0$ and so $\overline{X}_n \subseteq Y_n$;

- if $(c,z) \in Y_n - X_n$, then z is periodic for f_c, its period m is a proper divisor of n, and the multiplier of z as a fixed point of $f_c^{\circ m}$ is a primitive (n/m)th root of unity. In particular, $Q_m(c,z) = 0$. Since Q_n and Q_m do not have common factors, this only occurs for a finite set of points $(c,z) \in Y_n$. Thus, $Y_n \subseteq \overline{X}_n$. $\square$

REMARK 3. We have seen that $\overline{X}_n - X_n$ is finite. More generally, the set of points $(c_0, z_0) \in \overline{X}_n$ such that $f_{c_0}^{\circ n}(z_0) = z_0$ and $(f_{c_0}^{\circ n})'(z_0) = 1$ is finite. Indeed, for such a point, c_0 is a root of the discriminant of $P_n \in \mathbb{Z}[z]$. Since the roots of $P_n(0,z)$ are simple, this discriminant does not vanish at $c = 0$, so its roots form a finite set.

3   SMOOTHNESS OF THE DYNATOMIC CURVES

Our objective is now to give a proof of Theorem 1.1. We will prove the following more precise version. We shall denote by $\pi_c \colon \mathbb{C}^2 \to \mathbb{C}$ the projection $(c,z) \mapsto c$ and by $\pi_z \colon \mathbb{C}^2 \to \mathbb{C}$ the projection $(c,z) \mapsto z$.

THEOREM 3.1. *For every $n \geq 1$, the affine curve $\overline{X}_n$ is smooth. More precisely, for $(c_0, z_0) \in \overline{X}_n$, we have*

1. *if $z_0 \in X_n$ has multiplier different from 1, then $\pi_c \colon X_n \to \mathbb{C}$ is a local isomorphism;*

2. *if $z_0 \in X_n$ has multiplier 1, then $\pi_z \colon X_n \to \mathbb{C}$ is a local isomorphism; in addition, $\pi_c \colon X_n \to \mathbb{C}$ has local degree 2;*

3. *if $z_0 \in \overline{X}_n - X_n$ has multiplier a primitive sth root of unity, then $\pi_z \colon \overline{X}_n \to \mathbb{C}$ is a local isomorphism; in addition, $\pi_c \colon \overline{X}_n \to \mathbb{C}$ has local degree s.*

COROLLARY 3.2. *Any intersection between the curves $\overline{X}_m$ and $\overline{X}_n$ is transverse, and there are no threefold intersections.*

The idea for proving Theorem 3.1 is to apply the implicit function theorem. In particular, we will prove that $\partial Q_n / \partial z$ is nonzero at (c_0, z_0) in Case 1 (which is almost immediate) and in Cases 2 and 3, we show that that $\partial Q_n / \partial c$ is nonzero. This is where we have to work: following Epstein, we will first relate this partial derivative to the coefficient of a quadratic differential of the form $(f_{c_0})_* \mathbf{q} - \mathbf{q}$ (for a specific $\mathbf{q}$ in each case); we will then show that $(f_{c_0})_* \mathbf{q} \neq \mathbf{q}$ by using (a generalization of) Thurston's contraction principle. This approach is fundamentally different from Douady-Hubbard's original proof, where Fatou-Leau's flower theorem on parabolic periodic points as well as Douady-Hubbard's parabolic implosion theory play an essential role.

Once we know that in Cases 2 and 3 the projection $\pi_z \colon X_n \to \mathbb{C}$ is a local isomorphism, the local degree of the projection $\pi_c \colon X_n \to \mathbb{C}$ follows from Proposition 2.3. Indeed, if $Q_n(c_0, z) \in \mathbb{C}[z]$ has a root of order ν at z_0 and if $\partial Q_n / \partial c$ is nonzero at (c_0, z_0), then

$$Q_n(c_0 + \eta, z_0 + \varepsilon) = a\eta + b\varepsilon^\nu + o(\eta) + o(\varepsilon^\nu) \quad \text{with} \quad a \neq 0 \text{ and } b \neq 0,$$

and the solutions of $Q_n(c, z) = 0$ are locally of the form $(c(z), z)$, with

$$c(z_0 + \varepsilon) = c_0 - \frac{b}{a}\varepsilon^\nu + o(\varepsilon^\nu).$$

3.1 Case 1 of Theorem 3.1

Let $(c_0, z_0) \in X_n$ be such that the multiplier of z_0 as a fixed point of $f_{c_0}^{\circ n}$ is not 1. Then, $Q_n(c_0, z_0) = 0$, but $Q_k(c_0, z_0) \neq 0$ for $k < n$, and $(f_{c_0}^{\circ n})'(z_0) \neq 1$. Since

$$\prod_{k|n} Q_k(c, z) = P_n(c, z),$$

we have

$$\frac{\partial Q_n}{\partial z}(c_0, z_0) \cdot \prod_{k|n, k<n} Q_k(c_0, z_0) = \frac{\partial P_n}{\partial z}(c_0, z_0) = (f_{c_0}^{\circ n})'(z_0) - 1 \neq 0.$$

As a consequence, $\partial Q_n / \partial z$ is nonzero at (c_0, z_0). By the Implicit Function Theorem, X_n is smooth near (c_0, z_0) and the projection $\pi_c \colon X_n \to \mathbb{C}$ is a local isomorphism.

3.2 Case 2 of Theorem 3.1

Let $(c_0, z_0) \in X_n$ be such that the multiplier of z_0 as a fixed point of $f_{c_0}^{\circ n}$ is 1. As previously, since $Q_n(c_0, z_0) = 0$ and

$$\prod_{k|n} Q_k(c, z) = P_n(c, z),$$

we have

$$\frac{\partial Q_n}{\partial c}(c_0, z_0) \cdot \prod_{k|n, k<n} Q_k(c_0, z_0) = \frac{\partial P_n}{\partial c}(c_0, z_0).$$

Since for all $k < n$, $Q_k(c_0, z_0) \neq 0$, it is enough to prove that

$$\frac{\partial P_n}{\partial c}(c_0, z_0) \neq 0.$$

We shall use the following notations: for $n \geq 0$, we let $\zeta_n \colon \mathbb{C} \to \mathbb{C}$ be defined by

$$\zeta_n(c) := f_c^{\circ n}(z_0),$$

and we set

$$z_n := \zeta_n(c_0) = f_{c_0}^{\circ n}(z_0) \quad \text{and} \quad \delta_n := f_{c_0}'(z_n) = 2z_n.$$

Since $P_n(c, z_0) = f_c^{\circ n}(z_0) - z_0 = \zeta_n(c) - z_0$, we have

$$\frac{\partial P_n}{\partial c}(c_0, z_0) = \zeta_n'(c_0). \tag{3.1}$$

LEMMA 3.3 (Compare with [Mi2]). *We have*

$$\zeta_n'(c_0) = 1 + \delta_{n-1} + \delta_{n-1}\delta_{n-2} + \ldots + \delta_{n-1}\delta_{n-2}\cdots\delta_1.$$

PROOF. The function ζ_0 is constant (equal to z_0). From $\zeta_n(c) = (\zeta_{n-1}(c))^2 + c$, we obtain

$$\zeta_n'(c_0) = 1 + \delta_{n-1}\zeta_{n-1}'(c_0) \quad \text{with} \quad \zeta_0'(c_0) = 0.$$

The result follows by induction.   □

In order to prove that

$$1 + \delta_{n-1} + \delta_{n-1}\delta_{n-2} + \ldots + \delta_{n-1}\delta_{n-2} \cdots \delta_1 \neq 0,$$

we shall now work with meromorphic quadratic differentials.

3.2.1   Quadratic differentials

A meromorphic quadratic differential $\mathbf{q}$ on $\mathbb{C}$ takes the form $\mathbf{q} = q\,dz^2$ with q a meromorphic function on $\mathbb{C}$. We use $\mathcal{Q}(\mathbb{C})$ to denote the set of meromorphic quadratic differentials on $\mathbb{C}$ whose poles (if any) are all simple. If $\mathbf{q} = q\,dz^2 \in \mathcal{Q}(\mathbb{C})$ and U is a bounded open subset of $\mathbb{C}$, the norm

$$\|\mathbf{q}\|_U := \iint_U |q(x+iy)|\,dxdy$$

is well defined and finite.

EXAMPLE 3.4.

$$\left\| \frac{dz^2}{z} \right\|_{D(0,R)} = \int_0^{2\pi}\int_0^R \frac{1}{r}r\,drd\theta = 2\pi R\,.$$

3.2.2   Pushforward

For $f\colon \mathbb{C} \to \mathbb{C}$ a non-constant polynomial and $\mathbf{q} = q\,dz^2$ a meromorphic quadratic differential on $\mathbb{C}$, the pushforward $f_*\mathbf{q}$ is defined by

$$f_*\mathbf{q} := Tq\,dz^2 \quad \text{with} \quad Tq(z) := \sum_{f(w)=z} \frac{q(w)}{f'(w)^2}.$$

If $\mathbf{q} \in \mathcal{Q}(\mathbb{C})$, then $f_*\mathbf{q} \in \mathcal{Q}(\mathbb{C})$ also.

LEMMA 3.5 (Compare with [Mi2] or [L]). *For $f = f_c$, we have*

$$\begin{cases} f_*\left(\dfrac{dz^2}{z}\right) = 0 \\[2ex] f_*\left(\dfrac{dz^2}{z-a}\right) = \dfrac{1}{f'(a)}\left(\dfrac{dz^2}{z-f(a)} - \dfrac{dz^2}{z-c}\right) & \text{if } a \neq 0. \end{cases} \tag{3.2}$$

PROOF. If $f(w) = z$, then $w = \pm\sqrt{z-c}$ and

$$dw^2 = \frac{dz^2}{4(z-c)}.$$

We then have

$$f_* \left(\frac{dz^2}{z-a} \right) = \frac{dz^2}{4(z-c)} \left(\frac{1}{\sqrt{z-c} - a} + \frac{1}{-\sqrt{z-c} - a} \right)$$

$$= \frac{a \, dz^2}{2(z-c)(z - f(a))}.$$

If $a = 0$, we get the first equality in (3.2). Otherwise, we get the second equality using

$$\frac{1}{f'(a)} \left(\frac{1}{z - f(a)} - \frac{1}{z-c} \right) = \frac{f(a) - c}{2a(z-c)(z - f(a))} = \frac{a}{2(z-c)(z - f(a))}. \qquad \square$$

3.2.3   A particular quadratic differential

Let

$$\mathbf{q} := \sum_{k=0}^{n-1} \frac{\rho_k}{z - z_k} \, dz^2$$

be a quadratic differential in $\mathcal{Q}(\mathbb{C})$. Applying Lemma 3.5 and writing f for f_{c_0}, we obtain

$$f_* \mathbf{q} = \sum_{k=0}^{n-1} \frac{\rho_k}{\delta_k} \left(\frac{dz^2}{z - z_{k+1}} - \frac{dz^2}{z - c_0} \right).$$

We want to choose $\mathbf{q}$ so that $f_* \mathbf{q}$ and $\mathbf{q}$ differ only by a pole at c_0. It amounts then to solve the following linear system on the unknown coefficient vector $(\rho_0, \dots, \rho_{n-1})$:

$$\frac{\rho_{n-1}}{\delta_{n-1}} = \rho_0, \quad \frac{\rho_k}{\delta_k} = \rho_{k+1}, \quad k = 0, \dots, n-2$$

Notice that $\delta_0 \delta_1 \cdots \delta_{n-1}$ is the multiplier of z_0 as a periodic point of f_{c_0}, which is 1 by assumption. The preceding linear system has indeed a non-null solution:

$$\rho_0 = 1, \; \rho_{n-1} = \delta_{n-1}, \; \rho_{n-2} = \delta_{n-1} \delta_{n-2}, \; \dots, \; \rho_1 = \delta_{n-1} \delta_{n-2} \cdots \delta_1.$$

Therefore, for

$$\mathbf{q} := \sum_{k=0}^{n-1} \frac{\rho_k}{z - z_k} \, dz^2, \quad \rho_k = \delta_{n-1} \delta_{n-2} \cdots \delta_k, \quad k = n-1, n-2, \dots, 0 \qquad (3.3)$$

we have

$$f_* \mathbf{q} = \mathbf{q} - \left(\sum_{k=0}^{n-1} \rho_k \right) \cdot \frac{dz^2}{z - c_0}.$$

According to Equation (3.1), Lemma 3.3 and the definition of ρ_k,

$$\frac{\partial P_n}{\partial c}(c_0, z_0) = \zeta_n'(c_0) = \sum_{k=0}^{n-1} \rho_k.$$

It is therefore enough to prove that $f_* \mathbf{q} \neq \mathbf{q}$. This is done in the next paragraph using a contraction principle.

3.2.4   Contraction Principle

The following lemma is a weak version of Thurston's contraction principle (which applies in the setting of rational maps on $\mathbb{P}^1$).

LEMMA 3.6 (Contraction Principle). *For a non-constant polynomial f and a round disk V of radius large enough so that $U := f^{-1}(V)$ is relatively compact in V, we have*

$$\|f_*\mathbf{q}\|_V \leq \|\mathbf{q}\|_U < \|\mathbf{q}\|_V, \quad \forall\, \mathbf{q} \in \mathcal{Q}(\mathbb{C}).$$

PROOF. The strict inequality on the right is a consequence of the fact that U is relatively compact in V. The inequality on the left comes from

$$\|f_*\mathbf{q}\|_V = \iint_{x+iy\in V} \left| \sum_{f(w)=x+iy} \frac{q(w)}{f'(w)^2} \right| dxdy$$

$$\leq \iint_{x+iy\in V} \sum_{f(w)=x+iy} \left| \frac{q(w)}{f'(w)^2} \right| dxdy = \iint_{u+iv\in U} |q(u+iv)|\, dudv = \|\mathbf{q}\|_U.$$

$\square$

COROLLARY 3.7. *If $f\colon \mathbb{C} \to \mathbb{C}$ is a polynomial and if $\mathbf{q} \in \mathcal{Q}(\mathbb{C})$, then $f_*\mathbf{q} \neq \mathbf{q}$.*

This completes the proof in Case 2.

3.3   Case 3 of Theorem 3.1

Let $(c_0, z_0) \in \overline{X}_n - X_n$ be such that z_0 is periodic for f_{c_0} with period $m < n$ dividing n and multiplier ρ, a primitive sth root of unity with $s := n/m$.

According to Proposition 2.3(3), the polynomial $Q_n(c_0, z) \in \mathbb{C}[z]$ has a root of order $s \geq 2$ at z_0, so that

$$\frac{\partial Q_n}{\partial z}(c_0, z_0) = 0. \tag{3.4}$$

We want to show that

$$\frac{\partial Q_n}{\partial c}(c_0, z_0) \neq 0.$$

Let us write

$$P_n(c, z) = P_m(c, z) \cdot R(c, z) \quad \text{with} \quad R(c, z) = \prod_{k|n, k\nmid m} Q_k(c, z). \tag{3.5}$$

Since $Q_n(c_0, z_0) = 0$, we have

$$\frac{\partial R}{\partial z}(c_0, z_0) = \frac{\partial Q_n}{\partial z}(c_0, z_0) \cdot \prod_{k|n, k\nmid m, k<n} Q_k(c_0, z_0) = 0 \quad \text{and}$$

$$\frac{\partial R}{\partial c}(c_0, z_0) = \frac{\partial Q_n}{\partial c}(c_0, z_0) \cdot \prod_{k|n, k\nmid m, k<n} Q_k(c_0, z_0).$$

It is therefore enough to prove

$$\frac{\partial R}{\partial c}(c_0, z_0) \neq 0.$$

3.3.1   Variation along X_m

Note that $(c_0, z_0) \in X_m$ and the multiplier ρ of z_0 as a fixed point of $f_{c_0}^{om}$ is $\rho \neq 1$. Thus, according to Case 1, X_m is locally the graph of a function $\zeta(c)$ defined and holomorphic near c_0 with $\zeta(c_0) = z_0$. The point $\zeta(c)$ is periodic of period m for f_c. We denote by ρ_c its multiplier and set

$$\dot{\rho} := \frac{d\rho_c}{dc}\Big|_{c_0}.$$

LEMMA 3.8. *We have*

$$\frac{\partial R}{\partial c}(c_0, z_0) = \frac{s\dot{\rho}}{\rho(\rho - 1)}.$$

PROOF. Differentiating the first equation in (3.5) with respect to z, and evaluating at $(c, \zeta(c))$, we get

$$\underbrace{\frac{\partial P_n}{\partial z}(c, \zeta(c))}_{\rho_c^s - 1} = \underbrace{\frac{\partial P_m}{\partial z}(c, \zeta(c))}_{\rho_c - 1} \cdot R(c, \zeta(c)) + P_m(c, \zeta(c)) \cdot \underbrace{\frac{\partial R}{\partial z}(c, \zeta(c))}_{0},$$

whence

$$\rho_c^s - 1 = (\rho_c - 1) \cdot R(c, \zeta(c)).$$

Differentiating with respect to c and evaluating at c_0, we get

$$s\rho^{s-1}\dot{\rho} = \dot{\rho}\underbrace{R(c_0, z_0)}_{0} + (\rho - 1)\frac{\partial R}{\partial c}(c_0, z_0) + (\rho - 1)\underbrace{\frac{\partial R}{\partial z}(c_0, z_0)\zeta'(c_0)}_{0}$$

$$= (\rho - 1)\frac{\partial R}{\partial c}(c_0, z_0).$$

The result follows since $\rho^s = 1$ and so $\rho^{s-1} = 1/\rho$. $\qquad\qquad\square$

Thus, we are left with proving that $\dot{\rho} \neq 0$. This will be done by using a particular meromorphic quadratic differential having double poles along the cycle of z_0.

3.3.2   Quadratic differentials with double poles

LEMMA 3.9 (Compare with [L]). *For* $f = f_c$, *we have*

$$f_*\left(\frac{dz^2}{(z-a)^2}\right) = \frac{dz^2}{(z-f(a))^2} - \frac{1}{2a^2}\left(\frac{dz^2}{z-f(a)} - \frac{dz^2}{z-c}\right) \quad \text{if } a \neq 0.$$

PROOF. If $f(w) = z$, then $w = \pm\sqrt{z-c}$ and

$$dw^2 = \frac{dz^2}{4(z-c)}.$$

Then

$$f_*\left(\frac{dz^2}{(z-a)^2}\right) = \frac{dz^2}{4(z-c)}\left(\frac{1}{(\sqrt{z-c}-a)^2} + \frac{1}{(-\sqrt{z-c}-a)^2}\right)$$

$$= \frac{(z-c+a^2)\,dz^2}{2(z-c)(z-c-a^2)^2} = \frac{(z-c+a^2)\,dz^2}{2(z-c)(z-f(a))^2}.$$

Decomposing the last expression into partial fractions gives

$$\frac{z - c + a^2}{2(z - c)(z - f(a))^2} = \frac{A}{(z - f(a))^2} + \frac{B}{z - f(a)} + \frac{C}{z - c}$$

with

$$A = \frac{f(a) - c + a^2}{2(f(a) - c)} = \frac{2a^2}{2a^2} = 1, \quad C = \frac{c - c + a^2}{2(c - f(a))^2} = \frac{a^2}{2a^4} = \frac{1}{2a^2}$$

and

$$B = -C = -\frac{1}{2a^2}. \qquad \qquad \square$$

3.3.3 A particular quadratic differential with double poles

As in Case 2, we will try to find a quadratic differential $\mathbf{q}$ with double poles along the orbit of z_0 so that $(f_{c_0})_* \mathbf{q}$ and $\mathbf{q}$ differ by only a simple pole at c_0. Set

$$z_k := f_{c_0}^{\circ k}(z_0), \quad \delta_k := f'_{c_0}(z_k) = 2z_k.$$

Since

$$\delta_0 \delta_1 \cdots \delta_{m-1} = \rho \neq 1,$$

there is a unique m-tuple $\mu := (\mu_0, \ldots, \mu_{m-1}) \in \mathbb{C}^m$ such that

$$\mu_{k+1} = \frac{\mu_k}{2z_k} - \frac{1}{2z_k^2},$$

where the indices are considered to be modulo m. Indeed, this is a linear system in $\mathbb{C}^m$ of the form $\mu = A\mu + b$ with $A^m = (1/\rho)\mathbf{I}_m$. Thus

$$\mu = \frac{\rho}{\rho - 1}(\mathbf{I}_m + A + A^2 + \ldots + A^{m-1}) \cdot b.$$

Now consider the quadratic differential $\mathbf{q}$ (with double poles) defined by

$$\mathbf{q} := \sum_{k=0}^{m-1} \left(\frac{1}{(z - z_k)^2} + \frac{\mu_k}{z - z_k} \right) dz^2. \tag{3.6}$$

By the calculation of $(f_{c_0})_* \mathbf{q}$ in Lemmas 3.5 and 3.9, the polar parts of $\mathbf{q}$ and $(f_{c_0})_* \mathbf{q}$ along the cycle of z_0 are identical.

LEMMA 3.10 (Levin). Setting $f = f_{c_0}$, we have

$$f_* \mathbf{q} = \mathbf{q} - \frac{\dot{\rho}}{\rho} \cdot \frac{dz^2}{z - c_0}.$$

PROOF. Note that $f_* \mathbf{q}$ has an extra simple pole at the critical value c_0 with coefficient

$$\sum_{k=0}^{m-1} \left(-\frac{\mu_k}{2z_k} + \frac{1}{2z_k^2} \right) = -\sum_{k=0}^{m-1} \mu_{k+1}.$$

We need to show that this coefficient is equal to $-\dot{\rho}/\rho$.

Set
$$\zeta_k(c) := f_c^{\circ k}(\zeta(c)) \quad \text{and} \quad \dot{\zeta}_k := \zeta_k'(c_0).$$

Then
$$\zeta_{k+1}(c) = f_c(\zeta_k(c)) = (\zeta_k(c))^2 + c, \quad \zeta_n = \zeta_0 \quad \text{and} \quad \dot{\zeta}_{k+1} = 2z_k\dot{\zeta}_k + 1.$$

It follows that
$$\dot{\zeta}_{k+1}\mu_{k+1} - \mu_{k+1} = 2z_k\dot{\zeta}_k\mu_{k+1} = \dot{\zeta}_k\mu_k - \frac{\dot{\zeta}_k}{z_k}.$$

Therefore
$$\sum_{k=0}^{m-1} \mu_{k+1} = \sum_{k=0}^{m-1} \left(\dot{\zeta}_{k+1}\mu_{k+1} - \dot{\zeta}_k\mu_k + \frac{\dot{\zeta}_k}{z_k} \right) = \sum_{k=0}^{m-1} \frac{\dot{\zeta}_k}{z_k} = \frac{\dot{\rho}}{\rho},$$

where the last equality is obtained by evaluating the logarithmic derivative of
$$\rho_c := \prod_{k=0}^{m-1} 2\zeta_k(c)$$

at c_0. $\qquad \square$

To complete the proof that $\dot{\rho} \neq 0$, we will use a generalization of the contraction principle due to Epstein.

LEMMA 3.11 (Epstein). *We have $f_*\mathbf{q} \neq \mathbf{q}$.*

PROOF. The proof rests again on the contraction principle, but we can not apply Lemma 3.6 directly since $\mathbf{q}$ is not integrable near the cycle $\langle z_0, \ldots, z_{m-1} \rangle$. Consider a sufficiently large round disk V so that $U := f^{-1}(V)$ is relatively compact in V. Given $\varepsilon > 0$, we set
$$V_\varepsilon := \bigcup_{k=1}^{m} f^{\circ k}(D(z_0, \varepsilon)) \quad \text{and} \quad U_\varepsilon := f^{-1}(V_\varepsilon).$$

For ε sufficiently small, we have
$$\|f_*\mathbf{q}\|_{V-V_\varepsilon} \leq \|\mathbf{q}\|_{U-U_\varepsilon} = \|\mathbf{q}\|_{V-V_\varepsilon} - \|\mathbf{q}\|_{V-U} + \|\mathbf{q}\|_{V_\varepsilon-U_\varepsilon} - \|\mathbf{q}\|_{U_\varepsilon-V_\varepsilon}.$$

If we had $f_*\mathbf{q} = \mathbf{q}$, we would have
$$0 < \|\mathbf{q}\|_{V-U} \leq \|\mathbf{q}\|_{V_\varepsilon-U_\varepsilon} - \|\mathbf{q}\|_{U_\varepsilon-V_\varepsilon} \leq \|\mathbf{q}\|_{V_\varepsilon-U_\varepsilon}.$$

However, $\|\mathbf{q}\|_{V_\varepsilon-U_\varepsilon}$ tends to 0 as ε tends to 0, which is a contradiction. Indeed, $\mathbf{q} = q\, dz^2$, the meromorphic function q being equivalent to $1/(z - z_0)^2$ as z tends to z_0. In addition, since the multiplier of z_0 has modulus 1,
$$D(z_0, \varepsilon) \subset V_\varepsilon - U_\varepsilon \subset D(z_0, \varepsilon') \quad \text{with} \quad \frac{\varepsilon'}{\varepsilon} \xrightarrow[\varepsilon \to 0]{} 1.$$

Therefore,
$$\|\mathbf{q}\|_{V_\varepsilon-U_\varepsilon} \leq \int_0^{2\pi} \int_\varepsilon^{\varepsilon'} \frac{1 + o(1)}{r^2} r\, dr\, d\theta = 2\pi(1 + o(1)) \log \frac{\varepsilon'}{\varepsilon} \xrightarrow[\varepsilon \to 0]{} 0. \qquad \square$$

The proof of Theorem 1.1 is now completed.

4   IRREDUCIBILITY OF THE DYNATOMIC CURVES

Our objective is now to give a proof of Theorem 1.2, i.e., the irreducibility of the dynatomic curves X_n. Note that since the affine curve $\overline{X}_n$ is defined by a polynomial Q_n which has no repeated factors, this will prove that Q_n is irreducible. Since $\overline{X}_n$ is smooth, we may equivalently prove the following result.

THEOREM 4.1. *For every $n \geq 1$, the set $\overline{X}_n$ is connected.*

4.1   Kneading sequences

Set $\mathbb{T} = \mathbb{R}/\mathbb{Z}$ and let $\tau \colon \mathbb{T} \to \mathbb{T}$ be the angle-doubling map

$$\tau \colon \mathbb{T} \ni \theta \mapsto 2\theta \in \mathbb{T}.$$

By abuse of notation we shall often identify an angle $\theta \in \mathbb{T}$ with its representative in $[0, 1[$. In particular, the angle $\theta/2 \in \mathbb{T}$ is the element of $\tau^{-1}(\theta)$ with representative in $[0, 1/2[$ and the angle $(\theta + 1)/2$ is the element of $\tau^{-1}(\theta)$ with representative in $[1/2, 1[$.

Every angle $\theta \in \mathbb{T}$ has an associated kneading sequence $\nu(\theta) = \nu_1 \nu_2 \nu_3 \ldots$ defined by

$$\nu_k = \begin{cases} 1 & \text{if } \tau^{\circ(k-1)}(\theta) \in \left] \dfrac{\theta}{2}, \dfrac{\theta+1}{2} \right[, \\[2mm] 0 & \text{if } \tau^{\circ(k-1)}(\theta) \in \mathbb{T} - \left[\dfrac{\theta}{2}, \dfrac{\theta+1}{2} \right], \\[2mm] \star & \text{if } \tau^{\circ(k-1)}(\theta) \in \left\{ \dfrac{\theta}{2}, \dfrac{\theta+1}{2} \right\}. \end{cases}$$

For example, $\nu(1/7) = \overline{11\star}$ and $\nu(7/31) = \overline{1100\star}$.

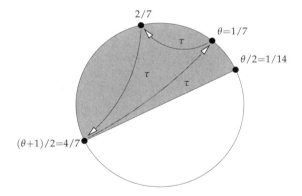

FIGURE 4.2: The kneading sequence of $\theta = 1/7$ is $\nu(1/7) = \overline{11\star}$.

We shall say that an angle $\theta \in \mathbb{T}$, periodic under τ, is *maximal in its orbit* if its representative in $[0, 1)$ is maximal among the representatives of $\tau^{\circ j}(\theta)$ in $[0, 1)$ for all $j \geq 1$. If the period is n and the binary expansion of θ is $.\overline{\varepsilon_1 \ldots \varepsilon_n}$, then θ is maximal in its orbit if and only if the periodic sequence $\overline{\varepsilon_1 \ldots \varepsilon_n}$ is maximal (in the lexicographic order) among its iterated shifts, where the shift of a sequence $\varepsilon_1 \varepsilon_2 \varepsilon_3$ indexed by $\mathbb{N}$ is

$$\sigma(\varepsilon_1 \varepsilon_2 \varepsilon_3 \ldots) = \varepsilon_2 \varepsilon_3 \varepsilon_4 \ldots.$$

EXAMPLE 4.3. The angle $7/31 = .\overline{00111}$ is not maximal in its orbit but the angle $28/31 = .\overline{11100}$ is maximal in the same orbit.

The following lemma indicates cases where the kneading sequence and the binary expansion coincide.

LEMMA 4.4. *Let $\theta \in \mathbb{T}$ be a periodic angle which is maximal in its orbit and let $.\overline{\varepsilon_1 \ldots \varepsilon_n}$ be its binary expansion. Then, $\varepsilon_n = 0$ and the kneading sequence $v(\theta)$ is $\overline{\varepsilon_1 \ldots \varepsilon_{n-1}\star}$.*

For example,

$$\frac{28}{31} = .\overline{11100} \quad \text{and} \quad v(\theta) = \overline{1110\star}.$$

PROOF. Since θ is maximal in its orbit under τ, the orbit of θ is disjoint from $]\theta/2, 1/2] \cup]\theta, 1]$. It follows that the orbit $\tau^{\circ j}(\theta)$, $j = 0, 1, \ldots, n - 2$ have the same itinerary relative to the two partitions

$$\mathbb{T} - \left\{0, \frac{1}{2}\right\} \quad \text{and} \quad \mathbb{T} - \left\{\frac{\theta}{2}, \frac{\theta+1}{2}\right\}.$$

The first one gives the binary expansion whereas the second gives the kneading sequence. Therefore, the kneading sequence of θ is $\overline{\varepsilon_1 \ldots \varepsilon_{n-1}\star}$. Since

$$\tau^{\circ(n-1)}(\theta) \in \tau^{-1}(\theta) = \left\{\frac{\theta}{2}, \frac{\theta+1}{2}\right\}$$

and since $(\theta + 1)/2 \in]\theta, 1]$, we must have $\tau^{\circ(n-1)}(\theta) = \theta/2 < 1/2$. So ε_n, as the first digit of $\tau^{\circ(n-1)}(\theta)$, must be equal to 0.   □

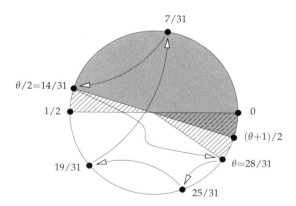

FIGURE 4.5: The kneading sequence of $\theta := 28/31 = .\overline{11100}$ is $v(28/31) = \overline{1110\star}$.

4.2   Filled-in Julia sets and the Mandelbrot set

We will use results proved by Douady and Hubbard in the Orsay notes [DH] that we now recall. Some simplified proofs of these results have been obtained by Schleicher in his Ph.D. thesis (see Schleicher [Sc1] and also Milnor [Mi4]) and by Petersen-Ryd [PR].

For $c \in \mathbb{C}$, we denote by K_c the filled-in Julia set of f_c, that is, the set of points $z \in \mathbb{C}$ whose orbit under f_c is bounded. We denote by M the Mandelbrot set, that is the set of parameters $c \in \mathbb{C}$ for which the critical point 0 belongs to K_c.

I. If $c \in M$, then K_c is connected. There is a conformal isomorphism ϕ_c between $\mathbb{C} - K_c$ and $\mathbb{C} - \overline{\mathbb{D}}$ which satisfies $\phi_c \circ f_c = f_0 \circ \phi_c$. The dynamical ray of angle $\theta \in \mathbb{T}$ is

$$R_c(\theta) := \{ z \in \mathbb{C} - K_c \mid \arg(\phi_c(z)) = 2\pi\theta \}.$$

If θ is rational, then as r tends to 1 from above, $\phi_c^{-1}(re^{2\pi i\theta})$ converges to a point $\gamma_c(\theta) \in K_c$. We say that $R_c(\theta)$ lands at $\gamma_c(\theta)$. We have $f_c \circ \gamma_c = \gamma_c \circ \tau$ on $\mathbb{Q}/\mathbb{Z}$. In particular, if θ is periodic under τ, then $\gamma_c(\theta)$ is periodic under f_c. In addition, $\gamma_c(\theta)$ is either repelling (its multiplier has modulus > 1) or parabolic (its multiplier is a root of unity).

II. If $c \notin M$, then K_c is a Cantor set. Here, there is a conformal isomorphism $\phi_c \colon U_c \to V_c$ between neighborhoods of ∞ in $\mathbb{C}$, which satisfies $\phi_c \circ f_c = f_0 \circ \phi_c$ on U_c. We may choose U_c so that U_c contains the critical value c and V_c is the complement of a closed disk. For each $\theta \in \mathbb{T}$, there is an infimum $r_c(\theta) \geq 1$ such that ϕ_c^{-1} extends analytically along $R_0(\theta) \cap \{ z \in \mathbb{C} \mid r_c(\theta) < |z| \}$. We denote by ψ_c this extension and by $R_c(\theta)$ the dynamical ray

$$R_c(\theta) := \psi_c \Big(R_0(\theta) \cap \{ z \in \mathbb{C} \mid r_c(\theta) < |z| \} \Big).$$

As r tends to $r_c(\theta)$ from above, $\psi_c(re^{2\pi i\theta})$ converges to a point $x \in \mathbb{C}$. If $r_c(\theta) > 1$, then $x \in \mathbb{C} - K_c$ is an iterated preimage of 0 and we say that $R_c(\theta)$ bifurcates at x. If $r_c(\theta) = 1$, then $\gamma_c(\theta) := x$ belongs to K_c and we say that $R_c(\theta)$ lands at $\gamma_c(\theta)$. Again, $f_c \circ \gamma_c = \gamma_c \circ \tau$ on the set of θ such that $R_c(\theta)$ does not bifurcate. In particular, if θ is periodic under τ and $R_c(\theta)$ does not bifurcate, then $\gamma_c(\theta)$ is periodic under f_c.

The Mandelbrot set is connected. The map

$$\phi_M \colon \mathbb{C} - M \ni c \mapsto \phi_c(c) \in \mathbb{C} - \overline{\mathbb{D}}$$

is a conformal isomorphism. For $\theta \in \mathbb{T}$, the parameter ray $R_M(\theta)$ is

$$R_M(\theta) := \{ c \in \mathbb{C} - M \mid \arg(\phi_M(c)) = 2\pi\theta \}.$$

It is known that if θ is rational, then as r tends to 1 from above, $\phi_M^{-1}(re^{2\pi i\theta})$ converges to a point $\gamma_M(\theta) \in M$. We say that $R_M(\theta)$ lands at $\gamma_M(\theta)$.

If θ is periodic for τ of exact period n and if $c = \gamma_M(\theta)$, then the point $\gamma_c(\theta)$ is periodic for f_c with period dividing n and multiplier a root of unity. If the period of $\gamma_c(\theta)$ for f_c is exactly n, then the multiplier is 1, $\gamma_c(\theta)$ disconnects K_c in exactly two connected components, and c is the root of a *primitive hyperbolic component* of M (as in Figure 4.6). Otherwise, c is the root of a *satellite hyperbolic component* of M (as in Figure 4.10).

The parameter ray $R_M(0)$ lands at $1/4$, and this is the only ray landing at $1/4$.

III. Let us now assume that $c \in \mathbb{C} - \{1/4\}$ is the root of a hyperbolic component of M, that is, f_c has a parabolic cycle. Then there are exactly two parameter rays $R_M(\theta)$ and $R_M(\eta)$ landing at c. We say that θ and η are *companion angles*. Both θ and η are periodic under τ with the same period. The hyperbolic component is primitive if and only if the orbits of θ and η under τ are distinct. Otherwise, the orbits are equal.

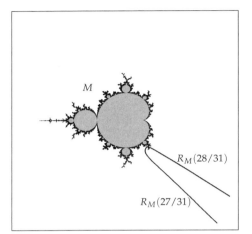
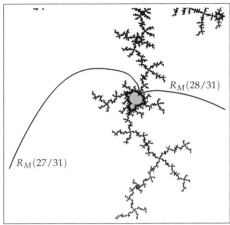

FIGURE 4.6: The parameter rays $R_M(27/31)$ and $R_M(28/31)$ land at a common root of a primitive hyperbolic component.

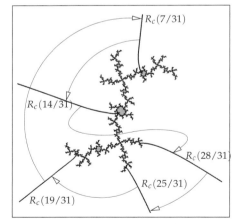
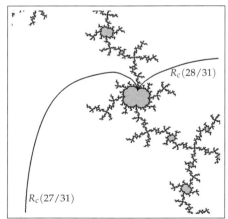

FIGURE 4.7: Filled-in Julia set K_c for the landing point $c := \gamma_M(28/31)$ illustrated in Figure 4.6, showing the orbit of the dynamical ray $R_c(28/31)$ on the left, and on the right showing that the rays $R_c(28/31)$ and $R_c(27/31)$ land at the same root point.

The dynamical rays $R_c(\theta)$ and $R_c(\eta)$ land on the Julia set at a common point $x_1 := \gamma_c(\theta) = \gamma_c(\eta)$. This point x_1 is the point of the parabolic cycle whose immediate basin contains the critical value c. The dynamical rays $R_c(\theta)$ and $R_c(\eta)$ are adjacent to the Fatou component containing c. The curve $R_c(\theta) \cup R_c(\eta) \cup \{x_1\}$ is a Jordan arc that cuts the plane in two connected components. One component, denoted V_0, contains the dynamical ray $R_c(0)$ and all the points of the parabolic cycle, except x_1. The other component, denoted V_1, contains the critical value c.

Since V_1 contains the critical value, its preimage $U_\star := f_c^{-1}(V_1)$ is connected and contains the critical point 0. It is bounded by the dynamical rays $R_c(\theta/2)$, $R_c(\eta/2)$, $R_c((\theta+1)/2)$ and $R_c((\eta+1)/2)$. Two of those dynamical rays land at the point x_0 of the parabolic cycle whose immediate basin contains the critical point 0. The two other dynamical rays land at $-x_0$. Since V_0 does not contain the critical value, its preimage has two connected components. One component, denoted U_0, contains the dynamical ray $R_c(0)$. The other component is denoted U_1 (see Figure 4.8).

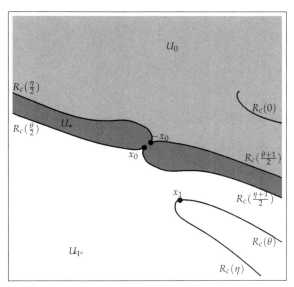

FIGURE 4.8: The regions for $c = \gamma_M(\theta)$

LEMMA 4.9. *Let $\theta \in \mathbb{T}$ be a periodic angle of period n which is maximal within its orbit. Then, $\gamma_M(\theta)$ is the root point of a hyperbolic component, which is*

- *of satellite type if $\theta = .\overline{11\ldots10} = \dfrac{2^n - 2}{2^n - 1}$ with $n \geq 2$,*

- *a primitive hyperbolic component in all other cases.*

PROOF. If $\theta = 0$, then $\gamma_M(0) = 1/4$ is the root of a hyperbolic component. So, without loss of generality, we may assume that $\theta \neq 0$. Let $n \geq 2$ be the period of θ under τ, let η be the companion angle of θ and let U_0 and U_1 and $U_\star$ be defined as before (see Figure 4.8).

Since θ is maximal in its orbit, $\tau^{\circ(n-1)}(\theta) = \theta/2$ (see Lemma 4.4). So, $R_c(\theta/2)$ lands on x_0. One of the two rays $R_c(\eta/2)$ and $R_c((\eta+1)/2)$ lands on x_0. Since $U_\star$

is connected and contains dynamical rays with angles in between $\eta/2$ and $\theta/2$ and dynamical rays with angles in between $(\eta+1)/2$ and $(\theta+1)/2$, the ray landing on x_0 has to be $R_c((\eta+1)/2)$. It follows that $(\eta+1)/2$ is in the orbit of η under τ.

Since $\theta/2 < \theta < (\theta+1)/2$ and since $R_c(\theta)$ avoids $U_\star$, we have $\theta \le (\eta+1)/2$. On the one hand, if $\theta < (\eta+1)/2$, then the orbit of θ under τ does not contain $(\eta+1)/2$ since otherwise θ would not be maximal in its orbit. In that case, the orbits of θ and η are disjoint and $\gamma_M(\theta)$ is the root of a primitive hyperbolic component. On the other hand, if $\theta = (\eta+1)/2$, then the rays $R_c((\eta+1)/2)$ and $R_c(\theta)$ are equal. In that case, their landing point is the same, so $x_0 = x_1 = f_c(x_0)$ is a fixed point of f_c. The rays landing at this fixed point are permuted cyclically. The dynamical rays $R_c(\theta)$ and $R_c(\eta)$ are consecutive among the rays landing at $x_0, \eta < (\eta+1)/2 = \theta$ and $R_c(\theta)$ is mapped to $R_c(2\theta) = R_c(\eta)$. It follows that each dynamical ray landing at x_0 is mapped to the one which is once further clockwise. Consequently, the kneading sequence of θ is $\overline{1\ldots1\star}$ and, according to Lemma 4.4, the binary expansion of θ is $.\overline{1\ldots10}$. See Figure 4.15. $\qquad\square$

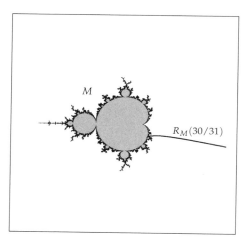
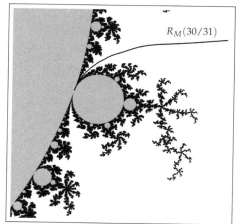

FIGURE 4.10: We have $30/31 = .\overline{11110}$, and the parameter ray $R_M(30/31)$ lands on the boundary of the main cardioid.

4.3   Outside the Mandelbrot set

The projection $\pi_c : \overline{X}_n \to \mathbb{C}$ is a ramified covering. According to Proposition 3.1, the critical points are the points $(c, z) \in \overline{X}_n$ such that $f_c^{on}(z) = z$ and $(f_c^{on})'(z) = 1$. So, the critical values are precisely the roots of the polynomial $\Delta_n \in \mathbb{Z}[c]$ which is the discriminant of $P_n \in \mathbb{Z}[z]$. Those critical values are contained in the Mandelbrot set since a parabolic cycle for f_c attracts the critical point of f_c.

The open set

$$W := \mathbb{C} - \left(M \cup R_M(0) \right)$$

is simply connected. This set avoids the critical values of the ramified covering $\pi_c : \overline{X}_n \to \mathbb{C}$. Let $W_n \subset X_n$ be the preimage of W by $\pi_c : \overline{X}_n \to \mathbb{C}$. It follows from the previous comment that $\pi_c : W_n \to W$ is a (unramified) cover, which is trivial since W is simply connected: each connected component of W_n maps isomorphically to W by π_c.

Note that each connected component of $\overline{X}_n$ is unbounded (because $\overline{X}_n$ is an affine curve) and so intersects W_n. Thus, in order to prove that $\overline{X}_n$ is connected, it is enough to prove that the closure $\overline{W}_n$ of W_n in $\overline{X}_n$ is connected. We shall say that two components of W_n are adjacent if they have a common boundary point in $\overline{X}_n$.

4.4   Labeling components of W_n

Here, we explain how the components of W_n may be labeled dynamically.

A parameter $c \in W$ belongs to a parameter ray $R_M(\theta)$ with $\theta \neq 0$ not necessarily periodic. The dynamical rays $R_c(\theta/2)$ and $R_c((\theta+1)/2)$ bifurcate on the critical point. The Jordan curve $R_c(\theta/2) \cup R_c((\theta+1)/2) \cup \{0\}$ separates the complex plane in two connected components. We denote by U_0 the component containing the dynamical ray $R_c(0)$ and by U_1 the other component (see Figure 4.11).

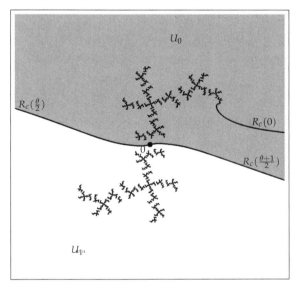

FIGURE 4.11: The regions U_0 and U_1 for a parameter c belonging to the ray $R_M(\theta)$, with $\theta = 28/31$.

The orbit of a point $z \in K_c$ has an itinerary with respect to this partition. In other words, to each $z \in K_c$, we can associate a sequence $\iota_c(z) \in \{0,1\}^{\mathbb{N}}$ whose jth term is equal to 0 if $f_c^{\circ(j-1)}(z) \in U_0$ and is equal to 1 if $f_c^{\circ(j-1)}(z) \in U_1$. A point $z \in K_c$ is periodic for f_c if and only if the itinerary $\iota_c(z)$ is periodic for the shift with the same period. The map $\iota_c \colon K_c \to \{0,1\}^{\mathbb{N}}$ is a bijection.

Let us define $\iota_n \colon W_n \to \{0,1\}^{\mathbb{N}}$ by

$$\iota_n(c,z) := \iota_c(z).$$

As c varies in W, the periodic points of f_c, the dynamical ray $R_c(0)$ and the Jordan curve $R_c(\theta/2) \cup R_c((\theta+1)/2) \cup \{0\}$ move continuously. As a consequence, the map $\iota_n \colon W_n \to \{0,1\}^{\mathbb{N}}$ is locally constant, whence constant on each connected component of W_n. So, each connected component V of W_n may be label-led by the itinerary $\iota_n(V)$. Since $\iota_c \colon K_c \to \{0,1\}^{\mathbb{N}}$ is injective, distinct components have distinct labels. Since $\iota_c \colon K_c \to \{0,1\}^{\mathbb{N}}$ is surjective, each periodic itinerary of period n is the label of a component of W_n. It follows that the number of connected components of W_n is equal to the number of n-periodic sequences in $\{0,1\}^{\mathbb{N}}$.

4.5 Turning around critical points

We now exhibit connected components of W_n which have common boundary points. The following statement is one of the key results in [LS].

PROPOSITION 4.12. *Let $\overline{\varepsilon_1 \ldots \varepsilon_{n-1}\star}$ be the kneading sequence of an angle $\theta \in \mathbb{T} - \{0\}$ which is periodic with period n. If $\gamma_M(\theta)$ is the root of a primitive hyperbolic component and if one follows continuously the periodic points of period n of f_c as c makes a small turn around $\gamma_M(\theta)$, then the periodic points with itineraries $\overline{\varepsilon_1 \ldots \varepsilon_{n-1}0}$ and $\overline{\varepsilon_1 \ldots \varepsilon_{n-1}1}$ get exchanged.*

PROOF. Set $c_0 := \gamma_M(\theta)$. Since c_0 is the root of a primitive hyperbolic component, the periodic point $x_1 := \gamma_{c_0}(\theta)$ has period n and multiplier 1. According to Case 2 of Theorem 3.1 (see also [DH, Exposé XIV, Proposition 3]), X_n is smooth at (c_0, x_1), the projection to the first coordinate has degree 2 and the projection to the second coordinate has degree 1. So, in a neighborhood of (c_0, x_1) in $\mathbb{C}^2$, X_n can be written as

$$\{(c_0 + \delta^2, x(\delta)), (c_0 + \delta^2, x(-\delta))\}$$

where $x \colon (\mathbb{C}, 0) \to (\mathbb{C}, x_1)$ is a holomorphic germ with $x'(0) \neq 0$. In particular, as c moves away from c_0, the periodic point x_1 of f_{c_0} splits into a pair of nearby periodic points $x(\pm\sqrt{c - c_0})$ for f_c, that get exchanged when c makes a small turn around c_0. So, it is enough to show that for $c \in \mathbb{C} - M$ close to c_0, those two periodic points have itineraries $\overline{\varepsilon_1 \ldots \varepsilon_{n-1}0}$ and $\overline{\varepsilon_1 \ldots \varepsilon_{n-1}1}$.

Let us denote by $V_0(c_0)$, $V_1(c_0)$, $U_0(c_0)$, $U_1(c_0)$ and $U_\star(c_0)$ the open sets V_0, V_1, U_0, U_1 and $U_\star$ defined in Section 4.2, part III (for $c_0 = c$). For $j \geq 0$, set $x_j := f_{c_0}^{\circ j}(x_0)$ and observe that for $j \in [1, n-1]$, we have $x_j \in U_{\varepsilon_j}(c_0)$.

For $c \in R_M(\theta)$, consider the following compact subsets of the Riemann sphere $\mathbb{C} \cup \{\infty\}$:

$$R(c) := R_c(\theta) \cup \{c, \infty\} \quad \text{and} \quad S(c) := R_c(\theta/2) \cup R_c((\theta+1)/2) \cup \{0, \infty\}.$$

Denote by $U_0(c)$ the component of $\mathbb{C} - S(c)$ containing $R_c(0)$ and by $U_1(c)$ the other component. From any sequence $c_j \in R_M(\theta)$ converging to c_0, we can extract a subsequence so that $R(c_j)$ and $S(c_j)$ converge, in the Hausdorff topology on compact subsets of $\mathbb{C} \cup \{\infty\}$, to connected compact sets R and S, respectively. Since $S(c) = f_c^{-1}(R(c))$, we have $S = f_{c_0}^{-1}(R)$. According to [PR, Sections 2 and 3], $R \cap (\mathbb{C} - K_{c_0}) = R_{c_0}(\theta)$, the intersection of R with the boundary of K_{c_0} is reduced to $\{x_1\}$, and the intersection of R with the interior of K_{c_0} is contained in the immediate basin of x_1, whence in $V_1(c_0)$. It follows that as $c \in R_M(\theta)$ tends to c_0, any Hausdorff accumulation value of the family of compact sets $R(c)$ is contained in $\overline{V}_1(c_0)$ and so any accumulation value of the family of compact sets $S(c)$ is contained in $\overline{U}_\star(c_0)$. In other words, any compact subset of $\mathbb{C} - \overline{U}_\star(c_0)$ is contained in $\mathbb{C} - S(c)$ for $c \in R_M(\theta)$ close enough to c_0. More precisely, every compact subset of $U_0(c_0)$ is contained in a connected compact set $L \subset U_0(c_0)$ whose interior intersects $R_{c_0}(0)$; for $c \in R_M(\theta)$ close enough to c_0, L intersects $R_c(0)$ and is contained in $\mathbb{C} - S(c)$, whence in $U_0(c)$. As a consequence, every compact subset of $U_0(c_0)$

is contained in $U_0(c)$ for $c \in R_M(\theta)$ close enough to c_0. Similarly, every compact subset of $U_1(c_0)$ is contained in $U_1(c)$ for $c \in R_M(\theta)$ close enough to c_0.

Fix $j \in [1, n-1]$ and let D_j be a sufficiently small disk around x_j so that

$$\overline{D}_j \subset U_{\varepsilon_j}(c_0) \subset \mathbb{C} - \overline{U}_*(c_0).$$

According to the previous discussion, if $c \in R_M(\theta)$ is close enough to c_0, we have

$$f_c^{\circ (j-1)}\big(x(\pm\sqrt{c-c_0})\big) \subset D_j \subset U_{\varepsilon_j}(c).$$

So, the itineraries of $x(\pm\sqrt{c-c_0})$ are of the form $\overline{\varepsilon_1 \ldots \varepsilon_{n-1}\varepsilon^{\pm}}$ with $\varepsilon^{\pm} \in \{0,1\}$ and $\varepsilon^+ \neq \varepsilon^-$ (each itinerary corresponds to a unique point in K_c). The result follows. $\qquad \square$

COROLLARY 4.13. *Let $\overline{\varepsilon_1 \ldots \varepsilon_{n-1}\star}$ be the kneading sequence of an angle $\theta \in \mathbb{T} - \{0\}$ which is periodic with period n. If $\gamma_M(\theta)$ is the root of a primitive hyperbolic component, then the components of W_n with labels $\overline{\varepsilon_1 \ldots \varepsilon_{n-1}0}$ and $\overline{\varepsilon_1 \ldots \varepsilon_{n-1}1}$ are adjacent.*

PROOF. According to the previous proposition, the closures of those components both contain the point (c_0, x_1) with $c_0 := \gamma_M(\theta)$ and $x_1 := \gamma_{c_0}(\theta)$. $\qquad \square$

PROPOSITION 4.14. *Let $\theta = (2^n - 2)/(2^n - 1) = .\overline{1 \ldots 10}$ be periodic with period $n \geq 2$. If one follows continuously the periodic points of period n of f_c as c makes a small turn around $\gamma_M(\theta)$, then the periodic points in the cycle of $\iota_c^{-1}(\overline{1 \ldots 10})$ are cyclically permuted.*

PROOF. Set $c_0 := \gamma_M(\theta)$. As mentioned earlier, all the dynamical rays $R_{c_0}\big(\tau^{\circ j}(\theta)\big)$ land on a common fixed point x_0. This fixed point is parabolic and each ray landing at x_0 is mapped to the one which is once further clockwise (see Figure 4.15). It follows that the multiplier of f_{c_0} at x_0 is $\omega := e^{-2\pi i/n}$.

According to Case 3 of Theorem 3.1, $\overline{X}_n$ is smooth at (c_0, x_0), the projection to the first coordinate has local degree n and the projection to the second coordinate has local degree 1. It follows that in a neighborhood of (c_0, x_0) in $\mathbb{C}^2$, $\overline{X}_n$ can be written as

$$\big\{(c_0 + \delta^n, x(\delta)), (c_0 + \delta^n, x(\omega\delta)), \ldots, (c_0 + \delta^n, x(\omega^{n-1}\delta))\big\}$$

where $x \colon (\mathbb{C}, 0) \to (\mathbb{C}, x_0)$ is a holomorphic germ satisfying $x'(0) \neq 0$. In addition,

$$f_{c_0 + \delta^n}\big(x(\delta)\big) = x(\omega\delta).$$

So, for c close to c_0, the set $x\{\sqrt[n]{c-c_0}\}$ is a cycle of period n of f_c, and when c makes a small turn around c_0, the periodic points in the cycle $x\{\sqrt[n]{c-c_0})\}$ get permuted cyclically. So, it is enough to show that for $c \in \mathbb{C} - M$ close enough to c_0, the point $\iota_c^{-1}(\overline{1 \ldots 10})$ belongs to $x\{\sqrt[n]{c-c_0}\}$.

Equivalently, we must show that there is a sequence $c_j \in \mathbb{C} - M$ converging to c_0, such that the sequence of periodic point $y_j = \iota_{c_j}^{-1}(\overline{1 \ldots 10})$ converges to x_0. Let $c_j \in R_M(\theta)$ converge to c_0. As in the previous proof, consider the following compact subsets of the Riemann sphere $\mathbb{C} \cup \{\infty\}$:

$$R(c_j) := R_{c_j}(\theta) \cup \{c_j, \infty\} \quad \text{and} \quad S(c_j) := R_{c_j}(\theta/2) \cup R_{c_j}\big((\theta+1)/2\big) \cup \{0, \infty\}.$$

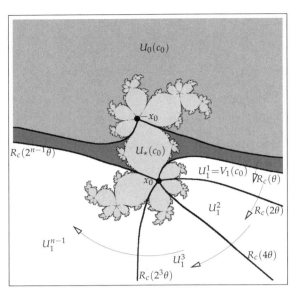

FIGURE 4.15: For $c_0 := \gamma_M(.\overline{11110})$, the dynamical rays $R_{c_0}\big(\tau^{\circ j}(\theta)\big)$ land on a common fixed point x_0.

Let $U_0(c_j)$ denote the component of $\mathbb{C} - S(c_j)$ containing $R_{c_j}(0)$, and let $U_1(c_j)$ be the other component. Without loss of generality, and extracting a subsequence if necessary, we may assume that the sequence y_j converges to a point y and that the sequence $R(c_j)$ and $S(c_j)$ have Hausdorff limits R and S. Passing to the limit on $f_{c_j}^{\circ n}(y_j) = y_j$, we see that $f_{c_0}^{\circ n}(y) = y$, and so y is periodic for f_{c_0} with period dividing n. In particular, it is contained in the boundary of K_{c_0}. We must show that $y = \{x_0\}$.

It follows from [PR, Sections 2 and 3] that $R \cap (\mathbb{C} - K_{c_0}) = R_{c_0}(\theta)$, the intersection of R with the boundary of K_{c_0} is reduced to $\{x_0\}$ and the intersection of R with the interior of K_{c_0} is contained in the immediate basin of x_0. We cannot quite conclude that L is contained in $\overline{U}_\star(c_0)$, but rather that it is contained in $\overline{U}_\star(c_0) \cup \mathring{K}_{c_0}$. As in the previous proof, it follows that the Hausdorff limit of $\overline{U}_1(c_j)$ is contained in $\overline{U}_\star(c_0) \cup U_1(c_0) \cup \mathring{K}_{c_0}$. Since $\iota_{c_j}(y_j) = \overline{1\dots 10}$, we know that $y_j, f_{c_j}(y_j), \dots, f_{c_j}^{\circ(n-2)}(y_j)$ belong to $U_1(c_j)$. So the points $y, f_{c_0}(y), \dots, f_{c_0}^{\circ(n-2)}(y)$ belong to $\overline{U}_\star(c_0) \cup U_1(c_0) \cup \mathring{K}_{c_0}$. Since y is in the boundary of K_{c_0}, we deduce that $y, f_{c_0}(y), \dots, f_{c_0}^{\circ(n-2)}(y)$ belong to $\overline{U}_\star(c_0) \cup U_1(c_0)$.

The dynamical rays landing at x_0 divide $U_1(c_0)$ in $n-1$ connected components U_1^j label-led clockwise so that

$$V_1(c_0) = U_1^1 \xrightarrow{f_{c_0}} U_1^2 \xrightarrow{f_{c_0}} \cdots \xrightarrow{f_{c_0}} U_1^{n-1} \xrightarrow{f_{c_0}} \overline{U}_\star(c_0) \cup U_0(c_0).$$

The component $U_\star(c_0)$ maps with degree 2 to $U_1^1 = V_1(c_0)$ (see Figure 4.15).

Now, we claim that the orbit of y intersects $\overline{U}_\star(c_0)$. Indeed, either y itself is in $\overline{U}_\star(c_0)$, or y is in U_1^j for some $j \geq 1$. Then, $f_{c_0}^{\circ(n-j)}(y) \in \overline{U}_\star(c_0) \cup U_0(c_0)$. Since it cannot be in $U_0(c_0)$, it belongs to $\overline{U}_\star(c_0)$.

The map $f_{c_0}^{\circ n}\colon U_\star(c_0) \to \mathbb{C} - \overline{U}_1(c_0)$ is a proper map of degree 2 and we have $U_\star(c_0) \subset \mathbb{C} - \overline{U}_1(c_0)$. Note that $x_0 \in \overline{U}_\star(c_0)$ is a multiple fixed point of $f_{c_0}^{\circ n}$ and that there is an attracting petal contained in $U_\star(c_0)$. It follows from a version of the Lefschetz fixed point formula (see [GM, Lemma 3.7]) that x_0 is the only fixed point of $f_{c_0}^{\circ n}$ contained in $\overline{U}_\star(c_0)$.

As a consequence, the orbit of the periodic point y contains x_0, and since x_0 is a fixed point, we have $y = x_0$ as required. □

COROLLARY 4.16. *The components of W_n whose labels contain a single 0 are adjacent.*

PROOF. Let $\theta := .\overline{1\ldots10}$ and $(c_0, x_0) := (\gamma_M(\theta), \gamma_{c_0}(\theta))$. By Proposition 4.14, every component of W_n whose label is a shift of $\overline{1\ldots10}$ contains (c_0, x_0) in its boundary. But every n-periodic label containing a single 0 is indeed a shift of $\overline{1\ldots10}$, so the result follows. □

4.6 Proof of Theorem 4.1

We will finally deduce that $\overline{W}_n$ is connected. According to Corollary 4.16, components of W_n whose label contain a single 0 have a common boundary point. So, it is enough to show that a component of W_n whose label has at least two 0 has a common boundary point with a component of W_n whose label has one less 0.

The map $F\colon \mathbb{C}^2 \to \mathbb{C}^2$ defined by

$$F(c, z) := (c, f_c(z))$$

restricts to an isomorphism $F\colon X_n \to X_n$. It permutes the components of W_n as follows: the label of $F(C)$ is the shift of the label of C. In addition, two components C_1 and C_2 of W_n are adjacent if and only if $F(C_1)$ and $F(C_2)$ are adjacent.

Let C be a connected component of W_n whose label ι contains at least two 0. Let $\overline{\varepsilon_1 \ldots \varepsilon_n} = \sigma^{\circ k}(\iota)$ be maximal (in the lexicographic order) among the iterated shifts of ι. Then, the angle $\theta := .\overline{\varepsilon_1 \ldots \varepsilon_n}$ is maximal in its orbit. According to Lemma 4.4, $\varepsilon_n = 0$ and the kneading sequence $\nu(\theta)$ is $\overline{\varepsilon_1 \ldots \varepsilon_{n-1}\star}$. According to Lemma 4.9, $\gamma_M(\theta)$ is the root of a primitive hyperbolic component. According to Corollary 4.13, the component $F^{\circ k}(C)$ which is labeled $\overline{\varepsilon_1 \ldots \varepsilon_{n-1}0}$ is adjacent to the component C' of W_n which is labeled $\overline{\varepsilon_1 \ldots \varepsilon_{n-1}1}$. Then, $F^{\circ(n-k)}(C')$ is a component of W_n adjacent to $F^{\circ(n-k)}(F^{\circ k}(C)) = C$, and its label contains one less 0 than the label of C.

This completes the proof of Theorem 4.1. □

5 Bibliography

[B]   T. BOUSCH, *Sur quelques problèmes de dynamique holomorphe*, Ph.D. thesis, Université de Paris-Sud, Orsay, 1992.

[BKM] A. BONIFANT, J. KIWI and J. MILNOR, *Cubic polynomial maps with periodic critical orbit. II. Escape regions*, Conform. Geom. Dyn. 14 (2010), 68–112.

[DH]   A. DOUADY and J. H. HUBBARD, *Etude dynamique des polynômes complexes (Deuxième partie)*, Publications Mathématiques d'Orsay, 1985.

[E]   A. L. EPSTEIN, *Transversality Principles in Holomorphic Dynamics*, Preprint. http://homepages.warwick.ac.uk/~mases/Transversality.pdf.

[GM]   L. R. GOLDBERG and J. MILNOR, *Fixed points of polynomial maps, Part II: Fixed point portraits.* Ann. Sci. Éc. Norm. Supér., IV. Sér. 26(1) (1993), 51–98.

[LS]   E. LAU and D. SCHLEICHER, *Internal addresses in the Mandelbrot set and irreducibility of polynomials,* arXiv:math/9411238v1. Stony Brook IMS Preprint 1994/19. http://www.math.sunysb.edu/preprints/ims94-19.pdf.

[L]   G. LEVIN, *On explicit connections between dynamical and parameter spaces,* Journal d'Analyse Mathematique, 91 (2003), 297–327.

[Ma]   M. MANES, *Moduli spaces for families of rational maps on* $\mathbb{P}^1$, J. Number Theory, 129(7) (2009), 1623–1663.

[Mi1]   J. MILNOR, *Geometry and dynamics of quadratic rational maps; With an appendix by the author and Tan Lei,* Experiment. Math., 2(1) (1993), 37–83.

[Mi2]   J. MILNOR, *Tsujii's monotonicity proof for real quadratic maps,* Preprint (2000). http://www.math.sunysb.edu/~jack/PREPRINTS/tsujii.ps.

[Mi3]   J. MILNOR, *Cubic polynomial maps with periodic critical orbit. I,* in "Complex dynamics, Families and friends," ed. D. Schleicher, A. K. Peters, Wellesley, MA (2009), 333–411.

[Mi4]   J. MILNOR, *Periodic orbits, external rays and the Mandelbrot set: An Expository Account; Géométrie complexe et systèmes dynamiques,* Astérisque 261 (2000), 277–333.

[Mo]   P. MORTON, *On certain algebraic curves related to polynomial maps,* Compositio Math. 103(3) (1996), 319–350.

[PR]   C. L. PETERSEN and G. RYD, *Convergence of rational rays in parameter-spaces,* in "The Mandelbrot set, Theme and Variations," ed. Tan Lei, London Mathematical Society, Lecture Note Series 274. Cambridge University Press, 2000.

[Sc1]   D. SCHLEICHER, *External Rays of the Mandelbrot Set,* Astérisque 261 (2000), 409–447.

[Sc2]   D. SCHLEICHER, *Internal addresses of the Mandelbrot set and Galois groups of polynomials* (2008). arXiv:math/9411238v2.

[Si]   J.H. SILVERMAN, "The arithmetic of dynamical systems." Graduate Texts in Math. 241, Springer, New York, 2007.

Multicorns are not path connected

John Hamal Hubbard and Dierk Schleicher

> Milnor, that intrepid explorer,
> Traveled cubics in hopes to discover
> Some exotic new beast:
> Northwest and southeast
> He found tricorns lurking there under cover.

ABSTRACT. The tricorn is the connectedness locus in the space of antiholomorphic quadratic polynomials $z \mapsto \bar{z}^2 + c$. We prove that the tricorn is not locally connected and not even pathwise connected, confirming an observation of John Milnor from 1992. We extend this discussion more generally for antiholomorphic unicritical polynomials of degrees $d \geq 2$ and their connectedness loci, known as multicorns.

1 INTRODUCTION

The **multicorn** $\mathcal{M}_d^*$ is the connectedness locus in the space of antiholomorphic unicritical polynomials $p_c(z) = \bar{z}^d + c$ of degree d, i.e., the set of parameters for which the Julia set is connected. The special case $d = 2$ is the **tricorn**, which is the formal antiholomorphic analog to the Mandelbrot set.

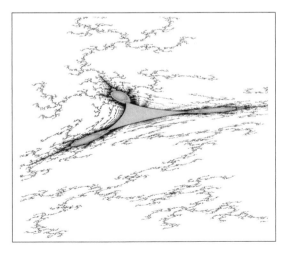

FIGURE 1.1: A "little tricorn" within the tricorn $\mathcal{M}_2^*$ illustrating that the "umbilical cord" converges to the little tricorn without landing at it. (See also Plate 2.)

The second iterate is

$$p_c^{\circ 2}(z) = \overline{\left(\overline{z^d} + c\right)}^d + c = (z^d + \overline{c})^d + c$$

and thus is holomorphic in the dynamical variable z but no longer complex analytic in the parameter c. Much of the dynamical theory of antiholomorphic polynomials (in short, antipolynomials) is thus in analogy to the theory of holomorphic polynomials, except for certain features near periodic points of odd periods. For instance, a periodic point of odd period k may be the simultaneous landing point of dynamic rays of periods k and $2k$ (which is invisible from the holomorphic second iterate of the first return map); see [NS, Lemma 3.1].

However, the theory of parameter space of multicorns is quite different from that of the Mandelbrot set and its higher-degree cousins, the multibrot sets of degree d, because the parameter dependence is only real analytic. Already the open mapping principle of the multiplier map fails, so it is not a priori clear that every indifferent orbit is on the boundary of a hyperbolic component and that bifurcations multiplying periods occur densely on boundaries of hyperbolic components. However, it turns out that many properties of parameter space are quite similar to that of the Mandelbrot set, except near hyperbolic components of odd period. For instance, there is a simple recursive relation for the number of hyperbolic components of period n for the multibrot set, given by $s_{d,n} = d^{n-1} - \sum_{k|n,\,k<n} s_{d,k}$: for multicorns, the same result holds, except if n is twice an odd number; in that case, the number of hyperbolic components equals $s_{d,n} + 2s_{d,n/2}$ [MNS]. Similarly, the multiplier map is an open map on the closure of any hyperbolic component of even period, except where it intersects the boundary of an odd-period-hyperbolic component.

However, boundaries of odd-period hyperbolic components have some quite interesting properties. The multiplier map is constant along their boundaries (all boundary points have parabolic orbits of multiplier $+1$); bifurcations only double the period (no higher factors), and these period doublings occur along arcs rather than at isolated points (see Corollary 3.9). Adjacent to these parabolic arcs, there are comb-like structures where the multicorn fails to be locally connected, and $(\sin 1/x)$-like structures accumulate on the centers of many parabolic arcs: even pathwise connectivity fails there. Nonetheless, some boundary arcs of odd-period hyperbolic components also feature "open beaches" with sub-arcs of positive length that form part of boundary of a hyperbolic component without any further decorations, so the hyperbolic component and the escape locus meet along a smooth arc. (We do not know whether the number of such arcs is finite or not.)

Overview of Paper and Results. In this paper we study the boundaries of hyperbolic components of $\mathcal{M}_d^*$ of odd period, focusing on local connectivity and pathwise connectivity. In Section 2, we investigate parabolic dynamics, especially of odd period, and review Ecalle cylinders and their special features in antiholomorphic dynamics: the existence of an invariant curve called the *equator*. We then discuss parabolic arcs on the boundary of hyperbolic components of odd period. In Section 3, we investigate these arcs from the point of view of the holomorphic fixed-point index, and we show that period-doubling bifurcations occur near both ends of all parabolic arcs. We then discuss, in Section 4, perturbations of parabolic periodic points and introduce continuous coordinates for the perturbed dynamics. In Section 5 we introduce an invariant tree in parabolic dynamics, similar to the Hubbard tree for postcritically finite polynomials, and discuss the dynamical properties of parabolic maps that we will later transfer into parameter space. This transfer is then done in Section 6, using perturbed Fatou coordinates: these are somewhat simplified in the antiholomorphic setting because of the existence of the invariant equator. Some concluding remarks and further results are discussed in Section 7.

Relations to Holomorphic Parameter Spaces. Much of the relevance of the tricorn (and the higher-dimensional multicorns) comes from the fact that it is related to natural holomorphic parameter spaces. Clearly, the tricorn space is the (real two-dimensional but not complex-analytic) slice $c = \bar{a} = b$ in the complex two-dimensional space of maps $z \mapsto (z^2 + a)^2 + b$, one of the natural complex two-dimensional spaces of polynomials. Perhaps more interestingly, the tricorn is naturally related to the space of *real* cubic polynomials: this space can be parametrized as $z \mapsto \pm z^3 - 3a^2 z + b$ with $a, b \in \mathbb{R}$. It was in the context of this space that Milnor discovered and explored the tricorn [M2, M3] as one of the prototypical local dynamical features in the presence of two active critical points; compare Figure 1.2. To see how antiholomorphic dynamics occurs naturally in the dynamics of a real cubic polynomial p, suppose there is an open bounded topological disk $U \subset \mathbb{C}$ containing one critical point so that $p(U)$ contains the closure of the complex conjugate of U. Denoting complex conjugation of p by p^* and the closure of U by $\overline{U}$, we have $p^*(U) \supset \overline{U}$. If, possibly by suitable restriction, the map $p \colon U \to p(U)$ is proper holomorphic, then $p^* \colon U \to p^*(U)$ is the antiholomorphic analogue of a polynomial-like map in the sense of Douady and Hubbard. Since p commutes with complex conjugation, the dynamics of p and of p^* are the same (the even iterates coincide), so the dynamics of p near one critical point is naturally described by the antiholomorphic polynomial p^* (and the other critical point is related by conjugation). The advantage of the antiholomorphic point of view is that, while U and $p(U)$ may be disjoint domains in $\mathbb{C}$ without obvious dynamical relation, there is a well-defined antiholomorphic dynamical system $p^* \colon U \to p^*(U)$. This is even more useful when $\overline{U}$ is a subset not of $p^*(U)$, but of a higher iterate: in this case, like for ordinary polynomial-like maps, the interesting dynamics of a high-degree polynomial is captured by a low-degree polynomial or, in this case, antipolynomial.

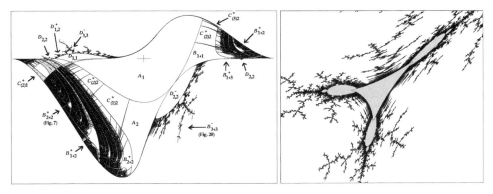

FIGURE 1.2: The connectedness locus of real cubic polynomials and a detail from the southeast quadrant, showing a tricorn-like structure. (See also Plate 4.) Pictures from Milnor [M2], reprinted by permission of Taylor & Francis.

Are There Embedded Tricorns? It was numerically "observed" by several people that the tricorn contains, around each hyperbolic component of even period, a small copy of the Mandelbrot set, and around each odd period component a small copy of the tricorn itself; and similar statements hold for certain regions of the real cubic connectedness locus — much as the well-known fact that the Mandelbrot set contains a small copy of itself around each hyperbolic component. A small tricorn within the big one is shown in Figure 1.3. However, we believe that most, if not all, "little tricorns" are not homeomorphic to the actual tricorn (both within the tricorn space and within the real cubic locus); quite possibly most little tricorns might not even be homeomorphic to each other. Indeed, a subset of the real axis connects the main hyperbolic component (of period 1) to the period 3 "airplane" component (along the real axis, the tricorn and the Mandelbrot set coincide obviously): we say that the "umbilical cord" of the period 3 tricorn lands. However, we prove for many little tricorns that their umbilical cords do not land but rather form some kind of $(\sin 1/x)$-structure. Our methods apply only to "prime" little tricorns: these are the ones not contained in larger "little tricorns," so we do not disprove continuity of the empirically observed embedding map given by the straightening theorem (even though this seems very likely). Two little tricorns could be homeomorphic to each other only if they have matching sizes of the wiggles of the umbilical cords of all the infinitely many little tricorns they contain, where the size of such a wiggle is measured in terms of Ecalle heights, as introduced later in this paper; see Definition 2.4.

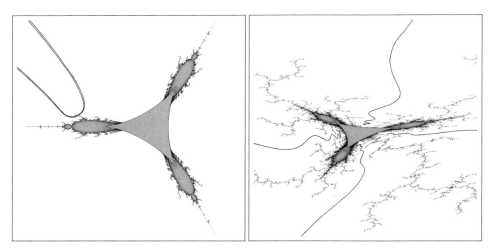

FIGURE 1.3: The tricorn and a blow-up showing a "small tricorn" of period 5. Shown in both pictures are the four parameter rays accumulating at the boundary of the period 5 hyperbolic component (at angles 371/1023, 12/33, 13/33, and 1004/1023). The wiggly features of these non-landing rays are clearly visible in the blow-up. (See also Plate 5.)

Failure of continuity of the straightening map was shown in other contexts, for instance by Epstein and by Inou. Failure of local connectivity and of pathwise connectivity was numerically observed by Milnor [M2] for the tricorn. For complex parameter spaces, failure of local connectivity was observed by Lavaurs for the cubic connectedness locus (a brief remark in his thesis) and by Epstein and

Yampolsky [EY] for real slices of cubic polynomials. Nakane and Komori [NK] showed that certain "stretching rays" in the space of real cubic polynomials do not land.

REMARK 1.4. This work was inspired by John Milnor in many ways: he was the first to have observed the tricorn and its relevance in the space of iterated (real) cubic maps, he made systematic studies about the local behavior of parameter spaces and under which conditions little tricorns appear there, he observed the loss of local connectivity and even of path connectivity of the tricorn, he introduced the term "tricorn"—and his home page shows non-landing rays of the kind that he observed and that we discuss here.

REMARK 1.5. In this paper, we need certain background results from the (still unpublished) earlier manuscript [MNS], which has more detailed results of bifurcations especially at hyperbolic components of odd period. In setting up notation and background, it seems more convenient to complete these proofs here rather than to strictly avoid overlap.

Acknowledgements. We would like to thank Adam Epstein for many inspiring and helpful discussions on tricorns, parabolic perturbations, and more. We would also like to thank Shizuo Nakane for numerous discussions, many years ago, about antiholomorphic dynamics. We are most grateful to two anonymous referees for numerous detailed and helpful comments. The second author would also like to thank Cornell University for its hospitality and the German Research Council DFG for its support during the time this work was carried out.

Finally, we wish to thank Jack Milnor, in gratitude for much inspiration, friendship, and generosity: mathematical and otherwise.

2 ANTIHOLOMORPHIC AND PARABOLIC DYNAMICS

In many ways, antiholomorphic maps have similar dynamical properties as holomorphic ones because the second iterate is holomorphic. There are a number of interesting features specific to antiholomorphic dynamics though, especially near periodic points of odd period k. The multiplier of a periodic point of odd period k is not a conformal invariant; only its absolute value is, and the multiplier of the $(2k)$th iterate (the second return map) is always non-negative real. This has interesting consequences on boundaries of hyperbolic components of odd period: all boundary parameters are parabolic with multiplier $+1$ (for the holomorphic second return map).

Another unusual feature is that dynamic rays landing at the same point of odd period k need not all have the same period. These rays can have period k or $2k$ (not higher), and both periods of rays can coexist: see [NS, Lemma 3.1].

We will also show that the number of periodic points of odd period k can change: but of course the number of periodic points of periods k and $2k$, which are both periodic points of period k for the holomorphic second iterate, must remain constant; the only thing that can happen is that two orbits of odd period k turn into one orbit of period $2k$, and this always occurs on boundaries of hyperbolic components of odd period k: see Lemma 3.5.

The straightening theorem [DH2] for polynomials has an antiholomorphic analogue. We state it here for easier reference; the proof is the same as in the holomorphic case.

THEOREM 2.1 (The Antiholomorphic Straightening Theorem). *Suppose that U and V are two bounded topological disks so that the closure of U is contained in V. Suppose also that $f: U \to V$ is an antiholomorphic proper map of degree d. Then $f|_U$ is hybrid equivalent to an antiholomorphic polynomial p of the same degree d. If the filled-in Julia set of $f: U \to V$ (the set of points that can be iterated infinitely often) is connected, then p is unique up to conformal conjugation.*

As usual, two maps are hybrid equivalent if they are quasiconformally conjugation so that the complex dilatation vanishes on the filled-in Julia set. In the rest of this section, we discuss the local dynamics of parabolic periodic points of odd period k specifically for antipolynomials $p_c(z) = \bar{z}^d + c$.

LEMMA 2.2 (Simple and Double Parabolics). *Every parabolic periodic point of p_c of odd period, when viewed as a fixed point of an even period iterate of p_c, has parabolic multiplicity 1 or 2.*

PROOF. The first return map of any parabolic periodic point of odd period is antiholomorphic, but the second iterate of the first return map is holomorphic and has multiplier $+1$. This second iterate can thus be written in local coordinates as $z \mapsto z + z^{q+1} + \ldots$, where $q \geq 1$ is the multiplicity of the parabolic orbit. There are then q attracting Fatou petals, and each must absorb an infinite critical orbit of $p_c^{\circ 2}$. But $p_c^{\circ 2}$ has two critical orbits (the single critical orbit of p_c splits up into two orbits of $p_c^{\circ 2}$, for even and odd iterates), hence $q \leq 2$. (Viewing this periodic point as a fixed point of a higher iterate of $p_c^{\circ 2}$ does not change q: in the same local coordinates as before, the higher iterate takes the form $z \mapsto z + az^{q+1} + \ldots$, where $a \in \mathbb{N}$ measures which higher iterate we are considering.)   □

A parabolic periodic point with multiplicity 1 (resp. 2) is called a *simple (resp. double) parabolic point*. A parameter c so that p_c has a double parabolic periodic point is called a *parabolic cusp*.

LEMMA 2.3 (Ecalle Cylinders). *Let z_0 be a simple parabolic periodic point of odd period k of an antiholomorphic map f and let V be the attracting basin of z_0. Then there is a neighborhood U of z_0 and an analytic map $\varphi: U \cap V \to \mathbb{C}$ that is an isomorphism to the half-plane $\operatorname{Re} w > 0$ such that*

$$\varphi \circ f^{\circ k} \circ \varphi^{-1}(w) = \bar{w} + \frac{1}{2}.$$

The map φ is unique up to an additive real constant.

It follows that the quotient of $V \cap U$ by $f^{\circ 2k}$ is isomorphic to $\mathbb{C}/\mathbb{Z}$, and on this quotient cylinder f induces the map $x + iy \mapsto x + 1/2 - iy$ with $x \in \mathbb{R}/\mathbb{Z}$, $y \in \mathbb{R}$.

PROOF. The second iterate $f^{\circ 2}$ is holomorphic, and for this map z_0 is parabolic with period k. Since the parabolic point is simple, we have the usual conformal Fatou coordinates $\varphi: V \cap U \to \mathbb{C}$ with $\varphi \circ f^{\circ 2k} \circ \varphi^{-1}(w) = w + 1$ for a certain neighborhood U of z_0, where $\varphi(V \cap U)$ covers some right half-plane and φ is unique up to addition of a complex constant. Adjusting this constant and restricting U (which will no longer be a neighborhood of z_0), we may assume that $\varphi(V \cap U)$ is

the right half-plane $\mathrm{Re}\, w > 0$. It follows that $(V \cap U)/(f^{\circ 2k})$ is conformally isomorphic to the bi-infinite $\mathbb{C}/\mathbb{Z} \simeq \mathbb{C}^*$, so that $f^{\circ k}$ has to send this cylinder to itself
in an antiholomorphic way. The only antiholomorphic automorphisms of $\mathbb{C}/\mathbb{Z}$
are $w \mapsto \pm\overline{w} + \alpha'$ with $\alpha' \in \mathbb{C}/\mathbb{Z}$ (depending on the sign, the two ends of $\mathbb{C}/\mathbb{Z}$
are either fixed or interchanged), and lifting this to the right half-plane we get
$\varphi \circ f^{\circ k} \circ \varphi^{-1}(w) = \pm\overline{w} + \alpha$ with $\alpha \in \mathbb{C}$, hence

$$\varphi \circ f^{\circ 2k} \circ \varphi^{-1}(w) = w + \pm\overline{\alpha} + \alpha \overset{!}{=} w + 1$$

so either $2\,\mathrm{Re}\,\alpha = 1$ or $2i\,\mathrm{Im}\,\alpha = 1$. The latter case is impossible, and in the former
case we get $\mathrm{Re}\,\alpha = 1/2$, as claimed. But φ is still unique up to addition of a complex constant, and the imaginary part of this constant can be adjusted uniquely so
that α becomes real, i.e., $\alpha = 1/2$. $\qquad\square$

DEFINITION 2.4 (Ecalle Cylinder, Ecalle Height, and Equator). *The quotient cylinder*
$(V \cap U)/(f^{\circ 2k})$ isomorphic to $\mathbb{C}/\mathbb{Z}$ is called the Ecalle cylinder *of the attracting basin.*
Its equator *is the unique simple closed (Euclidean) geodesic of $\mathbb{C}/\mathbb{Z}$ that is fixed by the*
action of f: in those coordinates in which f takes the form $w \mapsto \overline{w} + 1/2$, this equator is
the projection of $\mathbb{R}$ to the quotient. Finally, the Ecalle height *of a point $w \in \mathbb{C}/\mathbb{Z}$ in the*
Ecalle cylinder is defined as $\mathrm{Im}\,w$.

Similarly to the Ecalle cylinders in the attracting basin, one can also define them for a
local branch of f^{-1} fixing z_0; all this requires is the local parabolic dynamics in a neighbor-
hood of z_0. To distinguish these, they are called incoming *and* outgoing *Ecalle cylinders*
(for f and f^{-1}, respectively). Both have equators and Ecalle heights.

Note that the identification of an Ecalle cylinder with $\mathbb{C}/\mathbb{Z}$ for usual holomorphic maps is unique only up to translation by a complex constant; in our case, with

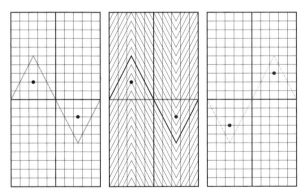

FIGURE 2.5: The Ecalle height of the critical value can be changed by putting a different complex structure onto the Ecalle cylinder and then by pull-backs onto the
entire parabolic basin. *Left*: the critical orbit (marked by heavy dots) in the Ecalle
cylinder, with a square grid indicating the complex structure; the equator is highlighted, and the critical value has Ecalle height 0.2. The grey zig-zag line will be
the new equator; it is invariant under $z \mapsto \overline{z} + 1/2$. *Center*: a grid of "distorted
squares" defines a new complex structure (in which each parallelogram should
become a rectangle); the dynamics is the same as before, and the new equator is
highlighted. *Right*: the new complex structure in the Ecalle cylinder after straightening; the Ecalle height of the critical value is now -0.3. The image of the old
equator is indicated in grey.

an antiholomorphic intermediate iterate and thus a preferred equator, this identification is unique up to a real constant. Therefore, there is no intrinsic meaning of $\operatorname{Re} w$ within the cylinder, or for $\operatorname{Re} \varphi(z)$ for $z \in V \cap U$. However, for two points $z, z' \in V \cap U$, the difference $\operatorname{Re} \varphi(z) - \operatorname{Re} \varphi(z')$ has a well-defined meaning in $\mathbb{R}$ called *phase difference*; this notion actually extends to the entire attracting basin V.

PROPOSITION 2.6 (Parabolic Arcs). *Every polynomial p_c with a simple parabolic periodic point of odd period is part of a real one-dimensional family of parabolic maps $p_{c(h)}$ with simple parabolic orbits. This family is real analytically parametrized by Ecalle height h of the critical value; more precisely, the map $h \mapsto p_{c(h)}$ is a real-analytic bijection from $\mathbb{R}$ onto a family of parabolic maps that we call a* parabolic arc.

We sketch the proof in Figure 2.5; see [MNS, Theorem 3.2] for details.

3   BIFURCATION ALONG ARCS AND THE FIXED-POINT INDEX

LEMMA 3.1 (Parabolic Arcs on Boundary of Odd-Period Components). *Near both ends, every limit point of every parabolic arc is a parabolic cusp.*

PROOF. Each limit point of parabolic parameters of period k must be parabolic of period k, so it could be a simple or double parabolic. But at simple parabolics, Ecalle height is finite, while it tends to ∞ at the ends of parabolic arcs. Therefore, each limit point of a parabolic arc is a parabolic cusp.   □

REMARK 3.2. In fact, the number of parabolic cusps of any given (odd) period is finite [MNS, Lemma 2.10], so each parabolic arc has two well-defined endpoints.

As the parameter tends to the end of a parabolic arc, the Ecalle height tends to $\pm\infty$, and the Ecalle cylinders (with first return map of period k, which permutes the two ends) becomes pinched; in the limit, the cylinder breaks up into two cylinders that are interchanged by the kth iterate, so each cylinder has a return map of period $2k$, which is holomorphic: the double parabolic dynamics in the limit is rigid and has no non-trivial deformations.

In the sequel, we will need the *holomorphic fixed-point index*: if f is a local holomorphic map with a fixed point z_0, then the index $\iota(z_0)$ is defined as the residue of $1/(z - f(z))$ at z_0. If the multiplier $\rho = f'(z_0)$ is different from 1, this index equals $1/(1 - \rho)$ and tends to ∞ as $\rho \to 1$. The most interesting situation occurs if several simple fixed points merge into one parabolic point. Each of their indices tends to ∞, but the sum of the indices tends to the index of the resulting parabolic fixed point, which is finite. Of course, analogous properties apply for the first return map of a periodic point.

If z_0 is a parabolic fixed point with multiplier 1, then in local holomorphic coordinates the map can be written as $f(w) = w + w^{q+1} + \alpha w^{2q+1} + \ldots$, and α is a conformal invariant (in fact, it is the unique formal invariant other than q: there is a formal, not necessarily convergent, power series that formally conjugates f to its first three terms). A simple calculation shows that α equals the parabolic fixed point index. The quantity $1 - \alpha$ is known as "résidu itératif" [BE]; its real part measures whether or not the parabolic fixed point of f in the given normal form can be perturbed into q or $q + 1$ attracting fixed points; Epstein introduced the notion "parabolic repelling" and "parabolic attracting" for these two situations, and in the latter case he obtains an extra count in his refined Fatou-Shishikura-inequality [E]. We will use these ideas in Theorems 3.8 and 7.1.

LEMMA 3.3 (Types of Perturbation of Odd-Period Parabolic Orbit). *Suppose p_{c_0} has a simple parabolic periodic point z_0 of odd period k. Then for any sequence $c_n \to c_0$ of parameters with $c_n \neq c_0$, the maps p_{c_n} have periodic points z_n and z'_n that both converge to z_0 as $c_n \to c_0$ and with multipliers $\rho_n := (p_{c_n}^{\circ 2k})'(z_n) \to 1$ and $\rho'_n := (p_{c_n}^{\circ 2k})'(z'_n) \to 1$, such that for large n either*

- *both z_n and z'_n have period k, we have $\rho_n, \rho'_n \in \mathbb{R}$, and one of the orbits is attracting, while the other one is repelling; or*

- *the points $z_n = z'_n$ are on a parabolic orbit of period k; or*

- *the points z_n and z'_n both have period $2k$, they are on the same orbit of p_{c_n}, and they satisfy $p_{c_n}^{\circ k}(z_n) = z'_n$ and $p_{c_n}^{\circ k}(z'_n) = z_n$. Their multipliers satisfy $\rho'_n = \overline{\rho_n} \notin \mathbb{R}$ and $\mathrm{Re}(\rho_n - 1) = O(\mathrm{Im}(\rho_n)^2)$.*

PROOF. The point z_0 is a simple parabolic fixed-point of the holomorphic map $p_{c_0}^{\circ 2k}$, so under small perturbations it must split up into exactly two fixed points of $p_{c_n}^{\circ 2k}$ (unless c_n is some other parameter on the parabolic arc, where $z_n = z'_n$ are still parabolic). As periodic points of p_{c_n}, these must both have period k or both period $2k$. They converge to the parabolic orbit, so their multipliers must tend to 1 and their fixed point indices $1/(1 - \rho_n)$ and $1/(1 - \rho'_n)$ must tend to ∞. However, the sum of these indices must tend to the finite fixed-point index of the parabolic periodic point.

If the period equals k, then the orbit of $p_c^{\circ 2k}(z_n) = z_n$ visits each of the k periodic points twice: once for an even (holomorphic) and once for an odd (antiholomorphic) iterate, and the chain rule implies that the multiplier ρ_n is real. The same argument applies to z'_n and ρ'_n. The two fixed-point indices are real and have large absolute values (once ρ and ρ' are close to 1), so their sum can be bounded only if one index is positive and the other one negative; hence one orbit must be attracting and the other one repelling.

If the period equals $2k$, then the periodic points z_n and z'_n that are near z_0 must be on the same orbit with $p_{c_n}^{\circ k}(z_n) = z'_n$ and $p_{c_n}^{\circ k}(z'_n) = z_n$. A similar argument as before shows $\rho'_n = \overline{\rho_n}$. For the sum of the fixed point indices, we obtain

$$\frac{1}{1 - \rho_n} + \frac{1}{1 - \overline{\rho_n}} = 2\,\mathrm{Re}\left(\frac{1}{1 - \rho_n}\right) = \frac{2(1 - \mathrm{Re}\,\rho_n)}{|1 - \rho_n|^2} . \tag{3.1}$$

Since his quantity must have a finite limit, the multipliers cannot be real. Writing $\varepsilon_n := 1 - \rho_n$, we have $\mathrm{Re}\,\varepsilon_n = O(\varepsilon_n^2)$; hence $\mathrm{Re}\,\varepsilon_n = O(\mathrm{Im}\,\varepsilon_n^2)$. □

In the following two results, we will show that both possibilities actually occur in every neighborhood of every simple parabolic parameter of odd period.

LEMMA 3.4 (Parabolics on Boundary of Hyperbolic Components). *If a map p_c has a parabolic periodic point of period k, then c is on the boundary of a hyperbolic component of period k.*

PROOF. We will employ a classical argument by Douady and Hubbard. Consider a map p_{c_0} with a parabolic orbit of odd period k. To see that it is on the boundary of a hyperbolic component W of period k, restrict the antipolynomial to an antipolynomial-like map of equal degree and perturb it slightly so as to make the indifferent orbit attracting: this can be achieved by adding a small complex multiple of an antipolynomial that vanishes on the periodic cycle but the derivative of which does not. Then apply the straightening theorem (Theorem 2.1) to bring it back into our family of maps p_c. This can be done with arbitrarily small Beltrami coefficients, so c is near c_0 (see also [MNS, Theorem 2.2]). $\square$

Of course, the indifferent orbit can also be made repelling by the same reasoning. However, one would expect that a perturbation of a simple parabolic periodic point, here of period k, creates two periodic points of period k. In our case, this is possible whenever the perturbation goes into the hyperbolic component W, and then one of the two orbits after perturbation is attracting and the other is repelling; a perturbation creating two repelling period k orbits is not possible within our family. It turns out, though, that if k is odd, then one can also perturb so that no nearby orbit of period k remains — and an orbit of period $2k$ is created. (The number of periodic orbits of given period must remain constant for perturbations of holomorphic maps such as $p_c^{\circ 2k}$, not for antiholomorphic maps such as $p_c^{\circ k}$.)

LEMMA 3.5 (Orbit Period Doubles in Bifurcation Along Arc). *Every parabolic arc with a parabolic orbit of period k (necessarily odd) is the locus of transition where two periodic orbits of period k (one attracting and one repelling near the arc) turn into one orbit of period $2k$ (attracting, repelling, or indifferent near the arc). Equivalently, every parameter c with a simple parabolic periodic orbit of odd period k is on the boundary of a hyperbolic component W of period k, and c has a neighborhood U so that for $c' \in W \cap U$, the parabolic orbit splits up into two orbits for $p_{c'}$ of period k (one attracting and one repelling), while it splits into one orbit of period $2k$ for $c' \in U \setminus \overline{W}$.*

PROOF. As in Lemma 3.4, consider a map p_{c_0} with a parabolic orbit of odd period k and restrict it to an antipolynomial-like map of equal degree. This time, we want to perturb it so as to make the period k orbit vanish altogether; therefore, we cannot just add a polynomial that takes the value zero along this orbit.

Let z_0 be one of the parabolic periodic points and change coordinates by translation so that $z_0 = 0$. By rescaling, we may assume that near 0, we have

$$p_{c_0}^{\circ k}(z) = \overline{z} + A\overline{z}^2 + o(\overline{z}^2)$$

(note that conjugation by complex scaling changes the coefficient in front of $\overline{z}$; see the following remark). The second iterate has the local form

$$p_{c_0}^{2k}(z) = z + (A + \overline{A})z^2 + o(z^2),$$

so the assumption that the parabolic orbit is simple means Re $A \neq 0$; conjugating if necessary by $z \mapsto -z$, we may assume that Re $A > 0$.

Let z_{k-1} be the periodic point with $p_{c_0}(z_{k-1}) = z_0$. Let f be an antipolynomial (presumably of large degree) that vanishes at the indifferent orbit except at z_{k-1}, where it takes the value $f(z_{k-1}) = 1$; assume further that the first and second derivatives of f vanish at the entire indifferent orbit, and that f vanishes to order d at the critical point. For sufficiently small $\varepsilon \in \mathbb{C}$, we will consider $f_\varepsilon(z) = p_{c_0}(z) + \varepsilon f(z)$; by slightly adjusting the domain boundaries, this will give an antipolynomial-like map of the same degree d as before, and it will continue to have a single critical point of maximal order. The map p_{c_0} has finitely many orbits of period k, all but one of which are repelling, and for sufficiently small ε these will remain repelling.

We claim there is a neighborhood U of $z_0 = 0$ so that for sufficiently small $\varepsilon > 0$, the map f_ε will not have a point of period k in this neighborhood. We use our local coordinates where $p_{c_0}^{\circ k}(z) = \bar{z} + A\bar{z}^2 + o(\bar{z}^2)$. We may assume that $|p_{c_0}^{\circ k}(z) - \bar{z} - A\bar{z}^2| \le |A z^2|$ in U, and also $|y| < 1/|8A|$.

We have

$$f_\varepsilon^{\circ k}(z) = p_{c_0}^{\circ k}(z) + \varepsilon + O(|z|^3)$$

because f vanishes to second order along the parabolic orbit. By restricting U and ε, we may assume that $|f_\varepsilon^{\circ k}(z) - p_{c_0}^{\circ k}(z) - \varepsilon| < (\operatorname{Re} A)|z|^2/2$.

Now suppose $z = x + iy \in U$. If $|y| > 8|A|x^2$, then

$$|p_{c_0}^{\circ k}(z) - \bar{z}| \le 2|A z^2| = 2|A|(x^2 + y^2) < 2|A| \left(\frac{|y|}{8|A|} + \frac{|y|}{8|A|} \right) = \frac{|y|}{2}$$

and

$$\begin{aligned}
|f_\varepsilon^{\circ k}(z) - \bar{z} - \varepsilon| &\le |f_\varepsilon^{\circ k}(z) - p_{c_0}^{\circ k}(z) - \varepsilon| + |p_{c_0}^{\circ k}(z) - \bar{z}| \\
&\le \frac{(\operatorname{Re} A)|z|^2}{2} + 2|A z^2| \\
&\le |y|,
\end{aligned}$$

so on this domain $p_{c_0}^{\circ k}$ and $f_\varepsilon^{\circ k}$ behave essentially like complex conjugation (plus an added real constant) and thus have no fixed points.

However, if $|y| \le 8|A|x^2$, that is, z is near the real axis, then

$$\operatorname{Re}(A\bar{z}^2) \ge (\operatorname{Re} A)\frac{|z|^2}{2}$$

and we have

$$\operatorname{Re} f_\varepsilon^{\circ k}(z) \ge \operatorname{Re} p_{c_0}^{\circ k}(z) + \varepsilon - |f_\varepsilon^{\circ k}(z) - p_{c_0}^{\circ k}(z) - \varepsilon| \ge \operatorname{Re} z + \operatorname{Re}(A\bar{z}^2) + \varepsilon + O(|z|^3)$$
$$\ge \operatorname{Re} z + \varepsilon > \operatorname{Re} z.$$

so that $f_\varepsilon^{\circ k}$ does not have a fixed point with $|y| \le 8|A|x^2$ either. The size of the neighborhood U is uniform for all sufficiently small ε.

Now apply the straightening theorem as in the Lemma 3.4: this yields antipolynomials p_{c_n} near c_0 with $c_n \to c_0$ for which there is one orbit of period k fewer than before perturbation, and these are all repelling, so we are outside of W. Since z_0 has period k for the *holomorphic* map $p_{c_0}^{\circ 2}$, there is a sequence (z_n) of periodic points of period k for $p_{c_n}^{\circ 2}$ with $z_n \to z_0$. These points must have period $2k$.

This shows that arbitrarily close to c_0 there are parameters for which the indifferent period k orbit has turned into an orbit of period $2k$; similarly, by Lemma 3.4 there are parameters near c_0 for which there is an attracting orbit of period k. Finally, by Lemma 3.3, any perturbation of p_{c_0} away from the parabolic arc either creates an attracting orbit of period k or an orbit of period $2k$ (which may be attracting, repelling, or indifferent; see Corollary 3.9 and Theorem 7.1). The transition between these two possibilities (attracting orbit of period k vs. orbit of period $2k$) can happen only when the period k orbit is indifferent, hence, on the boundary of a hyperbolic component of period k. This proves the claim. $\qquad\square$

REMARK 3.6. The local behavior of an antiholomorphic map at a fixed point (or a periodic point of odd period) is quite different from the holomorphic case. If the fixed point is at 0, such a map can be written $f(z) = a_1\bar{z} + a_2\bar{z}^2 + \ldots$; we will suppose $a_1 \neq 0$. The second iterate takes the form $f^{\circ 2}(z) = |a_1|^2 z + \ldots$, so indifferent orbits are always parabolic and $|a_1|$ is an invariant under conformal conjugations. However, a_1 itself is not: conjugating $u = \lambda z$ leads to

$$f_1(u) = \left(\frac{\bar{\lambda}}{\lambda}\right) a_1 \bar{u} + \left(\frac{\bar{\lambda}^2}{\lambda}\right) a_2 \bar{u}^2 + \ldots,$$

so $\arg a_1$ depends on the rotation of the coordinate system (the linear approximation df has eigenvalues $|a_1|$ and $-|a_1|$ with orthogonal eigenlines, and of course their orientation depends on the rotation of the coordinate system).

Specifically if $|a_1| = 1$, we can choose λ so that $f_1(u) = \bar{u} + A\bar{u}^2 + \ldots$ with $A \in \mathbb{C}$, and conjugation by scaling can change $|A|$. Note that we have

$$f_1^{\circ 2}(u) = u + (A + \bar{A})u^2 + \ldots.$$

If $\mathrm{Re}\, A \neq 0$, then there is a local *quadratic* conjugation $v = au + bu^2$ with $a \in \mathbb{R}$ that brings our map into the form $f_2(v) = \bar{v} + \bar{v}^2 + \ldots$, as can be checked easily. However, if

$$\mathrm{Re}\, A \neq 0,$$

there is no such change of coordinates because $f_1^{\circ 2}(u) = u + O(u^3)$, so the origin has a multiple parabolic fixed point.

PROPOSITION 3.7 (Fixed-Point Index on Parabolic Arc). *Along any parabolic arc of odd period, the fixed-point index is a real-valued real-analytic function that tends to $+\infty$ at both ends.*

PROOF. The fact that the fixed-point index is real valued follows for instance from (3.1). The Ecalle height parametrizes the arc real-analytically (Proposition 2.6), and as the Ecalle height is changed by a quasiconformal deformation, the residue integral that defines the fixed-point index depends analytically on the height (the integrand as well as the integration path). Therefore, the index depends real-analytically on Ecalle height.

The parabolic periodic point is simple for all parameters along the arc, but towards the end of a cusp the parabolic orbit merges with another repelling periodic point so as to form a double parabolic (Lemma 3.1). The orbit with which it merges is repelling, say with multiplier ρ, so its index $1/(1 - \rho)$ tends to ∞ in $\mathbb{C}$. In order for the limiting double parabolic to have finite index, the index $\iota(z_n)$ of the parabolic orbit of p_{c_n} must tend to ∞ as well as c_n tends to the end of a parabolic arc.

Note that the index $\iota(z_n)$ is real by (3.1), so it tends to $+\infty$ or to $-\infty$. Since $|\rho| > 1$, the index $1/(1 - \rho)$ always has real part less than $+1/2$. This implies that $\iota(z_n) \to +\infty$ (or the sum in the limit would not be finite). $\qquad\square$

THEOREM 3.8 (Odd-Even Bifurcation and Fixed-Point Index). *Every parabolic arc of period k intersects the boundary of a hyperbolic component of period $2k$ at the set of points where the fixed-point index is at least 1, except possibly at (necessarily isolated) points where the index has an isolated local maximum with value 1.*

PROOF. Consider a parameter p_{c_0} on a parabolic arc, and a sequence $c_n \to c_0$ so that all p_{c_n} have all periodic orbits of period k repelling. As in Lemma 3.5, let z_0 be a parabolic periodic point for p_{c_0}, let z_n be a periodic point of period $2k$ for p_{c_n} with $z_n \to z_0$, and let $z'_n := p_c^{\circ k}(z_n)$. Let ρ_n and $\rho'_n = \overline{\rho_n}$ be the multipliers of z_n and z'_n. The sum of the two fixed-point indices equals $2 \operatorname{Re}\left(1/\left(1 - \rho_n\right)\right)$. We have $|\rho_n| < 1$ if and only if $2 \operatorname{Re}\left(1/\left(1 - \rho_n\right)\right) > 1$.

Therefore, if c_0 is on the boundary of a period $2k$ component, we can choose c_n so that $|\rho_n| < 1$ and the fixed-point index at c_0 is at least 1. Conversely, if the fixed-point index is greater than 1, then we must have $|\rho_n| < 1$ for all large n, and the limit is on the boundary of a period $2k$ component. If the index equals 1, by real-analyticity of the index, either the index has an isolated local maximum there or the point is a limit point of points with index greater than 1 (note that the set of points with index 1 is isolated as the index is real-analytic and tends to ∞ at the ends). $\qquad\square$

COROLLARY 3.9 (Bifurcation Along Arcs). *Every parabolic arc has, at both ends, an interval of positive length at which a bifurcation from a hyperbolic component of odd period k to a hyperbolic component of period $2k$ occurs.* $\qquad\square$

COROLLARY 3.10 (Boundary of Bifurcating Component Lands). *Let W be a hyperbolic component of period $2k$ bifurcating from a hyperbolic component W_0 of odd period k, the set $\partial W \setminus \overline{W_0}$ accumulates only at isolated points in ∂W_0.* $\qquad\square$

This rules out the possibility that the boundary curve of W accumulates at ∂W_0, like a topologist's sine curve. The reason is that the set of limit points must have fixed point index 1, and the set of such points is discrete.

4   PARABOLIC PERTURBATIONS

In this section, we fix a hyperbolic component W of odd period k and a parameter $c_0 \in \partial W$ with a simple parabolic orbit $z_0, \ldots, z_{k-1}$. It is well known that there exists a neighborhood V of z_0 and a local coordinate $\varphi_{c_0} : V \to \mathbb{C}$ such that

$$\varphi_{c_0} \circ p_{c_0}^{\circ 2k} \circ \varphi_{c_0}^{-1}(\zeta) = \zeta + \zeta^2 h(\zeta),$$

with $\varphi_{c_0}(z_0) = 0$, $h(0) = 1$ and $|h(\zeta) - 1| < \varepsilon$ on V. We need to establish similar local coordinates for the local dynamics after perturbation.

PROPOSITION 4.1 (Perturbed Parabolic Dynamics). *For every $\varepsilon > 0$, one can choose neighborhoods V of c_0 and U of z_0 so that there is a $\varphi_c : V \to \mathbb{C}$ that satisfies*

$$f_c(\zeta) := \varphi_c \circ p_c^{\circ 2k} \circ \varphi_c^{-1}(\zeta) = \zeta + (\zeta^2 - a_c^2)h_c(\zeta),$$

with $|h_c(\zeta) - 1| < \varepsilon$ on V and $a_c \in \mathbb{C}$.

PROOF. Since $p_c^{\circ 2k}$ is holomorphic, after perturbation the parabolic fixed point splits up into two simple fixed points in the domain of φ_{c_0}. Let $\varphi_c := \varphi_{c_0} + b$, where b is chosen so the images of these fixed points are symmetric to 0, i.e., at some $\pm a_c \in \mathbb{C}$ (note that a_c may not be a continuous function in a neighborhood of c, but a_c^2 is). Then

$$h_c(\zeta) := \frac{\left(\varphi_c \circ p_c^{\circ 2k} \circ \varphi_c^{-1}\right)(\zeta) - \zeta}{(\zeta - a_c)(\zeta + a_c)}$$

must be holomorphic on V, and the map h_c is close to h at least for $\zeta \neq 0$, hence, also near 0. $\square$

Write $U^+ := U \cap W$ and $U^- := U \setminus \overline{W}$ (the parts inside and outside of the hyperbolic component W). By Lemma 3.3, for parameters $c \in U^-$ the parabolic orbit splits up into one orbit of period $2k$; denote it by $w_0(c), \dots, w_{2k-1}(c)$. By restricting U, we may assume that $|w_0(c) - z_0(c_0)| < \varepsilon$ and $|w_k(c) - z_0(c_0)| < \varepsilon$. Moreover, the multipliers $\rho_c := (p_c^{\circ 2k})'(z_0)$ and $\rho_c' := (p_c^{\circ 2k})'(z_k)$ are complex conjugate and $|\operatorname{Re}(\rho_c - 1)| \in O(|\operatorname{Im}(\rho_c)|^2)$. Since $f_c'(a_c) = 1 + 2a_c h_c(a_c) \in \{\rho_c, \rho_c'\}$ and h_c is close to 1, we see that a_c is almost purely imaginary.

Let L_c be the straight line through a_c and $-a_c$ when $c \in U^-$; when $c \in \partial W$, let L_c be the eigenline for eigenvalue -1 for the parabolic fixed point (every antiholomorphic fixed point with multiplier 1 has eigenvalues $+1$ and -1). This family of lines is continuous in c for $c \in \overline{U^-}$ (a one-sided neighborhood of c_0), and L_c is vertical (in ζ-coordinates) when $c \in \partial W$. For $c \in U^-$, let ℓ_c be the segment $[a_c, -a_c] \subset L_c$.

Choose $r > 0$, and consider the arc of circle K_c^{in} connecting a_c to $-a_c$ going through r, and K_c^{out} connecting the same two points through $-r$. If U is chosen so small that a_c is almost purely imaginary for $c \in U^-$, these arcs are well defined, and as $a_c \to 0$, each of these arcs has a limit, which is the circle through 0 and centered at $\pm r/2$; see Figure 4.2.

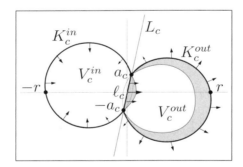

FIGURE 4.2: The line segment ℓ_c joining a_c to $-a_c$ with its image under f_c; the region between these (shaded) is a fundamental domain for the dynamics. The arcs K_c^{in} and K_c^{out} are also shown, as well as an inverse image of K_c^{out} and another fundamental domain bounded by K_c^{out} and its inverse image. This latter fundamental domain has a non-vanishing limit as $a_c \to 0$.

Denote by V_c^{in} the region bounded by K_c^{in} and ℓ_c, and let V_c^{out} denote the region bounded by K_c^{out} and ℓ_c. There is a fixed choice of $r > 0$ so that for all $c \in U^-$, iterates of $\zeta \in V_c^{\text{in}}$ under f_c will remain in V_c^{in} until they exit to V_c^{out} through ℓ_c; similarly iterates of $\zeta \in V_c^{\text{out}}$ under f_c^{-1} will remain in V_c^{out} (for $c \in U \cap \partial W$ these iterates never exit at all). The details of this argument are somewhat tedious but not difficult: as soon as $|a_c| \ll r$, the iterates in the first quadrant of K_c^{out} move upwards and to the right, while the iterates in the second quadrant (up to the point a_c) move upwards and to the left — except on a short piece of arc near the top of the circle, where the iteration moves essentially upwards. The argument is similar for the other parts of K_c^{out} and K_c^{in}.

PROPOSITION 4.3 (Ecalle Cylinders After Perturbation). *For every $c \in \overline{U^-}$, the quotients $C_c^{in} := V_c^{in}/p_c^{\circ 2k}$ and $C_c^{out} := V_c^{out}/p_c^{\circ 2k}$ (the quotients of V_c^{in} and V_c^{out} by the dynamics, identifying points that are on the same finite orbits entirely in V_c^{in} or in V_c^{out}) are complex annuli isomorphic to $\mathbb{C}/\mathbb{Z}$.*

PROOF. In the parabolic case of $c \in \partial W$, this is a standard result, proved using Fatou coordinates (see Milnor [M4, Sec. 10]). We will thus focus on the case $c \in U^-$.

Since the points $\pm a_c$ are almost purely imaginary and h_c is almost 1, it is easy to understand the dynamics of f_c, represented in Figure 4.2. In particular, the line segment $\ell = [-a_c, a_c]$ is sent by f_c to the arc $f_c(\ell) \subset \overline{V_c^{out}}$, still joining a_c to $-a_c$ but disjoint from ℓ (except at the endpoints). Let A^{out} be the domain bounded by ℓ and $f_c(\ell)$. Identifying the two boundary edges of A^{out} by f_c, we obtain a complex annulus that represents $C_c^{out} = A^{out}/p_c^{\circ 2k}$: every finite orbit in V^{out} enters $A^{out} \cup f(\ell)$ exactly once.

Finally, we must see that A^{out}/f_c is a bi-infinite annulus, i.e., isomorphic to $\mathbb{C}/\mathbb{Z}$ so that its ends are punctures.

This follows from the following lemma.

LEMMA 4.4. *Let λ be a non-real complex number, and let $g: u \mapsto \lambda u + O(u^2)$ be an analytic map defined in some neighborhood of 0. Let $\widetilde{Q} \subset \mathbb{C} \setminus \{0\}$ be the region bounded by $[0, r]$, $g([0, r])$ and $[r, g(r)]$, that we will take to include $(0, r)$ and $g((0, r))$ but not $(r, g(r))$. Then for r sufficiently small, $(0, r]$ and $g((0, r])$ are disjoint, so that the quotient of $\widetilde{Q}$ by the equivalence relation identifying $x \in (0, r)$ to $g(x)$ is homeomorphic to an annulus, and and it has infinite modulus.*

PROOF. The proof consists of passing to $\log u$ coordinates, where the corresponding annulus is bounded by $(-\infty, \log r) \subset \mathbb{R}$ and the image curve

$$\log g(u) = \log u + \log \lambda + O(u);$$

when setting $t = \log u$, the boundary identification relates $t \in \mathbb{R}^-$ to the image curve $t + \log \lambda + e^t$ (for $t \ll -1$). The claim follows. □

This lemma clearly applies to both ends of A^{out}, and the argument about A^{in} is the same. This proves the proposition. □

Recall the definition of Ecalle cylinders in the parabolic attracting and repelling petals of the holomorphic map $p_c^{\circ 2k}$ (Lemma 2.3), and the statement that the antiholomorphic iterate $p_c^{\circ k}$ introduces an antiholomorphic self-map of the Ecalle cylinders that interchanges the two ends. The situation is similar here: since the map $p_c^{\circ k}$ commutes with $p_c^{\circ 2k}$, it induces antiholomorphic self-maps from C_c^{in} to itself and from C_c^{out} to itself. As $p_c^{\circ k}$ interchanges the two periodic points at the end of the cylinders, it interchanges the ends of the cylinders, so it must fix a geodesic in the cylinders $\mathbb{C}/\mathbb{Z}$ that we call again the *equator*. Choosing complex coordinates in the cylinders for which the equator is at imaginary part 0, we can again define *Ecalle height* as the imaginary part in these coordinates. We will denote the Ecalle height of a point $z \in C_c^{in/out}$ by $E(z)$.

Since our arcs of circle $K_c^{in/out}$ depend continuously on c and have a finite non-zero limit as a_c tends to 0, the construction of the perturbed Ecalle cylinders depends continuously on $c \in \overline{U^-}$. We summarize this in the following proposition.

PROPOSITION 4.5 (Bundle of Ecalle Cylinders). *The disjoint unions*

$$\mathcal{C}^{\text{in}} := \bigsqcup_{c \in \overline{U^-}} \mathcal{C}_c^{\text{in}} \quad and \quad \mathcal{C}^{\text{out}} := \bigsqcup_{c \in \overline{U^-}} \mathcal{C}_c^{\text{out}}$$

form two-dimensional complex manifolds with boundary, and the natural maps

$$\mathcal{C}^{\text{in}} \to \overline{U^-} \quad and \quad \mathcal{C}^{\text{out}} \to \overline{U^-}$$

are smooth morphisms that make $\mathcal{C}^{\text{in}}$ and $\mathcal{C}^{\text{out}}$ into topologically trivial bundles with fibers isomorphic to $\mathbb{C}/\mathbb{Z}$.

The equators form subbundles of circles, and the Ecalle height of fixed points in $\mathbb{C}$ near 0 depends continuously on c.

REMARK 4.6. Here "smooth morphism" means that $\mathcal{C}^{\text{in}}$ and $\mathcal{C}^{\text{out}}$ are families of complex manifolds parametrized by $\overline{U^-}$ and that the fibers have analytic local coordinates that depend continuously on the parameter.

Of central importance to us is that above U^- (not the closure $\overline{U^-}$), the two bundles $\mathcal{C}^{\text{in}}$ and $\mathcal{C}^{\text{out}}$ are canonically isomorphic as follows.

DEFINITION 4.7 (The Transit Map). *The* transit map *is the conformal isomorphism*

$$T_c \colon \mathcal{C}_c^{\text{in}} \to \mathcal{C}_c^{\text{out}}$$

induced by the conformal isomorphism $p_c^{\circ 2k} \colon A^{\text{in}} \to A^{\text{out}}$.

This transit map clearly depends continuously on the parameter $c \in U^-$ and preserves the equators, hence Ecalle heights.

Finally, choose a smooth real curve $s \mapsto c(s)$ in $\overline{U^-}$ (in parameter space), parametrized by $s \in [0, \delta]$ for some $\delta > 0$, with $c(s) \in U^-$ for $s > 0$. Choose a smooth curve $s \mapsto \zeta(s)$ (in the dynamical planes, typically the critical value), also defined for $s \in [0, \delta]$ such that

$$\zeta(s) \in V^{\text{in}}(c(s))$$

for all $s \in [0, \delta]$. Then $s \mapsto \zeta(s)$ induces a map $\sigma : [0, \delta] \to \mathcal{C}^{\text{in}}$ with $\sigma(s) \in \mathcal{C}_{c(s)}^{\text{in}}$.

PROPOSITION 4.8 (Limit of Perturbed Fatou Coordinates). *The curve*

$$\gamma := s \mapsto T_{c(s)}(\sigma(s))$$

in $\mathcal{C}^{\text{out}}$, parametrized by $s \in (0, \delta]$, spirals as $s \searrow 0$ towards the circle on $\mathcal{C}_{c_0}^{\text{out}}$ at Ecalle height $E(\sigma(0))$.

Before proving this we need to say exactly what "spirals" means. We know that $\mathcal{C}^{\text{out}}$ is a trivial topological bundle of bi-infinite annuli $\mathbb{C}/\mathbb{Z}$ over $\overline{U^-}$; we can choose a trivialization

$$\Phi \colon \mathcal{C}^{\text{out}} \to \overline{U^-} \times \mathbb{C}/\mathbb{Z}$$

by deciding that the point r (see Figure 4.2) corresponds for all $c \in \overline{U^-}$ to the origin of $\mathbb{C}/\mathbb{Z}$. That allows us to define an *Ecalle phase* $\arg(\gamma(s))$ to be a continuous lift φ of

$$s \mapsto \text{Re}(pr_2(\Phi(\gamma(s)))) \in \mathbb{R}/\mathbb{Z}.$$

Spiralling will mean that the image of γ accumulates exactly on the circle on $\mathcal{C}_{c_0}^{\text{out}}$ at Ecalle height $E(\sigma(0))$, and that in the process the Ecalle phase tends to infinity.

PROOF. Since the transit map preserves Ecalle heights, the curve $t \mapsto \gamma(s)$ can only accumulate on the circle on $C_{c_0}^{\text{out}}$ at Ecalle height $E(\sigma(0))$. It remains to show that the Ecalle phase tends to infinity. The magnitude of the Ecalle phase essentially measures how many iterates of f_c it takes for $\zeta(s)$ to reach the fundamental domain in V_c^{out} shown in Figure 4.2.

This more or less obviously tends to infinity as $a_c \to 0$; to get from V_c^{in} to V_c^{out}, the orbit must cross ℓ_c, and near ℓ_c the map $f_c^{\circ k}$ moves points less and less as $a_c \to 0$. (In the language of Douady "it takes longer and longer to go through the egg-beater.") In fact, in the dynamics of the limit c_0 it takes infinitely many iterations of f_c^{-1} for r to get to the origin and, thus, arbitrarily many iterations to reach any small neighborhood X of the origin, and for sufficiently small perturbations the number of backwards iterations to go from r into X is almost the same.   □

REMARK 4.9. A computation in logarithmic coordinates shows that the Ecalle phase $\arg \gamma(s)$ (essentially the number or iterations required to get from $\zeta(s)$ to the fundamental domain in V_c^{out} containing r) satisfies

$$\arg \gamma(s) = \frac{\pi}{|a_c(s)|}(1 + o(1))$$

as $a_c \to 0$.

5   PARABOLIC TREES AND COMBINATORICS

DEFINITION 5.1 (Characteristic Parabolic Point and Principal Parabolic). *The char-acteristic point on a parabolic orbit is the unique parabolic periodic point on the boundary of the critical value Fatou component.*

A map with a parabolic orbit is called a principal parabolic *if the parabolic orbit is simple and each point on the parabolic orbit is the landing point of at least two periodic dynamic rays.*

REMARK 5.2. As proved in [MNS], every hyperbolic component W of odd period k in the multicorn $\mathcal{M}_d^*$ has a Jordan curve boundary consisting of exactly $d + 1$ parabolic arcs and $d + 1$ parabolic cusps where the arcs meet in pairs. Suppose $k \geq 3$. For each $c \in W$, each periodic bounded Fatou component has exactly $d + 1$ boundary points that are fixed under the first return map of the component, and these points together are the landing points of $d + 2$ periodic dynamic rays: one boundary fixed point is the landing point of two rays, both of period $2k$ and called the *dynamic root* of the component, and the other boundary fixed points are the *dynamic co-roots* and landing points of one ray each, of period k. Specifically for the critical value Fatou component, the two rays landing at the dynamic root separate this Fatou component and its co-roots from the entire critical orbit except the critical value (see Figure 5.3).

For each $c \in W$, the set of dynamic root and co-roots of the Fatou component containing the critical value is in natural bijection to the parabolic boundary arcs of W: at each of the $d + 1$ boundary arcs of W, a different one of the dynamic roots or co-roots becomes parabolic. The *parabolic root arc* is the arc at which the dynamic root becomes parabolic, while the d *co-root arcs* are those where one the d co-roots becomes parabolic. Principal parabolic maps p_c are thus maps from the root arc, and they exist on the boundary of each odd period component. At a parabolic

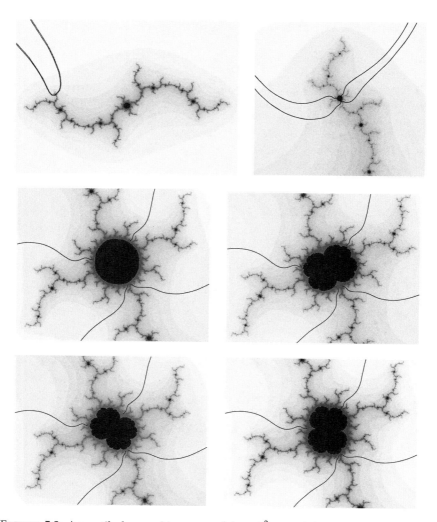

FIGURE 5.3: An antiholomorphic map $p_c(z) = \bar{z}^2 + c$ of degree $d = 2$ with an attracting cycle of period 5. The Fatou component containing the critical value has $d + 1 = 3$ boundary points that are fixed under $p_c^{\circ 5}$, and together these are the landing points of $d + 2 = 4$ dynamic rays: the dynamic root is the landing point of 2 rays (here, at angles 371/1023 and 404/1023 of period 10, and the two dynamic 2-roots are the landing points of one ray each (at angles 12/33 and 13/33 of period 5). *Upper left*: the entire Julia set with the four rays indicated. *Upper right*: blow-up of a neighborhood of the critical value Fatou component where the four rays can be distinguished. The hyperbolic component containing the parameter c is bounded by $d + 1$ parabolic arcs (see Figure 1.3): one arc contains the accumulation set of the parameter rays at angles 12/33 and 13/33 (the root arc), and the other two arcs contain the accumulation sets of one parameter ray each (at angles 12/33 and 13/33 respectively). The four remaining pictures show further blow-ups near the critical value, for parameters at the center (middle row, left), from the parabolic root arc (middle right) and from the two parabolic co-root arcs (bottom row). (See also Plates 7 and 8.)

cusp, a dynamic root or co-root merges with one of its adjacent dynamic (co-)roots. Specifically at a cusp at the end of the root arc, the dynamic root merges with a co-root: at such parameters, each parabolic periodic point is the landing point of two rays of period $2k$ and one ray of period k.

DEFINITION 5.4 (Parabolic Tree). *If p_c has a principal parabolic orbit of odd period k, we define its* parabolic tree *as the unique minimal tree within the filled-in Julia set that connects the parabolic orbit and the critical orbit, so that it intersects the critical value Fatou component along a simple $p_c^{\circ k}$-invariant curve connecting the critical value to the characteristic parabolic point, and it intersects any other Fatou component along a simple curve that is an iterated preimage of the curve in the critical value Fatou component. A* loose parabolic tree *is a tree that is homotopic to the parabolic tree, by a homotopy that fixes the Julia set (so it acts separately on bounded Fatou components). It is easy to see that the parabolic tree intersects the Julia set in a Cantor set, and these points of intersection are the same for any loose tree (not that for simple parabolics, any two periodic Fatou components have disjoint closures).*

This tree is defined in analogy to the Hubbard tree for postcritically finite polynomials. In our case, note first that the filled Julia set is locally connected hence path connected, so any minimal tree connecting the parabolic orbit is uniquely defined up to homotopies within bounded Fatou components. The parabolic tree is p_c-invariant (this is clear by construction separately in the Julia set and in the Fatou set). A simple standard argument (analogous to the postcritically finite case) shows that the critical value Fatou component has exactly one boundary point on the tree (the characteristic parabolic point), and all other bounded Fatou components have at most d such points (the preimages of the characteristic parabolic point). The critical value is an endpoint of the parabolic tree. All branch points of the parabolic tree are either in bounded Fatou components or repelling (pre)periodic points; in particular, no parabolic point (of odd period) is a branch point.

DEFINITION 5.5 (OPPPP: Odd Period Prime Principal Parabolic). *A principal parabolic map p_c with a parabolic orbit of odd period $k \geq 3$ is called* prime *if the parabolic tree does not have any proper connected subtree that connects at least two Fatou components and that is invariant under some iterate of p_c.*

Parabolics with these properties will be called OPPPP-parabolics, and these are the ones that we will work with.

REMARK 5.6. The condition of "prime" can be motivated informally as follows. Just like the Mandelbrot set contains countably many "little Mandelbrot set," it is experimentally observed (but not yet formally proved) that the multicorn $\mathcal{M}_d^*$ contains countably many "little multicorns," finitely many for each odd period $k \geq 3$ (these are combinatorial copies, not homeomorphic copies, for reasons we mentioned in the introduction); the period n means that periods of hyperbolic components in the combinatorial copy are n times the original periods. There is a natural map from the little multicorn onto $\mathcal{M}_d^*$ that is given by an antiholomorphic version of the straightening theorem, but it is not necessarily continuous. Each little multicorn, say, of period k, contains in turn countably many little multicorns, and all their periods are multiples of k. Under tuning (the inverse of straightening), the little multicorns thus form a semi-group (see [M1]), and a "prime" multicorn is one that cannot be written as a composition of other small multicorns. A map p_c

with an attracting or parabolic orbit of odd period k is prime if the parameter c is from the closure of the main hyperbolic component of a prime multicorn.

Formally speaking, any map p_c with a parabolic orbit of odd period $k \geq 3$ is clearly prime if the period k is prime (it may or many not be prime otherwise). This establishes the existence of infinitely many OPPPP parabolics, using the existence of hyperbolic components of all periods.

Concerning the latter, define a sequence $s_{d,k} := d^{k-1} - \sum_{m|k, m<n} s_{d,m}$ for each $d \geq 2$. Then $s_{d,k}$ is the number of hyperbolic components of period k for the "multibrot sets" of period d: each hyperbolic component of period k has a center parameter that satisfies $((c^d + c)^d + c + \ldots + c) = 0$, and dividing out solutions for periods k strictly dividing n we obtain the given recursive formula. It turns out that the number of hyperbolic components of the multicorns $\mathcal{M}_d^*$ of period k also equals $s_{d,k}$, except if k is twice an odd number: in the latter case, the number of hyperbolic components equals $s_{d,k} + 2s_{d,k/2}$ [MNS].

In order to reassure readers concerned that we might be talking about the empty set, here is a simple existence argument.

LEMMA 5.7 (Existence of Hyperbolic Components). *Every multicorn $\mathcal{M}_d^*$ has hyperbolic components of all odd periods.*

PROOF. Let k be an odd number and let $\varphi \in \mathbb{R}/\mathbb{Z}$ be an angle with period k under multiplication by $-d$ modulo 1; i.e., $\varphi = s/(d^k + 1)$ for some $s \in \mathbb{Z}$. The parameter ray $R(\varphi)$ at angle φ is defined as the set of parameters $c \in \mathbb{C} \setminus \mathcal{M}_d^*$ for which the critical value is on the dynamic ray at angle φ (and escapes to ∞). In [Na] it was shown that $\mathcal{M}_d^*$ is connected, and in particular that $R(\varphi)$ is a curve in $\mathbb{C} \setminus \mathcal{M}_d^*$ that converges to ∞ in one direction, and that accumulates at $\partial \mathcal{M}_d^*$ in the other direction. Let $c \in \partial \mathcal{M}_d^*$ be any accumulation point of $R(\varphi)$; note that we do *not* claim that $R(\varphi)$ has a well-defined limit point in $\partial \mathcal{M}_d^*$ (it usually will not), but its accumulation set is non-empty.

In the dynamics of p_c, the filled-in Julia set is connected, and the dynamic ray at angle φ lands at a periodic point that is repelling or parabolic. If the landing point is repelling, then stability under small perturbations assures that for parameters c' near c, the dynamic ray at angle φ lands at a repelling periodic point, and ray and landing point depend continuously on c'. But since the critical value has positive distance from ray and endpoint, this will remain so under perturbations, and this is a contradiction (compare [GM, Lemma B.1]).

Therefore, for p_c, the dynamic ray at angle φ lands at a parabolic periodic point. Let k' be the period of the parabolic orbit. Since k' is odd, the rays landing at this orbit must have period k' or $2k'$ [NS, Lemma 3.1]. This implies $k = k'$, so c is on the boundary of a hyperbolic component of period k (Lemma 3.4). □

LEMMA 5.8 (Analytic Arc Only for Real Parameters). *Suppose the filled-in Julia set of an OPPPP parabolic map p_c contains a simple analytic arc that connects two bounded Fatou components. If the critical value has Ecalle height zero, then p_c is conformally conjugate to a real map $p_{c'}$ (i.e., $c' \in \mathbb{R}$).*

PROOF. Let k be the period of the parabolic orbit, and let z_1 be the characteristic point on this orbit. Since the parabolic orbit is simple, any two bounded Fatou components have disjoint closures, so the analytic arc must traverse infinitely many bounded Fatou components. Iterating the analytic arc forward finitely many times and cutting at the critical point if necessary, we obtain a simple analytic arc connecting z_1 to some other bounded Fatou component that intersects the parabolic tree. Truncate if necessary so that the arc does not meet any branch point of the parabolic tree, nor any bounded Fatou component that contains a branch point, but so that it still connects z_1 to some other bounded Fatou component and so that all Fatou components that this arc intersects take more than k iterations to reach the critical value Fatou component. Call this piece of analytic arc J_1. Then $p_c^{\circ k} \colon J_1 \to p_c^{\circ k}(J_1) =: J_2$ is an analytic diffeomorphism between simple analytic arcs.

The arcs J_1 and J_2 are parts of the parabolic tree, except for homeomorphisms within bounded Fatou components (so they are part of a loose parabolic tree). They both start at z_1, which is not a branch point of the parabolic tree, so they must coincide at a Cantor set of points in the Julia set. As analytic arcs, they must thus coincide (except for truncation). It follows that one of the two arcs J_1 and J_2 is a sub-arc of the other. If $J_2 \subset J_1$, then p_c is not prime, so $J_2 \supset J_1$ and hence $J_{n+1} := p_c^{\circ k}(J_n) \supset J_n$ for all n. Again by definition of being prime, there is some N so that J_N covers the entire parabolic tree.

As long as $p_c^{\circ nk} \colon J_1 \to J_{n+1}$ is a homeomorphism, the image is a simple analytic arc. If during the iteration, the critical point is covered, the p_c-image will contain the critical value, but this cannot introduce any branching: suppose $J = p_c^{\circ m}(J_1)$ is a simple analytic arc that contains the critical point 0 in the interior and let J' and J'' be the components of $J \setminus \{0\}$. Then $p_c(J')$ and $p_c(J'')$ both start at the critical value and have z_1 as an interior point, so as above they must coincide in a neighborhood of z_1; hence $p_c(J') \cup p_c(J'') = p_c(J)$ is again a simple analytic arc. Therefore, all J_n are simple analytic arcs, and the same holds for the parabolic tree, which equals J_N. The parabolic tree thus is unbranched.

Now we claim that p_c is conformally conjugate to its complex conjugate $p_{\bar{c}}$ (they are obviously conjugate by an *anti*conformal homeomorphism, but we want a conformal conjugation). The condition of Ecalle height zero implies that $p_c^{\circ k}$ and $p_{\bar{c}}^{\circ k}$ are conformally conjugate on their incoming Ecalle cylinders, respecting the critical orbits. This conjugation can be pulled back to the incoming petal of the parabolic orbit and thus to their periodic Fatou components. It follows from local connectivity that this conformal conjugation on each individual Fatou component extends homeomorphically to the closure of the component.

The next step is to extend this conjugation homeomorphically to the filled-in Julia set, again using local connectivity. This is possible because the parabolic trees are unbranched, so their combinatorial structure is unaffected by complex conjugation.

Finally, we extend the conjugation to the basin of infinity, using the Riemann map between the basins of infinity so that ∞ is fixed. There are $d + 1$ choices for this conjugation near ∞, and one of them maps the rays landing at the parabolic orbit to the rays landing at the parabolic orbit (this is possible because we already know that the dynamics on the Julia set is conjugate); for this map, the extension to the boundary coincides with the conjugation on the Julia set we already have.

This way, we obtain a topological conjugation $h \colon \mathbb{C} \to \mathbb{C}$ between p_c and $p_{\bar{c}}$ that is conformal away from the Julia set. If we know that the Julia set is holomorphically removable, then we have a conformal conjugation on $\mathbb{C}$. This fact can be established directly without too much effort. Consider the equipotential E of p_c at some positive potential, and let $E_1, \ldots, E_k$ be piecewise analytic simple closed curves, one in each bounded periodic Fatou component of p_c, that surround the postcritical set in their Fatou components and that intersect the boundary of their

Fatou components in one point, which is on the parabolic orbit. Let V_0 be the domain bounded on the outside by E and on the inside by the E_i. Then there is a quasiconformal homeomorphism $h_0 \colon \mathbb{C} \to \mathbb{C}$ with $h_0 = h$ on $\mathbb{C} \setminus V$ (i.e., the homeomorphism h is modified on V_0 so as to become quasiconformal, possibly giving up on the condition that h_0 is a conjugation on V_0).

Now construct a sequence of quasiconformal homeomorphisms $h_n \colon \mathbb{C} \to \mathbb{C}$ as a sequence of pull-backs, satisfying $p_{\bar{c}} \circ h_{n+1} = h_n \circ p_c$: the construction assures that this is possible, and all h_n satisfy the same bounds on the quasiconformal dilatation as h_0. Moreover, the support of the quasiconformal dilatation shrinks to the Julia set, which has measure zero. By compactness of the space of quasiconformal maps with given dilatation, the h_n converge to a conformal conjugation between p_c and $p_{\bar{c}}$. This limiting conjugation must coincide with h on the Fatou set, so the Julia set is holomorphically removable, as claimed.

Finally, since p_c and $p_{\bar{c}}$ are conformally conjugate, we have $c = \zeta^s \bar{c}$, where ζ is a complex $(d+1)$st root of unity and $s \in \mathbb{Z}$, so writing $c = re^{2\pi i \varphi}$ it follows that $\varphi = -\varphi + s/(d+1)$ or $\varphi \in \mathbb{Z}/2(d+1)$. Since p_c is conformally conjugate to $p_{c'}$ with $c' = c\zeta^{s'}$ for $s' \in \mathbb{Z}$, we may add $s'/(d+1)$ to φ, so $\varphi \in \{0, 1/2\}$, and this means that p_c is conformally conjugate to a real map.   □

REMARK 5.9. From the statement that the parabolic tree is unbranched, there is an alternative argument that p_c is conformally conjugate to $p_{\bar{c}}$.

We can modify the parabolic tree topologically into a superattracting tree. The map on this tree extends to a postcritically finite orientation-reversing branched mapping whose combinatorial equivalence class is well defined and that is obviously combinatorially equivalent to its complex conjugate.

We can then apply Thurston's theorem (in fact, just Thurston rigidity) to claim that the corresponding superattracting antipolynomial $p_{c'}$ is conformally conjugate on $\mathbb{C}$ to its complex conjugate. This says that up to conjugacy, the tricorn with $p_{c'}$ at its center can be taken to be on the real axis, hence also the point on the principal boundary arc at Ecalle height 0.

We give this more elementary but somewhat tedious argument to avoid using Thurston's theorem for orientation-reversing branched maps, though the result is true and requires almost no modifications in the proof. Similarly, one can use the methods of "Posdronasvili" [DH1] or of Poirier [P] to prove that any two postcritically finite polynomials with unbranched Hubbard trees having identical combinatorics are conformally conjugate.

LEMMA 5.10 (Approximating Ray Pairs). *Every OPPPP parabolic map p_c with Ecalle height zero is either conformally conjugate to a map $p_{c'}$ with real parameter c', or the characteristic parabolic point z_1 is the limit of repelling preperiodic points w_n and $\tilde{w}_n$ on the parabolic tree so that all w_n have Ecalle heights $h > 0$ and all $\tilde{w}_n$ have Ecalle heights $-h$, with the following property: if φ and φ' are the external angles of the dynamic rays landing at z_1, then dynamic rays at angles φ_n and φ'_n land at w_n so that $\varphi_n \to \varphi$ and $\varphi'_n \to \varphi'$; similarly, dynamic rays at angles $\tilde{\varphi}_n$ and $\tilde{\varphi}'_n$ land at $\tilde{w}_n$ with $\tilde{\varphi}_n \to \varphi$ and $\tilde{\varphi}'_n \to \varphi'$.*

PROOF. Repelling preperiodic points are dense on the Cantor set where the parabolic tree intersects the Julia set (for instance by the condition of being prime), so choose one such point, say w_0, in the repelling petal of z_1 (repelling periodic points must accumulate at z_1 and cannot do this within the attracting petal; and some neighborhood of z_1 is covered by the union of attracting and repelling petals). If all repelling periodic points on the parabolic tree and near z_1 have Ecalle height 0, then the parabolic tree must intersect the Julia set entirely at Ecalle height zero, and then one can construct an analytic arc that satisfies the hypotheses of Lemma 5.8, so p_c is conformally conjugate to a real map.

If this is not the case, then we may assume that w_0 has non-zero Ecalle height h; to fix ideas, say, $h > 0$. Construct a sequence (w_n) so that $w_{n+1} := p_c^{\circ(-2k)}(w_n)$, choosing a local branch that fixes z_1 and so that all w_n are in the repelling petal of z_1; hence $w_n \to z_1$ as $k \to \infty$. All w_n have the same Ecalle height h.

As w_0 is on the parabolic tree, which is invariant, it follows that w_0 is accessible from outside of the filled Julia set on both sides of the tree, so w_0 is the landing point of (at least) two dynamic rays, "above" and "below" the tree. If φ_n and φ'_n are the corresponding angles of rays landing at w_n, then it follows that these sequences of angles converge to angles of rays landing at z_1 on both sides of the tree, and the claim follows.

Now let $w'_n := p_c^{\circ k}(w_n)$; then $w'_n \to z_1$ and all these points have Ecalle heights $-h$. The rays landing at z_1 have angles φ and φ' and their period is $2k$, so $p_c^{\circ k}$ permutes these and the claim about w'_n and its rays follows.   □

6   NON-PATHWISE CONNECTIVITY

We denote the dynamic ray at angle φ for the map p_c by $R_c(\varphi)$ and, as before, the parameter ray at angle φ by $R(\varphi)$.

THEOREM 6.1 (Rays Approximating at OPPPP Arc). *Let $\mathcal{A}$ be a prime parabolic root arc of odd period $k \geq 3$ that does not intersect the real axis or its images by a symmetry rotation of $\mathcal{M}_d^*$, and let $c \in \mathcal{A}$ be the parameter with Ecalle height zero. Let φ and φ' be the characteristic angles of the parabolic orbit for parameters $c \in \mathcal{A}$. Then there is a sub-arc $\mathcal{A}_\tau \subset \mathcal{A}$ of positive length and there are angles $\tilde{\varphi}_n \to \tilde{\varphi}$ and $\varphi'_n \to \varphi'$ so that $\mathcal{A}_\tau$ is contained in the limit of the parameter rays $R(\tilde{\varphi}_n)$, and also of $R(\varphi'_n)$ (this is the limit of the sequence of rays, not necessarily of individual rays: each $c \in \mathcal{A}_\tau$ is the limit of a sequence of points on the parameter rays $R(\tilde{\varphi}_n)$, and of another sequence on $R(\varphi'_n)$.)*

PROOF. In the dynamics of p_c, let z_1 be the characteristic parabolic point. The angles φ and φ' have period $2k$, so the rays $R_c(\varphi)$ and $R_c(\varphi')$ are interchanged by the first return map of z_1. In the outgoing Ecalle cylinders at z_1 of the holomorphic map $p_c^{\circ 2k}$, the rays $R_c(\varphi)$ and $R_c(\varphi')$ project to disjoint simple closed curves, not necessarily at constant Ecalle heights, but it makes sense to say which of the two rays has greater Ecalle heights (removing the projection of one ray from the Ecalle cylinder, the other ray is in the component with arbitrarily large positive or negative Ecalle heights). Without loss of generality, suppose that $R_c(\varphi)$ has greater heights than $R_c(\varphi')$. Let h^+ be the maximum of Ecalle heights of φ, and h^- be the minimum of Ecalle heights of φ'; since $p_c^{\circ k}$ interchanges $R_c(\varphi)$ and $R_c(\varphi')$, we have $h^+ = -h^- > 0$.

Consider the sequences of repelling preperiodic points w_n and $\tilde{w}_n$ converging to z_1 as provided by Lemma 5.10, and let $h > 0$ and $-h$ be their Ecalle heights. Then clearly $h < h^+$ (the points w_k are in the part of the Ecalle cylinder bounded by the rays $R_c(\varphi)$ and $R_c(\varphi')$). The rays $R_c(\tilde{\varphi}_n)$ terminate at the points $\tilde{w}_n$ with Ecalle heights $-h$, while they all project to the same ray in the Ecalle cylinder, in which they spiral upwards and converge towards the projection of the ray at

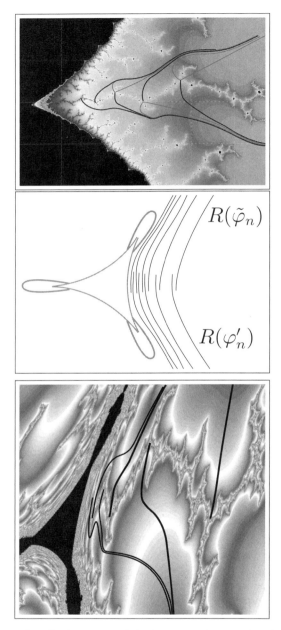

FIGURE 6.2: Loss of pathwise connectivity because of approximating overlapping
parameter rays. *Top*: Approximating preperiodic dynamic rays in the dynamic
plane with a parabolic orbit. Only the rays drawn by heavy lines are used in the
argument below; other rays landing at the same points are drawn in grey. *Middle*:
Symbolic sketch of the situation in the parameter space. *Bottom*: Actual parameter
rays accumulating in the same pattern, producing a double-comb-like structure.
(See also Plate 6.)

angle φ. Therefore, for any compact subinterval of $(-h, h)$, the rays $R_c(\tilde{\varphi}_n)$ have Ecalle heights in this entire compact interval. Similarly, the rays $R_c(\varphi'_n)$ terminate at the w_n with Ecalle height h and also have Ecalle heights within any compact subinterval of (h^-, h); see Figure 6.2.

Now let $c_t \in \mathcal{A}$ be the parameter where the critical value has Ecalle height $t \in \mathbb{R}$ (see Proposition 2.6). The points w_k depend real-analytically on t (like the entire Julia set); let $h(t)$ be their Ecalle heights; these too depend analytically on t. Therefore, there is a $\tau \in (0, h)$ so that $h(t) > t$ for all $t \in (-\tau, \tau)$. Choose $\varepsilon \in (0, \tau)$. Let $\mathcal{A}_\tau \subset \mathcal{A}$ be the sub-arc with Ecalle heights in $(-\tau + \varepsilon, \tau - \varepsilon)$.

To transfer these dynamic rays from the Ecalle cylinders to parameter space, we employ Proposition 4.8. Choose any smooth path $c \colon [0, \delta] \to \mathbb{C}$ with $c(0) = c_t \in \mathcal{A}_\tau$ but so that, except for $c(0)$, the path avoids closures of hyperbolic components of period k and so that the path is transverse to $\mathcal{A}$ at c_t.

In the outgoing cylinder of $c(0) = c_t \in \mathcal{A}$, all $R_c(\tilde{\varphi}_n)$ traverse Ecalle heights in $[-h + \varepsilon/2, h - \varepsilon/2]$. Since each ray $R_c(\tilde{\varphi}_n)$ and its landing point depend uniformly continuously on c, and since the projection into Ecalle cylinders is also continuous, there is a $\delta_\varepsilon > 0$ so that for all $c(s)$ with $s < \delta$ the projection of the rays $R_c(\tilde{\varphi}_n)$ into the Ecalle cylinders traverses heights $[-h + \varepsilon, h - \varepsilon]$, while the phase is uniformly continuous in s.

For $s \in [0, \delta]$, let $z(s)$ be the critical value. For $s > 0$, the critical orbit "transits" from the incoming Ecalle cylinder to the outgoing cylinder; as $s \searrow 0$, the image of the critical orbit in the outgoing Ecalle cylinder has Ecalle height tending to $t \in (-\tau + \varepsilon, \tau - \varepsilon) \subset (-h + \varepsilon, h - \varepsilon)$, while the phase tends to infinity. Therefore, there are $s \in (0, \delta_\varepsilon)$ arbitrarily close to 0 at which the critical value, projected into the incoming cylinder and sent by the transfer map to the outgoing cylinder, lands on the projection of the rays $R_c(\tilde{\varphi}_n)$. But in the dynamics of $p_{c(s)}$, this means that the critical value is on one of the dynamic rays $R_c(\tilde{\varphi}_n)$, so $c(s)$ is on the parameter ray $R(\tilde{\varphi}_n)$.

The analogous statement holds for φ'_n. $\square$

Now the proof of our main result is simple.

THEOREM 6.3 (Multicorns Are Not Path Connected). *For each $d \geq 2$, the multicorn $\mathcal{M}^*_d$ is not path connected.*

PROOF. Let W be any hyperbolic component of odd period not intersecting the real axis, let $\mathcal{A} \subset W$ be the parabolic root arc and suppose it is prime. By Theorem 6.1, there is a sub-arc $\mathcal{A}_\tau$ of positive length and there are two sequences of angles $\tilde{\varphi}_n$ and φ'_n converging to limits $\tilde{\varphi} \neq \varphi'$ so that the set

$$\bigcup_n R(\tilde{\varphi}_n) \cup \bigcup_n R(\varphi'_n) \cup \mathcal{A}_\tau$$

disconnects $\mathbb{C}$ into at least 2 path-components. If the angles are oriented so that $\varphi < \varphi'$, then W is in a different component from $R(0)$ and any hyperbolic component W' in the limit set of the angle $R(1/(2^n \pm 1))$ for sufficiently large n: any path connecting W to W' must accumulate at all points in $\mathcal{A}_\tau$, and this is impossible. $\square$

REMARK 6.4. We believe that the only hyperbolic components for which the umbilical cord lands are on the real axis, or symmetric to such components by a rotational symmetry of $\mathcal{M}_d^*$. For individual components, this can be verified numerically: all one needs to know is that the parabolic tree does not contain analytic arcs, even in the non-prime situation; for this it is good enough to know that the subtree of renormalizable points contains periodic points with non-real multipliers.

7   FURTHER RESULTS

The following proposition and its proof are inspired by more general results due to Bergweiler [Be] as well as Buff and Epstein [BE].

THEOREM 7.1 (No Bifurcation at Ecalle Height Zero).   *On every parabolic arc of period k, the point with Ecalle height zero has a neighborhood (along the arc) that does not intersect the boundary of a hyperbolic component of period 2k.*

PROOF. Suppose p_c is the center point of the parabolic arc (at Ecalle height 0). We will now discuss the local dynamics of the holomorphic first return map, i.e., the $(2k)$th iterate of p_c. The upper and lower endpoints of the Ecalle cylinders correspond to fixed points of $p_c^{\circ 2k}$; but they are interchanged by $p_c^{\circ k}$, so they are simultaneously attracting or repelling with complex conjugate multipliers.

Points in the outgoing petal at sufficiently large positive Ecalle heights will return to the incoming petal; this map is called the "horn map." This induces a conformal map from the upper end of the outgoing cylinder to the incoming cylinder which, by the Koebe compactness theorem, is close to a translation by a complex constant (writing the cylinders as $\mathbb{C}^*$, it is close to multiplication by a constant). Let η be the imaginary part of the translation constant. Therefore, for every $\varepsilon > 0$ there is an $H > 0$ so that points at Ecalle heights $h_o > H$ in the outgoing cylinder return to points in the incoming cylinder with height h_i so that $|h_i - h_o + \eta| < \varepsilon$, i.e. $h_i \in (h_o - \eta - \varepsilon, h_o - \eta + \varepsilon)$.

Cut the outgoing and incoming Ecalle cylinders at the equators into upper and lower half-cylinders and label the upper halves C_o and C_i, where i and o stand for "incoming" and "outgoing" (the whole discussion can be done analogously in the lower halves, with negative Ecalle heights, just as well; the antiholomorphic iteration assures that they are completely symmetric). Let $C_o' \subset C_o$ be the restriction to the parabolic basin. Let $f \colon C_o' \to C_o$ be a conformal isomorphism; it is unique up to addition of a real constant (a phase). There is a number $\delta \in \mathbb{C}$ so that asymptotically near the end, $f(z) = z - i\delta$. By adjusting the freedom in f, we can make δ purely real. The Schwarz lemma, together with the fact that $C_o \setminus C_o'$ contains open sets in the basin of infinity, implies $\delta > 0$. This number is called the *Grötzsch defect* of the outgoing cylinder (with respect to the parabolic basin).

For $h > 0$, let $C_i(h)$ and $C_o'(h)$ be C_i and C_o' restricted to Ecalle heights in $(0, h)$. Then $h = \mathrm{mod}\,(C_o'(h))$ and δ can be viewed as the limit, as $h \to \infty$, of $h - \mathrm{mod}\,(C_o'(h))$. The Grötzsch inequality implies that $\mathrm{mod}\,(C_o'(h)) \leq h - \delta$ for all h.

Choose $\varepsilon \in (0, \delta)$ and H depending on ε as before. Since by hypothesis the critical orbit is at Ecalle height 0, one can pull back $C_i(H)$ conformally into C_o; it must land within $C_o'(H + \eta + \varepsilon)$ (it must have Ecalle height less than $H + \eta + \varepsilon$ and it must be in the part within the attracting basin). Since $\mathrm{mod}\,(C_i(H)) = H$, while $\mathrm{mod}\,(C_o'(H + \eta + \varepsilon)) \leq H + \eta + \varepsilon - \delta$, this implies $\eta + \varepsilon - \delta \geq 0$ and thus $\eta > 0$.

By Proposition 4.8, the outgoing and incoming cylinders exist after small perturbations of the parameter outside of its hyperbolic component, the Ecalle heights depend continuously on the perturbation, and the same holds for the "horn maps" from the ends of the outgoing into the incoming cylinders.

If $\varepsilon < |\eta|$, then $\eta < 0$ means that points with great Ecalle heights the outgoing cylinder return into the incoming cylinder at greater heights. For such sufficiently small perturbations, $\eta < 0$ thus implies that points near the end of the cylinder will, after perturbation, converge to the end of the cylinder; the endpoint of the cylinder thus becomes an attracting fixed point. Similarly, if $\eta > 0$, then the end-points become repelling. Since these are the periodic points that bifurcate from the period k orbit, this shows that parameters c with $\eta > 0$ are not on the boundary of a period $2k$ hyperbolic component, and this is the case when the Ecalle height h is zero or sufficiently close to zero. $\qquad\square$

REMARK 7.2. This result can be strengthened in various ways. One can give an explicit lower bound on the Grötzsch defect δ: the basin at infinity alone occupies an annulus of modulus at least $1/(2k \log d)$, so $\delta > 1/(2k \log d)$ (compare [BE], Theorem B, and especially the first half of the proof). Moreover, one can deduce an inequality between the Ecalle height of the critical orbit and the fixed-point index: if the critical value has Ecalle height h, one can estimate the conformal modulus of the largest embedded annulus in $C_i(H)$ that separates the critical value from the upper boundary (this is a classical extremal length estimate; the modulus is $H - |h| + o(1)$), and this gives a correspondingly greater upper bound on the fixed-point index (after all, we know that for large Ecalle heights h, the fixed-point index must become greater than 1). Combining both facts, this implies a definite interval of Ecalle heights around 0, depending only on d, for which the parabolic arc does not meet bifurcating components.

THEOREM 7.3 (Decorations Along Parabolic Arc). *Every parabolic arc on a hyperbolic component of odd period k has Ecalle heights $h_1, h_2, h_3, h'_1, h'_2, h'_3 \in \mathbb{R}$ so that $h_3 > 0 > h'_3$, $h_3 > h_2, h'_2 > h'_3$, satisfying the following properties:*

- *the sub-arc with Ecalle heights $h > h_3$ is an arc of bifurcation to a component of period $2k$; and also for Ecalle heights $h < h'_3$;*

- *the sub-arc with Ecalle heights $h \in (h_2, h_3)$ is the limit of decorations (attached to the period $2k$ components bifurcating for large positive Ecalle heights); and also for Ecalle heights $h \in (h'_3, h'_2)$;*

- *if the arc is a root arc, then the sub-arc with Ecalle heights $h \in [h'_1, h_1]$ is the limit of the "umbilical cord."*

Note that by Theorem 7.1 we have $h_3 > 0 > h'_3$ (the two hyperbolic components near the end of the parabolic arc are disjoint), but we do not know whether it is always true that $h_2 > 0 > h'_2$ (if $h_2 < h'_2$, this would mean that the decorations from the period $2k$ components at both ends of the arc would overlap; this would imply $h_1 \geq |h'_2|$). We clearly have $h_3 > \max\{h_2, |h'_2|, h_1\}$. Loss of pathwise connectivity of the umbilical cord occurs whenever $h_1 > 0$.

If $h_1 < h_2$, then we have an "open beach," where the boundary of the multicorn locally equals just the parabolic arc without further decorations. We do not know whether there are infinitely many parabolic arcs for which this occurs.

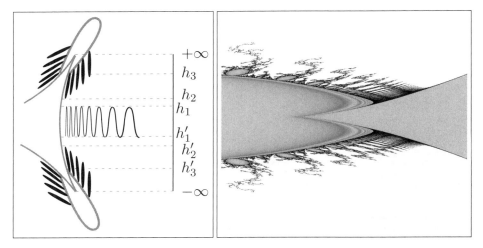

FIGURE 7.4: Illustration of Theorem 7.3. *Left*: Schematic illustration of the decorations along parabolic arcs, together with their threshold heights. *Right*: Decorations at a period 2 component that accumulate along arcs on the boundary of the
period 1 component. (See also Plate 3.)

SKETCH. For sufficiently large Ecalle heights, the parabolic arc is on the locus of
bifurcation from odd period k to period $2k$ (Corollary 3.9). Let $h_3 \in \mathbb{R}$ be the
infimum of such Ecalle heights, and consider p_c for c on this parabolic arc with
Ecalle height h_3. The boundary of the Fatou component, projected into the outgoing Ecalle cylinder, is not a geodesic in this cylinder (because this boundary is not
an analytic curve). It will thus project into the cylinder at an interval (h', h'') of
Ecalle heights with h'' strictly greater than h'. In fact, we have $h'' = h_3$.

For parameters $c(h)$ with h slightly less than h_3, the lower Ecalle height h' depends on h, but there is an interval (h_2, h_3) when $h'(h) < h$. For these, both escap-
ing points and the Julia set intersect the outgoing Ecalle cylinder at Ecalle height
in a neighborhood of h, so the parameter $c(h)$ can be approximated by parameters
inside and outside of $\mathcal{M}_d^*$.   □

REMARK 7.5. Each parabolic arc contains the accumulation set of one or two periodic parameter rays (two rays for root arcs, one for co-root arcs). By symmetry,
the rays accumulating at the boundary arcs of the period 1 hyperbolic component
actually land. We believe that all the other periodic parameter rays do not land, at
least those that accumulate at root arcs. Instead, we believe that they accumulate
at a sub-arc of positive length.

The reason is as follows. For a parameter c on a root arc of period $n \geq 3$, the
parabolic periodic point is the landing point of 2 dynamic rays. These form hyperbolic geodesics in the access that is bounded on one side by a periodic Fatou
component and on the other side by the parabolic tree, decorated by various structures of the Julia set. Even though the ray is an analytic curve, it would seem to
require a miracle that the boundaries of the access at the two sides are symmetric
enough so that the ray projects to an equator in the Ecalle cylinder. But if it does
not project to an equator, but has varying Ecalle height instead, then these wiggles

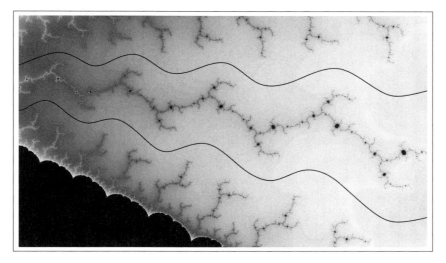

FIGURE 7.6: Heuristic argument why parameter rays should not land at parabolic root arcs but rather accumulate at a sub-arc of positive length. Sketch of the situation in the parabolic dynamics: near the top and bottom, there is the parabolic Fatou component (black, visible only in the bottom); in the middle there is the Julia set around the parabolic tree, and the two black curves are the two dynamic rays landing at the parabolic periodic point. They have no reason to have constant Ecalle height between two very different structures. While the parabolic basin is not stable under perturbations, the rays move continuously and keep their wiggles.

will transfer into parameter space to a ray that accumulates on the parabolic arc like a topologist's sine curve (see Figure 7.6).

REMARK 7.7. It is tempting to try to show for hyperbolic components of odd period k that the bifurcating period $2k$ component has wiggly boundary near the parabolic arc, transferring the wiggly boundary of the Fatou component to parameter space (using the fact that these Fatou components do not have analytic boundary arcs). However, this Fatou component is not stable under perturbation away from the parabolic arc, and we do not obtain wiggles in parameter space. In fact, it follows from Theorem 3.8 that the boundary of the period $2k$ component has a well-defined limit point on the parabolic arc: a simple parabolic with fixed point index $+1$, and those are isolated. What can be transferred into parameter space are the repelling periodic points on the boundary of the Fatou component, and the rays landing at them. These yield the decorations of the period $2k$ components that accumulate at parabolic arcs in a comb-like manner, as described in Theorem 7.3.

8 Bibliography

[Be] Walter Bergweiler, *On the number of critical points in parabolic basins*. Ergodic Theory Dynam. Systems **22** 3 (2002), 655–669.

[BE]   Xavier Buff and Adam Epstein, *A parabolic Pommerenke-Levin-Yoccoz inequality*. Fund. Math. **172** 3 (2002), 249–289.

[DH1]   Adrien Douady and John Hubbard, *Etude dynamique des polynômes Complexes* (The Orsay Notes). Publ. Math. Orsay (1984–85).

[DH2]   Adrien Douady and John Hubbard, *On the dynamics of polynomial-like maps*. Ann. Sci. Ec. Norm. Sup., 4e série, **18** 2 (1985), 277–343.

[E]   Adam Epstein, *Infinitesimal Thurston Rigidity and the Fatou-Shishikura Inequality*, Stony Brook IMS Preprint, **1** (1999). arXiv:math/9902158

[EY]   Adam Epstein and Mikhail Yampolsky, *Geography of the cubic connectedness locus: intertwining surgery*. Ann. Sci. Ec. Norm. Sup.,4e série, **32** 2 (1999), 151–185.

[GM]   Lisa Goldberg and John Milnor, *Fixed points of polynomial maps. Part II. Fixed point portraits*. Ann. Sci. Ec. Norm. Sup., 4e série, **26** 1 (1993), 51–98.

[M1]   John Milnor, *Self-similarity and hairiness in the Mandelbrot set*. In: Lecture Notes in Pure and Appl. Math. **114** (1989), Dekker, New York, 211–257.

[M2]   John Milnor, *Remarks on Iterated Cubic Maps*. Exper. Math. **1** 1 (1992), 5–25.

[M3]   John Milnor, *Hyperbolic Components* (with an Appendix by A. Poirier) Contemp. Math., **573**, Conformal dynamics and hyperbolic geometry, Amer. Math. Soc., (2012) 183–232. (Earlier version circulated as Stony Brook Preprint **3** (1992).)

[M4]   John Milnor, *Dynamics in one complex variable*, 3rd edition. Annals of Mathematics Studies **160**. Princeton University Press, Princeton, NJ, 2006.

[MNS]   Sabyasachi Mukherjee, Shizuo Nakane, and Dierk Schleicher, *On multicorns and unicorns II: Bifurcations in spaces of antiholomorphic polynomials*. Manuscript (in preparation).

[Na]   Shizuo Nakane, *Connectedness of the Tricorn*. Ergod. Th. & Dynam. Sys. **13** (1993), 349–356.

[NK]   Shizuo Nakane, Yohei Komori, *Landing property of stretching rays for real cubic polynomials*. Conform. Geom. Dyn. **8** (2004), 87–114.

[NS]   Shizuo Nakane and Dierk Schleicher, *On multicorns and unicorns I: Antiholomorphic dynamics, hyperbolic components and real cubic polynomials*. Int. J. Bif. Chaos. **13** 10 (2003), 2825–2844.

[P]   Alfredo Poirier, *Hubbard trees*. Fund. Math. **208** 3 (2010), 193–248.

[S]   Mitsuhiro Shishikura, *Bifurcation of parabolic fixed points*. In: The Mandelbrot set, theme and variations, ed. Tan Lei, London Math. Soc. Lecture Note Ser. **274**, Cambridge Univ. Press, Cambridge, 2000, 325–363.

Leading monomials of escape regions

Jan Kiwi

ABSTRACT. For each $p \geq 1$, we consider the affine algebraic curve $\mathcal{S}_p$ formed by all monic and centered cubic polynomials with a marked critical point which has period p under iterations. Each unbounded hyperbolic component of $\mathcal{S}_p$, called an escape region, has an associated vector of leading monomials which encodes the asymptotic behavior of the periodic critical orbit. We show that this vector determines the escape region, giving a positive answer to a question posed by Bonifant and Milnor.

1 INTRODUCTION

This note will use non-Archimedean methods to prove a result in classical holomorphic dynamics. In particular, we will prove a result in complex cubic polynomial dynamics using cubic polynomial dynamics over the field of formal Puiseux series.

The space of complex cubic polynomials, regarded as dynamical systems acting on the complex plane $\mathbb{C}$, has two complex dimensions. To understand how polynomials are organized in parameter space according to dynamics, it is of interest to study complex one-dimensional slices. Our focus in this paper is on slices formed by cubic polynomials with a periodic critical point. These slices have already received some attention in the literature (e.g., see [1], [2], [8], [10], [11], and [12]).

We will work in the parameter space of monic and centered cubic polynomials with a marked critical point. This parameter space is naturally identified with $\mathbb{C}^2$. More precisely, each $(a, v) \in \mathbb{C}^2$ corresponds to the map $F_{a,v} \colon \mathbb{C} \to \mathbb{C}$ with marked critical point $+a$, where

$$F_{a,v}(z) = (z - a)^2(z + 2a) + v.$$

Note that the critical points of $F_{a,v}$ are $\pm a$ and $F_{a,v}(a) = v = F_{a,v}(-2a)$.

For each integer $p \geq 1$, consider the set $\mathcal{S}_p$ consisting of all parameters (a, v) such that the marked critical point $+a$ of $F_{a,v}$ has period exactly p under iterations of $F_{a,v}$. The period p curve $\mathcal{S}_p \subset \mathbb{C}^2$ is a smooth algebraic curve [10, Theorem 5.2].

The emphasis of this paper is on the unbounded hyperbolic components of $\mathcal{S}_p$, called *escape regions*. Specifically, following Bonifant and Milnor, relevant information about the limiting behavior of the periodic critical orbit, as the parameters tend to infinity within an escape region $\mathcal{U}$, is encoded in an *associated leading monomial vector*. The objective of this note is to prove Theorem 1, which shows that this vector uniquely determines the escape region.

Partially supported by Research Network on Low Dimensional Dynamics ACT-17, and Fondecyt Grant 1110448, Conicyt, Chile.

The finite set of information encoded in the leading monomial vector of $\mathcal{U}$ can be described as follows. There exists a uniformizing parameter ρ on $\mathcal{U}$ so that $\rho^\mu = a$, where μ is a positive integer called the *multiplicity* of $\mathcal{U}$. For $1 \le j < p$, the periodic critical orbit element $a_j = F_{a,v}^j(a)$ satisfies the asymptotic equality

$$a_j = a + c_j \rho^{k_j} + o(|\rho|^{k_j}), \text{ as } \rho \to \infty,$$

with a complex coefficient $c_j \neq 0$ and an integer $k_j \le \mu$. Our main result implies that the escape region $\mathcal{U}$ is uniquely determined by these numbers c_j and k_j or, equivalently, by the associated leading monomial vector.

It is worth mentioning that it is an interesting open question, posed in [10], to determine whether $\mathcal{S}_p$ is connected for all $p \ge 1$. A more subtle issue is to understand the global topological type of $\mathcal{S}_p$. The main result of this note is one of the ingredients in the Euler characteristic formula for $\mathcal{S}_p$ given in [2, Theorem 7.2]. Thus, finding a formula for the Euler characteristic of the smooth compactification $\overline{\mathcal{S}_p}$ of $\mathcal{S}_p$ is equivalent to obtaining one for the number of escape regions of $\mathcal{S}_p$, which seems to be a difficult task (for an algorithm to compute this number see [6]). For further discussion of the interplay between escape regions and the structure of the bifurcation locus of $\mathcal{S}_p$, see [2, Section 8].

This paper is organized as follows. In Section 2 we state our main result: Theorem 1. In Section 3, we state an analogous result, Theorem 3.1, in the context of iteration of cubic polynomials over the field of formal Puiseux series, and proceed to show how Theorem 3.1 implies Theorem 1. In Section 4 we prove Theorem 3.1. The proof heavily relies on the results contained in [9]. Throughout, we recall the main definitions and results from [9] as we advance in the proof.

Acknowledgements. The main result of this paper was announced in [2]. Nevertheless, in a manuscript that circulated before, Bonifant and Milnor introduced leading monomial vectors associated to an escape region and asked whether they determine the escape region. I am grateful to them since their question is the main motivation for this work.

2   STATEMENT OF THE RESULTS

Throughout, we fix an integer $p \ge 1$ and consider the periodic curve $\mathcal{S}_p$. A polynomial $F_{a,v}$ in $\mathcal{S}_p$ has connected or disconnected Julia set according to whether the forward orbit of the critical point $-a$ is bounded or escapes to infinity. The polynomials in $\mathcal{S}_p$ with connected Julia set form a compact subset of $\mathcal{S}_p$ called the *connectedness locus* $\mathcal{C}(\mathcal{S}_p)$ *of* $\mathcal{S}_p$. The ones with disconnected Julia set form the *escape locus* $\mathcal{E}(\mathcal{S}_p)$.

An escape region is a connected component $\mathcal{U}$ of $\mathcal{E}(\mathcal{S}_p)$. Each escape region is an open and unbounded subset of $\mathcal{S}_p$ which is conformally isomorphic to a punctured disk. Thus, $\mathcal{S}_p$ is a punctured (possibly disconnected) complex curve with a one-to-one correspondence between punctures and escape regions. Adding to the curve $\mathcal{S}_p$ one *ideal point* $\infty_\mathcal{U}$ for each escape region $\mathcal{U}$ at the corresponding puncture, we obtain a compact and smooth complex curve $\overline{\mathcal{S}_p}$ (see [2]).

According to [10, Theorem 5.16], given an escape region $\mathcal{U} \subset \mathcal{S}_p$, as $(a, v) \in \mathcal{U}$ approaches $\infty_{\mathcal{U}}$, for each critical orbit element

$$a_j := F_{a,v}^j(a), \quad j = 1, \ldots, p - 1,$$

we have

$$a_j = \begin{cases} a + o(a), \text{ or} \\ -2a + o(a). \end{cases}$$

That is, the periodic critical orbit elements are "asymptotic" to the periodic critical point $+a$ or to the co-critical point $-2a$. The co-critical point receives its name from the property of also being a preimage of the marked critical value

$$F_{a,v}(-2a) = F_{a,v}(+a) = v.$$

An associated leading monomial vector $\vec{\mathbf{u}}$ will encode the asymptotic behavior of $a_j - a$, for $j = 1, \ldots, p - 1$. In order to be more precise we need to introduce the notion of multiplicity of an escape region $\mathcal{U}$ and auxiliary holomorphic functions u_j defined in a neighborhood of $\infty_{\mathcal{U}}$.

Following Milnor [10, Section 5A], there exists an integer $\mu \geq 1$, called the *multiplicity* of $\mathcal{U}$, and a local coordinate ζ of $\overline{\mathcal{S}_p}$ around $\infty_{\mathcal{U}}$ such that

$$\zeta^\mu = \frac{1}{3a},$$

where the ideal point $\infty_{\mathcal{U}}$ corresponds to $\zeta = 0$ and the preceding identity holds for $\zeta \neq 0$. In this coordinate,

$$u_j := \frac{a - a_j}{3a} = \sum_{k \geq 0} c_k \zeta^k$$

is holomorphic in a neighborhood of $\zeta = 0$.

REMARK 2.1. *The choice of the factor of 3 is for convenience. For example, with this choice, $c_0 \in \{0, 1\}$. If $a_j - a = o(a)$, then $c_0 = 0$ and if $a_j + 2a = o(a)$, then $c_0 = 1$.*

Passing to formal (Puiseux) power series, the complex variables ζ and

$$\xi := 1/3a$$

become transcendental elements $\boldsymbol{\zeta}$ and $\boldsymbol{\xi}$ over $\mathbb{C}$ such that $\boldsymbol{\zeta} = \boldsymbol{\xi}^{1/\mu}$. Thus, it is natural to associate with u_j the formal Puiseux power series:

$$\mathbf{u}_j = \sum_{k \geq k_0 \geq 0} c_k \boldsymbol{\xi}^{k/\mu} \in \mathbb{C}[[\boldsymbol{\xi}^{1/\mu}]].$$

Without loss of generality, we may assume that $c_{k_0} \neq 0$, and define the *leading monomial of $\mathbf{u}_j$* as

$$\mathbf{m}(\mathbf{u}_j) = c_{k_0} \boldsymbol{\xi}^{k_0/\mu}.$$

We say that

$$\vec{\mathbf{u}} = (\mathbf{u}_1, \ldots, \mathbf{u}_{p-1}, 0)$$

is a *vector of Puiseux series associated to the escape region $\mathcal{U}$.*

The asymptotic behavior of $a_j - a$ is encoded by the *vector of leading monomials associated to $\vec{\mathbf{u}}$*:

$$\mathbf{m}(\vec{\mathbf{u}}) = (\mathbf{m}(\mathbf{u}_1), \ldots, \mathbf{m}(\mathbf{u}_{p-1}), 0).$$

THEOREM 1. *Let $\vec{u}'$ and $\vec{u}''$ be vectors of Puiseux series associated to escape regions of $\mathcal{S}_p$.*

If $\mathbf{m}(\vec{u}') = \mathbf{m}(\vec{u}'')$, then $\vec{u}' = \vec{u}''$.

Vectors of Puiseux series associated to an escape region $\mathcal{U}$ of multiplicity μ are uniquely determined modulo the action of the group of (ring) automorphisms of $\mathbb{C}[[\xi^{1/\mu}]]$ fixing $\mathbb{C}[[\xi]]$ pointwise. This group is cyclic of order μ, generated by the unique automorphism γ of $\mathbb{C}[[\xi^{1/\mu}]]$ such that $\gamma(\xi^{1/\mu}) = \exp(2\pi i/\mu)\xi^{1/\mu}$. Thus, if all the coordinates of a monomial vector $\vec{\mathbf{m}}$ belong to $\mathbb{C}[[\xi^{1/\mu}]]$, then there is at most one escape region with an associated vector of leading monomials that agrees with $\gamma^j(\vec{\mathbf{m}})$ for some $j = 0, \ldots, \mu - 1$, where the action of γ on vectors is the one induced by the action on all coordinates (see Section 5 in [2]).

3   PUISEUX SERIES DYNAMICS STATEMENT

The aim of this section is to translate the statement of Theorem 1 into a question about iterations of cubic polynomials over the field of formal Puiseux series.

Consider the algebraic closure of $\mathbb{Q}$ contained in $\mathbb{C}$ and denote it by $\mathbb{Q}^a$. Let $\mathbb{Q}^a\langle\langle\xi\rangle\rangle$ be the field of formal Puiseux series with coefficients in $\mathbb{Q}^a$. That is, $\mathbb{Q}^a\langle\langle\xi\rangle\rangle$ is the injective limit of the fields of formal Laurent series $\mathbb{Q}^a((\xi^{1/m}))$, where $m \in \mathbb{N}$ and the limit is taken with respect to the obvious inclusions. The field $\mathbb{Q}^a\langle\langle\xi\rangle\rangle$ is algebraically closed (e.g. see [4]). It is naturally endowed with a non-Archimedean absolute value induced by the order of vanishing at $\xi = 0$. More precisely, for any $z \in \mathbb{Q}^a\langle\langle\xi\rangle\rangle$ there exists a positive integer m such that

$$z = \sum_{j \geq j_0} a_j \xi^{j/m}$$

for some $j_0 \in \mathbb{Z}$ and $a_j \in \mathbb{Q}^a$. Provided that $z \neq 0$, we let

$$\mathrm{ord}(z) = \min\left\{ \frac{j}{m} \mid a_j \neq 0 \right\}$$

and

$$|z|_o = \exp(-\mathrm{ord}(z)).$$

We will rely heavily on the results contained in [9], where the author studied cubic polynomial dynamics over the larger field $\mathbb{L}$ obtained as the completion of $\mathbb{Q}^a\langle\langle\xi\rangle\rangle$. Although it is not strictly necessary here, we will work over $\mathbb{L}$ so that we may refer to the aforementioned paper.

Any element z in $\mathbb{L}$ may be represented by a series

$$z = \sum_{k \geq 0} c_k \xi^{\lambda_k}$$

where $\lambda_k \in \mathbb{Q}$ is an increasing sequence such that $\lambda_k \to \infty$ as $k \to \infty$. Under the assumption that $c_0 \neq 0$, we have that $|z|_o = \exp(-\lambda_0)$, and we say that

$$\mathbf{m}(z) = c_0 \xi^{\lambda_0}$$

is the *leading monomial of z*.

For any given $\alpha \in \mathbb{L}$ with $|\alpha|_o > 1$, in [9] we considered the one parameter family of cubic polynomials

$$\psi_\nu(z) = \alpha^2(z-1)^2(z+2) + \nu \in \mathbb{L}[z],$$

where $\nu \in \mathbb{L}$. Note that the critical points of ψ_ν are $\omega^\pm = \pm 1$. When ± 1 play the role of critical points, we prefer to denote these numbers by $\omega^\pm$.

We say that a parameter ν is *periodic of period p* if $\psi_\nu^p(\omega^+) = \omega^+$ and p is the smallest positive integer with this property.

Given a period p parameter ν, for $j \geq 0$, let $\nu_j = \psi_\nu^j(\omega^+)$. We say that

$$\vec{\mathbf{m}}(\nu - \omega^+) = (\mathbf{m}(\nu_1 - \omega^+), \ldots, \mathbf{m}(\nu_{p-1} - \omega^+), 0)$$

is the *vector of leading monomials associated to ν*.

Now we may state Theorem 3.1 and show how Theorem 1 is a consequence of this result. The rest of the paper is devoted to establishing Theorem 3.1.

THEOREM 3.1. *Consider $\alpha \in \mathbb{L}$ such that $|\alpha|_o > 1$. Let*

$$\psi_\nu(z) = \alpha^2(z-1)^2(z+2) + \nu,$$

where $\nu \in \mathbb{L}$.

Assume that $\nu', \nu'' \in \mathbb{L}$ are periodic parameters such that the vectors of leading monomials associated to ν' and ν'' coincide. Then $\nu' = \nu''$.

PROOF OF THEOREM 1 ASSUMING THEOREM 3.1. According to equation (5) in [2, Section 4], if

$$\vec{\mathbf{u}} = (\mathbf{u}_1, \mathbf{u}_2, \ldots, \mathbf{u}_{p-1}, 0)$$

is a vector of Puiseux series associated to an escape region of $\mathcal{S}_p$, then

$$\mathbf{u}_{j+1} = \boldsymbol{\xi}^{-2}\mathbf{u}_j^2(\mathbf{u}_j - 1) + \mathbf{u}_1$$

for all $j = 1, \ldots, p-1$, with the understanding that $\mathbf{u}_p = 0$.

Consider the polynomial

$$\phi_\mathbf{u}(z) = \boldsymbol{\xi}^{-2}z^2(z-1) + \mathbf{u} \in \mathbb{Q}^a \langle\!\langle \boldsymbol{\xi} \rangle\!\rangle[z],$$

where $\mathbf{u} \in \mathbb{Q}^a \langle\!\langle \boldsymbol{\xi} \rangle\!\rangle$. If

$$\vec{\mathbf{u}} = (\mathbf{u}_1, \mathbf{u}_2, \ldots, \mathbf{u}_{p-1}, 0)$$

is a vector of Puiseux series associated to an escape region of $\mathcal{S}_p$, then $z = 0$ has period exactly p under $\phi_{\mathbf{u}_1}$ and $\mathbf{u}_j = \phi_{\mathbf{u}_1}^j(0)$.

After conjugacy by the affine isometry $z = -(w-1)/3$ of $\mathbb{Q}^a \langle\!\langle \boldsymbol{\xi} \rangle\!\rangle$, the map $\phi_\mathbf{u}$ becomes

$$\varphi_\mathbf{u}(w) = (3\boldsymbol{\xi})^{-2}(w-1)^2(w+2) - 3\mathbf{u} + 1.$$

We set $\alpha = (3\boldsymbol{\xi})^{-1}$. For $\nu = -3\mathbf{u} + 1$, we have that

$$\psi_\nu(z) = \alpha^2(z-1)^2(z+2) + \nu = \varphi_\mathbf{u}(z).$$

Thus, if $z = 0$ has period p under $\phi_\mathbf{u}$ and $\nu = -3\mathbf{u} + 1$, then $\omega^+ = +1$ has period p under ψ_ν. Moreover, $\mathbf{m}(\nu_j - \omega^+) = -3\mathbf{m}(\mathbf{u}_j)$ for $1 \leq j < p$, where $\nu_j = \psi_\nu^j(\omega^+)$.

Now consider two vectors $\vec{u}' = (\mathbf{u}'_j)$ and $\vec{u}'' = (\mathbf{u}''_j)$ of Puiseux series associated to some escape regions of $\mathcal{S}_p$ with the same associated vector of leading monomials. If $v' = -3\mathbf{u}'_1 + 1$ and $v'' = -3\mathbf{u}''_1 + 1$, then

$$\mathbf{m}(v'_j - \omega^+) = \mathbf{m}(v''_j - \omega^+)$$

for all $j = 1, \ldots, p - 1$, where $v'_j = \psi^j_{v'}(v')$ and $v''_j = \psi^j_{v''}(v'')$. From Theorem 3.1 it follows that $v' = v''$. Hence, $\mathbf{u}'_1 = \mathbf{u}''_1$ and $\vec{u}' = \vec{u}''$. □

4   ONE-PARAMETER FAMILIES

The aim of this section is to prove Theorem 3.1. We will simultaneously advance in the proof as we summarize the necessary results contained in [9]. Some of the following results overlap with [2] when $\alpha = (3\xi)^{-1}$. Since Theorem 1 was applied in [2], to avoid possible loss of rigor and to establish Theorem 3.1 for arbitrary α, we will not rely on the results of [2].

4.1   Closed balls, affine partitions, and action of polynomials

Although in the metric topology of $\mathbb{L}$ every ball is a clopen set, we say that

$$B_{\leq r}(z_0) = \{z \in \mathbb{L} \mid |z - z_0|_\circ \leq r\}$$

is a *closed ball* provided that $r \in \exp(\mathbb{Q})$. Similarly,

$$B_{<r}(z_0) = \{z \in \mathbb{L} \mid |z - z_0|_\circ < r\}$$

is called an *open ball* if $r \in \exp(\mathbb{Q})$.

4.1.1

Given a polynomial $\varphi \in \mathbb{L}[z]$ of degree at least 1, the image of an open (resp. closed) ball B under φ is again an open (resp. closed) ball. Moreover, $\varphi \colon B \to \varphi(B)$ has a well-defined degree. That is, for all $z' \in \varphi(B)$, the number of preimages of z' which lie in B, counting multiplicities, is independent of z'. The degree $\deg_B \varphi$ of $\varphi \colon B \to \varphi(B)$ may be computed from the number of critical points of φ in B as follows. Denote by $\deg_z \varphi$ the local degree of φ at z. Then

$$\deg_B \varphi = 1 + \sum_{z \in B} (\deg_z \varphi - 1).$$

For further details see [9, Proposition 2.1].

4.1.2   *Affine partitions*

Given a closed ball $B = B_{\leq r}(z_0)$ of radius r, the collection $\mathcal{P}_B$ formed by all the open balls of radius r contained in B will be called the *affine partition of B*. Affine partitions are naturally endowed with the structure of an affine line $\mathbb{A}^1_{\mathbb{Q}^a}$ over $\mathbb{Q}^a$. More precisely, for any choice of $z \in B$, the map

$$\begin{array}{ccc} \mathbb{Q}^a & \to & \mathcal{P}_B \\ w & \mapsto & B_{<r}(z + w\xi^{-\log r}) \end{array}$$

is a bijection which, modulo pre-composition by affine transformations, is independent of the choice of $z \in B$ (see [9, Section 2.3]).

To illustrate the preceding definition consider the *ring of integers* $\mathfrak{O}_{\mathbb{L}} = B_{\leq 1}(0)$. Observe that $\mathfrak{O}_{\mathbb{L}}$ has as unique maximal ideal $\mathfrak{M}_{\mathbb{L}} = B_{<1}(0)$. The field $\mathbb{Q}^a$ and the *residue field* $\mathfrak{O}_{\mathbb{L}}/\mathfrak{M}_{\mathbb{L}}$ are canonically identified via $w \mapsto w + \mathfrak{M}_{\mathbb{L}}$. This identification coincides with the above bijection for the choice of $z = 0$.

4.1.3

Polynomials not only map balls onto balls, but they also act on the corresponding affine partitions. Namely, consider a polynomial $\varphi \in \mathbb{L}[z]$ that maps the closed ball B onto the closed ball B' with degree $\deg_B \varphi$. If $D \in \mathcal{P}_B$, then $\varphi(D) \in \mathcal{P}_{B'}$. Moreover, the induced map $\varphi_* \colon \mathcal{P}_B \to \mathcal{P}_{B'}$ is a polynomial of degree $\deg_B \varphi$ in the corresponding affine structures (see [9, Proposition 2.5]).

For example, consider a polynomial

$$\varphi(z) = c_0 + c_1 z + \ldots + c_d z^d \in \mathfrak{O}_{\mathbb{L}}[z].$$

From the Newton polygon (e.g., see [5]), it is not difficult to check that

$$\varphi(\mathfrak{O}_{\mathbb{L}}) = \mathfrak{O}_{\mathbb{L}}$$

if and only if $\max\{|c_j|_o \mid 1 \leq j \leq d\} = 1$. In this case, denote by $\pi \colon \mathfrak{O}_{\mathbb{L}} \to \mathfrak{O}_{\mathbb{L}}/\mathfrak{M}_{\mathbb{L}}$ the quotient map. Then via the canonical identification of $\mathfrak{O}_{\mathbb{L}}/\mathfrak{M}_{\mathbb{L}}$ with $\mathbb{Q}^a$ we have that

$$\varphi_*(z) = \pi(c_0) + \pi(c_1)z + \ldots + \pi(c_d)z^d \in \mathbb{Q}^a[z].$$

4.2 Basic structure of dynamical space

Throughout the rest of the paper we fix $\alpha \in \mathbb{L}$ *with* $|\alpha|_o > 1$ *and let*

$$\psi_v(z) = \alpha^2(z-1)^2(z+2) + v,$$

where $v \in \mathbb{L}$. *Given parameters* $v, v', \ldots$ *in* $\mathbb{L}$, *we let* $v_j = \psi_v^j(\omega^+), v_j' = \psi_{v'}^j(\omega^+), \ldots$ *for all* $j \geq 0$.

4.2.1

The *filled Julia set* $K(\psi_v)$ consists of all $z \in \mathbb{L}$ such that $\{|\psi_v^j(z)|_o\}$ is bounded. The complement of $K(\psi_v)$ is the *basin of infinity*.

We want to study cases where the critical point ω^+ is periodic and, hence, belongs to $K(\psi_v)$. Since $\psi_v(\omega^+) = v$, this means that we want $v \in K(\psi_v)$, which implies easily that $|v|_o \leq 1$ (see [9, Lemma 5.1]).

Assume that $|v|_o \leq 1$; since $|\alpha|_o > 1$, it follows that every z with $|z|_o > 1$ belongs to the basin of infinity.

4.2.2

We say that

$$D_0 = \{z \in \mathbb{L} \mid |z|_o \leq 1\}$$

is the *level 0 dynamical ball*. For all $v \in \mathbb{L}$ such that $|v|_o \leq 1$, we have

$$\psi_v(D_0) = B_{\leq |\alpha|_o^2}(0).$$

Moreover, every element z of the annulus $B_{\leq|\alpha|_o^2}(0) \setminus D_0$ belongs to the basin of
infinity.

Note that $\psi_v(\omega^+) \in D_0$ if and only if $|v|_o \leq 1$. However, if $|v|_o \leq 1$, then
$\psi_v(\omega^-) \notin D_0$.

4.2.3

We can now imitate the Branner-Hubbard puzzle construction [3] in this non-
Archimedean context. Assume that $|v|_o \leq 1$. Let $L_0^v = D_0$ and for all integers
$\ell \geq 0$, recursively define
$$L_{\ell+1}^v = \psi_v^{-1}(L_\ell^v).$$
Thus,
$$L_0^v \supset L_1^v \supset L_2^v \supset \cdots.$$
Note that $z \in K(\psi_v)$ if and only if $z \in L_\ell^v$ for all $\ell \geq 0$.

We say that L_ℓ^v is the *level ℓ set* of ψ_v and if $z \in L_\ell^v$, then we say that z *is a level
ℓ point*. According to [9, Proposition 2.1], L_ℓ^v is the disjoint union of finitely many
closed balls called *level ℓ dynamical balls*. Each dynamical ball of level $\ell + 1$ maps
onto a dynamical ball of level ℓ. When z lies in the level ℓ set, we denote by $D_\ell^v(z)$
the level ℓ dynamical ball containing z.

4.2.4

Each level $\ell + 1$ dynamical ball $D_{\ell+1}$ is properly contained in a dynamical ball D_ℓ
of level ℓ. Thus, $D_{\ell+1}$ must be contained in one element of the affine partition of
D_ℓ. According to [9, Lemma 4.2(i)], assuming that v is a level ℓ point, we know
even more. That is, given an element P of the affine partition of D_ℓ there exists at
most one dynamical ball of level $\ell + 1$ contained in P.

4.2.5

The preceding is closely related to leading monomials. For example, assume z, ω^+
are level $\ell + 1$ points such that $D_\ell^v(z) = D_\ell^v(\omega^+)$ but $D_{\ell+1}^v(z) \neq D_{\ell+1}^v(\omega^+)$. By
the previous paragraph, z and ω^+ lie in different elements of the affine partition
of $D_\ell^v(\omega^+)$. Hence, $|\mathbf{m}(z - \omega^+)|_o$ is the radius of $D_\ell^v(\omega^+)$. Moreover, given an-
other level $\ell + 1$ point z', we have that $\mathbf{m}(z' - \omega^+) = \mathbf{m}(z - \omega^+)$ if and only if
$D_{\ell+1}^v(z') = D_{\ell+1}^v(z)$.

4.3 Marked grids

Assume that ω^+ is a level $\ell \geq 0$ point for ψ_v. Given any $z \in L_\ell^v$, we introduce the
level ℓ marked grid of z for ψ_v,
$$\mathbf{M} = \mathbf{M}_\ell^v(z)$$
as the array of 0s and 1s with entries $M(m,k)$ defined by
$$M(m,k) = \begin{cases} 1 & \text{if } \omega^+ \in D_m^v(\psi_v^k(z)), \\ 0 & \text{otherwise,} \end{cases}$$
where $0 \leq m + k \leq \ell$. In other words, $M(m,k) = 1$ if and only if the orbit point
$z_k = \psi_v^k(z)$ belongs to the same level ℓ dynamical ball as the critical point ω^+.

If $M(m,k) = 1$, then we say that (m,k) is a *marked position*. Observe that all positions $(0,k)$ are marked.

Marked grids were introduced by Branner and Hubbard [3] in the context of complex cubic polynomial dynamics.

4.3.1

The *level ℓ critical marked grid of ψ_v* is $\mathbf{M}_\ell^v(\omega^+)$ provided that $\omega^+ \in L_\ell^v$. When ω^+ and z lie in $K(\psi_v)$, the marked grid of z is defined for all positions (m,k) with non-negative entries. In this case $\mathbf{M}^v(z)$ denotes the infinite array and we say that $\mathbf{M}^v(\omega^+)$ is the *critical marked grid*.

4.3.2

Marked grids satisfy certain "rules." The first rule follows directly from the definition. That is, $M(m,k) = 1$ implies $M(n,k) = 1$ for all $0 \leq n \leq m$. For more about marked grid rules see [2, 9].

4.3.3

Assuming that $\omega^+ \in L_\ell^v$, the radius $r_\ell^v(z)$ of any dynamical ball $D_\ell^v(z)$ of level ℓ is uniquely determined by the marked grid $\mathbf{M}_\ell^v(z) = [M(m,k)]$. More precisely, if

$$S_\ell = \sum_{j=0}^{\ell-1} M(\ell - j, j),$$

then

$$-\log r_\ell^v(z) = \sum_{m=1}^{\ell} 2^{-S_m} \log |\alpha|_o^2,$$

according to [9, pp. 1354 and 1359].

4.3.4

The marked grid also determines the norm $|z_j - \omega^+|_o = |\mathbf{m}(z_j - \omega^+)|_o$ where $z_j = \psi_v^j(z)$. Consider $v \in \mathbb{L}$ and assume that both ω^+ and z are level $\ell \geq 0$ points. For $0 \leq j < \ell$ assume that ℓ_j is the "depth" of the jth column of $\mathbf{M}_\ell^v(z) = [M(m,k)]$. That is, $M(\ell_j, j) = 1$ and $M(\ell_j + 1, j) = 0$ (in particular, $\ell_j < \ell - j$). From 4.2.5, we have that

$$\operatorname{ord}(\mathbf{m}(z_j - \omega^+)) = -\log r_{\ell_j}^v(z_j),$$

where $r_{\ell_j}^v(z_j)$ is the radius of the level ℓ_j dynamical ball containing z_j. Moreover, this quantity may be computed using the formula given in 4.3.3.

4.3.5   *Leading monomials determine the marked grid*

LEMMA 4.1. *Let v and v' be period p parameters with the same associated leading monomial vector*

$$(\mathbf{m}_1, \ldots, \mathbf{m}_{p-1}, 0).$$

Then

$$\mathbf{M}^{\nu}(\omega^+) = \mathbf{M}^{\nu'}(\omega^+).$$

PROOF. Let $\mathbf{M}^{\nu}(\omega^+) = [M^{\nu}(m,k)]$ and $\mathbf{M}^{\nu'}(\omega^+) = [M^{\nu'}(m,k)]$.
We proceed by induction in $\ell \geq 0$ to show that

$$\mathbf{M}^{\nu}_{\ell}(\omega^+) = \mathbf{M}^{\nu'}_{\ell}(\omega^+)$$

for all ℓ. Recall that $\mathbf{M}^{\nu}_{\ell}(\omega^+)$ is the array of bits $M^{\nu}(j,k)$ with $0 \leq j,k$ and $j+k \leq \ell$. Thus our induction hypothesis will be that $M^{\nu}(j,k) = M^{\nu'}(j,k)$ for $j+k \leq \ell$.

Since the position $(0,0)$ of the grids under consideration are marked, the claim is true for $\ell = 0$. Hence we let $\ell \geq 0$ and assume that

$$\mathbf{M}^{\nu}_{\ell}(\omega^+) = \mathbf{M}^{\nu'}_{\ell}(\omega^+) = [M(j,k)].$$

For all $1 \leq k \leq \ell$, we have to show that $M^{\nu}(\ell-k+1,k) = M^{\nu'}(\ell-k+1,k)$.

In the case that $M(\ell-k,k) = 0$, from the first marked grid rule (see 4.3.2), we conclude that $M^{\nu}(\ell-k+1,k) = M^{\nu'}(\ell-k+1,k) = 0$.

In the case $M(j,k) = 1$, for $j = 1,\ldots,\ell-k$, set

$$e_j = \sum_{0 \leq m < j} M(j-m, k+m)$$

and let

$$s = \sum_{1 \leq j \leq \ell-k} 2^{-e_j} \log |\alpha|_o^2.$$

From the fact that ν and ν' have period p, it follows that

$$M^{\nu}(m,k) = M^{\nu}(m,k+p)$$

and $M^{\nu'}(m,k) = M^{\nu'}(m,k+p)$ for all (m,k). By definition, for all m,

$$M^{\nu}(m,0) = M^{\nu'}(m,0) = 1.$$

Thus, without loss of generality we may assume that $0 < k < p$. By 4.3.4, if $\mathrm{ord}(\mathbf{m}_k) > s$, then $M^{\nu}(\ell-k+1,k) = M^{\nu'}(\ell-k+1,k) = 1$; otherwise,

$$M^{\nu}(\ell-k+1,k) = M^{\nu'}(\ell-k+1,k) = 0.$$

$\square$

4.4 Basic structure of parameter space

Now we imitate the construction of parameter puzzles due to Branner-Hubbard.
For all integers $\ell \geq 0$,

$$\mathcal{L}_{\ell} = \{\nu \in \mathbb{L} \mid \psi^{\ell+1}_{\nu}(\omega^+) \in D_0\}$$

is the set of *level ℓ parameters*.

By [9, Proposition 2.1], $\mathcal{L}_{\ell}$ consists of a finite disjoint union of closed balls which we call *level ℓ parameter balls* and proceed to denote by $\mathcal{D}_{\ell}(\nu)$ the level ℓ parameter ball containing ν, when $\nu \in \mathcal{L}_{\ell}$. Note that there is a unique parameter ball $\mathcal{D}_0 = \{\nu \in \mathbb{L} \mid |\nu|_o \leq 1\}$ of level 0.

4.4.1

Parameters within a given level $\ell \geq 0$ parameter ball $\mathcal{D}_\ell$ share some dynamical balls and properties. More precisely, according to [9, Lemma 5.10], if $v, v' \in \mathcal{D}_\ell$, then

$$D_{\ell+1-k}^v(v_k) = D_{\ell+1-k}^{v'}(v_k')$$

for $1 \leq k \leq \ell$. Moreover, if $\mathcal{P}_{\ell+1-k}$ denotes the affine partition of $D_{\ell+1-k}^v(v_k)$, then

$$\psi_{v_*} = \psi_{v_*'} \colon \mathcal{P}_{\ell+1-k} \to \mathcal{P}_{\ell+1-(k+1)}.$$

4.4.2   Centers

Consider a parameter ball $\mathcal{D}_\ell$ of level $\ell \geq 0$. By [9, Proposition 5.5], the level $\ell+1$ critical marked grid of all parameters $v \in \mathcal{D}_\ell$ coincide. Denote such grid by $\mathbf{M}_{\ell+1} = [M(m,k)]$.

Let p_ℓ be such that $M(\ell+1-j, j)$ is unmarked for $j = 1, \ldots, p_\ell - 1$ and

$$M(\ell+1-p_\ell, p_\ell)$$

is marked. It follows that every periodic parameter in $\mathcal{D}_\ell$ has period p satisfying $p \geq p_\ell$.

According to [9, Proposition 5.5], there exists a unique parameter $\hat{v} \in \mathcal{D}_\ell$ such that $\hat{v}$ is periodic and the period of $\hat{v}$ is p_ℓ. We say that $\hat{v}$ *is the center of* $\mathcal{D}_\ell$.

4.4.3   Leading monomials of level ℓ centers

Although Lemma 4.2 is not needed in the proof of Theorem 3.1, it might be of independent interest, and it illustrates the relation between centers and leading monomials.

LEMMA 4.2. *Consider a period p parameter v with associated vector of leading monomials*

$$(\mathbf{m}_1, \ldots, \mathbf{m}_{p-1}, 0).$$

Let $\ell \geq 1$ and denote by v' the center of $\mathcal{D}_\ell(v)$. Then the vector of leading monomials associated to v' is

$$(\mathbf{m}_1, \ldots, \mathbf{m}_{p_\ell-1}, 0),$$

where $p_\ell \leq p$ is the period of v'.

PROOF. According to 4.4.1, for all $j \leq \ell$,

$$D_{\ell+1-j}^v(v_j) = D_{\ell+1-j}^{v'}(v_j').$$

From 4.4.2, it follows that, for $1 \leq j \leq p_\ell - 1$,

$$\omega^+ \notin D_{\ell+1-j}^v(v_j) = D_{\ell+1-j}^{v'}(v_j').$$

Thus, $\mathbf{m}(v_j - \omega^+) = \mathbf{m}_j$ and $\mathbf{m}(v_j' - \omega^+)$ agree, since

$$|v_j - v_j'|_o < |v_j - \omega^+|_o = |v_j' - \omega^+|_o = |\mathbf{m}_j|_o.$$

$\square$

4.5 Correspondence between parameter and dynamical balls

Given a periodic parameter v we would like to show that $\mathcal{D}_\ell(v)$ is uniquely determined, for all $\ell \geq 0$, by the associated leading monomial vector $\vec{\mathbf{m}}$. Our proof of this fact will be by induction. Provided that $\mathcal{D}_\ell(v)$ is determined by $\vec{\mathbf{m}}$, we will use the dynamical space of the center of $\mathcal{D}_\ell(v)$ to "compare" level $\ell + 1$ parameter balls compatible with $\vec{\mathbf{m}}$. This comparison relies on the following parameter-dynamical space correspondence results.

4.5.1

Propositions 5.5 and 5.7 in [9] establish the following.

Consider a parameter ball $\mathcal{D}_\ell$ of level $\ell \geq 0$ and let P be an element of the affine partition of $\mathcal{D}_\ell$. For any $v' \in \mathcal{D}_\ell$,

$$\mathcal{D}_\ell = D_\ell^{v'}(v'),$$

and the following holds:

P contains a dynamical ball $D_{\ell+1}^{v'}(z)$ of level $\ell + 1$ if and only if P contains a parameter ball $\mathcal{D}_{\ell+1}$ of level $\ell + 1$. In this case, the level $\ell + 1$ parameter ball is unique and

$$\mathbf{M}_{\ell+1}^{v'}(z) = \mathbf{M}_{\ell+1}^v(v)$$

for all $v \in \mathcal{D}_{\ell+1}$.

4.5.2   From parameter space to dynamical space

Here we establish the key lemmas to prove Theorem 3.1.

LEMMA 4.3. *Consider a period p parameter v with associated leading monomial vector*

$$(\mathbf{m}_1, \ldots, \mathbf{m}_{p-1}, 0).$$

Let v' be the center of the parameter ball $\mathcal{D}_\ell(v)$ of level ℓ. Denote by P the element of the affine partition of $\mathcal{D}_\ell(v)$ containing $\mathcal{D}_{\ell+1}(v)$ and by $p_{\ell+1}$ the period of the center of $\mathcal{D}_{\ell+1}(v)$.

Then there exists $z' \in P$ such that the following statements hold:

1. $\mathbf{M}_{\ell+1}^{v'}(z') = \mathbf{M}_{\ell+1}^v(v),$

2. $\psi_{v'}^{p_{\ell+1}-1}(z') = \omega^+,$

3. $\mathbf{m}_j = \mathbf{m}(\psi_{v'}^{j-1}(z') - \omega^+)$ *for all* $j = 1, \ldots, p_{\ell+1} - 1.$

PROOF. By 4.5.1, there exists a (unique) dynamical ball $D_{\ell+1}^{v'}(z_1) \subset P$. Moreover,

$$\mathbf{M}_{\ell+1}^{v'}(z_1) = \mathbf{M}_{\ell+1}^v(v).$$

Consequently, $\psi_{v'}^{p_{\ell+1}-1}(D_{\ell+1}^{v'}(z_1))$ contains ω^+, since the corresponding position in $\mathbf{M}_{\ell+1}^v(v)$ is marked. Thus, we may consider $z' \in D_{\ell+1}^{v'}(z_1)$ such that

$$\psi_{v'}^{p_{\ell+1}-1}(z') = \omega^+.$$

Therefore, $z' \in P$ and (1)–(2) hold.

By 4.4.1, we have that $\psi_{v'}^{j-1}(z') \in \psi_{v'}^{j-1}(P) = \psi_v^{j-1}(P)$. Moreover, $v_j \in \psi_v^{j-1}(P)$ and $\omega^+ \notin \psi_v^{j-1}(P)$, for $1 \le j \le p_{\ell+1} - 1$. Therefore,

$$\mathbf{m}(\psi_{v'}^{j-1}(z') - \omega^+) = \mathbf{m}(v_j - \omega^+) = \mathbf{m}_j.$$

That is, (3) also holds. □

In the dynamical space of the level ℓ center v', our previous lemma establishes that there exists a point z' "close" to the level $\ell + 1$ center with certain properties. Our next lemma establishes that this point z' is the unique point in the critical value dynamical ball of level ℓ with these properties.

LEMMA 4.4. *Consider a period p parameter v with associated leading monomial vector*

$$(\mathbf{m}_1, \ldots, \mathbf{m}_{p-1}, 0).$$

Let v' be the center of the parameter ball $\mathcal{D}_\ell(v)$ of level ℓ. Denote by $p_{\ell+1}$ the period of the center of $\mathcal{D}_{\ell+1}(v)$. Then there exists a unique $z' \in D_\ell^{v'}(v')$ such that the following statements hold:

1. $\mathbf{M}_{\ell+1}^{v'}(z') = \mathbf{M}_{\ell+1}^v(v)$,

2. $\psi_{v'}^{p_{\ell+1}-1}(z') = \omega^+$,

3. $\mathbf{m}_j = \mathbf{m}(\psi_{v'}^{j-1}(z') - \omega^+)$ *for all $j = 1, \ldots, p_{\ell+1} - 1$.*

PROOF. Existence follows from the previous lemma. To prove uniqueness let z'' be another value such that (1)–(3) hold. Let $z'_j = \psi_{v'}^{j-1}(z')$ and $z''_j = \psi_{v'}^{j-1}(z'')$. By (2),

$$\omega^+ = z'_{p_{\ell+1}-1} = z''_{p_{\ell+1}-1}.$$

Suppose, by descending induction on j, that

$$z'_{j+1} = z''_{j+1},$$

for some $1 \le j \le p_{\ell+1} - 2$. It is sufficient to prove that

$$z'_j = z''_j.$$

By (1) the maximal level m such that $D_m^{v'}(z'_j)$ contains ω^+ coincides with the maximal level m for which $D_m^{v'}(z''_j)$ contains the critical point. To simplify notation, let $D = D_m^{v'}(z'_j) = D_m^{v'}(z''_j)$.

Denote by P' (resp. P'') the element of the affine partition of D containing z'_j (resp. z''_j). According to 4.3.4, the radius of D is $|\mathbf{m}_j|_\circ$. From (3),

$$|z'_j - z''_j|_\circ \le \max\{|z'_j - \omega^+ - \mathbf{m}_j|_\circ, |z''_j - \omega^+ - \mathbf{m}_j|_\circ\} < |\mathbf{m}_j|_\circ.$$

Therefore, $P' = P''$.

From 4.2.4, $D_{m+1}^{v'}(z'_j) = D_{m+1}^{v'}(z''_j)$. But this dynamical ball D' of level $m + 1$ is critical point free. By 4.1.1, $\psi_{v'}: D' \to \psi_{v'}(D')$ is one-to-one. Since

$$\psi_{v'}(z'_j) = z'_{j+1} = z''_{j+1} = \psi_{v'}(w''_j),$$

we have that $z'_j = z''_j$. □

4.6   Leading monomials determine parameter balls

PROPOSITION 4.5. *If v' and v'' are periodic parameters with the same associated leading monomial vector, then*

$$\mathcal{D}_\ell(v') = \mathcal{D}_\ell(v'')$$

for all $\ell \geq 0$.

PROOF. Since there exists only one level 0 parameter ball, it is sufficient to assume that $\mathcal{D}_\ell(v') = \mathcal{D}_\ell(v'')$ and show that $\mathcal{D}_{\ell+1}(v') = \mathcal{D}_{\ell+1}(v'')$.

Let v be the center of $\mathcal{D}_\ell(v') = \mathcal{D}_\ell(v'')$. Denote by P' (resp. P'') the element of the affine partition of $\mathcal{D}_\ell(v') = \mathcal{D}_\ell(v'')$ containing v' (resp. v''). By Lemma 4.3 there exists $z' \in P'$ (resp. $z'' \in P''$) such that (1)–(3) of the lemma hold for z' (resp. for z''). By Lemma 4.4, $z' = z''$. Therefore, $P' = P''$. By 4.5.1, there is at most one parameter ball of level $\ell+1$ in each element of the affine partition of $\mathcal{D}_\ell(v') = \mathcal{D}_\ell(v'')$. Hence $\mathcal{D}_{\ell+1}(v') = \mathcal{D}_{\ell+1}(v'')$. $\square$

4.7   Leading monomials determine the parameter

We have established that the parameter balls of a periodic parameter are uniquely determined by the associated vector of leading monomials. However, different periodic parameters (of the same period) might share parameter balls at all levels. To finish the proof of Theorem 3.1 we have to deal with this situation.

4.7.1

Consider a periodic parameter $v \in \mathbb{L}$ of period p. For all $j \geq 0$, let

$$D_\infty^v(v_j) = \cap D_\ell^v(v_j).$$

By [9, Corollary 4.8], $D_\infty^v(v_j)$ is a closed ball. The closed balls $D_\infty^v(v_j)$ are periodic under the action of ψ_v, of some period q which divides p. Since only one of these balls contains the critical point ω^+, in view of 4.1.1, we have that

$$\psi_v^q \colon D_\infty^v(\omega^+) \to D_\infty^v(\omega^+)$$

has degree 2.

4.7.2

Let us continue under the preceding assumptions. Denote by $r_\ell^v(v_j)$ (resp. $r_\infty^v(v_j)$) the radius of $D_\ell^v(v_j)$ (resp. $D_\infty^v(v_j)$). According to [9, Lemma 5.9],

$$r_\ell^v(v) < r_{\ell+1-j}^v(v_j),$$

for all $j > 1$ and $\ell \geq 0$. Unfortunately this inequality is not sufficient for our purpose. However, from the proof of [9, Lemma 5.9], it is not difficult to conclude that there exists $\varepsilon_j > 0$ (independent of ℓ), such that $r_\ell^v(v) + \epsilon_j \leq r_{\ell+1-j}^v(v_j)$, for $j = 2, \ldots, q$. Hence, $r_\infty^v(v) < r_\infty^v(v_j)$, for all $j = 1, \ldots, q$.

4.7.3

Under the previous assumptions, in parameter space let $\mathcal{D}_\infty(\nu) = \cap \mathcal{D}_\ell(\nu)$. It follows that $\mathcal{D}_\infty(\nu) = D_\infty^\nu(\nu)$ is also a closed ball. Moreover, for all $\nu' \in \mathcal{D}_\infty(\nu)$ and all $j \geq 1$, $D_\infty^{\nu'}(\nu'_j) = D_\infty^\nu(\nu_j)$. Furthermore, $\psi_{\nu'}^q \colon D_\infty^{\nu'}(\omega^+) \to D_\infty^{\nu'}(\omega^+)$ is also a degree 2 map.

Observe that the period of the center of $\mathcal{D}_\ell(\nu)$ is eventually constant equal to q. Thus, $\mathcal{D}_\infty(\nu)$ contains a unique parameter of period q.

4.7.4

Before finishing the proof of Theorem 3.1, we need the following result.

LEMMA 4.6. *Consider two periodic parameters ν' and ν'' with the same associated vector of leading monomials, then $\mathcal{D}_\infty(\nu') = \mathcal{D}_\infty(\nu'')$ and both parameters ν' and ν'' are contained in the same element of the affine partition of $\mathcal{D}_\infty(\nu') = \mathcal{D}_\infty(\nu'')$.*

PROOF. The assertion that $\mathcal{D}_\infty(\nu') = \mathcal{D}_\infty(\nu'')$ follows from Proposition 4.5. Also, by 4.4.1, for all $j \geq 0$,

$$D_\infty^{\nu'}(\nu'_j) = D_\infty^{\nu''}(\nu''_j).$$

To simplify notation, let $B_j = D_\infty^{\nu'}(\nu'_j)$ and denote by r_j its radius. Since

$$D_\infty^{\nu'}(\nu') = \mathcal{D}_\infty(\nu') \ni \nu'',$$

it follows that $|\nu' - \nu''|_o \leq r_1$. Let q be the period of B_0 under $\psi_{\nu'}$. For all $2 \leq j < q$ and all $z \in B_j$,

$$|\psi_{\nu'}(z) - \psi_{\nu''}(z)|_o = |\nu' - \nu''|_o \leq r_1 < r_{j+1},$$

by 4.7.2. Thus, the actions of $\psi_{\nu'}$ and $\psi_{\nu''}$ on the affine partition of B_j coincide, for $2 \leq j < q$. Moreover, since B_j is critical point free for $2 \leq j < q$, these actions are injective.

We proceed by contradiction. If ν' and ν'' belong to different elements of the affine partition of B_1, then ν'_q and ν''_q belong to different elements of the affine partition of B_q. Thus,

$$|\mathbf{m}_q|_o \leq r_q = |\nu'_q - \nu''_q|_o < |\mathbf{m}_q|_o,$$

which gives us the desired contradiction.   □

4.7.5   Proof of Theorem 3.1

Consider a periodic parameter ν'. From Proposition 4.5 and the previous lemma, it is sufficient to show that each element of the affine partition of $\mathcal{D}_\infty = \mathcal{D}_\infty(\nu')$ contains at most one periodic parameter.

Let $\mathcal{D}_\infty = D_\infty^{\nu'}(\omega^+)$ and q be the period of $\mathcal{D}_\infty$ under $\psi_{\nu'}$. Denote by $\hat{\nu}$ the unique parameter of period q in $\mathcal{D}_\infty$. For all $\nu \in \mathcal{D}_\infty$, we have that $\psi_\nu^q \colon \mathcal{D}_\infty \to \mathcal{D}_\infty$ has degree 2 (see 4.7.1).

Recall from 4.1.2 that $\mathfrak{O}_{\mathbb{L}}$ denotes the (local) ring of integers of $\mathbb{L}$ with maximal ideal $\mathfrak{M}_{\mathbb{L}}$. Since the residue field $\mathfrak{O}_{\mathbb{L}}/\mathfrak{M}_{\mathbb{L}}$ is canonically identified with $\mathbb{Q}^a$, we denote by $\pi \colon \mathfrak{O}_{\mathbb{L}} \to \mathbb{Q}^a$ the quotient map.

Let $L: \mathcal{D}_\infty \to \mathfrak{O}_\mathbb{L}$ be (the restriction of) an affine map such that $L(\hat{v}) = 0$. Also, let $h: D_\infty \to \mathfrak{O}_\mathbb{L}$ be (the restriction of) an affine map such that $h(\omega^+) = 0$. For all $\lambda \in \mathfrak{O}_\mathbb{L}$,

$$\varphi_\lambda := h \circ \psi^q_{L^{-1}(\lambda)} \circ h^{-1}: \mathfrak{O}_\mathbb{L} \to \mathfrak{O}_\mathbb{L}$$

is a degree 2 map. Moreover, taking the coefficients of $\varphi_\lambda(z) \in \mathfrak{O}_\mathbb{L}[z, \lambda]$ modulo $\mathfrak{M}_\mathbb{L}$ we obtain a polynomial $Q_c(w) \in \mathbb{Q}^a[w, c]$ where $c = \pi(\lambda)$ and $w = \pi(z)$.

For each $c = \pi(\lambda)$, the polynomial $Q_c(w)$ is the action of φ_λ on the affine partition of $\mathfrak{O}_\mathbb{L}$ (see 4.1.3). Thus, $Q_c(w)$ has degree 2 on w and a critical point at $w = 0$, for all c. Therefore,

$$Q_c(w) = \eta w^2 + \beta(c)$$

for some non-zero constant $\eta \in \mathbb{Q}^a$ and some polynomial $\beta(c)$. Since the unique solution of $\varphi_\lambda(0) = 0$ is $\lambda = 0$, from the Newton polygon of $\varphi_\lambda(0)$ it follows that $\beta(c)$ must be a monomial, say bc^m. After a new change of coordinates in dynamical space ($w \mapsto w/\eta$) and a re-parametrization of $c \in \mathbb{Q}^a$, we may assume that $Q_c(w) = z^2 + c^m$.

For the standard complex quadratic family $f_c(z) = z^2 + c$, after Gleason (e.g., see [7]), it is known that all solutions of $f_c^n(0) = 0$ are simple, for all $n \geq 1$. It follows that all roots of $Q_c^n(0) = 0$ are simple, provided that $n \geq 2$. Hence, $\pi^{-1}(c)$ contains at most one periodic parameter of the family φ_λ, for all $c \in \mathbb{Q}^a$. Therefore, each element of the affine partition of $\mathcal{D}_\infty$ contains at most one periodic parameter. $\square$

5   Bibliography

[1]   Magnus Aspenberg and Michael Yampolsky.   Mating non-renormalizable quadratic polynomials. *Communications in Mathematical Physics*, 287:1–40, 2009.

[2]   Araceli Bonifant, Jan Kiwi, and John Milnor. Cubic polynomial maps with periodic critical orbit. II. Escape regions. *Conform. Geom. Dyn.*, 14:68–112, 2010.

[3]   Bodil Branner and John H. Hubbard. The iteration of cubic polynomials. Part 1: The global topology of parameter space, *Acta Math.*, 160(3–4):143–206, 1988.

[4]   Eduardo Casas-Alvero. *Singularities of Plane Curves*, **276** of *London Mathematical Society Lecture Note Series*. Cambridge University Press, Cambridge, 2000.

[5]   J. W. S. Cassels. *Local Fields*, **3** of *London Mathematical Society Student Texts*. Cambridge University Press, Cambridge, 1986.

[6]   Laura DeMarco and Aaron Schiff. Enumerating the basins of infinity of cubic polynomials. *J. Difference Equ. Appl.*, 16(5–6):451–461, 2010.

[7]   A. Douady and J. H. Hubbard. *Étude dynamique des polynômes complexes. I,II*, volume 85 of *Publications Mathématiques d'Orsay [Mathematical Publications of Orsay]*. Université de Paris-Sud, Département de Mathématiques, Orsay, 1985. With the collaboration of P. Lavaurs, Tan Lei, and P. Sentenac.

[8]   D. Faught. *Local connectivity in a family of cubic polynomials*. Ph.D. thesis, Cornell University, 1992.

[9] Jan Kiwi. Puiseux series polynomial dynamics and iteration of complex cubic polynomials. *Ann. Inst. Fourier (Grenoble)*, 56(5):1337–1404, 2006.

[10] John Milnor. Cubic polynomial maps with periodic critical orbit. I. In *Complex dynamics*, pages 333–411. A. K. Peters, Wellesley, MA, 2009.

[11] Mary Rees. Views of parameter space: Topographer and Resident. *Astérisque* (288):1–418, 2003.

[12] Pascale Roesch. Hyperbolic components of polynomials with a fixed critical point of maximal order. *Ann. Sci. École Norm. Sup. (4)*, 40:901–949, 2007.

Limiting behavior of Julia sets of singularly perturbed rational maps

Robert L. Devaney

ABSTRACT. We survey some recent results involving singular perturbations of the polynomial maps $z^n + c$ where $n \geq 2$, i.e., what happens when a pole is added to this family of maps. Here we consider the most interesting case where the pole lies at the origin and we have some natural symmetries, i.e., we restrict to the case

$$F_\lambda(z) = z^n + c + \frac{\lambda}{z^n}.$$

The case where $n = 2$ is by far the most interesting, as we show that as $\lambda \to 0$, the Julia sets of F_λ tend to the filled Julia set of the polynomial map $z^n + c$. This does not happen when $n > 2$, since there are always infinitely many annular components in the Fatou set when λ is close to 0.

1 INTRODUCTION

In recent years there have been a number of papers dealing with singular perturbations of complex dynamical systems. Most of these papers deal with maps of the form $z^n + c + \lambda/z^d$, where $n \geq 2$ and $d \geq 1$ and c is the center of a hyperbolic component of the multibrot set, i.e., the connectedness locus for the family $z^n + c$. These maps are called singular perturbations because, when $\lambda = 0$, the map is just the polynomial $z^n + c$ and the dynamical behavior for this map is completely understood. When $\lambda \neq 0$, the degree of the map changes and the dynamical behavior suddenly explodes.

Our goal in this paper is to give a survey of the behavior of these maps as the parameter λ tends to 0. By far the most interesting (and complicated) subfamily of these maps is the family $z^2 + c + \lambda/z^2$. The interesting fact here is that the Julia sets of these rational maps converge in the Hausdorff metric to the filled Julia set of the quadratic polynomial $z^2 + c$ as $\lambda \to 0$. This is somewhat surprising since it is known that if the Julia set of a rational map ever contains an open set, then that Julia set must in fact be the entire complex plane. Here the limiting set always contains an open set when c is the center of a hyperbolic component, but this set is never the entire complex plane. So, as $\lambda \to 0$, the Julia sets of these rational maps come arbitrarily close to subsets of $\mathbb{C}$ that contain open sets. The actual Julia set for $\lambda = 0$ is, of course, much different.

For example, in Figure 1.1, we display several Julia sets in the family $z^2 + \lambda/z^2$, where λ is small. The white regions lie in the complement of the Julia set. Note how these disks become smaller as λ moves closer to 0. The limiting set is the unit disk, which is the filled Julia set of z^2, but the actual Julia set when $\lambda = 0$ is just the unit circle.

This work was partially supported by grant #208780 from the Simons Foundation.

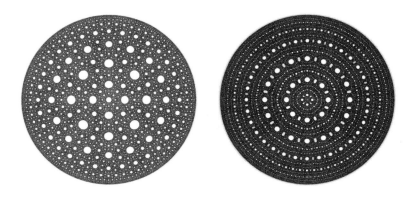

FIGURE 1.1: The Julia sets for $n = 2$ and $\lambda = -0.001$ and $\lambda = -0.00001$.

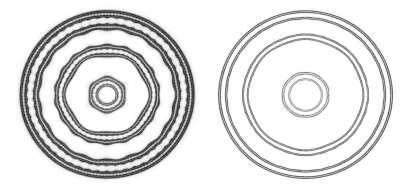

FIGURE 1.2: The Julia sets for $z^3 - 0.001/z^3$ and $z^4 - 0.001/z^4$ are both Cantor sets of circles.

In the more general case of the family $z^n + c + \lambda/z^d$, where $n, d \geq 2$ (but not both equal to 2), the situation is very different. For example, when $c = 0$, it is known that the Julia set is a Cantor set of simple closed curves, at least if $\lambda \neq 0$ is small enough. It is also known that there is always a round annulus of some definite width in the complement of the Julia set, so the Julia sets do not converge to the unit disk in this case (i.e., to the filled Julia set of z^n). When c is the center of some other hyperbolic component of the multibrot set, the Julia set again contains a Cantor set of simple closed curves, but now infinitely many of these curves are "decorated," so this situation is quite different.

In Figure 1.2, we display Julia sets from the family $z^n + \lambda/z^n$ where λ is small and $n = 3, 4$. Note that the complement of the Julia set in this case is a collection of annuli, and one of these annuli seems to have relatively large width. This is always the case as λ approaches the origin.

This paper is dedicated to Jack Milnor, whose books, papers, and lectures have been an inspiration to me from the very beginning of my mathematical career.

2   ELEMENTARY MAPPING PROPERTIES

For simplicity, for most of this paper, we will deal with the special case

$$F_\lambda(z) = z^n + \frac{\lambda}{z^n},$$

where $n \geq 2$. At the end of the paper we discuss the differences that occur when we add the parameter c.

In the dynamical plane, the object of principal interest is the *Julia set* of F_λ, which we denote by $J(F_\lambda)$. The Julia set is the set of points at which the family of iterates $\{F_\lambda^n\}$ fails to be a normal family in the sense of Montel. It is known that $J(F_\lambda)$ is also the closure of the set of repelling periodic points for F_λ as well as the boundary of the set of points whose orbits escape to ∞ under iteration of F_λ. See [13].

The point at ∞ is a superattracting fixed point for F_λ, and we denote the immediate basin of ∞ by B_λ. It is well known that F_λ is conjugate to $z \mapsto z^n$ in a neighborhood of ∞ in B_λ [16], [13]. There is also a pole of order n for F_λ at the origin, so there is a neighborhood of 0 that is mapped into B_λ by F_λ. If this preimage of B_λ is disjoint from B_λ (which it is when $|\lambda|$ is sufficiently small [6]), then we denote this preimage of B_λ by T_λ. So F_λ maps both B_λ and T_λ in n-to-one fashion onto B_λ. We call T_λ the *trap door* since any orbit that eventually enters the immediate basin of ∞ must pass through T_λ en route to B_λ.

The map F_λ has $2n$ free critical points given by $c_\lambda = \lambda^{1/2n}$. (We say "free" here since ∞ is also a critical point, but it is fixed, and 0 is also a critical point, but 0 is immediately mapped to ∞.) There are, however, only two critical values, and these are given by $v_\lambda = \pm 2\sqrt{\lambda}$. The map also has $2n$ prepoles given by $(-\lambda)^{1/2n}$. Note that all of the critical points and prepoles lie on the circle of radius $|\lambda|^{1/2n}$ centered at the origin. We call this circle the *critical circle*.

The map F_λ has some very special properties when restricted to circles centered at the origin. The following are straightforward computations:

1. F_λ takes the critical circle $2n$-to-one onto the straight line segment connecting the two critical values $\pm 2\sqrt{\lambda}$ and passing through 0.

2. F_λ takes any other circle centered at the origin to an ellipse whose foci are the two critical values.

We call the image of the critical circle the *critical segment*. Also, the straight ray connecting the origin to ∞ and passing through one of the critical points is called a *critical point ray*. Similar straightforward computations show that each of the critical point rays is mapped in two-to-one fashion onto one of the two straight line segments of the form tv_λ, where $t \geq 1$ and v_λ is the image of the critical point on this ray. So the image of a critical point ray is one of two straight rays connecting $\pm v_\lambda$ to ∞. Therefore, the critical segment, together with these two rays, forms a straight line through the origin.

We now turn to the symmetry properties of F_λ in both the dynamical and parameter planes. Let ν be the primitive $(2n)$th root of unity given by $\exp(\pi i/n)$. Then, for each j, we have $F_\lambda(\nu^j z) = (-1)^j F_\lambda(z)$. Hence, if n is even, we have $F_\lambda^2(\nu^j z) = F_\lambda(z)$. Therefore, the points z and $\nu^j z$ land on the same orbit after two iterations and so have the same eventual behavior for each j. If n is odd, the orbits of $F_\lambda(z)$ and $F_\lambda(\nu^j z)$ are either the same or else they are the negatives of each other.

In either case it follows that the orbits of $\nu^j z$ behave symmetrically under $z \mapsto -z$ for each j. Hence the Julia set of F_λ is symmetric under $z \mapsto \nu z$. In particular, each of the free critical points eventually maps onto the same orbit (in case n is even) or onto one of two symmetric orbits (in case n is odd). Thus these orbits all have the same behavior (up to the symmetry) and so the λ-plane is a natural parameter plane for each of these families. That is, like the well-studied quadratic family $z^2 + c$, there is only one free critical orbit for this family up to symmetry.

Let $H_\lambda(z)$ be one of the n involutions given by $H_\lambda(z) = \lambda^{1/n}/z$. Then we have $F_\lambda(H_\lambda(z)) = F_\lambda(z)$, so the Julia set is also preserved by each of these involutions. Note that each H_λ maps the critical circle to itself and also fixes a pair of critical points $\pm\sqrt{\lambda^{1/2n}}$. H_λ also maps circles centered at the origin outside the critical circle to similar circles inside the critical circle, and vice versa. It follows that two such circles, one inside and one outside the critical circle, are mapped onto the same ellipse by F_λ.

Since there is only one free critical orbit, we may use the orbit of any critical point to plot the picture of the parameter plane. In Figure 2.1 we have plotted the parameter planes in the cases $n = 3$ and $n = 4$. The parameter planes for F_λ also possess several symmetries. First of all, we have

$$\overline{F_\lambda(z)} = F_{\overline{\lambda}}(\overline{z}),$$

so that F_λ and $F_{\overline{\lambda}}$ are conjugate via the map $z \mapsto \overline{z}$. Therefore, the parameter plane is symmetric under complex conjugation.

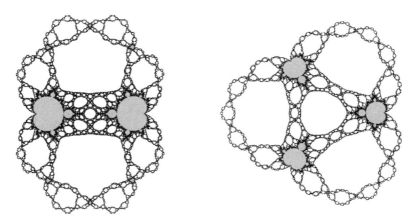

FIGURE 2.1: The parameter planes for the cases $n = 3$ and $n = 4$.

We also have $(n-1)$-fold symmetry in the parameter plane for F_λ. To see this, let ω be the primitive $(n-1)$st root of unity given by $\exp(2\pi i/(n-1))$. Then, if n is even, a straightforward computation shows that

$$F_{\lambda\omega}(\omega^{n/2}z) = \omega^{n/2}(F_\lambda(z)).$$

As a consequence, for each $\lambda \in \mathbb{C}$, the maps F_λ and $F_{\lambda\omega}$ are conjugate under the linear map $z \mapsto \omega^{n/2}z$. When n is odd, the situation is a little different. We now have

$$F_{\lambda\omega}(\omega^{n/2}z) = -\omega^{n/2}(F_\lambda(z)).$$

Since $F_\lambda(-z) = -F_\lambda(z)$, we therefore have that $F^2_{\lambda\omega}$ is conjugate to F^2_λ via the map $z \mapsto \omega^{n/2}z$. This means that the dynamics of F_λ and $F_{\lambda\omega}$ are "essentially" the same, though subtly different. For example, if F_λ has a fixed point, then, under this conjugacy, this fixed point and its negative are mapped to a 2-cycle for $F_{\lambda\omega}$. To summarize the symmetry properties of F_λ, we have the following:

PROPOSITION 2.2 (Symmetries in the dynamical and parameter plane). *The dynamical plane for F_λ is symmetric under the map $z \mapsto \nu z$, where ν is a primitive $(2n)$th root of unity, as well as the involution $z \mapsto \lambda^{1/n}/z$. The parameter plane is symmetric under both $z \mapsto \bar{z}$ and $z \mapsto \omega z$ where ω is a primitive $(n-1)$st root of unity.*

Recall that, for the quadratic family, if the critical orbit escapes to ∞, the Julia set is always a Cantor set. For F_λ, it turns out that there are three different possibilities for the Julia sets when the free critical orbit escapes. The following result is proved in [6].

THEOREM 2.3 (The Escape Trichotomy). *For the family of functions*

$$F_\lambda(z) = z^n + \frac{\lambda}{z^n}$$

with $n \geq 2$ and $\lambda \in \mathbb{C}$:

1. *If the critical values lie in B_λ, then the Julia set is a Cantor set.*

2. *If the critical values lie in T_λ (and, by assumption, T_λ is disjoint from B_λ), then the Julia set is a Cantor set of simple closed curves.*

3. *If the critical values lie in any other preimage of T_λ, then the Julia set is a Sierpinski curve.*

A *Sierpinski curve* is a planar set that is characterized by the following five properties: it is a compact, connected, locally connected and nowhere dense set whose complementary domains are bounded by simple closed curves that are pairwise disjoint. It is known from work of Whyburn [18] that any two Sierpinski curves are homeomorphic. In fact, they are homeomorphic to the well-known Sierpinski carpet fractal. From the point of view of topology, a Sierpinski curve is a universal set in the sense that it contains a homeomorphic copy of any planar, compact, connected, one-dimensional set. The first example of a Sierpinski curve Julia set was given by Milnor and Tan Lei [14].

Case 2 of the escape trichotomy was first observed by McMullen [12], who showed that this phenomenon occurs in each family provided that $n \neq 2$ and $|\lambda|$ is sufficiently small.

In the parameter plane pictures, the white regions consist of parameters for which the critical orbit escapes to ∞. The external white region is the set of parameters for which the Julia set is a Cantor set. The small central disk is the region containing parameters for which the Julia set is a Cantor set of simple closed curves. This is the McMullen domain, $\mathcal{M}$. And all the other white regions contain parameters whose Julia sets are Sierpinski curves. These are the Sierpinski holes.

In Figure 2.4 we display three Julia sets drawn from the family

$$F_\lambda(z) = z^4 + \lambda/z^4,$$

one corresponding to each of the three cases in the escape trichotomy.

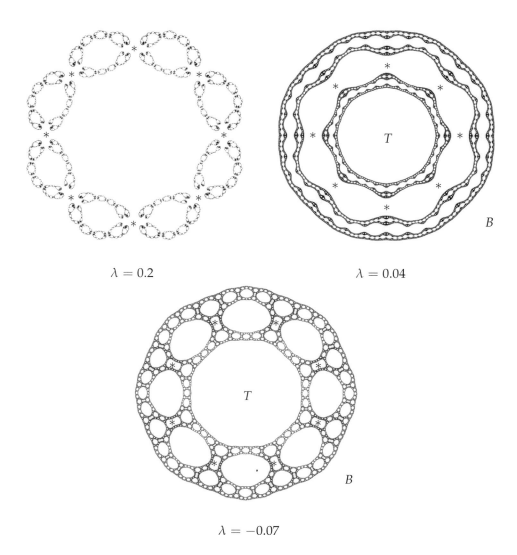

$\lambda = 0.2$ $\lambda = 0.04$

$\lambda = -0.07$

FIGURE 2.4: Some Julia sets for $z^4 + \lambda/z^4$: if $\lambda = 0.2$, $J(F_\lambda)$ is a Cantor set; if $\lambda = 0.04$, $J(F_\lambda)$ is a Cantor set of circles; and if $\lambda = -0.07$, $J(F_\lambda)$ is a Sierpinski curve. Asterisks indicate the location of critical points.

3 JULIA SETS CONVERGING TO THE UNIT DISK

In this section we describe the interesting limiting behavior of the family

$$F_\lambda(z) = z^2 + \frac{\lambda}{z^2}$$

as $\lambda \to 0$. In [5], the following result was proved:

THEOREM 3.1. *If λ_j is a sequence of parameters converging to 0, then the Julia sets of F_{λ_j} converge in the Hausdorff metric to the closed unit disk.*

Here is a sketch of the proof that the Julia sets of F_λ converge to the unit disk as $\lambda \to 0$. It is known that if c_λ does not lie in B_λ or T_λ, then $J(F_\lambda)$ is a connected set [4]. It has also been proved in that paper that if $|\lambda| < 1/16$, then the Julia set always contains an invariant *Cantor necklace*. A Cantor necklace is a set that is a continuous and one-to-one image of the following subset of the plane. Place the Cantor middle thirds set on the real axis. Then adjoin a circle of radius $1/3^j$ in place of each of the 2^j removed intervals at the jth level of the construction of the Cantor middle thirds set. The union of the Cantor set and the adjoined circles is the model for the Cantor necklace. See Figure 3.2. We remark that the existence of a Cantor necklace holds for any λ for which $J(F_\lambda)$ is connected, not just those with $|\lambda| < 1/16$ [4]. The only difference is that the boundaries of the open regions now may not be simple closed curves — they may just be the boundary of a bounded, simply connected, open set (which need not be a simple closed curve).

FIGURE 3.2: The Cantor middle-thirds necklace.

In the Julia set of F_λ, the invariant Cantor necklace has the following properties: the simple closed curve corresponding to the largest circle in the model is the boundary of the trap door. All the closed curves corresponding to the circles at level j correspond to the boundaries of preimages of ∂B_λ that map to this set after j iterations. The Cantor set portion of the necklace is an invariant set on which F_λ is conjugate to the one-sided shift map on two symbols. The two extreme points in this set correspond to a fixed point and its negative, both of which lie in ∂B_λ. Hence the Cantor necklace stretches completely "across" $J(F_\lambda)$. Moreover, it is known that the Cantor necklace is located in a particular subset of the Julia set. Specifically, let $c_0(\lambda)$ be the critical point of F_λ that lies in the sector $0 \leq \mathrm{Arg}\, z < \pi/2$ when $0 \leq \mathrm{Arg}\, \lambda < 2\pi$. Let c_j be the other critical points arranged in the clockwise direction around the origin as j increases. Let I_0 denote the sector bounded by the two critical point rays connecting the origin to ∞ and passing through c_0 and c_3. Let I_1 be the negative of this sector. Then, as shown in [4], the Cantor set portion of the necklace is the set of points in $J(F_\lambda)$ whose orbits remain in $I_0 \cup I_1$ for all λ with $0 \leq \mathrm{Arg}\, \lambda < 2\pi$. The appropriate preimages of T_λ all lie in $I_0 \cup I_1$ as well.

It is easy to check that, when λ is small, the boundary of B_λ is close to the unit circle, so $J(F_\lambda)$ is contained in a region close to the unit disk. We now show why the Julia sets of F_λ converge to the closed unit disk $\mathbb{D}$ as $\lambda \to 0$. Here convergence to the closed unit disk means convergence in the Hausdorff metric.

PROPOSITION 3.3. *Let $\epsilon > 0$ and denote the disk of radius ϵ centered at z by $B_\epsilon(z)$. There exists $\mu > 0$ such that, for any λ with $0 < |\lambda| \leq \mu$, $J(F_\lambda) \cap B_\epsilon(z) \neq \varnothing$ for all $z \in \mathbb{D}$.*

PROOF. Suppose that this is not the case. Then, given $\epsilon > 0$, we may find a sequence of parameters $\lambda_j \to 0$ and another sequence of points z_j in the unit disk $\mathbb{D}$ such that $J(F_{\lambda_j}) \cap B_{2\epsilon}(z_j) = \emptyset$ for each j. Since $\mathbb{D}$ is compact, there is a subsequence of the z_j that converges to some point $z^* \in \mathbb{D}$. This point z^* does not lie in T_λ since one checks easily that T_λ shrinks to the origin as $\lambda \to 0$. For each parameter in the corresponding subsequence, we then have $J(F_{\lambda_j}) \cap B_\epsilon(z^*) = \emptyset$ if j is sufficiently large. Hence we may assume at the outset that we are dealing with a sequence $\lambda_j \to 0$ such that $J(F_{\lambda_j}) \cap B_\epsilon(z^*) = \emptyset$.

Now consider the circle of radius $|z^*|$ centered at the origin. This circle meets $B_\epsilon(z^*)$ in an arc γ of length ℓ. Choose k so that $2^k \ell > 2\pi$. Since $\lambda_j \to 0$, we may choose j large enough so that $|F_{\lambda_j}^i(z) - z^{2^i}|$ is very small for $1 \leq i \leq k$, provided z lies outside the circle of radius $|z^*|/2$ centered at the origin. In particular, it follows that $F_{\lambda_j}^k(\gamma)$ is a curve whose argument increases by approximately 2π, i.e., the curve $F_{\lambda_j}^k(\gamma)$ wraps at least once around the origin. As a consequence, the curve $F_{\lambda_j}^k(\gamma)$ must meet the Cantor necklace in the dynamical plane. But this necklace lies in $J(F_{\lambda_j})$. Hence $J(F_{\lambda_j})$ must intersect this curve. Since the Julia set is backward invariant, it follows that $J(F_{\lambda_j})$ must intersect $B_\epsilon(z^*)$. This then yields a contradiction, and so the result follows. $\qquad \square$

REMARK 3.4. A similar result concerning the convergence to the unit disk occurs in the family of maps $G_\lambda(z) = z^n + \lambda/z$. See [15]. The difference here is that the Julia sets converge to the unit disk only if λ approaches the origin along the straight rays given by

$$\text{Arg } \lambda = \frac{(2k+1)\pi}{n-1},$$

where $k \in \mathbb{Z}$. In Figure 3.5 we display the parameter plane for the family $z^5 + \lambda/z$. Note that there are four accesses to the origin where the parameter plane is "interesting." It is along these rays that the Julia sets converge to the unit disk. On any other ray, G_λ always has attracting cycles whose basins extend from the boundary of T_λ to the boundary of B_λ.

4   THE CASE $n > 2$

In this section we show that the case $n > 2$ is quite different from the case $n = 2$. In particular, there is a McMullen domain whenever $n > 2$; moreover, the Julia sets no longer converge to the unit disk as $\lambda \to 0$.

Recall from the escape trichotomy that if the critical values lie in T_λ, then the Julia set of F_λ is a Cantor set of simple closed curves. This situation does not occur when $n = 2$. To see this, we need to specify the location of these critical values of $z^n + \lambda/z^n$. Let $\lambda^* = 4^{-n/(n-1)}$. Then one checks easily that if $|\lambda| = \lambda^*$, then $|v_\lambda| = |c_\lambda|$ so both the critical points and critical values lie on the critical circle. Then, if $|\lambda| < \lambda^*$, we have $|v_\lambda| < |c_\lambda|$, and so F_λ maps the critical circle strictly inside itself. So a slightly larger circle is mapped to an ellipse that lies strictly inside this circle. Then, using quasiconformal surgery, one can glue the map $z \mapsto z^n$ into the disk bounded on the outside by this circle. See [3] for details. This new map

FIGURE 3.5: The parameter plane for $z^5 + \lambda/z$.

is then conjugate to $z \mapsto z^n$, and the boundary of this map's basin of ∞ is then our original ∂B_λ. It then follows that B_λ is bounded by a simple closed curve lying strictly outside this disk. In particular, there is a preimage of B_λ surrounding the origin inside this circle. This is the trap door T_λ, which is therefore disjoint from B_λ.

Next we compute that

$$F_\lambda(v_\lambda) = 2^n \lambda^{n/2} + \frac{1}{2^n \lambda^{n/2-1}}.$$

When $n > 2$, as $\lambda \to 0$, we have $v_\lambda \to 0$ and so $F_\lambda(v_\lambda) \to \infty$. Thus, when $|\lambda|$ is small, v_λ does indeed lie in the trap door when $n > 2$. But when $n = 2$, $F_\lambda(v_\lambda) \to 1/4$ as $\lambda \to 0$. The point $1/4$ is not in B_λ for $|\lambda|$ small since the boundary of B_λ is close to the unit circle. Hence, v_λ does not lie in T_λ in this case.

There is another way to see why this is true. Suppose both critical values lie in T_λ. It is easy to see that T_λ is an open disk, so the question is: what is the preimage of T_λ? A natural first thought would be that the preimage of T_λ is a collection of open disks, one surrounding each preimage of $\pm v_\lambda$. But there are $2n$ such preimages, namely, the critical points, and so each of these disks would then necessarily be mapped two-to-one onto T_λ. But this would then mean that the map would have degree $4n$. But the degree of F_λ is $2n$, so the preimages of T_λ cannot be a collection of disjoint disks. Therefore, some of the preimages of T_λ must overlap. But then, by the symmetries discussed earlier, all these preimages must overlap, and so the preimage of T_λ is a connected set. By the Riemann-Hurwitz formula, we know that

$$\mathrm{conn}\,(F_\lambda^{-1}(T_\lambda)) - 2 = (\deg F_\lambda)(\mathrm{conn}\,(T_\lambda) - 2) + (\text{number of critical points})$$

where $\mathrm{conn}(X)$ denotes the number of boundary components of the set X. But both the degree of F_λ and the number of critical points in this formula is $2n$, and

$\mathrm{conn}(T_\lambda) = 1$. So it follows that the preimage of T_λ has two boundary components. That is, $F_\lambda^{-1}(T_\lambda)$ is an annulus.

This then is the beginning of McMullen's proof [12] that the Julia set in this case is a Cantor set of simple closed curves. We know that the complement of the Julia set contains the disks B_λ and T_λ as well as the annulus $F_\lambda^{-1}(T_\lambda)$. The entire preimage of B_λ is the union of B_λ and T_λ, while the entire preimage of T_λ is the annulus $F_\lambda^{-1}(T_\lambda)$. So what is the preimage of $F_\lambda^{-1}(T_\lambda)$? This preimage must lie in the two annular regions between $F_\lambda^{-1}(T_\lambda)$ and B_λ or T_λ. Call these annuli A_{in} and A_{out}. See Figure 4.1. Since the preimages of $F_\lambda^{-1}(T_\lambda)$ cannot contain a critical point, it follows that the preimages must be mapped as a covering onto $F_\lambda^{-1}(T_\lambda)$, in fact, as an n-to-one covering since F_λ is n-to-one on both B_λ and T_λ. So the preimage of $F_\lambda^{-1}(T_\lambda)$ consists of a pair of disjoint annuli, one in A_{in} and the other in A_{out}. Then the preimages of these annuli consist of four annuli, and so forth. What McMullen shows is that, when you remove all of these preimage annuli, what is left is a Cantor set of simple closed curves, each surrounding the origin.

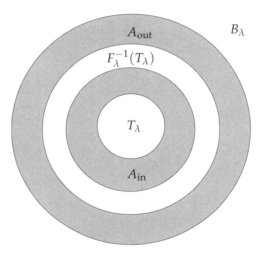

FIGURE 4.1: The annuli A_{in} and A_{out}.

Here, then, is the other reason why there is no McMullen domain when $n = 2$. From the preceding discussion, we have that each of the annuli A_{in} and A_{out} is mapped as an n-to-one covering onto the annulus A which is the union of $F_\lambda^{-1}(T_\lambda)$, A_{in}, and A_{out}. Then the modulus of A_{in} is equal to $\mathrm{mod}\,(A)/n$ and similarly for the modulus of A_{out}. But then, when $n = 2$, we have

$$\mathrm{mod}\,A_{\mathrm{in}} + \mathrm{mod}\,A_{\mathrm{out}} = \mathrm{mod}\,A.$$

This leaves no room for the intermediate annulus, $F_\lambda^{-1}(T_\lambda)$, so this picture cannot occur when $n = 2$.

So the question is: can these simple closed curves in the Julia set converge to the closed unit disk as $\lambda \to 0$. This, in fact, does not happen. The proof makes use of an important fact proved by Blé, Douady, and Henriksen concerning round annuli. We call an annulus of the form $0 < r_1 < |z| < r_2$ a round annulus. Then in [1] it is shown that any annulus in the plane that surrounds the origin and has modulus $\alpha > 1/2$ must contain a round annulus of modulus at least $\alpha - 1/2$.

As $\lambda \to 0$, we have that the annulus A stretches from ∂B_λ to ∂T_λ. Since ∂B_λ approaches the unit circle and ∂T_λ approaches 0 as $\lambda \to 0$, it follows that the modulus of A tends to ∞. So the moduli of A_{in} and A_{out} also tend to ∞. Then there is a subannulus, $A_1 \subset A_{\text{out}}$, that is mapped n-to-one onto A_{out}; consequently, we have mod $A_1 = $ mod A_{out}/n. Then A_1 contains a subannulus A_2 that is mapped n-to-one onto A_1, so mod $A_2 = $ mod $A_1/n = $ mod A_{out}/n^2. Continuing in this fashion, we find a sequence of annuli A_j whose moduli are given by mod A_{out}/n^j, and each of these annuli has one boundary in ∂B_λ. Adjacent to each A_j is another annulus E_j that eventually maps to T_λ and hence lie in the complement of the Julia set. One can estimate in similar fashion the moduli of these annuli, and note that they also lie "close" to ∂B_λ. Eventually we can find an annulus E_j whose modulus is larger than one and that lies outside the circle of some given radius centered at the origin. Then, as shown in [5], this annulus must contain a round annulus of modulus at least $1/2$ and so the Julia sets do not converge to the unit disk as $\lambda \to 0$.

5   OTHER c-VALUES

In this section we describe some other recent results involving the more general family

$$F_\lambda(z) = z^n + c + \frac{\lambda}{z^n}$$

where c is now the center of some other hyperbolic component of the multibrot set. As in the previous sections, the situation when $n = 2$ is quite different from that when $n > 2$. When $n = 2$ it has been shown in [10] that the Julia sets of F_λ converge to the filled Julia sets of $z^2 + c$ as $\lambda \to 0$. The proof here is a little more complicated since we no longer can show that Cantor necklaces lie in the Julia set. However, it can be shown that, for λ sufficiently small, $J(F_\lambda)$ is connected and ∂B_λ is homeomorphic to the Julia set of $z^2 + c$. The latter involves a holomorphic motions argument. In Figures 5.1 and 5.2 we display the quadratic Julia sets known as the basilica and the Douady rabbit together with small singular perturbations of these maps.

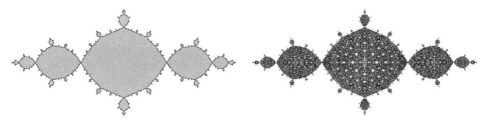

FIGURE 5.1: The Julia sets for $z^2 - 1 + \lambda/z^2$ where $\lambda = 0$ and $\lambda = -0.00001$. (See also Plate 9.)

When $n > 2$ for these families, the situation is a little different from the case when $c = 0$. The reason is that the interior of the filled Julia set of $z^n + c$ now consists of infinitely many disjoint disks. Only finitely many of these disks, say k,

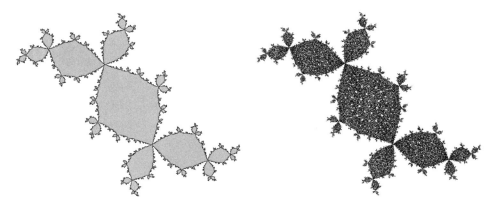

FIGURE 5.2: The Julia sets for $z^2 - 0.122 + 0.745i + \lambda/z^2$ where $\lambda = 0$ and $\lambda = -0.000001$. (See also Plate 10.)

contain the single superattracting cycle. When λ is small a similar holomorphic motions argument shows that ∂B_λ is again homeomorphic to the Julia set of the unperturbed polynomial, so we have k similar disks that surround the former superattracting cycle. If we consider just the points whose orbits remain in the union of these k closed disks, then similar arguments as in the case $c = 0$ show that this set consists of k different Cantor sets of simple closed curves, each surrounding points on the former superattracting cycle. Then all the infinitely many other preimages of these disks also contain Cantor sets of simple closed curves. However, none of these additional curves contain periodic points, as they all eventually map onto the original k Cantor sets of simple closed curves. So there must be more to the Julia sets than just these curves.

Indeed, in [2] it was shown that there are additional Cantor sets of point components in the Julia sets. These can be characterized by specifying how the points move around the disks that lie in the complement of ∂B_λ. In addition, countably many of the simple closed curves in the original k disks actually map onto the boundaries of the periodic disks. From the point of view of the entire Julia set, these boundaries are just a part of the entire set that makes up ∂B_λ. Hence these are no longer simple closed curves; rather, each of them has infinitely many "decorations" attached, i.e., preimages of the entire boundary of the basin of ∞. In Figures 5.3 and 5.3 we display the Julia set of the map $z^3 - i$ and its singular perturbation. Note that the annuli in the complement of the Julia set now have boundary curves with infinitely many attachments.

Finally, for most of this paper, we considered singular perturbations by which a pole was inserted in place of the critical point of $z^n + c$. There have been a number of papers that address other types of singluar perturbations. For example, in [7], maps of the form $z^n + \lambda/(z - a)^d$ were investigated. When a is nonzero but close to 0, the McMullen domain disappears. The Julia set now contains infinitely many closed curves, but they are no longer concentric. In fact, only one surrounds the origin. In addition, there are uncountably many point components in the Julia set. Similar phenomena occur in the family $z^2 + c + \lambda/z^2$ where c is in a hyperbolic component of the Mandelbrot set but not at its center. See [11].

We also remark that convergence of Julia sets to objects that are different from

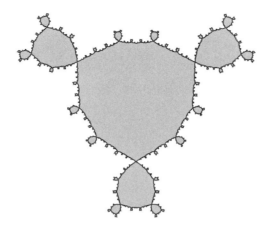

FIGURE 5.3: The Julia set for the unperturbed map $z^3 - i$. (See also Plate 11.)

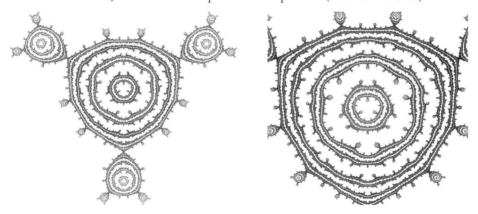

FIGURE 5.3: The Julia set for $z^3 - i + 0.0001/z^3$ and a magnification. (See also Plates 11 and 12.)

the Julia set of the limiting map is not restricted to singularly perturbed maps. Indeed, Douady [9] has shown that, when a family of polynomials approaches a map with a parabolic point, there are many possible limiting sets while the limiting polynomial's Julia set is quite different (and much tamer).

Acknowledgements. The author would like to thank the referee for pointing out many infelicities in the original version of this paper.

6 Bibliography

[1] Blé, G., Douady, A., and Henriksen, C. "Round Annuli." *Contemporary Mathematics* **355** (2004), 71–76.

[2]   Blanchard, P., Devaney, R. L., Garijo, A., and Russell, E. D. "A Generalized Version of the McMullen Domain." *Int'l J. Bifurcation and Chaos* **18** (2008), 2309–2318.

[3]   Blanchard, P., Devaney, R. L., Look, D. M., Moreno Rocha, M., Seal, P., Siegmund, S., and Uminsky, D. "Sierpinski Carpets and Gaskets as Julia Sets of Rational Maps." In *Dynamics on the Riemann Sphere*. European Math Society (2006), 97–119.

[4]   Devaney, R. L. "Cantor Necklaces and Structurally Unstable Sierpinski Curve Julia Sets for Rational Maps." *Qual. Theory Dynamical Systems* **5** (2006), 337–359.

[5]   Devaney, R. L., and Garijo, A. "Julia Sets Converging to the Unit Disk." *Proc. Amer. Math. Soc.* **136** (2008), 981–988.

[6]   Devaney, R. L., Look, D. M., and Uminsky, D. "The Escape Trichotomy for Singularly Perturbed Rational Maps." *Indiana University Mathematics Journal* **54** (2005), 1621–1634.

[7]   Devaney, R. L., and Marotta, S. "Evolution of the McMullen Domain for Singularly Perturbed Rational Maps." *Topology Proceedings* **32** (2008), 301–320.

[8]   Devaney, R. L., and Pilgrim, K. "Dynamic Classification of Escape Time Sierpinski Curve Julia Sets." *Fundamenta Mathematicae* **202** (2009), 181–198.

[9]   Douady, A. "Does the Julia Set Depend Continuously on the Polynomial?" *Proc. Symp. Appl. Math.* **49** (1994), 91–138.

[10] Kozma, R., and Devaney, R. L. "Julia Sets of Perturbed Quadratic Maps Converging to the Filled Quadratic Julia Sets." `http://math.bu.edu/people/bob/papers/kozma.pdf`

[11] Marotta, S. "Singular Perturbations in the Quadratic Family." *J. Difference Equations and Applications* **4** (2008), 581–595.

[12] McMullen, C. "The Classification of Conformal Dynamical Systems." *Current Developments in Mathematics*. International Press, Cambridge, MA (1995), 323–360.

[13] Milnor, J. *Dynamics in One Complex Variable.* 3rd ed. Annals of Mathematics Studies. Princeton University Press (2006).

[14] Milnor, J., and Tan Lei. "A 'Sierpinski Carpet' as Julia Set," Appendix F in "Geometry and Dynamics of Quadratic Rational Maps." *Experiment. Math.* **2** (1993), 37–83.

[15] Morabito, M., and Devaney, R. L. "Limiting Julia Sets for Singularly Perturbed Rational Maps." *International Journal of Bifurcation and Chaos* **18** (2008), 3175–3181.

[16] Petersen, C., and Ryd, G. *Convergence of Rational Rays in Parameter Spaces*, The Mandelbrot set: Theme and Variations, London Mathematical Society, Lecture Note Series 274, Cambridge University Press (2000), 161–172.

[17] Roesch, P. "On Capture Zones for the Family $f_\lambda(z) = z^2 + \lambda/z^2$." In *Dynamics on the Riemann Sphere*. European Mathematical Society (2006), 121–130.

[18] Whyburn, G. T. "Topological Characterization of the Sierpinski Curve." *Fundamenta Mathematicae* **45** (1958), 320–324.

On (non-)local connectivity of some Julia sets

Alexandre Dezotti and Pascale Roesch

ABSTRACT. This article deals with the question of local connectivity of the Julia set of polynomials and rational maps. It essentially presents conjectures and questions.

INTRODUCTION

In this note we discuss the following question: When is the Julia set of a rational map connected but not locally connected? We propose some conjectures and develop a model of non-locally connected Julia sets in the case of infinitely renormalizable quadratic polynomials, a situation where one hopes to find a precise answer.

The question of local connectivity of the Julia set has been studied extensively for quadratic polynomials, but there is still no complete characterization of when a quadratic polynomial has a connected and locally connected Julia set. In degree 2, the question reduces to the precise cases where the polynomial has a Siegel disk or is infinitely renormalizable. In his lecture [Mi3], J. Milnor proposed a quantitative condition to get a non-locally connected Julia set which is infinitely satellite renormalizable. This follows the work of A. Douady and D. E. K. Sørensen: in [So] a description of the topological nature of a non-locally connected Julia set is given, and some examples in the infinitely satellite renormalizable case are obtained. Nevertheless, the argument in [So] is by continuity and gives no explicit condition. Later, G. Levin gave such a condition in [Le] (see also Theorem 4.26). In section 4 of this note, we present a model (originally suggested by X. Buff) of what the structure of the post-critical set in that setting should be

In a previous work, we considered polynomials of higher degrees. Here we present an example (section 1.2) where the local connectivity can be deduced by renormalization.

The situation is even more complicated for rational maps. Indeed, there are examples of rational maps with Cremer points such that the Julia set is locally connected [Ro4]. It seems more difficult to find examples of non-locally connected Julia sets in the space of rational maps. Nevertheless, they exist and can be easily obtained by "tuning" from polynomials. From the way those rational maps are constructed, the natural question appears to be how much a rational map has to be related to a polynomial so that its Julia set is not locally connected? Are polynomials pathological rational maps? In all cases presented here, when the rational map or polynomial has a connected but not locally connected Julia set, a criterion is verified. We call it "Douady-Sullivan criterion" since it was first used by them.

Acknowledgements. The authors would like to thank the referee for his thorough reading of the manuscript and his comments, which helped improve the manuscript. Discussions with X. Buff, A. Chéritat, and Y. Yin, as well as remarks of

D. Cheraghi and H. Inou, inspired parts of this paper. This work was written when both authors were members of Institut de Mathématiques de Toulouse; we are grateful for their hospitality and support during our time there. We would also like to acknowledge the support of the grant ANR-08-JCJC-002-01.

1   LOCAL CONNECTIVITY

1.1   Generalities and first questions for polynomials

Recall that the *Julia set* of a rational map f is the minimal totally invariant (under f and f^{-1}) compact set containing at least 3 points. Its complement in the Riemann sphere, called the *Fatou set*, is an open set whose components are all eventually periodic by Sullivan's theorem. When the Julia set is connected, these components are all topological disks. Inside each of the periodic components the return map is conjugate near the boundary to some simple model (see [McM1]). If the boundary of the component is locally connected, the model extends to the boundary (by Carathéodory's theorem). One of the main reasons to consider the question of local connectivity for Julia sets is to get the model on the boundary.

Recall that the models are given by the following maps from the unit disc to itself:

- $Z_d(z) = z^d$, the *attracting case*,

- $B_d(z) = \frac{z^d + v}{1 + v z^d}$, where $v = \frac{d-1}{d+1}$, the *parabolic case*,

- $R_\theta(z) = e^{2i\pi\theta} z$, the *Siegel case* (the corresponding Fatou component is then called a *Siegel disk*).

In what follows we will always assume that the Julia sets considered are connected, even if it is not explicitly mentioned.

LEMMA 1.1 (Th. 4.4 of [W]). *The Julia set of a rational map is locally connected if and only if the boundary of each Fatou component is locally connected and for any $\epsilon > 0$ only finitely many Fatou components have diameter greater than ϵ.*

Hence, the question of whether Fatou components have a locally connected boundary is fundamental for a rational map. For a polynomial, the boundary of the unbounded Fatou component is the whole Julia set. Nevertheless, it is an interesting question to know if one can deduce some result looking only at the bounded Fatou components and their size. We will now give an answer to this question.

For polynomials let us recall the following result [RoYi]:

THEOREM 1.2 (R-Yin). *Any bounded periodic Fatou component of a polynomial containing a critical point is a Jordan domain.*

The following "classical" conjecture is the natural extension of this result to any bounded Fatou component. It has been proved recently in many cases by M. Shishikura.

CONJECTURE 1.3. *The boundary of a periodic Siegel disk of a polynomial is always a Jordan curve.*

A periodic point in the Julia set is called a *Cremer point* if the derivative of the return map at the fixed point is $e^{2i\pi t}$, with $t \in \mathbb{R} \setminus \mathbb{Q}$. Let us recall the following result (a proof will be sketched in section 3.1)

PROPOSITION 1.4. *If a polynomial has either a Cremer periodic point or a periodic Siegel disk with no critical point on the boundary (of the cycle generated by the disk), then its Julia set is not locally connected.*

In the case of Cremer quadratic polynomials, this settles the above question, since there are no bounded Fatou components. One can also construct non-locally connected Julia sets with Fatou components that are Jordan domains whose diameter tends to zero. Indeed, it is enough to take a polynomial containing both an attracting cycle and a Cremer point such that the orbit of the critical points does not accumulate on the boundary of the attracting basin. Then the attracting basin and all its pre-images are Jordan domains. Moreover, using the "shrinking lemma" (see [TY, Prop. A.3] or [LM, Section 11.1]) in the complement of the post-critical set $(\overline{P_f},$ where $P_f := \bigcup_{c \in crit} \bigcup_{n \geq 1} f^n(c))$, it is easy to see that the diameter of these Jordan domains goes to zero. Such examples are easy to find in cubic families with one attracting fixed point (see [Ro3]). There we can find copies of the Mandelbrot set (the connectedness locus for the quadratic family) in which we can choose a doubly renormalizable restriction containing a Cremer point (we will define renormalizable in Definition 1.7).

In the light of such examples, the previous question appears to be a naive one, but its original motivation leads to the following less naive question.

QUESTION 1.5. *Let P be a quadratic polynomial having a Siegel disk whose boundary is a Jordan curve containing the critical point. Is the Julia set locally connected?*

To our knowledge there is no known counterexample, and in higher degrees the situation is even more complicated. One could imagine building a cubic polynomial from a quadratic one that is non-locally connected the following way: One would have the Julia set of a quadratic polynomial with a Siegel disk without the critical point on its boundary sitting in the Julia set of a cubic polynomial. The critical point has to lie on some hairs around the Siegel disk. The idea is then to deform the cubic polynomial in the space of cubic polynomials in order to put the other critical point on the boundary of the Siegel disk. This kind of map has been considered in [BuHe], when the critical point belongs to strict pre-images of the Siegel disk. In our case, we get a polynomial with a Siegel disk containing one critical point on its boundary and another critical point on some hairs stemming from the Siegel disk. Nevertheless, it is not clear that having one critical point on the boundary of the Siegel disk will not force the hairs to disappear. Indeed, in the light of Douady-Sullivan criterion (see section 3), the non-local connectivity seems to appear when the map presents some injectivity, but here around the boundary of the Siegel disk the map is no longer injective, so there is no reason to expect the boundary to be topologically wild.

QUESTION 1.6. *Does there exist a non-renormalizable polynomial, of degree $d \geq 3$, with a Siegel disk containing at least one critical point on its boundary and whose Julia set is not locally connected?*

Recall the definition of renormalizable maps.

DEFINITION 1.7. *A polynomial is said to be* renormalizable *if some iterate admits a polynomial-like restriction whose filled Julia set is connected. A polynomial-like map is a proper holomorphic map $f : U \to V$, where U and V are topological disks with $\overline{U} \Subset V$; one defines its filled Julia set as $\bigcap f^{-n}(\overline{U})$. (A polynomial restricted to an appropriate disk U is an example of polynomial-like map).*

Notice that by the following connectedness principle (see [McM2]), in order for a renormalizable polynomial to have a locally connected Julia set, the Julia set of its renormalized map should be locally connected.

THEOREM 1.8 (Connectedness principle). *Let $f : \mathbb{C} \to \mathbb{C}$ be a polynomial with connected filled Julia set $K(f)$. Let $f^n : U \to V$ be a renormalization of f with filled Julia set K_n. Then $\partial K_n \subset J(f)$ and for any closed connected set $L \subset K(f)$, $L \cap K_n$ is also connected.*

Yoccoz proved that a quadratic polynomial that is finitely renormalizable and has only repelling periodic points has a locally connected Julia set. In higher degree, the following result appears to be the most general known result. See also [KL].

THEOREM 1.9 (Koslovski–van Strien). *The Julia set of a non-renormalizable polynomial without indifferent periodic points is locally connected, provided it is connected.*

One wonders if there is a way to combine previous results in order to justify the following:

QUESTION 1.10. *Is the Julia set of a polynomial locally connected provided that it is not infinitely renormalizable, it has no Cremer points, and there is a critical point on the boundary of any cycle of Siegel disks?*

Question 1.6 partially justifies Question 1.10. Here are some further justifications. Using the work done in [PR], one can construct a puzzle in the basins of parabolic cycles. Therefore, the proof of Theorem 1.9 (see [KovS]) will adapt to the case of parabolic cycles as soon as one can construct a "box mapping" with this puzzle. This can fail when the map is "parabolic-like" (see [Lo]). Nevertheless, the recent work of L. Lomonaco ([Lo]) on parabolic-like maps should take care of this case.

In the finitely renormalizable case, the proof proceeds by induction and uses the homeomorphism given by the straightening Theorem ([DoHu]). Theorem 1.11 presented in next section is an example of this method.

Finally, let us point out that the case of infinitely renormalizable polynomials is much more subtle. Kahn, Levin, Lyubich, McMullen, van Strien, and others gave conditions to obtain infinitely renormalizable quadratic polynomials with locally connected Julia sets. Douady and Sørensen gave examples of non-locally connected infinitely renormalizable quadratic polynomials. We will discuss infinitely renormalizable polynomials in section 3.2 and in section 4.

1.2   From Fatou components to the whole Julia set : an example

We would like to end this section with a concrete example. We prove local connectivity of the Julia set of a polynomial by knowing that it is renormalizable and that the small Julia set is locally connected.

THEOREM 1.11. *Let $f_a(z) = z^{d-1}(z + da/(d - 1))$, with $d \geq 3$ and $a \in \mathbb{C}$, be the family of polynomials of degree d, with one fixed critical point of maximal multiplicity (up to affine conjugacy). Assume that the Julia set $J(f_a)$ is connected. If f_a is renormalizable of lowest period k around the "free" critical point $-a$, we will denote by Q_c the unique quadratic polynomial conjugate to the restriction of f_a^k. Then the Julia set $J(f_a)$ is locally connected if and only if either f_a is not renormalizable or $J(Q_c)$ is locally connected.*

PROOF. First recall that if f_a or, equivalently, Q_c, is geometrically finite (i.e., if the post-critical set intersects the Julia set at finitely many points), then the Julia sets are locally connected by the result of [TY]. We recall in the following the construction of a graph "adapted" to the dynamics of the map f_a (as presented in [Ro1]). Given a graph Γ, the connected component of $\mathbb{C} \setminus f^{-n}(\Gamma)$ containing x, is called the puzzle piece of depth n containing x and is denoted by $P_n(x)$.

CLAIM 1.12. *There exists a graph Γ such that $\overline{P_n(x)} \cap J(f_a)$ is connected for all n. Moreover,*

1. *either the intersection $\bigcap \overline{P_n(x)}$ reduces to $\{x\}$,*

2. *or the end of the critical point is periodic: $\exists k > 0$ such that for all n large enough $f^k \colon P_{n+k}(-a) \to P_n(-a)$ is quadratic-like. It follows that the map is renormalizable: there exist $c \in \mathbb{C}$ and a quasiconformal homeomorphism $\phi \colon P_n(-a) \to V$, where V is a neighborhood of the filled Julia set of Q_c and ϕ conjugates the maps where it is defined. Moreover, for any $x \in J(f_a)$, either the impression $\bigcap \overline{P_n(x)}$ reduces to $\{x\}$ or it reduces to an iterated pre-image of the critical impression $\bigcap \overline{P_n(-a)} = I(-a)$.*

PROOF OF CLAIM 1.12. This result follows from the construction of the graph done in [Ro1], which we now recall. Denote by B the immediate basin of attraction of 0. The graph Γ under consideration is the union of two cycles of rays and two equipotentials. More precisely, in B we take the cycle generated by the internal ray of angle θ of the form $\frac{1}{d^l - 1}$ (for any l large enough), which has a repelling cycle as its landing point. We then take the cycle of external rays landing at this repelling cycle (on the boundary of B). For the equipotentials, we take any internal equipotential (in B) and any external equipotential.

It is not difficult to see (compare [Ro1]) that any point of the Julia set lying in some sector $U(\theta, \theta')$ (defined next) is surrounded by a non-degenerate annulus of the from $P_n(x) \setminus \overline{P_{n+1}(x)}$ (i.e., it lies in the central component of such an annulus). The sector $U(\theta, \theta')$ (with $\theta' < \theta$) is defined as follows. Consider the curve C formed by the internal rays in B of angles θ/d and $\theta' + 1/d$ and the external rays landing at the corresponding point; $U(\theta, \theta')$ is the connected component of the complement of this curve in $\mathbb{C}$ that contains the internal ray of angle 0.

We need to show that any point of the Julia set will fall under iteration in this domain; then using Yoccoz's theorem (see [Ro1]) we will get the announced claim. Any point of the filled Julia set belongs to a limb, and limbs are sent to limbs (except for the critical limb). Therefore, any point not in the critical limb will fall under iteration in $U(\theta, \theta')$. If the critical limb is fixed, then it is attached by a fixed ray and already belongs to $U(\theta, \theta')$. Otherwise, we look at the limb of the critical value and its orbit will fall in $U(\theta, \theta')$ since the angle of the critical limb is necessarily periodic (indeed, the sector of the wake containing the critical value has angular opening multiplied by d as long as it is not in the wake containing the critical point).

As a direct consequence of Yoccoz's result, if the map is not renormalizable, the whole Julia set is locally connected. We will now consider the case where f_a is renormalizable.

* *If the map f_a is renormalizable then we are in case 2 of the claim.*
Let K denote the filled Julia set of the renormalization f_a^k containing the critical point $-a$. We can choose θ such that the graph previously constructed does not cut K. Indeed, the graph is forward invariant, and any intersection point between Γ and K would be iterated to a point of the periodic cycle on the boundary of B. Hence, if K intersects ∂B under a cycle, it is enough to choose θ of a different period. Now, every puzzle piece $P_n(-a)$ contains the entire set K. Moreover, since K is periodic with some period k, the puzzle pieces $P_n(-a)$ are all mapped by f^k to $P_{n-k}(-a)$ as a quadratic-like map (since $\overline{P_{n+1}(-a)} \subset P_n(-a)$ for large n by the proof of the claim) so the critical point doesn't escape. Therefore, we are in case 2 of the claim.

* *Now we can assume that $K = I(-a)$ by taking the renormalization of lowest period. The filled-in Julia set $K(f_a)$ is the union of K and "limbs" of it.*
There are two rays landing at the non-separating fixed point (called p) of f_a^k in K (the fixed point corresponding by the conjugacy to the β fixed point) and exactly two rays landing at the pre-image of p by $f_{|K}^k$ (pre-image in K). Indeed, if there were more than two rays landing at p, then they define some new sector which is invariant by f_a^k and it then would contain some part of K, which gives the contradiction (the point p would be separating). These two rays separate $K(f_a)$ in three components, one containing B, denoted by L, and one containing neither B nor K, denoted by L'. The iterated pre-images of L' by f_a^k and L are called the limbs of K. Except for L, a connected component of $K(f_a) \setminus K$ will be mapped to a connected component of $K(f_a) \setminus K$. Therefore, any limb different from L is an (iterated) pre-image of L'.

* *For any point $x \in K$ and for any neighborhood U of x, there exists a sub-neighborhood $V \subset U$ such that $U \cap J(f_a)$ has finitely many connected components. Therefore, $J(f_a)$ is locally connected at the points of K.*
Since $K(Q_c)$ is locally connected, the image K is also locally connected. Therefore, there exists a neighborhood $V \subset U$ such that $V \cap K$ is connected. We prove now that the diameter of the limbs of K tends to 0, so that only finitely many of them enter V without being totally included in V. For this purpose, we prove that the diameter of $f_a^{-n}(L')$ tends to 0, meaning that for any $\epsilon > 0$, only finitely many of them have diameter greater than ϵ. For this, we shall use Yoccoz's puzzle for $K(Q_c)$. This puzzle is defined when the non-separating fixed point $\alpha(Q_c)$ is repelling (i.e., when both fixed points are repelling).

Therefore, we first consider the case when the fixed point α is not repelling. If it is attracting or parabolic, then the map f is geometrically finite and the result follows from [TY]. If the point is an irrationally indifferent fixed point for Q_c, then its image by the conjugacy will have the same rotation number (see [Na]). In the Cremer case, both Julia sets (are at the same time of Cremer type and) are non-locally connected. In the Siegel case, if the critical point is not on the boundary of the Siegel disk for one map, so it is for the other by the conjugacy, and both Julia sets are non-locally connected. If the critical points are on the boundary of the respective Siegel disks, then the post-critical sets stay in the boundary of the Siegel disks and remain away from the other fixed point and their first pre-images. Therefore, $\overline{L'} \cap \overline{\cup_{j \geq 0} f_a^j(-a)} = \varnothing$; then, using the so-called shrinking lemma (expansion in $\mathbb{C} \setminus \overline{\cup_{j \geq 0} (f_a^j(-a) \cup f_a^j(0))}$), the diameter of $f_a^{-n}(L')$ goes to 0.

Now we consider the case where the α fixed point of Q_c is repelling. The graph Γ_0 defining the Yoccoz puzzle for Q_c is the union of the external rays landing at the point α and some external equipotential. Let us define a new graph $\tilde{\Gamma}$ for f_a which is a combination of Γ and the cycle of external rays landing at the image of α by the conjugacy ϕ. The puzzle pieces have the same combinatorics for the map f_a and for the map Q_c using the conjugacy ϕ that allows identification of puzzle pieces.

Two cases appear in Yoccoz's result: either the map Q_c is non-renormalizable and then the nest of puzzle pieces shrink to points, or it is renormalizable and then it is easy to see that the orbit of 0 — the critical point of Q_c — is bounded away from the β fixed point and its pre-image $-\beta$. In the second case, using the conjugacy ϕ one obtains the result by applying the shrinking lemma, since the post-critical set of f_a will be disjoint from the limb L'. In the first case, we get a sufficiently small neighborhood of x such that the intersection with $J(f_a)$ is connected, since the diameter of the puzzle pieces in a nest for f_a also shrinks to 0.

$*$ *Conversely, we now assume that the Julia set of f_a is locally connected. We prove that the Julia set K of any renormalization of f_a around $-a$ is also locally connected.*

Let Φ denote the Riemann map of the complement of $K(f_a)$ (which in fact coincides with the Böttcher coordinate). Then $\Psi = \Phi^{-1}$ extends continuously to the boundary. The pre-image $K' = \Psi(K)$ is a compact subset of the unit circle. Therefore, its complement is a countable union of open intervals in the unit circle. Let Π be the projection from the unit circle to itself that collapses those open intervals to points, i.e., identifies the whole interval to one point. If t, t' are boundary points of such an open interval, then the external rays of angle t, t' land at the same point in K. Indeed, these two landing points are in K, and the landing point of any external ray in the interval between t and t' is not in K, which is a connected set. Then we can define a map from the unit circle to K as follows. For $\theta = \Pi(t)$, define $\overline{\Psi}(\theta) = \Psi(t)$. By the previous discussion, this map $\overline{\Psi}$ is well defined and continuous. Therefore, as the continuous image of the unit circle, K is locally connected. $\qquad\square$

2   RATIONAL MAPS

Which Fatou components of a rational map are Jordan domains?

The property of having a bounded Fatou component has no meaning for a rational map. The question is which properties of the bounded Fatou components of polynomials are used in the proofs of local connectivity results. Before we consider this issue, it is natural to ask if there exist rational maps with connected Julia sets but with Fatou components whose boundaries are not locally connected. We should consider only rational maps that are sufficiently far from polynomials.

DEFINITION 2.1.   *We say that a rational map is* veritable *if it is not topologically conjugate to a polynomial on its Julia set.*

Notice that in [Ro2] we introduced the notion of a *genuine rational map*, which by definition is a rational map that is not conjugate to a polynomial in a neighborhood of its Julia set. This condition is stronger than Definition 2.1. Indeed, rational maps of degree 2 with a fixed parabolic point at infinity of multiplier 1 are conjugate to quadratic polynomials on their Julia set (except in some special cases) but cannot be conjugate on a *neighborhood* of their Julia set simply because of the presence of a parabolic basin in the Fatou set (see [PR]).

2.1   Some rational maps as examples

∗ *Positive results* (see [Ro4])

Let us start with rational maps of low degree that fix a Fatou component containing a critical point. When the rational map is of degree 2 and the basin is attracting, the map is necessarily conjugate on its Julia set to a polynomial of degree 2; one can easily see it by using a surgery procedure. In the parabolic case, using McMullen's result ([McM1]) the rational map is conjugate to the one that fixes infinity with multiplier 1; as we mentioned before, those maps are conjugate on their Julia set to a quadratic polynomial if the Julia set does not contain a fixed Cremer point or a fixed Siegel disk (see [PR]). Therefore, the question of local connectivity for the boundary of the basin is almost equivalent to the same question for quadratic polynomials.

A rational map of degree 3 has four critical points in $\widehat{\mathbb{C}}$. First assume that three of them are fixed. It is then easy to see that the rational map is conjugate (by a Möbius transformation) to a *Newton method* associated to a polynomial of degree 3, i.e., to the rational map $N_P(z) = z - P(z)/P'(z)$, where P is a cubic polynomial with distinct roots. One should notice that if the fourth critical point is in the immediate basin of attraction of one of the three fixed points, then N_P is conjugate in a neighborhood of its Julia set to a polynomial of degree 3.

In other cases, the fixed Fatou components are always Jordan domains (see [Ro4]).

THEOREM 2.2. *Let N be a cubic Newton method that is a veritable rational map. Then the Fatou components containing a critical point are Jordan domains. Moreover, the Julia set is locally connected as soon as there is no "non-renormalizable" Cremer or Siegel point.*

Let us now consider the set of rational maps of degree 2 having a period 2 cycle of Fatou components containing a critical point. One example is the family of rational maps with a period 2 critical point studied in [AY]. In this article, the authors prove that if the map is non-renormalizable with only repelling periodic points, then it has a locally connected Julia set. It seems reasonable to believe that this result holds in general in this family and that it persists when the cycle becomes parabolic.

One may even wonder whether the critical Fatou components of a veritable rational map would always have locally connected boundaries.

∗ *Negative results* (see [Ro2])

In [Ro2], two families of examples that illustrate the following result are given.

THEOREM 2.3. *There exist veritable rational maps with connected Julia sets that posses a Fatou component with non-locally connected boundary.*

The first set of examples can be found in the works of Ghys [G] and Herman [He]; they are in the family

$$f_{a,t} = e^{2i\pi t} z^2 \frac{z - a}{1 - az} \quad (a > 3).$$

The second set of examples can be found in the family

$$g_a(z) = z^3 \frac{z - a}{1 - az}, \quad a \in \mathbb{C}.$$

They are obtained from perturbing a map g_{a_0} with a parabolic point in order to create a Cremer point.

We will briefly explain in next section why these maps have a non-locally connected Julia set. Nevertheless, such examples are not satisfactory since, roughly speaking, one can see the trace of a Julia set of a quadratic polynomial in them.

3 DOUADY-SULLIVAN CRITERION

A certain property is shared by all our examples of rational maps and polynomials with a Fatou component whose boundary is not locally connected. We will call it the *Douady-Sullivan criterion* since it was originally used by them to prove that the Julia set of a polynomial with a Cremer point is not locally connected.

First note that if a periodic Fatou component B is not simply connected, then ∂B is not connected and $J(f)$ is not connected (one can also deduce that ∂B is not locally connected at any point of its boundary).

DEFINITION 3.1 (Douady-Sullivan criterion). *A rational map f is said to satisfy the Douady-Sullivan criterion whenever f has a k-periodic Fatou component B that is simply connected and contains a critical point, and if there exists a compact set C in the boundary of B such that*

- *C does not contain any critical point of f^k, and*

- *the restriction $f^k \colon C \to C$ is a bijection.*

LEMMA 3.2. *Let f be a rational map that satisfies the Douady-Sullivan criterion. Let C denote the compact set and $B(p)$ the Fatou component appearing in the definition of the criterion, with $C \subset \partial B(p)$. Then*

- *either $\partial B(p)$ is not locally connected, or*

- *C is the finite union of parabolic or repelling cycles.*

PROOF. (Compare [Mi1]). The basin $B(p)$ is simply connected; assume that its boundary is locally connected. We assume that $k = 1$, replacing f^k by f. From Carathéodory's theorem, we know that the map is conjugate on the boundary to one of the models $Z_d(z) = z^d$ or $B_d(z) = \frac{z^d + v_d}{1 + v_d z^d}$ (where $v_d = \frac{d-1}{d+1}$) on the closed unit disk. Notice that the restriction of map B_d to the unit circle is topologically conjugate to the restriction of Z_d to the unit circle. Therefore, there exists a map $\gamma \colon \mathbb{S}^1 \to \partial B(p)$ that is a semi-conjugacy between $e^{2i\pi\theta} \mapsto e^{2i\pi d\theta}$ and f. One then considers the set $\Theta = \{\theta \in \mathbb{R}/\mathbb{Z} \mid \gamma(e^{2i\pi\theta}) \in C\}$. The map $m_d \colon \theta \mapsto d\theta$ is a bijection from Θ to itself. Indeed, if two rays landing at the same point are mapped onto the same ray, then they land at a critical point, which is excluded by our hypothesis in Definition 3.1. Moreover, every point of C has a pre-image in C and at least one ray lands at this point; therefore, $m_d \colon \Theta \to \Theta$ is surjective. It follows that $m_d \colon \Theta \to \Theta$ is a homeomorphism since Θ is compact.

Finally, notice that m_d is expanding and so Θ must be finite. Indeed, cover Θ by a finite number N of balls of radius ϵ sufficiently small; since m_d is a homeomorphism, the pre-image of a ball of radius ϵ is a ball of radius ϵ/d, so Θ is covered by the union of N balls of diameter ϵ/d, etc. It follows that Θ is the union of N points and those points are pre-periodic angles. Since m_d is a bijection, Θ is a union of periodic cycles.

One deduces that C is the union of cycles of f. These cycles are parabolic or repelling by the snail lemma (see [Mi1]). $\square$

3.1   Douady-Sullivan criterion in the previous examples

$*$ *Polynomial with a Cremer point*
Let f be a rational map. Recall that a *Cremer point* of f is a point of the Julia set $J(f)$ that is irrationally indifferent. The Julia set of a polynomial with a Cremer point is not locally connected. This follows from Lemma 3.2 since f satisfies the Douady-Sullivan criterion with $p = \infty$ and C being the cycle generated by the Cremer point.

$*$ *Polynomials with Siegel disks*
Let f be a polynomial and Δ a periodic Siegel disk for f such that no critical point is on the cycle generated by the boundary $\partial\Delta$. Then the Julia set $J(f)$ is not locally connected (see [Mi1] for instance). Indeed, f satisfies the Douady-Sullivan criterion, taking $p = \infty$ and C to be the cycle generated by boundary $\partial\Delta$.

$*$ *Rational maps: example of Ghys-Herman*
Now we consider the first family of examples studied in [Ro2], namely, the family

$$f_{a,t} = e^{2i\pi t} z^2 \frac{z - a}{1 - az}, \quad \text{with } a > 3, t \in \mathbb{R}.$$

The restriction of $f_{a,t}$ to $\mathbb{S}^1$ is an $\mathbb{R}$-analytic diffeomorphism. According to Denjoy's theorem, if the rotation number $\alpha = \rho(f_{a,t})$ is irrational, then $f_{a,t}$ is topologically conjugate on $\mathbb{S}^1$ to the rigid rotation R_α by some homeomorphism $h_{\alpha,t}$. E. Ghys shows (in [G]) that if $h_{\alpha,t}$ is quasisymmetric but not $\mathbb{R}$-analytic, then the polynomial $P_\alpha(z) = e^{2i\pi\alpha} z + z^2$ has a Siegel disk whose boundary is a quasicircle not containing the critical point. On the other hand, to compare $f_{a,t}$ with P_α, Ghys performs a surgery that provides a homeomorphism ψ such that $\psi(\mathbb{S}^1)$ is the boundary of the Siegel disk and such that the boundary of the immediate basin of ∞ for $f_{a,t}$ is the image by ψ of $J(P_\alpha)$.

As noticed previously, the Douady-Sullivan criterion implies that the Julia set of P_α is not locally connected, since the boundary of the Siegel disk contains no critical point. This implies that the boundary of the basin of ∞ for $f_{a,t}$ is not locally connected. To conclude, we use the following result of M. Herman: for any $a > 3$, there exists values of $t \in \mathbb{R}$ such that the conjugacy $h_{a,t}$ between $f_{a,t}$ and R_α is quasisymmetric but not C^2.

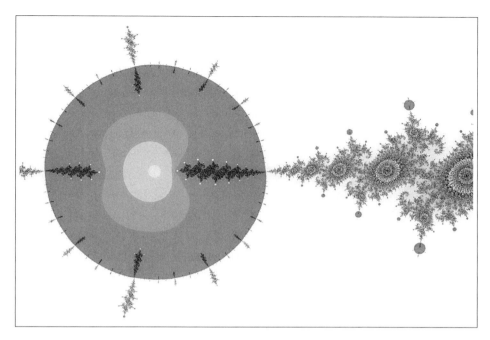

FIGURE 3.3: The Julia set of g_a for a value a next to 5. (See also Plate 13.)

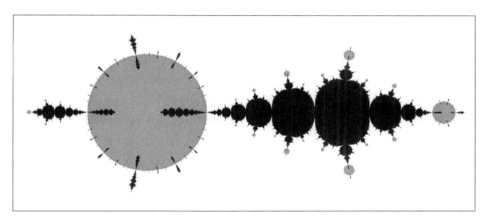

FIGURE 3.4: The Julia set of g_5. (See also Plate 14.)

∗ Rational maps: Perturbation of a fraction of Blaschke
In the second family of examples studied in [Ro2], namely,

$$g_a(z) = z^3 \frac{z-a}{1-az} \quad \text{with } a \in \mathbb{C},$$

the proof is much easier. One sees directly that the Blaschke product g_5 is renormalizable. Indeed, it admits a restriction that is quadratic-like in some open set bounded by rays in the immediate basin of the attracting fixed point 0. This restriction admits a parabolic point at $2 - \sqrt{3}$. There exists a neighborhood of $a_0 = 5$ in the parameter space such that for a in this neighborhood the map f_a has a fixed point $p(a)$ that is a holomorphic function satisfying $p(5) = 2 - \sqrt{3}$. Moreover, in this neighborhood, g_a admits a restriction that is polynomial-like in the neighborhood of $p(a)$. Therefore, one can find values of a near a_0 such that g_a is renormalizable, with the renormalized filled Julia set containing a Cremer point. Since the open sets defining the renormalization intersect the immediate basin of 0, the Cremer point thus obtained has to be on the boundary of the immediate basin. Those maps g_a verify the Douday-Sullivan criterion just by taking the Cremer point as the compact set C.

3.2   Infinitely renormalizable polynomials

Finally, there is a class of examples which we have not yet discussed. Indeed, in the class of infinitely renormalizable polynomials, one can find polynomials having connected but not locally connected Julia set. Several works have been devoted to their studies; see, for instance, [So],[Mi2],[Mi3], and [Le].

In this section, we consider particular infinitely renormalizable polynomials; these are the quadratic polynomials $Q_c(z) = z^2 + c$ where c is in a *limit of a sequence* (H_n) *of hyperbolic components such that H_{n+1} is attached to H_n with higher period*. Let us be more precise.

DEFINITION 3.5.

- *We will use the notation $\mathcal{M}$ for the classical Mandelbrot set, that is, the set of $c \in \mathbb{C}$ such that the orbit of the critical point of Q_c is bounded.*

- *Let H be a hyperbolic component of $\mathcal{M}$, i.e., a connected component of the interior of $\mathcal{M}$ such that Q_c has an attracting periodic point of some period k for every $c \in H$. In H, there exists a unique parameter $c \in H$ such that the critical point of the quadratic polynomial Q_c is periodic. We call the period of this point the period of H, and this parameter c is called the center of H.*

- *We say that a hyperbolic component H' of $\mathcal{M}$ is attached to H if its boundary intersects the boundary of H. In this case, their boundaries intersect at a unique point.*

  This intersection point is called the root of H' if the period of H' is greater than the period of H. Every hyperbolic component of $\mathcal{M}$ has at most one root.[1] We will write $r(H)$ for the root of the hyperbolic component H. At the parameter $c = r(H)$, the polynomial Q_c has a parabolic cycle of period less or equal to the period of H. If Q_c has a parabolic point of period equal to the period of H, with $c \in \partial H$, we call this the root of the hyperbolic component. As we will see later, it is the unique point where the multiplier function $\mu \colon \overline{H} \to \overline{\mathbb{D}}$ takes the value 1.

- *Let $(H_n)_{n\geq 0}$ be a sequence of hyperbolic components of $\mathcal{M}$. We say that $(H_n)_n$ is a* chain of components *arising from H_0 if for all $n \geq 0$, H_{n+1} is attached to H_n and the period of H_{n+1} is greater than the period of H_n.*

One should notice that when H' is attached to H at its root point $r(H')$, the period of H divides the period of H'.

The parameters c we will consider here are limits of the sequences of $(r(H_n))_n$ where $(H_n)_n$ is a chain of components.

---

[1] And they all have at least one root. But this root does not necessarily belong to the boundary of another hyperbolic component.

When a parameter c belongs to a hyperbolic component H of period k, the mapping Q_c has an attracting cycle $Z_H(c)$ of period k. The points in this cycle are holomorphic functions of $c \in H$. These holomorphic mappings can be extended to some regions containing the component H, but they do not extend to any neighborhood of the root $r(H)$.

Let H' be a hyperbolic component which is attached to H and whose period is $k' > k$. The cycle $Z_H(c)$ of Q_c, attracting when $c \in H$, becomes parabolic when $c = r(H')$. Then, when c enters H' a bifurcation occurs: the cycle $Z_H(c)$ becomes repelling while an attracting cycle $Z_{H'}(c)$ of period k' appears.

Let $(H_n)_n$ be a chain of components arising from H_0. Let k_n denote the period of H_n. It can be proved that, for all $n \geq 0$, the mappings Z_{H_n} have well-defined analytic continuations to some neighborhood of $\bigcup_{m \geq n} \overline{H_m} \setminus \{r(H_n)\}$.

We denote these continuations by the same Z_{H_n}. Note that the cycle $Z_{H_n}(c)$ is repelling for $c \in \bigcup_{m \geq n+1} \overline{H_m} \setminus \{r(H_{n+1})\}$.

LEMMA 3.6. *There exist chains of components $(H_n)_{n \geq 0}$ such that the sequence $(r(H_n))_n$ converges and such that the limit c_* has the following properties:*

1. *for all $n \geq 0$, the cycle $Z_{H_n}(c)$ converges to a repelling cycle $Z_{H_n}(c_*)$ as $c \to c_*$;*

2. *the closed set $Z(c_*) = \overline{\bigcup_{n \geq 1} Z_{H_n}(c_*)}$ does not contain 0;*

3. *Q_{c_*} is infinitely renormalizable; and*

4. *the Julia set $J(c_*)$ of Q_{c_*} is not locally connected.*

The difficulty in the choice of the sequence $(H_n)_n$ is in ensuring that the distance between the critical point 0 and the cycles $Z_{H_n}(c)$, for $c \in \bigcup_{m \geq n+1} \overline{H_m}$, is bounded below uniformly in n.

In [So],[Mi2],[Mi3], and [Le], one can find quantitative conditions in terms of the roots $r(H_n)$ which ensure that a chain of components $(H_n)_n$ converges to a unique parameter c_* with the forementioned properties (we will come back to this in section 4). But if one simply wants to show the existence of a sequence (H_n) satisfying the conditions of Lemma 3.6, one can proceed the following way.

We begin by choosing a component H_0 of period k_0 and a component H_1 of period $k_1 = rk_0$, attached to H_0. When the parameter c is the root of H_1, the cycle $Z_{H_0}(c)$ is parabolic (with multiplier $\neq 1$), and it can be followed in a neighborhood of $r(H_1)$ in $\mathbb{C}$ as a cycle of period k_0. By continuity, there exists $\varepsilon > 0$ and a neighborhood V_1 of $r(H_1)$ such that for $c \in V_1$ the cycle stays at a distance $\geq \varepsilon$ from the critical point 0 of Q_c. When c enters H_1, the cycle $Z_{H_0}(c)$ becomes repelling, and the attracting k_1-periodic cycle $Z_{H_1}(c)$ arises. This cycle appears in k_0 clusters of k_1/k_0 points (called *bifurcated cycles*) around the points of the cycle $Z_{H_0}(c)$.

This description is valid at all points of H_1 and remains valid on the boundary of the component H_1. It is easy to see that the distance between the cycle $Z_{H_1}(c)$ and the cycle $Z_{H_0}(c)$ tends to 0 as $c \in \overline{H_1}$ tends to $r(H_1)$. Let us choose a component H_2 which is attached to H_1 such that its root $r(H_2)$ belongs to V_1, and such that the distance between the cycle $Z_{H_1}(r(H_2))$ and $Z_{H_0}(r(H_2))$ is less than $\varepsilon/3$. By continuity, there exists a neighborhood V_2 of $r(H_2)$ contained in V_1 such that for $c \in V_2$, the distance between the cycles $Z_{H_1}(c)$ and $Z_{H_0}(c)$ is less than $\varepsilon/3$. Now we choose V_2 disjoint from $\overline{H_0}$, so that for all $c \in V_2$ the cycle $Z_{H_0}(c)$ is repelling. Repeating this argument, we can inductively build a chain of components $(H_n)_n$ and a decreasing sequence of neighborhoods V_n of $r(H_n)$, which are disjoint from $\overline{H_i}$ for all $i \leq n-2$ and such that for all $c \in V_n$ the distance between the cycles $Z_{H_{n-1}}(c)$ and $Z_{H_{n-2}}(c)$ is less than $\varepsilon/3^n$.

We can also choose the V_n such that their diameters tend to 0. Hence the sequence of the roots $r(H_n)$ converges to a point c_*. It is then easy to check that for all $n \geq 0$, the cycle $Z_{H_n}(c_*)$ is repelling and is at a distance of at least $\varepsilon/2$ from the critical point 0 of Q_{c_*}.

Now we give a proof of the fact that the Julia set of Q_{c_*} is not locally connected, only using the Douady-Sullivan criterion instead of more quantitative methods.

LEMMA 3.7. *Let c_* be a limit point of a chain of components $(H_n)_n$ such that for all $n \geq 0$, the cycle $Z_{H_n}(c)$ converges to a repelling cycle $Z_{H_n}(c_*)$ and such that the distance between the critical point 0 and the cycles $Z_{H_n}(c)$, when $c \in \bigcup_{m \geq n+1} H_m$, is bounded below uniformly in n. Then Q_{c_*} satisfies the Douady-Sullivan criterion.*

PROOF. For $q \geq 2$, we denote by $\mathbb{Z}_q$ the group of integers modulo q. In order to lighten notations, if $i \in \{0, \ldots, q-1\}$, we will use the number i for its residue class modulo q.

We consider the compact set $C = \overline{\bigcup_{n \geq 0} Z_{H_n}(c_*)}$, which by the assumption does not contain the critical point. Let us show that the mapping $Q_{c_*} : C \to C$ is bijective.

The surjectivity of Q_{c_*} onto C follows from the fact that Q_{c_*} is surjective onto $\bigcup_{n \geq 0} Z_{H_n}(c_*)$, since each $Z_{H_n}(c_*)$ is a cycle. The injectivity is more subtle, although it is obvious that Q_{c_*} is injective on $\bigcup_{n \geq 0} Z_{H_n}(c_*)$. We label the points of the cycles along clusters. Let k_i be the period of H_i. Define the numbers q_i for $i \geq 0$ by the relation $k_{i+1} = q_i k_i$. For later use we set $q_{-1} = k_0$.

Let z_i denote, for $i \in \mathbb{Z}_{k_0}$, the points of the first cycle $Z_{H_0}(c)$ in such a way that $Q_c(z_i) = z_{i+1}$ (where the indices are taken modulo k_0). Then, for $i \in \mathbb{Z}_{k_0}$ and $j \in \mathbb{Z}_{q_0}$, we let $z_{i,j}$ be the part of the cycle $Z_{H_1}(c)$ which bifurcates from the point z_i, with indices chosen so that if $i \neq k_0 - 1$, then $Q_c(z_{i,j}) = z_{i+1,j}$, and if $i = k_0 - 1$, then $Q_c(z_{i,j}) = z_{0,j+1}$.

For any $n \geq 0$, the cycle $Z_{H_{n+1}}(c)$ bifurcates from the cycle $Z_{H_n}(c)$. Because of this, we can label the points $z_{\varepsilon_0,\ldots,\varepsilon_n}$ of the cycle $Z_{H_n}(c)$ according to the dynamics. More precisely, if we set $k_0 = q_{-1}$, there is a map τ_n from $\mathbb{Z}_{q_{-1}} \times \mathbb{Z}_{q_0} \times \cdots \times \mathbb{Z}_{q_{n-1}}$ into itself such that $Q_c(z_{\varepsilon_0,\ldots,\varepsilon_n}) = z_{\tau(\varepsilon_0,\ldots,\varepsilon_n)}$. The mapping τ_n is defined in the following way. For $(\varepsilon_0,\ldots,\varepsilon_n) \in \mathbb{Z}_{q_{-1}} \times \mathbb{Z}_{q_0} \times \cdots \times \mathbb{Z}_{q_{n-1}}$, use $(\varepsilon_0',\ldots,\varepsilon_n')$ to denote $\tau(\varepsilon_0,\ldots,\varepsilon_n)$ and let $0 \leq j \leq n$ be the smallest integer such that $\varepsilon_j \neq k_{j-1} - 1$. Then, for $l < j$, set $\varepsilon_l' = 0$, $\varepsilon_j' = \varepsilon_j + 1$ and for $i > j$, set $\varepsilon_i' = \varepsilon_i$. Moreover the image of $(q_{-1},\ldots,q_{n-1})$ by τ is $(0,\ldots,0)$.

Taking the limit as $n \to \infty$, this definition yields a mapping τ from $\prod_{i=0}^{\infty} \mathbb{Z}_{q_{i-1}}$ into itself which is bijective. Now we need to check that the "parametrization" $\underline{\varepsilon} \in \prod_{i=-1}^{+\infty} \mathbb{Z}_{q_i} \mapsto z_{\underline{\varepsilon}} \in C$ is injective. Without loss of generality, we can assume that the sequence of hyperbolic components arises from the main cardioid of $\mathcal{M}$. In order to find neighborhoods which group clusters of bifurcated cycles together, we define disjoint graphs $\Gamma_1(c),\ldots,\Gamma_n(c),\ldots$ satisfying the following properties (see [Ro5]):

- $\Gamma_n(r(H_n))$ is made of external rays landing at the parabolic cycle of $Q_{r(H_n)}$;

- $\Gamma_n(c)$ exists in a neighborhood U_n of $\overline{\bigcup_{i \geq n} H_i} \setminus \{r(H_n)\}$;

- $\Gamma_n(c)$ depends continuously on c inside $U_n \cup \{r(H_n)\}$;

- $\bigcup_{k=1}^n \Gamma_k(c)$ separates the points of the cycles $Z_{H_n}(c)$ but not the points of the cycle $Z_{H_m}(c)$ where $m > n$;

- $\bigcup_{k=1}^n \Gamma_k(c)$ separates the points $z_{\varepsilon_0,\ldots,\varepsilon_n\ldots}$ which differ in at least one term ε_i for $0 \leq i \leq n$.

We obtain the graph $\Gamma_n(r(H_n))$ by considering the cycle of external rays landing at the parabolic cycle for the parameter $c = r(H_n)$. There exists a holomorphic motion of this graph defined in a region containing H_n and bounded by external rays in $\mathbb{C} \setminus \mathcal{M}$ landing at $r(H_n)$. At the parameter $r(H_n)$, the rays which $\Gamma_n(r(H_n))$ is made of separate the critical points of the iterate $Q_{r(H_n)}^{k_n}$. It follows that they also separate the point of the attracting cycle Z_{H_n}. Since the rays and the cycles cannot cross each other (the period being different), the graph $\bigcup_{k=1}^n \Gamma_k(c)$ separates those points $z_{\varepsilon_0,\ldots,\varepsilon_n\ldots}$ which differ in at least one term ε_i for $0 \leq i \leq n$. Hence if $\underline{\varepsilon} \neq \underline{\varepsilon}'$, then $z_{\underline{\varepsilon}} \neq z_{\underline{\varepsilon}'}$. The injectivity for finite sequences is obvious.

Assume now that two distinct sequences $z_{\varepsilon_0,\ldots,\varepsilon_n}$ and $z_{\varepsilon_0',\ldots,\varepsilon_n'}$ converge respectively to z and z' such that $z \neq z'$. Then $z = z_{\underline{\varepsilon}}$ and $z' = z_{\underline{\varepsilon}'}$ with $\underline{\varepsilon} \neq \underline{\varepsilon}'$. From $Q_{c_*}(z_{\underline{\varepsilon}}) = z_{\tau(\underline{\varepsilon})}$ and $Q_{c_*}(z_{\underline{\varepsilon}'}) = z_{\tau(\underline{\varepsilon}')}$ and from the fact that τ is injective, it follows that $Q_{c_*}(z)$ and $Q_{c_*}(z')$ are distinct. As a consequence, Q_{c_*} is injective on C. □

We can consider the same question with a sequence of primitive renormalizations. We say that a parameter is *primitive renormalizable* if it belongs to a primitive copy of $\mathcal{M}$ in $\mathcal{M}$, i.e., contained in a maximal copy of $\mathcal{M}$ in $\mathcal{M}$ which is not attached to the main cardioid of $\mathcal{M}$. We say that a parameter is *infinitely primitive renormalizable* if it belongs to an infinite sequence $\mathcal{M}_n$, each $\mathcal{M}_n$ being a primitive copy of $\mathcal{M}$ in $\mathcal{M}_{n-1}$.

QUESTION 3.8. *Do there exist infinitely primitive renormalizable quadratic polynomials having a connected but not locally connected Julia set?*

3.3 Conjectures for rational maps

With the preceding examples in mind, the following conjecture seems reasonable.

CONJECTURE 3.9. *Let f be a rational map whose Julia set is connected. If f has a periodic Fatou component which contains a critical point whose boundary is not locally connected, then f satisfies the Douady-Sullivan criterion.*

Notice that in this conjecture, we include polynomials among rational maps.

Let us return to the omnipresence of polynomials in our examples. We notice that in each of our examples, the boundary of the periodic critical Fatou component contains a copy of a non-locally connected quadratic Julia set. In the example of Ghys-Herman, the boundary of the immediate basin of ∞ is homeomorphic to the Julia set of a quadratic polynomial which is not locally connected. In the degree 4 Blaschke product example, the boundary of the immediate basin of 0 contains the image (by the straightening map of Douady-Hubbard) of a quadratic Julia set which is not locally connected. Motivated by these examples, we propose the following.

CONJECTURE 3.10. *Let f be a rational map whose Julia set is connected. If f has a periodic critical Fatou component U whose boundary is not locally connected, then ∂U contains the homeomorphic image of some non-locally connected polynomial Julia set.*

Notice that to be at the boundary of a Fatou component is crucial. Indeed, there exist cubic Newton maps N such that the Julia set $J(N)$ contains a quasiconformal copy of a non-locally connected quadratic Julia set even though $J(N)$ itself is locally connected (see [Ro4]).

Notice also that we do not ask that the homeomorphism conjugate the dynamics. Let us consider the map $f_t(z) = e^{2i\pi t}z^2(z-4)/(1-4z)$. This is an example in the class of Ghys-Herman studied in section 3.1. This map preserves the unit circle, it is of degree 1, and the critical points are not on the unit circle. Therefore, one can define a rotation number $\rho(f_t)$ of the restriction of the map on the circle. Since $\rho(f_t)$ is continuous in t, one can find some t such that $\rho(f_t)$ is not a Brjuno number. This implies, in particular, that there is no Herman ring around the unit circle. By the theory of Perez-Marco there is a "hedgehog" with hairs around the circle (see figure 3.11).

The boundary of the basin of ∞ (or 0) contains this "hairy circle" and its preimage touching at the critical point and all the iterated backward pre-images connected to this. Can we say that this compact connected set is homeomorphic to the Julia set of a quadratic polynomial? If that were the case, the quadratic polynomial would not be conjugate on its Julia set to f_t; indeed, the quadratic polynomial would have a Siegel disk, but here the rotation number is non-Brujno.

4   THE CASE OF INFINITELY SATELLITE-RENORMALIZABLE QUADRATIC POLYNOMIALS: A MODEL

Our aim in this section is to propose a conjectural condition on some combinatorial data related to an infinitely satellite-renormalizable quadratic polynomial that implies non-local connectedness its Julia set. We are interested in the combinatorial data consisting of the sequence of rotation numbers $(p_n/q_n)_n$ defined in section 4.1.

We are also interested in a description of the post-critical closure $\overline{P_f}$. There are some similarities with the hedgehogs and the Cantor bouquets.

This section contains a description and the beginning of the investigation of a geometric model of the sequence of straightenings of an infinitely satellite-renormalizable quadratic polynomial which provides such a conjectural condition. We owe the idea of this model to Xavier Buff.

In what follows, $\mathbb{N}$ will represents the set of non-negative integers.

4.1   Combinatorial data for satellite renormalizable polynomials

DEFINITION 4.1. *A hyperbolic component H of $\mathcal{M}$ is called* satellite *to another hyperbolic component L if it is attached to L at its root point $r(H)$.*

If H is a hyperbolic component of $\mathcal{M}$, the multiplier map λ_H of H will refer to the mapping that sends a parameter $c \in H$ to the complex number $\lambda_H(c)$ in $\mathbb{D}$ that is the multiplier of the unique attracting cycle of Q_c.

It is well known that the multiplier map can be extended to a homeomorphism from $\overline{H}$ onto the closed unit disk. Moreover, attached components meet only at parameters at which the multiplier map is a root of unity.

FIGURE 3.11: The Julia set of some f_t. In this image, the white region covers the Julia set. One can imagine the hairy circle. (See also Plate 15.) Image courtesy of H. Inou.

DEFINITION 4.2. *Let H and L be hyperbolic components of $\mathcal{M}$ such that H is satellite to L. The rational number p/q such that the multiplier map of L sends the root of H to $e^{2\pi i p/q}$ is called* the rotation number of H with respect to L; *i.e., $\lambda_L(r(H)) = e^{2\pi i p/q}$.*

DEFINITION 4.3. *For $c \in \mathbb{C}$, the quadratic polynomial Q_c is infinitely satellite renormalizable if it is a limit point of a sequence of hyperbolic components H_n of $\mathcal{M}$ such that H_{n+1} is satellite to H_n for all n.*

To each such c and H_0 we can associate a sequence of rotation numbers $(p_n/q_n)_n$; it is the sequence p_n/q_n of rotation numbers of H_{n+1} with respect to H_n.

Thanks to the connectedness principle (Theorem 1.8, compare [McM2]), if the Julia set of a renormalization of a polynomial is not locally connected then the Julia set of the original polynomial is not locally connected.

As a consequence, we are interested only in the tail of the sequence (p_n/q_n), which is independent of the choice of H_0 in the sense that if $(H_0, (p_n/q_n)_n)$ and

$(H'_0, (p'_n/q'_n)_n)$ are both associated to the same parameter c, then either there is n such that $H'_0 = H_n$ and $p'_k/q'_k = p_{n+k}/q_{n+k}$ (for all k) or such that $H_0 = H'_n$ and $p_k/q_k = p'_{n+k}/q'_{n+k}$.

Keeping this in mind, we will not mention H_0 when we talk about the sequence of rotation numbers of an infinitely satellite-renormalizable quadratic polynomial.

4.2   Definition of the model

Let $(p_n/q_n)_n$ be a sequence of reduced fractions in the interval $]0,1[$, where $q_n > 0$. We suppose that the sequence $(p_n/q_n)_n$ converges to 0.

Let $C > 1$ be a fixed constant and define t_n as

$$t_n = C\frac{p_n}{q_n}.$$

We refer to the Lemma 4.23 and the observation following the statement of Lemma 4.17 about the role of the constant C.

We denote by M_n the Möbius transformation

$$M_n(z) = \frac{1 - t_n/z}{1 - t_n}.$$

This mapping is characterized by the fact that it sends 0 to ∞, t_n to 0, and 1 to itself. We define the sequence of mappings $(\varphi_n)_n$ by

$$\varphi_n(z) = (M_n(z))^{q_n}.$$

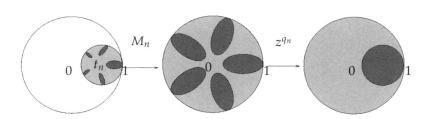

FIGURE 4.4: Schematic illustration of the mappings φ_n on the unit disk as a composition of the Möbius transformation M_n and the map $z \mapsto z^{q_n}$. The light grey part on the left is sent onto the light grey part on the right, as are the dark grey parts (the dark grey disk on the right is close to 0, so its pre-images are thin).

REMARK 4.5. We will always suppose that t_n belongs to the unit disk. Since we suppose $p_n/q_n \to 0$ this is true for n big enough.

Let

$$\Phi_n = \varphi_n \circ \cdots \circ \varphi_0.$$

We denote by K_∞ the set of points of $\overline{\mathbb{D}}$ which do not escape under Φ_n. That is,

$$K_\infty = \{z \in \overline{\mathbb{D}} : \forall n, \Phi_n(z) \in \overline{\mathbb{D}}\}.$$

Note that

$$K_\infty = \bigcap_n K_n, \quad \text{where} \quad K_n = \{z \in \overline{\mathbb{D}} : \varphi_n \circ \cdots \circ \varphi_0(z) \in \overline{\mathbb{D}}\}.$$

Also note that we have $K_n = \{z \in \overline{\mathbb{D}} : \forall k = 0, \ldots, n, \varphi_k \circ \cdots \circ \varphi_0(z) \in \overline{\mathbb{D}}\}$. Thus $(K_n)_n$ is a decreasing sequence of non-empty compact sets containing 1. In particular, K_∞ itself is non-empty and compact.

The model is defined by the sequence of mappings $(\varphi_n)_n$. The compact set K_∞ will play an important role in the study of the model and also as part of its realization.

Recall that satellite bifurcations correspond to cycle collisions. For example, it means that a small perturbation of a polynomial in the quadratic family having a parabolic fixed point with multiplier different from 1 has a cycle which belongs entirely to some small neighborhood of the perturbed fixed point.

If the multiplier of the fixed point of the former polynomial is $e^{2i\pi p_n/q_n}$ and if the rotation number of the cycle of the perturbed polynomial is p_{n+1}/q_{n+1}, then the displacement of the fixed point under perturbation is of the order of magnitude p_{n+1}/q_{n+1} while the explosion of the cycle happens at a speed whose order of magnitude is $\left(\frac{p_{n+1}}{q_{n+1}}\right)^{1/q_n}$.

The latter polynomial $f_n = z^2 + c_n$, which is a perturbation of a polynomial with a parabolic fixed point, is renormalizable. The renormalization replaces the q_nth iterate of the mapping f_n by a mapping $f_{n+1} = \mathcal{R}f_n$. The new map f_{n+1} has a fixed point which is the image of the exploding cycle by the renormalization map.

There exists a map $\tilde{\varphi}_n$ defined on the domain of renormalization such that $\mathcal{R}f_n \circ \tilde{\varphi}_n = \tilde{\varphi}_n \circ f_n^{\circ q_n}$. In the case where the quadratic polynomial f_n is infinitely satellite renormalizable, the quadratic polynomial f_{n+1} is again renormalizable. Thus an infinitely renormalizable f_0 yields a sequence of quadratic maps $(f_n)_n$.

The mapping φ_n is designed to be a geometric model of the straightening map of the nth satellite renormalization. In particular, it has the following properties:

- a fixed point at 1, which is also the critical point;

- the point t_n, which represents a fixed point, is sent to the center of the unit disk;

- the set of pre-images of t_{n+1} by φ_n represents the exploding cycle;

- the power map sends this cycle to a unique fixed point for the renormalized map.

When we consider the quadratic family, we can use the Douady-Sullivan criterion (compare 3.1) to show that if the set of accumulation points of the sequence of exploding cycles does not contain the critical point then the Julia set of the limit polynomial is not locally connected.

By its very construction, we know that K_∞ must contain these accumulation points. This fact allows us to determine a conjectural criterion for non-local connectedness of the Julia set.

4.3 The residual compact set

We begin with the study of the compact set $K_\infty = K_\infty\left(((p_n/q_n)_n, C\right)$ called the *residual compact set*. Recall that it is defined by

$$K_\infty = \bigcap_{n=0}^{\infty} K_n,$$

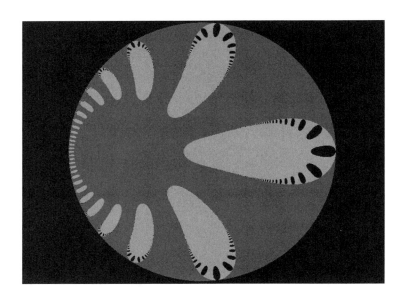

FIGURE 4.6: Example of the residual set K_n with $n = 2$ (it is a magnification of the right part of the unit disk). Here we have $\frac{p_0}{q_0} = \frac{1}{28}$, $\frac{p_1}{q_1} = \frac{1}{39670}$ (the value of p_2/q_2 plays no role in the shape of K_2). The black region surrounding everything is outside any K_n, the dark gray represents $K_0 \backslash K_1$, the light gray $K_1 \backslash K_2$, and the dark regions inside the light gray is K_2. The compact set K_n is symmetric with respect to the real axis and its intersection with the real axis contains a line segment bounded by 1 on the right (compare section 4.3.2).

where $K_n = \{z \in \overline{\mathbb{D}} : \forall k \leq n, \, \Phi_n(z) \in \overline{\mathbb{D}}\}$ and $\Phi_n = \varphi_n \circ \cdots \circ \varphi_0$.

In the next two sections we will label the connected components of K_∞ with an "odometer" and prove a result about the topological type of some of its components.

4.3.1 The address of a point in K_∞

LEMMA 4.7. *Let $t \in]0, 1[$ and $M(z) = \frac{1-t/z}{1-t}$. Then $M^{-1}\left(\overline{\mathbb{D}}\right)$ is the closed disk that has the line segment $\left[\frac{t}{2-t}, 1\right]$ as a diameter.*

PROOF. The Möbius transformation M commutes with $z \mapsto \bar{z}$, sends 1 to itself, and sends $\frac{t}{2-t}$ to -1. $\square$

LEMMA 4.8. *Let $n \in \mathbb{N}$ and E_n denote the mapping $z \mapsto z^{q_n}$. Then for each connected component of $E_n^{-1}\left(M_{n+1}^{-1}\left(\overline{\mathbb{D}}\right)\right)$, there exists a unique $k \in \{0, \ldots, q_n - 1\}$ such that $e^{2\pi i k p_n / q_n}$ belongs to this component.*

PROOF. From the previous lemma we know that $M_{n+1}^{-1}\left(\overline{\mathbb{D}}\right)$ is a disk which lies strictly to the right of 0. Then the connected components of its preimage by E_n are contained in sectors of angles π/q_n separated by sectors by sectors of the same angle. Moreover, $1 \in M_{n+1}^{-1}\left(\overline{\mathbb{D}}\right)$, so each component contains one and only one q_nth root of 1. Finally, note that since p_n/q_n is reduced, the sets $\{e^{2\pi i k p_n / q_n} \mid k \in \mathbb{N}\}$ and $\{e^{2\pi i k / q_n} \mid k \in \mathbb{N}\}$ coincide. $\square$

COROLLARY 4.9. *The number of connected components of K_n is*

$$N_n = \prod_{k=0}^{n-1} q_k, \quad with \ N_0 = 1.$$

PROOF. For $n \geq 2$ the mapping M_n is a homeomorphism between $E_{n-1}^{-1}(M_n^{-1}(\overline{\mathbb{D}}))$ and $\varphi_{n-2} \circ \cdots \circ \varphi_0(K_n)$ and, when $n = 1$, between K_1 and $E_0^{-1}(M_1^{-1}(\overline{\mathbb{D}}))$. $\square$

Thanks to the previous lemma, we can label the components of

$$\varphi_n^{-1}\left(\varphi_{n+1}^{-1}(\overline{\mathbb{D}})\right) = \varphi_n^{-1}\left(M_{n+1}^{-1}(\overline{\mathbb{D}})\right)$$

with the elements of $\mathbb{Z}_{q_n} = \mathbb{Z}/q_n\mathbb{Z}$. We label the component whose image under M_n contains $e^{2\pi i k p_n/q_n}$ by $k \in \mathbb{Z}_{q_n}$.

We can then define the *address* of $z \in K_n$ by $(k_0, \ldots, k_{n-1}) \in \mathbb{Z}_{q_0} \times \cdots \times \mathbb{Z}_{q_{n-1}}$, where the k_j are determined by the condition that $\varphi_{j-1} \circ \cdots \circ \varphi_0(z)$ belongs to the component of $\varphi_j^{-1}(\overline{\mathbb{D}})$ which has been labelled k_j.

The address of a point $z \in K_\infty$ is defined as the infinite sequence $(k_0, \ldots, k_n, \ldots)$ in $\prod_{n=0}^{\infty} \mathbb{Z}_{q_n}$, which is such that for each n, $(k_0, \ldots, k_{n-1})$ is the address of z in K_n. Every $z \in K_\infty$ has one and only one address, but the same address may correspond to several z.

DEFINITION 4.10. *Let $z \in K_\infty$. We say that $\alpha = (\alpha_0, \alpha_1, \ldots) \in \prod_{n \in \mathbb{N}} \mathbb{Z}_{q_n}$ is the address of z in K_∞ if for all $n \in \mathbb{N}$, the point $M_{n+1} \circ \varphi_n \circ \cdots \circ \varphi_0(z)$ belongs to the same connected component of $\varphi_n^{-1}(\overline{\mathbb{D}})$ as $e^{2\pi i \alpha_n p_n/q_n}$.*

In order to describe the structure of the compact set K_∞ we need to introduce odometers. Given a sequence of positive integers N_n such that N_n divides N_{n+1}, we call the *odometer with scale* $(N_n)_{n \in \mathbb{N}}$ the set

$$\mathcal{O} = \mathbb{Z}_{N_0} \times \prod_{n \in \mathbb{N}} \mathbb{Z}_{N_{n+1}/N_n}$$

equipped with the product topology of the discrete topology on each factor and with a continuous adding map $\sigma: \mathcal{O} \to \mathcal{O}$ defined by the following (compare [Dow]): for all $j = (j_n)_n \in \mathcal{O}$, $(\sigma(j))_0 = j_0 + 1$ and for $n > 0$,

$$(\sigma(j))_n = \begin{cases} j_n + 1 & \text{if } \forall k \leq n-1, \ j_k = \frac{N_{k+1}}{N_k} - 1, \\ j_n & \text{otherwise.} \end{cases}$$

Topologically, an odometer $\mathcal{O}$ is a Cantor set.

In the following we will identify the set of addresses

$$K_{addr} = \prod_{n \in \mathbb{N}} \mathbb{Z}_{q_n}$$

with the odometer with scale $(N_n)_{n \geq 1}$ where $N_n = \prod_{k=0}^{n-1} q_k$. We will refer to this odometer as the *addresses odometer*.

The adding map might be relevant from the dynamical perspective but not for the study of the topology K_∞; compare the proof of Lemma 3.7.

PROPOSITION 4.11. *Let $\mathcal{P}(K_\infty)$ be the set of subsets of K_∞. Let $\pi\colon K_\infty \to \mathcal{P}(K_\infty)$ be the mapping that sends a point to the connected component of K_∞ it belongs to. Consider the final topology on $\mathcal{P}(K_\infty)$ with respect to π, which is the finest topology that makes the map π continuous.*
Then the set of connected components of K_∞, when equipped with the final topology, is homeomorphic to the addresses odometer.

PROOF. Let $\alpha \in \prod_{n\in\mathbb{N}}\mathbb{Z}_{q_n}$. Then the set of points which have $(\alpha_0,\dots,\alpha_{n-1})$ as their address in K_n is a connected compact set homeomorphic to the closed unit disk. It follows that the set of points which have α as their address in K_∞ is a connected component of K_∞. Hence we have a one-to-one correspondence between the set of connected components of K_∞ and the set of addresses. We just need to show that the mapping that sends a point to its address is continuous.

Let $z, z' \in K_\infty$ and let α, α' be their respective addresses. Suppose that there exists n such that $\forall m \geq n$, $\alpha_m = \alpha'_m$. Then $\varphi_n \circ \cdots \circ \varphi_0(z)$ and $\varphi_n \circ \cdots \circ \varphi_0(z')$ belong to the same connected component of K_n. This component has a neighborhood in $\mathbb{C}$ which is disjoint from the other components. Hence, close addresses require the points to be close. $\qquad\square$

4.3.2   Topology of the critical component

In what follows, we denote the argument of a complex number in $]-\pi, \pi]$ by arg.

LEMMA 4.12. *Let $C > 1$, $q \in \mathbb{N}^*$, $t = C/q$ and let $\varphi\colon \overline{\mathbb{D}} \to \widehat{\mathbb{C}}$ be defined by*

$$\varphi(z) = \left(\frac{1 - t/z}{1 - t}\right)^q.$$

Then for all $z \in \overline{\mathbb{D}}$ such that $|\arg(z - t)| \leq \frac{\pi}{2}$, we have $|\arg\varphi(z)| > \dfrac{C|\operatorname{Im} z|}{2|z|^2}$.

PROOF. Because of the real symmetry of the mapping φ, it is sufficient to show the lemma for all $z = x + iy \in \overline{\mathbb{D}}$ such that $x \geq t$ and $y > 0$. Under these hypotheses, $\arg(z) = \arcsin(y/|z|)$ and $\arg(z - t) = \arcsin(y/|z - t|)$.
Since $\arg\varphi(z) = q(\arg(z - t) - \arg(z)) \geq 0$, we have

$$\arg\varphi(z) = q\left(\arcsin\left(\frac{y}{|z - t|}\right) - \arcsin\left(\frac{y}{|z|}\right)\right).$$

The function arcsin is convex on $]0, 1]$, so

$$\arcsin\left(\frac{y}{|z - t|}\right) \;\geq\; \arcsin\left(\frac{y}{|z|}\right) + \frac{|z|}{x}\left(\frac{y}{|z - t|} - \frac{y}{|z|}\right).$$

We estimate the difference $\frac{y}{|z-t|} - \frac{y}{|z|}$. Let $r = |z|$. Then

$$|z - t|^2 \;=\; r^2\left(1 + \frac{t^2}{r^2} - 2\frac{tx}{r^2}\right),$$

from which it follows that

$$|z - t| \;\leq\; r\left(1 + \frac{1}{2}\left(\frac{t^2}{r^2} - 2\frac{tx}{r^2}\right)\right).$$

Since $\frac{1}{2}\left(\frac{t^2}{r^2} - 2\frac{tx}{r^2}\right) \leq 0$, we have

$$\frac{1}{|z-t|} \geq \frac{1}{r} \cdot \frac{1}{1 + \frac{1}{2}\left(\frac{t^2}{r^2} - 2\frac{tx}{r^2}\right)}$$

$$\geq \frac{1}{r}\left(1 - \frac{1}{2}\left(\frac{t^2}{r^2} - 2\frac{tx}{r^2}\right)\right).$$

As a consequence,

$$\frac{1}{|z-t|} - \frac{1}{|z|} \geq \frac{xt}{r^3} - \frac{1}{2}\frac{t^2}{r^3}.$$

From the preceding, it follows that

$$\arg \varphi(z) \geq \frac{qty}{r^2}\left(1 - \frac{1}{2}\frac{t}{x}\right)$$

$$\geq \frac{Cy}{r^2}\left(1 - \frac{1}{2}\frac{t}{x}\right).$$

But $x \geq t$, so we have $\arg \varphi(z) \geq \frac{Cy}{2r^2}$. □

COROLLARY 4.13. *Under the assumptions of the previous lemma we have*

$$|\arg(z)| \leq \frac{\pi}{C}|\arg \varphi(z)||z|.$$

PROOF. From the previous lemma we have $|\arg \varphi(z)| \geq \frac{C}{2}\frac{|\operatorname{Im} y|}{|z|^2}$. Also note that $\frac{|\operatorname{Im} y|}{|z|} = |\sin(\arg(z))|$. Using the fact that $|\sin t| \geq \frac{2}{\pi}|t|$ for all $t \in [-\frac{\pi}{2}, \frac{\pi}{2}]$, we obtain $|\arg \varphi(z)| \geq \frac{C}{\pi|z|}|\arg z|$. □

DEFINITION 4.14. *Let K_∞ be the residual compact set of the model given by the data $C > 1$ and $(p_n/q_n)_n$. The critical component I_0 of K_∞ is the connected component of K_∞ which contains 1. The critical component $K_{n,0}$ of K_n is the connected component of K_n which contains 1.*

The critical component I_0 is the set of points in K_∞ which have $(0, \ldots, 0, \ldots)$ as their address. The set $K_{n,0}$ is the set of points in K_n which have $(0, \ldots, 0)$ as their address in K_n, or, equivalently, whose address in K_∞ starts with n noughts.

It follows from the definition that

$$I_0 = \bigcap_{n \in \mathbb{N}} K_{n,0}.$$

LEMMA 4.15. *The mapping $M_{n+1} \circ \varphi_n \circ \cdots \circ \varphi_0 \colon K_{n,0} \to \overline{\mathbb{D}}$ is the restriction of a biholomorphism defined in a neighborhood of $K_{n,0}$.*

PROOF. The mapping φ_k is the composition of a power map with a Möbius transformation. Then, for all $k < n$, there exists a holomorphic mapping ψ_k defined on $\mathbb{C} \backslash \mathbb{R}_-$ such that $\varphi_k \circ \psi_k = \operatorname{Id}_{\mathbb{C}\backslash\mathbb{R}_-}$. By definition, for all $k < n$, $\Phi_{k+1}(K_{n,0}) \subset \overline{\mathbb{D}}$, so $\varphi_k \circ \cdots \circ \varphi_0(K_{n,0}) \subset \{\operatorname{Re} > 0\}$. Hence $K_{n,0} \subset \psi_0 \circ \cdots \circ \psi_{n-1}(\mathbb{C}\backslash\mathbb{R}_-)$. □

LEMMA 4.16. *The intersection of $K_{n,0}$ with the real axis is a line segment containing 1.*

PROOF. The homeomorphism $(M_n \circ \varphi_{n-1} \circ \cdots \circ \varphi_0)|_{K_{n,0}}$ is a one-to-one mapping between the real points of $K_{n,0}$ and the real points of $\overline{\mathbb{D}}$. $\square$

LEMMA 4.17. *Suppose that $C \geq \pi$. Then for all $n \in \mathbb{N}$,*

$$K_{n,0} \subset \left\{ z \in \overline{\mathbb{D}} : |\arg(z)| \leq \frac{\pi}{2} \left(\frac{\pi}{C} \right)^n \right\}.$$

Assuming $C \geq \pi$ might not be optimal (compare Lemma 4.23). We do not know whether this hypothesis is necessary for the conclusion of Lemma 4.17.

PROOF. Let $z \in K_{n,0}$. From Lemma 4.7, it follows that $|\arg(z_k - t_{k+1})| \leq \frac{\pi}{2}$ for all $k = 0, \ldots, n-1$, where $z_k = \varphi_k \circ \cdots \circ \varphi_0(z) \in \overline{\mathbb{D}}$. Then Corollary 4.13 implies that for all $k \leq n-1$, $|\arg z_k| \leq \frac{\pi}{C} |\arg z_{k+1}|$. $\square$

COROLLARY 4.18. *If $C > \pi$, the critical component of K_∞ is either the point $\{1\}$ or a nontrivial line segment on the real axis.*

REMARK 4.19. The components of K_∞ that are sent to $\Phi_n(I_0)$ by some Φ_n are homeomorphic to I_0. As a consequence, we showed that a countable dense set of components of K_∞ are homeomorphic to the same line segment (possibly reduced to a point). These are dense because their images by the projection onto the odometer is the dense set of sequences $\alpha \in \prod_{n \in \mathbb{N}} \mathbb{Z}_{q_n}$ with finite support ($\alpha_n = 0$ for n big enough).

The next conjecture illustrates what a topological model for the compact space K_∞ could be. Notice that the definition of this model is close to those of a straight brush and hairy arcs (compare [AO], [Dev]).

Recall that K_{addr} is the addresses odometer, homeomorphic to the Cantor set of connected components of K_∞.

CONJECTURE 4.20. *There exists a closed subset B of $K_{addr} \times [0, 1]$ homeomorphic to K_∞ which satisfies*

1. $K_{addr} \times \{0\} \subset B$;

2. *for all $\alpha \in K_{addr}$ there exists $e_\alpha \in [0, 1]$ such that $(\alpha, t) \in B$ if and only if $0 \leq t \leq e_\alpha$. (Such e_α is called an upper endpoint);*

3. *the set of upper endpoints is dense in B.*

The definition of K_∞ might recall in some points one of the definitions of a Cantor bouquet (see the characterization of the Julia set of some maps of the exponential family as a Cantor bouquet in [Dev]):

- The set K_∞ is the set of non-escaping points under the compositions of a ordered countable family of holomorphic mappings $(\varphi_n)_n$.

- It is an intersection of a decreasing sequence of compact sets K_n (in the Riemann sphere).

- For every connected component ω of K_n the cardinality of the set of connected components of K_{n+1} which are included in ω are the same.

- The mapping φ_n sends $K_{n+1} \cap \omega$ homeomorphically onto K_n for all connected component ω of K_n.

- Connected components are ordered (vertical lines for the Julia sets of the exponential maps, circular ordering for the present object of our study). It allows to label the connected components of K_∞.

The above conjecture suggests how to unroll and how to straighten K_∞.

4.4   The conjectural non-local-connectedness criterion

LEMMA 4.21. *Let $x_0 = \inf\{x \in [0,1] : \forall y > x\, \forall n \in \mathbb{N}, \Phi_n(y) \geq t_{n+1}\}$. Then*

- $[x_0, 1] \subset K_\infty$;

- $x_0 = \lim_{n\to\infty} s_n$, *where, for all $n \in \mathbb{N}$, s_n is the unique pre-image of 0 by Φ_n which belongs to the component $K_{n,0}$;*

- *For all $\varepsilon > 0$, there exists $y \in [x_0 - \varepsilon, x_0[\setminus K_\infty$.*

PROOF. From Lemma 4.7 it follows that if a point $x \in K_n$ is such that $\operatorname{Re}\Phi_n(x) < 0$, then $x \notin K_\infty$. Since the mapping φ_n is increasing on $[t_n, 1]$ and $\Phi_{n-1}(s_n) = t_n$, the mapping Φ_n is also increasing on $[s_n, 1]$. As a consequence $[x_0, 1] \subset K_\infty$ and x_0 is the limit of the increasing sequence $(s_n)_n$.

Let $\varepsilon > 0$ and let n be such that $x_0 - s_n \leq \varepsilon/2$. If we take $y \in K_{n,0} \cap [s_n - \varepsilon/2, s_n[$, then we have $\Phi_n(y) < 0$ so $y \in [x_0 - \varepsilon, x_0[\setminus K_\infty$. $\square$

DEFINITION 4.22. *Let $C > 1$. We denote by $\mathcal{C}_C$ the set of all parameters $c \in \mathbb{C}$ such that the quadratic polynomial $z^2 + c$ is infinitely satellite renormalizable with the sequence of rotation numbers $(p_n/q_n)_n$ satisfying the following:*

1. *The sequence of positive numbers $(t_n)_n$ given by $t_n = C|p_n/q_n|$ is such that $\forall n \in \mathbb{N}^*$, $t_n \in]0,1[$.*

2. *Let φ_n be the mapping defined by $\varphi_n(z) = \left(\frac{1 - t_n/z}{1 - t_n}\right)^{q_n}$ and let $\Phi_n = \varphi_n \circ \cdots \circ \varphi_0$. Then*

$$\exists x_0 \in]0,1[, \forall n \in \mathbb{N}, \Phi_n(x_0) \geq t_{n+1}. \tag{4.1}$$

Because of the monotonicity of the mappings φ_n on $[t_n, 1]$, the second condition implies that the line segment $[x_0, 1]$ is included in K_∞ (compare Lemma 4.21). Conversely if there is $x_0 \in]0,1[$ such that $[x_0, 1] \subset K_\infty$, then x_0 satisfies the condition in (4.1).

LEMMA 4.23. *Let $C \geq C' > 1$. Then $\mathcal{C}_C \subset \mathcal{C}_{C'}$.*

PROOF. Let $c \in \mathcal{C}_C$ and let $\gamma = C'/C$. By assumption, the sequence of real numbers $t_n = C|p_n|/q_n$ is such that $t_n \in]0,1[$. Also, there exists $x_0 \in]0,1[$ such that for all n, $\Phi_n(x_0) \geq t_{n+1}$. We define the sequence of real numbers x_n by $x_{n+1} = \varphi_n(x_n)$.

Let $t'_n = \gamma t_n \in]0,1[$, $\varphi'_n(z) = \left(\frac{1 - t'_n/z}{1 - t'_n}\right)^{q_n}$, and $\Phi'_n = \varphi'_n \circ \cdots \circ \varphi'_0$. We show that the sequence $(x'_n)_n$ defined by $x'_0 = x_0$ and $x'_{n+1} = \varphi'_n(x'_n)$ satisfies $x'_n \geq x_n$ for all n.

By induction, suppose $x'_n \geq x_n$ and $x'_n \in [0,1]$. Since $t'_n \leq t_n$, we have

$$x'_{n+1} \geq \varphi_n(x'_n) \geq \varphi_n(x_n) = x_{n+1}.$$

Hence, $\Phi'_n(x_0) \geq \Phi_n(x_0) \geq t_{n+1} \geq t'_{n+1}$. $\square$

The main conjecture is the following (we purposely state it only in the case where $p_n = 1$).

CONJECTURE 4.24 (Non-local-connectedness criterion). *Let $c \in \mathbb{C}$ be such that the quadratic polynomial $Q_c(z) = z^2 + c$ is infinitely satellite renormalizable with the sequence of rotation numbers $(1/q_n)_n$. Then the Julia set of Q_c is not locally connected if and only if there exists $C > 1$ and a renormalization Q_{c*} of Q_c such that $c* \in \mathcal{C}_C$.*

We can justify this conjecture by the Douady-Sullivan criterion 3.1.

The cycles are modeled by the centers of the components of K_n, that is, by the preimages of 0 under Φ_n for some n. Thanks to Lemma 4.21, we know that the left boundary of the segment I_0 (the critical component) is a limit point of the set of centers of components of K_n, $n \in \mathbb{N}$. The other limit points belong to other components, which are at a positive distance from the point 1. As a consequence, if this segment is not reduced to a point, then the critical point is at a positive distance of the limit set of the cycles and we can apply the Douady-Sullivan criterion.

Another conjecture related to this model is the existence of an invariant compact set inside the Julia set which is homeomorphic to K_∞.

CONJECTURE 4.25. *Let $c \in \mathbb{C}$ be such that the quadratic polynomial $Q_c(z) = z^2 + c$ is infinitely satellite renormalizable with the sequence of rotation numbers $(1/q_n)_n$.*

Then there exists $C > 1$, a renormalization Q_{c*} of Q_c and an invariant compact subset of the Julia set of Q_{c*} homeomorphic to the residual compact set K_∞ of the model associated to the data $(C, (p_n/q_n)_n)$.

Moreover, the componentwise dynamics in this set is given by the adding map of the addresses odometer.

Non-locally connected quadratic Julia sets are still not well understood. Proving this conjecture may provide valuable information on the structure of and the dynamics on the Julia set in the case it is not locally connected.

It would be interesting to know if this homeomorphism extends to the whole plane and if it is even quasiconformal.

4.4.1   A test of the conjectural criterion

The following is an example of a situation where the Julia set is not locally connected and the conditions of the conjecture 4.25 are satisfied (that is $c_* \in \mathcal{C}_C$ for some renormalization Q_{c_*}).

As we mentioned earlier, Sørensen's article [So] does not contain explicit conditions for non-local connectedness, but Milnor has proposed such a condition in [Mi3]. G. Levin found an explicit condition which implies Milnor's condition. Indeed Levin's criterion is more general.

THEOREM 4.26 (Levin, [Le]). *Let $c \in \mathbb{C}$ be such that the polynomial $Q_c(z) = z^2 + c$ is infinitely satellite renormalizable with the sequence of rotation numbers denoted by $(p_n/q_n)_n$. Suppose that*

$$\limsup_{n\to\infty} \left| \frac{p_{n+1}}{q_{n+1}} \right|^{1/q_n} < 1.$$

Then the Julia set of Q_c is not locally connected.

Indeed, Levin's work yields a more general condition which is not easy to work with. No other explicit criterion based on the rotation numbers is known yet.

THEOREM 4.27. *Let $c \in \mathbb{C}$ be such that the polynomial $Q_c(z) = z^2 + c$ is infinitely satellite renormalizable with the sequence of rotation numbers $(p_n/q_n)_n$. Suppose that*

- *the sequence $(p_n)_n$ is bounded while $q_n \to \infty$,*

- $\limsup_{n \to \infty} \left| \frac{p_{n+1}}{q_{n+1}} \right|^{1/q_n} < 1.$

Then for all $C > 1$, there exists a renormalization Q_{c_*} of Q_c such that $c_* \in \mathcal{C}_C$.

PROOF. Since we can renormalize, we may suppose that there is an $\alpha \in [0, 1[$ such that $|p_{n+1}/q_{n+1}| \leq \alpha^{q_n}$. By hypothesis $\frac{p_n}{q_n} \to 0$, thus we can renormalize so that for all n, $t_n < 1$, where t_n is defined as $t_n = C|p_n/q_n|$.

Let $\beta \in]1, 1/\alpha[$ and $\eta = \frac{1}{1-\beta\alpha}$. For the same reason as before, we may also suppose that for all n, $\left(\frac{\beta}{1-t_n}\right)^{q_n} \geq C\eta$.

Define the sequence $(x_n)_n$ in the following way. We set $x_0 = \eta t_0$ and for all $n \geq 0$, define $x_{n+1} = \left(\frac{1-t_n/x_n}{1-t_n}\right)^{q_n}$. Then the sequence $(x_n)_n$ satisfies $x_n \geq \eta t_n$ for all n. In fact, by induction

$$
\begin{aligned}
x_{n+1} &\geq \left(\frac{1-1/\eta}{1-t_n}\right)^{q_n} = \left(\frac{\beta}{1-t_n}\right)^{q_n} \alpha^{q_n} \\
&\geq \eta t_{n+1}.
\end{aligned}
$$

$\square$

5 Bibliography

[AO]   J. M. AARTS and L. G. OVERSTEEGEN. *The geometry of Julia sets.* Trans. Amer. Math. Soc., **338**(2): 897–918, 1993.

[AY]   M. ASPENBERG and M. YAMPOLSKY. *Mating non-renormalizable quadratic polynomials.* Comm. Math. Phys. **287**(1): 1–40, 2009.

[BuHe] X. BUFF and C. HENRIKSEN. *Julia sets in parameter spaces.* Communications in Mathematical Physics **220**: 333–375, 2001.

[Dev]   R. L. DEVANEY. *Cantor bouquets, explosions, and Knaster continua: Dynamics of complex exponentials.* Publ. Mat., **43**(1): 27–54, 1999.

[DoHu] A. DOUADY and J. H. HUBBARD. *On the dynamics of polynomial-like mappings.* Ann. Sci. École Norm. Sup. **18**: 287–343, 1985.

[Dow]   T. DOWNAROWICZ. *Survey of odometers and Toeplitz flows.* In "Algebraic and topological dynamics," volume 385 of Contemp. Math., pp. 7–37. Amer. Math. Soc., Providence, RI, 2005.

[G]   É. GHYS. *Transformation holomorphe au voisinage d'une courbe de Jordan.* C. R. Acad. Sc. Paris, **298**: 385–388, 1984.

[He]   M. HERMAN. *Conjugaison quasi-symétrique des difféomorphisme du cercle à des rotations et applications aux disques singuliers de Siegel, I.* Manuscript, http://www.math.kyoto-u.ac.jp/~mitsu/Herman/qsconj2.

[KL]   J. KAHN and M. LYUBICH. *Local connectivity of Julia sets for unicritical polynomials.* Annals of Math. (2) **170**(1): 413–426, 2009.

[KovS] O. KOZLOVSKI and S. VAN STRIEN. *Local connectivity and quasi-conformal rigidity of non-renormalizable polynomials.* Proc. Lond. Math. Soc. (3), **99**(2): 275–296, 2009.

[Le]   G. LEVIN. *Multipliers of periodic orbits of quadratic polynomials and the parameter plane.* Israel J. Math., 170: 285–315, 2009.

[Lo]   L. LOMONACO. *Parabolic-like maps.* arXiv:1111.7150.

[LM]   M. LYUBICH and Y. MINSKY. *Laminations in holomorphic dynamics.* J. Differential Geom. **47**:17–94, 1997.

[McM1] C. MCMULLEN. *Automorphism of rational maps.* In Holomorphic Functions and Moduli I, pp. 31–60. Springer-Verlag, 1988.

[McM2] C. MCMULLEN. *Complex dynamics and renormalization.* volume 135 of Annals of Mathematics Studies. Princeton University Press, Princeton, NJ, 1994.

[Mi1]   J. MILNOR. *Dynamics in One Complex Variable.* Vieweg 1999, 2nd ed. 2000.

[Mi2]   J. MILNOR. *Local Connectivity of Julia Sets: Expository Lectures.* In "The Mandelbrot set, Theme and Variations," ed. Tan Lei, LMS Lecture Note Series **274**, pp. 67–116. Cambridge U. Press 2000.

[Mi3]   J. MILNOR. *Non-locally connected Julia sets constructed by iterated tuning.* Birthday lecture for "Douady 70." Revised May 26, 2006, Stony Brook. http://www.math.sunysb.edu/~jack/tune-b.pdf.

[Na]   V.A. NAĬSHUL'. *Topological invariants of analytic and area-preserving mappings and their application to analytic differential equations in C^2 and CP^2.* Trans. Moscow Math. Soc. **42**: 239–250, 1983.

[PR]   C. L. PETERSEN and P. ROESCH. *Parabolic tools.* Journ. of Diff. Equat. and Appl., **16**(05–06): 715–738, 2010.

[Ro1]   P. ROESCH. *Puzzles de Yoccoz pour les applications à allure rationnelle.* L'Enseignement Mathématique, **45**: 133–168, 1999.

[Ro2]   P. ROESCH. *Some rational maps whose Julia sets are not locally connected.* Conform. Geom. Dyn. **10**: 125–135, 2006.

[Ro3]   P. ROESCH. *Hyperbolic components of polynomials with a fixed critical point of maximal order.* Ann. Sci. École Norm. Sup. (4), **40**(6): 901-949, 2007.

[Ro4]   P. ROESCH. *Local connectivity for the Julia set of rational maps: Newton's famous example.* Annals of Math. (2), **168**(1): 127–174, 2008.

[Ro5]   P. ROESCH. *Quelques pas vers une l'étude de la frontière des domaines de stabilité en itération rationnelle.* Mémoire d'habilitation à diriger des recherches, Institut de Mathématiques de Toulouse, 2009.

[RoYi] P. ROESCH and Y. YIN. *The boundary of bounded polynomial Fatou components.* C. R. Acad. Sc. Paris, **346**(15–16): 877–880, 2008.

[So]   D. E. K. SØRENSEN. *Infinitely renormalizable quadratic polynomials, with non-locally connected Julia set.* J. Geom. Anal., **10**(1): 169–206, 2000.

[TY]   TAN LEI and YIN YONGCHENG. *Local connectivity of the Julia set for geometrically finite rational maps,* Science in China (Series A) **39**(1): 39–47, 1996.

[W]   G. WHYBURN, *Analytic Topology,* AMS Colloq. Publ. **28**, 1942.

Perturbations of weakly expanding critical orbits

Genadi Levin

ABSTRACT. Let f be a polynomial or a rational function which has r summable critical points. We prove that there exists an r-dimensional manifold Λ in an appropriate space containing f such that for every smooth curve in Λ through f, the ratio between parameter and dynamical derivatives along forward iterates of at least one of these summable points tends to a non-zero number.

1 INTRODUCTION

We say that a critical point c of a rational function f is *weakly expanding, or summable*, if, for the point $v = f(c)$ of the Riemann sphere $\bar{\mathbb{C}}$,

$$\sum_{n=0}^{\infty} \frac{1 + |f^n(v)|^2}{1 + |v|^2} \frac{1}{|(f^n)'(v)|} < \infty. \tag{1.1}$$

Throughout the paper, derivatives are standard derivatives of holomorphic maps; then the summand in (1.1) is a finite number for every $v \in \bar{\mathbb{C}}$ as soon as v is not a critical point of f^n.

In the present paper, we study perturbations of polynomials and rational functions with several (possibly, not all) summable critical points. This paper is a natural continuation of [Le02], [Le11], and, partly, [Le88] and [LSY]. Let us the state main result for rational functions (for a more complete account, see Theorem 3.6). We call a rational function *exceptional* if it is *double covered by an integral torus endomorphism* (another name: *a flexible Lattès map*): these form a family of explicitly described critically finite rational maps with Julia sets the Riemann sphere, see, e.g., [DH93], [McM], [Mi06]. Two rational functions are *equivalent* if they are conjugated by a Möbius transformation.

THEOREM 1.1. *Let f be an arbitrary non-exceptional rational function of degree $d \geq 2$. Suppose that $c_1, \ldots, c_r$ is a collection of r summable critical points of f, and suppose that the union $K = \cup_{j=1}^{r} \omega(c_j)$ of their ω-limit sets is a C-compact on the Riemann sphere (see Definition 3.5). For example, it is enough that K has zero Lebesgue measure on the plane. Replacing, if necessary, f by its equivalent, one can assume that the forward orbits of $c_1, \ldots, c_r$ avoid infinity. Consider the set X_f of all rational functions of degree d which are close enough to f and have the same number p' of different critical points with the same corresponding multiplicities.*

Then there is a p'-dimensional manifold Λ_f and its r-dimensional submanifold Λ, $f \in \Lambda \subset \Lambda_f \subset X_f$, with the following properties:

(a) *every $g \in X_f$ is equivalent to some $\hat{g} \in \Lambda_f$;*

Supported in part by ISF grant 799/08

(b) *for every one-dimensional family $f_t \in \Lambda$ through f such that*

$$f_t(z) = f(z) + tu(z) + O(|t|^2) \text{ as } t \to 0,$$

if $u \neq 0$, then, for some $1 \leq j \leq r$, the limit

$$\lim_{m \to \infty} \frac{\frac{d}{dt}|_{t=0} f_t^m(c_j(t))}{(f^{m-1})'(v_j)} = \sum_{n=0}^{\infty} \frac{u(f^n(c_j))}{(f^n)'(f(c_j))} \tag{1.2}$$

exists and is a non-zero *number. Here $c_j(t)$ is the critical point of f_t, such that $c_j(0) = c_j$, and $v_j = f(c_j)$.*

Furthermore, if f and all the critical points of f are real, the preceding maps and spaces can be taken real.

If $f_t(z) = z^d + t$, the preceding limit is a *similarity factor* between the dynamical and the parameter planes; see [RL01].

Let us list some cases when the set K defined previously has zero Lebesgue measure.

1. Suppose that every critical point c of f satisfies the Misiurewicz condition, i.e., c lies in the Julia set J of f and $\omega(c)$ contains no critical points and parabolic cycles. Then $\omega(c)$ is an invariant hyperbolic set of f (by Mane's hyperbolicity theorem [M]). By the bounded distortion property, it follows that $\omega(c)$ is of measure zero. Therefore, if $c_1, \ldots, c_{p'}$ satisfy the Misiurewicz condition, the Lebesgue measure $|K|$ of K is zero.

2. If all critical points satisfy the Collet-Eckmann condition (which clearly implies the summability) and the ω-limit set of each of them is not the whole sphere, then the measure of their union is zero [PR98]. See also [PR99] for a rigidity result for Collet-Eckmann holomorphic maps.

3. If all the critical points in the Julia set J of a rational function f are summable, f has no neutral cycles, and J is not the whole sphere, then the Lebesgue measure of J is equal to zero [BvS], [RL07]. In particular, $|K| = 0$. Moreover, as in case 2, if $J = \bar{\mathbb{C}}$, it is enough to assume that the ω-limit set of each of them is not the whole sphere, and then again $|K| = 0$ (the proof follows from [RLS], as explained in [RL]).

Theorem 1.1 being applied in case 1 of the preceding list yields the following corollary first obtained in [vS]. Let f be a non-exceptional rational function of degree d. Assume that every critical point c_j of f, $1 \leq j \leq 2d - 2$, is simple and satisfies the Misiurewicz condition. Replacing, if necessary, f by its equivalent, one can assume that the set $K = \cup_{j=1}^{2d-2} \omega(c_j)$ does not contain infinity. Let f_λ be a family of rational maps of degree d which depends (complex) analytically on $\lambda \in N$, where N is a $(2d - 2)$-dimensional complex manifold, and $f_{\lambda_0} = f$. In particular, there exist analytic functions $c_j(\lambda)$, $1 \leq j \leq 2d - 2$, such that $c_j(\lambda)$ are the critical points of f_λ, $c_j(\lambda_0) = c_j$. Furthermore, since the set $K \subset \mathbb{C}$ is hyperbolic for f, it persists for f_λ, for λ in some neighborhood W of λ_0: there exists a continuous map $(\lambda, z) \in W \times K \mapsto z(\lambda) \in \mathbb{C}$ (here $z(\lambda)$ is the "holomorphic motion" of the point $z \in K$), so that $z = z(\lambda_0) \in K$, the functions $z(\lambda)$ are analytic in W, and, moreover, for each $\lambda \in W$, the points $z(\lambda)$ are different for different initial points $z = z(\lambda_0)$, and the set $K(\lambda) = \{z(\lambda)\}$ is an invariant hyperbolic set for f_λ.

COROLLARY 1.2 ([vS]). *Assume the preceding hypotheses on the map f and the family f_λ and also that no two different maps from N are equivalent. Let $a_j = f^l(c_j) \in K$, where $l > 0$, and $a_j(\lambda)$ be the holomorphic motion of the point a_j. Then, for every l big enough, the map*

$$\mathbf{h}_l : \lambda \mapsto \left\{ f_\lambda^l(c_1(\lambda)) - a_1(\lambda), \dots, f_\lambda^l(c_{2d-2}(\lambda)) - a_{2d-2}(\lambda) \right\} \tag{1.3}$$

is (locally) invertible near λ_0.

See Subsection 3.3 for the details and for asymptotics of the inverse to the derivative of $\mathbf{h}_l$.

Let us make some further comments. In [Ts], Theorem 1.1 was shown for real quadratic polynomials which satisfy the Collet-Eckmann condition (to be precise, Tsujii proves that the limit (1.2) is positive for real Collet-Eckmann maps $t - x^2$). For all quadratic polynomials with the summable critical point, Theorem 1.1 was originally proven in [Le02] (see also [Av]). In turn, Corollary 1.2 had been known before for the quadratic polynomial family in the case of a strictly preperiodic critical point [DH85]. See also [RL01] for Corollary 1.2 for the family $z^d + c$ and [BE] for a proof in the particular case of Corollary 1.2 when each critical point is strictly preperiodic.

Here we state another corollary of Theorem 1.1, which is close to the main result of [Ma]; see Comment 1.4. Recall the following terminology. Let X be a complex manifold and $f_\lambda(z)$, $\lambda \in X$, be a holomorphic family of rational maps over X. The map f_{λ_0} is called structurally (respectively, quasiconformally) stable in X if there exists a neighborhood U of λ_0 such that: (a) f_{λ_0} and f_λ are topologically (respectively, quasiconformally) conjugate for every $\lambda \in U$, and (b) the conjugacy tends to the identity map as $\lambda \to \lambda_0$. By a fundamental result of [MSS] and [McSu], the sets of structurally stable and quasiconformally stable rational maps in X coincide; moreover, the quasiconformal conjugacies form a holomorphic motion. In particular, the condition (b) of the definition follows from (a). (We will not use these results though.) In the next statement, f is not necessarily non-exceptional. Note that the set of functions X_f and the manifold Λ_f that appeared in Theorem 1.1 are defined without any changes for all maps f (including the exceptional ones); see Subsection 3.2 and Proposition 3.4 in particular.

COROLLARY 1.3 (cf. [Ma]). *Suppose that a rational function f of degree $d \geq 2$ has a summable critical point c, and its ω-limit set $\omega(c)$ is a C-compact. Then f is not structurally stable in Λ_f.*

For the proof, see Subsection 3.3.

COMMENT 1.4. In this comment, we consider the case when f has only simple critical points. Then (and only then) X_f is the (local) space of all rational functions $\mathbf{Rat}_d$ of degree d which are close to f, and Λ_f is its $(2d - 2)$-dimensional submanifold. In this case, Corollary 1.3 was proved essentially in [Ma] (Theorem A). To be more precise, in Theorem A of [Ma], explicit conditions (1)–(4) are given, so that each of them implies that f is not quasiconformally stable in the space $\mathbf{Rat}_d$. In fact, each of these conditions (2)–(4) (but not (1)) implies that $\omega(c)$ is a C-compact; this is the only fact needed to show that f is not quasiconformally stable in this part of the proof from [Ma]. On the other hand, the condition (1) of Theorem A (which says that $c \notin \omega(c)$) requires somewhat different considerations and is not covered by Corollary 1.3.

For other results about transversality, see [Le11], the preprint [Ep10], and also [Gau]. For dynamical and statistical properties of one-dimensional and rational maps under different summability conditions, see [NvS], [GS], [BvS], [RL07], and [RLS].

The main results of the paper are contained in Theorem 2.3 (plus Comment 2.4) for polynomials and in Theorem 3.6 and Theorem 1.1 (stated previously) for rational functions. See also Comment 5.7 for a generalization which takes into account non-repelling cycles, and Propositions 2.10 and 5.4, which are of an independent interest. As usual, the polynomial case is more transparent and technically easier, so we consider it separately; see Section 2. Proposition 2.5 and Lemma 2.8 of this section are more general and are also used in Section 3, where, by the same method, we treat the case of rational functions.

One of the main tools of the paper is a Ruelle (or pushforward) operator T_f associated to a rational map f (see Subsection 2.4). It was introduced to the field of complex dynamics by Thurston (see [DH93]), and has been used widely since then. A fundamental property of the operator (observed by Thurston) is that it is contracting (see Subsection 5.1). The method of the proof of Theorem 2.3 and Theorem 3.6 consists in applying the following three basic components: **(i)** explicit (formal) identities involving the Ruelle operator, which are established in [LSY] (as a formula for the resolvent of T_f and in the case when f has simple critical points — see also [Le88]); this allows us to construct explicitly an integrable fixed point of the operator T_f, assuming that the determinant of some matrix vanishes (in other words, if some vectors of the coefficients in the identities are linearly dependent); **(ii)** the contraction property of the operator together with **(i)** lead to the conclusion that the vectors of the coefficients are linearly independent. The scheme **(i)**–**(ii)** appeared in [Le02], where the unicritical case was treated, and then was applied in [Ma] for rational functions with simple critical points and in [Le11].

The last component, which is the key new part of the paper, is as follows: **(iii)** a formula for the coefficients of the identities for an arbitrary polynomial or rational function via the derivative with respect to the canonical local coordinates in some functional spaces; see Propositions 2.10–2.12 and 5.4–5.5.

COMMENT 1.5. We will consider different subsets N of polynomials or rational functions equipped by the structure of a complex-analytic l-dimensional manifold and always having the following property: nearby maps in N have the same number of distinct critical points with the same corresponding multiplicities. In other words, given $f \in N$ with the distinct critical points $c_1, \ldots, c_{q'}$ and corresponding multiplicities $m_1, \ldots, m_{q'}$, there are q' functions $c_1(g), \ldots, c_{q'}(g)$ which are defined for g in a neighborhood of f in N such that $c_j(g) \to c_j$ as $g \to f$ and $c_j(g)$ is a critical point of g with multiplicity m_j. It is easy to see then that each $c_j(g)$ is a holomorphic function of $g \in N$ (provided $c_j \in \mathbb{C}$).

Throughout the paper we use the following convention about the notations. Let $f \in N$, and $\bar{x}(g) = \{x_1(g), \ldots, x_l(g)\}$ be a (local, near f) holomorphic coordinate of g in the space N. If $P \colon W \to \mathbb{C}$ is a function which is defined and analytic in a neighborhood W of f in the space N, we denote by $\frac{\partial P}{\partial x_k}$ the partial derivative of $P(g)$ w.r.t. $x_k(g)$ calculated at the point $\bar{x}(f)$, i.e., at $g = f$. For example, $\frac{\partial c_j}{\partial x_k}$ denotes $\frac{\partial c_j(g)}{\partial x_k}$ evaluated at $g = f$. Furthermore, if $P = g^m$, we denote $\frac{\partial f^m}{\partial x_k}(z)$ to be $\frac{\partial g^m}{\partial x_k}(z)$ evaluated at $g = f$. For a rational function $g(z)$, g' always means the derivative w.r.t. $z \in \mathbb{C}$. For instance, $\frac{\partial f'}{\partial v_k}$ means $\frac{\partial(\partial g/\partial z)}{\partial v_k}$ calculated at the point f. Note also that for a critical point $c(g) = c_j(g)$ of g, we have

$$\frac{\partial\big(g^m(c(g))\big)}{\partial x_k}\bigg|_{g=f} = \frac{\partial f^m}{\partial x_k}(c) \quad \text{where } c = c(f).$$

Acknowledgements. The paper was inspired by a recent question by Weixiao Shen to the author about a generalization of Corollary 1(b) of [Le02] to higher-degree polynomials. The answer is contained in Theorem 2.3. In turn, it has been used recently in [GaSh]. The author thanks Weixiao Shen for the above question and discussions, Feliks Przytycki for discussions, and Juan Rivera-Letelier for a few very helpful comments and for the reference [Gau]. Finally, the author thanks the referee for many comments that helped to improve the exposition.

2 POLYNOMIALS

2.1 Polynomial spaces

Let f be a monic centered polynomial of degree $d \geq 2$, i.e., it has the form

$$f(z) = z^d + a_1 z^{d-2} + \ldots + a_{d-1}.$$

Consider the space Π_d of all monic and centered polynomials of the same degree d. The vector of coefficients of $g \in \Pi_d$ defines a (global) coordinate in Π_d and identifies Π_d with $\mathbb{C}^{d-1}$.

Let $C = \{c_1, \ldots, c_p\}$ be the set of all *different* critical points of f, with the vector of multiplicities $\bar{p} = \{m_1, m_2, \ldots, m_p\}$, so that $f'(z) = d\Pi_{i=1}^{p}(z - c_i)^{m_i}$. Now, we consider a local subspace $\Pi_{d,\bar{p}}$ of Π_d associated to the polynomial f, as in [Le11]:

DEFINITION 2.1. *The space $\Pi_{d,\bar{p}}$ consists of all g from a neighborhood of f in Π_d with the same number p of different critical points $c_1(g), \ldots, c_p(g)$ and the same vector of multiplicities $\bar{p}$. Here $c_i(g)$ is close to $c_i = c_i(f)$ and has the multiplicity m_i, $1 \leq i \leq p$.*

In particular, $\Pi_{d,\overline{d-1}}$ consists of all monic centered polynomials of degree d close to f if and only if all the critical points of f are simple. At the other extreme case, the space $\Pi_{d,\bar{1}}$ consists of the unicritical family $z^d + v$.

Note that some of the critical values $v_1, \ldots, v_p$ of f may coincide.

Consider the vector of critical values $V(g) = \{v_1(g), \ldots, v_p(g)\}$, in which $v_i(g) = g(c_i(g))$. The set $\Pi_{d,\bar{p}}$ is an analytic subset of Π_d. The following fact is proved in Proposition 1 of [Le11]:

PROPOSITION 2.2. *$\Pi_{d,\bar{p}}$ is a p-dimensional complex analytic manifold, and the vector $V(g)$ is a local analytic coordinate in $\Pi_{d,\bar{p}}$.*

2.2 Main result

THEOREM 2.3. (a) *Let c be a weakly expanding critical point of f and $v = f(c)$. Then, for every $k = 1, \ldots, p$, the following limit exists:*

$$L(c, v_k) := \lim_{m \to \infty} \frac{\frac{\partial(f^m(c))}{\partial v_k}}{(f^{m-1})'(v)}. \tag{2.1}$$

(b) *Suppose that $c_1, \ldots, c_r$ are pairwise different weakly expanding critical points of f. Then the rank of the matrix*

$$\mathbf{L} = \left(L(c_j, v_k) \right)_{1 \leq j \leq r, 1 \leq k \leq p} \tag{2.2}$$

is equal to r, i.e., maximal.

COMMENT 2.4. Part (b) has the following geometric re-formulation. Denote by $1 \leq k_1 < \ldots < k_r \leq p$ the indexes for which the determinant of the square matrix $\left(L(c_j, v_{k_i}) \right)_{1 \leq j \leq r, 1 \leq i \leq r}$ is non-zero. We define a local r-dimensional submanifold Λ as the set of all $g \in \Pi_{d,\bar{p}}$ such that $v_i(g) = 0$ for every $i \neq k_1, k_2, \ldots, k_r$. Now consider a 1-dimensional family (curve) of maps $f_t \in \Lambda$ passing through f, such that $f_t(z) = f(z) + tu(z) + O(|t|^2)$ as $t \to 0$. If $u \neq 0$, then, for at least one weakly expanding critical point c_j of f,

$$\lim_{m \to \infty} \frac{\frac{d}{dt}|_{t=0} f_t^m(c_j(t))}{(f^{m-1})'(v_j)} \neq 0. \tag{2.3}$$

Here $c_j(t)$ is a holomorphic function with $c_j(0) = c_j$, so that $c_j(t)$ is a critical point of f_t, and $v_j = f(c_j)$.

In particular, if all critical points of f are *simple* and *weakly expanding*, then (2.3) holds for every curve f_t in Π_d through f with a non-degenerate tangent vector at f.

Furthermore, if $f(0)$ and all the critical points of f are real, the preceding maps and spaces can be taken as real.

Indeed, let $v_k(t)$ be the critical value of f_t, such that $v_k(0) = v_k$. Let $a_k = v_k'(0)$. As $f_t \in \Lambda$, $u(z) = \sum_{i=1}^r a_{k_i} \frac{\partial f}{\partial v_{k_i}}(z)$, where at least one of a_{k_i} must be non-zero. On the other hand, the limit R_j in (2.3) can be represented as $R_j = \sum_{k=1}^p a_k L(c_j, v_k)$. Therefore, $R_j = \sum_{i=1}^r a_{k_i} L(c_j, v_{k_i})$, where at least one of a_{k_i} is not zero. Now, if we assume that $R_j = 0$ for every $1 \leq j \leq r$, then the matrix $\left(L(c_j, v_{k_i}) \right)_{1 \leq j \leq r, 1 \leq i \leq r}$ degenerates, which is a contradiction.

2.3 Proof of Part (a) of Theorem 2.3

DEPENDENCE ON THE LOCAL COORDINATES

Suppose that a polynomial or a rational function f is included in a space N of polynomials or rational functions g with coordinates $\bar{x}(g) = \{x_1(g), \ldots, x_l(g)\}$, see Comment 1.5. Fix a critical point $c \in \mathbb{C}$ of f of multiplicity $m \geq 1$, and let $c(g)$ be the critical point of g of the multiplicity m which is close to c, if g is close to f. Consider also the corresponding critical value $v(g) = g(c(g))$. Recall that, according to our convention, $\frac{\partial f}{\partial x_k}$, $\frac{\partial f(a)}{\partial x_k}$, $\frac{\partial v}{\partial x_k}$, etc., mean, respectively, $\frac{\partial g}{\partial x_k}|_{g=f}$, $\frac{\partial g(a)}{\partial x_k}|_{g=f}$, $\frac{\partial v(g)}{\partial x_k}|_{g=f}$, etc.

PROPOSITION 2.5. *Assume that c and $v = f(c)$ lie in $\mathbb{C}$. Then the function*

$$\frac{\partial f}{\partial x_k}(z) - \frac{\partial v}{\partial x_k} \tag{2.4}$$

has at $z = c$ a zero of multiplicity at least m.

PROOF. Fix a point $a \in \mathbb{C}$ so that $f(a) \neq \infty$. For every g in the space N which is close to f, we may write

$$g(z) = g(a) + \int_a^z g'(w) dw, \tag{2.5}$$

and this holds for every z in the plane with $g(z) \neq \infty$. Hence,

$$\frac{\partial f}{\partial x_k}(z) = \frac{\partial f(a)}{\partial x_k} + \int_a^z \frac{\partial f'}{\partial x_k}(w) dw. \tag{2.6}$$

On the other hand,

$$v(g) = g(a) + \int_a^{c(g)} g'(w) dw. \tag{2.7}$$

Therefore,

$$\frac{\partial v}{\partial x_k} = \frac{\partial f(a)}{\partial x_k} + \frac{\partial c}{\partial x_k} f'(c) + \int_a^c \frac{\partial f'}{\partial x_k}(w) dw = \frac{\partial f(a)}{\partial x_k} + \int_a^c \frac{\partial f'}{\partial x_k}(w) dw. \tag{2.8}$$

Comparing this with (2.6), we have

$$\frac{\partial f}{\partial x_k}(z) - \frac{\partial v}{\partial x_k} = \int_c^z \frac{\partial f'}{\partial x_k}(w) dw. \tag{2.9}$$

As $c(g)$ is an m-multiple root of g', we get

$$\frac{\partial f'}{\partial x_k} = (z - c)^{m-1} r(z),$$

where $r(z)$ is a holomorphic function near c. Hence, as $z \to c$,

$$\frac{\partial f}{\partial x_k}(z) - \frac{\partial v}{\partial x_k} = (z - c)^m \frac{r(c)}{m} + O(z - c)^{m+1}. \tag{2.10}$$

This proves the statement. □

Now, we let $N = \Pi_{d,\bar{p}}$ and calculate the partial derivatives of a function $f \in \Pi_{d,\bar{p}}$ with respect to the local coordinates.

PROPOSITION 2.6. *For every $k = 1, \ldots, p$, the function $\frac{\partial f}{\partial v_k}(z)$ is a polynomial $p_k(z)$ of degree at most $d - 2$ which is uniquely characterized by the following condition: $p_k(z) - 1$ has zero at c_k of order at least m_k, while for every $j \neq k$, $p_k(z)$ has zero at c_j of order at least m_j. In particular, $\frac{\partial f}{\partial v_k}(c_j) = \delta_{j,k}$ (here and later on we use the notation $\delta_{j,k} = 1$ if $j = k$ and $\delta_{j,k} = 0$ if $j \neq k$); if c_k is simple (i.e., $m_k = 1$), then*

$$\frac{\partial f}{\partial v_k}(z) = \frac{f'(z)}{f''(c_k)(z - c_k)}. \tag{2.11}$$

PROOF. Since the coefficients of $g \in \Pi_{d,\bar{p}}$ are holomorphic functions of $V(g)$ and g is centered, the function $\frac{\partial f}{\partial v_k}(z)$ is indeed a polynomial in z of degree at most $d - 2$. Hence, it is enough to check that it satisfies the characteristic property of the polynomial $p_k(z)$. But since $\frac{\partial v_j(g)}{\partial v_k} = \delta_{j,k}$, this is a direct corollary of Proposition 2.5 applied for the coordinate $\bar{x} = V$ and the critical point $c(g) = c_j(g)$. □

PROOF OF THEOREM 2.3, PART(A). The following identity is easy to verify:

$$\frac{\partial f^m}{\partial v_k}(z) = (f^m)'(z) \sum_{n=0}^{m-1} \frac{\frac{\partial f}{\partial v_k}(f^n(z))}{(f^{n+1})'(z)}. \tag{2.12}$$

Letting $z \to c_j$ here, one gets

$$\frac{\partial f^m}{\partial v_k}(c_j) = (f^{m-1})'(v_j) \left\{ \frac{\partial f}{\partial v_k}(c_j) + \sum_{n=1}^{m-1} \frac{\frac{\partial f}{\partial v_k}(f^{n-1}(v_j))}{(f^n)'(v_j)} \right\}. \tag{2.13}$$

As we know, $\frac{\partial f}{\partial v_k}(c_j) = \delta_{j,k}$. Besides, $\frac{\partial f}{\partial v_k}(z) = f'(z)L_k(z) = p_k(z)$ is a polynomial of degree at most $d - 2$. Hence, for some constant C_k and all z,

$$\left| \frac{\partial f}{\partial v_k}(z) \right| \leq C_k(1 + |z|^{d-2}). \tag{2.14}$$

Now, assume that c_j is weakly expanding. As $c_j \in J$, the sequence $\{f^n(v_j)\}_{n \geq 0}$ is uniformly bounded. Then (2.14) and the summability condition (1.1) imply that the series $\sum_{n=1}^{\infty} \frac{\frac{\partial f}{\partial v_k}(f^{n-1}(v_j))}{(f^n)'(v_j)}$ converges absolutely. Thus we have

$$L(c_j, v_k) = \lim_{m \to \infty} \frac{\frac{\partial f^m}{\partial v_k}(c_j)}{(f^{m-1})'(v_j)} = \delta_{j,k} + \sum_{n=1}^{\infty} \frac{\frac{\partial f}{\partial v_k}(f^{n-1}(v_j))}{(f^n)'(v_j)}. \tag{2.15}$$

This ends the proof of Part (a) of Theorem 2.3. $\square$

The following corollary is immediate from (2.15).

COROLLARY 2.7.
$$L(c_j, v_k) = \delta_{j,k} + A(v_j, v_k), \tag{2.16}$$

for some function $A(x, y)$.

2.4   The Ruelle operator and an operator identity

THE OPERATOR

As in the proof of Corollary 1(b) of [Le02], the main tool for us is the following linear operator T_f associated with a rational function f, which acts on functions as follows:

$$T_f \psi(x) = \sum_{w: f(w)=x} \frac{\psi(w)}{(f'(w))^2},$$

provided x is not a critical value of f.

The next statement is about an arbitrary rational function which fixes infinity.

LEMMA 2.8. *Let f be any rational function so that $f(\infty) = \infty$. Let c_j, $j = 1, \ldots, p$ be all geometrically different critical points of f lying in the complex plane $\mathbb{C}$, such that the corresponding critical values $v_j = f(c_j)$, $j = 1, \ldots, p$ are also in $\mathbb{C}$. Denote by m_j the multiplicity of c_j, $j = 1, \ldots, p$. Then there are functions $L_1(z), \ldots, L_p(z)$ as follows. For* *every $z \in \mathbb{C}$ which is not a critical point of f, and for every $x \in \mathbb{C}$ which is not a critical value of f, and such that $x \neq f(z)$, we have*

$$T_f \frac{1}{z-x} := \sum_{y:f(y)=x} \frac{1}{f'(y)^2} \frac{1}{z-y} = \frac{1}{f'(z)} \frac{1}{f(z)-x} + \sum_{j=1}^{p} \frac{L_j(z)}{x-v_j}. \tag{2.17}$$

Furthermore, each function L_j obeys the following two properties:

(1) L_j *is a meromorphic function in the complex plane of the form*

$$L_j(z) = \sum_{i=1}^{m_j} \frac{q_{m_j-i}^{(j)}}{(z-c_j)^i}, \qquad q_{m_j-i}^{(j)} \in \mathbb{C}, \tag{2.18}$$

(2) *for every $k = 1, \ldots, p$, the function $f'(z)L_j(z) - \delta_{j,k}$ has zero at the point c_k of order at least m_j.*

PROOF. Fixing z, x as in the lemma, take R big enough and consider the integral

$$I = \frac{1}{2\pi i} \int_{|w|=R} \frac{dw}{f'(w)(f(w)-x)(w-z)}.$$

As the integrand is $O(1/w^2)$ at infinity, $I = 0$. On the other hand, applying the residue theorem,

$$I = -T_f \frac{1}{z-x} + \frac{1}{f'(z)} \frac{1}{f(z)-x} + \sum_{j=1}^{p} I_j(z,x).$$

Here

$$I_j(z,x) = \frac{1}{2\pi i} \int_{|w-c_j|=\epsilon} \frac{dw}{f'(w)(f(w)-x)(w-z)}.$$

Near $c = c_j$, $f'(w) = (w-c)^m r(w)$, where $m = m_j$ and $r = r_j$ is holomorphic with $r(c) \neq 0$. Denote

$$\frac{1}{r(w)} = \sum_{k=0}^{\infty} q_k(w-c)^k,$$

where $q_k = q_k^{(j)}$ and $q_0 = q_0^{(j)} = 1/r(c) \neq 0$. We can write

$$\frac{1}{f'(w)(f(w)-x)(w-z)}$$

$$= \frac{1}{(w-c)^m r(w)((f(c)-x) + O((w-c)^{m+1}))(w-z)}$$

$$= \frac{1}{r(w)(f(c)-x)(w-z)} \frac{1}{(w-c)^m} + O(w-c)$$

$$= \frac{1}{x-f(c)} \sum_{k=0}^{\infty} q_k(w-c)^{k-m} \sum_{n=0}^{\infty} \frac{(w-c)^n}{(z-c)^{n+1}} + O(w-c).$$

We see from here that $I_j(z,x) = L_j(z)/(x - f(c_j))$, where $L_j(z)$ has precisely the form (2.18). Now, consider $\tilde{L}(z) = f'(z)L_j(z)$. By (2.18), $\tilde{L}(z)$ has zero at every $c_k \neq c_j$ of order at least m_k. On the other hand, as $z \to c_j$, then

$$f'(z)L_j(z) = r_j(z)\left[q_0^{(j)} + q_1^{(j)}(z - c_j) + \ldots + q_{m_j-1}^{(j)}(z - c_j)^{m_j-1}\right]$$

$$= r_j(z)\left[\frac{1}{r_j(z)} - \sum_{k=m_j}^{\infty} q_k^{(j)}(z - c_j)^k\right] = 1 - (z - c_j)^{m_j}g(z),$$

where g is holomorphic near c_j. This finishes the proof of the property (2). □

As a simple corollary of the last two statements we have the following.

PROPOSITION 2.9. *Let $f \in \Pi_{d,\bar{p}}$. Then*

$$T_f \frac{1}{z - x} = \frac{1}{f'(z)}\frac{1}{f(z) - x} + \sum_{k=1}^{p}\frac{\frac{\partial f}{\partial v_k}(z)}{f'(z)}\frac{1}{x - v_k}. \tag{2.19}$$

PROOF. By property (1) of Lemma 2.8, we know $f'(z)L_k(z)$ is a polynomial of degree at most $d - 2$, which, by property (2) of Lemma 2.8, coincides with the polynomial $p_k(z) = \frac{\partial f}{\partial v_k}(z)$ introduced in Proposition 2.6. Therefore, indeed, we have $L_k(z) = \frac{\frac{\partial f}{\partial v_k}(z)}{f'(z)}$. □

THE OPERATOR IDENTITY AND ITS COROLLARY

PROPOSITION 2.10. *Let $f \in \Pi_{d,\bar{p}}$. We have (in formal series)*

$$\varphi_{z,\lambda}(x) - \lambda(T_f\varphi_{z,\lambda})(x) = \frac{1}{z - x} + \lambda\sum_{k=1}^{p}\frac{1}{v_k - x}\Phi_k(\lambda,z), \tag{2.20}$$

where

$$\varphi_{z,\lambda}(x) = \sum_{n=0}^{\infty}\frac{\lambda^n}{(f^n)'(z)}\frac{1}{f^n(z) - x}, \tag{2.21}$$

and

$$\Phi_k(\lambda,z) = \sum_{n=0}^{\infty}\lambda^{n+1}\frac{\frac{\partial f}{\partial v_k}(f^n(z))}{(f^{n+1})'(z)}, \tag{2.22}$$

for λ complex parameter and z, x complex variables.

PROOF. We use Proposition 2.9 and write (in formal series)

$\varphi_{z,\lambda}(x) - \lambda(T_f\varphi_{z,\lambda})(x)$

$$= \sum_{n=0}^{\infty}\frac{\lambda^n}{(f^n)'(z)}\frac{1}{f^n(z) - x}$$

$$- \lambda\sum_{n=0}^{\infty}\frac{\lambda^n}{(f^n)'(z)}\left\{\frac{1}{f'(f^n(z))}\frac{1}{f^{n+1}(z) - x} + \sum_{k=1}^{p}\frac{\frac{\partial f}{\partial v_k}(f^n(v_j))}{f'(f^n(v_j))}\frac{1}{x - v_k}\right\}$$

$$= \frac{1}{z - x} + \sum_{k=1}^{p}\frac{1}{v_k - x}\sum_{n=0}^{\infty}\lambda^{n+1}\frac{\frac{\partial f}{\partial v_k}(f^n(v_j))}{(f^{n+1})'(v_j)}.$$

□

COMMENT 2.11. If c_k is a simple critical point, $\Phi_k(\lambda, z) = \frac{\lambda}{f''(c_k)} \varphi_{z,\lambda}(c_k)$, and with $\Phi_k(\lambda, z)$ in such form, equations (2.20) and (5.10) from Proposition 5.4 of the next section appeared for the first time in [LSY] (where they were written via the resolvent $(1 - \lambda T_f)^{-1}$). The only new (and crucial) ingredient of Proposition 2.10 (as well as Proposition 5.4) is the representation of the coefficients $\Phi_k(\lambda, z)$ in (2.20) via derivatives with respect to the local coordinates in the appropriate space of maps.

We get Proposition 2.12 by setting $\lambda = 1$ and $z = v_j$ in Proposition 2.10 and combining it with (2.15); see the proof of Theorem 2.3(a) in Section 2.3.

PROPOSITION 2.12. *Let c_j be a summable critical point of $f \in \Pi_{d,\bar{p}}$. Then*

$$H_j(x) - (T_f H_j)(x) = \sum_{k=1}^{p} \frac{L(c_j, v_k)}{v_k - x}, \tag{2.23}$$

where

$$H_j(x) = \sum_{n=0}^{\infty} \frac{1}{(f^n)'(v_j)(f^n(v_j) - x)}. \tag{2.24}$$

2.5   Proof of Part (b) of Theorem 2.3

The proof is very similar to the one of Corollary 1(b) of [Le02], where the family of unicritical polynomials is considered (see also [Ma], [Le11] and references therein). But some additional considerations are needed if the number of summable critical points is more than 1 and their critical values coincide. Denote by S the set of indexes of given collection of r summable critical points of f. Assume the contrary, i.e., the rank of the matrix $\mathbf{L}$ is less than r. This holds if and only if there exist numbers a_j, for $j \in S$, which are not all zeros, such that, for every $1 \leq k \leq p$,

$$\sum_{j \in S} a_j L(c_j, v_k) = 0. \tag{2.25}$$

Then, by Proposition 2.12,

$$H = \sum_{j \in S} a_j H_j \tag{2.26}$$

is an integrable fixed point of T_f which is holomorphic in each component of the complement $\mathbb{C} \setminus J$. Let us show that $H = 0$ off J. We use that T_f is weakly contracting. Consider a component Ω of $\mathbb{C} \setminus J$. If Ω is not a Siegel disk, then considering the backward orbit $\cup_{n \geq 0} f^{-n}(\Omega)$, it is easy to find its open subset U such that $f^{-1}(U) \subset U$ and $U \setminus f^{-1}(U)$ is a non-empty open subset of Ω. Since $H = T_f H$, we then have (the integration is against the Lebesgue measure on the plane)

$$\int_U |H(x)| d\sigma_x = \int_U \left| \sum_{f(w)=x} \frac{H(w)}{f'(w)^2} \right| d\sigma_x$$

$$\leq \int_U \sum_{f(w)=x} \frac{|H(w)|}{|f'(w)|^2} d\sigma_x = \int_{f^{-1}(U)} |H(x)| d\sigma_x,$$

which is possible only if $H = 0$ in $U \setminus f^{-1}(U)$, hence, in Ω, because $U \subset \Omega$. And if Ω is a Siegel disk, we proceed as in the proof of Corollary 1(b) on p. 190 of [Le02] to show that $H = 0$ in Ω as well (see also Lemma 5.2). Thus, $H = 0$ off J. On the other hand, H can be represented as

$$H(x) = \sum_{k=0}^{\infty} \frac{\alpha_k}{x - b_k}, \tag{2.27}$$

where the points $b_k \in J$ are pairwise different and $|\alpha_k| < \infty$. Consider a measure with compact support $\mu = \sum_{k=0}^{\infty} \alpha_k \delta(b_k)$, where $\delta(z)$ is the Dirac measure at the point z. Then $H = 0$ off J implies that the measure μ annihilates any function which is holomorphic in a neighborhood of J. Indeed, for any such function $r(z)$ and an appropriate contour γ enclosing J,

$$\int r(z) d\mu(z) = \int \left(\frac{1}{2\pi i} \int_{\gamma} \frac{r(w)}{w - z} dw \right) d\mu(z)$$

$$= \frac{1}{2\pi i} \int_{\gamma} r(w) \left(\int \frac{d\mu(z)}{w - z} \right) dw = \frac{1}{2\pi i} \int_{\gamma} H(w) r(w) dw = 0.$$

As every point of J belongs also to the boundary of the basin of infinity, by a corollary from Vitushkin's theorem (see, e.g., [Ga]), every continuous function on J is uniformly approximated by rational functions. It follows that $\mu = 0$; in other words, the representation (2.27) is trivial:

$$\alpha_k = 0, \quad k = 0, 1, 2, \ldots. \tag{2.28}$$

That is to say, the left-hand side of (2.26) is zero:

$$\sum_{j \in S} a_j H_j = 0. \tag{2.29}$$

But, if the number of indexes in S is bigger than one, it does not imply immediately that all the numbers a_j in (2.29) must vanish. Indeed, some H_j can even coincide: by the definition of the function H_j (see Proposition 2.12), $H_{j_1} = H_{j_2}$ if $v_{j_1} = v_{j_2}$. So, we will use (2.25) along with (2.29) to prove Lemma 2.13.

LEMMA 2.13. $a_k = 0$, for $k \in S$.

PROOF. As the first step, we show that *different* functions H_j are linearly independent. Let us denote by V_i, $1 \leq i \leq q$, all the *different* critical values of f. For $1 \leq i \leq q$, introduce $\tilde{H}_i$ to be H_j, for every j, such that $v_j = V_i$. The set of indexes S is a disjoint union of subsets S_i, $1 \leq i \leq e$, so that $j \in S_i$ if and only if $v_j = V_i$. Then

$$0 = \sum_{i=1}^{e} \tilde{a}_i \tilde{H}_i, \quad \tilde{a}_i = \sum_{j \in S_i} a_j. \tag{2.30}$$

We show that this representation is trivial: $\tilde{a}_i = 0$, $1 \leq i \leq e$. By Proposition 2.12,

$$\tilde{H}_i(x) = \sum_{n=0}^{\infty} \frac{1}{(f^n)'(V_i)(f^n(V_i) - x)}, \tag{2.31}$$

where $\{V_i\}$ are pairwise different critical values of f. By contradiction, assume that some $\tilde{a}_i \neq 0$. Without loss of generality, one can assume further that $i = 1$, i.e., $\tilde{a}_1 \neq 0$.

Claim: V_1 is a point of a periodic orbit P of f, and, moreover, if P contains some $f^k(V_j)$, $k \geq 0$, with $\tilde{a}_j \neq 0$, then $V_j \in P$. By (2.30), the residue 1 of the first term in the sum (2.31) (with $i = 1$) must be cancelled. Hence, there is a critical value V_{i_1}, so that $V_1 = f^{k_1}(V_{i_1})$ with $\tilde{a}_{i_1} \neq 0$, for some minimal $k_1 > 0$. If $i_1 \neq 1$ here, then, by the same reason, there is V_{i_2} so that $V_{i_1} = f^{k_2}(V_{i_2})$ with $\tilde{a}_{i_2} \neq 0$, for some minimal $k_2 > 0$. We continue until we arrive at a critical value that has been met before. This proves the claim.

Denote by $V_{i_0} = V_1, V_{i_1}, \ldots, V_{i_s}$ all the different critical values which are points of the periodic orbit P and such that $\tilde{a}_{i_l} \neq 0$, $0 \leq l \leq s$. Since V_{i_l} are periodic points, each $\tilde{H}_{i_l}$, $0 \leq l \leq s$, is a rational function (see also (2.35)). We have

$$\sum_{l=0}^{s} \tilde{a}_{i_l} \tilde{H}_{i_l} = 0, \tag{2.32}$$

where every $\tilde{a}_{i_l}$ is non-zero. Denote by T the (minimal) period of P. Changing indexes, if necessary, there are some $0 < n_1 < n_2 < \ldots < n_s < T$, such that

$$f^{n_l}(V_1) = V_{i_l}, \quad 1 \leq l \leq s. \tag{2.33}$$

The rest is a direct, not difficult, calculation. From (2.31) and (2.33), for $1 \leq l \leq s$,

$$\tilde{H}_{i_l}(x) = (f^{n_l})'(V_1)\left(\tilde{H}_1(x) - \sum_{n=0}^{n_l-1} \frac{1}{(f^n)'(V_1)(f^n(V_1) - x)}\right), \tag{2.34}$$

while, if $\rho = (f^T)'(V_1)$,

$$\tilde{H}_1(x) = \frac{1}{1 - \rho^{-1}} \sum_{n=0}^{T-1} \frac{1}{(f^n)'(V_1)(f^n(V_1) - x)}. \tag{2.35}$$

(Note that $|\rho| > 1$, because V_1 is summable.) If $s = 0$ (i.e., V_1 is the only critical value $V_i \in P$ such that $\tilde{a}_i \neq 0$), then $\tilde{H}_1 = 0$, which is a contradiction with (2.35). Assume $s > 0$. By (2.34), the relation (2.32) turns into

$$\left(\tilde{a}_1 + \sum_{l=1}^{s} \tilde{a}_{i_l}(f^{n_l})'(V_1)\right)\tilde{H}_1(x) = \sum_{l=1}^{s} \tilde{a}_{i_l}(f^{n_l})'(V_1)\sum_{n=0}^{n_l-1} \frac{1}{(f^n)'(V_1)(f^n(V_1) - x)}, \tag{2.36}$$

where $\tilde{H}_1$ is given by (2.35). Since $0 \leq n_1 - 1 < \ldots < n_s - 1 \leq T - 2$ and the points $\{f^n(V_1)\}_{n=0}^{T-1}$ are pairwise different, the function on the right-hand side of (2.36) has no pole at $f^{T-1}(V_1)$ while $\tilde{H}_1$ has. Hence, $\tilde{a}_1 + \sum_{l=1}^{s} \tilde{a}_{i_l}(f^{n_l})'(V_1) = 0$. It means that the left-hand side in (2.36) is identically zero. Hence, the function in the right-hand side of (2.36) has no pole at $f^{n_s-1}(V_1)$, which is possible only if $\tilde{a}_{i_s} = 0$. This is already a contradiction, because, by our assumption, all $\tilde{a}_{i_l}$ are non-zero numbers.

The contradiction shows that $\tilde{a}_i = 0$, for $1 \leq i \leq e$. To get from this that all $a_k = 0$, we use Corollary 2.7 and rewrite (2.25) for every $k \in S$:

$$
\begin{aligned}
0 = \sum_{j \in S} a_j L(c_j, v_k) &= \sum_{i=1}^{e} \sum_{j \in S_i} a_j \left(\delta_{j,k} + A(V_i, v_k) \right) \\
&= \sum_{i=1}^{e} \left(\sum_{j \in S_i} a_j \delta_{j,k} + A(V_i, v_k) \sum_{j \in S_i} a_j \right) \\
&= \sum_{i=1}^{e} \left(\sum_{j \in S_i} a_j \delta_{j,k} + A(V_i, v_k) \tilde{a}_i \right) \\
&= \sum_{i=1}^{e} \sum_{j \in S_i} a_j \delta_{j,k} = \sum_{j \in S} a_j \delta_{j,k} = a_k. \quad \Box
\end{aligned}
$$

This ends the proof of Theorem 2.3(b), because this is a contradiction with the fact that at least one a_j, $j \in S$, is a non-zero number.

3   RATIONAL FUNCTIONS

3.1   Local spaces of rational maps

Let f be a rational function of degree $d \geq 2$ such that

$$
f(z) = \sigma z + b + \frac{P(z)}{Q(z)}, \tag{3.1}
$$

where $\sigma \neq 0, \infty$, and Q, P are polynomials of degrees $d - 1$ and at most $d - 2$, respectively, and which have no common roots. Without loss of generality, one can assume that $Q(z) = z^{d-1} + A_1 z^{d-2} + \ldots + A_{d-1}$ and $P(z) = B_0 z^{d-2} + \ldots + B_{d-2}$.

Let p' stand for the number of distinct critical points $c_1, \ldots, c_{p'}$ of f. Note that all of them are different from ∞. Denote by m_j the multiplicity of c_j, that is, the equation $f(w) = z$ has precisely $m_j + 1$ different solutions near c_j for z near $f(c_j)$ and $z \neq f(c_j)$, $j = 1, \ldots, p'$. Observe that it is equivalent to say that if $f = \frac{\hat{P}}{Q}$, where $\hat{P}(z) = Q(z)(\sigma z + b) + P(z)$, then c_j, for $1 \leq j \leq p'$, is a root of the polynomial $\hat{P}'Q - \hat{P}Q'$ of multiplicity m_j. Note that $\sum_{j=1}^{p'} m_j = 2d - 2$.

Let $\bar{p}' = \{m_j\}_{j=1}^{p'}$ denote the vector of multiplicities. Let $v_j = f(c_j)$, $1 \leq j \leq p'$, be the corresponding critical values. We assume that some of them can coincide as well as some can be ∞. By p we denote the number of critical points of f, so that their images (i.e., the corresponding critical values) avoid infinity.

As in [Le11], we define a local (near f) space $\Lambda_{d, \bar{p}'}$ of rational functions.

DEFINITION 3.1.   *A rational function g of degree d belongs to $\Lambda_{d, \bar{p}'}$ if and only if it has the form*

$$
g(z) = \sigma(g)z + b(g) + \frac{P_g(z)}{Q_g(z)}, \tag{3.2}
$$

where the numbers $\sigma(g)$, $b(g)$, and the polynomials

$$
Q_g = z^{d-1} + A_1(g)z^{d-2} + \ldots + A_{d-1}(g) \text{ and } P_g(z) = B_0(g)z^{d-2} + \ldots + B_{d-2}(g)
$$

are close to σ, b, Q, *and* P, *respectively. Moreover,* g *has* p' *different critical points* $c_1(g), \ldots, c_{p'}(g)$ *so that* $c_j(g)$ *is close to* c_j *and has the same multiplicity* m_j, $1 \leq j \leq p'$.

For g in a sufficiently small neighborhood of f in $\Lambda_{d,\bar{p}'}$, introduce a vector $\bar{v}(g) \in \mathbb{C}^{p'+2}$ as follows. Let us fix an order $c_1, \ldots, c_{p'}$ in the collection of all critical points of f. We will do it in such a way that the first p indexes correspond to finite critical values, i.e., $v_j \neq \infty$ for $1 \leq j \leq p$ and $v_j = \infty$ for $p < j \leq p'$ (if $p < p'$). Remark that $p \geq 1$: there always at least one critical value in the plane. There exist p' functions $c_1(g), \ldots, c_{p'}(g)$ which are defined and continuous in a small neighborhood of f in $\Lambda_{d,\bar{p}'}$ such that they constitute all different critical points of g. Moreover, $c_j(g)$ has the multiplicity m_j, $1 \leq j \leq p'$. Define now the vector $\bar{v}(g)$. If all critical values of f lie in the plane, then we set $\bar{v}(g) = \{\sigma(g), b(g), v_1(g), \ldots, v_{p'}(g)\}$. If some of the critical values v_j of f are infinity, that is, $v_j = \infty$ for $p < j \leq p'$, then we replace $v_j(g)$ by $1/v_j(g)$ in the definition of $\bar{v}(g)$:

$$\bar{v}(g) = \left\{ \sigma(g), b(g), v_1(g), \ldots, v_p(g), \frac{1}{v_{p+1}(g)}, \ldots, \frac{1}{v_{p'}(g)} \right\}.$$

In particular, $\bar{v} = \bar{v}(f) = \{\sigma, b, v_1, \ldots, v_p, 0, \ldots, 0\}$.

We can identify $g \in \Lambda_{d,\bar{p}'}$ as before with the point

$$\bar{g} = \{\sigma(g), b(g), A_1(g), \ldots, A_{d-1}(g), B_0(g), \ldots, B_{d-2}(g)\}$$

of $\mathbb{C}^{2d}$. Then $\Lambda_{d,\bar{p}'}$ is an analytic variaty in $\mathbb{C}^{2d}$. We again denote it by $\Lambda_{d,\bar{p}'}$. Every critical point c_j of f of multiplicity m_j is determined by $m_j - 1$ algebraic equations. Therefore, $\Lambda_{d,\bar{p}'}$ is defined by $\sum_{j=1}^{p'}(m_j - 1) = 2d - 2 - p'$ equations corresponding to the critical points. Thus the (complex) dimension of $\Lambda_{d,\bar{p}'}$ is at least $2d - \sum_{j=1}^{p'}(m_j - 1) = p' + 2$. In fact, the following is proved in [Le11], Sect. 7.

PROPOSITION 3.2. $\Lambda_{d,\bar{p}'}$ *is a complex-analytic manifold of dimension* $p' + 2$, *and* $\bar{v}(g)$ *defines a local holomorphic coordinate of* $g \in \Lambda_{d,\bar{p}'}$.

Two rational functions are called close if they are uniformly close in the Riemann metric on the sphere. We call two rational functions (*M*-)*equivalent* if there is a Möbius transformation M which conjugates them. Every rational function f is equivalent to some $\tilde{f}$ of degree $d \geq 2$ be of the form (3.1). Indeed, f has either a repelling fixed point or a fixed point with the multiplier 1 (see, e.g., [Mi91]). Hence, there exists a Möbius transformation P such that ∞ is a fixed non-attracting point of $\tilde{f} = P \circ f \circ P^{-1}$. See also Section 3.2.

If a critical value of f or its iterate is infinity, we consider also another space of rational maps which is biholomorphic to $\Lambda_{d,\bar{p}'}$. Let us fix a Möbius transformation M, such that $\alpha = M(\infty)$ lies outside of the critical orbits $\{f^n(c_k) : n \geq 0, 1 \leq k \leq p'\}$ of f. Then we make the same change of variable M for all maps from $\Lambda_{d,\bar{p}'}$ and obtain a new space with a natural coordinate.

DEFINITION 3.3. *Let* $\Lambda_{d,\bar{p}'}^M$ *be the set of maps*

$$\{M^{-1} \circ g \circ M : g \in \Lambda_{d,\bar{p}'}\}.$$

If $\tilde{g} \in \Lambda_{d,\bar{p}'}^M$, *the points* $c_k(\tilde{g}) = M^{-1}(c_k(g))$ *and* $v_k(\tilde{g}) = M^{-1}(v_k(g))$ *are the critical point (of multiplicity* m_k) *and the critical value of* $\tilde{g}$, $1 \leq k \leq p'$, *respectively. Then the vector*

$$\bar{v}^M(\tilde{g}) = \{\sigma(g), b(g), v_1(\tilde{g}), v_2(\tilde{g}), \ldots, v_{p'}(\tilde{g})\} \tag{3.3}$$

defines a holomorphic coordinate system in $\Lambda_{d,\bar{p}'}^M$.

The advantage of the new space is that the critical orbits of $\tilde{f} = M^{-1} \circ f \circ M$ lie in the plane (although they can be unbounded).

3.2 Subspaces

Suppose f is an arbitrary rational function of degree $d \geq 2$. Denote by p' the number of different critical points of f in the Riemann sphere and by $\bar{p}'$ the vector of multiplicities at the critical points. As it was mentioned, there is an alternative: either

(H): *f has a fixed point a, such that* $f'(a) \neq 0, 1$,

or

(N): *the multiplier of every fixed point of f is either 0 or 1, and there is a fixed point with the multiplier 1.*

Case **(N)** is degenerate. We consider each case separately.
(H). Let P be a Möbius transformation, such that $P(a) = \infty$. Then $\tilde{f} = P \circ f \circ P^{-1}$ belongs to $\Lambda_{d,\bar{p}'}$. Moreover, P can be chosen uniquely in such a way that $b(\tilde{f}) = 0$, and the critical value v_p of $\tilde{f}$ is equal to 1. Let us define a submanifold $\Lambda_{\tilde{f}}$ of $\Lambda_{d,\bar{p}'}$ consisting of $g \in \Lambda_{d,\bar{p}'}$ in a neighborhood of $\tilde{f}$, such that $b(g) = 0$, and $v_p(g) = 1$. The coordinate $\bar{v}(g)$ in $\Lambda_{d,\bar{p}'}$ restricted to $\Lambda_{\tilde{f}}$ is obviously a coordinate in that subspace, which turns it into a p'-dimensional complex manifold.

(N). There are two sub-cases to distinguish.
    (NN): f has a fixed point a such that $f'(a) = 1$ and $f''(a) \neq 0$. Let P be a Möbius transformation such that $P(a) = \infty$. Then $\tilde{f} = P \circ f \circ P^{-1}$ belongs to $\Lambda_{d,\bar{p}'}$. Moreover, P can be chosen uniquely in such a way that $v_p(\tilde{f}) = 1$ and $b(\tilde{f}) = 1$. Then we define $\Lambda_{\tilde{f}}$ to be the set of all $g \in \Lambda_{d,\bar{p}'}$ in a neighborhood of $\tilde{f}$ such that $b(g) = 1$ and $v_p(g) = 1$. Coordinates in $\Lambda_{\tilde{f}}$ are defined as in the previous case, and this turns $\Lambda_{\tilde{f}}$ into a p'-dimensional complex manifold.

    (ND): every fixed point with multiplier 1 is degenerate. Let a be one of them: $f'(a) = 1$ and $f''(a) = 0$. Then the Möbius map P can be chosen uniquely in such a way that $\tilde{f}(z) = P \circ f \circ P^{-1}(z) = z + O(1/z)$, and $\tilde{f}$ has a critical value equal to 1 in one attracting petal of ∞ and equal to 0 in another attracting petal of ∞. Then $\Lambda_{\tilde{f}}$ consists of $g \in \Lambda_{d,\bar{p}'}$ in a neighborhood of $\tilde{f}$ such that the critical value of g which is close to $v_{p-1}(\tilde{f}) = 1$ is identically equal to 1, and the critical value of g which is close to $v_p(\tilde{f}) = 0$ is identically equal to 0. Then $\Lambda_{\tilde{f}}$ is a p'-dimensional complex manifold.

It is easy to check that in any of these cases, every $\tilde{g} \in \Lambda_{d,\bar{p}'}$ is equivalent (by a linear conjugacy) to some $\tilde{g}_1 \in \Lambda_{\tilde{f}}$. Let us drop the condition that maps fix infinity and consider the set $X_{\tilde{f}}$ of all rational functions $\hat{g}$ of degree d which are close to $\tilde{f}$

and such that $\hat{g}$ has p' different critical points with the same corresponding multiplicities. Then, since any such $\hat{g}$ has a fixed point close to infinity, it is equivalent to some $\tilde{g} \in \Lambda_{d,\tilde{p}'}$ and, hence, to some $\tilde{g}_1 \in \Lambda_{\tilde{f}}$. In any of the cases **(H)**, **(NN)**, **(ND)**, we denote

$$X_f = \{g = P^{-1} \circ \hat{g} \circ P : \hat{g} \in X_{\tilde{f}}\}, \quad \Lambda_f = \{g = P^{-1} \circ \tilde{g} \circ P : \tilde{g} \in \Lambda_{\tilde{f}}\}, \qquad (3.4)$$

where the Möbius map P is taken as before. This shows the first part of the following statement.

PROPOSITION 3.4. (a) *Every rational function f of degree $d \geq 2$ is equivalent to some $\tilde{f} \in \Lambda_{d,\tilde{p}'}$, where $\tilde{f}$ is of one and only one type: either **(H)** or **(NN)** or **(ND)**. If X_f denotes the set of all rational functions g of degree d which are close to f and such that g have p' distinct critical points with the same corresponding multiplicities, then any $g \in X_f$ is equivalent to some $g_1 \in \Lambda_f$.*

(b) *In the case **(H)**, the set X_f is a complex-analytic manifold of dimension $p' + 3$, and the correspondence $\pi\colon g \in X_f \mapsto g_1 \in \Lambda_f$ is a well-defined holomorphic map.*

Let us show (b). In this case $\sigma(\tilde{f}) = \tilde{f}'(\infty) \neq 1$, hence, any $\hat{g} \in X_{\tilde{f}}$ has a unique fixed point β which is close to ∞. This defines a one-to-one map **B** between $X_{\tilde{f}}$ and $\Lambda_{d,\tilde{p}'}$ by **B**: $\hat{g} \mapsto \tilde{g} = M_\beta \circ \hat{g} \circ M_\beta^{-1}$, where $M(z) = \beta z/(\beta - z)$. It is also easy to see that in this case there is a natural map **A**: $\tilde{g} \in \Lambda_{d,\tilde{p}'} \mapsto \tilde{g}_1 \in \Lambda_{\tilde{f}}$ defined by $\tilde{g}_1 = A^{-1} \circ \tilde{g} \circ A$, where $A(z) = kz + e$ with $k = v_p(\tilde{g}) + \frac{b(\tilde{g})}{\sigma(\tilde{g})-1}$ and $e = -\frac{b(\tilde{g})}{\sigma(\tilde{g})-1}$. The vector $\bar{v}(\hat{g}) = \{\frac{1}{\beta}, \bar{v}(\tilde{g})\} \in \mathbb{C}^{p'+3}$ defines a complex-analytic structure on $X_{\tilde{f}}$, and the map $\mathbf{A} \circ \mathbf{B} \colon X_{\tilde{f}} \to \Lambda_{\tilde{f}}$ is holomorphic. Since, by (3.4), X_f, Λ_f are isomorphic to $X_{\tilde{f}}$, $\Lambda_{\tilde{f}}$, part (b) follows. Note in conclusion that the Möbius conjugacy between $g \in X_f$ and $g_1 \in \Lambda_f$, which we just constructed, tends to the identity map uniformly on the Riemann sphere as g tends to f.

3.3 Main result

DEFINITION 3.5. *A compact subset K of the Riemann sphere is called* C-compact *if there is a Möbius transformation M such that $M(K)$ is a compact subset of the complex plane with the property that every continuous function on $M(K)$ can be uniformly approximated by functions which are holomorphic in a neighborhood of $M(K)$.*

Clearly, C-compact sets must have empty interior. Vitushkin's theorem characterizes such compacts on the plane; see, e.g., [Ga]. As a simple corollary of this theorem, we have that each of the following conditions is sufficient for K to be a C-compact set:

1. K has Lebesgue measure zero,

2. every $z \in K$ belongs to the boundary of a component of the complement of K.

We call a rational function f *exceptional* if f is *double covered by an integral torus endomorphism* (also known as *a flexible Lattès map*): these form a family of explicitly described critically finite rational maps with Julia sets which are the whole Riemann sphere (see, for example, [DH93], [McM], and [Mi06]).

Let $f \in \Lambda_{d,\bar{p}'}$. Fix a Möbius transformation M, such that $M(\infty)$ is disjoint with the forward orbits of the critical points of f. (If $f^n(c_j) \neq \infty$ for all $1 \leq j \leq p'$ and all $n \geq 0$, one can let M be the identity map.) Consider $\tilde{f} = M^{-1} \circ f \circ M$ in the space $\Lambda_{d,\bar{p}'}^M$ with the coordinate $\bar{v}^M$ (see Definition 3.3). We denote by $\tilde{c}_k = M^{-1}(c_k)$, $\tilde{v}_k = M^{-1}(v_k)$, $1 \leq k \leq p'$, the corresponding critical points and critical values of $\tilde{f}$. As usual, $\frac{\partial(\tilde{f}^l(\tilde{c}_j))}{\partial \tilde{v}_k}$, etc., means $\frac{\partial(\tilde{g}^l(c_j(\tilde{g})))}{\partial v_k(\tilde{g})}$, etc., evaluated at $\tilde{g} = \tilde{f}$.

THEOREM 3.6. (a) *Let c_j be a weakly expanding critical point of f. Then, for every k between 1 and p', the following limits exist:*

$$L^M(c_j, v_k) := \lim_{l \to \infty} \frac{\frac{\partial(\tilde{f}^l(\tilde{c}_j))}{\partial \tilde{v}_k}}{(\tilde{f}^{l-1})'(\tilde{v}_j)}, \tag{3.5}$$

$$L^M(c_j, \sigma) := \lim_{l \to \infty} \frac{\frac{\partial(\tilde{f}^l(\tilde{c}_j))}{\partial \sigma}}{(\tilde{f}^{l-1})'(\tilde{v}_j)}, \quad L^M(c_j, b) := \lim_{l \to \infty} \frac{\frac{\partial(\tilde{f}^l(\tilde{c}_j))}{\partial b}}{(\tilde{f}^{l-1})'(\tilde{v}_j)}. \tag{3.6}$$

Moreover, if $p < p'$, i.e., f has a critical value at infinity), then, for $p + 1 \leq j \leq p'$ (i.e., when $v_j = \infty$) and $1 \leq k \leq p'$,

$$L^M(c_j, v_k) = \delta_{j,k}, \tag{3.7}$$

$$L^M(c_j, \sigma) = L^M(c_j, b) = 0. \tag{3.8}$$

(b) *Suppose that f is not exceptional and f has a collection of r weakly expanding critical points. Without loss of generality, one can assume that these points are $c_{j_1}, \ldots c_{j_r}$, with $1 \leq j_1 < \ldots < j_r \leq p'$, and they are ordered as follows. Denote by ν the number of such points with the corresponding critical values to be in the complex plane. Then denote by $1 \leq j_1 < \ldots < j_\nu \leq p$ the indexes of these points, and, if $\nu < r$, i.e., there are critical points from the collection with the corresponding critical values at infinity, then $r - \nu$ indexes of these points are the last ones: $j_{\nu+1} = p' - (r - \nu - 1)$, $j_{\nu+2} = p' - (r - \nu - 2)$, $\ldots, j_r = p'$.*

Denote by K the union of the ω-limit sets of these critical points. Assume that K is a C-compact set. Then the rank of the matrix $\mathbf{L}^M$ (defined shortly) is equal to r.

(H_∞). *Let f be of the type (H), i.e., $f(z) = \sigma z + O(1/z)$ as $z \to \infty$, where $\sigma \neq 1$, and also $v_p = 1$. Then*

$$\mathbf{L}^M = \left(L^M(c_{j_i}, \sigma), L^M(c_{j_i}, v_1), \ldots, L^M(c_{j_i}, v_{p-1}), L^M(c_{j_i}, v_{p+1}), \ldots, L^M(c_{j_i}, v_{p'}) \right)_{1 \leq i \leq r}.$$

(NN_∞). *If f is of the type (NN), i.e., $\sigma = 1$, $b \neq 0$, and $v_p = 1$, then*

$$\mathbf{L}^M = \left(L^M(c_{j_i}, v_1), \ldots, L^M(c_{j_i}, v_{p-1}), L^M(c_{j_i}, v_{p+1}), \ldots, L^M(c_{j_i}, v_{p'}) \right)_{1 \leq i \leq r}.$$

(ND_∞). *Finally, if f is of the type (ND): $\sigma = 1$, $b = 0$, $v_{p-1} = 1$, $v_p = 0$, then*

$$\mathbf{L}^M = \left(L^M(c_{j_i}, v_1), \ldots, L^M(c_{j_i}, v_{p-2}), L^M(c_{j_i}, v_{p+1}), \ldots, L^M(c_{j_i}, v_{p'}) \right)_{1 \leq i \leq r}.$$

PROOF OF THEOREM 1.1. We apply part (b) of Theorem 3.6 exactly as we apply Theorem 2.3 in Comment 2.4 and then apply Proposition 3.4. □

PROOF OF COROLLARY 1.2. Since $K = \cup_{j=1}^{2d-2} \omega(c_j)$ is a hyperbolic set for f and, hence, is of measure zero, Theorem 1.1 applies. As every critical point of f is simple and summable, $p' = r = 2d - 2$, i.e., $\Lambda = \Lambda_f$, and X_f is the (local) space of all close to f rational functions of degree d. Let $\bar{x} = (x_1, \ldots, x_{2d-2})$ be a local coordinate in Λ. Since all critical points of f are summable, f has no parabolic periodic orbits. Hence, f is of type $\mathbf{H}$, and the part (b) of Proposition 3.4 applies: X_f is a complex-analytic manifold, and the correspondence $\pi: g \in X_f \mapsto \hat{g} \in \Lambda_f$ (from Proposition 3.4 (a)) is a holomorphic map. Observe that the restriction of π on $N \subset X_f$ is one-to-one, because otherwise two different $g_1, g_2 \in N$ would be equivalent to the same $\pi(g_1) = \pi(g_2)$. Hence, the restriction $\pi|_N: N \to \Lambda_f$ is a local analytic isomorphism. This is equivalent to saying that, in the coordinates λ of N and $\bar{x}$ of Λ, the derivative of $\pi|_N$ at λ_0 is a non-degenerate $(2d - 2) \times (2d - 2)$ matrix π'_0. In turn, part (b) of Theorem 1.1 is equivalent to the statement that the matrix $\mathbf{L}_f = (L_{j,k})_{1 \le j,k \le 2d-2}$, where

$$L_{j,k} = \lim_{l \to \infty} \frac{\frac{\partial f^l(c_j)}{\partial x_k}}{(f^{l-1})'(v_j)}$$

is non-degenerate. The rest of the proof is a straightforward application of this fact. For every l big enough, we introduce the following objects:

1. an approximation matrix $\mathbf{L}_f^{(l)} = (L_{j,k}^{(l)})_{1 \le j,k \le 2d-2}$, where

$$L_{j,k}^{(l)} = \frac{\frac{\partial f^l(c_j)}{\partial x_k}}{(f^{l-1})'(v_j)},$$

2. two maps from the neighborhood W of λ_0 into $\mathbb{C}^{2d-2}$; these are the maps

$$C_l(\lambda) = \{f_\lambda^l(c_j(\lambda))\}_{1 \le j \le 2d-2} \quad \text{and} \quad \bar{a}_l(\lambda) = \{a_j(\lambda)\}_{1 \le j \le 2d-2}$$

(recall that $z(\lambda)$ is the holomorphic motion of the point $z = z(\lambda_0) \in K$, and $a_j = f^l(c_j) \in K$),

3. a diagonal $(2d - 2) \times (2d - 2)$ matrix D_l with the diagonal (j, j) element equal to $(f^{l-1})'(v_j)$, for $1 \le j \le 2d - 2$. Note that $(f^{l-1})'(v_j) \to \infty$ as $l \to \infty$.

Since the family of analytic functions $\{z(\lambda)\}$, where $\lambda \in W$, is bounded, then the norms of the derivative $\bar{a}'_l$ of $\bar{a}_l(\lambda)$ at λ_0 are uniformly (in l) bounded, too. By the definition, $\mathbf{h}_l(\lambda) = C_l(\lambda) - \bar{a}_l(\lambda)$; hence, for the derivatives at $\lambda = \lambda_0$, $\mathbf{h}'_l = C'_l - \bar{a}'_l$. By linear algebra, we have a representation:

$$C'_l = D_l \mathbf{L}_f^{(l)} \pi'_0.$$

Therefore,

$$\mathbf{h}'_l = D_l \mathbf{L}^{(l)}_f \pi'_0 - \bar{a}'_l = D_l \left(\mathbf{L}^{(l)}_f \pi'_0 - D_l^{-1} \bar{a}'_l \right).$$

As l tends to ∞, $\mathbf{L}^{(l)}_f$ tends to the invertible matrix $\mathbf{L}_f$, and D_l^{-1} tends to zero matrix. It follows, for every l big enough, $\mathbf{h_l}'$ is invertible. Moreover, asymptotically, $(\mathbf{h}'_l)^{-1}$ is $(\pi'_0)^{-1} \mathbf{L}_f^{-1} D_l^{-1}$. $\square$

PROOF OF COROLLARY 1.3. Let us first consider the case when f is exceptional [DH93]. Since all critical points of f are simple, then X_f is the (local) space of all rational functions $\mathbf{Rat}_d$ of degree d which are close to f, and Λ_f is its $(2d-2)$-dimensional submanifold. As f has no neutral cycles, by Proposition 3.4 (b), we have a well-defined map $g \in X_f \mapsto g_1 \in \Lambda_f$, where g is equivalent to g_1, and the Möbius conjugacy between them tends to the identity uniformly on the Riemann sphere as $g \to f$ (see the end of Subsection 3.2). It implies that f is structurally stable in X_f if and only if f is structurally stable in Λ_f. On the other hand, as f is critically finite, it is easy to see that the critical relations of f cannot be preserved for all maps in $X_f = \mathbf{Rat}_d$ which are close to f. Hence, f is not structurally stable in X_f and, then, in Λ_f.

Now, let f be a non-exceptional map. One can assume that $\omega(c)$ is a subset of the complex plane. By applying Theorem 1.1 to f and c, there exists a one-dimensional submanifold Λ (i.e., an analytic one-dimensional family f_t, $f_0 = f$) of Λ_f such that the limit

$$\lim_{m \to \infty} \frac{\frac{d}{dt}|_{t=0} f_t^m(c(t))}{(f^{m-1})'(f(c))} \tag{3.9}$$

exists and is a non-zero number. (Here $c(t)$ is the critical point of f_t which is close to $c(0) = c$.) This obviously implies that

$$\lim_{m \to \infty} \frac{d}{dt}\bigg|_{t=0} f_t^m(c(t)) = \infty. \tag{3.10}$$

This is enough to conclude that f is not structurally stable in the family f_t, hence, in Λ_f, too. Indeed, assuming, by a contradiction, that f is structurally stable in f_t, we get that the ω-limit set $\omega(c(t))$ of the critical point $c(t)$ of f_t tends to $\omega(c(0)) = \omega(c)$ as $t \to 0$ while the compact $\omega(c)$ lies in the plane. In particular, the sequence of holomorphic functions $\{f_t^m(c(t))\}_{m \geq 0}$ is uniformly bounded in a neigborhood of $t = 0$, which is a contradiction with (3.10). $\square$

4   PART (A) OF THEOREM 3.6

As in the polynomial case, we start by calculating partial derivatives of a function $g \in \Lambda_{d,\bar{p}'}$ w.r.t. the standard local coordinates of the space $\Lambda_{d,\bar{p}'}$, i.e., σ, b, and the critical values which are *not* infinity.

PROPOSITION 4.1. *Fix a function $f \in \Lambda_{d,\bar{p}'}$, where $f(z) = \sigma z + b + P(z)/Q(z)$.*

(a) *Let $v_k = f(c_k)$ be a critical value of f which is different from infinity (i.e., $1 \leq k \leq p$). Then $\frac{\partial f}{\partial v_k}(z)$ is a rational function $q_k(z)$ of degree $2d - 2$ of the form $\frac{\tilde{P}(z)}{(Q(z))^2}$, where $\tilde{P}$ is a polynomial of degree at most $2d - 3$. It is uniquely characterized by the following conditions:*

(i) *at $z = c_k$, the function $q_k(z) - 1$ has zero of order at least m_k;*

(ii) *at $z = c_j$, for every $j \neq k$, $1 \leq j \leq p$, $q_k(z)$ has zero of order at least m_j;*

(iii) *finally, at $z = c_j$, for every $p + 1 \leq j \leq p'$, the polynomial $\tilde{P}(z)$ has zero of order at least m_j.*

In particular, if c_k is simple (i.e., $m_k = 1$), then (2.11) holds:

$$\frac{\partial f}{\partial v_k}(z) = \frac{f'(z)}{f''(c_k)(z - c_k)}. \tag{4.1}$$

(b) *We have, as well,*

$$\frac{\partial f}{\partial \sigma}(z) = \frac{z}{\sigma}f'(z), \quad \frac{\partial f}{\partial b}(z) = \frac{1}{\sigma}f'(z). \tag{4.2}$$

PROOF. Let x be one of the following coordinates of the vector $\bar{v}(g)$, for $g \in \Lambda_{d,\bar{p}}$: x is either v_k, for $1 \leq k \leq p$, or σ, or b. Notice that, for $x = v_k$,

$$\frac{\partial v_j(g)}{\partial x} = \delta_{j,k}, \quad 1 \leq j \leq p, \tag{4.3}$$

and, for $x = \sigma$ or $x = b$,

$$\frac{\partial v_j(g)}{\partial x} = 0, \quad 1 \leq j \leq p. \tag{4.4}$$

As $g \in \Lambda_{d,\bar{p}'}$ is close to f, then $g(z) = \sigma(g)z + b(g) + \frac{P_g(z)}{Q_g(z)}$, where the polynomials $P_g(z) = B_0(g)z^{d-2} + \dots$ and $Q_g(z) = z^{d-1} + \dots$ have coefficients which are holomorphic functions of the vector $\bar{v}(g)$, and $P_f = P$, $Q_f = Q$. It follows that $\frac{f}{\partial v_k} = \frac{\tilde{P}}{Q^2}$, where $\tilde{P} = \frac{\partial P}{\partial v_k}Q - P\frac{\partial Q}{\partial v_k}$ is a polynomial of degree at most $2d - 3$. We apply Proposition 2.5 to the coordinate x as before and take into account (4.3)–(4.4). We conclude that, for every $1 \leq j \leq p$, c_j is at least a m_j-multiple zero of the functions $\frac{\partial f}{\partial v_k}(z) - \delta_{j,k}$, $\frac{\partial f}{\partial \sigma}(z)$, and $\frac{\partial f}{\partial b}(z)$. In particular, all this proves part (a) except for property (iii).

Let $\hat{P}(z) = (\sigma z + b)Q(z) + P(z)$, that is, $f = \frac{\hat{P}}{Q}$. Then $\frac{\partial f}{\partial x}(z)$ is a rational function of the form $\frac{P^x(z)}{(Q(z))^2}$, where

$$P^x = \frac{\partial \hat{P}}{\partial x}Q - \hat{P}\frac{\partial Q}{\partial x}. \tag{4.5}$$

Now, let $p + 1 \leq j \leq p'$, i.e., c_j is a root of $Q(z)$, such that $Q(z) = (z - c_j)^{m_j+1}\psi(z)$, where ψ is analytic near c_j and $\psi(c_j) \neq 0$. We cannot apply Proposition 2.5 directly in this case, because $v_j = f(c_j) = \infty$. To get around this, let us introduce a new (local) space N^* consisting of the functions $g^*(z) := 1/g(z)$, for $g \in \Lambda_{d,\bar{p}'}$ near f. Notice that the vector $\bar{v}(g)$ of coordinates in $\Lambda_{d,\bar{p}'}$ is also a local coordinate in N^*, the critical points of g^* and g (with the corresponding multiplicities) coincide, while for the critical values g^* and g, we have $v_j^*(g^*) = 1/v_j(g)$. In particular, $v_j^* = f^*(c_j) = 0$. Using the preceding notations, we have:

$$\frac{\partial f^*}{\partial x}(z) = -\frac{\frac{\partial f}{\partial x}(z)}{(f(z))^2} = -\frac{P^x(z)}{(\hat{P}(z))^2}. \tag{4.6}$$

On the other hand, we can apply Proposition 2.5 to the space N^*, the map f^*, its critical point c_j (with the corresponding critical value to be 0), and the coordinate x. Since $p < j \leq p'$, the critical value $v_j^*(g^*) = 1/v_j(g)$ of g^* is just the j-coordinate of the vector of coordinates $\bar{v}(g)$ in N^*, while x (which is either $v_k(g)$, for $1 \leq j \leq p$, or $\sigma(g)$, or $b(g)$) is a different coordinate of $\bar{v}(g)$. Hence, we also have

$$\frac{\partial v_j^*}{\partial x} = 0.$$

Then, by Proposition 2.5, c_j is at least an m_j-multiple root of $\frac{\partial f^*}{\partial x}(z)$; hence, by (4.6), c_j is at least an m_j-multiple root of $P^x(z)$, as well. In particular, this holds for $\tilde{P} = P^{v_k}$, which proves (iii) of part (a).

As for part (b), x is either σ or b. Let $x = \sigma$. Then

$$\frac{\partial f}{\partial \sigma}(z) = z + \frac{R(z)}{(Q(z))^2} = \frac{P^\sigma(z)}{(Q(z))^2},$$

where

$$R = \frac{\partial P}{\partial \sigma}Q - P\frac{\partial Q}{\partial \sigma}$$

is a polynomial of degree at most $2d - 3$ and $P^\sigma = zQ^2 + R$. In particular, we have $\frac{\partial f}{\partial \sigma}(z) = z + O(1/z)$. On the other hand, as we have seen, for every $1 \leq j \leq p$, c_j is at least an m_j-multiple zero of $\frac{\partial f}{\partial \sigma}(z)$, and, for $p < j \leq p'$, c_j is at least an m_j-multiple zero of P^σ. Therefore, $\frac{\partial f}{\partial \sigma}(z)/f'(z) = (z + O(1/z))/(\sigma + O(1/z^2))$, and, at the same time, it is a rational function without poles in the plane. Hence, it must be equal to z/σ. The proof for $\partial f/\partial b$ is very similar and is left to the reader. $\quad\square$

As in the polynomial case, we then have the following.

PROPOSITION 4.2. *Let $f \in \Lambda_{d,\bar{p}'}$. Then*

$$T_f \frac{1}{z - x} = \frac{1}{f'(z)} \frac{1}{f(z) - x} + \sum_{k=1}^{p} \frac{\frac{\partial f}{\partial v_k}(z)}{f'(z)} \frac{1}{x - v_k}. \tag{4.7}$$

PROOF. We use Lemma 2.8 and the part (a) of Proposition 4.1. By assertions (1) and (2) of Lemma 2.8, $f'(z)L_k(z)$ is a rational function of the form $\tilde{P}/Q^2$, where $\tilde{P}$ is a polynomial of degree at most $2d - 3$, such that $f'(z)L_k(z) - 1$ has a root at c_k with multiplicity at least m_k, and, for every $j \neq k$, $1 \leq j \leq p'$, the polynomial $\tilde{P}$ has a root at c_j with multiplicity at least m_j. Hence, $f'(z)L_k(z)$ coincides with the rational function $q_k(z)$. Therefore, indeed, $L_k(z) = \frac{\frac{\partial f}{\partial v_k}(z)}{f'(z)}$. $\quad\square$

If $v_k = \infty$ or $z = \infty$, Proposition 4.1 is not useful. In that case, we replace the space $\Lambda_{d,\bar{p}'}$ by the space

$$\Lambda_{d,\bar{p}'}^M = \{M^{-1} \circ g \circ M : g \in \Lambda_{d,\bar{p}'}\},$$

where M is a non-linear Möbius transformation, that is, $\beta = M^{-1}(\infty) \neq \infty$, $\alpha = M(\infty) \neq \infty$, and α is such that the forward orbits of critical points of f are different from α. Then the forward critical orbits of $\tilde{f} = M^{-1} \circ f \circ M$ lie in the complex plane. Hence, for any $l > 0$, if g is close enough to f, then the first l iterates of the critical points of $\tilde{g}$ also lie in the plane.

By Definition 3.3, $\tilde{g} = M^{-1} \circ g \circ M \in \Lambda_{d,\tilde{p}'}^M$ is a holomorphic function of $\sigma(g)$, $b(g)$, and the critical values $v_j(\tilde{g}) = M^{-1}(v_j(g))$ of $\tilde{g}$, $1 \leq j \leq p'$. By $\tilde{v}_j$, $\tilde{c}_j$ we denote $v_j(\tilde{f})$, $c_j(\tilde{f})$, i.e., the critical values and points of $\tilde{f}$. In particular, $\tilde{v}_j = \beta$ if and only if $p + 1 \leq j \leq p'$. As usual, $\{v_k\}$ denote the critical values of f, and $\partial \tilde{f}^n / \partial \tilde{v}_k$ means $\partial \tilde{g}^n / \partial v_k(\tilde{g})$ calculated at $\tilde{g} = \tilde{f}$.

PROPOSITION 4.3. (a) *Take* $l \geq 1$. *If* $f^i(v_j) \neq \infty$ *for* $0 \leq i \leq l - 1$, *then*

$$\frac{\frac{\partial \tilde{f}^l}{\partial \tilde{v}_k}(\tilde{c}_j)}{(\tilde{f}^{l-1})'(\tilde{v}_j)} = \frac{(M^{-1})'(v_j)}{(M^{-1})'(v_k)} \frac{\frac{\partial f^l}{\partial v_k}(c_j)}{(f^{l-1})'(v_j)}, \tag{4.8}$$

$$\frac{\frac{\partial \tilde{f}^l}{\partial \sigma}(\tilde{c}_j)}{(\tilde{f}^{l-1})'(\tilde{v}_j)} = (M^{-1})'(v_j) \frac{\frac{\partial f^l}{\partial \sigma}(c_j)}{(f^{l-1})'(v_j)}, \tag{4.9}$$

$$\frac{\frac{\partial \tilde{f}^l}{\partial b}(\tilde{c}_j)}{(\tilde{f}^{l-1})'(\tilde{v}_j)} = (M^{-1})'(v_j) \frac{\frac{\partial f^l}{\partial b}(c_j)}{(f^{l-1})'(v_j)}. \tag{4.10}$$

(b) *We have*

$$\frac{\partial \tilde{f}}{\partial \tilde{v}_k}(\beta) = 0, \ 1 \leq k \leq p', \qquad \frac{\partial \tilde{f}}{\partial \sigma}(\beta) = \frac{\partial \tilde{f}}{\partial b}(\beta) = 0, \tag{4.11}$$

$$\frac{\partial \tilde{f}}{\partial \tilde{v}_k}(\tilde{c}_j) = \delta_{j,k}, \qquad \frac{\partial \tilde{f}}{\partial \sigma}(\tilde{c}_j) = \frac{\partial \tilde{f}}{\partial b}(\tilde{c}_j) = 0, \ 1 \leq k, j \leq p'. \tag{4.12}$$

PROOF. (a) Since $\tilde{g}^l = M^{-1} \circ g^l \circ M$, $v_k(g) = M(v_k(\tilde{g}))$, where M is a fixed map, and by the conditions on v_k, c_j, the following calculations make sense.

$$\begin{aligned}
\frac{\frac{\partial \tilde{f}^l}{\partial \tilde{v}_k}(\tilde{c}_j)}{(\tilde{f}^{l-1})'(\tilde{v}_j)} &= \frac{(M^{-1})'\big(f^l(M(\tilde{c}_j))\frac{\partial f^l}{\partial v_k}(M(\tilde{c}_j))\frac{\partial v_k(g)}{\partial \tilde{v}_k}}{(M^{-1})'\big(f^{l-1}(M(\tilde{v}_j))\big)(f^{l-1})'(M(\tilde{v}_j))M'(\tilde{v}_j)} \\
&= \frac{(M^{-1})'(f^{l-1}(v_j))\frac{\partial f^l}{\partial v_k}(c_j)M'(\tilde{v}_k)}{(M^{-1})'(f^{l-1}(v_j))(f^{l-1})'(v_j)M'(\tilde{v}_j)} \\
&= \frac{(M^{-1})'(v_j)}{(M^{-1})'(v_k)} \frac{\frac{\partial f^l}{\partial v_k}(c_j)}{(f^{l-1})'(v_j)}.
\end{aligned} \tag{4.13}$$

The proof of (4.9)–(4.10) is similar.

(b) Since $\tilde{g}(\beta) = \beta \neq \infty$ for every $\tilde{g} \in \Lambda_{d,\tilde{p}'}^M$, we write

$$\tilde{g}(z) = \beta + \int_\beta^z \tilde{g}'(w)dw. \tag{4.14}$$

Since β is independent of g,

$$\frac{\partial \tilde{f}}{\partial \tilde{v}_k}(z) = \int_\beta^z \frac{\partial \tilde{f}'}{\partial \tilde{v}_k}(w)dw, \tag{4.15}$$

and similarly for the derivatives w.r.t. σ and b. To show (4.11), it remains to estimate (4.15) at $z = \beta$. To show (4.12), it is enough to apply Proposition 2.5 to the space $\Lambda_{d,\bar{p}'}^M$, the map $\tilde{f}$, the critical point $\tilde{c}_j$, and the coordinate in the vector $\bar{v}^M$, which is either $\tilde{v}_k$ or σ or b. $\qquad\square$

4.1 Proof of Part (a) of Theorem 3.6

Here we state and prove a refined version of Theorem 3.6(a), introducing notations $L(c_j, v_k)$, $L(c_j, \sigma)$, $L(c_j, b)$ to be used later on. Recall that the notations $L^M(c_j, v_k)$, $L^M(c_j, \sigma)$, $L^M(c_j, b)$ were introduced in Theorem 3.6.

THEOREM 4.4. *Let $f \in \Lambda_{d,\bar{p}'}$.*

1. *Assume $v_k \neq \infty$, c_j is summable and the whole orbit of c_j remains in the complex plane. Then the limits defined next exist and are expressed as the following series, which converge absolutely:*

$$L(c_j, v_k) := \lim_{l \to \infty} \frac{\frac{\partial(f^l(c))}{\partial v_k}}{(f^{l-1})'(v)} = \delta_{j,k} + \sum_{n=1}^{\infty} \frac{\frac{\partial f}{\partial v_k}(f^{n-1}(v_j))}{(f^n)'(v_j)}, \tag{4.16}$$

$$L(c_j, \sigma) := \lim_{l \to \infty} \frac{\frac{\partial(f^l(c_j))}{\partial \sigma}}{(f^{l-1})'(v_j)} = \frac{1}{\sigma} \sum_{n=0}^{\infty} \frac{f^n(v_j)}{(f^n)'(v_j)}, \tag{4.17}$$

$$L(c_j, b) := \lim_{l \to \infty} \frac{\frac{\partial(f^l(c_j))}{\partial b}}{(f^{l-1})'(v_j)} = \frac{1}{\sigma} \sum_{n=0}^{\infty} \frac{1}{(f^n)'(v)}. \tag{4.18}$$

Furthermore,

$$L^M(c_j, v_k) = \frac{(M^{-1})'(v_j)}{(M^{-1})'(v_k)} L(c_j, v_k), \tag{4.19}$$

$$L^M(c_j, \sigma) = (M^{-1})'(v_j) L(c_j, \sigma), \tag{4.20}$$

$$L^M(c_j, b) = (M^{-1})'(v_j) L(c_j, b). \tag{4.21}$$

2. *Assume $v_k \neq \infty$, c_j is summable and $f^l(v_j) = \infty$, for some minimal $l \geq 1$. Define, in this case,*

$$L(c_j, v_k) := \delta_{j,k} + \sum_{n=1}^{l} \frac{\frac{\partial f}{\partial v_k}(f^{n-1}(v_j))}{(f^n)'(v_j)}, \tag{4.22}$$

$$L(c_j, \sigma) := \frac{1}{\sigma} \sum_{n=0}^{l-1} \frac{f^n(v_j)}{(f^n)'(v_j)}, \tag{4.23}$$

$$L(c_j, b) := \frac{1}{\sigma} \sum_{n=0}^{l-1} \frac{1}{(f^n)'(v)}. \tag{4.24}$$

Then (4.19)–(4.21) hold.

3. *Assume $v_j = f(c_j) = \infty$. Then, for $1 \leq k \leq p'$,*

$$L^M(c_j, v_k) = \delta_{j,k}, \tag{4.25}$$

$$L^M(c_j, \sigma) = L^M(c_j, b) = 0. \tag{4.26}$$

PROOF. (1) It is enough to prove (4.16)–(4.18). Then, by (4.8)–(4.10) of Proposition 4.3, (4.19)–(4.21) follow. We can use identity (2.13) of Subsection 2.3: for every $l > 0$,

$$\frac{\frac{\partial f^l}{\partial v_k}(c_j)}{(f^{l-1})'(v_j)} = \frac{\partial f}{\partial v_k}(c_j) + \sum_{n=1}^{l-1} \frac{\frac{\partial f}{\partial v_k}(f^{n-1}(v_j))}{(f^n)'(v_j)}. \tag{4.27}$$

By part (a) of Proposition 4.1, $\frac{\partial f}{\partial v_k}(c_j) = \delta_{j,k}$, and, moreover, $\frac{\partial f}{\partial v_k}(z)$ is a rational function of the form $\tilde{P}/Q^2$, where degree of $\tilde{P}$ is less than the degree of Q^2. Hence, for some constant C_k and all z,

$$\left| \frac{\partial f}{\partial v_k}(z) \right| \leq C_k(1 + |f(z)|^2). \tag{4.28}$$

Now, assume that c_j is weakly expanding. Then

$$\sum_{n=1}^{\infty} \frac{|\frac{\partial f}{\partial v_k}(f^{n-1}(v_j))|}{|(f^n)'(v_j)|} \leq C_k \frac{1 + |f^n(v_j)|^2}{|(f^n)'(v_j)|} < \infty, \tag{4.29}$$

and (4.16) follows. The proof of the existence of $L(c_j, \sigma)$ and $L(c_j, b)$ is similar to the proof for the v_k-derivative. Indeed,

$$\frac{\partial f^l}{\partial \sigma}(z) = (f^l)'(z) \sum_{n=0}^{l-1} \frac{\frac{\partial f}{\partial \sigma}(f^n(z))}{(f^{n+1})'(z)}.$$

Here letting $z \to c_j$ and using part (b) of Proposition 4.1, we get

$$\begin{aligned}
\frac{\partial f^l}{\partial \sigma}(c_j) &= (f^{l-1})'(v_j) \left\{ \frac{\partial f}{\partial \sigma}(c_j) + \sum_{n=1}^{l-1} \frac{\frac{\partial f}{\partial \sigma}(f^{n-1}(v_j))}{(f^n)'(v_j)} \right\} \\
&= (f^{l-1})'(v_j) \frac{1}{\sigma} \left\{ c_j f'(c_j) + \sum_{n=1}^{l-1} \frac{f^{n-1}(v_j) f'(f^{n-1}(v_j))}{(f^n)'(v_j)} \right\} \\
&= \frac{(f^{l-1})'(v_j)}{\sigma} \sum_{n=1}^{l-1} \frac{f^{n-1}(v_j)}{(f^{n-1})'(v_j)},
\end{aligned}$$

and we get (4.17). Doing the same (with obvious changes) for the $\partial/\partial b$-derivative, we get (4.18).

(2) We use (4.27) with $\tilde{f}$ instead of f. Note that $f^j(v_j) = \infty$ if and only if $\tilde{f}^j(\tilde{v}_j) = \beta$. Then $\tilde{f}^j(\tilde{v}_j) = \beta$ for every $j \geq l$, and, hence, for $j \geq l$, by Proposition 4.3(b),

$$\begin{aligned}
\frac{\frac{\partial \tilde{f}^j}{\partial \tilde{v}_k}(\tilde{c}_j)}{(\tilde{f}^{j-1})'(\tilde{v}_j)} &= \frac{\partial \tilde{f}}{\partial \tilde{v}_k}(\tilde{c}_j) + \sum_{n=1}^{j-1} \frac{\frac{\partial \tilde{f}}{\partial \tilde{v}_k}(\tilde{f}^{n-1}(\tilde{v}_j))}{(\tilde{f}^n)'(\tilde{v}_j)} \\
&= \delta_{j,k} + \sum_{n=1}^{l-1} \frac{\frac{\partial \tilde{f}}{\partial \tilde{v}_k}(\tilde{f}^{n-1}(\tilde{v}_j))}{(\tilde{f}^n)'(\tilde{v}_j)}.
\end{aligned}$$

That is,

$$\frac{\frac{\partial f^j}{\partial \tilde{v}_k}(\tilde{c}_j)}{(\tilde{f}^{j-1})'(\tilde{v}_j)} = \frac{\frac{\partial f^l}{\partial \tilde{v}_k}(\tilde{c}_j)}{(\tilde{f}^{l-1})'(\tilde{v}_j)}, \tag{4.30}$$

while, by Proposition 4.3(a),

$$\frac{\frac{\partial f^l}{\partial \tilde{v}_k}(\tilde{c}_j)}{(\tilde{f}^{l-1})'(\tilde{v}_j)} = \frac{(M^{-1})'(v_j)}{(M^{-1})'(v_k)}\frac{\frac{\partial f^l}{\partial v_k}(c_j)}{(f^{l-1})'(v_j)} = \frac{(M^{-1})'(v_j)}{(M^{-1})'(v_k)}L(c_j, v_k).$$

This proves (4.19) in the considered case. The proof of (4.21) is similar.

(3) Relation (4.25) follows directly from (4.30) with $l = 1$ and Proposition 4.3. The proof of (4.26) is similar. $\square$

The next statement is a corollary from Theorem 4.4. It allows us to reduce evaluation of the rank of the matrix $\mathbf{L^M}$, which is defined in Theorem 3.6(b), to the evaluation of the rank of a simpler matrix $\mathbf{L}$ defined shortly. Recall that the integer number v between 0 and r is defined in Theorem 3.6(b) as the maximal number of points from the collection $c_{j_1}, \ldots c_{j_r}$, $1 \leq j_1 < \ldots < j_r \leq p'$, of r summable critical points of f such that the corresponding critical values are different from infinity. Moreover, if $v < r$, the last $r - v$ indexes $j_{v+1}, \ldots, j_r$ are $p' - (r - v - 1), \ldots, p'$.

COROLLARY 4.5. *The rank of the matrix $\mathbf{L^M}$ is bigger than or equal to*

$$r' := (r - v) + r_0, \tag{4.31}$$

where $r_0 \leq v$ is the rank of the matrix $\mathbf{L}$ obtained from $\mathbf{L^M}$ by the following three operations:

1. *Firstly, if $v < r$, then we cross out the last $r - v$ rows and the last $r - v$ columns of $\mathbf{L^M}$ (if $v = r$, we do nothing); let $\mathbf{L_1^M}$ be the resulting matrix.*

2. *Secondly, we cross out the last $(p' - p) - (r - v)$ columns of $\mathbf{L_1^M}$ (if $p' = p$, we again do nothing); let the resulting matrix be $\mathbf{L_2^M}$.*

3. *Finally, in $\mathbf{L_2^M}$ we replace the elements $L^M(c_{j_i}, v_k)$ and $L^M(c_{j_i}, \sigma)$ by $L(c_{j_i}, v_k)$ and $L(c_{j_i}, \sigma)$, respectively; the resulting matrix is called $\mathbf{L}$.*

In other words, in the case (H_∞),

$$\mathbf{L} = (L(c_{j_i}, \sigma), L(c_{j_i}, v_1), \ldots, L(c_{j_i}, v_{p-1}))_{1 \leq i \leq v}.$$

In the case (NN_∞),

$$\mathbf{L} = (L(c_{j_i}, v_1), \ldots, L(c_{j_i}, v_{p-1}))_{1 \leq i \leq v},$$

and in the case (ND_∞),

$$\mathbf{L} = (L(c_{j_i}, v_1), \ldots, L(c_{j_i}, v_{p-2})_{1 \leq i \leq v}.$$

PROOF. Let us consider last $r - v$ rows of $\mathbf{L^M}$, which correspond to the indices $j_i = p' - r + i$ of the summable critical points for $i = v + 1, \ldots, r$. By Theorem 4.4(3), on the i-row of $\mathbf{L^M}$, for $i = v + 1, \ldots, r$, all elements are 0, except for the "diagonal" one, which is equal to 1: $L^M(c_{p'-r+i}, v_{p'-r+i}) = 1$. In other words, the lower-right-hand corner of the matrix $\mathbf{L^M}$ contains the $r - v \times r - v$ identity matrix. Therefore, the rank of $\mathbf{L}^M$ is bigger than or equal to $r - \nu$ plus the rank of $\mathbf{L}_1^M$. If we cross out some of the columns of $\mathbf{L}_1^M$, then the rank can only decrease. Hence, the rank of $\mathbf{L}_1^M$ is bigger than or equal to the rank of $\mathbf{L}_2^M$. In turn, its elements are the corresponding elements of the matrix $\mathbf{L}$ defined before, but with the upper index M. Now, we employ the relations (4.19)–(4.21) of Theorem 4.4 and conclude that the ranks of $\mathbf{L}_2^M$ and $\mathbf{L}$ are actually equal. $\qquad\square$

5   PART (B) OF THEOREM 3.6

5.1   The Wolff-Denjoy series

Let $\{\alpha_k\}_{k=0}^{\infty}$, $\{b_k\}_{k=0}^{\infty}$ (where $b_i \neq b_j$, for $k \neq j$) be two sequences of complex numbers. Assume that

$$\sum_{k \geq 0} |\alpha_k|(1 + |b_k|^2) < \infty, \tag{5.1}$$

It implies that two series

$$A = \sum_{k \geq 0} \alpha_k, \quad B = \sum_{k \geq 0} \alpha_k b_k, \tag{5.2}$$

converge absolutely. Define

$$H(x) = \sum_{k=0}^{\infty} \frac{\alpha_k}{b_k - x}. \tag{5.3}$$

Then H is integrable in every disk $B_r = \{|x| < r\}$. Indeed, if σ_x denotes the element of the Lebesgue measure on the plane of the variable x, then

$$\int_{B_r} |H(x)| d\sigma_x \leq \sum_{k \geq 0} |\alpha_k| \int_{B_r} \frac{1}{|x - b_k|} d\sigma_x \leq 2\pi \sum_{k \geq 0} |\alpha_k|(r + |b_k|) < \infty.$$

We define a kind of regularization of H at infinity as

$$\hat{H}(x) = H(x) + \frac{A}{x} + \frac{B}{x^2} = \sum_{n \geq 0} \left(\frac{\alpha_n}{b_n - x} + \frac{\alpha_n}{x} + \frac{\alpha_n b_n}{x^2} \right).$$

The name is justified by the following claim (which must be known though).

LEMMA 5.1. $\hat{H}$ is integrable at infinity.

PROOF. For every n, the function $1/(b_n - x) + 1/x + b_n/x^2$ is integrable at infinity, and one can write, for $r > 0$:

$$\int_{|x| > r} \left| \frac{1}{b_n - x} + \frac{1}{x} + \frac{b_n}{x^2} \right| d\sigma_x = |b_n| \int_{|x| > r/|b_n|} \left| \frac{1}{1 - x} + \frac{1}{x} + \frac{1}{x^2} \right| d\sigma_x$$

$$\leq |b_n| \left(C_1 + \int_{2 > |x| > r/|b_n|} \frac{1}{|x|^2} d\sigma_x \right)$$

$$\leq |b_n| \left(C_2 + 2\pi \ln \frac{|b_n|}{r} \right) \leq C_3 |b_n|(1 + |b_n|),$$

where the constants C_i here and following depend only on r. Now, for every R big enough and using the condition (5.1), we have

$$
\int_{r<|x|<R} |\hat{H}(x)|d\sigma_x \;\leq\; \sum_{k=0}^{\infty} |\alpha_k| \int_{|x|>r} \left| \frac{1}{b_k - x} + \frac{1}{x} + \frac{b_k}{x^2} \right| d\sigma_x
$$

$$
\leq\; C_3 \sum_{k=0}^{\infty} |\alpha_k||b_k|(1 + |b_k|) \leq C_4,
$$

where C_4 does not depend on R. □

We need a contraction property of the operator T_f. This property is initially due to Thurston (see [DH93]). The proof of the following claim is a minor variation of [DH93], [Ep99], [Ma], [Le11].

LEMMA 5.2. *Let H be defined by the series (5.3) under condition (5.1), and let the numbers A, B be defined by (5.2). Denote by K the closure (on the Riemann sphere) of the set $\{b_k\}$:*

$$
K := \overline{\{b_k : k \geq 0\}} \subset \bar{\mathbb{C}}. \tag{5.4}
$$

Assume that K has no interior points. Let f be a rational function with the asymptotics at infinity $f(z) = \sigma z + b + O(1/z)$, such that H is a fixed point of the operator T_f associated with f.

1. *If $A = B = 0$, then either $H = 0$ on the complement K^c of K, or f is an exceptional map.*

2. *If either $|\sigma| \geq 1$ and $b = 0$, or $A = 0$ and $\sigma = 1$, then $H = 0$ on K^c, too.*

PROOF. Note that H is analytic in each component of K^c. Now, take R big enough and consider the disk $D(R) = \{|x| < R\}$. We claim that

$$
\lim_{R\to\infty} \left\{ \int_{f^{-1}(D(R))} |H(x)|d\sigma_x - \int_{D(R)} |H(x)|d\sigma_x \right\} \leq 0. \tag{5.5}
$$

Indeed, in case (1), this follows at once from the integrability of H at infinity. In case (2), the conditions on σ imply that there is $a > 0$ such that

$$
f^{-1}(D(R)) \subset D\left(R + |b| + \frac{a}{R}\right) \tag{5.6}
$$

(actually, $f^{-1}(D(R)) \subset D(R)$, if $|\sigma| > 1$). On the other hand,

$$
\lim_{R\to\infty} \int_{R<|x|<R+|b|+a/R} |H(x)|d\sigma_x = 0. \tag{5.7}
$$

Indeed, by Lemma 5.1, $H(x) = \hat{H}(x) - A/x - B/x^2$, where $\hat{H}$ is integrable at infinity. In particular, $\lim_{R\to\infty} \int_{R<|x|<R+|b|+a/R} |\hat{H}(x)|d\sigma_x = 0$. But an easy calculation shows that the conditions in the case (2) guarantee that

$$
\lim_{R\to\infty} \int_{R<|x|<R+|b|+a/R} \left| \frac{A}{x} + \frac{B}{x^2} \right| d\sigma_x = 0, \tag{5.8}
$$

so (5.7) is proved. This, along with (5.6), gives us (5.5) in the case (2).

Let us show that

$$|T_f H(x)| = \sum_{w:f(w)=x} \frac{|H(w)|}{|f'(w)|^2} \tag{5.9}$$

almost everywhere. Indeed, otherwise there is a set $A \subset D(R_0)$ of positive measure (for some R_0) and $\delta > 0$, such that $|T_f H(x)| < (1 - \delta) \sum_{w:f(w)=x} \frac{|H(w)|}{|f'(w)|^2}$ for $x \in A$. Then, for all $R > R_0$,

$$\begin{aligned}
\int_{D(R)} |H(x)| d\sigma_x &= \int_{D(R)\setminus A} |H(x)| d\sigma_x + \int_A |H(x)| d\sigma_x \\
&= \int_{D(R)\setminus A} |T_f H(x)| d\sigma_x + \int_A |T_f H(x)| d\sigma_x \\
&< \int_{f^{-1}(D(R)\setminus A)} |H(x)| d\sigma_x + (1 - \delta) \int_{f^{-1}(A)} |H(x)| d\sigma_x \\
&= \int_{f^{-1}(D(R))} |H(x)| d\sigma_x - \delta \int_{f^{-1}(A)} |H(x)| d\sigma_x,
\end{aligned}$$

which contradicts (5.5). With (5.9) holding almost everywhere, we proceed as in the previously cited papers. $\qquad\square$

LEMMA 5.3. *Let α_k and b_k ($k \geq 0$) be two sequences of complex numbers such that $\sum_{k\geq 0} |\alpha_k| < \infty$ and the closure K on the Riemann sphere of the set $\{b_k, k \geq 0\}$, where the points b_k are pairwise different, is a C-compact K. If $H(x) = \sum_{k\geq 0} \frac{\alpha_k}{b_k - x}$ is equal to 0 outside of K, then $\alpha_k = 0$ for every k.*

PROOF. (1) The case when K is a C-compact in the plane is classical; see, e.g., [BSZ] (and also the proof of Theorem 2.3).

(2) Now, assume that $\infty \in K$. Take $x_0 \in \mathbb{C} \setminus K$. Let $\epsilon > 0$ be so that $|b_k - x_0| > \epsilon$ for every k. Let $c_k = b_k - x_0$. Then the function $H_1(y) = \sum_{k\geq 0} \alpha_k / (c_k - y)$ is equal to 0 outside of the compact $K_1 = K - x_0$, which is also a C-compact but does not contain the origin. By the definition, the compact $K_2 = 1/K_1 = \{1/y : y \in K_1\}$ is a C-compact on the plane. Consider $H_2(z) = \sum_{k\geq 0}(\alpha_k/c_k)/(1/c_k - z)$. Since $|c_k| > \epsilon$, $\sum_{k\geq 0} |\alpha_k/c_k| < \infty$ still. But, for every $z = 1/y$ outside of K_2, so that y is outside of K_1, $H_2(z) = -yH_1(y) = 0$. Then we apply case (1). $\qquad\square$

5.2 The operator identity for rational functions

Since the proof of Proposition 2.10 is formal, we get the following using Proposition 4.2.

PROPOSITION 5.4. *Let $f \in \Lambda_{d,\bar{p}'}$. Given $z \in \mathbb{C}$, we define $l(z)$ to be the minimal l, such that $f^l(z) = \infty$. If there is no such l, then $l(z) = \infty$. We have*

$$\varphi_{z,\lambda}(x) - \lambda(T_f \varphi_{z,\lambda})(x) = \frac{1}{z - x} + \lambda \sum_{k=1}^{p} \frac{1}{v_k - x} \Phi_k(\lambda, z), \tag{5.10}$$

where

$$\varphi_{z,\lambda}(x) = \sum_{n=0}^{l(z)-1} \frac{\lambda^n}{(f^n)'(z)} \frac{1}{f^n(z) - x}, \tag{5.11}$$

$$\Phi_k(\lambda, z) = \sum_{n=0}^{l(z)-1} \lambda^{n+1} \frac{\frac{\partial f}{\partial v_k}(f^n(z))}{(f^{n+1})'(z)}. \tag{5.12}$$

Putting $\lambda = 1$ and $z = v_j$ in Proposition 5.4 and combining it with the definition of $L(\cdot, \cdot)$ from Theorem 4.4, we get Proposition 5.5.

PROPOSITION 5.5. *Let c_j, $1 \le j \le p$, be a summable critical point of $f \in \Lambda_{d,\bar{p}'}$, such that $v_j \ne \infty$. Then*

$$H_j(x) - (T_f H_j)(x) = \sum_{k=1}^{p} \frac{L(c_j, v_k)}{v_k - x}, \tag{5.13}$$

where

$$H_j(x) = \sum_{n=0}^{l(v_j)-1} \frac{1}{(f^n)'(v_j)(f^n(v_j) - x)}. \tag{5.14}$$

COMMENT 5.6. Note that the sum in (5.13) is over the critical values which are different from infinity. In particular, $l(v_j) \ge 1$ for $1 \le j \le p$.

5.3 Proof of Part (b) of Theorem 3.6

By Corollary 4.5, it is enough to show that the rank of the matrix $\mathbf{L}$ is equal to ν. Now, we follow closely the proof of Theorem 6 of [Le11]. Suppose first we are in the case (H_∞). Assume the rank of the matrix $\mathbf{L}$ is less than ν. Denote by S the set of indexes $j_1, \ldots, j_\nu$. Then the vectors

$$(L(c_j, \sigma), L(c_j, v_1), L(c_j, v_2), \ldots, L(c_j, v_{p-1})), \quad j \in S, \tag{5.15}$$

are linearly dependent: there exist ν numbers a_j, $j \in S$, not all zeros, such that:

$$\sum_{j \in S} a_j L(c_j, \sigma) = 0, \quad \sum_{j \in S} a_j L(c_j, v_k) = 0, \ 1 \le k \le p - 1. \tag{5.16}$$

On the other hand, for every $j \in S$,

$$H_j(x) - (T_f H_j)(x) = \sum_{k=1}^{p} \frac{L(c_j, v_k)}{v_k - x}, \tag{5.17}$$

where

$$H_j(x) = \sum_{n=0}^{l(v_j)-1} \frac{1}{(f^n)'(v_j)(f^n(v_j) - x)}, \tag{5.18}$$

By (5.16), we have that the linear combination $H = \sum_{j \in S} a_j H_j$ satisfies the relation

$$H(z) - (T_f H)(z) = \frac{L}{v_p - z}. \tag{5.19}$$

The function H is reduced to the form

$$H(z) = \sum_{k=0}^{\infty} \frac{\alpha_k}{b_k - z}, \quad b_k \neq b_j, \ k \neq j, \tag{5.20}$$

and since each c_j, $j \in S$, is summable, the sequences α_k, b_k satisfy condition (5.1). Recall the notations $A = \sum_{k \geq 0} \alpha_k$, $B = \sum_{k \geq 0} \alpha_k b_k$, and the regularization $\hat{H}$ of H is

$$\hat{H}(z) = H(z) + \frac{A}{z} + \frac{B}{z^2}.$$

We use the following asymptotics, as $z \to \infty$, which are easily checked (see, e.g., the proof of Lemma 8.2 of [Le11]):

$$T_f \frac{1}{z} = \frac{1}{\sigma z} + \frac{b}{\sigma z^2} + O\left(\frac{1}{z^3}\right), \quad T_f \frac{1}{z^2} = \frac{1}{z^2} + O\left(\frac{1}{z^3}\right). \tag{5.21}$$

Now, we can rewrite (5.19) as follows:

$$\hat{H}(z) - (T_f \hat{H})(z) = \frac{A}{z} - A\left(\frac{1}{\sigma z} + \frac{b}{\sigma z^2}\right) - \frac{L}{z} - \frac{L v_p}{z^2} + O\left(\frac{1}{z^3}\right),$$

or

$$\hat{H}(z) - (T_f \hat{H})(z) = \frac{A(1 - 1/\sigma) - L}{z} + \frac{-Ab/\sigma - L v_p}{z^2} + O\left(\frac{1}{z^3}\right). \tag{5.22}$$

But the function $\hat{H}$ is integrable at infinity; hence, so is $\hat{H} - T_f \hat{H}$, and we can write

$$I_R := \int_{|z|>R} |\hat{H}(z) - (T_f \hat{H})(z)| d\sigma_z \leq \int_{|z|>R} |\hat{H}(z)| d\sigma_z + \int_{f^{-1}(|z|>R)} |\hat{H}(z)| d\sigma_z.$$

It implies that $I_R \to 0$ as $R \to \infty$. But then $\hat{H}(z) - T_f \hat{H}(z) = O\left(\frac{1}{z^3}\right)$ at infinity. Hence, (5.22) implies

$$A\left(1 - \frac{1}{\sigma}\right) - L = 0, \quad -\frac{Ab}{\sigma} - L v_p = 0. \tag{5.23}$$

Since we are in the case (H_∞), then $b = 0$ and $v_p = 1$. Hence, by the second relation of (5.23), $L = 0$. That is, by (5.19), H is a fixed point of T_f. Furthermore, by another condition of the case (H_∞), $\sigma \neq 1$, which, along with $L = 0$ and the first relation of (5.23), imply that $A = 0$. Now we use that $\sum_{j \in S} a_j L(c_j, \sigma) = 0$. By assertions (4.17) and (4.24) of Theorem 4.4,

$$B = \sum_{k \geq 0} \alpha_k b_k = \sum_{j \in S} a_j \sum_{n \geq 0} \frac{f^n(v_j)}{(f^n)'(v_j)} = \sigma \sum_{j \in S} a_j L(c_j, \sigma) = 0. \tag{5.24}$$

Thus $A = B = 0$; hence, the regularization $\hat{H}$ of H takes the form

$$\hat{H}(z) = H(z) + \frac{A}{z} + \frac{B}{z^2} = H(z),$$

i.e., H is an integrable (on the plane) fixed point of T. By Lemma 5.2(1), either $H(z) = 0$ for every z outside of the set K or f is an exceptional rational function.

The latter is excluded; hence, the former holds. By Lemma 5.3 and the definition of H, we get that all α_k in (5.20) are zeros. In other words, $\sum_{j \in S} a_j H_j = 0$. Now we finish the proof as in the proof of Theorem 2.3(b), Subsection 2.5. Note that Corollary 2.7 remains true for rational functions (this follows directly from (4.16) and (4.22) of Theorem 4.4). This allows us to repeat the proof of Lemma 2.13 (with obvious changes) to conclude that all a_j, $j \in S$, are zeros. On the other hand, at least one a_j is not zero, which is a contradiction.

The remaining cases are quite similar.

(NN_∞). The relations (5.23) hold. Since $\sigma = 1$, then the first one gives us $L = 0$, and since $b \neq 0$, the second relation gives $A = 0$. Besides, (5.24) also holds. Then we apply Lemma 5.2(2) and end the proof as in the first case.

(ND_∞). That is, $\sigma = 1$ and $b = 0$. Now, assuming the contrary, we get a nontrivial linear combination $H = \sum_{j \in S} a_j H_j$ such that

$$H(z) - (T_f H)(z) = \frac{L_{p-1}}{v_{p-1} - z} + \frac{L_p}{v_p - z}. \tag{5.25}$$

Then

$$\hat{H}(z) - (T_f \hat{H})(z) = \frac{A}{z} - A\left(\frac{1}{\sigma z} + \frac{b}{\sigma z^2}\right) - \frac{L_{p-1} + L_p}{z} - \frac{L_{p-1} v_{p-1} + L_p v_p}{z^2} + O\left(\frac{1}{z^3}\right)$$

and since the function $\hat{H}$ is integrable at infinity, this implies that

$$A\left(1 - \frac{1}{\sigma}\right) - (L_{p-1} + L_p) = 0, \quad -\frac{Ab}{\sigma} - (L_{p-1} v_{p-1} + L_p v_p) = 0. \tag{5.26}$$

In the considered case, $\sigma = 1$, $b = 0$, and $v_{p-1} = 1 \neq 0 = v_p$; hence, by (5.26), $L_{p-1} = L_p = 0$. Thus H is a fixed point of T_f. We apply Lemma 5.2 (2) and end the proof as in the first case. $\qquad\qquad\square$

COMMENT 5.7. The main results of the paper — Theorem 2.3 and Theorem 3.6 — can be extended to also include non-repelling periodic orbits. Consider the case of rational functions (the polynomial case is very similar and simpler). Assume that $f \in \Lambda_{d,\bar{p}'}$ has r summable critical points c_j, for $j \in S$, and also r_a non-repelling periodic orbits O_j (different from infinity), for $j \in S_a$, where S and S_a are two disjoint subsets of $\{1, 2, \ldots, p'\}$. We assume that the multiplier ρ_j of each O_j is different from 1, and, if $\rho_j = 0$, O_j contains only a single and simple critical point. Assume for simplicity that the orbit of each c_j, $j \in S$, lies in the plane, and all critical values of f lie in the plane, too. Then the matrix $\mathbf{L^M}$ in Theorem 3.6 is, in fact, the matrix $\mathbf{L}$ introduced in Corollary 4.5. Let us extend the matrix $\mathbf{L}$ by the following r_a rows:

$$\left(\frac{\partial \rho_j}{\partial \sigma}, \frac{\partial \rho_j}{\partial v_1}, \ldots, \frac{\partial \rho_j}{\partial v_{p-1}}\right)_{j \in S_a} \quad \text{in the case } (H_\infty),$$

$$\left(\frac{\partial \rho_j}{\partial v_1}, \ldots, \frac{\partial \rho_j}{\partial v_{p-1}}\right)_{j \in S_a} \quad \text{in the case } (NN_\infty), \text{ and by}$$

$$\left(\frac{\partial \rho_j}{\partial v_1}, \ldots, \frac{\partial \rho_j}{\partial v_{p-2}}\right)_{j \in S_a} \quad \text{in the case } (ND_\infty).$$

Then the rank of the extended matrix is maximal, i.e., equal to $r + r_a$.

In the proof, we consider the system of $r + r_a$ relations as follows. For every $j \in S$, we have a relation of Proposition 5.5 of the present paper:

$$H_j(x) - (T_f H_j)(x) = \sum_{k=1}^{p} \frac{L(c_j, v_k)}{v_k - x}, \tag{5.27}$$

and for every $j \in S_a$, we have a similar relation (see Theorem 5 of [Le11]):

$$B_j(x) - (T_f B_j)(x) = \sum_{k=1}^{p} \frac{\partial \rho_j}{\partial v_k} \frac{1}{z - v_k}, \tag{5.28}$$

where $B_j(z) = \sum_{b \in O_j} \left(\frac{\rho_j}{(z-b)^2} + \frac{1}{1-\rho_j} \frac{(f^n)''(b)}{z-b} \right)$. Then we proceed precisely as in the proof of Theorem 6 of [Le11] (or the proof of Theorem 3.6(b) of this paper).

6 Bibliography

[Av]   Avila, A. *Infinitesimal perturbations of rational maps.* Nonlinearity 15 (2002), no. 3, 695–704.

[BE]   Buff, X. and Epstein, A. *Bifurcation measure and postcritically finite rational maps.* Complex dynamics, 491–512, A. K. Peters, Wellesley, MA, 2009.

[BSZ]   Brown, L., Shields, A. and Zeller, K. *On absolutely convergent exponential sums.* Trans. AMS 96 (1960), 162–182.

[BvS]   Bruin, H. and van Strien, S. *Expansion of derivatives in one dimensional dynamics.* Israel J. Math. 137 (2003), 223–263.

[DH85] Douady, A. and Hubbard, J. H. *On the dynamics of polynomial-like maps.* Ann. Scient. Ec. Norm. Sup., 4 serie, 18 (1985), 287–343.

[DH93] Douady, A. and Hubbard, J. H. *A proof of Thurston's topological characterization of rational maps.* Acta Math. 171 (1993), no. 2, 263–297.

[Ep99]   Epstein, A. *Infinitesimal Thurston rigidity and the Fatou-Shishikura inequality*, preprint Stony Brook IMS 1999/1, arXiv:9902158.

[Ep10]   Epstein, A. *Transversality in holomorphic dynamics*, preliminary version (see author's webpage, http://homepages.warwick.ac.uk/~mases/).

[Ga]   Gamelin, T. *Uniform Algebras*, Prentice Hall, 1969.

[GaSh] Gao, B., and Shen, W. *Summability implies Collet–Eckmann almost surely*, Ergod. Th. Dynam. Sys. (2013)   doi:10.1017/etds.2012.173.

[Gau]   Gauthier, T. *Strong-bifurcation loci of full Hausdorff dimension.* Ann. Sci. Ecole Norm. Sup. 45 (2012), 947–984. arXiv:1103.2656.

[GS]   Graczyk, J. and Smirnov, S. *Non-uniform hyperbolicity in complex dynamics.* Invent. Math. 175 (2009), 335–415.

[Le88]   Levin, G. M. *Polynomial Julia sets and Pade's approximations.* (in Russian) Proceedings of XIII Workshop on operator's theory in functional spaces, Kyubishev, USSR, 1988, 113–114.

[Le02]   Levin, G. *On an analytic approach to the Fatou conjecture.* Fundamenta Mathematicae 171 (2002), 177–196.

[Le11]   Levin, G. *Multipliers of periodic orbits in spaces of rational maps.* Ergod. Th. Dynam. Sys. 31 (2011), no. 1, 197–243.

[LSY]   Levin, G., Sodin, M. and Yuditski, P. *A Ruelle operator for a real Julia set.* Comm. Math. Phys. 141 (1991), 119–132.

[M]   Mañé, R. *On a theorem of Fatou.* Bol. Soc. Bras. Mat. 24 (1993), 1–11.

[MSS]   Mañé, R., Sad, P. and Sullivan, D. *On the dynamics of rational maps.* Ann. Sci. Ec. Norm. Sup. 16 (1983), 193–217.

[Mi91]   Milnor, J. *Dynamics in one complex variable*, 3rd ed. Annals of Mathematics Studies 160. Princeton University Press, Princeton, NJ, 2006.

[Mi06]   Milnor, J. *On Lattès maps.* Dynamics on the Riemann sphere, 9–43, Eur. Math. Soc., Zurich, 2006.

[Ma]   Makienko, P. *Remarks on Ruelle operator and invariant line field problem. II.* Ergod. Th. Dynam. Sys. 25 (2005), 1561–1581.

[McM]   McMullen, C. *Complex dynamics and renormalization*, Ann. of Math. Studies 135, Princeton University Press, Princeton, NJ, 1994.

[McSu]   McMullen, C. and Sullivan, D. *Quasiconformal homeomorphisms and dynamics III. The Teichmuller space of a holomorphic dynamical system.* Adv. Math. 135 (1998), 351–395.

[NvS]   Nowicki, T. and van Strien, S. *Absolutely continuous invariant measure under the summability condition.* Invent. Math. 105 (1991), 123–136.

[PR98]   Przytycki, F. and Rohde, S. *Porosity of Collet-Eckmann Julia sets.* Fund. Math. 155 (1998), 189–199.

[PR99]   Przytycki, F. and Rohde, S. *Rigidity of holomorphic Collet-Eckmann repellers.* Ark. Mat. 37 (1999), 357–371.

[RL01]   Rivera-Letelier, J. *On the continuity of Hausdorff dimension of Julia sets and similarity between the Mandelbrot set and Julia sets.* Fundamenta Mathematicae, v. 170 (2001), no. 3, 287–317.

[RL07]   Rivera-Letelier, J. *A connecting lemma for rational maps satisfying a no-growth condition.* Ergod. Th. Dynam. Sys. 27 (2007), 595–636.

[RL]   Rivera-Letelier, J. e-mail communication.

[RLS]   Rivera-Letelier, J. and Shen, W. *Statistical properties of one-dimensional maps under weak hyperbolicity assumptions.* arXiv:1004/0230, (2011).

[Ts]   Tsujii, M. *A simple proof for monotonicity of entropy in the quadratic family.* Erg. Th. Dyn. Syst. 20 (2000), no. 3, 925–933.

[vS]   Van Strien, S. *Misiurewicz maps unfold generically (even if they are critically non-finite).* Fund. Math. 163 (2000), 39–54.

Unmating of rational maps: Sufficient criteria and examples

Daniel Meyer

ABSTRACT. Douady and Hubbard introduced the operation of mating of polynomials. This identifies two filled Julia sets and the dynamics on them via external rays. In many cases one obtains a rational map. Here the opposite question is tackled. Namely, we ask when a given (postcritically finite) rational map f arises as a mating. A sufficient condition when this is possible is given. If this condition is satisfied, we present a simple explicit algorithm to unmate the rational map. This means we decompose f into polynomials that, when mated, yield f. Several examples of unmatings are presented.

1 INTRODUCTION

Douady and Hubbard observed that often "one can find" Julia sets of polynomials within Julia sets of rational maps. This prompted them to introduce the operation of *mating of polynomials*; see [Dou83]. It is a way to glue together two (connected and locally connected) filled Julia sets along their boundaries. The dynamics then descends to the quotient. Somewhat surprisingly, one often obtains a map that is topologically conjugate to a rational map.

Thurston's celebrated theorem on the classification of rational maps among (postcritically finite) topological rational maps (see [DH93]) can be applied to answer the question of when the mating of two polynomials results in (i.e., is topologically conjugate to) a rational map.

Here the opposite question is considered. Namely, we ask whether a given rational map f is a mating. This means to decide whether f arises as (i.e., is topologically conjugate to) a mating of polynomials P_w and P_b. If the answer is yes, one wishes to **unmate** the map f, i.e., to obtain the polynomials P_w and P_b.

The author has recently shown that in the case when f is postcritically finite and the Julia set of f is the whole Riemann sphere, every sufficiently high iterate indeed arises as a mating; see [Mey13] and [Mey]. Indeed, the result remains true for *expanding Thurston maps*.

The purpose of this paper is twofold. The first is to explain the methods used in [Mey13] and [Mey]. Namely, we present a sufficient condition that f arises as a mating of polynomials P_w and P_b. A simple algorithm to unmate f is presented, i.e., to obtain P_w and P_b in a simple combinatorial fashion. The presentation is largely expository, we do not attempt to provide full proofs (which can be found in [Mey13] and [Mey] and are somewhat technical).

The second purpose is to contrast the two opposite cases of (postcritically finite) rational maps. Namely, we contrast hyperbolic maps, where each critical point is contained in the Fatou set, with maps where each critical point is contained in the Julia set (equivalently the Julia set is the whole Riemann sphere $\hat{\mathbb{C}}$).

The following contrasting theorems are obtained.

THEOREM 4.10. *Let $f \colon \widehat{\mathbb{C}} \to \widehat{\mathbb{C}}$ be a hyperbolic, postcritically finite, rational map, with* $\#\mathrm{post}(f) = 3$, *and which is not (conjugate to) a polynomial. Then f does not arise as a mating.*

THEOREM 7.4. *Let $f \colon \widehat{\mathbb{C}} \to \widehat{\mathbb{C}}$ be a postcritically finite rational map, with Julia set* $\mathcal{J}(f) = \widehat{\mathbb{C}}$ *and* $\#\mathrm{post}(f) = 3$. *Then f or f^2 arises as a mating.*

In the case when the rational map f is hyperbolic (and postcritically finite) there is a necessary and sufficient condition that f arises as a mating. Namely, f arises as a mating if and only if f has an **equator** $\mathcal{E}$. An equator is a Jordan curve $\mathcal{E} \subset \widehat{\mathbb{C}} \setminus \mathrm{post}(f)$ such that $f^{-1}(\mathcal{E})$ is orientation-preserving isotopic to $\mathcal{E}$ rel. $\mathrm{post}(f)$. This theorem seems to be folklore, but it does not appear (to the knowledge of the author) in the literature. We give a proof here for the convenience of the reader; see Theorem 4.2. Equators were first introduced in this context by Wittner in his thesis [Wit88].

The existence of an equator, however, is not the right condition for f to arise as a mating in the non-hyperbolic case. There are many examples of maps, in particular, Lattès maps and maps where $\#\mathrm{post}(f) = 3$, that can be shown to have no equator (Proposition 4.3, Proposition 4.4, and Corollary 4.6). Nevertheless, it is known that many of these maps do indeed arise as a mating.

A sufficient condition that a postcritically finite rational map f with Julia set $\mathcal{J}(f) = \widehat{\mathbb{C}}$ arises as a mating is the existence of a **pseudo-equator**. This means there is a Jordan curve $\mathcal{C} \subset \widehat{\mathbb{C}}$ that *contains* all postcritical points and there is a pseudo-isotopy H rel. $\mathrm{post}(f)$ that deforms $\mathcal{C}$ to $\mathcal{C}^1 := f^{-1}(\mathcal{C})$ in an *elementary, orientation-preserving* way. Such a pseudo-equator not only is sufficient for f to arise as a mating, but it allows one to **unmate** f as well. More precisely, from the pseudo-equator (i.e., from the way $\mathcal{C}$ is deformed to $\mathcal{C}^1$) one obtains a matrix. From this matrix one can obtain the *critical portraits* of two polynomials P_{w} and P_{b}. These critical portraits determine P_{w} and P_{b} uniquely. The map f is (topologically conjugate to) the topological mating of P_{w} and P_{b}.

The question arises whether our sufficient condition is in fact necessary. The answer, however, turns out to be no, as an explicit example shows.

The organization of this paper is as follows. In Section 2 we review Moore's theorem. While this theorem is essential in the theory of matings, one usually encounters only a weak form of the theorem. We present a stronger version that deserves to be much better known.

In Section 3 we review the mating construction as well as relevant theorems.

In Section 4 we consider hyperbolic rational maps. We consider **equators**, existence and their connection to matings.

Section 5 introduces an example (it is a Lattès map), which is used in the following as an illustration.

The sufficient criterion for f to arise as a mating from [Mey13] and [Mey] for (postcritically finite) rational maps (whose Julia set is the whole sphere) is presented in Section 6. Namely, f arises as a mating if f has a pseudo-equator.

An equivalent formulation of the sufficient criterion is given in Section 7. Let $\mathcal{C} \subset \widehat{\mathbb{C}}$ be a Jordan curve with $\mathrm{post}(f) \subset \mathcal{C}$. We consider the preimages of the two components of $\widehat{\mathbb{C}} \setminus \mathcal{C}$. We color one component of $\widehat{\mathbb{C}} \setminus \mathcal{C}$ white, the other black. Each component of $\widehat{\mathbb{C}} \setminus f^{-1}(\mathcal{C})$ is a preimage of one component of $\widehat{\mathbb{C}} \setminus \mathcal{C}$. The closure of one such component is called a 1-**tile**. It it colored white/black if it is mapped by f to the white/black component of $\widehat{\mathbb{C}} \setminus \mathcal{C}$, respectively. The 1-tiles tile the sphere in a checkerboard fashion. At critical points several 1-tiles intersect. At each critical point we assign a **connection**. Such a connection formally assigns which 1-tiles are "connected" at this critical point.

The existence of a pseudo-equator is equivalent to the existence of connections at each critical point, such that the white 1-tiles form a spanning tree. Furthermore the "outline" of this spanning tree must be orientation-preserving isotopic to the Jordan curve $\mathcal{C}$.

We will need a description of polynomials that is adapted to matings, i.e., a description in terms of external angles. The one that best suits our needs is the description via **critical portraits**, introduced by Bielefeld-Fisher-Hubbard and Poirier (see [BFH92] and [Poi09]). Essentially, one records the external angles at critical points. Each such critical portrait yields a (unique up to affine conjugation) monic polynomial. We review the necessary material in Section 8.

In Section 9 we present the algorithm that unmates rational maps. Namely, one obtains from a pseudo-equator a matrix that encodes how one edge in $\mathcal{C}$ is deformed to several in $\mathcal{C}^1 := f^{-1}(\mathcal{C})$ by the corresponding pseudo-isotopy. From the Perron-Frobenius eigenvector one obtains the critical portraits of the two polynomials into which f unmates.

In Section 10 we present several examples of unmatings of rational maps.

In Section 11 we show that the existence of a pseudo-equator is not necessary for a rational map f to arise as a mating by giving an explicit example.

We conclude the paper in Section 12 with several open questions.

1.1   Notation

The Riemann sphere is denoted by $\widehat{\mathbb{C}} = \mathbb{C} \cup \{\infty\}$, and the 2-sphere is denoted by S^2 (which is a topological object, not equipped with a conformal structure). The unit circle is S^1, which will often be identified with $\mathbb{R}/\mathbb{Z}$. The circle at infinity is $\{\infty \cdot e^{2\pi i\theta} \mid \theta \in \mathbb{R}/\mathbb{Z}\}$, and the extension of $\mathbb{C}$ to the circle at infinity is $\overline{\mathbb{C}} = \mathbb{C} \cup \{\infty \cdot e^{2\pi i\theta} \mid \theta \in \mathbb{R}/\mathbb{Z}\}$. The Julia set is $\mathcal{J}$; the filled Julia set is $\mathcal{K}$.

We will often *color* objects (polynomials, sets) black and white. White objects will usually be equipped with a subscript w, black objects will be equipped with a subscript b.

If $f = a_0 + a_n(z - c)^n + \ldots$, then the local degree of f at c is $\deg_f(c) = n$.

2   MOORE'S THEOREM

Let X be a compact metric space. An equivalence relation $\sim$ on X is called **closed** if $\{(x,y) \in X \times X \mid x \sim y\}$ is a closed subset of $X \times X$. Equivalently, for any two convergent sequences $(x_n), (y_n)$ in X with $x_n \sim y_n$ for all $n \in \mathbb{N}$, it holds $\lim x_n \sim \lim y_n$.

A map $f \colon X \to Y$ *induces* an equivalence relation on X as follows. For all x and x' in X, let $x \sim x'$ if and only if $f(x) = f(x')$.

LEMMA 2.1. *Let X, Y be compact metric spaces and $f \colon X \to Y$ be a continuous surjection. Let $\sim$ be the equivalence relation on X induced by f. Then $\sim$ is closed and $X/\sim$ is homeomorphic to Y.*

PROOF. Let $x_n \to x$ and $y_n \to y$ be two convergent sequences in X such that $x_n \sim y_n$ for all $n \in \mathbb{N}$. Then

$$f(x) = \lim f(x_n) = \lim f(y_n) = f(y).$$

Thus $x \sim y$. This shows that $\sim$ is closed.

Define $h\colon X/\sim \to Y$ by $h([x]) = f(x)$. Clearly this is a well-defined bijective map. Let $\pi\colon X \to X/\sim$ be the quotient map. Consider an open set $V \subset Y$. Then $f^{-1}(V) = \pi^{-1}(h^{-1}(V))$ is open. Thus $h^{-1}(V)$ is open in $X/\sim$, by the definition of the quotient topology. Thus h is continuous, thus a homeomorphism.   □

Moore's theorem is of central importance in the theory of matings. However in the literature on matings, one usually encounters only a weak form of the theorem.

DEFINITION 2.2. *An equivalence relation $\sim$ on S^2 is of* Moore-type *if*

1. *$\sim$ is not trivial, i.e., there are at least two distinct equivalence classes;*

2. *$\sim$ is closed;*

3. *each equivalence class $[x]$ is connected;*

4. *no equivalence class separates S^2, i.e., $S^2 \setminus [x]$ is connected for each equivalence class $[x]$.*

The reason for the name "Moore-type" is the classical theorem of Moore that asserts that if $\sim$ is of Moore-type, then the quotient space $S^2/\sim$ is homeomorphic to S^2. This statement is, however, true in a stronger form, which we present next.

DEFINITION 2.3. *A homotopy $H\colon X \times [0,1] \to X$ is called a **pseudo-isotopy** if its restriction $H\colon X \times [0,1) \to X$ is an isotopy (i.e., $H(\cdot, t)$ is a homeomorphism for all $t \in [0,1)$). We will always assume that $H(x,0) = x$ for all $x \in X$.*

*Given a set $A \subset S^2$, we call H a pseudo-isotopy **rel.** A if H is a homotopy rel. A, i.e., if $H(a,t) = a$ for all $a \in A$, $t \in [0,1]$. We call the map $h := H(\cdot, 1)$ the **end** of the pseudo-isotopy H.*

We interchangeably write $H(\cdot, t) = H_t(\cdot)$ to unclutter notation.

LEMMA 2.4. *Let $H\colon S^2 \times I \to S^2$ be a pseudo-isotopy. We consider the equivalence relation induced by the end of the pseudo-isotopy, i.e., for $x, y \in S^2$,*

$$x \sim y \quad \Leftrightarrow \quad H_1(x) = H_1(y).$$

Then $\sim$ is of Moore-type.

PROOF. Since S^2 is not contractable, it follows that H_1 is surjective. Thus $\sim$ is not trivial, i.e., not all points in S^2 are equivalent. Lemma 2.1 shows that $\sim$ is closed. Furthermore $S^2/\sim = S^2$.

Consider an equivalence class $[x]$. Assume it is not connected. Then there is a Jordan curve C that separates $[x]$. Let $x_0 \in [x]$. We can assume that H keeps x_0 fixed; otherwise we can compose H with an isotopy such that the composition keeps x_0 fixed. There is a $\delta > 0$ such that $\text{dist}(H_t(C), x_0) > \delta$ for all $t \in [0,1]$.

We call the component of $S^2 \setminus C$ containing x_0 the interior of C, and the other one the exterior of C. Consider a point $y_0 \in [x]$ in the exterior of C. Let $y_t := H_t(y_0)$

for $t \in [0,1]$. By assumption $\lim_{t \to 1} y_t = x_0$. Thus there is a $t_0 \in [0,1)$ such that $|y_{t_0} - x_0| < \delta/2$. It follows that y_{t_0} is in the same component of $S^2 \setminus C_{t_0}$ as x_0, where $C_{t_0} := H_{t_0}(C)$. This is impossible.

Assume now that there is an equivalence class $[x]$ that separates S^2, i.e., $S^2 \setminus [x]$ has (at least) two components. We already know that each equivalence class $[y]$ is connected. Thus it follows that each equivalence class $[y] \neq [x]$ is contained in one of the components of $S^2 \setminus [x]$. It follows that if we remove the point $[x]$ from $S^2/\sim$ we obtain a disconnected set. This violates the fact that $S^2/\sim$ is homeomorphic to S^2. $\qquad\square$

THEOREM 2.5 (Moore, 1925). *Let $\sim$ be an equivalence relation on S^2. Then $\sim$ is of Moore-type if and only if the relation $\sim$ can be realized as the end of a pseudo-isotopy $H\colon S^2 \times [0,1] \to S^2$ (i.e., $x \sim y \Leftrightarrow H_1(x) = H_1(y)$).*

The "if direction" was shown in Lemma 2.4. A proof of the other direction can be found in [Dav86, Theorem 25.1 and Theorem 13.4]. From Lemma 2.1 we immediately recover the original form of Moore's theorem, i.e., that $S^2/\sim$ is homeomorphic to S^2.

The following theorem was originally proved by Baer (see [Bae27, Bae28]); a more modern treatment can be found in [Eps66, Theorem 2.1].

THEOREM 2.6. *Let $P \subset S^2$ be a finite set and $C, C'\colon S^1 \to S^2 \setminus P$ be Jordan curves that are homotopic in $S^2 \setminus P$, i.e., there is a homotopy $H\colon S^1 \times I \to S^2 \setminus P$ such that $H(S^1, 0) = C$ and $H(S^1, 1) = C'$. Furthermore, each component of $S^2 \setminus C$ contains at least a point in P.*

Then C, C' are isotopic rel. P, i.e., there is an isotopy $K\colon S^2 \times I \to S^2$ rel. P such that $K(\cdot, 0) = \mathrm{id}_{S^2}$ and $K(C, 1) = C'$.

3   MATING OF POLYNOMIALS

An excellent introduction to matings can be found in [Mil04]. In fact this paper was the main inspiration for the papers [Mey13] and [Mey]. See also [Wit88], [SL00], and [MP12].

We still give the basic definitions here, but we will be brief.

Recall that a **Thurston map** is an orientation-preserving branched covering map $f\colon S^2 \to S^2$ of degree at least 2 that is postcritically finite (see [BM11] for more background). A Thurston map f is called a **Thurston polynomial** if there is a totally invariant point, i.e., a point $\infty \in S^2$ such that $f(\infty) = f^{-1}(\infty) = \infty$.

3.1   Topological mating

Let P be a monic polynomial (i.e., the coefficient of the leading term is 1) with connected and locally connected filled Julia set $\mathcal{K}$. Let $\phi\colon \widehat{\mathbb{C}} \setminus \overline{\mathbb{D}} \to \widehat{\mathbb{C}} \setminus \mathcal{K}$ be the Riemann map normalized by $\phi(\infty) = \infty$ and $\phi'(\infty) = \lim_{z \to \infty} z/\phi(z) > 0$ (in fact, then $\phi'(\infty) = 1$). By Carathéodory's theorem (see for example [Mil99, Theorem 17.14]), ϕ extends continuously to

$$\sigma\colon S^1 = \partial\overline{\mathbb{D}} \to \partial\mathcal{K} = \mathcal{J}, \tag{3.1}$$

where $\mathcal{J}$ is the **Julia set** of P. We call the map σ the **Carathéodory semi-conjugacy** of $\mathcal{J}$. We remind the reader that every postcritically finite polynomial has connected and locally connected filled Julia set (see, for example, [Mil99, Theorem 19.7]).

Consider the equivalence relation on S^1 induced by the Carathéodory semi-conjugacy, namely,

$$s \sim t :\Leftrightarrow \sigma(s) = \sigma(t), \tag{3.2}$$

for all $s, t \in S^1$. Lemma 2.1 yields that $S^1/\sim$ is homeomorphic to $\mathcal{J}$, where the homeomorphism is given by $h: S^1/\sim \to \mathcal{J}$, $[s] \mapsto \sigma(s)$. Böttcher's theorem (see for example [Mil99, § 9]) says that the Riemann map ϕ conjugates z^d to the polynomial P on $\widehat{\mathbb{C}} \setminus \overline{\mathbb{D}}$, where $d = \deg P$. This means that the following diagram commutes:

$$
\begin{array}{ccc}
S^1/\sim & \xrightarrow{z^d/\sim} & S^1/\sim \\
\downarrow{\scriptstyle h} & & \downarrow{\scriptstyle h} \\
\mathcal{J} & \xrightarrow{P} & \mathcal{J}.
\end{array}
\tag{3.3}
$$

It will be convenient to identify the unit circle S^1 with $\mathbb{R}/\mathbb{Z}$. We still write

$$\sigma: \mathbb{R}/\mathbb{Z} \to \mathcal{J} = \partial\mathcal{K} \tag{3.4}$$

for the Carathéodory semi-conjugacy.

Consider two monic polynomials $P_{\mathtt{w}}$ and $P_{\mathtt{b}}$ (called the *white* and the *black* polynomial) of the same degree with connected and locally connected Julia sets. Let $\sigma_{\mathtt{w}}$ and $\sigma_{\mathtt{b}}$ be the Carathéodory semi-conjugacies of their Julia sets $\mathcal{J}_{\mathtt{w}}$ and $\mathcal{J}_{\mathtt{b}}$.

Glue the filled Julia sets $\mathcal{K}_{\mathtt{w}}$ and $\mathcal{K}_{\mathtt{b}}$ (of $P_{\mathtt{w}}$ and $P_{\mathtt{b}}$) together by identifying $\sigma_{\mathtt{w}}(t) \in \partial\mathcal{K}_{\mathtt{w}}$ with $\sigma_{\mathtt{b}}(-t) \in \partial\mathcal{K}_{\mathtt{b}}$. More precisely, we consider the disjoint union of $\mathcal{K}_{\mathtt{w}}$ and $\mathcal{K}_{\mathtt{b}}$, and let $\mathcal{K}_{\mathtt{w}} \amalg\!\!\!\perp \mathcal{K}_{\mathtt{b}}$ be the quotient obtained from the equivalence relation generated by $\sigma_{\mathtt{w}}(t) \sim \sigma_{\mathtt{b}}(-t)$ for all $t \in \mathbb{R}/\mathbb{Z}$. The minus sign is customary here, though not essential: identifying $\sigma_{\mathtt{w}}(t)$ with $\sigma_{\mathtt{b}}(t)$ amounts to the mating of $P_{\mathtt{w}}$ with $\overline{P_{\mathtt{b}}(\bar{z})}$. The **topological mating** of $P_{\mathtt{w}}$ and $P_{\mathtt{b}}$ is the map

$$P_{\mathtt{w}} \amalg\!\!\!\perp P_{\mathtt{b}}: \mathcal{K}_{\mathtt{w}} \amalg\!\!\!\perp \mathcal{K}_{\mathtt{b}} \to \mathcal{K}_{\mathtt{w}} \amalg\!\!\!\perp \mathcal{K}_{\mathtt{b}},$$

given by

$$P_{\mathtt{w}} \amalg\!\!\!\perp P_{\mathtt{b}}|_{\mathcal{K}_i} = P_i,$$

for $i = \mathtt{w}, \mathtt{b}$. It follows from (3.3) that it is well defined, namely, that

$$x_{\mathtt{w}} \sim x_{\mathtt{b}} \Rightarrow P_{\mathtt{w}}(x_{\mathtt{w}}) \sim P_{\mathtt{b}}(x_{\mathtt{b}}) \quad \text{for all } x_{\mathtt{w}} \in \mathcal{K}_{\mathtt{w}}, x_{\mathtt{b}} \in \mathcal{K}_{\mathtt{b}}.$$

DEFINITION 3.1. *A Thurston map* $f: S^2 \to S^2$ *is said to* **arise as a mating** *if* f *is topologically conjugate to* $P_{\mathtt{w}} \amalg\!\!\!\perp P_{\mathtt{b}}$, *the (topological) mating of polynomials* $P_{\mathtt{w}}$ *and* $P_{\mathtt{b}}$.

The Thurston map $g: S^2 \to S^2$ *is* **equivalent to a mating** *if* g *is Thurston equivalent to a Thurston map* $f: S^2 \to S^2$ *arising as a mating.*

The Rees-Shishikura-Tan theorem which follows is one of the most important results in the theory of matings.

THEOREM 3.2 ([TL92], [Ree92],[Shi00]). *Let $P_{\mathtt{w}} = z^2 + c_{\mathtt{w}}$ and $P_{\mathtt{b}} = z^2 + c_{\mathtt{b}}$ be two quadratic postcritically finite polynomials, where $c_{\mathtt{w}}$ and $c_{\mathtt{b}}$ are not in conjugate limbs of the Mandelbrot set. Then the topological mating of $P_{\mathtt{w}}$ and $P_{\mathtt{b}}$ exists and is topologically conjugate to a (postcritically finite) rational map (of degree 2).*

Recall that a *limb* of the Mandelbrot set is a component of the complement of the main cardioid.

3.2 Formal mating

We now define the **formal mating**. Its main purpose is to break up the topological mating into several steps.

We extend $\mathbb{C}$ to the **circle at infinity** $\{\infty \cdot e^{2\pi i\theta} \mid \theta \in \mathbb{R}/\mathbb{Z}\}$ and let

$$\overline{\mathbb{C}} = \mathbb{C} \cup \{\infty \cdot e^{2\pi i\theta} \mid \theta \in \mathbb{R}/\mathbb{Z}\}.$$

If P is a monic polynomial of degree d, set $P(\infty \cdot e^{2\pi i\theta}) = \infty \cdot e^{2\pi i d\theta}$ to obtain a continuous extension of P to $\overline{\mathbb{C}}$.

Consider now two monic polynomials $P_{\mathtt{w}} \colon \overline{\mathbb{C}}_{\mathtt{w}} \to \overline{\mathbb{C}}_{\mathtt{w}}, P_{\mathtt{b}} \colon \overline{\mathbb{C}}_{\mathtt{b}} \to \overline{\mathbb{C}}_{\mathtt{b}}$ of the same degree d. Here $\overline{\mathbb{C}}_{\mathtt{w}} = \overline{\mathbb{C}}_{\mathtt{b}} = \overline{\mathbb{C}}$, but we prefer that $P_{\mathtt{w}}$ and $P_{\mathtt{b}}$ are defined on disjoint domains. On the disjoint union $\overline{\mathbb{C}}_{\mathtt{w}} \sqcup \overline{\mathbb{C}}_{\mathtt{b}}$, we define the equivalence relation $\approx$ by identifying $\infty \cdot e^{2\pi i\theta} \in \overline{\mathbb{C}}_{\mathtt{w}}$ with $\infty \cdot e^{-2\pi i\theta} \in \overline{\mathbb{C}}_{\mathtt{b}}$. We call the quotient

$$\overline{\mathbb{C}}_{\mathtt{w}} \uplus \overline{\mathbb{C}}_{\mathtt{b}} := \overline{\mathbb{C}}_{\mathtt{w}} \sqcup \overline{\mathbb{C}}_{\mathtt{b}}/\approx,$$

the *formal mating sphere*. Clearly, this is a topological sphere. The map

$$P_{\mathtt{w}} \uplus P_{\mathtt{b}} \colon \overline{\mathbb{C}}_{\mathtt{w}} \uplus \overline{\mathbb{C}}_{\mathtt{b}} \to \overline{\mathbb{C}}_{\mathtt{w}} \uplus \overline{\mathbb{C}}_{\mathtt{b}}$$

given by

$$P_{\mathtt{w}} \uplus P_{\mathtt{b}}(x) = \begin{cases} P_{\mathtt{w}}(x), & x \in \overline{\mathbb{C}}_{\mathtt{w}} \\ P_{\mathtt{b}}(x), & x \in \overline{\mathbb{C}}_{\mathtt{b}} \end{cases}$$

is well defined. It is called the **formal mating** of $P_{\mathtt{w}}$ and $P_{\mathtt{b}}$. The **equator** of the formal mating sphere is the set $\{\infty \cdot e^{2\pi i\theta} \mid \theta \in \mathbb{R}/\mathbb{Z}\} \subset \overline{\mathbb{C}}_{\mathtt{w}} \subset \overline{\mathbb{C}}_{\mathtt{w}} \uplus \overline{\mathbb{C}}_{\mathtt{b}}$.

Assume now that the Julia sets of $P_{\mathtt{w}}$ and $P_{\mathtt{b}}$ are connected and locally connected. This means that every external ray lands (at a unique point). The *ray-equivalence* relation is the equivalence relation $\sim$ on the formal mating sphere generated by external rays. Here a **closed external ray** $R(\theta) \subset \overline{\mathbb{C}}_{\mathtt{w}}$ (where $\theta \in \mathbb{R}/\mathbb{Z}$) is one that includes its landing point as well as its point at infinity, that is, we have $\infty \cdot e^{2\pi i\theta} \in R(\theta)$. Then $\sim$ is the smallest equivalence relation on $\overline{\mathbb{C}}_{\mathtt{w}} \uplus \overline{\mathbb{C}}_{\mathtt{b}}$ such that all points in any closed external ray $R(\theta) \subset \overline{\mathbb{C}}_{\mathtt{w}}$, as well as all points in any closed external ray $R(\theta) \subset \overline{\mathbb{C}}_{\mathtt{b}}$, are equivalent. Note that the point $\infty \cdot e^{2\pi i\theta}$ (on the equator) is contained in both $R(\theta) \subset \overline{\mathbb{C}}_{\mathtt{w}}$ and $R(-\theta) \subset \overline{\mathbb{C}}_{\mathtt{b}}$. Thus all points on these two external rays are equivalent.

It is easy to show that $P_{\mathtt{w}} \uplus P_{\mathtt{b}}$ descends to the quotient $\overline{\mathbb{C}}_{\mathtt{w}} \uplus \overline{\mathbb{C}}_{\mathtt{b}}/\sim$. Moreover this quotient map is topologically conjugate to the topological mating $P_{\mathtt{w}} \perp\!\!\!\perp P_{\mathtt{b}}$ (see [MP12, Proposition 4.10]).

We will need the following theorem, originally proved by Rees [Ree92, §1.15]. A more general version was proved by Shishikura [Shi00, Theorem 1.7].

THEOREM 3.3. *Let P_w and P_b be two hyperbolic, postcritically finite, monic polynomials such that the formal mating is Thurston equivalent to a (postcritically finite) rational map $f\colon \widehat{\mathbb{C}} \to \widehat{\mathbb{C}}$. Then the topological mating $P_w \perp\!\!\!\perp P_b$ is topologically conjugate to f.*

It is well known that the Julia set of a postcritically finite rational map is locally connected. Furthermore each point in the Julia set of a polynomial is the landing point of only finitely many external rays (this is Douady's lemma; see [Shi00, Lemma 4.2]). Assume the quotient $\mathbb{C}_w \uplus \mathbb{C}_b /\sim$, equivalently $\mathcal{K}_w \perp\!\!\!\perp \mathcal{K}_b$, is a topological sphere. Then each ray equivalence class forms a tree. This tree is actually finite, as was proved by A. Epstein (see [MP12, Proposition 4.12]). If this tree does not contain a critical point, it is mapped homeomorphically by the formal mating $P_w \uplus P_b$. We thus have the following lemma.

LEMMA 3.4. *Assume the ray-equivalence relation $\sim$ on the formal mating sphere is as given before. Let $x \in S^1$ (i.e., in the equator of the formal mating sphere) be such that $[x]$ (i.e., the equivalence class of x) does not contain a postcritical point of the formal mating $P_w \uplus P_b$. Then the d preimages of $[x]$ by $P_w \uplus P_b$ are given by $[(x+k)/d]$ for $k = 0, \ldots, d-1$. The sets $[(x+k)/d]$ are disjoint.*

4   EQUATORS AND HYPERBOLIC RATIONAL MAPS

DEFINITION 4.1 (Equator). *Let $f\colon S^2 \to S^2$ be a Thurston map. Then a Jordan curve $\mathcal{E} \subset S^2 \setminus \mathrm{post}(f)$ is an **equator** for f if the following three conditions are satisfied.*

1. $\widetilde{\mathcal{E}} := f^{-1}(\mathcal{E})$ *consists of a single component.*

Since $\mathcal{E}$ contains no postcritical point, condition (1) implies that $\widetilde{\mathcal{E}}$ is a Jordan curve (in $S^2 \setminus f^{-1}(\mathrm{post}(f)) \subset S^2 \setminus \mathrm{post}(f)$). Furthermore the degree of the map $f\colon \widetilde{\mathcal{E}} \to \mathcal{E}$ is $d = \deg f$.

2. $\widetilde{\mathcal{E}}$ *is isotopic to $\mathcal{E}$ rel. $\mathrm{post}(f)$.*

Fix an orientation of $\mathcal{E}$. The curve $\widetilde{\mathcal{E}}$ then inherits an orientation in two distinct ways. First, the (covering) map $f\colon \widetilde{\mathcal{E}} \to \mathcal{E}$ induces an orientation on $\widetilde{\mathcal{E}}$.

Second, according to (2), let H_t be the isotopy rel. $\mathrm{post}(f)$ that deforms $\mathcal{E}$ to $\widetilde{\mathcal{E}}$, i.e., $H_0 = \mathrm{id}_{S^2}, H_1(\mathcal{E}) = \widetilde{\mathcal{E}}$. We choose the orientation on $\widetilde{\mathcal{E}}$ such that (the homeomorphism) $H_1\colon \widetilde{\mathcal{E}} \to \mathcal{E}$ is orientation-preserving.

3. $\widetilde{\mathcal{E}}$ *is **orientation-preserving** isotopic to $\mathcal{E}$ rel. $\mathrm{post}(f)$.*

This means that the two orientations on $\widetilde{\mathcal{E}}$ as given before agree.

If there is an equator $\mathcal{E}$ for the Thurston map f, we say that f has an equator. This notion was first introduced by Wittner in [Wit88]. The importance of equators for matings is shown by the following theorem.

THEOREM 4.2. *A hyperbolic postcritically finite rational map f arises as a mating if and only if f has an equator.*

We will need some preparation for the proof of this theorem. There are, however, many classes of Thurston maps that do not have an equator.

PROPOSITION 4.3. *Let $f \colon S^2 \to S^2$ be a Thurston map that is not a Thurston polynomial, with $\#\operatorname{post}(f) \leq 3$. Then f does not have an equator.*

PROOF. No Thurston map with $\#\operatorname{post}(f) = 0$ or $\#\operatorname{post}(f) = 1$ exists (see [BM11, Remark 5.16]).

If $\#\operatorname{post}(f) = 2$, then f is Thurston equivalent to $g = z^m \colon \widehat{\mathbb{C}} \to \widehat{\mathbb{C}}$, where $m \in \mathbb{Z} \setminus \{-1, 0, 1\}$ (see [BM11, Proposition 7.1]). The case $m \geq 2$ means that f is a Thurston polynomial. In the case $m \leq -2$ there is no equator, since each Jordan curve $\mathcal{C} \subset \widehat{\mathbb{C}} \setminus \operatorname{post}(g) = \mathbb{C} \setminus \{0\}$ which separates the postcritical points $0, \infty$ is isotopic (rel. $\operatorname{post}(g) = \{0, \infty\}$) to a small closed loop around 0. The preimage of such a loop is isotopic rel. $\{0, \infty\}$ to itself, but not orientation-preserving isotopic to itself.

Assume now that $\operatorname{post}(f)$ consists of three points, which are denoted for convenience by $0, 1, \infty$. Let $\mathcal{E} \subset S^2 \setminus \operatorname{post}(f)$ be a Jordan curve, which we assume to be not null-homotopic. Then $\mathcal{E}$ is isotopic in $S^2 \setminus \operatorname{post}(f)$ to a small loop $\mathcal{C}$ around a postcritical point, without loss of generality around ∞. Let $\widetilde{\mathcal{C}} := f^{-1}(\mathcal{C})$. For each $c \in f^{-1}(\infty)$ there is exactly one small loop around c in $\widetilde{\mathcal{C}}$. Thus $\widetilde{\mathcal{C}}$ is a single component if and only if ∞ has a single preimage c by f. If $c = \infty$ this means that f is a Thurston polynomial. If $c \neq \infty$, this means that $\widetilde{\mathcal{C}}$ is not isotopic rel. $\operatorname{post}(f)$ to $\mathcal{C}$. Thus if f is not a Thurston polynomial, it follows that $\mathcal{C}$, hence $\mathcal{E}$, is not an equator of f. $\square$

PROPOSITION 4.4. *Let $f \colon \widehat{\mathbb{C}} \to \widehat{\mathbb{C}}$ be a Lattès map. Then f does not have an equator.*

PROOF. A Lattès map f has three or four postcritical points. In the first case f has no equator, by Proposition 4.3.

In the second case we use the (well-known) explicit description of the map f; see [DH93, Proposition 9.3] and [Mil06, Theorem 3.1, Section 4, and Section 5].

Namely, there is a lattice $\Lambda = \mathbb{Z} \oplus \tau \mathbb{Z}$, where τ is in the upper half-plane. The group G is the subgroup of $\operatorname{Aut}(\mathbb{C})$ generated by the translations $z \mapsto z + 1$, $z \mapsto z + \tau$ and the involution $z \mapsto -z$. The quotient map $\wp \colon \mathbb{C} \to \mathbb{C}/G$ may be viewed as the Weierstraß $\wp$-function. Finally there is a map $L \colon \mathbb{C} \to \mathbb{C}$, given by $L(z) = az + b$, where $a, b \in \mathbb{C}$. The map f then is topologically conjugate to $L/G \colon \mathbb{C}/G \to \mathbb{C}/G$. Here $|a|^2 = \deg f$. There are two possible cases: either $a \in \mathbb{Z} \setminus \{-1, 0, 1\}$ or $a \notin \mathbb{R}$ (indeed, there is a more complete description).

Let $\mathcal{E} \subset S^2 \setminus \operatorname{post}(f)$ be an equator for f. Since f is not a Thurston polynomial it follows that $\mathcal{E}$ is not peripheral, i.e., each component of $S^2 \setminus \mathcal{E}$ contains two postcritical points. The curve $\mathcal{E}$ is isotopic rel. $\operatorname{post}(f)$ to a "straight curve" $\mathcal{C}$, i.e., every component of $\widetilde{\mathcal{C}} := \wp^{-1}(\mathcal{C}) \subset \mathbb{C}$ is a straight line. Note that if two such curves $\mathcal{C}, \mathcal{C}'$ lift to lines with distinct slopes, they are not isotopic rel. $\operatorname{post}(f)$.

Let $\mathcal{C}' := f^{-1}(\mathcal{C})$. Then the lift by $\wp$ to $\mathbb{C}$ satisfies $\widetilde{\mathcal{C}'} := \wp^{-1}(\mathcal{C}') = L^{-1}(\widetilde{\mathcal{C}})$.

If $a \in \mathbb{Z} \setminus \{-1, 0, 1\}$, then $\mathcal{C}$ has $|a|$ preimages by f, since $L^{-1}(\wp(\mathcal{C}))/G$ consist of $|a|$ distinct curves. In particular, $f^{-1}(\mathcal{C})$; hence $f^{-1}(\mathcal{E})$ does not consist of a single component. Thus $\mathcal{E}$ is not an equator.

If $a \notin \mathbb{R}$ then $\widetilde{\mathcal{C}'} = L^{-1}(\widetilde{\mathcal{C}})$ consist of parallel lines which intersect the lines in $\widetilde{\mathcal{C}}$ in the angle $\arg a$. In particular each component of $\widetilde{\mathcal{C}}$ is not isotopic rel. $\operatorname{post}(f)$ to $\mathcal{C}$. Thus $\mathcal{C}$, hence $\mathcal{E}$, is not an equator. $\square$

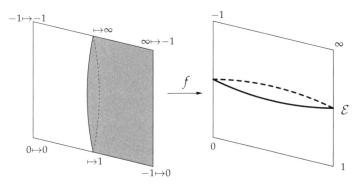

FIGURE 4.5: An obstructed parabolic map with an equator.

The previous result shows the following. While the existence of an equator is exactly the right condition to check whether a (postcritically finite) hyperbolic rational map arises as a mating (by Theorem 4.2), it is not the right condition in the case when f is not hyperbolic. Namely, there are many examples of Lattès maps known which arise as a mating (see [Mil04], [Mey13]), yet they do not have an equator by the previous theorem.

COROLLARY 4.6. *A rational map with parabolic orbifold that is not a (Thurston) polynomial does not have an equator.*

PROOF. A Thurston map f with parabolic orbifold has at most four postcritical points. If $\#\mathrm{post}(f) \le 3$, the result follows from Proposition 4.3. Every rational map with parabolic orbifold and $\#\mathrm{post}(f) = 4$ is a Lattès map, where the result follows from Proposition 4.4.   □

REMARK 4.7. It is possible to find obstructed Thurston maps with parabolic orbifold that have an equator. Namely, the map $z = x + yi \mapsto 2x + yi$ descends to the quotient by the Weierstraß $\wp$-function (to the lattice $L = \mathbb{Z} \oplus i\mathbb{Z}$). A topological model of this map f is given as follows. Glue two squares $[0,1]^2$ together along their boundaries to form a topological sphere. We label the vertices of the resulting "pillow" with $0, 1, -1, \infty$ for convenience. Each square (i.e., each side of the pillow) is divided into the two rectangles $[0, 1/2] \times [0,1]$ and $[1/2, 1] \times [0,1]$. These rectangles are now affinely mapped to the squares as indicated in Figure 4.5 (essentially by $x + yi \mapsto 2x + yi$). The resulting map is Thurston equivalent (indeed, topologically conjugate) to f. The map is obstructed and has parabolic orbifold. An equator $\mathcal{E}$ is indicated to the right in Figure 4.5.

The following theorem first appeared in Wittner's thesis [Wit88, Theorem 7.2.1], where he credits Thurston. Since Wittner's thesis is not easily available, we provide the proof here.

THEOREM 4.8. *Let $f\colon \widehat{\mathbb{C}} \to \widehat{\mathbb{C}}$ be a postcritically finite rational map that has an equator $\mathcal{E} \subset S^2 \setminus \mathrm{post}(f)$. Then f is Thurston equivalent to the formal mating of two polynomials.*

PROOF. Let f be a postcritically finite rational map that has an equator.

If f has parabolic orbifold, it follows from Corollary 4.6 that f is (conjugate to) a monic polynomial P. In this case the statement is trivial, since f is Thurston equivalent to the formal mating of P with z^d (where $d = \deg f$). Thus we can assume from now on that f has hyperbolic orbifold.

Let $\mathcal{E}$ be an equator for f, and $\widetilde{\mathcal{E}} := f^{-1}(\mathcal{E})$. Let $\widetilde{U}_{\mathsf{w}}$ and $\widetilde{U}_{\mathsf{b}}$ be the two components of $\widehat{\mathbb{C}} \setminus \widetilde{\mathcal{E}}$. The map f maps $\widetilde{U}_{\mathsf{w}}$, as well as $\widetilde{U}_{\mathsf{b}}$, (properly) to one component of $\widehat{\mathbb{C}} \setminus \mathcal{E}$. Thus we denote the two components U_{w} and U_{b} of $\widehat{\mathbb{C}} \setminus \mathcal{E}$ such that $f(\widetilde{U}_{\mathsf{w}}) = U_{\mathsf{w}}, f(\widetilde{U}_{\mathsf{b}}) = U_{\mathsf{b}}$.

Let $H \colon \widehat{\mathbb{C}} \times [0,1]$ be an isotopy rel. post(f) that deforms $\mathcal{E}$ to $\widetilde{\mathcal{E}}$, i.e., $H_0 = \mathrm{id}_{\widehat{\mathbb{C}}}$ and $H_1(\mathcal{E}) = \widetilde{\mathcal{E}}$. Then $\widetilde{\mathcal{E}}$ is (forward and backward) invariant for the map $\widetilde{f} \colon \widehat{\mathbb{C}} \to \widehat{\mathbb{C}}$ given by $\widetilde{f} = H_1 \circ f$, i.e., $\widetilde{f}(\widetilde{\mathcal{E}}) = \widetilde{\mathcal{E}} = \widetilde{f}^{-1}(\widetilde{\mathcal{E}})$. We can choose the isotopy H such that $\widetilde{f} \colon \widetilde{\mathcal{E}} \to \widetilde{\mathcal{E}}$ is topologically conjugate to $z^d \colon S^1 \to S^1$ (where $d = \deg \widetilde{f} = \deg f$). We will assume this from now on.

Since $\widetilde{\mathcal{E}}$ is orientation-preserving isotopic (rel. post(f)) to $\mathcal{E}$, it follows that $H_1(U_{\mathsf{w}}) = \widetilde{U}_{\mathsf{w}}$ and $H_1(U_{\mathsf{b}}) = \widetilde{U}_{\mathsf{b}}$.

On $\widetilde{X}_{\mathsf{w}} := \widetilde{U}_{\mathsf{w}} \cup \widetilde{\mathcal{E}}$ we consider the equivalence relation which has the following equivalence classes. The set $\widetilde{\mathcal{E}}$ is an equivalence class and every singleton $\{x\} \subset U_{\mathsf{w}}$ is an equivalence class.

Consider the quotient space $S^2_{\mathsf{w}} := \mathrm{clos}\,\widetilde{U}_{\mathsf{w}}/\widetilde{\mathcal{E}}$. Certainly S^2_{w} is a topological sphere.

The map $H_1 \circ f$ maps $\widetilde{\mathcal{E}}$ to $\widetilde{\mathcal{E}}$ and each $x \in \widetilde{U}_{\mathsf{w}}$ into $\widetilde{U}_{\mathsf{w}}$. Thus this map descends naturally to the quotient S^2_{w}, i.e., to a map $p_{\mathsf{w}} \colon S^2_{\mathsf{w}} \to S^2_{\mathsf{w}}$. Call the point $[\widetilde{\mathcal{E}}] \in S^2_{\mathsf{w}}$ for convenience ∞. We note that $p_{\mathsf{w}}(\infty) = p_{\mathsf{w}}^{-1}(\infty) = \infty$. Clearly p_{w} is a postcritically finite branched covering. Thus p_{w} is a Thurston polynomial.

It is well known that a Thurston polynomial is obstructed if and only if it has a *Levy cycle* (see [BFH92, Section 5]). Assume that p_{w} is obstructed, i.e., has a Levy cycle. Then f would have a Levy cycle as well. Since we assume that f is a rational map with hyperbolic orbifold, this cannot happen. Thus p_{w} is Thurston equivalent to a polynomial P_{w}, which we assume to be monic and centered. We extend this polynomial to the circle at infinity $S^1_{\infty} = \{\infty \cdot e^{2\pi i\theta} \mid \theta \in \mathbb{R}/\mathbb{Z}\}$; this extension is still denoted by $P_{\mathsf{w}} \colon \overline{\mathbb{C}}_{\mathsf{w}} \to \overline{\mathbb{C}}_{\mathsf{w}}$.

CLAIM 4.9. *There are homeomorphisms $h_0, h_1 \colon \overline{\mathbb{C}}_{\mathsf{w}} \to \widetilde{X}_{\mathsf{w}}$ with the following properties:*

- $h_0 \circ P_{\mathsf{w}}(z) = \widetilde{f} \circ h_1(z)$ for all $z \in \overline{\mathbb{C}}_{\mathsf{w}}$, i.e., the following diagram commutes

$$
\begin{array}{ccc}
\overline{\mathbb{C}}_{\mathsf{w}} & \xrightarrow{\ h_1\ } & \widetilde{X}_{\mathsf{w}} \\
{\scriptstyle P_{\mathsf{w}}}\downarrow & & \downarrow{\scriptstyle \widetilde{f}} \\
\overline{\mathbb{C}}_{\mathsf{w}} & \xrightarrow[\ h_0\ ]{} & \widetilde{X}_{\mathsf{w}}\,;
\end{array}
$$

- h_0, h_1 are isotopic rel. post$(P_{\mathsf{w}}) \cup S^1_{\infty}$, in particular $h_0 = h_1$ on S^1_{∞}.

PROOF OF CLAIM. We will deform $P_{\mathsf{w}} \colon \overline{\mathbb{C}}_{\mathsf{w}} \to \overline{\mathbb{C}}_{\mathsf{w}}$ by an isotopy rel. post$(P_{\mathsf{w}}) \cup S^1_{\infty}$ and $\widetilde{f} \colon \widetilde{X}_{\mathsf{w}} \to \widetilde{X}_{\mathsf{w}}$ by an isotopy rel. post$(\widetilde{f}) \cup \widetilde{\mathcal{E}}$. It is enough to prove the statement with these deformed maps.

More precisely, we can postcompose P_{w} with an isotopy rel. post$(P_{\mathsf{w}}) \cup S^1_{\infty}$ such that the resulting map is topologically conjugate to $\varphi \colon S^1 \times [0,1] \to S^1 \times [0,1]$,

$\varphi(\theta, t) = (d\theta, t)$ in an annulus A containing $S^1_\infty \subset \overline{\mathbf{C}}_\mathbf{w}$. We call this deformed map $\widetilde{P}_\mathbf{w}$. The boundary component of A distinct from S^1_∞ is called $\mathcal{C}$. Note that $\widetilde{P}_\mathbf{w}(\mathcal{C}) = \mathcal{C}$.

Since $P_\mathbf{w}$, hence $\widetilde{P}_\mathbf{w}$, is Thurston equivalent to $p_\mathbf{w}$, it follows that there are homeomorphisms $k_0, k_1 \colon \overline{\mathbf{C}}_\mathbf{w} \to \widetilde{U}_\mathbf{w}$ such that $k_0 \circ \widetilde{P}_\mathbf{w} = \widetilde{f} \circ k_1$ on $\overline{\mathbf{C}}_\mathbf{w}$, that are isotopic rel. post$(P_\mathbf{w})$.

Let $\widetilde{\mathcal{C}} := k_0(\mathcal{C}) \subset \widetilde{U}_\mathbf{w}$ and $\widetilde{\mathcal{C}}^1 := \widetilde{f}^{-1}(\mathcal{C}) = k_1(\mathcal{C})$. We can precompose $\widetilde{f}$ with a pseudo-isotopy I rel. post$(\widetilde{f}) \cup \widetilde{\mathcal{E}}$, so that the resulting map $\widetilde{f} \circ I_1$ leaves $\widetilde{\mathcal{C}}$ invariant, and $\widetilde{f} \circ I_1$ is topologically conjugate to φ as above. Let $\widetilde{A}$ be the annulus between $\mathcal{E}$ and $\widetilde{\mathcal{C}}$.

We define $h_0 = k_1$ on $\overline{\mathbf{C}}_\mathbf{w} \setminus A$ and $h_1 = I_1 \circ k_1$ on $\overline{\mathbf{C}}_\mathbf{w} \setminus A$. Since both $\widetilde{P}_\mathbf{w}$ and $\widetilde{f} \circ I_1$ are topologically conjugate to φ on A, respectively $\widetilde{A}$, we can extend h_0, h_1 to A (mapping to $\widetilde{A}$) such that the claim holds. $\square$

In the same fashion, we define $S^2_\mathbf{b} := \text{clos}\,\widetilde{U}_\mathbf{b}/\widetilde{\mathcal{E}}$ and the map $p_\mathbf{b} \colon S^2_\mathbf{b} \to S^2_\mathbf{b}$. Again this is a Thurston polynomial, which is Thurston equivalent to a (monic, centered) polynomial $P_\mathbf{b} \colon \overline{\mathbf{C}}_\mathbf{b} \to \overline{\mathbf{C}}_\mathbf{b}$ (i.e., is not obstructed). The analogous statement to the preceding claim holds for $P_\mathbf{b}$, that is, there are homeomorphisms $k_0, k_1 \colon \overline{\mathbf{C}}_\mathbf{b} \to \widetilde{X}_\mathbf{b}$ ($\widetilde{X}_\mathbf{b} = \widetilde{U}_\mathbf{b} \cup \widetilde{E}$) which are isotopic rel. post$(P_\mathbf{b}) \cup S^1_\infty$ and satisfy $k_0 \circ P_\mathbf{b} = \widetilde{f} \circ k_1$ on $\overline{\mathbf{C}}_\mathbf{b}$.

It is not necessarily true that $\widetilde{f}$ (hence f) is Thurston equivalent to the formal mating of $P_\mathbf{w}$ and $P_\mathbf{b}$. Recall however that $h_0 = h_1$ on $S^1_\infty \subset \overline{\mathbf{C}}_\mathbf{w}$ and $k_0 = k_1$ on $S^1_\infty \subset \overline{\mathbf{C}}_\mathbf{b}$. Thus the map $\tau = k_0^{-1} \circ h_0 \colon S^1_\infty \subset \overline{\mathbf{C}}_\mathbf{w} \to S^1_\infty \subset \overline{\mathbf{C}}_\mathbf{b}$ is a topological conjugacy from z^d to z^d, i.e., $\tau(z)^d = \tau(z^d)$. It is elementary that such a map must be of the form $\tau(z) = cz$, or $\tau(z) = c\bar{z}$, where $c = e^{2\pi i j/(d-1)}$, for some $j = 0, \ldots, d-2$.

If $\tau(z) = \bar{z} = -z$ we are done, since the maps h_0 and k_0, as well as h_1 and k_1 can be glued together to yield maps H_0, H_1 from the formal mating sphere $\overline{\mathbf{C}}_\mathbf{w} \uplus \overline{\mathbf{C}}_\mathbf{b}$ to $\widehat{\mathbf{C}}$ ($= \widetilde{U}_\mathbf{w} \cup \widetilde{\mathcal{E}} \cup \widetilde{U}_\mathbf{b}$), isotopic rel. post$(P_\mathbf{w}) \cup \text{post}(P_\mathbf{b}) \cup \widetilde{\mathcal{E}}$, such that for the formal mating $P_\mathbf{w} \uplus P_\mathbf{b}$ it holds $H_0 \circ P_\mathbf{w} \uplus P_\mathbf{b} = \widetilde{f} \circ H_1$. This means $\widetilde{f}$, hence f, is Thurston equivalent to the formal mating of $P_\mathbf{w}$ and $P_\mathbf{b}$.

If $\tau(z) = c\bar{z}$ we consider the polynomial $\widetilde{P}_\mathbf{w}(z) = cP_\mathbf{b}(c^{-1}\bar{z})$. Then $\widetilde{f}$ is Thurston equivalent to the formal mating of $P_\mathbf{w}$ and $\widetilde{P}_\mathbf{b}$, by the same argument as before.

If $\tau(z) = cz$, $\widetilde{f}$ is Thurston equivalent to the polynomial $\widetilde{P}_\mathbf{b} = \overline{cP_\mathbf{w}(c^{-1}\bar{z})}$, using the same argument again. $\square$

PROOF OF THEOREM 4.2. Let $f \colon \widehat{\mathbf{C}} \to \widehat{\mathbf{C}}$ be a hyperbolic postcritically finite rational map.

Assume f arises as a mating. Let $S^2 = \overline{\mathbf{C}}_\mathbf{w} \uplus \overline{\mathbf{C}}_\mathbf{b}$ be the formal mating sphere with equator $S^1 := \{\infty \cdot e^{2\pi i \theta} \mid \theta \in [0, 2\pi]\} \subset \overline{\mathbf{C}}_\mathbf{w} \subset S^2$ and $\sim$ be the ray-equivalence on S^2. Let $H \colon S^2 \times I \to S^2$ be the pseudo-isotopy realizing $\sim$ according to Moore's theorem.

We fix an appropriate $\epsilon > 0$ such that for $t \in [1 - \epsilon, 1]$ the curve $H(S^1, t)$ does not meet post(f), i.e., $H(S^1 \times [1 - \epsilon, 1]) \subset S^2 \setminus \text{post}(f)$. Consider the two parametrized curves $\gamma_1, \gamma_{1-\epsilon} \colon S^1 \to S^2$ given by $\gamma_1(\infty \cdot e^{2\pi i \theta}) = H_1(\infty \cdot e^{2\pi i d\theta})$ and, more generally, $\gamma_t(\infty \cdot e^{2\pi i \theta}) = H_t(\infty \cdot e^{2\pi i d\theta})$ for each $t \in [1 - \epsilon, 1]$, where $d = \deg f$.

Note that they cover the image (at least) d-fold. Clearly $H: S^1 \times [1 - \epsilon, 1] \to S^2$ is a homotopy deforming $\gamma_{1-\epsilon}$ to γ_1.

Consider now $\widetilde{\gamma}_1: S^1 \to S^2$ given by $\widetilde{\gamma}_1(\infty \cdot e^{2\pi i\theta}) := H_1(e^{2\pi i\theta})$. By (3.3) $\widetilde{\gamma}_1$ is a lift of γ_1 by f, i.e., $f \circ \widetilde{\gamma}(\infty \cdot e^{2\pi i\theta}) = \gamma(\infty \cdot e^{2\pi i\theta})$. By the standard lifting theorem of homotopies by covering maps, it follows that the homotopy $H: S^1 \times [1 - \epsilon, 1] \to S^2$ can be lifted by f to a homotopy $\widetilde{H}: S^1 \times [1 - \epsilon, 1] \to S^2$ with $\widetilde{H}_1 = \widetilde{\gamma}_1$, that is, $f \circ \widetilde{H} = H$. Let $\widetilde{\gamma}_t: S^1 \to S^2$ be given by $\widetilde{\gamma}_t(\infty \cdot e^{2\pi i\theta}) := \widetilde{H}_t(\infty \cdot e^{2\pi i\theta})$. Thus $f \circ \widetilde{\gamma}_t(\infty \cdot e^{2\pi i\theta}) = \gamma_t(\infty \cdot e^{2\pi i\theta})$. Thus $\widetilde{\gamma}_t$ is a component of $f^{-1}(\gamma_t)$.

It remains to show that there is no other component, that is, showing $\widetilde{\gamma}_t$ is the whole preimage of γ_t by f. This can be seen by using Lemma 3.4. Indeed, for any $\gamma_1(\infty \cdot e^{2\pi i\theta}) = H_1(\infty \cdot e^{2\pi i d\theta})$, the points $H_1(\infty \cdot e^{2\pi i\theta + \frac{k}{d}})$, $k = 0, \ldots, d - 1$ are the d *distinct* preimages by f. By continuity $\widetilde{\gamma}_t(\infty \cdot e^{2\pi i\theta + \frac{k}{d}})$, $k = 0, \ldots, d - 1$ are distinct for t sufficiently close to 1. Thus $\widetilde{\gamma}_t$ is the whole preimage of γ_t by f.

Since $\mathcal{C}_t := \gamma_t(S^1)$ contains no postcritical point, $\widetilde{\mathcal{C}}_t := \widetilde{\gamma}_t(S^1)$ is a simple curve. Finally, the concatenation of H and $\widetilde{H}$ deforms $\mathcal{C}_t$ to $\widetilde{\mathcal{C}}_t$; thus they are homotopic rel. post(f). Clearly each component of $S^2 \setminus \mathcal{C}_t$ contains at least a postcritical point. Thus $\mathcal{C}_t, \widetilde{\mathcal{C}}_t$ are orientation-preserving isotopic rel. post(f) by Theorem 2.6.

Assume now that f has an equator. Then f is Thurston equivalent to the formal mating of two polynomials P_w and P_b, by Theorem 4.8. From the Rees-Shishikura theorem, i.e., Theorem 3.3, it follows that f is topologically conjugate to (the topological mating) $P_w \perp\!\!\!\perp P_b$. $\qquad\square$

Theorem 4.2 together with Proposition 4.3 immediately yields the following.

THEOREM 4.10. *Let $f: \widehat{\mathbb{C}} \to \widehat{\mathbb{C}}$ be a hyperbolic, postcritically finite, rational map, with $\#\mathrm{post}(f) = 3$, that is not (conjugate to) a polynomial. Then f does not arise as a mating.*

In the case when f is a polynomial it arises trivially as the mating of itself with z^d (where $d = \deg f$). Note that each iterate f^n has the same postcritical set. Thus in the case of the previous theorem no iterate f^n arises as mating. This, of course, is in contrast to the results from [Mey13] and [Mey].

5   AN EXAMPLE

We provide an example of a rational map to be able to illustrate the following. We give several descriptions (naturally equivalent) of the same map.

Consider two equilateral triangles. We glue them together along their boundaries. This yields a topological sphere denoted by Δ. We color one side (i.e., one of the equilateral triangles) white and call it T_w; the other one is colored black and called T_b. We call T_w and T_b the 0-**triangles**.

Again glue two equilateral triangles (of the same size as before) together to form a topological sphere Δ^1, as before. Each face is now divided into four equilateral triangles of half the side-length. We color these small triangles black and white in a *checkerboard pattern*. This means that two small triangles which share an edge have different colors. These small triangles are called 1-**triangles**.

Consider a small white triangle $T_1 \subset \Delta^1$. The map f is given on T_1 as follows. Scale T_1 by the factor 2 and map it to the big white triangle in Δ. Consider now a small black triangle $T_2 \subset \Delta^1$ that intersects T_1 in an edge. The map f can be

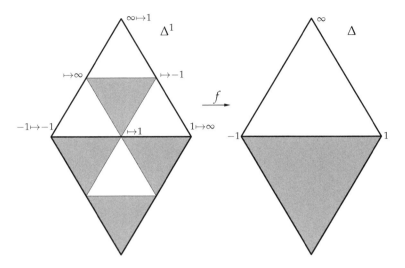

FIGURE 5.1: A Lattès map with signature $(3,3,3)$.

extended continuously to T_2 by scaling by the factor 2 and mapping it to the black triangle in Δ.

Continuing in this fashion we obtain a map $\Delta^1 \to \Delta$. Identifying Δ^1 with Δ we obtain a branched covering map from a topological sphere to itself. Note that the critical points (i.e., the points where f is not locally injective) are the vertices of the small triangles. These are mapped to the vertices of Δ; each such vertex is mapped to (possibly) another vertex of Δ. Thus the map is postcritically finite.

The map may not look like a rational map to many readers. To the reader familiar with Thurston's classification of rational maps among Thurston maps (see [DH93]) we remark that f has only 3 postcritical points, thus it has no Thurston obstruction (the orbifold of f is parabolic, though).

There is, however, a standard way to view every polyhedral surface as a Riemann surface (see [Bea84]). We outline the construction in the case at hand.

Consider first a point $p \in \Delta$ that lies in the interior of either one of the two triangles from which Δ is built, say, $p \in \text{int } T_w$. Then (any) orientation-preserving, isometric map from int T_w to (the interior of) an equilateral triangle in the plane is a chart.

Assume now that p lies on a common edge E of T_w and T_b but is not a vertex of Δ. Then the map to (the interior of) the union of two equilateral triangles in the plane is a chart. Here, of course, we demand that this chart is orientation preserving, and the interiors of T_w and T_b, E are mapped isometrically.

Finally, if p is a vertex of Δ we note that the (Euclidean) total angle around p is $2\pi/3$. Thus we can map a small open neighborhood of p in Δ essentially by the map $z \mapsto z^3$ to a planar domain.

Changes of coordinates are conformal; thus we have defined a Riemann surface. Clearly this is compact and simply connected, thus conformally equivalent to the Riemann sphere $\widehat{\mathbb{C}}$. Note that the map $f\colon \Delta \to \Delta$ is holomorphic with respect to these charts. Thus changing the coordinate system yields a holomorphic, and hence rational, map $f\colon \widehat{\mathbb{C}} \to \widehat{\mathbb{C}}$.

A slightly different way to construct the map may be thought of as constructing the uniformizing map above explicitly. Namely, map the equilateral triangle T_w to the upper half-plane conformally, i.e., by a Riemann map φ normalized such that the three vertices are mapped to $-1, 1, \infty$. Recall that T_w is subdivided into four triangles of half the side-length. Consider one (of the three) such triangles which is colored white T_1. We consider the image in the upper half-plane, i.e., $T_1' := \varphi(T_1) \subset \operatorname{clos} \mathbb{H}^+$.

We map T_1' conformally to the upper half-plane such that the images of the vertices by φ are mapped to $-1, 1, \infty$. Figure 5.1 indicates which vertices are mapped to which of the points $-1, 1, \infty$.

Consider now the black 1-triangle $T_2 \subset T_w$. It shares an edge E with T_1. Its image is $T_2' := \varphi(T_2)$.

Note that T_2 is the reflection of T_1 along E. Thus T_2' is the **conformal reflection** of T_1' along $E' := \varphi(E)$. By the Schwarz reflection principle it follows that the map $f \colon T_1' \to \operatorname{clos} \mathbb{H}^+$ extends conformally to T_2' and maps this set to the lower half-plane (vertices are mapped to $-1, 1, \infty$).

Continuing in this fashion we construct a map $f \colon \operatorname{clos} \mathbb{H}^+ = \varphi(T_w) \to \widehat{\mathbb{C}}$. There are now two ways to proceed. Either we map T_b conformally to the lower half-plane and proceed as before, or we note that the map $f \colon \operatorname{clos} \mathbb{H}^+ = \varphi(T_w) \to \widehat{\mathbb{C}}$ maps the extended real line $\widehat{\mathbb{R}} = \mathbb{R} \cup \{\infty\}$ to itself; thus we can use the Schwarz reflection principle again to extend f to the lower half-plane via $f(z) := \overline{f(\bar{z})}$.

A third way to construct the map f is as follows. Consider the triangular lattice $\Lambda := \mathbb{Z} \oplus \omega \mathbb{Z}$, where $\omega = \exp(\pi i / 3)$. Map the equilateral triangle with vertices $0, 1, \omega$ by a Riemann map $\wp$ to the upper half-plane, such that $\wp(0) = -1$, $\wp(1) = 1$, $\wp(\omega) = \infty$. This map extends by reflection to a holomorphic map $\wp \colon \mathbb{C} \to \widehat{\mathbb{C}}$. It is the **Weierstraß $\wp$-function** to the lattice Λ (slightly differently normalized than usual). Consider now the map $z \mapsto 2z$ on $\mathbb{C}$. It is straightforward to check that if $z, w \in \mathbb{C}$ are mapped by $\wp$ to the same point, then the same is true for $2z, 2w$. Thus there is a well-defined map $f \colon \widehat{\mathbb{C}} \to \widehat{\mathbb{C}}$ (which is the same map as before) such that the diagram

$$
\begin{array}{ccc}
\mathbb{C} & \xrightarrow{z \mapsto 2z} & \mathbb{C} \\
\wp \downarrow & & \downarrow \wp \\
\widehat{\mathbb{C}} & \xrightarrow{f} & \widehat{\mathbb{C}}
\end{array}
$$

commutes. Finally we note that the map f is given by

$$
f = \frac{2(z+1)(z-3)^3}{(z-1)(z+3)^3} - 1.
$$

6   A SUFFICIENT CRITERION FOR MATING

In Section 4 we saw a necessary and sufficient criterion for a hyperbolic postcritically finite rational map to arise as a mating.

Here we present another sufficient criterion for a map to arise as a mating from [Mey13] and [Mey].

Recall that for an equator we required the existence of a Jordan curve in $\widehat{\mathbb{C}}$ which avoids $\mathrm{post}(f)$ and that can be deformed by a isotopy rel. $\mathrm{post}(f)$ orientation preserving to its preimage.

The existence of an equator is the right condition to check whether f arises as a mating for *hyperbolic maps*. It is not the right condition however, for maps that are not hyperbolic. Here we concentrate on the case that the map f is as far away as possible from being hyperbolic, i.e., on the case where all critical points, and hence, all postcritical points, are in the Julia set.

Whereas in the hyperbolic case, we considered Jordan curves *avoiding* the post-critical set, we consider here Jordan curves *containing* the postcritical set. So let $\mathcal{C} \subset \widehat{\mathbb{C}}$ be a Jordan curve with $\mathrm{post}(f) \subset \mathcal{C}$. The postcritical points divide $\mathcal{C}$ into closed Jordan arcs, which are called 0-**edges**.

We consider the preimage

$$\mathcal{C}^1 := f^{-1}(\mathcal{C}).$$

This set can be naturally viewed as a graph embedded in the sphere. Namely, the set $\mathbf{V}^1 := f^{-1}(\mathrm{post}(f))$ is the set of vertices of this graph, the closure of one component of $f^{-1}(\mathcal{C}) \setminus \mathbf{V}^1$ is an edge (called a 1-**edge**) of this graph. All edges arise in this form. We call each point $p \in \mathbf{V}^1$ a 1-**vertex**. Note that every postcritical point, as well as every critical point, is a 1-vertex.

There may be multiple edges connecting two vertices, but there can be no loops. For each 1-edge $E^1 \subset f^{-1}(\mathcal{C})$ there is a 0-edge $E^0 \subset \mathcal{C}$ such that the map $f \colon E^1 \to E^0$ is a homeomorphism.

Note that each critical point c is incident to $2 \deg_f(c)$ 1-edges. In particular $\mathcal{C}^1$ is not a Jordan curve and thus cannot be isotopic to $\mathcal{C}$.

Roughly speaking we demand that there is a **pseudo-isotopy** rel. $\mathrm{post}(f)$ that deforms $\mathcal{C}$ to $f^{-1}(\mathcal{C})$.

To complete the picture, we let X_{w}^0 and X_{b}^0 be the closures of the two components of $\widehat{\mathbb{C}} \setminus \mathcal{C}$. Then X_{w}^0 will be colored white, and X_{b}^0 is colored black. They are called the two 0-*tiles*. We orient $\mathcal{C}$, so that it is positively oriented as boundary of the white 0-tile X_{w}^0.

The closure of each component of $\widehat{\mathbb{C}} \setminus \mathcal{C}^1$ is called a 1-*tile*. It is not very hard to show that each 1-tile X is mapped by f homeomorphically to either X_{w}^0 or X_{b}^0, see [BM11, Proposition 5.17]. In the first case we color X white, and in the second, black.

Recall the definition of a pseudo-isotopy from Definition 2.3. There are many "bad" ways in which $\mathcal{C}$ may be deformed by a pseudo-isotopy to $f^{-1}(\mathcal{C})$. For example a whole interval or a Cantor set may be deformed to a postcritical point. Also some arc $A \subset \mathcal{C}$ may be deformed to some edge in $f^{-1}(\mathcal{C})$ in a highly nontrivial fashion.

DEFINITION 6.1. *An* **elementary, orientation-preserving, pseudo-isotopic** *deformation of $\mathcal{C}$ to $\mathcal{C}^1 = f^{-1}(\mathcal{C})$ rel. $\mathrm{post}(f)$ is a pseudo-isotopy $H \colon S^2 \times [0,1]$ rel. $\mathrm{post}(f)$ (with $H_0 = \mathrm{id}_{S^2}$) such that the following holds:*

1. $H_1(\mathcal{C}) = \mathcal{C}^1$;

2. *the set of points $w \in \mathcal{C}$ such that $H_1(w)$ is a 1-vertex is* finite. *The set of all such points is denoted by* $\mathbf{W} := (H_1)^{-1}(\mathbf{V}^1) \cap \mathcal{C}$. *Note that* $\mathrm{post}(f) \subset \mathbf{W}$;

3. *restricted to* $C \setminus \mathbf{W}$ *the homotopy* H *is an isotopy, i.e.,*

$$H_1 \colon C \setminus \mathbf{W} \to C^1 \setminus \mathbf{V}^1 \text{ is a homeomorphism.}$$

This means there is a bijection between 1-*edges and closures of components of* $C \setminus \mathbf{W}$. *Furthermore if* $E^1 \subset C^1$ *is a* 1-*edge and* $A \subset C$ *is the corresponding arc (i.e., closure of a component of* $C \setminus \mathbf{W}$*), then* $H_1 \colon A \to E^1$ *is a homeomorphism.*

4. *we assign an* orientation *to* C. *For each* 1-*edge* $E^1 \subset C^1$, *there are two ways to map it homeomorphically to (a part of)* C; *namely, by*

$$H_1 \colon A \to E^1 \quad \text{and by} \quad f \colon E^1 \to E^0.$$

Here $A \subset C$ *is a (closed) arc corresponding to* E^1 *as before,* $E^0 \subset C$ *is a* 0-*edge. We demand that* H *deforms* C orientation-preserving *to* C^1, *meaning that the orientations on* E^1 *induced by the two homeomorphisms above agree for each* 1-*edge.*

If there is a pseudo-isotopy H *for* C *as before, we say that* f *has a* **pseudo-equator**.

REMARK 6.2. Let the pseudo-isotopy $H \colon S^2 \times [0,1] \to S^2$ be as before. Then for any point $p \in C^1 \setminus \mathbf{V}^1$, there is exactly one point $q \in C$ that is mapped by H_1 to p. For a point $c \in \mathbf{V}^1$, there are exactly $\deg_f(c)$ points in C that are mapped by H_1 to c. In this sense the curve C is deformed by H_1 to C^1 as simply as possible.

The following two theorems are proved in [Mey13] and [Mey].

THEOREM 6.3. *Let* $f \colon \widehat{\mathbb{C}} \to \widehat{\mathbb{C}}$ *be a postcritically finite rational map with Julia set* $\mathcal{J}(f) = \widehat{\mathbb{C}}$. *Assume* f *has a pseudo-equator as in Definition* 6.1. *Then* f *arises as a (topological) mating of two (postcritically finite, monic) polynomials* $P_{\mathbf{w}}$ *and* $P_{\mathbf{b}}$.

THEOREM 6.4. *Let* $f \colon \widehat{\mathbb{C}} \to \widehat{\mathbb{C}}$ *be a postcritically finite rational map with Julia set* $\mathcal{J}(f) = \widehat{\mathbb{C}}$. *Then every sufficiently high iterate* f^n *of* f *has a pseudo-equator.*

REMARK 6.5. The preceding two theorems remain true in the case when the map f is not a rational map. Namely, they hold for *expanding Thurston maps*. The statement of Theorem 6.3 has to be slightly modified in the presence of periodic critical points.

EXAMPLE 6.6. We consider the example discussed in Section 5. Recall that post(f) is $\{-1, 1, \infty\}$. These points are the vertices of the pillow in Figure 5.1. We choose $C = \widehat{\mathbb{R}}$. In the model of the map indicated in Figure 5.1, C is the common boundary of the two triangles which form the pillow. The preimage $C^1 := f^{-1}(C)$ is the union of all the edges of the small triangles to the left in Figure 5.1.

The elementary, orientation-preserving, pseudo-isotopic deformation H rel. post(f) that deforms C to $C^1 = f^{-1}(C)$ is indicated (to the left) in Figure 6.8.

EXAMPLE 6.7. We consider again the same example from Section 5. Again $C = \widehat{\mathbb{R}}$. We have chosen to draw the curve C on the right of Figure 6.9 as the boundary of the black 0-tile, however. In Figure 6.9 we show a pseudo-isotopy that deforms C to C^1 in an *orientation-reversing* way. By this we mean that the shown pseudo-isotopy H satisfies all properties from Definition 6.1 except (4). This is seen as follows.

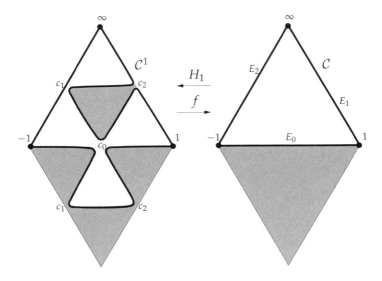

FIGURE 6.8: Deforming $\mathcal{C}$ to $\mathcal{C}^1$.

If we traverse $\mathcal{C}$ positively as boundary of the white 0-tile, we go through the postcritical points in the cyclic order $-1 \to 1 \to \infty \to -1$. However if we traverse $\mathcal{C}^1 := f^{-1}(\mathcal{C})$ in the orientation given by f (i.e., each 1-edge is traversed positively as boundary of the white 1-tile in which it is contained), then the postcritical points are traversed in the order $-1, \infty, 1$. More precisely, we see that the orientations on each 1-edge E^1 induced by f and by H_1 are opposite.

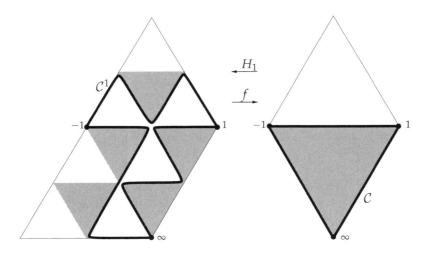

FIGURE 6.9: Orientation reversing pseudo-isotopy.

If we compare Figure 6.8 with Figure 6.9, we see one difference. Namely, in the pseudo-isotopy indicated in Figure 6.8 the interior of $\mathcal{C}$, more precisely the interior of the white 0-tile X_{w}^0, is deformed to all white 1-tiles. On the other hand, the pseudo-isotopy indicated in Figure 6.9 deforms the interior of the black 0-tile to the white 1-tiles. This is a general phenomenon.

LEMMA 6.10. *Let H be an elementary, pseudo-isotopic deformation of $\mathcal{C}$ to $\mathcal{C}^1$, i.e., a pseudo-isotopy rel. $\mathrm{post}(f)$ that satisfies (1), (2), and (3) (but not necessarily (4)) of Definition 6.1. Then the following hold:*

1. *Either H deforms $\mathcal{C}$ to $\mathcal{C}^1$ orientation-preserving (i.e., satisfies (4) of Definition 6.1), or H deforms $\mathcal{C}$ to $\mathcal{C}^1$ orientation-reversing. The latter means that for any 1-edge the orientations induced by the homeomorphisms*

$$f\colon E^1 \to E^0 \text{ and } H_1\colon A \to E^1$$

disagree. Here $E^0 \subset \mathcal{C}$ is a 0-edge and $A \subset \mathcal{C}$ is an arc as in Definition 6.1 (3).

Let $x, y \in S^2 \setminus (H_1)^{-1}(\mathcal{C}^1)$, i.e., two points that are mapped by H_1 to the interior of (possibly distinct) 1-tiles. Let $X^0 \ni x$, $Y^0 \ni y$ be the 0-tiles containing x, y, and $X^1 \ni H_1(x)$, $Y^1 \ni H_1(y)$ be the 1-tiles containing $H_1(x), H_1(y)$. Then

2. *X^0, Y^0 have the same color $\iff X^1, Y^1$ have the same color.*

3. *X^0 and X^1 have the same color $\iff H$ is orientation-preserving, i.e., H satisfies (4) of Definition 6.1.*

PROOF. (1) Consider a critical point, i.e., a 1-vertex c. This point is contained in the 1-edges $E_0, \ldots, E_{n-1}$ (where $n = 2\deg_f(c)$), which are labeled mathematically positively around c.

We first note that the orientation on these 1-edges induced by f alternates. This is seen as follows. There are n black as well as n white 1-tiles around c, the colors of the 1-tiles around c alternate. The endpoint of the edge E_j is c (via the orientation given by f) if and only if the sector between E_{j-1}, E_j contains a white 1-tile (and the sector between E_j, E_{j+1} contains a black 1-tile). Conversely, the initial point of E_j is c if and only if the sector between E_{j-1}, E_j contains a black 1-tile (and the sector between E_j, E_{j+1} contains a white 1-tile).

We now consider the orientation on 1-edges induced by the pseudo-isotopy H.

Let $A, A' \subset \mathcal{C}$ be two adjacent arcs that are deformed to 1-edges E, E' as in Definition 6.1 (3), where $A \cap A'$ is deformed to c by H. We assume that A' succeeds A with respect to the orientation of $\mathcal{C}$. Then (with respect to the orientation induced by H) the endpoint of E is c, while c is the initial point of E'.

Since H is a pseudo-isotopy, $H_{1-\epsilon}$ is a homeomorphism for all $0 < \epsilon < 1$. It follows that in the sector between E, E' there is an even number of 1-edges.

Thus from the first claim it follows that the orientation on E induced by H agrees with the orientation on E induced by f if and only if the orientations on E' induced by H and f agree.

Consider the arc A'' on $\mathcal{C}$ that succeeds A'. This is deformed by H to a 1-edge E''. The orientations induced on E'' by f and H agree if and only if the respective orientations on E' agree. Continuing in this fashion we obtain the desired statement.

(2) Let $x^1 := H_1(x), y^1 := H_1(y)$. We want to consider the winding number. To be able to do that we assume that ∞ is contained in the interior of the black 0-tile; furthermore we assume that H deforms ∞ neither to $\mathcal{C}^1$ nor to x^1 or y^1. Then we can define the winding number $N_{\mathcal{C}}(x), N_{\mathcal{C}}(y)$ of $\mathcal{C}$ for x and y.

Then $N_{\mathcal{C}}(x), N_{\mathcal{C}}(y)$ agree if and only if X^0, Y^0 have the same color. This happens if and only if $N_{\mathcal{C}^1}(x^1)$ and $N_{\mathcal{C}^1}(y^1)$ agree. These winding numbers are computed by mapping $\widehat{\mathbb{C}} \setminus \{H_1(\infty)\}$ (orientation preserving) to $\mathbb{C}$, i.e., identifying $\infty^1 := H_1(\infty)$ with ∞. The curve $\mathcal{C}^1 = H_1(\mathcal{C})$ is traversed in the direction induced by H_1 as well as the orientation of $\mathcal{C}$.

Assume that $H_1(\infty)$ is contained in a black 1-tile. Each 1-edge is contained in exactly one white 1-tile. Consider the 1-edges $E_1, \ldots, E_k$ contained in the boundary of a white 1-tile Z. By (1) all these 1-edges are either positively or negatively oriented as boundary of Z. In the first case the winding number of $E_1 \cup \ldots \cup E_k$ is 1, in the second case -1, for all points in the interior of Z. For all points in the complement of Z, the winding number is 0. The same argument applies to all white 1-tiles. Note that by (1) the boundaries of all white 1-tiles have the same orientation induced by H_1, i.e., are either all positively or all negatively oriented. Since taking the union of all 1-edges in the boundaries of all white 1-tiles yields all 1-edges, it follows that the winding number of $\mathcal{C}^1$ for all points in the interior of some white 1-tile is either 1 or -1, while the winding number for all points in the interior of black 1-tiles is 0. This finishes the claim. The argument in the case when $H_1(\infty)$ is contained in a white 1-tiles is completely analogous.

(3) We use the setting as before. Assume X^0 is white; then the winding number of $\mathcal{C}^1$ for x^1 is 1 if and only if the orientation induced by H_1 of the 1-edges in the boundary of X^1 agrees with the orientation of them as boundary of X^1. This happens if and only if H is orientation-preserving by (1). If X^0 is white, the argument is completely analogous. $\square$

Kameyama considers curves that are similar to pseudo-equators as considered here. We give a brief overview, but refer to [Kam03] for complete statements as well as proofs. Let a Thurston map $f \colon S^2 \to S^2$ be given. Let $\mathrm{post}_a(f)$ be the set of postcritical points that eventually land in a critical periodic cycle and $\mathrm{post}_r(f) = \mathrm{post}(f) \setminus \mathrm{post}_r(f)$. In the case when f is a rational map it holds that $\mathrm{post}_a(f)$ is the set of all postcritical points contained in the Fatou set of f and $\mathrm{post}_r(f)$ is the set of postcritical points contained in its Julia set.

Kameyama considers curves $\gamma \subset S^2 \setminus \mathrm{post}_a(f)$ with finitely many self-intersections that are *oriented*, i.e., are images of a Jordan curve $\widetilde{\gamma} \subset S^2$ by the end of some pseudo-isotopy on $S^2 \setminus \mathrm{post}_a(f)$. He now assumes that $f^{-1}(\gamma)$ is homotopic in $S^2 \setminus \mathrm{post}_a(f)$ via a homotopy rel. $\mathrm{post}_r(f)$. For a rational Thurston map f he proves the following. Such a curve γ as before exists if and only if some iterate f^n is Thurston equivalent to the *degenerate mating* of two polynomials. See [Shi00] for the definition (as well as background) of the degenerate mating.

7 CONNECTIONS

There is an equivalent way to describe the existence of a pseudo-equator as in Definition 6.1. Intuitively the description is most easily explained via a picture, as in Figure 6.8.

Namely, consider the left picture in Figure 6.8. Each of the critical points c_j is contained in exactly 3 white, as well as 3 black, 1-tiles. Consider first the critical point c_0. Here all white 1-tiles are *connected* at c_0, while none of the black 1-tiles are connected at c_0.

At the critical point c_1 there are two white 1-tiles connected, and two black 1-tiles that are connected. Also there is one white, as well as one black, 1-tile that is not connected to any other 1-tile at c_1.

At the critical point c_2 all three black 1-tiles containing c_2 are connected; all three white 1-tiles containing c_2 are not connected to any other 1-tile.

The connection of white 1-tiles at each critical point c_j is *complementary* to the connection of black 1-tiles at c_j. The white, as well as the black, 1-tiles are connected in such a way that the resulting *white connection graph* is a *spanning tree*.

The connection of white 1-tiles may be *represented geometrically* as in Figure 6.8. Taking the boundary of this geometric representation of the white cluster results in a curve that is isotopic to C rel. $post(f)$.

We now proceed to make the preceding description precise.

Let $X_0, \ldots X_{2n-1}$ be the 1-tiles intersecting in a 1-vertex v, ordered mathematically positively around v. The white 1-tiles have even index, the black ones odd index. We consider a decomposition $\pi_w = \pi_w(v)$ of $\{0, 2, \ldots, 2n - 2\}$ (i.e., of indices corresponding to white 1-tiles around v); and a decomposition $\pi_b = \pi_b(v)$ of $\{1, 3, \ldots, 2n - 1\}$ (i.e., of indices corresponding to black 1-tiles around v). They satisfy the following:

- They are *decompositions*. This means $\pi_w = \{b_1, \ldots, b_N\}$, where each **block** b_i is a subset of $\{0, 2, \ldots, 2n - 2\}$ with $b_i \cap b_j = \emptyset$ for $i \neq j$, and such that $\bigcup b_i = \{0, 2, \ldots, 2n - 2\}$. Similarly for π_b.

- The decompositions π_w and π_b are **non-crossing**. This means the following. Two distinct blocks $b_i, b_j \in \pi_w$ are *crossing* if there are numbers $a, c \in b_i$, $b, d \in b_j$ and

$$a < b < c < d.$$

Each partition π_w, π_b does not contain any (pair of) crossing blocks.

- The partitions π_w and π_b are **complementary**. This means the following. Given π_w, the partition π_b is the unique, biggest partition of $\{1, 3, \ldots, 2n - 1\}$ such that $\pi_w \cup \pi_b$ is a non-crossing partition of $\{0, 1, \ldots, 2n - 1\}$.

Such a partition $\pi_w \cup \pi_b$ is called a **complementary non-crossing partition**, or **cnc-partition**. We may *represent* a cnc-partition *geometrically* as follows. Let $i, j \in b \in \pi_w$. We call these indices *succeeding* (in b), if $i + 1, i + 2, \ldots, j - 1 \pmod{2n} \notin b$. Let $e_k := \exp(2\pi k i / 2n)$, $k = 0, \ldots, 2n - 1$ be the $2n$ unit roots. Consider the closed unit disk $\overline{\mathbb{D}}$. For each pair of succeeding indices i, j in any white block $b \in \pi_w$, we draw a Jordan arc g_m ($m = 1, \ldots, n$) in $\overline{\mathbb{D}}$ connecting e_{i+1}, e_j. Each such arc intersects $\partial\mathbb{D}$ only in its endpoints (i.e., in e_{i+1}, e_j). Furthermore, two distinct such arcs are disjoint. Figure 7.1 shows a geometric representation of the cnc-partition $\pi_w \cup \pi_b$ given by $\pi_w = \{\{0, 4, 6\}, \{2\}\}$, $\pi_b = \{\{1, 3\}, \{5\}, \{7\}\}$.

The arcs g_m divide $\overline{\mathbb{D}}$ into $n + 1$ components. We color those components in a black and white checkerboard fashion, such that components which share an arc g_m as its boundary have different colors. Furthermore we color the component having the circular arc between e_0, e_1 on $\partial\mathbb{D}$ white.

A geometric representation of a cnc-partition is to be thought of as a blow-up of the 1-edges incident to a 1-vertex v. We will sometimes need to know where in the geometric representation the original vertex is located. This is achieved by *marking* the cnc-partition. This means we mark one of the arcs g_m. Equivalently we

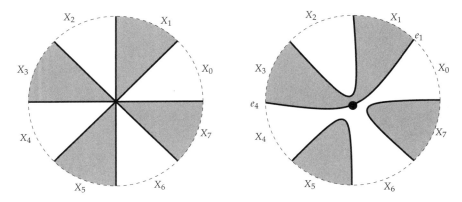

FIGURE 7.1: Connection at a vertex.

may mark a pair of succeeding indices i, j contained in some white block $b \in \pi_{\mathbf{w}}$. If a cnc-partition is marked, we always let the marked arc g_m contain the origin.

In Figure 7.1 the arc connecting e_{0+1} and e_4, or, equivalently, the succeeding indices 0 and 4 of the block $\{0, 4, 6\}$, is marked.

A **connection** (of 1-tiles) assigns to each 1-vertex v a cnc-partition $\pi_{\mathbf{w}}(v) \cup \pi_{\mathbf{b}}(v)$ as before. Furthermore in the case when $v = p$ is a postcritical point, the cnc-partition is marked. This marking is mostly relevant in the case when the postcritical point is at the same time a critical point, since then there are several choices of which arc to mark. Two 1-tiles $X_i, X_j \ni v$ are said to be **connected at** v if the indices i, j are contained in the *same* block of $\pi_{\mathbf{w}}(v) \cup \pi_{\mathbf{b}}(v)$. Note that tiles of different color are never connected. The 1-tile X_i is **incident** (at v) to the block $b \ni i$ of $\pi_{\mathbf{w}}(v) \cup \pi_{\mathbf{b}}(v)$.

The **white connection graph** is defined as follows. For each white 1-tile there is a vertex and for each block $b \in \pi_{\mathbf{w}}(v)$ (for any 1-vertex v) there is a vertex. There is an edge in the white connections graph between $b \in \pi_{\mathbf{w}}(v)$ and the white 1-tile $X \ni v$ if and only if X is incident to b at v.

A **geometric representation** of a connection is achieved as follows. For each 1-vertex v a neighborhood of v "looks like" the picture on the left in Figure 7.1. This is replaced by the picture on the right in Figure 7.1, i.e., by a geometric representation of the cnc-partition that represents the connection at v. More precisely, there is a neighborhood U_v of v and a homeomorphism $\varphi = \varphi_v \colon U_v \to \mathbb{D}$ with the following properties. The point v is mapped to the origin. For any 1-edge $E \ni v$, the map φ maps $E \cap U_v$ to some ray $R_k = \{r \exp(2\pi k i / 2n) \mid 0 \leq r < 1\}$, where $k = 0, \dots, 2n - 1$. If $X \ni v$ is a white 1-tile, then $X \cap U_v$ is mapped by φ to the sector between two rays R_{2k}, R_{2k+1}. Finally, distinct 1-vertices v, w have disjoint neighborhoods U_v, U_w. Draw a geometric representation of the cnc-partition $\pi_{\mathbf{w}}(v) \cup \pi_{\mathbf{b}}(v)$ in $\mathbb{D}$ as above. Replace U_v by the preimage (of this geometric representation) by φ_v (for each 1-vertex v).

Given a geometric representation of a connection we obtain several black and white components. The following is [Mey13, Lemma 6.23].

LEMMA 7.2. *Assume a connection (of 1-tiles) is given. Then the following are equivalent.*

- *The white connection graph is a spanning tree.*

- *In each geometric representation (of the connection) there is a single white component V_w, which is a Jordan domain.*

The following is [Mey13, Lemma 7.2].

LEMMA 7.3. *Let $f: \widehat{\mathbb{C}} \to \widehat{\mathbb{C}}$ be a postcritically finite rational map with Julia set $\mathcal{J}(f) = \widehat{\mathbb{C}}$ and $\mathcal{C} \subset \widehat{\mathbb{C}}$ be a Jordan curve with $\mathrm{post}(f) \subset \mathcal{C}$. Then the following are equivalent.*

- *There is a pseudo-isotopy $H: \widehat{\mathbb{C}} \times [0,1] \to \widehat{\mathbb{C}}$ as in Definition 6.1 (i.e., f has a pseudo-equator).*

- *There is a connection of 1-tiles such that the white connection graph is a spanning tree and the boundary ∂V_w (of the white component V_w from Lemma 7.2) is orientation-preserving isotopic to $\mathcal{C}$ rel. $\mathrm{post}(f)$.*

Here ∂V_w is positively oriented as boundary of the white component V_w (and $\mathcal{C}$ is positively oriented as boundary of the white 0-tile X_w^0 as before).

THEOREM 7.4. *Let $f: \widehat{\mathbb{C}} \to \widehat{\mathbb{C}}$ be a postcritically finite rational map, with Julia set $\mathcal{J}(f) = \widehat{\mathbb{C}}$ and $\#\mathrm{post}(f) = 3$. Then f or f^2 arises as a mating.*

PROOF. Let $f: \widehat{\mathbb{C}} \to \widehat{\mathbb{C}}$ be as in the statement. Let $\mathcal{C} \subset \widehat{\mathbb{C}}$ be a Jordan curve with $\mathrm{post}(f) \subset \mathcal{C}$. We choose the closure of one component of $\widehat{\mathbb{C}} \setminus \mathcal{C}$ to be X_w^0. Then $\mathcal{C}$ is positively oriented as boundary of the white 0-tile X_w^0. It is easy to construct a connection of white 1-tiles such that the resulting white connection graph is a spanning tree, see [Mey13, Corollary 6.20]. Indeed, one starts with a connection where no white 1-tiles are connected at any 1-vertex, successively "adding" other 1-tiles until one obtains a spanning tree as desired.

From now on the connection (of white 1-tiles), which results in the white connection graph being a spanning tree, is fixed. Let V_w be the white component, according to Lemma 7.2. Let ∂V_w be its boundary, oriented positively as boundary of V_w.

The following fact is well known (a proof can, however, be found in [BM11, Lemma 11.10]). Let $P \subset S^2$ be a set with $\#P \leq 3$ and $\gamma, \gamma' \subset S^2$ be two Jordan curves, which both contain P. Then γ, γ' are isotopic rel. P. However if $\gamma, \gamma' \subset S^2$ are oriented, it is not necessarily true that γ, γ' are *orientation-preserving* isotopic rel. P.

If ∂V_w is orientation-preserving isotopic to $\mathcal{C}$ rel. $\mathrm{post}(f)$ we are done by Lemma 7.3 and Theorem 6.3, i.e., f arises as a mating.

Assume ∂V_w is not orientation-preserving isotopic to $\mathcal{C}$ rel. $\mathrm{post}(f)$. Then ∂V_w is orientation-reversing isotopic to $\mathcal{C}$ rel. $\mathrm{post}(f)$. We obtain an elementary, orientation-reversing pseudo-isotopy H that deforms $\mathcal{C}$ to $\mathcal{C}^1$, as in Lemma 6.10 (1).

Let $\widetilde{H}: \widehat{\mathbb{C}} \times [0,1] \to \widehat{\mathbb{C}}$ be the lift of H by f with $\widetilde{H}_0 = \mathrm{id}_{\widehat{\mathbb{C}}}$. We note some properties of the lift $\widetilde{H}$; proofs may be found in [Mey13, Lemma 3.5 and Lemma 3.6]. Namely, $\widetilde{H}$ is a pseudo-isotopy rel. $\mathbf{V}^1$ (which is the set of 1-vertices) such that $f \circ \widetilde{H}(x,t) = H(f(x),t)$ for all $x \in \widehat{\mathbb{C}}$, $t \in [0,1]$. This pseudo-isotopy deforms $\mathcal{C}^1$ to $\mathcal{C}^2 := f^{-1}(\mathcal{C}^1) = f^{-2}(\mathcal{C})$, i.e., $\widetilde{H}_1(\mathcal{C}^1) = \mathcal{C}^2$. We call $\mathbf{V}^2 := f^{-2}(\mathrm{post}(f))$ the set of 2-vertices, the closure of one component of $\mathcal{C}^2 \setminus \mathbf{V}^2$ is called a 2-edge. Only finitely many points $x \in \mathcal{C}^1$ are deformed to any 2-vertex (i.e., to any point $v \in f^{-2}(\mathrm{post}(f))$). The points x as before divide $\mathcal{C}^1$ into closed arcs. There is a bijection between such closed arcs $\widetilde{A} \subset \mathcal{C}^1$ and 2-edges. Finally, for each such arc $\widetilde{A} \subset \mathcal{C}^1$, there is a 1-edge E and the map $\widetilde{H}_1 \colon \widetilde{A} \to E$ is a homeomorphism.

CLAIM 7.5. $\widetilde{H}$ *deforms* $\mathcal{C}^1$ *to* $\mathcal{C}^2$ *in an orientation-reversing way.*

   As before, this means that the orientations induced on $\mathcal{C}^2$ *by* $f \colon \mathcal{C}^1 \to \mathcal{C}^2$ *and* $\widetilde{H}_1 \colon \mathcal{C}^1 \to \mathcal{C}^2$ *disagree.*

   Each 2-edge $\widetilde{E}$ *is mapped by* f *homeomorphically to a 1-edge* $E \subset \mathcal{C}^1$ *(see [BM11, Prop 5.17]). Recall from Definition 6.1 (2) that there is a closed arc* $A \subset \mathcal{C}$ *that is deformed by* H *to* E. *Since* $\widetilde{H}$ *is the lift of* H *by* f, *there is an arc* $\widetilde{A} \subset \mathcal{C}^1$ *that is mapped homeomorphically to* $\widetilde{E}$ *by* $\widetilde{H}_1$, *and* $f(\widetilde{A}) = A$. *Note that* $f \colon \widetilde{A} \to A$, *as well as* $f \colon \widetilde{E} \to E$, *is orientation-preserving. Thus it follows that* H *is orientation-reversing if and only if* $\widetilde{H}$ *is orientation-reversing, proving the claim. The argument is worked out in more detail in [Mey13, Lemma 3.12].*

   Consider $K \colon \widehat{\mathbb{C}} \times [0,1] \to \widehat{\mathbb{C}}$ given by $K(x,t) := \widetilde{H}(H(x,t),t)$. This is a pseudo-isotopy rel. $\mathrm{post}(f)$ that deforms $\mathcal{C}$ in an elementary, orientation-preserving way to $\mathcal{C}^2 = f^{-2}(\mathcal{C})$. Thus it follows from Theorem 6.3 that f^2 arises as a mating. $\qquad \square$

REMARK 7.6. We do not know whether it is necessary to take the second iterate f^2 in the previous theorem.

8   CRITICAL PORTRAITS

In the next section we will describe an algorithm to unmate a rational map f. More precisely, from a pseudo-equator as in Definition 6.1 we can recover the white and black polynomials P_{w} and P_{b} that yield f as their mating. We need, however, a description of polynomials that is adapted to the situation. Namely, we need a description in terms of external angles.

   Fix an integer $d \geq 2$ (d will be the degree of the rational map f as well as the polynomials P_{w} and P_{b}). The map $\mu \colon \mathbb{R}/\mathbb{Z} \to \mathbb{R}/\mathbb{Z}$ is $\mu(t) := dt \bmod 1$. We will somehow abuse notation by identifying a point $x \in \mathbb{R}$ with the corresponding equivalence class $[x] \in \mathbb{R}/\mathbb{Z}$; similarly, we identify $q \in \mathbb{Q}$ with the corresponding $[q] \in \mathbb{Q}/\mathbb{Z}$.

DEFINITION 8.1. *A list* $\mathcal{A} = A_1, \ldots, A_m$ *is called a* **critical portrait** *if conditions (CP 1)–(CP 7) are satisfied.*

(CP 1) *Each* $A_j \subset \mathbb{Q}/\mathbb{Z} \subset \mathbb{R}/\mathbb{Z}$ *is a finite set, and distinct sets* A_i, A_j *are disjoint.*

(CP 2) μ *maps each set* A_j *to a single point,*

$$\mu(A_j) = \{a_j\},$$

   for all $j = 1, \ldots, m$.

(CP 3) $\sum_j \left(\#A_j - 1 \right) = d - 1$.

(CP 4) *The sets are* non-crossing. *This means the following. Two distinct sets* A_i, A_j *are called* crossing, *if there are (representatives)* $s, u \in A_i, t, v \in A_j$ *such that* $0 \leq s < t < u < v \leq 1$, *otherwise* non-crossing. *All distinct sets* A_i, A_j *of the critical portrait are non-crossing.*

(CP 5) *No $a \in A_j$ is periodic under μ (for any $j = 0, \ldots, m - 1$).*

The set $\mathbf{A} := \bigcup \{\mu^k(A_j) \mid j = 1, \ldots, m, \ k \geq 1\}$, *i.e., the union of forward orbits of all angles in any of the sets A_j, then is a finite set.*

(CP 6) *No set A_j contains more than one point of $\mathbf{A}$.*

The following definition is somewhat technical. The reader should think of $\mathbb{R}/\mathbb{Z} = S^1$ as being the boundary of the unit disk $\mathbb{D}$. We form the convex hull of each set A_j (with respect to the hyperbolic metric on $\mathbb{D}$). We remove all these hulls. The closure of one remaining component is called a 1-gap. Points in the same gap should be thought of as being not separated by the sets A_j. We are interested only in points on the unit circle.

Here is the formal definition. The points $\bigcup A_j$ divide the circle into closed intervals $[a, b]$, i.e., $a, b \in \bigcup A_j$ and (a, b) does not contain any point from $\bigcup A_j$. A 1-gap is a union of such intervals. Two intervals $[a_1, b_1], [a_2, b_2]$ belong to the same 1-gap if and only if for any two points $c_1 \in (a_1, b_1), c_2 \in (a_2, b_2)$, the sets $\{c_1, c_2\}, A_j$ are non-crossing for all $j = 1, \ldots, n$.

Two points $x, y \in \mathbb{R}/\mathbb{Z}$ belong to the same n-gap ($n \geq 1$) if $\mu^k(x), \mu^k(y)$ belong to the same 1-gap for all $k = 0, \ldots, n - 1$.

(CP 7) *There is a constant $n_0 \in \mathbb{N}$ such that the following holds. Two distinct points $a, b \in \mathbf{A}$ are not contained in the same n_0-gap.*

THEOREM 8.2 (Bielefeld-Fisher-Hubbard [BFH92]). *Let $\mathcal{A}$ be a critical portrait as in Definition 8.1. Then there is a (unique up to affine conjugacy) monic polynomial P of degree d realizing it. The polynomial P is postcritically finite, and each critical point of P is strictly preperiodic.*

That a polynomial P realizes a critical portrait means that for each A_j there is a critical point c_j of P. The degree of P at c_j is $\#A_j$. If $a \in A_j$, then the external ray $R(a)$ lands at c_j. There is a generalization of the preceding theorem due to Poirier [Poi09].

9   UNMATING THE MAP

A pseudo-equator for f as in Definition 6.1 not only guarantees that f arises as a mating, but it is possible to explicitly find the polynomials that when mating give the map f. This is described here.

Let $\mathcal{C} \supset \text{post}(f)$ be a pseudo-equator for the (rational, postcritically finite) map $f \colon \widehat{\mathbb{C}} \to \widehat{\mathbb{C}}$, whose Julia set is the whole sphere.

Let X_w^0, X_b^0 be the two 0-tiles (defined in terms of $\mathcal{C}$) which are colored white and black. We orient $\mathcal{C}$ so that it is positively oriented as boundary of (the white 0-tile) X_w^0.

Recall that the postcritical points divide $\mathcal{C}$ into (closed) 0-edges $E_0, \ldots, E_{k-1}$, which we label positively on $\mathcal{C}$.

Consider now a 1-edge E^1. We say it is of *type j* if $f(E^1) = E_j$. Each 0-edge is deformed by H_1 into several 1-edges. We record how many 1-edges of each type are contained in such a deformed 0-edge in the matrix $M = (m_{ij})$ defined as follows:

$$m_{ij} := \text{number of 1-edges of type } j \text{ contained in } H_1(E_i).$$

This matrix is called the *edge replacement matrix* of the pseudo-isotopy H. Since there are exactly d 1-edges of each type, it follows that $\sum_i m_{ij} = d = \deg f$. It is relatively easy to show that the matrix M is *primitive*, i.e., that $M^n > 0$ for some $n \in \mathbb{N}$ (see [Mey13, Lemma 4.3]). Thus it follows from the Perron-Frobenius theorem that d is a simple eigenvalue (which is in fact the spectral radius of M), with eigenvector $l = (l_j) > 0$. We normalize l by $\sum l_j = 1$; this makes l unique.

To illustrate we consider the pseudo-isotopy shown in Figure 6.8. Here the edge replacement matrix and the corresponding eigenvector is

$$M = \begin{pmatrix} 2 & 2 & 1 \\ 2 & 1 & 2 \\ 0 & 1 & 1 \end{pmatrix}, \quad l = \frac{1}{15} \begin{pmatrix} 7 \\ 6 \\ 2 \end{pmatrix}.$$

The vector l describes the *lengths* of the 0-edges. This in turn will be used to find the external angles at the 0-vertices (= postcritical points). More precisely, for each $p \in \mathrm{post}(f)$ we will define an external angle $\theta(p)$.

The 0-edges $E_j \subset C$ inherit the orientation of C. Let $p_0, \dots, p_{k-1}$ be the postcritical points, labeled in positively cyclical order on C, such that p_0 is the initial point of (the first) 0-edge E_0.

Assume first that p_0 is a fixed point of f (as in the example from Figure 6.8). Then we set the external angle of p_0 equal to 0, i.e., $\theta(p_0) = 0$. The external angle of the other postcritical points is now given by

$$\theta(p_1) = l_0, \ \theta(p_2) = l_0 + l_1, \ \dots, \ \theta(p_j) = l_0 + \dots + l_{j-1}.$$

Thus the difference between the external angles of p_j and p_{j+1} is always given by l_j (here indices are taken mod k).

Assume now that p_0 is mapped by f to p_j. Let $l(p_0, p_j) := l_0 + \dots + l_{j-1}$, i.e., the total length of all 0-edges between p_0 and p_j (in the positive direction on C). We now desire that $d\,\theta(p_0) = \theta(p_j) = \theta(p_0) + l(p_0, p_j)$. Thus we define

$$\theta(p_0) = \frac{l(p_0, p_j)}{d - 1},$$

and $\theta(p_1) = \theta(p_0) + l_0, \dots, \theta(p_i) = \theta(p_0) + l_0 + \dots + l_{i-1}$. We will, however, need only $\theta(p_0)$ in the following.

We now assign external angles at the critical points. More precisely, we want to find the external angles at critical points of polynomials P_{w} and P_{b} (called the white/black polynomials). These are the polynomials into which f is unmated, i.e., f will be (topologically equivalent to) the mating of P_{w} and P_{b}. The external angles will give the critical portraits of the polynomials P_{w} and P_{b}.

First we define the length of a 1-edge E^1 of type j by $l(E^1) = l_j/d$. Note that there are d 1-edges of each type; thus $\sum l(E^1) = 1$ (where the sum is taken over all 1-edges).

Denote by $\mathbf{E}^1$ the set of all 1-edges. Since H deforms C to $C^1 = \bigcup \mathbf{E}^1$, it follows that the orientation of C together with H induces a cyclical ordering $E_0^1, \dots, E_{kd-1}^1$ on the 1-edges as well as an orientation on each 1-edge. Here we start the labeling at p_0, i.e., the initial point of E_0^1 is p_0. Note that the type of 1-edges in this cyclical ordering is changing cyclically, i.e., if E_i^1 is of type j, then E_{i+1}^1 is of type $j + 1$ (here the lower index is taken mod kd, the type is taken mod k).

Assume the 1-edge E_j^1 ends at the critical point c. Then an external angle associated with c is

$$\theta(E_j^1) := \theta(p_0) + l(E_0^1) + \ldots + l(E_j^1).$$

There are $\deg_f(c)$ such (oriented) 1-edges ending at c, and hence, different external angles associated to c. We will have to decide which belong to critical points of the white polynomial and which belong to critical points of the black polynomial. However, the situation is more complicated: c might be associated to several distinct critical points of the white polynomial, as well as several distinct critical points of the black polynomial.

Recall that the set $\mathbf{W}$ was the set of preimages of the 1-vertices by H_1 located on $\mathcal{C}$ (see Definition 6.1(2)). The points in $\mathbf{W}$ divide $\mathcal{C}$ into closed arcs, each of which is mapped by H_1 homeomorphically to a 1-edge by H_1 (see Definition 6.1(3)). Since there are as many such arcs as 1-edges (i.e., kd), there are kd points in $\mathbf{W}$. Let the points in $\mathbf{W}$ be $w_0, \ldots, w_{kd-1}$ labeled positively on $\mathcal{C}$, such that $w_0 = p_0$.

Consider $[c] := (H_1)^{-1}(c)$, i.e., the set of points that are deformed by H to c. This set contains $\deg_f(c)$ points of $\mathbf{W} = \{w_j\} \subset \mathcal{C}$ (see Remark 6.2). Recall from Lemma 2.4 that this set is connected. To clarify: the set of points in $\mathcal{C}$ that is deformed by H to c (i.e., $[c] \cap \mathcal{C}$) is finite, while the set of *all* points in $\widehat{\mathbf{C}}$ that is deformed by H to c (i.e., $[c]$) is infinite.

Consider the components of $[c] \setminus \mathcal{C}$. Such a component C is called white/black if it is contained in the white/black 0-tile (i.e., in X_w^0 or X_b^0). Furthermore we call C non-trivial if it contains at least two distinct points $w_i, w_j \in \mathbf{W}$ in its boundary. Each non-trivial white component C corresponds to a critical point of the white polynomial, and each non-trivial black component C corresponds to a critical point of the black polynomial.

Let C be a white non-trivial component. The set of external angles associated to C is now the set

$$\{\theta(E_j^1) \mid w_j \in \mathbf{W} \text{ contained in the closure of } C\}.$$

The list of all these sets of external angles (for all critical points c and all white components of $[c] \setminus \mathcal{C}$) forms the critical portrait of the white polynomial P_w.

The white critical portrait for the pseudo-equator indicated in Figure 6.8 is

$$\left\{\frac{7}{60}, \frac{22}{60}, \frac{37}{60}\right\}, \left\{\frac{43}{60}, \frac{58}{60}\right\}.$$

The critical portrait of the black polynomial is constructed in almost the same fashion. There is a slight difference, however. This difference appears because in the construction of mating, a point with external angle θ in the Julia set of one polynomial is identified with a point with external angle $-\theta$ from the Julia set of the other polynomial.

Thus let C be a non-trivial black component of $[c] \setminus \mathcal{C}$ (for some critical point c). Then the set of external angles associated to C is

$$\{1 - \theta(E_j^1) \mid w_j \in \mathbf{W} \text{ contained in the closure of } C\}.$$

The list of all these sets (for all critical points c and all non-trivial black components of $[c] \setminus \mathcal{C}$) form the critical portrait of the black polynomial P_b.

The black critical portrait of the pseudo-equator in Figure 6.8 is

$$\left\{\frac{15}{60}, \frac{30}{60}, \frac{45}{60}\right\}, \left\{\frac{2}{60}, \frac{47}{60}\right\}.$$

Thus the critical portraits of the white and black polynomials can be read off from the pseudo-isotopy in an elementary combinatorial way. These determine the white and black polynomials P_w and P_b uniquely. In [Mey13] and [Mey] it is shown that $P_w \perp\!\!\!\perp P_b$ is topologically conjugate to f. Thus the existence of a pseudo-equator not only shows that f arises as a mating, but it allows us to recover the polynomials into which f unmates in an elementary fashion.

10   EXAMPLES OF UNMATINGS

Here we show several examples of unmatings.

We first show that shared matings are ubiquitous. We list all shared matings of the example from Section 5 that can be found with the sufficient condition from Section 6.

The critical portraits of the polynomials into which the map f unmates are as follows. They are obtained from the pseudo-equators shown in Figure 10.1. It is convenient to write the common denominator of the angles of a critical portrait outside the parentheses, i.e., instead of $\{n/N, m/N\}, \{i/N, j/N\}$, we write $\frac{1}{N}\{n, m\}, \{i, j\}$, and so on. By w we denote the critical portrait of the white polynomial, and by b the critical portrait of the (corresponding) black polynomial into which f unmates.

$$w : \frac{1}{60}\{7, 22, 37\}, \{43, 58\} \qquad b : \frac{1}{60}\{2, 47\}, \{15, 30, 45\} \qquad (10.1)$$

$$w : \frac{1}{20}\{2, 7, 17\}, \{10, 15\} \qquad b : \frac{1}{20}\{2, 7, 17\}, \{10, 15\} \qquad (10.2)$$

$$w : \frac{1}{60}\{11, 26\}, \{29, 59\}, \{30, 45\} \qquad b : \frac{1}{60}\{1, 46\}, \{19, 34\}, \{15, 45\} \qquad (10.3)$$

$$w : \frac{1}{60}\{14, 59\}, \{26, 41\}, \{15, 45\} \qquad b : \frac{1}{60}\{1, 31\}, \{15, 30\}, \{34, 49\} \qquad (10.4)$$

$$w : \frac{1}{20}\{3, 18\}, \{7, 17\}, \{10, 15\} \qquad b : \frac{1}{20}\{3, 18\}, \{7, 17\}, \{10, 15\} \qquad (10.5)$$

$$w : \frac{1}{60}\{7, 37\}, \{15, 30\}, \{43, 58\}, \qquad b : \frac{1}{60}\{2, 47\}, \{15, 45\}, \{23, 38\} \qquad (10.6)$$

The pseudo-equators in Figure 10.1 are not all possible; all others, however, are obtained from these by rotation (by $2\pi/3$ and $4\pi/3$). In the preceding, the point labeled p in Figure 10.1 is the point -1. We now rotate each pseudo-equator by $2\pi/3$, meaning the point labeled p is 1. From these new pseudo-equators we obtain unmatings of f into polynomials with the following critical portraits.

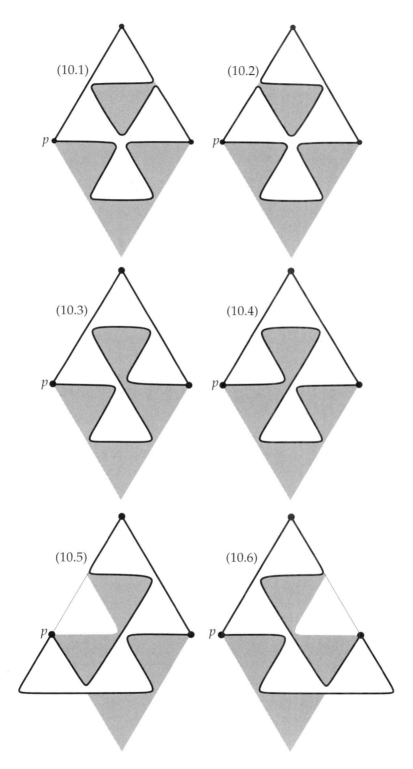

FIGURE 10.1: Different pseudo-equators.

$$\text{w}: \frac{1}{60}\{2,47\},\{15,30,45\} \qquad\qquad \text{b}: \frac{1}{60}\{7,22,37\},\{43,58\} \qquad (10.1')$$

$$\text{w}: \frac{1}{60}\{13,58\},\{15,30,45\} \qquad\qquad \text{b}: \frac{1}{60}\{2,17\},\{23,38,53\} \qquad (10.2')$$

$$\text{w}: \frac{1}{20}\{1,11\},\{5,10\},\{14,19\} \qquad\qquad \text{b}: \frac{1}{20}\{1,11\},\{5,10\},\{14,19\} \qquad (10.3')$$

$$\text{w}: \frac{1}{20}\{1,6\},\{9,19\},\{10,15\} \qquad\qquad \text{b}: \frac{1}{20}\{1,6\},\{9,19\},\{10,15\} \qquad (10.4')$$

$$\text{w}: \frac{1}{60}\{13,58\},\{15,45\},\{22,37\} \qquad\qquad \text{b}: \frac{1}{60}\{2,17\},\{23,53\},\{30,45\} \qquad (10.5')$$

$$\text{w}: \frac{1}{60}\{2,47\},\{15,45\},\{23,38\} \qquad\qquad \text{b}: \frac{1}{60}\{7,37\},\{15,30\},\{43,58\} \qquad (10.6')$$

Finally, we rotate the pseudo-equators in Figure 10.1 by $4\pi/3$, meaning that the point labeled p is ∞. We obtain unmatings of the map f into polynomials with the following critical portraits.

$$\text{w}: \frac{1}{20}\{3,13,18\},\{5,10\} \qquad\qquad \text{b}: \frac{1}{20}\{3,13,18\},\{5,10\} \qquad (10.1'')$$

$$\text{w}: \frac{1}{60}\{2,17\},\{23,38,53\} \qquad\qquad \text{b}: \frac{1}{60}\{13,58\},\{15,30,45\} \qquad (10.2'')$$

$$\text{w}: \frac{1}{60}\{1,46\},\{15,45\},\{19,34\} \qquad\qquad \text{b}: \frac{1}{60}\{11,26\},\{29,59\},\{30,45\} \qquad (10.3'')$$

$$\text{w}: \frac{1}{60}\{1,31\},\{15,30\},\{34,49\} \qquad\qquad \text{b}: \frac{1}{60}\{14,59\},\{15,45\},\{26,41\} \qquad (10.4'')$$

$$\text{w}: \frac{1}{60}\{2,17\},\{23,53\},\{30,45\} \qquad\qquad \text{b}: \frac{1}{60}\{13,58\},\{15,45\},\{22,37\} \qquad (10.5'')$$

$$\text{w}: \frac{1}{20}\{2,17\},\{3,13\},\{5,10\} \qquad\qquad \text{b}: \frac{1}{20}\{2,17\},\{3,13\},\{5,10\} \qquad (10.6'')$$

Here are some observations from these examples. Several, namely, (10.2), (10.5), (10.3'), (10.4'), (10.1''), (10.6''), are obtained by mating the same white polynomial to the same black polynomial.

Somewhat more interesting (and possibly surprising) is the following phenomenon. Consider the pairs (10.1) and (10.1'), (10.6) and (10.6'), (10.3) and (10.3''), (10.4) and (10.4''), as well as (10.2') and (10.2''). In each of these cases the map f is unmated into the same two polynomials. However, the pseudo-equators by which these unmatings are achieved are distinct. Thus the corresponding invariant Peano curves are distinct. Put differently, f can be obtained as the mating of two polynomials P_w and P_b in two distinct ways. Note, however, that the role of the black and white polynomials is interchanged in each case.

We next consider some unmatings of the map $g\colon \widehat{\mathbb{C}} \to \widehat{\mathbb{C}}$ given by

$$g(z) = 1 + \frac{\omega - 1}{z^3}, \qquad (10.7)$$

where $\omega = \exp(4\pi i/3)$. Note that $g(z) = \tau(z^3)$, where $\tau(w) = 1 + (\omega - 1)/w$ is a Möbius transformation that maps the upper half-plane to the half-plane above

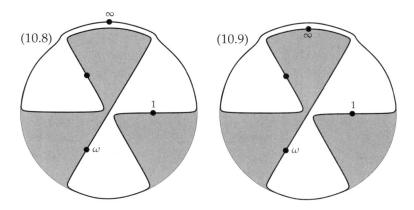

FIGURE 10.2: Almost the same connection.

the line through ω and 1 (more precisely $\tau : 0 \mapsto \infty, 1 \mapsto \omega, \infty \mapsto 1$). The critical points of g are 0 and ∞, which are mapped as follows

$$0 \xrightarrow{\ 3:1\ } \infty \xrightarrow{\ 3:1\ } 1 \longrightarrow \omega \ .$$

Thus $\text{post}(g) = \{\infty, 1, \omega\}$, i.e., g is postcritically finite. Since each critical point is strictly preperiodic, it follows that the Julia set of g is all of $\widehat{\mathbb{C}}$. The crucial property for our purposes is that ∞ is at the same time a critical as well as a postcritical point.

As $\mathcal{C} \supset \text{post}(g)$ we choose the extended line through ω and 1, i.e., the circle on $\widehat{\mathbb{C}}$ through ω, 1, and ∞, oriented positively as the boundary of the half-plane above this line. Then

$$g^{-1}(\mathcal{C}) = \bigcup_{j=0,\ldots,5} R_j,$$

where $R_j = \{r \exp(2\pi i j/6) \mid 0 \le r \le \infty\}$.

We show two pseudo-equators for g in Figure 10.2. In both cases the points on the circle all represent the point ∞. The narrow white channel indicates that two white 1-tiles are connected at ∞ (in both cases). The connection of 1-tiles is the same in (10.8) and (10.9), with one exception: the marking of the connection at ∞ (more precisely the cnc-partition representing the connection at ∞) differ. This means that in the geometric representation of Figure 10.2 the point ∞ is in different positions. From these pseudo-equators we obtain unmatings of g into polynomials with the following critical portraits.

$$\text{w}: \frac{1}{27}\{4, 22\}, \{12, 21\} \qquad \text{b}: \frac{1}{27}\{5, 14\}, \{15, 24\} \tag{10.8}$$

$$\text{w}: \frac{1}{27}\{7, 25\}, \{12, 21\} \qquad \text{b}: \frac{1}{27}\{2, 11\}, \{15, 24\} \tag{10.9}$$

REMARK 10.3. At first the author thought that this example would result in two distinct pairs of polynomials that would not only yield the same rational map when mated, but also give the same invariant Peano curve. This, however, is not true. We do not give the precise argument here. This can, however, be found in [Mey13, Section 8].

11   A MATING NOT ARISING FROM A PSEUDO-EQUATOR

In this section we give an example of a mating that does not arise via a pseudo-equator. Thus we show that the sufficient condition that a rational map arises as a mating given in Theorem 6.3 is not necessary.

More precisely, we will mate two postcritically finite polynomials and show that the resulting rational map f has no pseudo-equator. The Julia set of this map will be the whole sphere.

Each sufficiently high iterate f^n of f has a pseudo-equator by Theorem 6.4, but we do not know how many iterates are necessary.

In the construction of the topological mating, a point in the white Julia set where the external ray R_θ lands is identified with a point in the black Julia set where the external ray $R_{-\theta}$ lands. To make the following more readable, we define for any $\theta \in [0,1)$ $\theta^* := 1 - \theta$. Thus in the topological mating, the landing point of R_θ in the white Julia set is identified with the landing point of R_{θ^*} in the black polynomial.

The white polynomial will be $P_w := z^2 + i$. It is well-known that the critical portrait of P_w is $\{1/12, 7/12\}$. The external angles are mapped under the angle doubling map in the following way:

$$\frac{1}{12}, \frac{7}{12} \longrightarrow \frac{1}{6} \longrightarrow \frac{1}{3} \overset{\frown}{\longrightarrow} \frac{2}{3}\ .$$

Furthermore, the external rays with angles $\{1/7, 2/7, 4/7\}$ land at the α-fixed point of P_w.

As the black polynomial P_b we pick the (quadratic, monic, postcritically finite) polynomial with the critical portrait $\{(1/28)^*, (15/28)^*\}$. Thus $P_b = z^2 + c_b$, where c_b is the landing point of the external ray $R_{(1/14)^*}$ in the Mandelbrot set. The external angles are mapped under angle doubling in the following way:

$$\left(\frac{1}{28}\right)^*, \left(\frac{15}{28}\right)^* \longrightarrow \left(\frac{1}{14}\right)^* \longrightarrow \left(\frac{1}{7}\right)^* \overset{\frown}{\longrightarrow} \left(\frac{2}{7}\right)^* \longrightarrow \left(\frac{4}{7}\right)^*\ .$$

LEMMA 11.1. *The topological mating of P_w and P_b exists and is topologically conjugate to a rational map $f \colon \widehat{\mathbb{C}} \to \widehat{\mathbb{C}}$ of degree 2. The two critical points a, b of f are mapped by f in the following way:*

$$a \overset{2:1}{\longrightarrow} p_1 \longrightarrow p_2 \overset{\frown}{\circlearrowright}$$

$$b \overset{2:1}{\longrightarrow} q_1 \longrightarrow q_2 \overset{\frown}{\longrightarrow} q_3\ .$$

Thus f has 5 postcritical points, namely, p_1, p_2, q_1, q_2, q_3. The Julia set of f is the whole sphere, since each critical point is strictly preperiodic.

PROOF. It is well known that i, i.e., the critical value of P_w, is in the $1/3$-limb of the Mandelbrot set. Indeed the external angle at the critical value i is $1/6$. The external rays $R_{1/7}$ and $R_{2/7}$ disconnect the $1/3$-limb from the main cardioid. Since $1/7 < 1/6 < 2/7$, it follows that i lies in this limb.

The conjugate limb is the $2/3$-limb; it is disconnected from the main cardioid by the external rays $R_{5/7}, R_{6/7}$. Since $(1/14)^* = (13/14) > 6/7$, it follows that i, c_b do not lie in conjugate limbs of the Mandelbrot set. From the Rees-Shishikura-Tan theorem (Theorem 3.2) it follows that the topological mating of P_w and P_b exists and is topologically conjugate to a rational map (which is postcritically finite and of degree 2).

Clearly the three postcritical points of P_b with external angles $(1/7)^*$, $(2/7)^*$, $(4/7)^*$ are all identified with the α-fixed point of P_w under the topological mating, i.e., correspond to a single postcritical point in the mating. It remains to show that there are not more identifications of postcritical points.

Consider first the external angle $1/3$. Let $z_{(1/3)^*}$ be the landing point of $R_{(1/3)^*}$ in the Julia set $\mathcal{J}_b$ of P_b. We want to show that $R_{(1/3)^*}$ is the only external ray landing at $z_{(1/3)^*}$. Indeed $(1/3)^*$ is fixed under the second iterate of the angle-doubling map. Thus if R_{θ^*} lands at $z_{(1/3)^*}$ it has to be fixed under the second iterate of the angle doubling map as well. Thus either $\theta^* = (2/3)^*$ or $\theta^* = 0$. In the first case, the external rays $R_{1/3}$, $R_{2/3}$ in the dynamical plane of P_w together with the external rays $R_{(1/3)^*}$, $R_{(2/3)^*}$ disconnect the formal mating sphere. This cannot happen by the Rees-Shishikura-Tan theorem. The second case cannot happen, since for quadratic polynomials the landing point of R_0, i.e., the β-fixed point, is not the landing point of any other external ray.

By the same argument it follows that the landing point of $R_{(2/3)^*}$ (in the dynamical plane of P_b) is not the landing point of any other external ray.

Consider now the landing point $z_{(1/6)^*}$ of $R_{(1/6)^*}$, in the dynamical plane of P_b. This point is not a critical point which is mapped to $z_{(1/3)^*}$ by P_b. Since by the preceding there is a single external ray landing at $z_{(1/3)^*}$, it follows that there is a single external ray landing at $z_{(1/6)^*}$.

Thus it follows that the three postcritical points of P_w descend to three distinct postcritical points q_1, q_2, q_3 of f, and their dynamics is as described in the statement.

It remains to show that the postcritical point of P_b at which the external ray $R_{(1/14)^*}$ (in the dynamical plane of P_b) lands is not identified with other postcritical or critical points.

We first note that at the three postcritical points of P_b at which the external rays $R_{(1/7)^*}$, $R_{(2/7)^*}$, $R_{(4/7)^*}$ land, there is no other external ray which lands. Indeed, any other such external ray would be a fixed point of the third iterate of the angle-doubling map, i.e., it has to be an external ray of the form $R_{(k/7)^*}$, where $k = 3, 5, 6$. Assume first that $R_{(1/7)^*}$ and $R_{(6/7)^*}$ land at the same point. This is impossible, since the external rays $R_{(1/28)^*}$ and $R_{(15/28)^*}$ (which both land at the critical point of P_b) are not contained in one component of $\mathbb{C} \setminus (R_{(1/7)^*} \cup R_{(6/7)^*}$. The same argument shows that $R_{(1/7)^*}$ and $R_{(5/7)^*}$ cannot land at the same point. Finally, assume that $R_{(1/7)^*}$ and $R_{(3/7)^*}$ land at the same point. Mapping these external rays by P_b yields that $R_{(2/7)^*}$ and $R_{(6/7)^*}$ land at the same point. This is again impossible, since $R_{(1/28)^*} \cup R_{(15/28)^*}$ do not lie in the same component of $\mathbb{C} \setminus (R_{(2/7)^*} \cup R_{(6/7)^*})$.

Consider the landing point $z_{1/14}$ of the external ray $R_{1/14}$ in the dynamical plane of P_w. This is the preimage of the α-fixed point of P_w. Thus the external rays $R_{9/14}$ and $R_{11/14}$ (in the dynamical plane of P_w) land at $z_{1/14}$ as well.

Consider now the corresponding external rays $R_{(9/14)^*}$ and $R_{(11/14)^*}$ in the dynamical plane of R_b. These are mapped by P_b to $R_{(2/7)^*}$ and $R_{(4/7)^*}$. By the preceding discussion, there is no other external ray landing at these endpoints. Thus there are no other external rays landing at the same points as $R_{(9/14)^*}$ and $R_{(11/14)^*}$.

Thus the 4 postcritical points of P_b descend to 2 postcritical points p_1, p_2 of f, which are mapped as stated. $\square$

THEOREM 11.2. *The rational map* $f: \widehat{\mathbb{C}} \to \widehat{\mathbb{C}}$ *from Lemma 11.1, which arises as the mating of* P_w *and* P_b, *does not have a pseudo-equator (as in Definition 6.1).*

PROOF. To show that no pseudo-equator as in Definition 6.1 exists, we will consider several cases. Namely, we consider the different cyclical orders in which $\mathcal{C}$ may traverse the postcritical points. Assume a pseudo-isotopy $H: \widehat{\mathbb{C}} \times [0,1] \to \widehat{\mathbb{C}}$ deforms $\mathcal{C}$ to $\mathcal{C}^1 = f^{-1}(\mathcal{C})$, as in Definition 6.1. Then H induces an order in which $\mathcal{C}^1$ is traversed. The cyclical order in which $\mathcal{C}^1$ traverses the postcritical points has to agree with the cyclical order of the postcritical points on $\mathcal{C}$, since H is a pseudo-isotopy rel. post(f). We will show that this is impossible.

Thus let $\mathcal{C} \subset \widehat{\mathbb{C}}$ be a Jordan curve with post(f) $\subset \mathcal{C}$, which is (arbitrarily) oriented. The white and black 0-tiles X_w^0, X_b^0, as well as the white and black 1-tiles, are defined in terms of $\mathcal{C}$ as usual ($\mathcal{C}$ is positively oriented as the boundary of X_w^0). There are two white (as well as two black) 1-tiles, which intersect at the critical points a, b. Recall from Lemma 7.3 that if $\mathcal{C}$ is a pseudo-equator, there has to be a connection of 1-tiles such that the white connection graph is a spanning tree.

To obtain such a connection of white 1-tiles that yields a spanning tree, we have to connect the two white 1-tiles either at a or at b. Recall that each white 1-tile is mapped homeomorphically to X_w^0 by f. Thus the two preimages of any point p_2, q_2, q_3, i.e., the postcritical points which are not critical values, have to lie in distinct white 1-tiles. Note that $\mathcal{C}^1$ has to traverse the boundary of each of the white 1-tiles positively (see Definition 6.1 and Lemma 6.10).

Let q_2' be the preimage of q_3 by f distinct from q_2. Adding this point to the diagram in Lemma 11.1, we obtain that the points $a, b, p_1, p_2, q_1, q_2, q_2', q_3$ are mapped by f as follows:

$$
a \xrightarrow{2:1} p_1 \longrightarrow p_2 \righthookarrow
$$
$$
b \xrightarrow{2:1} q_1 \longrightarrow q_2 \xrightarrow{} q_3 \longleftarrow q_2' \, .
$$

Note that the set $\{a, b, p_1, p_2, q_1, q_2, q_2', q_3\}$ contains all preimages of the postcritical points.

CASE 1 (p_1, p_2 *are adjacent on* $\mathcal{C}$).

This means that one arc on $\mathcal{C}$ between p_1, p_2 does not contain any other postcritical point. One such situation is pictured in Figure 11.3. Note that p_1, p_2 are both preimages of p_2. Thus p_1, p_2 are contained in distinct white 1-tiles.

Assume that the two white 1-tiles are connected at either a or b, and traverse $\mathcal{C}^1 := f^{-1}(\mathcal{C})$ in the order induced by this connection. Each edge in $\mathcal{C}^1$ is traversed positively as boundary of a white 1-tile (see Definition 6.1 and Lemma 6.10). Mark each postcritical point in $\mathcal{C}$ by a different symbol as to the right in Figure 11.3. Similarly we mark each point (which is a preimage of a postcritical point) by the same symbol as its image. See the left side of Figure 11.3. Note that $\mathcal{C}^1$ traverses the symbols in the same cyclical order as $\mathcal{C}$ (twice). Thus on each of the two paths on $\mathcal{C}^1$ between p_1 and p_2 (which are marked by the same symbol), every other symbol has to appear. In particular the symbol marking the preimages of q_2 have to appear on each such path. Since these mark the postcritical points q_1 and q_3, it

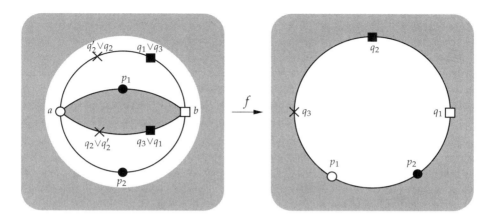

FIGURE 11.3: Case 1.

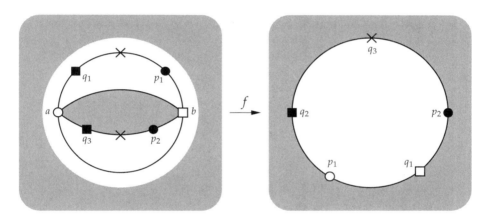

FIGURE 11.4: Case 3.

follows that on each path of $\mathcal{C}^1$ between p_1 and p_2 there is a postcritical point. This shows that Case 1 is impossible.

By exactly the same argument as before we can rule out that q_1, q_3 are adjacent on $\mathcal{C}$. Indeed they obtain the same marking (in $\mathcal{C}^1$), each path on $\mathcal{C}^1$ between them has to contain either p_1 or p_2 (which have the same marking).

CASE 2 (p_2, q_2 are adjacent on $\mathcal{C}$).

We choose the orientation on $\mathcal{C}$ such that q_2 succeeds p_2 on $\mathcal{C}$. Consider the white 1-tile X containing p_2 in its boundary ∂X. Then p_2 is succeeded on ∂X, hence on $\mathcal{C}^1$, by a 1-vertex which is mapped to q_2 by f. This 1-vertex must be q_1 or q_3. Hence on $\mathcal{C}^1$ the postcritical point p_2 is succeeded by either q_1 or q_3. Hence $\mathcal{C}^1$ is not obtained as a pseudo-isotopic deformation of $\mathcal{C}$ rel. $\mathrm{post}(f)$ (in which case p_2 would be succeeded by q_2 on $\mathcal{C}^1$). Hence Case 2 does not happen.

From the preceding it follows that the cyclical order of the postcritical points on $\mathcal{C}$ is either $p_1, q_1, p_2, q_3, q_2, p_1$ or $p_1, q_3, p_2, q_1, q_2, p_1$. Here we are using the fact that we can choose the orientation on $\mathcal{C}$ arbitrarily.

CASE 3 (The cyclical order of the postcritical points on $\mathcal{C}$ is $p_1, q_1, p_2, q_3, q_2, p_1$).
The situation is illustrated in Figure 11.4. We mark the points in $\mathcal{C}$ and $\mathcal{C}^1$ as before.
Since on $\mathcal{C}$ the point p_1 is succeeded by q_1 and p_2 is succeeded by q_3, the same has
to hold on $\mathcal{C}^1$. Note however, that on $\mathcal{C}^1$ between p_1, q_1 and between p_2, q_3, there is
a preimage of q_3. One of these preimages (marked with an × in the figure) of q_3 has
to be q_2. Thus the cyclical ordering of the postcritical points on $\mathcal{C}^1$ does not agree
with the one on $\mathcal{C}$.

CASE 4 (The cyclical order of the postcritical points on $\mathcal{C}$ is $p_1, q_3, p_2, q_1, q_2, p_1$).

The situation is exactly analogous to Case 3. In the right of Figure 11.4 we have
to interchange q_1 and q_3. Since p_1 is succeeded by q_3 and p_2 is succeeded by q_1 on $\mathcal{C}$,
the same has to hold in $\mathcal{C}^1$. Thus in the right of Figure 11.4 we have to interchange
q_1 and q_3 for the situation at hand. One of the points marked by × has to be q_2,
which gives a contradiction as before.   □

12   OPEN QUESTIONS

Let $f \colon \widehat{\mathbb{C}} \to \widehat{\mathbb{C}}$ be a rational map. Is the existence of an equator for f *sufficient* for
f to arise as a mating? When f is postcritically finite, this is widely expected, but
there does not seem to be a proof in the literature.

A (pseudo-) equator $\mathcal{E}$ has to be *orientation-preserving* (pseudo-) isotopic to
$f^{-1}(\mathcal{E})$. *Orientation-reversing* (pseudo-) equators seem to be as common as orientation-preserving ones. From such a pseudo-equator it is possible to construct a
semi-conjugacy from $z^{-d} \colon S^1 \to S^1$ to the map f, where $d = \deg f$ (in the case
when f is postcritically finite and $\mathcal{J}(f) = \widehat{\mathbb{C}}$). There should be some sort of
"orientation-reversing mating" associated to such orientation-reversing (pseudo-)
equators.

The form of Moore's theorem presented here (see Theorem 2.5) allows one to
shrink each ray-equivalence class to a point by a pseudo-isotopy H. Is it possible
to put further smoothness assumptions on H? In particular, can one *embed H in a
holomorphic motion*?

13   Bibliography

[Bae27]   R. Baer. Kurventypen auf Flächen. *J. Reine Angew. Math.*, 156:231–246,
          1927.

[Bae28]   R. Baer. Isotopie von Kurven auf orientierbaren, geschlossen Flächen
          und ihr Zusammenhang mit der topologischen Deformation der
          Flächen. *Reine Angew. Math.*, 159:101–116, 1928.

[Bea84]   A. F. Beardon. *A primer on Riemann surfaces*, volume 78 of *London Mathe-
          matical Society Lecture Note Series*. Cambridge University Press, 1984.

[BFH92]   Ben Bielefeld, Yuval Fisher, and John Hubbard. The classification of criti-
          cally preperiodic polynomials as dynamical systems. *J. Amer. Math. Soc.*,
          5(4):721–762, 1992.

[BM11]   M. Bonk and D. Meyer. *Expanding Thurston Maps*. Mathematical Surveys
          and Monographs. AMS. to appear. arXiv:1009.3647.

[Dav86] R. J. Daverman. *Decompositions of manifolds*, volume 124 of *Pure and Applied Mathematics*. Academic Press, Orlando, FL, 1986.

[DH93] A. Douady and J. H. Hubbard. A proof of Thurston's topological characterization of rational functions. *Acta Math.*, 171(2):263–297, 1993.

[Dou83] A. Douady. Systèmes dynamiques holomorphes. In *Bourbaki seminar, Vol. 1982/83*, volume 105 of *Astérisque*, pages 39–63. Soc. Math. France, 1983.

[Eps66] D. B. A. Epstein. Curves on 2-manifolds and isotopies. *Acta Math.*, 115:83–107, 1966.

[Kam03] A. Kameyama. On Julia sets of postcritically finite branched coverings, II. S^1-parametrization of Julia sets. *J. Math. Soc. Japan*, 55(2):455–468, 2003.

[Mey] D. Meyer. Expanding Thurston maps as quotients. Preprint. arXiv:0910.2003.

[Mey13] D. Meyer. Invariant Peano curves of expanding Thurston maps. *Acta Math.*, 210:95–171, 2013. arXiv:0907.1536.

[Mil99] John Milnor. *Dynamics in one complex variable. Introductory lectures.* Friedr. Vieweg & Sohn, Braunschweig, 1999.

[Mil04] J. Milnor. Pasting together Julia sets: a worked out example of mating. *Experiment. Math.*, 13(1):55–92, 2004.

[Mil06] J. Milnor. On Lattès maps. In *Dynamics on the Riemann Sphere, A Bodil Branner Festschrift*, Eur. Math. Soc. Hjorth and Petersen, 2006.

[MP12] D. Meyer and C. L. Petersen. On the notions of mating. *Ann. Fac. Sci. Toulouse Math.*, 21(5):839-876, 2012. arXiv:1307.7934.

[Poi09] A. Poirier. Critical portraits for postcritically finite polynomials. *Fund. Math.*, 203(2):107–163, 2009.

[Ree92] M. Rees. A partial description of parameter space of rational maps of degree two, I. *Acta Math.*, 168(1-2):11–87, 1992.

[Shi00] M. Shishikura. On a theorem of M. Rees for matings of polynomials. In *The Mandelbrot set, theme and variations*, volume 274 of *London Mathematical Society Lecture Note Series*, pages 289–305. Cambridge University Press, Cambridge, 2000.

[SL00] M. Shishikura and Tan Lei. A family of cubic rational maps and matings of cubic polynomials. *Experiment. Math.*, 9(1):29–53, 2000.

[TL92] Tan L. Matings of quadratic polynomials. *Ergodic Theory Dynam. Systems*, 12(3):589–620, 1992.

[Wit88] B. S. Wittner. *On the bifurcation loci of rational maps of degree two.* Ph.D. thesis, Cornell University, 1988.

A framework toward understanding the characterization of holomorphic dynamics

Yunping Jiang

with an appendix by Tao Chen,
Yunping Jiang, and Linda Keen

ABSTRACT. This paper gives a review of our work on the characterization of geometrically finite rational maps and then outlines a framework for characterizing holomorphic maps. Whereas Thurston's methods are based on estimates of hyperbolic distortion in hyperbolic geometry, the framework suggested here is based on controlling conformal distortion in spherical geometry. The new framework enables one to relax two of Thurston's assumptions: first, that the iterated map has finite degree and, second, that its post-critical set is finite. Thus, it makes possible to characterize certain rational maps for which the post-critical set is not finite as well as certain classes of entire and meromorphic coverings for which the iterated map has infinite degree.

1 CHARACTERIZATION

Suppose $f\colon S^2 \to S^2$ is an orientation-preserving branched covering of the two-sphere S^2, often simply referred to as a "branched covering." We use the notation $f^n = \underbrace{f \circ \cdots \circ f}_{n}$ for the n-fold composition. Let $d = \deg f > 1$ be the degree of f and let

$$\Omega_f = \{c \mid \deg_c f \geq 2\}$$

be the set of all branched points, where $\deg_c f$ means the local degree of f at c. Let

$$P_f = \overline{\cup_{n\geq 1} f^n(\Omega_f)}$$

be the post-critical set. A rational map which is a quotient of two polynomials is a holomorphic branched covering.

DEFINITION 1.1 (Combinatorial equivalence). *Two branched coverings f and g are said to be combinatorially equivalent if there are two homeomorphisms $\phi, \psi\colon S^2 \to S^2$ such that*

$$\begin{CD} (S^2, P_f) @>{\phi}>> (S^2, P_g) \\ @V{f}VV @VV{g}V \\ (S^2, P_f) @>{\psi}>> (S^2, P_g) \end{CD}$$

commutes and $\phi \overset{homotopy}{\sim} \psi$ rel. P_f. (The term "rel. P_f" means that $\psi|P_f = \phi|P_f$ and the homotopy preserves this equation, refer to Definition 4.1.)

The following is a basic problem in complex dynamics.

PROBLEM 1. *When is a branched covering f combinatorially equivalent to a rational map?*

Consider the complex plane $\mathbb{C}$ and consider the Riemann sphere $\widehat{\mathbb{C}} = \mathbb{C} \cup \{\infty\}$ which is the two-sphere equipped with the standard conformal structure. Let

$$M(\mathbb{C}) = \{\mu \mid \mu \text{ is a measurable function on } \mathbb{C} \text{ such that } \|\mu\|_\infty < 1\}$$

be the open unit ball of the complex Banach space $\mathcal{L}^\infty(\mathbb{C})$. Each μ in $M(\mathbb{C})$ is called a Beltrami coefficient and can be viewed as a conformal structure on $\widehat{\mathbb{C}}$. The Beltrami equation

$$w_{\bar{z}} = \mu w_z \tag{1.1}$$

has a unique quasiconformal homeomorphism solution fixing $0, 1, \infty$ (see, for example, [Ahlfors-Bers]). We always use w^μ to denote this solution.

Assume f is locally quasiconformal except on Ω_f. In other words, we assume that f is quasiregular, that is, $f = R \circ g$, where g is a quasiconformal homeomorphism and R is a rational map. Given a Beltrami coefficient μ (viewed as a conformal structure on $\widehat{\mathbb{C}}$), the pull-back $f^*\mu$ of μ by f can be calculated as

$$f^*\mu(z) = \frac{\mu_f(z) + \mu(f(z))\theta(z)}{1 + \overline{\mu_f(z)}\mu(f(z))\theta(z)},$$

where $\mu_f(z) = f_{\bar{z}}(z)/f_z(z)$ is a Beltrami coefficient and where $\theta(z) = \overline{f_z(z)}/f_z(z)$.

Let $E \supseteq P_f$ be a closed subset of $\widehat{\mathbb{C}}$ such that $f(E) \subseteq E$. Suppose $0, 1, \infty \in E$. Then $f^*\mu \sim_E \mu$ means that $(w^{f^*\mu})^{-1} \circ w^\mu \overset{homotopy}{\sim} id$ rel. E (the term "rel. E" means that $(w^{f^*\mu})^{-1} \circ w^\mu | E = id$ and the homotopy preserves this equation; refer to Definition 4.1).

PROBLEM 2. *Can we view $f: (\widehat{\mathbb{C}}, E) \to (\widehat{\mathbb{C}}, E)$ as a holomorphic map? More precisely, can we find a Beltrami coefficient μ such that $f^*\mu \sim_E \mu$ rel. E?*

A special case is when $E = \widehat{\mathbb{C}}$. The answer to Problem 2 in this special case is the uniform quasiconformality.

THEOREM 1.2 ([Sullivan;2]). *There is a Beltrami coefficient μ (in other words, a conformal structure on $\widehat{\mathbb{C}}$) such that*

$$f^*\mu = \mu \text{ if and only if } \sup_{n \geq 0} K(f^n) \leq K_0 < \infty,$$

where $K(f^n)$ means the maximal quasiconformal dilatation of f^n.

REMARK 1.3. Sullivan proved Theorem 1.2 for two-dimensional quasiconformal semi-groups acting on the Riemann sphere. Hinkkanen [Hinkkanen] showed that a similar method can also be applied to prove the theorem for any quasi-entire function, $f = e \circ g$, and any quasi-meromorphic function, $f = m \circ g$, where g is a quasiconformal homeomorphism and e and m are an entire function and a meromorphic function on the complex plane, as a two-dimensional semi-group acting on the complex plane. Sullivan [Sullivan;1] and Tukia [Tukia] also proved the theorem for two-dimensional quasiconformal groups acting on the Riemann sphere.

2 OBSTRUCTION

In general, answers to Problems 1 and 2 are not easy. There is a topological obstruction which we call a Thurston obstruction in this study.

Suppose $E \supseteq P_f$ is a closed subset of the Riemann sphere (it is not necessary to assume that $0, 1, \infty \in E$). A simple closed curve γ in $S^2 \setminus E$ is called non-peripheral if every component of $S^2 \setminus \gamma$ contains at least two points from E. A *multi-curve* $\Gamma = \{\gamma_1, \ldots, \gamma_n\}$ is a set of finitely many pairwise disjoint, non-homotopic, and non-peripheral curves in $S^2 \setminus E$.

Since $E \supseteq P_f$, $\Omega_f \subseteq f^{-1}(E)$ and $f: S^2 \setminus f^{-1}(E) \to S^2 \setminus E$ is a covering map of finite degree and every component of $f^{-1}(\gamma)$ is a simple closed curve. A multi-curve Γ is said to be *f-stable* if for any $\gamma \in \Gamma$, every non-peripheral component of $f^{-1}(\gamma)$ is homotopic to an element of Γ rel. E.

For a stable multi-curve Γ, define f_Γ as follows: For each $\gamma_j \in \Gamma$, let $\gamma_{i,j,\alpha}$ denote the components of $f^{-1}(\gamma_j)$ homotopic to γ_i in $S^2 \setminus E$ and $d_{i,j,\alpha}$ be the degree of $f|_{\gamma_{i,j,\alpha}}: \gamma_{i,j,\alpha} \to \gamma_j$. Define

$$f_\Gamma(\gamma_j) = \sum_i \Big(\sum_\alpha \frac{1}{d_{i,j,\alpha}} \Big) \gamma_i.$$

Then $f_\Gamma: \mathbb{R}^\Gamma \to \mathbb{R}^\Gamma$ is a linear transformation with a non-negative matrix $A = (f_\Gamma)$. There exists a largest non-negative eigenvalue $\lambda(A)$ of A with a non-negative vector.

DEFINITION 2.1 (Thurston obstruction).   *A stable multi-curve Γ is called a* Thurston obstruction *if the largest eigenvalue $\lambda(A)$ is greater than or equal to 1.*

DEFINITION 2.2 (Levy cycle).   *If there is a f-invariant finite set of non-peripheral curves $\Gamma = \{\gamma_0, \gamma_1, \gamma_2, \ldots, \gamma_n\}$ then Γ is called a* Levy cycle. *Here, f-invariant means that there is a component γ'_i of $f^{-1}(\gamma_{i+1})$ such that $\gamma'_i \sim \gamma_i$ rel P_f and $f: \gamma'_i \to \gamma_{i+1}$, with i taken mod n, is a homeomorphism.*

REMARK 2.3. Usually a Levy cycle is only a multi-curve but it can be extended to a stable multi-curve. A stable Levy cycle is a special kind of Thurston obstruction. On the other hand, for a topological polynomial (not necessarily post-critically finite) or a post-critically finite branched covering of degree two, a Thurston obstruction always contains a Levy cycle (see [Levy] and [Rees]; also refer to [Tan] and [Godillon]). Among all Thurston obstructions (if they exist), the canonical obstruction is the most interesting one (see Definition 4.7).

3 REVIEW

A branched covering f is called post-critically finite if P_f is a finite subset. In this case, an answer to Problems 1 and 2 is Thurston's theorem, as follows.

PROPOSITION 3.1.   *Any post-critically finite branched covering f is combinatorially equivalent to a quasi-regular post-critically finite branched covering $g = R \circ h$, where R is a rational map and h is a quasiconformal homeomorphism; in other words, g is locally quasiconformal except on Ω_g.*

PROOF. Recall that Ω_f is the set of branched points of f in $\widehat{\mathbb{C}}$. Consider the topological surface $S = \widehat{\mathbb{C}} \setminus \Omega_f$. For every $p \in S$, let U_p be a small neighborhood about p such that $\phi_p = f|_U \colon U \to f(U) \subset \widehat{\mathbb{C}}$ is injective. Then $\alpha = \{(U_p, \phi_p)\}_{p \in S}$ defines an atlas on S with charts (U_p, ϕ_p). If $U_p \cap U_q \neq \varnothing$, then

$$\phi_p \circ \phi_q^{-1}(z) = z \colon \phi_q(U_p \cap U_q) \to \phi_p(U_p \cap U_q).$$

Thus all transition maps are conformal (one-to-one and analytic) and the atlas α defines a Riemann surface structure on S which we again denote by α. Denote this Riemann surface by (S, α). From the uniformization theorem, (S, α) is conformally equivalent to the Riemann surface $\widehat{\mathbb{C}} \setminus A$ with the standard complex structure induced by $\widehat{\mathbb{C}}$, where A consists of $\#(\Omega_f)$ points. Thus we have a homeomorphism $\widehat{h} \colon \widehat{\mathbb{C}} \to \widehat{\mathbb{C}}$ such that $\widehat{h} \colon (S, \alpha) \to \widehat{\mathbb{C}} \setminus A$ is conformal. Thus $R = f \circ \widehat{h}^{-1} \colon \widehat{\mathbb{C}} \to \widehat{\mathbb{C}}$ is holomorphic with critical points at $\widehat{h}(\Omega_f)$. Since the set P_f is finite, following the standard procedure in quasiconformal mapping theory, there is a K-quasiconformal homeomorphism $h \colon \widehat{\mathbb{C}} \to \widehat{\mathbb{C}}$ such that h is isotopic to $\widehat{h}$ rel. P_f. The map $g = R \circ h$ is a quasi-regular map and combinatorially equivalent to f. This completes the proof of the proposition. $\qquad\square$

Associated to the finite set $E = P_f$, one can introduce an orbifold structure as follows:

Define the signature $\nu_f \colon S^2 \to \mathbb{N} \cup \{\infty\}$ as

$$\nu_f(x) = \begin{cases} 1, & x \notin \cup_{n \geq 1} f^{\circ n}(\Omega_f); \\ \mathrm{lcm}\{\deg_y f^n \mid n > 0, f^n(y) = x\}, & \text{otherwise,} \end{cases}$$

where lcm means the least common multiple and $\deg_y f^n$ means the local degree of f^n at y. The orbifold associated to f is $\mathcal{O}_f = (S^2, \nu_f)$ and the Euler characteristic of $\mathcal{O}_f$ is defined as

$$\chi(\mathcal{O}_f) = 2 - \sum_{x \in S^2} \left(1 - \frac{1}{\nu_f(x)}\right).$$

It is known that $\chi(\mathcal{O}_f) \leq 0$ for any post-critically finite branched covering f. The orbifold $\mathcal{O}_f$ is called hyperbolic if $\chi(\mathcal{O}_f) < 0$ and parabolic if $\chi(\mathcal{O}_f) = 0$.

THEOREM 3.2 ([Thurston], in [Douady-Hubbard]). *A post-critically finite branched covering f with hyperbolic orbifold is combinatorially equivalent to a rational map if and only if it has no Thurston obstruction. Moreover, the rational map is unique up to conjugation by an automorphism of the Riemann sphere.*

REMARK 3.3. The paper [Douady-Hubbard] is a good source to learn Thurston's theorem, where a complete, detailed, and comprehensive proof is presented. The video recordings of Sullivan's lectures given at the CUNY Graduate Center in 1981–1986 [Sullivan;3] are another source for learning this theorem.

Furthermore, we have the following.

THEOREM 3.4 ([Pilgrim]). *If a post-critically finite branched covering f is not combinatorially equivalent to a rational map (i.e., a negative answer to Problem 1), then it must have the canonical obstruction (see Definition 4.7). The canonical obstruction is a Thurston obstruction.*

Things become very different when one turns to geometrically finite branched coverings. We started this study in [Cui-Jiang-Sullivan;M] and then summarized it in [Cui-Jiang-Sullivan]. The latter divides our study into two parts; the first part concentrates on a local combinatorial theory, and the second part focuses on a global combinatorial theory. A branched covering f is called geometrically finite if the cardinality of P_f is infinite but the accumulation set P'_f is finite. In this case, $P'_f = \{a_1, \ldots, a_k\}$ consists of finitely many periodic orbits. In [Cui-Jiang], it is proved that this definition of a geometrically finite branched covering is equivalent to the traditional definition of a geometrically finite rational map when the branched covering is holomorphic. Recall that the traditional definition of a geometrically finite rational map is that a rational map is geometrically finite if the intersection of the post-critical set and the Julia set is a finite set. There was little progress towards the understanding of the characterization of geometrically finite rational maps (for example, [Brown]) until our work, which we will discuss shortly. Note that for a general rational map R, we have the following.

THEOREM 3.5 ([McMullen]). *Suppose R is a rational map. Take $E = P_R$. Let Γ be a f-stable multi-curve on $\widehat{\mathbb{C}} \setminus E$. Then $\lambda(\Gamma) \leq 1$. Only in the following two cases, $\lambda(\Gamma)$ may be 1:*

1. *R is post-critically finite with $\#(E) = 4$ and the orbifold $\mathcal{O}_f$ is $(2,2,2,2)$. Moreover, R is double covered by an integral torus endomorphism.*

2. *E is an infinite set and Γ includes the essential curves in a finite system of annuli permuted by R. These annuli lie in Siegel disks or Herman rings (refer to [Milnor]) for R and each annulus is a connected component of $\widehat{\mathbb{C}} \setminus E$.*

A geometrically finite rational map has no Thurston obstruction (refer to the proof of [Douady-Hubbard, Theorem 4.1] or refer to Theorem 3.5 since it has no Siegel disks and Herman rings). Furthermore, we found a counterexample in the geometrically finite case which prevents a similar theorem to Thurston's theorem under the combinatorial equivalence.

THEOREM 3.6 (Counterexample [Cui-Jiang]). *There is a geometrically finite branched covering f having no Thurston obstruction and giving a negative answer to Problem 1 (i.e., it is not combinatorially equivalent to a rational map).*

The idea to construct this counterexample is to start with $q(z) = \lambda z + z^2$, $0 < |\lambda| < 1$. Then 0 is its attractive fixed point such that the critical point $c = -\lambda/2$ is in the immediate basin of 0. The post-critical orbit $\{q(c), \ldots, q^n(c), \ldots\}$ has a limiting point 0. Since $q(z)$ is a geometrically finite rational map, it has no Thurston obstruction. Apply Dehn twists along this post-critical orbit such that the topological structure near this limiting point has infinite complexity and such that there is no new Thurston obstruction being introduced. Remember that according to Kœnig's theorem (see [Milnor]), the topological structure of a holomorphic map near an attractive fixed point must be very simple. Thus the newly constructed geometrically finite branched covering can not be combinatorially equivalent to a rational map. Pictures in [Cui-Jiang, Figures 1–3] provide further ideas about this construction.

Based on this counterexample, we define a semi-rational branched covering and a sub-hyperbolic semi-rational branched covering.

DEFINITION 3.7 (Semi-rational and sub-hyperbolic [Cui-Jiang]). *Suppose* $f: \hat{C} \to \hat{C}$ *is a geometrically finite branched covering of degree* $d \geq 2$. *We say* f *is* semi-rational *if*

1. f *is holomorphic in a neighborhood of* P'_f;

2. *each cycle* $< p_0, \ldots, p_{k-1} >$ *of period* $k \geq 1$ *in* P'_f *is either attracting (that is,* $0 < |(f^k)'(p_0)| < 1$), *super-attracting (that is,* $(f^k)'(p_0) = 0$), *or parabolic (that is,* $[(f^k)'(p_0)]^q = 1$ *for some integer* $1 \leq q < \infty$);

3. *for each parabolic cycle in* P'_f, *every attracting petal associated to this cycle contains at least one point in a critical orbit.*

Furthermore, if in addition P'_f *contains only attracting and super-attracting periodic cycles, then we call* f sub-hyperbolic semi-rational.

Furthermore, we proved the following.

THEOREM 3.8 ([Cui-Jiang]). *Any semi-rational branched covering is always combinatorially equivalent to a sub-hyperbolic semi-rational branched covering.*

Therefore, we concentrated our study on the sub-hyperbolic semi-rational case. In the same paper [Cui-Jiang], we gave the following definition of the *CLH-equivalence* (combinatorial and locally holomorphic equivalence) among all sub-hyperbolic semi-rational branched coverings.

DEFINITION 3.9 (The CLH-equivalence [Cui-Jiang]). *Suppose* f *and* g *are two sub-hyperbolic semi-rational branched coverings. We say that they are* CLH-equivalent *if there are two homeomorphisms* $\phi, \varphi: \hat{C} \to \hat{C}$ *such that*

1. ϕ *is isotopic to* φ *rel.* P_f;

2. $\phi \circ f = g \circ \varphi$; *and*

3. $\phi|U_f = \varphi|U_f$ *is holomorphic on some open set* $U_f \supset P'_f$.

REMARK 3.10. In the study of the family $x \mapsto x + a + \frac{b}{2\pi} \sin(2\pi x)$, a similar concept called strong combinatorial equivalence is defined in [Epstein-Keen-Tresser], but the meaning is a little different from the preceding definition.

Similar to Proposition 3.1, we have the following.

PROPOSITION 3.11. *Any sub-hyperbolic semi-rational branched covering* f *is CLH-equivalent to a quasi-regular sub-hyperbolic semi-rational branched covering* $g = R \circ h$, *where* R *is a rational map and* h *is a quasiconformal homeomorphism; in other words,* g *is locally quasiconformal except on* Ω_g.

PROOF. Suppose $P'_f = \{a_1, \ldots, a_n\}$. Let D_i be a round disk of radius $r_i > 0$ centered at a_i such that there is no point P_f on ∂D_i and such that $f|_{D_i}$ is holomorphic. Let $U = \cup_{i=1}^n D_i$. Let $P = \hat{C} \setminus U$. Then $\#(P) < \infty$.

Recall that Ω_f is the set of branched points of f in $\hat{C}$. Consider the topological surface $S = \hat{C} \setminus \Omega_f$. For every $p \in S$, let U_p be a small neighborhood about p such that $\phi_p = f|_U: U \to f(U) \subset \hat{C}$ is injective. When $p \in D_i$, we pick $U_p \subset D_i$. Then $\alpha = \{(U_p, \phi_p)\}_{p \in S}$ defines an atlas on S with charts (U_p, ϕ_p). If $U_p \cap U_q \neq \varnothing$, then $\phi_p \circ \phi_q^{-1}(z) = z \colon \phi_q(U_p \cap U_q) \to \phi_p(U_p \cap U_q)$. Thus all transition maps are conformal (one-to-one and analytic) and the atlas α defines a Riemann surface structure on S which we again denote by α. Denote this Riemann surface by (S, α). From the uniformization theorem, (S, α) is conformally equivalent to the Riemann surface $\widehat{\mathbb{C}} \setminus A$ with the standard complex structure induced by $\widehat{\mathbb{C}}$, where A consists of $\#(\Omega_f)$ points. Thus we have a homeomorphism $\widehat{h} \colon \widehat{\mathbb{C}} \to \widehat{\mathbb{C}}$ such that $\widehat{h} \colon (S, \alpha) \to \widehat{\mathbb{C}} \setminus A$ is conformal. Moreover, $\widehat{h}|_U \colon U \to h(U)$ is also conformal. Thus $R = f \circ \widehat{h}^{-1} \colon \widehat{\mathbb{C}} \to \widehat{\mathbb{C}}$ is holomorphic with critical points at $\widehat{h}(\Omega_f)$. Since the set P is finite, following the standard procedure in quasiconformal mapping theory, there is a K-quasiconformal homeomorphism $h \colon \widehat{\mathbb{C}} \to \widehat{\mathbb{C}}$ such that h is isotopic to $\widehat{h}$ rel. $P \cup U$. The map $g = R \circ h$ is a quasi-regular map and CLH-equivalent to f. This completes the proof of the proposition. $\qquad\square$

Furthermore, we proved the following theorem.

THEOREM 3.12 ([Cui-Jiang-Sullivan] ([Cui-Tan]), [Zhang-Jiang]). *Suppose f is a sub-hyperbolic semi-rational branched covering. Then it is CLH-equivalent to a rational map if and only if f has no Thurston obstruction. Moreover, the rational map is unique up to the conjugation by an automorphism of the Riemann sphere.*

REMARK 3.13. The idea in the proof given in [Cui-Jiang-Sullivan] and the idea in the proof given in [Zhang-Jiang] are very different. The former is to decompose a sub-hyperbolic semi-rational branched covering along a stable multi-curve into finitely many post-critically finite branched coverings and to check they have no Thurston obstruction if the original map does not have a Thurston obstruction, and then to apply Thurston's theorem. A good source to learn this proof is [Cui-Tan], where the proof was rewritten with more detailed and comprehensive explanations.

The main idea in the latter proof is first to develop Thurston's idea, working on an induced map from the associated Teichmüller space into itself. In this case, we have an infinite dimensional Teichmüller space. Similar to Thurston's idea, we study an iterated sequence by the induced map. Each element in this iterated sequence represents a punctured Riemann sphere with infinitely many punctures. We then compare the geometry of this punctured Riemann sphere with the geometry of another Riemann sphere with finitely many punctures (see [Zhang-Jiang, Lemma5.6]). Here, the shielding ring lemma (Lemma 4.4) guarantees this comparison. Note that the punctured Riemann sphere with finitely many punctures, which is used in the comparison, has no dynamical relationship with the punctured Riemann sphere with infinitely many punctures. This comparison enables us to eventually prove that the induced map strongly contracts the Teichmüller distance along this iterated sequence.

Both methods in [Cui-Jiang-Sullivan] (see also [Cui-Tan]) and in [Zhang-Jiang] can be further developed. The former method of "decomposition along a stable multi-curve" can be used to study the renormalization of rational maps as well as their combinatorics (for example: trees, wandering continuum) and branched coverings with "Siegel disks" and "Herman rings," for example, the recent work in [Zhang;1], [Zhang;2], and [Wang] for the characterization of rational maps with Siegel disks and Herman rings. The latter method of "iterations in the Teichmüller space" can be used to characterize several kinds of holomorphic maps beyond post-critically finite rational maps, including hyperbolic maps, exponential maps, and some other meromorphic maps, and maps with "Siegel disks" and "Herman rings" (see §4).

Furthermore, we proved the following in a later paper.

THEOREM 3.14 ([Chen-Jiang]). *If a sub-hyperbolic semi-rational branched covering f is not CLH-equivalent to a rational map (i.e., a negative answer to Problem 1 under the CLH-equivalence), then it must have the canonical Thurston obstruction (see Definition 4.7). The canonical obstruction is a Thurston obstruction.*

4   GEOMETRY

In this section, a framework is outlined for the further study of Problems 1 and 2 (under either the combinatorial equivalence or the CLH-equivalence). We divide this study into three steps. The first step is to introduce the important concept of *bounded geometry*, and to study Problems 1 and 2 under this condition. The second step is to prove the equivalence between the bounded geometry condition and Thurston's topological condition. The third step is to prove that the equivalence between the bounded geometry condition and the canonical topological condition.

The bounded geometry condition is useful in the study of this direction and should be fully developed. It is an analytic condition but can be very well connected with Thurston's topological condition and the canonical topological condition in the critically finite case and in the sub-hyperbolic semi-rational case, as we will show. With the bounded geometry condition we will be able to work in this direction based on calculations in *spherical geometry*. This enables us to work directly on Teichmüller spaces $T(E)$ of closed subsets E on the Riemann sphere. This is different from Thurston's original work based on calculations in hyperbolic geometry (refer to [Douady-Hubbard]), which first works on moduli spaces and then lifts results obtained on moduli spaces to Teichmüller spaces (this process needs assumptions that finite degree and finiteness on the post-critical set (refer to [Douady-Hubbard, Lemma 5.2]). The bounded geometry condition can be also defined for quasi-entire functions or quasi-meromorphic functions and used in the study of Problems 1 and 2 (under either the combinatorial equivalence or the CLH-equivalence) for these maps without involving moduli spaces and [Douady-Hubbard, Lemma 5.2] directly.

The idea of bounded geometry is very important in complex dynamics. Besides its application to single dynamics, it could be applied to the parameter space (for example, infinitely renormalizable quadratic polynomials and hyperbolic rational maps with Sierpinsky curve Julia sets).

Since our definition of bounded geometry relates to Teichmüller spaces $T(E)$ of closed subsets E on the Riemann sphere, let us briefly mention the definition of $T(E)$ and some related properties.

Suppose E is a closed subset of $\widehat{\mathbb{C}}$. Suppose $0, 1, \infty \in E$ (which allows us to consider it is as a subset in spherical geometry rather than consider it as a subset in hyperbolic geometry which treats two subsets are the same if one is the image of another one under a Möbius transformation). Recall that we use w^μ to denote the unique quasiconformal homeomorphism with $0, 1, \infty$ fixed and with Beltrami coefficient μ in $M(\mathbb{C})$ (refer to the Beltrami equation (1.1)).

DEFINITION 4.1 (Teichmüller space of a closed subset).   *Two elements $\mu, \nu \in M(\mathbb{C})$ are said to be E-equivalent, denoted as $\mu \sim_E \nu$, if*

$$(w^\mu)^{-1} \circ w^\nu \overset{homotopy}{\sim} id \quad rel.\ E.$$

(The term "rel. E" means that we have a continuous map $H(t,z)\colon [0,1] \times \widehat{\mathbb{C}} \to \widehat{\mathbb{C}}$ such that (a) $H(0,z) = z$ and $H(1,z) = (w^\mu)^{-1} \circ w^\nu(z)$ for all $z \in \widehat{\mathbb{C}}$ and (b) $H(t,z) = z$ for all $0 \le t \le 1$ and all $z \in E$.) The Teichmüller space of E is defined as

$$T(E) = \{[\mu]_E\},$$

the space of all E-equivalence classes.

The space $T(E)$ is very useful for people working on complex dynamics, and the study of $T(E)$ is closely related to the study of holomorphic motions of E. Much is currently known about $T(E)$ (see [Earle-McMullen], [Lieb], [Epstein], [Earle-Mitra], [Mitra], [Gardiner-Lakic], [Jiang-Mitra], [Beck-Jiang-Mitra-Shiga], [Gardiner-Jiang-Wang], [Jiang-Mitra-Shiga], and [Jiang-Mitra-Wang]). The reader may find many similarities and differences between $T(E)$ and classical Teichmüller spaces of Riemann surfaces. An excellent book on the classical Teichmüller theory which is comprehensive for people working on complex dynamics is [Hubbard]. We now give a very brief summary of some basic properties of $T(E)$.

i. $T(E)$ has a complex manifold structure such that the projection

$$P_E(\mu) = [\mu]_E \colon M(\mathbb{C}) \to T(E)$$

is holomorphic.

ii. The projection P_E is a holomorphic split submersion, that is, for any $\tau \in T(E)$, $\exists U$ (neighborhood) about τ and a holomorphic section $s_\tau \colon U \to M(\mathbb{C})$ such that $P_E \circ s_\tau = id$.

iii. $\exists$ a continuous global section $S \colon T(E) \to M(\mathbb{C})$ such that $P_E \circ S = id$. Therefore, $T(E)$ is contractible.

iv. If $\dim T(E) \ge 2$, there is no global holomorphic section.

v. $T(E)$ is biholomorphically equivalent to $\prod_i T(S_i, \partial S_i) \times M(E)$, where S_i are components of $\widehat{\mathbb{C}} \setminus E$ and $T(S_i, \partial S_i)$ is the Teichmüller space of the hyperbolic Riemann surface S_i with the ideal boundary ∂S_i.

vi. We have the lifting property for $T(E)$. Let Δ be the open disk $\{z \mid |z| < 1\}$. Then any holomorphic map $f \colon \Delta \to T(E)$ can be lifted to a holomorphic map $\tilde{f} \colon \Delta \to M(\mathbb{C})$ such that $P_E \circ \tilde{f} = f$, that is, the following diagram commutes.

vii. Teichmüller's metric d_T on $T(E)$ coincides with Kobayashi's pseudo-metric d_K on $T(E)$.

PROPOSITION 4.2. *Suppose $f\colon \widehat{\mathbb{C}} \to \widehat{\mathbb{C}}$ is quasiregular (or a quasi-entire or a quasi-meromorphic). Suppose E is a closed subset of $\widehat{\mathbb{C}}$ such that $E \supseteq V_f = f(\Omega_f)$ and $E \subseteq f^{-1}(E)$ (in the quasi-entire and quasi-meromorphic case, V_f must contain all asymptotic values and essential singularities). If $\mu \sim_E \nu$, then $f^*\mu \sim_E f^*\nu$.*

PROOF. By the assumption $\mu \sim_E \nu$, we can find

$$H(t,z)\colon [0,1] \times \widehat{\mathbb{C}} \to \widehat{\mathbb{C}}$$

such that

1. $H(0,z) = z$ for all $z \in \widehat{\mathbb{C}}$;

2. $H(1,z) = (w^\mu)^{-1} \circ w^\nu(z)$ for all $z \in \widehat{\mathbb{C}}$; and

3. $H(t,z) = z$ for all $z \in E$ and all $0 \le t \le 1$.

Since $V_f \subseteq E$, we have that $\Omega_f \subseteq f^{-1}(E)$. This implies $f\colon \widehat{\mathbb{C}} \setminus f^{-1}(E) \to \widehat{\mathbb{C}} \setminus E$ is a covering map. The homotopy $H(t,z)\colon [0,1] \times (\widehat{\mathbb{C}} \setminus E) \to \widehat{\mathbb{C}} \setminus E$ rel. ∂E can be lift to a homotopy $\widetilde{H}(t,z)\colon [0,1] \times (\widehat{\mathbb{C}} \setminus f^{-1}(E)) \to \widehat{\mathbb{C}} \setminus f^{-1}(E)$ rel. $\partial f^{-1}(E)$ such that

$$H(t,f(z)) = f(\widetilde{H}(t,z)), \quad \forall z \in \widehat{\mathbb{C}} \setminus f^{-1}(E), \ 0 \le t \le 1$$

and

$$\widetilde{H}(t,z) = z, \quad \forall z \in \partial f^{-1}(E), \ 0 \le t \le 1.$$

Define $\widetilde{H}(t,z) = z$ for all $z \in f^{-1}(E)$ and $0 \le t \le 1$. Then the new defined map, which we still denote as $\widetilde{H}$, is a continuous map $\widetilde{H}(t,z)\colon [0,1] \times \widehat{\mathbb{C}} \to \widehat{\mathbb{C}}$ such that

a. $\widetilde{H}(z,0) = z$ for all $z \in \widehat{\mathbb{C}}$; and

b. $\widetilde{H}(z,t) = z$ for all $z \in E$ and all $0 \le t \le 1$.

Therefore, it is a homotopy from the identity to $\widetilde{H}_1(z) = \widetilde{H}(1,z)$.

Let $H_1(z) = H(1,z)$. Since $H_1 \circ f = f \circ \widetilde{H}_1$ and $H_1 = (w^\mu)^{-1} \circ w^\nu$ is quasiconformal, $\widetilde{H}_1$ is quasiconformal.

Now by using two commuting equations,

$$(w^\mu)^{-1} \circ w^\nu(z) \circ f = f \circ \widetilde{H}_1 \quad \text{and} \quad g \circ w^{f^*\mu} = w^\mu \circ f,$$

where g is holomorphic, we have that

$$g \circ w^{f^*\mu} \circ \widetilde{H}_1 = w^\mu \circ f \circ \widetilde{H}_1 = w^\mu \circ (w^\mu)^{-1} \circ w^\nu(z) \circ f = w^\nu(z) \circ f.$$

Since g is holomorphic,

$$\mu_{w^{f^*\mu} \circ \widetilde{H}_1} = \mu_{g \circ w^{f^*\mu} \circ \widetilde{H}_1} = \mu_{w^\nu(z) \circ f} = f^*\nu.$$

Since both $w^{f^*\mu} \circ \widetilde{H}_1$ and $w^{f^*\nu}$ fix $0, 1, \infty$, we get

$$w^{f^*\mu} \circ \widetilde{H}_1 = w^{f^*\nu}.$$

In other words,

$$\widetilde{H}_1 = (w^{f^*\mu})^{-1} \circ w^{f^*\nu}.$$

Thus $\widetilde{H}(t,z)\colon [0,1] \times \widehat{\mathbb{C}} \to \widehat{\mathbb{C}}$ is a homotopy from the identity to $(w^{f^*\mu})^{-1} \circ w^{f^*\nu}$ rel. $f^{-1}(E)$. But $E \subseteq f^{-1}(E)$, so the last statement implies that $f^*\mu \sim_E f^*\nu$. This completes the proof. $\qquad\qquad\qquad\qquad\qquad\qquad\qquad\qquad\qquad\qquad\qquad\quad$ $\square$

From the previous proposition, we can induce a map

$$\sigma_f(\tau) = [f^*\mu]_E \colon T(E) \to T(E), \quad \mu \in \tau.$$

Since

$$\sigma_f(\tau) = P_E \circ f^* \circ s_\tau,$$

σ_f is holomorphic. Thus it is weakly contracting due to the property that the Kobayashi metric and the Teichmüller metric are the same. That is,

$$d_T(\sigma_f(\tau), \sigma_f(\tau')) \leq d_T(\tau, \tau'), \quad \forall \tau, \tau' \in T(E).$$

Suppose f is holomorphic on E and $\{0, 1, \infty\} \subset E \subseteq f^{-1}(E)$. For any μ with $\mu|E = 0$, we have that $f^*\mu|E = 0$.

PROBLEM 3. *Can we find a (unique) fixed point $\tau = [\mu]_E$ with $\mu|E = 0$ of σ_f?*

A positive answer to Problem 2 (under either combinatorially equivalence or CLH-equivalence) is equivalent to a positive answer to Problem 3 (by recalling Theorem 1.2).

For any $\tau_0 = [\mu_0]_E \in T(E)$ with $\mu_0|E = 0$, let $\tau_n = \sigma_f(\tau_{n-1}) = [\mu_n]_E \in T(E)$, where $\mu_n = f^*\mu_{n-1}$ with $\mu_n|E = 0$ for all $n \geq 1$. Let w^{μ_n} be the corresponding quasiconformal homeomorphism solution of the Beltrami equation (1.1) with $0, 1, \infty$ fixed. Then we can define a sequence of subsets $E_n = w^{\mu_n}(E)$ such that $0, 1, \infty \in E_n$. Roughly speaking, f has bounded geometry if we have a subsequence E_{n_i} "converging" to E_∞ as i goes to ∞ such that E_∞ is homeomorphic to E. We will give a more precise definition in the critically finite case and in the sub-hyperbolic semi-rational case. We use d_{SP} to mean the spherical distance on $\widehat{\mathbb{C}}$.

DEFINITION 4.3 (Bounded geometry in the critically finite case). *Suppose f is a critically finite branched covering. Suppose $E = P_f = \{0, 1, \infty, p_1, \ldots, p_{k-3}\}$. We say f has bounded geometry if there are a constant $C > 0$ and a point $\tau_0 \in T(E)$ such that*

$$d_{SP}(a, b) \geq C, \quad \forall a, b \in E_n = w^{\mu_n}(E), \quad n \geq 0.$$

LEMMA 4.4 (Shielding ring lemma [Zhang-Jiang]). *Suppose f is a sub-hyperbolic semi-rational branched covering. Let $P'_f = \{a_i\}$ be the set of accumulation points of P_f. There exists a collection of finite number of open disks $\{D_i\}$ centered at a_i and a collection of finite number of annuli $\{A_i\}$ (we call them the shielding rings) such that*

i. $\overline{A_i} \cap P_f = \varnothing$;

ii. $A_i \cap D_i = \varnothing$, but one components of ∂A_i is the boundary of D_i;

iii. $(\overline{D_i \cup A_i}) \cap (\overline{D_j \cup A_j}) = \varnothing$ for $i \neq j$;

iv. f is holomorphic on $\overline{D_i} \cup A_i$; and

v. *every $f(\overline{D_i} \cup A_i)$ is contained in D_{i+1} for $1 \leq i \leq k-1$ and $f(\overline{D_k} \cup A_k)$ is contained in D_1, where k is the period of a_i.*

Suppose $\{(D_i, A_i)\}$ are domains and annuli constructed in the shielding ring lemma. Take

$$\mathcal{D}_0 = \cup_{i=1}^k \overline{D_i} \quad \text{and} \quad P_0 = P_f \setminus \mathcal{D}_0.$$

Then $\#(P_0) < \infty$. Take

$$E = P_0 \cup \mathcal{D}_0. \tag{4.1}$$

For any $\tau_0 = [\mu_0]_E$ such that $\mu_0|\mathcal{D}_0 = 0$, we have that $\mu_n|\mathcal{D}_0 = 0$ for all $n \geq 1$. Then $E_n = w^{\mu_n}(E)$ for $n = 0, 1, \ldots$. Note that

$$E_n = P_n \cup \mathcal{D}_n,$$

where

$$P_n = w^{\mu_n}(P_0) \quad \text{and} \quad \mathcal{D}_n = w^{\mu_n}(\mathcal{D}_0).$$

Every component D in $\mathcal{D}_0$ is a round disk with center a. The image $D_n = w^{\mu_n}(D)$ is a topological disk in $\mathcal{D}_n$. We call $a_n = w^{\mu_n}(a)$ the center of D_n. Then $\mathcal{D}_n$ consists of finitely many disjoint topological disks. Suppose $0, 1, \infty \in E$. Then $0, 1, \infty \in E_n$ for all $n > 0$.

DEFINITION 4.5 (Bounded geometry in the sub-hyperbolic semi-rational case). *We say f has bounded geometry if there are a constant $C > 0$ and a point $\tau_0 = [\mu_0]_E \in T(E)$ with $\mu_0|\mathcal{D}_0 = 0$ such that*

  i. *$d_{SP}(a, b) \geq C$, for any two points $a, b \in P_n = w^{\mu_n}(P_0)$ and for all $n \geq 0$;*

  ii. *$d_{SP}(A, B) \geq C$, for any two different components A and B of $\mathcal{D}_n = w^{\mu_n}(\mathcal{D}_0)$ and for all $n \geq 0$;*

  iii. *$d_{SP}(a, A) \geq C$, for any point $a \in P_n$ and any component A of $\mathcal{D}_n$ and for all $n \geq 0$;*

  iv. *each component D_n of $\mathcal{D}_n$ contains a round disk of radius C centered at the center point a_n of D_n.*

REMARK 4.6. Both Definition 4.3 and Definition 4.5 are not preserved under actions of Möbius transformations.

Consider $\widehat{\mathbb{C}} \setminus E$ with the standard conformal structure $\tau_0 = [0]_E$ as the hyperbolic Riemann surface R_0 and consider $\widehat{\mathbb{C}} \setminus E$ with the conformal structure $\tau_n = [\mu_n]_E$ as the hyperbolic Riemann surface R_n. Then R_n is Teichmüller equivalent to $\widehat{\mathbb{C}} \setminus E_n$.

For any non-peripheral simple closed curve γ in $\widehat{\mathbb{C}} \setminus E$, $w^{\mu_n}(\gamma)$ is a non-peripheral simple closed curve in $\widehat{\mathbb{C}} \setminus E_n$. We use $l_n(\gamma)$ to denote the hyperbolic length of a unique geodesic which is homotopic to $w^{\mu_n}(\gamma)$ in $\widehat{\mathbb{C}} \setminus E_n$. Under the condition of bounded geometry, we can find a positive constant $b > 0$ such that

$$l_n(\gamma) \geq b, \quad \forall n \geq 0, \quad \forall \gamma \text{ (non-peripheral simple closed curve).} \tag{4.2}$$

However, to get from (4.2) to bounded geometry, we need the assumption that $0, 1, \infty \in E$ (consequently, $0, 1, \infty \in E_n$).

In both the critically finite and the sub-hyperbolic semi-rational cases, *non-bounded geometry* is equivalent to saying that there is a sequence of non-peripheral simple closed curves γ_{n_i} in $\widehat{\mathbb{C}} \setminus E$ such that

$$l_{n_i}(\gamma_{n_i}) \to 0, \quad \text{as} \quad i \to \infty.$$

In particular, we have the following definition. (Refer to [Pilgrim] for the post-critically finite case and to [Chen-Jiang] for the sub-hyperbolic semi-rational case.)

DEFINITION 4.7 (Canonical obstruction). *Let*

$$\Gamma_c = \{\gamma \mid l_n(\gamma) \to 0, \quad n \to \infty\}$$

where γ is a non-peripheral simple closed curve in $\widehat{\mathbb{C}} \setminus E$. If $\Gamma_c \neq \emptyset$, then it is called the canonical obstruction for f.

Clearly, if $\Gamma_c \neq \emptyset$, then f has no bounded geometry. Moreover, we have the following.

THEOREM 4.8 (Equivalent statements for the critically finite case). *Suppose f is a critically finite branched covering with $\#(P_f) \geq 4$ and has a hyperbolic orbifold. Then the following are equivalent:*

1. *f can be viewed as a holomorphic map (i.e, it is combinatorially equivalent to a unique rational map).*

2. *f has bounded geometry.*

3. *$\Gamma_c = \emptyset$.*

4. *f has no Thurston obstruction.*

Similarly, we have the following.

THEOREM 4.9 (Equivalent statements for the sub-hyperbolic semi-rational case). *Suppose f is a sub-hyperbolic semi-rational branched covering. Then the following are equivalent:*

1. *f can be viewed as a holomorphic map (i.e, it is CLH-equivalent to a unique rational map).*

2. *f has bounded geometry.*

3. *$\Gamma_c = \emptyset$.*

4. *f has no Thurston obstruction.*

Proofs of these two theorems can now be put into one framework through bounded geometry. Let $E = P_f$ in the post-critically finite case and let $E = P_0 \cup \mathcal{D}_0$ as defined in Equation (4.1) in the sub-hyperbolic semi-rational case.

First, prove that $\sigma_f^k \colon T(E) \to T(E)$ is contracting for some $k \geq 1$, i.e., that we have $d_T(\sigma_f^k(\tau), \sigma_f^k(\tau')) < d_T(\tau, \tau')$ for any $\tau, \tau' \in T(E)$. In the post-critically finite case, $k = 2$, and in the sub-hyperbolic semi-rational case, $k = 1$. Secondly, prove the family of rational maps $\{g_n = w^{\mu_n} \circ f \circ (w^{\mu_{n+1}})^{-1}\}$ forms a compact subset in the space of all rational maps of degree $d = \deg(f)$ with the uniform convergence topology under the assumption of bounded geometry. Thirdly, consider the operator $d\sigma_f^k$ from the cotangent space $T^*_{\tau_{n+k}} T(E)$ to the cotangent space $T^*_{\tau_n} T(E)$ and show that it is strictly contracting by using bounded geometry and the formula for $d\sigma_f \colon T^*_{\sigma_f(\tau)} T(E) \to T^*_\tau T(E)$ defined as a transfer operator

$$\mathcal{L}q(z) = \sum_{g_n(w)=z} \frac{q(w)}{(g_n'(w))^2},$$

where $q(w)$ is the coefficient of an integrable holomorphic quadratic differential on the hyperbolic Riemann surface $\widehat{\mathbb{C}} \setminus E_{n+1}$. All the preceding steps imply that σ_f^k is strongly contracting on the sequence $\{\tau_n\}_{n=0}^\infty$ in $T(E)$; that is, there is a constant $0 < c < 1$ such that

$$d_T(\sigma_f^k(\tau_{n+1}), \sigma_f^k(\tau_n)) \le c d_T(\tau_{n+1}, \tau_n).$$

Then prove that the bounded geometry is equivalent to the no Thurston obstruction condition. And finally, by using some properties of Thurston obstructions (if they exist), prove that Γ_c is not empty.

In this framework, we first prove the equivalence (1) $\Leftrightarrow$ (2) and then prove (2) $\Leftrightarrow$ (3) and (2) $\Leftrightarrow$ (4). A detailed explanation will be given in a later expository paper.

The existing proof of (3) $\Rightarrow$ (2) (see [Pilgrim], [Chen-Jiang]) uses a Thurston obstruction as a bridge to argue that if f has no bounded geometry, then f has a Thurston obstruction; furthermore, by using this Thurston obstruction, prove that $\Gamma_c \ne \emptyset$. However, we are more interested in a proof without using those properties of Thurston obstructions.

PROBLEM 4. *Find a direct proof of* (3) $\Rightarrow$ (2) *in both Theorem 4.8 and Theorem 4.9. In other words, without using a Thurston obstruction as a bridge, show that if f has no bounded geometry, then* $\Gamma_c \ne \emptyset$.

This is an interesting problem because, for a quasi-entire or a quasi-meromorphic function, it is not clear to us that the equivalence between the bounded geometry condition and the no Thurston obstruction condition due to infinite degree and the existence of asymptotic values and essential singularities (refer to section 6).

5 GEOMETRIZATION

Given the canonical obstruction $\Gamma_c = \{\gamma_1, \ldots, \gamma_n\} \ne \emptyset$ for a map f which is either (i) a critically finite branched covering or (ii) a sub-hyperbolic semi-rational branched covering, we can use curves in Γ_c (up to homotopy) to divide the sphere S^2 into finitely many topological surfaces, which are topological two-spheres removing a finite number of disks as follows.

Let E be the post-critical set P_f in case (i) or the set defined in Equation (4.1) in case (ii). Consider the Riemann surface $\widehat{\mathbb{C}} \setminus E$. Suppose $A_{0,i}$ $(i = 1, \ldots, n)$ are a collection of disjoint annuli whose core curves are γ_i $(i = 1, \ldots, n)$, respectively. Set

$$A_0 = \bigcup_{i=1}^{n_0} A_{0,i} \quad \text{and} \quad n_0 = n.$$

Let

$$A_1 = \bigcup_{i=1}^{n_1} A_{1,i}$$

be the union of preimage of elements of A_0 such that every element of A_1 is homotopic to some element in A_0 rel. E since Γ_c is a stable and full multi-curve. Up to homotopy, we can assume that

1. every curve γ_i is the core curve of the annulus $A_{0,i}$;

2. every $A_{1,k}$ is a component of the preimage of some $A_{0,j}$ and homotopic to some $A_{0,i}$, denote by $A_{1,ji,\alpha}$;

3. the union $\cup_{j,\alpha} A_{1,ji,\alpha}$ of elements of A_1 which are homotopic to γ_i, denote by $B_{1,i,i}$, is contained inside $A_{0,i}$;

4. the two outmost annuli of $B_{1,i,i}$ share their outer boundary curves with $A_{0,i}$; and

5. restricted to a boundary curve χ of $A_{0,i}$, the map $f\colon \chi \to f(\chi)$ is conjugated to $z \mapsto z^d \colon \mathbb{S}^1 \to \mathbb{S}^1$ for some d. Note that $f(\chi)$ is a boundary of $A_{0,j}$.

Each component $A_{0,i}$ of A_0 is called a thin part, and each component of $\overline{\mathbb{C}} \backslash A_0$ is called a thick part.

Let $\mathfrak{P}_0 = \{P_i^0\}_{i=1}^{n_0}$ be the collection of all thick parts. The pull-back of all thick parts by f^k is denoted as $\mathfrak{P}_k = f^{-k}(\mathfrak{P}_0)$. Each element of $\mathfrak{P}_k$ belongs to one and only one of the following four classes:

  i. Disk component: if it is a topological disk D and $D \cap P_f = \varnothing$.

  ii. Punctured disk component: if it is a topological disk P and $\sharp(P \cap P_f) = 1$.

  iii. Annulus component: if it is an annulus A and $A \cap P_f = \varnothing$.

  iv. Complex component: if it is not in (i), (ii), and (iii).

Since all elements of Γ_c are non-peripheral and non-homotopic to each other, the thick parts $P_0^0, \ldots, P_n^0$ are all complex components.

For each $P_i^0 \in \mathfrak{P}_0$ and any k, since P_i^0 is a component of $\widehat{\mathbb{C}} \setminus A_0$ satisfying assumptions (1)–(5), there exists an unique component of $\mathfrak{P}_k$, denoted by P_i^k, such that each component of ∂P_i^k is either peripheral or some component of ∂P_i^0 and such that each component of ∂P_i^0 is some component of ∂P_i^k. This gives a relationship between P_i^0 and P_i^k. We use $P_i^0 \approx P_i^k$ to denote this relation. Therefore, P_i^k and $f(P_i^k)$ are complex components and, furthermore, if $f(P_i^k) = P_j^{k-1}$, then $f(P_i^l) = P_j^{l-1}$ for any l. By using the action of the branched covering f, for each P_i^0, we have a unique P_i^k such that $P_i^0 \approx P_i^k$; then $f(P_i^k) = P_j^{k-1}$ and so there is a unique P_j^0 such that $P_j^0 \approx P_j^k$. Let $n = n_0$. This defines a map $\tau\colon \{0, 1, \ldots, n\} \to \{0, 1, \ldots, n\}$ such that $\tau(i) = j$ if $f(P_i^k) = P_j^{k-1}$. Under the relationship $\approx$, each component of $\mathfrak{P}_0$ is eventually periodic and at least one is periodic. Thus we conclude the following.

Decomposition. *Suppose f is a post-critically finite or a sub-hyperbolic semi-rational branched covering not combinatorially equivalent or CLH-equivalent to a rational map. Then f can be decomposed into thin parts and thick parts according to the canonical obstruction. Furthermore, an equivalence relation can be introduced on thick parts such that every thick part is eventually periodic and at least one is periodic.*

Suppose $P_0^0 \in \mathfrak{P}_0$ is a periodic component of periodic $k > 0$. Suppose that $\gamma_1, \ldots, \gamma_p$ are boundary curves of P_0^0 and that $\gamma_1, \ldots, \gamma_p, \beta_1, \ldots, \beta_q$ are boundary curves of P_0^k, where $\beta_j (j = 1, \ldots, q)$ are peripheral curves. For any β_j, it must be a component of $f^{-k}(\gamma_i)$ for some γ_i. For any γ_i, it must be a component of $f^{-k}(\gamma_l)$ for some γ_l. Denote

$$P_0^0 \setminus P_0^k = \bigcup_{j=1}^{q} D(\beta_j), \quad \overline{\mathbb{C}} \setminus P_0^0 = \bigcup_{i=1}^{p} D(\gamma_i).$$

Let

$$d_{\beta_j} = deg(f^k \colon \beta_j \to \gamma_i), \quad d_{\gamma_i} = deg(f^k \colon \gamma_i \to \gamma_l = f^k(\gamma_i)).$$

Define a new branched covering by

$$\widetilde{f} = \widetilde{f}_{P_0^0} = \begin{cases} f^k, & z \in P_0^k, \\ \varphi_j \circ z^{d_{\beta_j}} \circ \psi_j, & z \in D(\beta_j) \ (j = 1, \dots, q), \\ \varphi_i \circ z^{d_{\gamma_i}} \circ \psi_i, & z \in D(\gamma_i) \ (i = 1, \dots, p), \end{cases} \tag{5.1}$$

where ψ_j, φ_j^{-1} are homeomorphisms from $D(\beta_j)$ and $D(\gamma_i)$ to unit disk $\mathbb{D}$, respectively, and ψ_i, φ_i^{-1} are homeomorphisms from $D(\gamma_i)$ and $D(\gamma_l) = D(f^k(\gamma_i))$ to unit disk $\mathbb{D}$, respectively, such that $\widetilde{f}$ is continuous. Marking a point in each $D(\gamma_i)$, say z_i, we set

$$P_{\widetilde{f}} = P_f|_{P_0^0} \cup \left(\bigcup_{j=1}^{q} z_j \right).$$

If $D(\beta_j)$ contains a point, say z^*, belonging to P_f and $f^k(\beta_j) = \gamma_i$, we can select φ_j, φ_i and ψ_j, ψ_i such that $\widetilde{f}(z^*) = z_i$. Also, if $f^k(\gamma_i) = \gamma_l$, we can select $\varphi_i, \varphi_l, \psi_i, \psi_l$ such that $\widetilde{f}(z_i) = z_l$. So $\widetilde{f}(P_{\widetilde{f}}) \subseteq P_{\widetilde{f}}$.

Extension. *Any periodic thick part P_i^0 of period k can be extended to a topological two-sphere with finitely many marked points, and the restriction $f^k|P_i^k$ can be extended to either a critically finite–type branched covering or a sub-hyperbolic semi-rational–type branched covering $\widetilde{f} = \widetilde{f}_{P_i^0}$ such that*

$$\widetilde{f}|P_i^k = f^k|P_i^k.$$

THEOREM 5.1 ([Selinger]). *Suppose f is a post-critically finite branched covering which is not combinatorially equivalent to a rational map. Then every $\widetilde{f}$ associated to a hyperbolic orbifold is combinatorially equivalent to a unique rational map (up to conjugation of an automorphism of the Riemann sphere).*

REMARK 5.2. Actually, the paper [Selinger] contains more. In the post-critically finite case, let $E = P_f$, then the Riemann surface $\widehat{\mathbb{C}} \setminus E$ is a punctured sphere with finitely many punctures. The Teichmüller space $T(E)$ is the same as the Teichmüller space $T(\widehat{\mathbb{C}} \setminus E)$ of Riemann surfaces with basepoint $\widehat{\mathbb{C}} \setminus E$. Thus, one can define its augmented Teichmüller space $\overline{T}(E)$, which is the space of all Teichmüller equivalent classes but allows an equivalent class with a representation of continuous maps from $\widehat{\mathbb{C}} \setminus E$ to a Riemann surface with nodes that are allowed to send a whole non-trivial closed annulus or a whole non-trivial closed curve in the complement of marked points to a node. From [Masur], we know that the closure of the Teichmüller space $T(E)$ under the Weil-Petersson metric is its augmented Teichmüller space. Selinger proved that σ_f from $T(E)$ into itself can be extended to a continuous self-map of $\overline{T}(E)$ such that it either has a fixed point in $T(E)$ or a fixed point in the boundary $\partial \overline{T}(E) = \overline{T}(E) \setminus T(E)$.

REMARK 5.3. In [Bonnot-Yampolsky], one can find an alternative proof of Theorem 5.1, which applies a result in [Minsky] and a result in [Hassinsky].

In our latest study, we worked on the geometrization of a sub-hyperbolic semi-rational type branched covering as follows.

THEOREM 5.4 ([Cheng-Jiang]). *Suppose f is a sub-hyperbolic semi-rational branched covering which is not CLH-equivalent to a rational map. Then every $\widetilde{f} = \widetilde{f}_{P_i^0}$ is either a post-critically finite type branched covering or a sub-hyperbolic semi-rational type branched covering. In the post-critically finite type case, if the orbifold associated to $\widetilde{f}$ is hyperbolic, then $\widetilde{f}$ is combinatorially equivalent to a rational map; in the sub-hyperbolic semi-rational type case, $\widetilde{f}$ is CLH-equivalent to a rational map. Moreover, in the both cases, the realized rational map is unique up to conjugation of an automorphism of the Riemann sphere.*

Unlike in the post-critically finite case, in the sub-hyperbolic semi-rational case, we work on an infinite-dimensional Teichmüller space of the Riemann sphere minus a set of finitely many points and a set of finitely many topological disks. A major work in [Cheng-Jiang] is to overcome this difficulty. First, we proved the following proposition in [Cheng-Jiang], which is useful in general.

PROPOSITION 5.5. *Suppose R_1 is a Riemann surface, which is the Riemann sphere minus a set E consisting of finite number of points and finite number of disks, with complex structure $\tau_1 = [\mu_1]$. Suppose $\gamma_1, \ldots, \gamma_n$ are non-peripheral non-homotopic simple closed curves on R_1 with $l_{\tau_1}(\gamma_i) < \varepsilon$ (ε sufficiently small). Let R_2 be a Riemann surface with complex structure $\tau_2 = [\mu_2]$ obtained from R_1 by cutting along γ_i and capping every hole by a puncture disk. If there exists a constant $K > 0$ such that for every non-peripheral simple closed curve β other than $\gamma_1, \ldots, \gamma_n$, $l_{\tau_1}(\beta) \geq K$, then there exists a constant $\widetilde{K} = \widetilde{K}(K, \varepsilon) > 0$ such that for every non-peripheral simple closed curve $\widetilde{\beta}$ of R_2, $l_{\tau_2}(\widetilde{\beta}) \geq \widetilde{K}$.*

Secondly, for any initial conformal structure $\tau_0 = [\mu_0]_E$ on the Riemann sphere, by using the decomposition and the extension, we have a conformal structure $\widetilde{\tau}_0 = [\widetilde{\mu}_0]$ which is μ_0 on P_i^k and 0 on other places. Then we have two iterated sequences $\{\tau_n = \sigma_f^n(\tau_0)\}$ and $\{\widetilde{\tau}_n = \sigma_{\widetilde{f}}^n(\tau_0)\}$. Compare these two iterated sequences and other three modified sequences of conformal structures (refer to [Cheng-Jiang, Proposition 5]) and then compare the geometry of punctured Riemann spheres with infinitely many punctures with the geometry of punctured Riemann spheres with finitely many punctures (refer to [Zhang-Jiang, Lemma 5.6] and [Cheng-Jiang, Lemma 5 and Proposition 4]). After we constructed these five different, but related, punctured Riemann spheres, we then used some of the methods and ideas from [Douady-Hubbard] and [Selinger] to complete the proof.

REMARK 5.6. From Theorem 5.4, we know $\widetilde{f}$ has bounded geometry as defined for $\{\widetilde{\tau}_n\}$. However, we are still interested in a direct proof of the bounded geometry property for $\widetilde{f}$. With this direct proof, we can have a proof of Theorem 5.4 using our framework in the previous section.

REMARK 5.7. In the sub-hyperbolic semi-rational case, E defined in Equation (4.1) has finitely many points and finitely many disks. The Riemann surface $\widehat{\mathbb{C}} \setminus E$ is the Riemann sphere with finitely many points and finitely many disks removed. The Teichmüller space $T(E)$ is biholomorphically equivalent to the Teichmüller space $T(\widehat{\mathbb{C}} \setminus E)$ of Riemann surfaces with basepoint $\widehat{\mathbb{C}} \setminus E$ times the open unit ball $M(E)$ of the complex Banach space $\mathcal{L}^\infty(E)$. It is an interesting problem to understand how to study the augmented Teichmüller space and how to generalize the result in [Masur] for this case. After that we could study a full version of the result in [Selinger] in the sub-hyperbolic semi-rational case. One possibility for us is to define $T_0(E)$ as follows. For each disk component D_i of E, consider a ring A_i attaching to it such that $D_i \cup A_i$ is a bigger disk and such that $\{D_i \cup A_i\}$ and all point components of E are pairwise disjoint. Let $\widetilde{E} = E \cup \cup_i A_i$. Consider $M_0(\mathbb{C}) = \{\mu \in M(\mathbb{C}) \mid \mu|\widetilde{E} = 0\}$. Define $T_0(E) = \{[\mu]_E \mid \mu \in M_0(\mathbb{C})\}$ as the space of all E-equivalence classes of elements in $M_0(\mathbb{C})$. We would like to understand the augmented Teichmüller space $\overline{T}_0(E)$. It is still an interesting project for us.

Acknowledgements. I would like to thank referees for their very useful comments, which I have included in this paper. I would like to thank all my collaborators mentioned in this paper who spent a lot of time with me to discuss this interesting subject. This research is partially supported by the collaboration grant (#199837) from the Simons Foundation, the CUNY collaborative incentive research grant (#1861), and awards from PSC-CUNY. This research is also partially supported by the collaboration grant (#11171121) from the NSF of China and a collaboration grant from Academy of Mathematics and Systems Science and the Morningside Center of Mathematics at the Chinese Academy of Sciences.

6   APPENDIX ON TRANSCENDENTAL FUNCTIONS BY TAO CHEN, YUNPING JIANG, AND LINDA KEEN

We believe that the framework described in the preceding sections also applies to the characterization problem for families of quasi-entire and quasi-meromorphic functions with certain finiteness properties. At present, we are interested in the characterization of meromorphic functions with a finite number of asymptotic values but no critical points and the characterization of general exponential functions of the form $P(z)e^{Q(z)}$, where $P(z)$ and $Q(z)$ are two polynomials.

More precisely, in [Chen-Jiang-Keen;1], we define two model spaces of maps we call $\mathcal{AV}2$ and $\mathcal{AV}3$. These are universal covering maps of the two-sphere with two or three removed points that we model on the space of meromorphic functions with two or three asymptotic values and no critical points. We call the spaces of meromorphic functions $\mathcal{M}2$ and $\mathcal{M}3$. Examples of elements in $\mathcal{M}2$ are the tangent map $\lambda \tan z$ and the exponential map $e^{\beta z}$. Nevanlinna [Nevanlinna] characterized meromorphic maps with p asymptotic values and no critical values as functions whose Schwarzian derivative is a polynomial of degree $p - 2$. Thus, a meromorphic map in $\mathcal{M}2$ has constant Schwarzian derivative and a meromorphic map in $\mathcal{M}3$ has a linear function as its Schwarzian derivative. For a more detailed description, the reader is referred to our paper [Chen-Jiang-Keen;1].

The bounded geometry and canonical obstruction conditions can also be defined for post-singularly finite maps in $\mathcal{AV}2$ and $\mathcal{AV}3$. We are now working on a proof of the following statement that characterizes post-singularly finite maps in $\mathcal{M}2$ and $\mathcal{M}3$:

A post-singularly finite map in $\mathcal{AV}2$ or in $\mathcal{AV}3$ is combinatorially equivalent to a post-singularly finite transcendental meromorphic function in $\mathcal{M}2$ or $\mathcal{M}3$ if and only if it has bounded geometry. The realization is unique up to conjugation by an affine map of the plane.

The proof follows the framework described in the previous sections. The new ingredient here is the addition of a topological constraint, which we define in order to control the holomorphic maps obtained by the Thurston iteration process and show they remain in a compact subset of the parameter space.

In [Chen-Jiang-Keen;2], we study the characterization of entire post-singularly finite (p, q)-*exponential maps* of the form $E(z) = P(z)e^{Q(z)}$ with P and Q polynomials of degrees $p \geq 0$ and $q \geq 0$, respectively, and $p + q \geq 1$. We use the notation $\mathcal{E}_{p,q}$ for this family. Again, we define a model space $\mathcal{T}E_{p,q}$ of *topological exponential maps of type* p, q. These are infinite-degree branched coverings with a single finite asymptotic value, normalized to be at the origin, and modeled on the functions $E(z) = P(z)e^{Q(z)}$. For a more detailed description, the reader is referred to our paper [Chen-Jiang-Keen;2].

The bounded geometry and canonical obstruction conditions can also be defined for post-singularly finite maps in $\mathcal{T}E_{p,q}$. We are now working on a proof of the following statement that characterizes post-singularly finite maps in $\mathcal{T}E_{p,q}$:

A post-singularly finite map in $\mathcal{T}E_{p,q}$ is combinatorially equivalent to a post-singularly finite entire map of the form Pe^Q if and only if it has bounded geometry.

Again, the new ingredient is the addition of a topological constraint which we need in order to control the holomorphic maps obtained by the Thurston iteration process and show they remain in a compact subset of the parameter space.

The framework described in Sections 1–5 can also be extended to transcendental maps in another direction. We can define sub-hyperbolic semi-holomorphic topological exponential maps and sub-hyperbolic semi-holomorphic topological meromorphic maps with two asymptotic values (recall Definition 3.7). Some results have straightforward extensions to these maps and conjectures similar to those stated before also make sense in this context.

Our framework for rational functions involved the proof of the equivalence of statements (1)–(4) in the preceding sections. At present, for a post-singularly finite and post-critically finite map or a general map in $\mathcal{AV}2$, $\mathcal{AV}3$, or $\mathcal{T}E_{p,q}$, it is premature for us to make any statement about the equivalence between (1) with f as a transcendental function and (4), that f has no Thurston obstruction. It is also premature for us to make any statement about the equivalence between (2) for a transcendental f with bounded geometry and (4). The reason is the following: If E has infinitely many components, the definition of a Thurston obstruction in Section 2 may not make sense because when the degree of f is infinite, a non-peripheral closed curve γ on $\widehat{\mathbb{C}} \setminus E$, $f^{-1}(\gamma)$ may contain infinitely many non-peripheral closed curves. If E has finitely many components in the complex plane and ∞ is an isolated component, one may be able to define a Thurston obstruction formally, in a manner similar to that in Section 2, because even if the map f has infinite degree, all but finitely many preimages of a non-peripheral curve γ on $\widehat{\mathbb{C}} \setminus E$ are peripheral or null homotopic rel. E. Even in this case, however, if one would like to prove (4) $\Rightarrow$ (1) by following the idea of the proof in [Douady-Hubbard], one needs first to find a new proof of Lemma 5.2 in [Douady-Hubbard], which is the key to the proof of Thurston's theorem (Theorem 3.2) and whose proof crucially depends on the finiteness of the degree of f. Furthermore, in order to prove (4) $\Rightarrow$ (2) by following the ideas in this framework, one needs first to find a new proof of Theorem 7.1 in [Douady-Hubbard], which only works for maps of finite degree and which plays an important role in our proof of (4) $\Rightarrow$ (2) in Theorems 4.8 and 4.9. Note that when a Thurston obstruction can be appropriately defined, (1) $\Rightarrow$ (4) can be proved just by following the proof in [Douady-Hubbard, Theorem 4.1] and the proof of Theorem 3.5 in [McMullen]; (2) $\Rightarrow$ (4) is currently being studied in [Chen-Jiang-Keen;1] and [Chen-Jiang-Keen;2].

In [Hubbard-Schleicher-Shishikura], the authors use a Levy cycle condition rather than a Thurston obstruction to characterize the non-existence of convergence. This is because, by the universal covering property of the map, every Thurston obstruction necessarily contains a Levy cycle. In fact, for general quasi-meromorphic and quasi-entire functions, a "canonical Thurston obstruction" is more natural than a "Levy cycle." We are working on a characterization problem for f a post-singularly finite function in either $\mathcal{AV}2$, $\mathcal{AV}3$ or $\mathcal{TE}_{p,q}$ in terms of canonical Thurston obstructions. The delicate part of the problem is the proof that (3) $\Rightarrow$ (2) (see Problem 4).

7 Bibliography

[Ahlfors-Bers] L. V. Ahlfors and L. Bers, Riemann's mapping theorem for variable metrics. *Annals of Math. (2)* **72** (1960), 385–404. MR 0115006 (22 #5813).

[Beck-Jiang-Mitra-Shiga] M. Beck, Y. Jiang, S. Mitra, and H. Shiga, Extending holomorphic motions and monodromy. *Annales Academiæ Scientiarum Fennicæ Mathematica* **37** (2012), 53–67. MR 2920423.

[Bonnot-Yampolsky] S. Bonnot and M. Yampolsky, Geometrization of post-critically finite branched coverings. arXiv:1102.2247.

[Brown] D. Brown, Using spider theory to explore parameter spaces. Ph.D. Dissertation, Cornell University, 2001. MR 2702308.

[Chen-Jiang] T. Chen and Y. Jiang, Canonical Thurston obstructions for sub-hyperbolic semi-rational branched coverings. *Conformal Geometry and Dynamics*, **17** (2013), 6–25.

[Chen-Jiang-Keen;1] T. Chen, Y. Jiang and L. Keen, Bounded geometry and families of meromorphic functions with two and three asymptotic values. arXiv:1112.2557 [math.DS].

[Chen-Jiang-Keen;2] T. Chen, Y. Jiang, and L. Keen, Bounded geometry and characterization of post-singularly finite (p,q)-exponential maps. Work in progress.

[Cheng-Jiang] T. Cheng and Y. Jiang, Geometrization of sub-hyperbolic semi-rational branched coverings. arXiv:1207.1292

[Cui-Jiang] G. Cui and Y. Jiang, Geometrically finite and semi-rational branched coverings of the two-sphere. *Transactions of the American Mathematical Society* **363**, (May 2011), 2701-2714. MR 2763733 (2012e:37090).

[Cui-Jiang-Sullivan;M] G. Cui, Y. Jiang and D. Sullivan, On geometrically finite branched covering maps. Manuscript, 1996.

[Cui-Jiang-Sullivan] G. Cui, Y. Jiang and D. Sullivan, On geometrically finite branched covering maps-I: Locally combinatorial attracting and II. Realization of rational maps (refer to its rewritten version [Cui-Tan]). *Complex Dynamics and Related Topics*, New Studies in Advanced Mathematics, 2004, The International Press, 1–14 and 15–29. MR 2504307. MR 2504308.

[Cui-Tan] G. Cui and L. Tan, A characterization of hyperbolic rational maps. *Invent. Math.* **183** (2011), 451–516. MR 2772086 (2012c:37088).

[Douady-Hubbard] A. Douady and J. Hubbard, A proof of Thurston's topological characterization of rational functions. *Acta. Math.* **171**, (1993), 263–297. MR 1251582 (94j:58143).

[Earle-McMullen] C. J. Earle and C. McMullen, *Quasiconformal isotopies. Holomorphic Functions and Moduli*, Vol. I (Berkeley, CA, 1986), 143–154, Math. Sci. Res. Inst. Publ. **10**, Springer, New York, 1988. MR 0955816 (89h:30028).

[Earle-Mitra] C.J. Earle and S. Mitra, Variation of moduli by holomorphic motions. *Contemp. Math.* **256** (2000), 39–67. MR 1759669 (2001f:30031).

[Epstein] A. Epstein, Towers of finite type complex analytic maps. Ph.D. Dissertation, The CUNY Graduate Center, 1993. MR 2690048.

[Epstein-Keen-Tresser] A. Epstein, L. Keen, and C. Tresser, The set of maps $F_{a,b} : \mapsto x + a + \frac{b}{2\pi}\sin(2\pi x)$ with any given rotation interval is contractible. *Commun. Math. Phys.* **173** (1995), 313–333. MR 1355627 (97e:58129).

[Gardiner-Jiang-Wang] F. Gardiner, Y. Jiang, and Z. Wang, Holomorphic motions and related topics. *Geometry of Riemann Surfaces*, London Mathematical Society Lecture Note Series **368**, 2010, 156–193. MR 2665009 (2011j:37087).

[Gardiner-Lakic] F. Gardiner and N. Lakic, *Quasiconformal Teichmüller Theory*. AMS, Providence, RI, 2000. MR 1730906 (2001d:32016).

[Godillon] S. Godillon, Construction de fractions rationnelles à dynamique prescrite. Ph.D. Dissertation, Université de Cergy-Pontoise, Mai 2010.

[Hassinsky] P. Haissinsky, Deformation loaclisee de surfaces de Riemann. *Publ. Math.* **49** (2005), 249–255. MR 2140209 (2006a:30047).

[Hinkkanen] A. Hinkkanen, Uniformly quasiregular semigroups in two dimensions. *Annales Academiæ Scientiarum Fennicæ Mathematica* **21**, 1996, 205–222. MR 1375517 (96m:30029).

[Hubbard] J. Hubbard, *Teichmüler Theory and Applications to Geometry, Topology, and Dynamics*, Volume I: *Teichmüller theory*. Matrix Edition, Ithaca, NY 14850. 2006. MR 2245223 (2008k:30055).

[Hubbard-Schleicher-Shishikura] J. Hubbard, D. Schleicher, and M. Shishikura, Exponential Thurston maps and limits of quadratic differentials. *Journal of the American Mathematical Society* **22** (2009), 77–117. MR 2449055 (2010c:37100).

[Jiang-Mitra] Y. Jiang and S. Mitra, Some applications of universal holomorphic motions. *Kodai mathematical Journal* **30**, 85–96. MR 2319079 (2008c:32019).

[Jiang-Mitra-Shiga] Y. Jiang, S. Mitra, H. Shiga, Quasiconformal motions and iso-morphisms of continuous families of Mobius groups. *Israel Journal of Mathematics* **188** (2012), 177–194.

[Jiang-Mitra-Wang] Y. Jiang, S. Mitra, Z. Wang, Liftings of holomorphic maps into Teichmülller space. *Kodai Mathematical Journal* **32**, 544–560. MR 2582017 (2010m:32015).

[Levy] S. Levy, Critically finite rational maps. Ph.D. Dissertation, Princeton University, 1985. MR 2634168.

[Lieb] G. Lieb, Holomorphic motions and Teichmüller space, Ph.D. Dissertation, Cornell University 1990. MR 2638376.

[Masur] H. Masur, Extension of the Weil-Petersson metric to the boundary of Teichmüller space, *Duke Math. J.* **43**, (1976), 623–635. MR 0417456.

[McMullen] C. McMullen, *Complex dynamics and Renormalization*. Ann. of Math. Studies **79**, Princeton University Press, Princeton 1994. MR 1312365 (96b:58097).

[Milnor] J. Milnor, *Dynamics in One Complex Variable, Introductory Lectures*. Vieweg, 2nd ed., 2000. MR 1721240 (2002i:37057).

[Minsky] Y. Minsky, Extremal length estimates and product regions in Teichmüller space. *Duke Math. J.* **83** (1996), 249–286. MR 1390649.

[Mitra] S. Mitra, Extensions of holomorphic motions. *Israel Journal of Mathematics* **159** (2007), 277–288. MR 2342482 (2009f:32025).

[Nevanlinna] R. Nevanlinna, *Analytic Functions*. Translated from the second German edition by Phillip Emig. Die Grundlehren der mathematischen Wissenschaften, Band **162**, Springer-Verlag, New York–Berlin 1970. MR 0279280 (43 #5003).

[Pilgrim] K. M. Pilgrim, Cannonical Thurston obstruction. *Advances in Mathematics* **158** (2001), 154–168. MR 1822682 (2001m:57004).

[Rees] M. Rees, A partial description of the parameter space of rational maps of degree two, Part 2. *Proc. London Math. Soc.* (1995) 644–690. MR 1317518 (96g:58161).

[Selinger] N. Selinger, Thurston's pullback map on the argmented Teichmüller space and applications. *Invent Math.* **189** (2012), 111–142.

[Sullivan;1] D. Sullivan, The ergodic theory at infinity of an arbitrary discrete group of hyperbolic motions. *Riemann Surfaces and Related Topics, Proceedings of the 1978 Stony Brook Conference*. Ann. of Math. Stud. **97**, Princeton University Press, Princeton, 1981, 465–496. MR 0624833 (83f:58052).

[Sullivan;2] D. Sullivan, *Conformal dynamical systems. Geometric dynamics*, Lecture notes in Math. **1007**, Springer, Berlin, 1983, 725–752. MR 0730296.

[Sullivan;3] D. Sullivan, 11 lectures on Thurston's geometric theorem and Thurston's critical finite theorem #1–6 and #26 from 1981 to 1986. Einstein Chair Seminar Video Tape Collections, CUNY Graduate Center. http://www.math.sunysb.edu/Videos/Einstein/

[Tan] L. Tan, Matings of quadratic polynomials. *Ergod. Th. Dynam. Sys.* **12** (1992), 589–620. MR 1182664 (93h:58129).

[Thurston] W. Thurston, The combinatorics of iterated rational maps (1985), published in: *Complex dynamics: Families and Friends*, ed. by D. Schleicher, A. K. Peters (2008), 1–108.

[Tukia] P. Tukia, On two-dimensional quasiconformal groups. *Ann. Acad. Sci. Fenn. Ser. A. I. Math.* **5**, 1980, 73–78. MR 0595178 (82c:30031).

[Wang] X. Wang, A decomposition theorem for Herman maps. arXiv:1203.5563.

[Zhang;1] G. Zhang, Topological models of polynomials of simple Siegel disk type. Ph.D. Dissertation, CUNY Graduate Center 2002.

[Zhang;2] G. Zhang, Dynamics of Siegel rational maps with prescribed combinatorics. arXiv:0811.3043.

[Zhang-Jiang] G. Zhang and Y. Jiang, Combinatorial characterization of sub-hyperbolic rational maps. *Advances in Mathematics*, **221** (2009), 1990–2018. MR 2522834 (2010h:37095).

Part II

One Real Variable

Metric stability for random walks
(with applications in renormalization theory)

Carlos Gustavo Moreira and Daniel Smania

ABSTRACT. Consider deterministic random walks $F\colon I \times \mathbb{Z} \to I \times \mathbb{Z}$, defined by

$$F(x,n) = (f(x), \psi(x) + n),$$

where f is an expanding Markov map on the interval I and $\psi\colon I \to \mathbb{Z}$. We study the universality (i.e., stability) of ergodic (for instance, recurrence and transience), geometric, and multifractal properties in the class of perturbations of the type $\tilde{F}(x,n) = (f_n(x), \tilde{\psi}(x,n) + n)$, which are topologically conjugate with F and where the f_n are expanding Markov maps exponentially close to f as $|n| \to \infty$. We give applications of these results in the study of the regularity of conjugacies between (generalized) infinitely renormalizable maps of the interval and the existence of wild attractors for one-dimensional maps.

1 INTRODUCTION

1.1 Metric stability for random walks

In the study of a dynamical system, some of the most important questions concerns the stability of their dynamical properties under (most of the) perturbations: how robust are they?

Here we are mainly interested in the stability of metric (measure-theoretic) properties of dynamical systems. A well-known example is that of (C^2) expanding maps on the circle; this class is structurally stable, and all such maps have an absolutely continuous and ergodic invariant probability satisfying certain decay of correlations estimates. In particular, in the measure theoretic sense, most of the orbits are dense in the phase space.

Now let us study a slightly more complicated situation: consider a C^2 Markov almost-onto expanding map of the interval $f\colon I \to I$ with bounded distortion and large images (see Section 2 for details) and let $\psi\colon I \to \mathbb{Z}$ be a function which is constant on each interval of the Markov partition of f. We can define $F\colon I \times \mathbb{Z} \to I \times \mathbb{Z}$ as

$$F(x,n) := (f(x), \psi(x) + n).$$

The second entry of (x,n) will be called its **state**. We also assume that

$$\inf \psi > -\infty \tag{1.1}$$

and that F is topologically mixing.

The map F is refered to in the literature in many ways: as a "skew-product of f and the translation on the group $\mathbb{Z}$," a "group extension of f," or even a "deterministic random walk generated by f," and its metric behavior is very well

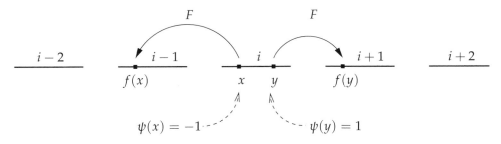

FIGURE 1.1: A deterministic random walk.

studied: for instance, are most the orbits recurrent? Everything depends on the **mean drift**

$$M = \int \psi d\mu,$$

where μ is the absolutely continuous invariant probability of f (the function ψ will be called the **drift function**). Indeed, note that

$$F^n(x,i) = \left(f^n(x), \; i + \sum_{k=0}^{n-1} \psi(f^k(x)) \right).$$

By the Birkhoff ergodic theorem,

$$\lim_{n\to\infty} \frac{\pi_2(F^n(x,i)) - \pi_2(x,i)}{n} = \lim_{n\to\infty} \frac{1}{n} \sum_{k=0}^{n-1} \psi(f^k(x)) = M$$

for almost every $x \in I$ (here $\pi_2(x,n) := n$). In particular if $M \neq 0$, then almost every point $(x,i) \in I \times \mathbb{Z}$ is **transient**; in other words we have

$$\lim_{n\to\infty} |\pi_2(F^n(x,i))| = \infty,$$

and so most of the points are not recurrent.

On the other hand, if $M = 0$, most of points are recurrent (see Guivarc'h [G]): by the central limit theorem for expanding maps (here we need to assume that ψ is not constant and $f \in On$: see Section 2) of the interval

$$\sup_{\epsilon \in \mathbb{R}} \left| \mu\left(x \in I : \frac{\sum_{k=0}^{n-1} \psi(f^k(x))}{\sigma \sqrt{n}} \leq \epsilon \right) - \frac{1}{\sqrt{2\pi}} \int_{-\infty}^{\epsilon} e^{-\frac{u^2}{2}} \, du \right| \leq \frac{C}{\sqrt{n}}.$$

Given $\delta > 0$ we can easily obtain, taking $\epsilon = n^{-1/4}$ and applying Borel-Cantelli lemma, that

$$\mu(A_+) := \mu\left(x \in I : \; \limsup_{n\to\infty} \frac{\sum_{k=0}^{n-1} \psi(f^k(x))}{\sqrt[2+\delta]{n}} = \infty \right) \geq \frac{1}{2}, \tag{1.2}$$

$$\mu(A_-) := \mu\left(x \in I : \; \liminf_{n\to\infty} \frac{\sum_{k=0}^{n-1} \psi(f^k(x))}{\sqrt[2+\delta]{n}} = -\infty \right) \geq \frac{1}{2}. \tag{1.3}$$

Clearly A_+ and A_- are invariant sets: the ergodicity of f implies that

$$\mu(A_+ \cap A_-) = 1. \tag{1.4}$$

By the conditions on ψ in Eq. (1.1), since f is expanding with distortion control and that F is transitive, we can easily conclude that almost every point in $I \times \mathbb{Z}$ is an F-recurrent point.

Note that the random walk F is a dynamical system quite similar to expanding circle maps: F is an expanding map, with good bounded distortion properties, but F does not have an invariant probability measure which is absolutely continuous with respect to the Lebesgue measure on $I \times \mathbb{Z}$. Moreover, in general the random walk is not even recurrent and the recurrence property loses its stability: given a recurrent random walk (f, ψ), it is possible to obtain a transient random walk by changing f and ψ a little bit.

Since the non-compactness of the phase space seems to be the origin of the lack of stability of recurrence and transience properties, a natural question is to ask if such properties are stable by compact perturbations. The answer is yes. Indeed, as we are going to see in Theorems 3.1–3.4, the transience and recurrence are preserved even by non-compact perturbations which decrease fast away from state 0. For instance, we can choose perturbations like

$$\tilde{F}(x, n) = (f_n(x), \psi(x) + n),$$

where, for some $\lambda \in [0, 1)$,

$$|f_n - f|_{C^3} \leq \lambda^{|n|}. \tag{1.5}$$

The notations and conventions are more or less obvious: we have postponed the rigorous definitions to the next section.

With respect to the stability of transience and recurrence, there is a previous quite elegant result by R. L. Tweedie [T]: if p_{ij} are the transition probabilities of a Markov chain on $\mathbb{Z}$, then any perturbation $\tilde{p}_{ij}$ so that

$$(1 + \epsilon_i)^{-1} p_{ij} \leq \tilde{p}_{ij} \leq p_{ij}(1 + \epsilon_i), \ j \neq i,$$

and

$$\prod_{i=0}^{\infty}(1 + \epsilon_i) < \infty$$

preserves the recurrence or transience of the original Markov chain. But Tweedie's argument does not appear to work in our setting. Our result coincides with that of Tweedie in the very special case where f and f_n are linear Markov maps and $\epsilon_i \sim C\lambda^{|i|}$.

In the transient case we can tell a little more: there will be a conjugacy between the original random walk f and its perturbation which is a martingale strongly quasisymmetric map (for short, mSQS-map; see Section 3.1) with respect to certain dynamically defined set of partitions. Unlike the usual class of one-dimensional quasisymmetric functions, which does not share many of most interesting properties of higher-dimensional quasisymmetric maps, the one-dimensional mSQS-maps are much closer to their high-dimensional cousins, as quasiconformal maps in dimension 2. For instance, they are absolutely continuous.

We also study the behavior of the Hausdorff dimension of dynamically defined sets. Denote by $\Omega_+(F)$ the set of points which have non-negative states along the positive orbit by F. We prove that $\Omega_+(F)$ has Hausdorff dimension strictly smaller than one if and only if $\Omega_+(\tilde{F})$ has dimension less than one for all perturbation satisfying Eq. (1.5). Furthermore we give a variational characterization for the Hausdorff dimension $HD(\Omega_+(F))$ as the minimum of $HD(\Omega_+(\tilde{F}))$, where $\tilde{F}$ runs over the set of such perturbations. For these results we study of the stability of the multifractal spectrum of the random walk F under perturbations.

1.2   Applications to (generalized) renormalization theory

A unimodal map is a map with a unique critical point. Under reasonable conditions (real-analytic maps with negative Schwarzian derivative and non-flat critical point) two non-renormalizable unimodal maps with the same topological entropy are indeed topologically conjugate. A key question in one-dimensional dynamics is about the regularity of the conjugacy: Is it Hölder? Is it absolutely continuous? Since Dennis Sullivan's work in the 1980s, the quasisymmetry of the conjugacy became a very useful tool to obtain deep results in one-dimensional dynamics. Later on, the density of the hyperbolic maps in the real quadratic family was proved verifying the quasisymmetry of the conjugacies for all combinatorics, including infinitely renormalizable ones (see [L2], [GS]).

Note that quasisymmetric maps are not, in general, absolutely continuous: they do not even preserve (in general) sets of Hausdorff dimension one. Is the conjugacy between unimodal maps absolutely continuous? The answer is no: M. Martens and W. de Melo [MdM] proved that under the forementioned reasonable conditions, an absolutely continuous conjugacy is actually C^∞, provided the unimodal maps

(i)   do not have a periodic attractor,

(ii)   are not infinitely renormalizable,

(iii)   do not have a wild attractor (the topological and measure-theoretic attractor must coincide).

Since we can change the eigenvalues of the periodic points of maps preserving its topological class and the eigenvalues are preserved by C^1 conjugacies, we conclude that in general a conjugacy between unimodal maps is not absolutely continuous.

Condition (i) is clearly necessary. This work (Theorem 3.11) shows that Condition (ii) is necessary, proving that the conjugacy between two arbitrary Feigenbaum unimodal maps with same critical order is *always* absolutely continuous. Actually the conjugacy is martingale strongly quasisymmetric with respect to a set of dynamically defined partitions.

Condition (iii) is never violated when the critical point is quadratic (see [L1]). But for certain topological classes of unimodal maps wild attractor appears when the order of the critical point increases: Fibonacci maps are the simplest kind of such maps [BKNvS, Br]. We are going to prove (Theorem 3.16) that a Fibonacci map with even order has a wild attractor if and only if all Fibonacci maps with the same even order are conjugate to each other by an absolutely continuous mapping (in particular all these Fibonacci maps have a wild attractor). So Condition (iii) is necessary.

To show that Conditions (ii) and (iii) are necessary, the (generalized) renormalization theory for unimodal maps and the study of perturbations of transient and recurrent random walks are going to be crucial. Feigenbaum and Fibonacci unimodal maps admit induced maps which are essentially perturbations of deterministic random walks (Section 8). In the Fibonacci case the transience of this random walk is equivalent to the existence of a wild attractor. Random walks associated to a Feigenbaum map will always be transient.

For both Feigenbaum and Fibonacci maps there are infinitely many periodic points (indeed in the Fibonacci case the periodic points are also dense in an interval). It is well known that the conjugacy between critical circle maps with same irrational rotation number and satisfying a certain Diophantine condition is absolutely continuous, but we think that these are the first interesting examples of a similar phenomena for maps with many periodic points.

1.3 Acknowledgements

We thank the referee for the very useful, detailed and precise comments and suggestions. D. S. was partially supported by CNPq 470957/2006-9, 310964/2006-7, 472316/03-6, 303669/2009-8, 305537/2012-1 and FAPESP 03/03107-9, 2008/02841-4, 2010/08654-1, and C. G. M. was partially supported by CNPq and by the Balzan Research Project of J. Palis. The authors would like to thank the hospitality of ICMC-USP and IMPA.

2 EXPANDING MARKOV MAPS, RANDOM WALKS, AND THEIR PERTURBATIONS

In this article we will deal with maps

$$F \colon I \times \mathbb{Z} \to I \times \mathbb{Z}$$

which are piecewise C^2 diffeomorphisms, which means that there is a partition $\mathcal{P}^0$ of $I \times \mathbb{Z}$ so that each element $J \in \mathcal{P}^0$ is an open interval where $F|_{\overline{J}}$ is a C^2 diffeomorphism. Denote $I_n = I \times \{n\}$. Denote by m the Lebesgue measure in the in $I \times \mathbb{Z}$, that is, if $A \subset I \times \mathbb{Z}$ is a Borel set, then

$$m(A) = \sum_n m_I(\pi(A \cap I_n)),$$

where m_I is the Lebesgue measure in the interval I and $\pi(n, x) = x$.

If A_J denotes the unique affine transformation which maps the interval J to $[0, 1]$ and preserves orientation, then define, for each $J \in \mathcal{P}^0$,

$$\tau^F_J := A_J \circ F^{-1} \circ A_{F(J)}^{-1}.$$

Throughout this article we will assume that F satisfies some of the following properties:

(Mk) Markovian: For each $J \in \mathcal{P}^0$, $F(J)$ is a connected union of elements in $\mathcal{P}^0$. In particular, we can write $F(x, n) = (f_n(x), n + \psi(x, n))$, where $f_n \colon I \to I$ is a piecewise C^2 diffeomorphism relative to the partition $\mathcal{P}^0_n := \{J \in \mathcal{P}^0 : J \subset I_n\}$ and $\psi \colon I \times \mathbb{Z} \to \mathbb{Z}$, called the **drift function**, is constant on each element of $\mathcal{P}^0$.

(LBD) Lower Bounded Drift: F is Markovian and $\min \psi > -\infty$.

(LI) Large Image: F is Markovian and there exists $\delta > 0$ so that for each $J \in \mathcal{P}^0$ we have $|F(J)| \geq \delta$.

(On) Onto: F is Markovian and for each $J \in \mathcal{P}^0$ we have $F(J) = I^n$, for some $n \in \mathbb{Z}$.

(BD) Bounded Distortion: There exists $C > 0$ so that every $J \in \mathcal{P}_n^0$ and map τ_J is a C^2 function satisfying

$$\sup_J \left| \frac{D^2 \tau_J}{(D\tau_J)^2} \right| \leq C.$$

(sBD) Strong Bounded Distortion: There exists $C > 0$ so that every $J \in \mathcal{P}_n^0$ and map τ_J is a C^2 function satisfying

$$\sup_J \left| \frac{D^2 \tau_J}{(D\tau_J)^2} \right| \leq C|J|.$$

(Ex) Expansivity: If $J \in \mathcal{P}_n^0 := \{J \in \mathcal{P}^0 : J \subset I_n\}$, denote $\phi_J := f_n^{-1}|_{f_n(J)}$. Then either ϕ_J can be extended to a function in a δ-neighborhood of J so that

$$S\phi_J > 0,$$

where $S\phi_J$ denotes the Schwarzian derivative of ϕ_J, or there exists $\theta \in (0,1)$ so that for all n and $J \in \mathcal{P}_n^0$ we have

$$|\phi_J'| < \theta$$

on $f_n(J)$.

(Ra) Regularity a: There exists $N \in \mathbb{N}$, $\delta > 0$ and $C > 0$ with the following properties: the intervals in $\mathcal{P}_n^0$ are positioned in $I_n = [a,b]$ in such way that the complement of

$$\bigcup_{J \in \mathcal{P}_n^0} int\ J$$

contains at most N accumulation points

$$c_1^n < c_2^n < \cdots < c_{i_n}^n,$$

with $i_n \leq N$, which are in the interior of I_n. Furthermore, $|c_{i+1}^n - c_i^n| \geq \delta$ and $|a - c_1^n|, |b - c_{i_n}^n| \geq \delta$. Moreover, given P and $Q \in \mathcal{P}_n^0$ so that $\overline{P} \cap \overline{Q} \neq \phi$, then

$$\frac{1}{C} \leq \frac{|P|}{|Q|} \leq C. \tag{2.1}$$

(Rb) Regularity b: Assume Ra. There exists $C > 0$, $\lambda \in (0,1)$, $\delta > 0$ so that for each $1 < i < i_n$ we can find a point

$$d_i^n \in (c_i^n, c_{i+1}^n),$$

which does not belong to any $P \in \mathcal{P}_n^0$, and

$$\min\{|c_{i+1}^n - d_i^n|, |d_i^n - c_i^n|\} \geq \delta$$

with the following property: If J is a connected component of

$$I_n \setminus \{d_i^n, c_j^n\}_{i,j}$$

then we can enumerate the set

$$\{P\}_{P \in \mathcal{P}_n^0,\ P \subset J} = \{J_i\}_{i \in \mathbb{N}}$$

in such way that $\partial J_i \cap \partial J_{i+1} \neq \phi$ for each i and

$$\frac{|J_{i+j}|}{|J_i|} \leq C\lambda^j$$

for $i \geq 0, j > 0$.

(GD) Good Drift: If ψ is the drift function of the random walk, then for each $n \in \mathbb{Z}$ there exists x such that $\psi(x, n) > 0$. Moreover there exists $\gamma \in (0, 1)$ and $C > 0$ so that for every $k \geq 0$,

$$m(\{(x, n) \text{ s.t. } \psi(x, n) \geq k\}) \leq C\gamma^k.$$

(T) Transitive: F has a dense orbit.

For convenience of notation, if, for instance, F is Markovian and it has bounded distortion, we will write $F \in Mk + BD$.

A **deterministic random walk** (or, simply, **random walk**) is a map

$$F \in Mk + LBD + LI + Ex + BD + GD.$$

It is generated by the pair $(\{f_n\}, \psi)$ if

$$F(x, n) := (f_n(x), \psi(x, n) + n).$$

When $f_n = f \in Mk$ and $\psi(x, n) = \psi(x)$, we say that F is the **homogeneous deterministic random walk** generated by the pair (f, ψ). There is a large literature about such random walks. We will sometimes assume the following property:

(aO) Almost Onto: For every $i, j \in \Lambda$, there exists a finite sequence

$$i = i_0, i_1, i_2, \ldots, i_{n-1}, i_n = j \in \Lambda \quad \text{so that} \quad f(I_{i_k}) \cap f(I_{i_{k+1}}) \neq \varnothing$$

for each $k < j$.

Denote $\pi(x, n) := \pi_2(x, n) := n$. A random walk is called **transient** if, for almost every $(x, n) \in I \times \mathbb{Z}$,

$$\lim_{k \to \infty} |\pi_2(F^k(x, n))| = \infty,$$

and it is **recurrent** if for almost every $(x, n) \in I \times \mathbb{Z}$

$$\#\{k\colon \pi_2(F^k(x,n)) = n\} = \infty.$$

Making use of usual bounded distortion tricks, it is easy to show that every $F \in Mk + LI + Ex + BD + T$ is either recurrent or transient.

A (topological) **perturbation** of a random walk is a random walk $\tilde{F}$, generated by a pair $(\{\tilde{f}_i\}, \tilde{\psi})$, so that $H \circ F = \tilde{F} \circ H$ for some homeomorphism

$$H\colon I \times \mathbb{Z} \to I \times \mathbb{Z}$$

which preserves states: $\pi_2(H(x, i)) = i$.

Define $\mathcal{P}^n(F) := \vee_{i=0}^{n-1} F^{-i} \mathcal{P}^0(F)$. If F and $\tilde{F}$ are random walks and H is a topological conjugacy that preserves states between F and $\tilde{F}$, then for each interval L such that $L \subset J \in \mathcal{P}^{n-1}(F)$, define

$$\text{dist}_n(L) := \sup_{y \in L} \left| \ln \frac{D\tilde{F}^n(H(y))}{DF^n(y)} \right|,$$

Similarly, define

$$\text{dist}_n(x) := \left| \ln \frac{D\tilde{F}^n(H(x))}{DF^n(x)} \right|$$

and

$$\text{dist}_\infty(x) := \sup_n \text{dist}_n(x).$$

Another kind of random walk which will have a central role in our results are those which are **asymptotically small** perturbations: these are perturbations $(\{\tilde{f}_i\}, \tilde{\psi})$ of a deterministic random walk $(\{f_i\}, \psi)$ such that there exists $\lambda \in (0, 1)$ and $C > 0$ satisfying either

$$\left| \log \frac{D\tilde{F}(H(p))}{DF(p)} \right| \leq C\lambda^{|\pi_2(p)|}, \tag{2.2}$$

if ψ is bounded, or

$$\left| \log \frac{D\tilde{F}(H(p))}{DF(p)} \right| \leq C\lambda^{\pi_2(p)}, \tag{2.3}$$

for $\pi_2(p) \geq 0$ and $D\tilde{F}(H(p)) = DF(p)$ otherwise, if ψ has only a lower bound.

It is easy to see that Properties Ra, Rb and GD are invariant by asymptotically small perturbations (if we allow to change the constants described in these properties).

Let $F = (\{f_i\}, \psi)$ be a random walk, where ψ is Lebesgue integrable on compact subsets of $I \times \mathbb{Z}$. We say that F is **strongly transient** if $K > 0$ and

$$\mathbb{E}(\psi \circ F^n | \mathcal{P}^{n-1}(F)) > K$$

for every $n \geq 1$. We will also say that F is K-strongly transient. Here we are considering conditional expectations relative to the Lebesgue measure. As the notation suggests, every strongly transient random walk is transient. Moreover, we have the following large deviations result.

PROPOSITION 2.1. *Let $F = (f, \psi) \in On + sBD + Ra + Rb$ be a homogeneous random walk with positive mean drift. Let $K := \int \psi \, dm > 0$. Then F is transient and for every small $\epsilon > 0$ there exist $\lambda \in [0, 1)$ and $C > 0$ so that for each $P \in \mathcal{P}^0$, we have*

$$m\left(p \in P \colon \ \pi_2(F^n(p)) - \pi_2(p) < (K - \epsilon)n \right) \leq C\lambda^n |P|.$$

PROPOSITION 2.2. *Every K-strongly transient random walk $F \in Ra + Rb$ is transient. Furthermore for every small $\epsilon > 0$ there exist $\lambda \in [0, 1)$ and $C > 0$ so that for each $P \in \mathcal{P}^0$ we have*

$$m\left(p \in P \colon \ \pi_2(F^n(p)) - \pi_2(p) < (K - \epsilon)n \right) \leq C\lambda^n |P|.$$

We will postpone the proof of Propositions 2.2 and 2.1 until Section 5.

REMARK 2.3. By the Birkhoff ergodic theorem it is easy to see that a sufficiently high iteration of a homogeneous random walk with positive mean drift is strongly transient (see the proof of Proposition 5.1 for details).

3 STATEMENTS OF RESULTS

3.1 Stability of transience

THEOREM 3.1 (Stability of Transience I). *Assume that the random walk F defined by the pair $(\{f_i\}, \psi)$ is strongly transient. Then every asymptotically small perturbation G of F is also transient. Indeed there is a topological conjugacy between F and G which is an absolutely continuous map and preserves the states.*

We have a similar theorem for all transient homogeneous random walks.

THEOREM 3.2 (Stability of Transience II). *Suppose that the homogeneous random walk F defined by the pair (f, ψ) has positive mean drift. Then every asymptotically small perturbation of F is topologically conjugated to F by an absolutely continuous map which preserves the states.*

We can be more precise regarding the regularity of the conjugacy if the drift is non-negative.

Let $\mathcal{A}_0, \mathcal{A}_1, \ldots, \mathcal{A}_n, \mathcal{A}_{n+1}, \ldots$ be a succession of partitions by intervals of $I \times \mathbb{Z}$ such that $\mathcal{A}_{n+1}$ refines $\mathcal{A}_n$ and whose union generates the Borel algebra of $\sqcup_n I_n$. We say that $h \colon \sqcup_n I_n \to \sqcup_n I_n$ is a **martingale strongly quasisymmetric (mSQS)** map with respect to the **stochastic basis** $\cup_n \mathcal{A}_n$ if there exist $C > 0$ and $\alpha \in (0, 1]$ so that

$$\frac{m(h(B))}{|h(J)|} \leq C \left(\frac{m(B)}{|J|} \right)^\alpha$$

for all Borel $B \subset J \in \cup_n \mathcal{A}_n$, and the same inequality holds replacing h by h^{-1} and $\cup_n \mathcal{A}_n$ by $\cup_n h(\mathcal{A}_n)$.

THEOREM 3.3 (Strongly Quasisymmetric Rigidity). *Let F be either a strongly transient random walk or a transient homogeneous random walk with positive mean drift. Moreover, assume in both cases that $\psi \geq 0$. Then every asymptotically small perturbation G of F is topologically conjugated to F by an absolutely continuous map h which preserves the states. Furthermore, h on $\cup_{i \geq 0} I_i$ is a martingale strongly quasisymmetric mapping with respect to the stochastic basis $\cup_i \mathcal{P}^i$.*

3.2 Stability of recurrence

In the recurrent case, we are going to restrict ourselves to the stability of the metric properties of homogeneous random walks under asymptotically small perturbations: it is easy to see that the recurrence is not stable by perturbations which are not asymptotically small. Nevertheless, we have the following.

THEOREM 3.4 (Stability of Recurrence). *Suppose that $F \in On + T$ is a recurrent homogeneous random walk generated by the pair (f, ψ). Then every asymptotically small perturbation of F is also recurrent.*

If p is a periodic point with prime period n, then $DF^n(p)$ is called the spectrum of the periodic point p. Note that we cannot expect, as in the transient case, an absolutely continuous conjugacy which preserves states between F and G, once asymptotically small perturbations do not preserve (in general) the spectrum of the periodic points.

PROPOSITION 3.5 (Rigidity). *Suppose that the random walk $F \in On$ generated by a pair $(\{f_i\}_i, \psi)$ is recurrent. If there is an absolutely continuous conjugacy which preserves states H between F and a random walk G, then H is C^1 in each state. In particular, the spectrum of the corresponding periodic points of F and G are the same.*

The reader should compare this result with similar results by Shub and Sullivan [ShSu] for expanding maps on the circle and de Melo and Martens [MdM] for unimodal maps.

3.3 Stability of the multifractal spectrum

Let F be a random walk and denote

$$\Omega_+(F) := \{p \colon \pi_2(F^j p) \geq 0, \text{ for } j \geq 0\},$$

$$\Omega_+^k(F) := \{(x, k) \colon \pi_2(F^j(x, k)) \geq 0, \text{ for } j \geq 0\}, \text{ and}$$

$$\Omega_{+\beta}^k(F) := \{(x, k) \in \Omega_+^k \text{ s.t } \lim_n \frac{\pi_2(F^n(x, k))}{n} \geq \beta\}.$$

THEOREM 3.6. *Let $F \in Ra + Rb + On$ be a random walk. Then, for all $k \in \mathbb{Z}$ and $\beta > 0$, the Hausdorff dimension $HD(\Omega_{+\beta}^k)$ is invariant by asymptotically small perturbations.*

We will need the following.

PROPOSITION 3.7. *Let $F \in Ra + Rb + On$ be a homogeneous random walk. Then*

$$HD(\Omega_+^k(F)) = \lim_{\beta \to 0^+} HD(\Omega_{+\beta}^k(F)).$$

As a consequence of Theorem 3.6 and Proposition 3.7, we have Theorem 3.8.

THEOREM 3.8. *Let $F \in Ra + Rb + On$ be a homogeneous random walk. If G is an asymptotically small perturbation of F then*

$$HD(\Omega_+^k(G)) \geq HD(\Omega_+^k(F)). \tag{3.1}$$

We cannot replace the inequality in Eq. (3.1) by an equality. Indeed, even if $HD(\Omega_+^k(F)) < 1$, we have that $\sup HD(\Omega_+^k(G)) = 1$, where the supremum is taken on all asymptotically small perturbations G of F. Nevertheless, consider Theorem 3.9.

THEOREM 3.9. *Let $F \in Ra + Rb + On + T$ be the homogeneous random walk generated by the pair (f, ψ). Consider $M = \int \psi d\mu$, where μ is the unique absolutely continuous invariant measure of f.*

- *If $M > 0$, then for all asymptotically small perturbations G of F, we have $m(\Omega_+(G)) > 0$.*

- *If $M = 0$, then for all asymptotically small perturbations G of F, we have $HD(\Omega_+(G)) = 1$ but $m(\Omega_+(G)) = 0$.*

- *If $M < 0$, then for all asymptotically small perturbations G of F, we have $HD(\Omega_+(G)) < 1$.*

REMARK 3.10. Since the authors are more familiar with deterministic rather than stochastic terminology, we stated and proved the results in this work for determinist random walks. However, we believe that these results could be easily translated to the theory of chains with complete connections (g-measures, chains of infinite order) and one-sided shifts on an infinite alphabet.

3.4   Applications to renormalization theory of one-dimensional maps

THEOREM 3.11. *Let f and g be unimodal maps which are infinitely renormalizable with the same bounded combinatorial type and even critical order. Then the continuous conjugacy h between f and g is a strongly quasisymmetric mapping with respect to a certain stochastic basis of intervals $\mathcal{P}$.*

The set of intervals $\mathcal{P}$ is defined using a map induced by f. See the details in Section 8.1.

REMARK 3.12. D. Sullivan [Su, dMvS] shows that, under the assumptions of Theorem 3.11, the conjugacy h is a quasisymmetric map. However, it is known that quasisymmetric maps on the real line are not in general absolutely continuous maps.

Let $\mathcal{F}_d$ be the class of analytic maps with negative Schwarzian derivative which are infinitely renormalizable in the Fibonacci sense, and with even critical order d (see Section 8.2 for definitions). If f is a Fibonacci map, denote by $J_{\mathbb{R}}(f)$ the maximal invariant set of f. Let $\mathcal{F}_d^{uni}$ be the class of Fibonacci *unimodal* maps with negative Schwarzian derivative.

THEOREM 3.13 (Metric Universality). *For each even critical order d, $d \geq 4$, one of the following statements holds:*

- $HD(J_{\mathbb{R}}(f)) < 1$, *for all $f \in \mathcal{F}_d$.*

- $HD(J_{\mathbb{R}}(f)) = 1$ *and $m(J_{\mathbb{R}}) = 0$ for all $f \in \mathcal{F}_d$.*

- $HD(J_{\mathbb{R}}(f)) = 1$ *and f has a wild attractor (in particular, $m(J_{\mathbb{R}}(f)) > 0$) for all $f \in \mathcal{F}_d$*

THEOREM 3.14 (Measurable Deep Point). *Let $f \in \mathcal{F}_d$, where $d \geq 4$ is an even integer, and assume that 0 is its critical point. If $J_{\mathbb{R}}(f)$ has positive Lebesgue measure then there exists $\alpha > 0$ and $C > 0$ so that*

$$m\left(x \in (-\delta, \delta): \ x \notin J_{\mathbb{R}}(f)\right) \leq C\delta^{1+\alpha}.$$

REMARK 3.15. Indeed α can be taken depending only on d.

THEOREM 3.16. *For each even critical order d, $d \geq 4$, the following statements are equivalent:*

1. *There exists $f \in \mathcal{F}_d$ such that $m(J_{\mathbb{R}}(F)) > 0$.*

2. *There exists $f \in \mathcal{F}_d$ with a wild attractor.*

3. *There exist maps $f, g \in \mathcal{F}_d^{uni}$ which are conjugated by a continuous absolutely continuous maps h, but f has a periodic point p whose eigenvalue is different from the eigenvalue of the periodic point $h(p)$ of g.*

4. *All maps in $\mathcal{F}_d$ have wild attractors.*

5. *All maps in $\mathcal{F}_d^{uni}$ can be conjugated with each other by an absolutely continuous conjugacy.*

4   PRELIMINARIES

4.1   Probabilistic tools

We are going to collect here a handful of probabilistic tools which are going to be useful along the article. A good reference for these results is [B].

Most of the probabilistic results in dynamical systems (large deviation, central limit theorem) assume that the observables are quite regular; usual regularity assumptions are either Hölder continuity or bounded variation. Fix a random walk $(f, \psi) \in On + sBD + Ra + Rb + GD$. Then f has a unique absolutely continuous invariant probability μ. Moreover, this invariant measure is ergodic (see [B, page 29]). We are interested in $\mathcal{P}^0$-measurable observables with integer values which do not have such regularity. Fortunately, this is almost true: denote by $\mathcal{O}(f)$ the class of $\mathcal{P}^0$-measurable functions $\phi \colon I \to \mathbb{Z}$ so that

- $\phi \in L^2(\mu)$;

- if P denotes the Perron-Frobenius-Ruelle operator of f, then $P\phi$ has bounded variation.

Then $\psi \in \mathcal{O}(f)$. Up to simple modifications in the proofs in [B], we have the following.

PROPOSITION 4.1 (Large Deviations Theorem [B]). *Suppose $(f, \psi) \in On + sBD + Ra + Rb + GD$. Then for every $\psi \in \mathcal{O}(f)$ and $\epsilon > 0$, there exists $\gamma \in (0, 1)$ and $C \geq 0$ so that*

$$\mu\left(\left\{x \in I \colon \left|\frac{1}{n}\sum_{i=0}^{n-1}\psi(f^i(x)) - \int \psi d\mu\right| \geq \epsilon\right\}\right) \leq C\gamma^n.$$

PROPOSITION 4.2 (Proposition 6.1 of [B]). *If $(f, \psi) \in On + sBD + Ra + Rb + GD$, then for every $\psi \in \mathcal{O}(f)$, the limit*

$$\sigma^2 := \lim_{n \to \infty} \int \left(\frac{1}{\sqrt{n}} \sum_{k=0}^{n-1} \psi(f^k(x)) \right)^2 d\mu$$

exists. Furthermore, $\sigma^2 = 0$ if and only if there exist a function $\alpha \in L^2(\mu)$ so that $\psi = \alpha \circ f - \alpha$.

PROPOSITION 4.3 (Central Limit Theorem: Theorem 8.1 in [B]). *Suppose we have $(f, \psi) \in On + Mk + sBD + Ra + Rb + GD$. Then for every $\psi \in \mathcal{O}(f)$ with $\sigma^2 \neq 0$,*

$$\sup_{\epsilon \in \mathbb{R}} \left| \mu \left(x \in I: \frac{\sum_{k=0}^{n-1} \psi(f^k(x))}{\sigma \sqrt{n}} \leq \epsilon \right) - \frac{1}{\sqrt{2\pi}} \int_{-\infty}^{\epsilon} e^{-\frac{u^2}{2}} \, du \right| \leq \frac{C}{\sqrt{n}}, \qquad (4.1)$$

Indeed we are going to see that the assumption $\sigma^2 \neq 0$ is very weak: to this end we need the following result.

PROPOSITION 4.4 (Theorem 3.1 in [AD]). *Let $f: \cup_i I_i \to I$ be a map in $Mk + BD + Ex + Ra + Rb$. Let $\psi: \cup_i I_i \to \mathbb{S}^1$ be a $\mathcal{P}_0$-measurable function. If*

$$\psi = \frac{\alpha \circ f}{\alpha},$$

where α is measurable, then α is $\mathcal{P}^\star$-measurable, where $\mathcal{P}^\star$ is the finest partition of I so that $f(I_i)$ is included in an atom of $\mathcal{P}^\star$ for each $i \in \Lambda$.

PROPOSITION 4.5. *Let $\psi: \cup_i I_i \to \mathbb{Z}$ be a $\mathcal{P}^0$-measurable function. If $\psi = \alpha \circ f - \alpha$, where α is measurable, then α is constant on $f(I_i)$, for each $i \in \Lambda$.*

PROOF. Note that we can assume that $\alpha(x) \in \mathbb{Z}$, for every x. Indeed, the relation $\psi = \alpha \circ f - \alpha$ implies that the function $\beta(x) = \alpha(x) \mod 1$ is f-invariant, so we can replace α by $\alpha - \beta$, if necessary. Fix an irrational number γ. Then

$$e^{2\pi\gamma\psi(x)i} = \frac{e^{2\pi\gamma\alpha(f(x))i}}{e^{2\pi\gamma\alpha(x)i}},$$

so by Proposition 4.4 we have that $e^{2\pi\gamma\alpha(x)i}$ is a $\mathcal{P}^\star$-measurable function. Since $j \in \mathbb{Z} \to e^{2\pi\gamma ji} \in \mathbb{S}^1$ is one-to-one, we get that α is $\mathcal{P}^\star$-measurable. □

A Markov map f is almost onto if and only if $\mathcal{P}_0^\star = \{I\}$, so we have the following corollaries.

COROLLARY 4.6. *Under the conditions of Proposition 4.5, if f is almost onto, then α is constant.*

COROLLARY 4.7. *For every nonconstant $\psi \in \mathcal{O}(f)$ we have that $\sigma^2 \neq 0$. In particular, the central limit theorem as given in Eq. (4.1) holds for every non-constant ψ.*

Let $\mathcal{A}_0 \subset \mathcal{A}_1 \subset \mathcal{A}_2 \subset \cdots$ be an increasing sequence of σ-subalgebras of a probability space $(\Omega, \mathcal{A}, \mu)$. A **martingale difference sequence** is a sequence of functions $\psi_n: \Omega \to \mathbb{R}$, where ψ_n is $\mathcal{A}_n$-measurable for $n \geq 1$, so that

$$\mathbb{E}(\psi_n | \mathcal{A}_{n-1}) = 0$$

for every n. Here $\mathbb{E}(\psi|\mathcal{B})$ denotes de conditional expectation of ψ relative to the sub-algebra $\mathcal{B}$. When $\mathcal{B}$ is generated by atoms $\{J_i\}_i$ then $\mathbb{E}(\psi|\mathcal{B})$ is the function defined as

$$\mathbb{E}(\psi|\mathcal{B})(x) = \frac{1}{\mu(J_i)} \int_{J_i} \psi \, d\mu$$

for every $x \in J_i$.

The following proposition is the classic Azuma-Hoeffding inequality: see, for instance, Exercise E14.2 in [W].

PROPOSITION 4.8 (Azuma-Hoeffding Inequality). *Let ψ_n be a martingale difference sequence and furthermore assume that $\|\psi_i\|_\infty = c_i < \infty$. Define*

$$\psi := \sum_{i=1}^{n} \psi_i.$$

Then

$$\mu\big(x \in \Omega\colon |\psi - \mathbb{E}(\psi)| > t\big) \leq 2 \exp\left(-\frac{t^2}{2 \sum_{i=1}^{n} c_i^2}\right).$$

4.2   How to construct asymptotically small perturbations

As we will see in the next proposition, it is easy to construct asymptotically small perturbations of a random walk.

PROPOSITION 4.9. *Let F and G be random walks satisfying the Properties LI, Ex, sBD, Ra and Rb, where G is a topological perturbation of F. Assume that there exist $C > 0$ and $\lambda \in (0,1)$ with the following properties: if I_j^n is as in Properties Ra and Rb, then*

i. *for every $I_j^n \in \mathcal{P}_n^0$ we have*

$$\left| \log \frac{|I_{j+1}^n|}{|I_j^n|} \frac{|H(I_j^n)|}{|H(I_{j+1}^n)|} \right| \leq C\lambda^{|n|+|j|};$$

ii. *for every $J \in \mathcal{P}_n^0$ we have*

$$\left| \tau_J^F - \tau_{H(J)}^G \right|_{C^2} \leq C\lambda^{|n|};$$

iii. *if $I_i^n = [a_i^n, b_i^n]$, then*

$$\max_i \max\{|a_i^n - H(a_i^n)|, |b_i^n - H(b_i^n)|\} \leq C\lambda^{|n|};$$

iv. *either ψ is a bounded funtion or ψ has a lower bound and $F = G$ on $\bigcup_{n<0} I_n$.*

Then G is an asymptotically small perturbation of F. Furthermore, there exist $\beta \in [0,1)$ and $C > 0$ so that

$$|H(p) - p| \leq C\beta^{|\pi_2(p)|}.$$

PROOF. We will assume that ψ is bounded: the other case is analogous. Consider $(x, n) \in I \times \mathbb{Z}$ and $(y, n) = H(x, n)$. Denote $(x_i, n_i) := F^i(x, n)$, $(y_i, n_i) := G^i(y, n)$. Let $\delta_i = |y_i - x_i|$ and $\tilde{\delta}_i = |A_{G(H(J_{i-1}))}(y_i) - A_{F(J_{i-1})}(x_i)|$. Here $(x_i, n_i) \in J_i \in \mathcal{P}^0$. It is easy to conclude, using (iii) and Property LI, that

$$\tilde{\delta}_i \leq \frac{\delta_i}{|F(J_{i-1})|} + C\lambda^{|n_i|} \tag{4.2}$$

and, making use of (ii), to get

$$\left| \tau^G_{H(J_{i-1})}(A_{G(H(J_{i-1}))}(y_i)) - \tau^F_{J_{i-1}}(A_{F(J_{i-1})}(x_i)) \right| \leq D\tau^F_{J_{i-1}}(z_i) \frac{\delta_i}{|F(J_{i-1})|} + C\lambda^{|n_i|}.$$

Here $z_i \in [0, 1]$. Since $D\tau^F_{J_{i-1}}(z_i)|F(J_{i-1})|/|J_{i-1}| \leq \lambda$ (Property Ex), we get, using (iii) again,

$$\delta_{i-1} \leq \lambda \delta_i + C\lambda^{|n_i|}. \tag{4.3}$$

Because ψ is bounded, $|n_{i+1} - n_i| \leq B = \max |\psi|$. This means that if $i < n/2B$ then $|n_i| > |n_0|/2$. Since $\delta_{[\frac{n}{2B}]} \leq 1$, Eq. (4.3) implies

$$|H(x, n) - (x, n)| = |y_0 - x_0| \leq C\lambda^{|n|/2}. \tag{4.4}$$

In particular, by Eq. (4.2) and Property (ii), we have

$$\left| D\tau^G_{H(J_0)}(A_{G(H(J_0))}(y_1)) - D\tau^F_J(A_{F(J_0)}(x_1)) \right| \leq C\lambda^{|n|/2}. \tag{4.5}$$

By $Ra + Rb$ there exists $\theta \in (0, 1)$ so that

$$\theta^{|i|} \leq |I^n_i|. \tag{4.6}$$

Let i be so that $J = I^n_i$.
Case A. $|i| \geq |n/2|(\log \lambda / \log \theta)$: Due to (i) and (iii) and Property Ra, there exists $C > 0$ so that

$$\left| \log \frac{|H(I^n_i)|}{|I^n_i|} \right| \leq C\lambda^n.$$

Combining this with $sBD + LI$ and (iii), we can conclude that for every $p \in I^n_i$ with $|i| \geq |n/2|(\log \lambda / \log \theta)$, we have

$$\left| \log \frac{DG(H(p))}{DF(p)} \right| \leq C\lambda^{(|n| \log \lambda)/(2 \log \theta)}.$$

Case B. $|i| < |n/2|(\log \lambda / \log \theta)$: In this case, by (iii) and Eq. (4.6) we have

$$\log \frac{|H(I^n_i)|}{|I^n_i|} \leq C \frac{|H(b^n_i) - b^n_i| + |H(a^n_i) - a^n_i|}{|b^n_i - a^n_i|} \leq C\lambda^{|n|/2}.$$

Now using Eq. (4.4) and Eq. (4.5) we can easily obtain

$$\left| \log \frac{DG(H(p))}{DF(p)} \right| \leq C\lambda^{|n|/2}.$$

$\square$

5 STABILITY OF TRANSIENCE

We will begin this section with the large deviations result to transient homogeneous random walks and strongly transient random walks:

PROOF OF PROPOSITION 2.1. Let $P \in \mathcal{P}^0(F)$ be such that $F(P) = I_\ell$. By Proposition 4.1 we have that for every $\epsilon > 0$, there exist $\gamma < 1$ such that

$$m\left(p \in I_\ell\colon \ \pi_2(F^{n-1}p) - \pi_2(p) < (K - \epsilon/2)\,(n-1)\right) \leq C\gamma^{n-1},$$

for every ℓ. By Property BD we have

$$m\left(p \in P\colon \ \pi_2(F^n p) - \pi_2(F(p)) < (K - \epsilon/2)\,(n-1)\right) \leq C\gamma^{n-1}|P|. \tag{5.1}$$

Denote by Λ_P^n the set in the left-hand side of Eq. (5.1). Now let n_0 be such that $\min \psi > -\epsilon(n_0 - 1)/2 - \epsilon + K$. Then for $n \geq n_0$ we have

$$\tilde{\Lambda}_P^n := \left\{p \in P\colon \ \pi_2(F^n p) - \pi_2(p) < (K - \epsilon)\,n\right\} \subset \Lambda_P^n. \tag{5.2}$$

Indeed,

$$\pi_2(F^n p) - \pi_2(p) < (K - \epsilon)n$$

implies

$$\pi_2(F^n p) - \pi_2(F(p)) < (K - \epsilon/2)\,(n-1).$$

So

$$m\left(p \in P\colon \ \pi_2(F^n p) - \pi_2(p) < (K - \epsilon)\,n\right) \leq C_2 \gamma^n |P|$$

for every n. This completes the proof.   □

PROOF OF PROPOSITION 2.2. Fix $\epsilon > 0$ small. We intend to apply the Azuma-Hoeffding inequality, but since ψ is not necessarily bounded, we need to make some adjustments first: fix $P \in \mathcal{P}^0(F)$ and define

$$\mathcal{F}_0 := \{P\} \quad \text{and} \quad \mathcal{F}_n := \{Q\}_{Q \subset P,\ Q \in \mathcal{P}^n(F)}.$$

Since $F \in GD$, by the usual distortion control tricks for F, we can find $M > \min \psi$ such that $\alpha(x) := \min\{\psi(x), M\}$ satisfies

$$\mathbb{E}(\alpha \circ F^n | \mathcal{F}_{n-1}) \geq K - \frac{\epsilon}{4} \tag{5.3}$$

for every $n \geq 1$. Here we are considering conditional expectations relative to the probability

$$\mu_P(A) := \frac{m(A)}{|P|},$$

where m is the Lebesgue measure.

Define the martingale difference sequence

$$\Psi_n := \alpha \circ F^n - \mathbb{E}(\alpha \circ F^n | \mathcal{F}_{n-1}).$$

Of course $\|\Psi_n\|_\infty \leq M$, if M is large enough. By the Azuma-Hoeffding inequality we have

$$m\left(p \in P: \left|\sum_{i=1}^{n} \Psi_i(p)\right| > t\right) \le 2\exp\left(-\frac{t^2}{2nM^2}\right)|P|.$$

Taking $t = \epsilon n/4$ we obtain

$$m\left(p \in P: \left|\sum_{i=1}^{n} \Psi_i(p)\right| > \frac{\epsilon}{4} n\right) \le 2\exp\left(-\frac{\epsilon^2 n}{32M^2}\right)|P|. \tag{5.4}$$

Since

$$\pi_2(F^{n+1}p) - \pi_2(F(p)) = \sum_{i=1}^{n} \psi(F^i(p))$$

$$\ge \sum_{i=1}^{n} \alpha(F^i(p)) = \sum_{i=1}^{n} \Psi_i(p) + \sum_{i=1}^{n} \mathbb{E}(\alpha \circ F^i | \mathcal{F}_{i-1})(x)$$

$$\ge \sum_{i=1}^{n} \Psi_i(p) + \left(K - \frac{\epsilon}{4}\right)n.$$

Due to Eq. (5.4), this implies that

$$m\left(p \in P: \ \pi_2(F^n p) - \pi_2(F(p)) = \sum_{i=1}^{n-1} \psi(F^i(p))\right)$$

$$< \left(K - \frac{\epsilon}{2}\right)(n-1) \le C_1 \exp\left(-\frac{\epsilon^2 n}{32M^2}\right)|P|.$$

Let n_0 be such that $\min \psi > -\epsilon(n_0 - 1)/2 - \epsilon + K$. Then for $n \ge n_0$ we have that

$$\pi_2(F^n p) - \pi_2(p) < (K - \epsilon)n$$

implies

$$\pi_2(F^n p) - \pi_2(F(p)) < \left(K - \frac{\epsilon}{2}\right)(n-1).$$

So

$$m\left(p \in P: \ \pi_2(F^n p) - \pi_2(p) < (K - \epsilon)\, n\right) \le C_2 \exp\left(-\frac{\epsilon^2 n}{32M^2}\right)|P|$$

for every n. This completes the proof. $\qquad\square$

PROPOSITION 5.1. *Let F be either a homogeneous random walk with positive mean drift or a strongly transient random walk. Then any asymptotically small perturbation G of F has the following property: there exists $\lambda \in [0,1)$, $C > 0$ and $\tilde{K} > 0$ so that for every $P \in \mathcal{P}^0(G)$*

$$m\left(p \in P: \ \sum_{i=0}^{n-1} \psi(G^i(p)) < \tilde{K}n\right) \le C\lambda^n |P|.$$

In particular, G is also transient.

PROOF. We will carry out the proof assuming the strong transience: the homogeneous case with positive mean drift is analogous: Fix $\epsilon > 0$. Let $\tilde{\delta}_1 > 0$ be small enough such that

$$(1 - \tilde{\delta}_1)(K - \epsilon) + \tilde{\delta}_1 \min \psi > K - 2\epsilon.$$

Due the bounded distortion of G, there exists $\delta_1 > 0$ such that for every $n \geq 1$ and every $P \in \mathcal{P}^{n-1}(G)$, interval $Q \subset G^n(P)$, and set $A \subset Q$ satisfying

$$\frac{m(A)}{m(Q)} \geq 1 - \delta_1,$$

we have

$$\frac{m(P \cap G^{-n}A)}{m(P \cap G^{-n}Q)} \geq 1 - \tilde{\delta}_1. \tag{5.5}$$

By Proposition 2.2 we have

$$m\left(p \in P\colon \sum_{i=0}^{n-1} \psi(F^i(p)) < (K - \epsilon)n \text{ for some } n \geq n_0\right) \leq C_1 \exp(-C_2 n_0)|P|, \tag{5.6}$$

for every $P \in \mathcal{P}_j^0(F)$. Since G is an asymptotically small perturbation, Eq. (2.2) implies that

$$m\left(p \in H(P)\colon \sum_{i=0}^{n-1} \psi(G^i(p)) < (K - \epsilon)n \text{ for some } n \geq n_0\right) \leq C_3 \exp(-C_4 n_0)|H(P)| \tag{5.7}$$

provided that $P \in \mathcal{P}_j^0(F)$, $j \geq 2 \, |\min \psi| \, n_0$. Indeed, the set in the l.h.s. of Eq. (5.6) can the written as the pairwise disjoint union of the sets Δ_j, $j \geq n_0$, where Δ_j is defined as

$$\Delta_j = \left\{ p \in P\colon \sum_{i=0}^{n-1} \psi(F^i(p)) \geq (K - \epsilon)n \ \text{ for all } n_0 \leq n < k \right.$$
$$\left. \text{ and } \ \sum_{i=0}^{k-1} \psi(F^i(p)) < (K - \epsilon)k \right\}.$$

So by Eq. (2.2) and Eq. (2.3) we have that

$$\text{dist}_k(p) \leq C n_0 \lambda^{|\min \psi| \, n_0} + \sum_{i=n_0+1}^{\infty} C \lambda^{(K-\epsilon)i} \leq \tilde{C} < \infty$$

for every $p \in \Delta_k$, $k \geq n_0$, and $j \geq 2 \, |\min \psi| \, n_0$. In particular,

$$m(H(\Delta_k)) \leq \tilde{C} m(\Delta_k). \tag{5.8}$$

Note that the set in the l.h.s. of Eq. (5.7) is the pairwise disjoint union of $H(D_j)$. Since $P \in \mathcal{P}_j^0$ we have $m(P) \leq C m(H(P))$, so from Eq. (5.8) we obtain Eq. (5.7).

In particular, there exists $n_0 = n_0(\delta_1)$ such that for every $P \in \mathcal{P}_j^0(G)$ and all $j \geq 2 \, |\min \psi| \, n_0$, we have

$$m(\tilde{\Omega}_P) \geq (1 - \delta_1)|P|, \tag{5.9}$$

where $\tilde{\Omega}_P$ is the set of points $p \in P$ such that $\pi_2(G^n(p)) \geq |\min \psi| n_0$ for all $n \geq 0$ and $\pi_2(G^n(p)) - \pi_2(p) \geq (K - \epsilon)n$ for all $n \geq n_0$.

By the GD condition, there exists n_1 such that for $n \geq n_1$ we have

$$m\left(p \in P: \text{ there exists } i \leq n \text{ such that } \psi(F^i(p)) \geq n\right) \leq \frac{\delta_1}{4}.$$

By Eq (5.6) there exists $n_2 > n_1$ such that

$$m\left(p \in P: \sum_{i=0}^{n_2-1} \psi(F^i(p)) > (K - \epsilon)n_2\right) \geq \left(1 - \frac{\delta_1}{4}\right)|P|. \tag{5.10}$$

So

$$m\left(p \in P: \sum_{i=0}^{n_2-1} \psi(F^i(p)) > (K - \epsilon)n_2 \text{ and } \psi(F^i(p)) < n_2 \text{ for every } i \leq n_2\right)$$
$$\geq \left(1 - \frac{\delta_1}{2}\right)|P|. \tag{5.11}$$

Note that for p in the set in Eq. (5.11) we have $\pi_2(G^i(p)) - \pi_2(p) \leq (n_2)^2$ for every $i \leq n_2$. Since G is an asymptotically small perturbation of F, this observation and Eq. (5.11) implies that there exists $n_3 \gg (n_2)^2$ such that for $P \in \mathcal{P}_j^0(G)$, with $j \leq -n_3$, we have

$$m\left(p \in P: \sum_{i=0}^{n_2-1} \psi(G^i(p)) > (K - \epsilon)n_2 \text{ and } \psi(G^i(p)) < n_2 \text{ for every } i \leq n_2\right)$$
$$\geq (1 - \delta_1)|P|. \tag{5.12}$$

So for $P \in \mathcal{P}_j^0(G)$, with $j \leq -n_3$, we have

$$m\left(p \in P: \sum_{i=0}^{n_2-1} \psi(G^i(p)) > (K - \epsilon)n_2\right) \geq (1 - \delta_1)|P|. \tag{5.13}$$

Claim A: Almost every point $x \in I \times \{j\}, j \leq -n_3$, visits at least once (and consequently infinitely many times) the set

$$\bigcup_{j \geq -n_3} I \times \{j\}. \tag{5.14}$$

Indeed, define a new random walk $\tilde{G}: I \times \mathbb{Z} \to I \times \mathbb{Z}$,

$$\tilde{G}(x, n) := (\tilde{g}_n(x), n + \tilde{\psi}(x, n)),$$

in the following way. Let T be an integer larger than $n_2(K - \epsilon)$. If $n \geq -n_3$, then define $\tilde{g}_n: I \to I$ as an affine expanding map, onto on each element of $\mathcal{P}_n^{n_2}$, and $\tilde{\psi}(x, n) = T$.

For (x, n), with $n < -n_3$, define $\tilde{G}(x, n) = G^{n_2}(x, n)$. In this case

$$\tilde{\psi}(x, n) = \sum_{i=0}^{n_2-1} \psi(G^i(x, n)).$$

It is not difficult to see that if the $\tilde{G}$-orbit of a point (x, n), with $n < -n_3$, visits the set in Eq. (5.14) at least once, then the G-orbit of (x, n) visits the same set at least once.

To prove the claim, it is enough to show that $\tilde{G}$ is strongly transient. Indeed, let P be an element of the Markov partition $\mathcal{P}_j^{k-1}(\tilde{G})$. If $\pi_2(\tilde{G}^i(P)) \geq -n_3$, for some $i \leq k$ then $\pi_2(\tilde{G}^k(P)) \geq -n_3$, so

$$\frac{1}{|P|} \int_P \tilde{\psi} \circ \tilde{G}^k \, dm = \frac{1}{|P|} \int_P T \, dm \geq (K - \epsilon)n_2. \tag{5.15}$$

Otherwise, $\pi_2(\tilde{G}^i(P)) < -n_3$ for every $i \leq k$. In particular, $\tilde{G}^i = G^{in_2}$ on P, for every $i \leq k$. Note that

$$\tilde{G}^k P = \bigcup_i Q_i,$$

where $\{Q_i\}_i$ is the family of all interval Q such that $Q \in \mathcal{P}_j^0(G)$ for some $j < -n_3$ and $Q \cap \tilde{G}^k P \neq \varnothing$ (this is a consequence of the Markovian property of G). By Eq. (5.13) we have

$$m\left(q \in Q_i \colon \tilde{\psi}(q) \geq (K - \epsilon)n_2 \right) \geq (1 - \delta_1)|Q_i|,$$

so by the distortion control in Eq. (5.5), we obtain

$$m\left(p \in P \cap \tilde{G}^{-k}Q_i \colon \tilde{\psi}(\tilde{G}^k p) \geq (K - \epsilon)n_2 \right) \geq (1 - \tilde{\delta}_1)|P \cap \tilde{G}^{-k}Q_i|;$$

consequently,

$$\begin{aligned}
\int_P \tilde{\psi} \circ \tilde{G}^k \, dm &= \sum_i \int_{P \cap \tilde{G}^{-k}Q_i} \tilde{\psi} \circ \tilde{G}^k \, dm \\
&\geq \sum_i ((1 - \tilde{\delta}_1)(K - \epsilon)n_2 + \tilde{\delta}_1 n_2 \min \psi)|P \cap \tilde{G}^{-k}Q_i| \\
&\geq \sum_i (K - 2\epsilon)n_2|P \cap \tilde{G}^{-k}Q_i| \\
&= (K - 2\epsilon)n_2|P|.
\end{aligned} \tag{5.16}$$

Equations (5.15) and (5.16) imply that $\tilde{G}$ is strongly transient, so by Proposition 2.2, $\tilde{G}$ is transient. This concludes the proof of the claim.

Claim B: The G-orbit of almost every point of $I \times \mathbb{Z}$ eventually arrives at $\tilde{\Omega}_P$, for some $P \in \mathcal{P}_j^0$, with $j > 2|\min \psi|n_0$.

Since F is transient and G is topologically conjugate to F, the set

$$\Omega := \{p \colon -n_3 \leq \pi_2(p) \leq 2|\min \psi|n_0 \text{ and } \lim_n \pi_2(G^n(p)) = +\infty\}$$

is dense on

$$\bigcup_{j=-n_3}^{2|\min \psi|n_0} I \times \{j\}.$$

This implies that for every non-empty open set $O \subset I_j$, with $-n_3 \leq j \leq 2|\min \psi|n_0$, we have

$$m\Big((x,j) \in O: \exists\, k \geq 0 \text{ s.t. } G^k(x,j) \in \tilde{\Omega}_P, \text{ with } P \in \mathcal{P}_q^0(G), q > 2|\min \psi|n_0\Big) > 0,$$
(5.17)

where $\tilde{\Omega}_P$ is as in Eq. (5.9). Indeed, pick a point $p \in O \cap \Omega$. By property Ex and the definition of Ω, there exists k and $Q \in \mathcal{P}_j^k(G)$ such that $Q \subset O$, $P = G^k(Q) \in \mathcal{P}_q^0$, with $q > 2|\min \psi|n_0$. By Eq. (5.9) we have $m(\tilde{\Omega}_P) > 0$, so

$$m(O \cap G^{-k}\tilde{\Omega}_P) \geq m(Q \cap G^{-k}\tilde{\Omega}_P) > 0.$$

In particular, there exists $\tilde{\delta} > 0$ such that for every interval $J \subset I_j$, with $-n_3 \leq j \leq 2|\min \psi|n_0$ and $|J| \geq \delta$, where δ is as in the LI property, we have

$$m\Big((x,j) \in J: \exists\, k \geq 0 \text{ s.t. } G^k(x,j) \in \tilde{\Omega}_P, \text{ with } P \in \mathcal{P}_q^0(G), q > 2|\min \psi|n_0\Big) > \tilde{\delta}|J|,$$
(5.18)

It follows that there exists $\delta_3 > 0$ such that for every i and every $Q \in \mathcal{P}^{i-1}(G)$ such that $\pi_2(G^i Q) \geq -n_3$, we have that

$$m\Big(p \in Q: \exists k \geq 0 \text{ s.t. } G^k p \in \tilde{\Omega}_P, \text{ with } P \in \mathcal{P}_q^0(G), q > 2|\min \psi|n_0\Big) \geq \delta_3 |Q|.$$
(5.19)

Indeed, if $\pi_2(G^i Q) \leq 2|\min \psi|n_0$ we can apply Eq. (5.18), and the BD and LI properties. Otherwise, apply Eq. (5.9) and the BD property.

We will show Claim B by contradiction. Suppose that it does not hold. Then there is a set W of positive measure whose G-orbit of its elements never hits $\tilde{\Omega}_P$ for any $P \in \mathcal{P}_j^0$, with $j > 2|\min \psi|n_0$. Pick a Lebesgue density point p of W whose G-orbit visits

$$\bigcup_{j \geq -n_3} I \times \{j\}$$

infinitely many times, which is possible due Claim A. In particular there exists a sequence $Q_k \in \mathcal{P}^{n_k-1}(G)$ such that $|Q_k| \to_n 0$, $p \in Q_k$, $\pi_2(G^{n_k} Q_k) \geq -n_3$ and

$$\lim_k \frac{m(Q_k \cap W)}{|Q_k|} = 1.$$

That contradicts Eq. (5.19). This concludes the proof of Claim B.

Note that Claim B implies the following: almost every point in $I \times \{j\}$ belongs to the set

$$\Lambda_j := \bigcup_{k \geq 0} \Lambda_j^k,$$

where

$$\Lambda_j^k := \Big\{p \in I \times \{j\}: \pi_2(G^n(p)) - \pi_2(G^k(p)) \geq (K - \epsilon)(n - k) \text{ for all } n \geq k + n_0\Big\}$$

Let k_0 be large enough such that for every $-n_3 \leq j \leq 2|\min \psi|n_0$, we have

$$m(A \cap \bigcup_{k \leq k_0} \Lambda_j^k) \geq (1 - \delta_1)|A|$$

for every interval $A \subset I \times \{j\}$ satisfying $|A| \geq \delta$, where $\delta > 0$ is as in the LI property. Pick n_4 satisfying $n_4 \geq k_0 + n_0$ and

$$n_4 > \frac{-k_0 \min \psi}{\epsilon} - k_0.$$

It is easy to see that if $p \in \bigcup_{k \leq k_0} \Lambda_j^k$, then

$$\pi_2(G^{n_4} p) - \pi_2(p) = \sum_{i=0}^{n_4 - 1} \psi(G^i p) \geq (K - 2\epsilon) n_4.$$

In a argument similar to the proof of Claim A, consider the random walk $\hat{G}$ which is defined in the following way: if $\pi_2(p) \leq -n_3$, define $\hat{G}(p) = G^{n_2}$. If $\pi_2(p) \geq 2|\min \psi| n_0$ define $\hat{G}(p) = G^{n_0}$. Finally if $-n_3 < \pi_2(p) < 2|\min \psi| n_0$, define $\hat{G}(p) = G^{n_4}$. The random walk $\hat{G}$ is $3\hat{K}$-strongly transient, for some $\hat{K} > 0$. The proof is quite similar to the proof of the strong transience of $\tilde{G}$, so we leave it to the reader. So $\hat{G}$ is transient. It is easy to see that this implies that G is transient. Finally, Proposition 2.2 implies that

$$m(p \in P \colon \ \pi_2(\hat{G}^n(p)) - \pi_2(p) < 2\hat{K}n) \leq C\hat{\lambda}^n |P|,$$

for some $\hat{\lambda} \in (0, 1)$, which implies

$$m(Y_P^n) \leq C\hat{\lambda}^n |P|,$$

where

$$Y_P^n := \{p \in P \colon \ \exists \, m \geq n \ s.t. \ \pi_2(\hat{G}^m(p)) - \pi_2(p) < 2\hat{K}m\}.$$

Let $n_5 = \max\{n_0, n_4, n_2\}$. Let $p \in P$ be such that

$$\pi_2(G^i(p)) - \pi_2(p) < \frac{\hat{K}}{n_5} i.$$

There exists m and j such that $\hat{G}^m(p) = G^j(p)$, with $i \geq j$, $|i - j| \leq n_5$. Note that

$$m \leq i \leq j + n_5 \leq (m + 1) n_5,$$

so we can find i_0 such that for every $i \geq i_0$ we have

$$\frac{-n_5 \min \psi}{m} + \hat{K} \frac{m + 1}{m} < 2\hat{K}.$$

So

$$\pi_2(\hat{G}^m(p)) - \pi_2(p) = \pi_2(G^j(p)) - \pi_2(G^i(p)) + \pi_2(G^i(p)) - \pi_2(p)$$

$$\leq -n_5 \min \psi + \frac{\hat{K}}{n_5} i \leq -n_5 \min \psi + \hat{K}(m + 1) < 2\hat{K}m,$$

where $m \geq \frac{i}{n_5} - 1$. This implies

$$\left\{ p \in P \colon \pi_2(G^i(p)) - \pi_2(p) < \frac{\hat{K}}{n_5} i \right\} \subset Y_P^{(i/n_5) - 1},$$

and so

$$m\left(p \in P \colon \pi_2(G^i(p)) - \pi_2(p) < \frac{\hat{K}}{n_5} i \right) \leq C\hat{\lambda}^{i/n_5} |P|.$$

This completes the proof.   □

Let $n > 0$ and j be integers and F be a deterministic random walk. Then any connected component C of $F^{-n} \int I_j$ is called a **cylinder**. It follows from the Markovian property of F that a cylinder is a disjoint union of intervals in $\mathcal{P}^{n-1}$. The **length** $\ell(C)$ of the cylinder C is n. If C is a cylinder of length n so that $F^i(C) \subset I_{j_i}$, for $i < n$, we will denote $C = C(j_0, j_1, \ldots, j_n)$.

PROPOSITION 5.2. *Let $F = (\{f_i\}, \psi) \in Mk + LBD + LI + Ex + BD$. Assume that there exists $\epsilon > 0$ so that for $K > 0$, we have*

$$m\left(\left\{p \in I_n : \psi(p) < -K\right\}\right) \leq \frac{1}{K^{2+\epsilon}}, \tag{5.20}$$

provided $n \geq n_0$. Then

$$\lim_k m\left(\left\{p \in I_{n_k} : \text{ there exists } i \leq k^2 \text{ so that } \psi(F^i(p)) < -k\right\}\right) = 0, \tag{5.21}$$

uniformly for all sequences satisfying $n_k > k^3 + n_0$.

PROOF. For each k and $i \leq k^2$, denote

$$\Lambda_{n_k}^i = \{p \in I_{n_k} : \ \psi(F^j(p)) \geq -k \text{ for every } j < i \text{ and } \psi(F^i(p)) < -k\}.$$

The set in the l.h.s. of Eq. (5.21) is the union of the sets $\Lambda_{n_k}^i$. The interval I_{n_k} is the union of the cylinders in $\mathcal{P}_{n_k}^{i-1}$. Let $Q \in \mathcal{P}_{n_k}^{i-1}$ and suppose that $Q \cap \Lambda_{n_k}^i \neq \emptyset$. Then $\pi_2(F^i(Q)) \geq n_0$. By the LI property and Eq. (5.20), we get

$$\frac{m(p \in F^i(Q) : \ \psi(p) < -k)}{m(F^i(Q))} \leq C \frac{1}{k^{2+\epsilon}}.$$

By the BD property

$$m(Q \cap \Lambda_{n_k}^i) \leq \frac{C}{k^{2+\epsilon}} m(Q).$$

As a consequence,

$$m(I_{n_k} \cap \Lambda_{n_k}^i) \leq \frac{C}{k^{2+\epsilon}}.$$

So

$$m(I_{n_k} \cap \cup_{i \leq k^2} \Lambda_{n_k}^i) \leq \frac{C}{k^{\epsilon}}.$$

$\square$

REMARK 5.3. For a homogeneous random walk, the condition on ψ is equivalent to $1_{I_0} \cdot \psi \in L^{2+\epsilon}(m)$.

Let F and G be random walks which are topologically conjugated by a homeomorphism h that preserves states. For any $p \in I \times \mathbb{Z}$, define

$$C_p := \sup_{i \geq 0} \text{dist}_i(p).$$

For each $n_0 \in \mathbb{Z} \cup \{-\infty\}$, define

$$\Omega_{n_0+}(F) := \{p : \pi_2(F^n(p)) \geq n_0, \text{ for all } n \geq n_0\}.$$

In particular, $\Omega_{-\infty+}(F) = I \times \mathbb{Z}$.

PROPOSITION 5.4. *Let F and G be random walks which are conjugated by a homeomorphism h which preserves states. Suppose that there exists a F-forward invariant set Λ so that*

(H1) $C_p := \sup_{i \geq 0} dist_i(p) < \infty$, *for each $p \in \Lambda$.*

Then we have that h is absolutely continuous on $\cup_i F^{-i}\Lambda$ and h^{-1} is absolutely continuous on $\cup_i G^{-i}h(\Lambda)$.

Furthermore, if also

(H2) *there exists $C > 0$, $M > 0$ and $n_0 \in \mathbb{Z} \cup \{-\infty\}$ so that for every $n \geq n_0$ with $n \in \mathbb{Z}$ and $P \in \mathcal{P}_n^0$, we have $m(p \in P \cap \Lambda\colon C_p \leq C) \geq M|P|$,*

then h is absolutely continuous on $\cup_i F^{-i}(\Omega_{n_0+}(F))$ and h^{-1} is absolutely continuous on $\cup_i G^{-i}(\Omega_{n_0+}(G))$. In particular, when $n_0 = -\infty$ we have that h and h^{-1} are absolutely continuous on $I \times \mathbb{Z}$.

PROOF. For each $j \in \mathbb{N}$ denote

$$\Lambda_j := \{p \in \Lambda\colon \sup_i dist_i(p) \leq j\}.$$

Note that Λ_i is forward invariant.

We claim that h is absolutely continuous on Λ_j and h^{-1} is absolutely continuous on $h(\Lambda_j)$. Indeed, for each $p \in \Lambda_j$ and $k \in \mathbb{N}$, denote $F^k p = (x_k, n_k)$. Denote by $J_k(x) \in \mathcal{P}^k$ the unique interval which contains x so that F^k maps $J_k(x)$ diffeomorphically onto $Q_k \subset I_{n_k}$. There is some ambiguity here if x is in the boundary of $J_k(x)$, but these points are countable, so they are irrelevant for us.

If we use the analogous notation to $h(x)$ and G, we have $h(J_k(x)) = J_k(h(x))$ and, due the $BD + LI$ property of the random walks F and G, there exist $C_1, C_2 > 0$ such that

$$C_1 e^{-dist_k(p)} \leq \frac{|h(J_k(x))|}{|J_k(x)|} \leq C_2 e^{dist_k(p)}. \tag{5.22}$$

So, if $p \in \Lambda_j$ then

$$C_1 e^{-j} \leq \frac{|h(J_k(x))|}{|J_k(x)|} \leq C_2 e^j, \quad \text{for all } k \in \mathbb{N}. \tag{5.23}$$

Let $A \subset \Lambda_j$ be a set with positive Lebesgue measure. We claim that $h(A)$ also has positive Lebesgue measure. Indeed, choose a compact set $K \subset A$ with positive Lebesgue measure. Denote $U_k := \cup_{x \in K} J_k(x)$. Since $|J_k(x)| \leq \lambda^k$, we have that $\lim_k m(U_k) = m(K)$ and $\lim_k m(h(U_k)) = m(h(K))$. Since U_k is a countable disjoint union of intervals of the type $J_k(x)$, by Eq. (5.23)

$$C_1 e^{-j} \leq \frac{m(h(U_k))}{m(U_k)} \leq C_2 e^j, \quad \text{so} \quad C_1 e^{-j} \leq \frac{m(h(K))}{m(K)} \leq C_2 e^j, \tag{5.24}$$

and we conclude that $h(K)$ also has positive Lebesgue measure. An identical argument shows that, if $A \in \Lambda_j$ has positive Lebesgue measure, then $h^{-1}A$ also has positive Lebesgue measure. The proof of the claim is finished and so h and h^{-1} are absolutely continuous on $\Lambda = \cup_j \Lambda_j$ and $h(\Lambda) = \cup_j h(\Lambda_j)$.

Now it is easy to conclude that h and h^{-1} are absolutely continuous on $\cup_i F^{-i}\Lambda$ and $\cup_i G^{-i}h(\Lambda)$.

Now assume **H2**. We claim $\cup_i F^{-i}\Lambda$ has full Lebesgue measure on $\Omega_{n_0+}(F)$. Indeed, Assume that $m(\Omega_{n_0+}(F) \setminus \cup_i F^{-i}\Lambda) > 0$ and choose a Lebesgue density point p of this set. Then

$$\lim_k \frac{m(J_k(p) \cap \Omega_{n_0+}(F) \setminus \cup_i F^{-i}\Lambda)}{|J_k(x)|} = 1.$$

Due the bounded distortion of F, if $F^k(p) = (x_k, n_k)$ and $F^k(J_k(x)) = Q_k \subset I_{n_k}$, with $n_k \geq n_0$, where Q_k is a union of intervals in $\mathcal{P}^0_{n_k}$, then

$$\limsup_k \frac{m(Q_k \cap \Lambda)}{|Q_k|} \leq C\Big(1 - \liminf_k \frac{m(J_k(x) \cap \Omega_{n_0+}(F) \setminus \cup_i F^{-i}\Lambda)}{|J_k(x)|}\Big) = 0,$$

which contradicts **H2**.

Since on the set $\{p \in P \cap \Lambda:\ C_p \leq C\}$, $\text{dist}_k(p)$ is uniformly bounded with respect to k and p, we can use an argument identical to the proof of Eq. (5.24) to conclude that

$$\frac{m(p \in P \cap \Lambda:\ C_p \leq C)}{m\big(h(p) \in h(P) \cap h(\Lambda):\ C_p \leq C\big)} \leq C_1,$$

so $m(h(P \cap \Lambda:\ C_p \leq C)) \geq \tilde{C}M|h(P)|$, for all $P \in \mathcal{P}^0_n$, $n \geq n_o$ and using an argument as before, we conclude that $\cup_i G^{-i}h(\Lambda)$ has full Lebesgue measure on $\Omega_{n_0+}(G)$. Since $h\ (h^{-1})$ is absolutely continuous on $\cup_i F^{-i}\Lambda\ (\cup_i G^{-i}h(\Lambda))$ and

$$m\big(\Omega_{n_0+}(F) \setminus \cup_i F^{-i}\Lambda\big) = m\big(h(\Omega_{n_0+}(F) \setminus \cup_i F^{-i}\Lambda)\big)$$
$$= m\big(\Omega_{n_0+}(G) \setminus \cup_i G^{-i}h(\Lambda)\big) = 0,$$

we have that h and h^{-1} are absolutely continuous on $\Omega_{n_0+}(F)$ and $\Omega_{n_0+}(G)$. Now it is easy to prove that h is absolutely continuous on $\cup_i F^{-i}\Omega_{n_0+}(F)$ and h^{-1} is absolutely continuous on $\cup_i G^{-i}\Omega_{n_0+}(G)$. $\square$

PROOF OF THEOREM 3.1. By Proposition 5.1, G is transient. In particular, for all $n_0 \in \mathbb{Z}$ the sets

$$\cup_i F^{-i}\Omega_{n_0+}(F) \text{ and } \cup_i G^{-i}\Omega_{n_0+}(G)$$

have full Lebesgue measure. So by Proposition 5.4, to prove that h and h^{-1} are absolutely continuous, it is enough to find a forward invariant set satisfying the assumptions **H1** and **H2** for some $n_0 \in \mathbb{Z}$. Indeed, fix $\delta > 0$ (we will choose δ later). Consider the F-forward invariant set

$$\Lambda = \Lambda_\delta := \left\{ p:\ \liminf_k \frac{\pi_2(F^k(p)) - \pi_2(p)}{k} \geq \frac{\delta}{3} \right\}.$$

We claim that Λ satisfies **H1**. Indeed take $p \in \Lambda$. Then, for $k \geq k_0(p)$ we have $n_k := \pi_2(F^k(p)) \geq k\delta/4$. So

$$
\begin{aligned}
\text{dist}_k(p) &\leq \sum_{i=0}^{k-1} \left| \log \frac{DF(F^{i+1}(p))}{DG(h(F^{i+1}(p)))} \right| \\
&\leq \sum_{i=0}^{k_0-1} \left| \log \frac{DF(F^{i+1}(p))}{DG(h(F^{i+1}(p)))} \right| + \sum_{i=k_0}^{k-1} \left| \log \frac{DF(F^{i+1}(p))}{DG(h(F^{i+1}(p)))} \right| \\
&\leq \sum_{i=0}^{k_0-1} \left| \log \frac{DF(F^{i+1}(p))}{DG(h(F^{i+1}(p)))} \right| + \sum_{i=k_0}^{k-1} \lambda^{n_i} \\
&\leq \sum_{i=0}^{k_0-1} \left| \log \frac{DF(F^{i+1}(p))}{DG(h(F^{i+1}(p)))} \right| + \sum_{i=k_0}^{\infty} \lambda^{i\delta/4} \\
&\leq K_p + C(\delta).
\end{aligned}
\tag{5.25}
$$

Here λ is as in Eq. (2.2) and Eq. (2.3). To prove that Λ satisfies **H2**, by Proposition 2.2 for each $P \in \mathcal{P}_i^0$ we have

$$
m(p \in P : \ \pi_2(F^k(p)) - \pi_2(p) < \delta k) \leq C\lambda^k |P|,
\tag{5.26}
$$

provided δ is small enough. From Eq. (5.26) we obtain

$$
\mu(p \in P : \ \pi_2(F^n(p)) - \pi_2(p) \geq \delta n \text{ for all } n \geq n_0) \geq (1 - C\lambda^{n_0})|P|.
\tag{5.27}
$$

In particular, we have that, for every n,

$$
\pi_2(F^n(p)) \geq \delta(n - n_0) + \pi_2(p) + n_0 \min \psi.
\tag{5.28}
$$

in the set in Eq. (5.27). Using the same argument as in Eq. (5.25) we can easily obtain **H2** from Eq. (5.28) and Eq. (5.27), choosing n_0 large enough. $\qquad \square$

PROOF OF THEOREM 3.2. Since the mean drift is positive, by the Birkhoff ergodic theorem F is transient. By Proposition 5.1, G is also transient. Now the proof goes exactly as the Theorem 3.1, except that to obtain Eq. (5.26) we use Proposition 2.1 instead of Proposition 2.2. $\qquad \square$

PROOF OF THEOREM 3.3. Let F be either a K-strongly recurrent random walk or a homogeneous random walk with mean drift $K = \int \psi \, dm$. Let $\epsilon < K$. By Proposition 5.1 there exists $\theta < 1$ such that for every i we have

$$
m\left(p \in I_i : \ \frac{\pi_2(F^k(p)) - \pi_2(p)}{k} \leq \epsilon \right) \leq C\theta^k.
\tag{5.29}
$$

Using an argument as in the proof of Theorem 3.1, we can conclude that

$$
m\left(p \in I_i : \ \frac{\pi_2(F^k(p)) - \pi_2(p)}{k} \geq \epsilon \text{ for } k \geq k_0 \right) \geq 1 - C\theta^{k_0}
\tag{5.30}
$$

for every $i \geq 0$. By Theorem 3.1 and Theorem 3.2 the conjugacy h is absolutely continuous. Let $\delta = \sup_p \text{dist}_1(p)$. Then Eq. (5.30) implies that there exist $C > 0$ such that

$$
m(p \in I_i : \ \text{dist}_k(p) \geq \delta n + C \text{ for some } k) \leq C\theta^n,
\tag{5.31}
$$

for $i \geq 0$. Denote $\Lambda_1 := \{p \in I_i : h'(p) \leq 1\}$ and, for $n \geq 1$,

$$\Lambda_n := \{p \in I_i : e^{\delta(n-1)} < h'(p) \leq e^{\delta n}\}.$$

By Eq. (5.31) we have $m(\Lambda_n) \leq C\theta^n$. Indeed, Let $J_k(p)$ be as in the proof of Proposition 5.4. By the Lebesgue differentiation theorem, for almost every p we have

$$\lim_k \frac{|h(J_k(p))|}{|J_k(p)|} = h'(p).$$

On the other hand, by Eq. (5.22) we have that for almost every $p \in I_i$ outside the set in Eq. (5.31)

$$C_1 e^{-(n\delta+C)} \leq \frac{|h(J_k(p))|}{|J_k(p)|} \leq C_2 e^{n\delta+C},$$

so

$$C_1 e^{-(n\delta+C)} \leq h'(p) \leq C_2 e^{n\delta+C}, \tag{5.32}$$

in a subset of I_i with measure that is larger than $1 - C\theta^n$. Of course, this implies $m(\Lambda_n) \leq C\theta^n$. Let $B \subset I_i$ be an arbitrary Lebesgue measurable set. Let k_1 be so that

$$\theta^{k_1+1} < |B| \leq \theta^{k_1}.$$

First we prove Theorem 3.3 assuming that $e^{\delta}\theta < 1$. Since h is absolutely continuous we have

$$\begin{aligned}
|h(B)| &= \int_B h' \, dm \\
&= \sum_{n=0}^{k_1} \int_{B \cap \Lambda_n} h' \, dm + \sum_{n=k_1+1}^{\infty} \int_{B \cap \Lambda_n} h' \, dm \\
&\leq \sum_{n=0}^{k_1} \theta^{k_1} e^{\delta n} + \sum_{n=k_1+1}^{\infty} C(e^{\delta}\theta)^n \\
&\leq C(e^{\delta}\theta)^{k_1} \leq C|B|^{1+\delta/\ln\theta}.
\end{aligned}$$

Now if $B \subset J \in \mathcal{P}^n$ and $F^n(J) = Q \subset I_i$, with $|Q|, |h(Q)| \geq C$ (due to Property LI for F and G), then due to the bounded distortion of F and G

$$\frac{|h(B)|}{|h(J)|} \leq C\frac{|h(F^n(B))|}{|h(Q)|} \leq C\left(\frac{|F^n(B)|}{|Q|}\right)^{1+\delta/\ln\theta} \leq C\left(\frac{|B|}{|J|}\right)^{1+\delta/\ln\theta}.$$

To prove a similar inequality to h^{-1}, define

$$\tilde{\Lambda}_n := \{p \in I_i : e^{\delta(n-1)} < (h^{-1})'(p) \leq e^{\delta n}\}.$$

Of course,

$$h^{-1}\tilde{\Lambda}_n = \{p \in I_i : e^{-\delta n} < h'(p) \leq e^{-\delta(n-1)}\},$$

so by Eq. (5.31) and Eq. (5.32) we obtain

$$m(h^{-1}\tilde{\Lambda}_n) \leq C\theta^n.$$

In particular,

$$m(\tilde{\Lambda}_n) = \int_{h^{-1}\tilde{\Lambda}_n} h'(x)\,dm \le C(e^{-\delta}\theta)^n.$$

Note that this argument gives us an exponential upper bound even if δ is large.

Now we can switch the roles of F and G to obtain the inequality to h^{-1}, which shows that h is a mSQS-homeomorphism relative to the stochastic basis $\cup_n \mathcal{P}^n$.

To complete the proof when $e^{\delta}\theta \ge 1$ we do the following: find a continuous path of random walks F_t with $F_0 = F$ and $F_1 = G$ so that for every $t \in [0,1]$ we have that F_t is an asymptotically small perturbation of F and moreover there exist $\epsilon > 0$ and $\theta < 1$ such that Eq. (5.29) holds for every random walk in this family. Using the compactness of $[0,1]$ we can find a finite sequence of random walks $F_{t_0} = F, F_{t_1}, F_{t_2}, \ldots, F_{t_n} = G$ so that F_{t_i} and $F_{t_{i+1}}$ are conjugated by a map H_i such that

$$\delta_i := \sup_p \left| \frac{DF_{t_{i+1}}(H_i(p))}{DF_{t_i}(p)} \right|$$

satisfy $e^{\delta_i}\theta < 1$. So the conjugacy H_i is mSQS with respect to some dynamically defined stochastic basis. Composing these conjugacies we find a mSQS-conjugacy between F and G.   □

6 STABILITY OF RECURRENCE

Let $F = (f, \psi)$ be a homogeneous random walk and let G be an asymptotically small perturbation of F. To avoid a cumbersome notation, in this section we make the convention that all inequalities holds only for large n. Moreover, in this section we assume that ψ is unbounded. Recall that in this case we assume that asymptotically small perturbations G coincide with F on negative states. The case where ψ is bounded is similar.

The following is an easy consequence of the central limit theorem for Birkhoff sums (Proposition 4.3).

COROLLARY 6.1. *Let a_n be a positive increasing sequence. Then*

$$\mu\left(\frac{|S_n|}{\sqrt{n}} > a_n\right) \le Ce^{-\sigma^2 a_n^2/2} + C\frac{1}{\sqrt{n}}.$$

Here

$$S_n(x) = \sum_{k=0}^{n-1} \psi(f^k(x)).$$

PROOF. Use Proposition 4.3 and and note that the estimate

$$\int_{-\infty}^{v} e^{-\frac{u^2}{2}}\,du \le Ce^{-v^2/2}$$

holds for $v \ll 0$.   □

Given $n \in \mathbb{N}$, split $[0, 2n] \cap \mathbb{N}$ in $\sqrt{\log n}$ blocks (called main blocks), denoted B_j, with length

$$\frac{n}{\log^{8j} n}, \ j = 1, \ldots, \sqrt{\log n},$$

and between the main blocks we put little blocks H_j, called holes, of length $\log^4 n$. These holes will warranty the independence between the events in distinct main blocks. Put these blocks in the following order:

$$\cdots < B_{j+1} < H_{j+1} < B_j < H_j < \cdots ,$$

with $\min B_{\sqrt{\log n}} = 0$. Note that we left most of the second half of the interval $[0, 2n] \cap \mathbb{N}$ uncovered.

Define

$$S(j) = \sum_{i \in B_j} \psi \circ f^i \quad \text{and} \quad H(j) = \sum_{i \in H_j} \psi \circ f^i .$$

Denote $|B_j| := \max B_j - \min B_j$.

LEMMA 6.2. *We have*

$$\mu\left(\sum_{i=0}^{|B_j|} \psi \circ f^i \geq \frac{\sqrt{n}}{\log^{4j} n} \log^3 n\right) \leq C\frac{\log^{4j} n}{\sqrt{n}} .$$

PROOF. This follows from Corollary 6.1. $\qquad\qquad\square$

PROPOSITION 6.3. *For every $\epsilon > 0$ we have*

$$\mu\left(S(j) > \frac{\sqrt{n}}{\log^{4j} n} \log^3 n, \text{ for some } j \leq \sqrt{\log n}\right) \leq C\frac{1}{\sqrt[2+\epsilon]{n}},$$

provided n is large enough.

PROOF. For $j \leq \sqrt{\log n}$ define

$$\Lambda_j := \left\{x \in I : \ S(j)(x) > \frac{\sqrt{n}}{\log^{4j} n} \log^3 n\right\}$$

$$= \left\{x \in I : \ \sum_{i < |B_j|} \psi \circ f^{i + \min B_j}(x) > \frac{\sqrt{n}}{\log^{4j} n} \log^3 n\right\}$$

and for each $P \in \mathcal{P}^{\min B_j}$ denote $\Lambda_j(P) := \Lambda_j \cap P$.

Due to Lemma 6.2 and the bounded distortion of $f^{\min B_j}$ on P, we have

$$m\left(\Lambda_j(P)\right) \leq C\frac{\log^{4j} n}{\sqrt{n}} |P| .$$

Summing on j and P

$$m\left(\bigcup_j \bigcup_P \Lambda_j(P)\right) \leq \sqrt{\log n} \, \frac{\log^{4j} n}{\sqrt{n}} \ll C\frac{1}{\sqrt[2+\epsilon]{n}} .$$

$\qquad\qquad\square$

PROPOSITION 6.4. *For every $\epsilon > 0$ and $d > 0$ we have*

$$\mu\left(\left|\sum_{i\in H_j} \psi(f^i(x))\right| > \log^8 n, \text{ for some } j \leq \sqrt{\log n}\right) \leq C\frac{1}{n^d}, \qquad (6.1)$$

provided n is large enough.

PROOF. For $i \in H_j - 1$, with $j \leq \sqrt{\log n}$, define

$$\Lambda_i^j := \{x \in I\colon |\psi(f^i(x))| > \log^4 n.\}.$$

By expanding and bounded distortion properties of f and condition GD, we have that

$$\mu(\Lambda_i^j) \leq C\lambda^{\log^4 n}.$$

Since $|H_j| = \log^4 n$, if x belongs to the set in Eq. (6.1), then $x \in \Lambda_i^j$, for some $i \in H_j - 1$, with $j \leq \sqrt{\log n}$. So

$$\mu\left(\left|\sum_{i\in H_j} \psi(f^i(x))\right| > \log^8 n, \text{ for some } j \leq \sqrt{\log n}\right) \leq \mu\left(\bigcup_{j\leq\sqrt{\log n}} \bigcup_{i\in H_j-1} \Lambda_i^j\right)$$

$$\leq \sqrt{\log n} \, \log^4 n \, n^{\log \lambda \log^3 n}$$

$$\ll \frac{1}{n^d},$$

where the last inequality holds for n large enough. □

PROPOSITION 6.5 (Independence between Distant Events). *There exists $\lambda < 1$ so that the following holds: Let C_1 be a disjoint union of elements of $\mathcal{P}^{n-1}$ and let C_2 be a disjoint union of elements of $\mathcal{P}^{k-1}$. We have*

$$\mu(C_1 \cap f^{-(n+d)}C_2) = \mu(C_1)\mu(C_2)(1 + O(\lambda^d)).$$

Here $n = \ell(C_1)$.

PROOF. Let $J \in \mathcal{P}^{n-1}$. Since $F \in On$ we have $f^n(J) = I$. Define the measure $\rho(A) := \mu(f^{-n}A \cap J)/\mu(J)$. Note that by the bounded distortion property of f, we have that $\log d\rho/dm$ is uniformly α-Hölder, that is,

$$\left|\log \frac{d\rho}{dm}(x) - \log \frac{d\rho}{dm}(y)\right| \leq C|x - y|^\alpha, \qquad (6.2)$$

where C and α do not depend on n and C_1. Furthermore, it is bounded above by a constant which does not depend on n. By the well-know theory of Ruelle-Perron-Frobenius operators for Markov expanding maps (see, for instance, [V]), if P is the Perron-Frobenius-Ruelle operator of f, then there exists $\lambda < 1$ so that

$$P^d \frac{d\rho}{dm} = (1 + O(\lambda^d))\frac{d\mu}{dm}.$$

So

$$\frac{\mu(J \cap f^{-(n+d)}C_2)}{\mu(J)} = \rho(f^{-d}C_2) = \int 1_{C_2} \circ f^d \frac{d\rho}{dm} dm$$

$$= \int 1_{C_2} P^d \frac{d\rho}{dm} dm$$

$$= (1 + O(\lambda^d)) \int 1_{C_2} \frac{d\mu}{dm} dm$$

$$= (1 + O(\lambda^d))\mu(C_2).$$

The constant λ is the contraction of the Ruelle-Perron-Frobenious operator in certain cone of positive functions and whose logarithm is α-Hölder continuous (see [V]). Since all functions $\log \frac{d\rho}{dm}$ belong to the very same cone (due to Eq. (6.2)), λ does not depend on C_1. Since C_1 is a disjoint union of intervals $J \in \mathcal{P}^{n-1}$, we have finished the proof. $\qquad\square$

COROLLARY 6.6. *There exists $M > 0$ so that*

$$\mu\left(S_j < \frac{\sqrt{n}}{\log^{4j} n} M \text{ for all } j \le \sqrt{\log n}\right) \le C \left(\frac{2}{3}\right)^{\sqrt{\log n}}.$$

PROOF. Choose $M > 0$ so that

$$\frac{1}{\sqrt{2\pi}} \int_{-\infty}^{M} e^{-u^2/2} du < \frac{2}{3}.$$

Consider

$$C_j := \left\{x \text{ s.t.} \sum_{i=0}^{|B_j|} \psi \circ f^i(x) < \frac{\sqrt{n}}{\log^{4j} n} M\right\}.$$

Note that C_j is a the disjoint union of elements of $\mathcal{P}^{|B_j|-1}$. The central limit theorem tells us that if n is large enough, then

$$\mu(C_j) < \frac{2}{3}$$

for every $j \le \sqrt{\log n}$.

Recall that between B_j and B_{j+1}, there is a hole H_{j+1} with length $\log^4 n$. Denote

$$\Lambda_j := \bigcap_{i=1}^{j} f^{-\sum_{k=i+1}^{j}(|B_k|+|H_k|)} C_i.$$

Note that Λ_j is a disjoint union of elements of $\mathcal{P}^{|B_1|+\sum_{k=2}^{j}(|B_k|+|H_k|)-1}$ and

$$\Lambda_j = C_j \cap f^{-|B_j|-|H_j|}\Lambda_{j-1}.$$

Moreover,

$$\Lambda_{\sqrt{\log n}} = \left\{x \text{ s.t. } S_j < \frac{\sqrt{n}}{\log^{4j} n} M \text{ for all } j \le \sqrt{\log n}\right\}. \tag{6.3}$$

By Proposition 6.5 , we obtain

$$\mu(\Lambda_j) = \left(1 + O(\lambda^{|H_j|})\right) \mu(C_j)\mu(\Lambda_{j-1}).$$

So by Eq. (6.3),

$$\mu\left(S_j < \frac{\sqrt{n}}{\log^{4j} n} \; M \text{ for all } j \leq \sqrt{\log n}\right) \leq \left(\frac{2}{3}\right)^{\sqrt{\log n}} \left(1 + O(\lambda^{\log^4 n})\right)^{\sqrt{\log n}}$$

$$\leq C \left(\frac{2}{3}\right)^{\sqrt{\log n}}.$$

□

PROPOSITION 6.7. *There exists $C > 0$ so that*

$$\mu\left(x \in I: \text{ there exists } i < \ell^3 \text{ so that } \sum_{k=0}^{i} \psi \circ f^k(x) > \frac{\ell}{2}\right) \geq 1 - C\left(\frac{2}{3}\right)^{\sqrt{3\log \ell}}$$

PROOF. Let M be as in Corollary 6.6. Denote $n = \ell^3$ and define

$$\mathcal{A}_\ell := \left\{x: \text{ there exists } i < \ell^3 \text{ so that } \sum_{k=0}^{i} \psi \circ f^k(x) > \frac{\ell}{2}\right\},$$

$$\mathcal{B}_\ell := \left\{x: |S_j| < \frac{\sqrt{n}}{\log^{4j} n} \log^3 n, \text{ for all } j \leq \sqrt{\log n}\right\},$$

$$\mathcal{C}_\ell := \left\{x: S_j \geq \frac{\sqrt{n}}{\log^{4j} n} M, \text{ for some } j \leq \sqrt{\log n}\right\},$$

$$\mathcal{D}_\ell := \left\{x: |H_j(x)| \leq \log^8 n, \text{ for all } j \leq \sqrt{\log n}\right\}.$$

We claim that if ℓ is large, then $\mathcal{B}_\ell \cap \mathcal{C}_\ell \cap \mathcal{D}_\ell \subset \mathcal{A}_\ell$. Indeed, let $x \in \mathcal{B}_\ell \cap \mathcal{C}_\ell \cap \mathcal{D}_\ell$. Then for some $j_0 \leq \sqrt{\log n}$,

$$S_{j_0}(x) \geq \frac{\sqrt{n}}{\log^{4j_0} n} \; M.$$

We claim that, if $m = max \, B_{j_0}$, then

$$\sum_{0}^{m} \psi \circ f^i(x) > \frac{\ell}{2}.$$

Indeed, since $x \in \mathcal{D}_\ell$,

$$\left| \sum_{i \in H_j, \, j > j_0} \psi \circ f^i(x) \right| \leq \sqrt{\log n} \log^8 n = o(\ell).$$

Moreover, since $x \in \mathcal{B}_\ell$,

$$\left| \sum_{i \in B_j, \, j > j_0} \psi \circ f^i(x) \right| \leq \sum_{j > j_0} \frac{\sqrt{n}}{\log^{4j} n} \log^3 n \leq C \frac{\sqrt{n}}{\log^{4j_0+1} n}.$$

So

$$\sum_0^m \psi \circ f^i(x) = \sum_{i \in B_{j_0}} \psi \circ f^i(x) + \sum_{i \in B_j,\, j > j_0} \psi \circ f^i(x) + \sum_{i \in H_j,\, j > j_0} \psi \circ f^i(x)$$

$$\geq \left(M - \frac{C}{\log n} \right) \frac{\sqrt{n}}{\log^{4j_0} n} + o(\ell) > C\ell^{\frac{6}{5}} - o(\ell) > \frac{\ell}{2},$$

and we finished the proof of the claim. To finish the proof, note that by Proposition 6.3, Corollary 6.6, and Proposition 6.4,

$$\mu(\mathcal{A}_\ell) \geq \mu(\mathcal{B}_\ell \cap \mathcal{C}_\ell \cap \mathcal{D}_\ell) \geq 1 - C\frac{1}{\sqrt[2+\epsilon]{n}} - C\left(\frac{2}{3}\right)^{\sqrt{\log n}} - C\frac{1}{n^d} \geq 1 - C\left(\frac{2}{3}\right)^{\sqrt{\log n}}.$$

$\square$

Let $C > 0$ and $\lambda \in (0,1)$ be as in Eq. (2.2) and Eq. (2.3). Define

$$\mathrm{Dist}_n(p) := \sum_{i=0}^{n-1} C\lambda^{\pi_2(F^i p)}.$$

Of couse $\mathrm{dist}_n(p) \leq \mathrm{Dist}_n(p)$.

PROPOSITION 6.8. *There exist ϵ and D so that for every $\ell \geq 0$,*

$$\mu\left(\left\{ p \in I_\ell : \text{ there exists } i \text{ so that } F^i(p) \in \bigcup_{t \in [\min \psi, -\min \psi]} I_t \text{ and } \mathrm{dist}_i(p) \leq D \right\} \right) \geq \epsilon.$$

PROOF. For $\ell \geq 0$ and k, define B_k^ℓ as the set of all $p \in I_\ell$ such that there exists

$$j \leq \sum_{i=0}^{k-1} \frac{\ell^3}{2^{3i}}$$

satisfying

$$\pi_2(F^j(p)) \leq \frac{\ell}{2^k} \text{ and } \mathrm{Dist}_j(p) \leq \sum_{i=0}^{k-1} C\frac{\ell^3}{2^{3i}} \lambda^{\ell/2^i + \min \psi}.$$

We are going to prove by ascending induction on $k \geq 0$ that there is $C > 0$ so that for every $\ell \geq 0$, we have

$$\mu(B_k^\ell) \geq \prod_{i=0}^{k-1} \left(1 - C\left(\frac{2}{3}\right)^{\sqrt{\log(\ell/2^i)}} \right), \tag{6.4}$$

for all $k \geq 1$ and $\mu(B_0^\ell) = 1$.

Note that $B_0^\ell = I_\ell$, so $\mu(B_0^\ell) = 1$. Now assume the induction hypothesis for some $k \geq 0$. Take $p \in B_k^\ell$. Let $p \in L = C(i_0, i_1, \ldots, i_{j-1})$, where j is the smallest integer as in the definition of B_k^ℓ. In particular

$$\frac{\ell}{2^k} + \min \psi \leq \pi_2(F^j(p)) \leq \frac{\ell}{2^k}. \tag{6.5}$$

Note that $L \subset B_k^\ell$ and $F^j(L) = I_r$, with $r := \pi_2(F^j(p))$. Applying Proposition 6.7 to $-\psi$ we get

$$\mu\left(x \in I_r\colon \text{ there exists } i < \frac{\ell^3}{2^{3k}} \text{ so that } \sum_{n=0}^{i} \psi \circ f^n(x) < -\frac{\ell}{2^{k+1}}\right) \quad (6.6)$$

$$\geq 1 - C\left(\frac{2}{3}\right)^{\sqrt{\log(\ell/2^k)}}.$$

Denote

$$D_L := \left\{x \in L\colon \text{ there exists } i < \frac{\ell^3}{2^{3k}} \text{ so that } \sum_{n=0}^{i} \psi \circ f^n(f^j(x)) < -\frac{\ell}{2^{k+1}}\right\}$$

Due to the bounded distortion property for F, the estimate in Eq. (6.6) implies

$$\frac{\mu(D_L)}{|L|} \geq 1 - C\left(\frac{2}{3}\right)^{\sqrt{\log(\ell/2^k)}}. \quad (6.7)$$

We claim that $D_L \subset B_{k+1}^\ell$. Indeed, let $x \in D_L$. Take the smallest i so that

$$\sum_{n=0}^{i} \psi \circ f^n(f^j(x)) < -\ell/2^{k+1}.$$

Then by Eq. (6.5) we have $\pi_2(F^{j+h}(p)) \geq \frac{\ell}{2^{k+1}} + \min \psi$, for every $0 \leq h < i$, so

$$\mathrm{Dist}_i(F^j(p)) \leq \sum_{h=0}^{i} C\lambda^{\pi_2(F^{j+h}(p))} \leq C\frac{\ell^3}{2^{3k}}\lambda^{\frac{\ell}{2^{k+1}}+\min \psi}.$$

So $D_L \subset B_{k+1}^\ell$. Since B_k^ℓ is a disjoint union of cylinders L, the estimate in Eq. (6.7) implies that Eq. (6.4) holds, replacing k by $k+1$. This concludes the induction step.

Define

$$D := \sum_{i=0}^{\infty} C\frac{\ell^3}{2^{3i}}\lambda^{\ell/2^i+\min \psi} < \infty.$$

Let k be so that $2^k \leq \ell \leq 2^{k+1}$. Now it is easy to check that

$$\mu\left(\left\{x \in I_\ell\colon \text{ there exists } i \text{ so that } F^i(p) \in I_0 \text{ and } \mathrm{dist}_i(p) \leq D\right\}\right)$$

$$\geq C\mu(B_k^\ell) \geq \prod_{i=0}^{k-1}\left(1 - C\left(\frac{2}{3}\right)^{\sqrt{\log(\ell/2^i)}}\right) \geq C\prod_{i=0}^{k-1}\left(1 - C\left(\frac{2}{3}\right)^{\sqrt{\log(2^k/2^i)}}\right)$$

$$\geq \exp\left(-C\sum_{i=1}^{\infty}\left(\frac{2}{3}\right)^{\sqrt{i \log 2}}\right) > \tilde{C} > 0,$$

which finishes the proof.   □

PROOF OF THE STABILITY OF RECURRENCE (THEOREM 3.4). Since F is recurrent, its mean drift is zero. By Corollary 4.7 we can apply the central limit theorem as in the introduction to conclude Eq. (1.2) and Eq. (1.4). Because G coincides with F on negative states, the orbit by G of almost every point p satisfying $\pi_2(p) < 0$ will enter

$$\cup_{i \geq 0} I^i.$$

As a consequence the orbit by G of almost every point p visits this set infinitely many times. Let $\ell \geq 0$.

By Proposition 6.8, there exist $D > 0$ and $\epsilon > 0$ so that

$$A_\ell := \left\{ p \in I_\ell \colon \text{ there exists } i \text{ so that } F^i(p) \in \bigcup_{t=\min \psi}^{-\min \psi} I_t \text{ and } \mathrm{Dist}_i(p) < D \right\}$$

satisfies $\mu(A_\ell) > \epsilon$, for all $\ell \geq 0$.

Consider a cylinder $C_F = C_F(\ell, k_1, \ldots, k_{i-1}, k_i) \subset A_\ell$, with $C_F \neq \varnothing$, satisfying $|k_j| > -\min \psi$ for $0 < j < i$, $\min \psi \leq k_i \leq -\min \psi$ and $\mathrm{Dist}_i(x) < D$, for every $x \in C_F$. We claim that the corresponding cylinder $C_G = C_G(\ell, k_1, \ldots, k_{i-1}, k_i)$ for the perturbed random walk G satisfies

$$\frac{1}{C} \leq \frac{|C_G|}{|C_F|} \leq C,$$

where C depends only on D. Because we used $\mathrm{Dist}_i(p)$ instead of $\mathrm{dist}_i(p)$ in the definition of A_ℓ, the set A_ℓ is a disjoint union of cylinders of this type, so we obtain that $B_\ell = H(A_\ell)$ satisfies $m(B_\ell) > C\epsilon > 0$, for all $\ell \geq 0$.

To prove that the set of points whose orbits returns infinitely many times to

$$\bigcup_{t=\min \psi}^{-\min \psi} I_t$$

has full Lebesgue measure, it is enough to prove that $\Lambda := \cup_{j \geq 0, \ell} G^{-j} B_\ell$ has full Lebesgue measure.

Indeed, assume by contradiction that Λ is not full. Choose a Lebesgue density point p of the complement of Λ and also satisfying $\limsup_k \pi_2(G^k(p)) \geq 0$. Then there exist a sequence of cylinders $C_k \in \mathcal{P}^{k-1}$ so that $p \in C_k$ and

$$\frac{m(C_k \setminus \Lambda)}{|C_k|} \to_k 1. \tag{6.8}$$

But $G^k(C_k) = I_{\ell_k}$, with $\ell_k = \pi_2(G^k(C_k))$, and $m(I_{\ell_k} \cap B_{\ell_k}) \geq C\epsilon |I_{\ell_k}|$. By the bounded distortion property,

$$\frac{m(\Lambda \cap C_k)}{|C_k|} > \frac{m(G^{-k} B_{\ell_k} \cap C_k)}{|C_k|} > \tilde{C}\epsilon,$$

which contradicts Eq. (6.8). Now we can use that G is transitive and has bounded distortion to prove that G is recurrent. $\square$

PROOF OF PROPOSITION 3.5. Since F is recurrent, almost every point of I^0 returns to I^0 at least once. So the first return map $R_F \colon I^0 \to I^0$ is defined almost everywhere is I^0, and the same can be said about R_G. Of course, the absolutely continuous conjugacy H also conjugates the expanding Markovian maps R_F and R_G. Using the same argument used in Shub and Sullivan [ShSu] and Martens and de Melo [MdM], we can prove that H is actually C^1 on I^0. Using the dynamics, it is easy to prove that H is C^1 everywhere. $\square$

7   STABILITY OF THE MULTIFRACTAL SPECTRUM

7.1   Dynamically defined intervals and root cylinders

When we are dealing with Markov expanding maps with *finite* Markov partitions, for each arbitrary interval J we can find an element of $\cup_j \mathcal{P}^j$ which covers J and has more or less the same size that J. Note that this is no longer true when the Markov partition is infinite. Since coverings by intervals are crucial in the study of the Hausdorff dimension of a one-dimensional set, this trick is very useful to estimate the dimension of dynamically defined sets, once we can replace an arbitrary covering by intervals by another one with essentially the same metric properties but whose elements are themselves *dynamically defined* sets (cylinders).

Consider $j \geq 0$ and let $\{C_i\}_{i \in \Theta} \subset \mathcal{P}^j$ be a finite or countable family of cylinders such that $W := \bigcup_i \overline{C_i}$ is connected, $W \subset J \in \mathcal{P}^{j-1}$ and $F^j(int\, W)$ does not contain any point d_i^n (as defined in property Rb). Then W is called a dynamically defined interval (dd-interval, for short) of level j. Define the root cylinder of W as the unique cylinder C_{i_0} with the following property: if $\#\Theta = \infty$, then W is a semi-open interval and C_{i_0} will be the cylinder so that $\partial C_{i_0} \cap \partial W \neq \emptyset$. Otherwise W is closed, and let C_{i_0} be the unique cylinder such that $F = \partial C_{i_0} \cap \partial W$ is the boundary of a semi-open dd-interval which contains W. The following lemmas are an easy consequence of the regularity properties $Ra + Rb$ and it will be useful to recover the trick described above for (certain) infinite Markov partitions. The proof is very simple.

LEMMA 7.1. *For every $d \in (0,1)$ there exists $K > 1$ so that for every dd-interval $W := \cup_i \overline{C_i}$ with root cylinder C_{i_0} we have*

$$\frac{1}{K} \leq \frac{|W|^\alpha}{\sum_i |C_i|^\alpha} \leq K \tag{7.1}$$

$$\frac{1}{K} \leq \frac{|C_{i_0}|^\alpha}{\sum_i |C_i|^\alpha} \leq K \tag{7.2}$$

for every $1 \geq \alpha \geq d$. Indeed the constant K depends only on d and constants in the properties $Ra + Rb + Ex + BD$.

PROOF. Due to Property Ra, we can enumerate C_i in such way that C_0 is the root cylinder of W and $\partial C_{i+1} \cap \partial C_i \neq \emptyset$. Moreover if j is the level of W, then $W \subset J \in \mathcal{P}^{j-1}$. Let $F^j(J) = I_n$. In particular, $F^j(C_i) \in \mathcal{P}_n^0$ and $F^j(W) = \cup_i F^j(C_i)$ is a dd-interval of level 0, with root cylinder $F^j(C_0)$ and $\partial F^j(C_{i+1}) \cap F^j(\partial C_i) \neq \emptyset$. By Property Rb we have

$$\frac{|F^j(C_i)|}{|F^j(C_0)|} \leq C\lambda^i.$$

By the BD property, we have that

$$\frac{|C_i|}{|C_0|} \leq C\lambda^i,$$

so we obtain Eq. (7.2) since

$$|C_0|^\alpha \leq \sum_i |C_i|^\alpha \leq |C_0|^\alpha \sum_{i=0}^\infty C\lambda^{di}.$$

In particular, for $\alpha = 1$ we have

$$1 \leq \frac{|W|}{|C_0|} \leq C, \tag{7.3}$$

From Eq. (7.3) and Eq. (7.2) we can easily get Eq. (7.1) for every $d \leq \alpha \leq 1$. $\qquad\square$

LEMMA 7.2. *Let N be as in Properties $Ra + Rb$. For every $d \in (0,1)$ there exists $K > 1$ so that the following holds: For every interval $J \subset I \times \mathbb{Z}$ there exist m dd-intervals W_j, all of same level, with $m \leq 2N$, satisfying the following properties:*

- *The interior of these dd-intervals are pairwise disjoint.*

- *The closure of the union of W_j covers J, i.e.,* $\quad J \subset \overline{\bigcup_j W_j}.$

- *We have*
$$\frac{1}{K} \leq \frac{\sum_{i=1}^m |W_i|^\alpha}{|J|^\alpha} \leq K$$

    for every $1 \geq \alpha > d$.

Indeed, the constant K depends only on d and constants in the $Ra + Rb + Ex + BD$ properties.

PROOF. Let $\mathcal{P}^{-1} = \{I_n\}_n$. Define the sequence of partitions $\mathcal{Q}^j$, $j \geq 0$, of $I \times \mathbb{Z}$ in the following way: $\mathcal{Q}^0$ is the family of the connected components of

$$I \times \mathbb{Z} \setminus \{c_i^n, d_i^n\}_{i,n}$$

and an interval Q belongs to $\mathcal{Q}^j$, $j \geq 1$, if there exists $P \in \mathcal{P}^{j-1}$ such that Q is one of the connected components of

$$P \cap F^{-j}\{c_i^n, d_i^n\}_i,$$

where $n = \pi_2(F^j P)$. Here c_i^n, d_i^n are as in Properties Ra and Rb. Note that each $P \in \mathcal{P}^{j-1}$ contains at most $2N$ intervals in $\mathcal{Q}^j$ and each $Q \in \mathcal{Q}^j$ is a dd-interval of level j. First we consider the case

$$\left\{ j \geq 0 \text{ such that } \#\{Q \in \mathcal{Q}^j : J \cap Q \neq \varnothing\} \leq 2 \right\} = \varnothing. \tag{7.4}$$

Let $n = \pi_2(J)$. Then J intersects at least three connected components of

$$I_n \setminus \{c_i^n, d_i^n\}_i,$$

so it contains one of the connected components of this set. In particular if

$$\{W_j\}_j := \{Q \in \mathcal{Q}^0 \text{ such that } Q \cap J \neq \varnothing\},$$

then by Properties $Ra + Rb$ we have $\max_j |W_j| \geq \delta$, so

$$1 \leq \frac{\sum_{i=1}^m |W_i|^\alpha}{|J|^\alpha} \leq \frac{2N}{\delta}.$$

If Eq. (7.4) does not hold, let

$$j_0 = \max\{j \geq 0 \text{ such that } \#\{Q \in \mathcal{Q}^j : J \cap Q \neq \varnothing\} \leq 2\}.$$

Let $Q, R \in \mathcal{Q}^{j_0}$ be such that $J \subset \overline{Q \cup R}$. Then

$$Q = \cup_i \overline{D_i}, \ R = \cup_i \overline{E_i},$$

with $D_i, E_i \in \mathcal{P}^{j_0}$, $\partial D_i \cap \partial D_{i+1} \neq \varnothing$, $\partial E_i \cap \partial E_{i+1} \neq \varnothing$ and D_0 and E_0 are the root intervals of Q and R. Let

$$i_Q := \min\{i \geq 0 \text{ such that } D_i \cap J \neq \varnothing\},$$
$$i_R := \min\{i \geq 0 \text{ such that } E_i \cap J \neq \varnothing\},$$
$$Q_{i_Q} = \cup_{i \geq i_Q} \overline{D_i}, \qquad R_{i_R} = \cup_{i \geq i_R} \overline{E_i}.$$

Note that Q_{i_Q} and R_{i_R} are dd-intervals of level j_0. Without lost of generality, suppose that $|D_{i_Q}| \geq |E_{i_R}|$. Then $\overline{Q_{i_Q} \cup R_{i_R}} \supset J$ is an interval. Let K be the constant given by Proposition 7.1 for $\alpha > d$. Then

$$(1 + C_1)|D_{i_Q}| \leq |D_{i_Q}| + |D_{i_Q+1}| \leq |Q_{i_Q}| \leq |Q_{i_Q}| + |R_{i_R}| \leq 2K|D_{i_Q}|, \qquad (7.5)$$

where the first inequality follows from Eq. (2.1). We have three cases.

Case 1. Suppose that $Q \neq R$ and the intervals are in the order

$$D_{i_Q} < D_{i_Q+1} < \cdots < E_{i_R+1} < E_{i_R}.$$

Then $|J| \geq C_1|D_{i_Q}|$; otherwise

$$J \subset D_{i_Q} \cup D_{i_Q+1},$$

which contradicts $J \cap E_{i_R} \neq \varnothing$. So

$$1 \leq \frac{|Q_{i_Q}|^\alpha + |R_{i_R}|^\alpha}{|J|^\alpha} \leq \frac{(4K|D_{i_Q}|)^\alpha}{C_1^\alpha |D_{i_Q}|^\alpha} \leq \frac{4K}{C_1}.$$

Case 2. Suppose that $Q \neq R$ and the intervals are in the order

$$\cdots < D_{i_Q+1} < D_{i_Q} < E_{i_R} < E_{i_R+1} < \cdots$$

Then $i_Q = i_R = 0$ and there exists $y \in \partial D_0 \cap \partial E_0$. By Properties $Ra + Rb + BD$ there exist $[d, y], [y, e] \in \mathcal{P}^{j_0+1}$, with $[d, y] \subset D_0$, $[y, e] \subset E_0$ such that

$$C_3|D_0| \leq |d - y|, |e - y| \leq C_2|D_0|.$$

since J intersects D_0 and Q_0 and at least three intervals in $\mathcal{Q}^{j_0+1}$ intersect J, we have that either $[d, y]$ or $[y, e]$ is contained on J. So $|J| \geq C_2|D_0|$. We conclude

$$1 \leq \frac{|Q_{i_Q}|^\alpha + |R_{i_R}|^\alpha}{|J|^\alpha} \leq \frac{(4K|D_{i_Q}|)^\alpha}{C_2^\alpha |D_{i_Q}|^\alpha} \leq \frac{4K}{C_2}.$$

Case 3. Suppose that $R = Q$, that is $J \subset Q_{i_Q}$. By Properties $Ra + Rb + BD$,

$$C_5|D_{i_Q}| \leq |D_{i_Q+1}| \leq C_4|D_{i_Q}|.$$

Using $Ra + Rb + BD$ again, for every interval $S \subset D_{i_Q} \cup D_{i_Q+1}$ such that $S \in \mathcal{Q}^{j_0+1}$ we have

$$|S| \geq C_6|D_{i_Q}|.$$

Since at least three intervals in $\mathcal{Q}^{j_0+1}$ intersect J, there is $S \subset D_{i_Q} \cup D_{i_Q+1}$ with $S \in \mathcal{Q}^{j_0+1}$ such that $S \subset J$. So $|J| \geq C_6|D_{i_Q}|$. We conclude that

$$1 \leq \frac{|Q_{i_Q}|^\alpha}{|J|^\alpha} \leq \frac{(4K|D_{i_Q}|)^\alpha}{C_6^\alpha |D_{i_Q}|^\alpha} \leq \frac{4K}{C_6}.$$

$\square$

7.2   Dimension of dynamically defined sets

Let $f \in Mk + BD + Ex$ and denote by $\mathcal{P}^0$ its Markov partition. Let

$$\mathcal{I} := \{C_i\}_i \subset \cup_n \mathcal{P}^n$$

be a finite or countable family of disjoint cylinders. Define the induced Markov map $f_\mathcal{I} \colon \cup_i C_i \to I$ by

$$f_\mathcal{I}(x) = f^{\ell(C_i)-1}(x), \text{ if } x \in C_i.$$

We can also define an induced drift function $\Psi \colon \cup_i C_i \to \mathbb{Z}$ in the following way: Define, for $x \in C \in \mathcal{P}_0^n$,

$$\Psi_\mathcal{I}(x) := \sum_{i=0}^{n-1} \psi(f^i(x)).$$

Under the same conditions on x, define $N_\mathcal{I}(x) = n$. The maximal invariant set of $f_\mathcal{I}$ is

$$\Lambda(\mathcal{I}) := \{x \in I \colon f^j(x) \in \bigcup_i C_i, \text{ for all } j \geq 0\}.$$

Denote by $HD(\mathcal{I})$ the Hausdorff dimension of the maximal invariant set of $f_\mathcal{I}$.
   We are going to use the following result.

PROPOSITION 7.3 (Theorem 1.1 in [MU]). *We have*

$$HD(\mathcal{J}) = \sup\{HD(\mathcal{I}) \colon \mathcal{I} \subset \mathcal{J}, \mathcal{I} \text{ finite}\}.$$

Before we give the proof of Proposition 3.7, we need to introduce some tools which are useful to estimate the Hausdorff dimension.

Let $\mathcal{I}$ be a finite collection of disjoint cylinders. Then there exists β such that

$$\sum_{C \in \mathcal{I}} |C|^\beta = 1;$$

we will call β the **virtual Hausdorff dimension** of $f_{\mathcal{I}}$, denoted $VHD(\mathcal{I})$. The virtual Hausdorff dimension is a nice way to estimate $HD(\mathcal{I})$: indeed if $f_{\mathcal{I}}$ is linear on each interval of the Markov partition, then these values coincide. When the distortion is positive, these values remain related, as expressed in the following result (which is included, for instance, in the proof of Theorem 3, Section 4.2 of [PT]).

PROPOSITION 7.4. *Let $\mathcal{I}$ be a finite family of disjoint cylinders. Then*

$$|HD(\mathcal{I}) - VHD(\mathcal{I})| \le \frac{d}{\log \lambda - d'}$$

where

$$d := \sup_{C \in \mathcal{I}} \sup_{x,y \in C} \log \frac{Df_{\mathcal{I}}(y)}{Df_{\mathcal{I}}(x)} \qquad and \qquad \lambda := \inf_{C \in \mathcal{I}} \inf_{x \in C} |Df_{\mathcal{I}}(x)|.$$

Recall that if $\mathcal{I}$ is finite, then $f_{\mathcal{I}}$ has an invariant probability measure $\mu_{\mathcal{I}}$ supported on its maximal invariant set $\Lambda(\mathcal{I})$ such that for any subset $S \subset \Lambda(\mathcal{I})$ satisfying $\mu_{\mathcal{I}}(S) = 1$, we have $HD(S) = HD(\mathcal{I})$ (see, for instance, [PU]).

Note that for a homogeneous random walk F,

$$\Omega_+^k(F) = \{k\} \times \left\{ x \in I \text{ s.t.} \sum_{i=0}^{j} \psi(f^i(x)) + k \ge 0, \text{ for } j \ge 0 \right\}, \qquad and$$

$$\Omega_{+\beta}^k(F) = \{k\} \times \left\{ x \in I \text{ s.t.} \sum_{j=0}^{n-1} \psi(f^j(x)) + k \ge 0 \text{ for all } n \ge 0 \right.$$

$$\left. and \underline{\lim}_n \frac{1}{n} \sum_{j=0}^{n-1} \psi(f^j(x)) \ge \beta \right\}.$$

Define $\pi_1(x, n) := x$. The following is an easy consequence of this observation.

LEMMA 7.5. *If F is a homogeneous random walk, then $\pi_1(\Omega_+^0(F)) \subset \pi_1(\Omega_+^k(F))$ and $\pi_1(\Omega_{+\beta}^0(F)) \subset \pi_1(\Omega_{+\beta}^k(F))$, for all $k \ge 0$. Furthermore,*

$$HD(\Omega_+^0(F)) = HD(\Omega_+^k(F)) \qquad and \qquad HD(\Omega_{+\beta}^0(F)) = HD(\Omega_{+\beta}^k(F)).$$

PROPOSITION 7.6. *Let F be a homogeneous random walk. Then there exists a sequence of finite families of cylinders*

$$\mathcal{F}_s \subset \cup_i \mathcal{P}_0^i$$

so that the following hold:

- *$\Lambda(\mathcal{F}_s) \subset \Omega_+^0(F)$,*

- *Denote $\beta_n := \int \Psi_{\mathcal{F}_s} d\mu_{\mathcal{F}_s}$. Then $\beta_n > 0$.*

- $\lim_{s \to \infty} HD(\mathcal{F}_s) = HD(\Omega^0_+(F))$.

PROOF. Denote $d = HD\,\Omega^0_+(F) \leq 1$. Given any $s \in \mathbb{N}^\star$, $m_{d_s}(\Omega_+(F)) = \infty$, where $d_s := d(1 - 1/s) < 1$. Here m_D denotes the D-dimensional Hausdorff measure. By Theorem 5.4 in [F], for each positive number M we can find a compact subset $\Lambda_s \subset \Omega^0_+(F)$ satisfying $m_{d_s}(\Lambda_s) = M$. We may assume that Λ_s does not have isolated points. We will specify M later.

In particular, for each ϵ small enough the following hold:

i. *For every* family of intervals $\{J_i\}_i$ which covers Λ_s, with $|J_i| < \epsilon$, we have

$$\frac{M}{2} \leq \sum_i |J_i|^{d_s}.$$

ii. *There exists* a family of intervals $\{J_i\}_i$, with $|J_i| \leq \epsilon$, which covers Λ_s and

$$\sum_i |J_i|^{d_s} \leq 2M.$$

Furthermore, we can assume that $\partial J_i \subset \Lambda_s$.

Assume that $d_s \geq d/2$. By Lemma 7.1 and Lemma 7.2, there exists some K such that we can replace the special covering $\{J_i\}$ in (ii) by a new covering by dd-intervals $\{W_i^\ell\}_{i,\,\ell}$, with root cylinders R_i^ℓ, where

$$J_i \cap \Lambda_s \subset \overline{\bigcup_\ell W_i^\ell}, \tag{7.6}$$

$$W_i^\ell := \bigcup_k \overline{C_k^{i\ell}}, \quad \text{for each } \ell \leq m_{i\ell} \leq 2N, \tag{7.7}$$

$$\frac{1}{K} \leq \frac{\sum_\ell |R_i^\ell|^{d_s}}{|J_i|^{d_s}} \leq K, \tag{7.8}$$

$$\frac{1}{K} \leq \frac{\sum_k |C_k^{i\ell}|^{d_s}}{|R_i^\ell|^{d_s}} \leq K. \tag{7.9}$$

Indeed, we can replace W_i^ℓ by a dd-subinterval of it, if necessary, in such way that $R_i^\ell \cap \Lambda_s \neq \phi$ and Eq. (7.6), Eq. (7.7), Eq. (7.8) and Eq. (7.9) hold, except perhaps the lower bound in Eq. (7.8), since the new root cylinder could be smaller than the original one. The above estimates, together with the fact that $\{W_i^\ell\}$ covers Λ_s (up to a countable set) gives

$$\frac{M}{2K^2} \leq \sum_{i,\ell,k} |C_k^{i\ell}|^{d_s} \leq 2K^2 M. \tag{7.10}$$

The lower bound in Eq. (7.10) follows from (i). Since these intervals are cylinders, if necessary we can replace this family of cylinders by a subfamily of disjoint cylinders which covers Λ_s up to a countable number of points and such that each cylinder intersects Λ_s. Indeed, we can choose a finite subfamily $\mathcal{F}_s := \{C_r\}_r$ satisfying

$$\frac{M}{3K^2} \leq \sum_r |C_r|^{d_s} \leq 2K^2 M. \tag{7.11}$$

Let's call this finite subfamily $\mathcal{F}_s$. Note that since $C_r \cap \Lambda_s \neq \varnothing$, we have

$$\sum_{t=0}^{\ell} \psi(f^t(x)) \geq 0$$

for every $x \in C_r$ and $\ell \leq \ell(C_r)$. If

$$\sum_{t=0}^{\ell(C_r)} \psi(f^t(x)) = 0$$

for every C_r, choose a very small cylinder $\tilde{C}$ satisfying

$$\tilde{C} \cap \bigcup_r C_r = \varnothing$$

and such that

$$\sum_{t=0}^{\ell} \psi(f^t(x)) \geq 0$$

for every $x \in \tilde{C}$ and $\ell < \ell(\tilde{C})$, and

$$\sum_{t=0}^{\ell(\tilde{C})} \psi(f^t(x)) > 0$$

on $\tilde{C}$, and moreover

$$\frac{M}{3K^2} \leq |\tilde{C}|^{d_s} + \sum_r |C_r|^{d_s} \leq 3K^2 M. \tag{7.12}$$

Add $\tilde{C}$ to the family $\mathcal{F}_s$. Then, if μ_s is the geometric invariant measure of $f_{\mathcal{F}_s}$, we have

$$\int \Psi_{\mathcal{F}_s} \, d\mu_s > 0.$$

We can find such $\tilde{C}$ because $F \in On + GD$ implies that there is at least a point x_0 such that

$$\min_{k \geq 0} \sum_{i=0}^{k} \psi(f^i(x_0)) > 0.$$

and $x_0 \notin \Lambda_s$. By Proposition 7.4 and Eq. (7.12)

$$|HD(\Lambda(f_{\mathcal{F}_s})) - d_s| \leq -\frac{C}{\log \epsilon}.$$

Since ϵ can be taken arbitrary, we can choose $\mathcal{F}_s$ such that

$$HD(\Lambda(f_{\mathcal{F}_s})) \to_s d.$$

$\square$

COROLLARY 7.7. *If F is a homogeneous random walk, we have that*

$$HD(\Omega_+(F)) = \lim_{\beta \to 0^+} HD(\Omega_{+\beta}(F)) = \sup_{\beta > 0} HD(\Omega_{+\beta}(F)).$$

PROOF. Due Lemma 7.5, it is enough to prove the corollary for $k = 0$. Of course, $\Omega^0_{+\beta}(F) \subset \Omega^0_+(F)$ and $\beta_0 \leq \beta_1$ implies $\Omega^0_{+\beta_1}(F) \subset \Omega^0_{+\beta_0}(F)$, so

$$\lim_{\beta \to 0^+} HD(\Omega^0_{+\beta}(F)) = \sup_{\beta > 0} HD(\Omega^0_{+\beta}(F)) \leq HD(\Omega^0_+(F)).$$

To obtain the opposite inequality, let $\mathcal{F}_s$ be as in Proposition 7.6. Denote

$$\gamma_s := \int \Psi_{\mathcal{F}_s} \, d\mu_{\mathcal{F}_s}, \text{ and } W_n := \int N_{\mathcal{F}_s} \, d\mu_{\mathcal{I}}$$

and $\beta_s := \gamma_s / W_s$. By the Birkhoff ergodic theorem there is subset $T_s \subset \Lambda(\mathcal{I}_n)$ such that $\mu_{\mathcal{F}_s}(T_s) = 1$ and

$$\lim_k \frac{1}{k} \sum_{i=0}^{k-1} \psi(f^i(x)) = \lim_k \frac{\sum_{j=0}^{k-1} \Psi_{\mathcal{I}_n}(f^j_{\mathcal{F}_s}(x))}{\sum_{j=0}^{k-1} N_{\mathcal{I}_n}(f^j_{\mathcal{F}_s}(x))} = \frac{\gamma_s}{W_s} = \beta_s > 0$$

for every $x \in T_s$. Since the Hausdorff dimension of $\mu_{\mathcal{F}_s}$ is equal to $HD(\mathcal{F}_s)$, we have that $HD(T_s) = HD(\mathcal{F}_s)$. Note also that

$$T_s \subset \Omega^0_{+\beta_s},$$

which implies $HD(\mathcal{F}_s) \leq HD(\Omega^0_{+\beta_s})$, so by the choice of $\mathcal{F}_s$, we conclude that

$$HD(\Omega^0_+) = \lim_s HD(\mathcal{F}_s) \leq \overline{\lim}_s HD(\Omega^0_{+\beta_s}) \leq \sup_{\beta > 0} HD(\Omega^0_{+\beta}).$$

$\square$

PROOF OF THEOREM 3.6. Define

$$\Gamma_n(F) := \left\{ x \in \Omega^k_{+\beta}(F) \text{ s.t. } \pi_2(F^i(x,k)) \geq \frac{\beta}{2} i, \text{ for all } i \geq n \right\}.$$

Of course

$$\Omega^k_{+\beta}(F) = \bigcup_n \Gamma_n(F).$$

To prove the theorem, it is enough to verify that $HD(\Gamma_n(F)) = HD(\Gamma_n(G))$. Indeed, for every $\epsilon > 0$ and $\alpha \in (HD(\Gamma_n(F)), 1)$ there exists a covering of $\Gamma_n(F)$ by intervals A_i so that

$$\sum_j |A_j|^\alpha \leq \epsilon.$$

Note that we can assume that $\partial A_j \subset \Gamma_n(F)$. Since G is an asymptotically small perturbation of F, it is easy to see that G also satisfies the properties $Ra + Rb$, replacing the points c_i^n and d_i^n by $h(c_i^n)$ and $h(d_i^n)$ and modifying the constant. Indeed, we can choose constants in the definitions of the Properties $Ex + BD + Ra + Rb$

which works for both random walks, so we can take $K > 0$ in the statements of Lemma 7.2 and Lemma 7.1 in such way that it works for both random walks.

In particular (as in the proof of Proposition 7.6), for each A_j we can find at most $2N$ dd-intervals

$$W_j^\ell := \bigcup_k \overline{C_k^{j\ell}}, \text{ with } \ell \le m_j \le 2N$$

which satisfy

$$A_i \cap \Gamma_n(F) \subset \overline{\bigcup_\ell W_i^\ell}$$

and

$$\sum_{k,\ell} |C_k^{j\ell}|^\alpha \le K|A_j|^\alpha.$$

Furthermore, we can assume that the root R_j^ℓ of W_j^ℓ satisfies

$$\frac{1}{K} \le \frac{|R_j^\ell|^\alpha}{\sum_k |C_k^{j\ell}|^\alpha} \le K \tag{7.13}$$

and $R_\ell^j \cap \Gamma_n(F) \ne \varnothing$.

The constant K does not depend on α, j, or ℓ. In particular, the union of all cylinders $C_k^{j\ell}$ covers $\Gamma_n(F)$ up to a countable set and

$$\sum_{j,k,\ell} |C_k^{j\ell}|^\alpha \le K\epsilon. \tag{7.14}$$

Note that if $x \in \Gamma_n(F)$, then

$$\text{dist}_i(x) \le r_n := Cn + C\lambda^n$$

for every $i \in \mathbb{N}$. So

$$e^{-r_n} \le \frac{|\mathcal{P}_F^i(x)|}{|\mathcal{P}_G^i(h(x))|} \le e^{r_n}.$$

There is a point in the cylinder R_j^ℓ which belongs to $\Gamma_n(F)$, so

$$e^{-\alpha r_n} \le \frac{|R_j^\ell|^\alpha}{|h(R_j^\ell)|^\alpha} \le e^{\alpha r_n}. \tag{7.15}$$

Note that $h(W_j^\ell) = \overline{\bigcup_k h(C_k^{j\ell})}$ is a dd-interval for G and $h(R_j^\ell)$ is its root cylinder. So, using Eq. (7.13)

$$\frac{1}{K} \le \frac{|h(R_j^\ell)|^\alpha}{\sum_i |h(C_i^{j\ell})|^\alpha} \le K. \tag{7.16}$$

But the union of the cylinders $h(C_k^{j\ell})$ covers $\Gamma_n(G)$ up to a countable set and so Eq. (7.13), Eq. (7.14), Eq. (7.15), and Eq. (7.16) give

$$\sum_{j,k,\ell} |h(C_k^{j\ell})|^\alpha \le K^3 e^{\alpha r_n}\epsilon.$$

Since $\alpha > HD(\Gamma_n(F))$ and ϵ is arbitrary, we obtain that $HD(\Gamma_n(G)) \leq HD(\Gamma_n(F))$. Switching the roles of F and G in the preceding argument gives the opposite inequality. □

LEMMA 7.8. *Let $G \in On + Ra + Rb$ be a random walk. For every $\alpha > 0$ there exist ϵ and C so that*

$$\sum_{P \in \mathcal{P}_\ell^n} |P|^{1-\epsilon} \leq C(1+\alpha)^n, \tag{7.17}$$

for all n and ℓ.

PROOF. Indeed, denote

$$\mathcal{P}_\ell^n = \{Q^j\}_j \text{ and } \mathcal{P}_\ell^{n+1} = \{Q_k^j\}_{j,k}, \tag{7.18}$$

in such way that $Q_k^j \subset Q_j$. To avoid cumbersome notation we are omitting explicit indexing on n and ℓ. Since $G \in BD + Ra + Rb$, it is possible to order Q_k^j so that there exist C and $\lambda < 1$ satisfying

$$\frac{|Q_k^j|}{|Q^j|} \leq C\lambda^k, \tag{7.19}$$

for every j, k, n. As a consequence the family of functions

$$h_{j,\ell,n}(\epsilon) = \sum_k \frac{|Q_k^j|^{1-\epsilon}}{|Q^j|^{1-\epsilon}}$$

is an equicontinuous set of functions in a small neighborhood of 0. In particular, since $h_{j,\ell,n}(0) = 1$, there exists ϵ_0 so that, for every $\epsilon < \epsilon_0$ and every j, ℓ, and n

$$\sum_k \frac{|Q_k^j|^{1-\epsilon}}{|Q^j|^{1-\epsilon}} \leq 1 + \alpha. \tag{7.20}$$

So

$$\sum_{P \in \mathcal{P}_\ell^{n+1}} |P|^{1-\epsilon} = \sum_{j,k} |Q_k^j|^{1-\epsilon} \leq (1+\alpha) \sum_j |Q_j|^{1-\epsilon} = (1+\alpha) \sum_{P \in \mathcal{P}_\ell^n} |P|^{1-\epsilon}.$$

□

From now on we are going to assume that $F = (f, \psi) \in On$ is a homogeneous random walk with negative mean drift and G is an asymptotically small perturbation of F.

LEMMA 7.9. *Let $G \in On + Ra + Rb$ be a random walk that is an asymptotically small perturbation of a homogeneous random walk $F \in On + Ra + Rb$ with negative mean drift. Then for every $\alpha > \int \psi \, d\mu$, there exists $C > 0$, $\sigma < 1$ so that for any $n_1 \geq n_0$, with n_0 large enough,*

$$m\{p \in I_{n_1} : \pi_2(G^k(p)) \geq n_0, \text{ for } k \leq n, \text{ and } \pi_2(G^n(p)) - n_1 \geq \alpha n\} \leq C\sigma^n. \tag{7.21}$$

PROOF. Denote

$$\Lambda_{n_0,n_1}^n(G) := \{p \in I_{n_1}: \ \pi_2(G^k(p)) \geq n_0 \text{ for all } k \leq n \text{ and } \pi_2(G^n(p)) - n_1 \geq \alpha n\}.$$

The statement for F is consequence of the large deviations estimate (see, for instance, [B]) for every $K > 0$ there exists $C_K > 0$, $\gamma_K \in (0,1)$ such that

$$m\left\{p \in I: \ \left|\frac{\sum_{k=0}^{n-1} \psi(f^k(p))}{n} - \int \psi \, d\mu\right| \geq K\right\} \leq C_K \gamma_K^n.$$

Pick $K = \alpha - \int \psi \, d\mu$ and $\tilde{\sigma} = \gamma_K$. Then for every n_1

$$m\left\{p \in I_{n_1}: \ \frac{\sum_{k=0}^{n-1} \psi(f^k(\pi_1(p)))}{n} = \pi_2(F^n(p)) - n_1 \geq \alpha n\right\} \leq C\tilde{\sigma}^n,$$

which implies (of course)

$$m(\Lambda_{n_0,n_1}^n(F)) \leq C\tilde{\sigma}^n. \tag{7.22}$$

We are going to use this estimate to obtain Eq. (7.21) for the perturbation of F.

Indeed, for every $\delta > 0$, there is n_0 so that if $\pi_2(x) \geq n_0$, then

$$1 - \delta \leq \frac{|DF(x)|}{|DG(H(x))|} \leq 1 + \delta. \tag{7.23}$$

Here H is the topological conjugacy between F and G which preserves states. Note that $\Lambda_{n_0,n_1}^n(F)$ is a disjoint union of elements $Q_i \in \mathcal{P}^n(F)$, so $\Lambda_{n_0,n_1}^n(G)$ is a disjoint union of the intervals $H(Q_i)$. Due to the BD property for F and G, Eq. (7.22), and Eq. (7.23), we have

$$m(\Lambda_{n_0,n_1}^n(G)) = \sum_i |H(Q_i)| \leq \sum_i C(1+\delta)^n |Q_i| \leq C(1+\delta)^n \tilde{\sigma}^n. \tag{7.24}$$

Choose n_0 large enough such that $\sigma := (1+\delta)\tilde{\sigma} < 1$. $\square$

We would like to replace n_0 by an arbitrary state in Eq. (7.21). The following lemma will be useful for this task:

LEMMA 7.10. *Let p_n and q_n sequences of non-negative real numbers such that*

1. $p_0 + q_0 \leq 1$,

2. *there exists $\epsilon > 0$ and $\ell \geq 1$ such that $s_n := p_n + q_n \leq (1-\epsilon)^\ell p_{n-\ell} + q_{n-\ell}$ for every $n \geq \ell$ and $q_n \leq C(1-\epsilon)^n + \sum_{k=1}^n (1-\epsilon)^k p_{n-k}$, for every n.*

Then there exists $C > 0$ and $\delta = \delta(\epsilon) > 0$ such that $s_n \leq C(1-\delta)^n$, for every $n \in \mathbb{N}$.

PROOF. If $n \geq \ell$, we have $s_n \leq (1-\epsilon)p_{n-\ell} + q_{n-\ell} = (1-\epsilon)s_{n-\ell} + \epsilon q_{n-\ell}$. It follows by induction that if $n = i\ell + r$, with $r < \ell$, then

$$s_n \leq (1-\epsilon)^i s_r + \sum_{k=0}^{i-1} \epsilon(1-\epsilon)^{k\ell} q_{n-(k+1)\ell}$$

$$\leq C(1-\epsilon)^{n/\ell} s_r + \sum_{k=0}^{n-\ell} \epsilon(1-\epsilon)^k q_{n-\ell-k}.$$

Since $q_{n-\ell} \leq C(1-\epsilon)^{n-\ell} + \sum_{k=1}^{n-\ell}(1-\epsilon)^k p_{n-\ell-k}$, we obtain

$$s_n \leq C(1-\epsilon)^{n/\ell} s_r + C\epsilon(1-\epsilon)^{n/\ell} + \sum_{k=1}^{n-\ell} \epsilon(1-\epsilon)^k (p_{n-\ell-k} + q_{n-\ell-k})$$

$$\leq (1-\epsilon)^{n/\ell} C(s_r + \epsilon) + \sum_{k=1}^{n-\ell} \epsilon(1-\epsilon)^k s_{n-\ell-k},$$

for every $n \geq \ell$.

We claim that there exist $\delta < 1$ and K so that $s_n \leq K(1-\delta)^n$ for every n. Indeed, fix $\delta < 1$, For each n, define $K_n := s_n/(1-\delta)^n$. Note that

$$s_n \leq (1-\epsilon)^{n/\ell} C(s_r + \epsilon) + \sum_{k=1}^{n-1} \epsilon(1-\epsilon)^k s_{n-\ell-k}$$

$$\leq (1-\epsilon)^{n/\ell} C(s_r + \epsilon) + \sum_{k=1}^{n-\ell} \epsilon(1-\epsilon)^k K_{n-\ell-k}(1-\delta)^{n-\ell-k}$$

$$\leq \left[\left(\frac{(1-\epsilon)^{1/\ell}}{1-\delta} \right)^n C(\max_{j<\ell} s_j + \epsilon) + \max_{i < n-\ell} K_i \frac{\epsilon}{(1-\delta)^\ell} \sum_{k=1}^{n-\ell} \left(\frac{1-\epsilon}{1-\delta} \right)^k \right] (1-\delta)^n.$$

$$(7.25)$$

Choose $\delta > 0$ close enough to 0 so that

$$\sigma_1 := \frac{(1-\epsilon)^{1/\ell}}{1-\delta} < 1, \text{ and}$$

$$\sigma_2 := \frac{\epsilon}{(1-\delta)^\ell} \sum_{k=1}^{\infty} \left(\frac{1-\epsilon}{1-\delta} \right)^k < 1.$$

Then by Eq. (7.25) we have $K_n \leq \sigma_2 \max_{i< n-\ell} K_i + C\sigma_1^n$, for every $n > \ell$, which easily implies that $\max_i K_i < \infty$. $\square$

Define
$$\Omega_+^{n_1,n} := \{p \in I_{n_1} \colon \pi_2(G^k(p)) \geq 0, \text{ for } 0 \leq k \leq n\}.$$

LEMMA 7.11. *Let $G \in On + Ra + Rb$ be a random walk that is an asymptotically small perturbation of a homogeneous random walk $F \in On + Ra + Rb$ with negative mean drift. Then there exists $\delta < 1$ so that for every $n_1 \geq 0$ there exists $C = C(n_1)$ satisfying*

$$m(\Omega_+^{n_1,n}(G)) \leq C(1-\delta)^n.$$

PROOF. Take n_0 as in Lemma 7.9 and fix $n_1 \geq 0$. Define the sets and sequences

$$s_n := m(\Omega_+^{n_1,n})$$
$$p_n := m(B^n), \text{ where } B^n := \{p \in \Omega_+^{n_1,n} \colon \pi_2(G^n(p)) \in [0, n_0]\}, \text{ and}$$
$$q_n := m(C^n), \text{ where } C^n := \{p \in \Omega_+^{n_1,n} \colon \pi_2(G^n(p)) > n_0\}.$$

To prove Lemma 7.11, it is enough to verify that these sequences satisfy the assumptions of Lemma 7.10. Indeed, of course, $p_0 + q_0 \leq 1$. To prove the other assumptions, take $i \in [0, n_0]$. Since G is topologically transitive, there are $\ell_i \in \mathbb{N}$ and intervals $J_i \subset I_i$ so that $\pi_2(G^{\ell_i}(J_i)) < 0$. Use the notation $\ell = max_{0 \leq i \leq n_0} \ell_i$ and $r = min_{0 \leq i \leq n_0} |J_i| / |I_i|$.

Clearly $\Omega_+^{n_1, n} = B^n \cup C^n \subset B^{n-\ell} \cup C^{n-\ell}$. Let $J \subset B^{n-\ell}$ be an interval so that $G^{n-\ell}(J) = I_i$, with $0 \leq i \leq n_0$. Note that $B^{n-\ell}$ is a disjoint union of such intervals. By the bounded distortion control for G,

$$\frac{m(J \cap \Omega_+^{n_1, n})}{m(J)} \leq 1 - \frac{m(J \cap G^{-(n-\ell)} J_i)}{m(J)} \leq \left(1 - \frac{r}{c}\right) \quad (7.26)$$

Choose ϵ_0 satisfying $(1 - r/c) \leq (1 - \epsilon_0)^\ell$. Then Eq. (7.26) implies

$$m(B^{n-\ell} \cap \Omega_+^{n_1, n}) \leq (1 - \epsilon_0)^\ell m(B^{n-\ell})$$

and we obtain

$$s_n = m(B^{n-\ell} \cap \Omega_+^{n_1, n}) + m(C^{n-\ell} \cap \Omega_+^{n_1, n}) \leq (1 - \epsilon_0)^\ell p_{n-\ell} + q_{n-\ell}.$$

It remains to prove that $q_n \leq \sum_{k=1}^{n} (1 - \epsilon)^k p_{n-k}$. There are two kind of points p in C^n:

Type 1. For every $j \leq n$ we have $\pi_2(G^j(p)) \geq n_0$ (in particular, $n_1 \geq n_0$). We are going to estimate the measure of the set of these points, denoted Θ_1^n. It follows from Lemma 7.9, choosing $\alpha = \int \psi \, d\mu / 2 < 0$, that

$$m(\{p \in I_{n_1}: \ \pi_2(G^k(p)) \geq n_0, \text{ for } k \leq n \text{ and } \pi_2(G^n(p)) \geq n_1 + \alpha n\}) \leq C\sigma^n. \quad (7.27)$$

Note that if $n \geq (n_0 - n_1)/\alpha$, then $n_1 + \alpha n \leq n_0$. Then the set in the l.h.s. of Eq. (7.27) contains Θ_1^n. In particular,

$$m(\Theta_1^n) \leq C_{n_1} \sigma^n$$

for some $\sigma < 1$ which does not depend on n_1.

Type 2. For some $j \leq n$ we have $\pi_2(G^j(p)) \leq n_0$. Denote the set of these points by Θ_2^n. Denote by $\Theta_{2,k}^n$ the set of points p so that $k \geq 1$ is the smallest natural satisfying $\pi_2(G^{n-k} p) \leq n_0$. Clearly Θ_2^n is a disjoint union of these sets. We are going to estimate their measure. Note that $\Theta_{2,k}^n \subset B^{n-k}$. The set B^{n-k} is a disjoint union of intervals L so that $\pi_2(G^{n-k} L) = I_i$, for some $i \leq n_0$. To estimate

$$\frac{m(\Theta_{2,k}^n \cap L)}{|L|},$$

we note that $L \subset B^{n-k}$, and that $\Theta_{2,k}^n \cap L$ is the set of points $p \in L$ for which $\pi_2(G^{n-k+j} p) > n_0$ for every $0 < j \leq k$. Define

$$L_y := \{p \in L: \psi(G^{n-k} p) = y\}.$$

First, note that for $y \leq n_0 - i$ we have

$$|L_y \cap \Theta_{2,k}^n| = 0, \quad (7.28)$$

since $p \in L_y \cap \Theta_{2,k}^n$ satisfies $\pi_2(G^{n-k+1}p) = i + \psi(G^{n-k}p) = i + y > n_0$. In particular, for $y < 0$ we have $|L_y \cap \Theta_{2,k}^n| = 0$, which implies, due the bounded distortion control,

$$\frac{m(L \cap \Theta_{2,k}^n)}{|L|} \leq \frac{\sum_{y \geq 0} |L_y|}{|L|} \leq (1 - \delta),$$

for some $\delta < 1$ which does not depend on k, L or n_1. This implies

$$m(\Theta_{2,k}^n) \leq (1 - \delta)m(B^{n-k}) = (1 - \delta)p_{n-k}. \tag{7.29}$$

Furthermore, again using the distortion control and the regularity condition GD (big jumps are rare), we have

$$\frac{\sum_{y > -\alpha(k-1)} |L_y \cap \Theta_{2,k}^n|}{|L|} \leq \frac{\sum_{y > -\alpha(k-1)} |L_y|}{|L|} \leq C\gamma^k, \tag{7.30}$$

for some $C \geq 0$ and $\gamma < 1$.

To estimate $|L_y \cap \Theta_{2,k}^n|/|L_y|$, in the case $n_0 - i \leq y \leq -\alpha(k-1)$, recall that $G^{n-k+1}L_y = I_{i+y}$, with $i + y > n_0$. By Lemma 7.9, we have

$$m\{p \in I_{i+y} : \pi_2(G^m(p)) \geq n_0 \text{ for } m \leq k-1, \text{ and } \pi_2(G^{k-1}(p)) \geq i + y + \alpha(k-1)\}$$
$$\leq C\sigma^k.$$

Since $i + y + \alpha(k-1) \leq n_0$, this implies that

$$m\{p \in I_{i+y} : \pi_2(G^m(p)) \geq n_0, \text{ for every } m \leq k-1\} \leq C\sigma^k.$$

The points in $L_y \cap \Theta_{2,k}^n$ are exactly the points whose $(n - k + 1)$th iteration belongs to the set in the preceding estimate. Using the bound distortion control we have

$$\frac{|L_y \cap \Theta_{2,k}^n|}{|L_y|} \leq C\sigma^k,$$

so

$$\frac{|\sum_{n_0-i \leq y \leq -\alpha(k-1)} L_y \cap \Theta_{2,k}^n|}{|L|} \leq C\frac{|\sum_{n_0-i \leq y \leq -\alpha(k-1)} L_y \cap \Theta_{2,k}^n|}{\sum_{n_0-i \leq y \leq -\alpha(k-1)} |L_y|} \leq C\sigma^k. \tag{7.31}$$

Choose $\epsilon < \epsilon_0$ so that $\min\{\max\{C\sigma^k, C\gamma^k\}, 1 - \delta\} \leq (1 - \epsilon)^k$, for every $k \geq 0$, and put together Eq. (7.28), Eq. (7.29), Eq. (7.30), and Eq. (7.31) to get the bound $m(L \cap \Theta_{2,k}^n) \leq (1 - \epsilon)^k|L|$. Since B^{n-k} is a disjoint union of such intervals L, we obtain

$$m(\Theta_{2,k}^n) \leq (1 - \epsilon)^k m(B^{n-k}) = (1 - \epsilon)^k p_{n-k}$$

and now we can conclude with

$$q_n = m(\Theta_1^n) + \sum_k m(\Theta_{2,k}^n) \leq C_{n_1}\sigma^n + \sum_k (1 - \epsilon)^k p_{n-k}.$$

$\square$

Now we are ready to prove Theorem 3.9.

PROOF OF THEOREM 3.9. There are three cases:

F is transient with M > 0. If $M > 0$, then the random walk F is transient and it is easy to see (using for instance Proposition 4.1) that $m(\Omega_+(F)) > 0$. Since the conjugacy with an asymptotically small perturbation G is absolutely continuous (Theorem 3.2), we conclude that $m(\Omega_+(G)) > 0$.

F is recurrent (M = 0). If $M = 0$, then F is recurrent [G] and its asymptotically small perturbations are recurrent by Theorem 3.4. In particular, almost every point visits negative states infinitely many times, so $m(\Omega_+(G)) = 0$. It remains to prove that $HD \, \Omega_+(G) = 1$. By Theorem 3.8 it is enough to verify that $HD \, \Omega_+(F) = 1$. Indeed, it is easy to show using the central limit theorem that if

$$\int \psi \, d\mu = 0,$$

then there exist $C > 0$ and for each n, subsets $\mathcal{A}_n \subset \mathcal{P}_0^n$ so that

$$\sum_{i=0}^{n-1} \psi(f^i(x)) > 0$$

for all $x \in J \in \mathcal{A}_n$ and

$$1 \geq m(\bigcup_{J \in \mathcal{A}_n} J) > C > 0. \tag{7.32}$$

Here C does not depend on n. Of course, we can assume that $\mathcal{A}_n$ is finite. Property Ex implies that there exists $\theta \in (0, 1)$ such that

$$\sup_{J \in \mathcal{A}_n} |J| \leq \theta^n.$$

Consider the function

$$h(\epsilon) := \sum_{J \in \mathcal{A}_n} |J|^{1-\epsilon}.$$

Then by Eq. (7.32) if $0 \leq \epsilon < 1$, we have

$$h'(\epsilon) := \sum_{J \in \mathcal{A}_n} -\log |J| |J|^{1-\epsilon} \geq -Cn \log \theta.$$

In particular, if

$$\tilde{\epsilon} := \frac{C - 1}{Cn \log \theta}$$

then $h(\tilde{\epsilon}) \geq 1$. Since $h(0) \leq 1$, there exist $\epsilon_n = 1 - O(1/n)$ such that $h(\epsilon_n) = 1$. But $VHD(\mathcal{A}_n) = \epsilon_n$, so

$$|VHD(\mathcal{A}_n) - 1| \leq \frac{C}{n}.$$

By Property BD that there exists $C_1 > 0$ such that for every n

$$d_n := \sup_{C \in \mathcal{A}_n} \sup_{x, y \in C} \log \frac{Df_{\mathcal{A}_n}(y)}{Df_{\mathcal{A}_n}(x)} \leq C_1$$

and since $\mathcal{A}_n \subset \mathcal{P}^n$, by Property Ex we have that there exists $\theta \in (0,1)$ such that for every n,

$$\lambda_n := \inf_{C \in \mathcal{I}} \inf_{x \in C} |Df_{\mathcal{I}}(x)| \geq \frac{1}{\theta^n}.$$

We can apply Proposition 7.4 to obtain

$$\left| HD\, \Lambda(\mathcal{A}_n) - VHD(\mathcal{A}_n) \right| = O\left(\frac{1}{n}\right),$$

so

$$HD(\mathcal{A}_n) = 1 - O\left(\frac{1}{n}\right).$$

If $\mu_{\mathcal{A}_n}$ is the geometric invariant measure of $f_{\mathcal{A}_n}$, then

$$\int \psi_{\mathcal{A}_n}\, d\mu_{\mathcal{A}_n} > 0.$$

So by the Birkhoff ergodic theorem,

$$\lim_{n \to \infty} \sum_{i=0}^{n-1} \psi(f^i(x)) = +\infty \tag{7.33}$$

in a set $S_n \subset \Lambda(\mathcal{A}_n)$ satisfying $\mu_{\mathcal{A}_n}(S_n) = 1$, so $HD\, S_n = 1 - O(1/n)$. In particular, the set S of points satisfying Eq. (7.33) has Hausdorff dimension 1. We can decompose S in subsets B_j defined by

$$B_j := \left\{ x \in S \colon \min_n \sum_{i=0}^{n-1} \psi(f^i(x)) \geq -j \right\}.$$

Clearly, $\sup_j HD\, B_j = 1$.

By Properties $GD + On$, for each j there are k_j and $J_j \neq \emptyset \in \mathcal{P}^{k_j}$ so that for all $x \in J_j$ we have

$$\sum_{i=0}^{\ell-1} \psi(f^i(x)) \geq 0$$

for every $\ell \leq k_j$ and

$$\sum_{i=0}^{k_j} \psi(f^i(x)) \geq j.$$

Then

$$(J_j \cap f^{-k_j} B_j) \times \{0\}$$

belongs to $\Omega_+(F)$ for every j. This implies $HD\, \Omega_+(F) \geq HD\, B_j$, so

$$HD\, \Omega_+(F) \geq \sup_j HD\, B_j = 1.$$

F is transient with M < 0. By Lemma 7.11, there is some $\delta \in (0,1)$, which does not depend on n_1, so that

$$m(\Omega_+^{n_1,n}) \leq C(1-\delta)^n. \tag{7.34}$$

By Lemma 7.8, there exists ϵ so that

$$\sum_{P \in \mathcal{P}^n,\, P \subset I_k} |P|^{1-\epsilon} \leq C(1-\delta)^{-n/2}. \tag{7.35}$$

Denote by $\{J_i^n\}_i \subset \mathcal{P}^n$ the family of disjoint intervals so that $\Omega_+^{n_1,n} = \cup_i J_i^n$. We claim that there exists $C > 0$ satisfying

$$\sum_i |J_i^n|^{1-\epsilon/4} \leq C(1-\delta)^n. \tag{7.36}$$

Since $\sup_i |J_i^n| \to_n 0$, this proves that $HD\, \Omega_+^{n_1,\infty} \leq 1 - \epsilon/4$.

Indeed,

$$\sum_i |J_i^n|^{1-\epsilon/4} = \sum_{|J_i| > (1-\delta)^{2n/\epsilon}} |J_i^n|^{1-\epsilon/4} + \sum_{|J_i| \leq (1-\delta)^{2n/\epsilon}} |J_i^n|^{1-\epsilon/4}$$

$$\leq (1-\delta)^{n/2} \sum_i |J_i^n| + (1-\delta)^{3n/2} \sum_i |J_i^n|^{1-\epsilon}$$

$$\leq C(1-\delta)^{n/2},$$

where in the last line we made use of Eq. (7.34) and Eq. (7.35). The proof is complete. □

8   APPLICATIONS TO ONE-DIMENSIONAL RENORMALIZATION THEORY

8.1   (Classic) infinitely renormalizable maps

Let I denote the interval $[-1,1]$. Consider a real analytic unimodal map $f \colon I \to I$, with negative Schwarzian derivative and even-order critical point at 0. The map f is called **infinitely renormalizable** if there exists an sequence of natural numbers $n_0 < n_1 < n_2 < \cdots$ and a nested sequence of intervals $I = I_0 \supset I_1 \supset I_2 \supset \cdots$ so that

- $f^{n_k} \partial I_k \subset \partial I_k$,

- $f^{n_k} I_k \subset I_k$,

- $f^{n_k} \colon I_k \to I_k$ is a unimodal map.

We say that f has **bounded combinatorics** if there exists a positive C so that $n_{k+1}/n_k \leq C$, for all k. Two infinitely renormalizable maps f and g have the same combinatorics if there exists a homeomorphism $h \colon I \to I$ such that $f \circ h = h \circ g$.

The following result is a deep result in renormalization theory.

PROPOSITION 8.1 ([McM96]). *Let f and g be two infinitely renormalizable unimodal maps with the same bounded combinatorics and same even order. Then for every $r > 0$, there exist $C > 0$ and $\lambda < 1$ so that*

$$\left\| \frac{1}{|I_k^f|} f^{n_k}(|I_k^f| \cdot) - \frac{1}{|I_k^g|} g^{n_k}(|I_k^g| \cdot) \right\|_{C^r} \leq C\lambda^k.$$

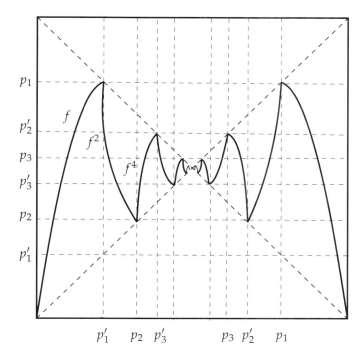

FIGURE 8.2: The "Bat" map: the induced map F for a Feigenbaum unimodal map.

Here $|I_k^f|$ denotes the length of I_k^f.

PROOF OF THEOREM 3.11. Let f be an infinitely renormalizable map with bounded combinatorics. We are going to define an induced map $F: I \to I$, following Y. Jiang (see [J1, J2]): Let p_k be the periodic point in ∂I_k. Define E as the set

$$\left\{1, -1, -p_k, p_k, f(p_k), -f(p_k), \ldots, f^{n_k-1}(p_k), -f^{n_k-1}(p_k)\right\} - \left\{f(p_k), -f(p_k)\right\}.$$

The set E cuts $I_{k-1} \setminus I_k$ in m_k intervals. Denote these intervals by $M_{k-1,i}$, with $i = 1, \ldots, m_k$. For each $x \in M_{k-1,i}$, define $n(x) \geq 1$ as the minimal positive integer so that

$$I_k \subset f^{n(x)n_{k-1}} M_{k-1,1}.$$

Note that $f^{n(x)n_{k-1}}$ does not have critical points on $M_{k-1,i}$. Define the induced map F, which is defined everywhere in I, except for a countable set of points:

$$F(x) := f^{n(x)}(x), \quad \text{for } x \in I_k \setminus I_{k+1}.$$

Figure 8.2 shows the induced map for an infinitely renormalizable map satisfying $n_{i+1} = 2n_i$ for all i (the so-called Feigenbaum maps). The map F is Markovian with respect to the partition

$$\mathcal{P} := \{M_{k,i}\}_{k \in \mathbb{N}, i \leq m_k}.$$

Furthermore, if f and g have the same bounded combinatorics and even order, then by Proposition 8.1, the corresponding induced maps F and G satisfy

$$\left\| \frac{1}{|I_k^f|} F(|M_{k,i}^f| \cdot + |I_k^f| - |M_{k,i}^f|) - \frac{1}{|I_k^g|} G(|M_{k,i}^g| \cdot + |I_k^g| - |M_{k,i}^g|) \right\|_{C^r([0,1])} \leq C\lambda^k.$$

Define L_k as, say, the right component of $I_k \setminus I_{k+1}$ and $\gamma_k \colon I \to L_k$ as the unique bijective order preserving affine map between this two intervals. We are going to define a random walk $\mathcal{F} \colon I \times \mathbb{N} \to I \times \mathbb{N}$ from the map F in the following way:

$$\mathcal{F}(x,k) := \begin{cases} (\gamma_i^{-1} \circ F \circ \gamma_k(x), i) & \text{if } F \circ \gamma_k(x) \in L_i; \\ (\gamma_i^{-1} \circ (-F) \circ \gamma_k(x), i) & \text{if } F \circ \gamma_k(x) \in -L_i. \end{cases} \qquad (8.1)$$

It is easy to see that we can extend $\mathcal{F} \colon I \times \mathbb{Z} \to I \times \mathbb{Z}$ to a strongly transient deterministic random walk with non-negative drift. Indeed, if $k < 0$, define

$$\mathcal{F}(x,k) := \begin{cases} (\gamma_i^{-1} \circ F \circ \gamma_0(x), k+i) & \text{if } F \circ \gamma_k(x) \in L_i; \\ (\gamma_i^{-1} \circ (-F) \circ \gamma_0(x), k+i) & \text{if } F \circ \gamma_k(x) \in -L_i. \end{cases}$$

Furthermore, if g is another infinitely renormalizable map with the combinatorics of f, then by Proposition 8.1 and Proposition 4.9 we can define the corresponding random walk $\mathcal{G} \colon I \times \mathbb{N} \to I \times \mathbb{N}$ and extend this to a random walk $\mathcal{G}$ on $I \times \mathbb{Z}$ by defining $\mathcal{G}(x,k) = \mathcal{F}(x,k)$ if $k < 0$. Then $\mathcal{G}$ is an asymptotically small perturbation of $\mathcal{F}$. So we can apply Theorem 3.3 to conclude that there is a conjugacy between F and G which is strongly quasisymmetric with respect to the nested sequence of partitions defined by the random walk $\mathcal{F}$. We can now easily translate this result in terms of the original unimodal maps f and g, saying that the continuous conjugacy h between f and g is a strongly quasisymmetric mapping with respect to $\mathcal{P}$. $\qquad \square$

REMARK 8.3. An interesting case is when the unimodal map f is a periodic point to the renormalization operator: there exists n_0 and λ, with $|\lambda| < 1$ so that

$$\frac{1}{\lambda} f^{n_0}(\lambda x) = f(x).$$

In this case, if we take $n_k = k n_0$, then the induced map F will satisfy the functional equation

$$F(\lambda x) = \lambda F(x). \qquad (8.2)$$

Define the relation $\sim$ in the following way:

$$x \sim y \text{ iff there exists } i \in \mathbb{Z} \text{ so that } x = \pm \lambda^i y.$$

By Eq. (8.2), F preserves this relation, so we can take the quotient of F by the relation $\sim$. Note that

$$L_0 = \mathbb{R}^\star / \sim.$$

It is easy to see that $q = F/\sim \colon L_0 \to L_0$ is a Markov expanding map. Now define $\psi \colon L_0 \to \mathbb{Z}$ as $\psi(x) = k$, if $f(x) \in I_k \setminus I_{k+1}$. Then $\mathcal{F}$ is exactly the homogeneous random walk defined by the pair (q, ψ).

8.2   Fibonacci maps

The Fibonacci renormalization is the simplest way to generalize the concept of classical renormalization as described in Section 8.1. Actually, we could prove all the results stated for Fibonacci maps to a wider class of maps: maps which are infinitely renormalizable in the generalized sense and with periodic combinatorics and bounded geometry, but we will keep ourselves in the simplest case to avoid more technical definitions and auxiliary results with its long proofs.

Consider the class of real analytic maps f with $Sf < 0$ and defined in a disjoint union of intervals $I_1^0 \sqcup I_1^1$, where $-I_1^1 = I_1^0$, so that the following hold:

- The map $f \colon I_1^1 \to I_0^0 := f(I_1^1)$ is a diffeomorphism. Furthermore, I_1^1 is compactly contained in I_0^0.

- The map $f \colon I_1^0 \to I_0^0$ is an even map which has as 0 as its unique critical point of even order.

We say that f is **Fibonacci renormalizable** if

$$f(0) \in I_1^1, \ f^2(0) \in I_0^1 \text{ and } f^3(0) \in I_0^1.$$

In this case, the Fibonacci renormalization of f is defined as the first return map to the interval I_1^0 restricted to the connected components of its domain which contain the points $f(0)$ and $f^2(0)$. This new map is denoted $\mathcal{R}f$: it could be Fibonacci renormalizable again, and so on, obtaining an infinite sequence of renormalizations $\mathcal{R}f, \mathcal{R}^2f, \mathcal{R}^3f, \ldots$.

We will denote the set of infinitely renormalizable maps in the Fibonacci sense with a critical point of order d by $\mathcal{F}_d$. A map $f \in \mathcal{F}_d$ will be called a **Fibonacci map**.

As in the original map f, the nth renormalization $f_n := \mathcal{R}^n f$ of f is a map defined in two disjoint intervals, denoted I_n^0 and I_n^1, where $-I_0^n = I_0^n$. Indeed f_n on I_n^0 is a unimodal restriction of the S_nth iteration of f, where $\{S_n\}$ is the Fibonacci sequence

$$S_0 = 1, \ S_1 = 2, \ S_2 = 3, \ S_3 = 5, \ \ldots, S_{k+2} = S_{k+1} + S_k, \ldots$$

and f_n on I_n^1 is the restriction of the (S_{n-1})th iteration of f.

Denote by p_k the sequence of points $p_k \in \partial I_0^k$ so that

$$f_k(p_{k+1}) = p_k$$

and denote $I_0^k = [p_k, p_k']$.

It is possible to define a sequence u_k of points satisfying

1. $\cdots < \ p_{k+1} < u_k < p_k < \cdots < p_0,$

2. f^{S_k} is monotone on $[0, u_k]$,

3. $f^{S_k}(u_{k+1}) = u_k,$

4. $f^{S_k}(u_k) = u_{k-2}.$

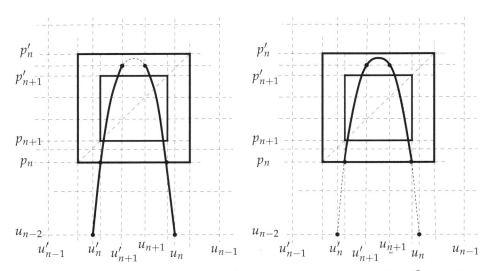

FIGURE 8.4: On the left figure the solid curves represent the part of the f^{S_n} used in the definition of the induced map. On the right figure the solid curve is the part of f^{S_n} which coincides with the nth Fibonacci renormalization on its central domain.

We are going to define an induced map for an infinitely renormalizable map in the Fibonacci sense in the following way: Firstly, define $f_{-1}: I_0^0 \setminus I_0^1$ as an C^3 monotone extension of f_0 on I_1^1 which has negative Schwarzian derivative and bounded distortion. Define $F: I_0^0 \to \mathbb{R}$ as

$$F(x) := f^{S_i}(x) \ if \ x \in [u_i, -u_i] \setminus [u_{i+1}, -u_{i+1}]$$

for each $i \geq 0$.

Define L_i as, say, the right component of $[u_i, -u_i] \setminus [u_{i+1}, -u_{i+1}]$ and $\gamma_i: I \to L_i$ as the unique bijective order preserving affine map between these two intervals.

We are ready to define the map $\mathcal{F}: I \times (\mathbb{N} \setminus \{0\}) \to I \times \mathbb{N}$ as

$$\mathcal{F}(x,k) := \begin{cases} (\gamma_i^{-1} \circ F \circ \gamma_k(x), i) & \text{if } F \circ \gamma_k(x) \in L_i, \\ (\gamma_i^{-1} \circ (-F) \circ \gamma_k(x), i) & \text{if } F \circ \gamma_k(x) \in -L_i. \end{cases}$$

If the order of the critical point is even and larger than two, then there is a very special Fibonacci map $f^\star$, called the Fibonacci fixed point (see, for instance [Sm]), whose induced map $F^\star$ satisfies (choosing a good u_0)

$$F^\star(\lambda x) = \pm \lambda F^\star(x) \tag{8.3}$$

for some $\lambda \in (0,1)$. In this case we can use the argument in Remark 8.3 to conclude that the corresponding map $\mathcal{F}^\star: I \times (\mathbb{N} \setminus \{0\}) \to I \times \mathbb{N}$ can be extended to a homogeneous random walk $\mathcal{F}^\star: I \times \mathbb{Z} \to I \times \mathbb{Z}$. For an arbitrary Fibonacci map f, we can extend $\mathcal{F}: I \times (\mathbb{N} \setminus \{0\}) \to I \times \mathbb{N}$ to a random walk $\mathcal{F}: I \times \mathbb{Z} \to I \times \mathbb{Z}$ defining $\mathcal{F}(x,k) = \mathcal{F}^\star(x,k)$ for $k \leq 0$. Then $\mathcal{F}$ is not homogeneous; however, due to Proposition 4.9 and the following result, $\mathcal{F}$ is an asymptotically small perturbation of $\mathcal{F}^\star$:

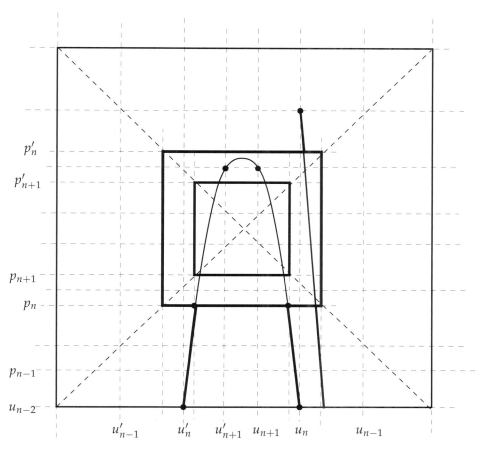

FIGURE 8.5: The solid curves inside the medium square represent the graph of the nth Fibonacci renormalization f_n. The solid curves inside the largest square form the graph of an extension of f_n which has the same maximal invariant set.

PROPOSITION 8.6 (See [Sm]). *For each even integer larger than two the following holds: for every Fibonacci map f, denote*

$$g_i = \alpha_i^{-1} \circ f^{S_i} \circ \alpha_{i+1} \colon I \to I,$$

where $\alpha_i \colon I \to [u_i^f, -u_i^f]$ is an bijective affine map so that $\alpha_i^{-1}(f_{i+1}(0)) > 0$ and consider the correspondent maps $g_i^\star$ for $f^\star$. Then

$$\|g_i - g_i^\star\|_{C^r} \le K_r \rho^i$$

for some $\rho < 1$ and every $r \in \mathbb{N}$.

The **real Julia set** of f, denoted $J_{\mathbb{R}}(f)$, is the maximal invariant of the map

$$f \colon I_1^0 \sqcup I_1^1 \to I_0^0,$$

in other words,

$$J_{\mathbb{R}}(f_j) := \cap_i f_j^{-i} I_j^0.$$

Denote
$$\Omega_+^j(F) := \{(x,i) \text{ s.t. } \pi_2(F^n(x,i)) \geq j \text{ for all } n \geq 0\}.$$

PROPOSITION 8.7. *There exists some k_0 so that*

$$\Omega_+^{j+1}(F) \subset J_\mathbb{R}(f_j) \subset \Omega_+^{j-1}(F).$$

In particular

$$HD\ \Omega_+^{j+1}(F) \leq HD\ J_\mathbb{R}(f_j) \leq HD\ \Omega_+^{j-1}(F), \qquad (8.4)$$

and, for the Fibonacci fixed point, since $\Omega_+^{j+1}(F)$ is an affine copy of $\Omega_+^{j-1}(F)$, we have

$$HD\ \Omega_+^j(F) = HD\ J_\mathbb{R}(f) \qquad (8.5)$$

for all $j \geq 0$.

PROOF. Denote by F_ℓ the restriction of F to $\cup_{i \geq \ell} L_i$. Then the maximal invariant set of F_ℓ,
$$\Lambda(F_\ell) := \cap_{i \in \mathbb{N}} F^{-i}\mathbb{R},$$
is $\Omega_+^\ell(F)$. Consider the extension of f_j described in Figure (8.5). Let's call this extension $\tilde{f}_j$. An easy analysis of its graph shows that f_j and $\tilde{f}_j$ have the same maximal invariant set. We claim that $\tilde{f}_{j+1}$ is just a map induced by $\tilde{f}_j$. Indeed, the restriction of $\tilde{f}_{j+1}$ to $[u_{j+1}, u'_{j+1}]$ coincides with $\tilde{f}_j^2$ on the same interval. On the rest of $\tilde{f}_{j+1}$-domain $\tilde{f}_{j+1}$ coincides with $\tilde{f}_j$.

By consequence, for $i \geq j$ the map $\tilde{f}_i$ is induced by $\tilde{f}_j$ and, since F_{j+1} restricted to L_i is equal to $\tilde{f}_i$, we obtain that F_{j+1} is a map induced by $\tilde{f}_j$. In particular,

$$\Lambda(F_{j+1}) \subset \Lambda(\tilde{f}_j) = J_\mathbb{R}(f_j).$$

To prove that $\Lambda(\tilde{f}_j) \subset \Lambda(F_{j-1})$, we are going to prove that

$$x \in \Lambda(\tilde{f}_j) \text{ implies } F_{j-1}(x) \in \Lambda(\tilde{f}_j). \qquad (8.6)$$

If x belongs to the interval $I_j^1 \subset L_{j-1}$, where $\tilde{f}_j$ coincides with F_{j-1}, then we have $F_{j-1}(x) \in \Lambda(\tilde{f}_j)$. Otherwise $x \in I_j^0 \subset \cup_{i \geq j} L_i$, so $x \in \Lambda(\tilde{f}_j) \cap L_i$, for some $i \geq j$; then F_{j-1} is an iteration of $\tilde{f}_j$ on L_i, so $F_{j-1}(x) \in \Lambda(\tilde{f}_j)$. This finishes the proof of Eq. (8.6). Since $\Lambda(\tilde{f}_j)$ is invariant by the action of F_{j-1} we have $\Lambda(\tilde{f}_j) \subset \Lambda(F_{j-1})$.
□

PROOF OF THEOREM 3.13. Consider the homogeneous random walk $F^\star = (g, \psi)$ induced by $f^\star$. Denote
$$M = \int \psi\, d\mu,$$

where μ is the absolutely continuous invariant measure of g. Using Thorem 3.9, there are three cases:

1. M < 0. In this case $\mathbf{F}^\star$ is transient and we have that $HD\ \Omega_+(F) < 1$ for every asymptotically small perturbation of $F^\star$, in particular, when F is a random walk induced by a Fibonacci map f. By Proposition 8.7, $HD\ J_\mathbb{R}(f) < 1$.

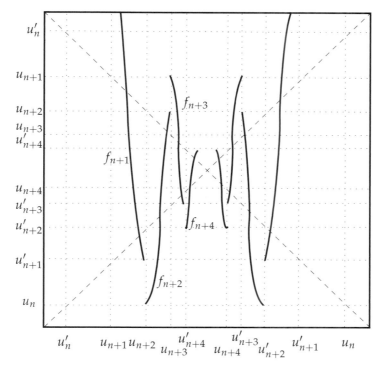

FIGURE 8.8: Induced map F for a Fibonacci map.

2. M = 0. Then $F^\star$ is recurrent [G] so every asymptotically small perturbation G of $F^\star$ is recurrent and $m(\Omega_+(G)) = 0$ but $HD \, \Omega_+(G) = 1$. By Proposition 8.7 we obtain $m(J_{\mathbb{R}}(f)) = 0$ and $HD \, J_{\mathbb{R}}(f) = 1$.

3. M > 0. In this case $\mathbf{F}^\star$ is transient with $m(\Omega_+(F^\star)) > 0$ and the conjugacy between $F^\star$ and any asymptotically small perturbation of it is absolutely continuous on $\Omega_+^i(F^\star)$. In particular, $m(\Omega_+(F)) > 0$ for every random walk F induced by a Fibonacci map f so $m(J_{\mathbb{R}}(f)) > 0$ by Proposition 8.7. $\qquad\square$

A map $f\colon I \to I$ is called a unimodal map if f has a unique critical point, with even order d, which is a maximum, and $f(\partial I) \subset \partial I$. We will assume that f is real analytic, symmetric with respect the critical point and $Sf < 0$. If the critical value is high enough, then f has a reversing fixed point p. Let $I_0^0 := [-p, p]$. Consider the map of first return R to f: if $x \in I_0^0$ and $f^r(x) \in I_0^0$, but $f^n(x) \notin I_0^0$ for $i < r$, define

$$R(x) := f^r(x).$$

If there exists exactly two connected components I_1^0 and I_1^1 of the domain of R containing points in the orbit of the critical point, and furthermore the map

$$R\colon I_1^0 \cup I_1^1 \to I_0^0$$

is a Fibonacci map, then we will call f an **unimodal Fibonacci map**. The class of all unimodal Fibonacci maps will be denoted $\mathcal{F}_d^{uni}$.

PROOF OF THEOREM 3.14. We will use the notation in the proof of Theorem 3.13. Since $m(J_{\mathbb{R}}(f)) > 0$, we conclude that the mean drift M of $F^\star$ is positive. By Proposition 5.1 any asymptotically small perturbation G of $F^\star$ has the following property: there exist $\lambda \in [0,1)$, $C > 0$ and $K > 0$ so that for every $P \in \mathcal{P}^0(G)$,

$$m\left(p \in P \colon \sum_{i=0}^{n-1} \psi(G^i(p)) < Kn\right) \leq C\lambda^n |P|.$$

This implies that

$$m\left(p \in I_j \colon \sum_{i=0}^{\ell} \psi(G^i(p)) \geq K\ell \text{ for every } \ell \geq n\right) \geq (1 - C\lambda^n).$$

So if $j = n|\min \psi|$ we obtain

$$m(\Omega_+^j(G)) \geq 1 - C\lambda^{C_1 j}.$$

Here $C_1 > 0$. If G is a random walk induced by a Fibonacci map g, then this implies that for j large,

$$m\left(L_j \setminus J_{\mathbb{R}}(g)\right) = m\left((-L_j) \setminus J_{\mathbb{R}}(g)\right) \leq C\lambda^{C_1 j} |L_j|.$$

Since

$$[-u_{j+1}, u_{j+1}] = \bigcup_{i \geq j} L_i \cup (-L_i),$$

we conclude that

$$m\left([u_{j+1}, -u_{j+1}] \setminus J_{\mathbb{R}}(g)\right) \leq C\lambda^{C_1 j} |u_{j+1}|. \tag{8.7}$$

For every δ, choose j so that $|u_{j+2}| \leq \delta \leq |u_{j+1}|$. Because $|u_{j+2}| > \theta |u_{j+1}|$, where $\theta \in (0,1)$ does not depend on j, we have that $|u_j| \geq C\theta^j$. Together with Eq. (8.7) this implies

$$m\left([-\delta, \delta] \setminus J_{\mathbb{R}}(g)\right) \leq C\lambda^{C_1 j} |u_{j+1}| \leq C|u_{j+1}|^{1+\alpha} \leq C|\delta|^{1+\alpha}.$$

$\square$

PROOF OF THEOREM 3.16. We will prove each one of the following implications:

(1) implies (2): From the proof of Theorem 3.13, if $m(J_{\mathbb{R}}(f)) > 0$ for some $f \in \mathcal{F}_d$ the mean drift M of the homogeneous random walk $\mathcal{F}^\star$ of $f^\star$ is positive. So $\mathcal{F}^\star$ (and all its asymptotically small perturbations) is transient (to $+\infty$). In terms of the original Fibonacci map f, this means that almost every orbit in $J_{\mathbb{R}}(f)$ accumulates in the post-critical set: So f has a wild attractor.

(2) implies (3): If there exists a wild attractor for f, then $m(J_{\mathbb{R}}(f)) > 0$. From the proof of Theorem 3.13 we obtain that the mean drift M of $\mathcal{F}^\star$ is positive. So there exists an absolutely continuous conjugacy between $\mathcal{F}^\star$ and any asymptotically small perturbation of $\mathcal{F}^\star$. This implies that any two maps $f_1, f_2 \in \mathcal{F}_d$ admits a continuous and absolutely continuous conjugacy

$$h \colon J_{\mathbb{R}}(f_1) \to J_{\mathbb{R}}(f_2).$$

Now consider two arbitrary maps $g_1, g_2 \in \mathcal{F}_d^{uni}$. Then we already know that there exists an absolutely continuous conjugacy

$$h \colon J_{\mathbb{R}}(R_{g_1}) \to J_{\mathbb{R}}(R_{g_2})$$

between the induced Fibonacci maps R_{g_1} and R_{g_2} associated with g_1 and g_2. Of course, h is just the restriction of a topological conjugacy between g_1 and g_2. By a Blokh and Lyubich result [BL] (see also page 332 in [dMvS]), every map of $\mathcal{F}_d^{uni}$ is ergodic with respect the Lebesgue measure. Since g_1 and g_2 have wild attractors, this implies that the orbit of almost every point $x \in I$ hits $J_{\mathbb{R}}(R_{g_1})$ at least once. Let $n(x)$ be a time when this happens.

So consider a arbitrary measurable set $B \subset I$ so that $m(B) > 0$. Then, for at least one $n_0 \in \mathbb{N}$, the set

$$B_{n_0} := \{x \in B \colon n(x) = n_0\}$$

has positive Lebesgue measure. This implies that $f^{n_0} B_{n_0}$ has positive Lebesgue measure, so $m(h(f^{n_0} B_{n_0})) > 0$. Now it is easy to conclude that $m(h(B_{n_0})) > 0$ and $h(B) > 0$. Switching the places of g_1 and g_2 in this argument, we can conclude that h is absolutely continuous on I.

Finally, note that the eigenvalues of the periodic points are not constant on the class $\mathcal{F}_d^{uni}$.

(3) implies (4): By the argument in Martens and de Melo [MdM], if a Fibonacci map does not have a wild attractor, then any continuous absolutely continuous conjugacy with other Fibonacci map is C^1: in particular, the conjugacy preserves the eigenvalues of the periodic points. So if (3) holds, then we can use the same argument in the proof of the previous implication to conclude that every Fibonacci map has a wild attractor.

(4) implies (5): The proof goes exactly as the proof of (2)$\Rightarrow$ (3).

(5) implies (1): The proof goes exactly as the proof of (3)$\Rightarrow$ (4).

$\square$

9   Bibliography

[AD]   J. Aaronson and M. Denker. Local limit theorems for partial sums of stationary sequences generated by Gibbs-Markov maps. *Stoch. Dyn.* 1, 193–237, 2001.

[BL]   A. Blokh and M. Lyubich. Measurable dynamics of S-unimodal maps of the interval. *Ann. Sci. École Norm. Sup. (4)* 24, no. 5, 545–573, 1991.

[B]   A. Broise. Transformations dilatantes de l'intervalle et théorèmes limites. in *Études spectrales d'operateurs de transfert et applications, Astérisque*, 238, 1996.

[Br]   H. Bruin. Topological conditions for the existence of absorbing Cantor sets. *Trans. Amer. Math. Soc.* 350, no. 6, 2229–2263, 1998.

[BKNvS] H. Bruin, G. Keller, T. Nowicki, S. van Strien. Wild Cantor attractors exist. *Ann. of Math. (2)* 143 no. 1, 97–130, 1996.

[F]   K. Falconer. The geometry of fractal sets. Cambridge University Press, Cambridge, 1985.

[GS]   J. Graczyk and G. Świątek. Generic hyperbolicity in the logistic family. *Ann. of Math. (2)* 146, no. 1, 1–52, 1997.

[G]   Y. Guivarc'h.   Propriétés ergodiques, en mesure infinie, de certains systèmes dynamiques fibrés. *Ergodic Theory Dynam. Systems*, 9, no. 3, 433–453, 1989.

[J1]   Y. Jiang. On the quasisymmetrical classification of infinitely renormalizable maps I. Maps with Feigeinbaum's topology. *Preprint ims91-19a*, IMS-SUNY at Stony Brook, 1991. arXiv:math/9201294

[J2]   Y. Jiang. On the quasisymmetrical classification of infinitely renormalizable maps II. Remarks on maps with a bounded type topology. *Preprint ims91-19b*, IMS-SUNY at Stony Brook, 1991. arXiv:math/9201295

[L1]   M. Lyubich. Combinatorics, geometry and attractors of quasi-quadratic maps. *Ann. of Math. (2)*, 148 (1994), no. 2, 347–404.

[L2]   M. Lyubich. Dynamics of quadratic polynomials I, II. *Acta Math.* 178, no. 2, 185–247, 247–297, 1997.

[MdM]   M. Martens and W. de Melo. The multipliers of periodic points in one-dimensional dynamics. *Nonlinearity* 12, 217–227, 1999.

[MU]   R. Mauldin and M. Urbański. Thermodynamic formalism and multifractal analysis of conformal infinite iterated functions systems. *Acta Math. Hungar.*, 96, 27–98, 2002.

[dMvS]   W. de Melo and S. van Strien. One-dimensional dynamics. Ergebnisse der Mathematik und ihrer Grenzgebiete (3), 25, Springer-Verlag, Berlin, 1993.

[McM96]   C. McMullen. Renormalization and 3-manifolds which fiber over the circle. Annals of Mathematics Studies, 142, Princeton University Press, Princeton, 1996.

[PT]   J. Palis and F. Takens. Hyperbolicity and sensitive chaotic dynamics at homoclinic bifurcations. Cambridge University Press, Cambridge, 1993.

[PU]   F. Przytycki and M. Urbański. Conformal fractals: ergodic theory methods. London Mathematical Society Lecture Note Series, 371, Cambridge University Press, Cambridge, 2010.

[ShSu]   M. Shub and D. Sullivan. Expanding endomorphisms of the circle revisited. *Ergodic Theory Dynam. Systems* 5, no. 2, 285–289, 1985.

[Sm]   D. Smania. Puzzle geometry and rigidity: The Fibonacci cycle is hyperbolic. *J. Amer. Math. Soc.* 20, no. 3, 629–673, 2007.

[Su]   D. Sullivan. Bounds, quadratic differentials, and renormalization conjectures. American Mathematical Society centennial publications, Vol. II (Providence, RI, 1988), 417–466, Amer. Math. Soc., Providence, RI, 1992.

[T]   R. L. Tweedie. Perturbations of countable Markov chains and processes. *Ann. Inst. Statist. Math.* 32, 283–290, 1980.

[V]   M. Viana. Stochastic dynamics of deterministic systems. 21º Colóquio Brasileiro de Matemática, IMPA, 1997. http://www.impa.br/~viana/

[W]   D. Williams. Probability with Martingales. Cambridge Mathematical Textbooks, 1991.

Milnor's conjecture on monotonicity of topological entropy: Results and questions

Sebastian van Strien

ABSTRACT. This note discusses Milnor's conjecture on monotonicity of entropy and gives a short exposition of the ideas used in its proof, which was obtained in joint work with Henk Bruin, see [BvS09]. At the end of this note we explore some related conjectures and questions.

1 MOTIVATION

In their seminal and widely circulated 1977 preprint, "On iterated maps of the interval: I,II," Milnor and Thurston proved the following.

THEOREM (Milnor and Thurston [MT77], see also [MT88]). *The function $C^{2,b} \to \mathbb{R}$ which associates to a mapping $g \in C^{2,b}$ its topological entropy $h_{top}(g)$ is continuous.*

Here $C^{2,b}$ stands for C^2 maps of the interval with b non-degenerate critical points (non-degenerate means second derivative non-zero). This theorem relies on a result of Misiurewicz and Szlenk [MS77, MS80], who had previously shown that $\gamma(f) := \exp(h_{top}(f))$ is equal to the growth rate of the number of laps (i.e., intervals of monotonicity) of f^n. The crucial new ingredient in the proof of Milnor and Thurston's theorem is a formula which shows that $\gamma(f)$ is also the largest zero of a certain meromorphic function.

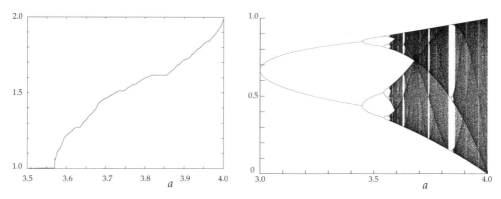

FIGURE 1.1: On the left, $\exp(h_{top}(f_a))$ as a function $a \in [3.5, 4]$ for $f_a(x) = ax(1-x)$. On the right is the well-known bifurcation diagram.

For the quadratic (logistic) family $q_a(x) = ax(1-x)$, the entropy function $a \mapsto h(q_a)$ appears monotone (in the weak sense); see Figure 1.1. Monotonicity indeed holds.

THEOREM. *The topological entropy of $x \mapsto ax(1-x)$ increases with $a \in \mathbb{R}$.*

This theorem was proven in a later version of Milnor and Thurston's preprint, which was published in 1988 [MT88]. Their proof, together with other proofs of this theorem which appeared in the mid-1980s (see Douady and Hubbard [DH85, Dou95] and Sullivan [dMvS93]) all rely on ideas from holomorphic dynamics. In the late 1990s, Tsujii gave a different proof, which does not rely on holomorphic dynamics [Tsu00] but still requires the family of maps to be quadratic (or of the form $z \mapsto z^d + c$ with d an even integer). It is worth emphasizing that it is not known whether the topological entropy is monotone for the family $x \mapsto |x|^r + c$ with $c \in \mathbb{R}$ and $r > 0$ not an integer.

These proofs all show a stronger statement: periodic orbits never disappear when a increases; see the bifurcation diagram on the right of Figure 1.1. In this way, we have an instance of the following.

HEURISTIC PRINCIPLE. *Families of real polynomial maps undergo bifurcations in the simplest possible way.*

Later on it was shown (independently by Graczyk and Świątek [GŚ97] and Lyubich [Lyu97]) that, within the quadratic family, hyperbolic maps are dense and so the periodic windows are dense.

In this survey, we will discuss a generalization of the Milnor-Thurston theorem which solves a conjecture due to Milnor on monotonicity of entropy for more general families of one-dimensional maps.

Acknowledgements. The author would like to thank Charles Tresser and the anonymous referee for helpful comments on an earlier version.

2   MILNOR'S MONOTONICITY OF ENTROPY CONJECTURE

Milnor proposed a conjecture which makes the heuristic principle precise in the case of polynomials of higher degree. Indeed, given $\epsilon \in \{-, +\}$, consider the space P_ϵ^b of real polynomials f with

1. precisely b distinct critical points, all of which are real, non-degenerate, and contained in $(-1, 1)$;

2. $f\{\pm 1\} \subset \{\pm 1\}$;

3. shape $\epsilon = \epsilon(f)$, where

$$\epsilon(f) = \begin{cases} +1 & \text{if } f \text{ is increasing at the left endpoint of } [-1,1], \\ -1 & \text{otherwise.} \end{cases}$$

Note that P_ϵ^b consists of polynomials of degree $d = b + 1$. In particular, P_ϵ^1 corresponds to quadratic maps q_c. Taking $\epsilon = +$, the required normalization in P_ϵ^1 gives $q_c(\pm 1) = 1$, from which it follows that the quadratic family takes the form $x \mapsto q_c(x) = -(c+1)x^2 + c$, where $c \in [-1, 1]$ is the critical value.

P_ϵ^b forms a b-dimensional space which can be parametrized by the critical values of the maps f. So for each choice of

$$(\zeta_1, \ldots, \zeta_b) \in [-1, 1]^b \text{ with } (\zeta_{i-1} - \zeta_i)(\zeta_i - \zeta_{i+1}) < 0 \text{ for each } i = 2, \ldots, b-1,$$

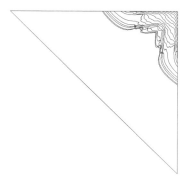

FIGURE 2.1: Level sets of the topological entropy within the space P_ϵ^3, where the axes correspond to the position of the two critical values ζ_1, ζ_2 in $[-1, 1]$, where $\zeta_1 < \zeta_2$.

there exists a unique map $f_{\zeta_1,\dots,\zeta_b} \in P^d$ with critical values $\zeta_1,\dots,\zeta_b$ and the map $(\zeta_1,\dots,\zeta_b) \mapsto f_{\zeta_1,\dots,\zeta_b}$ is continuous ([dMvS93, page 120] and [MTr00, page 132 and appendix]).

To formalize the heuristic principle in higher dimensions, one may hope that the topological entropy of f depends monotonically on the position of its critical values, i.e., that the map $(\zeta_1,\dots,\zeta_b) \mapsto h_{top}(f_{\zeta_1,\dots,\zeta_b})$ is monotone in each of its components. This turns out *not* to be true, as is clear by looking at the level sets of the entropy function in Figure 2.1 but as can also be proved rigorously see [BvS11].

Instead, Milnor suggested that one should investigate the level sets of topological entropy, noting that monotonicity of entropy within the quadratic family q_c is equivalent to the statement that for each h_0, the level set

$$I(h_0) := \{c \in \mathbb{R}; h_{top}(q_c) = h_0\}$$

is connected. Milnor coined these level sets **isentropes** (following the terminology used in thermodynamics for sets with a given entropy) and proposed the following conjecture.

MILNOR'S MONOTONICITY OF ENTROPY CONJECTURE. *Given $\epsilon \in \{-1, 1\}$, the set of $f \in P_\epsilon^b$ with topological entropy equal to h_0 is connected.*

Another approach to monotonicity using smooth one-parameter families of maps has been introduced by Yorke and co-workers; we briefly review this line of work in Section 4.

Milnor first posed this conjecture in the cubic case in [DGMT95, Mil92] and subsequently in joint work with Tresser in the previous setting in [MTr00, page 125]. At the end of this note, we state some further conjectures in this direction, due to Milnor and others.

Milnor and Tresser realized that the treatment of this conjecture in the cubic family required only a weak generalization of the rigidity results for unimodal maps with quadratic critical points (which had been obtained independently by Lyubich [Lyu97] and Graczyk and Świątek [GŚ97]) rather than a full rigidity statement for cubic maps. Subsequently, they posed the precise analytical question which needed to be answered to experts. This question was solved successfully by Heckman, a student of Świątek, as his PhD [Hec96]. Using Heckman's work, Milnor and Tresser then solved this conjecture in the cubic case (when $b = 2$).

THEOREM (Milnor and Tresser [MTr00]). *The entropy conjecture is true for real cubic maps.*

Note that Milnor and Tresser did not make any assumptions on whether the critical points were real, because any real cubic map which is not bimodal is, in fact, monotone (and so has zero topological entropy). We should also remark that in the cubic case, the parameter space is two-dimensional. Accordingly, Milnor and Tresser's proof considers certain curves corresponding to the existence of a critical point belonging to a periodic orbit, uses that certain conjugacy classes are unique, and then relies on planar topology to obtain connectedness of isentropes. As a model for the parameter space, Milnor and Tresser use stunted sawtooth maps; see Figure 3.7.

Another particular case of Milnor's conjecture was solved by Anca Radulescu (a student of Jack Milnor) [Ra07]. In this work the conjecture was proved within the space of real quartic polynomials which are compositions of two real quadratic maps.

Some time ago, in joint work with Henk Bruin, we proved Milnor's entropy conjecture for arbitrary degree.

MAIN THEOREM (Bruin and van Strien [BvS09]). *The entropy conjecture is true in general.*

In other words, fix $\epsilon \in \{-1, 1\}$, given any integer $d \geq 1$ and any $h_0 \in \mathbb{R}$. Then the set of $f \in P_\epsilon^b$ with topological entropy equal to h_0, is connected.

The purpose of this note is to present some of the ideas of the proof of the latter theorem. At the end of this note we will state some open problems.

3   IDEA OF THE PROOF

Our proof of the main theorem has four steps:

1. A generalization of the notion of hyperbolic component, namely, the space $\mathcal{PH}(f)$ of maps which are **partial conjugate** to f (as defined later in this section). A key point is showing that these sets are **cells** (i.e., homeomorphic images of balls in some $\mathbb{R}^n$). Moreover, we give a full description of the **bifurcations** which occur at the boundary of these sets.

2. The introduction of a more suitable **parameter space**. Following Milnor and Tresser, this is done by using the space $\mathcal{S}^b$ of **stunted sawtooth maps** as a model for the space P^b. Since one can assign to each polynomial $f \in P^b$ a stunted sawtooth map $\Psi(f)$, the spaces P^b and $\mathcal{S}^b$ are naturally related. As mentioned, the space $\mathcal{S}^b$ was already used by Milnor and Tresser.

3. To have a more faithful description of the parameter space, we restrict to the class of **admissible sawtooth maps** $\mathcal{S}_*^b$. In the real case, the admissibility condition corresponds to the absence of wandering intervals; for general polynomials this would correspond to the "absence of Levy cycles." A nontrivial result is that isentropes within the space $\mathcal{S}_*^b$ of admissible sawtooth maps are **contractible**.

4. A proof that $\Psi\colon P^b \to \mathcal{S}_*^b$ is "almost" a **homeomorphism**. This statement (which we will make precisely shortly) is the main rationale for introducing the space $\mathcal{S}_*^b$ rather than the much more pleasant space $\mathcal{S}^b$.

First ingredient: Rigidity and the partial conjugacy class

The first step in the proof is to introduce the notion of **partial conjugacy class** and to show that one has generic bifurcation at the boundary of these sets.

Let $B(f)$ consists of all points x so that $f^n(x)$ tends to a (possibly one-sided) periodic attractor. Let $f, g\colon [-1,1] \to [-1,1]$ be two d-modal maps.

DEFINITION 3.1. *We say that two f, g are **partially conjugate** if there is a homeomorphism $h\colon [-1,1] \to [-1,1]$ such that*

- *h maps $B(f)$ onto $B(g)$;*
- *h maps the ith critical point of f to the ith critical point of g;*
- *$h \circ f(x) = g \circ h(x)$ for all $x \notin B(f)$.*

DEFINITION 3.2. *We denote by $\mathcal{PH}(f)$ the class of maps which are partially conjugate to f, also called the **partial conjugacy class** associated to f. Furthermore, we denote by $\mathcal{PH}^o(f)$ the set of maps $g \in \mathcal{PH}(f)$ with*

- *only hyperbolic periodic points, and*
- *no critical point of g maps to the boundary of a component of $B(g)$.*

EXAMPLE 3.3. Consider the quadratic family $q_c(x) = -(c+1)x^2 + c, c \in [-1,1]$, and let a_n be the first period-doubling bifurcation creating a periodic orbit of period 2^{n+1}. Then all the maps corresponding to $c \in (-1, a_0]$ are partially conjugate. Similarly, all the maps corresponding to $c \in (a_n, a_{n+1}]$ are partially conjugate. If a polynomial $f \in P^b$ has no periodic attractors, then $g \in P^b$ is partially conjugate to f only if these maps are topologically conjugate. Hence, in this case, $\mathcal{PH}^o(f) = \mathcal{PH}(f)$ and, by Theorem 3.4, $g = f$ and $\mathcal{PH}(f) = \{f\}$.

If all critical points of f are in the basin of hyperbolic periodic attractors, then $\mathcal{PH}^o(f)$ agrees with the usual hyperbolic component, but the previous definition also makes sense if not all critical points are in basins of periodic attractors. The fact that hyperbolic components are cells was shown by Douady and Hubbard in the case of quadratic maps and by Milnor in the case of polynomials of higher degree [Mil92b]. The following three theorems from [BvS09] generalize these results to the space $\mathcal{PH}^o(f)$.

THEOREM 3.4. *Let $f \in P_\varepsilon^b$. Then*

- *$\mathcal{PH}^o(f)$ is a submanifold with dimension equal to the number of critical points in $B(f)$;*
- *$\mathcal{PH}(f) \subset \overline{\mathcal{PH}^o(f)}$.*

In particular, if no critical point of f is in the basin of a periodic attractor, then $\mathcal{PH}(f)$ is a single point. In fact, the description of $\mathcal{PH}^o(f)$ is more detailed.

THEOREM 3.5. *$\mathcal{PH}^o(f)$ is parametrized by (finite) **Blaschke** products and, for example, critical relations unfold transversally.*

Here Blaschke products are maps of the $\mathbb{D}$ of the type $z \mapsto z \prod_{i=1}^{n-1} \frac{z-a_i}{1-\bar{a}_i z}$. If each periodic attractor has precisely one critical point in its basin, then this description simplifies: $\mathcal{PH}^o(f)$ is parametrized by **multipliers** at the periodic attractors (so this corresponds to Douady and Hubbard's result).

That critical relations unfold transversally in special cases (when a critical point is eventually periodic) was previously proved in [vS00] and [BE09]. In addition to the preceding theorem, we need the following additional transversality properties at the boundary of partial conjugacy classes (and consequently at the "boundary" of the space of real Blaschke products):

THEOREM 3.6. *For each $g \in P_\varepsilon^b$ and each $f \in \overline{\mathcal{PH}(g)} \setminus \mathcal{PH}^o(g)$, there exists a continuous family f_t, $t \in [0,1]$ with $f_0 = f$, $f_1 = g$, $f_t \in \mathcal{PH}^o(g)$ for $t \in (0,1]$ and so that at $t = 0$, one has one or more of the following bifurcations:*

- *saddle-node: creation of one-sided attractor, which then becomes becomes an attracting + repelling pair;*

- *pitchfork: a two-sided attractor, which becomes repelling and spins off a pair of attracting orbits;*

- *period-doubling: multiplier -1 with a creation/destruction of a attractor of double the period;*

- *homoclinic bifurcation: a critical point in the basin is (eventually) mapped to the boundary of the basin.*

In other words, near $f \in \overline{\mathcal{PH}(g)} \setminus \mathcal{PH}^o(g)$ several of these bifurcations can occur at the same time, but each of these is generic. So, for example, in the saddle-node bifurcation case, $(f^n)''$ is non-zero at a periodic point of period n. In [BvS09] a more precisely formulated and slightly weaker version of Theorem 3.6 is stated and proved, as this suffices for the proof of Milnor's conjecture. Theorem 3.6 will be proved elsewhere (and is related to transversality results due to Epstein [BE09] and also due to Levin [Lev11], but in Theorem 3.6 we also ensure that all the maps f_t, $t \in (0,1]$ are partially conjugate).

The proofs of Theorems 3.4, 3.5, and 3.6 rely on complex methods, in particular, quasiconformal rigidity for maps within the space P^b: two topologically conjugate maps in P^b are quasiconformally conjugate. This result was proved by Kozlovski, Shen, and van Strien in [KSvS07a]. As was shown in [KSvS07b], it implies density of hyperbolicity within one-dimensional systems.

Second ingredient: The space of stunted sawtooth maps as a model for the parameter space

One can assign to each map $f \in P^b$ a so-called kneading map in the following way:

- Given a piecewise monotone d-modal map f with turning points $c_1, \ldots, c_b$, associate to $x \in [-1,1]$ its **itinerary** $i_f(x)$ consisting of a sequence of symbols $i_{f,n}(x)$, $n \geq 0$ from the alphabet $\mathcal{A} = \{I_0, c_1, I_1, c_2, \ldots, c_b, I_b\}$ (where $i_{f,n}(x)$ is the symbol s in $\mathcal{A}$ iff $f^n(x)$ belongs to the corresponding interval or singleton).

- As is well-known, $x \mapsto i_f(x)$ is **monotone** with respect to a variant of the lexicographic ordering (the signed lexicographical ordering).
- So, the following is well-defined:

$$\nu_i := \lim_{x \downarrow c_i} i_f(x).$$

- The **kneading invariant** $\nu(f)$ of f is defined as

$$\nu(f) := (\nu_1, \dots, \nu_b).$$

Any kneading sequence which is realized by some piecewise monotone d-modal map is called *admissible*.

Since the space of kneadings with the natural topology is not connected, following Milnor and Tresser, we find it easier to work in another space, namely, the space of stunted sawtooth maps. These are stunted versions of some fixed sawtooth map S with slope $\pm\lambda$, where $\lambda > 1$, as drawn in Figure 3.7. Stunted sawtooth maps T are modifications of S with each peak "stunted" (i.e., replaced by a plateau). In other words, T has slopes $\pm\lambda, 0$. The space of **stunted sawtooth maps** is denoted by $\mathcal{S}^b$ and can be parametrized by the parameters ζ_i, as in Figure 3.7.

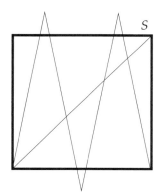 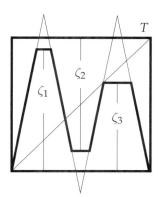

FIGURE 3.7: The sawtooth map S on the left and a stunted sawtooth map $T \in \mathcal{S}^3$ on the right (drawn in bold) with 3 plateaus. Adjacent plateaus of maps in $\mathcal{S}^b$ are allowed to touch.

To each map $f \in P^b$ we will assign a *unique* stunted sawtooth map $\Psi(f) \in \mathcal{S}^b$. Let $\nu(f) = (\nu_1, \dots, \nu_b)$ be the kneading invariant of f, and let s_i be the *unique point* in the $(i+1)$th lap I_i of S such that

$$\lim_{y \downarrow s_i} i_S(y) = \nu_i := \lim_{x \downarrow c_i} i_f(x).$$

Such a point s_i exists because all kneading sequences are realized by S. It is *unique* since S is expanding and so distinct points have different different kneading sequences. Next we associate to $f \in P^b$ the stunted sawtooth map $\Psi(f) \in \mathcal{S}^b$, which

- is constant on a plateau Z_i with right endpoint s_i, and
- agrees with S outside $\cup Z_i$.

Although all critical points of any map $f \in P^b$ are distinct, several plateaus of $\Psi(f) \in \mathcal{S}^b$ can touch (and so the number of genuine plateaus in $\Psi(f) \in \mathcal{S}^b$ can be less than b). We should emphasize that the map

$$P^b \ni f \mapsto \Psi(f) \in \mathcal{S}^b$$

is non-continuous, non-surjective, and also non-injective. Nevertheless, as we will see, weaker versions of the properties hold. Moreover, the space $\mathcal{S}^b$ has the following useful property: Let ζ_i describe the **height** of the ith plateau of T as in Figure 3.7; then $T \mapsto h_{top}(T)$ is monotone increasing in each parameter ζ_i. Using this, one can show *isentropes within $\mathcal{S}^b$ are connected* (and even contractible). We should emphasize that the approach we mentioned in this subsection goes back to [DGMT95, MTr00]. As our proof exploits the map $\Psi \colon P^b \to \mathcal{S}^b$, our next ingredient is to consider a subspace of $\mathcal{S}^b$.

Third ingredient: Addressing non-surjectivity of Ψ by introducing the space $\mathcal{S}^b_*$ of non-degenerate sawtooth maps

Polynomial maps have **no wandering intervals**. Hence, if the endpoints of an interval containing two distinct critical points have the same itineraries, then the interval is contained in the basin of a periodic attractor. Analogously, $\mathcal{S}^b_* \subset \mathcal{S}^b$ consists of maps T so that if

- an interval J contains two plateaus *and*

- $n > 0$ is so that $T^n(J)$ is a point,

- *then J is contained in the basin of a periodic attractor of T.*

(This corresponds to absence of a Levy-cycle obstruction.) This space $\mathcal{S}^b_*$ will be crucial in our discussion. Note that $\mathcal{S}^b_* = \mathcal{S}^b$ when $b = 1, 2$. The space $\mathcal{S}^b_*$ is messier than the original space $\mathcal{S}^b$ but still has the following (rather non-trivial) property.

THEOREM 3.8. *Isentropes within $\mathcal{S}^b_*$ are connected and even contractible.*

The proof of the analogous statement for the space $\mathcal{S}^b$ is much simpler and was already given in Milnor and Tresser [MTr00]. Even though the map

$$P^b \ni f \mapsto \Psi(f) \in \mathcal{S}^b_*$$

still is not surjective, it turns out that there is an interpretation (making this map set valued) in which it does become surjective. Indeed, define the *plateau-basin* $W(T)$:

$$W(T) = \{y; T^k(y) \in \text{interior}(\cup_{i=1}^b Z_{i,T}) \text{ for some } k \geq 0\}.$$

In order to ignore what happens within the basins of periodic attractors, define the equivalence class

$$\langle T \rangle = \{\tilde{T} \in \mathcal{S}^b; W(\tilde{T}) = W(T)\}$$

and also define

$$[T] = \text{closure}(\langle T \rangle).$$

Using this, we get surjectivity of Ψ.

THEOREM 3.9 ("Surjectivity" of Ψ). *For each $T \in \mathcal{S}^b_*$ there exists $f \in P^b$ so that $T \in [\Psi(f)]$.*

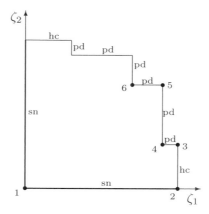

FIGURE 3.10: The case of a periodic component W of $W(T)$ of period $s_1 + s_2$ so that W and the component W' of $W(T)$ containing $T^{s_1}(W)$ both contain a plateau.

Fourth ingredient: Ψ is almost injective and almost continuous

To prove Milnor's conjecture we need to show that isentropes are connected within the space P^b. Since, by Theorem 3.8, the corresponding statement is true within the space $\mathcal{S}_*^b$ of non-degenerate sawtooth maps, we want to show that the spaces P^b and $\mathcal{S}_*^b$ are essentially homeomorphic. As we have shown in Theorem 3.9, $f \mapsto [\Psi(f)]$ is surjective. The next two propositions show that this map is essentially homeomorphic.

THEOREM 3.11 (Injectivity of Ψ). *If $f_1, f_2 \in P^b$ and $[\Psi(f_1)] \cap [\Psi(f_2)] \neq \emptyset$, then $\overline{\mathcal{PH}(f_1)} \cap \overline{\mathcal{PH}(f_2)} \neq \emptyset$.*

THEOREM 3.12 (Continuity of Ψ). *Suppose $f_n \in P^b$ converges to $f \in P^b$. Then any limit of $\Psi(f_n)$ is contained in $[\Psi(f)]$.*

The proof of Theorems 3.11 and 3.12 relies strongly on the fact that on the boundary of a set $\mathcal{PH}^o(f)$ one has generic bifurcations; see Theorems 3.5 and 3.6. Basically, this involves a description of the boundary of the set $[\Psi(f)]$, showing what bifurcations occur at this boundary; see Figure 3.10.

From the previous four theorems it easily follows that isentropes in P^b are connected, thus proving Milnor's conjecture.

4   OPEN PROBLEMS

In this section we will pose some further conjectures and questions on monotonicity of entropy. For other questions and a broader survey on one-dimensional dynamics, see [vS10].

Two questions: Are isentropes contractible? Are hyperbolic maps dense within "most" isentropes?

The following questions are due to Milnor [MTr00, page 125]:

QUESTION 4.1 (Milnor). *Fix $\epsilon \in \{-1, 1\}$. Are isentropes in P_ϵ^b contractible and cellular?*

Note that the isentropes in the space of admissible stunted sawtooth maps $\mathcal{S}_*^b$ are contractible (as mentioned, this is by no means a trivial result). Obviously, even though the map $\Psi \colon P_\epsilon^b \to \mathcal{S}_*^b$ is "almost" a homeomorphism, it is not easy to use the subtle deformation in $\mathcal{S}_*^b$ to construct one in the space P_ϵ^b. We believe that this conjecture is true, but this is work in progress.

A somewhat different conjecture was posed (in an email) by Thurston.

QUESTION 4.2 (Thurston). *Does there exist a dense set of level sets $H \subset [0, \log(d)]$ (where $d = b + 1$) so that for any $h_0 \in H$, the isentrope $I(h_0)$ in P_ϵ^b contains a dense set of hyperbolic maps?*

As usual, we say that a map is **hyperbolic** if each critical point is in the basin of a periodic attractor. As mentioned, it is known that hyperbolic maps are dense within P_ϵ^b [KSvS07a]. Solving Thurston's question most probably requires an ability to perturb a map to a hyperbolic map while staying inside an isentrope.

Question: Are isentropes always non-locally connected?

Even though all isentropes are connected, we have the following.

THEOREM 4.3 ([BvS11]). *When $b \geq 4$, not all isentropes within P^b are non-locally connected.*

A related theorem is due to [FT86] for maps of the circle. The previous theorem works only when $b \geq 4$, and it is possible that all isentropes are locally connected within P^b when $b = 3$. The previous theorem is in some sense the analogue of Milnor's theorem stating that the connectedness locus for cubic maps (in the complex plane) is not locally connected.

CONJECTURE 4.4. *When $b \geq 4$, no isentrope within P^b is locally connected. The boundary of the isentropy corresponding to zero-entropy is non-locally connected.*

Conjecture: Isentropes within the space of real polynomials are connected

In another direction, Tresser posed the following conjecture. Consider the space Pol_ϵ^d of real polynomials f of degree d, not necessarily with all critical points on the real line but still with $f(\{\pm 1\}) \subset \{\pm 1\}$ and $\epsilon(f) = \epsilon$ as in the definition of P^b.

CONJECTURE 4.5 (Tresser). *Fix $\epsilon \in \{-1, 1\}$. Isentropes in Pol_ϵ^d are connected.*

To explain this conjecture, let us take $d = 4$ and consider the subspace $Pol_\epsilon^{4,1}$ of real degree 4 polynomials $f \in Pol_\epsilon^4$, with one critical point c_1 on the real and the other two critical points c_2, c_3 (with $c_2 = \bar{c}_3$) off the real line. Maps in $Pol_\epsilon^{4,1}$ are unimodal. So consider the question of whether isentropes within $Pol_\epsilon^{4,1}$ are connected. To be specific, consider the set $\Sigma_{\log(2)}$ of maps $f \in Pol_\epsilon^{4,1}$ with $f(c_1) = 1$. Such maps, restricted to the real line, are conjugate to $x \mapsto 4x(1 - x)$, and have topology entropy $\log(2)$. The situation seems good, because we can prove that any two maps $f, \tilde{f} \in \Sigma_{\log(2)}$ are quasisymmetrically conjugate on the real line. However this does not imply that one can connect $f, \tilde{f}$ by an arc within $\Sigma_{\log(2)}$. Indeed, the dynamics of $f, \tilde{f}$ on the complex plane are entirely unrelated, and so

f and $\tilde{f}$ are in general certainly not quasiconformally conjugate. It seems therefore hopeless to use quasiconformal surgery to prove that $f, \tilde{f}$ can be connected by a path in $\Sigma_{\log(2)}$. On the other hand, the set $\Sigma_{\log(2)}$ forms a codimension-one algebraic subset of the (real) three-dimensional parameter space. By a somewhat tedious explicit calculation this algebraic set can be shown to be connected.

In spite of these difficulties, we recently proved the following, in joint work with Cheraghi.

THEOREM 4.6 ([CvS]). *Isentropes within the space $Pol_c^{4,1}$ are connected.*

The main ingredient in our proof is the property that critically finite polynomials with a given combinatorial type are unique.

Conjecture: Isentropes within more general unimodal families are connected.

Of course, one can ask what happens if one considers wider classes of functions. In recent work with Lasse Rempe, we have recently been able to prove results of the following type:

THEOREM 4.7 ([RvS10]). *The topological entropy of the map $f_a\colon [0,1] \to [0,1]$ defined by $f_a(x) = a \cdot \sin(\pi x)$ depends monotonically on a.*

In the proof of this theorem it is heavily used that $x \mapsto \sin(x)$ is a transcendental map with some additional geometric properties; for a proof and a much more general theorem [RvS10], where the following theorem is proven.

THEOREM 4.8 ([RvS10]). *Each isentrope within the space of trigonometric polynomials is connected.*

However, it is far from clear how to obtain results without relying on complex tools. For example, the following well-known conjecture has been open for the last 30 years.

CONJECTURE 4.9. *Take $\ell > 1$ not an integer. Then the topological entropy of the map $x \mapsto -(c+1)|x|^\ell + c$ depends monotonically on c.*

When ℓ is an integer, the corresponding statement can be proved as in the quadratic case. More generally, the following conjecture was posed by Nusse and Yorke.

CONJECTURE 4.10. *Let f be S-unimodal of the unit interval and symmetric, i.e.,*

$$f(1-x) = f(x).$$

Does the topological entropy of the map $f_a = a \cdot f$ depend monotonically on a?

Note that if one drops the assumption that f is symmetric, then this the conjecture definitely does not hold, as was shown by Zdunik, Nusse and Yorke, Kolyada and others (for references, see [dMvS93]). In fact, to prove this conjecture it is enough to show that, under the previous assumptions, periodic orbits of

$$[0,1] \ni x \mapsto a \cdot f(x) \in [0,1]$$

can never be destroyed as a increases. It should be noted that some partial results toward this can be obtained by applying the notion of rotation number; see [GT92],

[Blo94], and [BM97]; under mild conditions periodic orbits with particular types of combinatorics (and uncountably many types of aperiodic behavior) do not disappear as a increases. The previous conjecture is subtle: there are C^3 close maps $f, g: [0, 1] \to [0, 1]$ of this type for which $f \leq g$ and $h_{top}(f) > h_{top}(g)$ [Bru95].

Question: Is antimonotonicity common?

We should note that even though isentropes within the space of cubic polynomials are connected, isentropes are complicated non-locally connected topological sets [BvS11]. Related to this, one has the following.

THEOREM 4.11 ([BvS09]). *It is well known that cubic polynomials can be parametrized by their critical values. However, the entropy of a cubic polynomial does not depend monotonically on these parameters separately.*

Dawson, Grebogi, Kan, Koçak, and Yorke proposed that this phenomenon holds more generally by stating the following general conjecture.

ANTIMONOTONICITY CONJECTURE ([DG91],[DGK93],[DGMT95]). *A smooth one-dimensional map depending on one parameter has antimonotone parameter values whenever two critical points have disjoint orbits and are contained in the interior of a chaotic attractor.*

A further discussion about the relation between connectedness of isentropes and the above antimonotony conjecture can be found in [MTr00].

Question: Are isentropes within families of higher-dimensional maps connected?

As we have seen, even though isentropes for cubic maps are connected, one has antimonotonicity. Motivated by this, we pose the following question.

QUESTION 4.12 (Hénon maps). *Let $H(x, y) = (1 - ax^2 + by, y)$ be the family of Hénon maps. It is known ([KKY92] and [DGY$^+$92]) that for fixed b, the set of parameters*

$$\{a; h_{top}(H_{a,b}) = h_0\}$$

is not connected. However, is it possible that isentropes $I(c) = \{(a, b); h_{top}(H_{a,b}) = h_0\}$ are connected? For all h_0? For some h_0? For $h_0 = 0$?

For preliminary results in this direction and further references, see [GT91]. A positive answer for the case when $h_0 = 0$ would mean that the boundary of chaos (as defined by positivity of topological entropy) is connected. If so, a decent picture emerges of how one can move from simple (i.e., zero entropy dynamics) to chaotic dynamics.

Of course, the difficulty in resolving the last question is that one can no longer rely on holomorphic dynamics.

5 Bibliography

[BE09] Xavier Buff and Adam Epstein, *Bifurcation measure and postcritically finite rational maps*, Complex dynamics, A K Peters, Wellesley, MA, 2009, pp. 491–512. MR 2508266 (2010g:37072).

[Blo94]   Alexander Blokh, *Rotation Number for Unimodal Maps*, MSRI (1994), Preprint #058-94.

[BM97]   Alexander Blokh and Michał Misiurewicz, *Entropy of twist interval maps*, Israel J. Math. **102** (1997), 61–99. MR 1489101 (99b:58139).

[Bru95]   Henk Bruin, *Non-monotonicity of entropy of interval maps*, Phys. Lett. A **202** (1995), no. 5–6, 359–362. MR 1336987 (96g:58051).

[BvS09]   Henk Bruin and Sebastian van Strien, *Monotonicity of entropy for real multimodal maps*, preprint, 2009. arXiv:0905.3377

[BvS11]   Henk Bruin and Sebastian van Strien, *Non-local connectedness of isentropes*, Dynamical Systems, an International Journal **38** (2013), 381–392.

[CvS]   Davoud Cheraghi and Sebastian van Strien, *In preparation*.

[DG91]   Silvina Ponce Dawson and Celso Grebogi, *Cubic maps as models of two-dimensional antimonotonicity*, Chaos Solitons Fractals **1** (1991), no. 2, 137–144. MR 1295901.

[DGK93]   Silvina Ponce Dawson, Celso Grebogi, and Hüseyin Koçak, *Geometric mechanism for antimonotonicity in scalar maps with two critical points*, Phys. Rev. E (3) **48** (1993), no. 3, 1676–1682. MR 1377914 (96j:58110).

[DGMT95]   Silvina P. Dawson, Roza Galeeva, John Milnor, and Charles Tresser, *A monotonicity conjecture for real cubic maps*, Real and complex dynamical systems (Hillerød, 1993), NATO Adv. Sci. Inst. Ser. C Math. Phys. Sci., vol. 464, Kluwer Acad. Publ., Dordrecht, 1995, pp. 165–183. MR 1351522 (96g:58132).

[DGY+92]   Silvina P. Dawson, Celso Grebogi, James A. Yorke, Ittai Kan, and Hüseyin Koçak, *Antimonotonicity: inevitable reversals of period-doubling cascades*, Phys. Lett. A **162** (1992), no. 3, 249–254. MR 1153356 (92k:58182).

[DH85]   Adrien Douady and John Hamal Hubbard, *Étude dynamique des polynômes complexes. Partie II*, Publications Mathématiques d'Orsay [Mathematical Publications of Orsay], vol. 85, Université de Paris-Sud, Département de Mathématiques, Orsay, 1985, With the collaboration of P. Lavaurs, Tan Lei and P. Sentenac. MR 87f:58072b.

[dMvS93]   Welington de Melo and Sebastian van Strien, *One-dimensional dynamics*, Ergebnisse der Mathematik und ihrer Grenzgebiete (3) [Results in Mathematics and Related Areas (3)], vol. 25, Springer-Verlag, Berlin, 1993. MR 95a:58035.

[Dou95]   Adrien Douady, *Topological entropy of unimodal maps: monotonicity for quadratic polynomials*, Real and complex dynamical systems (Hillerød, 1993), NATO Adv. Sci. Inst. Ser. C Math. Phys. Sci., vol. 464, Kluwer Acad. Publ., Dordrecht, 1995, pp. 65–87. MR 1351519 (96g:58106).

[FT86]   B. Friedman and C. Tresser, *Comb structure in hairy boundaries: Some transition problems for circle maps*, Phys. Lett. A **117** (1986), no. 1, 15–22. MR 851335 (87j:58061).

[GŚ97]   Jacek Graczyk and Grzegorz Świątek, *Generic hyperbolicity in the lo-
            gistic family*, Ann. of Math. (2) **146** (1997), no. 1, 1–52. MR 1469316
            (99b:58079).

[GT91]   Jean-Marc Gambaudo and Charles Tresser, *How horseshoes are cre-
            ated*, Instabilities and nonequilibrium structures, III (Valparaíso, 1989),
            Math. Appl., vol. 64, Kluwer Acad. Publ., Dordrecht, 1991, pp. 13–25.
            MR 1177837 (93m:58083).

[GT92]   Jean-Marc Gambaudo and Charles Tresser, *A monotonicity property
            in one-dimensional dynamics*, Symbolic dynamics and its applications
            (New Haven, CT, 1991), Contemp. Math., vol. 135, Amer. Math. Soc.,
            Providence, RI, 1992, pp. 213–222. MR 1185089 (93i:58087).

[Hec96]   Christopher Heckman, *Monotonicity and the construction of quasiconfor-
            mal conjugacies in the real cubic family.*, Ph.D. thesis, Stony Brook, 1996.

[KKY92]   Ittai Kan, Hüseyin Koçak, and James A. Yorke, *Antimonotonicity: Con-
            current creation and annihilation of periodic orbits*, Ann. of Math. (2) **136**
            (1992), no. 2, 219–252. MR 1185119 (94c:58135).

[KSvS07a]   Oleg Kozlovski, Weixiao Shen, and Sebastian van Strien, *Rigidity for
            real polynomials*, Ann. of Math. (2) **165** (2007), no. 3, 749–841.

[KSvS07b]   Oleg Kozlovski, Weixiao Shen, and Sebastian van Strien, *Density of hy-
            perbolicity in dimension one*, Ann. of Math. (2) **166** (2007), no. 1, 145–182.
            MR 2342693 (2008j:37081).

[Lev11]   Genadi Levin, *Multipliers of periodic orbits in spaces of rational maps*, Er-
            godic Theory Dynam. Systems **31** (2011), no. 1, 197–243. MR 2755929
            (2012a:37095).

[Lyu97]   Mikhail Lyubich, *Dynamics of quadratic polynomials. I, II*, Acta Math. **178**
            (1997), no. 2, 185–247, 247–297. MR 98e:58145.

[Mil92]   John Milnor, *Remarks on iterated cubic maps*, Experiment. Math. **1** (1992),
            no. 1, 5–24. MR 1181083 (94c:58096).

[Mil92b]   John Milnor (with an appendix by Alfredo Poirier), *Hyperbolic com-
            ponents*, in Conformal dynamics and hyperbolic geometry, Con-
            temp. Math. **573**, (ed. by F. Bonahon, R. L. Devaney, F. P. Gardiner
            and D. Šarić), 183–232. Amer. Math. Soc., Providence, RI, 2012.
            arXiv:1205.2668

[MT77]   John Milnor and William Thurston, *On iterated maps of the interval, I+
            II*, (An expanded version was published as [MT88]), 1977.

[MT88]   John Milnor and William Thurston, *On iterated maps of the interval*, Dy-
            namical systems (College Park, MD, 1986–87), Lecture Notes in Math.,
            vol. 1342, Springer, Berlin, 1988, pp. 465–563. MR 970571 (90a:58083).

[MTr00]   John Milnor and Charles Tresser, *On entropy and monotonicity for real
            cubic maps*, Comm. Math. Phys. **209** (2000), no. 1, 123–178, With an
            appendix by Adrien Douady and Pierrette Sentenac. MR 1736945
            (2001e:37048).

[MS77]   Michał Misiurewicz and Wiesław Szlenk, *Entropy of piecewise monotone mappings*, Dynamical systems, Vol. II—Warsaw, Soc. Math. France, Paris, 1977, pp. 299–310. Astérisque, No. 50. MR 0487998 (58 #7577).

[MS80]   Michał Misiurewicz and Wiesław Szlenk, *Entropy of piecewise monotone mappings*, Studia Math. **67** (1980), no. 1, 45–63. MR 579440 (82a:58030).

[Ra07]   Anca Radulescu, *The connected isentropes conjecture in a space of quartic polynomials*, Discrete Contin. Dyn. Syst. **19** (2007), no 1, 139–175. MR 2318278 (2009j:37022).

[RvS10] Lasse Rempe and Sebastian van Strien, *Density of hyperbolicity for real transcendental entire functions with real singular values*, In preparation, 2010. arXiv:1005.4627

[Tsu00] Masato Tsujii, *A simple proof for monotonicity of entropy in the quadratic family*, Ergodic Theory Dynam. Systems **20** (2000), no. 3, 925–933. MR 1764936 (2001g:37048).

[vS00]   Sebastian van Strien, *Misiurewicz maps unfold generically (even if they are critically non-finite)*, Fund. Math. **163** (2000), no. 1, 39–54. MR 2001g:37064.

[vS10]   Sebastian van Strien, *One-dimensional dynamics in the new millennium*, Discrete Contin. Dyn. Syst. **27** (2010), no. 2, 557–588. MR 2600680 (2011j:37071).

Entropy in dimension one

William P. Thurston

Bill Thurston fell ill during the process of completing a draft version of this manuscript. During his illness, Sarah Koch assisted him in making revisions. Sadly, Bill passed away in August of 2012. The editors received help from several people in preparing this paper for publication, including Mladen Bestvina, Michael Handel, Wolf Jung, Sarah Koch, Doug Lind, Curt McMullen, John Milnor, Lee Mosher, and Tan Lei. Footnotes and a section of additional remarks (§13) have been added by John Milnor.

1 INTRODUCTION

The topological entropy $h(f)$ of a map from a compact topological space to itself, $f\colon X \to X$, is a numerical measure of the unpredictability of trajectories $x, f(x), f^2(x), \ldots$ of points under f: it is the limiting upper bound for exponential growth rate of ϵ-distinguishable orbits, as $\epsilon \to 0$. Here ϵ can be measured with respect to an arbitrary metric on X, or one can merely think of it as a neighborhood of the diagonal in $X \times X$, since all that matters is whether or not two points are within ϵ of each other. The number of ϵ-distinguishable orbits of length n is the maximum cardinality of a set of orbits of f such that no two are always within ϵ.

More formally, given a metric d on X, define the ϵ-count of X, $N(X, \epsilon, d)$ to be the maximum cardinality of a set $S \subset X$ such that no point is within ϵ of any other. Given a continuous function $f\colon X \to X$, define a metric $d_{f,n}$ on X as $d_{f,n}(x, y) = \sup_{0 \le i < n} \{d(f^i(x), f^i(y))\}$. Then

$$h(f) = \lim_{\epsilon \to 0} \limsup_{n \to \infty} \frac{1}{n} \log\left(N(X, \epsilon, d_{f,n})\right).$$

If f is a Lipschitz self-map of a compact m-manifold with Lipschitz constant $K \ge 1$, then it's easy to see that

$$h(f) \le m \log(K) \tag{1.1}$$

(compare [Mis]). The upper bound is attained in cases such as $x \mapsto Kx$ acting on the torus $\mathbb{R}^m/\mathbb{Z}^m$, where K is an integer. On the other hand, for a continuous map f that is not Lipschitz, $h(f)$ need not be finite, even for simple situations such as homeomorphisms of S^2 or continuous maps of intervals.

A differentiable map f of an interval to itself is *postcritically finite* or *critically finite* if the union of forward orbits of the critical points is finite. In particular, the set of critical points for f must be finite.

For a map f that is not differentiable, we can define any interior point that is a local maximum or local minimum to be a turning point, or topological critical point. (Note that with this definition, not all smooth critical points are topological critical points.) Let $c(f)$ denote the *modality*, or number of turning points for such

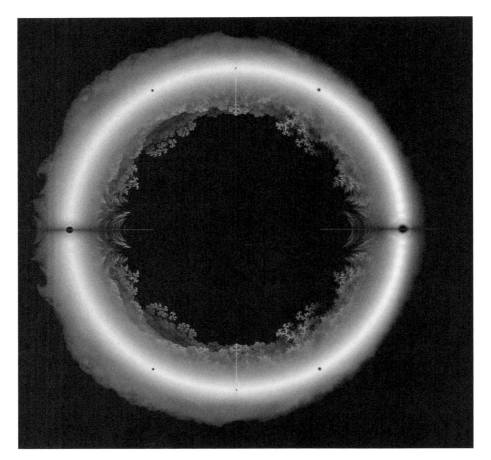

FIGURE 1.1: This is a plot of roughly 8×10^8 roots of defining polynomials for $\exp(h(f))$, where f ranges over a sample of about 10^7 postcritically finite quadratic maps of the interval with postcritical orbit of length ≤ 80. The brightness is proportional to the log of the density; the highest concentration is at the unit circle. (See also Plate 16.)

a map, and let $\mathrm{Var}(f)$ denote the total variation of f, i.e., its arclength considered as a path. For maps of the interval to itself with finitely many critical points, there are two simple ways to characterize the topological entropy.

THEOREM 1.2 (Misiurewicz-Szlenk, [MS1], [MS2]). *For a continuous map f of an interval to itself with finitely many turning points, the topological entropy equals*

$$h(f) = \lim_{n \to \infty} \frac{1}{n} \log(\mathrm{Var}(f^n)) = \lim_{n \to \infty} \frac{1}{n} \log c(f^n).$$

In particular, these are actual limits, not just limits of lim sups. There are good algorithms to actually compute the entropy [MT].

The first main goal of this paper is to characterize what values of entropy can occur for postcritically finite maps.

THEOREM 1.3. *A positive real number h is the topological entropy of a postcritically finite self-map of the unit interval if and only if* $\exp(h)$ *is an algebraic integer that is at least*

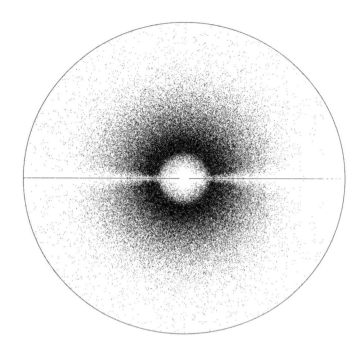

FIGURE 1.4: Shown are the roots of the minimal polynomials for 5,932 degree 21 Perron numbers, obtained by sampling 20,000 monic degree 21 polynomials with integer coefficients between 5 and -5, and keeping those that have a root in $[1, 2]$ larger than all other roots. The indicated circle has radius 2.

as large as the absolute value of any conjugate of $\exp(h)$. *The map may be chosen to be a polynomial, all of whose critical points are in* $(0, 1)$.

Two maps $f_1, f_2\colon X \to X$ are *conjugate* if there is a homeomorphism g of X conjugating one to the other, i.e., $g \circ f_1 = f_2 \circ g$. They are *semiconjugate* if there is a map g satisfying the condition that is continuous and surjective but not necessarily a homeomorphism. Basically, a semiconjugacy can collapse out certain kinds of subsidiary behavior of the dynamics.

Many phenomena of 1-dimensional dynamics are irrelevant for the study of entropy; there is a relatively simple family of non-smooth examples that has central importance when f is PL (piecewise-linear) continuous and $|f'| = \lambda > 1$ is constant, wherever f' exists. We call such an f a *uniform expander*, or a *uniform λ-expander* if we want to be more specific. These maps are often called maps with *constant slope*. The importance of the uniform expanders is indicated by this theorem.

THEOREM 1.5 ([MT]). *Every continuous self-map g of an interval with finitely many turning points and with $h > 0$ is semi-conjugate to a uniform λ-expander* $\mathrm{PL}(g)$ *with the same topological entropy* $h = \log(\lambda)$. *If g is postcritically finite, so is* $\mathrm{PL}(g)$. *(But if* $\mathrm{PL}(g)$ *is postcritically finite, it does not imply that g is postcritically finite.)*

In [ALM], a more general version of Theorem 1.5 is proven, which applies in more circumstances, including, for instance, piecewise continuous maps and self-maps of graphs.

In other words, Theorem 1.3 reduces to the study of expansion constants for 1-dimensional uniform expanders.

In the case of a postcritically finite map f, a uniform expander model is easily computed. If necessary, first trim the domain interval until it maps to itself surjectively, by taking the intersection of its forward images. If this is a point, then the entropy is 0. Otherwise, the two endpoints are either images of an endpoint, or images of a turning point. In this case, conjugate by an affine transformation to make the interval $[0, 1]$.

If we now subdivide $[0, 1]$ by cutting at the union of postcritical orbits (including the critical points) into intervals J_i, each J_i maps homeomorphically to a finite union of other subintervals. If there is a uniform expander F with the same qualitative behavior, that is, having an isomorphic subdivision into subintervals each mapped homeomorphically to the corresponding union of other subintervals, then the lengths of the intervals satisfy a linear condition: The sum of the lengths of intervals J_j hits is λ times the length of J_j. In other words, the lengths of the intervals of the subdivision define a positive eigenvector for a non-negative matrix, with eigenvalue λ.

The Perron-Frobenius theorem gives necessary and sufficient conditions for this to exist. Here is some terminology: a non-negative matrix is *ergodic* if the sum of some number of its positive powers is strictly positive, and it is *mixing* if some power (and hence, all subsequent powers) is strictly positive. The incidence matrix for f is ergodic if and only if for each pair of intervals J_i and J_j, some $f^n(J_i)$ contains J_j. The incidence matrix is mixing if and only if for each J_i, some $f^n(J_i)$ image covers all intervals.

The Perron-Frobenius theorem says that any non-negative matrix has at least one non-negative eigenvector with non-negative eigenvalue $\geq$ the absolute value of any other eigenvalue. If the matrix is ergodic, there is a unique strictly positive eigenvector; its eigenvalue is automatically [strictly] positive.[1]

From any non-negative eigenvector for the incidence matrix, we can make a uniformly expanding model by subdividing the unit interval into subintervals whose lengths equal the corresponding coordinate of the eigenvector normalized to have L^1 norm $= 1$. We thus obtain a PL map; its topological entropy is the log of the eigenvalue.

If all the entries of the matrix are integers, then its characteristic polynomial has integer coefficients, so it's an immediate corollary that the expansion constant for a postcritically finite uniform λ-expander is at least as large as the absolute value of its Galois conjugates.

In [MT, Lemma 7.9] there is a more general formula (very quick on a computer) for a semiconjugacy to a uniform expander for a general map with finitely many critical points.

Doug Lind proved a converse to the integer Perron-Frobenius theorem.

THEOREM 1.6 ([Li]). *For any real algebraic integer $\lambda > 0$ that is strictly larger than its Galois conjugates in absolute value, there exists a non-negative integer matrix that is mixing (some power strictly positive) and has λ as an eigenvalue.*

We will prove Lind's theorem, on our way to other results, in Section 3.

---

[1]For an elegant description of the entire non-zero spectrum of a nonnegative mixing integer matrix, see [KOR].

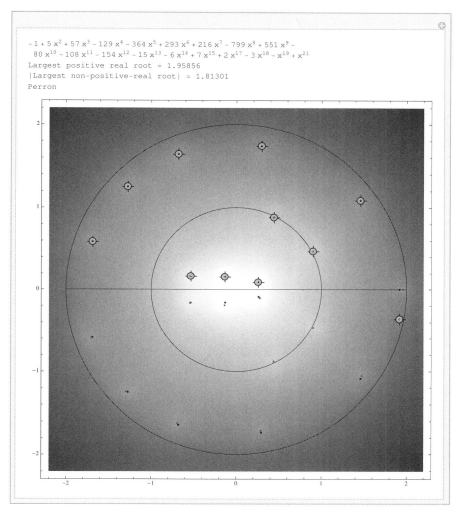

$-1 + 5\,x^2 + 57\,x^3 - 129\,x^4 - 364\,x^5 + 293\,x^6 + 216\,x^7 - 799\,x^8 + 551\,x^9 -$
$80\,x^{10} - 108\,x^{11} - 154\,x^{12} - 15\,x^{13} - 6\,x^{14} + 7\,x^{15} + 2\,x^{17} - 3\,x^{18} - x^{19} + x^{21}$
Largest positive real root = 1.95856
|Largest non-positive-real root| = 1.81301
Perron

FIGURE 1.7: This is a plot of the roots of an irreducible polynomial P of degree 21, as printed above the picture, defining a Perron number, $\lambda \approx 1.95856$. Although λ is an algebraic unit < 2, the map $x \mapsto \lambda|x| - 1$ cannot be postcritically finite since P has (several) roots not in (far away from) the region shown in Figure 1.1. To construct P, first a monic real polynomial P_0 was defined by its roots with a mouse (the black dots). The coefficients of P_0 were rounded to define the integer polynomial P, whose roots are indicated by a lighter gray dot (typically close to a root of P_0), sometimes obscuring the roots of P_0. The constant term was made small by balancing roots inside and outside the unit circle. When roots for P_0 are chosen so that those away from the unit circle are spaced well apart (relative to the sizes of coefficients), they tend to be fairly stable under rounding. The shading, proportional to $\log(1 + |P|)$, is a guide to stability: clicking new roots into darker areas is typically stabilizing. (See also Note 13.1.)

In view of Lind's theorem together with the Perron-Frobenius theorem, these numbers are called *Perron* numbers. An algebraic integer λ that satisfies the weak inequality $\lambda \geq |\lambda^\alpha|$, where α ranges over the Galois group of λ, is a *weak Perron number*.

There are two important (and better-known) special cases of Perron numbers. A positive real algebraic integer is a *Pisot number*, or *Pisot-Vijayaraghavan number*, or *PV number* if all its Galois conjugates are in the open unit disk. Since the product of all Galois conjugates is a nonzero integer, λ is bigger than its conjugates. It turns out that the set of Pisot numbers is a closed subset of $\mathbb{R}$. If $\lambda > 1$ is a real algebraic integer that has at least one conjugate on the unit circle and all conjugates are in the closed unit disk, then λ is called a *Salem number*. Since the complex conjugate of a point on the unit circle is its inverse, every element of the Galois conjugacy class of a point on the unit circle is also Galois conjugate to its inverse; in particular λ is Galois conjugate to $1/\lambda$. Therefore, $1/\lambda$ is the only Galois conjugate of λ in the open unit disk, since the inverse of any other conjugate in the open unit disk would give another conjugate of λ outside the unit disk.

Note: The dimension of the matrix in Lind's theorem is not bounded by the degree of λ: for any integer $n > 0$, there are, in fact, cubic Perron numbers λ that are not eigenvalues for nonnegative matrices smaller than $n \times n$.[2]

The proof of Theorem 1.3 uses techniques motivated by Doug Lind's methods.

It is also interesting to investigate what happens for postcritically finite maps subject to a bound on the number of turning points, in particular, a single turning point (i.e., for quadratic polynomials). The situation is very different. Figure 1.1 shows the Galois conjugates of $\exp(h)$ for postcritically finite real quadratic maps. This is a path-connected set, with much structure visible.[3]

Most Perron numbers between 1 and 2 do not have roots in this set. For example, Figure 1.4 shows a sampling of degree 21 polynomials with coefficients between -5 and 5 that happen to define a Perron number between 1 and 2 (out of 20000 random polynomials, 5937 fit the condition). These however are not random Perron numbers of degree 21: more typically, many of the coefficients are much larger. Figure 1.7 shows a degree 21 example constructed by hand, first specifying a collection of real points and pairs of complex conjugate points in the disk of radius 2, expanding the monic polynomial with those points as roots, and rounding the coefficients to the nearest integers. With care in spacing and placement of roots, the integer polynomial has roots near the given choices. (When the points away from the unit circle cluster too much, their positions become unstable with respect to rounding).

More generally, if Γ is a finite graph and $f: \Gamma \to \Gamma$ is a continuous map which is an embedding when restricted to any edge, we will say that f is *postcritically finite* if the forward orbit of every vertex is finite. If f has the additional property that for all k, f^k restricted to any edge is an immersion (it is an embedding on a sufficiently small neighborhood of any point), then f is a *traintrack map*.

Entropy for graph maps behaves similarly to entropy for intervals.

---

[2]Lind points out an easy family of examples where the degree of the dominant eigenvalue λ for a nonnegative $n \times n$ matrix must be strictly less than n. When the degree is equal to n, the trace of λ must equal the trace of the matrix, which is non-negative. But a Perron number may well have negative trace. The dominant root of $p(t) = t^3 + 3t^2 - 15t - 46$, with trace -3, provides an example.

[3]See Note 13.2, as well as the further discussion in §7.

THEOREM 1.8 (Alsedá-Llibre-Misiurewicz [ALM]). *Let $f\colon \Gamma \to \Gamma$ be a continuous self-map of a finite graph which has finitely many exceptional points $x(f)$, where f is not a local homeomorphism. Then*

$$h(f) = \lim_{n\to\infty} \frac{1}{n} \log \mathrm{Var}(f^n).$$

If f is a degree d covering map $S^1 \to S^1$, this yields $h(f) = \log(d)$; otherwise, the entropy also satisfies

$$h(f) = \lim_{n\to\infty} \frac{1}{n} \log(x(f^n)).$$

A self-map of a graph that is a homotopy equivalence defines an outer automorphism of its fundamental group, that is, an automorphism up to conjugacy (since we're not specifying a base point that must be preserved).

Handel and Bestvina [BH] showed that for any outer automorphism that is irreducible in the sense that no free factor is preserved up to conjugacy, there is a graph Γ and a traintrack map of Γ to itself representing the outer automorphism. They also developed a theory of relative traintracks that addresses outer automorphisms that are reducible. Traintrack theory is a powerful tool, parallel in many ways to pseudo-Anosov theory for self-homotopy-equivalences of surfaces. Algebraically, you can look at the action of an outer automorphism on conjugacy classes, represented by cyclically reduced words in a free group. The lengths of images of cyclically reduced words have a limiting exponential growth rate. There is a well-understood situation when traintrack automorphisms[4] can have conjugacy classes that are fixed under an automorphism: for instance, any automorphism of the free group on $\{a, b\}$ either fixes the conjugacy class of the commutator $[a, b]$ or carries it to its inverse. (See [LS, p. 44].) Apart from these, the lengths of the sequence of images of any conjugacy classes under iterates of a traintrack automorphism have exponent of growth equal to the topological entropy of the traintrack map, as measured in any generating set.

Peter Brinkmann wrote a very handy java application Xtrain that implements the Bestvina-Handel algorithm, http://math.sci.ccny.cuny.edu/pages?name= XTrain. I used this program extensively to work out and check examples for the next theorem, which is the second main goal of this paper.

THEOREM 1.9. *A positive real number h is the topological entropy for an ergodic traintrack representative of an outer automorphism of a free group if and only if its expansion constant $\exp(h)$ is an algebraic integer that is at least as large as the absolute value of any conjugate of $\exp(h)$.*

Note: Even though automorphisms are invertible, the expansion constant need not be an algebraic unit.

The relationship between the expansion constant of an automorphism and the expansion constant of its inverse is mysterious, but there is one special case where it's possible to control the expansion constant for both an automorphism ϕ and its inverse ϕ^{-1}. An automorphism ϕ is *positive* with respect to a set G of free generators if ϕ of any generator is a positive word in the generators, that is, it maps the semigroup they generate into itself. This implies that ϕ comes from a traintrack map of the bouquet of oriented circles defined by G.

---

[4]Presumably the term "traintrack automorphism" refers to the outer automorphism of a free group which is induced by some given traintrack map.

DEFINITION 1.10. *A linear transformation A is* bipositive[5] *with respect to a basis B if B can be expressed as the disjoint union $B = P \cup N$ such that A is non-negative with respect to B, and its inverse is non-negative with respect to the basis $P \cup (-N)$. An automorphism ϕ is* bipositive *if it is positive with respect to a set G of free generators, and its inverse is positive with respect to a set of generators obtained by replacing some subset of elements of G by their inverses.*

At one point, I hoped that the criteria in the following theorem would characterize all pairs of expansion constants for a free group automorphism and its inverse. This turned out to be false (Theorem 1.9), but the characterization of such pairs in this special case is still interesting. The proof of the following statement will be sketched in Section 12.

THEOREM 1.11. *A pair (λ_1, λ_2) of positive real numbers is the pair of expansion constants for ϕ and ϕ^{-1}, where ϕ is bipositive, if and only if it is the pair of positive eigenvalues for a bipositive element of $GL(n, \mathbb{Z})$ for some n, if and only if λ_1 and λ_2 are real algebraic units such that the Galois conjugates of λ_1 and λ_2^{-1} are contained in the closed annulus $\lambda_2^{-1} \leq |z| \leq \lambda_1$.*

Theorem 1.11 does not extend in an immediate way to the general case. Classification of the set of pairs of expansion constants that can occur in general remains mysterious. As already noted, these expansion constants need not be units. Moreover, there are examples of traintrack automorphisms where the Galois conjugates of λ_1 and λ_2^{-1} are *not* contained in the annulus $\lambda_2^{-1} \leq |z| \leq \lambda_1$. It is consistent with what I currently know that *every* pair of weak Perron numbers greater than 1 is the pair of expansion constants for ϕ and ϕ^{-1}.

2   SPECIAL CASE: PISOT NUMBERS

The special case that λ is a Pisot number has a particularly easy theory, so we will look at that first.

It is easy to see that the topological entropy of a map $f : I \to I$ with $d - 1$ topological critical points can be at most $\log(d)$: each point has at most d preimages, so the total variation of f^n is at most d^n.

THEOREM 2.1. *For any integer $d > 1$ and any Pisot number $\lambda \leq d$, there is a postcritically finite map $f : I \to I$ of degree d (that is, having $d - 1$ critical points) with entropy $\log(\lambda)$.*

In fact, when λ is Pisot, every λ-uniformly expanding map whose critical points are in $\mathbb{Q}(\lambda)$ is postcritically finite.

PROOF. One way to construct uniform λ-expanders is to create their graphs by folding. Start with the graph of the linear function $x \mapsto \lambda x$ on the unit interval. Now reflect the portion of the graph above the line $y = 1$ through that line. Reflect the portion of the new function that is below the line $y = 0$ through that line. Continue until the entire graph is folded into the strip $0 \leq y \leq 1$.

---

[5]The following example was suggested by D. A. Lind: Let

$$A = \begin{pmatrix} 1 & 1 \\ 1 & 2 \end{pmatrix}, \quad N = \left\{ \begin{pmatrix} 1 \\ 0 \end{pmatrix} \right\} = \{\mathbf{e}_1\}, \quad \text{and} \quad P = \left\{ \begin{pmatrix} 0 \\ 1 \end{pmatrix} \right\} = \{\mathbf{e}_2\}.$$

Then A is non-negative with respect to $\{\mathbf{e}_1, \mathbf{e}_2\}$, and A^{-1} is non-negative with respect to $\{-\mathbf{e}_1, \mathbf{e}_2\}$.

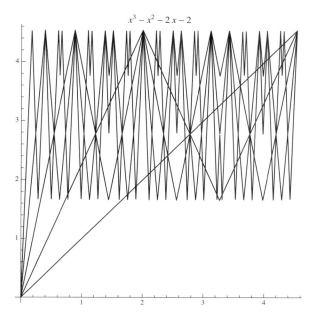

$$x^3 - x^2 - 2x - 2$$

FIGURE 2.2: This is the graph of the first 4 iterates of the function described in the proof, for the Pisot number satisfying $x^3 - x^2 - 2x - 2 = 0$. The Pisot root is $\lambda = 2.2695308$, and the other roots are $-0.63476542 + 0.69160123\,i$ and its conjugate, of modulus 0.938743. The first critical point, 2, maps to the endpoint 2λ, which is fixed. The other critical point, $1 + \lambda$, maps to a fixed point on the third iterate.

This results in a function that may have fewer than $d - 1$ critical points, so repeatedly reflect segments of the graph through horizontal segments $y = \alpha$ at heights $\alpha \in \mathbb{Q}(\lambda)$ until the function has the desired number $d - 1$ of critical points.

For convenience, rescale by some integer n to clear all denominators, so that all critical points are algebraic integers. Therefore, the postcritical orbits are contained in $\mathbb{Z}[\lambda]$.

For each Galois conjugate λ^α of λ, there is an embedding of $\mathbb{Z}[\lambda]$ in $\mathbb{C}$. In each such embedding, the postcritical orbits remain bounded: the action of f_{λ^α} on any point is some composition of functions of the form $f_{\lambda^\alpha, i}(x) = \pm \lambda^\alpha(x) + a_i$, which act as contractions, so there is a compact subset $K \subset \mathbb{C}$ that the Galois conjugates of every piece, $f_{\lambda^\alpha, i}$ takes inside itself.

By construction, the orbit of the critical points under f_λ is also bounded, since f_λ is a map of an interval to itself.

For any fixed bound B, there is only a finite number of algebraic integers of $\mathbb{Q}(\lambda)$ satisfying the bound $|\lambda^\alpha| < B$, since in the embedding into the product of real embeddings and a selection of one from each pair of complex conjugate embeddings, the algebraic integers in $\mathbb{Q}(\lambda)$ form a lattice.[6] The postcritical set is contained in such a set, so it is finite.   □

---

[6]This argument appears in the paper by K. Schmidt [Sch]. It was obtained independently by A. Bertrand (see [Ber] and [Ber1]). It actually goes back to an idea of Gelfond (see [Ge]). These papers concern the β-transformation $T_\beta : [0, 1) \to [0, 1)$, $T_\beta(x) = \beta x \mod 1$, which is not continuous, but the arguments are the same. In particular Schmidt's paper shows that if β is Pisot, then the set of points (pre-)periodic under T_β is exactly $\mathbb{Q}(\beta) \cap [0, 1)$.

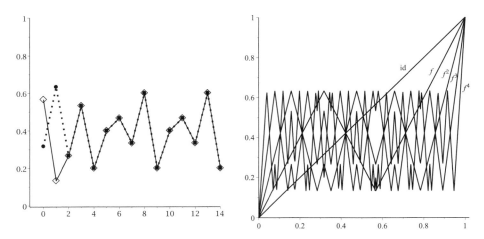

FIGURE 2.3: The diagram on the left shows the postcritical orbits for the Pisot construction where $\lambda = 2$, $d = 3$, with the critical points chosen as 19/60 and 17/30. Here the two critical orbits come together after two iterations, and land on a period five orbit after four iterations. On the right is the plot of the first four iterates of this piecewise linear function of entropy $\log 2$. (See also Plate 17.)

This phenomenon is closely related to why decimal representations of rational numbers are eventually periodic. There is a theory of β-expansions, similar to decimal expansions but with β a real number; when β is Pisot, the digits of the β-expansion of any element of $\mathbb{Q}(\lambda)$ are eventually periodic (by an almost identical proof).

The details of the preceding construction of f_λ are not important. *Any λ-expander whose critical points are in $\mathbb{Q}(\lambda)$ will do.* As long as $\lambda < d > 2$, there are infinitely many. To make the proof work as phrased, rescale the unit interval to clear all denominators, so that all critical points become algebraic integers in $\mathbb{Q}(\lambda)$.

Even in the case that λ is an integer less than d, this gives countably many different examples provided $d > 2$. For instance, Figure 2.3 shows the critical point orbits when $\lambda = 2$, $d = 3$ and the critical points are chosen as 19/60 and 17/30. On the fourth iterate, they settle into a single periodic orbit of period 4. There is a unique cubic polynomial, up to affine conjugacy, having the same order structure for the postcritical orbits, with entropy $\log(2)$.

In general, there is a non-empty convex $(d - 2)$-dimensional space of λ-uniform-expanders for every $1 < \lambda < d$. If λ is Pisot, then postcritically finite examples are dense in this set.

There are many Pisot numbers: for any real algebraic number y, it is easy to see that there are infinitely many Pisot numbers in $\mathbb{Q}(y)$. In fact, the intersection of the lattice of algebraic integers with the unit cylinder centered around any line corresponding to an embedding of $\mathbb{Q}(y)$ in $\mathbb{R}$ consists of Pisot numbers, except for those in a closed ball containing the origin. However, in the geometric sense, Pisot numbers are rare: in [Sa] Salem proved that the set of Pisot number is a countable closed subset of $\mathbb{R}$, making use of a theorem of Pisot that a real number x is Pisot if and only if sequence of minimum differences of x^n from the nearest integer is square-summable. The golden ratio is the smallest accumulation point of Pisot numbers, and the *plastic number* 1.3247..., a root of $x^3 - x - 1$, is the smallest Pisot number.

It is elementary and well-known that postcritically finite maps are dense among uniform expanders, but the preceding construction raises a question that does not seem obvious for $d > 2$:

QUESTION 2.4. *For fixed d, is there a dense set of numbers $1 < \lambda < d$ for which the set of postcritically finite maps is dense among λ-uniform expanders? For which λ are there infinitely many non-affinely equivalent postcritically finite maps? For which λ are postcritically finite maps dense?*

One way to get infinite families of postcritically finite maps with the same λ is to take dynamical extensions of maps with fewer critical points (*c.f. Section 10*), taking care only to introduce new critical points that map to positions whose forward orbit is finite. But this construction cannot work when $\lambda > d - 1$.

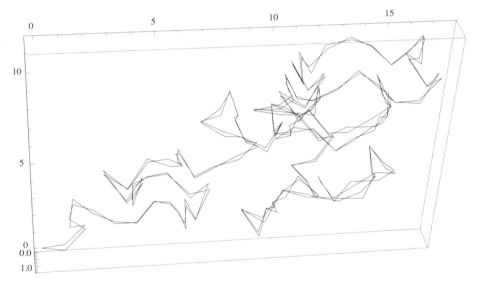

FIGURE 2.5: The tent map $x \mapsto \lambda|x| - 1$ for the degree 6 Salem number $\lambda = 1.4012683679\ldots$ that satisfies $\lambda^6 - \lambda^4 - \lambda^3 - \lambda^2 + 1 = 0$ is postcritically finite, with the critical point having period 270, quite large compared to the degree of λ. This figure shows the absolute value of the projection to the λ-line (the thin direction) as well as to the two complex places of λ. The trajectory resembles a random walk in the plane. Since random walks in $\mathbb{E}^2$ are recurrent (they have probability 1 of visiting any set of positive measure infinitely often), one would expect it to eventually return. It does, but as random walks often do (the mean return time is ∞), it takes a long time.

Salem numbers are closely related to Pisot numbers: some people conjecture that the union of Salem numbers and Pisot numbers is a closed subset of $\mathbb{R}$. However, the construction that worked for Pisot numbers is inadequate for Salem numbers. The Galois conjugates of the linear pieces of f_λ are isometries of $\mathbb{C}$ for any conjugate on the unit circle. Let n be the degree of the Salem number, so for each linear piece of f_λ there are $n/2 - 1$ Galois conjugate complex isometries, one for each complex place. If we take the product of these isometries over all complex places of $\mathbb{Q}(\lambda)$ we get an action by isometries on $\mathbb{C}^{n/2-1}$, where the first derivative of the action of each linear piece is $\pm U_\lambda$, where U_λ is a unitary transformation. In the unitary group, the orbit is dense on an $n/2 - 1$ torus, acting as an irrational rotation of the quotient of the torus by $\pm I$ (thus factoring out the complication of the variable sign of f_λ). The action of the sequence of linear functions on the $(n/2 - 1)$-tuple of moduli appears to behave like a random walk in $\mathbb{R}^{(n-2)/2}$. Sometimes they are periodic, sometimes with fairly large periods, but Brownian motion in dimension bigger than 2 is not recurrent, and a few experiments for $n \geq 8$ indicate that they typically drift slowly toward infinity and thus are not postcritically finite. However, there could be some reason (opaque to me) why they might not act randomly in the long run, and they could eventually cycle. At least it seems hard to prove any particular example is not postcritically finite.

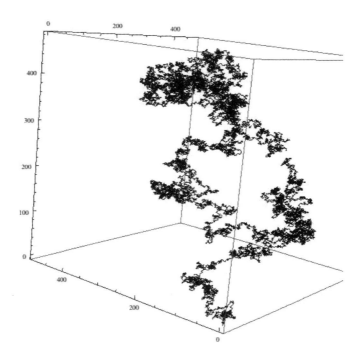

FIGURE 2.6: This diagram is a 3-dimensional projection of the first 200,000 iterates of the critical point for $x \mapsto \lambda|x| - 1$ where $\lambda = 1.17628\ldots$ is Lehmer's constant with minimal polynomial $1 + x - x^3 - x^4 - x^5 - x^6 - x^7 + x^9 + x^{10}$. There are 4 complex places; this shows a 3-dimensional projection for the quadruple of absolute values. It resembles a Brownian path in 3 dimensions. The map is postcritically finite if and only if the path closes. Its values are always algebraic integers, so if it comes sufficiently close to the start it's fairly likely to close (become eventually periodic). The quadruple of radii determines a 4-torus in $\mathbb{C}^4 \times \mathbb{R}^2$, with the current lattice point a bounded distance from that torus. When the radii are on the order of 200, this bounded neighborhood has volume on the order of $(2\pi \times 200)^4$, with roughly 10^{12} lattice points, so the chances of returning seem remote.

For example, for the Salem number $1.7220838\ldots$ of degree 4 which satisfies $x^4 - x^3 - x^2 - x + 1 = 0$, the critical point of the tent map is periodic of period 5.

For the Salem number $1.401268367939\ldots$ satisfying $x^6 - x^4 - x^3 - x^2 + 1 = 0$, the critical point has period 270. Note that its square is also a Salem number, for which the period is $135 = 270/2$.

One of the most famous Salem numbers is the Lehmer constant, defined by the polynomial $1 + x - x^3 - x^4 - x^5 - x^6 - x^7 + x^9 + x^{10} = 0$. The single root outside the unit circle is approximately 1.17628. This is the smallest-known Salem number and, in fact, the smallest-known Mahler measure for any algebraic integer. (Mahler measure is the product of the absolute value of all Galois conjugates outside the unit circle.) Figure 2.6 is a diagram of the first 200,000 elements of the critical point in $x \mapsto \lambda|x| - 1$, projected from $\mathbb{C}^4 \times \mathbb{R}^2$ to the quadruple of radii in the complex factors and from there a projection into 3 dimensions. The map is postcritically finite if and only if the path closes. Its values are always algebraic integers, so if it comes sufficiently close to the start it's fairly likely to close. The quadruple of radii determines a 4-torus in $\mathbb{C}^4 \times \mathbb{R}^2$, with the current lattice point a bounded distance from that torus. When the radii are on the order of 200, as in this case, this bounded neighborhood has volume on the order of $(2\pi \times 200)^4$, with roughly 10^{12} lattice points, so the chances of looping appear remote unless the path wanders close to the origin, where the tori are smaller. For comparison, the variance of a random walk with step size 1 in $\mathbb{R}^n$ equals the number of steps, so for a random walk of length 200,000, the standard deviation is $\sqrt{200,000} \approx 447$; projected from 8 dimensions to 3, the standard deviation would be ≈ 274, in line with the picture. An experiment with a selection of small Salem numbers of moderate degree greater than 6 showed similar results, with none of them exhibiting a cycle within $500,000$ iterates.[7]

3   CONSTRUCTING INTERVAL MAPS: FIRST STEPS

For a Perron number that is not Pisot, the situation is much more delicate. To develop a strategy, first we'll discuss incidence matrices. There are two reasonable versions for an incidence matrix that are transposes of each other. We will use the version whose columns each represent the image of a subinterval of the domain and whose rows represent a subinterval in the range, so that each entry counts how many times the subinterval that indexes its column crosses the subinterval that indexes its row.

Suppose the unit interval I is subdivided into n subintervals $J_1, \ldots, J_n$ (in order). Let $V = \{0 = v_0, v_1, \ldots, v_n = 1\}$ be the vertex set. Every function $g \colon V \to V$ that takes adjacent vertices to distinct vertices extends to a postcritically finite piecewise linear map. In this way, a large but finite and specialized set of $n \times n$ matrices can be realized as incidence matrices. Incidence matrices are easy to recognize. Each column consists of a consecutive sequence of 1s, and is otherwise 0. There is a matching between the ends of the blocks of consecutive 1s in adjacent columns, with every column except the first and last having one end of the block of 1s matched to the left and the other end of the block of 1s matched to the right.

Here is a generalization of this concept. Consider a map $f \colon I \to I$ with finitely many critical points such that $f(V) = V$ and f of the critical set is contained in V, that is, V contains all critical values. Under these conditions, an extended incidence matrix is still defined for f, whose entries a_{ij} count how many times $f(J_j)$ crosses interval J_i.

---

[7]For further discussion, see Note 13.3.

PROPOSITION 3.1. *An $n \times n$ non-negative integer matrix A is an extended incidence matrix if and only if*

1. *the nonzero entries in each column form a consecutive block, and*

2. *there is a map $\phi \colon \{0, 1, \ldots, n\} \to \{0, 1, \ldots, n\}$ such that in column i, the entries in rows greater than $\phi(i-1)$ and not greater than $\phi(i)$, and no other entries, are odd.*[8]

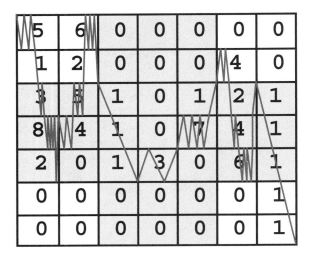
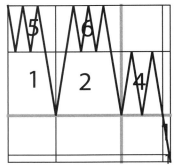

FIGURE 3.2: The matrix M on the left satisfies the necessary and sufficient conditions of proposition 3.1 to be an incidence matrix for a self-map of the interval: in each column, the positive entries in each column form a connected block as do the odd entries, and the odd blocks in the columns can be matched together end to end, from the left column to the right, to form a chain. The jagged line defines a PL function realizing the given matrix, since it crosses each square the specified number of times. If you reflect the line across the top edge of the matrix, it matches the usual convention for drawing a graph (since the convention for matrices that the row numbers increase going downward is opposite the convention for graphs of functions). On the right, the widths of rows and columns of the matrix have been adjusted in proportion to the positive eigenvector for the transpose M^t. The matrix M^t is not ergodic, entries 3,4,5 of the positive eigenvector are 0, and the shaded blocks at left have collapsed to the gray lines at right. Now the graph can be drawn with constant absolute slope.

PROOF. The necessity of the conditions is easy. The map ϕ represents the map f restricted to V. The image of any interval J_i is necessarily the union of a consecutive block of intervals. The subinterval between the images of the first and last endpoints, $[f(v_{i-1}), f(v_i)]$ (which could be a degenerate interval) is traversed an odd number of times, and the rest of the image is traversed an even number of times.

Sufficiency of the conditions is also easy. To map J_i, start from $v_{\phi(i-1)}$, go to the lowest vertex in the image, zigzag across the lowest interval until its degree is

---

[8]Condition (2) is not quite right as stated. See Note 13.4.

used up, then the next lowest, etc. until you get to $v_{\phi(i)}$, at which point proceed to the highest vertex in the image and work back. $\qquad\square$

Note: Sufficiency can also be reduced to the familiar condition that a graph admits a Eulerian[9] path from vertex a to vertex b if and only if it is connected and either $a = b$ and all vertices have even valence, or a and b have odd valence and all other vertices have even valence. For each i, apply this to the graph Γ_i that has a_{ij} edges connecting v_{j-1} to v_j. The entire map f is really a Eulerian path in the graphs Γ_i connected in a chain by joining vertex $\phi(i)$ of Γ_i to that of Γ_{i+1}, followed by the natural projection to $[0, 1]$.

PROPOSITION 3.3. *The topological entropy of any map with extended incidence matrix A is the log of the largest positive eigenvalue of A.*

PROOF. The total variation of f^n is a positive linear combination of matrix entries of A^n (if all intervals have equal length, it is their sum). By 1.2, $h(f)$ is the exponent of growth of total variation, so this equals the log of the largest positive eigenvalue of A. $\qquad\square$

Suppose we are given a (strict) Perron number λ. Our strategy is to first construct a strictly positive extended incidence matrix that has positive eigenvalue λ^N, for a large power N of λ. We will promote this to an example with entropy λ by implanting it as the return map replacing a periodic cycle of a map with entropy less than λ.

Afterward, we will deal with additional issues involving questions of mixing and weak Perron numbers.

Given λ, let O_λ be the ring of algebraic integers in the field $\mathbb{Q}(\lambda)$, and let V_λ be the real vector space $V_\lambda = \mathbb{Q}(\lambda) \otimes_\mathbb{Q} \mathbb{R}$. Another way to think of it is that V_λ is the product of the real and complex places of $\mathbb{Q}(\lambda)$, that is, the product of a copy of $\mathbb{R}$ for each real root of the minimal polynomial for λ and a copy of $\mathbb{C}$ for each pair of complex conjugate roots. The ring operations of $\mathbb{Q}(\lambda)$ extend continuously to V_λ, but division is discontinuous where the projection to any of the places is 0. Yet another way to think of V_λ is in terms of the companion matrix C_λ for the minimal polynomial P_λ of λ. We can identify $\mathbb{Q}(\lambda)$ with the set of all polynomials in C_λ with rational coefficients, and V_λ with the vector space on which the companion matrix acts. The real and complex places of $\mathbb{Q}(\lambda)$ correspond to the minimal invariant subspaces of C_λ. When P_λ is factored into polynomials that are irreducible over $\mathbb{R}$, the terms are linear and positive quadratic; the subspaces are in one-to-one correspondence with these factors.

We are assuming that λ is a Perron number, so for multiplication of V_λ by λ, the λ-eigenvector is dominant. If we consider the convex cone K_λ consisting of all points where the projection to the λ eigenspace is larger than the projection to any of the other invariant subspaces, then $\lambda \times K_\lambda$ is contained in the interior of K_λ except at the origin.

PROPOSITION 3.4. *There is a rational polyhedral convex cone KR_λ contained in K_λ and containing $\lambda \times K_\lambda$.*

---

[9]An *Eulerian path* is one that passes through each edge exactly once. Thurston's manuscript uses the term *Hamiltonian path*, but this would be historically incorrect. Euler first studied such paths, while Hamilton studied paths which pass through each *vertex* exactly once.

PROOF. It is easiest to think of this projectively, in $\mathbb{P}(V_\lambda)$. The projective image $\mathbb{P}(K_\lambda)$ of K_λ is a convex set (specifically, a product of intervals and disks), with the image of $\mathbb{P}(\lambda \times K_\lambda)$ contained in its interior. Since rational points $\mathbb{P}(\mathbb{Q}(\lambda))$ are dense in $\mathbb{P}(V_\lambda)$, we can readily find a set of rational points in the interior of $\mathbb{P}(K_\lambda)$ whose convex hull contains $\mathbb{P}(\lambda \times K_\lambda)$. This gives us the desired rational polyhedral convex cone.   □

Let S_λ be the additive semigroup $O_\lambda \cap KR_\lambda \setminus \{0\}$.

PROPOSITION 3.5. S_λ is finitely generated as a semigroup.

PROOF. This is a standard fact. Here's how to see it using elementary topology: We can complete S_λ by adding on the set of projective limits $\mathbb{P}(KR_\lambda)$. The completion is compact. It's easy to see that the set of closures U_s of the "ideals" $s + S_\lambda$ for $s \in S_\lambda$ form a basis for this topology.[10] By compactness, the cover by basis elements has a finite subcover $U_{s_1}, \ldots, U_{s_k}$. For any such cover, the set $G = \{s_1, \ldots, s_k\}$ form a generating set: given any element, write it as $s_i + s$, and continue until the remainder term s is 0.   □

PROOF OF CONVERSE PERRON-FROBENIUS (THEOREM 1.6). Equipped with this picture, it is now easy to prove the converse Perron-Frobenius theorem of Lind. Consider the free abelian group $\mathbb{Z}^G$ on the set G of semigroup generators. The positive semigroup in $\mathbb{Z}^G$ maps surjectively to S_λ. Call this map p. We can lift the action of multiplication by λ to an endomorphism of the positive semigroup by sending each generator g to an arbitrary element of $p^{-1}(\lambda\, p(g))$. In coordinate form, this is described by a non-negative integer matrix. In S_λ, projection to the λ-space is a dual eigenvector, that is, a linear functional multiplied by λ under the transformation. It is strictly positive[11] on S_λ. Therefore, the pullback of this function is a strictly positive λ-eigenvector of the transpose matrix, proving Lind's theorem.   □

Note that the minimum size of a generating set might be much larger than the dimension of V_λ. For instance, if λ is a quadratic algebraic integer, V_λ is the plane, and KR_λ could be bounded by any pair of rational rays that make slightly less than a 45° angle to the λ eigenvector. Minimal generating sets can be determined using continued fraction expansions of the slopes; they can be arbitrarily large.

As we shall presently see, it can happen in higher dimensions that the minimum size of a generating set for S_λ can be very large, no matter how we choose a semigroup S_λ on which multiplication by λ acts as an endomorphism.

---

[10]This is not quite right as stated, since these sets are not open and do not form a basis for the topology. However, each point in the limit set $\mathbb{P}(KR_\lambda) \subset \overline{S}_\lambda$ has an open neighborhood contained in some U_s. Choosing finitely many U_{s_j} which cover a neighborhood of the limit set, and adding the finitely many s_k which are not included in these U_{s_j}, we obtain a finite generating set.

[11]In the diagram

$$
\begin{array}{ccc}
\mathbb{Z}_+^G & \xrightarrow{L} & \mathbb{Z}_+^G \\
p \downarrow & & p \downarrow \\
S_\lambda & \xrightarrow{\lambda\cdot} & S_\lambda \quad \overset{\iota}{\subset} \quad \mathbb{R}
\end{array}
$$

the composition $\iota \circ p \in \mathrm{Hom}(\mathbb{Z}_+^G, \mathbb{R})$ is a dual eigenvector which is strictly positive, in the sense that every basis element in $\mathbb{Z}_+^G$ maps to a positive real number.

4   SECOND STEP: CONSTRUCTING A MAP FOR λ^N

Now we need to address the special requirements for an incidence matrix for a map having a finite invariant set as the set of critical values. Given any Perron number λ of degree d, we will construct such a matrix for some power, probably large, of λ. From Section 3, we assume we have a set G of semigroup generators for a semigroup in $S_\lambda \subset V_\lambda = \mathbb{Q}(\lambda) \otimes_{\mathbb{Q}} \mathbb{R}$ invariant under multiplication by λ.

Choose a finite sequence of generators, including each generator at least once, such that the partial sums of the sequence contain all 2^d mod 2 congruence classes, that is, the partial sums map surjectively to $O_\lambda/(2O_\lambda)$. If necessary, adjoin additional elements to the sequence so that the sum T of the entire sequence is 0 (mod $2O_\lambda$).

The action of λ (by multiplication) on the projective completion of S_λ has a unique attracting fixed point. For any $s \in S_\lambda$, the closure of $s + S_\lambda$ contains[12] the fixed point, so there is some power N such that $\lambda^N \times S_\lambda \subset 3T + S_\lambda$.

Now mark off an interval of length T into segments whose lengths are given by the chosen sequence $g_1, \ldots, g_k$ of generators (in the embedding of O_λ in $\mathbb{R}$ where λ goes to λ). We'll construct a λ^N-expander map by induction, going from left to right, starting with $0 \to 0$. There is at least one point $q = g_1 + g_2 + \cdots + g_k$ in this subdivision that has the same value mod $2O_\lambda$ as $\lambda^N \times g_1$. Since $\lambda^N \times g_1$ can be expressed as $3T$ plus a sum of generators, it can also be expressed as $q + 2T$ plus a sum of generators. Since $(\lambda^N \times g_1 - q - 2T)$ is in S_λ and congruent to zero mod 2, it can be expressed as 2α, where $\alpha \in S_\lambda$. We can write 2α as a linear combination of generators with even coefficients, so we can write $\lambda^N \times g_1$ as a strictly positive sum of generators, where each generator between 0 and q occurs an odd number of times, and each other generator occurs an even number of times.

We can continue in exactly the same way for λ^N times each of the points in the subdivision. We first pick where each point g_i goes based on the congruence class of $\lambda^N \times g_i$ mod 2, then express the difference as an even and strictly positive linear combination of all generators in the sequence. Finally, we end with T going to either 0 or T, as we choose. The incidence matrix satisfies the conditions of 3.1, so we have constructed a λ^N uniform expander.

5   POWERS AND ROOTS: COMPLETION OF PROOF OF THEOREM 1.3

Given a Perron number λ, we'll first construct a map of S^1 to itself that is a λ uniform expander, because the construction is a little nicer for S^1. From the preceding section, for some N we construct a λ^N uniform expanding map f_{λ^N} of an interval that takes each endpoint to itself. Let ρ be any rotation of the circle of order N. The circle can be subdivided into N intervals that are cyclically permuted by the rotation. Define a metric on the circle so that in this cyclic order, the N intervals have length $1, \lambda, \lambda^2, \ldots, \lambda^{N-1}$. Now define $g_\lambda \colon S^1 \to S^1$ by mapping each of the first $N-1$ intervals affinely to the next and mapping the last interval to the first using f_{λ^N} with affine adjustments in the domain and range to send the last interval exactly to the first.[13] Since the first interval has length $1/\lambda^{N-1}$ times the last, with this affine adjustment f_{λ^N} also expands uniformly by λ.

---

[12]In fact this closure is a neighborhood of the fixed point. For any generator s_j of S_λ, the succesive products $\lambda^n \times s_j$ converge to the fixed point, and it follows that $\lambda^N \times S_\lambda \subset s + S_\lambda$ for large N.

[13]For a figure to illustrate this construction, see Note 13.5.

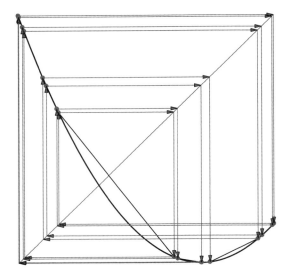

FIGURE 5.1: This is the graph of a quadratic map in the initial period-doubling cascade, with critical point of period 16. This, and all other quadratic maps in this family, have entropy 0. If the critical orbit is blown up and replaced with a sequence of intervals where the return map has is a λ^{16}-uniform expander, then the resulting map is semiconjugate to a λ-uniform expander.

In the case of the circle, we can easily modify the construction of g_λ to make the incidence matrix mixing: this will happen if we change any small piece of f_{λ^N} to stray into a neighboring interval and back, when it gets to the upper endpoint. If the rotation is chosen as a rotation by $1/N$ and if the first generator is chosen to be a "small" element in S_λ, these intervals are small, and straying is probably possible with ease, but in any case, by taking N to be a somewhat higher power, it can be readily guaranteed.

In the case of the unit interval, we need a substitute for a rotation of the circle. For this, we can use the well-known period-doubling cascade for quadratic self-maps of an interval (see Figure 5.1 for an illustrative example). In this period-doubling family, there is a quadratic map q_n with entropy 0 in which the critical point has period 2^n. We can blow up the forward and backward orbit of the critical point, replacing each point x in the orbit by a small interval I_x of any length l_x such that the set of lengths is summable. Extend the map to these intervals by affine homeomorphisms, with the exception of the interval for the critical point; for that, we can use any affine map that takes both endpoints to the right hand endpoint.

Since the entropy of the quadratic map is 0, the number of critical points of q_n^k grows subexponentially in k, so if we assign length λ^{-k} to each interval for a point that is critical for q_n^k but not for q_n^{k-1}, the set of lengths is summable. After blowing up the full orbit of the critical point in this way, we can define a pseudo-metric on the interval where the length of an interval is the measure of its intersection with the blown-up orbit; everything else collapses to measure 0.

By the preceding section, we can find an f_{λ^N} of the form $N = 2^n$ that maps the unit interval to itself, taking both endpoints to 0. Implant this, using affine the blown-up orbit; everything else collapses to measure 0.

By the preceding section, we can find an $f_{\lambda N}$ of the form $N = 2^n$ that maps the unit interval to itself, taking both endpoints to 0. Implant this, using affine adjustments in the domain and the range, for the map from the interval for the critical point of q_n to its image. The image of the critical interval has length $\lambda^{2^n - 1}$ since the original critical point had period 2^n, so the implanted map is a uniform λ-expander.

We now have a map which is a local homeomorphism with Radon derivative λ in the complement of the critical interval, where it is a uniform λ-expander. Therefore, the entire map is a piecewise-linear uniform λ-expander, with entropy $\log(\lambda)$. This completes the proof of Theorem 1.3 for strong Perron numbers.[14]

Now we'll address weak Perron numbers.

PROPOSITION 5.2. *A positive real number λ is a weak Perron number if and only if some power of λ is a [strong] Perron number.*

PROOF. In one direction this is pretty obvious: the nth roots of any algebraic integer are algebraic integers, and their ratios to each other are nth roots of unity. Thus the positive real nth root of a Perron number is a weak Perron number.

In the other direction, suppose λ is a weak Perron number. Let B be the set of Galois conjugates of λ with absolute value λ, and let b be their product. This product b is a real number equal to $\lambda^{\#(B)}$. The Galois conjugates of b are products of $\#(B)$ Galois conjugates of λ; for any subset of conjugates of this cardinality other than B, the product is strictly smaller, so b is a Perron number. □

(Since $\lambda^n = b$, the other elements of B also satisfy this equation,[15] so they are just the products of λ with an nth root of unity.)

Now given a weak Perron λ, we first find a power k so that λ^k is a Perron number. In the family of degree 2 uniform expanders, functions with periodic critical point are dense, and functions with critical point having period a multiple of k are also dense. Choose such a function g whose entropy is less than λ, where the critical point has period a sufficiently high multiple kn of k. Blow up the full (backwards and forward) orbit of the critical point, and implant a map of the form $f_{\lambda^{kn}}$, as constructed in Section 4, in the interval replacing the critical point. Since the growth rate of critical points for powers of g is less than the growth rate of powers of λ, we can construct a metric, just as before, that is uniformly expanded by λ.

Note that the λ-uniform expanders we have constructed are very far from mixing. This is, of course, impossible if λ is only a weak Perron number, by the Perron-Frobenius theorem, but for a Perron number λ it is tempting to try to generalize the straying technique that worked earlier.

There are two difficulties. The first is an essential problem.

PROPOSITION 5.3. *For any self-map f of the interval with entropy strictly between 0 and $\log \sqrt{2}$, there are two disjoint subintervals that are interchanged by the map, and, in particular, f is not mixing.*

---

[14]For a simplified version of this argument, see Note 13.6; and for a proof that the required map can be realized as a polynomial, see [dMvS, Theorem 6.4].

[15]In fact, any Galois automorphism α which takes λ to an element of B must permute the elememts of B, since otherwise the element $(\lambda^\alpha)^n = (\lambda^n)^\alpha = (\prod_{\beta \in B} \lambda^\beta)^\alpha$ would have a smaller absolute value. It follows that $(\lambda^\alpha)^n = \lambda^n$.

PROOF.[16] Let $D(f)$ be the set of points that have more than one preimage. Note that $D(f) \subset D(f^2)$ and also $f(D(f)) \subset D(f^2)$; in fact, $D(f^2) = D(f) \cup f(D(f))$. Let $a = \inf(D(f^2))$, and $b = \sup(D(f^2))$. Then $[a, b]$ is mapped into itself and has the same entropy as f.   □

In particular, they never mix. By induction, if the entropy is less than $2^{-n} \log 2$, there are 2^n disjoint intervals cyclically permuted by the map.

The second problem is that the piecewise-linear map we constructed before has coefficients in $\mathbb{Q}(\lambda)$ (either because the formulas from kneading theory for the infinite sums of intervals are rational functions of λ or because they are determined by linear functions with coefficients in O_λ), but it is not obvious how to get coordinates to be in O_λ. Maybe it's possible to analyze and control the algebra, but if so it's beyond the scope of this paper.

6   MAPS OF ASTERISKS

Postcritically finite self-maps of graphs, including in particular Hubbard trees, give another interesting collection of one-dimensional postcritically finite maps.

REMARK 6.1.   Any expanding self-map of a tree can be promoted to a self-map of a Hubbard tree by adding extra information as to a planar embedding and choosing a branched covering map for the planar neighborhood of each vertex that acts on its link in the given way. Each such promotion is the Hubbard tree for a unique polynomial up to affine automorphism.

For use later in constructing automorphisms of free groups, we will look at a special case, the *asterisks* $*$. An *n-pointed asterisk* is the cone on a set of n points, which we'll refer to as the *tips* of the asterisk.

THEOREM 6.2.   *For every Perron number λ there is a postcritically finite λ-uniform expanding self-map f of some asterisk such that*

- *f fixes the center vertex,*

- *f permutes the n tips,[17] and*

- *the incidence matrix for f is mixing.*

PROOF. This is similar to the proof for self-maps of the interval. In principal it is easier because there is no order information to worry about, but we will use a very similar method.

An incidence matrix for a self-map of an asterisk of the given form is a nonnegative integer matrix that has exactly one odd entry in each row and each column. Given any such matrix, we can construct a corresponding self-map of an asterisk by permuting the tips according to the matrix mod 2, which is a permutation matrix, and running each edge out and back various edges to generate the even part of the matrix.

As before, find a subsemigroup S_λ of O_λ that excludes 0 and is invariant under multiplication by λ.

---

[16]This proof is seriously incomplete. Compare Note 13.7.
[17]This condition has been simplified by the editors

Let G be a finite set of generators for S_λ that maps surjectively to $O_\lambda/(2O_\lambda)$. Let $T = \sum_{g \in G} g$.

Let n be an integer such that $\lambda^n - 1$ is congruent to 0 mod 2, and for each $g \in G$, $\lambda^n \times g$ is contained in $g + 2(T + \lambda \times g) + S_\lambda$.

Now make an asterisk whose points are indexed by $G \times \{1, 2, \ldots, n-1\}$. Map the edge to point (g, i) homeomorphically to the edge to point $(g, i+1)$ when $i < n - 2$.

The final set of tips $(g, n-1)$ will map to tips $(g, 1)$. For each g, $\lambda^n \times g - g$ is congruent to 0 mod $2O_\lambda$ and so can be expressed as a strictly positive even linear combination of $G \cup \{\lambda \times g\}$.

Use these linear combinations to construct an asterisk map. Each edge to the first set of tips maps homeomorphically to a new edge for the first $n - 1$ iterates. On the nth iterate, it maps to a path that makes at least one round trip to all the first-layer points as well as one second-layer point, finally ending back where it started. We can easily arrange the order of traversal so that every edge is represented in the first segment of one of these edge paths.

If the asterisk is given a metric where edge (g, i) has length equal to the value of $\lambda^i g$ in the standard embedding of O_λ in $\mathbb{R}$, where λ is the Perron number, this map is a λ-uniform expander. It is mixing: by the nth iterate, the image of an edge with index (g, i) contains all generators of the form $(*, i)$; by the $2n$th iterate, the image of an edge contains all generators of the form $(*, i)$ and $(*, i+1)$. After n^2 iterates, the edge maps surjectively to the entire asterisk. $\qquad\square$

This construction was intended to avoid the need for complicated conditions and bookkeeping. It's clear that a more careful construction could produce suitable asterisk maps for a typical Perron number λ that are much smaller (but still might be quite large).

7   ENTROPY IN BOUNDED DEGREE

The constructions for maps of given entropy have been very uneconomical with the complexity of the maps. First, there is a potentially dramatic (but sometimes unavoidable) blowup in going from a Perron number λ to a finite set of generators for a subsemigroup S_λ of algebraic integers invariant by multiplication by λ. Even then, there is another possibly large blowup in finding a power of λ such that $\lambda^N \times S_\lambda$ is sufficiently deep inside S_λ to guarantee an easy construction of a suitable incidence matrix. In other words: unlike Pisot numbers, a typical Perron number is probably unlikely to be the growth constant for a typical postcritically finite map of an interval to itself.

In some sense, expansion constants for bounded degree systems are almost Pisot: most of their Galois conjugates don't seem to wander very far outside the unit circle. Figure 1.1 illustrate this point: most of the Galois conjugates of $\exp(h(f))$ for postcritically finite quadratic maps are in or near the unit circle. Since their minimal polynomials have constant term ± 1 or ± 2, the inside and outside roots are approximately balanced. If they don't wander outside the circle, they can't wander very far inside the circle, and most roots are near the unit circle. In contrast, Figure 1.4 illustrates that Perron numbers less than 2 can have roots spread the disk of the their radius.

DEFINITION 7.1. *To get some understanding of what's going on, we'll translate questions about the distribution of roots into questions about dynamics of a semigroup of affine maps, elaborating on the point of view taken in the discussion of Pisot numbers in Section 2. We'll consider the sets of piecewise linear uniform expander functions $F(d, \epsilon)$ that take ∂I onto ∂I, where $d > 1$ is the number intervals on which the function is linear and $\epsilon = \pm 1$ determines the sequence of slopes. $\epsilon(-1)^i \lambda$ is the sequence of slopes in subintervals $i = 0$ through $i = d - 1$. The constant terms for the linear function f_i in the first and last interval are determined by the condition that $f(\partial I) = \partial I$, implying that $f_0(0) = 0$ if $\epsilon = 1$ and $f_0(0) = 1$ when $\epsilon = -1$, with a similar equation for the last interval. In all other subintervals, the constant term is a free variable C_i subject to linear inequalities. That is, the ith critical point c_i, determined by $\epsilon(-1)^{i-1}\lambda c_i + C_{i-1} = \epsilon(-1)^i \lambda c_i + C_i$ must be inside the unit interval, so we have the inequalities*[18]

$$0 \le \epsilon(-1)^{i-1}\frac{C_i - C_{i-1}}{2\lambda} \le 1.$$

Now for any other field embedding $\sigma \colon \mathbb{Q}(\lambda, C_1, \ldots, C_{i-1}) \to \mathbb{C}$, we can look at the collection of image functions f_i^σ. Since the choice of which f_i to apply is determined by inequalities, we will look at the product action. Define $f^{1+\sigma}$ to act on $I \times \mathbb{C}_\sigma$ by $f^{1+\sigma} \colon (x, z) \mapsto (f(x), f_i^\sigma(z))$ where f_i is a linear piece that f applies to x. In the ambiguous case where x is one of the critical points, this definition is discontinuous, so we will look at both images, which are the limit from the left and the limit from the right.

Define the *boundedness set* $B(f^{1+\sigma})$ to be the set of (x, z) such that its (forward) orbit stays bounded. If f is postcritically finite, then the critical points[19] in particular have bounded orbits, so in addition the full orbit of the critical points (under taking inverse images as well as forward images) are contained in the boundedness locus. Define the limit set $L(f^{1+\sigma})$ to be the smallest closed set containing all ω-limit sets for (x, z).

There are three qualitatively different cases, depending on whether $|\sigma(\lambda)|$ is less than 1, equal to 1, or greater than 1.

In the first case, $|\sigma(\lambda)| < 1$, every orbit remains bounded in the $\mathbb{C}_\sigma$ direction, since the map is the composition of a sequence of contractions. Everything outside a certain radius is contracted, so the limit set[20] is compact. Some special examples of this are illustrated in Figures 7.2, 7.3, and 7.4.

The second case, when $|\sigma(\lambda)| = 1$, seems hardest to understand, since the f_i^σ act as isometries. Perhaps the limit set in these cases is all of $\mathbb{C}$.

In the third case, when $|\sigma(\lambda)| > 1$, the maps f_i^σ are expansions, so there is a radius R such that everything outside the disk $D_R(0)$ of radius R centered at the origin escapes to ∞. Given $x \in [0, 1]$, the only z such that $f^{1+\sigma}(x, z)$ has second coordinate inside $D_R(0)$ are inside a disk of radius $R/\sigma(\lambda)$ about the preimage of 0. Starting far along in the sequence of iterates and working backward to the beginning, we find a sequence of disks, shrinking geometrically by the factor $1/\sigma(\lambda)$, that contain all bounded orbits. In the end, there's a unique point $b(x) \in \mathbb{C}$ such that $(x, b(x))$ remains bounded. The point $b(x)$ can also be expressed as the sum of a power series in $\sigma(\lambda)^{-1}$ with bounded coefficients that depend on the C_i and the kneading data for x, matching with formulas from [MT].

---

[18]Actually the C_i must satisfy the stronger inequalities $0 \le f(c_i) = (C_{i-1} + C_i)/2 \le 1$; and also $C_0 > C_2 > C_4 > \cdots$ and $C_1 < C_3 < \cdots$ if $\epsilon = +1$, or the opposite inequalites if $\epsilon = -1$.

[19]By a critical point for $f^{1+\sigma}$ is meant a pair $(x, \sigma(x))$, where x is a critical point for f.

[20]The original manuscript stated incorrectly that the boundedness set is compact.

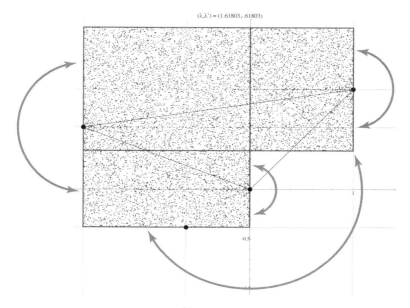

FIGURE 7.2: This diagram shows $L(f^{1+\sigma})$, where $\lambda = 1.61803\ldots$, the golden ratio, is a Pisot number, and $\sigma(\lambda) = -1/\lambda$. It was drawn by taking a random point near the origin, first iterating it 15,000 times and then plotting the next 30,000 images. The dynamics is hyperbolic and ergodic, so almost any point would give a very similar picture. The critical point is periodic of period 3, and the limit set is a finite union of rectangles with a critical point on each vertical side. The dynamics reflects the rectangle on the upper right in a horizontal line and arranges it as the rectangle on the lower left. The big rectangle formed by the two stacked rectangles at left is reflected in a vertical line, stretched horizontally and squeezed vertically, and arranged as the two side-by-side rectangles on the top. Although the dynamics is discontinuous, the sides of the figure can be identified, as indicated (partially) by the arrows to form a tetrahedron to make it continuous: see Figure 7.3. (See also Plate 18.) For further discussion, see Notes 13.8 and 13.9.

FIGURE 7.3: This is the diagram from Figure 7.2 taped together into a tetrahedron, where the dynamics acts continuously (but reverses orientation). The map $f^{1+\sigma}$ acts on the tetrahedron as an Anosov diffeomorphism. There are coordinates in which it becomes the Fibonacci recursion $(s, t) \mapsto (t, s + t)$, modulo a $(2, 2, 2, 2)$ symmetry group generated by 180° rotations about lattice points. Compare Note 13.9. (See also Plate 19.)

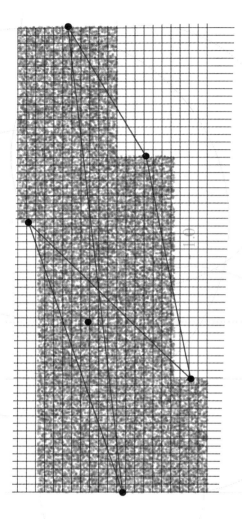

FIGURE 7.4: This diagram shows $L(f^{1+\sigma})$ where $\lambda = 1.7220838\ldots$ is a Salem number of degree 4 satisfying $\lambda^4 - \lambda^3 - \lambda^2 - \lambda + 1 = 0$, and $\sigma(\lambda) = 1/\lambda$. The critical point is periodic of period 5, and the limit set is a finite union of rectangles with a critical point on each vertical side. The dynamics multiplies the left half (to the left of the vertical line through the uppermost black dot) by the diagonal matrix with entries $(-\lambda, -1/\lambda)$, and then translates until the lowermost black dot goes to the uppermost black dot. The right half of the figure is multiplied by the diagonal matrix $(\lambda, 1/\lambda)$, and translated to fit in the lower left. As with the example in Figure 7.2, it can be folded up, starting by folding the vertical sides at the black dots, to form (topologically) an S^2 on which the dynamics acts continuously, a pseudo-Anosov map of the $(2, 2, 2, 2, 2)$-orbifold. This phenomenon has been explored, in greater generality, by André de Carvalho and Toby Hall [CH], and it is also related to the concept of the dual of a hyperbolic groupoid developed by Nekrashevych [Ne]. (See also Plate 21.)

PROPOSITION 7.5. *If f is postcritically finite and if $|\sigma(\lambda)| > 1$, then $b(x)$ depends continuously on x.*

PROOF. When f is postcritically finite, the preperiodicity of f at any critical point c_i is equivalent to an identity among compositions of the f_i. Therefore, $f^{1+\sigma}$ satisfies the same identity when applied to $(c_i, \sigma(c_i))$; therefore, its orbit is bounded so $b(c_i) = \sigma(c_i)$.

Since $b(x)$ is the sum of a geometrically convergent series, the value depends continuously on the coefficients. The coefficients change continuously except for where x is a precritical point; but the limits from the two sides coincide at precritical points, since they coincide for critical points. Therefore, $b(x)$ is continuous. $\square$

When f is not postcritically finite, $b(x)$ might not be continuous; there could well be different limits from the left and from the right at critical points and, therefore, at precritical points: the natural domain for $b(x)$ in general is a Cantor set obtained by cutting the interval at the countable dense set of precritical points. For example, if λ is transcendental, then σ could send it to any other transcendental number, and it's obvious that the generically $b(x)$ would not be continuous.

This phenomenon points to the inadequacy of considering only the algebraic properties of σ when our real goal is to control the geometry. Here is a formulation of the appropriate condition, in a slightly bigger context.

DEFINITION 7.6. *Suppose that $f \in F(d, \epsilon)$ (see Definition 7.1). Let V be a complex vector space, and let $g\colon [0,1] \times V \to [0,1] \times V$ have the form*

$$g\colon (x, v) \mapsto (f(x), \epsilon(-1)^i A(v) + C_i)$$

for $c_i < x < c_{i+1}$, where $A\colon V \to V$ is a linear expanding map, and each C_i is a constant vector. Then g is a friendly extension of f if there is a (necessarily unique) continuous map $x \mapsto b(x)$ such that the orbit of $g(x, b(x))$ is bounded.

Since postcritically finite systems are dense in $F(d, \epsilon)$, there are many postcritically finite examples, and many have Galois conjugates of λ outside the unit circle. Since boundedness depends continuously on the kneading data, the coefficient λ and the constant terms C_i, we can pass to limits of postcritically finite systems; if we take a convergent sequence of postcritically finite systems having a convergent sequence of friendly extensions, then the limit is also a friendly extension.

In the case $d = 2$, these correspond to the kneading roots which were collected in Figure 1.1. Note that even for a λ for which $x \mapsto \lambda|x| - 1$ is postcritically finite, it is a limit of λ_i of much higher degree for which $x \mapsto \lambda_i|x| - 1$ is also postcritically finite; in fact, it is a limit of cases where the critical point is periodic. The minimal polynomials for these nearby polynomials can be of much higher degree, so there can be many more friendly extensions than just the ones associated with the roots of the minimal polynomial for λ. Figure 7.7 is a 3-dimensional figure of friendly extensions, where the vertical axis is the λ direction, and the horizontal direction is $\mathbb{C}$; the plotted points outside the unit cylinder are expansion factors for friendly extensions.[21]

Figure 7.8 is a very thin slice of the set depicted in Figure 7.7, halfway up and of thickness 10^{-9}. A movie made from frames of this thickness, at 30 frames per second, would last a year. This slab is thin enough to freeze the motion of 7 isolated

---

[21]Here Thurston is considering the case $V = \mathbb{C}$ with $A(v) = uv$, where the coefficient u is called the expansion factor.

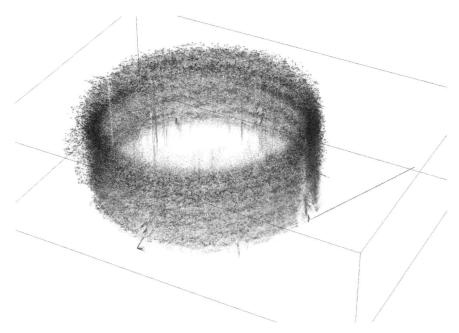

FIGURE 7.7: This picture shows the parameter values for friendly extensions of tent maps (the space $F(2, +1)$). The vertical direction is the λ-axis. The horizontal plane is $\mathbb{C}$, and the points shown are roots of minimal polynomials for postcritically finite examples. With a finer resolution and expanded vertical scale, the friendly extensions would appear as a network of very frizzy hairs (usually of Hausdorff dimension greater than 1), sometimes joining and splitting, but always transverse to the horizontal planes. The diagonal line on the right is λ itself. (See also Plate 22.)

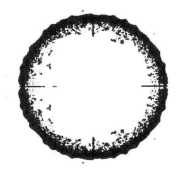

FIGURE 7.8: This image shows roots of 20,000 postcritically finite polynomials of degree about 80 in a thin slab, $1.500000059 \leq \lambda < 1.500000060$. There are 7 isolated spots visible outside the unit circle, with the smallest at 1.5. This slice of expansion factors of length 10^{-9} is thin enough to confine one friendly extension to stay within each spot as λ changes. Roots closer to the unit circle move quite fast with λ, so that even in this short interval the closer roots wander over large areas that overlap and sometimes perhaps collide and split. Roots in the closed unit disk do not depend continuously on λ, but they are confined to (and dense in) closed sets that include the unit circle and increases monotonically with λ, converging at $\lambda = 2$ to the inside portion of Figure 1.1.

friendly parameters, which you can see around the periphery. One of the 7, at position 1.5, is the original controlling expansion constant λ. Closer to the unit circle, the friendly parameters are packed closer together, and they move so quickly that they blur together into a big cloud.

REMARK 7.9. The "set of friends" changes continuously, regarded as an atomic measure on the parameter space for candidate systems, in the weak topology.

8   TRAINTRACKS

We have defined a traintrack map of a graph to be a map such that each edge is mapped by a local embedding under all iterates; now we'll define a traintrack structure.

DEFINITION 8.1. *A traintrack structure τ for a graph Γ is a collection of 2-element subsets of the link of each vertex, called the set of legal turns.*

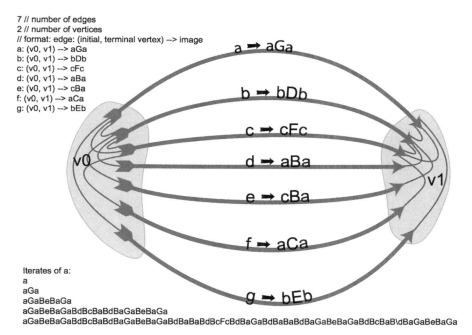

```
7 // number of edges
2 // number of vertices
// format: edge: (initial, terminal vertex) --> image
a: (v0, v1) --> aGa
b: (v0, v1) --> bDb
c: (v0, v1) --> cFc
d: (v0, v1) --> aBa
e: (v0, v1) --> cBa
f: (v0, v1) --> aCa
g: (v0, v1) --> bEb
```

a ⇒ aGa

b ⇒ bDb

c ⇒ cFc

d ⇒ aBa

e ⇒ cBa

f ⇒ aCa

g ⇒ bEb

```
Iterates of a:
a
aGa
aGaBeBaGa
aGaBeBaGaBdBcBaBdBaGaBeBaGa
aGaBeBaGaBdBcBaBdBaGaBeBaGaBdBaBaBdBcFcBdBaGaBdBaBaBdBaGaBeBaGaBdBcBaB\dBaGaBeBaGa
```

An automorphism of F(6) with λ = 3.

FIGURE 8.2: This is a traintrack structure preserved by an automorphism of the free group of rank 6 with expansion constant 3. The edges are labeled with letters, where a capital letter indicates the reverse direction. One set of generators is[22] $\{bA, cA, dA, eA, fA, gA\}$. The two vertices v_0 and v_1 are "exploded" to show legal turns, depicted as thin lines connecting the heads or tails of the arrows. This example will be used in §9 as a template for constructing traintrack maps with any possible Perron number as expansion constant.

The mental image is that of a railroad switch, or more generally a switchyard, where for each incoming direction there is a set of possible outgoing directions where trains can be diverted without reversing course. A path on Γ is *legal* if it is a local embedding, and at each vertex it takes a legal turn.

In describing edge paths, we must first choose an orientation for each edge. We use the convention that a lowercase letter denotes the forward direction on the edge, and the corresponding capital letter to denote the backward direction on the edge. This convention also applies to generators of groups (which we can think of as edges in a cell complex having the given group as its fundamental group).

A map $f: \Gamma \to \Gamma$ *preserves* τ if f maps every τ-legal path to a τ-legal path. It follows that f is a traintrack map: since the forward images of edges must always be legal paths, in particular, they are mapped by local embeddings.

For any traintrack map f, there is always at least one traintrack structure that f preserves. The *minimal invariant traintrack structure* allows turns only if they are ever taken by the forward image of some edge of the graph. The *maximal invariant traintrack structure* allows all turns that are never folded by iterates of f, that is, it allows any turn that is always mapped to be locally embedded.

There is an interesting special case of traintrack structure for a bouquet of circles: the positive traintrack structure, where the legal paths correspond to paths that are positive words in the generators of the free group. These paths follow a consistent orientation along the circles. More generally, an *orientation* for a traintrack structure is a choice of direction for each edge such that every legal path maintains a consistent orientation, either always positive or always negative. A traintrack structure is *orientable* it admits such an orientation. This implies the weaker condition that we can choose a consistent orientation along any legal path.[23]

An oriented traintrack on a graph Γ defines a convex cone C in $H^1(\Gamma, \mathbb{Z})$. For an automorphism ϕ that preserves such a structure, $H^1(\phi, \mathbb{R})$ takes C to C or to $-C$. Therefore, $\pm\phi$ has an eigenvector inside C. Just as in the Perron-Frobenius theorem, an eigenvector strictly inside C has the largest eigenvalue, and in any case, there is an eigenvector in the cone (possibly on its boundary) whose eigenvalue is largest. This eigenvalue is the same as the expansion constant for ϕ, since the incidence matrix for ϕ is the matrix for $\pm\phi$ acting on (simplicial) 1-chains of Γ. As an eigenvector for an element of $GL(n, \mathbb{Z})$, such an eigenvalue is always an algebraic unit.

As illustrated by the example of Figure 8.2, the expansion constant for a traintrack map need not be an algebraic unit: the figure describes a map ϕ_3 which clearly has expansion constant 3. In fact, if we assign each edge the length one, then every edge maps to a path of length three.

To check that this map induces a group automorphism, first write down the images of the generators $\{bA, cA, dA, eA, fA, gA\}$ and collapse the a edge to a point[24] (thus striking out all a's):

---

[22]Note that in discussing the fundamental group of a graph, there is no need to consider only legal paths.

[23]This definition has been changed by the editors. Let us call the weaker condition "weak orientability." We are grateful to Lee Mosher for the following example of a traintrack structure which is weakly orientable but not orientable. Let Γ be the graph with four vertices v_0, v_1, v_2, v_3, together with a loop at v_i for $i > 0$ and an edge joining v_i to v_0 for $i > 0$; and let τ be the traintrack structure which contains all three turns at v_0, together with exactly two of the three turns at v_i including the turn of the loop at v_i.

[24]As an example, we must first expand, $bA \mapsto (bDb)(AgA)$ and only then delete each a or A to obtain $b \mapsto bDbg$.

$$\phi_3: b \mapsto bDbg$$
$$c \mapsto cFcg$$
$$d \mapsto Bg$$
$$e \mapsto cBg$$
$$f \mapsto Cg$$
$$g \mapsto bEbg \qquad (8.1)$$

Every surjective selfmap of a free group is an automorphism,[25] so it's enough to check that all 6 generators are in the image. We get c from $eD \mapsto c$, and given c we get g from $f \mapsto Cg$. Given c and g, we get b from the image of E and f from the image of C. Given these four generators we get the remaining two, d from the image of B and e from the image of G. Therefore, ϕ_3 is a self-homotopy-equivalence of the graph, so it gives a traintrack map. To see that the map preserves the traintrack structure, first note that every edge maps to a legal path. Now note that the three edges a, b, c each map to a path starting and ending in the same way, with a, b or c, respectively. All words involving only these three edges are legal for this traintrack. The other four edges map to words beginning and ending with a, b or c, so it is easy to check that legal turns are mapped to legal turns.

Since the expansion constant is not an algebraic unit, there can be no orientable traintrack structure invariant by the automorphism up to homotopy. For this particular example, the structure has an anti-orientation, such that every legal path reverses orientation at every vertex.

For $m \geq 0$, a map ϕ_{3+2m} with expansion factor $3 + 2m$ can be defined, preserving the same traintrack:

$$\phi_{3+2m}: a \mapsto aG(aB)^m a$$
$$b \mapsto bD(bC)^m b$$
$$c \mapsto cF(cA)^m c$$
$$d \mapsto aB(aB)^m a$$
$$e \mapsto cB(aB)^m a$$
$$f \mapsto aC(aB)^m a$$
$$g \mapsto bE(bA)^m b \qquad (8.2)$$

A sequence of steps identical to that for ϕ_3 shows that ϕ_{3+2m} is a traintrack map. For completeness, we can define ϕ_1 as the identity map of this traintrack; ϕ_1 is also a traintrack map. There are similar constructions for even expansion factors, but we do not need them.

9   SPLITTING HAIRS

Let $f: \Gamma_0 \to \Gamma_0$ be any purely expanding self-map of a tree. Every tree is bipartite; that is, we can partition the vertices into to sets V_0 and V_1 such that every edge of Γ_0 has one endpoint in each of the two sets.

---

[25]See, for example, [LS, p. 14].

The map f may not respect this partition; if not, let Γ be the barycentric sub-division of Γ_0, with one new vertex $v(e)$ in the middle of each edge e of Γ_0. For each edge e of the original Γ_0, choose one of the edges e' in its image, and adjust f so that $v(e) \mapsto v(e')$. The new map still has a non-negative incidence matrix. Any positive linear function that is a dual eigenvector for its transpose pulls back to a linear function that is an eigenvalue for the original f, so its eigenvalue is the same.[26] Therefore, there are positions for the new vertices $v(e)$ so all vertices of Γ are mapped (by the original f) to vertices of Γ. The graph Γ has a bipartite parti-tion, where V_0 consists of original vertices and V_1 consists of new vertices, where each partition element is invariant by f.

Given any graph Γ and an integer n, define a new graph $\mathrm{Split}_n(\Gamma)$ to be ob-tained by replacing each edge of Γ with n edges having its same pair of endpoints. For each edge e, label the new edges with subscripts $e_a, e_b, \ldots$.

When Γ has a bipartite structure $\{V_0, V_1\}$, define a traintrack structure on $\mathrm{Split}_7(\Gamma)$ using the prototype of Figure 8.2: that is, a path is legal if and only if the sequence of subscripts defines a legal path in the prototype.

When f is a map of Γ to itself that preserves the partition elements and maps each edge to an edge-path of length at least one, define

$$\mathrm{Split}_7(f)\colon\ \mathrm{Split}_7(\Gamma) \to \mathrm{Split}_7(\Gamma)$$

using the prototypes ϕ_i defined in Section 8. That is, for an edge e_x of $\mathrm{Split}_7(\Gamma)$, if $f(e_x)$ has combinatorial length k, lift f to the map $\mathrm{Split}_7(f)$ by applying the sequence of subscripts $\phi_k(x)$ to the sequence of edges $f(e_x)$.

Note that the ϕ_{2m+1} themselves are defined by this process, starting with the self-maps of the unit interval that fold it over itself an odd number of times.

PROPOSITION 9.1. *For any bipartite structure $\{V_0, V_1\}$ on a graph Γ and any map $f\colon \Gamma \to \Gamma$ that preserves the partition elements and maps each edge to an edge-path of length at least one, $\mathrm{Split}_7(f)$ is a traintrack map.*

PROOF. Notice that in the prototype traintrack, every legal turn has at least one of its ends among the edges a, b, c. Any turns among these three edges are legal, and every ϕ_{2m+1} preserves their beginnings and end. Every ϕ_{2m+1} with $m > 1$ maps beginnings and ends in the same way, so a turn between edges whose com-binatorial image length is more than 1 maps to a legal turn. Similarly, a legal turn between edges one or both of which have combinatorial image length 1 maps to a legal turn. $\qquad\square$

PROPOSITION 9.2. *If Γ is a tree with bipartite structure $\{V_0, V_1\}$, and if $f\colon \Gamma \to \Gamma$ is a self-map such that*

- *f preserves V_0 and V_1, and*

- *f is a local embedding on each edge,*

- *the map $f_1\colon e \mapsto f(e)_1$, that is, e goes to the first element of its image edgepath, is a permutation, and*

---

[26]The incidence matrix M' for the subdivided traintrack map is made up out of 2×2 blocks $\begin{bmatrix} a & b \\ c & d \end{bmatrix}$ such that $a + b = c + d$ is equal to the corresponding entry for the original incidence matrix M. If $(x_1, x_2, \ldots, x_{2n})$ is a left eigenvector for M', then $(x_1 + x_2, \ldots, x_{2n-1} + x_{2n})$ is a left eigenvector for M with the same eigenvalue.

- *the map f_2: $e \mapsto f(e)_2$ if $f(e)$ has length more than 1, and $e \mapsto f(e)_1$ otherwise, is also a permutation,*

then $\mathrm{Split}_7(\Gamma)$ *is a homotopy equivalence.*

PROOF. The set of edges with subscript a forms a spanning tree for $\mathrm{Split}_7(\Gamma)$. If we collapse the spanning tree to a point, we obtain a bouquet of circles, so the remaining edges give a free set of generators for the fundamental group.

Pick a basepoint $*$ on the graph $\mathrm{Split}_7(\Gamma)$. As an edge path, the generator corresponding to e_x is obtained by prefixing e_x by the subscript a path from $*$ to its first vertex, and appending the A path from its end vertex back to $*$, then striking out all as and As.

As before, we just need to show that all generators are in the image of $\mathrm{Split}_7(f)$. Consider the 6 generators lying over any particular edge x of Γ. Let x' be the first edge in the edgepath $f(x)$. We will follow the same outline that showed ϕ_3 is a homotopy equivalence. The image of $f(x_e X_D)$ is x_c'[27] Applying this to all edges x of Γ, we obtain all generators x_c in the image. Modulo the x_c generators, from $\mathrm{Split}_7(f)(x_f)$ we get all generators of the form $f_2(x)_g$. Since f_2 is a permutation, this gives all subscript g generators. From $\mathrm{Split}_7(f)(X_D)$ modulo the g generators, we get the $f_1(x)_b$ generators. Continue in the same sequence that was used to show ϕ_3 is surjective: modulo previous generators, each edge has a payload generator in the first or second slot of its image, so we get all generators. $\square$

Given any Perron number λ, we can now apply Proposition 9.2 to the asterisk map f_λ constructed by Theorem 6.2, where we take V_0 to consist of the center vertex and V_1 to consist of all the tips. We obtain a traintrack map $\mathrm{Split}_7(f_\lambda)$. The maps were constructed so that every edge occurs as the first segment of some image edge-path. The same edge occurs as the second segment, if the path has length more than 1, so the maps $f_{\lambda,1}$ and $f_{\lambda,2}$ are both permutations. Therefore, $\mathrm{Split}_7(f_\lambda)$ is traintrack homotopy equivalence of $\mathrm{Split}_7(\Gamma)$, which uniformly expands with expansion factor λ.

The maps ϕ_{2m+1} are mixing when $m > 1$, from which it follows that $\mathrm{Split}_7(f_\lambda)$ is mixing.

If λ is a weak Perron number, let n be the least integer such that λ^n is a Perron number. Make an asterisk from n copies of an asterisk for λ^n, and map it to itself by permuting the factors except for the final map, which is a copy of f_{λ^n}. From this asterisk map f_λ, we obtain a traintrack automorphism $\mathrm{Split}_7(f_\lambda)$ with expansion constant λ. This completes the proof of Theorem 1.9.

10   DYNAMIC EXTENSIONS

If f: $X \to X$ is a continuous map, then an *extension* of f is a space Y with a continuous surjective map F: $Y \to Y$ and a semiconjugacy p: $Y \to X$ of F to f, that is, it satisfies $p \circ F = f \circ p$.

The hair-splitting construction of section 9 is a special case, which we'll call a *graph extension*, where X and Y are graphs and edges of Y are mapped to nontrivial edge-paths of X. If f is a λ-uniform expander, then so is F. Given two graph extensions F_1 and F_2, they have a fiber product F_3, consisting of all pairs of points in Γ_1 and Γ_2 that map to the same point in Γ. We can pick a connected component

---

[27]The image of $x_e X_D$ under $\mathrm{Split}_7(f)$ is $x_c' X_A'$, which becomes x_c' after the collapse.

of the fiber product to obtain a connected graph that is a common extension of both; thus the set of connected graph extensions is a partially ordered set where any two elements have an upper bound (but not necessarily a least upper bound).

It seems natural to ask which uniformly expanding self-maps of graphs have graph extensions that are traintrack self-homotopy-equivalences. It would also be interesting to strengthen the condition, to require that the fundamental group of the extension graph maps surjectively to the fundamental group of the base; in that case, a necessary condition is that f itself be a homotopy equivalence, otherwise neither it nor its extension would be surjective on π_1. If necessary, we could also weaken the condition by allowing a subdvision of Γ before taking the extension.

Although Proposition 9.2 required that the first and second elements in the edgepaths for f are partitions, this condition does not seem essential. One trick would be to modify the formula of 8.2 to lift f, we could use lifts in patterns such as $b \mapsto b(Cb)^{m1}D(bC)^{m2}g$ and $e \mapsto (aB)^{m1}c(Ba)^{m2}$ to adjust the payload edges to be somewhere else along the edgepath, taking care that the payload edges map surjectively, and that they are chosen so that there is an order that will unlock them inductively.

It seems plausible that by a combination of duplicating edges, subdividing edges, and lifting with adjustable payloads, any expanding self-homotopy-equivalence of a graph would have an extension that is a traintrack map of a graph whose fundamental group maps surjectively to the base. However, it is beyond the scope of the current paper to pursue this.

It would also seem interesting to understand a theory of minimal uniformly expanding maps, ones that cannot be expressed as non-trivial extensions. Perhaps extensions that merely identify vertices should be factored out. This should be related to a theory of finitely-generated abelian semigroups that exclude 0 and are invariant by a transformation and which, tensored with $\mathbb{R}$, have λ as dominant eigenvalue. *They can be thought of as positive λ-modules.*

11   BIPOSITIVE MATRICES

DEFINITION 11.1. *A pair of real numbers (a, b) is a* conjugate pinching pair *if a and b^{-1} are algebraic units whose Galois conjugates are contained in the open annulus of inner radius b^{-1} and outer radius a. They are a* weak conjugate pinching pair *if their conjugates are contained in the closed annulus.*

Here is an elementary fact.

PROPOSITION 11.2. *A real number is the expansion constant for an element of $\mathrm{GL}(n, \mathbb{Z})$ for some n if and only if it is a weak Perron number.*

A pair (a, b) of algebraic units occurs as the pair of expansion constants for A and A^{-1} if and only if the pair is a weak conjugate pinching pair.

PROOF. Sufficiency is easy. Given a weak Perron number a, the companion matrix for the minimal polynomial for a has expansion constant a. Given a conjugate pinching pair (a, b), take the sum of the companion matrix for a minimal polynomial for a with the inverse of a companion matrix for the minimal polynomial for b.

Necessity is pretty obvious: a must be the maximum absolute value, and b the minimum, among all roots of χ_A. □

DEFINITION 11.3. *An invertible matrix A is* bipositive *with respect to a basis B if B admits a partition into two parts P and Q such that A maps the orthant spanned by $P \cup Q$ to itself, and A^{-1} maps the orthant spanned by $P \cup -Q$ to itself.*

If either of the two positive matrices associated with a bipositive matrix is mixing, the other is as well, so in this case we also say the bipositive matrix is mixing. Similarly, we will call a bipositive matrix *mixing* if all sufficiently high powers have all non-zero entries. More geometrically, given a positive matrix, we can make a graph whose vertices are the basis elements and whose edges correspond to nonzero entries. The matrix is primitive if there are edge-paths of every sufficiently long length from each vertex to each others. The set of lengths of edge-paths from one vertex to another is a semigroup, and it is easy to see that its complement is finite if and only if the elements have a common divisor larger than one. It follows that an irreducible positive or bipositive matrix A is mixing if and only if every positive power of A is ergodic.

THEOREM 11.4. *A pair (a, b) is the pair of positive eigenvalues for an invertible bipositive integral matrix of some dimension if and only if it is a conjugate pinching pair, or a weak conjugate pinching pair such that all conjugates of a and b^{-1} of modulus a or b^{-1} are a or b^{-1} times roots of unity.*

REMARK 11.5. The clause about roots of unity addresses the issue that although all Galois conjugates of a weak Perron number a that have *maximum* modulus are at angles that are roots of unity, the Galois conjugates of *minimum* modulus need not be.

For instance, if a is the plastic number $1.32472\ldots$ which is a root of $x^3 - x - 1$, its other two conjugates, $-0.662359\ldots \pm i0.56228\ldots$, have modulus $1/\sqrt{a}$, so $(a, \sqrt{a})$ is a weak conjugate pinching pair. However, the two complex Galois conjugates of a are at angles that are irrational multiples of π. Therefore, $(a, \sqrt{a})$ is not the pair of expansion constants for any bipositive matrix.

PROOF. We will first establish the theorem in the case (a, b) is a [strict] conjugate pinching pair.

First, let $A \in \mathrm{GL}(n, \mathbb{Z})$ have eigenvectors S and T of eigenvalues a and b^{-1}. By restricting to the smallest rationally defined subspace of $\mathbb{R}^n$ containing S and T, we may assume that all other characteristic roots of A are in the interior of the annulus $b^{-1} < |x| < a$. Let H be the real subspace spanned by S and T and $p \colon \mathbb{R}^n \to H$ be the projection that commutes with A.

Let L be the line through the origin of slope $a \times b$ in the ST-plane. The linear transformation A multiplies slopes by $1/a \times b$, so $A(L)$ has slope 1. Let L' be the reflection of L in the S-axis; thus $A(L')$ is the reflection of $A(L)$.

Choose a set P_0 of lattice points such that $p(P_0)$ lies in the interior of the first quadrant of the ST-plane, P_0 generates the lattice $\mathbb{Z}^n$ as a group, and the convex cone generated by P_0 contains the angle between L and $A(L)$, except 0, in its interior. Similarly, choose Q_0 to be a set of lattice vectors that generate $\mathbb{Z}^n$ as a group, that are mapped by p to the interior of the fourth quadrant, generates $\mathbb{Z}^n$ as a group, and contains the angle between L' and $A(L')$.

Let σ_P be the semigroup generated by P_0 and let σ_Q be the semigroup generated by Q_0.

We claim that for sufficiently large $k > 0$ and for any two elements $p_1, p_2 \in P_0$, there is a $j > 0$ such that

$$A^j(p_2) - A^{-k}(p_1) \in \sigma_Q,$$

and, similarly, for any $q_1, q_2 \in Q_0$, there is a $j > 0$ such that

$$A^j(q_2) - A^{-k}(q_1) \in \sigma_P.$$

To see this, we will make use of a basic fact about the semigroups of lattice elements:[28]

LEMMA 11.6. *Let U be a finite set of elements of the lattice $\mathbb{Z}^n \subset \mathbb{R}^n$, and let C_U be the convex cone generated by U. There is some constant R such that any element g of the group G_U generated by U whose radius R ball $B_R(g)$ is contained in C_U is also contained in the semigroup S_U generated by U.*

PROOF. The convex cone C_U is the set of all non-negative real linear combinations of elements of U. If we express an element of $G_U \cap C_U$ as a non-negative real linear combination of elements of U and round the coefficients to the nearest integer, we see that g is within a bounded distance R_1 of some element of S_U.

Let F be the set $G_U \cap B_{R_1}(0)$. Express each element of $f \in F$ as an integer linear combination $\sum_{u \in U} k_{u,f} u$. Let R_2 be the maximum, among these linear combinations, of the norm of the set of negative coefficients, and set $R = R_1 + R_2$.

Now for any element $g \in G_U$ such that $B_R(g) \subset C_U$, there are elements $s \in S_U$ and $f \in F$ such that $g = s + f$. $\square$

Now for k large, the image in projective space of $A^{-k}(p_1)$ is close to the image of the eigenspace T. The images $A^j(p_2)$ march in the direction of the S axis, projectively converging to the S eigenspace, with the slope in the ST-plane of the line from $A^{-k}(p_1)$ to $A^j(p_2)$ decreasing by a factor of $a \times b$ at each iterate. If k is suitably large, at least one of these iterates is captured in the interior of the cone $A^{-k}(p_1) + C_Q$ and so, by the lemma, in the semigroup σ_Q.[29]

Now we can describe an irreducible bipositive matrix. Take the free abelian group FA generated by

$$\bigcup_{g \in P_0 \cup Q_0} \left\{ A^{-k}(g), \ldots, A^{j(g)-1}(g) \right\}.$$

We will choose a bipositive map $\tilde{A}$ of FA to itself that commutes with evaluation in $\mathbb{Z}^n$. The generators $A^h(g)$ pass off from one to the next until $A^{j(g)-1}(g)$. Choose a cyclic permutation of p of P_0 and a cyclic permutation q of Q_0. For each $g \in P_0$ choose an expression $A^{j(g)}(g) = A^{-k}(p(g)) + sq$ where sq is an element of σ_Q, and similarly for $g \in Q_0$ express $A^{j(g)}(g) = A^{-k}(q(g)) + sp$, with $sp \in \sigma_P$. Use this to give the final links in the chain, to define $\tilde{A}$.

The matrix for the linear transformation can be expressed as an upper triangular matrix followed by a permutation; hence, it is invertible. In block form, the

---

[28]See for example Exercise 7.15 of [MSt].

[29]If k is suitably large, at least one of these iterates is the center of a large ball captured in the interior of the cone $A^{-k}(p_1) + C_Q$ and so, by the lemma, in the semigroup σ_Q.

P generators are each expressed as another P generator plus an element of the σ_Q semigroup, and vice versa. The inverse has the same form, but with semigroup elements subtracted; reversing the sign of the Q generators turns the inverse into a positive matrix.

The matrix for FA is irreducible because of the cyclic permutations: the images of each generator eventually involves each other generator, and (except in the trivial case $a = b = 1$) the powers of the matrix are eventually strictly positive. The positive eigenvalue for FA is a, since projection to the S-axis gives a linear function, positive on the positive orthant for FA, that is a dual eigenvector of eigenvalue a. Similarly, the positive eigenvalue for FA^{-1} is b.

If (a, b) is a weak conjugate pinching pair such that all conjugates of a or b^{-1} on the circle of maximal or minimal modulus have arguments that are rational multiples of 2π, let h be a common multiple of the denominators. Then (a^h, b^h) is a strict conjugate pinching pair; let A_0 be a matrix realizing this pair of pinching constants. Take the direct sum of h copies of the underlying vector space, permute them cyclically, with return map A_0. This is a bipositive matrix realizing (a, b).

For the converse: consider any bipositive matrix A with positive eigenvalue a and positive eigenvalue b for A^{-1}, where one or both are not the unique characteristic roots of maximum modulus *say there are k roots of modulus a*. Then there is a k-dimensional subspace with a metric where A acts as a similarity, expanding by a factor of a. Hence, the projective image of this subspace is mapped isometrically. Its intersection with the image of the positive orthant is a polyhedron mapped isometrically to itself; hence, its vertices are permuted. It follows that all characteristic roots of modulus a have the form of a times an mth root of unity. $\qquad\square$

A simple example of the construction for a weak conjugate pinching pair is the Fibonacci transformation $x \mapsto y$ and $y \mapsto x + y$, with eigenvalues $\phi = (1 + \sqrt{5})/2$, the golden ratio and $-1/\phi$. The square of this transformation is $x \mapsto x + y$ and $y \mapsto x + 2y$, which is bipositive: its inverse maps the second quadrant to itself. This translates into a bipositive map for $(\phi, 1/\phi)$ in dimension 4, that expresses a bipositive linear recurrence for four successive terms of the Fibonacci sequence,

$$x_1 \mapsto x_2, \quad x_2 \mapsto x_3, \quad x_3 \mapsto x_2 + x_3, \quad x_3 \mapsto x_1 + 2x_2.$$

As another example, $a = (1 + \sqrt{2})^{1/3}$ and $b = (\sqrt{2} - 1)^{1/3})$ can be realized by a bipositive transformation in dimension 12.

QUESTION 11.7 (suggested by Martin Kassabov). *Suppose $A \in \mathrm{GL}(n, \mathbb{Z})$ has dominant eigenvalue a and A^{-1} has dominant eigenvalue b, where (a, b) is a [strict] conjugate pinching pair. Is there a basis for which some power of A is bipositive?*

REMARK 11.8. Although elementary matrices generate $\mathrm{SL}(n, \mathbb{Z})$, and together with permutations generate $\mathrm{GL}(n, \mathbb{Z})$, the semigroup is a very different matter: for $n \geq 3$, the elementary positive semigroup is not even finitely generated.

To see the gap between the elementary positive semigroup and the full positive semigroup of $\mathrm{GL}(n, \mathbb{Z})$, let's focus on the case $n = 3$ (the embedding in $\mathrm{GL}(3 + n, \mathbb{Z})$ that fixes all but the first 3 basis elements gives examples, albeit atypical, for arbitrary dimension ≥ 3).

Let's look at the action of these semigroups on the basis triangle B in $\mathbb{RP}^2$. The image of the triangle by a word in the generators gives a sequence of subtriangles of this triangle, where at each step you bisect one of the sides and throw one half away. In particular, each possible proper image is contained in one of six half-triangles of B.

Now consider a general positive element of $\mathrm{GL}(3, \mathbb{Z})$. It maps the tetrahedron spanned by 0, together with the basis elements to a "clean" lattice tetrahedron, that intersects lattice points only at its vertices. This property (clean) characterizes the possible images. Furthermore, any clean triangle in the positive orthant with one vertex at the origin can be extended (in many ways) to a clean tetrahedron: just add any vertex in one of the two lattice planes neighboring the plane containing the triangle.

But there are many clean triangles; in fact, the set of lattice points in the positive orthant that are primitive (i.e., the 1-simplex from the origin to the point is clean) have density $(1 - 2^{-3}) \times (1 - 3^{-3}) \times (1 - 5^{-3}) \cdots = 1/\zeta(3) \approx 0.831907 \ldots$, and given a primitive lattice point p, the density of lattice points q such that the triangle $\Delta(0, p, q)$ is clean is $1/\zeta(2) = 6/\pi^2 \approx 0.607927 \ldots$. It follows that every line segment contained in B can be approximated in the Hausdorff topology by an image of B under the positive semigroup of $\mathrm{GL}(3, \mathbb{Z})$. Many such line segments cross all 3 altitudes of B, so a positive element of $\mathrm{GL}(3, \mathbb{Z})$ that maps B to a nearby triangle is not in the positive elementary semigroup.

Nonetheless, images of B under the positive semigroup of $\mathrm{GL}(3, \mathbb{Z})$ are quite restricted. For instance, it's easy to see that the centroid of B, corresponding to the line $x = y = z$, cannot be in the interior of any image of B. For any finite collection of triangles not containing the centroid in their interior, most lines through the centroid are not contained in any one of them. Therefore, no finite set of positive $\mathrm{GL}(3, \mathbb{Z})$ images of B cover all possible images.

Note that the proof of Theorem 11.4 actually gave something stronger. If we have a basis B that is partitioned into two parts P and Q, then any elementary transform that replaces an element of P by its sum with an element of Q, or an element of Q by its sum with an element of P is bipositive. The *elementary bipositive semigroup* [with respect to (P, Q)] is the semigroup generated by these cross-type elementary transformations, together with permutations that preserve P and Q.

THEOREM 11.9. *A pair of real algebraic units (a, b) is the pair of expansion constants for an elementary bipositive matrix and its inverse if and only if it is a conjugate pinching pair such that all Galois conjugates of a or b^{-1} of maximal or minimal modulus have arguments that are roots of unity.*

PROOF. From the proof of Theorem 11.4, the condition on arguments is an equivalent form of the hypothesis in the case that (a, b) does not strictly pinch. □

12   TRACKS, DOUBLETRACKS, ZIPPING AND A SKETCH OF THE PROOF OF THEOREM 1.11

Continuous maps are often inconvenient for representing homotopy equivalences between graphs, because a self homotopy equivalence cannot be made into a self-homeomorphism of any graph in the homotopy class unless it has finite order up to homotopy.

Continuous maps of graphs can be inconvenient as geometric representatives of group automorphisms of the free group, since they are usually not invertible. As we have seen, it is not easy to see at a glance whether a given map of graphs is a homotopy equivalence. There are algorithms to check, but they can be tedious.

As an alternative, we can represent self-homotopy equivalences by continuous 1-parameter families of graphs. These have the advantage of being reversible. If we restrict to graphs that have no vertices of valence 1, these can be locally described by moving attachment points of edges along paths in the complement of the edge.

A *zipping* of a traintrack τ to a traintrack σ is a 1-parameter family of structures $(\Gamma_t, \tau_t) | \tau \in [0, 1]$ that may be thought of as squeezing together legal paths. It's elementary to see that for any traintrack map f, there is a zipping that yields the homotopy class of f: just progressively and locally zip together the identifications that will be made by the map.

A zipping can be translated into a sequence of reversible steps, consisting of a motion of an attachment point of one edge along a legal path on its complement, starting in a direction in its linkgroup. When (Γ, τ) is zipped to (Γ', τ'), every bi-infinite τ-legal path becomes a bi-infinite τ'-legal path. The inverse of a zipping is an *unzipping*, or *splitting*.

A *doubletrack* structure for Γ is a pair (σ, τ) of traintrack structures on Γ. A graph Γ equipped with a pair of traintrack structures is a *doubletrack*.

A *bizipping* between doubletracks is a 1-parameter family of doubletracks that is a zipping of the first traintrack structure and an unzipping of the second. This yields a traintrack map of one structure whose inverse is a traintrack map for the other structure.

REMARK 12.1. It seems likely that invariant foliations could provide a good alternative to traintracks. Bestvina and Handel introduced a concept of traintracks relative to an invariant filtration of a graph and showed that relative traintrack maps exist for every outer automorphism of a free group (not just in the irreducible case). Instead, one could look at foliations of finite depth on a manifolds of sufficiently high dimension (as a function of the rank of the free group), with singularities having links based on polyhedra, and satisfying the condition (used to great effect by Novikov) that there are no null-homotopic closed transversals. Such a foliation picks out a class of bi-infinite words in the free group. Bestvina-Handel's theorem on existence of relative traintracks would appear to translate to the existence of a homeomorphism of some open manifold homotopy-equivalent to a bouquet of circles that preserves such foliation. Pairs of foliations could substitute for double-tracks.

We will not take the detour of trying to develop this point of view here.

We now sketch the proof of Theorem 1.11.

PROOF. We will now analyze the [main] case when there are strict inequalities: let (a, b) be a strictly pinching pair of algebraic units.

From Theorem 11.4, let A be a bipositive matrix with positive eigenvectors S for A and T for A^{-1} having eigenvalues a and b. Let p be the invariant projection of the vector space for A to the plane spanned by S and T. The basis is partitioned into two sets, P and Q, with $p(P)$ in the first quadrant and $p(Q)$ in the fourth quadrant. The negations of these two sets give vectors that are mapped into the other two quadrants.

We claim that, for a suitable choice of P and Q, there is a permutation α of the basis such that for any $g \in P \cup Q$ the difference $A(\alpha g) - g$ is a non-negative linear combination of $\pm$ basis vectors that map to the quadrant neighboring the quadrant of $p(g)$ across the S axis. (To help with visualization: for most basis elements, α will be chosen so that $A(g) = \alpha^{-1}(g)$.)

Suppose this claim is true. Let Γ be a graph with a single vertex whose edges correspond to $P \cup Q$. Let E be the homomorphism $\pi_1(\Gamma) \to \mathbb{R}^n$ that maps the loop of an edge labeled by a vector to that vector, and let $\tilde{\Gamma}$ be the corresponding covering space. (Γ can be visualized as an embedded graph in the torus $\mathbb{R}^n / \mathbb{Z}^n$, and $\tilde{\Gamma}$ is lift to a graph in $\mathbb{R}^n$). Define a doubletrack structure (σ, τ) on Γ where a σ-legal path lifted to $\tilde{\Gamma}$ has monotone projection to the S-axis and a τ-legal path lifts to have monotone projection to the T-axis. The linear transformation A maps Γ to a new graph $A(\Gamma)$ mapped to $\mathbb{R}^n$. Using the preceding claim we can construct a bizipping that slides the image edges back to the originals.[30]   □

13   SUPPLEMENTARY NOTES (MOSTLY BY JOHN MILNOR)

NOTE 13.1 (See Figure 1.7). The following email exchange between Curt Mc-Mullen and Thurston took place in February 2012.

> CTM: *What proof do you have in mind that this Perron number does not give a critically finite tent map?*

> WPT: *I check these things with a simple computer program that computes in the field, represented as polynomials modulo the defining polynomial for the field. The field can be mapped to $\mathbb{C}$ according to various roots, but it's computationally simpler just to look at several of the coefficients. As you iterate the self-map of the interval, if the image in one of the real or complex places gets outside some easily computed bound, it escapes to infinity. This becomes very obvious (without worrying about precise bounds) when you actually do it, because in the non-Perron cases, if they escape, you quickly get numbers on the order of 10^4 or 10^6 or so which are way out of bounds. Anyway, with the computer, it's immediate (for non-Perron cases).*

Presumably, in the Perron case it takes a bit longer, but still becomes obvious. Note that this procedure makes sense only if the polynomial is known to be irreducible. However, this is easy to check using Maple or Mathematica.

NOTE 13.2. Thurston states on page 344:

> *Figure 1.1 shows the Galois conjugates of $\exp(h)$ for postcritically finite real quadratic maps. This is a path-connected[31] set, with much structure visible.*

The same message from February 2012 clarifies this statement. But first let us provide some necessary background from kneading theory. Let f be a unimodal map with topological entropy h. According to [MT], the kneading data for f gives rise to a function of the form

$$k_f(t) = 1 \pm t \pm t^2 \pm \ldots ,$$

---

[30]The inverse of the permutation α is implicit in the description of the basis for the free abelian group FA in the proof of Lemma 11.6.

[31]Or, more precisely, the topological closure of this set is path-connected.

which is defined and holomorphic throughout the open unit disk, which has no roots within the disk $|t| < e^{-h}$, and which has e^{-h} as a root provided that $h > 0$. In the postcritically finite case, the coefficients of this power series are eventually periodic, say, of period p. It follows that k_f extends to a rational function with poles only at pth roots of unity. If the expansion constant $\lambda = e^h$ has defining polynomial $P(t)$ of degree n, it follows that the numerator of the rational function $k_f(t)$ is a multiple of the "reversed" polynomial $t^n P(1/t)$.

Here is Thurston's commentary (mildly edited).

> *For the portion of Figure 1.1 outside the unit circle, the roots depend continuously on the entropy. This comes from a general principle — first of all, among power series with ± 1 coefficients, there are a bounded number of roots inside a circle of radius $r < 1$. (See the original paper that Milnor and I wrote.) Further, the roots change continuously with the coefficients. Since the set of possible coefficients is totally disconnected, this doesn't yet logically say they move continuously, but in fact they do change continuously with entropy via the maps*
>
> $$ entropy \;\mapsto\; power\ series\ (from\ kneading\ theory) \;\mapsto\; roots. $$
>
> *However, the roots move very fast with entropy.*
>
> *For the roots inside the unit circle, if you look at the case when they are roots of a finite power series, they are approximately determined by the last bunch of terms of the series (the reciprocal is a root of the reverse series) — they don't depend continuously on entropy. What actually happens is that in a small interval of entropy, there is a set of length N tails that can occur; in the usual topology on power series they are traced over in a discontinuous, dense way. The portion "blacked out" inside the unit disk gradually increases, and it always remains connected.[32] The unit circle remains in the set as well.*

The statement that the roots of k_t within the unit disk depend continuously on entropy depends on the observation that the only discontinuities of the correspondence $f \mapsto k_f(t)$ occur when f is critically periodic. (Here we assume that f varies within a well-behaved family, for example, quadratic maps, or uniformly expanding maps.) As we pass through a critically periodic f, there is a jump discontinuity in which $k_f(t)$ is multiplied by a function which has no zeros in the unit disk. Thus the roots of k_t within the unit disk always depend continuously on entropy. (More precisely, in order to keep track of roots near the boundary, it would be clearer to say that the correspondence $h \mapsto \{roots\ in\ \mathbb{D}\} \cup \partial\mathbb{D}$ is continuous, using the Hausdorff topology.)

Note that in an example such as that of Figure 1.7, it would be perfectly possible to compute the roots of $k_f(t)$ within any disk $|t| < 1 - \epsilon$, and to compare them with the roots of the defining polynomial for $1/\lambda$. There will always be at least one root $1/\lambda$ in common; but in the postcritically finite case every Galois conjugate of $1/\lambda$ must also be a root of k_f.

---

[32]Here he does not say path-connected. We are not sure whether this is just an oversight or whether there is a problem with path-connectedness. See [Ti] for a proof of path connectedness for the closure of the possibly larger set consisting of reciprocals of zeros of the kneading determinant for critically periodic unimodal maps.

Evidently the conjugates of λ inside the unit disk (or the roots of $k_f(t)$ outside the unit disk), form a much more difficult object of study, well worth a more detailed exposition and further investigation. As one example, it would be curious to compare this with Oylyzko and Poonen study [OP] of polynomials with 0, 1 coefficients. For other similar studies, see http://math.ucr.edu/home/baez/roots.

NOTE 13.3 (D. Lind).   Here are further remarks on the discussion in §2 (page 351). Let $T_\beta : [0,1) \to [0,1)$ defined by $T_\beta(x) = \beta x \mod 1$ be the β-transformation. A key property is whether 1 has a finite orbit under T_β. This is the case iff the corresponding subshift has finite type (a result due to Parry). Such a β is called a *Parry number* by some. Every Pisot number is Parry, and Schmidt suggested that every Salem number is Parry.

David Boyd (see [Bo1]) confirmed that Salem numbers of degree 4 are Parry. In [Bo2], he investigated the degree 6 case. In this paper Boyd describes in detail the heuristic based on random walks discussed in this paper, and in particular the predictions that *almost every* degree 6 Salem number should be Parry, while in degree 8 and higher both a positive proportion of Salem numbers should be Parry and a positive proportion should be *not* Parry.

He considers the 11,836 Salem numbers of degree 6 and trace at most 15. For all but a handful, he computes a (finite) orbit of 1. However, there are few examples ($x^6 - 3x^5 - x^4 - 7x^3 - x^2 - 3x + 1$ being one of them) for which the orbit size is at least 10^9 and possibly infinite. This is left unresolved in Boyd's paper.

NOTE 13.4 (See page 352).   Given an arbitrary function $\phi \colon \{0,\dots,n\} \to \{0,\dots,n\}$, we can extend uniquely to a function $f \colon [0, n] \to [0, n]$ which is linear on each subinterval $I_j = [j-1, j]$. The associated incidence matrix $A = [a_{i,j}]$ is defined by

$$a_{i,j} = \begin{cases} 1 & \text{if } f(I_i) \supset I_j, \\ 0 & \text{otherwise}. \end{cases}$$

It follows easily that $a_{i,j} = 1$ if and only if

$$\min\left(\phi(i-1), \phi(i)\right) < j \leq \max\left(\phi(i-1), \phi(i)\right).$$

If the leading eigenvalue λ of this matrix is non-zero, then the topological entropy of f is given by $h = \log(\lambda)$.

More generally, consider a piecewise monotone map $f \colon [0, n] \to [0, n]$ which maps the finite set $\{0, \dots, n\}$ into itself, and such that every critical value lies in this finite set. We can then define $\widehat{a}_{i,j} \geq 0$ to be the number of times that the image of I_i covers I_j. The topological entropy of f is the logarithm of the leading eigenvalue $\lambda \geq 1$.

Let ϕ be the restriction of f to $\{0, \dots, n\}$, and let $[a_{i,j}]$ be the matrix of zeros and ones associated with this function ϕ. Then it is not hard to see that $\widehat{a}_{i,j} \equiv a_{i,j} \pmod 2$. Furthermore, for each j, the set of i for which $\widehat{a}_{i,j} > 0$ must form a connected block. Conversely, given a matrix $\widehat{a}_{i,j}$ satisfying these conditions for an appropriately chosen ϕ, it is not hard to construct a corresponding piecewise montone map.

NOTE 13.5 (See page 355). The figure below illustrates Thurston's procedure for passing from a postcritically finite λ^N-expanding map of the interval to a postcritically finite λ-expanding map of the circle as discussed at the beginning of Section 5. In this example, $N = 3$ and $\lambda^3 = 2$, so that the subintervals on the right have lengths $1, 2^{1/3}, 2^{2/3}$. Critical orbits have been marked.

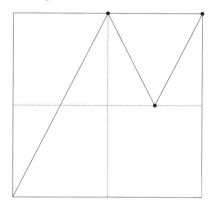 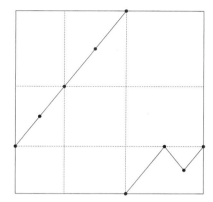

NOTE 13.6 (See page 357). The proof that existence of a postcritically finite λ^N-expanding interval map implies the existence of a postcritically finite λ-expanding interval map can be simplified as follows. First consider the special case $N = 2$. Divide the unit interval into four subintervals by the points

$$b = \lambda a, \quad c = \lambda b, \quad \text{and} \quad b = \frac{1}{1+\lambda}.$$

The following figure illustrates the case $\lambda^2 = 5$. A uniform 5-expander is graphed on the left. Reflecting in the y-axis and squeezing the unit square onto the rectangle $[b, c] \times [a, b]$ we obtain the graph on the right, with slope $\pm\sqrt{5}$ everywhere.

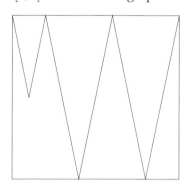 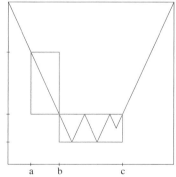

A similar construction works for any $\lambda > 1$ and any uniform λ^2-expander which maps both endpoints of the unit interval to 1. Note that the number of critical points is preserved. In the critically periodic case, the period is doubled.

As an example, consider the family of inverted tent maps

$$V_\mu(x) = \begin{cases} 1 - \mu x & \text{for } 0 \le x \le \frac{1}{2}, \\ \mu x + 1 - \mu & \text{for } \frac{1}{2} \le x \le 1, \end{cases}$$

with $1 < \mu \leq 2$. This construction applied to V_μ yields the map $V_{\sqrt{\mu}}$. If we choose μ_0 so that V_{μ_0} is critically periodic of period p, then by iterating we obtain a sequence of maps V_{μ_k} of critical period $2^k p$, where the expansion constant $\mu_k = \sqrt[2^k]{\mu_0} > 1$ converges to 1 as $k \to \infty$.

Now let λ be a Perron number or weak Perron number, and let λ^N be a power such that there exists an uniformly λ^N-expanding map f which is postcritically finite. After replacing N with some multiple if necessary, we can construct an inverted tent map V_μ which is critically periodic of period N, with $\mu < \lambda$. Let

$$\frac{1}{2} \mapsto x_1 \mapsto \cdots \mapsto x_N = \frac{1}{2}$$

be the critical orbit for V_μ. For each $1 \leq i \leq N$ let L_i be a short interval of length $\epsilon \lambda^i$ centered at x_i. Construct a new piecewise linear map g as follows. Each L_i with $i < N$ is to map linearly onto L_{i+1}, with slope either λ or $-\lambda$ according as $x_i > 1/2$ or $x_i < 1/2$. On the other hand, map L_N into L_1 by a composition $a' \circ f \circ a''$, where a' and a'' are carefully chosen affine maps, so the the iterate $g^{\circ N}$ maps L_N onto itself by a map which is affinely conjugate to f. Finally, extend g over the $N+1$ complementary intervals by affine maps. If ϵ is sufficiently small, then the slope of this map g will have absolute value at most λ everywhere. Hence it will be a Lipschitz map with Lipschitz constant λ. Hence it follows from the inequality (1.1) that $h(g) \leq \log(\lambda)$. Since the inequality $h(g) \geq \log(\lambda)$ can be checked by looking at $g^{\circ n}|_{L_n}$ and since g is postcritically finite, this completes the proof. $\square$

NOTE 13.7 (See Proposition 5.3). For any interval map f with entropy

$$0 < h(f) < \sqrt{2},$$

there exists a pair of disjoint subintervals which are mapped to each other by f. Thurston's argument is incomplete. However, the statement can be proved as follows. It suffices to consider the case of a uniform λ-expander. The proof will be by induction on the number of critical points. If f has only one critical point, the argument is straightforward. In fact, if $c_0 \mapsto c_1 \mapsto \cdots$ is the critical orbit, then the two intervals $[c_2, c_4] \leftrightarrow [c_3, c_1]$ are disjoint whenever $\lambda < \sqrt{2}$. For the general case, replacing I by a smaller interval if necessary, we may assume that $f : I \to I$ is onto.

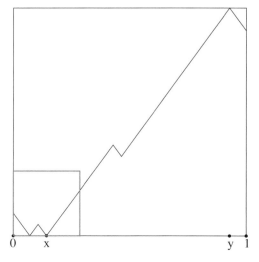

Let $[x, y] \subset I$ be the unique shortest subinterval which maps onto $I = [0,1]$. If $f(x) = 0$ with $x > 0$, as in the diagram, then the interval $[0, 2x]$ maps into itself and has fewer critical points; hence, the conclusion follows by induction. If $x = f(x) = 0$, then the interval $[\epsilon, 1]$ maps into itself for small $\epsilon > 0$. As ϵ increases, we must either reach a value where the map $[\epsilon, 1] \to [\epsilon, 1]$ has ϵ as a critical value, so that the preceding argument applies, or a value such that f has only one critical point in $[\epsilon, 1]$. In either case we are finished. Now suppose that $f(x) = 1$. In this case consider the map $f^2 = f \circ f$ from $[0, 1]$ to itself. Then there is a shortest interval $[x', y']$ which maps onto $[0, 1]$ under f^2, where now $f^2(x')$ is necessarily equal to 0. If $x' > 0$, then the interval $[0, 2x']$ is mapped into itself by f^2. (Here the inequality $\lambda^2 < 2$ is essential.) There are now two possibilities. If the intervals $[0, 2x']$ and $f[0, 2x']$ intersect in at most a single endpoint, then (replacing $[0, 2x']$ by $[0, 2x' - \epsilon]$ if necessary), we have found the required pair of disjoint intervals. On the other hand, if these two intervals overlap, then their intersection is f-invariant and has fewer critical points, and the conclusion follows by induction. Finally, if $x' = 0$, then again we can pass to the interval $[\epsilon, 1]$ for carefully chosen $\epsilon > 0$, and continue as before. $\qquad\square$

NOTE 13.8 (See Figure 7.2 and Plates 18–20). It will be convenient to work with the map $f(x) = \lambda|x| - 1$, where $\lambda = (1 + \sqrt{5})/2$ with conjugate

$$\lambda^\sigma = 1 - \lambda = -1/\lambda.$$

The corresponding map $F = f^{1+\sigma}$ then has the form

$$F(x, y) = \begin{cases} (\lambda x - 1, \ \lambda^\sigma y - 1) & \text{if} \quad x > 0, \\ (-\lambda x - 1, \ -\lambda^\sigma y - 1) & \text{if} \quad x < 0. \end{cases}$$

This is piecewise linear (with a discontinuity along the line $x = 0$), and with Jacobian determinant -1. If we place three boxes

$$A = [-1, 0] \times [1 - \lambda, \lambda - 1], \quad B = [-1, 0] \times [-1 - \lambda, 1 - \lambda],$$
$$C = [0, \lambda - 1] \times [-1 - \lambda, 1 - \lambda]$$

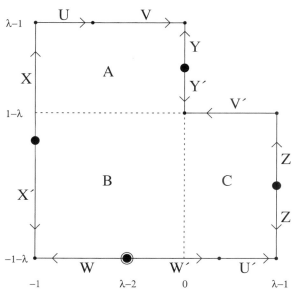

in the (x, y)-plane, as shown in the figure, then it is not hard to check that

$$A \cup B \cup C$$

maps bijectively onto itself with

$$C \xrightarrow{\cong} A \quad \text{and} \quad A \cup B \xrightarrow{\cong} B \cup C.$$

Here the fundamental period three orbit has been indicated by heavy dots, and the fixed point by a circled dot. Every point in $[-1, \lambda - 1] \times \mathbb{R}$ eventually maps into this union. (Here the horizontal scale has been exagerated by a factor of two. To compare this with Thurston's picture, it is necessary to reflect, interchanging top and bottom.)

NOTE 13.9 (See Figure 7.3 and Plate 19). Thurston points out that the discontinuity along the line $x = 0$ (and even the discontinuity in derivatives) can be eliminated by identifying the top and bottom halves of each of the outer vertical lines. This will yield a topological 2-disk bounded by the four top and bottom edges. In fact, he further identifies boundary points so as to obtain a closed surface of genus zero with a flat Euclidean structure except at four cone points. (The cone points correspond to the the large dots in the figures: the period three orbit together with the fixed point.) The boundary of $A \cup B \cup C$ is divided into twelve intervals which are identified in pairs, as indicated in the figure of Note 13.8. The dynamics remains well defined and continuous under these identifications, with

$$U \xrightarrow{\cong} V' \xrightarrow{\cong} A \cap B, \qquad U' \xrightarrow{\cong} V \xrightarrow{\cong} A \cap B,$$

and with

$$W \xrightarrow{\cong} W' \cup U', \qquad W' \xrightarrow{\cong} W \cup U.$$

The map is a linear area preserving Anosov map which preserves the horizontal and vertical foliations, except at the cone points.

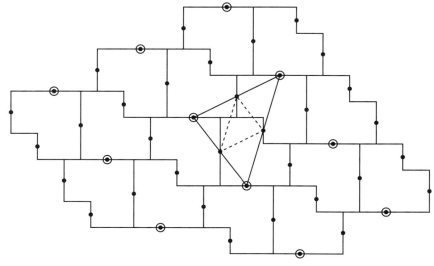

For a different way of describing this construction, consider the group of Euclidian motions of the plane generated by $180°$ rotation about the four marked periodic points. The images of $A \cup B \cup C$ under the action of this group forms a tiling of the plane, as shown in the figure. (See also Plate 20.) The map F on $A \cup B$ now extends to a linear automorphism of the plane. If we translate coordinates so that the fixed point $W \cap W'$ is at the origin, then this extended map is given simply by $(x, y) \mapsto \left(-\lambda x, \; (\lambda - 1)y \right)$.

The triangle in the figure forms an alternate fundamental domain for this action. If we fold this triangle along the dotted lines, then we obtain the tetrahedron which Thurston described as a model for the quotient space. (In fact there is a one parameter family of such examples, since the ratio between horizontal scale and vertical scale can be chosen arbitrarily—the ratio 2 was used for convenience in these figures.)

14 Bibliography

[ALM] L. Alsedà, J. Llibre, and M. Misiurewicz. *Combinatorial dynamics and entropy in dimension one*, in *Advanced Series in Nonlinear Dynamics*. **5** World Scientific Publishing Co. Inc., River Edge, NJ, second edition, 2000.

[Ber] A. Bertrand. Développements en base de Pisot et répartition modulo 1. *C. R. Acad. Sci. Paris Sér. A-B*, **285**(6):A419–A421, 1977.[33]

[Ber1] A. Bertrand-Mathis. Développement en base θ, répartition modulo un de la suite $(x\theta^n)_n > 0$, langages, codes et θ-shift. *Bull. Soc. Math. France* **114** (1986), 271–323, 1986.[33]

[BH] M. Bestvina and M. Handel. Train tracks and automorphisms of free groups. *Ann. of Math. (2)*, **135**(1):1–51, 1992.[33]

[Bo1] D. Boyd. Salem numbers of degree four have periodic expansions. In *Théorie des nombres (Quebec, PQ, 1987)*, pages 57–64. de Gruyter, Berlin, 1989.[33]

[Bo2] D. Boyd. On the beta expansion for Salem numbers of degree 6. *Math. Comp.*, **65**(214):861–875, S29–S31, 1996.[33]

[CH] A. de Carvalho and T. Hall. Unimodal generalized pseudo-Anosov maps. *Geometry & Topology*, **8**:1127–1188, 2004.

[Ge] A. O. Gelfond. A common property of number systems. *Izv. Akad. Nauk SSSR. Ser. Mat.*, **23**:809–814, 1959.[33]

[KOR] K. Kim, N. Ormes and F. Rousch. The spectra of nonnegative integer matrices via formal power series. *J. AMS* **13**:773–806, 2000.[33]

[Li] D. A. Lind. The entropies of topological Markov shifts and a related class of algebraic integers. *Ergodic Theory Dynam. Systems*, **4**(2):283–300, 1984.

[LS] R. Lyndon and P. Schupp *Combinatorial Group Theory*. Springer, 1977, 2001.[33]

[dMvS] W. de Melo and S. van Strien, *One-Dimensional Dynamics*. Springer, 1993.[33]

---

[33]These references have been added by the editors.

[MSt]   E. Miller and B. Sturmfels. *Combinatorial Commutative Algebra*. Graduate Texts in Mathematics. Springer, New York; 2005 edition.[33]

[MT]   J. Milnor and W. Thurston. On iterated maps of the interval. In *Dynamical systems (College Park, MD, 1986–87)*, **1342** of *Lecture Notes in Math.*, pages 465–563. Springer, Berlin, 1988.

[Mis]   M. Misiurewicz. On Bowen's definition of topological entropy. *Discrete and Continuous Dynamical Systems* **10**(3):827–833, 2004.[33]

[MS1]   M. Misiurewicz and W. Szlenk. Entropy of piecewise monotone mappings. In *Dynamical systems, Vol. II—Warsaw*, pages 299–310. Astérisque, **50** Soc. Math. France, Paris, 1977.

[MS2]   M. Misiurewicz and W. Szlenk. Entropy of piecewise monotone mappings. *Studia Math.*, **67**(1):45–63, 1980.

[Ne]   V. Nekrashevych. Hyperbolic groupoids: Definitions and duality. 2011.

[OP]   A. M. Odylyzko and B. Poonen. Zeros of polynomials with 0, 1 coefficients. *L'Enseign. Math.* **39**: 317–348, 1993.[33]

[Pa]   W. Parry On the β-expansions of real numbers, *Acta Math. Acad. Sci. Hungar.*, **11**: 269–278 (1960).[33]

[Sa]   R. Salem. A remarkable class of algebraic numbers. Proof of a conjecture of Vijayaraghavan. *Duke Math. Journal*, **11**:103–108, 1944.[33]

[Sch]   K. Schmidt. On periodic expansions of Pisot numbers and Salem numbers. *Bull. London Math. Soc.*, **12**(4):269–278, 1980.[33]

[Ti]   G. Tiozzo. On Thurston's entropy spectrum for unimodal maps. Preprint, Harvard University, 2013

Part III

Several Complex Variables

On Écalle-Hakim's theorems in holomorphic dynamics

Marco Arizzi and Jasmin Raissy

ABSTRACT. In this survey we provide detailed proofs for the results by Hakim regarding the dynamics of germs of biholomorphisms tangent to the identity of order $k + 1 \geq 2$ and fixing the origin.

1 INTRODUCTION

One of the main questions in the study of local discrete holomorphic dynamics, i.e., in the study of the iterates of a germ of a holomorphic map of $\mathbb{C}^p$ at a fixed point, which can be assumed to be the origin, is when it is possible to holomorphically conjugate it to a "simple" form, possibly its linear term. It turns out (see [Ab3], [Ab4], [Br], [CC], [IY], [Yo] and Chapter 1 of [Ra] for general surveys on this topic) that the answer to this question strongly depends on the arithmetical properties of the eigenvalues of the linear term of the germ.

It is not that useful to search for a holomorphic conjugacy in a full neighborhood of the origin in the so-called *tangent to the identity case,* that is, when the linear part of the germ coincides with the identity, but the germ is not the identity. Nevertheless, it is possible to study the dynamics of such germs, which is indeed very interesting and rich, using the conjugacy approach in smaller domains having the origin on their boundaries. The one-dimensional case, was first studied by Leau [Le] and Fatou [Fa], who provided a complete description of the dynamics in a pointed neighbourhood of the origin. More precisely, in dimension 1, a tangent to the identity germ can be written as

$$f(z) := z + az^{k+1} + O(z^{k+2}), \tag{1.1}$$

where the number $k + 1 \geq 2$ is usually called the *order* of f. We define the *attracting directions* $\{v_1, \dots, v_k\}$ for f as the kth roots of $-|a|/a$, and these are precisely the directions v such that the term av^{k+1} points in the direction opposite to v. An *attracting petal* P for f is a simply-connected domain such that $0 \in \partial P$, $f(P) \subseteq P$ and $\lim_{n \to \infty} f^n(z) = 0$ for all $z \in P$, where f^n denotes the nth iterate of f. The attracting directions for f^{-1} are called *repelling directions* for f and the attracting petals for f^{-1} are *repelling petals* for h. Then the Leau-Fatou flower theorem is the following result (see, *e.g.,* [Ab4], [Br], [Mi]). We write $a \approx b$ whenever there exist constants $0 < c < C$ such that $ca \leq b \leq Ca$.

THEOREM 1.1 (Leau-Fatou, [Le], [Fa]). *Let f be as in (1.1). Then for each attracting direction v of h, there exists an attracting petal P for f (said centered at v) such that for each $z \in P$ the following hold:*

J. Raissy supported in part by FSE, Regione Lombardia.

1. $f^n(z) \neq 0$ for all n and $\lim_{n \to \infty} \frac{f^n(z)}{|f^n(z)|} = v$,

2. $|f^n(z)|^k \approx \frac{1}{n}$.

Moreover, the union of all k attracting petals and k repelling petals for f forms a punctured open neighborhood of 0.

By Property (1), attracting [resp. repelling] petals centered at different attracting [resp. repelling] directions must be disjoint.

For dimension $p \geq 2$ the situation is more complicated and a general complete description of the dynamics in a full neighborhood of the origin is still unknown (see [AT] for some interesting partial results). Analogously to the one-dimensional case, we can write our germ as sum of homogeneous polynomials,

$$F(z) = z + P_{k+1}(z) + O(\|z\|^{k+2}),$$

where $k + 1 \geq 2$ is the *order* of F.

Very roughly, Écalle using his resurgence theory [Ec], and Hakim with classical tools [Ha] proved that generically, given a tangent to the identity germ of order $k + 1$, it is possible to find one-dimensional "petals," called *parabolic curves*, that is, one-dimensional F-invariant analytic discs having the origin on the boundary and where the dynamics is of *parabolic type*. That is, the orbits converge to the origin tangentially to a particular direction, called *characteristic* (see Definition 4.1). Abate, in [Ab2], then proved that in dimension 2 such parabolic curves always exist. Hakim also gave sufficient conditions, that here we call *attracting* (see Definition 4.8) for the existence of basins of attraction along *non-degenerate characteristic directions* (see Definition 4.1) modeled on such parabolic curves, proving the following result.

THEOREM 1.2 (Hakim [Ha]). *Let F be a tangent to the identity germ fixing the origin of order $k + 1 \geq 2$, and let $[v]$ be a non-degenerate characteristic direction. If $[v]$ is attracting, then there exist k parabolic invariant domains, where each point is attracted by the origin along a trajectory tangential to $[v]$.*

Hakim's techniques have been largely used in the study of the existence of parabolic curves (see [Ab2], [BM], [Mo], [R1]), basins of attraction and Fatou-Bieberbach domains, i.e., proper open subset of $\mathbb{C}^p$ biholomorphic to $\mathbb{C}^p$, (see [BRZ], [Ri], [R2], [V1]).

The aim of this survey is to make available important results and very useful techniques that were included, up to now, only in [Ha2], a preprint which is not easily retrievable, and where the case $k > 1$ was stated with no detailed proofs.

We shall provide, from Section 3 up to Section 7, the reformulations for any order $k + 1 \geq 2$, with detailed proofs, of the results published by Hakim in [Ha] (Hakim gave detailed proofs of her results for $k = 1$ only), and, in the last three sections, again reformulating definitions, lemmas, propositions and theorems for any order $k + 1 \geq 2$, we shall provide detailed proofs for the unpublished results, including her construction of Fatou-Bieberbach domains, obtained by Hakim in [Ha2].

ACKNOWLEDGMENTS. We would like to thank the anonymous referee for useful comments and remarks, which improved the presentation of the paper. The second named author would like to thank the Dipartimento di Matematica e Applicazioni of the Università degli Studi di Milano Bicocca, where this work was carried out.

2 NOTATION

In the following we shall work in $\mathbb{C}^p$, $p \geq 2$ with the usual Euclidean norm

$$\|z\| = \left(\sum_{i=1}^{p} |z_i|^2 \right)^{1/2}.$$

We shall denote by $\mathbb{D}_{r,k}$ the following subset of $\mathbb{C}$

$$\mathbb{D}_{r,k} = \left\{ z \in \mathbb{C} \mid |z^k - r| < r \right\},$$

which has exactly k connected components, that will be denoted by $\Pi_{r,k}^1, \ldots, \Pi_{r,k}^k$.

Let $F \colon \mathbb{C}^p \to \mathbb{C}^p$ be a holomorphic map. We shall denote with $F'(z_0)$ the Jacobian matrix of F in z_0. If, moreover, we write $\mathbb{C}^p = \mathbb{C}^s \times \mathbb{C}^t$, then $\frac{\partial F}{\partial x}$ and $\frac{\partial F}{\partial y}$ will be the Jacobian matrices of $F(\cdot, y)$ and $F(x, \cdot)$.

Given $f, g_1, \ldots, g_s \colon \mathbb{C}^m \to \mathbb{C}^k$, we shall write

$$f = O(g_1, \ldots, g_s),$$

if there exist $C_1, \ldots, C_s > 0$ so that

$$\|f(w)\| \leq C_1 \|g_1(w)\| + \cdots + C_s \|g_s(w)\|;$$

and, moreover, with $f = o(g)$ we mean

$$\frac{\|f(w)\|}{\|g(w)\|} \to 0 \text{ as } w \to 0.$$

Similarly, given a sequence $w_n \in \mathbb{C}^p$, we shall write

$$w = O\left(\frac{1}{n}\right) \iff \exists C > 0 : |w_n| \leq \frac{C}{n};$$

$$w = o\left(\frac{1}{n}\right) \iff \frac{w_n}{1/n} \to 0 \text{ as } n \to \infty.$$

Given $\{x_n\}$ a sequence in a metric space(M, d), by $x_n \tilde{x}$ we mean that, for n sufficiently large, $d(x_n, x) \to 0$.

Finally, we shall denote with $\text{Diff}(\mathbb{C}^p, 0)$ the space of germs of biholomorphisms of $\mathbb{C}^p$ fixing the origin.

3 PRELIMINARIES

One of the main tools in the study of the dynamics for tangent to the identity germs is the blow-up of the origin. In our case, it will suffice one blow-up to simplify our germ.

DEFINITION 3.1. *Let $F \in \text{Diff}(\mathbb{C}^p, O)$ be tangent to the identity. The order $v_0(F)$ of F is the minimum $v \geq 2$ so that $P_v \not\equiv 0$, where we consider the expansion of as sum of homogeneous polynomials*

$$F(z) = \sum_{k \geq 1} P_k(z),$$

where P_k is homogeneous of degree k ($P_1(z) = z$). We say that F is non-degenerate *if $P_{v_0(F)}(z) = 0$ if and only if $z = 0$.*

Let $\widetilde{\mathbb{C}}^p \subset \mathbb{C}^p \times \mathbb{CP}^{p-1}$ be defined by

$$\widetilde{\mathbb{C}}^p \{(v,[l]) \in \mathbb{C}^p \times \mathbb{CP}^{p-1} : v \in [l]\}.$$

Using coordinates $(z_1,\ldots,z_p) \in \mathbb{C}^p$ and $[S_1 : \cdots : S_p] \in \mathbb{CP}^{p-1}$, we obtain that $\widetilde{\mathbb{C}}^p$ is determined by the relations

$$z_h S_k = z_k S_h$$

for $h,k \in \{1,\ldots,p\}$. It is well known that $\widetilde{\mathbb{C}}^p$ is a complex manifold of the same dimension as $\mathbb{C}^p$. Given $\sigma \colon \widetilde{\mathbb{C}}^p \to \mathbb{C}^p$ the projection, the *exceptional divisor* $E := \sigma^{-1}(0)$ is a complex submanifold of $\widetilde{\mathbb{C}}^p$ and $\sigma|_{\widetilde{\mathbb{C}}^p \setminus E} \colon \widetilde{\mathbb{C}}^p \setminus E \to \mathbb{C}^p \setminus \{0\}$ is a biholomorphism. The datum $(\widetilde{\mathbb{C}}^p, \sigma)$ is usually called *blow-up of $\mathbb{C}^p$ at the origin*.

Note that an atlas of $\widetilde{\mathbb{C}}^p$ is given by $\{(V_j, \varphi_j)\}_{1 \le j \le p}$, where

$$V_i = \{(z,[S]) \in \widetilde{\mathbb{C}}^p \mid S_j \ne 0\},$$

and $\varphi_j \colon V_j \to \mathbb{C}^p$ is given by

$$\varphi_j(z_1,\ldots,z_p, [S_1 : \cdots : 1 : \cdots : S_p]) = \left(S_1,\ldots,z_j,\ldots,S_p\right),$$

since the points in $\{S_j = 1\}$ satisfy $z_k = z_j S_k$ for $k \in \{1,\ldots,p\} \setminus \{j\}$. Moreover, we have

$$\varphi_j^{-1}(z_1,\ldots,z_p) = (z_1 z_j,\ldots,z_j,\ldots,z_p z_j, [z_1 : \cdots : 1 : \cdots : z_p]) \in V_j.$$

The projection $\sigma \colon \widetilde{\mathbb{C}}^p \to \mathbb{C}^p$ is given by $\sigma(z,[S]) = z$, and in the charts (V_j, φ_j) it is given by

$$\sigma \circ \varphi_j^{-1}(z_1,\ldots,z_p) = (z_1 z_j,\ldots,z_j,\ldots,z_p z_j).$$

PROPOSITION 3.2. *Let $F \in \mathrm{Diff}(\mathbb{C}^p, 0)$ be tangent to the identity, and let $(\widetilde{\mathbb{C}}^p, \sigma)$ be the blow-up of $\mathbb{C}^p$ at the origin. Then there exists a unique lift $\tilde{F} \in \mathrm{Diff}(\widetilde{\mathbb{C}}^p, E)$ so that*

$$F \circ \sigma = \sigma \circ \tilde{F}.$$

Moreover, F acts as the identity on the points of the exceptional divisor , i.e.,

$$\tilde{F}(0, [S]) = (0, [S]).$$

We omit the proof of the previous result, which can be found in [Ab1]. It is also possible to prove that there exists a unique lift for any endomorphism G of $(\mathbb{C}^p, 0)$ so that $G(z) = \sum_{k \ge h} P_k(z)$, where h is the minimum integer such that $P_h \not\equiv 0$ and so that $P_h(z) = 0$ if and only if $z = 0$, and in such a case the action on the exceptional divisor is $\tilde{G}(0, [S]) = (0, [P_h(S)])$.

4   CHARACTERISTIC DIRECTIONS

We shall use the following reformulation of Definition 2.1 and Definition 2.2 of [Ha] for the case $k + 1 \ge 2$.

DEFINITION 4.1. *Let $F \in \mathrm{Diff}(\mathbb{C}^p, 0)$ be a tangent to the identity germ of order $k + 1$, and let P_{k+1} be the homogeneous polynomial of degree $k + 1$ in the expansion of F as sum of homogeneous polynomials (that is, the first non-linear term of the series). We shall say that $v \in \mathbb{C}^p \setminus \{0\}$ is a* characteristic direction *if $P_{k+1}(v) = \lambda v$ for some $\lambda \in \mathbb{C}$. Moreover, if $P_{k+1}(v) \neq 0$, we shall say that the characteristic direction is* non-degenerate, *otherwise, we shall call it* degenerate.

Since characteristic direction are well defined only as elements in $\mathbb{CP}^{p-1}$, we shall use the notation $[v] \in \mathbb{CP}^{p-1}$.

DEFINITION 4.2. *Let $F \in \mathrm{Diff}(\mathbb{C}^p, 0)$ be a tangent to the identity germ. A* characteristic trajectory *for F is an orbit $\{X_n\} := \{F^n(X)\}$ of a point X in the domain of F, such that $\{X_n\}$ converges to the origin tangentially to a complex direction $[v] \in \mathbb{CP}^{p-1}$, that is,*

$$\begin{cases} \lim_{n \to \infty} X_n = 0, \\ \lim_{n \to \infty} [X_n] = [v]. \end{cases}$$

The concepts of characteristic direction and characteristic trajectory are indeed linked as next result shows. We shall use coordinates, following Hakim [Ha], $z = (x, y) \in \mathbb{C} \times \mathbb{C}^{p-1}$ and $(x_n, y_n) := (f_1^n(x, y), f_2^n(x, y)) \in \mathbb{C} \times \mathbb{C}^{p-1}$ for the n-tuple iterate of F. We have the following generalization of Proposition 2.3 of [Ha] for the case $k + 1 \geq 2$.

PROPOSITION 4.3. *Let $F \in \mathrm{Diff}(\mathbb{C}^p, 0)$ be a tangent to the identity germ, and let $\{X_n\}$ be a characteristic trajectory tangent to $[v]$ at the origin. Then v is a characteristic direction. Moreover, if $[v]$ is non-degenerate, choosing coordinates so that $[v] = [1 : u_0]$, writing $P_{k+1}(z) = (p_{k+1}(z), q_{k+1}(z)) \in \mathbb{C} \times \mathbb{C}^{p-1}$, we have*

$$x_n^k \approx -\frac{1}{nk p_{k+1}(1, u_0)}, \quad \text{as } n \to \infty, \tag{4.1}$$

where $X_n = (x_n, y_n)$.

PROOF. If $P_{k+1}([v]) = 0$, then $[v]$ is a degenerate characteristic direction and there is nothing to prove. Hence we may assume $P_{k+1}([v]) \neq 0$, and, up to reordering the coordinates, we may assume that $[v] = [1 : u_0]$ and F is of the form

$$\begin{cases} x_1 = x + p_{k+1}(x, y) + p_{k+2}(x, y) + \cdots, \\ y_1 = y + q_{k+1}(x, y) + q_{k+2}(x, y) + \cdots, \end{cases} \tag{4.2}$$

where $x_1, x, p_j(x, y) \in \mathbb{C}$ and $y_1, y, q_j(x, y) \in \mathbb{C}^{p-1}$. Since $\{X_n\}$ is a characteristic trajectory tangent to $[v]$, we have

$$\lim_{n \to \infty} \frac{y_n}{x_n} = u_0.$$

Now we blow-up the origin and we consider a neighborhood of $[v]$. If the blow-up is $y = ux$, with $u \in \mathbb{C}^{p-1}$, then the first coordinate of our map becomes

$$x_1 = x(1 + p_{k+1}(1, u)x^k + p_{k+2}(1, u)x^{k+1} + \cdots), \tag{4.3}$$

whereas the other coordinates become

$$u_1 = \frac{y_1}{x_1} = u + r(u)x^k + O(x^{k+1}), \tag{4.4}$$

where
$$r(u) := q_{k+1}(1,u) - p_{k+1}(1,u)u.$$

As a consequence, the non-degenerate characteristic directions of F of the form $[1 : u]$ coincide with the ones so that u is a zero of the polynomial map $r(u)$:

$$\begin{cases} p_{k+1}(1,u) = \lambda \\ q_{k+1}(1,u) = \lambda u \end{cases} \iff r(u) = q_{k+1}(1,u) - p_{k+1}(1,u)u = 0.$$

It remains to prove that if $u_n = \frac{y_n}{x_n}$ converges to u_0, then $r(u_0) = 0$. Since $u_n \to u_0$, the series

$$\sum_{n=0}^{\infty} (u_{n+1} - u_n) \tag{4.5}$$

is convergent. Thanks to (4.4), assuming $r(u_n) \neq 0$, we obtain

$$u_{n+1} - u_n = r(u_n)x_n^k + O\big(x_n^{k+1}\big) \approx r(u_0)x_n^k.$$

We can now prove (4.1). In fact from

$$\frac{1}{x_1} = \frac{1}{x}\left(1 - p_{k+1}(1,u)x^k + O(x^{k+1})\right),$$

we deduce

$$\frac{1}{x_1^k} = \frac{1}{x^k} - kp_{k+1}(1,u) + O(x),$$

and, hence,

$$\frac{1}{nx_n^k} = \frac{1}{nx^k} - \frac{k}{n}\sum_{j=0}^{n-1}\left(p_{k+1}(1,u_j) + O\big(x_j\big)\right).$$

Setting $a_j := p_{k+1}(1,u_j) + O(x_j)$, since $a_j \to p_{k+1}(1,u_0)$, the average $\frac{1}{n}\sum_{j=0}^{n-1} a_j$ converges to the same limit. It follows that, as $n \to \infty$, $\frac{1}{nx_n^k}$ converges to $-kp_{k+1}(1,u_0)$ and

$$x_n^k \approx -\frac{1}{nkp_{k+1}(1,u_0)}.$$

If $r(u_0) \neq 0$, then we could find $C \neq 0$ such that

$$u_{n+1} - u_n \approx \frac{C}{n}r(u_0),$$

and the series $\sum_{n=0}^{\infty} (u_{n+1} - u_n)$ would not converge, contradicting (4.5); hence, $r(u_0) = 0$, and this concludes the proof. $\qquad\square$

Unless specified, thanks to the previous results, without loss of generality, we shall assume that any given $F \in \mathrm{Diff}(\mathbb{C}^p, 0)$ tangent to the identity germ of order $k + 1 \geq 2$, with a non-degenerate characteristic direction $[v]$ is of the form

$$\begin{cases} x_1 = x(1 + p_{k+1}(1,u)x^k + O(x^{k+1})), \\ u_1 = u + (q_{k+1}(1,u) - p_{k+1}(1,u)u)x^k + O(x^{k+1}). \end{cases} \tag{4.6}$$

LEMMA 4.4. *Let $F \in \text{Diff}(\mathbb{C}^p, 0)$ be a tangent to the identity germ of order $k + 1 \geq 2$, of the form (4.6), with a non-degenerate characteristic direction $[v] = [1 : u_0]$. Then there exists a polynomial change of coordinates holomorphically conjugating F to a germ with first component of the form*

$$x_1 = x - \frac{1}{k}x^{k+1} + O\left(x^{k+1}\|u\|, x^{2k+1}\right).$$

PROOF. We shall first prove that it is possible to polynomially conjugate F to a germ whose first coordinate has no terms in x^h for $h = k + 2, \ldots, 2k$. Thanks to (4.3), expanding $p_{k+1}(1, u)$ in u_0, we obtain

$$x_1 = f(x, u) = x + p_{k+1}(1, u_0)x^{k+1} + O\left(\|u\|x^{k+1}, x^{k+2}\right).$$

Now we use the same argument one can find in [Be, Theorem 6.5.7, p. 122], conjugating f to polynomials f_h, for $1 \leq h < k$, of the form

$$f_h(x, u) = x + p_{k+1}(1, u_0)x^{k+1} + b_h x^{k+h+1} + O\left(\|u\|x^{k+1}, x^{k+h+2}\right),$$

that is, changing polynomially the first coordinate x and leaving the others invariant, in order to get

$$f_k(x, u) = x + p_{k+1}(1, u_0)x^{k+1} + O\left(\|u\|x^{k+1}, x^{2k+1}\right).$$

Let us consider $g(x) = x + \beta x^{h+1}$, with $\beta := \frac{b_h}{(k-h)p_{k+1}(1, u_0)}$, and set

$$\Phi = (g, \text{id}_{p-1}) : (x, u) \mapsto (g(x), u).$$

Then, conjugating $F_h = (f_h, \Psi_h)$ via Φ, we have $F_{h+1} \circ \Phi = \Phi \circ F_h$, which is equivalent to

$$\begin{cases} f_{h+1}(g(x), u) = g(f_h(x, u)), \\ \Psi_{h+1}(g(x), u) = \Psi_h(x, u). \end{cases} \tag{4.7}$$

Since $\Phi(0) = 0$ and the Taylor expansion of Φ up to order $k + 1$ depends only on $d\Phi_0$, we must have

$$\begin{cases} f_{h+1}(x, u) = x + \sum_{m=k+1}^{\infty} A_m x^m + O\left(\|u\|x^{k+1}\right), \\ \Psi_{h+1}(x, u) = u + r(u)x^k + O\left(x^{k+1}\right), \end{cases}$$

and in particular these changes of coordinates do not interfere on Ψ in the order that we are considering.

Let us consider the terms up to order $k + h + 2$ in the first equation of (4.7). We obtain

$$g(f_h(x, u)) = x + p_{k+1}(1, u_0)x^{k+1} + b_h x^{k+h+1}$$
$$+ \beta(x^{h+1} + (h+1)x^{k+h+1}) + O\left(\|u|x^{k+1}, x^{k+h+2}\right),$$

and

$$f_{h+1}(g(x), u) = x + \beta x^{k+1} + A_{k+1}x^{k+1} + \cdots$$
$$+ A_{k+h+1}x^{k+h+1} + A_{k+1}\beta(k+1)x^{k+h+1} + O\left(x^{k+h+2}, \|u\|x^{k+1}\right).$$

Hence, the coefficients A_m satisfy

$$\begin{cases} A_{k+1} = p_{k+1}(1, u_0), \ A_{k+2} = 0, \ \ldots, \ A_{k+h} = 0, \\ b_h + (h+1)p_{k+1}(1, u_0)\beta = \beta(k+1)A_{k+1} + A_{k+h+1}, \end{cases}$$

yielding $A_{k+h+1} = 0$. In particular, there exists b_{h+1} such that

$$f_{h+1}(x, u) = x + p_{k+1}(1, u_0)x^{k+1} + b_{h+1}x^{k+h+2} + O\big(\|u\|x^{k+1}, x^{k+h+3}\big).$$

Repeating inductively this procedure up to $h = k - 1$, we conjugate with a polynomial (and hence holomorphic) change of coordinates our original F to a germ with no terms in x^h for $h = k + 2, \ldots, 2k$, i.e.,

$$x_1 = f(x, u) = x + p_{k+1}(1, u_0)x^{k+1} + O\big(\|u\|x^{k+1}, x^{2k+1}\big). \tag{4.8}$$

Finally, using the change of coordinates acting as $x \mapsto X = \sqrt[k]{-p_{k+1}(1, u_0)k}\, x$ on the first coordinate and as the identity on the other coordinates, the germ (4.8) is transformed into

$$X_1 = X - \frac{1}{k}X^{k+1} + O\big(\|u\|X^{k+1}, X^{2k+1}\big),$$

in the first component, whereas the other components become

$$U_1 = U - r(U)\,\frac{X^k}{k\,p_{k+1}(1, u_0)} + O(X^{k+1}). \qquad \square$$

Up to now, we simply acted on the first component of F, mainly focusing on the characteristic direction $[v]$. We shall now introduce a class of $(p - 1) \times (p - 1)$ complex matrices, which takes care of the remaining $p - 1$ components of F. We consider the Taylor expansion of r in u_0, and we have

$$u_1 = u - \frac{x^k}{k\,p_{k+1}(1, u_0)}r'(u_0)(u - u_0) + O\big(\|u - u_0\|^2 x^k, x^{k+1}\big),$$

where $r'(u_0) = \mathrm{Jac}(r)(u_0)$. It is then possible to associate to the characteristic direction $[v] = [1 : u_0]$ the matrix

$$A(v) = \frac{1}{k\,p_{k+1}(1, u_0)}r'(u_0),$$

and hence, assuming without loss of generality $u_0 = 0$, after the previous reductions, the germ F has the form

$$\begin{cases} x_1 = x - \dfrac{x^{k+1}}{k} + O\big(\|u\|x^{k+1}, x^{2k+1}\big), \\ u_1 = (I - x^k A)u + O\big(\|u\|^2 x^k, x^{k+1}\big). \end{cases} \tag{4.9}$$

The next result gives us a more geometric interpretation of this matrix.

LEMMA 4.5. *Let $F \in \mathrm{Diff}(\mathbf{C}^p, 0)$ be a tangent to the identity germ of order $k + 1 \geq 2$ and let $[v] \in \mathbf{CP}^{p-1}$ be a non-degenerate characteristic direction for F with associate matrix $A(v)$. Then the projection $\widetilde{P}_{k+1}$ in $\mathbf{CP}^{p-1}$ of the homogeneous polynomial P_{k+1}*

of degree $k + 1$ in the expansion of F as sum of homogeneous polynomials induces $\widetilde{P}_{k+1} \colon \mathbb{CP}^{p-1} \to \mathbb{CP}^{p-1}$, defined by

$$\widetilde{P}_{k+1} \colon [x] \mapsto [P_{k+1}(x)],$$

which is well-defined in a neighborhood of v. Moreover, $[v]$ is a fixed point of $\widetilde{P}_{k+1}$ and $A(v)$ is the matrix associated to the linear operator

$$\frac{1}{k} \left(d(\widetilde{P}_{k+1})_{[v]} - \mathrm{id} \right).$$

PROOF. The germ F can be written as

$$F(z) = z + P_{k+1}(z) + P_{k+1}(z) + \cdots,$$

where P_h is homogeneous of degree h. Let $[v]$ be a non-degenerate characteristic direction for F. The p-tuple P_{k+1} of homogeneous polynomials of degree $k + 1$ induces a meromorphic map $\widetilde{P}_{k+1} \colon \mathbb{CP}^{p-1} \to \mathbb{CP}^{p-1}$ given by

$$\widetilde{P}_{k+1} \colon [x] \mapsto [P_{k+1}(x)],$$

and it is clear that the non-degenerate characteristic directions correspond to the fixed points of such a map, and the degenerate characteristic directions correspond to the indeterminacy points.

We may assume, without loss of generality, $v = (1, u_0)$. Then

$$U = \left\{ [x_1 : \cdots : x_p] \in \mathbb{CP}^{p-1} \mid x_1 \neq 0 \right\}$$

is an open neighborhood of $[v]$ and the map $\varphi_1 \colon U \to \mathbb{C}^{p-1}$ defined as

$$[x_1 : \cdots x_p] \mapsto \left(\frac{x_2}{x_1}, \cdots, \frac{x_p}{x_1} \right) = (u_1, \ldots, u_{p-1}),$$

is a chart of $\mathbb{CP}^{p-1}$ around $[v]$.

The differential $d(\widetilde{P}_{k+1})_{[v]} \colon T_{[v]}\mathbb{CP}^{p-1} \to T_{[v]}\mathbb{CP}^{p-1}$ is a linear map, and it is represented, in $u_0 = \varphi_1([v])$, by the Jacobian matrix of the map

$$g := \varphi_1 \circ \widetilde{P}_{k+1} \circ \varphi_1^{-1} \colon \varphi_1(U) \to \varphi_1(\widetilde{P}_{k+1}(U))$$

given by

$$u = (u_1, \ldots, u_{p-1}) \mapsto \left(\frac{q_{k+1,1}(1, u_1, \ldots, u_{p-1})}{p_{k+1}(1, u_1, \ldots, u_{p-1})}, \ldots, \frac{q_{k+1,p-1}(1, u_1, \ldots, u_{p-1})}{p_{k+1}(1, u_1, \ldots, u_{p-1})} \right).$$

We can associate to $[v]$ the linear endomorphism

$$\mathcal{A}_F([v]) = \frac{1}{k} \left(d(\widetilde{P}_{k+1})_{[v]} - \mathrm{id} \right) \colon T_{[v]}\mathbb{CP}^{p-1} \to T_{[v]}\mathbb{CP}^{p-1},$$

and we can then prove that the matrix of $\mathcal{A}_F([v])$ coincides with $A(v)$. In fact, let $g_1, \ldots, g_{p-1}$ be the components of g. Since $g(u_0) = u_0$, we have

$$\frac{\partial g_i}{\partial u_j}(u_0) = \frac{1}{p_{k+1}(1, u_0)} \left[\frac{\partial q_{k+1,i}}{\partial u_j}(1, u_0) - \frac{\partial p_{k+1}}{\partial u_j}(1, u_0)u_{0,i} \right],$$

for $i, j = 1, \ldots, p - 1$. Therefore, it follows from $r_i(u) = q_{k+1,i}(1, u) - p_{k+1}(1, u)u_i$ that

$$\frac{\partial r_i}{\partial u_j}(u_0) = \frac{\partial q_{k+1,i}}{\partial u_j}(1, u_0) - \frac{\partial p_{k+1}}{\partial u_j} u_{0,i} u_{0,i} - p_{k+1}(1, u_0)\delta_{i,j},$$

and, hence,

$$A(v) = \frac{1}{k}(g'(u_0) - \mathrm{id}),$$

concluding the proof. $\qquad\square$

LEMMA 4.6. *Let $F \in \mathrm{Diff}(\mathbb{C}^p, 0)$ be a tangent to the identity germ and let $\varphi \in \mathbb{C}[\![X]\!]^p$ be an invertible formal transformation of $\mathbb{C}^p$. If $F = I + \sum_{h \geq k+1} P_h$ and $\varphi = Q_1 + \sum_{j \geq 2} Q_j$ are the expansion of F an φ as sums of homogeneous polynomials, then the expansion of $F^* = \varphi^{-1} \circ F \circ \varphi$ is of the form $I + \sum_{h \geq k+1} P_h^*$, and*

$$P_{k+1}^* = Q_1^{-1} \circ P_{k+1} \circ Q_1. \tag{4.10}$$

PROOF. It is obvious that the linear term of F^* is the identity. It then suffices to consider the equivalent condition $F \circ \varphi = \varphi \circ F^*$ and to compare homogeneous terms up to order $k + 1$, writing $F^* = \sum_{h \geq 1} P_h^*$. $\qquad\square$

We are now able to prove, as in Proposition 2.4 of [Ha], that we can associate to $[v]$ the class of similarity of $A(v)$.

PROPOSITION 4.7. *Let $F \in \mathrm{Diff}(\mathbb{C}^p, 0)$ be a tangent to the identity germ of order $k + 1 \geq 2$ and let $[v] = [1 : u_0] \in \mathbb{CP}^{p-1}$ be a non-degenerate characteristic direction for F. Then the class of similarity of $A(v)$ is invariant under formal changes of the coordinates.*

PROOF. We may assume without loss of generality $[v] = [1 : 0]$, and, hence, $r(0) = 0$. Up to a linear change of the coordinate we have

$$u_1 = u + x^k r'(0)u + O\big(\|u\|^2 x^k, x^{k+1}\big).$$

It suffices to consider linear changes of the coordinates. Indeed, writing F in its expansion as sum of homogeneous polynomials $F = I + P_{k+1} + \sum_{j \geq k+2} P_j$, if F is conjugated by $\varphi \in \mathrm{Diff}(\mathbb{C}^p, 0)$ of the form $\varphi = L + \sum_{j \geq 2} Q_j$, by Lemma 4.6 we have

$$F^* = \varphi^{-1} \circ F \circ \varphi = I + L^{-1} \circ P_{k+1} \circ L + \cdots,$$

and hence the expansion of F^* up to order $k + 1$ depends only on $d\varphi_0$.

The projection of P_{k+1}^* on $\mathbb{CP}^{p-1}$ is, with the notation of Lemma 4.5,

$$\widetilde{P}_{k+1}^* = \tilde{L}^{-1} \circ \widetilde{P}_{k+1} \circ \tilde{L},$$

where $\tilde{L}$ is just the linear transformation of $\mathbb{CP}^{p-1}$ induced by L and $\widetilde{P}_{k+1}$ is the projection of P_{k+1}. Note that $[v^*]$ is a characteristic direction for F^* if and only if $[Lv^*]$ is a characteristic direction for F. Since we have

$$d(\widetilde{P}_{k+1}^*)_{[v^*]} = \tilde{L}^{-1} \circ d(\widetilde{P}_{k+1})_{[v]} \circ \tilde{L},$$

we obtain

$$\frac{1}{k}\big[d(\tilde{P}_{k+1}^*)_{[v^*]} - I\big] = \tilde{L}^{-1} \circ \frac{1}{k}\big(d(\tilde{P}_{k+1})_{[v]} - I\big) \circ \tilde{L},$$

yielding, by Lemma 4.5,

$$A^*(v^*) = L^{-1}A(v)L,$$

which is the statement.   □

As a corollary, we obtain that the eigenvalues of $A(v)$ are holomorphic (and formal) invariants associated to $[v]$, and so the following definition is well posed.

DEFINITION 4.8. *Let $F \in \text{Diff}(\mathbb{C}^p, 0)$ be a tangent to the identity germ of order $k+1 \geq 2$ and let $[v] \in \mathbb{CP}^{p-1}$ be a non-degenerate characteristic direction for F. The class of similarity of the matrix $A(v)$ is called (with a slight abuse of notation) the* matrix associated *to $[v]$ and it is denoted by $A(v)$. The eigenvalues of the matrix $A(v)$ associated to $[v]$ are called* directors *of v. The direction $[v]$ is called* attracting *if all the real parts of its directors are strictly positive.*

5   CHANGES OF COORDINATES

We proved in the previous section that in studying germs $F \in \text{Diff}(\mathbb{C}^p, 0)$ tangent to the identity in a neighborhood of a non-degenerate characteristic direction $[v]$, we can reduce ourselves to the case $v = (1, 0)$ and F of the form:

$$\begin{cases} x_1 = f(x, u) = x - \frac{1}{k}x^{k+1} + O(\|u\|x^{k+1}, x^{2k+1}), \\ u_1 = \Psi(x, u) = (I - x^k A)u + O(\|u\|^2 x^k, \|u\|x^{k+1}) + x^{k+1}\psi_1(x), \end{cases} \tag{5.1}$$

where $A = A(v)$ is the $(p-1) \times (p-1)$ matrix associated to v, and ψ_1 is a holomorphic function. Moreover, we may assume A to be in Jordan normal form.

In this section we shall perform changes of coordinates to find F-invariant holomorphic curves, tangent to the direction $u = 0$, that is, we want to find a function u holomorphic in an open set U having the origin on its boundary, and such that

$$\begin{cases} u : U \to \mathbb{C}^{p-1}, \\ u(0) = 0, \, u'(0) = 0, \\ u(f(x, u(x))) = \Psi(x, u(x)). \end{cases}$$

If we have such a function, the F invariant curve will just be $\phi(x) = (x, u(x))$.

We now give precise definitions that generalize Definition 1.2 of [Ha] and Definition 1.5 of [Ha2] for the case $k + 1 \geq 2$.

DEFINITION 5.1. *Let $F \in \text{Diff}(\mathbb{C}^p, 0)$ be a tangent to the identity germ. A subset $M \subset \mathbb{C}^p$ is a* parabolic manifold *of dimension d at the origin tangent to a direction V if*

1. *there exists a domain S in $\mathbb{C}^d$, with $0 \in \partial S$, and an injective map $\psi : S \to \mathbb{C}^p$ such that $\psi(S) = M$ and $\lim_{z \to 0} \psi(z) = 0$;*

2. *for any sequence $\{X_h\} \subset S$ so that $X_h \to 0$, we have $[\psi(X_h)] \to [V]$;*

3. *M is F-invariant and for each $p \in M$ the orbit of p under F converges to 0.*

A parabolic manifold of dimension 1 will we called parabolic curve.

We shall search for a function $\psi = (\mathrm{id}_\mathbb{C}, u)$, defined on the k connected components of $\mathbb{D}_r = \{x \in \mathbb{C} \mid |x^k - r| < r\}$, and taking values in $\mathbb{C}^p$, verifying

$$u(f(x, u(x))) = \Psi(x, u(x)),$$

and, taking r sufficiently small, we shall obtain parabolic curves.

The idea is to first search for a formal transformation, and then to show its convergence in a sectorial neighborhood of the origin. The general obstruction to this kind of procedure is given by the impossibility of proving directly the convergence of the formal series.

As we said, in this section we shall change coordinates to further simplify F, by means of changes defined in domains of $\mathbb{C}^p$, with 0 on the boundary, and involving square roots and logarithms in the first variable x.

Following Hakim [Ha], we shall first deal with the 2-dimensional case ($p = 2$), generalizing Propositions 3.1 and 3.5 of [Ha] for the case $k + 1 \geq 2$, to better understand the changes of coordinates that we are going to use. The equations (5.1) for $p = 2$ are the following:

$$\begin{cases} x_1 = f(x, u) = x - \frac{1}{k} x^{k+1} + O(u x^{k+1}, x^{2k+1}), \\ u_1 = \Psi(x, u) = (1 - x^k \alpha) u + x^{k+1} \psi_1(x) + O(u^2 x^k, u x^{k+1}), \end{cases} \tag{5.2}$$

where $\alpha \in \mathbb{C}$ is the director, and we shall need to consider separately the case $k\alpha \in \mathbb{N}$ and the case $k\alpha \notin \mathbb{N}$.

5.1 Case $p = 2$ and $k\alpha \notin \mathbb{N}^*$

PROPOSITION 5.2. *Let $F = (f, \Psi) \in \mathrm{Diff}(\mathbb{C}^2, 0)$ be of the form (5.2). If $k\alpha \notin \mathbb{N}$, then there exists a unique sequence $\{P_h\}_{h \in \mathbb{N}} \subset \mathbb{C}[x]$ of polynomials with $\deg(P_h) = h$ for each $h \in \mathbb{N}$, such that*

$$\begin{cases} P_h(0) = 0, \\ \Psi(x, P_h(x)) = P_h(f(x, P_h(x))) + x^{h+k+1} \psi_{h+1}(x). \end{cases} \tag{5.3}$$

Moreover, $P_{h+1}(x) = P_h(x) + c_{h+1} x^{h+1}$, where $c_{h+1} = \frac{k \psi_{h+1}(0)}{k\alpha - (h+1)}$.

PROOF. We shall argue by induction on h.

If $h = 1$, we have to search for $P_1 = c_1 x$ satisfying (5.3). We have

$$\Psi(x, P_1(x)) = c_1 x \left(1 - \alpha x^k + O(x^{k+1})\right) + x^{k+1} \psi_1(x)$$

and

$$P_1(f(x, P_1(x))) = c_1 \left(x - \frac{1}{k} x^{k+1} + O(x^{k+2})\right).$$

Hence,

$$\Psi(x, P_1(x)) - P_1(f(x, P_1(x))) = c_1 x^{k+1} \left(\frac{1}{k} - \alpha + \frac{\psi_1(0)}{c_1}\right) + O(x^{k+2}).$$

To delete the terms of order less than $k + 2$, we must set $c_1 = \frac{k \psi_1(0)}{k\alpha - 1}$, which is possible since $k\alpha \notin \mathbb{N}^*$.

Let us now assume that we have a unique polynomial P_h of degree h satisying (5.3). We search for a polynomial P_{h+1} of degree $h+1$ and such that

$$\Psi\left(x, P_{h+1}(x)\right) = P_{h+1}\left(f(x, P_{h+1}(x))\right) + x^{h+k+2}\psi_{h+2}(x).$$

We can write P_{h+1} as $P_{h+1}(x) = p_h(x) + c_{h+1}x^{h+1}$, where p_h is a polynomial of degree $\leq h$ and $p_h(0) = 0$. In particular,

$$P_{h+1}\left(f(x, P_{h+1}(x))\right) = p_h\left(f(x, P_{h+1}(x))\right) + c_{h+1}(f(x, P_{h+1}(x)))^{h+1}.$$

Let $x_1 = f(x, u) = x - \frac{1}{k}x^{k+1} + x^{k+1}\varphi(x, u)$, with $\varphi(x, u) \in O(x, u)$. We have

$$p_h\left(f(x, P_{h+1}(x))\right) = p_h\left(x - \frac{1}{k}x^{k+1} + x^{k+1}\varphi(x, p_h(x))\right) + O(x^{k+h+2})$$

$$= p_h\left(f(x, p_h(x))\right) + O(x^{k+h+2}),$$

and

$$(f(x, P_{h+1}(x)))^{h+1} = x^{h+1}\left[1 - \frac{h+1}{k}x^k + O(x^{k+1})\right]$$

$$= x^{h+1}\left[1 - \frac{h+1}{k}x^k\right] + O(x^{h+k+2}).$$

It thus follows that

$$P_{h+1}\left(f(x, P_{h+1}(x))\right)$$
$$= p_h\left(f(x, p_h(x))\right) + c_{h+1}x^{h+1} - c_{h+1}\frac{h+1}{k}x^{h+k+1} + O(x^{h+k+2}).$$

By the second equation of (5.2), $u_1 = u[1 - \alpha x^k + x^k\phi(x, u)] + x^{k+1}\psi_1(x)$, with $\phi(x, u) \in O(x, u)$, and hence

$$\Psi\left(x, P_{h+1}(x)\right) = \left[p_h(x) + c_{h+1}x^{h+1}\right] \cdot \left[1 - \alpha x^k + x^k\phi(x, P_{h+1}(x))\right] + x^{k+1}\psi_1(x)$$

$$= \Psi\left(f(x, p_h(x))\right) + c_{h+1}x^{h+1} - \alpha c_{h+1}x^{h+k+1} + O(x^{h+k+2}).$$

$$(5.4)$$

Therefore,

$$\Psi\left(x, P_{h+1}(x)\right) - P_{h+1}\left(f(x, P_{h+1}(x))\right)$$
$$= \Psi\left(f(x, p_h(x))\right) - p_h\left(f(x, p_h(x))\right) + c_{h+1}\left(\frac{h+1}{k} - \alpha\right)x^{h+k+1} + O(x^{h+k+2}).$$

$$(5.5)$$

To have P_{h+1} satisying (5.3), we need

$$\Psi\left(f(x, p_h(x))\right) - p_h\left(f(x, p_h(x))\right) + c_{h+1}x^{h+k+1}\left(\frac{h+1}{k} - \alpha\right) = O(x^{h+k+2}),$$

that is, p_h has to solve (5.3); and this implies, by our induction hypothesis, $p_h = P_h$. Substituting P_h to p_h in (5.5) and expanding ψ_{h+1} in a neighborhood of 0 we get

$$\Psi\left(x, P_{h+1}(x)\right) - P_{h+1}\left(f(x, P_{h+1}(x))\right)$$
$$= x^{h+k+1}\left[\psi_{h+1}(0) + c_{h+1}\left(\frac{h+1}{k} - \alpha\right)\right] + O(x^{h+k+2}),$$

and so we have to set the leading coefficient of P_{h+1} to be

$$c_{h+1} = \frac{k\psi_{h+1}(0)}{k\alpha - (h+1)},$$

which is possible since $k\alpha \notin \mathbb{N}^*$, and then we are done. $\qquad\square$

The following reformulation of Corollary 3.2 of [Ha] for the case $k+1 \geq 2$, shows that we can rewrite the equations of F in a more useful way, with a suitable change of coordinates.

COROLLARY 5.3. *Let $F = (f, \Psi) \in \text{Diff}(\mathbb{C}^2, 0)$ be of the form (5.2), with $k\alpha \notin \mathbb{N}$. Then, for any $h \in \mathbb{N}$, there exists a holomorphic change of coordinates conjugating F to*

$$\begin{cases} x_1 = \tilde{f}(x, u) = x - \frac{1}{k}x^{k+1} + O(ux^{k+1}, x^{2k+1}), \\ u_1 = \widetilde{\Psi}(x, u) = (1 - \alpha x^k)u + x^{h+k}\psi_h(x) + O(u^2 x^k, ux^{k+1}). \end{cases} \tag{5.6}$$

PROOF. It is clear that the change of coordinates will involve only u. Let $h \in \mathbb{N}$, and let P_{h-1} be the polynomial of degree $h - 1$ of Proposition 5.2 and consider the change of coordinates

$$\begin{cases} X = x, \\ U = u - P_{h-1}(x). \end{cases}$$

The first equation of (5.2) does not change, whereas the second one becomes

$$U_1 = u_1 - P_{h-1}(x_1)$$
$$= \Psi(X, U + P_{h-1}(X)) - P_{h-1}(f(X, U + P_{h-1}(X))),$$

where we have

$$\Psi(X, U + P_{h-1}(X)) = U[1 - \alpha X^k] + \Psi(X, P_{h-1}(X)) + O(U^2 X^k, U X^{k+1}).$$

Analogously to the previous proof, we can expand $P_{h-1}(f(X, U + P_{h-1}(X)))$ at the first order in U, obtaining

$$P_{h-1}(f(X, U + P_{h-1}(X)))$$
$$= P_{h-1}\left(X - \frac{1}{k}X^{k+1} + X^{k+1}\varphi_1(X, P_{h-1}(X)) + O(U X^{k+1})\right)$$
$$= P_{h-1}(f(X, P_{h-1}(X))) + O(U X^{k+1}).$$

Therefore, we have

$$U_1 = X^{h+k}\psi_h(X) + U\left(1 - \alpha X^k + O(U X^k, X^{k+1})\right),$$

and this concludes the proof. $\qquad\square$

5.2   Case $p = 2$ and $k\alpha \in \mathbb{N}^*$

We now consider the case $k\alpha \in \mathbb{N}^*$, $k\alpha \geq 1$. Proposition 3.3 of [Ha] becomes the following.

PROPOSITION 5.4. *Let $F = (f, \Psi) \in \text{Diff}(\mathbb{C}^2, 0)$ be of the form (5.2), with $k\alpha \in \mathbb{N}$. Then there exists a sequence $\{P_h(x, t)\}_{h \in \mathbb{N}}$ of polynomials in two variables (x, t) such that*

$$\tilde{u}_h(x) := P_h \left(x, x^{k\alpha} \log x \right)$$

has degree $\leq h$ in x (where consider as constant the terms in $\log x$). Moreover,

$$\Psi \left(x, \tilde{u}_h(x) \right) - \tilde{u}_h \left(f(x, \tilde{u}_h(x)) \right) = x^{h+k+1} \psi_{h+1}(x), \tag{5.7}$$

where ψ_{h+1} satisfies

1. *$x^{h+k} \psi_{h+1}$ is holomorphic in x and $x^{k\alpha} \log x$;*

2. *$\psi_{h+1}(x) = R_{h+1}(\log x) + O(x)$, with R_{h+1} a polynomial of degree $p_{h+1} \in \mathbb{N}$, $p_{h+1} \leq h + 1$.*

PROOF. The proof is done by induction on h.

If $h < k\alpha$, then the same argument of Proposition 5.2 holds, since the polynomials P_h are still well defined. As a consequence, also the change of variables $u \mapsto u - P_{k\alpha-1}(x)$ is well defined and hence we can assume that the second component of F is of the form

$$u_1 = u \left(1 - \alpha x^k + O(ux^k, x^{k+1}) \right) + x^{k\alpha+k} \psi_{k\alpha}(x).$$

It is clear that, for $h < k\alpha$, the functions ψ_h are holomorphic in x and thus they satisfy the conditions (1) and (2) of the statement.

We can then assume that F is of the form

$$\begin{cases} x_1 = f(x, u) = x - \frac{1}{k} x^{k+1} + x^{k+1} \varphi_1(x, u), \\ u_1 = \Psi(x, u) = u \left(1 - \alpha x^k + x^k \varphi_2(x, u) \right) + x^{k\alpha+k} \psi_{k\alpha}(x), \end{cases} \tag{5.8}$$

where φ_1 and φ_2 are holomorphic functions or order at least 1 in x and u.

If $h = k\alpha$, it suffices to consider $P_{k\alpha}(x, t) = ct$, where $c = -k\psi_{k\alpha}(0)$. In fact $\tilde{u}_{k\alpha}(x) = cx^{k\alpha} \log x$ verifies (5.7) if

$$\Psi \left(x, \tilde{u}_{k\alpha}(x) \right) - \tilde{u}_{k\alpha} \left(f(x, \tilde{u}_{k\alpha}(x)) \right)$$
$$= \tilde{u}_{k\alpha}(x) \left[1 - \alpha x^k + x^k \varphi_2(x, \tilde{u}_{k\alpha}(x)) \right] + x^{k\alpha+k} \psi_{k\alpha}(x)$$
$$\qquad - \tilde{u}_{k\alpha} \left(x - \frac{1}{k} x^{k+1} + x^{k+1} \varphi_1(x, \tilde{u}_{k\alpha}(x)) \right)$$
$$= O \left(x^{k\alpha+k+1} (\log x)^{p_h} \right),$$

for some $p_h \in \mathbb{N}$. Recall that

$$\begin{cases} \frac{\partial f}{\partial u} = x^{k+1} \frac{\partial \varphi_1}{\partial u} = O(x^{k+1}), \\ \frac{\partial \Psi}{\partial u} = 1 - \alpha x^k + x^k \left(\varphi_2(x, u) + u \frac{\partial \varphi_2(x, u)}{\partial u} \right) = 1 - \alpha x^k + O(x^{k+1}, ux^k). \end{cases} \tag{5.9}$$

We have

$$\tilde{u}_{k\alpha}(x) \left[1 - \alpha x^k + x^k \varphi_2(x, \tilde{u}_{k\alpha}(x)) \right] + x^{k\alpha+k} \psi_{k\alpha}(x)$$
$$= cx^{k\alpha} \log x - \alpha c x^{k\alpha+k} \log x + x^{k\alpha+k} \log x \cdot \varphi_2 \left(x, x^{k\alpha} \log x \right) + x^{k\alpha+k} \psi_{k\alpha}(x)$$

and

$$\tilde{u}_{k\alpha}\left(x - \frac{1}{k}x^{k+1} + x^{k+1}\varphi_1(x, \tilde{u}_{k\alpha}(x))\right) =$$

$$c\left(x - \frac{x^{k+1}}{k} + O(x^{k\alpha+k+1}\log x, x^{2k+1})\right)^{k\alpha}\log\left(x - \frac{x^{k+1}}{k} + O(x^{k\alpha+k+1}\log x, x^{2k+1})\right)$$

$$= cx^{k\alpha}\log x - c\alpha x^{k\alpha+k}\log x - \frac{c}{k}x^{k\alpha+k} + O(x^{2k\alpha+k}(\log x)^2, x^{k\alpha+2k}\log x).$$

Therefore,

$$\Psi\left(x, \tilde{u}_{k\alpha}(x)\right) - \tilde{u}_{k\alpha}\left(f(x, \tilde{u}_{k\alpha}(x))\right)$$
$$= x^{k\alpha+k}\psi_{k\alpha}(0) + \frac{c}{k}x^{k\alpha+k} + x^{k\alpha+k+1}O\left(x^{k\alpha-1}(\log x)^2, \log x\right).$$

If $c = -k\psi_{k\alpha}(0)$, then

$$\Psi\left(x, \tilde{u}_{k\alpha}(x)\right) - \tilde{u}_{k\alpha}\left(f(x, \tilde{u}_{k\alpha}(x))\right) = x^{k\alpha+k+1}\psi_{k\alpha+1}(x) = O(x^{k\alpha+k+1}(\log x)^2).$$

In particular, note that

$$\psi_{k\alpha+1}(x) = R_{k\alpha+1}(\log x) + O(x),$$

where $R_{k\alpha+1}$ is a polynomial of degree 1 or 2, depending on whether $k\alpha = 1$ or $k\alpha > 1$. Also, in this case $\psi_{k\alpha+1}$ satisfies Conditions (1) and (2) of the statement. Indeed, since $k\alpha + k \geq 2$, we have that $x^{k\alpha+k}R_{k\alpha+1}(\log x)$ is holomorphic in $x^{k\alpha}\log x$ and x.

We are left with the case $h > k\alpha$. The inductive hypothesis ensures that (5.7) holds for $h - 1$ and there exists a polynomial $R_h(t)$ of degree $\leq h$ so that $\psi_h(x) = R_h(\log x) + O(x)$. We search for $\tilde{u}_h$ of the form

$$\tilde{u}_h(x) = \tilde{u}_{h-1}(x) + x^h Q_h(\log x), \tag{5.10}$$

where Q_h is a polynomial, and we shall prove that $\tilde{u}_h$, of the form (5.10), satisfies (5.7) if and only if Q_h is the unique polynomial solution of the following differential equation

$$(h - k\alpha)Q_h(t) - Q_h'(t) = kR_h(t).$$

In fact, we have

$$\Psi(x, \tilde{u}_h(x)) - \tilde{u}_h\left(f(x, \tilde{u}_h(x))\right) = \Psi\left(x, \tilde{u}_{h-1}(x) + x^h Q_h(\log x)\right) - \tilde{u}_{h-1}(f(x, \tilde{u}_h(x)))$$
$$- \left(f(x, \tilde{u}_h(x))\right)^h Q_h(\log(f(x, \tilde{u}_h(x)))).$$

Thanks to the inductive hypothesis, in $\tilde{u}_h$ for $h \geq k\alpha$, the term of lower degree is $cx^{k\alpha}\log x$. We have

$$\Psi(x, \tilde{u}_{h-1}(x) + x^h Q_h(\log x))$$
$$= \Psi\left(x, \tilde{u}_{h-1}(x) + x^h Q_h(\log x)\right)$$
$$= \Psi(x, \tilde{u}_{h-1}(x)) + \frac{\partial\Psi}{\partial u}(x, \tilde{u}_{h-1}(x))x^h Q_h(\log x)$$
$$+ \sum_{n\geq 2}\frac{1}{n!}\frac{\partial^n\Psi}{\partial u^n}(x, \tilde{u}_{h-1}(x))\left(x^h Q_h(\log x)\right)^n$$
$$= \Psi(x, \tilde{u}_{h-1}(x)) + x^h Q_h(\log x) - \alpha x^{k+h}Q_h(\log x)$$
$$+ O\left(x^{h+k+k\alpha}(\log x)^{\deg Q_h + 1}, x^{h+k+1}(\log x)^{\deg Q_h}\right).$$

Analogously to the previous proof, using the first equation in (5.9), we have

$$
f\left(x, \tilde{u}_{h-1}(x) + x^h Q_h(\log x)\right)
$$
$$
= f(x, \tilde{u}_{h-1}(x)) + \sum_{n \geq 1} \frac{1}{n!} \frac{\partial^n f}{\partial x^n}(f(x, \tilde{u}_{h-1}(x))) \left(x^h Q_h(\log x)\right)^n
$$
$$
= f(x, \tilde{u}_{h-1}(x)) + O\left(x^{h+k+1}(\log x)^{\deg Q_h}\right).
$$

Therefore,

$$
\tilde{u}_{h-1}(f(x, \tilde{u}_h(x))) = \tilde{u}_{h-1}\left(f(x, \tilde{u}_{h-1}(x)) + O\left(x^{h+k+1}(\log x)^{\deg Q_h}\right)\right)
$$
$$
= \tilde{u}_{h-1}\left(f(x, \tilde{u}_{h-1}(x))\right) + O\left(x^{h+k+k\alpha}(\log x)^{\deg Q_h + 1}\right).
$$

Finally, expanding Q_h in a neighborhood of $\log x$, and considering the terms of degree $h + k$, we obtain

$$
[f(x, \tilde{u}_h(x))]^h Q_h\left(\log(f(x, \tilde{u}_h(x)))\right)
$$
$$
= \left[x - \frac{x^{k+1}}{k} + O(x^{k+1}\tilde{u}_h(x), x^{k+2})\right]^h Q_h\left(\log\left(x - \frac{x^{k+1}}{k} + O(x^{k+1}\tilde{u}_h(x), x^{k+2})\right)\right)
$$
$$
= \left[x^h - \frac{h}{k}x^{h+k} + O\left(x^{k+h}\tilde{u}_h(x), x^{k+h+1}\right)\right]
$$
$$
\times \left[Q_h(\log x) - \frac{x^k}{k}Q_h'(\log x) + O\left(x^k\tilde{u}_h(x)(\log x)^{\deg Q_h - 1}, x^{k+1}(\log x)^{\deg Q_h - 1}\right)\right]
$$
$$
= x^h Q_h(\log x) - \frac{x^{h+k}}{k}Q_h'(\log x) - \frac{h}{k}x^{h+k}Q_h(\log x)
$$
$$
+ O\left(x^{h+k+k\alpha}(\log x)^{\deg Q_h + 1}, x^{h+k+1}(\log x)^{\deg Q_h}\right).
$$

The inductive hypothesis implies

$$
\Psi(x, \tilde{u}_{h-1}(x)) - \tilde{u}_{h-1}(f(x, \tilde{u}_{h-1}(x))) = x^{k+h}\psi_h(x),
$$

with $\psi_h(x) = R_h(\log x) + o(x)$. Reordering the terms, we then obtain

$$
\Psi(x, \tilde{u}_h(x)) - \tilde{u}_h(f(x, \tilde{u}_h(x)))
$$
$$
= x^{h+k}\left[R_h(\log x) + \left(\frac{h}{k} - \alpha\right)Q_h(\log x) + \frac{1}{k}Q_h'(\log x)\right] \tag{5.11}
$$
$$
+ O\left(x^{h+k+k\alpha}(\log x)^{\deg Q_h + 1}, x^{h+k+1}(\log x)^{\deg Q_h}\right),
$$

where $R_h(t)$ is the polynomial of degree $p_h \leq h$. Hence, $\tilde{u}_h$ satisfies (5.7) if and only if Q_h is the unique solution of

$$
(k\alpha - h)Q_h(t) - Q_h'(t) = kR_h(t). \tag{5.12}
$$

Therefore, R_{h+1} is a polynomial so that $\deg R_{h+1} \leq h + 1$, and we can have $\deg R_{h+1} = h + 1$ only if $k\alpha = 1$. Moreover, if $k\alpha = 1$, $\deg R_{h+1}$ can be more that $h + 1$.

We finally have to verify that ψ_{h+1} is holomorphic and that $\tilde{u}_h$ is a polynomial in x and $x^{k\alpha} \log x$ of degree $\leq h$ in x. Since Q_h solves the differential equation (5.12), it has to be a polynomial of the same degree as R_h. Moreover, since $x^h R_h(\log x)$ is a polynomial in x and $x^{k\alpha} \log x$, we have $p_h \leq \frac{h}{k\alpha}$. We thus conclude that $\tilde{u}_h$ is a polynomial in x and $x^{k\alpha} \log x$ of degree $\leq h$. Thanks to (5.11), $x^{h+k}\psi_{h+1}(x)$ is holomorphic in x and $x^{k\alpha} \log x$.

Summarizing, the sequence of polynomials is the following

$$P_h(x,t) = \begin{cases} \sum\limits_{i=1}^{h} c_i x^i, c_i = \dfrac{k\psi_i(0)}{k\alpha - (i+1)} & \text{if } h < k\alpha, \\[2ex] \psi_{k\alpha}(0)t & \text{if } h = k\alpha, \\[1ex] P_{h-1}(x,t) + x^h Q_h(\log x) & \text{if } h > k\alpha. \end{cases}$$

$\square$

Similarly to the case $k\alpha \notin \mathbb{N}^*$, we deduce the following reformulation of Corollary 3.4 of [Ha] for the case $k + 1 \geq 2$.

COROLLARY 5.5. *Let $F = (f, \Psi) \in \text{Diff}(\mathbb{C}^2, 0)$ be of the form (5.2), with $k\alpha \in \mathbb{N}$. Then for any $h \in \mathbb{N}$, so that $h \geq \max\{k, k\alpha\}$, it is possible to choose local coordinates in which F has the form*

$$\begin{cases} x_1 = \tilde{f}(x, u) = x - \frac{1}{k}x^{k+1} + O(ux^{k+1}, x^{2k+1} \log x), \\ u_1 = \widetilde{\Psi}(x, u) = u\left(1 - \alpha x^k + O(ux^k, x^{k+1} \log x)\right) + x^{h+k}\psi_h(x), \end{cases}$$

where $\tilde{f}$, $\widetilde{\Psi}$ and $x^{h+k-1}\psi_h(x)$ are holomorphic in x, $x^{k\alpha} \log x$ and u.

PROOF. Consider $h \geq \max\{k, k\alpha\}$, and let $\tilde{u}_{h-1}$ be the polynomial map in x and $x^{k\alpha} \log x$ given by the previous result. With the change of coordinates

$$\begin{cases} X = x, \\ U = u - \tilde{u}_{h-1}, \end{cases}$$

the first equation becomes

$$X_1 = X - \frac{1}{k}X^{k+1} + O\left(UX^{k+1}, X^{2k+1} \log X\right).$$

In particular, the term $x^{2k+1} \log x$ appears only if $k\alpha = 1$. The second equation becomes

$$\begin{aligned} U_1 &= u_1 - \tilde{u}_{h-1}(x_1) \\ &= U\left(1 - \alpha X^k\right) + O\left(U^2 X^k, UX^{k+1} \log x\right) + X^{k+h}\psi_h(X). \end{aligned}$$

Again, the term $UX^{k+1} \log x$ appears only if $k\alpha = 1$; otherwise we have UX^{k+1}. $\square$

REMARK 5.6. Note that if $k\alpha \in \mathbb{N}^*$, due to the presence of the logarithms, all the changes of coordinates used are not defined in a full neighborhood of the origin, but in an open set having the origin on its boundary.

5.3   General case: $p > 2$

Now we deal with the general case of dimension $p > 2$. Also, in this case, the allowed changes of coordinates will depend on the arithmetic properties of the directors associated to the characteristic direction.

PROPOSITION 5.7. *Let $F = (f, \Psi) \in \mathrm{Diff}(\mathbb{C}^p, 0)$ be of the form (5.1), let $[v] = [1 : 0]$ be a non-degenerate characteristic direction, and let $\{a_1, \dots, a_s\}$ be the directors of $[v]$ so that $ka_j \in \mathbb{N}$. Then, for all $h \in \mathbb{N}$, there exists $\tilde{u}_h : \mathbb{C} \to \mathbb{C}^{p-1}$ so that its components are polynomials in $x, x^{ka_1} \log x, \dots, x^{ka_s} \log x$ of degree $\le h$ in x, and the change of coordinates $u \mapsto u - \tilde{u}_h(x)$ conjugates F to*

$$
\begin{cases}
x_1 = \tilde{f}(x, U) = x - \dfrac{1}{k} x^{k+1} + O\big(\|U\| x^{k+1}, x^{2k+1} \log x\big), \\
U_1 = \widetilde{\Psi}(x, U) = (I - A x^k) U + O\big(\|U\|^2 x^k, \|U\| x^{k+1} \log x\big) + x^{k+h} \psi_h(x),
\end{cases}
$$
$$(5.13)$$

with $\psi_h(x) = R_h(\log x) + O(x)$, where $R_h(t) = (R_h^1(t), \dots, R_h^{p-1}(t))$ is a polynomial map with $\deg R_h^i = p_h^i \le h$, for each $i = 1, \dots, p - 1$.

PROOF. We may assume without loss of generality that A is in Jordan normal form. For each fixed h, the jth component of $\tilde{u}_h$ is determined by the components from $p - 1$ to $j + 1$, and each of them is determined with the results proved in dimension 2.

It suffices to prove the statement when A is a unique Jordan block of dimension $p - 1$ with eigenvalue α and with elements out of the diagonal equal to α. The equations of F are

$$
\begin{cases}
x_1 \quad = f(x, u) = x - \frac{1}{k} x^{k+1} + O\big(x^{2k+1}, \|(u,v)\| x^{k+1}\big), \\
u_{1,j} \quad = \Psi_j(x, u) = (1 - x^k \alpha) u_j - x^k \alpha u_{j+1} + O\big(\|u\|^2 x^k, \|u\| x^{k+1}\big) + x^{k+1} \psi_j(x) \\
u_{1,p-1} = \Psi_{p-1}(x, u) = (1 - x^k \alpha) u_{p-1} + O\big(\|u\|^2 x^k, \|u\| x^{k+1}\big) + x^{k+1} \psi_{p-1}(x),
\end{cases}
$$

for $j = 1, \dots, p - 2$ and where $\psi_1, \dots, \psi_{p-1}$ are holomorphic bounded functions.

We proceed by induction on h. If $h = 0$, it suffices to consider $\tilde{u}_0 \equiv 0$. In fact,

$$
\Psi_j(x, \tilde{u}_0) - \tilde{u}_{0,j}(f(x, \tilde{u}_0)) = x^{k+1} \psi_j(x), \quad \text{for } j = 1, \dots, p - 1.
$$

Let us then assume by inductive hypothesis that there exist $\tilde{u}_{h-1}$ such that

$$
\Psi_j(x, \tilde{u}_{h-1}) - \tilde{u}_{h-1,j}(f(x, \tilde{u}_{h-1})) = x^{k+h} \psi_{h,j}(x), \quad \text{for } j = 1, \dots, p - 1. \quad (5.14)
$$

As in the 2-dimensional case, we want to find polynomials $Q_{h,1}, \dots, Q_{h,p-1}$ so that

$$
\tilde{u}_{h,j}(x) = \tilde{u}_{h-1,j}(x) + Q_{h,j}(\log x) x^h, \quad \text{for } j = 1, \dots, p - 1,
$$

verify (5.14) for h. Proposition 5.4 gives us that $\tilde{u}_{h,p-1}$ is a solution if and only if $Q_{h,p-1}$ verifies

$$
(k\alpha - h) Q_{h,p-1}(t) - (Q_{h,p-1}(t))'(t) = k R_{h,p-1}(t).
$$

Moreover, we have $\deg R_{h,p-1} = p_{h,p-1} \le h$. We proceed in the same way for the remaining $\tilde{u}_{h,j}$'s, except for the fact that the equations are a bit different from the ones used before. In particular

$$
\begin{cases}
\dfrac{\partial \Psi_j}{\partial u_j}(x,u) = 1 - \alpha x^k + O\big(x^{k+1}, \|u\|x^k\big), \\[4mm]
\dfrac{\partial \Psi_j}{\partial u_{j-1}}(x,u) = -\alpha x^k + O\big(x^{k+1}, \|u\|x^k\big).
\end{cases}
$$

Hence,

$$
\begin{aligned}
\Psi_j(x, \tilde u_h) &= \Psi_j(x, \tilde u_{h-1} + x^h Q_h(\log x)) \\
&= \Psi_j(x, \tilde u_{h-1}) + (1 - \alpha x^k) x^h Q_{h,j}(\log x) - \alpha x^{k+h} Q_{h,j+1}(\log x) \\
&\quad + O\big(x^{k+h+1}(\log x)^{p_h}, \|\tilde u_{h-1}\| x^{k+h}(\log x)^{p_h}\big),
\end{aligned}
$$

where $p_h = \max \deg Q_{h,j}$, and

$$
\tilde u_h \left(f(x, \tilde u_h) \right) = \tilde u_{h-1} \left(f(x, \tilde u_h) \right) + [f(x, \tilde u_h)]^h \, Q_h \left(\log(f(x, \tilde u_h)) \right).
$$

We have

$$
\tilde u_{h-1} \left(f(x, \tilde u_h) \right) = \tilde u_{h-1} \left(f(x, \tilde u_{h-1}) \right) + O\big(x^{h+k+1} \log x, \|\tilde u_h\| x^{h+k} \log x\big),
$$

and

$$
\begin{aligned}
[f(x, \tilde u_h)]^h \, Q_h \left(\log(f(x, \tilde u_h)) \right) &= \left[x \left(1 - \frac{1}{k} x^k + O\big(x^{2k}, \|\tilde u_h\| x^k\big) \right) \right]^h \\
&\quad \times Q_h \left(\log x + \log\Big(1 - \frac{1}{k} x^k + O\big(x^{2k}, \|\tilde u_h\| x^k\big)\Big) \right) \\
&= x^h Q_h(\log x) - x^{k+h} \left(\frac{1}{k} Q_h'(\log x) + \frac{h}{k} Q_h(\log x) \right) \\
&\quad \times O\big(x^{2k+h}(\log x)^{l_1}, \|\tilde u_h\| x^{k+h}(\log x)^{l_2}\big),
\end{aligned}
$$

for some integer l_1 and l_2. It follows that

$$
\begin{aligned}
&\Psi_j(x, \tilde u_h) - \tilde u_{h,j}(f(x, \tilde u_h)) \\
&= x^{k+h} \left[\psi_{h,j}(x) + \frac{1}{k} Q_{h,j}'(\log x) + \frac{h}{k} Q_{h,j}(\log x) - \alpha Q_{h,j}(\log x) - \alpha Q_{h,j+1}(\log x) \right] \\
&\quad + O\big(x^{2k+h}(\log x)^{l_1}, \|\tilde u_h\| x^{k+h}(\log x)^{l_2}\big).
\end{aligned}
$$

Hence, $\tilde u_h$ solves the equations if and only if $Q_{h,j}$ solves

$$
[h - k\alpha] Q_{h,j}(t) + Q_{h,j}'(t) = k\alpha Q_{h,j+1}(t) - k R_{h,j}(t), \quad \text{for } j = 1, \ldots, p-2
$$

and, moreover, $\deg R_{h,j} \le h$. $\qquad\square$

REMARK 5.8. It is clear that in the previous proposition that we have no restrictions on h, and, hence, we can choose $h = k\bar h$, obtaining F of the form

$$
\begin{cases}
x_1 = f(x,u) = x - \frac{1}{k} x^{k+1} + O(\|u\| x^{k+1}, x^{2k+1} \log x) \\
u_1 = \Psi(x,u) = (I - Ax^k)u + O\big(\|u\|^2 x^k, \|u\| x^{k+1} \log x\big) + x^{k(\bar h+1)} \psi_h(x),
\end{cases}
$$

where $\psi_h(x) = R_{k\bar{h}}(x) + O(x)$. Then, up to changing the degree of the polynomials in $\log x$, for any $h \in \mathbb{N}$ we can write

$$u_1 = \Psi(x, u) = (I - Ax^k)u + O\big(\|u\|^2 x^k, \|u\| x^{k+1} \log x\big) + x^{k(h+1)} \psi_h(x).$$

6   EXISTENCE OF PARABOLIC CURVES

From now on, without loss of generality, we shall assume that non-degenerate characteristic direction is $[1 : 0] \in \mathbb{CP}^{p-1}$. Moreover, thanks to Proposition 5.7 and to Remark 5.8, after blowing up the origin, it is possible to change coordinates, in a domain having the origin on its boundary, such that F, in the coordinates $(x, u) \in \mathbb{C} \times \mathbb{C}^{p-1}$, has the form

$$\begin{cases} x_1 = f(x, u) = x - \frac{1}{k}x^{k+1} + O(\|u\| x^{k+1}, x^{2k+1} \log x) \\ u_1 = \Psi(x, u) = (I - Ax^k)u + O\big(\|u\|^2 x^k, \|u\| x^{k+1}(\log x)^{p_h}\big) + x^{k(h+1)} \psi_h(x), \end{cases}$$
(6.1)

for an arbitrarily chosen $h \in \mathbb{N}$, and with $p_h \in \mathbb{N} \setminus \{0\}$ depending on h.

REMARK 6.1. The existence of parabolic curves $S_1, \ldots, S_k$ tangent to a given direction $[v]$ at 0 is equivalent to finding u defined and holomorphic on the k connected components $\Pi^1_r, \ldots, \Pi^k_r$ of $\mathbb{D}_r := \{x \in \mathbb{C} \mid |x^k - r| < r\}$ and such that

$$\begin{cases} u(f(x, u(x))) = \Psi(x, u(x)) \\ \lim_{x \to 0} u(x) = \lim_{x \to 0} u'(x) = 0. \end{cases}$$
(6.2)

We are going to prove the existence of such curves finding a fixed point of a suitable operator between Banach spaces. We shall then need to further simplify our equations via a change of coordinates holomorphic that will be holomorphic on $\mathrm{Re}\,(x^k) > 0$. Let us consider the new coordinates $(x, w) \in \mathbb{C} \times \mathbb{C}^{p-1}$, where $w \in \mathbb{C}^{p-1}$ is defined, on $\mathrm{Re}\,(x^k) > 0$, by

$$u = x^{kA}w := \exp(kA \log x)w.$$

Hence $u_1 = x_1^{kA}w_1$. Starting from (6.1) we obtain

$$x_1 - x = -\frac{1}{k}x^{k+1} + O\big(\|u\| x^{k+1}, x^{2k+1} \log x\big)$$

and

$$u_1 - (I - x^k A)u = O\big(\|u\|^2 x^k, \|u\| x^{k+1} \log x, x^{k(h+1)}(\log x)^{p_h}\big).$$
(6.3)

Moreover, we have

$$\begin{aligned} x_1^{kA} &= \exp\left(kA\left(\log x + \log\left(1 - \frac{1}{k}x^k + O\big(\|u\| x^k, x^{2k} \log x\big)\right)\right)\right) \\ &= x^{kA}\left[\big(I - x^k A\big) + O\big(\|u\| x^k, x^{2k} \log x\big)\right]. \end{aligned}$$
(6.4)

Using $x^{kA}w = u$, we have $x^{kA}w_1 = x^{kA}x_1^{-kA}x_1^{kA}w_1 = x^{kA}x_1^{-kA}u_1$. Set

$$H(x, u) := x^{kA}(w - w_1) = u - x^{kA}x_1^{-kA}u_1.$$
(6.5)

Thanks to (6.4), we have

$$
\begin{aligned}
H(x,u) &= u - \left[(I - x^k A) + O\left(\|u\| x^k, x^{2k} \log x \right) \right]^{-1} u_1 \\
&= - \left[(I - x^k A) + O\left(\|u\| x^k, x^{2k} \log x \right) \right]^{-1} \\
&\quad \times \left[u_1 - \left[(I - x^k A) + O\left(\|u\| x^k, x^{2k} \log x \right) \right] u \right] \\
&= O\left(\|u\|^2 x^k, \|u\| x^{k+1} \log x, x^{k(h+1)} (\log x)^{p_h} \right).
\end{aligned}
\tag{6.6}
$$

Therefore, we can write

$$
w_1 = w - x^{-kA} H(x,u).
$$

Now we have all the ingredients to search for parabolic curves tangent to the direction $[v]$. For the moment, we impose only that u is at least of order $k+1$. We have the following generalization of Lemma 4.2 of [Ha] .

LEMMA 6.2. *Let f be a holomorphic function defined as in the first equation of (6.1). For any u so that $u(x) = x^{k+1} \ell(x)$, for some bounded holomorphic map $\ell \colon \Pi_r^i \to \mathbf{C}^{p-1}$, let $\{x_n\}$ be the sequence of the iterates of x via*

$$
x_1 = f_u(x) := f\left(x, u(x) \right).
$$

Then, for r small enough, for any ℓ so that $\|\ell\|_\infty \leq 1$, and any $n \in \mathbf{N}$, if $x \in \Pi_r^i$ then $x_n \in \Pi_r^i$, and, moreover,

$$
|x_n| \leq 2^{1/k} \frac{|x|}{\left(|1 + nx^k| \right)^{\frac{1}{k}}}.
$$

PROOF. Thanks to the hypothesis on u we can rewrite the first equation of (6.1), obtaining

$$
x_1 = x - \frac{1}{k} x^{k+1} + a x^{2k+1} + b x^{2k+1} \log x + O\left(x^{2k+2} (\log x)^l, x^{2k+2} \right).
$$

By Proposition 5.7, we have the term $b x^{2k+1} \log x$ only if 1 is an eigenvalue of kA. Moreover, we have

$$
x_1^k = x^k \left[1 - x^k + ka x^{2k} + kb x^{2k} \log x + \frac{1}{k^2} \binom{k}{2} x^{2k} + O\left(x^{2k+1} (\log x)^l \right) \right].
$$

Hence,

$$
\frac{1}{x_1^k} = \frac{1}{x^k} + 1 + (1 - a_1) x^k - b_1 x^k \log x + O(x^{k+1} (\log x)^l),
$$

where $O(x^{k+1} (\log x)^l)$ represents a function bounded by $K |x|^{k+1} |\log x|^l$, with K not depending on u, because $\|\ell\|_\infty \leq 1$. It is thus possible to write

$$
\frac{1}{x_1^k} = \frac{1}{x^k} \left(1 + x^k + x^{2k} \left(\frac{a}{k} + \frac{2b}{k} \log x \right) + O(x^{2k+1} (\log x)^l) \right)
\tag{6.7}
$$

where the same considerations hold as before. We can now define the following change of variable on $\operatorname{Re} x^k > 0$

$$
\frac{1}{z} = \frac{1}{x^k} + a \log x + b (\log x)^2.
$$

Therefore, (6.7) becomes

$$\frac{1}{z_1} = \frac{1}{z} + 1 + O(x^{k+1}(\log x)^l),$$

where we used

$$\log x_1 = \log x - \frac{1}{k}x^k + O(x^{2k}(\log x)^2)$$

and

$$(\log x_1)^2 = (\log x)^2 - \frac{2}{k}x^k \log x + O(x^{2k}(\log x)^2).$$

We then deduce

$$\frac{1}{z_n} = \frac{1}{z_{n-1}} + 1 + O(x_{n-1}^k(\log x_{n-1})^l) = \cdots = \frac{1}{z} + n + O(1).$$

On the other hand,

$$\frac{1}{z_n} = \frac{1}{x_n^k} + a \log x_n + b \log^2 x_n = \frac{1}{x_n^k}\left[1 + ax_n^k \log x_n + bx_n^k \log^2 x_n\right],$$

and, hence,

$$\frac{1}{z} + n + O(1) = \frac{1}{x^k}\left(1 + nx^k\right)\left[1 + \frac{ax^k \log x + bx^k \log^2 x + O(x^k)}{1 + nx^k}\right].$$

If r is small enough, f_u is an attracting map from Π_r^i in itself, and hence for any $\varepsilon > 0$ there exists $\bar{n}$ so that, for each $n > \bar{n}$,

$$\left|ax_n^k \log x_n + bx_n^k \log^2 x_n\right| < \varepsilon$$

and

$$\left|\frac{ax^k \log x + bx^k \log^2 x + O(x^k)}{1 + nx^k}\right| < \varepsilon.$$

Therefore, for $n > \bar{n}$ and r small enough

$$|x_n|^k = \left|\frac{x^k}{1 + nx^k}\right|\left|1 + ax_n^k \log x_n + bx_n^k \log^2 x_n\right|\left|\frac{1}{1 + \frac{ax^k \log x + bx^k \log^2 x + O(x^k)}{1 + nx^k}}\right|$$

$$\leq 2\frac{|x|^k}{|1 + nx^k|},$$

and hence we obtain the statement.   □

Analogously to Corollary 4.3 in [Ha], for the case $k + 1 \geq 2$ we have the following very useful inequality.

COROLLARY 6.3. *Let f be a holomorphic function defined as in the first equation of (6.1). For any u so that $u(x) = x^{k+1}\ell(x)$, for some bounded holomorphic map $\ell: \Pi_r^i \to \mathbb{C}^{p-1}$ with $\|\ell\|_\infty \leq 1$, let $\{x_n\}$ be the sequence of the iterates of x via*

$$x_1 = f_u(x) := f(x, u(x)),$$

and let r be sufficiently small. Then for any $\mu > k$ ($\mu \in \mathbb{R}$) and for any $q \in \mathbb{N}$, there exists a constant $C_{\mu,q}$ such that, for any $x \in \Pi_r^i$, we have

$$\sum_{n=0}^{\infty} |x_n|^{\mu} |\log x_n|^q \leq C_{\mu,q} |x|^{\mu-k} |\log |x||^q.$$

PROOF. If $x \in \Pi_r^i$, then $\operatorname{Re} x^k > 0$, and hence

$$\left|1 + nx^k\right|^2 = 1 + nx^k + n\bar{x}^k + n^2 \left|x^k\right|^2 \geq 1 + \left|nx^k\right|^2.$$

Then the inequality of the previous lemma becomes

$$|x_n| \leq 2^{1/k} \frac{|x|}{|1 + nx|^{\frac{1}{k}}} \leq 2^{1/k} \frac{|x|}{\sqrt[2k]{1 + |nx^k|^2}}.$$

Recalling that, for x sufficiently small, $|\log x| \leq K_1 |\log |x||$, for each $\mu > k$ and each $q \in \mathbb{N}$, we have

$$|x_n|^{\mu} |\log x_n|^q \leq K_1 |x_n|^{\mu} |\log |x_n||^q \leq K_2 \frac{|x|^{\mu}}{\sqrt[2k]{(1 + |nx^k|^2)^{\mu}}} \left| \log \frac{2^{1/k} |x|}{\sqrt[2k]{1 + |nx^k|^2}} \right|^q,$$

where $K_2 = K_1 2^{\mu/k}$. We then have that there exists K so that

$$\sum_{n=0}^{\infty} |x_n|^{\mu} |\log x_n|^q \leq K \int_0^{\infty} \frac{|x|^{\mu}}{(1 + |tx^k|^2)^{\mu/2k}} \left| \log \frac{2^{1/k} |x|}{\sqrt[2k]{1 + |tx^k|^2}} \right|^q dt \tag{6.8}$$

$$= K |x|^{\mu-k} \int_0^{\infty} \frac{1}{(1 + s^2)^{\mu/2k}} \left| \log \frac{2^{1/k} |x|}{\sqrt[2k]{1 + s^2}} \right|^q ds.$$

To conclude, the following estimate suffices:

$$\left| \log \frac{2^{1/k} |x|}{\sqrt[2k]{1 + s^2}} \right|^q \leq |\log |x||^q \sum_{j=0}^{q} \binom{q}{j} \left| \log \frac{\sqrt[2k]{1 + s^2}}{2^{1/k}} \right|^j.$$

In fact, we have

$$\int_0^{\infty} \frac{1}{(1 + s^2)^{\mu/2k}} \left| \log \frac{2^{1/k} |x|}{\sqrt[2k]{1 + s^2}} \right|^q ds$$

$$\leq |\log |x||^q \sum_{j=0}^{q} \binom{q}{j} \int_0^{\infty} \left| \log \frac{\sqrt[2k]{1 + s^2}}{2^{1/k}} \right|^j \frac{1}{(1 + s^2)^{\mu/2k}} ds.$$

that, together with (6.8), yields

$$\sum_{n=0}^{\infty} |x_n|^{\mu} |\log x_n|^q \leq C_{\mu,q} |x|^{\mu-k} |\log |x||^q,$$

where

$$C_{\mu,q} := K \sum_{j=0}^{q} \binom{q}{j} \int_0^{\infty} \left| \log \frac{\sqrt[2k]{1 + s^2}}{2^{1/k}} \right|^j \frac{1}{(1 + s^2)^{\mu/2k}} ds,$$

concluding the proof. $\square$

We have the following analogous of Lemma 4.4 of [Ha].

LEMMA 6.4. *Let f be a holomorphic function defined as in the first equation of* (6.1). *For any u so that $u(x) = x^{k+1}\ell(x)$, for some bounded holomorphic map $\ell\colon \Pi_r^i \to \mathbb{C}^{p-1}$, let $\{x_n\}$ be the sequence of the iterates of x via*

$$x_1 = f_u(x) := f\left(x, u(x)\right).$$

Then, if r is sufficiently small, for any ℓ so that $\|\ell\|_\infty \leq 1$ and $\|x\ell'\|_\infty \leq 1$, for each $n \in \mathbb{N}$ and each $x \in \overline{\Pi_r^i}$, we have

$$\left|\frac{dx_n}{dx}\right| \leq 2\frac{|x_n|^{k+1}}{|x|^{k+1}}.$$

PROOF. Arguing as in the proof of Lemma 6.2, we have

$$\frac{1}{x_1^k} + a\log x_1 + b(\log x_1)^2 = \frac{1}{x^k} + 1 + a\log x + b(\log x)^2 + \varphi(x, u), \qquad (6.9)$$

where φ is holomorphic in $x, u, x^{\ell_j}\log x$ and

$$\varphi(x, u) = O\left(x^{2k}(\log x)^l, \|u\|\right) = O\left(x^{2k}(\log x)^l, x^{k+1}\|\ell\|\right).$$

By (6.9) we therefore have

$$\frac{1}{x_n^k} + a\log x_n + b(\log x_n)^2 = \frac{1}{x^k} + n + a\log x + b(\log x)^2 + \sum_{p=0}^{n-1}\varphi(x_p, u(x_p)).$$

Differentiating, we obtain

$$-\left[\frac{k - ax_n^k - 2bx_n^k\log x_n}{x_n^{k+1}}\right]\frac{dx_n}{dx} = -\left[\frac{k - ax^k - 2bx^k\log x}{x^{k+1}}\right]$$

$$+ \sum_{p=0}^{n-1}\frac{d}{dx_p}\left[\varphi(x_p, u(x_p))\right]\frac{dx_p}{dx}. \qquad (6.10)$$

We shall now proceed by induction on n. We first have to estimate the sum of the remainders $\varphi(x_p, u(x_p))$. From the hypotheses for ℓ and the form of φ, we deduce the existence of a constant K so that

$$\left|\frac{d}{dx}\varphi(x, u(x))\right| \leq K\left(|\log|x|| + \|\ell\| + \|x\ell'\|\right)|x|.$$

For $n = 1$ we have

$$\left|\frac{dx_1}{dx}\right| = \left|\frac{k - ax^k - 2bx^k\log x + x^{k+1}\frac{d}{dx}\varphi(x, u(x))}{k - ax_1^k - 2bx_1^k\log x_1} \cdot \frac{x_1^{k+1}}{x^{k+1}}\right| \leq D\frac{|x_1|^{k+1}}{|x|^{k+1}},$$

for a constant $D \in \mathbb{R}$, that can be chosen to be $D = 2$, if r is small enough.

Let us assume, by inductive hypothesis, $\left|\frac{dx_p}{dx}\right| \le 2\frac{|x_p|^{k+1}}{|x|^{k+1}}$ for any $p < n$. Then, by the previous corollary, we have

$$\left|\sum_{p=0}^{n-1} \frac{d}{dx_p}\left[\varphi(x_p, u(x_p))\right] \frac{dx_p}{dx}\right|$$

$$\le 2K(1 + \|\ell\| + \|x\ell'\|) \sum_{p=0}^{\infty} \frac{|x_p|^{k+2}}{|x|^{k+1}} + \sum_{p=0}^{\infty} \frac{|x_p|^{k+2}|\log|x_p||}{|x|^{k+1}}$$

$$\le 2K(1 + \|\ell\| + \|x\ell'\|)\frac{C_{k+2,0}}{|x|^{k-1}} + \frac{C_{k+2,1}}{|x|^{k-1}} = \frac{K_1}{|x|^{k-1}}.$$

Therefore, we obtain

$$\left|\frac{dx_n}{dx}\right| \le \frac{\left|k - ax^k - 2bx^k \log x\right| + K_1 |x|^2}{|k - ax_n^k - 2bx_n^k \log x_n|} \cdot \frac{|x_n|^{k+1}}{|x|^{k+1}} \le 2\frac{|x_n|^{k+1}}{|x|^{k+1}},$$

for r small enough, and we are done. $\square$

6.1   The operator T

To find our desired holomorphic curve, we shall use, as announced, a certain operator acting on the space of maps u of order $k + 1 \ge 2$. We saw that, given a map $u(\cdot) = x^{k+1}\ell(\cdot)$, with $\ell\colon \Pi_r^i \to \mathbb{C}^{p-1}$, the iterates $\{x_n\}$ of $x_0 \in \Pi_r^i$ defined via

$$x_{j+1} = f_u(x_j) := f(x_j, u(x_j))$$

are well-defined for r sufficiently small. With this choice for u, the operator

$$Tu(x) = x^{kA} \sum_{n=0}^{\infty} x_n^{-kA} H(x_n, u(x_n)),$$

where A is the matrix associated to the non-degenerate characteristic direction we are studying,

$$H(x, u) := x^{kA}(w - w_1) = u - x^{kA}x_1^{-kA}u_1,$$

and $\{x_n\}$ is the sequence of the iterates of x under f_u, is well defined, since the series converges normally. We shall now restrict the space of definition of T, to obtain a contracting operator. In particular, we are going to search for positive constants r, C_0 and C_1 so that T is well defined on a closed subset of the Banach space of the maps of order $k + 1 \ge 2$.

We have the following analogous of Definition 4.7 of [Ha].

DEFINITION 6.5. *Let $k \in \mathbb{N} \setminus \{0\}$. Let $h, q \in \mathbb{N}$ be such that $hk \ge 3$ and $h \ge 1$, and let $r > 0$. For any $i = 1, \dots, k$, let $B_{h,q,r}^i$ be the space of maps $u\colon \Pi_r^i \to \mathbb{C}^{p-1}$, of the form $u(\cdot) = x^{kh-1}(\log x)^q t(\cdot)$ with t holomorphic and bounded. The space $B_{h,q,r}^i$ endowed with the norm $\|u\| = \|t\|_\infty$ is a Banach space.*

DEFINITION 6.6. *Let $k \in \mathbb{N} \setminus \{0\}$, and let $h, q \in \mathbb{N}$ be such that $hk \ge 3$ and $h \ge 1$. Let r, C_0 and C_1 be positive real constants and let $E_T^i(r, C_0, C_1) \subset B_{h,q,r}^i$ be the closed subset of $B_{h,q,r}^i$ given by the maps so that*

1. $\|u(x)\| \leq C_0 \, |x|^{kh-1} \, |\log |x||^q$, for any $x \in \Pi_r^i$;

2. $\|u'(x)\| \leq C_1 \, |x|^{kh-2} \, |\log |x||^q$, for any $x \in \Pi_r^i$.

Let T *be the operator defined as*

$$\mathrm{T}u(x) = x^{kA} \sum_{n=0}^{\infty} x_n^{-kA} H(x_n, u(x_n)), \qquad (6.11)$$

where A *is the matrix associated to the non-degenerate characteristic direction we are studying; as in (6.5) we have*

$$H(x, u) = x^{kA}(w - w_1) = u - x^{kA} x_1^{-kA} u_1,$$

and $\{x_n\}$ *is the sequence of the iterates of* x *under* f_u.

We shall devote the rest of the section to proving that the restriction of T to $E_T^i(r, C_0, C_1)$ is a continuous operator and a contraction. It will thus admit a unique fixed point u, and we shall prove that the unique fixed point is a solution of the functional equation (6.2).

We shall need the following reformulation of Lemma 4.1 of [Ha] for the case $k + 1 \geq 2$.

LEMMA 6.7. *Let* $\{\alpha_1, \alpha_2, \ldots, \alpha_{p-1}\}$ *be the directors of* A, *and let* $\lambda = \max_j \{\mathrm{Re}\, \alpha_j\}$. *If* $\varepsilon > 0$, *then for any* $x \in \Pi_r^i$, *with* r *small enough, we have*

$$\|x^{-kA}\| \leq |x|^{-k(\lambda+\varepsilon)}.$$

PROOF. We may assume without loss of generality that A is in Jordan normal form, i.e., $A = D + N$, where

$$D = \mathrm{Diag}(\alpha_1, \alpha_2, \ldots, \alpha_{p-1}), \ DN = ND, \ N^{p-1} = 0.$$

Since D and N commute, we have $x^{-kA} = x^{-k(D+N)} = x^{-kD} \exp(-kN \log x)$, and so we have the following estimate

$$\|x^{-kA}\| \leq \|x^{-kD}\| \|\exp(-kN \log x)\| \leq K \, |x|^{-k\lambda} \, |\log x|^{p-2} \leq |x|^{-k(\lambda+\varepsilon)}$$

for r small enough, and we are done. $\qquad \square$

REMARK 6.8. It follows from (6.10) that if $u \in B_{h,q,r}^i$, then the operator H verifies

$$H(x, u(x)) = O\left(x^{k(h+1)}(\log x)^{q+1}, x^{k(h+1)}(\log x)^{p_h}\right),$$

mapping $B_{h,q,r}^i$ into intself. We shall see that

$$(\mathrm{T}u)(x) = O\left(x^{kh-1}(\log x)^q\right),$$

for $q \geq p_h$.

We have the following generalization of Lemma 4.5 of [Ha] for the case $k + 1 \geq 2$.

LEMMA 6.9. *Let* T *be the operator defined in Definition 6.6. Let* $\lambda = \max_j\{\operatorname{Re}\alpha_j\}$, *where* $\alpha_1, \ldots, \alpha_{p-1}$ *are the directors of the non-degenerate characteristic direction* $[v]$, *and let* h *be an integer so that* $h > \lambda + \varepsilon$. *Let* p_h *be as in (6.6). Then, for* r *sufficiently small, there exists a constant* C_0 *so that, for any* u *satisfying*

$$\|u(x)\| \le C_0\,|x|^{hk-1}\,|\log|x||^{p_h}\,, \tag{6.12}$$

for each $x \in \Pi_r^i$, *we have that* Tu *satisfies the same inequality in* Π_r^i.

PROOF. By the definition, we have

$$Tu(x) = \sum_{n=0}^{\infty} \left(\frac{x_n}{x}\right)^{-kA} H(x_n, u(x_n)).$$

Thanks to Equation (6.6) we know that

$$H(x, u) = O\Big(\|u\|^2 x^k, \|u\| x^{k+1} \log x, x^{k(h+1)} (\log x)^{p_h}\Big)\,.$$

Therefore, there exist K_1, K_2, K_3 such that

$$\|H(x, u)\| \le K_1 \|u\|^2\,|x|^k + K_2 \|u\|\,|x|^{k+1}\,|\log x| + K_3\,|x|^{k(h+1)}\,|\log x|^{p_h}$$

in a neighborhood of 0. From the hypothesis $\|u(x)\| \le C_0\,|x|^{hk-1}\,|\log|x||^{p_h}$, it follows that for all $x \in \Pi_r^i$,

$$\|H(x, u(x))\| \le K\,|x|^{k(h+1)}\,|\log|x||^{p_h}\,,$$

with K not depending on C_0 provided that r is sufficiently small. Then we have

$$\|H(x_n, u(x_n))\| \le K\,|x_n|^{k(h+1)}\,|\log|x_n||^{p_h}$$

for $x \in \Pi_r^i$ and r small. By Lemma 6.7 we have

$$\left\|\left(\frac{x_n}{x}\right)^{-kA}\right\| \le \left|\frac{x_n}{x}\right|^{-k(\lambda+\varepsilon)}\,.$$

Applying all these inequalities to $Tu(x)$, and using Corollary 6.3 (note that $h > \lambda + \varepsilon$), we obtain

$$\|Tu(x)\| \le K \sum_{n=0}^{\infty} \left|\frac{x_n}{x}\right|^{-k(\lambda+\varepsilon)}\,|x_n|^{k(h+1)}\,|\log|x_n||^{p_h} \le K'\,|x|^{kh}\,|\log|x||^{p_h}$$

$$\le K''\,|x|^{kh-1}\,|\log|x||^{p_h}\,,$$

and we are done. $\square$

For our estimates we shall need the following technical result, generalizing Lemma 4.6 of [Ha] for the case $k + 1 \ge 2$.

LEMMA 6.10. *Let* T *be the operator defined as in Definition 6.6. Let* h, p_h, *and* C_0 *be as in Lemma 6.9. Then, for* r *sufficiently small, there exists a constant* C_1 *such that for any* u *satisfying (6.12) and*

$$\|u'(x)\| \le C_1\,|x|^{hk-2}\,|\log|x||^{p_h}\,, \tag{6.13}$$

for each $x \in \Pi_r^i$, *then* $(Tu)'$ *satisfies the same inequality in* Π_r^i.

PROOF. By the definition of T we have

$$Tu(x) = x^{kA} \sum_{n=0}^{\infty} (x_n)^{-kA} H(x_n, u(x_n)).$$

Then, differentiating, we obtain

$$\frac{d}{dx} Tu(x)$$

$$= \underbrace{\frac{d}{dx} x^{kA} \left(\sum_{n=0}^{\infty} x_n^{-kA} H(x_n, u(x_n)) \right)}_{S_1} + \underbrace{x^{kA} \sum_{n=0}^{\infty} \frac{\partial}{\partial u} \left(x_n^{-kA} H(x_n, u(x_n)) \right) \frac{du}{dx_n} \frac{dx_n}{dx}}_{S_2}$$

$$+ \underbrace{x^{kA} \sum_{n=0}^{\infty} \frac{\partial}{\partial x_n} \left(x_n^{-kA} H(x_n, u(x_n)) \right) \frac{dx_n}{dx}}_{S_3}.$$

We then have to estimate S_1, S_2, and S_3. Since

$$\frac{dx^{kA}}{dx} = kAx^{-1} x^{kA},$$

we have

$$S_1 = kAx^{-1} x^{kA} \left(\sum_{n=0}^{\infty} x_n^{-kA} H(x_n, u(x_n)) \right),$$

and thus, using the same inequalities as in the previous proof, we obtain

$$\|S_1\| \le \frac{k\|A\|}{|x|} C_0 |x|^{kh-1} |\log |x||^{p_h} = D_1 |x|^{kh-2} |\log |x||^{p_h},$$

where $D_1 = k\|A\|C_0$. For the second term, we have

$$S_2 = x^{kA} \sum_{n=0}^{\infty} x_n^{-kA} \frac{\partial H}{\partial u}(x_n, u(x_n)) \frac{du}{dx_n} \frac{dx_n}{dx}.$$

Since $kh \ge 3$, the hypotheses of Lemma 6.4 are satisfied; hence,

$$\left| \frac{dx_n}{dx} \right| \le 2 \left| \frac{x_n}{x} \right|^{k+1}.$$

Moreover, $H(x, u) = O\left(\|u\|^2 x^k, \|u\| x^{k+1} \log x, x^{k(h+1)} (\log x)^{p_h} \right)$ implies that there exist constants K_1 and K_2 so that

$$\left\| \frac{\partial}{\partial u} (H(x, u)) \right\| \le K_1 \|u\| |x|^k + K_2 |x|^{k+1} |\log |x||,$$

and our hypothesis gives that there is C_0 so that $\|u(x)\| \le C_0 |x|^{kh-1} |\log |x||^{p_h}$. Therefore,

$$\left\| \frac{\partial H}{\partial u}(x, u(x)) \right\| \le K_1 C_0 |x|^{kh+k-1} |\log |x||^{p_h} + K_2 |x|^{k+1} |\log |x|| \le C |x|^{k+1} |\log |x||$$

for some constant C, not depending on C_0. If C_1 is so that

$$\|u'(x)\| \leq C_1 |x|^{kh-2} |\log|x||^{p_h},$$

then

$$\left\|\frac{\partial H}{\partial u}(x_n, u(x_n))\frac{du(x_n)}{dx_n}\frac{dx_n}{dx}\right\| = \left\|\frac{\partial H}{\partial u}(x_n, u(x_n))\right\|\left\|\frac{du(x_n)}{dx_n}\right\|\left|\frac{dx_n}{dx}\right|$$

$$\leq 2CC_1 |x|^{-(k+1)} |x_n|^{2k+kh} |\log|x_n||^{p_h}.$$

Analogously to the proof of the previous result, $\left\|\left(\frac{x_n}{x}\right)^{-kA}\right\| \leq \left|\frac{x_n}{x}\right|^{-k(\lambda+\varepsilon)}$ and, by Corollary 6.3, we have

$$\|S_2\| \leq \sum_{n=0}^{\infty} 2CC_1 \left|\frac{x_n}{x}\right|^{-k(\lambda+\varepsilon)}\left|\frac{x_n}{x}\right|^{k+1} |x_n|^{kh+k-1} |\log|x_n||^{p_h+1}$$

$$\leq D_2 |x|^{kh-2} |\log|x||^{p_h},$$

with D_2 not depending on C_0 and C_1.

We are left with the third term

$$S_3 = x^{kA} \sum_{n=0}^{\infty} \frac{\partial G}{\partial x}(x_n, u(x_n))\frac{dx_n}{dx},$$

where $G(x, u) = x^{-kA}H(x, u)$, and hence

$$\frac{\partial G}{\partial x} = -\frac{kA}{x}x^{-kA}H(x, u) + x^{-kA}\frac{\partial H}{\partial x}(x, u).$$

With the same computations as before, using

$$H(x, u) = O\left(\|u\|^2 x^k, \|u\| x^{k+1}\log x, x^{k(h+1)}(\log x)^{p_h}\right)$$

and $\|u(x)\| \leq C_0 |x|^{kh-1} |\log|x||^{p_h}$, we have that there exist constants K_1, K_2 and K_3 so that

$$\left\|\frac{\partial H}{\partial x}\right\| \leq K_1\|u\|^2 |x|^{k-1} + K_2\|u\| |x|^k |\log x| + K_3 |x|^{k(h+1)-1} |\log|x||^{p_h}$$

and thus there exists C, depending of C_0, so that

$$\left\|x^{kA}\frac{\partial G}{\partial x}(x, u(x))\right\| \leq C |x|^{k(h+1)-1} |\log|x||^{p_h+1}.$$

Again using Corollary 6.3, we obtain

$$\|S_3\| \leq K_4 \sum_{n=0}^{\infty} \left|\frac{x_n}{x}\right|^{-k(\lambda+\varepsilon)+k+1} |x_n|^{k(h+1)-1} |\log|x_n||^{p_h+1}$$

$$\leq D_3 |x|^{kh-2} |\log|x||^{p_h},$$

with D_3 independent of C_0. Summing up, we obtain

$$\left\|\frac{d}{dx}Tu(x)\right\| \leq \|S_1\| + \|S_2\| + \|S_3\| \leq (D_1 + D_2 + D_2) |x|^{kh-2} |\log|x||^{p_h},$$

and setting $C_1 = D_1 + D_2 + D_3$, we conclude the proof. $\qquad\square$

The previous two lemmas prove that T is an endomorphism of $E_T^i(r, C_0, C_1)$. Now we have to prove that T is a contraction. We shall need the following reformulation of Lemma 4.9 of [Ha] for the case $k + 1 \geq 2$.

LEMMA 6.11. *Let* $u(\cdot) = x^{kh-1}(\log x)^{p_h} \ell_1(\cdot)$ *and* $v(\cdot) = x^{kh-1}(\log x)^{p_h} \ell_2(\cdot)$ *be in* $E_T^i(r, C_0, C_1)$ *and let* $\{x_n\}$ *and* $\{x_n'\}$ *be the iterates of x via* f_u *and* f_v. *Then there exists a constant K so that*

$$|x_n' - x_n| \leq K |x|^{kh} |\log |x||^{p_h} \|\ell_2 - \ell_1\|_\infty$$

for any n and r small enough.

PROOF. Let x and x' be in Π_r^i. We estimate

$$f_v(x') - f_u(x) = f(x', v(x')) - f(x, u(x)).$$

Thanks to (6.1), we can find constants a, b, c and $m(x, u)$ so that

$$\begin{cases} f_v(x') = x' - \frac{1}{k}(x')^{k+1} + (x')^{2k+1}(a + b \log x') + c(x')^{k+1} v(x') + m(x', v), \\ f_u(x) = x - \frac{1}{k}x^{k+1} + x^{2k+1}(a + b \log x) + cx^{k+1} u(x) + m(x, u). \end{cases}$$

Therefore, we have

$$f_v(x') - f_u(x) = (x' - x)\left[1 + \frac{1}{k}\sum_{i=0}^{k}(x')^i x^{k-i} + O\left(|(x'')|^{2k}\left|\log |x|''\right|\right)\right]$$
$$+ (v(x') - u(x))O\left(|x''|^{k+1}\right), \tag{6.14}$$

where $x'' = \max\{|x'|, |x|\}$. Lemma 6.2 implies $x_n^k \approx (x_n')^k \approx \frac{1}{n}$ as $n \to \infty$; then we can replace $|x''|^k$ with $|x|^k$. Moreover, since

$$v(x') = v(x) + O\left(x^{kh-2}(\log x)^{p_h}\right)(x' - x), \tag{6.15}$$

we obtain

$$v(x') - u(x) = v(x') - v(x) + v(x) - u(x)$$
$$= (x' - x)O\left(|x|^{kh-2}|\log |x||^{p_h}\right) + O\left(|x|^{kh-1}|\log |x||^{p_h}\right)\|\ell_2 - \ell_1\|_\infty.$$

Then, substituting in (6.14), we have

$$f_v(x') - f_u(x) = (x' - x)\left[1 - \frac{1}{k}\sum_{i=0}^{k}(x')^i x^{k-i} + O\left(|x|^{kh+k-1}|\log |x||, |x|^{2k}|\log |x||\right)\right]$$
$$+ O\left(|x|^{kh+k}|\log |x||^{p_h}\right)\|\ell_2 - \ell_1\|_\infty.$$

We are left with estimating $f_v(x') - f_u(x)$. For x and x' in Π_r^i and r small enough, we have

$$\left|1 - \frac{1}{k}\sum_{i=0}^{k}(x')^i x^{k-i} + O\left(|x|^{2k}|\log |x||\right)\right| = 1 + O\left(x^k\right) \leq 1.$$

Moreover, there exists a constant K such that

$$\left|f_v(x') - f_u(x)\right| \leq |x' - x| + K|x|^{kh+k}|\log|x||^{p_h}\|\ell_2 - \ell_1\|_\infty.$$

Iterating, we obtain

$$\left|f_v^n(x') - f_u^n(x)\right| \leq |x' - x| + K\sum_{i=0}^{n-1}|x_i|^{kh+k}|\log|x_i||^{p_h}\|\ell_2 - \ell_1\|_\infty,$$

for any n.

In particular, if $x = x'$, we have

$$\left|x'_n - x_n\right| \leq K\sum_{i=0}^{n-1}|x_i|^{kh+k}|\log|x_i||^{p_h}\|\ell_2 - \ell_1\|_\infty$$

$$\leq K'|x|^{kh}|\log|x||^{p_h}\|\ell_2 - \ell_1\|_\infty,$$

where we used Corollary 6.3 to deduce the last inequality, and we put

$$K' = KC_{k(h+1),p_h'}. \qquad\qquad\qquad\qquad \square$$

We now have all the ingredients to prove, as in [Ha, Proposition 4.8], that $T|_{E_T^i(r,C_0,C_1)}$ is a contraction.

PROPOSITION 6.12. *Let* T *be the operator defined in Definition 6.6. Then for* r *small enough,*

$$T|_{E_T^i(r,C_0,C_1)}\colon E_T^i(r,C_0,C_1) \to E_T^i(r,C_0,C_1)$$

is a contraction.

PROOF. We have to prove that given

$$u(\cdot) = x^{kh-1}(\log x)^{p_h}\ell_1(\cdot) \quad \text{and} \quad v(\cdot) = x^{kh-1}(\log x)^{p_h}\ell_2(\cdot)$$

in $E_T^i(r,C_0,C_1)$, we have

$$\|Tu - Tv\| \leq C\|u - v\|$$

with $C < 1$.

We have

$$Tu(x) - Tv(x) = x^{kA}\sum_{n=0}^{\infty}\left[x_n^{-kA}H(x_n,u(x_n)) - x_n'^{-kA}H(x_n',v(x_n'))\right];$$

hence,

$$Tu(x) - Tv(x) = \underbrace{x^{kA}\sum_{n=0}^{\infty}x_n^{-kA}\left[H(x_n,u(x_n)) - H(x_n',v(x_n'))\right]}_{S_1}$$

$$+ \underbrace{x^{kA}\sum_{n=0}^{\infty}\left[x_n^{-kA} - x_n'^{-kA}\right]H(x_n',v(x_n'))}_{S_2}.$$

For S_1, since $H(x, u) = O\big(\|u\|^2 x^k, \|u\|x^{k+1}\log x, x^{k(h+1)}(\log x)^{p_h}\big)$, for

$$u(x) = x^{kh-1}(\log x)^{p_h}\ell_1(x),$$

there exist $\alpha(x, u)$ and $\beta(x, u)$ holomorphic in the variables x, u and $x^k(\log x)^{p_h}$, so that

$$H(x, u) = ux^{k+1}(\log x)\alpha(x, u) + x^{k(h+1)}(\log x)^{p_h}\beta(x, u).$$

Therefore, by the inequalities in the proof of Lemma 6.10, we obtain

$$\big\|H(x_n, u(x_n)) - H(x'_n, v(x'_n))\big\|$$
$$\leq K\left[\left\|\frac{\partial H}{\partial x}(x_n, u(x_n))\right\||x_n - x'_n| + \left\|\frac{\partial H}{\partial u}(x_n, u(x_n))\right\|\,\|u(x_n) - v(x'_n)\|\right]$$
$$\leq K_1\left[\|u(x_n) - v(x'_n)\|\,|x_n|^{k+1}\,|\log|x_n|| + |x_n - x'_n|\,|x_n|^{k(h+1)-1}\,|\log|x_n||^{p_h}\right].$$

Arguing as in the proof of Lemma 6.11, thanks to (6.15), there exist constants A', B', and K_2 such that

$$\|v(x'_n) - u(x_n)\|$$
$$\leq A'\,|x'_n - x_n|\,|x_n|^{kh-2}\,|\log|x_n||^{p_h} + B'\,|x_n|^{kh-1}\,|\log|x_n||^{p_h}\,\|\ell_2 - \ell_1\|_\infty$$
$$\leq K_2\,|x_n|^{kh-2}\,|\log|x_n||^{p_h}\left[|x|^{kh}\,|\log|x||^{p_h} + |x_n|\right]\|\ell_2 - \ell_1\|_\infty,$$

where the last inequality follows from the previous lemma. Then

$$\|S_1\| \leq K_1\sum_{n=0}^{\infty}\left|\frac{x_n}{x}\right|^{-k(\lambda+\varepsilon)}\Big\{|x_n - x'_n|\,|x_n|^{k(h+1)-1}\,|\log|x_n||^{p_h}$$
$$+ K_2\,|x_n|^{k(h+1)-1}\,|\log|x_n||^{p_h+1}\left[|x|^{kh}\,|\log|x||^{p_h} + |x_n|\right]\|\ell_2 - \ell_1\|_\infty\Big\}.$$

Moreover, setting

$$\tilde{S} := |x_n - x'_n|\,|x_n|^{k(h+1)-1}\,|\log|x_n||^{p_h}$$
$$+ K_2\,|x_n|^{k(h+1)-1}\,|\log|x_n||^{p_h+1}\left[|x|^{kh}\,|\log|x||^{p_h} + |x_n|\right]\|\ell_2 - \ell_1\|_\infty,$$

we have

$$\tilde{S} \leq K'\,|x_n|^{k(h+1)-1}\,|\log|x_n||^{p_h}\,|x|^{kh}\,|\log|x||^{p_h}\,\|\ell_2 - \ell_1\|_\infty$$
$$+ K_2\,|x_n|^{k(h+1)-1}\,|\log|x_n||^{p_h+1}\,|x|^{kh}\,|\log|x||^{p_h}\,\|\ell_2 - \ell_1\|_\infty$$
$$+ K_2\,|x_n|^{k(h+1)}\,|\log|x_n||^{p_h+1}\,\|\ell_2 - \ell_1\|_\infty,$$

and applying Corollary 6.3,

$$\|S_1\| \leq C_1\,|x|^{2kh-1}\,|\log|x||^{2p_h}\,\|\ell_2 - \ell_1\|_\infty + C_2\,|x|^{2kh-1}\,|\log|x||^{2p_h+1}\,\|\ell_2 - \ell_1\|_\infty$$
$$+ C_3\,|x|^{kh}\,|\log|x||^{p_h+1}\,\|\ell_2 - \ell_1\|_\infty$$
$$\leq K_1\,|x|^{kh}\,|\log|x||^{p_h+1}\,\|\ell_2 - \ell_1\|_\infty.$$

We now consider S_2. We can write

$$x_n^{-kA} - x_n'^{-kA} = x_n^{-kA}\left(I - \exp\left(-A\log\frac{x'_n}{x_n}\right)\right).$$

Therefore,

$$\left\| \left(I - \exp\left(-kA \log \frac{x_n'}{x_n} \right) \right) H(x_n', v(x_n')) \right\|$$

$$\leq C \left\| kA \log \frac{x_n'}{x_n} \right\| \|H(x_n', v(x_n'))\|$$

$$\leq C' \frac{|x_n' - x_n|}{|x_n|} |x_n|^{k(h+1)} |\log |x_n||^{p_h}$$

$$\leq C'' x^{k(h+1)-1} |\log |x_n||^{p_h} |x|^{kh} |\log |x||^{p_h} \|\ell_2 - \ell_1\|_\infty.$$

By Corollary 6.3 we have

$$\|S_2\| \leq K_2 |x|^{2kh-1} |\log |x||^{2p_h} \|\ell_2 - \ell_1\|_\infty.$$

Thus, for r small enough, there exists K such that

$$\|Tu(x) - Tv(x)\| \leq K |x|^{kh} |\log |x||^{p_h} \|\ell_2 - \ell_1\|_\infty.$$

From the definition of the norm in $E_T^i(r, C_0, C_1)$, we have then that for r small enough, there is $c < 1$ such that

$$\|Tu - Tv\| \leq c \|u - v\|,$$

proving that $T|_{E_T^i(r, C_0, C_1)}$ is a contraction. $\square$

COROLLARY 6.13. *Let* T *be the operator defined in Definition 6.6. Then there exists* $u \colon \Pi_r^i \to \mathbb{C}^{p-1}$ *holomorphic and satisfying* (6.2).

PROOF. Thanks to the previous proposition, T is a contraction, and hence it has a unique fixed point $u \in E_T^i(r, C_0, C_1)$. It suffices to prove that this u satisfies (6.2). The definition of H gives us that $f(x,u)^{-kA} \Psi(x,u) = x^{-kA} u - x^{-kA} H(x,u)$, and, hence,

$$H(x, u(x)) = u(x) - x^{kA} x_1^{-kA} \Psi(x, u(x)).$$

We therefore obtain

$$Tu(x) = x^{kA} \sum_{n=0}^{\infty} x_n^{-kA} H(x_n, u(x_n))$$

$$= u(x) - x^{kA} x_1^{-kA} \Psi(x, u(x)) + x^{kA} x_1^{-kA} [u(x_1) - x_1^{kA} x_2^{-kA} \Psi(x_1, u(x_1))]$$

$$+ \cdots.$$

This implies that $Tu = u$ if and only if

$$-x^{kA} x_1^{-kA} [\Psi(x, u(x)) - u(x_1)] - x^{kA} x_2^{-kA} [\Psi(x_1, u(x_1)) - u(x_2)] + \cdots = 0,$$

i.e.,

$$\Psi(x_n, u(x_n)) = u(f(x_n, u(x_n))) \text{ for any } n \geq 0,$$

and this concludes the proof. $\square$

7 EXISTENCE OF ATTRACTING DOMAINS

In this section, we shall prove that given non-degenerate attracting characteristic direction $[v]$, it is possible to find not only a curve tangent to $[v]$, but also a open connected set, containing the origin on its boundary and so that each of its points is attracted by the origin tangentially to $[v]$, that is, the following generalization of Theorem 5.1 of [Ha] for the case $k + 1 \geq 2$.

THEOREM 7.1. *Let $F \in \mathrm{Diff}(\mathbb{C}^p, 0)$ be a tangent to the identity germ of order $k + 1 \geq 2$, and let $[v]$ be a non-degenerate characteristic direction. If $[v]$ is attracting, then there exist k parabolic invariant domains, where each point is attracted by the origin along a trajectory tangential to $[v]$.*

PROOF. Since $[v]$ is a non-degenerate characteristic direction, we can find $r, c > 0$ so that we can choose coordinates $(x, y) \in \mathbb{C} \times \mathbb{C}^{p-1}$ holomorphic in the sector

$$S_{r,c}^i = \left\{ (x, y) \in \mathbb{C} \times \mathbb{C}^{p-1} \mid x \in \Pi_r^i, \|y\| \leq c |x| \right\},$$

where Π_r^i is one of the connected components of $\mathbb{D}_r = \left\{ \left| x^k - r \right| < r \right\}$, so that, after the blow-up $y = ux$, F is of the form

$$\begin{cases} x_1 = f(x, u) = x - \frac{1}{k} x^{k+1} + O\left(\|u\| x^{k+1}, x^{2k+1} \log x \right), \\ u_1 = \Psi(x, u) = (I - x^k A) u + O\left(\|u\| x^{k+1} \log x, \|u\|^2 x^k \right). \end{cases}$$

In particular, after the blow-up, $\|u\| \leq c$.

Without loss of generality, we may assume that A is in Jordan normal form. Let $\{\alpha_1, \ldots, \alpha_{p-1}\}$ be the eigenvalues of A. Thanks to the hyptheseis, we have

$$\mathrm{Re}\, \alpha_j > 0, \quad j = 1, \ldots, p - 1,$$

and hence there exists a constant $\lambda > 0$ so that $\mathrm{Re}\, \alpha_j > \lambda$ for all $j = 1, \ldots, p - 1$. We can also assume that the elements off the diagonal in the Jordan blocks are all equal ε, with $\varepsilon < \lambda$.

We shall now restrict our sectorial domain to obtain good estimates for x_1 and u_1. We define, for $j = 1, \ldots, p - 1$,

$$\Delta_j := \{ x \in \mathbb{C} \mid \left| 1 - \alpha_j x^k \right| \leq 1 \}.$$

Consider the sector

$$S_{\gamma, \rho} := \{ x \in \mathbb{C} \mid |\mathrm{Im}\, x| \leq \gamma \mathrm{Re}\, x, |x| \leq \rho \};$$

Since $\mathrm{Re}\, \alpha_j > 0$, there exist positive constants γ and ρ so that, setting for each $i = 1, \ldots, k$,

$$S_{\gamma, \rho}^i := \{ x \in \Pi_r^i \mid x^k \in S_{\gamma, \rho} \},$$

we have

$$S_{\gamma, \rho}^i \subset \bigcap_{j=1}^{p-1} \Delta_j \cap \overline{\mathbb{D}}_r \subset \Pi_r^i.$$

We want to check that, for any $i = 1, \ldots, k$, the k sets

$$A_{\gamma, \rho, c}^i := \{ (x, u) \in \mathbb{C} \times \mathbb{C}^{p-1} \mid x \in S_{\gamma, \rho}^i, \|u\| \leq c \}$$

are invariant attractive domains.

Recalling that there is K so that

$$\|u_1 - (I - x^k A)u\| \leq K(\|u\|\,|x|^{k+1}\,|\log|x|| + \|u\|^2\,|x|^k),$$

for $(x, u) \in A^i_{\gamma,\rho,c}$, we have

$$\|u_1\| \leq \|(I - x^k A)u\| + K\|u\|\,|x|^k\,(|x|\,|\log|x|| + \|u\|),$$

and, provided that γ, ρ and c are small enough,

$$\|u_1\| \leq \|u\|\|I - x^k A\| \leq \|u\|(1 - \lambda\,|x|^k) \leq \|u\|, \tag{7.1}$$

where we used that

$$\|I - x^k A\| \leq \max_j \left|1 - \alpha_j x^k\right| + |x|^k \varepsilon \leq 1 - (\lambda + \varepsilon')|x^k| + \varepsilon|x^k|.$$

Therefore, $\|u_1\| \leq c$.

To estimate x_1, since we know that $x_1 = x - \frac{1}{k}x^{k+1} + O\big(\|u\|x^{k+1}, x^{2k+1}\log x\big)$, we have

$$\frac{1}{x_1^k} = \frac{1}{x^k} + 1 + O\big(\|u\|, x^k \log x\big). \tag{7.2}$$

Therefore, there is $\tilde{C}$, not depending on u, so that

$$\left|\frac{1}{x_1^k} - \frac{1}{x^k} - 1\right| \leq \tilde{C}\|u\| + K\,|x|^k\,|\log|x|| \leq \tilde{C}c + K\,|x|^k\,|\log|x||. \tag{7.3}$$

We shall use this last inequality to prove that $A^i_{\gamma,\rho,c}$ is an invariant domain. In particular, it suffices to check

$$\begin{cases} \|u_1\| \leq c, \\ \left|\mathrm{Im}\,x_1^k\right| \leq \gamma\mathrm{Re}\,x_1^k, \\ \left|x_1^k\right| \leq \rho. \end{cases}$$

We already estimated u_1 in (7.1). On the other hand, to prove that $S^i_{\gamma,\rho,c}$ is f-invariant, it suffices to prove that, for u small enough,

$$(S^i_{\gamma,\rho,c})^* = \{x \in \mathbb{C} \mid \frac{1}{x} \in S_{\gamma,\rho}\}$$

is $1/(f)^k$-invariant, which follows from (7.3) using the same argument as in the proof of Leau-Fatou flower theorem.

To finish, it remains to check that, given a point $(x, u) \in S^i_{\gamma,\rho,c}$ its iterates converge to the origin along the direction $[1 : 0]$. We shall first show that $x_n^k \approx \frac{1}{n}$ and $\|u_n\| \leq C\frac{1}{n^\lambda}$, for any fixed $0 < \lambda < \max_j \mathrm{Re}\,\alpha_j$. It follows from (7.2) that

$$\frac{1}{x_n^k} = \frac{1}{x^k} + n + \sum_{i=0}^{n-1} O\big(\|u_i\|, x_i^k \log x_i\big)$$

and, hence,

$$\frac{1}{nx_n^k} = \frac{1}{nx^k} + 1 + \frac{1}{n}\sum_{i=0}^{n-1} O\Big(\|u_i\|, x_i^k \log x_i\Big),$$

where the sum is bounded. Therefore,

$$\frac{1}{nx_n^k} = O(1),$$

yielding

$$x_n^k \approx \frac{1}{n}.$$

Finally, take $\mu < \lambda$ (where λ is the positive constant so that $\max_j \operatorname{Re} \alpha_j > \lambda$). Then

$$x_1^{-k\mu} = x^{-k\mu}\left[1 - \frac{1}{k}x^k + O\Big(x^{2k}\log x, \|u\|x^k\Big)\right]^{-k\mu}$$
$$= x^{-k\mu}\left[1 + \mu x^k + O\Big(x^{2k}\log x, \|u\|x^k\Big)\right]$$

and, hence,

$$|x_1|^{-k\mu} \leq |x|^{-k\mu}\left|1 + \mu x^k + O\Big(\|u\|x^k, x^{2k}\log x\Big)\right| \leq |x|^{-k\mu}\left(1 + \lambda |x|^k\right).$$

It thus follows that

$$\|u_1\|\,|x_1|^{-k\mu} \leq \|u\|(1 - \lambda |x|^k)\,|x|^{-k\mu}\left(1 + \lambda |x|^k\right) = \|u\|\,|x|^{-k\mu}\left(1 - \lambda^2 |x|^{2k}\right)$$
$$< \|u\|\,|x|^{-k\mu}.$$

Therefore, there exists C so that

$$\|u_n\|\,|x_n|^{-k\mu} < \|u\|\,|x|^{-k\mu} \leq C,$$

implying

$$\|u_n\| \leq C\,|x_n|^{k\mu}.$$

Then, $\|u_n\| = O\big(1/n^{k\lambda}\big)$. This shows that each $(x,u) \in A^i_{\gamma,\rho,c}$ converges to the origin along the direction $[1:0]$. $\qquad\square$

8 PARABOLIC MANIFOLDS

Let $\Phi \in \mathrm{Diff}(\mathbb{C}^p, 0)$ be a tangent to the identity germ of order $k+1 \geq 2$, and let $[V] = [1:0]$ be a non-degenerate characteristic direction. We can divide the set of the directors of $[V]$ into two sets: the *attracting directors*, i.e., the set $\{\lambda_1,\ldots,\lambda_a\}$ with $\operatorname{Re}\lambda_j > 0$ for $j = 1,\ldots,a$, and the *non-attracting directors*, i.e., the set $\{\mu,\ldots,\mu_b\}$ with $\operatorname{Re}\mu_h \leq 0$ for $h = 1,\ldots,b$. Let d_j be the multiplicity of λ_j for $j = 1,\ldots,a$ and let $d := d_1 + \cdots + d_a$. We know that, after the blow-up, we can assume that Φ is of the form

$$\begin{cases} x_1 = f(x,u,v) = x - \frac{1}{k}x^{k+1} + F(x,u,v), \\ u_1 = g(x,u,v) = (I_d - x^k A)u + G(x,u,v), \\ v_1 = h(x,u,v) = (I_l - x^k B)v + H(x,u,v), \end{cases} \qquad (8.1)$$

where A is the $d \times d$ matrix in Jordan normal form associated to the attracting directors, B is the $l \times l$ matrix in Jordan normal form associated to the non-attracting directors (where $l := p - d - 1$), and with F, G, H so that

$$
\begin{cases}
F(x,u,v) = O\big(\|(u,v)\|x^{k+1}, x^{2k+1}\log x\big), \\
G(x,u,v) = O\big(\|(u,v)\|x^{k+1}\log x, \|(u,v)\|^2 x^k\big), \\
H(x,u,v) = O\big(\|(u,v)\|x^{k+1}\log x, \|(u,v)\|^2 x^k\big).
\end{cases}
\tag{8.2}
$$

Moreover F, G, H are holomorphic in an open set of the form

$$
\Delta_{r,\rho} = \Big\{ (x,u,v) \in \mathbb{C} \times \mathbb{C}^d \times \mathbb{C}^{p-d-1} \ \Big|\ |x^k - r| < r,\ \|(u,v)\| < \rho \Big\},
$$

and, therefore, also in the set

$$
S_{\gamma,s,\rho} := \Big\{ (x,U) \in \mathbb{C} \times \mathbb{C}^{p-1} \ \Big|\ |\mathrm{Im}\, x^k| \le \gamma \,\mathrm{Re}\, x^k,\ |x^k| < s,\ \|U\| < \rho \Big\} \subset \Delta_{r,\rho}.
$$

In the next result, the analogous of Proposition 2.2 of [Ha2], we shall see that it is possible to further modify the last $p - d - 1$ components of Φ.

PROPOSITION 8.1. *Let* $\Phi \in \mathrm{Diff}(\mathbb{C}^p, 0)$ *be a tangent to the identity germ of order* $k + 1 \ge 2$ *as in* (8.1), *with* $[V] = [1 : 0]$ *non-degenerate characteristic direction so that the matrix* $A(v) = \mathrm{Diag}(A,B)$ *satisfies*

$$
\begin{aligned}
&\mathrm{Re}\,\lambda_j > \alpha > 0, \quad \text{for any } \lambda_j \text{ eigenvalue of } A, \\
&\mathrm{Re}\,\mu_j \le 0, \qquad\;\; \text{for any } \mu_j \text{ eigenvalue of } B.
\end{aligned}
$$

Then, for any choice of $N, m \ge 2$, *it is possible to choose coordinates* (x,u,v) *in* $\Delta_{r,\rho}$, *with* H *satisfying*

$$
H(x,u,0) = O\big(|x|^k \|u\|^m + |x|^N \|u\|\big).
$$

PROOF. Thanks to (8.2), it is possible to write $H(x,u,v)$ in a more convenient form. Indeed, for any $N \in \mathbb{N}$, we have

$$
H(x,u,0) = \sum_{k \le s \le N,\, t \in E_s} c_{s,t}(u) x^s (\log x)^t + O\big(\|u\|\, |x|^N |\log|x||^{h_N}\big),
\tag{8.3}
$$

for some $h_N \in \mathbb{N}$ depending on N, where for any s we define E_s as the (finite) set of integers t so that the series or Equation (8.3) contains the term $x^s(\log x)^t$ and where $c_{s,t}(u)$ are holomorphic in $\|u\| \le \rho$ and $c_{s,t}(0) \equiv 0$. We shall prove by induction on s, t and the order of $c_{s,t}(u)$ that, if $s \le N$, using changes of coordinates of the form $\tilde{v} = v - \varphi(x,u)$, it is possible to obtain $c_{s,t}$ of order at least m. We shall need the following reformulation of Lemma 2.3 of [Ha2] for the case $k + 1 \ge 2$.

LEMMA 8.2. *Let* $\Phi \in \mathrm{Diff}(\mathbb{C}^p, 0)$ *be a tangent to the identity germ of order* $k + 1 \ge 2$ *as in* (8.1), *with* $[V] = [1 : 0]$ *so that* $A(v) = \mathrm{Diag}(A,B)$ *satisfies*

$$
\begin{aligned}
&\mathrm{Re}\,\lambda_j > \alpha > 0, \quad \text{for any } \lambda_j \text{ eigenvalue of } A, \\
&\mathrm{Re}\,\mu_j \le 0, \qquad\;\; \text{for any } \mu_j \text{ eigenvalue of } B.
\end{aligned}
$$

Let H *be so that* (8.3) *holds, let* $\bar{s}$ *be the smallest integer in* (8.3), *and let* $m \ge 2$; *for such an* $\bar{s}$, *let* $\bar{t}$ *be the greatest integer in* $E_{\bar{s}}$ *so that* $c_{\bar{s},\bar{t}}$ *has order* $\bar{d}$ *less than* m. *Then there* *exists a polynomial map $P(u)$, homogeneous of degree $\bar{d}$, with values in $\mathbf{C}^l$, such that, after changing v in*

$$\tilde{v} = v - x^{\bar{s}-k}(\log x)^{\bar{t}}P(u),$$

$c_{\bar{s},\bar{t}}(u)$ *has order greater than* $\bar{d}$.

PROOF. Since $c_{\bar{s},\bar{t}}(u)$ has order $\bar{d}$, we can write

$$c_{\bar{s},\bar{t}}(u) = Q(u) + O\left(\|u\|^{\bar{d}+1}\right),$$

where $Q(u)$ is a homogeneous polynomial of degree $\bar{d}$ and takes values in $\mathbf{C}^l$. Moreover, the term $c_{\bar{s},\bar{t}}(u)x^{\bar{s}}(\log x)^{\bar{t}}$ in (8.3) is

$$H(x,u,0) = c_{\bar{s},\bar{t}}(u)x^{\bar{s}}(\log x)^{\bar{t}} + \sum_{\substack{k \le s \le N, \\ t \in E_s, (s,t) \ne (\bar{s},\bar{t})}} c_{s,t}(u)x^s(\log x)^t + O\left(\|u\|\,|x|^N\,|\log|x||^{h_N}\right).$$

(8.4)

Using a change of coordinates of the form

$$\tilde{v} = v - x^{\bar{s}-k}(\log x)^{\bar{t}}P(u),$$

with $P(u)$ homogeneous polynomial, we have

$$\begin{aligned}
\tilde{v}_1 &= v_1 - x_1^{\bar{s}-k}(\log x_1)^{\bar{t}}P(u_1) \\
&= (I_l - x^k B)(\tilde{v} + x^{\bar{s}-k}(\log x)^{\bar{t}}P(u)) + H(x,u,v) - x_1^{\bar{s}-k}(\log x_1)^{\bar{t}}P(u_1) \\
&= (I_l - x^k B)\tilde{v} + \tilde{H}(x,u,\tilde{v}),
\end{aligned}$$

where

$$\begin{aligned}
\tilde{H}(x,u,\tilde{v}) &:= (I_l - x^k B)x^{\bar{s}-k}(\log x)^{\bar{t}}P(u) + H(x,u,\tilde{v} + x^{\bar{s}-k}(\log x)^{\bar{t}}P(u)) \\
&\quad - x_1^{\bar{s}-k}(\log x_1)^{\bar{t}}P(u_1).
\end{aligned}$$

Expanding $\tilde{H}(x,u,0)$ we obtain

$$\begin{aligned}
\tilde{H}(x,u,0) &= x^{\bar{s}-k}(\log x)^{\bar{t}}P(u) - Bx^{\bar{s}}(\log x)^{\bar{t}}P(u) + H(x,u,x^{\bar{s}-k}(\log x)^{\bar{t}}P(u)) \\
&\quad - x_1^{\bar{s}-k}(\log x_1)^{\bar{t}}P(u_1).
\end{aligned}$$

(8.5)

We have

$$\begin{aligned}
&H(x,u,x^{\bar{s}-k}(\log x)^{\bar{t}}P(u)) \\
&= Q(u)x^{\bar{s}}(\log x)^{\bar{t}} + O\left(\|u\|^{\bar{d}+1}x^{\bar{s}}(\log x)^{\bar{t}}, \|u\|^{\bar{d}}x^{\bar{s}}(\log x)^{\bar{t}-1}, \|u\|\,|x|^N\,|\log|x||^{h_N}\right),
\end{aligned}$$

and

$$\begin{aligned}
&x_1^{\bar{s}-k}(\log x_1)^{\bar{t}}P(u_1) \\
&= \left[x^{\bar{s}-k} - \frac{\bar{s}-k}{k}x^{\bar{s}} + O\left(\|u\|x^{\bar{s}}, x^{\bar{s}+k}\log x\right)\right](\log x)^{\bar{t}}P(u_1) \\
&\quad + O\left(x^{\bar{s}}(\log x)^{\bar{t}}\right)P(u_1) \\
&= x^{\bar{s}-k}(\log x)^{\bar{t}}P(u) - x^{\bar{s}-k}(\log x)^{\bar{t}}\langle\operatorname{grad} P; x^k Au\rangle - \frac{\bar{s}-k}{k}x^{\bar{s}}(\log x)^{\bar{t}}P(u) \\
&\quad + O\left(x^s(\log x)^{\bar{t}}, \|u\|x^{\bar{s}}(\log x)^t, x^{\bar{s}+k}(\log x)P(u_1)\right),
\end{aligned}$$

where we used

$$P(u_1) = P((I_d - x^k A)u) + O(x^{\bar{s}})$$
$$= P(u) + \langle \text{grad } P, -x^k Au \rangle + O(x^{2k}, x^{\bar{s}}).$$

It is then clear that the terms of order $\bar{s} - k$ in (8.5) cancel each other, whereas we can put in evidence the terms of order $\bar{s}$ in x and of order $\bar{d}$ in u. In particular, the l homogeneous polynomials of degree $\bar{d}$ of $\tilde{c}_{\bar{s},\bar{t}}$ in (8.5) vanish identically if and only if P satisfies the following l equations

$$\langle \text{grad } P_i, Au \rangle - \left(\left(B - \frac{\bar{s} - k}{k} I_l \right) P(u) \right)_i = -Q_i(u) \quad i = 1, \dots, l. \qquad (8.6)$$

These equations form a square linear system in the coefficients of P. Therefore, to prove that such a system has a solution it suffices to prove that

$$\langle \text{grad } P_i, Au \rangle - \left(\left(B - \frac{\bar{s} - k}{k} I_l \right) P(u) \right)_i = 0 \quad i = 1, \dots, l \implies P = 0. \qquad (8.7)$$

Moreover, since B is in Jordan normal form, if we denote by $\varepsilon_{i,i+1}$ the elements out of the diagonal, we can rewrite the previous equation as

$$\frac{\partial P_i}{\partial u_1}(Au)_1 + \cdots + \frac{\partial P_i}{\partial u_q}(Au)_q - \left(\mu_i - \frac{\bar{s} - k}{k} \right) P_i - \varepsilon_{i,i+1} P_{i+1} = 0, \qquad (8.8)$$

recalling that, for any $1 \le i < l$, we have $\varepsilon_{i,i+1} = 0$ or 1, and $\varepsilon_{l,l+1} = 0$. Therefore, arguing by decreasing induction over i from l to 1, we reduce ourselves to solve

$$\frac{\partial R}{\partial u_1}(Au)_1 + \cdots + \frac{\partial R}{\partial u_d}(Au)_d - \left(\mu_i - \frac{\bar{s} - k}{k} \right) R = 0 \implies R = 0, \qquad (8.9)$$

for a homogeneous polynomial R of degree $\bar{d}$. By Euler formula, we know that

$$R = \bar{d}^{-1} \left[\frac{\partial R}{\partial u_1} u_1 + \cdots + \frac{\partial R}{\partial u_d} u_d \right].$$

We can, therefore, reduce ourselves to solve

$$\frac{\partial R}{\partial u_1}(C_i u)_1 + \cdots + \frac{\partial R}{\partial u_d}(C_i u)_d = 0 \implies R = 0, \qquad (8.10)$$

where $C_i = A - (\mu_i - \bar{s} + k)\bar{d}^{-1} I_d$ is invertible, since from our hypotheses $\text{Re}\,(\alpha - \frac{\mu_i - \bar{s} + k}{\bar{d}}) > 0$. We prove (8.10) with a double induction, on the dimension d and on the degree $\bar{d}$ of R. For any degree $\bar{d}$, if $d = 1$, then there exists a constant K_i so that $R = K_i u_1^{\bar{d}}$; then, since $\alpha - \frac{\mu_i - \bar{s} + k}{\bar{d}} \neq 0$, we have

$$\frac{\partial R}{\partial u_1} \left(\alpha - \frac{\mu_i - \bar{s} + k}{\bar{d}} \right) u_1 = 0 \implies \bar{d} K_i u_1^{\bar{d}} = 0 \implies K_i = 0,$$

implying $R = 0$. Similarly, for any dimension d, if $\bar{d} = 1$, then there exist constants $a_1, \dots, a_d$ so that $R = a_1 u_1 + \cdots + a_d u_d$; hence,

$$a_1(C_i u)_1 + \cdots + a_d(C_i u)_d = 0 \implies a_1 = \cdots = a_d = 0 \implies R = 0.$$

Assume, by inductive hypothesis, that (8.10) holds for any pair $(d-1, \bar{d})$ and $(d, \bar{d}-1)$, with $d > 1$ and $\bar{d} > 1$, and we shall prove that (8.10) holds also for $(d, \bar{d})$. Assume that $\langle \text{grad}\, R, Cu \rangle = 0$ for a certain homogeneous polynomial R of degree $\bar{d}$ in q variables. By inductive hypothesis, setting

$$\tilde{R}(u_1, \ldots, u_{d-1}) := R(u_1, \ldots, u_{d-1}, 0),$$

we have

$$\langle \text{grad}\, \tilde{R}, C \cdot (u_1, \ldots, u_{d-1}, 0) \rangle = 0 \Longrightarrow \tilde{R} = 0,$$

and so $R(u) = u_d S(u)$, with S homogeneous polynomial of degree $\bar{d} - 1$ in d variables. Therefore,

$$\frac{\langle \text{grad}\, R, Cu \rangle}{u_d} = 0 \Longrightarrow \langle \text{grad}\, S, Cu \rangle + \left(\lambda_d - \frac{\mu_i - \bar{s} + k}{\bar{d}} \right) S = 0.$$

Again, by Euler's formula, we can then write

$$\langle \text{grad}\, S, C'u \rangle = 0,$$

with $C' = C + \frac{\lambda_d - (\mu_i - \bar{s} + k)/\bar{d}}{\bar{d}} I_d$, and applying the inductive hypothesis, we obtain $S = 0$, and thus $R = 0$. □

We shall now apply the previous lemma, for the integers s and t, until $c_{s,t}(u)$ has order at least m. Then either $E_s = \varnothing$, or the greatest t' in E_s is less than t. In this last case, we can again apply the lemma, with integers s and t', until $E_s = \varnothing$. We can then apply the lemma with $s + 1$ instead of s, until we have $s + 1 = N$. This proves the proposition. □

We shall prove, analogously to the way we found a parabolic curve, that we can find parabolic manifolds as fixed points of a certain operator between spaces of functions, proving the following generalization of Theorem 1.6 of [Ha2] for the case $k + 1 \geq 2$.

THEOREM 8.3. *Let $\Phi \in \text{Diff}(\mathbb{C}^p, 0)$ be a tangent to the identity germ of order $k + 1 \geq 2$. Let $[V]$ be a non-degenerate characteristic direction and let $A = A(V)$ be its associated matrix. If A has exactly d eigenvalues, counted with multiplicity, with strictly positive real parts, then there exists a parabolic manifold of dimension $d + 1$, with 0 on its boundary and tangent to $\mathbb{C}V \oplus E$ in 0, where E is the eigenspace associated to the attracting directors, and so that each of its points is attracted to the origin along the direction $[V]$. Moreover, it is possible to find coordinates (x, u, v) in a sector of $\mathbb{C} \times \mathbb{C}^d \times \mathbb{C}^{p-d-1}$ so that the parabolic manifold is locally defined by $\{v = 0\}$.*

PROOF. We may assume that Φ is of the form (8.1), with $[V] = [1 : 0]$ so that $A(v) = \text{Diag}(A, B)$ satisfies

$$\text{Re}\, \lambda_j > \alpha > 0, \quad \text{for any } \lambda_j \text{ eigenvalue of } A,$$
$$\text{Re}\, \mu_j \leq 0, \qquad \text{for any } \mu_j \text{ eigenvalue of } B,$$

and

$$H(x, u, 0) = O(|x|^k \|u\|^m + |x|^N \|u\|),$$

with $m, N > 0$.

We shall search for $\phi(x, u)$, holomorphic in a sector

$$S_{\gamma, s, \rho} = \{(x, u) \in \mathbb{C} \times \mathbb{C}^d \mid |\mathrm{Im}\, x^k| \leq \gamma \mathrm{Re}\, x^k, |x| \leq s, \|u\| \leq \rho\}, \qquad (8.11)$$

so that, for

$$\begin{cases} x_1^\phi = f(x, u, \phi(x, u)), \\ u_1^\phi = g(x, u, \phi(x, u)), \end{cases}$$

we have

$$\phi(x_1^\phi, u_1^\phi) = h(x, u, \phi(x, u)). \qquad (8.12)$$

Repeating the same changes of coordinates performed in the Section 6, we first transform v_1, for $\mathrm{Re}\, x > 0$ by setting

$$w = x^{-kB} v,$$

and we define H_1 as

$$w - w_1 = x^{-kB} H_1(x, u, v).$$

From the definitions of x_1 and u_1 in (8.1), we have

$$x_1^{-kB} = x^{-kB} \left[\left(I + x^k B \right) + O\left(\|u\| x^k, x^{2k} \log x \right) \right],$$

and

$$\begin{aligned} w_1 &= x^{-kB} \left[\left(I + x^k B \right) + O\left(\|u\| x^k, x^{2k} \log x \right) \right] \left[(I - x^k B) x^{kB} w + H(x, u, v) \right] \\ &= \left(I + O\left(\|u\| x^k, x^{2k} \log x \right) \right) w + x^{-kB} \left(I + O(x^k) \right) H(x, u, v). \end{aligned}$$

Hence, $H_1(x, u, v)$ satisfies the same estimates as $H(x, u, v)$:

$$H_1(x, u, v) = O\left(\|u\|^2 x^k, \|u\| x^{k+1} \log x \right)$$

and

$$H_1(x, u, 0) = O\left(|x|^k \|u\|^m + |x|^N \|u\| \right). \qquad (8.13)$$

Therefore, (8.12) is equivalent to

$$x^{-kB} \phi(x, u) - x_1^{-kB} \phi(x_1^\phi, u_1^\phi) = x^{-kB} H_1(x, u, \phi(x, u)). \qquad (8.14)$$

Operator T. Let $\{(x_n, u_n)\}$ be the iterates defined by

$$\begin{cases} x_1^\phi = f(x, u, \phi(x, u)) = x - \dfrac{x^{k+1}}{k} + F(x, u, \phi(x, u)), \\ u_1^\phi = g(x, u, \phi(x, u)) = \left(I_d - x^k A \right) u + G(x, u, \phi(x, u)), \end{cases}$$

with f and g as in (8.1), and ϕ holomorphic from the sector $S_{\gamma, s, \rho}$, defined in (8.11), to $\mathbb{C}^{p-d-1}$. Now we consider the operator

$$T\phi(x, u) := x^{kB} \sum_{n=0}^\infty x_n^{-kB} H_1(x_n, u_n, \phi(x_n, u_n)).$$

We shall prove that this operator, restricted to a suitable closed subset $\mathcal{F}$ of the Banach space of bounded holomorphic maps $\phi\colon S_{\gamma,s,\rho} \to \mathbb{C}^{p-d-1}$, is a contraction. Then there exists a unique fixed point in $\mathcal{F}$, and, by the definition of T, such a fixed point will be a solution of (8.14).

We shall proceed as follows:

1. We shall prove that there exists a constant $K_0 > 0$ such that

$$\|\phi(x,u)\| \leq K_0 \left(\|u\|^m + |x|^{N-k} \|u\| \right)$$
$$\implies \|T\phi(x,u)\| \leq K_0 \left(\|u\|^m + |x|^{N-k} \|u\| \right). \tag{8.15}$$

2. We shall prove that if $K_0 > 0$ satisfies (8.15), then there exist positive constants K_1 and K_2 such that

$$\begin{cases} \left| \frac{\partial \phi}{\partial x} \right| \leq K_1 (\|u\|^m |x|^{-1} + \|u\| |x|^{N-k-1}), \\ \left| \frac{\partial \phi}{\partial u} \right| \leq K_2 (\|u\|^{m-1} + |x|^{N-k}), \end{cases}$$
$$\implies \begin{cases} \left| \frac{\partial T\phi}{\partial x} \right| \leq K_1 (\|u\|^m |x|^{-1} + \|u\| |x|^{N-k-1}), \\ \left| \frac{\partial T\phi}{\partial u} \right| \leq K_2 (\|u\|^{m-1} + |x|^{N-k}). \end{cases} \tag{8.16}$$

3. Considering the Banach space $(F_0, \|\cdot\|_0)$ defined as

$$F_0 = \{ \phi\colon S_{\gamma,s,\rho} \to \mathbb{C}^{p-d-1} \mid \|\phi\|_0 < +\infty \},$$

with the norm

$$\|\phi\|_0 := \sup_{x,u} \left\{ \frac{\|\phi(x,u)\|}{\|u\|^m + |x|^{N-k} \|u\|} \right\},$$

we shall prove that the subset $\mathcal{F}$ of F_0, given by the maps ϕ satisfying (8.15) and (8.16) with the constants K_0, K_1 and K_2 we found in (1) and (2) is closed.

4. We shall finally show that T is a contraction.

We first prove the following analogous of Proposition 3.2 of [Ha2].

PROPOSITION 8.4. *If m and N are integers so that H_1 satisfies (8.13), then there exists a positive constant K_0 such that, if*

$$\|\phi(x,u)\| \leq K_0 \left(\|u\|^m + |x|^{N-k} \|u\| \right), \tag{8.17}$$

then

1. *the series defining the operator T is uniformly convergent in*

$$S_{\gamma,s,\rho} \cap \{ (x,u) \in \mathbb{C}^p \mid \|u\| \, |x|^{-k\alpha} \leq 1 \};$$

2. *also $\|T\phi(x,u)\|$ satisfies the same inequality*

$$\|T\phi(x,u)\| \leq K_0 \left(\|u\|^m + |x|^{N-k} \|u\| \right).$$

PROOF. Since all the eigenvalues of A have strictly positive real parts, as we saw in Theorem 7.1, for any $(x, u) \in S_{\gamma, s, \rho}$ we have

$$\lim_{n \to \infty} \|u_n\| \, |x_n|^{-k\alpha} = 0,$$

where $\alpha > 0$ is strictly less then the real parts of the eigenvalues of A. Therefore, without loss of generality, we may assume that $\|u\| \, |x|^{-k\alpha}$ is bounded by 1. Let $\beta < k\alpha$ be a positive real number so that each eigenvalue μ_j of B satisfies $\operatorname{Re}\mu_j < \beta$. By Lemma 6.7, this implies that there exists a constant $C_1 > 0$ so that

$$\|x^{kB} x_n^{-kB}\| \leq C_1 \left| \frac{x_n}{x} \right|^{-k\beta}.$$

Moreover, choosing γ, s, ρ small enough, if $(x, u) \in S_{\gamma, s, \rho}$, then

$$\left| x_n^k \right| \leq \frac{2}{n}, \quad \|u_n\| \leq \|u\| \, |x|^{-k\alpha} \, |x_n|^{k\alpha}.$$

By the hypotheses on $H_1(x, u, v)$, there exist positive constants K_1 and K_2 so that

$$\begin{aligned}
\|H_1(x, u, v)\| \leq & K_1 \left(\|u\|^m |x|^k + |x|^N \|u\| \right) \\
& + K_2 \left(\|v\| \, |x|^{k+1} \, |\log|x||^q + \|v\|^2 \, |x|^k + \|u\| \|v\| \, |x|^k \right)
\end{aligned} \tag{8.18}$$

for a certain $q \in \mathbb{N}$.

Let us assume that $\|\phi(x, u)\| \leq K \left(\|u\|^m + |x|^{N-k} \|u\| \right)$ for a constant $K > 0$. For $v = \phi(x, u)$, we have

$$\begin{aligned}
\|v\| \, |x|^{k+1} & |\log|x||^q + \|v\|^2 \, |x|^k + \|u\| \|v\| \, |x|^k \\
& = O\left(\|u\| + |x| \, |\log|x||^q \right) \left(|x|^k \|u\|^m + |x|^N \|u\| \right).
\end{aligned}$$

Hence, taking s and ρ small enough, we have

$$\|H_1(x, u, \phi(x, u))\| \leq (K_1 + 1) \left(|x|^k \|u\|^m + |x|^N \|u\| \right),$$

and, therefore,

$$\begin{aligned}
\|T\phi(x, u)\| \leq & \sum_{n=0}^{\infty} \left\| \left(\frac{x_n}{x} \right)^{-kB} H_1(x_n, u_n, \phi(x_n, u_n)) \right\| \\
\leq & (K_1 + 1) \sum_{n=0}^{\infty} \left| \frac{x_n}{x} \right|^{-k\beta} \left(|x_n|^k \|u_n\|^m + |x_n|^N \|u_n\| \right) \\
\leq & (K_1 + 1) \sum_{n=0}^{\infty} \left| \frac{x_n}{x} \right|^{-k\beta} \left(|x_n|^{k+k\alpha m} \|u\|^m |x|^{-k\alpha m} + |x_n|^{N+k\alpha} \|u\| \, |x|^{-k\alpha} \right).
\end{aligned}$$

Since $k\alpha > \beta$, the series is normally convergent in the set $\{ \|u\| \, |x|^{-k\alpha} \leq 1 \}$. By Corollary 6.3, there exists a positive constant K_0, depending only on H_1, so that

$$\|T\phi(x, u)\| \leq K_0 \left(\|u\|^m + |x|^{N-k} \|u\| \right).$$

Then, to conclude the proof it suffices to take $K = K_0$. $\qquad\square$

Let $\mathcal{F}_0$ be the set of holomorphic maps from $S_{\gamma,s,\rho}$ to $\mathbb{C}^p$, satisfying (8.17) with the constant K_0 of Proposition 8.4. We just proved that T maps $\mathcal{F}_0$ into itself. Since we want T to be a contraction, we need to restrict this set. We first do it by restricting the domain of definition of the maps in $\mathcal{F}_0$.

Choice of the domain of definition $\mathcal{D}$. In the following, instead of $S_{\gamma,s,\rho}$, we shall use the following domain of definition for the maps ϕ

$$\mathcal{D} := S_{\gamma,s,\rho} \cap \{(x,u) \in \mathbb{C}^r \mid \|u\| \, |x|^{-k\alpha} \leq 1\},$$

and we shall denote with $\mathcal{F}_0$ the set of maps $\phi \colon \mathcal{D} \to \mathbb{C}^r$ satisfying (8.17). We shall prove a result analogous to Proposition 8.4 for the partial derivatives of ϕ. To do so, we shall need bounds for the series

$$\sum_{n=0}^{\infty} \left\| \frac{\partial}{\partial x} \left\{ \left(\frac{x_n}{x} \right)^{-kB} H_1(x_n, u_n, \phi(x_n, u_n)) \right\} \right\|$$

and

$$\sum_{n=0}^{\infty} \left\| \frac{\partial}{\partial u} \left\{ \left(\frac{x_n}{x} \right)^{-kB} H_1(x_n, u_n, \phi(x_n, u_n)) \right\} \right\|.$$

We thus have to control the partial derivatives $\left| \frac{\partial x_n}{\partial x} \right|$, $\left\| \frac{\partial u_n}{\partial x} \right\|$, $\left\| \frac{\partial x_n}{\partial u} \right\|$, and $\left\| \frac{\partial u_n}{\partial u} \right\|$.

Following Lemma 3.5 of [Ha2], we have the following estimates.

LEMMA 8.5. *Let $\delta = min\{k\alpha, k\}$, and let $\varepsilon > 0$, with $\varepsilon < \delta$. Then, for γ, s and ρ small enough, we have the following inequalities in $\mathcal{D}$:*

$$\left| \frac{\partial x_n}{\partial x} \right| \leq \left| \frac{x_n}{x} \right|^{1+\delta-2\varepsilon}, \qquad \left\| \frac{\partial u_n}{\partial x} \right\| \leq \frac{\|u\| \, |x_n|^{\delta-\varepsilon}}{|x|^{1+\delta-2\varepsilon}},$$

$$\left\| \frac{\partial x_n}{\partial u} \right\| \leq \frac{|x_n|^{1+\delta-2\varepsilon}}{|x|^{\delta-\varepsilon}}, \quad and \quad \left\| \frac{\partial u_n}{\partial u} \right\| \leq \left| \frac{x_n}{x} \right|^{\delta-\varepsilon}.$$

PROOF. We argue by induction over n. If $n = 1$, deriving

$$x_1 = x - \frac{1}{k}x^{k+1} + O\big(x^{2k+1}, \|u\|x^{k+1} \log x\big)$$

and $u_1 = (I - x^k A)u + O\big(\|u\|^2 x^k, \|u\|x^{k+1} \log x\big)$ with respect to x and u, we obtain

$$\left| \frac{\partial x_1}{\partial x} \right| = \left| 1 - \frac{k+1}{k}x^k + o(x^k) \right| \leq \left| \frac{x_1}{x} \right|^{k+1-2\varepsilon},$$

because

$$\left| \frac{|x_1|}{|x|} \right|^{k+1-2\varepsilon} = \left| 1 - \frac{k+1-\varepsilon}{k}x^k + o(x^k) \right|$$

and

$$\left\| \frac{\partial u_1}{\partial x} \right\| \leq K\|u\| \, |x|^{k-1}.$$

Moreover,

$$\left\| \frac{\partial x_1}{\partial u} \right\| \leq K \, |x|^{k+1} \leq \left| \frac{|x_1|^{1+\delta-2\varepsilon}}{|x|^{\delta-\varepsilon}} \right| \quad and \quad \left\| \frac{\partial u_1}{\partial u} \right\| \leq \left| 1 - \alpha x^k \right| \leq \left| \frac{x_1}{x} \right|^{\delta-\varepsilon},$$

for γ, s and ρ small enough. By the definition of Φ we deduce

$$\left|\frac{\partial x_{n+1}}{\partial x}\right| \leq \left|1 - \frac{k+1}{k}x_n^k + o(x_n^k)\right|\left|\frac{\partial x_n}{\partial x}\right| + K|x_n|^{k+1}\left\|\frac{\partial u_n}{\partial x}\right\|$$

and

$$\left\|\frac{\partial u_{n+1}}{\partial x}\right\| \leq K\|u_n\|\left|\frac{\partial x_n}{\partial x}\right| + \left|1 - \alpha x_n^k\right|\left\|\frac{\partial u_n}{\partial x}\right\|.$$

Hence, by inductive hypothesis,

$$\left|\frac{\partial x_{n+1}}{\partial x}\right| \leq \left|\frac{x_n}{x}\right|^{1+\delta-2\varepsilon}\left(1 - \frac{k+1}{k}\operatorname{Re} x_n^k + o(x_n^k) + K\|u\||x_n|^{1+\varepsilon}\right)$$

$$\leq \left|\frac{x_{n+1}}{x}\right|^{1+\delta-2\varepsilon} = \left|1 - \frac{1+\delta-2\varepsilon}{k}x_n^k + o(x_n^k)\right|,$$

because $1 + \delta - 2\varepsilon < \frac{k+1}{k}$. On the other side, using the inductive hypothesis and the inequality $\|u_n\| \leq \|u\||x|^{-k\alpha}|x_n|^{k\alpha}$, we obtain

$$\left\|\frac{\partial u_{n+1}}{\partial x}\right\| \leq \frac{\|u\|}{|x|^{1+\delta-2\varepsilon}}|x_n|^{\delta-\varepsilon}\left(1 - \alpha\operatorname{Re} x_n^k + o(x_n^k) + K|x|^{-k\alpha}|x_n|^{1+k\alpha-\varepsilon}\right),$$

which is less than $\frac{\|u\|}{|x|^{1+\delta-2\varepsilon}}|x_{n+1}|^{\delta-\varepsilon}$, because $\delta - \varepsilon < k\alpha$. Arguing analogously by induction, we prove also the inequalities for the partial derivatives with respect to u. In fact,

$$\left\|\frac{\partial x_{n+1}}{\partial u}\right\| \leq \left|1 - \frac{k+1}{k}x_n^k + o(x_n^k)\right|\left\|\frac{\partial x_n}{\partial u}\right\| + K|x_n|^{k+1}\left\|\frac{\partial u_n}{\partial u}\right\|$$

$$\leq \left|\frac{|x_n|^{k+1-2\varepsilon}}{|x|^{\delta-\varepsilon}}\right|\left[1 - \frac{k+1}{k}x_n^k + o(x_n^k) + K|x_n|^{\delta+\varepsilon}\right]$$

$$\leq \frac{|x_n|^{k+1-2\varepsilon}}{|x|^{\delta-\varepsilon}},$$

because $\delta + 1 - \varepsilon < k + 1$ and

$$\left\|\frac{\partial u_{n+1}}{\partial u}\right\| \leq K\|u_n\|\left\|\frac{\partial x_n}{\partial u}\right\| + \left|1 - \alpha x_n^k\right|\left\|\frac{\partial u_n}{\partial u}\right\|$$

$$\leq \left|\frac{x_n}{x}\right|^{\delta-\varepsilon}\left[1 - \alpha\operatorname{Re} x_n^k + o(x_n^k)K\|u\|\frac{|x_n|^{2k+k\alpha-\delta}}{|x|^{k\alpha}}\right]$$

$$\leq \left|\frac{x_{n+1}}{x}\right|^{\delta-\varepsilon},$$

because $\delta - \varepsilon < k\alpha$. This concludes the proof. $\qquad\square$

We can now prove the following reformulation of Proposition 3.4 of [Ha2] for the case $k + 1 \geq 2$.

PROPOSITION 8.6. *Let ϕ be in $\mathcal{F}_0$. There exist positive constants K_1 and K_2 so that, if we have*

$$\begin{cases} \left| \frac{\partial \phi}{\partial x} \right| \leq K_1 \left(\|u\|^m |x|^{-1} + |x|^{N-k-1} \|u\| \right), \\ \left| \frac{\partial \phi}{\partial u} \right| \leq K_2 \left(\|u\|^{m-1} + |x|^{N-k} \right), \end{cases} \tag{8.19}$$

than the same inequalities hold for $\left\| \frac{\partial T\phi}{\partial x} \right\|$ and $\left\| \frac{\partial T\phi}{\partial u} \right\|$.

PROOF. We first deal with the partial derivative of H_1. There exist positive constants C_1 and C_2 so that

$$\|H_1(x,u,v)\| \leq C_1 \left(\|u\|^m |x|^k + |x|^N \|u\| \right)$$
$$+ C_2 \left(\|v\| |x|^{k+1} |\log |x||^q + \|v\|^2 |x|^k + \|u\| \|v\| |x|^k \right).$$

Then there exist positive constants C_3 and C_4 such that

$$\left\| \frac{\partial H_1}{\partial x} \right\| \leq C_3 \left(\|u\|^m |x|^{k-1} + |x|^{N-1} \|u\| \right)$$
$$+ C_4 \left(\|v\| |x|^k |\log |x||^q + \|v\|^2 |x|^{k-1} + \|u\| \|v\| |x|^{k-1} \right).$$

On the other side,

$$\left\| \frac{\partial H_1}{\partial u} \right\| \leq C_5 \left(\|u\|^{m-1} |x|^k + |x|^N \right) + C_6 \|v\| |x|^k,$$

for some positive constants C_5 and C_6. Finally, there exist positive constants C_7 and C_8 such that

$$\left\| \frac{\partial H_1}{\partial v} \right\| \leq C_7 \left(|x|^{k+1} |\log |x||^q + \|v\| |x|^k + \|u\| |x|^k \right) \leq C_8 |x|^k.$$

Let us assume that there exist positive constants K and K' such that

$$\begin{cases} \left\| \frac{\partial \phi}{\partial x} \right\| \leq K \left(\|u\|^m |x|^{-1} + |x|^{N-k-1} \|u\| \right), \\ \left\| \frac{\partial \phi}{\partial u} \right\| \leq K' \left(\|u\|^{m-1} + |x|^{N-k} \right). \end{cases}$$

Then we have

$$\left\| \frac{\partial T\phi}{\partial x} \right\| \leq \sum_{n=0}^{\infty} \left\| \frac{\partial}{\partial x} \left\{ \left(\frac{x_n}{x} \right)^{-kB} H_1(x_n, u_n, \phi(x_n, u_n)) \right\} \right\|$$
$$\leq \sum_{n=0}^{\infty} \left[\left\| \frac{\partial}{\partial x} \left(\left(\frac{x_n}{x} \right)^{-kB} \right) \right\| \|H_1(x_n, u_n, \phi(x_n, u_n))\| \right.$$
$$+ \left\| \left(\frac{x_n}{x} \right)^{-kB} \right\| \left(\left\| \frac{\partial H_1}{\partial x} \right\| \left\| \frac{\partial x_n}{\partial x} \right\| + \left\| \frac{\partial H_1}{\partial u} \right\| \left\| \frac{\partial u_n}{\partial x} \right\| \right)$$
$$+ \left. \left\| \left(\frac{x_n}{x} \right)^{-kB} \right\| \left\| \frac{\partial H_1}{\partial v} \right\| \left(\left| \frac{\partial \phi}{\partial x} \right| \left| \frac{\partial x_n}{\partial x} \right| + \left| \frac{\partial \phi}{\partial u} \right| \left| \frac{\partial u_n}{\partial x} \right| \right) \right].$$

We now use Lemma 8.5 to give estimates. We have

$$\left\|\frac{\partial}{\partial x}\left(\left(\frac{x_n}{x}\right)^{-kB}\right)\right\| \|H_1(x_n, u_n, \phi(x_n, u_n))\|$$

$$\leq \|-kB\| \left|\frac{x_n}{x}\right|^{-k\beta} \left[\frac{1}{|x_n|} \left|\frac{x_n}{x}\right|^{1+\delta-2\varepsilon} + \frac{1}{|x|}\right]$$

$$\times \left(\|u_n\|^m |x_n|^{-1} + |x_n|^{N-k-1} \|u_n\|\right)$$

$$\times |x_n|^{k+1} \left[C_1 + C_2 K_0 \left(|x_n| \,|\log|x_n||^q + \|\phi(x_n, u_n)\| + \|u_n\|\right)\right].$$

Similarly,

$$\left\|\frac{\partial H_1}{\partial x}\right\| \left|\frac{\partial x_n}{\partial x}\right| \leq \left(\|u_n\|^m |x_n|^{-1} + |x_n|^{N-k-1} \|u_n\|\right) |x_n|^k$$

$$\times \left[C_3 + C_4 K_0 \left(|x_n| \,|\log|x_n||^q + \|\phi(x_n, u_n)\| + \|u_n\|\right)\right] \left|\frac{x_n}{x}\right|^{1+\delta-2\varepsilon}$$

and

$$\left\|\frac{\partial H_1}{\partial u}\right\| \left|\frac{\partial u_n}{\partial x}\right| \leq \left(\|u_n\|^m |x_n|^{-1} + |x_n|^{N-k-1} \|u_n\|\right) |x_n|^k$$

$$\times \left[C_5 \frac{|x_n|}{\|u_n\|} + C_6 |x_n|\right] \frac{\|u\| \,|x_n|^{\delta-\varepsilon}}{x^{1+\delta-2\varepsilon}}.$$

Finally,

$$\left\|\frac{\partial H_1}{\partial v}\right\| \left[\left|\frac{\partial\phi}{\partial x}\right| \left|\frac{\partial x_n}{\partial x}\right| + \left|\frac{\partial\phi}{\partial u}\right| \left|\frac{\partial u_n}{\partial x}\right|\right] \leq C_8 \left(\|u_n\|^m |x_n|^{-1} + |x_n|^{N-k-1} \|u_n\|\right) |x_n|^k$$

$$\times \left[K \left|\frac{x_n}{x}\right|^{1+\delta-2\varepsilon} + K' \frac{\|u\| \,|x_n|^{1+\delta-\varepsilon}}{\|u_n\| \,|x|^{1+\delta-2\varepsilon}}\right].$$

By Corollary 6.3, we have the estimate

$$\sum_{n=0}^{\infty} |x_n|^\mu \,|\log x_n|^q \leq C_{\mu,q} |x|^{\mu-k} \,|\log x|^q,$$

for a constant $C_{\mu,q} > 0$, and, hence, there exists a positive constant K_1, depending only on H_1, so that

$$\left\|\frac{\partial T\phi}{\partial x}\right\| \leq K_1 \left(\|u\|^m |x|^{-1} + |x|^{N-k-1} \|u\|\right).$$

Setting $K = K_1$, we proved the first inequality. In a similar way, we estimate $\left\|\frac{\partial T\phi}{\partial u}\right\|$, obtaining

$$\left\|\frac{\partial T\phi}{\partial u}\right\| \leq \sum_{n=0}^{\infty} \left\|\frac{\partial}{\partial u}\left(\left(\frac{x_n}{x}\right)^{-kB}\right)\right\| \|H_1(x_n, u_n, \phi(x_n, u_n))\|$$

$$+ \left\|\left(\frac{x_n}{x}\right)^{-kB}\right\| \left\|\frac{\partial}{\partial u}\left(H_1(x_n, u_n, \phi(x_n, u_n))\right)\right\|.$$

For the first term , we have

$$\left\| \frac{\partial}{\partial u} \left(\left(\frac{x_n}{x} \right)^{-kB} \right) \right\| \left\| H_1(x_n, u_n, \phi(x_n, u_n)) \right\|$$

$$\leq \left\| -kB \frac{1}{x_n} \frac{\partial x_n}{\partial u} \left(\frac{x_n}{x} \right)^{-kB} \right\| \left(\|u_n\|^{m-1} + |x_n|^{N-k} \right) |x_n|^k$$

$$\times \underbrace{\left[C_1 \|u_n\| + C_2 \|u_n\| \left(|x_n| \, |\log |x_n||^q + \|\phi(x_n, u_n)\| + \|u_n\| \right) \right]}_{\tilde{K}(x_n, u_n)}$$

$$\leq \tilde{C} \left| \frac{1}{x_n} \right| \left| \frac{\partial x_n}{\partial u} \right| \left| \frac{x_n}{x} \right|^{-k\beta} \left(\|u_n\|^{m-1} + |x_n|^{N-k} \right) |x_n|^k \, \tilde{K}(x_n, u_n).$$

The second term contains the partial derivatives of H_1 with respect to x, u and v, and we have

$$\left\| \frac{\partial}{\partial u} \left(H_1(x_n, u_n, \phi(x_n, u_n)) \right) \right\|$$

$$\leq \left(\|u_n\|^{m-1} + |x_n|^{N-k} \right) |x_n|^k$$

$$\times \left(\left[C_3 \|u_n\| + C_4 K_0 \|u_n\| \left(|x_n| \, |\log |x_n||^q + \|\phi(x_n, u_n)\| + \|u_n\| \right) \right] \frac{|x_n|^{\delta-\varepsilon}}{|x|^{\delta-\varepsilon}} \right.$$

$$\left. + \left[C_5 + C_6 K_0 \|u_n\| \right] \left| \frac{x_n}{x} \right|^{\delta-\varepsilon} + C_8 \left[K \|u_n\| \, |x_n|^{-1} \frac{|x_n|^{1+\delta-2\varepsilon}}{|x|^{\delta-\varepsilon}} + K' \left| \frac{x_n}{x} \right|^{\delta-\varepsilon} \right] \right)$$

$$\leq \overline{K}(x_n, u_n, x) \left(\|u_n\|^{m-1} + |x_n|^{N-k} \right) |x_n|^k .$$

Therefore,

$$\left\| \frac{\partial T \phi}{\partial u} \right\| \leq K_2 \left(\|u\|^{m-1} + |x|^{N-k} \right),$$

and the constant K_2 depends only on H_1. Taking $K' = K_2$, we conclude the proof.
□

Definition of $\mathcal{F}$. We are left with finding a suitable subset of maps such that T is a contraction. Let m and N be integers satisfying (8.13). Let F_0 be the Banach space of the holomorphic maps ϕ, defined on $S_{\gamma, s, \rho}$, such that

$$\|\phi\|_0 := \sup_{x,u} \left\{ \frac{\|\phi(x, u)\|}{\|u\|^m + |x|^{N-k} \|u\|} \right\}$$

is bounded, endowed with the norm $\|\phi\|_0$. Define $\mathcal{F}$ as the closed subset of F_0 given by the maps satisfying (8.17) and (8.19), with the constants K_0, K_1 and K_2 given by Propositions 8.4 and 8.6.

PROPOSITION 8.7. *If $\mathcal{F}$ is the subset defined before, then $T|_{\mathcal{F}}$ is a contraction.*

PROOF. Let ϕ and ψ be in $\mathcal{F}$. We need to control

$$S := \left\| \sum_{n=0}^{\infty} \left(\frac{x_n}{x} \right)^{-kB} H_1(x_n, u_n, \phi(x_n, u_n)) - \sum_{n=0}^{\infty} \left(\frac{x_n'}{x} \right)^{-kB} H_1(x_n', u_n', \psi(x_n', u_n')) \right\|,$$

where (x_n, u_n) and (x'_n, u'_n) are the iterates of (x, u) via (8.1), respectively, with ϕ and ψ. We can bound S with the sum of S_1 and S_2, where

$$S_1 := \left\| \sum_{n=0}^{\infty} \left(\frac{x_n}{x} \right)^{-kB} H_1(x_n, u_n, \phi(x_n, u_n)) - \sum_{n=0}^{\infty} \left(\frac{x_n}{x} \right)^{-kB} H_1(x_n, u_n, \psi(x_n, u_n)) \right\|$$

and

$$S_2 := \left\| \sum_{n=0}^{\infty} \left(\frac{x_n}{x} \right)^{-kB} H_1(x_n, u_n, \psi(x_n, u_n)) - \sum_{n=0}^{\infty} \left(\frac{x'_n}{x} \right)^{-kB} H_1(x'_n, u'_n, \psi(x'_n, u'_n)) \right\|.$$

It is easy to control the term S_1. From (8.18), we have S_1 bounded above by

$$C \sum_{n=0}^{\infty} \left| \frac{x_n}{x} \right|^{-k\beta} \left(\|u_n\|^m + |x_n|^{N-k} \|u_n\| \right) \left(|x_n|^{k+1} |\log x_n|^q + |x_n|^k \|u_n\| \right) \|\phi - \psi\|_0,$$

for some integer q. By Corollary 6.3, since $\|u_n\| \le \|u\| |x_n|^{k\alpha} |x|^{-k\alpha}$, we obtain

$$S_1 \le C' \left(\|u\|^m + |x|^{N-k} \|u\| \right) \left(|x| |\log x|^q + \|u\| \right) \|\phi - \psi\|_0.$$

To estimate S_2, we have to estimate the dependence of $\{(x_n, u_n)\}$ on ϕ in (8.1). We have the following reformulation of Lemma 3.7 of [Ha2] .

LEMMA 8.8. *Let $\delta = \min\{k\alpha, k\}$. Let ε be a positive real number, with $\varepsilon < \delta$, and $\mathrm{Re}\,\lambda_j > \alpha + \varepsilon$ for each eigenvalue λ_j of A. Let ϕ and ψ be in $\mathcal{F}$, and let $\{(x_n, u_n)\}$ and $\{(x'_n, u_n)'\}$ be the iterates via (8.1) associated to ϕ and ψ. Then for γ, s and ρ small enough, the following estimates hold in $S_{\gamma, s, \rho}$:*

$$|x_n - x'_n| \le |x_n|^{1+\delta-\varepsilon} |x|^{-\delta} \left(\|u\|^m + |x|^{N-k-1} \|u\| \right) \|\phi - \psi\|_0,$$

and

$$\|u_n - u'_n\| \le |x_n|^{\delta} |x|^{-\delta} \left(\|u\|^m + |x|^{N-k-1} \|u\| \right) \|\phi - \psi\|_0.$$

PROOF. We use the following notation: $\Delta x_n := |x_n - x'_n|$, $\Delta u_n := \|u_n - u'_n\|$, and $\Delta\phi(x, u) := \left(\|u\|^m + |x|^{N-k-1} \|u\| \right) \|\phi - \psi\|_0$. We argue by induction over n. If $n = 1$, thanks to (8.1), there exists $K > 0$ such that

$$\begin{cases} \Delta x_1 \le K |x|^{k+1} \Delta\phi(x, u), \\ \Delta u_1 \le K \left(|x|^{k+1} + |x|^k \|u\| \right) \Delta\phi(x, u), \end{cases}$$

for γ and s small enough. Let us assume that the inequalities hold for n, and we prove that they hold also for $n + 1$. Since x_n^k and $(x'_n)^k$ are equivalent to $\frac{1}{n}$, we have $(x'_n)^k = x_n^k + o(x_n^k)$. From the definition of Φ it follows

$$\Delta x_{n+1}$$
$$= \left| x_n^k \left(1 - x_n^k + O(x_n^{2k}, \|u_n\| x_n^k) \right) - (x'_n)^k \left(1 - (x'_n)^k + O((x'_n)^{2k}, \|u'_n\| (x'_n)^k) \right) \right|$$
$$\le \Delta x_n \left| 1 - x_n^k + o(x_n^k) \right| + K |x_n|^{k+1} \Delta u_n + K |x_n|^{k+1} \Delta\phi(x_n, u_n)$$

and

$$\Delta u_{n+1} \leq K \|u_n\| \Delta x_n + |1 - (\alpha + \varepsilon) x_n + o(x_n)| \, \Delta u_n$$
$$+ K \left(|x_n|^{k+1} + \|u_n\| \, |x_n|^k \right) \Delta \phi(x_n, u_n).$$

Thanks to $\|u_n\| \leq \|u\| \, \left| \frac{x_n}{x} \right|^{k\alpha}$, we have

$$\Delta \phi(x_n, u_n) \leq \left(\|u\|^m + |x|^{N-k-1} \|u\| \right) \left(\left| \frac{x_n}{x} \right|^{mk\alpha} + \left| \frac{x_n}{x} \right|^{k\alpha + N - k - 1} \right) \|\phi - \psi\|_0,$$

and, since $\left| \frac{x_n}{x} \right|^\gamma \leq \left| \frac{x_n}{x} \right|^\delta$ when $\gamma > \delta$, we have

$$|x|^\delta \Delta \phi(x_n, u_n) \leq 2 \, |x_n|^\delta \Delta \phi(x, u).$$

By inductive hypothesis, we may bound $|x|^\delta \frac{\Delta x_{n+1}}{\Delta \phi}$ and $|x|^\delta \frac{\Delta u_{n+1}}{\Delta \phi}$. We thus obtain

$$|x|^\delta \frac{\Delta x_{n+1}}{\Delta \phi} \leq \left| 1 - \varepsilon x_n^k + o(x_n^k) \right| |x_{n+1}|^{1+\delta-\varepsilon} + K |x_n|^{2+\delta} + 2K |x_n|^{2+\delta}$$
$$\leq \left| 1 - \varepsilon x_n^k + o(x_n^k) \right| |x_{n+1}|^{1+\delta-\varepsilon} \leq |x_{n+1}|^{1+\delta-2\varepsilon}$$

and

$$|x|^\delta \frac{\Delta u_{n+1}}{\Delta \phi} \leq K \|u_n\| \, |x_n|^{1+\delta-\varepsilon} + \left| 1 - \varepsilon x_n^k + o(x_n^k) \right| |x_{n+1}|^\delta$$
$$+ K \left(|x_n| + \|u_n\| \right) |x_n|^{1+\delta}$$
$$\leq \left| 1 - \varepsilon x_n^k + o(x_n^k) \right| |x_{n+1}|^\delta \leq |x_{n+1}|^\delta.$$

Since $\|u_n\| \, |x_n|^{-k\alpha} = o(1)$, we can now prove the last inequality

$$\|u_n\| \, |x_n|^{1+\delta-\varepsilon} = o(|x_n|^{1+k\alpha+\delta-\varepsilon}) = |x_n|^\delta o(|x_n|),$$

for ε small enough. $\qquad \square$

We can now estimate S_2 as follows.

$$S_2 \leq \sum_{n=0}^\infty K_1 \left| \frac{x_n}{x} \right|^{-k\beta-\varepsilon} \left(\|u_n\|^m + |x_n|^{N-k-1} \|u_n\| \right) \Delta x_n$$
$$+ K_2 \left| \frac{x_n}{x} \right|^{-k\beta-\varepsilon} \left(\|u_n\|^{m-1} |x_n|^k + |x_n|^{N-k-1} \right) \Delta u_n.$$

By Lemma 8.8, Corollary 6.3 and the fact that $\|u_n\| \, |x_n|^{-\varepsilon} = o(1)$, we thus obtain

$$S_2 \leq K \left(\|u\|^{m-1} + |x_n|^{N-k-1} \right) \left(\|u\|^m + |x|^{N-k-1} \|u\| \right) \|\phi - \psi\|_0.$$

Therefore, T is a contraction. $\qquad \square$

Taking ϕ the unique fixed point of T, we can use the following change of coordinates: $\tilde{v} = v - \phi(x, u)$. Then

$$
\begin{aligned}
\tilde{v}_1 &= v_1 - \phi(x_1, u_1) = (I - x^k B)v + H(x, u, v) - \phi(x_1, u_1) \\
&= (I - x^k B)(\tilde{v} + \phi(x, u)) + H(x, u, \tilde{v} + \phi(x, u)) - \phi(x_1, u_1) \\
&= (I - x^k B)\tilde{v} + \underbrace{(I - x^k B)\phi(x, u) + H(x, u, \phi(x, u))}_{=\phi(x_1^\phi, u_1^\phi)} \\
&\quad + \sum_{n \geq 1} \left[\frac{1}{n!} \frac{\partial^n H}{\partial v^n}(x, u, \phi(x, u))\tilde{v}^n \right] - \phi(x_1, u_1) \\
&= (I - x^k B)\tilde{v} + \phi(x_1^\phi, u_1^\phi) - \phi(x_1^\phi, u_1^\phi) + \cdots \\
&= (I - x^k B)\tilde{v} + \tilde{H}(x, u, \tilde{v}),
\end{aligned}
$$

with $\tilde{H}(x, u, \tilde{v}) = O(\|\tilde{v}\|)$, and hence $H(x, u, 0) = 0$. Therefore, we can apply Theorem 7.1 to $\Phi|_{\{\tilde{v}=0\}}$, and this concludes the proof of Theorem 8.3. $\qquad \square$

We then deduce the following reformulation of Corollary 3.8 of [Ha2] .

COROLLARY 8.9. *Let $\Phi \in \mathrm{Diff}(\mathbb{C}^p, 0)$ be a tangent to the identity germ of order $k + 1 \geq 2$ and let $[V]$ be a non-degenerate characteristic direction. Let $\{\lambda_1, \ldots, \lambda_h\}$ be the directors associated to $[V]$ with strictly positive real parts and assume that*

$$
\alpha_1 > \alpha_2 > \cdots > \alpha_h > 0,
$$

where $\alpha_j = \mathrm{Re}\,\lambda_j$. Then there exists an increasing sequence

$$
M_1 \subset M_2 \subset \cdots \subset M_h
$$

of parabolic manifolds, defined in a sector, attracted by the origin along the direction $[V]$. Moreover, for any $1 \leq i \leq h$, the dimension of M_i is $1 + \sum_{\mathrm{Re}\,\lambda_j \geq \alpha_i} m_{alg}(\lambda_j)$ and M_i is tangent at the origin to $\mathbb{C}V \bigoplus_{\mathrm{Re}\,\lambda_j \geq \alpha_i} E_{\lambda_j}$, where E_{λ_j} is the eigenspace associated to the eigenvalue λ_j.

We can also deduce a partial converse of Theorem 7.1, using the following result, which holds for germs of biholomorphisms and, hence, also for global biholomorphisms.

LEMMA 8.10. *Let $\Phi \in \mathrm{Diff}(\mathbb{C}^p, 0)$ be a tangent to the identity germ of order $k + 1 \geq 2$. If $X = (x, y) \in \mathbb{C}^p \setminus \{(0, 0)\}$ is so that $X_n = \Phi^n(x, y)$ converges to the origin and $[X_n]$ converges to $[1 : 0]$, then there exist constants γ, s, and ρ so that, for any $n > n_0$, with n_0 large enough, we have $x_n \neq 0$ and $X_n = (x_n, y_n) \in S_{\gamma, s, \rho}$, where, for $x \neq 0$ and $U = \frac{y}{x}$, we set*

$$
S_{\gamma, s, \rho} = \left\{ (x, U) \in \mathbb{C} \times \mathbb{C}^{p-1} \;\middle|\; \left|\mathrm{Im}\,x^k\right| \leq \gamma\,\mathrm{Re}\,x^k, \left|x^k\right| < s, \|U\| < \rho \right\}.
$$

PROOF. Since $X_n = (x_n, y_n)$ converges to 0 and $[X_n]$ converges to $[1 : 0]$, we have that x_n is definitively different from 0. Moreover, X_n definitively lies in $D_{s,\rho} := \{(x, U) \mid |x| \leq s, \|U\| \leq \rho\}$. Thanks to Proposition 4.3, the first component of Φ is of the form $x_1 = x - \frac{1}{k}x^{k+1} + O(\|U\|x^{k+1}, x^{2k+1})$, and $x_n^k \approx \frac{1}{n}$. Therefore, for any γ arbitrarily small and any n large enough, we have $\left|\mathrm{Im}\,x_n^k\right| \leq \gamma\,\mathrm{Re}\,x_n^k$, and hence X_n definitively lies in $S_{\gamma, s, \rho}$. $\qquad \square$

COROLLARY 8.11. *Let* $\Phi \in \mathrm{Diff}(\mathbb{C}^p, 0)$ *be a tangent to the identity germ of order* $k+1 \geq 2$, *and let* $[V]$ *be a non-degenerate characteristic direction. If there exists an attracting domain* Ω *where all the orbits converge to the origin along* $[V]$, *then all the directors of* $[V]$ *have non-negative real parts.*

REMARK 8.12. It is not true that if $[V]$ is a non-degenerate characteristic direction and there exists an attracting domain Ω where all the orbits converge to the origin along $[V]$, then all the directors of $[V]$ have strictly positive real parts. In fact, as shown by Vivas in [V2], it is possible to find examples of germs having attracting domains along non-degenerate characteristic direction even when the directors have vanishing real parts.

9   FATOU COORDINATES

We have the following analogous of Theorem 1.9 of [Ha2].

THEOREM 9.1. *Let* $\Phi \in \mathrm{Diff}(\mathbb{C}^p, 0)$ *be a tangent to the identity germ. Let* $[V]$ *be an attracting non-degenerate characteristic direction. Then there is an invariant domain* D, *with* $0 \in \partial D$, *so that every point of* D *is attracted to the origin along the direction* $[V]$ *and such that* $\Phi|_D$ *is holomorphically conjugated to the translation*

$$
\begin{cases}
\dfrac{1}{x_1} = \dfrac{1}{x} + 1, \\
U_1 = U,
\end{cases}
$$

with $(x, U) \in \mathbb{C} \times \mathbb{C}^{p-1}$.

We may assume that $[V] = [1 : 0]$ and that its associated matrix A is in Jordan normal form, with the non-zero elements out of the diagonal equal to $\varepsilon > 0$ small.

Let $\lambda_1, \dots, \lambda_h$ be the distinct eigenvalues of A, and up to reordering, we may assume that, setting $\alpha_j = \mathrm{Re}\,(\lambda_j)$, we have

$$
\alpha_1 \geq \alpha_2 \geq \cdots \geq \alpha_h > \alpha > 0.
$$

Let $J_1, \dots, J_h$ be the Jordan blocks of A, where J_l is the block relative to λ_l for $1 \leq l \leq h$, and let $u = (u^1, \dots, u^h) \in \mathbb{C}^{p-1}$ be the splitting of the coordinates of $\mathbb{C}^{p-1}$ associated to the splitting of A in Jordan blocks. Therefore, we can write

$$
\Phi(x, u) =
\begin{cases}
x_1 = x - \frac{1}{k}x^{k+1} + F(x, u), \\
u_1^1 = (I^1 - x^k J_1)u^1 + G^1(x, u), \\
u_1^2 = (I^2 - x^k J_2)u^2 + G^2(x, u), \\
\quad \vdots \\
u_1^h = (I^h - x^k J_h)u^h + G^h(x, u),
\end{cases}
\tag{9.1}
$$

where I^l is an identity matrix of same dimension of the block J_l for $1 \leq l \leq h$, and

$$
G^j(x, u) = O(\|u\|^2 x^k, \|u\|x^{k+1}|\log x|).
$$

Set
$$
u^{\leq j} := (u^1, \dots, u^j) \quad \text{and} \quad u^{>j} := (u^{j+1}, \dots, u^h),
$$

and analogous definitions for $u^{<j}$ and $u^{\geq j}$.

Given $N \in \mathbb{N}$ with $N \geq k+1$, thanks to (8.3), for every $1 \leq j \leq h$, we can write

$$G^j(x,u) = P_N^j(x,u) + O\big(\|u\| \, |x|^N \, |\log|x||^{h_N}\big), \qquad (9.2)$$

with

$$P_N^j(x,u) = \sum_{k \leq s \leq N, \, t \in E_s} c_{s,t}^j(u) x^s (\log x)^t,$$

for some $h_N \in \mathbb{N}$ depending on N, where for any s we defined E_s as the (finite) set of integers t so that the series above contains the term $x^s(\log x)^t$, and where $c_{s,t}(u)$ are holomorphic in $\|u\| \leq \rho$ and $c_{s,t}(0) \equiv 0$.

The following result is analogous to Proposition 4.1 of [Ha2].

PROPOSITION 9.2. *Let* $\Phi \in \mathrm{Diff}(\mathbb{C}^p, 0)$ *be a tangent to the identity germ of order* $k+1 \geq 2$ *as in (9.1), with* $[V] = [1:0]$ *attracting non-degenerate characteristic direction. For any positive integers* $N \geq k+1$ *and* m, *there exist local holomorphic coordinates (defined in a sector) such that (9.2) holds, and moreover*

$$P_N^j(x, (0, u^{>j})) = O(x^k \|u^{>j}\|^m) \qquad (9.3)$$

for $1 \leq j \leq h$.

PROOF. We want to change coordinates holomorphically in order to remove the terms in $u^{>j}$ with degree less than m from $P_N^j(x, (0, u^{>j}))$. We use holomorphic changes of coordinates of the form $\tilde{u}^j = u^j - q_j(x, u^{>j})$, where the q_j's are polynomials in x, $\log x$ and $u^{>j}$ with $q_j(x,0) = 0$; if we obtain (9.3) for $j = 1, \ldots, j_0$, then changing the variables u^j for $j > j_0$ will provide no effect on the first j_0 variables. We shall then perform the construction by induction on j, considering only changes on $u^{\geq j}$ with $u^{<j} = 0$, which allow us to forget about the first $j-1$ coordinates.

Let $v = u^{>j}$, and let us consider the matrix B_j defined as

$$B_j = \mathrm{Diag}(A_{j+1}, \ldots, A_h).$$

We now have to prove a statement similar to the one in Proposition 8.1 but with the opposite notation, i.e., we look for changes of coordinates of the form $\tilde{u} = u - \varphi(x, v)$ such that

$$G^j(x, 0, v) = O(|x|^k \|v\|^m + |x|^N \|v\|),$$

and hence the rôles of $A(=J_j)$ and $B(=B_j)$ are here exchanged.

Note that, if $1, \lambda_1, \ldots, \lambda_h$ are rationally independent, then we can prove the statement exactly as in the proof of Proposition 8.1.

Otherwise, let $Q(v) x^s (\log x)^t$ be the lower degree term in $P_N^j(x, (0, v))$, with $Q(v)$ homogeneous polynomial of degree d in v. The change of coordinates

$$\tilde{u}^j = u^j - x^{s-k} (\log x)^t P(v)$$

deletes the term $Q(v) x^s (\log x)^t$ if P solves

$$P((I_q - x^k B)v) - (I_r - x^k J_j)P(v) - \frac{s-k}{k} x^k P(v) = x^k Q(v) + O(\|v\| x^{k+1}),$$

where r and q are the dimensions, respectively, of u^j and v. Therefore, by decreasing induction on the indices of the components of u^j, we may reduce ourselves to solve, for the r components P_i of P, equations of the form

$$\frac{\partial P_i}{\partial v_1}(Bv)_1 + \cdots + \frac{\partial P_i}{\partial v_q}(Bv)_q - \left(\lambda_j - \frac{s-k}{k}\right) P_i = \tilde{Q}_i(v), \qquad (9.4)$$

that is, by Euler formula, of the form

$$\frac{\partial P_i}{\partial v_1}(Cv)_1 + \cdots + \frac{\partial P_i}{\partial v_q}(Cv)_q = \tilde{Q}_i(v), \qquad (9.5)$$

where $C = B - \frac{\lambda_j - (s-k)/k}{d} I_q$. We solve these equations component by component, by comparing the coefficients of the monomials v^T, where $T \in \mathbb{N}^q$ in both sides. For any $T := (t_1, \ldots, t_q) \in \mathbb{N}^q$, we define the weight of T as

$$w(T) = t_1 + 2t_2 + \cdots + qt_q.$$

If $P_i(v) = av^T$ and $\tilde{Q}_i(v) = cv^T$, Equation (9.5) is reduced, modulo terms of greater weight, to

$$a(\nu_1 t_1 + \cdots + \nu_q t_q) = c,$$

where $\nu_1, \ldots, \nu_q$ are the eigenvalues of C. Now, if $\nu_1 t_1 + \cdots + \nu_q t_q \neq 0$, then we can solve the equation; otherwise, we can consider the change of coordinates

$$\tilde{u}^j = u^j - av^T x^{s-k}(\log x)^{t+1},$$

under which the terms in $v^T x^{s-k}(\log x)^{t+1}$ and in $v^T x^s(\log x)^{t+1}(\log x)^{t+1}$ in the left-hand side vanish and the equation is reduced to $a(t+1) = -c$; this change introduces new terms in $x^s(\log x)^{t+1}$, but it can happen only finitely many times and, hence is not a problem. Iterating this procedure on the weight of T, given a degree d, we have to solve the case of T of maximal weight, i.e., $v^T + v_q^d$. In this case, the equation is simply $av_q d = c$, and it is solvable if $\nu_q \neq 0$; if $\nu_q = 0$, as before, we can consider the change

$$\tilde{u}^j = u^j - av_q^d x^{s-k}(\log x)^{t+1},$$

and we are done. $\qquad\square$

We can then deduce the following reformulations of Corollaries 4.2, 4.3, and 4.4 of [Ha2] for the case $k+1 \geq 2$.

COROLLARY 9.3. *Let $\Phi \in \text{Diff}(\mathbb{C}^p, 0)$ be a tangent to the identity germ of order $k+1 \geq 2$ as in (9.1), with $[V] = [1:0]$ attracting non-degenerate characteristic direction. For any positive integers $N \geq k+1$ and m, there exist local holomorphic coordinates such that*

$$G^{\leq j}\left(x, (0, u^{>j})\right) = O\left(|x|^k \|u^{>j}\|^m + |x|^N \|u^{>j}\|\right), \qquad (9.6)$$

for $1 \leq j \leq h$.

COROLLARY 9.4. *Let $\Phi \in \text{Diff}(\mathbb{C}^p, 0)$ be a tangent to the identity germ of order $k+1 \geq 2$ as in (9.1), with $[V] = [1:0]$ attracting non-degenerate characteristic direction. Let $0 < \varepsilon < \alpha$, and assume that the local coordinates are chosen so that the non-zero* *coefficients out of the diagonal in A are equal to $\varepsilon_0 > 0$ small enough, and* (9.6) *is satisfied with m and N such that*

$$m\alpha_h - \alpha_1 \geq 1 \quad \text{and} \quad N + k(\alpha_h - \alpha_1) \geq k + 1. \tag{9.7}$$

Then, for every j and for each $(x, u) \in S_{\gamma, s, \rho}$ with γ, s, ρ small enough, there exists a constant $K > 0$ such that

$$\|u_n^{\leq j}\| \leq |x_n|^{k(\alpha_j - \varepsilon)}(\|u^{\leq j}\| |x|^{-k(\alpha_j - \varepsilon)} + K|x|^k), \tag{9.8}$$

and, moreover, $\|u_n^{\leq j}\| |x_n|^{-k(\alpha_j - \varepsilon)}$ converges to zero as n tends to infinity.

PROOF. From the proof of Theorem 7.1, we know that, taking $\alpha = \alpha_h - \varepsilon$ we have

$$\|u_n\| \leq |x_n|^{k\alpha} \|u\| |x|^{-k\alpha}.$$

Hence, from (9.6) and (9.7) we obtain

$$|x_{n+1}|^{-k(\alpha_j - \varepsilon)} \left\| G^{\leq j}\left(x, (0, u^{>j})\right) \right\| \leq K_1 |x_n|^{k+1}.$$

Arguing as in the proof of Theorem 7.1, choosing ε and ε_0 small enough, since the eigenvalues of $J_j - k(\alpha_j - \varepsilon)I^j$ have positive real parts, for each $x \in S_{\gamma, s, \rho}$ with γ, s, ρ small enough, we have

$$\|I^j - (J_j - k(\alpha_j - \varepsilon)I^j)x^k + o(x^k)\| \leq 1.$$

We have

$$|x_{n+1}|^{-k(\alpha_j - \varepsilon)} \|u_{n+1}^{\leq j}\| \leq \|I^j - (J_j - k(\alpha_j - \varepsilon)I^j)x^k + o(x^k)\| |x_n|^{-k(\alpha_j - \varepsilon)} \|u_n^{\leq j}\|$$
$$+ |x_{n+1}|^{-k(\alpha_j - \varepsilon)} \|G^{\leq j}(x, (0, u^{>j}))\|.$$

Hence, setting $V_n^{\leq j} := |x_n|^{-k(\alpha_j - \varepsilon)} \|u_n^{\leq j}\|$, we obtain

$$V_{n+1}^{\leq j} \leq V_n^{\leq j} + K_1 |x_n|^{k+1}.$$

Since for any $(x, u) \in S_{\gamma, s, \rho}$, for γ, s, ρ small enough, there exists $0 < c < 1$ such that $|x_{n+1}|^k \leq |x_n|^k(1 - c|x_n|)$, and we have

$$|x_n|^{k+1} \leq \frac{|x_n|^k - |x_{n+1}|^k}{c},$$

implying that there exists a positive constant $K > 0$ such that

$$V_{n+1}^{\leq j} + K|x_{n+1}|^k \leq V_n^{\leq j} + K|x_n|^k \leq \cdots \leq V^{\leq j} + K|x|^k,$$

proving (9.8). Moreover, this proves that there exists $\varepsilon_0 \ll \varepsilon_1 < \varepsilon$ such that $\|u_n^{\leq j}\| |x_n|^{-k(\alpha_j - \varepsilon)} \leq \varepsilon_1$, and so $\|u_n^{\leq j}\| |x_n|^{-k(\alpha_j - \varepsilon)}$ converges to zero as $n \to +\infty$. $\square$

Thanks to the last corollary, we may assume without loss of generality that our germ Φ is of the form (9.1) and in the hypotheses of Corollary 9.4. Define the set

$$\mathcal{D} := \{(x, u) \in S_{\gamma, s, \rho} : \|u_n^{\leq j}\| |x_n|^{-k(\alpha_j - \varepsilon)} \leq 1\}. \tag{9.9}$$

To prove Theorem 9.1, we shall need this reformulation of Lemma 4.5 of [Ha2].

LEMMA 9.5. *Let $\Phi \in \mathrm{Diff}(\mathbb{C}^p, 0)$ be a tangent to the identity germ. Let $[V]$ be an attracting non-degenerate characteristic direction. Consider local holomorphic coordinates where Φ satisfies the hypotheses of Corollary 9.4 with ε such that $3\varepsilon < \min(\alpha_1, 1)$. Then the sequence $\{x_n^{-kA} u_n\}$ converges normally on the set $\mathcal{D}$ defined by (9.9).*

PROOF. Given $(x, u) \in \mathcal{D}$, we shall bound $\|x_{n+1}^{-kJ_j} u_{n+1}^j - x_n^{-kJ_j} u_n^j\|$, for each $j = 1, \ldots, h$. We have

$$\|x_{n+1}^{-kJ_j} u_{n+1}^j - x_n^{-kJ_j} u_n^j\| \le K \|x_n^{-kJ_j}\| \|u_n^{\le j}\| \left(\|u_n\| |x_n|^k + |x_n|^{k+1} |\log x_n|^q \right)$$
$$+ K' \|x_n^{-kJ_j}\| \|G^j(x, (0, u^{>j}))\|,$$

for some positive integer q. Since $(x, u) \in \mathcal{D}$, we have $\|u_n^{\le j}\| \le |x_n|^{k(\alpha_j - \varepsilon)}$, and, hence, using the inequality $\|x_n^{-kJ_j}\| \le |x_n|^{-k\alpha_j - k\varepsilon}$, we obtain

$$\|x_{n+1}^{-kJ_j} u_{n+1}^j - x_n^{-kJ_j} u_n^j\|$$
$$\le K_1 |x_n|^{-k\alpha_j - k\varepsilon} \left(|x_n|^{k(1 + \alpha + \alpha_j - 2\varepsilon)} + |x_n|^{k+1+k\alpha_j - k\varepsilon} |\log x_n|^q \right) + K_2 |x_n|^{k+1},$$

and, hence, there exists $K > 0$ such that

$$\|x_{n+1}^{-kJ_j} u_{n+1}^j - x_n^{-kJ_j} u_n^j\| \le K \left(|x_n|^{k(1+\alpha_1 - 3\varepsilon)} + |x_n|^{k+1-2k\varepsilon} |\log x_n|^q \right).$$

Thus, we are done.   □

We now have all the ingredients to prove Theorem 9.1.

PROOF OF THEOREM 9.1. Thanks to the previous lemma, we can define in $\mathcal{D}$ the following holomorphic bounded map

$$H(x, u) := \sum_{n=0}^{\infty} \left(x_{n+1}^{-kA} u_{n+1} - x_n^{-kA} u_n \right), \tag{9.10}$$

which satisfies

$$\|H(x, u)\| \le K \left(|x|^{k(\alpha_1 - 3\varepsilon)} + |x_n|^{k-2k\varepsilon} |\log x_n|^q \right) \le K \left(|x|^{k(\alpha_1 - 3\varepsilon)} + |x_n|^{k-3k\varepsilon} \right). \tag{9.11}$$

Therefore, the holomorphic map

$$U(x, u) := x^{-kA} u + H(x, u) = \lim_{n \to +\infty} x_n^{-kA} u_n \tag{9.12}$$

is invariant. The main term near to the origin is $x^{-kA} u$, and the level sets $\{U(x, u) = c\}$ with $c \in \mathbb{C}$ are complex invariant analytic curves. Therefore, taking (x, U) as new coordinates, Φ becomes

$$\begin{cases} x_1 = x - \frac{1}{k} x^{k+1} + \tilde{F}(x, U), \\ U_1 = U, \end{cases} \tag{9.13}$$

where $\tilde{F}$ is a holomorphic function of order at least $k+2$ in x, and U behaves as a parameter. We can thus argue as in Fatou [Fa], and change coordinates, in $\mathcal{D}$, in the first coordinate x, with a change depending on U, to obtain Φ of the form

$$\begin{cases} \dfrac{1}{z_1} = \dfrac{1}{z} + 1, \\ U_1 = U, \end{cases}$$

and this concludes the proof. $\square$

We thus deduce the following reformulation of Corollary 4.6 of [Ha2] .

COROLLARY 9.6. *Let* $\Phi \in \mathrm{Diff}(\mathbb{C}^p, 0)$ *be a tangent to the identity germ of order* $k+1 \geq 2$ *and let* $[V]$ *be a non-degenerate characteristic direction. Assume that* $[V]$ *has exactly d (counted with multiplicity) directors with positive real parts. Let M be the parabolic manifold of dimension $d+1$ provided by Theorem 8.3. Then there exist local holomorphic coordinates (x, u, v) such that $M = \{v = 0\}$, and $\Phi|_M$ is holomorphically conjugated to*

$$\begin{cases} \dfrac{1}{z_1} = \dfrac{1}{z} + 1, \\ U_1 = U. \end{cases}$$

PROOF. Thanks to Theorem 8.3 there exist local holomorphic coordinates (x, u, v) defined in a sector $S_{\gamma, s, \rho}$ such that the parabolic manifold M is defined by $M = \{v = 0\}$, and Φ is defined by (8.1) with F, G, and H satisfying (8.2), and $H(x, u, 0) = 0$. Then $\Phi|_M$ is given by

$$\begin{cases} x_1 = x - \dfrac{x^{k+1}}{k} + F(x, u, 0), \\ u_1 = (I_d - x^k A)u + G(x, u, 0), \end{cases}$$

where all the eigenvalues of A have positive real parts. Let $\lambda_1, \ldots, \lambda_h$ be the distinct eigenvalues of A, and let $\alpha_j = \mathrm{Re}\,(\lambda_j)$. Up to reordering, we may assume $\alpha_1 > \cdots > \alpha_h > \alpha > 0$.

Let m and $N \geq k+1$ be positive integers such that $m\alpha_h - \alpha_1 \geq 1$ and $N + k(\alpha_h - \alpha_1) \geq k+1$. We can thus write the Taylor expansion of G as

$$G(x, u, 0) = \sum_{\substack{1 \leq s \leq N \\ t \in E_s}} c_{s,t}(u) x^s (\log x)^t + O(|x|^k \|u\|^m + |x|^N \|u\|),$$

where $c_{s,t}(u)$ is a polynomial and $\deg(c_{s,t}(u)) \leq m$. Therefore, we can apply Theorem 9.1 to $\Phi(x, u, 0)$ and we are done. $\square$

10   FATOU-BIEBERBACH DOMAINS

In this section we shall assume that Φ is a global biholomorphism of $\mathbb{C}^p$ fixing the origin and tangent to the identity of order $k+1 \geq 2$.

DEFINITION 10.1. *Let Φ be a global biholomorphism of $\mathbb{C}^p$ fixing the origin and tangent to the identity of order $k+1 \geq 2$. Let $[V]$ be a non-degenerate characteristic direction of Φ at 0. The* attractive basin to $(0, [V])$ *is the set*

$$\Omega_{(0, [V])} := \{X \in \mathbb{C}^p \setminus \{0\} : \Phi^n(X) \to 0, [\Phi^n(X)] \to [V]\}. \tag{10.1}$$

We shall study the attractive basin $\Omega_{(0,[V])}$ when some of the directors of $[V]$ have positive real parts.

We can assume that, writing $X = (x, y) \in \mathbb{C} \times \mathbb{C}^{p-1}$, $[V] = [1 : 0]$ and Φ is of the form

$$\begin{cases} x_1 = x + p_{k+1}(x, y) + p_{k+2}(x, y) + \cdots, \\ y_1 = y + q_{k+1}(x, y) + q_{k+2}(x, y) + \cdots, \end{cases}$$

with $p_{k+1}(1, 0) = -1/k$ and $q_{k+1}(1, 0) = 0$.

Thanks to Lemma 8.10, we have

$$\Omega_{(0,[V])} = \bigcup_{n \geq 0} \Phi^{-n} \left(\Omega_{(0,[V])} \cap S_{\gamma, s, \rho} \right), \tag{10.2}$$

and we can restrict ourselves to study $\Omega_{(0,[V])} \cap S_{\gamma, s, \rho}$.

Since $S_{\gamma, s, \rho} \cap \{x = 0\} = \varnothing$, we can use the blow-up $y = xu$ and we can assume that, in the sector, Φ has the form

$$\begin{cases} x_1 = x - \frac{1}{k} x^{k+1} + O(\|u\| x^{k+1}, x^{k+2}), \\ u_1 = (I_{p-1} - x^k A) u + O(\|u\| x^k, \|u\| x^{k+1}), \end{cases} \tag{10.3}$$

where $A = A([V])$ is the matrix associated to $[V]$, and we can perform all the changes of coordinates used to prove Theorem 8.3 and Theorem 9.1.

We thus can prove the following generalization of Theorem 5.2 of [Ha2] for the case $k + 1 \geq 2$.

THEOREM 10.2. *Let Φ be a global biholomorphism of $\mathbb{C}^p$ fixing the origin and tangent to the identity of order $k + 1 \geq 2$ and let $[V]$ be a non-degenerate characteristic direction of Φ at 0. If $[V]$ is attracting, then the attractive basin $\Omega_{(0,[V])} \subset \mathbb{C}^p$ is a domain isomorphic to $\mathbb{C}^p$, that is, it is a Fatou-Bieberbach domain.*

PROOF. We can reduce ourselves to consider Φ as in (10.3), with A in Jordan normal form. Let $\lambda_1, \ldots, \lambda_h$ be the distinct eigenvalues of A, and let $\alpha_j = \text{Re}(\lambda_j)$. Up to reordering, we may assume $\alpha_1 > \cdots > \alpha_h > \alpha > 0$. Let $\varepsilon > 0$ be small and such that

$$\alpha_1 > \alpha_1 - \varepsilon > \alpha_2 > \alpha_2 - \varepsilon > \cdots > \alpha_h > \alpha_h - \varepsilon > 0.$$

Thanks to Theorem 9.1 and Corollary 9.4, the coordinates $u = (u^1, \ldots, u^h)$ adapted to the structure in blocks of A can be chosen such that, for n large enough, we have

$$\|u_n^j\| \leq |x_n|^{k(\alpha_j - \varepsilon)}, \tag{10.4}$$

and we know that on

$$\mathcal{D} = \{(x, u) \in S_{\gamma, s, \rho} : \|u_n^j\| \leq |x_n|^{k(\alpha_j - \varepsilon)}, \text{ for } j = 1, \ldots, h\}, \tag{10.5}$$

we can conjugate holomorphically Φ to the translation

$$\begin{cases} \dfrac{1}{z_1} = \dfrac{1}{z} + 1, \\ U_1 = U, \end{cases}$$

with a change of the form $(z(x, u), U(x, u))$ such that

$$U(x, u) = x^{-kA} u + O(x^\eta), \tag{10.6}$$

for some positive η, and $z(x,u) \approx x^k$ as $x \to 0$.

Let $\psi \colon \mathcal{D} \to \mathbb{C}^p$ be defined by

$$\psi(x,u) := (Z(x,u), U(x,u)) = \left(\frac{1}{z(x,u)}, U(x,u) \right),$$

and let $\tau \colon \mathbb{C}^p \to \mathbb{C}^p$ be the translation $\tau(Z,U) := (Z+1, U)$. We know that $\mathcal{D}$ is Φ-invariant

$$\tau \circ \psi = \psi \circ \Phi. \tag{10.7}$$

Let us consider $W := \psi(\mathcal{D})$. For γ small enough and $R > 0$ big enough, the projection $Z(W)$ of W on $\mathbb{C}$ contains the set

$$\Sigma_{\gamma,R} := \{ Z \in \mathbb{C} : |\operatorname{Im} Z| < \gamma \operatorname{Re} Z, |Z| > R \}. \tag{10.8}$$

For any fixed $Z \in \mathbb{C}$ and $r > 0$, consider the generalized polydisc

$$P_{(Z,r)} := \{ (Z,U) \in \mathbb{C}^p : \|U^j\| \le r \text{ for } j = 1, \dots, h \}.$$

The definition (10.5) of $\mathcal{D}$ and the form (10.6) of $U(x,u)$ imply that for $Z \in \Sigma_{\gamma,R}$ and R big enough, W contains the generalized polydisc $P_{(Z, |Z|^{\epsilon/2})}$. For $|Z|$ tending to infinity, the fiber of W above Z contains generalized polydisc $P_{(Z,r)}$ of radius arbitrarily large. Hence, we have

$$\bigcup_{n \ge 0} \tau^{-n}(W) = \mathbb{C}^p. \tag{10.9}$$

The end of the argument is then as in Fatou [Fa2, Fa3], as follows.

Since $\mathcal{D} \subset \Omega_{(0,[V])}$, and, thanks to (10.4), for n large enough, for every $X \in \Omega_{(0,[V])}$, $X_n \in \mathcal{D}$, we also have

$$\Omega_{(0,[V])} = \bigcup_{n \ge 0} \Phi^{-n}(\mathcal{D}). \tag{10.10}$$

Therefore, we can extend the isomorphism $\psi \colon \mathcal{D} \to W$, to

$$\tilde{\psi} \colon \Omega_{(0,[V])} \to \mathbb{C}^p$$

as follows: given $X \in \Omega_{(0,[V])}$, consider n_0 such that $\Phi^{n_0}(X) \in \mathcal{D}$, and define

$$\tilde{\psi}(X) := \tau^{-n_0} \circ \psi \circ \Phi^{n_0}(X).$$

Thanks to (10.7), the definition does not depend on n. It is immediate to check that $\tilde{\psi}$ is injective, whereas its surjectivity follows from (10.10). $\qquad\square$

This last result is the generalization of Theorems 1.10 and 1.11 of [Ha2] for the case $k + 1 \ge 2$.

THEOREM 10.3. *Let $\Phi \in \operatorname{Diff}(\mathbb{C}^p, 0)$ be a tangent to the identity germ. Let $[V]$ be a nondegenerate characteristic direction, and assume it has exactly d directors, counted with multiplicities, with strictly positive real parts, greater than $\alpha > 0$. Then*

1. *if the remaining directors have strictly negative real parts, the attractive basin $\Omega_{(0,[V])}$ is biholomorphic to $\mathbb{C}^{d+1}$;*

2. *otherwise, considering coordinates such that* $[V] = [1 : 0]$, *the set*

$$\widetilde{\Omega}_{(0,[V])} := \{X \in \Omega_{(0,[V])} : \lim_{n \to +\infty} X_n^{-k\alpha} u_n = 0\}$$

is biholomorphic to $\mathbb{C}^{d+1}$, *and moreover its definition does not depend on* α.

PROOF. Thanks to the previous results we can apply Lemma 8.10 and Property (10.2). We can thus choose local holomorphic coordinates in a sector, such that, after the blow-up, Φ has the form

$$\begin{cases} x_1 = f(x, u, v) = x - \dfrac{1}{k} x^{k+1} + F(x, u, v), \\ \\ u_1 = g(x, u, v) = (I_d - x^k A)u + G(x, u, v), \\ v_1 = h(x, u, v) = (I_{p-d-1} - x^k B)v + H(x, u, v), \end{cases}$$

where A, and B are in Jordan normal form, A has eigenvalues with strictly positive real parts, B has eigenvalues with non-positive real parts, and F, G, and H satisfying (8.2). Moreover, thanks to Theorem 8.3, we may assume $H(x, u, 0) = 0$.

If $X \in \Omega_{(0,[V])}$, for γ, s, ρ arbitrarily small positive numbers, then $X_n \in S_{\gamma,s,\rho}$, for n big enough.

Assume that B has only eigenvalues with strictly negative real parts. Therefore, thanks to the previous equations, we have $\|v_{n+1}\| > \|v_n\|$ for n big enough, so v_n cannot converge to 0 unless we have $v_n = 0$. Hence

$$\Omega_{(0,[V])} \cap S_{\gamma,s,\rho} \subset \{v = 0\},$$

and we can apply the same argument as in Theorem 10.2 to $\Phi|_{S_{\gamma,s,\rho} \cap \{v=0\}}$.

If B has eigenvalues with non-positive real parts, since in $\widetilde{\Omega}_{(0,[V])}$, for n big enough, we have $\|x_{n+1}^{-k\alpha} v_{n+1}\| > \|x_n^{-k\alpha} v_n\|$, we cannot have $x_n^{-k\alpha} v_n$ converging to 0 unless $v_n = 0$. Therefore we argue as before, but considering $\widetilde{\Omega}_{(0,[V])}$. $\square$

11 Bibliography

[Ab1] M. Abate, *Diagonalization of non-diagonalizable discrete holomorphic dynamical systems*. Amer. J. Math. **122** (2000), 757–781.

[Ab2] M. Abate, *The residual index and the dynamics of holomorphic maps tangent to the identity*, Duke Math. J., **107**, (2001), 173–207.

[Ab3] M. Abate, *Discrete local holomorphic dynamics*. Proceedings of 13th Seminar on Analysis and Its Applications, Isfahan 2003, Eds. S. Azam et al., University of Isfahan, Iran (2005), 1–32.

[Ab4] M. Abate, *Discrete local holomorphic dynamics*. In "Holomorphic dynamical systems," Eds. G. Gentili, J. Guenot, G. Patrizio, Lect. Notes in Math. 1998, Springer, Berlin (2010), 1–55.

[AT] M. Abate and F. Tovena: *Poincaré-Bendixson theorems for meromorphic connections and holomorphic homogeneous vector fields*. J. Diff. Eq. **251**, (2011), no. 9, 2612–2684. MR 2825343 (2012h:32026).

[Be]   A. Beardon. "Iteration of rational functions." Springer, New York, 1995.

[Br]   F. Bracci, *Local dynamics of holomorphic diffeomorphisms*. Boll. UMI (8) 7-B (2004), 609–636.

[BM]   F. Bracci and L. Molino, *The dynamics near quasi-parabolic fixed points of holomorphic diffeomorphisms in* $\mathbb{C}^2$. Amer. J. Math. **126** (2004), 671–686.

[BRZ] F. Bracci and J. Raissy and D. Zaitsev, *Dynamics of multi-resonant biholomorphisms*. Int. Math. Res. Notices, first published online August 27, 2012. `http://imrn.oxfordjournals.org/content/early/2012/08/27/imrn.rns192.abstract`

[CC]   A. Candel and L. Conlon, "Foliations, I." Graduate Studies in Math. 23, American Mathematical Society, Providence, RI, 2000. MR 1732868.

[Ec]   J. Écalle, *Les fonctions résurgentes, Tome III: L'équation du pont et la classification analytiques des objects locaux*. Publ. Math. Orsay, 85-5, Université de Paris-Sud, Orsay, 1985.

[Fa]   P. Fatou, *Substitutions analytiques et equations fonctionelles de deux variables*. Ann. Sc. Ec. Norm. Sup. (1924), 67–142.

[Fa2] P. Fatou, *Sur les fonctions uniformes de deux variables*. C. R. Acad. Sci. Paris **175** (1922), 862–865.

[Fa3] P. Fatou, *Sur certaines fonctions uniformes de deux variables*. C. R. Acad. Sci. Paris **175** (1922), 1030–1033.

[Ha]   M. Hakim, *Analytic transformations of* $(\mathbb{C}^p, 0)$ *tangent to the identity*, Duke Math. J. **92** (1998), 403–428.

[Ha2] M. Hakim, *Transformations tangent to identity. Stable pieces of manifolds*. Preprint.

[IY]   Y. Ilyashenko and S. Yakovenko, "Lectures on analytic differential equations." Graduate Studies in Math. 86. American Mathematical Society, Providence, RI, 2008. MR 2363178.

[Le]   L. Leau, *Étude sur les equations fonctionelles à une ou plusieurs variables*, Ann. Fac. Sci. Toulouse, **11** (1897) E1–E110.

[Mi]   J. Milnor, "Dynamics in one complex variable," 3rd edition. Annals of Mathematics Studies **160**. Princeton University Press, Princeton, NJ, 2006.

[Mo]   L. Molino, *The dynamics of maps tangent to the identity and with nonvanishing index*, Trans. Amer. Math. Soc., **361** (2009), 1597–1623.

[Ra]   J. Raissy, "Geometrical methods in the normalization of germs of biholomorphisms," Ph.D. Thesis, Univeristy of Pisa, (2010). `http://etd.adm.unipi.it/theses/available/etd-02112010-094712/`

[Ri]   M. Rivi, *Parabolic manifolds for semi-attractive holomorphic germs*. Michigan Math. J., **49** (2001), no. 2, 211–241.

[R1]   F. Rong, *Quasi-parabolic analytic transformations of* $\mathbb{C}^n$. J. Math. Anal. Appl. **343** (2008), no. 1, 99–109.

[R2]   F. Rong, *Quasi-parabolic analytic transformations of* $\mathbb{C}^n$. *Parabolic manifolds.*, Ark. Mat., **48** (2010), 361–370.

[V1]   L. Vivas, *Fatou-Bieberbach domains as basins of attraction of automorphisms tangent to the identity*, J. Geom. Anal. **22** (2012), no. 2, 352–382. MR 2891730 (2012m:32017).   arXiv:0907.2061.

[V2]   L. Vivas, *Degenerate characteristic directions for maps tangent to the Identity*, to appear, Indiana J. Math. (2013). arXiv:1106.1471.

[Yo]   J-C. Yoccoz, *Centralisateurs et conjugaison différentiable des difféomorphismes du cercle. Petits diviseurs en dimension 1.* Astérisque **231** (1995), 89–242. MR 1367354.

Index theorems for meromorphic self-maps of the projective space

Marco Abate

ABSTRACT. We study global meromorphic self-maps of complex projective space using techniques from the theory of local dynamics of holomorphic germs tangent to the identity, More precisely, we prove three index theorems, relating suitably defined local residues at the fixed and indeterminacy points of a meromorphic map $f \colon \mathbb{P}^n \dashrightarrow \mathbb{P}^n$ with Chern classes of $\mathbb{P}^n$.

1 INTRODUCTION

In this short note we would like to show how the techniques introduced in [ABT1] (see also [ABT2], [ABT3], [Br], [BT] and [AT2]) for studying the local dynamics of holomorphic germs tangent to the identity can be used to study global meromorphic self-maps of the complex projective space $\mathbb{P}^n$. More precisely, we shall prove the following index theorem.

THEOREM 1.1. *Let $f \colon \mathbb{P}^n \dashrightarrow \mathbb{P}^n$ be a meromorphic self-map of degree $\nu + 1 \geq 2$ of the complex n-dimensional projective space. Let $\Sigma(f) = \mathrm{Fix}(f) \cup I(f)$ be the union of the indeterminacy set $I(f)$ of f and the fixed points set $\mathrm{Fix}(f)$ of f. Let $\Sigma(f) = \sqcup_\alpha \Sigma_\alpha$ be the decompositon of Σ in connected components, and denote by N the tautological line bundle of $\mathbb{P}^n$. Then*

(i) *we can associate to each Σ_α a complex number $\mathrm{Res}^1(f, \Sigma_\alpha) \in \mathbb{C}$, depending only on the local behavior of f nearby Σ_α, so that*

$$\sum_\alpha \mathrm{Res}^1(f, \Sigma_\alpha) = \int_{\mathbb{P}^n} c_1(N)^n = (-1)^n \, ;$$

(ii) *given a homogeneous symmetric polynomial $\varphi \in \mathbb{C}[z_1, \dots, z_n]$ of degree n, we can associate to each Σ_α a complex number $\mathrm{Res}^2_\varphi(f, \Sigma_\alpha) \in \mathbb{C}$, depending only on the local behavior of f nearby Σ_α, such that*

$$\sum_\alpha \mathrm{Res}^2_\varphi(f, \Sigma_\alpha) = \int_{\mathbb{P}^n} \varphi(T\mathbb{P}^n - N^{\otimes \nu}) \, ;$$

(iii) *if $\nu > 1$, given a homogeneous symmetric polynomial $\psi \in \mathbb{C}[z_0, \dots, z_n]$ of degree n we can associate to each Σ_α a complex number $\mathrm{Res}^3_\psi(f, \Sigma_\alpha) \in \mathbb{C}$, depending only on the local behavior of f nearby Σ_α, such that*

$$\sum_\alpha \mathrm{Res}^3_\psi(f, \Sigma_\alpha) = \int_{\mathbb{P}^n} \psi\big((T\mathbb{P}^n \oplus N) - N^{\otimes \nu}\big) \, .$$

I would like to thank Francesca Tovena, Jasmin Raissy, and Matteo Ruggiero for many useful conversations, Nurìa Fagella and the Institut de Matemàtica de la Universitat de Barcelona for their wonderful hospitality during the preparation of this note, and Jack Milnor for creating so much beautiful mathematics and for being an inspiration to us all.

In this statement, if $\varphi \in \mathbb{C}[z_1, \ldots, z_n]$ is a homogeneous symmetric polynomial and E and F are vector bundles over $\mathbb{P}^n$, we put

$$\varphi(E - F) = \tilde{\varphi}\big(c_1(E - F), \ldots, c_r(E - F)\big) ,$$

where the $c_j(E - F)$ are the Chern classes of the virtual bundle $E - F$, and $\tilde{\varphi}$ is the unique polynomial in $\mathbb{C}[z_1, \ldots, z_r]$ such that

$$\varphi = \tilde{\varphi}(\sigma_1, \ldots, \sigma_r) ,$$

where $\sigma_1, \ldots, \sigma_r \in \mathbb{C}[z_1, \ldots, z_n]$ are the elementary symmetric functions on n variables.

REMARK 1.2. If $\Sigma(f)$ is finite, then the number of points in $\Sigma(f)$, counted with respect to a suitable multiplicity, is

$$\frac{1}{\nu}[(\nu + 1)^{n+1} - 1] = \sum_{k=0}^{n} \binom{n+1}{k+1} \nu^k ;$$

see, e.g., [AT1].

REMARK 1.3. Since the total Chern class of $T\mathbb{P}^n$ in terms of the first Chern class of N is given by $c(T\mathbb{P}^n) = \big(1 - c_1(N)\big)^{n+1}$, the total Chern class of $T\mathbb{P}^n - N^{\otimes \nu}$ is given by

$$c(T\mathbb{P}^n - N^{\otimes \nu}) = \frac{\big(1 - c_1(N)\big)^{n+1}}{1 + \nu c_1(N)} = \big(1 - c_1(N)\big)^{n+1} \sum_{j=0}^{n} (-1)^j \nu^j c_1^j(N)$$

and thus the Chern classes of $T\mathbb{P}^n - N^{\otimes \nu}$ are given by

$$c_j(T\mathbb{P}^n - N^{\otimes \nu}) = (-1)^j \sum_{k=0}^{j} \binom{n+1}{k} \nu^{j-k} c_1^j(N) .$$

In particular,

$$\int_{\mathbb{P}^n} c_n(T\mathbb{P}^n - N^{\otimes \nu}) = \sum_{k=0}^{n} \binom{n+1}{k} \nu^{n-k} \quad \text{and} \quad \int_{\mathbb{P}^n} c_1^n(T\mathbb{P}^n - N^{\otimes \nu}) = (\nu + 1 + n)^n .$$

$$(1.1)$$

Arguing in a similar way we get that the Chern classes of $(T\mathbb{P}^n \oplus N) - N^{\otimes \nu}$ are given by

$$c_j\big((T\mathbb{P}^n \oplus N) - N^{\otimes \nu}\big) = (-1)^j \sum_{k=0}^{j} \left[\binom{n+1}{k} - \binom{n+1}{k-1} \right] \nu^{j-k} c_1^j(N) ,$$

with the convention $\binom{n+1}{-1} = 0$.

REMARK 1.4. As the proof of Theorem 1.1 presented at the end of this note will make clear, this index theorem is a direct consequence of the index theorems proved in [ABT1] on holomorphic self-maps pointwise fixing a hypersurface S of a complex manifold M. In turn, as shown in a very general setting in [ABT2], these index theorems follow from the existence of a holomorphic action (or partial connection) of a suitable tensor power $N_S^{\otimes \nu}$ of the normal bundle N_S on a (possibly virtual) vector bundle E on S. The embedding of S into M yields three natural vector bundles to consider: N_S, TS, and $TM|_S$; accordingly, we get three index theorems. Indeed, case (i) of Theorem 1.1 corresponds to the existence of a holomorphic action on N_S (a *Camacho-Sad action*); case (ii) corresponds to the existence of a holomorphic action on $TS - N_S^{\otimes \nu}$ (a *Baum-Bott action*); and case (iii) corresponds to the existence of a holomorphic action on $TM|_S - N_S^{\otimes \nu}$ (a *Lehmann-Suwa action*).

REMARK 1.5. As a personal aside, in my opinion index theorems, giving quantitative and explicit links between local and global objects, rank among the most beautiful results in mathematics and, as such, are an end in themselves. For this reason, since I was a budding young mathematician I hoped to be able to discover at least one new index theorem during my career. So, as you can imagine, I was pretty excited when in [A2] I realized that I was proving such a theorem; and now that thanks to the work done in [ABT1] and [ABT2] (and in this note) I can say that I actually proved not one but several new index theorems, I consider myself (mathematically) realized and content (and thus ready to explore new avenues of research in the future, of course). But index theorems, besides being beautiful, are also useful; for instance, they can be of help in classifying objects or behaviors, imposing constraints that should be satisfied. We shall not discuss applications of Theorem 1.1 in this note; however similar (though less general) theorems already present in the literature have been, for instance, applied to show that generic holomorphic self-maps of $\mathbb{P}^n(\mathbb{C})$ have infinitely many non-attracting periodic points (see [U]); to classify complex homogenous vector fields (see [G]); to classify particular classes of Liouville integrable complex Hamiltonian systems (see [P1], [P2]); and to study the (non-)existence of compact positive dimensional fixed points set of holomorphic self-maps of complex surfaces (see [ABT1]).

2 THE PROOF

The usefulness of a theorem like Theorem 1.1 is, of course, related to the possibility of computing the residues involved, and indeed we have explicit formulas for the (generic) case of point components, expressed in terms of Grothendieck residues (the residues at components of positive dimension can be expressed by using Lehmann-Suwa theory; see [LS], [S, Chapter IV], [ABT1, (6.1)], and [BS]; see also [B]). To state the formulas we recall three definitions.

DEFINITION 2.1. *Let $\mathcal{O}_{w^o}^n$ be the ring of germs of holomorphic functions at $w^o \in \mathbb{C}^n$. Take $g_1, \ldots, g_n \in \mathcal{O}_{w^o}^n$ such that w^o is an isolated zero of $g = (g_1, \ldots, g_n)$. Then the Grothendieck residue of $h \in \mathcal{O}_{w^o}^n$ along $g_1, \ldots, g_n$ is defined (see [H], [L, Section 5], [LS, Section 4], [S, pp.105–107]) by the formula*

$$\mathrm{Res}_{w^o} \left[\frac{h(w)\, dw_1 \wedge \cdots \wedge dw_n}{g_1, \ldots, g_n} \right] = \left(\frac{1}{2\pi i} \right)^n \int_{\Gamma} \frac{h}{g_1 \cdots g_n}\, dw_1 \wedge \cdots \wedge dw_n \,,$$

where $\Gamma = \{|g_j(w)| = \varepsilon \mid j = 1, \ldots, n\}$ for $\varepsilon > 0$ small enough, oriented so that $d\arg(g_1) \wedge \cdots \wedge d\arg(g_n) > 0$.

DEFINITION 2.2. *Let $\varphi \in \mathbb{C}[z_1, \ldots, z_n]$ be a homogeneous symmetric polynomial in n variables, and $L\colon V \to V$ an endomorphism of an n-dimensional complex vector space V. Then we set*

$$\varphi(L) = \varphi(\lambda_1, \ldots, \lambda_n),$$

where $\lambda_1, \ldots, \lambda_n \in \mathbb{C}$ are the eigenvalues of L.

DEFINITION 2.3. *We shall say that a homogeneous polynomial self-map*

$$F = (F_0, \ldots, F_n)\colon \mathbb{C}^{n+1} \to \mathbb{C}^{n+1}$$

of degree $\nu + 1$ induces the meromorphic self-map $f\colon \mathbb{P}^n \dashrightarrow \mathbb{P}^n$ if

$$f([z_0 : \cdots : z_n]) = [F_0(z_0, \ldots, z_n) : \cdots : F_n(z_0, \ldots, z_n)]$$

for all $[z_0 : \cdots : z_n] \in \mathbb{P}^n$. It is well known (see, e.g., [FS1]) that every meromorphic self-map of $\mathbb{P}^n$ of degree $\nu + 1$ is induced by a unique homogeneous polynomial self-map of $\mathbb{C}^{n+1}$ of degree $\nu + 1$.

We can now state our next theorem.

THEOREM 2.4. *Let $f\colon \mathbb{P}^n \dashrightarrow \mathbb{P}^n$ be a meromorphic self-map of degree $\nu + 1 \geq 2$ of the complex n-dimensional projective space. Let $\Sigma(f) = \mathrm{Fix}(f) \cup I(f)$ be the union of the indeterminacy set $I(f)$ of f and the fixed points set $\mathrm{Fix}(f)$ of f, and assume that $p = [1 : w_1^o : \cdots : w_n^o]$ is an isolated point in $\Sigma(f)$. Let w^o denote $(w_1^o, \ldots, w_n^o)$, let $F = (F_0, \ldots, F_n)\colon \mathbb{C}^{n+1} \to \mathbb{C}^{n+1}$ be the homogeneous polynomial self-map of degree $\nu + 1$ inducing f, and define $g\colon \mathbb{C}^n \to \mathbb{C}^n$ by setting*

$$g_j(w) = F_j(1, w) - w_j F_0(1, w) \tag{2.1}$$

for all $w \in \mathbb{C}^n$ and $j = 1, \ldots, n$. Then

(i) *we have*

$$\mathrm{Res}^1(f, p) = \mathrm{Res}_{w^o} \left[\frac{F_0(1, w)^n \, dw_1 \wedge \cdots \wedge dw_n}{g_1, \ldots, g_n} \right];$$

(ii) *if $\varphi \in \mathbb{C}[z_1, \ldots, z_n]$ is a homogeneous symmetric polynomial of degree n, then*

$$\mathrm{Res}^2_\varphi(f, p) = \mathrm{Res}_{w^o} \left[\frac{\varphi(dg_w) \, dw_1 \wedge \cdots \wedge dw_n}{g_1, \ldots, g_n} \right];$$

(iii) *if $\psi \in \mathbb{C}[z_0, \ldots, z_n]$ is a homogeneous symmetric polynomial of degree n, then*

$$\mathrm{Res}^3_\psi(f, p) = \mathrm{Res}_{w^o} \left[\frac{\psi\big(F_0(1, w), \mu_1(w), \ldots, \mu_n(w)\big) \, dw_1 \wedge \cdots \wedge dw_n}{g_1, \ldots, g_n} \right],$$

where $\mu_1(w), \ldots, \mu_n(w)$ are the eigenvalues of dg_w.

This statement is effective because we can explicitly compute a Grothendieck residue using an algorithm suggested by Hartshorne (see [BB] and [H]). Without loss of generality, we can assume $w^o = O$. Since the origin is an isolated zero of g, there exist minimal positive integers $\alpha_1, \ldots, \alpha_n$ such that $w_1^{\alpha_1}, \ldots, w_n^{\alpha_n}$ belong to the ideal generated by $g_1, \ldots, g_n$ in $\mathcal{O}^n = \mathcal{O}_O^n$. Hence, there exist holomorphic functions $b_{ij} \in \mathcal{O}^n$ such that

$$w_i^{\alpha_i} = \sum_{j=1}^{n} b_{ij} g_j .$$

The properties of the Grothendieck residue (see [H]) then imply

$$\mathrm{Res}_O \begin{bmatrix} h \, dw_1 \wedge \cdots \wedge dw_n \\ g_1, \cdots, g_n \end{bmatrix} = \mathrm{Res}_O \begin{bmatrix} h \det(b_{ij}) \, dw_1 \wedge \cdots \wedge dw_n \\ w_1^{\alpha_1}, \cdots, w_n^{\alpha_n} \end{bmatrix} .$$

The right-hand side is now evaluated by expanding $h \det(b_{ij})$ in a power series in the w_i; the residue is given by the coefficient of $w_1^{\alpha_1-1} \cdots w_n^{\alpha_n-1}$.

The easiest case is when $\det(dg_O) \neq 0$. Indeed, since O is an isolated zero of g we can write

$$g_j = \sum_{i=1}^{k} c_{ji} w_i .$$

Differentiating this and evaluating in O we get

$$\frac{\partial g_j}{\partial w_k}(O) = c_{jk}(O) ;$$

therefore, if $\det(dg_O) \neq 0$ we can invert the matrix (c_{ji}) in a neighborhood of O and write

$$w_i = \sum_{j=1}^{n} b_{ij} g_j ,$$

where (b_{ij}) is the inverse matrix of (c_{ji}). So we have $\alpha_1 = \cdots = \alpha_n = 1$ and also $\det(b_{ij})(O) = 1/\det(dg_O)$.

It thus follows that

$$\det(dg_{w^o}) \neq 0 \implies \mathrm{Res}_{w^o} \begin{bmatrix} h(w) \, dw_1 \wedge \cdots \wedge dw_n \\ g_1, \ldots, g_n \end{bmatrix} = \frac{h(w^o)}{\mu_1 \cdots \mu_n} , \qquad (2.2)$$

where $\mu_1, \ldots, \mu_n \in \mathbb{C}$ are the eigenvalues of dg_{w^o}.

In our situation, $g_j(w) = F_j(1, w) - w_j F_0(1, w)$; therefore, we have

$$\frac{\partial g_j}{\partial w_k}(w) = \frac{\partial F_j}{\partial w_k}(1, w) - w_j \frac{\partial F_0}{\partial w_k}(1, w) - \delta_k^j F_0(1, w) ,$$

where δ_k^j is Kronecker's delta.

So if $p = [1 : w_1^o : \cdots : w_n^o]$ is an indeterminacy point, we have $F_0(1, w^o) = 0$ and

$$\mathrm{Jac}(g)(w^o) = \left(\frac{\partial F_j}{\partial w_k}(1, w^o) - w_j^o \frac{\partial F_0}{\partial w_k}(1, w^o) \right)_{j,k=1,\ldots,n} .$$

If instead p is a fixed point, we can consider the differential of f at p. Let χ be the usual chart of $\mathbb{P}^n$ centered at $[1 : 0 : \cdots : 0]$. Then

$$\tilde{f}(w) = \chi \circ f \circ \chi^{-1}(w) = \left(\frac{F_1(1, w)}{F_0(1, w)}, \ldots, \frac{F_n(1, w)}{F_0(1, w)} \right) ,$$

and so

$$\frac{\partial \tilde{f}_j}{\partial w_k}(w) = \frac{1}{F_0(1,w)} \left[\frac{\partial F_j}{\partial w_k}(1,w) - \frac{F_j(1,w)}{F_0(1,w)} \frac{\partial F_0}{\partial w_k}(1,w) \right] .$$

In particular,

$$\frac{\partial \tilde{f}_h}{\partial w_k}(w^o) = \frac{1}{F_0(1,w^o)} \left[\frac{\partial F_j}{\partial w_k}(1,w^o) - w_j^o \frac{\partial F_0}{\partial w_k}(1,w^o) \right] ; \qquad (2.3)$$

therefore,

$$\mathrm{Jac}(g)(w^o) = F_0(1,w^o) \left[\mathrm{Jac}(\tilde{f})(w^o) - I \right] .$$

It follows that the eigenvalues $\mu_1, \ldots, \mu_n$ of dg_{w^o} are related to the eigenvalues $\lambda_1, \ldots, \lambda_n$ of df_p by the formula

$$\forall j = 1, \ldots, n \quad \mu_j = F_0(1,w^o)(\lambda_j - 1) .$$

In particular, $\det(dg_{w^o}) \neq 0$ if and only if 1 is not an eigenvalue of df_p, that is, if and only if p is a simple fixed point of f.

Summing up, these computations give the following particular case of Theorem 2.4.

COROLLARY 2.5. *Let* $f \colon \mathbb{P}^n \dashrightarrow \mathbb{P}^n$ *be a meromorphic self-map of degree* $\nu + 1 \geq 2$ *of the complex* n-*dimensional projective space.*

(a) *Assume that* $p \in \mathbb{P}^n$ *is a simple (necessarily isolated) fixed point of* f, *and let* $\lambda_1, \ldots, \lambda_n \neq 1$ *be the eigenvalues of* df_p. *Then*

    (i) *we have*

$$\mathrm{Res}^1(f,p) = \frac{(-1)^n}{(1-\lambda_1) \cdots (1-\lambda_n)} ;$$

    (ii) *if* $\varphi \in \mathbb{C}[z_1, \ldots, z_n]$ *is a homogeneous symmetric polynomial of degree* n, *then*

$$\mathrm{Res}_\varphi^2(f,p) = \frac{\varphi(1-\lambda_1, \ldots, 1-\lambda_n)}{(1-\lambda_1) \cdots (1-\lambda_n)} ;$$

    (iii) *if* $\psi \in \mathbb{C}[z_0, \ldots, z_n]$ *is a homogeneous symmetric polynomial of degree* n, *then*

$$\mathrm{Res}_\psi^3(f,p) = (-1)^n \frac{\psi(1, \lambda_1 - 1, \ldots, \lambda_n - 1)}{(1-\lambda_1) \cdots (1-\lambda_n)} .$$

(b) *Assume that* $p = [1 : w_1^o : \cdots : w_n^o]$ *is an isolated indeterminacy point, and that* $\det G \neq 0$, *where*

$$G = \left(\frac{\partial F_j}{\partial w_k}(1,w^o) - w_j^o \frac{\partial F_0}{\partial w_k}(1,w^o) \right)_{j,k=1,\ldots,n} ,$$

and $F = (F_0, \ldots, F_n) \colon \mathbb{C}^{n+1} \to \mathbb{C}^{n+1}$ *is the homogeneous polynomial self-map of degree* $\nu + 1$ *inducing* f. *Denote by* $\mu_1, \ldots, \mu_n \neq 0$ *the eigenvalues of* G. *Then*

    (i) *we have*

$$\mathrm{Res}^1(f,p) = 0 ;$$

(ii) *if $\varphi \in \mathbb{C}[z_1, \ldots, z_n]$ is a homogeneous symmetric polynomial of degree n, then*

$$\text{Res}_\varphi^2(f, p) = \frac{\varphi(\mu_1, \ldots, \mu_n)}{\mu_1 \cdots \mu_n} \ ;$$

(iii) *if $\psi \in \mathbb{C}[z_0, \ldots, z_n]$ is a homogeneous symmetric polynomial of degree n, then*

$$\text{Res}_\psi^3(f, p) = \frac{\psi(0, \mu_1, \ldots, \mu_n)}{\mu_1 \cdots \mu_n} \ .$$

This corollary shows that our Theorem 1.1 is related to Ueda's index theorem, which, however, applies only to holomorphic self-maps having only simple fixed points:

THEOREM 2.6 ([U]). *Let $f : \mathbb{P}^n \to \mathbb{P}^n$ be holomorphic of degree $\nu + 1 \geq 2$, and assume that all fixed points of f are simple (and thus isolated). For $p \in \text{Fix}(f)$, let $\lambda_1(p), \ldots, \lambda_n(p) \neq 1$ denote the eigenvalues of df_p. Then*

$$\forall k = 0, \ldots, n \sum_{p \in \text{Fix}(f)} \frac{\sigma_k(df_p)}{(1 - \lambda_1(p)) \cdots (1 - \lambda_n(p))} = (-1)^k (\nu + 1)^k \ .$$

For instance, the case $k = 0$ is a consequence of Theorem 1.1(i), and the cases $k > 0$ can be linked to Theorem 1.1(ii) and (iii) by using the formula

$$\sigma_j(I - L) = \sum_{\ell=0}^{j} \binom{n - \ell}{n - j} (-1)^\ell \sigma_\ell(L) \ ,$$

valid for every endomorphism L of an n-dimensional complex vector space.

REMARK 2.7. Corollary 2.5 shows that our Theorem 1.1 generalizes to the case of non-simple or non-isolated fixed (or indeterminacy) points the classical holomorphic Lefschetz formula (see, e.g., [GH, p. 426]), as well as results obtained by Guillot [G] and, in a different context, by Przybylska [P1, P2].

EXAMPLE 2.8. *Let us consider the map $f([z]) = [z_0^{\nu+1} : \cdots : z_n^{\nu+1}]$. This map is holomorphic, and all its fixed points are simple. More precisely, for each $\ell = 0, \ldots, n$ the set $\text{Fix}(f)$ contains exactly $\binom{n+1}{\ell+1} \nu^\ell$ fixed points of the form $[\zeta_0 : \cdots : \zeta_n]$, where $\ell + 1$ of the ζ_j are ν th roots of unity and the remaining ζ_js are equal to 0. If $p \in \text{Fix}(f)$ has exactly $\ell + 1$ non-zero homogeneous coordinates, we shall say that p is a fixed point of level ℓ.*

Using Equation (2.3) it is easy to see that if $p \in \text{Fix}(f)$ is a fixed point of level ℓ then the eigenvalues of df_p are $\nu + 1$ with multiplicity ℓ and 0 with multiplicity $n - \ell$. In particular, $(1 - \lambda_1(p)) \cdots (1 - \lambda_n(p)) = (-1)^\ell \nu^\ell$, and so Theorem 1.1(i) becomes

$$(-1)^n \sum_{\ell=0}^{n} \binom{n + 1}{\ell + 1} (-1)^\ell = (-1)^n \ ,$$

while Theorem 1.1(ii) becomes

$$(-1)^n \sum_{\ell=0}^{n} \binom{n + 1}{\ell + 1} (-1)^\ell \varphi(\underbrace{\nu, \ldots, \nu}_{\ell \text{ times}}, -1, \ldots, -1) = \int_{\mathbb{P}^n} \varphi(T\mathbb{P}^n - N^{\otimes \nu}) \ ,$$

and Theorem 1.1(iii) becomes

$$(-1)^n \sum_{\ell=0}^{n} \binom{n+1}{\ell+1}(-1)^\ell \psi(1,\underbrace{\nu,\ldots,\nu}_{\ell \text{ times}},-1,\ldots,-1) = \int_{\mathbb{P}^n} \psi((T\mathbb{P}^n \oplus N) - N^{\otimes \nu}) \,.$$

For instance, taking $\varphi = \sigma_1^n$ and recalling Equation (1.1) we get

$$(\nu+1+n)^n = \int_{\mathbb{P}^n} c_1^n(T\mathbb{P}^n - N^{\otimes\nu}) = (-1)^n \sum_{\ell=0}^{n} \binom{n+1}{\ell+1}(-1)^\ell ((\nu+1)\ell - n)^n \,.$$

The (non-trivial) equality of the left and right-hand sides in this formula can also be proved directly by using Abel's formula,

$$(x+y)^r = \sum_{k=0}^{r} \binom{r}{k} x(x-kz)^{k-1}(y+kz)^{r-k}\,, \tag{2.4}$$

valid for all $r \in \mathbb{N}$ and $x,\, y,\, z \in \mathbb{C}$ with $x \neq 0$ (see, e.g., [K, p. 56]). Indeed we have

$$\sum_{\ell=0}^{n} \binom{n+1}{\ell+1}(-1)^\ell ((\nu+1)\ell - n)^n$$

$$= \sum_{\ell=0}^{n} \binom{n+1}{\ell+1}(n-(\nu+1)\ell)^\ell (-n+(\nu+1)\ell)^{n-\ell}$$

$$= \sum_{k=1}^{n+1} \binom{n+1}{k}(n+\nu+1-k(\nu+1))^{k-1}(-n-\nu-1+k(\nu+1))^{n+1-k}$$

$$= \frac{1}{n+\nu+1}\left[\sum_{k=0}^{n+1} \binom{n+1}{k}(n+\nu+1)(n+\nu+1-k(\nu+1))^{k-1}\right.$$

$$\left. (-n-\nu-1+k(\nu+1))^{n+1-k} + (-1)^n(n+\nu+1)^{n+1}\right]$$

$$= (-1)^n(\nu+n+1)^n\,,$$

thanks to Equation (2.4) applied with $r = n+1$, $x = -y = n+\nu+1$ and $z = \nu+1$.

Analogously, taking $\varphi = \sigma_n$ we get

$$\sum_{k=0}^{n} \binom{n+1}{k}\nu^{n-k} = \int_{\mathbb{P}^n} c_n(T\mathbb{P}^n - N^{\otimes\nu})$$

$$= \sum_{\ell=0}^{n} \binom{n+1}{\ell+1}\nu^\ell = \frac{(\nu+1)^{n+1}-1}{\nu} = \sum_{\ell=0}^{n}(\nu+1)^\ell\,,$$

in agreement with both Equation (1.1) and Theorem 2.6.

Let us finally show how to use the local theory developed in [ABT1] and [AT2] to prove Theorems 1.1 and 2.4.

Given a meromorphic map $f \colon \mathbb{P}^n \dashrightarrow \mathbb{P}^n$ of degree $\nu + 1 \geq 2$, we again define $F = (F_0, \ldots, F_n) \colon \mathbb{C}^{n+1} \to \mathbb{C}^{n+1}$ to be the homogeneous polynomial map of degree $\nu + 1$ inducing f. First of all, we associate to F the homogeneous vector field

$$Q = \sum_{j=0}^{n} F_j \frac{\partial}{\partial z_j} \, .$$

It is well known that the time-1 map f_Q of Q is a germ of holomorphic self-map of $\mathbb{C}^{n+1}$ tangent to the identity; furthermore, we can write

$$f_Q(z) = z + F(z) + O(\|z\|^{\nu+2}).$$

In particular, by definition a direction $v \in \mathbb{C}^{n+1} \setminus \{O\}$ is a non-degenerate characteristic direction for f_Q if and only if $[v] \in \mathbb{P}^n$ is a fixed point of f; and it is a degenerate characteristic direction for f_Q if and only if $[v]$ is an indeterminacy point of f. Furthermore, since f has degree at least 2 then Q is non-dicritical, that is, not all directions are characteristic.

Now let $\pi \colon M \to \mathbb{C}^{n+1}$ be the blow-up of the origin, and denote the exceptional divisor by $E = \pi^{-1}(O)$. By construction, E is canonically biholomorphic to $\mathbb{P}^n$; furthermore, the blow-up M can be identified with the total space of the normal bundle N_E of E in M, and N_E is isomorphic to the tautological line bundle N on $\mathbb{P}^n$.

We can now lift f_Q to the blow-up, obtaining (see, e.g., [A1]) a germ $\hat{f}_Q$ about E of holomorphic self-map of M fixing E pointwise, and we may apply all the machinery developed in [ABT1]. First of all, since Q is non-dicritical, then $\hat{f}_Q$ is tangential and has order of contact ν with E ([ABT1, Proposition 1.4]). Then we can define the canonical section X_f (see [ABT1, Proposition 3.1 and Corollary 3.2]), which is a global holomorphic section of $TE \otimes (N_E^*)^{\otimes \nu}$. In local coordinates in M centered at $[1 : 0 : \cdots : 0] \in E$, we can write

$$X_f = \sum_{j=1}^{n} g_j \frac{\partial}{\partial w_j} \otimes (dw_0)^{\otimes \nu} \, , \tag{2.5}$$

where $g_1, \ldots, g_n$ are given by Equation (2.1), and E in these coordinates is given by $\{w_0 = 0\}$.

In particular, we can think of X_f as a bundle morphism $X_f \colon N_E^{\otimes \nu} \to TE$, vanishing exactly on the characteristic directions of f_Q, and thus we have proved the following (already noted, for instance, in [FS2, p. 409]).

PROPOSITION 2.9. *Let $f \colon \mathbb{P}^n \dashrightarrow \mathbb{P}^n$ be a meromorphic self-map of degree $\nu + 1 \geq 2$. Then the canonical section X_f, given locally by Equation (2.5), defines a global singular holomorphic foliation of $\mathbb{P}^n$ in Riemann surfaces, singular exactly at the fixed and indeterminacy points of f.*

We can now apply the index theorems proved in [ABT1]. Theorem 1.1(i) follows immediately from the Camacho-Sad-like index Theorem 6.2 in [ABT1], recalling that $\int_{\mathbb{P}^n} c_1(N)^n = (-1)^n$. Theorem 1.1(ii) follows from the Baum-Bott index theorem (see [S, Theorem III.7.6] and [ABT1, Theorem 6.4]); and Theorem 1.1(iii) follows from the Lehmann-Suwa-like index theorem 6.3 in [ABT1], that can be applied to this situation because E is automatically comfortably embedded in M (see [ABT1, Example 2.4]), and recalling that the exact sequence

$$O \longrightarrow TE \longrightarrow TM|_E \longrightarrow N_E \longrightarrow O$$

implies that $c(TM|_E) = c(T\mathbb{P}^n \oplus N_E)$.

Theorem 2.4 also follows from [ABT1]. Indeed, Theorem 2.4(i) is an immediate consequence of [ABT1, Theorem 6.5(i)] and the computations in [AT2, Section 5]. The same computations and [ABT1, Theorem 6.6] yield Theorem 2.4(iii); and Theorem 2.4(ii) follows from [BB] and [S, Theorem III.5.5].

3   Bibliography

[A1]   M. Abate, *Diagonalization of non-diagonalizable discrete holomorphic dynamical systems*, Amer. J. Math. **122** (2000) 757–781.

[A2]   M. Abate, *The residual index and the dynamics of holomorphic maps tangent to the identity*, Duke Math. J. **107** (2001) 173–207.

[ABT1] M. Abate, F. Bracci, and F. Tovena, *Index theorems for holomorphic self-maps*, Ann. of Math. **159** (2004) 819–864.

[ABT2] M. Abate, F. Bracci, and F. Tovena, *Index theorems for holomorphic maps and foliations*, Indiana Univ. Math. J. **57** (2008) 2999–3048.

[ABT3] M. Abate, F. Bracci, and F. Tovena, *Embeddings of submanifolds and normal bundles*, Adv. Math. **220** (2009) 620–656.

[AT1]   M. Abate and F. Tovena, *Parabolic curves in* $\mathbb{C}^3$, Abstr. Appl. Anal. **2003** (2003) 275–294.

[AT2]   M. Abate and F. Tovena, *Poincaré-Bendixson theorems for meromorphic connections and homogeneous vector fields*, J. Diff. Eq. **251** (2011) 2612–2684.

[BB]   P. Baum and R. Bott, *Singularities of holomorphic foliations*, J. Diff. Geom. **7** (1972) 279–342.

[B]   G. Biernat, *La représentation paramétrique d'un résidu multidimensionnel*, Rev. Roumaine Math. Pures Appl. **36** (1991) 207–211.

[Br]   F. Bracci, *The dynamics of holomorphic maps near curves of fixed points*, Ann. Scuola Norm. Sup. Pisa **2** (2003) 493–520.

[BS]   F. Bracci and T. Suwa, *Perturbation of Baum-Bott residues*, Preprint, arXiv:1006.3706 (2010).

[BT]   F. Bracci and F. Tovena, *Residual indices of holomorphic maps relative to singular curves of fixed points on surfaces*, Math. Z. **242** (2002) 481–490.

[FS1]   J.-E. Fornæss and N. Sibony, *Complex dynamics in higher dimension. I*, Astérisque **222** (1994) 201–231.

[FS2]   J.-E. Fornæss and N. Sibony, *Riemann surface laminations with singularities*, J. Geom. Anal. **18** (2008) 400–442.

[GH] P. Griffiths and J. Harris, "Principles of algebraic geometry," Wiley, New York, 1978.

[G]   A. Guillot, *Un théorème de point fixe pour les endomorphismes de l'espace projectif avec des applications aux feuilletages algébriques,* Bull. Braz. Math. Soc. **35** (2004) 345–362.

[H]   R. Hartshorne, "Residues and duality," Lecture Notes in Math. **20** Springer, Berlin, 1966.

[K]   D. E. Knuth, "The Art of Computer Programming, Vol. 1: Fundamental Algorithms," Addison-Wesley, Reading, MA, 1973.

[L]   D. Lehmann, *Résidues des sous-variétés invariants d'un feuilletage singulier,* Ann. Inst. Fourier, Grenoble **41** (1991) 211–258.

[LS] D. Lehmann and T. Suwa, *Residues of holomorphic vector fields relative to singular invariant subvarieties,* J. Diff. Geom. **42** (1995) 165–192.

[P1] M. Przybylska, *Darboux points and integrability of homogeneous Hamiltonian systems with three and more degrees of freedom,* Regul. Chaotic Dyn. **14** (2009) 263–311.

[P2] M. Przybylska, *Darboux points and integrability of homogeneous Hamiltonian systems with three and more degrees of freedom. Nongeneric cases,* Regul. Chaotic Dyn. **14** (2009) 349–388.

[S]   T. Suwa, "Indices of vector fields and residues of singular holomorphic foliations," Hermann, Paris, 1998.

[U]   T. Ueda, "Complex dynamics on projective spaces—index formula for fixed points," Dynamical systems and chaos, **1** (Hachioji, 1994) World Sci. Publ., River Edge, NJ, (1995) 252–259.

Dynamics of automorphisms of compact complex surfaces

Serge Cantat

ABSTRACT. Recent results concerning the dynamics of holomorphic diffeomorphisms of compact complex surfaces are described that require a nice interplay between algebraic geometry, complex analysis, and dynamical systems.

1 INTRODUCTION

1.1 Automorphisms

1.1.1 Automorphisms

Let M be a compact complex manifold. By definition, holomorphic diffeomorphisms $f \colon M \to M$ are called **automorphisms**; they form a group, the group $\mathrm{Aut}(M)$ of automorphisms of M. Endowed with the topology of uniform convergence, $\mathrm{Aut}(M)$ is a topological group, and a theorem due to Bochner and Montgomery shows that this topological group is a complex Lie group, whose Lie algebra is the algebra of holomorphic vector fields on M (see [18]). The connected component of the identity in $\mathrm{Aut}(M)$ is denoted $\mathrm{Aut}(M)^0$, and the group of its connected components is

$$\mathrm{Aut}(M)^\sharp = \mathrm{Aut}(M)/\mathrm{Aut}(M)^0.$$

1.1.2 Curves

If $M = \mathbb{P}^1(\mathbb{C})$, then $\mathrm{Aut}(M)$ is $\mathrm{PGL}_2(\mathbb{C})$, the group of linear projective transformations. In particular, $\mathrm{Aut}(M)$ is connected, and the dynamics of all elements $f \in \mathrm{Aut}(M)$ is easily described. (However, the theory of Kleinian groups shows the richness of the dynamics of subgroups of $\mathrm{Aut}(\mathbb{P}^1(\mathbb{C}))$).

If $M = \mathbb{C}/\Lambda$ is an elliptic curve, the connected component of the identity $\mathrm{Aut}(M)^0$ coincides with $\mathbb{C}/\Lambda$, acting by translations. The group $\mathrm{Aut}(M)$ is the semi-direct product of $\mathrm{Aut}(M)^0$ by the finite group F of similarities of $\mathbb{C}$ preserving Λ. The group F contains $z \mapsto -z$ and is generated by this involution in all cases except when Λ is similar to $\mathbb{Z}[\sqrt{-1}]$, and then $|F| = 4$, or when Λ is similar to $\mathbb{Z}[\omega]$, where ω is a cubic root of 1, and then $|F| = 6$.

If M is a connected curve of genus $g \geq 2$, Hurwitz's theorem shows that $\mathrm{Aut}(M)$ is finite, with at most $84(g-1)$ elements.

1.1.3 Connected components

Starting with $\dim(M) = 2$, the group $\mathrm{Aut}(M)$ may have an infinite number of connected components.

As an example, let $E = \mathbb{C}/\Lambda$ be an elliptic curve and $M = E^n$ be the product of n copies of E; in other words, M is the torus $\mathbb{C}^n/\Lambda^n$. The group $\mathrm{GL}_n(\mathbb{Z})$ acts linearly on $\mathbb{C}^n$, preserves the lattice $\Lambda^n \subset \mathbb{C}^n$, and, therefore, embeds into the group $\mathrm{Aut}(M)$. All non-trivial elements B in $\mathrm{GL}_n(\mathbb{Z})$ act non-trivially on the homology of M, so that distinct matrices fall in distinct connected components of $\mathrm{Aut}(M)$. Thus, $\mathrm{Aut}(M)^\sharp$ is infinite if $n \geq 2$.

As a specific example, one can take

$$B = \begin{pmatrix} 2 & 1 \\ 1 & 1 \end{pmatrix}.$$

The automorphism induced by B on $E \times E$ is an Anosov diffeomorphism: it expands a holomorphic linear foliation of $E \times E$ by a factor $(3 + \sqrt{5})/2 > 1$ and contracts another transverse linear foliation by $(3 - \sqrt{5})/2 < 1$.

This example shows that there are compact complex surfaces X for which $\mathrm{Aut}(X)$ has an infinite number of connected components and $\mathrm{Aut}(X)$ contains elements $f \colon X \to X$ that exhibit a very nice dynamics; as we shall explain, these two properties are intimately linked together. The following section provides another example that will be used all througout this survey.

1.2 An example

Consider the affine space of dimension 3, with coordinates (x_1, x_2, x_3), and compactify it as $\mathbb{P}^1 \times \mathbb{P}^1 \times \mathbb{P}^1$. Denote by π_i the projection onto $\mathbb{P}^1 \times \mathbb{P}^1$ that forgets the ith factor; for example, in affine coordinates,

$$\pi_2(x_1, x_2, x_3) = (x_1, x_3).$$

Let $X \subset \mathbb{P}^1 \times \mathbb{P}^1 \times \mathbb{P}^1$ be a smooth surface such that all three projections π_i induce ramified covers of degree 2, still denoted π_i, from X to $\mathbb{P}^1 \times \mathbb{P}^1$. Equivalently, X is defined in the affine space by a polynomial equation

$$P(x_1, x_2, x_3) = 0$$

that has degree 2 with respect to each variable. For instance, one can take

$$P(x_1, x_2, x_3) = (1 + x_1^2)(1 + x_2^2)(1 + x_3^2) + Ax_1x_2x_3 - 2,$$

for all parameters $A \neq 0$, as in [86]. Since $\pi_i \colon X \to \mathbb{P}^1 \times \mathbb{P}^1$ is a 2-to-1 cover, there is an involutive automorphism s_i of X such that $\pi_i \circ s_i = \pi_i$. For example, if (x_1, x_2, x_3) is a point of X, then

$$s_2(x_1, x_2, x_3) = (x_1, x_2', x_3)$$

where x_2 and x_2' are the roots of the equation $P(x_1, t, x_3) = 0$ (with x_1 and x_2 fixed).

As we shall see in the following pages, there are no non-trivial relations between these involutions. In other words, the subgroup of $\mathrm{Aut}(X)$ generated by s_1, s_2, and s_3 is isomorphic to the free product $\mathbb{Z}/2\mathbb{Z} * \mathbb{Z}/2\mathbb{Z} * \mathbb{Z}/2\mathbb{Z}$. Moreover, if f is a non-trivial element of this group, then

- either f is conjugate to one of the s_i, and then f is an involution;

- or f is conjugate to an iterate $(s_i \circ s_j)^n$ of one of the compositions $s_i \circ s_j, i \neq j$, and the dynamics of f is easily described since the closure of typical orbits are elliptic curves;

- or f has a rich dynamics, with positive topological entropy, an infinite number of saddle periodic points, etc.

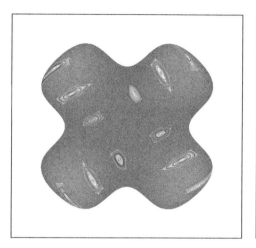 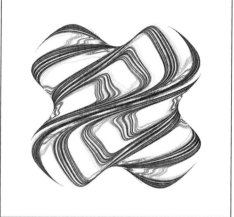

FIGURE 1.1: Here, X is defined by a polynomial with real coefficients. The automorphism $f = s_1 \circ s_2 \circ s_3$ preserves the real part $X(\mathbb{R})$. On the left, several orbits are plotted, while on the right, an approximation of a stable manifold of a saddle fixed point is drawn. (See also Plate 23; picture realized by V. Pit, based on a program by C. T. McMullen [85].)

1.3   Aims and scope

This text describes the dynamics of automorphisms of compact complex surfaces when it is rich, as in the previous example $f = s_1 \circ s_2 \circ s_3$. We restrict the study to compact Kähler surfaces. This is justified by the fact that the topological entropy of all automorphisms vanishes on compact complex surfaces which are not Kähler, as explained in Section 2.5 and the appendix.

Not much is known, but a nice interplay between algebraic geometry, complex analysis, and dynamical systems provides a few interesting results. This leads to a precise description of the main stochastic properties of the dynamics of automorphisms, whereas topological properties seem more difficult to obtain.

Our goal is to present the main results of the subject to specialists in algebraic geometry and to specialists in dynamical systems as well; this implies that several definitions and elementary explanations need to be given that are common knowledge for a large proportion of potential readers. No proof is detailed, but a few arguments are sketched in order to enlighten the interplay between algebraic geometry, complex analysis, and dynamical systems. When a result holds for automorphisms of projective surfaces over any algebraically closed field **k**, we mention it.

We tried as much as possible to focus on topics which are not covered by other recent surveys on holomorphic dynamics in several complex variables; we recommend [96], [31], [67], [5], and [49] for complementary material.

1.4 Acknowledgement

Thanks to Eric Bedford, Charles Favre, and Curtis T. McMullen for interesting comments and references. A large part of the work on this paper was done while the author was a CNRS researcher at the Département de Mathématiques et Applications at the Ecole Normale Supérieure de Paris; I am very grateful to my former colleagues for the wonderful atmosphere in this institute.

2 HODGE THEORY AND AUTOMORPHISMS

Let X be a connected, compact, Kähler surface. Our goal in this section is to describe the action of automorphisms $f \in \mathrm{Aut}(X)$ on the cohomology groups of X. The Hodge structure plays an important role; we refer to the four books [64], [100], [2], and [82] as general references for this topic and to [31] for details concerning the action of $\mathrm{Aut}(X)$ on the cohomology of X.

2.1 Hodge decomposition, intersection form, and Kähler cone

2.1.1 Cohomology groups

Hodge theory implies that the de Rham cohomology groups $H^k(X, \mathbb{C})$ split into direct sums

$$H^k(X, \mathbb{C}) = \bigoplus_{p+q=k} H^{p,q}(X, \mathbb{C}),$$

where classes in the Dolbeault cohomology groups $H^{p,q}(X, \mathbb{C})$ are represented by closed forms of type (p, q). For example, $H^{1,0}(X, \mathbb{C})$ and $H^{2,0}(X, \mathbb{C})$ correspond, respectively, to holomorphic 1-forms and holomorphic 2-forms. This bigraded structure is compatible with the cup product. $H^{p,q}(X, \mathbb{C})$ is permuted with $H^{q,p}(X, \mathbb{C})$ by complex conjugation; it defines a real structure on the complex vector space $H^{1,1}(X, \mathbb{C})$, for which the real part is

$$H^{1,1}(X, \mathbb{R}) = H^{1,1}(X, \mathbb{C}) \cap H^2(X, \mathbb{R}),$$

and on the space $H^{2,0}(X, \mathbb{C}) \oplus H^{0,2}(X, \mathbb{C})$. We denote by $h^{p,q}(X)$ the dimension of $H^{p,q}(X, \mathbb{C})$.

2.1.2 Intersection form

Since X is canonically oriented by its complex structure, it admits a natural fundamental class $[X] \in H_4(X, \mathbb{Z})$; this provides an identification of $H^4(X, \mathbb{Z})$ with $\mathbb{Z}$. Hence, the **intersection form** defines an integral bilinear form on $H^2(X, \mathbb{Z})$. We denote by $\langle \cdot | \cdot \rangle$ the bilinear form which is induced on $H^{1,1}(X, \mathbb{R})$ by the intersection form[1]

$$\forall u, v \in H^{1,1}(X, \mathbb{R}), \quad \langle u | v \rangle = \int_X u \wedge v.$$

---

[1]Here u and v are implicitly represented by $(1, 1)$-forms and the evaluation of the cup product of u and v on the fundamental class $[X]$ is identified to the integral of $u \wedge v$ on X.

THEOREM 2.1 (Hodge index theorem). *Let X be a connected compact Kähler surface. On the space $H^{1,1}(X, \mathbb{R})$, the intersection form $\langle \cdot | \cdot \rangle$ is non-degenerate and of signature $(1, h^{1,1}(X) - 1)$.*

In particular, $\langle \cdot | \cdot \rangle$ endows $H^{1,1}(X, \mathbb{R})$ with the structure of a Minkowski space that will play an important role in the following sections.

REMARK 2.2. If Ω and Ω' are holomorphic 2-forms, then

$$\int_X \Omega \wedge \overline{\Omega'} > 0, \tag{2.1}$$

where $\overline{\Omega'}$ is the complex conjugate. As a consequence, the intersection form is positive definite on the real part of $H^{2,0}(X, \mathbb{C}) \oplus H^{0,2}(X, \mathbb{C})$, and the signature of the intersection form on $H^2(X, \mathbb{R})$ is $(2h^{2,0}(X) + 1, h^{1,1}(X) - 1)$.

EXAMPLE 2.3. Let X be a smooth surface of degree $(2, 2, 2)$ in $\mathbb{P}^1 \times \mathbb{P}^1 \times \mathbb{P}^1$, as in Section 1.2. In the affine space $\mathbb{C} \times \mathbb{C} \times \mathbb{C}$, with coordinates (x_1, x_2, x_3), X is defined by a polynomial equation $P(x_1, x_2, x_3) = 0$. Since X is smooth, every point of X is contained in an open set where one of the partial derivatives of P does not vanish; hence, we can define a holomorphic 2-form Ω on the affine part of X by

$$\Omega = \frac{dx_1 \wedge dx_2}{\partial P / \partial x_3} = \frac{dx_2 \wedge dx_3}{\partial P / \partial x_1} = \frac{dx_3 \wedge dx_1}{\partial P / \partial x_2}.$$

As the reader can check, this form extends to X as a non-vanishing holomorphic 2-form Ω_X, because P has degree 2 with respect to each variable (an instance of the "adjunction formula" [64]). Now, if Ω' is another holomorphic 2-form, then $\Omega' = \psi \Omega_X$ for some holomorphic, and therefore constant, function $\psi \colon X \to \mathbb{C}$. Thus, $H^{2,0}(X, \mathbb{C})$ is generated by $[\Omega_X]$ and so $h^{2,0}(X) = 1$; by conjugation, we have $h^{0,2}(X) = 1$.

The generic fibers of the projection

$$\sigma_1 \colon (x_1, x_2, x_3) \in X \to x_1 \in \mathbb{P}^1$$

are elliptic, with a finite number of singular fibers — for a generic choice of P, one sees that π has exactly 24 singular fibers which are isomorphic to a rational curve with a double point. This implies that X is simply connected with Euler characteristic 24. Thus, $H^1(X, \mathbb{Z})$, $H^{1,0}(X, \mathbb{C})$, and $H^{0,1}(X, \mathbb{C})$ vanish, and $H^{1,1}(X, \mathbb{C})$ has dimension 20. Consequently, the signature of the intersection form on $H^2(X, \mathbb{R})$ is $(3, 19)$.

2.1.3   *Kähler and nef cones*

Classes $[\kappa] \in H^{1,1}(X; \mathbb{R})$ of Kähler forms are called **Kähler classes**. The **Kähler cone** of X is the subset $\mathcal{K}(X) \subset H^{1,1}(X, \mathbb{R})$ of all Kähler classes. This cone is convex and is contained in one of the two connected components of the cone

$$\{u \in H^{1,1}(X, \mathbb{R}) \mid \langle u | u \rangle > 0\}.$$

Its closure $\overline{\mathcal{K}}(X) \subset H^{1,1}(X, \mathbb{R})$ is the **nef cone** (where "nef" simultaneously stands for "numerically eventually free" and "numerically effective").

2.1.4 The Néron-Severi group

The **Néron-Severi group** of X is the discrete subgroup of $H^{1,1}(X, \mathbb{R})$ defined by

$$NS(X) = H^{1,1}(X, \mathbb{R}) \cap H^2(X, \mathbb{Z}).$$

Lefschetz's theorem on $(1,1)$-classes asserts that this space coincides with the group of Chern classes of holomorphic line bundles on X. The dimension $\rho(X)$ of $NS(X)$ is the **Picard number** of X; by definition $\rho(X) \leq h^{1,1}(X)$. Similarly, we denote by $NS(X, \mathbf{A})$ the space $NS(X) \otimes_{\mathbb{Z}} \mathbf{A}$ for $\mathbf{A} = \mathbb{Q}, \mathbb{R}, \mathbb{C}$.

When Y is a projective surface defined over an algebraically closed field $\mathbf{k}$, the Néron-Severi group $NS(Y)$ is defined as the group of classes of curves modulo numerical equivalence; this definition coincides with the definition just given when $\mathbf{k} = \mathbb{C}$ and $Y = X$ is a complex projective surface.

REMARK 2.4. Let X be a projective surface, embedded as a degree d surface in some projective space $\mathbb{P}^n(\mathbb{C})$. Recall that the line bundle $\mathcal{O}(1)$ on $\mathbb{P}^n(\mathbb{C})$ is the inverse of the tautological line bundle: holomorphic sections of $\mathcal{O}(1)$ are given by linear functions in homogeneous coordinates and their zero-sets are hyperplanes of $\mathbb{P}^n(\mathbb{C})$; the Chern class of $\mathcal{O}(1)$ is a Kähler class (represented by the Fubini-Study form). Restricting $\mathcal{O}(1)$ to X, we obtain a line bundle, the Chern class of which is an integral Kähler class with self-intersection d. Equivalently, intersecting X with two generic hyperplanes, one gets exactly d points. This shows that $\langle \cdot | \cdot \rangle$ restricts to a non-degenerate quadratic form of signature $(1, \rho(X) - 1)$ on $NS(X, \mathbb{R})$ when X is projective.

In the other direction, the description of Kähler cones for surfaces (see [80]) and Kodaira's embedding theorem imply that X is a projective surface as soon as $NS(X)$ contains classes with positive self-intersection.

EXAMPLE 2.5. Let X be a smooth, generic surface of degree $(2, 2, 2)$ in $\mathbb{P}^1 \times \mathbb{P}^1 \times \mathbb{P}^1$; it can be shown that $NS(X)$ is isomorphic to $\mathbb{Z}^3$, with generators given by the classes $[C_i]$ of the fibers of the projection $\sigma_i \colon X \to \mathbb{P}^1$ onto the ith factor (an instance of Noether-Lefschetz theorem [100, 101]). When X is a torus of $\mathbb{C}^2 / \Lambda$, its Picard number $\rho(X)$ is at most 4. For example, generic tori have Picard number 0 and $\rho(\mathbb{C}/\mathbb{Z}[\sqrt{-1}] \times \mathbb{C}/\mathbb{Z}[\sqrt{-1}]) = 4$.

2.2 Automorphisms

2.2.1 Action on cohomology groups

The group $\mathsf{Aut}(X)$ acts by pull-back on $H^*(X, \mathbb{Z})$, where $H^*(X, \mathbb{Z})$ stands for the graded direct sum of the cohomology groups $H^k(X, \mathbb{Z})$. This action provides a morphism

$$f \in \mathsf{Aut}(X) \mapsto f^* \in \mathsf{GL}\,(H^*(X, \mathbb{Z})), \qquad (2.2)$$

the image of which preserves

1. the graded structure, i.e., each subspace $H^k(X, \mathbb{Z})$, acting trivially on $H^0(X, \mathbb{Z})$ and $H^4(X, \mathbb{Z})$;

2. the Poincaré duality;

3. the Hodge decomposition, commuting with complex conjugation;

4. the Kähler cone $\mathcal{K}(X)$.

Moreover,

5. the cup product is equivariant with respect to the action of $\text{Aut}(X)$; in particular, $\text{Aut}(X)$ preserves the intersection form $\langle \cdot | \cdot \rangle$ on $H^{1,1}(X, \mathbb{R})$.

The connected component of the identity $\text{Aut}(X)^0 \subset \text{Aut}(X)$ acts trivially on the cohomology of X; the following theorem shows that this group has finite index in the kernel of the morphism (2.2).

THEOREM 2.6 (Fujiki [61], Lieberman [83]). *Let M be a compact Kähler manifold. If $[\kappa]$ is a Kähler class on M, the connected component of the identity $\text{Aut}(X)^0$ has finite index in the group of automorphisms of M fixing $[\kappa]$.*

In other words, the group of connected components $\text{Aut}(M)^\sharp$ almost embeds into $\text{GL}(H^*(M, \mathbb{Z}))$.

2.2.2 Eigenvalues and dynamical degree

On the space $H^{2,0}(X, \mathbb{C})$ (resp. on $H^{0,2}(X, \mathbb{C})$), the group $\text{Aut}(X)$ preserves the positive hermitian product

$$([\Omega], [\Omega']) \mapsto \int_X [\Omega] \wedge [\overline{\Omega'}]$$

from Equation (2.1). This shows that the image of $\text{Aut}(X)$ in $\text{GL}(H^{2,0}(X, \mathbb{C}))$ (resp. in $\text{GL}(H^{0,2}(X, \mathbb{C}))$) is contained in a unitary group.

Since $\text{Aut}(X)$ preserves the Hodge decomposition and the integral structure of the cohomology, it preserves the Néron-Severi group. When X is projective, $NS(X)$ intersects the Kähler cone (see Remark 2.4) and by Hodge index theorem, the intersection form is negative definite on its orthogonal complement $NS(X)^\perp \subset H^{1,1}(X, \mathbb{R})$.

These facts imply the following lemma.

LEMMA 2.7. *Let X be a compact Kähler surface, and f be an automorphism of X. Let $u \in H^2(X, \mathbb{C})$ be a non-zero eigenvector of f^*, with eigenvalue λ. If $|\lambda| > 1$, then u is contained in $H^{1,1}(X, \mathbb{C})$ and is contained in $NS(X, \mathbb{C})$ when X is projective.*

Now let u be an eigenvector of f^* in $H^1(X, \mathbb{C})$ with eigenvalue β. Its $(1, 0)$ and $(0, 1)$ parts $u_{1,0}$ and $u_{0,1}$ are also eigenvectors of f^*, with the same eigenvalue β. Since $u_{1,0}$ is represented by a holomorphic 1-form, we have

$$u_{1,0} \wedge \overline{u_{1,0}} \neq 0$$

as soon as $u_{1,0} \neq 0$. Thus, $\beta\overline{\beta}$ is an eigenvector of f^* in $H^{1,1}(X, \mathbb{R})$.

LEMMA 2.8. *The square of the spectral radius of f^* on $H^1(X, \mathbb{C})$ is bounded from above by the largest eigenvalue of f^* on $H^{1,1}(X, \mathbb{R})$. The spectral radius of f^* on $H^*(X, \mathbb{C})$ is equal to the spectral radius of f^* on $H^{1,1}(X, \mathbb{R})$.*

We shall denote by $\lambda(f)$, or simply λ, the spectral radius of f^*; this number is the **dynamical degree** of f. As we shall see in Sections 2.4.3 and 4.4.2, $\lambda(f)$ is an eigenvalue of f^*, $\lambda(f)$ is and algebraic integer, and its logarithm is equal to the topological entropy of f as a transformation of the complex surface X.

EXAMPLE 2.9 (see Section 7.3 for explicit examples). Let f_0 be a birational transformation of the plane $\mathbb{P}^2(\mathbb{C})$. By definition, the degree $\deg(f_0)$ of f_0 is the degree of the pre-image of a generic line by f_0. Equivalently, there are homogeneous polynomials P, Q, and R of the same degree d and without common factors of degree > 1 such that $f[x : y : z] = [P : Q : R]$ in homogeneous coordinates; this number d is equal to $\deg(f_0)$.

Suppose there is a birational map $\varphi\colon X \dashrightarrow \mathbb{P}^2(\mathbb{C})$ such that $f := \varphi^{-1} \circ f_0 \circ \varphi$ is an automorphism of X. Then $\lambda(f) = \lim_n \deg(f_0^n)^{1/n}$. This formula justifies the term "dynamical degree."

2.3   Isometries of Minkowski spaces

This paragraph is a parenthesis on the geometry of Minkowski spaces and their isometries.

2.3.1   Standard Minkowski spaces

The standard Minkowski space $\mathbb{R}^{1,m}$ is the real vector space $\mathbb{R}^{1+m}$ together with the quadratic form

$$x_0^2 - x_1^2 - x_2^2 - \ldots - x_m^2.$$

Let $\langle \cdot | \cdot \rangle_m$ be the bilinear form which is associated with this quadratic form. Let w be the vector $(1, 0, \ldots, 0)$; it is contained in the hyperboloid of vectors u with $\langle u | u \rangle_m = 1$. Define $\mathbb{H}_m$ to be the connected component of this hyperboloid that contains w, and let dist_m be the distance on $\mathbb{H}_m$ defined by (see [14], [72], and [98])

$$\cosh(\mathrm{dist}_m(u, u')) = \langle u | u' \rangle_m.$$

The metric space $(\mathbb{H}_m, \mathrm{dist}_m)$ is a Riemannian, simply connected, and complete space of dimension m with constant sectional curvature -1; these properties uniquely characterize it up to isometry.[2]

The projection of $\mathbb{H}_m$ into the projective space $\mathrm{P}(\mathbb{R}^{1,m})$ is one-to-one onto its image. In homogeneous coordinates, its image is the ball $x_0^2 > x_1^2 + \ldots + x_m^2$, and the sphere obtained by projection of the isotropic cone $x_0^2 = x_1^2 + \ldots + x_m^2$ is its boundary. In what follows, $\mathbb{H}_m$ is identified with its image in $\mathrm{P}(\mathbb{R}^{1,m})$ and its boundary is denoted by $\partial \mathbb{H}_m$; hence, boundary points correspond to isotropic lines in $\mathbb{R}^{1,m}$.

2.3.2   Isometries

Let $\mathrm{O}_{1,m}(\mathbb{R})$ be the group of linear transformations of $\mathbb{R}^{1,m}$ preserving the bilinear form $\langle \cdot | \cdot \rangle_m$. The group of isometries $\mathrm{Isom}\,(\mathbb{H}_m)$ coincides with the subgroup of $\mathrm{O}_{1,m}(\mathbb{R})$ that preserves the chosen sheet $\mathbb{H}_m$ of the hyperboloid consisting of all $u \in \mathbb{R}^{1,m}$ for which $\langle u | u \rangle_m = 1$. This group acts transitively on $\mathbb{H}_m$ and on its unit tangent bundle.

---

[2] The Riemannian structure is defined as follows. If u is an element of $\mathbb{H}_m$, the tangent space $T_u\mathbb{H}_m$ is the affine space through u that is parallel to $u^\perp$, where $u^\perp$ is the orthogonal complement of $\mathbb{R}u$ with respect to $\langle \cdot | \cdot \rangle_m$; since $\langle u | u \rangle_m = 1$, the form $\langle \cdot | \cdot \rangle_m$ is negative definite on $u^\perp$, and its opposite defines a positive scalar product on $T_u\mathbb{H}_m$; this family of scalar products determines a Riemannian metric, and the associated distance coincides with dist_m (see [14]).

If $h \in O_{1,m}(\mathbb{R})$ is an isometry of $\mathbb{H}_m$ and $v \in \mathbb{R}^{1,m}$ is an eigenvector of h with eigenvalue λ, then either $\lambda^2 = 1$ or v is isotropic. Moreover, since $\mathbb{H}_m$ is homeomorphic to a ball, h has at least one eigenvector v in $\mathbb{H}_m \cap \partial\mathbb{H}_m$. Thus, there are three types of isometries: **elliptic** isometries, with a fixed point u in $\mathbb{H}_m$; **parabolic** isometries, with no fixed point in $\mathbb{H}_m$ but a fixed vector v in the isotropic cone; **loxodromic** (or **hyperbolic**) isometries, with an isotropic eigenvector v corresponding to an eigenvalue $\lambda > 1$. They satisfy the following additional properties (see [14]).

1. An isometry h is elliptic if and only if it fixes a point u in $\mathbb{H}_m$. Since $\langle \cdot | \cdot \rangle_m$ is negative definite on the orthogonal complement $u^\perp$, the linear transformation h fixes pointwise the line $\mathbb{R}u$ and acts by rotation on $u^\perp$ with respect to $\langle \cdot | \cdot \rangle_m$.

2. An isometry h is parabolic if it is not elliptic and fixes a vector v in the isotropic cone. The line $\mathbb{R}v$ is uniquely determined by the parabolic isometry h. For all points u in $\mathbb{H}_m$, the sequence $h^n(u)$ converges toward the boundary point $\mathbb{R}v$ in the projective space $\mathbb{P}(\mathbb{R}^{1,m})$ as n goes to $+\infty$ and $-\infty$.

3. An isometry h is hyperbolic if and only if h has an eigenvector v_h^+ with eigenvalue $\lambda > 1$. Such an eigenvector is unique up to scalar multiplication, and there is another, unique, isotropic eigenline $\mathbb{R}v_h^-$ corresponding to an eigenvalue < 1; this eigenvalue is equal to $1/\lambda$. If u is an element of $\mathbb{H}_m$,

$$\frac{1}{\lambda^n} h^n(u) \longrightarrow \frac{\langle u | v_h^- \rangle_m}{\langle v_h^+ | v_h^- \rangle_m} v_h^+$$

as n goes to $+\infty$, and

$$\frac{1}{\lambda^n} h^n(u) \longrightarrow \frac{\langle u | v_h^+ \rangle_m}{\langle v_h^+ | v_h^- \rangle_m} v_h^-$$

as n goes to $-\infty$. On the orthogonal complement of $\mathbb{R}v_h^+ \oplus \mathbb{R}v_h^-$, h acts as a rotation with respect to $\langle \cdot | \cdot \rangle_m$.

The type of h is also characterized by the growth of the iterates h^n: For any norm $\| \cdot \|$ on the space $\text{End}(\mathbb{R}^{1,m})$, the sequence $\|h^n\|$ is bounded if h is elliptic, grows like $C^{\text{ste}}n^2$ if h is parabolic, and grows like λ^n if h is hyperbolic with $\lambda > 1$ as in (3).

2.4 Types of automorphisms and geometry

Let X be a connected, compact, Kähler surface.

2.4.1 The hyperbolic space $\mathbb{H}_X$

The intersection form on $H^{1,1}(X, \mathbb{R})$ is non-degenerate of signature $(1, h^{1,1}(X) - 1)$; as such, it is isometric to the standard Minkowski form in dimension $h^{1,1}(X)$. One, and only one, sheet of the hyperboloid $\{u \in H^{1,1}(X, \mathbb{R}) \mid \langle u | u \rangle = 1\}$ intersects the Kähler cone $\mathcal{K}(X)$: we denote by $\mathbb{H}_X$ this hyperboloid sheet; as in Section 2.3, the intersection form endows $\mathbb{H}_X$ with the structure of a hyperbolic space $\mathbb{H}_m$ of dimension $m = h^{1,1}(X) - 1$.

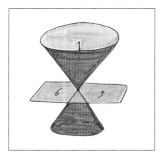 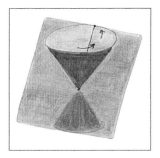 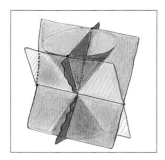

FIGURE 2.10: Three types of automorphisms (from left to right): elliptic, parabolic, and loxodromic. Elliptic isometries preserve a point in $\mathbb{H}_m$ and act as a rotation on the orthogonal complement. Parabolic isometries fix an isotropic vector v; all positive and negative orbits in $\mathbb{H}_m$ converge toward the line $\mathbb{R}v$; the orthogonal complement of $\mathbb{R}v$ contains it and is tangent to the isotropic cone. Loxodromic isometries dilate an isotropic line, contract another one, and act as a rotation on the intersection of the planes tangents to the isotropic cone along those lines (see also Figure 2.12 , as well as Plates 24 and 25).

2.4.2 *Isometries induced by automorphisms*

Since automorphisms of X act by isometries with respect to the intersection form and preserve the Kähler cone, they preserve the hyperbolic space $\mathbb{H}_X$. This provides a morphism

$$\mathsf{Aut}(X) \to \mathsf{Isom}\,(\mathbb{H}_X).$$

By definition, an automorphism f is either **elliptic**, **parabolic**, or **loxodromic**,[3] according to the type of $f^* \in \mathsf{Isom}\,(\mathbb{H}_X)$.

2.4.3 *Loxodromic automorphisms*

Let f be a loxodromic automorphism, and let $\lambda(f)$ be its dynamical degree. We know from Sections 2.2.2 and 2.3.2 that $\lambda(f)$ is the unique eigenvalue of f^* on $H^2(X, \mathbb{C})$ with modulus > 1. It is real, positive, and its eigenspace is a line: this line is defined over $\mathbb{R}$, is contained in $H^{1,1}(X, \mathbb{C})$, and is isotropic with respect to the intersection form. Moreover, $\lambda(f)$ is an algebraic number because $\lambda(f)$ is an eigenvalue of f^* and f^* preserves the lattice $H^2(X, \mathbb{Z})$. Since the other eigenvalues of f^* on $H^2(X, \mathbb{C})$, beside $\lambda(f)$ and its inverse $1/\lambda(f)$, have modulus 1, this implies that $\lambda(f)$ *is either a reciprocal quadratic integer or a Salem number.*[4]

REMARK 2.11. The set of Salem numbers is not well understood. In particular, its infimum is unknown. However, dynamical degrees of automorphisms provide only a small subset of the set of Salem numbers, and McMullen proved in [87] that the minimum of all dynamical degrees $\lambda(f)$, for f describing the set of all loxodromic automorphisms of compact Kähler surfaces, is equal to Lehmer's number

---

[3]In the literature, the terminology used for "loxodromic" is either "hyperbolic," "hyperbolic on the cohomology," or "with positive entropy," depending on the authors.

[4]By definition, an algebraic number λ is a **Salem number** if λ is real, $\lambda > 1$, its degree is ≥ 4, and the conjugates of λ are $1/\lambda$ and complex numbers of modulus 1. In particular, quadratic integers are not considered as Salem numbers here.

$\lambda_{10} \simeq 1.17628$, the largest root of

$$x^{10} + x^9 - x^7 - x^6 - x^5 - x^4 - x^3 + x + 1.$$

This is the smallest known Salem number; for comparison, the smallest quadratic integer is the golden mean, $\lambda_G \simeq 1.61803$. Lehmer's number is realized as the dynamical degree of automorphisms on some rational surfaces (see [7] and [87]) and K3 surfaces (both on projective and non-projective K3 surfaces; see [88] and [86] respectively).

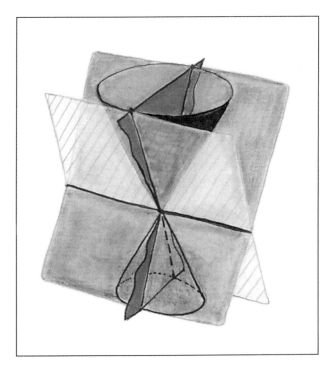

FIGURE 2.12: Action of a loxodromic automorphism on $H^{1,1}(X, \mathbb{R})$. The two invariant isotropic lines $\mathbb{R}v_f^+$ and $\mathbb{R}v_f^-$ generate a dark gray plane; the vector spaces $(\mathbb{R}v_f^+)^\perp$ and $(\mathbb{R}v_f^-)^\perp$ are tangent to the isotropic cone, and their intersection N_f is the orthogonal complement to the dark gray plane. (See also Plate 25.)

The Kähler cone $\mathcal{K}(X)$ is contained in the convex cone $\mathbb{R}_+ \mathbb{H}_X$. Let $[\kappa]$ be an element of $\mathcal{K}(X)$. Then, from Section 2.3.2, $(1/\lambda(f)^n)(f^*)^n[\kappa]$ converges toward a non-zero eigenvector of f^* for the eigenvalue $\lambda(f)$. Thus, there exists a non-zero nef vector $v_f^+ \in H^{1,1}(X, \mathbb{R})$ such that

$$f^* v_f^+ = \lambda(f) v_f^+.$$

We fix such an eigenvector v_f^+ in what follows; this choice is unique up to a positive scalar factor, because the eigenspace for $\lambda(f)$ is a line. The same argument, applied to f^{-1}, provides a nef vector v_f^- such that $f^* v_f^- = (1/\lambda(f)) v_f^-$. Changing

v_f^- in a scalar multiple, we assume that

$$\langle v_f^+ | v_f^- \rangle = 1.$$

The orthogonal complements of v_f^+ and of v_f^- intersect along a codimension 2 subspace

$$N_f := (v_f^+)^\perp \cap (v_f^-)^\perp \subset H^{1,1}(X, \mathbb{R});$$

the intersection form $\langle \cdot | \cdot \rangle$ is negative definite on N_f.

2.4.4   Elliptic and parabolic automorphisms

The following result provides a link between this classification in types and the geometry of the transformation $f: X \to X$.

THEOREM 2.13 (Gizatullin, Cantat [63, 27]). *Let X be a connected, compact, Kähler surface. Let f be an automorphism of X.*

(i) *If f is elliptic, a positive iterate f^k of f is contained in the connected component of the identity* Aut$(X)^0$; *in particular, $f^* \in$ GL$(H^*(X, \mathbb{Z}))$ has finite order.*

(ii) *If f is parabolic, there is an elliptic fibration $\pi_f: X \to B$ and an automorphism $\bar{f}$ of the curve B such that $\pi_f \circ f = \bar{f}$. If C is a fiber of the fibration, its class $[C]$ is contained in the unique isotropic line which is fixed by f^*; in particular, this line intersects $NS(X) \setminus \{0\}$. Moreover, if $\bar{f}$ does not have finite order, then X is isomorphic to a torus $\mathbb{C}^2/\Lambda$.*

Moreover, f is elliptic if and only if $\|(f^n)^*\|$ is a bounded sequence, f is parabolic if and only if $\|(f^n)^*\|$ grows quadratically, and f is loxodromic if and only if $\|(f^n)^*\|$ grows exponentially fast, like $\lambda(f)^n$.

2.4.5   Projective surfaces over other fields

Suppose X is a complex projective surface. Since $NS(X, \mathbb{R})$ intersects the Kähler cone, it also intersects $\mathbb{H}_X$ on an Aut(X)-invariant, totally isometric subspace. Thus, *the type of every automorphism f is the same as the type of f^* as an isometry of the hyperbolic subspace $\mathbb{H}_X \cap NS(X, \mathbb{R})$.* In particular, if f is loxodromic, the two isotropic eigenlines are contained in $NS(X, \mathbb{R})$. On the other hand, there is no vector u in $NS(X)$ such that $f^* u = \lambda u$ with $\lambda > 1$, because f^* determines an automorphism of the lattice $NS(X)$. Hence, when f is loxodromic, the eigenline corresponding to the eigenvalue $\lambda(f)$ is irrational with respect to the lattice $NS(X)$.

Now let Y be a smooth projective surface defined over an algebraically closed field **k**. Let f be an automorphism of Y. Then f acts on the Néron-Severi group $NS(Y)$ by isometries with respect to the intersection form, where $NS(Y)$ is defined as the group of numerical classes of divisors (see [68] for Néron-Severi groups). Hodge index theorem applies and shows that $NS(Y, \mathbb{R})$ is a Minkowski space with respect to its intersection form. Consequently, automorphisms of Y can also be classified in three categories in accordance with the type of the isometry f^* of $NS(Y, \mathbb{R})$; as said earlier, this definition is compatible with the previous one — which depends on the action of f^* on $H^{1,1}(X, \mathbb{R})$ — when X is a smooth complex projective surface.

Theorem 2.13 also applies to this setting, as shown by Gizatullin in [63].

2.4.6 Two examples

A family of complex tori. Consider an elliptic curve $E = \mathbb{C}/\Lambda$, and the abelian surface $X = E \times E$, as in Section 1.1.3. The group $\mathsf{SL}_2(\mathbb{Z})$ acts linearly on $\mathbb{C}^2$ and this action preserves the lattice $\Lambda \times \Lambda$, so that $\mathsf{SL}_2(\mathbb{Z})$ embeds as a subgroup of $\mathrm{Aut}(X)$. Let B be an element of $\mathsf{SL}_2(\mathbb{Z})$ and $\mathrm{tr}(B)$ be the trace of B. Then, the automorphism f_B of X induced by B is

- elliptic if and only if $B = \pm \mathrm{Id}$ or $\mathrm{tr}(B) = -1, 0$, or 1;

- parabolic if and only if $\mathrm{tr}(B) = -2$ or 2 and $B \neq \pm \mathrm{Id}$;

- loxodromic if and only if $|\mathrm{tr}(B)| > 2$; in this case, the dynamical degree of f_B is the square of the largest eigenvalue of B.

Since the type of an automorphism depends only on its action on the cohomology of X, all automorphisms of the form $t \circ f_B$ where $t \in \mathrm{Aut}(X)^0$ is a translation have the same type as f_B.

REMARK 2.14. The appendix of [62] lists all 2-dimensional tori with a loxodromic automorphism.

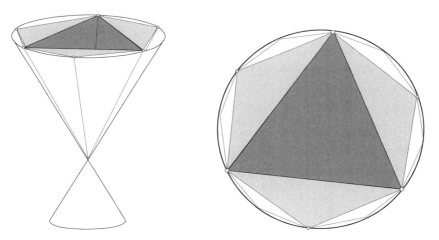

FIGURE 2.15: Action of involutions: on the left, a picture of N_X with the triangular cone $\mathbb{R}^+[C_1] + \mathbb{R}^+[C_2] + \mathbb{R}^+[C_3]$ and its images under the three involutions. On the right, a projective view of the same picture: the triangular cone becomes a gray ideal triangle Δ.

Surfaces of degree (2,2,2) in $\mathbb{P}^1(\mathbb{C}) \times \mathbb{P}^1(\mathbb{C}) \times \mathbb{P}^1(\mathbb{C})$. Let X be a smooth surface of degree $(2,2,2)$ in $\mathbb{P}^1 \times \mathbb{P}^1 \times \mathbb{P}^1$. Let $N_X \subset NS(X)$ be the subgroup of the Néron-Severi group which is generated by the three classes $[C_i]$, $i = 1, 2, 3$, where $[C_i]$ is the class of the fibers of the projection $\sigma_i \colon X \to \mathbb{P}^1$ defined by $\sigma_i(x_1, x_2, x_3) = x_i$. One easily checks that the three involutions s_i^* preserve the space N_X; on N_X, the matrix of s_1^* in the basis $([C_1], [C_2], [C_3])$ is equal to

$$\begin{pmatrix} -1 & 0 & 0 \\ 2 & 1 & 0 \\ 2 & 0 & 1 \end{pmatrix},$$

and the matrices of s_2^* and s_3^* are obtained from it by permutation of the coordinates. Thus, on N_X, the map s_i is the orthogonal reflection with respect to the plane $\mathsf{Span}([C_j], [C_k])$ (for $\{i, j, k\} = \{1, 2, 3\}$). The space $\mathbb{H}_X \cap (N_X \otimes \mathbb{R})$ is isometric to the Poincaré disk. Denote by Δ the ideal triangle of the disk with vertices $[C_1]$, $[C_2]$, $[C_3]$. Then Δ is a fundamental domain for the action of the group generated by the s_i^*, as shown on Figure 2.15. The group generated by the involution s_i^* acts by symmetries of the tessellation of the disks by ideal triangle. This proves that there are no non-obvious relations between the involutions, as stated in the introduction.

The transformation $s_i \circ s_j$, for $i \neq j$, is parabolic, and all parabolic elements in the group $\langle s_1, s_2, s_3 \rangle$ are conjugate to some iterate of one of these parabolic automorphisms. A prototypical example of a loxodromic automorphism is the composition $g = s_3 \circ s_2 \circ s_1$. Its action on N_X is given by the matrix

$$\begin{pmatrix} -1 & -2 & -6 \\ 2 & 3 & 10 \\ 2 & 6 & 15 \end{pmatrix},$$

and the eigenvalues of this matrix are

$$\lambda = 9 + 4\sqrt{5}, \quad \frac{1}{\lambda} = 9 - 4\sqrt{5}, \quad \text{and} \; -1.$$

Thus, the dynamical degree of g is $\lambda(g) = 9 + 4\sqrt{5}$.

2.5 Classification of surfaces

Compact complex surfaces have been classified (see [2]), and this classification, known as Enriques-Kodaira classification, has been extended to projective surfaces over algebraically closed fields by Mumford and Bombieri. This classification can be used to list all types of surfaces that may admit a loxodromic automorphism. Since this classification is not used in the sequel, we postpone the statement to an appendix of this survey.

All we need to know is that after (several successive) contractions of smooth periodic curves with self-intersection -1, there are four main types of surfaces with loxodromic automorphisms: rational surfaces obtained from $\mathbb{P}^2$ by a finite sequence of at least ten blow-ups, tori, K3 surfaces, and Enriques surfaces. Complex Enriques surfaces are quotients of K3 surfaces by a fixed-point-free involution, so that the main examples, beside the well-known case of tori, are given by rational surfaces and K3 surfaces.

Surfaces of degree $(2, 2, 2)$ are examples of K3 surfaces; Section 7.3 provides examples on rational surfaces.

3 GROUPS OF AUTOMORPHISMS

In order to illustrate the strength of our knowledge of $\mathsf{Isom}\,(\mathbb{H}_X)$, let us study the structure of subgroups of $\mathsf{Aut}(X)^\sharp$. In this section, we denote by $\mathsf{Aut}(X)^*$ the image of $\mathsf{Aut}(X)$ in $\mathsf{GL}\,(H^*(M, \mathbb{Z}))$; up to finite index, $\mathsf{Aut}(X)^*$ coincides with $\mathsf{Aut}(X)^\sharp$.

3.1 Torsion

Let us start with a remark concerning torsion in $\mathsf{Aut}(X)^\sharp$. Let A be a subgroup of $\mathsf{Aut}(X)$, and A^* be its image in $\mathsf{GL}\,(H^*(X, \mathbb{Z}))$. The subgroup G_3 of all elements g

in GL $(H^*(X, \mathbb{Z}))$ such that

$$g = \mathrm{Id} \bmod(3)$$

is a finite-index, torsion-free subgroup of GL $(H^*(X, \mathbb{Z}))$. Denote by A_0^* its intersection with A^* and by A_0 its pre-image in A. Then A_0 is a finite index subgroup of A and A_0^* is torsion free.

LEMMA 3.1. *Let X be a connected, compact Kähler surface. Up to finite index in the group* Aut(X), *every elliptic element of* Aut(X) *acts trivially on the cohomology of X.*

The same statement holds for arbitrary compact Kähler manifolds M if "elliptic" is replaced by "with finite order on $H^{1,1}(M, \mathbb{R})$."

3.2 Free subgroups and dynamical degrees

We can now prove the following result that provides a strong form of the Tits alternative.

THEOREM 3.2 (Strong Tits Alternative [22], [24], [94], [104]). *Let X be a connected compact Kähler surface. If A is a subgroup of* Aut$(X)^*$, *there is a finite index subgroup A_0 of A which satisfies one of the following properties*

- *A_0 contains a non-abelian free group, all of whose elements $g^* \neq \mathrm{Id}$ are loxodromic isometries of $H^{1,1}(X, \mathbb{R})$;*

- *A_0 is cyclic and acts by loxodromic isometries on $H^{1,1}(X, \mathbb{R})$;*

- *A_0 is a free abelian group of rank at most $h^{1,1}(X, \mathbb{R}) - 2$ whose elements $g^* \neq \mathrm{Id}$ are parabolic isometries of $H^{1,1}(X, \mathbb{R})$ (fixing a common isotropic line).*

REMARK 3.3. From the classification of compact Kähler surfaces and of holomorphic vector fields on surfaces, one easily proves the following: if Aut$(X)^\sharp$ is infinite, either X is a torus, or Aut$(X)^0$ is trivial [32, 20]. Consequently, Theorem 3.2 applies directly to subgroups of Aut(X) (instead of Aut$(X)^\sharp$) when X is not a torus.

PROOF. By Section 3.1, we can assume that A is torsion free, so that it does not contain any elliptic element. Thus, either A contains a loxodromic element, or all elements of $A \setminus \{\mathrm{Id}\}$ are parabolic.

Assume A contains a loxodromic element h^*. If A does not fix any isotropic line of $H^{1,1}(X, \mathbb{R})$, then the ping-pong lemma (see [35] and [24]) implies that A contains a free non-abelian subgroup, all of whose elements $f \neq \mathrm{Id}$ are loxodromic. Otherwise, A fixes an isotropic line $\mathbb{R}v$. Denote by

$$\alpha \colon A \to \mathbb{R}_+$$

the morphism defined by $g^*v = \alpha(g^*)v$ for all g^* in A. Since h^* fixes $\mathbb{R}v$ and h^* is loxodromic, $\alpha(h^*) = \lambda(h)^{\pm 1}$ and $\mathbb{R}v$ is an irrational line with respect to $H^2(X, \mathbb{Z})$. Since this line is A-invariant and irrational, we obtain: A contains no parabolic element, $\alpha(g^*) = \lambda(g)^{\pm 1}$ for all g in A, and α is injective. Moreover, all values of α in an interval $[a, b] \subset \mathbb{R}_+^*$ are algebraic integers of degree at most $\dim(H^2(X, \mathbb{Z}))$ whose conjugates are bounded by $\max(b, 1/a)$. Consequently, there are finitely many possible values in compact intervals, and the image of α is discrete, and hence, cyclic. Thus, either A contains a non-abelian free group or A is cyclic.

Assume A does not contain any loxodromic element. Then all elements of $A \setminus \{\mathrm{Id}\}$ are parabolic. As in [24], this implies that A preserves a unique isotropic line $\mathbb{R}u \subset H^{1,1}(X, \mathbb{R})$. If g^* is an element of A, its eigenvalues in $H^2(X, \mathbb{Z}) \otimes \mathbb{C}$ are algebraic integers, and all of them have modulus 1. By the Kronecker lemma, all of them are roots of 1. This implies that a finite index subgroup of A acts trivially on $u^\perp/(\mathbb{R}u)$. From this, it follows easily that, up to finite index, A is abelian of rank at most $h^{1,1}(X) - 2 = \dim(u^\perp) - 1$. $\square$

3.3 Mapping class groups

Let S be a connected, closed, and oriented surface of genus $g \geq 2$. The modular group, or mapping class group, of S is the group $\mathrm{Mod}\,(S)$ of isotopy classes of homeomorphisms of S; thus, $\mathrm{Mod}\,(S)$ is the group of connected components of the group of homeomorphisms of S, and is a natural analogue of the group $\mathrm{Aut}(X)^\sharp$. Let us list a few useful analogies between modular groups $\mathrm{Mod}\,(S)$ and groups of automorphisms $\mathrm{Aut}(X)$.

On one hand, $\mathrm{Aut}(X)^\sharp$ acts almost faithfully on the cohomology of X; on the other, $\mathrm{Mod}\,(S)$ coincides with the group of outer automorphisms of the fundamental group $\pi_1(S)$. Thus, both $\mathrm{Aut}(X)^\sharp$ and $\mathrm{Mod}\,(S)$ are determined by their respective action on the algebraic topology of the surface.

For instance,

- $\mathrm{Aut}(X)^\sharp$ acts by isometries on the hyperbolic space $\mathbb{H}_X$ and we derived from this action a strong form of the Tits alternative for subgroups of $\mathrm{Aut}(X)^\sharp$ (see Theorem 3.2);

- similarly, $\mathrm{Mod}\,(S)$ acts on the complex of curves of S, a Gromov hyperbolic space (see [74, 84]), and $\mathrm{Mod}\,(S)$ also satisfies a strong form of Tits alternative (see [73, 15]). For example, solvable subgroups are almost abelian, and torsion-free abelian subgroups have rank at most $3g$ (see [16]).

Thus, subgroups of $\mathrm{Mod}\,(S)$ satisfy properties which are similar to those listed in Theorem 3.2.

If f^* is an element of $\mathrm{Aut}(X)^*$, we know that f^* is either elliptic, parabolic, or loxodromic; this classification parallels the Nielsen-Thurston classification of mapping classes $g \in \mathrm{Mod}\,(S)$: elliptic automorphisms correspond to finite-order elements of $\mathrm{Mod}\,(S)$, parabolic to composition of Dehn twists along pairwise disjoint simple closed curves, and loxodromic to pseudo-Anosov classes (there are no "reducible" automorphisms beside "Dehn twists"). As we shall see, when f is a loxodromic automorphism, the classes $[v_f^\pm]$ are represented by laminar currents on the surface X that will play a role similar to the stable and unstable foliations for pseudo-Anosov homeomorphisms.

As explained in [24], this analogy is even more fruitful for birational transformations of X.

4 PERIODIC CURVES, PERIODIC POINTS, AND TOPOLOGICAL ENTROPY

We now focus on the dynamics of loxodromic automorphisms on connected compact Kähler surfaces.

The main goal of this section is to explain how ideas of algebraic geometry, including geometry over finite fields, of topology, and of dynamical systems can be used to study periodic curves and periodic points of loxodromic automorphisms.

Compact Kähler surface X	Higher genus, closed surface S
f acts on the hyperbolic space $\mathbb{H}_X$	h acts on Teichmüller space $T(S)$ (resp. on the complex of curves)
f is loxodromic	h is pseudo-Anosov
dynamical degree $\lambda(f)$	dilatation factor $\lambda(h)$
cohomology classes v_f^+ and v_f^-	fixed points of h on $\partial T(S)$
$\mathrm{h}_{top}(f) = \log \lambda(f)$	$\mathrm{h}_{top}(h) = \log \lambda(h)$
closed laminar currents T_f^+ and T_f^-	measured stable, unstable foliations of h

TABLE 3.4: Automorphisms versus mapping classes. Here, f is an automorphism of a connected, compact Kähler surface X, and h is a pseudo Anosov homeomorphism of a closed oriented surface S. (See the following sections for topological entropy and the laminar currents $T_f^\pm$.)

4.1 Periodic curves

Let $E \subset X$ be a curve which is invariant under the loxodromic automorphism f. We denote by $[E]$ its class[5] in $H^{1,1}(X, \mathbb{R})$. Since $f^*[E] = [E]$, $[E]$ is contained in N_f. Thus, the intersection form is negative definite on the subspace of $H^{1,1}(X, \mathbb{R})$ generated by the classes of all f-invariant or f-periodic curves. The Grauert-Mumford contraction theorem (see [2]) can, therefore, be applied to this set of curves and provides the following result.

PROPOSITION 4.1 (Cantat, Kawaguchi [27], [30], [79]). *Let f be a loxodromic automorphism of a connected, compact, Kähler surface X. There exist a (singular) surface X_0, a birational morphism $\pi \colon X \to X_0$, and an automorphism f_0 of X_0 such that*

1. *$\pi \circ f = f_0 \circ \pi$;*

2. *a curve $E \subset X$ is contracted by π if and only if E is f-periodic, if and only if $[E]$ is contained in N_f.*

This implies that the number of f-periodic curves is finite when f is loxodromic. Moreover, we can assume that f does not have any periodic curve if we admit singular models X_0 for the surface X. When f is an automorphism of a projective surface Y defined over an algebraically closed field $\mathbf{k}$, as in Section 2.4.5, the same result holds.

---

[5]This class is the dual of the homology class of E. Equivalently, $[E]$ is the Chern class of the line bundle $\mathcal{O}_X(E)$.

THEOREM 4.2 (Castelnuovo [17], [34], [44]). *Let f be a loxodromic automorphism of a connected, compact, Kähler surface X. If E is a periodic curve of f, then (the normalization of) E has genus 0 or 1.*

REMARK 4.3. Loxodromic automorphisms of complex tori have no periodic curve. On a K3 surface (resp. on an Enriques surface), the genus formula shows that all irreducible periodic curves are smooth rational curves. There are examples of rational surfaces X with an automorphism f such that f is loxodromic and f fixes an elliptic curve point-wise (see Example 3.1 and Remark 3.2 in [30]).

4.2   Fixed-point formulae

Lefschetz's formula (see [64]) provides a link between fixed points of f and its action on the cohomology of X.

4.2.1   Lefschetz's formula

Let M be a smooth oriented manifold and g be a smooth diffeomorphism of M. Let p be an isolated fixed point of g and U be a chart around p. One defines the index $\mathsf{Ind}(g; p)$ of g at p as the local degree of the map $\mathrm{Id}_M - g$.

The graph $\Gamma_g \subset M \times M$ of g intersects the diagonal Δ at (p, p). This intersection is transversal if and only if 1 is not an eigenvalue of the tangent map Dg_p and if and only if $\det(Dg_p - \mathrm{Id}) \neq 0$. In this case, the index of g at p satisfies

$$\mathsf{Ind}(g; p) = \mathrm{sign}(\det(Dg_p - \mathrm{Id})).$$

Another equivalent definition of $\mathsf{Ind}(g; p)$ is as follows. Orient Γ_g around (p, p) in such a way that the map $x \mapsto (x, g(x))$ preserves the orientation; then $\mathsf{Ind}(g; p)$ is the intersection number of Γ_g with the diagonal Δ at (p, p). Thus, one gets

$$\sum_{g(p)=p} \mathsf{Ind}(g; p) = \Delta \cdot \Gamma_g,$$

where $\Delta \cdot \Gamma_g$ denotes the intersection number of Δ and Γ_g, a quantity which can be computed in terms of the action of g^* on the cohomology of M. One obtains $\Delta \cdot \Gamma_g = L(g)$, where $L(g)$ denotes the Lefschetz number

$$L(g) := \sum_{k=0}^{\dim M} (-1)^k \mathrm{tr}(g^*_{|H^k(M,\mathbb{R})}).$$

Thus,

$$\sum_{g(p)=p} \mathsf{Ind}(g; p) = L(g)$$

when all fixed points of M are isolated. If all fixed points of f are non-degenerate, one gets the estimate $|\mathrm{Fix}(f)| \geq |L(f)|$.

4.2.2   Shub-Sullivan theorem and automorphisms

In order to apply Lefschetz's fixed-point formula to count periodic points, one needs to control the indices of the iterates g^n. This is exactly what the following result does.

THEOREM 4.4 (Shub-Sullivan [95]). *Let* $g\colon U \to \mathbb{R}^m$ *be a map of class* C^1, *where* U *is an open subset of* $\mathbb{R}^m$ *that contains the origin* 0. *Assume that* 0 *is an isolated fixed point of all positive iterates* g^n, $n > 0$. *Then* $\mathrm{Ind}(g^n; 0)$ *is bounded as a function of* n.

Let f be a loxodromic automorphism of a compact Kähler surface X. Suppose that f does not have any curve of periodic points; then all periodic points are isolated, because the set of periodic points of period $n > 0$ is an analytic subset of X without components of positive dimension. From Lefschetz's formula and the Shub-Sullivan theorem, there is an infinite number of periodic points, because $L(f^n)$ grows like $\lambda(f)^n$ as n goes to ∞. As a simple corollary, we obtain the following.

COROLLARY 4.5. *If* f *is a loxodromic automorphism of a compact Kähler surface, the set* $\mathrm{Per}(f)$ *of periodic points of* f *is infinite.*

To prove the existence of an infinite number of isolated periodic points (i.e., of periodic points that are not contained in curves of periodic points), one needs (i) a version of Lefschetz's formula that would take into account curves of fixed points and (ii) control of the indices along such curves; this is done in [75] for area preserving automorphisms (see [75] for examples showing that indices of f^k along curves of fixed points are not always bounded).

4.2.3 Holomorphic fixed-point formulae

In the holomorphic setting, one can derive more precise formulae. Let f be a holomorphic endomorphism of a compact complex manifold M. For each integer $r \in \{0, \ldots, \dim_{\mathbb{C}}(M)\}$, define Lefschetz's number of index r by

$$L^r(f) = \sum_{s=0}^{s=\dim_{\mathbb{C}}(M)} (-1)^s \mathrm{tr}(f^*_{|H^{r,s}(M,\mathbb{C})}).$$

For example, when M is a complex surface, Poincaré duality implies

$$L^0(f) = \overline{L^2(f)} = 1 - \mathrm{tr}(f^*_{|H^{0,1}}) + \mathrm{tr}(f^*_{|H^{0,2}}),$$
$$L^1(f) = 2\mathrm{tr}(f^*_{|H^{1,0}}) - \mathrm{tr}(f^*_{|H^{1,1}}).$$

THEOREM 4.6 (Atiyah-Bott fixed-point theorem [1]). *Let* f *be a holomorphic endomorphism of a compact complex manifold* M. *If all fixed points of* f *are non-degenerate, then*

$$L^r(f) = \sum_{f(p)=p} \frac{\mathrm{tr}(\wedge^r Df_p)}{\det(\mathrm{Id} - Df_p)}.$$

As a consequence, on a compact Kähler surface with no non-zero holomorphic form, every endomorphism has at least one fixed point. This remark applies, for example, to surfaces obtained from the projective plane by a finite sequence of blow-ups.

4.3 Periodic points are Zariski dense

As explained in the previous paragraph, every loxodromic automorphism of a compact Kähler surface has an infinite number of periodic points. Here is a stronger result for projective surfaces.

THEOREM 4.7 (Fakhruddin [59], Junyi [102]). *Let* **k** *be an algebraically closed field. Let X be an irreducible projective surface and f be an automorphism of X, both defined over* **k**. *If f is loxodromic, the set* $\mathrm{Per}(f) \subset X(\mathbf{k})$ *is Zariski dense in X. Moreover, for every curve* $Z \subset X$ *there is a periodic orbit of f in* $X \setminus Z$.

Finer results hold when f is an automorphism of a connected compact Kähler surface (see below §4.4.3).

Let us try to convey some of the ideas that lead to a proof of this theorem. First, recall that loxodromic automorphisms have a finite number of periodic curves, as shown by Section 4.1.

4.3.1   Finite fields

Let us first assume that both X and f are defined over a finite field $\mathbf{F}_q$, with q elements. Pick a point x in X and choose a finite extension $\mathbf{F}_{q^l}$ of $\mathbf{F}_q$ such that $x \in X(\mathbf{F}_{q^l})$. Since f is defined over $\mathbf{F}_q$, f permutes the points of the finite set $X(\mathbf{F}_{q^l})$, so that the orbit of x is finite. This shows that all points are periodic !

There is another, more powerful, technique to construct periodic points over finite fields. Let Z be any Zariski closed proper subset of X. We shall construct a periodic orbit of f which is entirely contained in the complement of Z, a result that is stronger than the existence of a periodic point in $X \setminus Z$.

Let $\overline{\mathbf{F}}_q$ be an algebraic closure of $\mathbf{F}_q$, and let $\Phi_q \colon X \to X$ be the geometric Frobenius automorphism (on $\overline{\mathbf{F}}_q$, Φ_q raises numbers t to the power t^q). First, note that the orbit of Z under the action of the Frobenius morphism is Zariski closed: There is an integer $k \geq 0$ such that

$$\cup_n \Phi_q^n(Z) = Z \cup \Phi_q(Z) \cup \ldots \cup \Phi_q^k(Z)$$

because Z is defined over a finite extension of $\mathbf{F}_q$. Denote by Z' this proper, Φ_q-invariant, Zariski closed subset of X. Then, fix an affine Zariski open subset $U \subset X$ that does not intersect Z'. Denote by $\Gamma_f \subset X \times X$ the graph of f, and by $\Gamma_f(U)$ its intersection with $U \times U$. We can apply the following theorem to $S = \Gamma_f(U)$.

THEOREM 4.8 (Hrushovski). *Let U be an irreducible affine variety over* $\mathbf{F}_q$, *and let* $S \subset U \times U$ *be an irreducible variety over* $\overline{\mathbf{F}}_q$, *with* Φ_q *the Frobenius automorphism on U. If the two projections of S on U are dominant, the set of points of S of the form* $(x, \Phi_q^m(x))$, *for x in U and* $m \geq 1$, *is Zariski dense in S.*

Thus, there exists a positive integer m and a point $x \in U$ such that $(x, \Phi_q^m(x))$ is contained in $\Gamma_f(U)$; in other words,

$$f(x) = \Phi_q^m(x).$$

Since f is defined over $\mathbf{F}_q$, it commutes to Φ_q, and

$$f^n(x) = \Phi_q^{mn}(x) \in X \setminus Z'$$

for all $n \geq 1$. But x is periodic under Φ_q, because its coordinates live in a finite extension of $\mathbf{F}_q$; hence, $f^n(x) = x$ for some positive integer n. This provides a periodic orbit in the complement of Z', as desired.

4.3.2 Arbitrary fields

Assume now that X and f are defined over the field of rational numbers $\mathbb{Q}$. After reduction modulo a sufficiently large prime power $q = p^l$, one gets an automorphism

$$f_q \colon X_{\mathbf{F}_q} \to X_{\mathbf{F}_q}.$$

The Néron-Severi group of $X(\mathbb{C})$ is generated by classes of curves which are defined on a finite extension K of $\mathbb{Q}$. Thus, if p and l are large enough, the action of f_q on $NS(X_{\mathbf{F}_q})$ is loxodromic, with the same dynamical degree as $f \colon X_{\mathbb{C}} \to X_{\mathbb{C}}$. In particular, f_q has a finite number of periodic curves and $\mathrm{Per}(f_q) \subset X_{\mathbf{F}_q}(\overline{\mathbf{F}}_q)$ is Zariski dense in $X_{\mathbf{F}_q}$. Pick an isolated periodic point m of some period n, i.e., a periodic point $m \in X_{\mathbf{F}_q}(\overline{\mathbf{F}}_q)$ which is not contained in a curve of periodic points. Then one can lift m to a periodic point $\hat{m} \in X(\overline{\mathbb{Q}})$ (roughly speaking, the equation $f^n(m) = m$ determines a scheme of dimension 1, which is not contained in the special fiber $X_{\mathbf{F}_q}$ because m is an isolated fixed point of f^n; thus, its intersection with the generic fiber provides a periodic point). Since the set of such points m is Zariski dense, the lifts $\hat{m}$ form a Zariski dense subset of $X(\overline{\mathbb{Q}})$.

When $\mathbf{k}$ is an arbitrary, algebraically closed field, one first replaces it by a finitely generated subring over which X and f are defined. Then standard techniques show that the same strategy — reduction plus lift — can be applied.

4.4 Topological entropy and saddle periodic points

Let us come back to the dynamics of loxodromic automorphisms on compact Kähler surfaces and apply tools from dynamical systems to understand periodic points.

4.4.1 Entropy

Let g be a continuous transformation of a compact metric space (Z, dist). The **topological entropy** $\mathsf{h}_{top}(g)$ is defined as follows. Let ϵ be a positive number and n a positive integer. One says that a finite subset A of Z is separated at scale ϵ during the first n iterations, or simply that A is (ϵ, n)-separated, if and only if, for any pair of distinct points a and b in A, there exists a time $0 \le k < n$ such that

$$\mathrm{dist}(g^k(a), g^k(b)) \ge \epsilon.$$

Let $N(\epsilon, n)$ denote the maximum number of elements in (ϵ, n)-separated subsets. Then, one defines successively

$$\mathsf{h}_{top}(g; \epsilon) = \limsup_{n \to \infty} \frac{1}{n} \log N(\epsilon, n)$$

and, taking finer and finer scales of observation of the dynamics,

$$\mathsf{h}_{top}(g) = \lim_{\epsilon \to 0} \mathsf{h}_{top}(g; \epsilon).$$

So, topological entropy counts the rate at which the dynamics of g creates distinct orbits, when observed with an arbitrarily small, but positive, scale. As an example, the transformation $z \mapsto z^d$ of the unit circle $\{z \in \mathbb{C}; |z| = 1\}$, has entropy $\log(d)$.

4.4.2   Gromov-Yomdin formula

Computing topological entropy is a difficult problem in practice, but for holomorphic transformations $f \colon M \to M$ of compact Kähler manifolds, entropy coincides with the logarithm of the spectral radius of $f^* \in \mathsf{GL}\,(H^*(M, \mathbb{C}))$.

THEOREM 4.9 (Gromov [66], Yomdin [103, 65]). *Let f be a diffeomorphism of a compact manifold M, and let $\lambda(f)$ denote the spectral radius of the linear transformation $f^* \colon H^*(M, \mathbb{C}) \to H^*(M, \mathbb{C})$.*

- *If M and f are of class $\mathcal{C}^\infty$, then $\mathsf{h}_{top}(f) \geq \log \lambda(f)$.*

- *If M is a Kähler manifold and f is holomorphic, $\mathsf{h}_{top}(f) = \log \lambda(f)$.*

For automorphisms of compact Kähler surfaces, one gets

$$\mathsf{h}_{top}(f) = \log \lambda(f),$$

where $\lambda(f)$ is the dynamical degree of f^*.

REMARK 4.10. When f is an automorphism of a compact Kähler manifold M, one can replace $\lambda(f)$ by the largest eigenvalue of f^* on the sum $\bigoplus_p H^{p,p}(M, \mathbb{R})$ in Gromov-Yomdin theorem.

EXAMPLE 4.11. Let $M = \mathsf{SL}_2(\mathbb{C})/\Gamma$ where Γ is a co-compact lattice in $\mathsf{SL}_2(\mathbb{C})$. Let t be a positive real number. The automorphism f_t of M defined by left multiplication by

$$\left(\begin{array}{cc} \exp(t) & 0 \\ 0 & \exp(-t) \end{array} \right)$$

is isotopic to the identity (let t go to 0) but has positive entropy. This does not contradict Gromov's theorem because M is not Kähler.

4.4.3   Saddle periodic points

Let p be a periodic point of the automorphism f and let k be its period. One says that p is a saddle (or hyperbolic) periodic point if one eigenvalue of the tangent map $D(f^k)_p$ has modulus > 1 and the other has modulus < 1. Since f has topological entropy $\log(\lambda(f))$ and X has dimension 2, one can apply a result due to Katok.

THEOREM 4.12 (Katok [77]). *Let f be a loxodromic automorphism of a compact Kähler surface. The set of saddle periodic points of f is Zariski dense in X. The number $N(f, k)$ of saddle periodic points of f of period at most k grows like $\lambda(f)^k$: for all $\epsilon > 0$,*

$$\limsup \frac{1}{k} \log(N(f, k)) \geq \log(\lambda(f) - \epsilon).$$

The same result holds for isolated periodic points in place of saddle periodic points.

Katok's proof requires several non-trivial dynamical constructions, including the full strength of Pesin theory. It provides f-invariant subsets $\Lambda_l \subset X$, $l \geq 1$, and numbers ϵ_l going to 0 with l, such that (i) the restriction of f to Λ_l is conjugate to a horse-shoe map (and is therefore well understood [78]), and (ii) the number of periodic points of period n in these sets grows like $(\lambda(f) - \epsilon_l)^n$ with n.

5 INVARIANT CURRENTS

5.1 Currents

5.1.1 Definitions

Let X be a compact Kähler surface, and $\wedge^{1,1}(X, \mathbb{R})$ be the space of smooth real-valued $(1,1)$-forms on X with its usual Fréchet topology. By definition, a $(1,1)$-current is a continuous linear functional on $\wedge^{1,1}(X, \mathbb{R})$. For simplicity, $(1,1)$-currents are called **currents** in this text. The value of a current T on a form ω is denoted by $(T|\omega)$.

EXAMPLE 5.1. **a.** Let α be a continuous $(1,1)$-form, or, more generally, a $(1,1)$-form with distribution coefficients. Then α defines a current $\{\alpha\}$ (also denoted by α in what follows):

$$(\{\alpha\}|\omega) = \int_X \alpha \wedge \omega.$$

b. Let $C \subset X$ be a curve. The current of integration on C is defined by

$$(\{C\}|\omega) = \int_C \omega.$$

This is well defined even if C is singular; moreover, $\{C\}$ extends to a linear functional on the space of continuous forms for the topology of uniform convergence.

Recall that a $(1,1)$-form ω is positive if $\omega(u, \sqrt{-1}u) \geq 0$ for all tangent vectors u. A current T is **positive** if it takes non-negative values on the convex cone of positive forms. When positive, T extends as a continuous linear functional on the space of continuous $(1,1)$-forms with the topology of uniform convergence. Given two currents T and T', one says that T is larger than T', written $T \geq T'$, if the difference $T - T'$ is a positive current.

A current is **closed** if it vanishes on the space of exact forms. For example, the current associated to a smooth $(1,1)$-form α is positive (resp. closed) if and only if α is a positive (resp. closed) form. The current of integration on a curve $C \subset X$ is positive and closed (because C has empty boundary).

5.1.2 Cohomology classes

Let T be a closed current. Then T defines a linear form on the space $H^{1,1}(X, \mathbb{R})$, and there is a unique cohomology class $[T]$ such that

$$(T|\omega) = \langle [T] | [\omega] \rangle$$

for all closed forms ω of type $(1,1)$. By definition, $[T]$ is the cohomology class of T.

5.1.3 Mass and compact sets of currents

Let T be a positive current on a Kähler surface X. Let κ be a Kähler form on X. The trace measure of T is the positive measure $\|T\|$ defined by

$$\int_X \xi \|T\| = (T|\xi\kappa)$$

for all smooth functions ξ; it depends on the choice of the Kähler form κ. The mass $M(T)$ of T is the total mass of the trace measure $\|T\|$. When T is closed, we obtain

$$M(T) = \langle [T] | [\kappa] \rangle,$$

so that the mass depends only on the cohomology class $[T]$.

The space of currents is endowed with the weak topology: a sequence of currents (T_i) converges toward a current T if $(T_i | \omega)$ converges towards $(T | \omega)$ for all smooth forms. The set of positive currents with mass at most B (B any positive real number) is a compact convex set for this topology. In particular, if T_i is a sequence of closed positive currents with uniformly bounded cohomology classes, one can extract a converging subsequence.

5.1.4 *Potentials (see [64, 38])*

The differential operator d decomposes as $d = \partial + \bar{\partial}$ where, in local coordinates $z_i = x_i + \sqrt{-1} y_i$, the operators ∂ and $\bar{\partial}$ are given by

$$\partial = \sum_i \frac{1}{2} \left(\frac{\partial}{\partial x_i} - \sqrt{-1} \frac{\partial}{\partial y_i} \right) dz_i, \quad \bar{\partial} = \sum_i \frac{1}{2} \left(\frac{\partial}{\partial x_i} + \sqrt{-1} \frac{\partial}{\partial y_i} \right) d\bar{z}_i.$$

Denote by d^c the operator $\frac{1}{2\pi}(\bar{\partial} - \partial)$; then

$$dd^c = \frac{\sqrt{-1}}{\pi} \partial \bar{\partial}.$$

Let T be a closed and positive current. Locally, T can be written as

$$T = dd^c u$$

for some function u, called a local **potential** of T (see [64], §3.2). The positivity of T is equivalent to the **pluri-subharmonicity** of u, which means that (i) u is upper semi-continuous with values into $\{-\infty\} \cup \mathbb{R}$ and (ii) u is subharmonic along all holomorphic disks $\varphi \colon \mathbb{D} \to X$ (i.e., $u \circ \varphi$ is either identically $-\infty$ or subharmonic on $\mathbb{D}$). Pluri-subharmonic functions are locally integrable, and the equation $T = dd^c u$ means that

$$(T | \omega) = \int_X u \, dd^c \omega$$

for all smooth forms ω with support in the open set where the equality $T = dd^c u$ is valid. When the local potentials of T are continuous (resp. smooth, Hölder continuous, etc.), one says that T has continuous (resp. smooth, Hölder continuous, etc.) potentials.

5.1.5 *Multiplication (see [4, 38])*

In general, distributions, and currents as well, cannot be multiplied, but Bedford and Taylor introduced a pertinent way to multiply two closed positive currents T_1 and T_2 when one of them, say T_2, has continuous potentials. The product, a positive measure $T_1 \wedge T_2$, is defined by the following local formula:

$$(T_1 \wedge T_2 | \psi) = (T_1 | u_2 dd^c (\psi))$$

for all smooth functions ψ with support on open sets where $T_2 = dd^c(u_2)$; when both T_1 and T_2 have continuous potentials, this definition is symmetric in T_1 and T_2.

Cohomology classes and products of currents are compatible, which means that the total mass of the measure $T_1 \wedge T_2$ is equal to the intersection of the classes $[T_1]$ and $[T_2]$ (for closed positive currents with continuous potentials).

5.1.6 Automorphisms

Let f be an automorphism of X and T be a current. Define $f_* T$ by

$$(f_* T | \omega) = (T | f^* \omega), \quad \forall \, \omega \in \wedge^{1,1}(X, \mathbb{R}).$$

The operator f_* maps closed (resp. positive) currents to closed (resp. positive) currents. Define f^* by $f^* = (f^{-1})_*$; it satisfies $[f^* T] = f^*[T]$, where the right-hand side corresponds to the action of f on the cohomology group $H^{1,1}(X, \mathbb{R})$.

EXAMPLE 5.2. If $C \subset X$ is a curve, then $f_*\{C\}$ is the current of integration on the curve $f(C)$. If α is a $(1,1)$-form, then $f^*\{\alpha\} = \{f^* \alpha\}$.

5.2 The currents T_f^+ and T_f^- and the probability measure μ_f

THEOREM 5.3 (see [27], [48], [50], and [90]). *Let f be a loxodromic automorphism of a compact Kähler surface X. There is a unique closed positive current T_f^+ such that $[T_f^+] = v_f^+$. The local potentials of T_f^+ are Hölder continuous,*

$$f^* T_f^+ = \lambda(f) T_f^+,$$

and $\mathbb{R}^+ T_f^+$ is an extremal ray in the convex cone of closed positive currents.

The extremality means that a convex combination $sT + (1-s)T'$ of two closed positive currents T and T' is proportional to T_f^+ if and only if both T and T' are proportional to T_f^+.

Applied to f^{-1}, this result shows that there is a unique closed positive current T_f^- such that $[T_f^-] = v_f^-$. This current has Hölder continuous potentials; it satisfies

$$f^* T_f^- = \frac{1}{\lambda(f)} T_f^-,$$

and the ray $\mathbb{R}^+ T_f^-$ is also extremal.

COROLLARY 5.4. *Let f be a loxodromic automorphism of a compact Kähler surface X. Let $C \subset X$ be a curve, and $\{C\}$ be the current of integration on C. Then*

$$\frac{1}{\lambda(f)^n} (f^n)^* \{C\} \to \langle [C] | v_f^- \rangle \, T_f^+$$

as n goes to $+\infty$. Let κ be a Kähler form on X and let $\{\kappa\}$ be the current determined by this form. Then

$$\frac{1}{\lambda(f)^n} (f^n)^* \kappa^n \to \langle [\kappa] | v_f^- \rangle \, T_f^+$$

as n goes to $+\infty$.

PROOF OF COROLLARY 5.4. Let C be a curve and $[C]$ be its cohomology class. Decompose $[C]$ as

$$[C] = [C]_+ + [C]_- + [C]_N,$$

where $[C]_{\pm}$ is contained in $\mathbb{R} v_f^{\pm}$ and $[C]_N$ is in the orthogonal complement N_f. Since $\langle [C] | v_f^+ \rangle = \langle [C_-] | v_f^+ \rangle$ and $\langle v_f^+ | v_f^- \rangle = 1$, we have $[C_+] = \langle [C] | v_f^- \rangle v_f^+$. When n goes to $+\infty$, the sequence $(f^*)^n [C]_N$ is bounded and $(f^n)^* [C]_-$ goes to 0. Thus,

$$\frac{1}{\lambda(f)^n} (f^n)^* [C] \to \langle [C] | v_f^- \rangle v_f^+.$$

In particular, the sequence of currents $(f^n)^* \{C\} / \lambda(f)^n$ has bounded mass, and all limits of convergent subsequences are currents with cohomology class $\langle [C] | v_f^- \rangle v_f^+$. Since $\langle [C] | v_f^- \rangle T_f^+$ is the unique closed positive current with cohomology class $\langle [C] | v_f^- \rangle v_f^+$, the sequence $(f^*)^n \{C\} / \lambda(f)^n$ converges toward $\langle [C] | v_f^- \rangle T_f^+$.

The same proof applies to Kähler forms κ.   □

5.2.1   The measure μ_f

Since T_f^+ and T_f^- have continuous potentials, we can multiply them: this defines a probability measure

$$\mu_f = T_f^+ \wedge T_f^-;$$

the total mass of μ_f is 1 because our choice for the cohomology classes v_f^+ and v_f^- implies

$$\langle [T_f^+] | [T_f^-] \rangle = \langle v_f^+ | v_f^- \rangle = 1.$$

The probability measure μ_f is f-invariant because T_f^+ is multiplied by $\lambda(f)$, while T_f^- is divided by the same quantity.

Note that μ_f is uniquely determined by the diffeomorphism f and the f-invariant complex structure on X: both T_f^+ and T_f^- are uniquely determined by the equation $f^* T_f^{\pm} = \lambda(f)^{\pm} T_f^{\pm}$ up to scalar multiplication, so that the product $\mu_f = T_f^+ \wedge T_f^-$ is uniquely determined once one imposes $\mu_f(X) = 1$.

REMARK 5.5. Since T_f^+ and T_f^- have Hölder continuous potentials, one can show that the Hausdorff dimension of μ is strictly positive. We refer to [58] for a discussion of this topic for endomorphisms of projective spaces.

5.3   The measure μ_f is mixing

The following statement is deeper than Corollary 5.4. It applies, for example, when S is the current of integration over a disk Δ contained in a stable manifold of f (see Section 6.1.2 below).

THEOREM 5.6 (Bedford-Smillie, Fornaess-Sibony [10], [27], [60]). *Let f be a loxodromic automorphism of a connected compact Kähler surface X. Let S be a positive current and $\psi \colon X \to \mathbb{R}_+$ be a smooth function which vanishes in a neighborhood of the support of ∂S. Then*

$$\frac{1}{\lambda(f)^n} (f^n)^* (\psi S)$$

converges toward

$$(T_f^- | \psi S) T_f^+$$

in the weak topology as n goes to $+\infty$.

The number $(T_f^- | \psi S)$ is the total mass of the positive measure $T_f^- \wedge (\psi S)$. One drawback of this statement resides in the difficulty of proving that $T_f^- \wedge (\psi S)$ is not zero, but there is at least one interesting and easily accessible corollary.

COROLLARY 5.7 ([10], [27], [60]). *If f is a loxodromic automorphism of a connected compact Kähler surface, the measure μ_f is ergodic and mixing.*

Ergodicity means that all f-invariant measurable subsets have measure 0 or 1. The mixing property is stronger, and says that $\mu_f(f^n(A) \cap B)$ converges towards $\mu_f(A)\mu_f(B)$ as n goes to ∞ for all pairs (A, B) of measurable subsets of X. Equivalently, μ_f is mixing if and only if

$$\int_X (\phi \circ f^n)\, \psi \, d\mu_f \to \int_X \phi \, d\mu_f \int_X \psi \, d\mu_f$$

for all pairs of smooth (resp. smooth and non-negative) functions (ϕ, ψ) on X.

To prove the corollary, start with two smooth functions ϕ and ψ with non-negative values. By definition of μ_f we have

$$
\begin{aligned}
\int_X (\phi \circ f^n)\, \psi \, d\mu_f &= (T_f^+ \wedge T_f^- \,|\, (\phi \circ f^n)\psi) \\
&= ((\phi T_f^+) \wedge T_f^- \,|\, (\psi \circ f^{-n})) \\
&= \left(\frac{1}{\lambda(f)^n}(f^n)^*(\phi T_f^+) \wedge T_f^- \,\Big|\, \psi \right)
\end{aligned}
$$

because $f^* T_f^- = \lambda(f)^{-1} T_f^-$. From Theorem 5.6, we obtain

$$\frac{1}{\lambda(f)^n}(f^n)^*(\phi T_f^+) \to c T_f^+, \quad \text{with } c = \int_X \phi \, d\mu_f.$$

Since products of currents are compatible with weak convergence, the sequence $\int_X (\phi \circ f^n)\psi \, d\mu_f$ converges toward

$$c\,(T_f^+ \wedge T_f^- \,|\, \phi) = \int_X \phi \, d\mu_f \int_X \psi \, d\mu_f,$$

as desired.

6   ENTIRE CURVES, STABLE MANIFOLDS, AND LAMINARITY

6.1   Entire curves and stable manifolds

6.1.1   *Entire curves*

By definition, an **entire curve** on X is a non-constant holomorphic map $\xi \colon \mathbf{C} \to X$. Fix such a curve and a Kähler form κ on X. For each real number $r \geq 0$, denote by $\mathbb{D}_r \subset \mathbb{C}$ the open disk of radius r, and denote by $A(r)$ and $L(r)$ the area and perimeter of $\xi(\mathbb{D}_r)$:

$$A(r) = \int_{t=0}^{r} \int_{\theta=0}^{2\pi} \|\xi'(te^{i\theta})\|_\kappa^2 \, t dt d\theta, \qquad (6.1)$$

$$L(r) = \int_{\theta=0}^{2\pi} \|\xi'(te^{i\theta})\|_\kappa \, t d\theta, \qquad (6.2)$$

where $\|\xi'(te^{i\theta})\|_\kappa$ is the norm of the velocity vector $\xi'(te^{i}\theta)$ with respect to the Kähler metric defined by κ. By the Cauchy-Schwartz inequality, one gets Ahlfors's inequality,

$$L(r)^2 \leq 2\pi r \frac{dA}{dr}(r).$$

It implies that the infimum limit of the ratio $L(r)/A(r)$ vanishes. As a consequence, there are sequences of radii (r_n), going to $+\infty$ with n, such that

$$\frac{1}{A(r_n)}\{\xi(\mathbb{D}_{r_n})\}$$

converges toward a closed positive current.

Another useful family of currents is defined by

$$N(r) = \frac{1}{T(r)} \int_0^r \{\xi(\mathbb{D}_t)\}\frac{dt}{t}$$

with

$$T(r) = \int_0^r A(t)\frac{dt}{t}.$$

Then, Ahlfors and Jensen inequalities, together with the Nakai-Moishezon theorem, imply the following result.

THEOREM 6.1 (Positivity of Ahlfors currents [19], [27], [39], [90]). *Let X be a compact Kähler surface with a Kähler form κ. Let $\xi: \mathbb{C} \to X$ be an entire curve. There exist sequences of radii (r_n) going to ∞ such that $(N(r_n))$ converges toward a closed positive current. Let T be such a current.*

1. *If $\xi(\mathbb{C})$ is contained in a compact curve E, this curve has genus 0 or 1 (it may be singular), and T is equal to the current $\langle \kappa|[E]\rangle^{-1}\{E\}$.*

2. *If $A(r)$ is bounded, then $\xi(\mathbb{C})$ is contained in a rational curve E.*

If $\xi(\mathbb{C})$ is not contained in a compact curve, then

3. *$[T]$ intersects all classes of curves positively, i.e. $\langle[T]|[C]\rangle \geq 0$ for all curves $C \subset X$;*

4. *$[T]$ is in the nef cone and $\langle[T]|[T]\rangle \geq 0$.*

6.1.2 Stable manifolds

Let now p be a saddle periodic point of the loxodromic automorphism f, and let k be its period (there is an infinite number of such points by Theorem 4.12). Locally,

f^k is linearizable: there is an open subset U of X containing p, and a holomorphic diffeomorphism $\psi\colon U \to V$ onto a ball $V \subset \mathbf{C}^2$ such that $\psi(p) = (0,0)$ and

$$\psi \circ (f^k) \circ \psi^{-1}(x,y) = (\alpha x, \beta y),$$

where α, β are the eigenvalues of $D(f^k)_p$, $|\alpha| < 1$ and $|\beta| > 1$. Thus, locally, the set of points q near p such that $f^{kn}(q)$ stays in U and converges toward p as n goes to $+\infty$ is the image of the horizontal axis by ψ^{-1}. This set is the **local stable manifold** $W^s_{loc}(p)$ of p. The **stable manifold** $W^s(p)$ is the set of points q in X such that $f^{kn}(q)$ converges towards p as n goes to $+\infty$. This set coincides with the increasing union of all $f^{-kn}(W^s_{loc}(p))$, $n \geq 1$. Unstable and local unstable manifolds are defined similarly.

By construction, every stable manifold $W^s(p)$ is the holomorphic image of a Riemann surface which is homeomorphic to the plane $\mathbf{R}^2$, and f induces an automorphism of this Riemann surface which fixes p and acts as a contraction around p. This implies that $W^s(p)$ is not isomorphic to the unit disk, because all automorphisms of $\mathbf{D}$ are isometries with respect to the Poincaré metric. From Riemann uniformization theorem, we deduce that $W^s(p)$ is parametrized by an entire curve: there exists an entire curve $\xi^s_p \colon \mathbf{C} \to X$ such that ξ^s_p is injective, $\xi^s_p(0) = p$ and $\xi^s_p(\mathbf{C}) = W^s(p)$. The map ξ^s_p is unique up to composition with similitudes $z \mapsto \gamma z$. This implies that

$$f^k \circ \xi^s_p(z) = \xi^s_p(\alpha z).$$

THEOREM 6.2 (see [9], [10], and [27]). *Let f be a loxodromic automorphism of a compact Kähler surface X. Let p be a saddle periodic point of f of period k, and $\xi^s_p \colon \mathbf{C} \to X$ be a parametrization of its stable manifold. If $\xi^s_p(\mathbf{C})$ has finite area, it is contained in a periodic rational curve of period k. If $\xi^s_p(\mathbf{C})$ has infinite area, then all closed limits of sequences $N_{\xi^s_p}(r_n)$ coincide with a positive multiple of T^+_f.*

An important feature of this theorem is that T^+_f can be recovered from every saddle periodic point (once periodic curves have been contracted). This will lead to a strong relationship between saddle periodic points and the invariant probability measure μ_f.

REMARK 6.3. **a.** Let $\varphi \colon \mathbf{D} \to \mathbf{R}^+$ be a smooth function with compact support such that $\varphi(0) > 0$. Denote by $\{(\xi^s_p)_*(\varphi\mathbf{D})\}$ the current that is defined by

$$(\{(\xi^s_p)_*(\varphi\mathbf{D})\}|\omega) = \int_{\mathbf{D}} \varphi \, (\xi^s_p)^*\omega.$$

Theorem 5.6 implies that the sequence

$$\frac{1}{\lambda(f)^{kn}} (f^{kn})^* \{(\xi^s_p)_*(\varphi\mathbf{D})\}$$

converges toward a multiple of T^+_f; by Theorem 6.2, the limit cannot be zero and is therefore a positive multiple of T^+_f. Similarly, T^-_f is an Ahlfors current for unstable manifolds of saddle periodic points of f. Since $T^+_f \wedge T^-_f$ does not vanish, this suggests that the stable and unstable manifolds of saddle periodic points p and q always intersect, except when p or q is contained in a periodic curve, a fact which is proved in Section 6.4.2.

b. One can also deduce the following asymptotic behavior from Theorem 6.2, in which $A(r)$ denotes the area of $\xi_p^s(\mathbb{D}_r)$,

$$\limsup_{r \to \infty} \frac{\log(A(r))}{\log(r)} = \frac{\log(\lambda(f))}{|\log(\alpha^{1/k})|}.$$

This means that the growth rate of ξ_p^s is the ratio between the topological entropy of f and the Lyapunov exponent of $\hat{f}$ at p (see [28]).

PROOF OF THE FIRST ASSERTION OF THEOREM 6.2. Replacing f by f^k, we assume that p is a fixed point. Let T be a closed current for which exists a sequence of radii (r_n) such that $N_{\xi_p^s}(r_n)$ converges toward T. Since $f^*\{\xi_p^s(\mathbb{D}_t)\} = \{\mathbb{D}_{t/\alpha}\} \geq \{\mathbb{D}_t\}$ for all $t \geq 0$, we get

$$f^* N_{\xi_p^s}(r_n) = \frac{1}{T_{\xi_p^s}(r_n)} \int_{t=0}^{r_n} f^*\{\xi_p^s(\mathbb{D}_t)\} \frac{dt}{t} \geq N_{\xi_p^s}(r_n)$$

and $f^*T \geq T$. This implies that

$$\langle (f^m)^*[T] | [\kappa] \rangle \geq \langle [T] | [\kappa] \rangle$$

for all Kähler classes $[\kappa]$ and all positive integers m. Since f^* preserves intersections,

$$\frac{1}{\lambda(f)^m} \langle [T] | [\kappa] \rangle = \langle (f^m)^*[T] | \frac{1}{\lambda(f)^m} (f^m)^*[\kappa] \rangle \geq \langle [T] | \frac{1}{\lambda(f)^m} (f^m)^*[\kappa] \rangle,$$

and taking limits on both sides, $0 \geq \langle [T] | [T_f^+] \rangle$. Moreover, Theorem 6.1(4) implies $\langle [T] | [T] \rangle \geq 0$. By the Hodge index theorem, $[T]$ is proportional to $[T_f^+]$ and, by Theorem 5.3, T is a positive multiple of T_f^+.   □

6.2   Laminarity

Since T_f^+ is obtained as Ahlfors currents for stable manifolds, i.e., for injective entire curves $\xi_p^s \colon \mathbb{C} \to X$, one expects that the structure of T_f^+ retains some information from the injectivity of the entire curves ξ_p^s (see Figure 1.1, right, and the close-up in Figure 6.5). This leads to the theory of laminar currents.

6.2.1   *Uniformly laminar currents*

Let Γ be a family of disjoint horizontal graphs in the bidisk $\mathbb{D} \times \mathbb{D}$: each element of Γ intersects the vertical disk $\{0\} \times \mathbb{D}$ in a unique point. If a is such a point of intersection and Γ_a is the graph of the family Γ containing a, there exists a holomorphic mapping $\varphi_a \colon \mathbb{D} \to \mathbb{D}$ such that $\Gamma_a = \{(x, \varphi_a(x)); x \in \mathbb{D}\}$.

REMARK 6.4. Let A be the set of intersection points a of the graphs $\Gamma_a \in \Gamma$ with the vertical disk $\{0\} \times \mathbb{D}$. Then, Γ determines a holomorphic motion of the set A parametrized by the disk $\mathbb{D}$ (see [52]): A moves along the graphs from its initial position in $\{0\} \times \mathbb{D}$ to nearby disks $\{z\} \times \mathbb{D}$ by $a = \varphi_a(0) \in A \mapsto \varphi_a(z)$. By the so-called Λ-lemma, (i) this motion extends to a motion of its closure $\bar{A}$ and (ii) the motion from $\{0\} \times \mathbb{D}$ to $\{z\} \times \mathbb{D}$ is Hölder continuous.

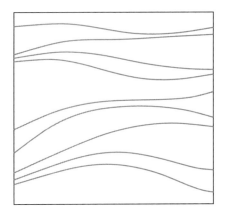

FIGURE 6.5: Stable lamination. This is a close-up of Figure 1.1 (right).

FIGURE 6.6: Uniform lamination. A family of horizontal graphs in a bidisk.

Given a finite positive measure ν on $\{0\} \times \mathbb{D}$ (or on $\overline{A}$), one obtains a measured family of disjoint graphs (Γ, ν); it determines a closed positive current $T_{\Gamma,\nu}$ in $\mathbb{D} \times \mathbb{D}$, which is defined by

$$(T_{\Gamma,\nu}|\omega) = \int_{a\in\mathbb{D}} \int_{\Gamma_a} \omega \, d\nu(a) = \int_{a\in\mathbb{D}} \int_{\mathbb{D}} \varphi_a^* \omega \, d\nu(a).$$

One says that $T_{\Gamma,\nu}$ is the current of integration over (Γ, ν).

By definition, a **flow box** Γ of a complex surface Z is a closed set of disjoint horizontal graphs in a bidisk $\mathbb{D} \times \mathbb{D} \simeq U \subset Z$; a **measured flow box** is a flow box Γ, together with a transverse measure ν. Thus, every measured flow box (Γ, ν) defines a current of integration $T_{|\Gamma,\nu}$ on U. A current T on the surface Z is **uniformly laminar** if T is locally given by integration over a measured flow box. If T is uniformly laminar, those flow boxes can be glued together to define a lamination of the support of T.

EXAMPLE 6.7. Let $X = \mathbb{C}^2/\Lambda$ be a complex torus. Let α be a non-zero holomorphic 1-form on X; such a form is induced by a constant 1-form $adx + bdy$ on $\mathbb{C}^2$, and the kernel of α determines a holomorphic foliation $\mathcal{F}_\alpha$ on X, whose leaves are projections of parallel lines $ax + by = C^{ste}$. Let T be the current $\{\alpha \wedge \overline{\alpha}\}$. Then T is uniformly laminar: locally, T is given by integration over disks in the leaves of the foliation $\mathcal{F}_\alpha$ with respect to Lebesgue measure on the transversal.

6.2.2 Laminar currents [9], [27], [57]

A positive current T on a complex surface Z is **laminar** if there is an increasing sequence of open subsets $\Omega_i \subset Z$, $i \in \mathbb{N}$ and an increasing sequence of currents T_i supported on Ω_i such that

   (i) for each i, $\|T\|(\partial\Omega_i) = 0$, i.e., the boundary of Ω_i does not support any mass of T;

  (ii) each T_i is uniformly laminar in its domain of definition Ω_i;

 (iii) the sequence of currents $(T_i)_{i\geq1}$ weakly converges toward T.

Equivalently, T is laminar if there is a family of disjoint measured flow boxes (Γ_i, ν_i) such that

$$T = \sum_i T_{|\Gamma, \nu_i}.$$

Thus, every laminar current has a representation

$$T = \int_{\mathcal{A}} \{\Delta_a\} \, d\mu(a) \qquad (6.3)$$

as a current of integration over a measured family of disjoint disks (each disk Δ_a is the image of $\mathbb{D}$ by an injective holomorphic map $\varphi_a \colon \mathbb{D} \to \mathbb{C}$ which extends to a neighborhood of $\mathbb{D}$ in $\mathbb{C}$). In general, those disks Δ_a cannot be glued together into a lamination.

EXAMPLE 6.8. Let p be a point of the projective plane $\mathbb{P}^2(\mathbb{C})$. Identify the set of lines through p with the projective line $\mathbb{P}^1(\mathbb{C})$; each line $L_x, x \in \mathbb{P}^1(\mathbb{C})$, determines a current of integration $\{L_x\}$. Let ν_p be a probability measure on this set of lines. Then

$$T_p = \int_{\mathbb{P}^1(\mathbb{C})} \{L_x\} \, d\nu_p(x)$$

is a laminar current. Let $q \in \mathbb{P}^2(\mathbb{C})$ be a second point and ν_q a probability measure on the space of lines through q. This provides a second laminar current T_q. Suppose that (i) the supports of ν_p and ν_q have zero Lebesgue measure, but (ii) both ν_p and ν_q are given by continuous potentials (i.e., $d\nu = dd^c(u)$ locally on $\mathbb{P}^1(\mathbb{C})$, with u continuous). Then $T_p + T_q$ is laminar, $T_p + T_q$ has continuous potentials, and $T_p \wedge T_q$ is a well-defined positive (non-zero) measure (see [54] for details). This example shows that the constitutive disks of $T_p + T_q$ cannot be glued together to form a lamination of the support of $T_p + T_q$.

6.3   The current T_f^+ is laminar

THEOREM 6.9 (de Thélin, [36]). *Let Ω be a bounded open subset of $\mathbb{C}^2$. Let (C_n) be a sequence of curves, defined in neighborhoods of Ω, and let (d_n) be a sequence of positive real numbers such that $(1/d_n)\{C_n\}$ converges toward a closed positive current T on Ω. If genus$(C_n) \leq \mathrm{C}^{\mathrm{ste}} d_n$, the current T is laminar.*

REMARK 6.10. Let κ be a Kähler form on $\mathbb{C}^2$, and let $Area(C_n)$ denote the area of the curve C_n with respect to κ. If T is not zero, then

$$\frac{Area(C_n)}{d_n} = \frac{\langle \{C_n\} | \kappa \rangle}{d_n} \to \langle T | \kappa \rangle,$$

so that d_n is asymptotically proportional to, and can be replaced by, $Area(C_n)$.

Being laminar is a local property. To explain how the proof of Theorem 6.9 starts, one can, therefore, choose a linear projection $\pi \colon \mathbb{C}^2 \to \mathbb{C}$ and assume that Ω is a bidisk $\mathbb{D} \times \mathbb{D}$ on which π coincides with the projection on the first factor. Let $r_m = 2^{-m}$. For each $m \geq 1$, tessellate the complex line $\mathbb{C}$ into the open squares of size r_m defined by

$$Q_m(i, j) = r_m Q_0 + r_m(i, j), \quad (i, j) \in \mathbb{Z}^2,$$

where $Q_0 = \{(x, y) \in \mathbb{R}^2 = \mathbb{C}; |x| < 1, |y| < 1\}$. This induces a tessellation $\mathcal{Q}_m$ of the unit disk $\mathbb{D}$ into pieces $Q_m(i, j) \cap \mathbb{D}$. Fix an index n and consider the curve C_n. For each element Q of $\mathcal{Q}_m$, organize the connected components D of $\pi^{-1}(Q) \cap C_n$ in two families:

- D is a good component if $\pi \colon D \to Q$ is an isomorphism,

- D is a bad component otherwise (e.g., $\pi \colon D \to Q$ has degree > 1).

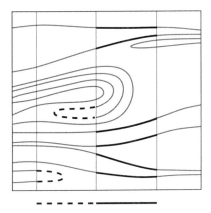

FIGURE 6.11: Good and bad components. Two bad components drawn above the center left tile (dashed black) and five good components above the center right tile (solid black).

Denote by $\mathcal{G}_{n,m}$ the set of all good components when Q runs over all tiles of the tessellation $\mathcal{Q}_m$, and define the currents

$$\{\mathcal{G}_{n,m}\} = \sum_{D \in \mathcal{G}_{n,m}} \{D\}.$$

By construction, each $\{\mathcal{G}_{n,m}\}/d_n$ is uniformly laminar on an open subset Ω_m of Ω (Ω_m is the union of the $\pi^{-1}(Q_m(i, j) \cap \mathbb{D}))$. Moreover, when m is fixed, these currents are laminar with respect to the same bidisks and projections; this implies that the limit of these currents as n goes to infinity, m being fixed, is uniformly laminar on Ω_m. For all (n, m) the total mass of $\{\mathcal{G}_{n,m}\}$ is bounded from above by d_n. If the total area of bad components becomes small with respect to d_n when n and m become large, this will show that T is laminar. The control of bad components is precisely what de Thélin obtains, using ideas from Ahlfors and Nevanlinna theory, under the assumption $\mathrm{genus}(C_n) = O(d_n)$ and for a generic projection π (see [36]). In our context, one can apply this strategy to

$$C_n = f^{-n}(C), \quad d_n = \lambda(f)^n, \quad T = T_f^+,$$

where C is as in Corollary 5.4. For those examples, and with carefully chosen bidisks $\Omega \subset X$ and projections $\pi \colon \Omega \to \mathbb{D}$, one obtains the estimate (see [53])

$$0 \leq (T - \frac{1}{d_n}\{\mathcal{G}_{n,m}\}|\kappa) \leq C^{\mathrm{ste}} r_m^2.$$

Hence, T is laminar, with an explicit rate of convergence of order $O(r_m^2)$ in the proof. Such currents are said to be strongly approximated by algebraic curves (see [54] and [57], Prop. 5.1); if T is such a current — i.e., closed, positive, laminar and strongly approximated by algebraic curves — and T does not charge any analytic set, we say that T is a **good laminar current**.

THEOREM 6.12 (Bedford-Lyubich-Smillie [3],[27], [53]). *Let f be a loxodromic automorphism of a compact Kähler surface X. Both T_f^+ and T_f^- are laminar, and are good laminar currents when X is projective.*

Note, however, that the currents T_f^+ and T_f^- are rarely uniformly laminar, as Figure 6.5 suggests (see [27], §7, and [31]).

REMARK 6.13. Assume that X contains a curve C such that $\lambda(f)^{-n}(f^n)^*\{C\}$ converges toward T_f^+. Then the Néron-Severi group $NS(X, \mathbb{R})$ contains the classes $(f^n)^*[C]$ and, therefore, $[T_f^+]$ as well; it also contains $[T_f^-]$. Since $[T_f^+] + [T_f^-]$ has positive self-intersection, this implies that $NS(X)$ contains classes with positive self-intersection, so that X is projective. This is the reason why X is assumed to be projective in the last part of Theorem 6.12.

6.4   Good laminar currents, and contraction properties for $T_f^{\pm}$

When a current is laminar, its building disks may intersect transversally with positive probability, as in Example 6.8. Good laminar currents inherit a stronger laminar structure; see [53],[54],[55], and [56].

6.4.1   *Analytic continuation and weak lamination*

Let T be a laminar current. One says that a disk $\Delta = \varphi(\mathbb{D})$ is subordinate to T if there is an open set $\Omega \subset X$ containing Δ and a uniformly laminar current S on Ω such that $S \leq T$, Δ is contained in the support of S, and Δ lies inside one of the leaves of the lamination associated to S. If Γ is a flow box, one defines $T_{|\Gamma}$ by Formula (6.3), but restricted to disks Δ_a contained in disks of Γ.
    For good laminar currents T, Dujardin proved that

1. If two disks Δ_1 and Δ_2 are subordinate to T, they are **compatible**: their intersection is an open subset of Δ_1 and Δ_2;

2. If Γ is any flow box, the restriction $T_{|\Gamma}$ is uniformly laminar: there is a measure ν_T such that $T_{|\Gamma}$ is the current of integration over the measured family of graphs (Γ, ν_T).

As a corollary, if Δ is subordinate to T, then all disks contained in the analytic continuation of Δ are subordinate to T. Thus, disks subordinate to T, or, more generally, flow boxes whose constitutive disks are subordinate to T, can be glued together in a compatible way. This provides a "weak lamination" for the support of T, and T is determined by a holonomy invariant transverse measure for this weak lamination.

REMARK 6.14. Consider a current T_f^+, where f is a loxodromic automorphism. Since $\mathbb{R}^+ T_f^+$ is extremal in the convex cone of closed positive currents, the transverse invariant measure for T_f^+ is ergodic: If it decomposes into a sum of two non-trivial, holonomy invariant, transverse measures ν_1 and ν_2, then ν_1 is proportional to ν_2 (see [55]).

6.4.2 Geometric product

The second crucial fact concerning good laminar currents is a geometric definition of their intersection. Let T_1 and T_2 be two good laminar currents with continuous potentials. The product $T_1 \wedge T_2$ has been defined in Section 5.1.5 using Bedford-Taylor technique. One may also be tempted to represent each T_i in the form

$$T_i = \int_{\mathcal{A}_i} \{\Delta_a\} \, d\mu_i(a),$$

as in Formula 6.3, and define

$$T_1 \cap T_2 = \int_{\mathcal{A}_1} \int_{\mathcal{A}_2} \{\Delta_{a_1} \cap \Delta_{a_2}\} \, d\mu_1(a_1) d\mu_2(a_2),$$

where $\{\Delta_{a_1} \cap \Delta_{a_2}\}$ is the sum of the Dirac masses on the set $\Delta_{a_1} \cap \Delta_{a_2}$ if this set is finite and is zero otherwise. If both T_1 and T_2 have continuous potential, one may expect that[6]

$$T_1 \wedge T_2 = T_1 \cap T_2;$$

if this equality holds, one says that the product of T_1 and T_2 is geometric. Good laminar currents with continuous potentials satisfy this formula (see [55]). The intersection of such currents is, therefore, a sum of geometric intersections of uniformly laminar currents in flow boxes. Since T_f^+ and T_f^- are good laminar currents with continuouours potentials, the product $T_f^+ \wedge T_f^-$ is geometric. Since T_f^+ and $T^- f$ are Ahlfors currents with respect to stable and unstable manifolds, one obtains the following statement (see Remark 6.3(a)).

THEOREM 6.15 (Bedford-Lyubich-Smillie [9]). *Let f be a loxodromic automorphism of a connected compact Kähler surface. Let p and q be saddle periodic points of f. Assume that the stable manifold W_p^s and the unstable manifold W_q^u are not contained in periodic algebraic curves. Then*

- *the set of transverse intersections of W_p^s and W_q^u is dense in the support of μ_f;*

- *every intersection of W_p^s and W_q^u is contained in the support of μ_f.*

6.4.3 Contraction properties along T_f^+

Thanks to Section 6.4.1, we now have a well behaved notion of disks subordinate to the good laminar current T_f^+. Since T_f^+ is multiplied by $\lambda(f)$ under the action of f^*, most of those constitutive disks must be contracted by f. This is precisely what Dujardin proved in [57].

To state the result, consider a flow box Γ in some bidisk $U \simeq \mathbb{D} \times \mathbb{D}$. Denote by A the intersection of Γ with the vertical disk $\{0\} \times \mathbb{D}$, as in Section 6.2.1, and

---

[6]If $T_1 = T_2$ is the current of integration over a compact curve C with non-zero self-intersection, one cannot hope to define the product of T_1 with T_2 in such a simple way. But good laminar currents do not charge compact curves.

by Δ_a the unique element of Γ containing $a \in A$. Restricting T_f^+ to Γ, one obtains a uniformly laminar current $T_{f|\Gamma}^+$ given by a measure ν^+ on A:

$$T_{f|\Gamma}^+ = \int_{a \in \Gamma} \{\Delta_a\} \, d\nu^+(a).$$

Dujardin's theorem controls the diameter of images $f^n(\Delta_a)$ (for any Riemannian metric on X):

THEOREM 6.16 (Dujardin [57]). *Let f be a loxodromic automorphism of a connected complex algebraic surface X. Let Γ be a flow box and ν^+ be the transverse measure associated to $T_{f|\Gamma}^+$. For all $\epsilon \in (0,1)$ there is a constant $C(\epsilon)$ and a subset A_ϵ of measure $\nu^+(A_\epsilon) > \nu^+(A)(1-\epsilon)$ such that*

$$\mathrm{diam}(f^n(\Delta_a))^2 \leq C(\epsilon) \frac{n^2}{\lambda(f)^n}, \quad \forall \, a \in A_\epsilon.$$

7   FATOU AND JULIA SETS

7.1   Definition

Let g be an automorphism of a compact complex manifold M. A point $x \in M$ is in the **Fatou set** $\mathrm{Fat}(g)$ of g if there exists an open neighborhood U of p on which the sequence $(g^n)_{n \in \mathbb{Z}}$ forms a normal family of holomorphic mappings from U to X. Taking only positive (resp. negative) iterates, one can also define the forward Fatou set $\mathrm{Fat}_+(g)$ (resp. backward Fatou set $\mathrm{Fat}_-(g)$).

REMARK 7.1. Let d be the dimension of M. If p is a fixed point in $\mathrm{Fat}(g)$, then (g^n) is a normal family in a neighborhood of p, so that g is locally linearizable near p: there is a germ of holomorphic diffeomorphism $\psi \colon (U, p) \to (\mathbb{C}^d, 0)$ and a unitary transformation D such that

$$\psi \circ g \circ \psi^{-1} = D.$$

The linear transformation D is conjugate to Dg_p by $D\psi_p$. In the other direction, if Dg_p is a unitary transformation and g is locally conjugate to Dg_p, then p is in the Fatou set. Section 7.3 provides examples of this type for automorphisms of projective surfaces.

7.2   Kobayashi hyperbolicity and pseudo-convexity

7.2.1   *Kobayashi pseudo-distance*

Let M be a complex manifold. A chain of disks Ψ between two points x and y is a finite family of marked disks

$$\psi_i \colon (\mathbb{D}; z_{i,1}, z_{i,2}) \to (M; x_{i,1}, x_{i,2}), \quad 1 \leq i \leq l,$$

such that

- $\psi_i(z_{i,1}) = x_{i,1}$ and $\psi_i(z_{i,2}) = x_{i,2}$;

- $x_{1,1} = x$, $x_{l,2} = x_{i+1,1}$ for all $1 \leq i \leq l-1$, and $x_{l,2} = y$.

The hyperbolic length of such a chain is

$$\mathsf{hl}(\Psi) = \sum_{i=1}^{i=l} \mathsf{dist}_{\mathbb{D}}(z_{i,1}, z_{i,2}).$$

The **Kobayashi pseudo-distance** $\mathsf{dist}_M(x, y)$ between two points x and y is the infimum of the hyperbolic length $\mathsf{hl}(\Psi)$ over all chains of disks Ψ joining x to y (see [81]). The Kobayashi pseudo-distance satisfies all axioms of a distance, except that it can take the value $+\infty$ (exactly when x and y are in two distinct connected components of M), and it may vanish for pairs of distinct points. One says that M is **Kobayashi hyperbolic** if $\mathsf{dist}_M(x, y) > 0$ for all $x \neq y$ in M.

REMARK 7.2. **a.** When $M = \mathbb{D}$, the Kobayahsi pseudo-distance coincides with the Poincaré metric $\mathsf{dist}_{\mathbb{D}}$. When $M = \mathbb{C}$, the Kobayashi pseudo-distance $\mathsf{dist}_{\mathbb{C}}$ vanishes identically.

b. Holomorphic mappings between complex manifolds are distance decreasing: $\mathsf{dist}_N(f(x), f(y)) \leq \mathsf{dist}_M(x, y)$ if $f \colon M \to N$ is holomorphic. In particular, M is not hyperbolic when it contains an entire curve.

c. When M is Kobayashi hyperbolic, the topology induced by dist_M is the same as the topology of M as a complex manifold.

7.2.2   Brody re-parametrization and hyperbolicity

Fix a hermitian metric on the manifold M. Assume that there exists a sequence of holomorphic mappings $\psi_m \colon \mathbb{D} \to M$ such that $\|\psi_m'(0)\|$ goes to ∞ with m, where $\psi'(0)$ is the velocity vector of the curve at $z = 0$, the center of the unit disk, and $\|\psi_m'(0)\|$ is its norm with respect to the fixed hermitian metric.

LEMMA 7.3 (Brody lemma (see [81], Chapter III)). *There exists a sequence of real numbers r_m and automorphisms $h_m \in \mathsf{Aut}(\mathbb{D})$ such that*

1. *r_m goes to $+\infty$ with m;*

2. *$\varphi_m(z) = \psi_m \circ h_m(z/r_m)$ is holomorphic on $\mathbb{D}_{r_m}$, and its velocity at the origin has norm 1;*

3. *the norm of the derivative of φ_m satisfies $\limsup_m \max_{z \in K} \|\varphi_m'(z)\| \leq 1$ for all compact subsets $K \subset \mathbb{C}$.*

Suppose now that M is compact. By (3), the sequence (φ_m) is equicontinuous. Hence, a subsequence of (φ_m) converges toward an entire curve $\varphi \colon \mathbb{C} \to M$ such that

$$\|\varphi'(0)\| = 1, \quad \text{and} \quad \|\varphi'(z)\| \leq 1, \quad \forall z \in \mathbb{C}.$$

Such an entire curve is called a **Brody curve**. As a corollary of this construction, one gets the following theorem.

THEOREM 7.4 (Brody [81]). *Let M be a compact complex manifold, with a fixed hermitian metric $\| \cdot \|$. Then M is Kobayashi hyperbolic if and only if there is a uniform upper bound on $\|\psi'(0)\|$ for all holomorphic disks $\psi \colon \mathbb{D} \to M$, if and only if there is no Brody curve $\varphi \colon \mathbb{C} \to M$.*

7.2.3   Fatou sets are almost hyperbolic

We are now in a position to explain the following result.

THEOREM 7.5 (Dinh-Sibony [47], Moncet [90]). *Let f be a loxodromic automorphism of a compact Kähler surface X. The Fatou set $\mathsf{Fat}(f)$*

1. *coincides with the complement of the supports of T_f^+ and T_f^-, i.e.,*

$$\mathsf{Fat}(f) = X \setminus \mathsf{Support}(T_f^+ + T_f^-);$$

2. *is Kobayashi hyperbolic modulo periodic curves;*

3. *is pseudo-convex.*

To be more precise, the second assertion says that the Kobayashi pseudo-distance $\mathrm{dist}_{\mathsf{Fat}(f)}$ vanishes exactly along the set of algebraic periodic curves C of the following types:

- C is elliptic and contained in $\mathsf{Fat}(f)$;

- C is rational and $C \cap \mathsf{Fat}(f)$ is equal to C minus 0, 1, or 2 points.

To prove this theorem, one studies the complement

$$\Omega = X \setminus \mathsf{Support}(T_f^+ + T_f^-)$$

of the support of T_f^+ and T^-. This set is f-invariant and contains $\mathsf{Fat}(f)$. If it is not Kobayashi hyperbolic, there exists a Brody curve $\varphi\colon \mathbb{C} \to \overline{\Omega}$. Let A be an Ahlfors current associated to φ, as in Section 6.1.1. Suppose that φ is not contained in a compact curve; then $\langle [A]|[A] \rangle \geq 0$, by Theorem 6.1(4). Dinh and Sibony prove that A does not intersect T_f^+ and T_f^-, a fact which is not obvious because the supports of A, of T_f^+, and of T_f^- could very well be contained in $\overline{\Omega} \setminus \Omega$ and could have non-trivial intersection. Therefore,

$$\langle [A]|[T_f^+] \rangle = \langle [A]|[T_f^-] \rangle = 0,$$

By the Hodge index theorem, these equalities imply $\langle [A]|[A] \rangle < 0$, a contradiction. Thus, $\varphi(\mathbb{C})$ is contained in a compact curve C. This curve is either elliptic or rational, $A = \{C\}$, and $[A]$ does not intersect $[T_f^+]$ and $[T_f^-]$. According to Proposition 4.1, C is periodic. This shows that Ω is Kobayashi hyperbolic modulo rational or elliptic periodic curves.

7.3   Examples

Consider the birational map $f\colon \mathbb{P}^2 \dashrightarrow \mathbb{P}^2$ given in affine and homogeneous coordinates by

$$
\begin{aligned}
f(x,y) &= (a+y, b+y/x), \\
f[x:y:z] &= [axz+yx : bxz+yz : zx],
\end{aligned}
$$

for some parameter $(a,b) \in \mathbb{C}^2$. It has three indeterminacy points, namely

$$p_1 = [1:0:0], \ p_2 = [0:1:0], \ \text{and} \ p_3 = [0:0:1].$$

Let Δ be the triangle whose edges are the three coordinate axes $\{x = 0\}$, $\{y = 0\}$, and $\{z = 0\}$. Each axis of Δ is blown down to a point by f: the first on p_2, the third on p_1, and the second axis $\{y = 0\}$ on the point

$$p_4 = [a : b : 0].$$

Define $p_{4+m} = f^m(p_4)$, for $m \geq 1$, pick an integer $n \geq 1$, and suppose that the parameter (a, b) has been chosen in such a way that

1. $p_j \notin \Delta$ for all $4 \leq j \leq n$, (in particular, f is well defined at p_j);

2. $p_{n+1} = p_3$.

Blowing up all points $(p_j)_{j=1}^n$, one gets a rational surface $X \to \mathbb{P}^2$ on which f lifts to a well-defined automorphism $\hat{f}\colon X \to X$. As shown by Bedford and Kim and by McMullen, this leads to an infinite family of automorphisms on rational surfaces (the number of possible parameters (a, b) and the number of blow-ups increase with n). The dynamical degree of $\hat{f}$ depends only on n, is equal to 1 if and only if $n \leq 9$, and is equal to the unique root $\lambda_n > 1$ of the polynomial equation

$$t^{n-2}(t^3 - t - 1) + (t^3 + t^2 - 1) = 0$$

when $n \geq 10$. The sequence (λ_n) is increasing and converges toward the smallest Pisot number, i.e., the root $\lambda_P > 1$ of $t^3 = t + 1$. When $n = 10$ the dynamical degree is equal to Lehmer's number (see Section 2.4.3).

For $n = 10$, there is a parameter $(a, b) = (a, \bar{a})$ with

$$a \simeq 0.04443 - 0.44223\sqrt{-1}$$

that satisfies Properties (1) and (2). The dynamical degree of the corresponding automorphism $\hat{f}$ is equal to Lehmer's number. This automorphism has a fixed point q such that the tangent map $D\hat{f}_q$ is (conjugate to) a unitary transformation with eigenvalues α and β. Moreover, α and β are algebraic numbers, and are multiplicatively independent. Results from Diophantine approximation imply that products like $\alpha^k \beta^l$ are not well approximated by roots of 1, so that Siegel's linearization theorem can be applied (see [86], [87]). In a neighborhood of the fixed point q, $\hat{f}$ is conjugate to its linear part $D\hat{f}_q$; this neighborhood is, therefore, contained in the Fatou set of $\hat{f}$.

In [8], a similar construction leads to examples of Fatou components containing several fixed points and invariant rational curves.

7.4   Julia sets

There are three Julia sets for each loxodromic automorphism f. The forward Julia set $J^+(f)$ is the complement of the forward Fatou set $\mathrm{Fat}^+(f)$. The backward Julia set $J^-(f)$ is the complement of $\mathrm{Fat}^-(f)$. The Julia set $J(f)$ is the intersection $J^+(f) \cap J^-(f)$. The support of T_f^+ (resp. T_f^-) is contained in $J^-(f)$ (resp. $J^+(f)$); the support of μ_f is contained in $J(f)$.

In the previous paragraph, examples of loxodromic automorphisms with nonempty Fatou set have been described. One can construct examples for which the

Lebesgue measure of $J^+(f)$ vanishes and the forward orbit of every point in $\text{Fat}^+(f)$ goes to an attracting fixed point (see [87]). Beside these examples, not much is known. For instance, the following questions remain open. Does there exist a loxodromic automorphism of a projective K3 surface — for example, a smooth surface of degree $(2,2,2)$ in $\mathbb{P}^1 \times \mathbb{P}^1 \times \mathbb{P}^1$ — with non-empty Fatou set? Does there exist a loxodromic automorphism f of a K3 surface for which μ_f is singular with respect to Lebesgue measure ? (Based on Figure 1.1, one may expect a positive answer to the second question). We refer to [31] for other open problems of this type.

8   THE MEASURE OF MAXIMAL ENTROPY AND PERIODIC POINTS

In this section, two important characterizations of the measure μ_f are described. Both show that μ_f, a measure which is uniquely determined by the f-invariant complex structure on the manifold X, is also uniquely determined by its dynamical properties.

8.1   Entropy, Pesin theory, and laminarity

8.1.1   *Entropy of invariant measures*

Let $(Z, \mathcal{T}, \mu)$ be a probability space, with σ-algebra $\mathcal{T}$ and probability measure μ. Let g be a measure-preserving transformation of $(Z, \mathcal{T}, \mu)$. Let $\mathcal{P} = \{P_i\,; 1 \le i \le l\}$ be a partition of Z into a finite number of measurable subsets P_i: the P_i are disjoint, have positive measure, and cover a subset of full measure in Z. One defines the entropy of $\mathcal{P}$ with respect to μ by

$$\mathsf{h}(\mathcal{P}, \mu) = -\sum_i \mu(P_i) \log(\mu(P_i)).$$

By pullback, g transforms $\mathcal{P}$ into a new partition $g^*\mathcal{P}$ of Z, whose elements are the subsets $g^{-1}(P_i)$. Iterating, we get a sequence of partitions $g^{-k}(\mathcal{P})$, and we denote by

$$\mathcal{P}_n = \mathcal{P} \vee g^{-1}(\mathcal{P}) \vee \ldots \vee g^{-n+1}(\mathcal{P})$$

the partition generated by the first n elements of this sequence. The entropy of g with respect to μ is then defined as the supremum

$$\mathsf{h}(g, \mu) = \sup_{\mathcal{P}} \left\{ \limsup_n \frac{1}{n} \mathsf{h}(\mathcal{P}_n, \mu) \right\}$$

over all measurable partitions of Z in a finite number of pieces (see [78]).

Assume, now, that g is a continuous transformation of a compact space Z. Let $\mathcal{T}$ be the σ-algebra of Borel subsets of Z. Then g has at least one invariant probability measure μ on $(Z, \mathcal{T})$. By the so called variational principle [78], it turns out that the supremum of $\mathsf{h}(g, \mu)$ over all g-invariant probability measures is equal to the topological entropy of g:

$$\mathsf{h}_{top}(g) = \sup_{\mu} \mathsf{h}(g, \mu).$$

Newhouse's theorem asserts that this supremum is a maximum when g is a diffeomorphism of class $\mathcal{C}^\infty$ on a compact manifold (see [93]).

8.1.2 Pesin Theory [78]

Since f has positive topological entropy, there exist invariant, ergodic, probability measures ν with $h(f;\nu) > 0$. The Lyapunov exponents of f with respect to such a measure ν are defined point-wise by

$$
\begin{aligned}
\chi^+(x) &= \limsup_{n \to +\infty} \frac{1}{n} \log \|D(f^n)_x\|, \\
\chi^-(x) &= \limsup_{n \to -\infty} \frac{1}{n} \log \|D(f^{-n})_x\|.
\end{aligned}
$$

Since ν is ergodic, both χ^+ and χ^- are constant on a set of full measure. Ruelle's inequality implies that χ^+ is positive and χ^- is negative (both count with multiplicity two because f preserves the complex structure).

By Osseledet's theorem, the tangent space $T_x X$ splits ν almost everywhere into the direct sum $E^u(x) \oplus E^s(x)$ of two lines such that $f_* E^s(x) = E^s(f(x))$, the derivative of f^n along $E^s(x)$ decreases as $\exp(-n\chi^-(x))$ with n, and the lines $E^u(x)$ satisfy similar properties for f^{-1}. By Pesin theory, there are stable and unstable manifolds

$$
\xi_x^{s/u} \colon \mathbb{C} \to X
$$

through ν-almost every point x. The image of the stable manifold ξ_x^s is the set of points y in X such that the distance between $f^n(x)$ and $f^n(y)$ goes to 0 with n. It is tangent to $E^s(x)$ at x; f maps ξ_x^s to $\xi_{f(x)}^s$ (with a different parametrization).

In what follows, we denote by $\Lambda_f(\nu)$ a measurable set of full measure such that every $x \in \Lambda_f(\nu)$ has non-zero Lyapunov exponents and stable and unstable manifolds, as before. By definition, the union Λ_f of those sets $\Lambda_f(\nu)$, where ν describes the set of invariant and ergodic probability measures with positive entropy, is the set of **hyperbolic (or saddle) points**.

One can show (see [27]) that the sequence of currents $\lambda(f)^{-n}(f^n)^* \{\xi_x^s(\psi\mathbb{D})\}$ converges toward a *positive* multiple of T_f^+ almost surely (here ψ is any non-negative smooth function with compact support in $\mathbb{D}$ such that $\psi(0) > 0$). Thus, ν-almost every point determines T_f^+ and T_f^- through its stable and unstable manifolds. Since $\mu_f = T_f^+ \wedge T_f^-$, it follows that μ_f takes every invariant measure with positive entropy "into account."

8.1.3 Laminar versus dynamical structures

Let us compare the local structure of the dynamics of f to the local structure of the currents T_f^+ and T_f^-, as given by the flow boxes from Section 6.4.1.

Let $U \subset X$ be a bidisk, $U \simeq \mathbb{D} \times \mathbb{D}$. Let $x \in \Lambda_f \cap U$ be a hyperbolic point. The connected component of $\xi_x^s(\mathbb{C}) \cap U$ that contains x is the **local stable manifold** of x in U, denoted $W_{loc}^s(x)$. A similar definition applies for local unstable manifolds. A **Pesin box** (U, K) is a pair of a bidisk $U \simeq \mathbb{D} \times \mathbb{D}$ in X and a compact subset K of U such that

- every point x in K is a hyperbolic point and its local stable and unstable manifolds are horizontal and vertical graphs in U;

- for all pairs of distinct points (x, y) in K, $W_{loc}^s(x) \cap W_{loc}^u(y)$ is a singleton and is contained in K.

In particular, the local stable (resp. unstable) manifolds determine a lamination K^s (resp. K^u) in U. From the second property, one may identify K to the product of the transversal $A^u = K \cap W^u_{loc}(x)$ and $A^s = K \cap W^s_{loc}(y)$ (for all pairs $(x, y) \in K^2$).

Let (U, K) be a Pesin box. Restricting T^+_f to the lamination K^s, as in Section 6.4.1, one obtains a current of integration $T^+_{f|K^s}$ on K^s with respect to a transverse measure μ^+_K. Similarly, T^-_f determines a transverse measure μ^-_K for K^u. Since T^+_f and T^-_f are Ahlfors currents with respect to stable and unstable manifolds and the constitutive disks of $T^\pm_f$ are contracted by $f^\pm$ (see Theorem 6.16), it can be shown that

1. Pesin boxes provide a complete family of flow boxes for T^+_f; in other words, T^+_f is a sum of $T^+_{f|K^s}$ over a family of Pesin boxes;

2. the restriction $\mu_{f|K}$ is equal to the product $\mu^+_K \otimes \mu^-_K$ (on $K \simeq A^u \times A^s$);

3. the Lyapunov exponents $\chi^+(\mu_f)$ and $\chi^-(\mu_f)$ satisfy

$$\chi^+(\mu_f) \geq \frac{1}{2}\log\lambda(f) > 0 > -\frac{1}{2}\log(\lambda(f) \geq \chi^-(f).$$

The second property shows that μ_f has a product structure in Pesin boxes. This is a strong property, from which it follows that *the dynamical system (X, μ, f) is measurably equivalent to a Bernouilli shift*. In other words, the dynamics of f with respect to μ_f is equivalent to the dynamics of a random coin flip.

8.2   Two characterizations of μ_f

This interplay between Pesin theory and the laminar structure of T^+_f and T^-_f leads to two characterizations of the invariant measure μ_f.

THEOREM 8.1 ([9], [27], [57]). *Let f be a loxodromic automorphism of a compact Kähler surface X. The measure μ_f has maximal entropy, i.e.,*

$$\mathsf{h}(f, \mu_f) = \mathsf{h}_{top}(f) = \log(\lambda(f)),$$

and is the unique invariant probability measure with maximal entropy.

Thus, the measure μ_f is the unique point on the compact convex set of f-invariant probability measures at which the entropy $\mathsf{h}(f, \cdot)$ is maximal. This is an implicit characterization of μ_f.

THEOREM 8.2 (Bedford-Lyubich-Smillie [9], [3], [27]). *Let f be a loxodromic automorphism of a complex projective surface X. Let $\mathrm{Per}(f, k)$ be the set of isolated periodic points of f with period at most k. Then*

$$\frac{1}{\lambda(f)^k} \sum_{p \in \mathrm{Per}(f,k)} \delta_p$$

converges toward μ_f as k goes to $+\infty$. The same result holds if $\mathrm{Per}(f, k)$ is replaced by the set $\mathrm{Per}_{sad}(f, k)$ of saddle periodic points of period at most k.

If p is a saddle periodic point, either p is contained in the support of μ_f, or p is contained in a cycle of periodic rational curves.

Hence, μ_f is determined by the repartition of periodic points of f. This provides a simple characterization of μ_f, using only the simplest dynamical objects (periodic points) associated to $f\colon X \to X$.

8.3 Application: Complex versus real dynamics

COROLLARY 8.3. *Let X be a smooth projective surface defined over the real numbers $\mathbb{R}$. Let f be an automorphism of X defined over $\mathbb{R}$. The entropy of $f\colon X(\mathbb{R}) \to X(\mathbb{R})$ is equal to the entropy of $f\colon X(\mathbb{C}) \to X(\mathbb{C})$ if and only if all saddle periodic points of f which are not contained in rational periodic curves are contained in $X(\mathbb{R})$.*

PROOF. The topological entropy of f on $X(\mathbb{R})$ is at most equal to the entropy on $X(\mathbb{C})$. If f is not loxodromic, the entropy of f on $X(\mathbb{R})$ and on $X(\mathbb{C})$ vanish. We can now assume that f is loxodromic.

Assume that almost all saddle periodic points are contained in $X(\mathbb{R})$. Then $\mu_f(X(\mathbb{R})) = 1$, and, by the variational principle, the topological entropy of f on $X(\mathbb{R})$ is equal to the entropy
on $X(\mathbb{C})$.

Assume now that the entropy of f on $X(\mathbb{R})$ is equal to the entropy on $X(\mathbb{C})$. Then, by Newhouse's theorem and the uniqueness of the measure of maximal entropy, μ_f is supported on $X(\mathbb{R})$. Let p be a saddle periodic point of f. If p is not contained in a periodic curve, Theorem 6.15 shows that p is contained in the support of μ_f. Thus, p is contained in $X(\mathbb{R})$. $\qquad\square$

There are examples of automorphisms of rational surfaces, defined over $\mathbb{R}$, for which the entropy on $X(\mathbb{C})$ and $X(\mathbb{R})$ coincide (see [8]). This phenomenon is not possible on tori, because all automorphisms are induced by complex affine transformations of the universal cover $\mathbb{C}^2$. It is an open problem to decide whether such examples exist on K3 surfaces. See [89, 90] for a discussion of this problem.

9 COMPLEMENTS

Since we focussed on the dynamics of automorphisms of surfaces, we didn't address several interesting questions. What about higher dimension, and birational transformations? We list a few references that may help the reader.

9.1 Birational transformations

Let X be a connected compact Kähler surface, with a Kähler form κ. Denote by $\mathrm{Bir}(X)$ the group of its birational (or bimeromorphic) transformations. Let f be an element of $\mathrm{Bir}(X)$. If f is not an automorphism, the indeterminacy sets $\mathrm{Ind}(f)$ and $\mathrm{Ind}(f^{-1})$ are not empty; iterating f, it may happen that the union of indeterminacy points of iterates of f is dense in $X(\mathbb{C})$.

The sequence

$$d_n = \int_X (f^n)^* \kappa \wedge \kappa$$

controls several features of the dynamics of f. One can classify birational transformations f such that the dynamical degree

$$\lambda(f) = \lim_{n \to +\infty} (d_n)^{1/n}$$

is equal to 1; up to conjugation by birational maps, the list is the same as in Theorem 2.13, with one more case: It may happen that d_n grows linearly with n, in which case f preserves a pencil of rational curves (see [63, 40]). Moreover, the group $\mathrm{Bir}(X)$ acts on an infinite dimensional hyperbolic space $\mathbb{H}_\infty$, and this classification into types can be explained in terms of elliptic, parabolic and loxodromic isometries; such a classification can then be used to study groups of birational transformations (see [24] and [33]).

When f is a birational transformation with $\lambda(f) > 1$, the dynamics of f should be similar to the dynamics of loxodromic automorphisms. Unfortunately, the techniques described in the previous paragraphs do not apply directly when there are indeterminacy points. Assume, however, that

$$\sum_{n \geq 0} \frac{1}{\lambda(f)^n} \log \mathrm{dist}(f^n(\mathrm{Ind}(f^{-1})), \mathrm{Ind}(f)) < +\infty$$

(this is the so-called Bedford-Diller condition). Then pluri-potential theory can be applied successfully to construct good f-invariant laminar currents and the unique measure of maximal entropy; this measure describes the distribution of periodic points. See [6] and [57], as well as [41], [42], and [43]. On the other hand, there are families of birational mappings f_t for which the strategy through pluri-potential analysis fails for a dense set of parameters t (see [21]).

9.2   Higher dimension

We did not mention any result concerning the dynamics of automorphisms on higher dimensional complex manifolds, but many recent results have been obtained recently. Here is a short list of relevant references

- [48] and [50] consider invariant currents and measures for automorphisms in any dimension; they prove uniqueness results, mixing properties, etc.

- [76] shows that most stable manifolds are uniformized by $\mathbb{C}^k$, where k is the complex dimension of the stable manifold; [37] provides estimates for the Lyapunov exponents.

- [50] and [46] use all these results to study naturally invariant measures similar to the measure μ_f in the case of automorphisms of surfaces.

Groups of automorphisms of compact Kähler manifolds are not well understood yet, but Hodge theory provides a powerful tool in any dimension, which can be used in a spirit similar to Section 3.2: see [45], [23], [104], and [25], for example.

9.3   Other topics

Similar tools can be applied to describe the dynamics of (non-invertible) endomorphisms of compact complex manifolds; see [96] and [49] for two surveys on this topic.

In the case of automorphisms of the plane $\mathbb{C}^2$ (and certain affine surfaces [29]), the currents T_f^+ and T_f^- have global potentials: There are Green functions G_f^+ and G_f^- such that $T_f^\pm = dd^c G_f^\pm$ on $\mathbb{C}^2$ and $G_f^\pm \circ f = \lambda(f)^\pm G_f^\pm$. Those invariant functions provide new tools that can be used to obtain deeper results. See [97] and [5] for two surveys. While most results described in the previous paragraphs concern the stochastic properties of the dynamics of automorphisms, there are important results concerning topological aspects of the dynamics of automorphisms of $\mathbb{C}^2$ (see [70], [71], [69], [11], [12], and [13] for example).

10 APPENDIX: CLASSIFICATION OF SURFACES

Thanks to Enriques-Kodaira classification of surfaces (see [2]), we describe the geometry of surfaces that admit a loxodromic automorphism.

10.1 Kodaira dimension

Let X be a connected compact complex surface. Consider the canonical bundle $K_X = \det(T^*X)$. Its holomorphic sections are holomorphic 2-forms, and the holomorphic sections of its tensor powers $K_X^{\otimes n}$ can be expressed in local coordinates in the form $a(z_1, z_2)(dz_1 \wedge dz_2)^n$ for some holomorphic function a. Fix a positive integer n, fix a point $x \in X$, and consider the evaluation map

$$ev_x \colon \Omega \in H^0(X, K_X^{\otimes n}) \mapsto \Omega_x \in \det(T_x^*X).$$

Since $\det(T_x^*X)$ has dimension 1, it can be identified with $\mathbb{C}$, and this identification is unique up to a non-zero scalar multiple. Thus, either ev_x vanishes identically, or it defines an element $[ev_x]$ of $\mathbb{P}(H^0(X, K_X^{\otimes n})^*)$. If $H^0(X, K_X^{\otimes n})$ is not reduced to $\{0\}$, this construction provides a meromorphic mapping

$$\Phi_n \colon x \in X \dashrightarrow [ev_x] \in \mathbb{P}(H^0(X, K_X^{\otimes n})^*).$$

By definition, the Kodaira dimension $\mathrm{kod}(X)$ is equal to $-\infty$ if $H^0(X, K_X^{\otimes n})$ is reduced to $\{0\}$ for all $n \geq 1$ and is equal to the maximum of $\dim(\Phi_n(X))$, $n \geq 1$, otherwise.

10.2 Automorphisms

The group $\mathrm{Aut}(X)$ acts linearly by pullback on the sections of $K_X^{\otimes n}$ and preserves the positive homogeneous function

$$\omega \in H^0(X, K_X^{\otimes n}) \mapsto \int_X (\Omega \wedge \overline{\Omega})^{1/n}.$$

Hence, the image of $\mathrm{Aut}(X)$ in $\mathsf{GL}\,(H^0(X, K_X^{\otimes n})^*)$ is relatively compact and, in fact, is a finite group [99, 92].

The meromorphic mapping Φ_n is equivariant with respect to the natural action of $\mathrm{Aut}(X)$ on X and its projective linear action on $\mathbb{P}(H^0(X, K_X^{\otimes n})^*)$. Consequently, when $\dim(\Phi_n(X)) > 0$, one obtains a non-trivial factorization of the dynamics, with a projective linear action of a finite group on the image $\Phi_n(X)$.

As an easy consequence, *a connected compact complex surface with a loxodromic automorphism has Kodaira dimension 0 or $-\infty$.*

10.3   Classification of surfaces with loxodromic automorphisms

Among compact complex surfaces with $\mathrm{kod}(X) = 0$, there are three important types.

- complex tori of dimension 2, i.e., quotients of $\mathbb{C}^2$ by a lattice Λ;

- K3 surfaces, i.e., simply connected surfaces with trivial canonical bundle (or equivalently, surfaces with a non-vanishing holomorphic 2-form and trivial first Betti number);

- Enriques surfaces, i.e., quotients of K3 surfaces by fixed-point-free involutions.

This does not exhaust the list of surfaces with $\mathrm{kod}(X) = 0$; there are also bi-elliptic surfaces, which are quotients of tori, Kodaira surfaces, which are not Kähler, and blow-ups of all these five types of surfaces. Surfaces of degree $(2,2,2)$ in $\mathbb{P}^1 \times \mathbb{P}^1 \times \mathbb{P}^1$ are examples of K3 surfaces.

There are three types of surfaces with negative Kodaira dimension. The first type is given by rational surfaces, i.e., surfaces which are birationally equivalent to the projective plane $\mathbb{P}^2$. The second type is made of ruled surfaces $\pi\colon X \to B$, where the generic fibers of π are rational curves and the basis B has genus ≥ 1. The third type is given by VII_0-surfaces (those surfaces are not Kähler). Section 7.3 provides examples of rational surfaces with loxodromic automorphisms.

The following result classifies surfaces with interesting automorphisms and explains why we focussed on compact Kähler surfaces.

THEOREM 10.1 (Cantat [26], Nagata [91]). *Let X be a connected compact complex surface. Assume that $\mathrm{Aut}(X)$ contains an automorphism f with positive topological entropy (resp. assume that X is Kähler and $\mathrm{Aut}(X)$ contains a loxodromic automorphism f). Then X is a Kähler surface, and*

- *either X is obtained from the plane $\mathbb{P}^2(\mathbb{C})$ by a finite sequence of at least ten blow-ups;*

- *or there is a holomorphic birational map $\pi\colon X \to X_0$ such that $\pi \circ f \circ \pi^{-1}$ is an automorphism of X_0 and X_0 is a torus, a K3 surface, or an Enriques surface (X_0 is the "minimal model of X").*

10.4   Positive characteristic

Let us now consider projective surfaces defined on an algebraically closed field $\mathbf{k}$. Enriques-Kodaira's classification has been extended to this context by Bombieri and Mumford. New phenomena appear when both $\mathrm{kod}(Y) = 0$ and the characteristic of $\mathbf{k}$ is positive, the main cases being $\mathrm{char}(\mathbf{k}) = 2$, or 3; for example, there are K3 surfaces which are unirational in characteristic 2, a fact which is impossible for complex surfaces.

Nevertheless, with appropriate definitions,[7] the previous theorem still remains valid: *if $\mathrm{Aut}(Y)$ contains a loxodromic automorphism f, then either Y is obtained from*

---

[7] A K3 surface Y is a surface with Kodaira dimension $\mathrm{kod}(Y) = 0$, Betti numbers $b_1(Y) = 0$ and $b_2(Y) = 22$, and characteristic $\chi(\mathcal{O}_Y) = 2$. An Enriques surface is a surface with Kodaira dimension $\mathrm{kod}(Y) = 0$, Betti numbers $b_1(Y) = 0$ and $b_2(Y) = 10$, and characteristic $\chi(\mathcal{O}_Y) = 1$.

$\mathbb{P}_\mathbf{k}^2$ *by a sequence of at least ten blow-ups, or there is a birational morphism* $\pi\colon Y \to Y_0$ *such that* $\pi \circ f \circ \pi^{-1}$ *is an automorphism of* Y_0 *and* Y_0 *is an abelian surface, a K3 surface, or an Enriques surface.*

More interestingly, one can construct surfaces in positive characteristic with automorphisms groups which are surprisingly large compared to the case of complex surfaces (see the theory of complex multiplication for tori, and [51] for an example on a K3 surface).

11   Bibliography

[1]   M. F. Atiyah and R. Bott. A Lefschetz fixed point formula for elliptic complexes. II. Applications. *Ann. of Math. (2)*, 88:451–491, 1968.

[2]   W. Barth, Peters, and van de Ven. *Compact complex surfaces.* Springer-Verlag, 1984.

[3]   E. Bedford, M. Lyubich, and J. Smillie. Distribution of periodic points of polynomial diffeomorphisms of $\mathbb{C}^2$. *Invent. Math.*, 114(2):277–288, 1993.

[4]   E. Bedford and B. A. Taylor. The Dirichlet problem for a complex Monge-Ampère equation. *Invent. Math.*, 37(1):1–44, 1976.

[5]   Eric Bedford. Dynamics of rational surface automorphisms. In *Holomorphic dynamical systems*, volume 1998 of *Lecture Notes in Math.*, pages 57–104. Springer, Berlin, 2010.

[6]   Eric Bedford and Jeffrey Diller. Energy and invariant measures for birational surface maps. *Duke Math. J.*, 128(2):331–368, 2005.

[7]   Eric Bedford and Kyounghee Kim. Periodicities in linear fractional recurrences: degree growth of birational surface maps. *Michigan Math. J.*, 54(3):647–670, 2006.

[8]   Eric Bedford and Kyounghee Kim. Dynamics of rational surface automorphisms: Rotation domains. *Amer. J. Math*, 134(2):379–405, 2012.

[9]   Eric Bedford, Mikhail Lyubich, and John Smillie. Polynomial diffeomorphisms of $\mathbb{C}^2$. IV. The measure of maximal entropy and laminar currents. *Invent. Math.*, 112(1):77–125, 1993.

[10]   Eric Bedford and John Smillie. Polynomial diffeomorphisms of $\mathbb{C}^2$. III. Ergodicity, exponents and entropy of the equilibrium measure. *Math. Ann.*, 294(3):395–420, 1992.

[11]   Eric Bedford and John Smillie. Polynomial diffeomorphisms of $\mathbb{C}^2$. VI. Connectivity of *J. Ann. of Math. (2)*, 148(2):695–735, 1998.

[12]   Eric Bedford and John Smillie. Polynomial diffeomorphisms of $\mathbb{C}^2$. VIII. Quasi-expansion. *Amer. J. Math.*, 124(2):221–271, 2002.

[13]   Eric Bedford and John Smillie. Real polynomial diffeomorphisms with maximal entropy: Tangencies. *Ann. of Math. (2)*, 160(1):1–26, 2004.

[14]   Riccardo Benedetti and Carlo Petronio. *Lectures on hyperbolic geometry*. Universitext. Springer-Verlag, Berlin, 1992.

[15]   Mladen Bestvina, Mark Feighn, and Michael Handel. The Tits alternative for Out(F_n). I. Dynamics of exponentially-growing automorphisms. *Ann. of Math. (2)*, 151(2):517–623, 2000.

[16]   Joan S. Birman, Alex Lubotzky, and John McCarthy. Abelian and solvable subgroups of the mapping class groups. *Duke Math. J.*, 50(4):1107–1120, 1983.

[17]   Jérémy Blanc, Ivan Pan, and Thierry Vust. On birational transformations of pairs in the complex plane. *Geom. Dedicata*, 139:57–73, 2009.

[18]   S. Bochner and D. Montgomery. Locally compact groups of differentiable transformations. *Ann. of Math. (2)*, 47:639–653, 1946.

[19]   Marco Brunella. Courbes entières et feuilletages holomorphes. *Enseign. Math. (2)*, 45(1-2):195–216, 1999.

[20]   Marco Brunella. *Birational geometry of foliations*. Publicações Matemáticas do IMPA. [IMPA Mathematical Publications]. Instituto de Matemática Pura e Aplicada (IMPA), Rio de Janeiro, 2004.

[21]   Xavier Buff. Courants dynamiques pluripolaires. *Ann. Fac. Sci. Toulouse Math. (6)*, 20(1):203–214, 2011.

[22]   S. Cantat. Sur la dynamique du groupe d'automorphismes des surfaces $K3$. *Transform. Groups*, 6(3):201–214, 2001.

[23]   S. Cantat. Version kählérienne d'une conjecture de Robert J. Zimmer. *Ann. Sci. École Norm. Sup. (4)*, 37(5):759–768, 2004.

[24]   S. Cantat. Groupes de transformations birationnelles du plan. *Annals of Math.*, 174(1):299–340, 2011.

[25]   S. Cantat and A. Zeghib. Holomorphic actions, Kummer examples, and Zimmer program. *Ann. Sci. École Norm. Sup. (4)*, 45(3):447–489, 2012.

[26]   Serge Cantat. Dynamique des automorphismes des surfaces projectives complexes. *C. R. Acad. Sci. Paris Sér. I Math.*, 328(10):901–906, 1999.

[27]   Serge Cantat. Dynamique des automorphismes des surfaces $K3$. *Acta Math.*, 187(1):1–57, 2001.

[28]   Serge Cantat. Croissance des variétés instables. *Ergodic Theory Dynam. Systems*, 23(4):1025–1042, 2003.

[29]   Serge Cantat. Bers and Hénon, Painlevé and Schrödinger. *Duke Math. J.*, 149(3):411–460, 2009.

[30]   Serge Cantat. Invariant hypersurfaces in holomorphic dynamics. *Math. Res. Lett.*, 17(5):833–841, 2010.

[31]   Serge Cantat. Quelques aspects des systèmes dynamiques polynomiaux: existence, exemples et rigidité. *Panorama et Synthèse*, 30:13–87, 2010.

[32]   Serge Cantat and Charles Favre. Symétries birationnelles des surfaces feuilletées. *J. Reine Angew. Math.*, 561:199–235, 2003.

[33]   Serge Cantat and Stéphane Lamy. Normal subgroups in the Cremona group, with an appendix by Yves de Cornulier. *Acta Math.*, 210(1): 31–94, 2013.

[34]   Julian Lowell Coolidge. *A treatise on algebraic plane curves*. Dover Publications Inc., New York, 1959.

[35] Pierre de la Harpe. *Topics in geometric group theory*. Chicago Lectures in Mathematics. University of Chicago Press, Chicago, 2000.

[36] Henry de Thélin. Sur la laminarité de certains courants. *Ann. Sci. École Norm. Sup. (4)*, 37(2):304–311, 2004.

[37] Henry de Thélin. Sur les exposants de Lyapounov des applications méromorphes. *Invent. Math.*, 172(1):89–116, 2008.

[38] J.-P. Demailly. *Complex analytic and differential geometry*, volume 1 of *Open-Content Book*. Grenoble, 2009.

[39] Jean-Pierre Demailly. Variétés hyperboliques et équations différentielles algébriques. *Gaz. Math.*, (73):3–23, 1997.

[40] J. Diller and C. Favre. Dynamics of bimeromorphic maps of surfaces. *Amer. J. Math.*, 123(6):1135–1169, 2001.

[41] Jeffrey Diller, Romain Dujardin, and Vincent Guedj. Dynamics of meromorphic maps with small topological degree I: from cohomology to currents. *Indiana Univ. Math. J.*, 59(2):521–561, 2010.

[42] Jeffrey Diller, Romain Dujardin, and Vincent Guedj. Dynamics of meromorphic maps with small topological degree III: geometric currents and ergodic theory. *Ann. Sci. Éc. Norm. Supér. (4)*, 43(2):235–278, 2010.

[43] Jeffrey Diller, Romain Dujardin, and Vincent Guedj. Dynamics of meromorphic mappings with small topological degree II: energy and invariant measure. *Comment. Math. Helv.*, 86(2):277–316, 2011.

[44] Jeffrey Diller, Daniel Jackson, and Andrew Sommese. Invariant curves for birational surface maps. *Trans. Amer. Math. Soc.*, 359(6):2793–2991 (electronic), 2007.

[45] T.-C. Dinh and N. Sibony. Groupes commutatifs d'automorphismes d'une variété kählérienne compacte. *Duke Math. J.*, 123(2):311–328, 2004.

[46] Tien-Cuong Dinh and Henry de Thélin. Dynamics of automorphisms on compact Kähler manifolds. *Adv. Math.*, 229(5):2640–2655, 2012.

[47] Tien-Cuong Dinh and Nessim Sibony. Green currents for holomorphic automorphisms of compact Kähler manifolds. prepublication Orsay, arXiv:math/0311322, pages 1–41. 2003.

[48] Tien-Cuong Dinh and Nessim Sibony. Green currents for holomorphic automorphisms of compact Kähler manifolds. *J. Amer. Math. Soc.*, 18(2):291–312 (electronic), 2005.

[49] Tien-Cuong Dinh and Nessim Sibony. Dynamics in several complex variables: endomorphisms of projective spaces and polynomial-like mappings. In *Holomorphic dynamical systems*, volume 1998 of *Lecture Notes in Math.*, pages 165–294. Springer, Berlin, 2010.

[50] Tien-Cuong Dinh and Nessim Sibony. Super-potentials for for currents on compact Kähler manifolds and dynamics of automorphisms. *J. Algebraic Geom.*, 19(3):473–529, 2010.

[51] I. Dolgachev and S. Kondō. A supersingular K3 surface in characteristic 2 and the Leech lattice. *Int. Math. Res. Not.*, (1):1–23, 2003.

[52]   Adrien Douady. Prolongement de mouvements holomorphes (d'après Słodkowski et autres). *Astérisque*, (227):Exp. No. 775, 3, 7–20, 1995. Séminaire Bourbaki, Vol. 1993/94.

[53]   Romain Dujardin. Laminar currents in $\mathbb{P}^2$. *Math. Ann.*, 325(4):745–765, 2003.

[54]   Romain Dujardin. Sur l'intersection des courants laminaires. *Publ. Mat.*, 48(1):107–125, 2004.

[55]   Romain Dujardin. Structure properties of laminar currents on $\mathbb{P}^2$. *J. Geom. Anal.*, 15(1):25–47, 2005.

[56]   Romain Dujardin. Approximation des fonctions lisses sur certaines laminations. *Indiana Univ. Math. J.*, 55(2):579–592, 2006.

[57]   Romain Dujardin. Laminar currents and birational dynamics. *Duke Math. J.*, 131(2):219–247, 2006.

[58]   Christophe Dupont. On the dimension of invariant measures of endomorphisms of $\mathbb{CP}^k$. *Math. Ann.*, 349(3):509–528, 2011.

[59]   Najmuddin Fakhruddin. Questions on self maps of algebraic varieties. *J. Ramanujan Math. Soc.*, 18(2):109–122, 2003.

[60]   John Erik Fornæss and Nessim Sibony. Complex dynamics in higher dimensions. In *Complex potential theory (Montreal, PQ, 1993)*, volume 439 of *NATO Adv. Sci. Inst. Ser. C Math. Phys. Sci.*, pages 131–186. Kluwer Acad. Publ., Dordrecht, 1994. Notes partially written by Estela A. Gavosto.

[61]   A. Fujiki. On automorphism groups of compact Kähler manifolds. *Invent. Math.*, 44(3):225–258, 1978.

[62]   Étienne Ghys and Alberto Verjovsky. Locally free holomorphic actions of the complex affine group. In *Geometric study of foliations (Tokyo, 1993)*, pages 201–217. World Sci. Publ., River Edge, NJ, 1994.

[63]   M. H. Gizatullin. Rational *G*-surfaces. *Izv. Akad. Nauk SSSR Ser. Mat.*, 44(1):110–144, 239, 1980.

[64]   P. Griffiths and J. Harris. *Principles of algebraic geometry*. Wiley Classics Library. John Wiley & Sons, Inc., New York, 1994. Reprint of the 1978 original.

[65]   M. Gromov. Entropy, homology and semialgebraic geometry. *Astérisque*, (145-146):5, 225–240, 1987. Séminaire Bourbaki, Vol. 1985/86.

[66]   M. Gromov. On the entropy of holomorphic maps. *Enseign. Math. (2)*, 49(3-4):217–235, 2003.

[67]   V. Guedj. Propriétés ergodiques des applications rationnelles. *Panorama et Synthèse*, 30:90-202, 2010.

[68]   Robin Hartshorne. *Algebraic geometry*. Springer-Verlag, New York, 1977. Graduate Texts in Mathematics, No. 52.

[69]   John Hubbard, Peter Papadopol, and Vladimir Veselov. A compactification of Hénon mappings in $\mathbb{C}^2$ as dynamical systems. *Acta Math.*, 184(2):203–270, 2000.

[70]   John H. Hubbard and Ralph W. Oberste-Vorth. Hénon mappings in the complex domain. I. The global topology of dynamical space. *Inst. Hautes Études Sci. Publ. Math.*, (79):5–46, 1994.

[71] John H. Hubbard and Ralph W. Oberste-Vorth. Hénon mappings in the complex domain. II. Projective and inductive limits of polynomials. In *Real and complex dynamical systems (Hillerød, 1993)*, volume 464 of *NATO Adv. Sci. Inst. Ser. C Math. Phys. Sci.*, pages 89–132. Kluwer Acad. Publ., Dordrecht, 1995.

[72] John Hamal Hubbard. *Teichmüller theory and applications to geometry, topology, and dynamics. Vol. 1: Teichmüller theory.* With contributions by Adrien Douady, William Dunbar, Roland Roeder, Sylvain Bonnot, David Brown, Allen Hatcher, Chris Hruska and Sudeb Mitra, With forewords by William Thurston and Clifford Earle. Matrix Editions, Ithaca, 2006.

[73] Nikolai V. Ivanov. *Subgroups of Teichmüller modular groups*, volume 115 of *Translations of Mathematical Monographs.* American Mathematical Society, Providence, RI, 1992. Translated from the Russian by E. J. F. Primrose and revised by the author.

[74] Nikolai V. Ivanov. Automorphism of complexes of curves and of Teichmüller spaces. *Internat. Math. Res. Notices*, (14):651–666, 1997.

[75] Katsunori Iwasaki and Takato Uehara. Periodic points for area-preserving birational maps of surfaces. *Math. Z.*, 266(2):289–318, 2010.

[76] Mattias Jonsson and Dror Varolin. Stable manifolds of holomorphic diffeomorphisms. *Invent. Math.*, 149(2):409–430, 2002.

[77] A. Katok. Lyapunov exponents, entropy and periodic orbits for diffeomorphisms. *Inst. Hautes Études Sci. Publ. Math.*, (51):137–173, 1980.

[78] Anatole Katok and Boris Hasselblatt. *Introduction to the modern theory of dynamical systems*, volume 54 of *Encyclopedia of Mathematics and its Applications.* Cambridge University Press, Cambridge, 1995. With a supplementary chapter by Katok and Leonardo Mendoza.

[79] Shu Kawaguchi. Projective surface automorphisms of positive topological entropy from an arithmetic viewpoint. *Amer. J. Math.*, 130(1):159–186, 2008.

[80] A. Lamari. Le cône kählérien d'une surface. *J. Math. Pures Appl. (9)*, 78(3):249–263, 1999.

[81] Serge Lang. *Introduction to complex hyperbolic spaces.* Springer-Verlag, New York, 1987.

[82] R. Lazarsfeld. *Positivity in algebraic geometry. I and II*, volume 48–49 of *Ergebnisse der Mathematik und ihrer Grenzgebiete. 3. Folge. A Series of Modern Surveys in Mathematics [Results in Mathematics and Related Areas. 3rd Series. A Series of Modern Surveys in Mathematics].* Springer-Verlag, Berlin, 2004.

[83] D. I. Lieberman. Compactness of the Chow scheme: Applications to automorphisms and deformations of Kähler manifolds. In *Fonctions de plusieurs variables complexes, III (Sém. François Norguet, 1975–1977)*, pages 140–186. Springer, Berlin, 1978.

[84] Howard A. Masur and Yair N. Minsky. Geometry of the complex of curves. I. Hyperbolicity. *Invent. Math.*, 138(1):103–149, 1999.

[85] Curtis T. McMullen. http://www.math.harvard.edu/~ctm/programs.html and http://www.math.harvard.edu/~ctm/gallery/index.html.

[86] Curtis T. McMullen. Dynamics on *K*3 surfaces: Salem numbers and Siegel disks. *J. Reine Angew. Math.*, 545:201–233, 2002.

[87] Curtis T. McMullen. Dynamics on blowups of the projective plane. *Publ. Math. Inst. Hautes Études Sci.*, (105):49–89, 2007.

[88] Curtis T. McMullen. Dynamics with small entropy on projective K3 surfaces. Preprint, pages 1–39, 2011.

[89] Arnaud Moncet. Real versus complex volumes on real algebraic surfaces. *Int. Math. Res. Notices*, 16:3723–3762, 2011.

[90] Arnaud Moncet. Thèse de doctorat de mathématique. Univ. Rennes 1, 2012.

[91] Masayoshi Nagata. On rational surfaces. II. *Mem. Coll. Sci. Univ. Kyoto Ser. A Math.*, 33:271–293, 1960/1961.

[92] Noboru Nakayama and De-Qi Zhang. Building blocks of étale endomorphisms of complex projective manifolds. *Proc. Lond. Math. Soc. (3)*, 99(3):725–756, 2009.

[93] Sheldon E. Newhouse. Continuity properties of entropy. *Ann. of Math. (2)*, 129(2):215–235, 1989.

[94] Keiji Oguiso. Bimeromorphic automorphism groups of non-projective hyperkähler manifolds—a note inspired by C. T. McMullen. *J. Differential Geom.*, 78(1):163–191, 2008.

[95] M. Shub and D. Sullivan. A remark on the Lefschetz fixed point formula for differentiable maps. *Topology*, 13:189–191, 1974.

[96] Nessim Sibony. Dynamique des applications rationnelles de $\mathbb{P}^k$. In *Dynamique et géométrie complexes (Lyon, 1997)*, volume 8 of *Panor. Synthèses*, pages ix–x, xi–xii, 97–185. Soc. Math. France, Paris, 1999.

[97] J. Smillie. Dynamics in two complex dimensions. In *Proceedings of the International Congress of Mathematicians, Vol. III (Beijing, 2002)*, pages 373–382, Higher Ed. Press, Beijing, 2012.

[98] William P. Thurston. *Three-dimensional geometry and topology. Vol. 1*, Edited by Silvio Levy. volume 35 of *Princeton Mathematical Series*. Princeton University Press, Princeton, NJ, 1997.

[99] Kenji Ueno. *Classification theory of algebraic varieties and compact complex spaces*. Lecture Notes in Mathematics, Vol. 439. Springer-Verlag, Berlin, 1975. Notes written in collaboration with P. Cherenack.

[100] C. Voisin. *Théorie de Hodge et géométrie algébrique complexe*, volume 10 of *Cours Spécialisés [Specialized Courses]*. Société Mathématique de France, Paris, 2002.

[101] Joachim Wehler. *K*3-surfaces with Picard number 2. *Arch. Math. (Basel)*, 50(1):73–82, 1988.

[102] Junyi Xie. Periodic points of birational maps on the projective plane. *preprint*, arXiv:1106.1825:1–26, 2011.

[103] Y. Yomdin. Volume growth and entropy. *Israel J. Math.*, 57(3):285–300, 1987.

[104] D.-Q. Zhang. A theorem of Tits type for compact Kähler manifolds. *Invent. Math.*, 176(3):449–459, 2009.

Bifurcation currents and equidistribution in parameter space

Romain Dujardin

ABSTRACT. We review the use of techniques of positive currents for the study of parameter spaces of one-dimensional holomorphic dynamical systems (rational mappings on $\mathbb{P}^1$ or subgroups of the Möbius group $PSL(2, \mathbb{C})$). The topics covered include the construction of bifurcation currents and the characterization of their supports, the equidistribution properties of dynamically defined subvarieties of the parameter space.

INTRODUCTION

Let $(f_\lambda)_{\lambda \in \Lambda}$ be a holomorphic family of dynamical systems acting on the Riemann sphere $\mathbb{P}^1$, parameterized by a complex manifold Λ. The "dynamical systems" in consideration here can be polynomial or rational mappings on $\mathbb{P}^1$, as well as groups of Möbius transformations. It is a very basic idea that the product dynamics $\widehat{f}$ acting on $\Lambda \times \mathbb{P}^1$ by $\widehat{f}(\lambda, z) = (\lambda, f_\lambda(z))$ is an important source of information on the bifurcation theory of the family. The input of techniques from higher-dimensional holomorphic dynamics into this problem recently led to a number of interesting new results in this area, especially when the parameter space Λ is multidimensional. Our purpose in this paper is to review these recent developments.

The main new idea that has arisen from this interaction is the use of positive closed currents. We will see that the consideration of the dynamics of $\widehat{f}$ gives rise to a number of interesting currents on $\Lambda \times \mathbb{P}^1$ and Λ. Positive currents have an underlying measurable structure, so it would be fair to say that we are studying these parameter spaces at a measurable level, somehow in the spirit of the ergodic theoretic approach to dynamics.[1]

In general, a basic way in general to construct and study dynamical currents is to view them as limits of sequences of dynamically defined subvarieties. This will be another major theme in this paper.

We will try as much as possible to emphasize the similarities between methods of higher-dimensional dynamics, of the study of families of rational maps, and that of Möbius subgroups. We will also state a number of open questions to foster further developments of this theory.

An interesting outcome of these methods is the possibility of studying "higher codimensional" phenomena — like the property for a rational map of having several periodic critical points. These phenomena are difficult to grasp using elementary complex analysis techniques because of the failure of Montel's theorem in

---

[1] In this respect it is instructive to compare this with the more topological point of view of Branner and Hubbard, which was summarized 20 years ago by Branner in [Bra].

higher dimension. In the same vein, we will see that when $\dim(\Lambda) > 1$, the bifurcation locus of a family of rational maps possesses a hierarchical structure, which may conveniently be formalized using bifurcation currents. When $(f_\lambda)_{\lambda \in \Lambda}$ is the family of polynomials of degree $d \geq 3$, the smallest of these successive bifurcation loci is the right analogue of the Mandelbrot set in higher degree, with whom it shares many important properties.

◇

Contents. Let us now outline the contents of this article. Section 1 is of general nature. We explain how the non-normality of a sequence of holomorphic mappings $f_n \colon \Lambda \to X$ between complex manifolds is related to certain closed positive currents of bidegree (1,1) on Λ. We also show that the preimages under f_n of hypersurfaces of X tend to be equidistributed. This will provide — at least at the conceptual level — a uniform framework for many of the subsequent results.

Since these facts are not so easy to extract from the literature, we explain them in full detail; therefore, the presentation is a bit technical. The reader who wants to dive directly into holomorphic dynamics is advised to skip this section on a first reading.

Sections 2 and 3 are devoted to the study of bifurcation currents for polynomials and rational maps on $\mathbb{P}^1$, which is the most developed part of the theory. In Section 2 we present two (related) constructions of bifurcation currents of bidegree (1,1): the "absolute" bifurcation current T_{bif} and the bifurcation current associated with a marked critical point. In both cases, the support of the bifurcation current is equal to the corresponding bifurcation locus. We also show that these currents describe the asymptotic distribution of families of dynamically defined codimension 1 subsets of parameter space. More precisely we will be interested in the families of hypersurfaces $\mathrm{PerCrit}(n,k)$ (resp. $\mathrm{Per}(n,w)$) defined by the condition that a critical point satisfies $f^n(c) = f^k(c)$ (resp. f possesses a periodic n-cycle of multiplier w).

In Section 3, we study "higher" bifurcation currents, which are obtained by taking exterior products of the previous ones. We will develop the idea that the supports of these currents define a dynamically meaningful filtration of the bifurcation locus and then characterize them precisely. We also explain why bifurcation currents should display some laminar structure in parts of parameter space, and give some results in this direction.

Many of the proofs will be sketched, the reader being referred to the original papers for complete arguments. Let us also mention a recent set of lecture notes by Berteloot [Bt2], which covers most of this material with greater detail (and complete proofs).

In Section 4 we introduce currents associated to bifurcations of holomorphic families of subgroups of $\mathrm{PSL}(2, \mathbb{C})$, which is in a sense the counterpart of Section 2 in the Kleinian groups setting. The existence of such a counterpart is in accordance with the so-called *Sullivan dictionary* between rational and Kleinian group dynamics; nevertheless, its practical implementation requires a number of new ideas. To be specific, let (ρ_λ) be a holomorphic family of representations of a given finitely generated group into $\mathrm{PSL}(2, \mathbb{C})$ (satisfying certain natural assumptions). We construct a bifurcation current on Λ associated to a random walk on G. As before, this current is supported precisely on the "bifurcation locus" of the family, and it describes the asymptotic distribution of natural codimension 1 subsets of parameter space. We will see that the key technical ingredient here is the ergodic theory of random products of matrices.

In Section 5 we outline some possible extensions of the theory. An obvious generalization would be to consider rational mappings in higher dimension. A basic difficulty is that in that setting the understanding of bifurcation phenomena is still rather poor.

We do not include a general discussion about plurisubharmonic (psh for short) functions and positive currents. Good reference sources for this are the books by Demailly [De, Chap. I and III] and Hörmander [Hö, Chap. 4]. See also [Ca2] in this volume for a short presentation. We do not require much knowledge in holomorphic dynamics, except for the basic properties of the maximal entropy measure [Ly1, FLM].

◇

Bibliographical overview. Let us briefly review the main references that we will be considering in the paper. Bifurcation currents were introduced by DeMarco in [DeM1]. In this paper she constructs a current T_{bif} on any holomorphic family of rational maps, whose support is the bifurcation locus. This current is defined in terms of the critical points. In [DeM2], she proves a formula for the Lyapunov exponent of a rational map on $\mathbb{P}^1$, relative to its maximal entropy measure, which extends prior results due to Manning [Man] and Przytycki [Pr]. From this she deduces that T_{bif} is the dd^c of the Lyapunov exponent function.

In [BB1], Bassanelli and Berteloot generalize DeMarco's formula to higher-dimensional rational maps and initiate the study of the higher exterior powers of the bifurcation current (associated with rational mappings on $\mathbb{P}^1$), by showing that $\text{Supp}(T_{\text{bif}}^k)$ is accumulated by parameters possessing k indifferent cycles. In [DF], Favre and the author study the asymptotic distribution of the family of hypersurfaces PerCrit(n, k). The structure of the space of polynomials of degree d is also investigated, with emphasis on the higher-dimensional analogue of the boundary of the Mandelbrot set. A finer description is given in the particular case of cubic polynomials in [Du2].

Several equidistribution theorems for the family of hypersurfaces Per(n, w) are obtained by Bassanelli and Berteloot in [BB2], [BB3]. [BB2] also discusses the laminarity properties of bifurcation currents in the space of quadratic rational maps.

In [BE], Buff and Epstein develop a method based on transversality ideas to characterize the supports of certain "higher" bifurcation currents. This was recently generalized by Gauthier [Ga], leading in particular to Hausdorff dimension estimates for the supports of these currents, which generalize Shishikura's famous result that the boundary of the Mandelbrot set has dimension 2 [Sh].

Bifurcation currents for families of subgroups of PSL$(2, \mathbb{C})$, satisfying properties similar to those previously mentioned, were designed by Deroin and the author in [DD1].

Related developments in higher-dimensional holomorphic dynamics can be found in [DS1], [Ph].

◇

We close this introduction with a few words on the connections between these ideas and the work of Milnor. Alone or with coauthors, he wrote a number of papers, most quite influential, on parameter spaces of polynomials and rational functions, such as [Mi1], [Mi2], [Mi3], [Mi4], [Mi5], and [BKM]. These articles share an emphasis on multidimensional issues and the role played by subvarieties of parameter space such as $\mathrm{PerCrit}(n, k)$ and $\mathrm{Per}(n, w)$. I hope he will appreciate the way in which these ideas reappear here.

Many thanks to Serge Cantat, Charles Favre, and Thomas Gauthier for their useful comments.

1   PROLOGUE: NORMAL FAMILIES, CURRENTS AND EQUIDISTRIBUTION

Let Λ be a complex manifold of dimension d and let X be a compact Kähler manifold of dimension k, endowed with Kähler form ω. Let (f_n) be a sequence of holomorphic mappings from Λ to X. In this section we explain a basic construction relating the non-normality of the sequence (f_n) and certain positive (1,1) currents on Λ. When applied to particular situations it will give rise to various bifurcation currents. This construction is also related to higher-dimensional holomorphic dynamics since we may take $\Lambda = X$ and f_n be the family of iterates of a given self-map on X.

The problems we consider are local on Λ so without loss of generality we assume that Λ is an open ball in $\mathbb{C}^d$. We say that the family (f_n) is *quasi-normal* if for every subsequence of (f_n) (still denoted by (f_n)) there exists a further subsequence (f_{n_j}) and an analytic subvariety $E \subset \Lambda$ such that (f_{n_j}) is a normal family on $\Lambda \setminus E$ (see [IN] for a discussion of this and other related notions).

1.1   A normality criterion

The following result is a variation on well-known ideas, but it is apparently new.

THEOREM 1.1. *Let Λ and X be as described earlier, and (f_n) be a sequence of holomorphic mappings from Λ to X. If the sequence of bidegree (1,1) currents $f_n^* \omega$ has locally uniformly bounded mass on Λ, then the family (f_n) is quasi-normal on Λ.*

Recall that the mass of a positive current of bidegree (1,1) in an open set $\Omega \subset \mathbb{C}^d$ is defined by $\mathbf{M}_\Omega(T) = \sup |\langle T, \varphi \rangle|$, where φ ranges among test $(d-1, d-1)$ forms $\sum \varphi_{I,J} dz_I \wedge d\bar{z}_J$ with $\left\| \varphi_{I,J} \right\|_{L^\infty} \leq 1$.

A few comments are in order here. First, if $d = 1$, the result is a well-known consequence of Bishop's criterion for the normality of a sequence of analytic sets [Ch, §15.5] (see Lemma 1.2 for the proof). The point here is that if $d > 1$ our assumption does *not* imply that the volumes of the graphs of the f_n are locally bounded (see [IN, Example 5.1]). Secondly, it is clear that the converse of Theorem 1.1 is false, i.e., there exist quasi-normal families such that $f_n^* \omega$ has unbounded mass. For this, take any sequence of holomorphic mappings $\mathbb{D} \to \mathbb{P}^1$, converging on compact subsets of $\mathbb{D}^*$ to $z \mapsto \exp(1/z)$.

LEMMA 1.2. *Theorem 1.1 holds when $\dim(\Lambda) = 1$.*

PROOF. Let $\Gamma(f_n) \subset \Lambda \times X$ be the graph of f_n. Let π_1, π_2 be the first and second projections on $\Lambda \times X$. Let β be the standard Kähler form on $\Lambda \subset \mathbb{C}$. Then if $U \Subset \Lambda$, the volume of $\Gamma(f_n) \cap \pi_1^{-1}(U)$ relative to the product Hermitian structure equals

$$\int_{\Gamma(f_n) \cap \pi_1^{-1}(U)} \pi_1^* \beta + \pi_2^* \omega = \text{vol}_\Lambda(U) + \int_U \left(\pi_2 \circ \left(\pi_1|_{\Gamma(f_n)}^{-1} \right) \right)^* \omega$$

$$= \text{vol}_\Lambda(U) + \int_U f_n^* \omega.$$

Therefore, our assumption implies that there is a uniform bound on the volumes[2] of the analytic sets $\Gamma(f_n) \cap \pi_1^{-1}(U)$. By Bishop's theorem one can extract a subsequence n_j such that the $\Gamma(f_{n_j}) \cap \pi_1^{-1}(U)$ converge in the Hausdorff topology to a one-dimensional analytic set Γ of $\pi_1^{-1}(U)$.

We claim that Γ is the union of a graph and finitely many vertical curves. Here vertical means that it projects to a point on Λ. Indeed, note that by Lelong's lower bound on the volume of an analytic set [Ch, §15.3], the volume of any analytic subset of X is uniformly bounded below. Since $\text{vol}_{\Lambda \times X}(\Gamma) \leq \liminf \text{vol}_{\Lambda \times X}(\Gamma(f_{n_j}))$, this implies that Γ contains only finitely many vertical components. Let $E \subset \Lambda$ be the projection of these components. We claim that the f_{n_j} converge locally uniformly outside E. Indeed let $V \subset \Lambda$ be a connected open subset disjoint from E. Since the $\Gamma(f_{n_j})$ converge in the Hausdorff topology, we see that $\Gamma \cap \pi_1^{-1}(V)$ is non-empty. Now, since π_1 is proper, we infer that $\pi_1|_{\Gamma \cap \pi_1^{-1}(V)}$ is a branched covering, which must be of degree 1 (for if not, generic fibers of π_1 would intersect $\Gamma(f_{n_j})$ in several points for large j). We conclude that Γ is a graph over V, of some $f: V \to X$, and that the f_{n_j} converge uniformly to f there. $\square$

The possibility of vertical components of Γ over a locally finite set is known as the *bubbling phenomenon*. An important consequence of the proof is that there exists a constant δ_0 (any number smaller than the infimum of the volumes of 1-dimensional subvarieties of X will do) such that if $\limsup \int_V f_n^* \omega \leq \delta_0$, then no bubbling occurs in V. An easy compactness argument yields the following:

COROLLARY 1.3. *Let V be a one-dimensional disk and let $\delta_0 > 0$ be as described previously. For every $K \Subset V$, there exists a constant $C(V, K, \delta_0)$ so that if $f: V \to X$ is a holomorphic mapping satisfying $\int_V f^* \omega \leq \delta_0$, then $\|df\|_{L^\infty(K)} \leq C$.*

We now prove the theorem. The idea, based on an argument from [DS1] (see also [Du3, Prop. 5.7]), is to use a slicing argument together with a theorem due to Sibony and Wong [SW]. For convenience let us state this result first.

THEOREM 1.4 (Sibony-Wong). *Let g be a holomorphic function defined in the neighborhood of the origin in $\mathbb{C}^d$, which admits a holomorphic continuation to a neighborhood of $\bigcup_{L \in E} L \cap B(0, R)$, where $E \subset \mathbb{P}^{d-1}$ is a set of lines through the origin, of measure $\geq 1/2$ (relative to the Fubini-Study volume on $\mathbb{P}^{d-1}$).*

Then there exists a constant $C_{SW} > 0$ such that g extends to a holomorphic function of $B(0, C_{SW}R)$, and furthermore

$$\sup_{B(0, C_{SW}R)} |g| \leq \sup_{\bigcup_{L \in E} L \cap B(0,R)} |g|. \tag{1.1}$$

---

[2]This is where we use the assumption that $\dim(\Lambda) = 1$. In higher dimension, to estimate this volume one needs to integrate the exterior power $(\pi_1^* \beta + \pi_2^* \omega)^d$ where $d = \dim(\Lambda)$.

PROOF OF THEOREM 1.1. Recall that Λ was supposed to be an open subset in $\mathbb{C}^d$. Denote by β the standard Kähler form on $\mathbb{C}^d$. Let $T_n = f_n^* \omega$ and consider a subsequence (still denoted by n) such that T_n converges to a current T on Λ. Let $\sigma_T = T \wedge \beta^{d-1}$ be the trace measure of T. For every $p \in \Lambda, \frac{1}{c_{2d-2} r^{2d-2}} \sigma_T(B(p,r))$ converges as $r \to 0$ to the Lelong number of T at p, denoted by $\nu(T, p)$ (here c_{2d-2} is the volume of the unit ball in $\mathbb{C}^{2d-2}$). By Siu's semi-continuity theorem [De], for each $c > 0$, $E_c(T) = \{p, \nu(T,c) \geq c\}$ is a proper analytic subvariety of Λ. Fix $c = \delta_0/4$, where δ_0 is as before. We show that (f_n) is a normal family on $\Lambda \setminus E_c(T)$.

Indeed, let $p \notin E_c(T)$. Then for $r < r(\delta_0)$ (which will be fixed from now on),

$$\frac{1}{c_{2d-2} r^{2d-2}} \int_{B(p,r)} T \wedge \beta^{d-1} \leq \frac{\delta_0}{3};$$

hence, for large n,

$$\frac{1}{c_{2d-2} r^{2d-2}} \int_{B(p,r)} T_n \wedge \beta^{d-1} \leq \frac{\delta_0}{2}.$$

Now let $\alpha_p = dd^c \log \|z - p\|$. By Crofton's formula [De, Cor. III.7.11], we know $\alpha_p^{d-1} = \int_{\mathbb{P}^{d-1}} [L] dL$ is the average of the currents of integrations along the complex lines through p (w.r.t. the unitary invariant probability measure on $\mathbb{P}^{d-1}$). By a well-known formula due to Lelong [Ch, §15.1], for every $r_1 < r$,

$$\frac{1}{c_{2d-2} r^{2d-2}} \sigma_{T_n}(B(p,r)) - \frac{1}{c_{2d-2} r_1^{2d-2}} \sigma_{T_n}(B(p,r_1)) = \int_{r_1 < \|z-p\| < r} T_n \wedge \alpha_p^{d-1}.$$

Since T_n is a smooth form, it has zero Lelong number at p and we can let r_1 tend to zero. We conclude that for every large-enough n, $\int_{B(p,r)} T_n \wedge \alpha_p^{d-1} \leq \delta_0/2$. Applying Crofton's formula we see that

$$\int_{\mathbb{P}^{d-1}} \left(\int_{L \cap B(p,r)} f_n^* \omega \right) \leq \frac{\delta_0}{2};$$

therefore there exists a set E_n of lines of measure at least $1/2$ such that if $L \in E_n$ then $\int_{L \cap B(p,r)} f_n^* \omega < \delta_0$. By Corollary 1.3, for each such line $L \in E_n$, the derivative of $f_n|_{L \cap B(p,r)}$ is locally uniformly bounded. Extract a further subsequence so that $f_n(p)$ converges to some $x \in X$. Thus, reducing r if necessary, we can assume that for $L \in E_n$, $f_n|_{L \cap B(p,r)}$ takes its values in a fixed coordinate patch containing x, which may be identified to a ball in $\mathbb{C}^k$. By Theorem 1.4, there exists a constant C_{SW} such that $f_n|_{B(p,C_{SW}r)}$ takes its values in the chart, with the same bound on the derivative. This implies that (f_n) is a normal family in $B(p, C_{SW}r)$, thereby concluding the proof. $\qquad \square$

REMARK 1.5. The proof shows that if it can be shown that the Lelong numbers of the cluster values of $(f_n)^* \omega$ are smaller than δ_0 (for instance if the potentials are uniformly bounded), then the family is actually normal.

1.2 Equidistribution in codimension 1

We now turn to the case where the mass of $f_n^* \omega$ tends to infinity and show that the preimages of hypersurfaces under f_n of X tend to equidistribute in the sense of currents. This idea goes back to the work of Russakovskii, Shiffmann and Sodin [RSo, RSh]. Dinh and Sibony [DS2] later gave a wide generalization of these results. Here we present a simple instance of this phenomenon, which is inspired by (and can be deduced from) [DS2].

Let Λ, X and (f_n) be as before and set $d_n = \int_\Lambda f_n^*\omega \wedge \beta^{d-1}$, so that $d_n^{-1}f_n^*\omega$ is a sequence of currents of bounded mass on Λ. A first remark is that if ω' is another Kähler form, there exists a constant $C \geq 1$ such that $C^{-1}\omega \leq \omega' \leq C\omega$; hence, $d_n^{-1}f_n^*\omega'$ also has bounded mass.

By definition, a holomorphic family of subvarieties $(H_a)_{a\in A}$ of dimension l parameterized by a complex manifold A is the data of a subvariety H in $A \times X$, of dimension $\dim(A) + l$, such that for every $a \in A$, $\pi_1^{-1}(a) \cap H =: H_a$ has dimension l. Of course here we are identifying every fiber $\pi_1^{-1}(a)$ with X, using the second projection. We will need only to consider the case $l = k - 1$.

We need a notion of a "sufficiently mobile" family of hypersurfaces. For this, let us assume for simplicity that X is a projective manifold. We say that $(H_a)_{a\in A}$ is a *substantial* family of hypersurfaces on X if

the hypersurfaces (H_a) are hyperplane sections relative to some embedding $\iota\colon X \hookrightarrow \mathbb{P}^N$ and there exists a positive measure ν on A such that the current $\int [H_a]d\nu(a)$ has locally bounded potentials.

Notice that the family of *all* hyperplane sections relative to some projective embedding of X is substantial. Indeed by Crofton's formula there exists a natural smooth measure dL on the dual projective space $\check{\mathbb{P}}^N$ (i.e., the space of hyperplanes) such that the average current of integration is the Fubini-Study form, i.e., $\int [L]dL = \omega_{FS}$. Therefore, on X we get that $\int [\iota^{-1}(L)]dL = \iota^*\omega_{FS}$, and the family is substantial. From this it follows, for instance, that the family of hypersurfaces of given degree in $\mathbb{P}^k$ is substantial.

Here is the equidistribution statement. We do not strive for maximal generality here and it is likely that some of the assumptions could be relaxed. For instance, in view of applications to random walks on groups it is of interest to obtain similar results for non-compact X (to deal with examples like $X = \mathrm{SL}(n, \mathbb{C})$, etc.). One might also obtain equidistribution statements in higher codimension by introducing appropriate dynamical degrees.

THEOREM 1.6. *Let Λ be a complex manifold of dimension d and X be a projective manifold. Let $f_n\colon \Lambda \to X$ be a family of holomorphic mappings such that $d_n = \int_\Lambda f_n^*\omega \wedge \beta^{d-1}$ tends to infinity. Let $(H_a)_{a\in A}$ be an substantial holomorphic family of hypersurfaces in X.*

Let $\mathcal{E}$ be the set of $a \in A$ such that

$$\frac{1}{d_n}(f_n^*[H_a] - f_n^*\omega)$$

does not converge to zero in the sense of currents. Then

   i. *if the series $\sum d_n^{-1}$ converges, then $\mathcal{E}$ is pluripolar;*

   ii. *if for every $t > 0$ the series $\sum \exp(-td_n)$ converges, then $\mathcal{E}$ has zero Lebesgue measure.*

As the proof easily shows, in case (ii) Lebesgue measure can actually be replaced by any *moderate measure*, that is, a measure satisfying an inequality of the form $m(\{u < -t\}) \leq Ce^{-\alpha t}$ on any compact class of psh functions. This is wide class of measures which contains, for instance, the area measure on totally real submanifolds of maximal dimension. We refer to [DS2] for details.

COROLLARY 1.7. *Under the assumptions of the theorem, if the sequence $\frac{1}{d_n} f_n^* \omega$ converges to some current T, then for $a \notin \mathcal{E}$, $\frac{1}{d_n} f_n^*[H_a]$ converges to T.*

PROOF OF THEOREM 1.6. Without loss of generality, assume that A is an open ball in $\mathbb{C}^{\dim(A)}$.

LEMMA 1.8. *Under the assumptions of the theorem, there exists a Kähler form ω on X and a negative function $(a, x) \mapsto u(a, x)$ on $A \times X$ such that*

i. *for every $a \in A$, $dd_x^c u(a, \cdot) = [H_a] - \omega$;*

ii. *the L^1 norm $\|u(a, \cdot)\|_{L^1(X)}$ is locally uniformly bounded with respect to $a \in A$;*

iii. *for every x, $u(\cdot, x)$ is psh on A.*

Assuming this result for the moment, let us continue with the proof of the theorem. Suppose first that the series $\sum d_n^{-1}$ converges. Let m be a positive measure with compact support on A such that psh functions are m-integrable. We claim that for m-a.e. a, $\frac{1}{d_n}(f_n^*[H_a] - f_n^*\omega)$ converges to zero. By Lemma 1.8(i), for this it is enough to prove that $\frac{1}{d_n} \int u_a \circ f_n$ tends to zero in $L^1_{\mathrm{loc}}(\Lambda)$ (here and in what follows we denote $u(a, \cdot)$ by u_a).

Let us admit the following lemma for the moment.

LEMMA 1.9. *The function defined by $x \mapsto \int u_a(x)dm(a)$ is ω-psh, that is,*

$$dd^c \int u_a(x)dm(a) \geq -\omega,$$

and it is bounded on X.

Fix $\Lambda' \Subset \Lambda$. By Fubini's theorem and Lemma 1.9, setting $\widetilde{u}(x) = \int u_a(x)dm(a)$ we have that

$$\int \left(\frac{1}{d_n} \int_{\Lambda'} (-u_a \circ f_n(\lambda))d\lambda \right) dm(a) = \frac{1}{d_n} \int_{\Lambda'} (-\widetilde{u} \circ f_n(\lambda))d\lambda \leq \frac{C}{d_n},$$

so

$$m\left(\left\{ a, \frac{1}{d_n} \int_{\Lambda'} |u_a \circ f_n(\lambda)| \, d\lambda \geq \varepsilon \right\} \right) \leq \frac{C}{\varepsilon d_n},$$

and the result follows from the Borel-Cantelli lemma.

To conclude the proof of case (i) in the theorem, we argue that if the exceptional $\mathcal{E}$ set was not pluripolar, then it would contain a non-pluripolar compact set K. By the work of Bedford and Taylor [BT] there exists a Monge-Ampère measure $m = (dd^c v)^N$ supported on K, with v bounded. It is well known that for such a measure psh functions are integrable (see [BT] or [De, Prop. III.3.11]), so we are in contradiction with the previous claim.

Assume now that $\sum \exp(-td_n)$ converges for all t. By Lemma 1.9 applied to m the Lebesgue measure (cut off to any compact subset of A) the family of negative psh functions

$$\left\{ a \mapsto \int_{\Lambda'} u_a \circ f_n(\lambda)d\lambda \right\}_{n \geq 1}$$

is bounded in $L^1_{\mathrm{loc}}(A)$. Let $A' \Subset A'' \Subset A$. It follows from an inequality due to Hörmander [Hö, Prop 4.2.9] that there exist constants (C_0, α_0) such that if φ is any negative psh function on A such that $\|\varphi\|_{L^1(A'')} \le 1$, then

$$\mathrm{vol}\left(\{a \in A',\ \varphi(a) < -t\}\right) \le C_0 \exp(-\alpha_0 t).$$

It follows that there exist constants (C, α) independent of n such that

$$\mathrm{vol}\left(\left\{a \in A',\ \frac{1}{d_n}\int_{\Lambda'}|u_a \circ f_n(\lambda)|\,d\lambda > \varepsilon\right\}\right) \le C\exp(-\varepsilon\alpha d_n),$$

and again the Borel-Cantelli implies that for (Lebesgue) a.e. a, $\frac{1}{d_n}u_a \circ f_n$ converges to zero in $L^1_{\mathrm{loc}}(A)$. $\square$

PROOF OF LEMMA 1.8. By definition there is an embedding $\iota : X \hookrightarrow \mathbb{P}^N$ and a holomorphic family $(L_a)_{a \in A}$ of hyperplanes such that $H_a = \iota^{-1}(L_a)$. There exists a holomorphic family of linear forms $(\ell_a)_{a \in A}$ on $\mathbb{C}^{N+1}$ such that $\{\ell_a = 0\}$ is an equation of L_a. We normalize so that $|\ell_a| \le 1$ on the unit ball. Now define $\varphi(a, \cdot)$ on $\mathbb{P}^N$ by

$$\varphi(a, z) = \frac{\log |\ell_a(Z)|}{\log \|Z\|},$$

where Z is any lift of z and $\|\cdot\|$ is the Hermitian norm. Then φ satisfies (i)–(iii) relative to the family L_a on $\mathbb{P}^N$, i.e., $dd^c_z\varphi(a, \cdot) = [L_a] - \omega_{FS}$, etc.

We now put $u(a, x) = \varphi(a, \iota(x))$ and claim that it satisfies the desired requirements (with $\omega = \iota^*\omega_{FS}$). Properties (i) and (iii) are immediate. Suppose, for contradiction, that (ii) does not hold; then by the Hartogs Lemma [Hö, pp. 149–151] we would get a sequence $a_n \to a_0 \in A$ such that u_{a_n} diverges uniformly to $-\infty$. But if $x_0 \notin H_{a_0}$, it is clear that u is locally uniformly bounded near (a_0, x_0), hence the contradiction. $\square$

PROOF OF LEMMA 1.9. Let $A' \Subset A$ be an open set containing $\mathrm{Supp}(m)$. According to [DS1, Prop. 3.9.2] there exists a constant C such that for every psh function φ on A, $\|\varphi\|_{L^1(m)} \le C\|\varphi\|_{L^1(A')}$. From this we infer that for every x in X, $\int |u(a, x)|\,dm(a) \le C\int_{A'}|u(a, x)|\,da$, where da denotes the Lebesgue measure.

Now we claim that there exists a constant C' such that for any negative psh function on A, $\|\varphi\|_{L^1(A')} \le C'\|\varphi\|_{L^1(\nu)}$, where ν is the measure from the definition of substantial families. Indeed, by the Hartogs lemma (see [Hö, pp.149-151]) the set

$$\left\{\varphi \text{ negative psh s.t. } \int |\varphi|\,d\nu \le 1\right\}$$

is relatively compact in L^1_{loc}, hence bounded.

From these two facts we infer that for every x,

$$\int |u(a, x)|\,dm(a) \le CC'\int |u(a, x)|\,d\nu(a).$$

We now show that $x \mapsto \int u_a(x)d\nu(a)$ is uniformly bounded. For this, note first that this function is integrable, because by Lemma 1.8, $\|u(a, \cdot)\|_{L^1(X)}$ is locally uniformly bounded. So we can take the dd^c in x, and we obtain that

$$dd^c_x\left(\int u_a(\cdot)d\nu(a)\right) = \int [H_a]d\nu(a) - \omega,$$

and we conclude by using our assumption that the local potentials of $\int [H_a] d\nu(a)$ are bounded. Finally, the same argument shows that $x \mapsto \int u_a(x) dm(a)$ is ω-psh, since by uniform boundedness of $\|u(a, \cdot)\|_{L^1(X)}$, we can permute the dd^c in x and integration with respect to m. $\square$

We close this section by highlighting the following direct consequence of Theorem 1.1, which appears as a key step in the characterization of the supports of certain bifurcation currents.

It can also be used to obtain a coordinate-free proof of the fact that the support of the Green current of a holomorphic self-map of $\mathbb{P}^k$ coincides with the Julia set (a result originally due to Fornæss-Sibony [FS2] and Ueda [U], and generalized to other contexts in [Gue] and [Dil], for example). More precisely, it implies that if $\Omega \subset \mathbb{P}^k$ is an open set disjoint from the support of the Green current T, then the sequence of iterates f^n is normal in Ω (see Remark 1.5 for an explanation how to obtain normality rather than quasi-normality, which applies in this case).

PROPOSITION 1.10. *Let Λ be a complex manifold of dimension d and X be a compact Kähler manifold of dimension k, endowed with a Kähler form ω. Let (f_n) be a sequence of holomorphic mappings $\Lambda \to X$, and assume that the sequence $\frac{1}{d_n} f_n^* \omega$ converges to a current T.*

Assume furthermore that for every test function φ, one has the following estimate

$$\left\langle \frac{1}{d_n} f_n^* \omega - T, \varphi \right\rangle = O\left(\frac{1}{d_n} \right). \tag{1.2}$$

Then the sequence (f_n) is quasi-normal outside $\operatorname{Supp}(T)$.

PROOF. If U is an open set disjoint where $T \equiv 0$, then (1.2) implies that the sequence $(f_n^* \omega)$ has locally uniformly bounded mass on U. It then follows from Theorem 1.1 that (f_n) is quasi-normal on U. $\square$

Notice that in the context of endomorphisms of $\mathbb{P}^k$, the converse inclusion $\operatorname{Supp}(T) \subset J$ is an easy consequence of the definitions.

Conversely, if $d_n \to \infty$, and U is an open set such that $U \cap \operatorname{Supp}(T) \neq \emptyset$, then the sequence (f_n) is *not* normal on U. Indeed it follows from the explicit expression of $f_n^* \omega$ in local coordinates that the L^2 norm of the derivative of f_n tend to infinity in U.

2   BIFURCATION CURRENTS FOR FAMILIES OF RATIONAL MAPPINGS ON $\mathbb{P}^1$

2.1   Generalities on bifurcations

Let us first review a number well-known facts on holomorphic families of rational maps. Let $(f_\lambda)_{\lambda \in \Lambda}$ be a holomorphic family of rational maps $f_\lambda : \mathbb{P}^1 \to \mathbb{P}^1$ of degree $d \geq 2$ parameterized by a connected complex manifold Λ. By definition, a *marked critical point* is a holomorphic map $c : \Lambda \to \mathbb{P}^1$ such that $f_\lambda'(c(\lambda)) = 0$ for all $\lambda \in \Lambda$.

Given any family (f_λ) if c_0 is a given critical point at parameter λ_0, there exists a branched cover $\pi : \widetilde{\Lambda} \to \Lambda$ such that the family of rational mappings $(\widetilde{f}_{\widetilde{\lambda}})_{\widetilde{\lambda} \in \widetilde{\Lambda}}$

defined for $\tilde{\lambda} \in \tilde{\Lambda}$ by $\tilde{f}_{\tilde{\lambda}} = f_{\pi(\tilde{\lambda})}$ has a marked critical point $\tilde{c}(\tilde{\lambda})$ with $\tilde{c}(\tilde{\lambda}_0) = c_0$. Specifically, it is enough to parameterize the family by

$$\tilde{\Lambda} = \hat{\mathcal{C}} = \left\{ (\lambda, z) \in \Lambda \times \mathbb{P}^1, \ f'_\lambda(z) = 0 \right\}$$

(or its desingularization if $\hat{\mathcal{C}}$ is not smooth). Then the first projection $\pi_1 \colon \hat{\mathcal{C}} \to \Lambda$ makes it a branched cover over Λ, and for any $\tilde{\lambda} = (\lambda, z) \in \hat{\mathcal{C}}$, we set $\tilde{c}(\tilde{\lambda}) = z$, which is a critical point for $\tilde{f}_{\tilde{\lambda}} := f_{\pi_1(\tilde{\lambda})}$.

Therefore, taking a branched cover of Λ if necessary, it is always possible to assume that all critical points are marked.

We always denote with a subscript λ the dynamical objects associated to f_λ: Julia set, maximal entropy measure, etc.

In a celebrated paper, Mañé, Sad, and Sullivan [MSS], and independently Lyubich [Ly2], showed the existence of a decomposition

$$\Lambda = \mathrm{Bif} \cup \mathrm{Stab}$$

of the parameter space Λ into a (closed) bifurcation locus and a (open) stability locus, which is similar to the Fatou-Julia decomposition of dynamical space.

Stability is defined by a number of equivalent properties, according to the following theorem [MSS, Ly2].

THEOREM 2.1 (Mañé-Sad-Sullivan, Lyubich). *Let $(f_\lambda)_{\lambda \in \Lambda}$ be a holomorphic family of rational maps of degree $d \geq 2$ on $\mathbb{P}^1$. Let $\Omega \subset \Lambda$ be a connected open subset. The following conditions are equivalent:*

  i. *the periodic points of (f_λ) do not change nature (attracting, repelling, indifferent) in Ω;*

  ii. *the Julia set J_λ moves under a holomorphic motion for $\lambda \in \Omega$;*

  iii. *$\lambda \mapsto J_\lambda$ is continuous for the Hausdorff topology;*

  iv. *for any two parameters λ, λ' in Ω, $f_\lambda|_{J_\lambda}$ is conjugate to $f_{\lambda'}|_{J_{\lambda'}}$.*

If, in addition, the critical points $\{c_1, \ldots, c_{2d-2}\}$ are marked, these conditions are equivalent to

  v. *the family of meromorphic functions $(f_\lambda^n(c_i(\lambda)))_{n \geq 0} \colon \Lambda \to \mathbb{P}^1$ is normal for any $1 \leq i \leq 2d - 2$.*

If these conditions are satisfied, we say that (f_λ) is *J-stable* in Ω (which we simply abbreviate as *stable* in this paper). The stability locus is the union of all such Ω, and Bif is by definition its complement.

Another famous result is the following [MSS], [Ly2].

THEOREM 2.2. *Let $(f_\lambda)_{\lambda \in \Lambda}$ be as in Theorem 2.1. The stability locus Stab is dense in Λ.*

If (f_λ) is the family of all polynomials or all rational functions of degree d, it is conjectured that $\lambda \in \mathrm{Stab}$ if and only if all critical points converge to attracting cycles (the *hyperbolicity conjecture*). More generally, the work of McMullen and Sullivan [McMS] leads to a conjectural description of the components of the stability locus in any holomorphic family of rational maps (relying on the the so-called *no invariant line fields conjecture*).

We now explain how the bifurcation locus can be seen in a number of ways as the limit set (in the topological sense) of countable families of analytic subsets of codimension 1. This is a basic motivation for a description of the bifurcation locus in terms of positive closed currents of bidegree (1,1).

A marked critical point c is said to be *passive* in Ω if the family $(f_\lambda^n(c(\lambda)))_{n\geq 0}$ is normal and *active* at λ_0 if, for every neighborhood $V \ni \lambda_0$, c is not passive in V (this convenient terminology is due to McMullen). The characterization v. of stability in Theorem 2.1 then rephrases as "a critically marked family is stable iff all critical points are passive in Ω."

A typical example of a passive critical point is that of a critical point converging to an attracting cycle, for this property is robust under perturbations. If Λ is the family of all polynomials or all rational functions of degree d, according to the hyperbolicity conjecture, all passive critical points should be of this type. Indeed, any component of passivity would intersect the hyperbolicity locus. On the other hand, in a family of rational mappings with a persistent parabolic point (resp. a persistent Siegel disk), a critical point attracted by this parabolic point (resp. eventually falling in this Siegel disk) is passive.

Let (f_λ) be the space of all polynomials or rational mappings of degree d with a marked critical point c. It seems to be an interesting problem to study the geometry of *hyperbolic passivity components*,

that is, components of the passivity locus associated to c, where c converges to an attracting cycle. Does there exist a "center" in this component, that is a subvariety where c is periodic? How does the topology of the component related to that of its center?

The following result is an easy consequence of Montel's theorem (see, e.g., [Lev1] or [DF] for the proof).

THEOREM 2.3. *Let $(f_\lambda)_{\lambda\in\Lambda}$ be a holomorphic family of rational maps of degree d on $\mathbb{P}^1$, with a marked critical point c. If c is active at λ_0, then $\lambda_0 = \lim_n \lambda_n$, where for every n, $c(\lambda_n)$ is periodic (resp. falls onto a repelling cycle).*

Combined with Theorem 2.1 this implies that in any (not necessarily critically marked) holomorphic family, the family of hypersurfaces, defined by the condition that a critical point is periodic (resp. preperiodic) cluster on the bifurcation locus.

COROLLARY 2.4. *Let $(f_\lambda)_{\lambda\in\Lambda}$ be a holomorphic family of rational maps of degree d on $\mathbb{P}^1$. Then*

$$\mathrm{Bif} \subset \overline{\{\lambda, \exists\, c(\lambda)(pre)periodic\ critical\ point\}}$$

and, more precisely,

$$\mathrm{Bif} = \overline{\{\lambda, \exists\, c(\lambda)critical\ point\ falling\ non\text{-}persistently\ on\ a\ repelling\ cycle\}}.$$

At this point a natural question arises: assume that several marked critical points are active at λ_0. Is it possible to perturb λ_0 so that these critical points become simultaneously (pre)periodic? It turns out that the answer to the question is no, which is a manifestation of the failure of Montel's theorem in higher dimension (see Example 3.3). An important idea in higher-dimensional holomorphic dynamics is that the use of currents and pluripotential theory is a way to get around this difficulty. As it turns out, the theory of bifurcation currents will indeed provide a reasonable understanding of this problem.

The following simple consequence of item (i) of Theorem 2.1 provides yet another dense codimension 1 phenomenon in the bifurcation locus.

COROLLARY 2.5. *Let* $(f_\lambda)_{\lambda \in \Lambda}$ *be a holomorphic family of rational maps of degree d on* $\mathbb{P}^1$. *Then for every* $\theta \in \mathbb{R}/2\pi\mathbb{Z}$,

$$\text{Bif} = \overline{\{\lambda, \ f_\lambda \text{ admits a non-persistent periodic point of multiplier } e^{i\theta}\}}.$$

Again one might ask: what is the set of parameters possessing *several* non-persistent indifferent periodic points?

2.2 The bifurcation current

It is a classical observation that in a holomorphic family of dynamical systems, Lyapunov exponents often depend subharmonically on parameters (for instance, this idea plays a key role in [He]). Also, since the 1980's, potential theoretic methods appear to play an important role in the study the quadratic family and the Mandelbrot set (see, e.g., [DH] and [Sib]).

DeMarco made this idea more systematic by putting forward the following definition [DeM2].

PROPOSITION-DEFINITION 2.6. *Let* $(f_\lambda)_{\lambda \in \Lambda}$ *be a holomorphic family of rational maps of degree* $d \geq 2$. *For each* $\lambda \in \Lambda$, *let* μ_λ *be the unique measure of maximal entropy of* f_λ. *Then*

$$\chi \colon \lambda \longmapsto \chi(f_\lambda) = \int \log \|f_\lambda'\| \, d\mu_\lambda$$

is a continuous psh function on Λ *(here the norm of the differential is relative to any Riemannian metric on* $\mathbb{P}^1$).

The bifurcation current X *of the family* (f_λ) *is by definition* $T_{\text{bif}} = dd^c \chi$.

The continuity of χ was originally proven by Mañé [Mñ]. Actually χ is Hölder continuous, as can easily be seen from DeMarco's formula for χ (see Footnote 3), and the joint Hölder continuity in (λ, z) of the dynamical Green's function (see also [DS3, §2.5] for another approach).

The significance of this definition is justified by the following result [DeM1, DeM2].

THEOREM 2.7 (DeMarco). *The support of* T_{bif} *is equal to* Bif.

PROOF (SKETCH). The easier inclusion is the fact that $\text{Supp}(T_{\text{bif}}) \subset \text{Bif}$, or equivalently, that χ is pluriharmonic on the stability locus. A neat way to see this is to use the following approximation formula, showing that the Lyapunov exponent of f_λ can be read on periodic orbits: for every rational map f of degree d,

$$\chi(f) = \lim_{n \to \infty} \frac{1}{d^n} \sum_{p \text{ repelling of period } n} \frac{1}{n} \log^+ |(f^n)'(p)| \tag{2.1}$$

(see Berteloot [Bt1] for the proof). It thus follows from characterization (i) of Theorem 2.1 that if (f_λ) is stable on some open set Ω, then χ is pluriharmonic there (notice that by Hartogs' Lemma the pointwise limit of a uniformly bounded sequence of pluriharmonic functions is pluriharmonic).

Let us sketch DeMarco's argument for the converse inclusion. It is no loss of generality to assume that all critical points are marked. The main ingredient is a formula relating χ and the value of the Green's function at critical points (see also Proposition 2.10). We do not state this formula precisely here[3] and note only that it generalizes the well-known formula due to Przytycki [Pr] (see also Manning [Man]) for the Lyapunov exponent of a monic polynomial P:

$$\chi(P) = \log d + \sum_{c \text{ critical}} G_P(c). \qquad (2.2)$$

Here G_P denotes the dynamical Green's function of P in $\mathbb{C}$, defined by

$$G_P(z) = \lim_{n \to \infty} d^{-n} \log^+ |P^n(z)|.$$

From this one deduces that if χ is pluriharmonic in some open set Ω, then all critical points are passive in Ω; hence, the family is stable by Theorem 2.1. □

A first consequence of this result, which was a source of motivation in [DeM1], is that if Λ is a Stein manifold (e.g. an affine algebraic manifold), then the components of the stability locus are also Stein.

In the most studied quadratic family $(P_\lambda(z)) = (z^2 + \lambda)_{\lambda \in \mathbb{C}}$, we see from (2.2) that the bifurcation "current" (which is simply a measure in this case) is defined by the formula $\mu_{\text{bif}} = dd^c G_{P_\lambda}(0)$, where G_{P_λ} is the Green's function. As expected, we recover the usual parameter space measure, that is, the harmonic measure of the Mandelbrot set.

2.3 Marked critical points

In this paragraph we present a construction due to Favre and the author [DF] of a current associated to the bifurcations of a marked critical point. It would also be possible to define this current by lifting the dynamics to $\mathbb{C}^2 \setminus \{0\}$ and evaluating appropriate dynamical Green's function at the lifted critical points, in the spirit of [DeM1]. However our construction is more intrinsic and generalizes to other situations.

Let $(f_\lambda, c(\lambda))$ be a holomorphic family of rational maps of degree $d \geq 2$ with a marked critical point. Let $f_n \colon \Lambda \to \mathbb{P}^1$ be defined by $f_n(\lambda) = f_\lambda^n(c(\lambda))$. Let ω be a Fubini-Study form on $\mathbb{P}^1$.

In the spirit of Section 1 we have the following result.

THEOREM 2.8. *Let (f, c) be a holomorphic family of rational maps of degree $d \geq 2$ with a marked critical point, and set $f_n \colon \lambda \mapsto f_\lambda^n(c(\lambda))$. Then the sequence of currents $(d^{-n} f_n^* \omega)$ converges to a current T_c on Λ. The support of T_c is the activity locus of c.*

By definition, T_c is *the bifurcation current* (also referred to as the *activity current*) associated to (f, c).

PROOF (SKETCH). The convergence of the sequence of currents $(d^{-n} f_n^* \omega)$ does not follow from the general formalism of Section 1. The proof relies on equidistribution results for preimages of points under f^n instead. For this, it is convenient to consider the product dynamics on $\Lambda \times \mathbb{P}^1$. Let $\widehat{\Lambda} = \Lambda \times \mathbb{P}^1$. The family f_λ

---

[3] The expression of DeMarco's formula for a rational map f is $\chi(f) = \sum_{i=1}^{2d-2} G_F(c_j) + H(f)$, where G_F is the dynamical Green's function of a homogeneous lift $F \colon \mathbb{C}^2 \to \mathbb{C}^2$ of f, the c_j are certain lifts of the critical points, and $H(f)$ depends pluriharmonically on f.

lifts to a holomorphic map $\widehat{f} \colon \widehat{\Lambda} \to \widehat{\Lambda}$ mapping (λ, z) to $(\lambda, f_\lambda(z))$. We denote by $\pi_1 \colon \widehat{\Lambda} \to \Lambda$ and $\pi_2 \colon \widehat{\Lambda} \to \mathbb{P}^1$ the natural projections, and let $\widehat{\omega} = \pi_2^* \omega$.

The following proposition follows from classical techniques in higher-dimensional holomorphic dynamics.

PROPOSITION 2.9. *Let (f_λ) be a holomorphic family of rational maps of degree $d \geq 2$ and $\widehat{f}$ be its lift to $\Lambda \times \mathbb{P}^1$, as just described. Then the sequence of currents $d^{-n} \widehat{f}^{n*} \widehat{\omega}$ converges to a limit $\widehat{T}$ in $\Lambda \times \mathbb{P}^1$.*

The current $\widehat{T}$ should be understood as "interpolating" the family of maximal measures μ_λ.

Let $\widehat{c} = \{(\lambda, c(\lambda)), \lambda \in \Lambda\} \subset \widehat{\Lambda}$ (resp. $\widehat{f}^n(\widehat{c})$) be the graph of c (resp. $f^n(c)$). As observed in Lemma 1.2, $d^{-n} f_n^* \omega = (\pi_1)_* \left(d^{-n} \widehat{\omega}|_{\widehat{f}^n(c)} \right)$. Now $\widehat{f}^n$ induces a biholomorphism $\widehat{c} \to \widehat{f}^n(\widehat{c})$, so $d^{-n} \widehat{\omega}|_{\widehat{f}^n(c)} = d^{-n}((\widehat{f}^n)^* \widehat{\omega})|_{\widehat{c}}$. Thus we see that the bifurcation current T_c is obtained by slicing $\widehat{T}$ by the hypersurface $\widehat{c}$ and projecting down to Λ: $T_c = (\pi_1)_* \left(\widehat{T}|_{\widehat{c}} \right)$.

Making this precise actually requires a careful control on the convergence of the sequence $(d^{-n}(\widehat{f}^n)^* \widehat{\omega})$ to $\widehat{T}$. We obtain this via a classical computation. Write $d^{-1} \widehat{f}^* \widehat{\omega} - \widehat{\omega} = dd^c g_1$ so then $d^{-n}(\widehat{f}^n)^* \widehat{\omega} - \widehat{\omega} = dd^c g_n$, where $g_n = \sum_{j=0}^{n-1} d^{-j} g_0 \circ \widehat{f}^j$. Therefore, (g_n) converges uniformly to g_∞, with $\widehat{\omega} + dd^c g_\infty = \widehat{T}$.

In particular, we have that $|g_n - g_\infty| = O(d^{-n})$, which implies that the assumption (1.2) in Proposition 1.10 is satisfied. Hence the family $(f_\lambda^n(c(\lambda))$ is quasinormal outside T_c. To see that it is actually normal, we notice that the uniform control on the potentials allows to apply Remark 1.5.

Conversely, $\mathrm{Supp}(T_c)$ is contained in the activity locus. Indeed it follows from the explicit expression of $d^{-n} f_n^* \omega$ in local coordinates on Λ that if $U \subset \Lambda$ is an open set intersecting $\mathrm{Supp}(T_c)$, the L^2 norm of the derivative of f_n (relative to the spherical metric on $\mathbb{P}^1$) grows exponentially in U. The result follows. $\square$

Observe that the fact that c is a critical point does not play any role here. We might as well associate activity/passivity loci and a bifurcation current to any holomorphically moving point $(a(\lambda))$, and Theorem 2.8 holds in this case (this type of consideration appears in [BaDeM], for example).

For the quadratic family $(P_\lambda(z)) = (z^2 + \lambda)_{\lambda \in \mathbb{C}}$, which has a marked critical point at 0, one easily checks that the current $\widehat{T}$ is defined in $\mathbb{C} \times \mathbb{C}$ by the formula $\widehat{T} = dd^c_{(\lambda, z)} G_{P_\lambda}(z)$, and that the bifurcation current associated to the critical point is again $dd^c_\lambda G_\lambda(0) = \mu_{\mathrm{bif}}$.

More generally, DeMarco's formula for the Lyapunov exponent of a rational map gives the relationship between these currents and the bifurcation current T_{bif} defned in §2.2.

PROPOSITION 2.10. *Let (f_λ) be a family of rational maps with all critical points marked $\{c_1, \ldots, c_{2d-2}\}$, then $T_{\mathrm{bif}} = sum T_i$, where T_i is the bifurcation current associated to c_i.*

2.4   Equidistribution of critically preperiodic parameters

We keep hypotheses as before, that is, we work with a family of rational maps with a marked critical point (f, c). Our goal is to show that the bifurcation current T_c is the limit in the sense of currents of sequences of dynamically defined codimension 1 subvarieties.

The first result follows directly from Theorem 1.6. It is a quantitative version of the fact that near an activity point, $f_\lambda^n(c(\lambda))$ assumes almost every value in $\mathbb{P}^1$.

THEOREM 2.11.   *Let $(f_\lambda, c(\lambda))_{\lambda \in \Lambda}$ be a holomorphic family of rational maps of degree $d \geq 2$ with a marked critical point, and T_c be the associated bifurcation current.*
   There exists a pluripolar exceptional set $\mathcal{E} \subset \mathbb{P}^1$ such that if a $\notin \mathcal{E}$, then

$$\lim_{n \to \infty} \frac{1}{d^n} [H_{n,a}] \to T_c, \text{ where } H_{n,a} = \{\lambda, \ f_\lambda^n(c(\lambda)) = a\}.$$

Note that $H_{n,a}$ is defined not only as a set, but as an analytic subvariety, with a possible multiplicity. It is likely that the size of the exceptional set can be estimated more precisely.

QUESTION 2.12.   *Is the exceptional set in Theorem 2.11 finite, as in the case of a single mapping?*

It is dynamically more significant to study the distribution of parameters for which c becomes periodic (resp. preperiodic), that is, to try to make Theorem 2.3 an equidistribution result. This is expected to be more difficult because in this case the set of targets that $f_\lambda^n(c(\lambda))$ is supposed to hit (the set of periodic points, say) is both countable and moving with λ.

Let $e \in \{0, 1\}$ be the cardinality of the exceptional set of f_λ for generic λ. If $e = 2$, then the family is trivial. If $e = 1$, it is conjugate to a family of polynomials. Given two integers $n > m \geq 0$ denote by $\mathrm{PerCrit}(n, m)$ the subvariety of Λ defined by the (non necessarily reduced) equation $f_\lambda^n(c(\lambda)) = f_\lambda^m(c(\lambda))$.

It is convenient to adopt the convention that $[\Lambda] = 0$. This means that if some subvariety V like $\mathrm{PerCrit}(n, k)$ turns out to be equal to Λ, then we declare that $[V] = 0$.

The following equidistribution theorem was obtained in [DF].

THEOREM 2.13 (Dujardin-Favre).   *Let (f, c) be a non-trivial holomorphic family of rational maps on $\mathbb{P}^1$ of degree $d \geq 2$, with a marked critical point, and denote by e the generic cardinality of the exceptional set. Assume furthermore that the following technical assumption is satisfied:*

(H) *For every $\lambda \in \Lambda$, there exists an immersed curve $\Gamma \subset \Lambda$ through λ such that the complement of the set $\{\lambda, c(\lambda) \text{ is attracted by a periodic cycle}\}$ is relatively compact in Γ.*

Then for every sequence $0 \leq k(n) < n$, we have that

$$\lim_{n \to \infty} \frac{[\mathrm{PerCrit}(n, k(n))]}{d^n + d^{(1-e)k(n)}} = T_c.$$

Notice that with our convention, if $c(\lambda)$ is periodic throughout the family, then both sides of the equidistribution equation vanish.

QUESTION 2.14. *Is assumption (H) really necessary?*

One might at least try to replace it by a more tractable condition like Λ being an algebraic family — compare with Theorem 4.13. It is easy to see that (H) holds, e.g., in the space of all polynomial or rational maps of degree d (see [DF]).

In the 1-parameter family $(z^d + \lambda)_{\lambda \in \mathbb{C}}$ of unicritical polynomials of degree d, the theorem implies the equidistribution of the centers of components of the degree d Mandelbrot set

$$\lim_{n \to \infty} \frac{1}{d^n} [\text{PerCrit}(n, 0)] = \lim_{n \to \infty} \frac{1}{d^n} \sum_{f_c^n(0) = 0} \delta_c = \mu_{\text{bif}}. \qquad (2.3)$$

This result had previously been proven by Levin in [Lev2] (see also McMullen [McM1]).

Another interesting approach to this type of equidistribution statements is to use arithmetic methods based on height theory (following work of Zhang, Autissier, Chambert-Loir, Thuillier, Baker-Rumely, and others). In particular a proof of (2.3) along these lines was obtained, prior to [DF], by Baker and Hsia in [BaH, Theorem 8.15]).

Since the varieties $\text{PerCrit}(n, 0)$ and $\text{PerCrit}(n, k)$ have generally many irreducible components (e.g., $\text{PerCrit}(n - k, 0) \subset \text{PerCrit}(n, k)$) and since it is difficult to control multiplicities, the theorem does not directly imply that T_c is approximated by parameters where c is genuinely preperiodic. To ensure this, we use a little trick based on the fact that T_c gives no mass to subvarieties (since it has local continuous potentials).

Denote by $\text{PreperCrit}(n, k) \subset \text{PerCrit}(n, k)$ the union of irreducible components of $\text{PerCrit}(n, k)$ (with their multiplicities) along which c is strictly preperiodic at generic parameters. As sets we have that

$$\text{PreperCrit}(n, k) = \overline{\text{PerCrit}(n, k) \setminus \text{PerCrit}(n - k, 0)}.$$

COROLLARY 2.15. *Under the assumptions of the theorem, if k is fixed, then*

$$\frac{1}{d^n + d^{(1-e)(n-k)}} [\text{PreperCrit}(n, n - k)] \to T_c \, .$$

PROOF. $[\text{PerCrit}(n, n - k)] - [\text{PreperCrit}(n, n - k)] = [D_n]$ is a sequence of effective divisors supported on $\text{PerCrit}(k, 0)$. Assume by contradiction that the conclusion of the corollary does not hold. In this case, T_c would give positive mass to $\text{PerCrit}(k, 0)$, which cannot happen. $\qquad \square$

Here is a heuristic geometric argument justifying the validity of Theorem 2.13. For each λ, (pre)periodic points equidistribute towards the maximal measure μ_λ [Ly1]. For this one "deduces" that in $\Lambda \times \mathbb{P}^1$, the sequence of integration currents on the hypersurfaces

$$\left\{ (\lambda, z) \in \Lambda \times \mathbb{P}^1, \ f_\lambda^n(z) = f_\lambda^{k(n)}(z) \right\},$$

conveniently normalized, converge to $\widehat{T}$. By "restricting" this convergence to the graph $\widehat{c}$, one gets the desired result. The trouble here is that there is no general result showing that the slices $\widehat{T}_n|_{\widehat{c}}$ converge to $\widehat{T}|_{\widehat{c}}$.

PROOF OF THEOREM 2.13 (SKETCH). Assume for simplicity that (f_λ) is a family of polynomials of degree d. A psh potential of $d^{-n}\mathrm{PerCrit}(n, k(n))$ is given by $d^{-n} \log \left| f_\lambda^n(c(\lambda)) - f_\lambda^{k(n)}(c(\lambda)) \right|$. We need to show that this sequence converges to $G_\lambda(c(\lambda))$, or equivalently that

$$\frac{1}{d^n} \log \left| f_\lambda^n(c(\lambda)) - f_\lambda^{k(n)}(c(\lambda)) \right| - \frac{1}{d^n} \log^+ \left| f_\lambda^n(c(\lambda)) \right| \xrightarrow[n \to \infty]{} 0 \text{ in } L^1_{\mathrm{loc}}(\Lambda) \qquad (2.4)$$

converges to zero. We argue by case-by-case analysis, depending on the behavior of c. For instance it is clear that (2.4) holds when c escapes to infinity or is attracted to an attracting cycle. On the other hand, there are parts of parameter space where the convergence is delicate to obtain directly. So instead we apply some potential-theoretic ideas (slightly reminiscent of the proof of Brolin's theorem [Bro]). One of these arguments is based on the maximum principle, and requires a certain compactness property leading to assumption (H).   □

The speed of convergence in Theorem 2.13 is unknown in general. Furthermore, the proof is ultimately based on compactness properties of the space of psh functions, so it is not well suited to obtain such an estimate. The only positive result in this direction is due to Favre and Rivera-Letelier [FRL], based on the method of [BaH], and concerns the unicritical family.

THEOREM 2.16 (Favre-Rivera Letelier). *Consider the unicritical family of polynomials $(z^d + \lambda)_{\lambda \in \mathbb{C}}$. Then for any any compactly supported C^1 function φ, if $0 \le k(n) < n$ is any sequence, as $n \to \infty$ we have*

$$\left| \langle \mathrm{PerCrit}(n, k(n)) - T_{\mathrm{bif}}, \varphi \rangle \right| \le C \left(\frac{n}{d^n} \right)^{1/2} \|\varphi\|_{C^1}.$$

2.5 Equidistribution of parameters with periodic orbits of a given multiplier

In [BB2] and [BB3], Bassanelli and Berteloot studied the distribution of parameters for which there exists a periodic cycle of a given multiplier. For this, given any holomorphic family of rational maps (f_λ), we need to define the subvariety $\mathrm{Per}(n, w)$ of parameter space defined by the condition that f_λ admits a cycle of exact period n and multiplier w. Doing this consistently as w crosses the value 1 requires a little bit of care. The following result, borrowed from [BB2], originates from the work of Milnor [Mi2] and Silverman [Sil].

THEOREM 2.17. *Let (f_λ) be a holomorphic family of rational maps of degree $d \ge 2$. Then for every integer n there exists a holomorphic function p_n on $\Lambda \times \mathbb{C}$, which is polynomial on $\mathbb{C}$, such that*

i. *for any $w \in \mathbb{C} \setminus \{1\}$, $p_n(\lambda, w) = 0$ if and only if f_λ admits a cycle of exact period n and of multiplier w;*

ii. *for $w = 1$, $p_n(\lambda, 1) = 0$ if and only if f admits a cycle of exact period n and of multiplier 1 or a cycle of exact period m whose multiplier is a primitive rth root of unity, and $n = mr$.*

We now put $\mathrm{Per}(n, w) = \{\lambda, \ p_n(\lambda, w) = 0\}$. The equidistribution result is the following [BB2, BB3] (recall our convention that $[\Lambda] = 0$).

THEOREM 2.18 (Bassanelli-Berteloot). *Let $(f_\lambda)_{\lambda \in \Lambda}$ be a holomorphic family of rational maps of degree $d \geq 2$.*

i. *For any $w \in \mathbb{C}$ such that $|w| < 1$, $\frac{1}{d^n}\left[\mathrm{Per}(n, w)\right] \to T_{\mathrm{bif}}$.*

ii. *Let $d\theta$ denote the normalized Lebesgue measure on $\mathbb{R}/2\pi\mathbb{Z}$. Then for every $r > 0$,*

$$\frac{1}{d^n} \int_{\mathbb{R}/2\pi\mathbb{Z}} \left[\mathrm{Per}(n, re^{i\theta})\right] d\theta \to T_{\mathrm{bif}}.$$

If, moreover, $(f_\lambda)_{\lambda \in \Lambda}$ is the family of all polynomials of degree d, then

iii. *for any w such that $|w| \leq 1$, $\frac{1}{d^n}\left[\mathrm{Per}(n, w)\right] \to T_{\mathrm{bif}}$.*

PROOF (EXCERPT). For fixed λ, the polynomial $w \mapsto p_n(\lambda, w)$ can be decomposed into

$$p_n(\lambda, w) = \prod_{i=1}^{N_d(n)} (w - w_j(\lambda)),$$

where the degree $N_d(n)$ satisfies $N_d(n) \sim \frac{d^n}{n}$ and the $w_j(\lambda)$ are the multipliers of the periodic cycles of period n of f_λ. For simplicity, in (i) let us only discuss the case where the multiplier w equals 0. We can write

$$\frac{1}{d^n}[\mathrm{Per}(n, 0)] = \frac{1}{d^n} dd^c \log |p_n(\lambda, 0)| = dd^c \left(\frac{1}{d^n} \sum_{i=1}^{N_d(n)} \log \left| w_j(\lambda) \right| \right).$$

We see that the potential of $\frac{1}{d^n}[\mathrm{Per}(n, 0)]$ is just the average value of the logarithms of the multipliers of repelling orbits. Now for n large enough, all cycles of exact period n are repelling, so from (2.1) we see[4] that the sequence of potentials converges pointwise to $\chi(f_\lambda)$. Furthermore, it is easy to see that this sequence is locally uniformly bounded from above, so by Hartogs' lemma it converges in L^1_{loc}. By taking dd^c we see that $\frac{1}{d^n}[\mathrm{Per}(n, 0)]$ converges to T_{bif}.

The argument for (ii) is similar. For simplicity assume that $r = 1$. We write

$$\frac{1}{d^n} \int \left[\mathrm{Per}(n, e^{i\theta})\right] d\theta = \frac{1}{d^n} dd^c \left(\int \log \left| p_n(\lambda, e^{i\theta}) \right| d\theta \right)$$

$$= dd^c \left(\frac{1}{d^n} \sum_{i=1}^{N_d(n)} \int \log \left| e^{i\theta} - w_j(\lambda) \right| d\theta \right)$$

$$= dd^c \left(\frac{1}{d^n} \sum_{i=1}^{N_d(n)} \log^+ \left| w_j(\lambda) \right| \right),$$

where the well-known formula $\log^+ |z| = \int \log \left| z - e^{i\theta} \right| d\theta$ gives the last equality. As before for each λ, when n is large enough all points of period n are repelling, so the potentials converge pointwise to χ. We conclude as before.

---

[4]The additional $\frac{1}{n}$ in that formula follows from the fact that in (2.1) the sum is over periodic points, while here we sum over periodic cycles.

The proof of (iii) is more involved since for $|w| = 1$, in the estimation of the potentials we have to deal with the possibility of cycles of large period with multipliers close to w. To overcome this difficulty, the authors use a global argument (somewhat in the spirit of the use of (H) in Theorem 2.13), requiring the additional assumption that (f_λ) is the family of polynomials. $\qquad\qquad\square$

It is a useful fact that in assertions (i) and (ii) of Theorem 2.18, no global assumption on Λ is required (see the comment following the proof of Theorem 3.2). On the other hand in (iii), it is expected that the convergence of $d^{-n}[\mathrm{Per}(n, w)]$ to T_{bif} (even for $|w| > 1$) holds in any family of rational mappings. Note that by using techniques similar to those of Theorem 1.6, it can be shown that in any family of rational maps, the set of $w \in \mathbb{C}$ violating the convergence in (iii) is polar.

3   HIGHER BIFURCATION CURRENTS AND THE BIFURCATION MEASURE

A crucial difference between one- and higher-dimensional families of rational mappings is the presence of a hierarchy of bifurcations according to the number of bifurcating critical points. These "higher bifurcation loci" are rather delicate to define precisely, and the main thesis in this section is that the formalism of bifurcation currents is well suited to deal with these questions.

In this respect, let us start by suggesting a certain "dictionary" of analogies between these issues and the dynamics of holomorphic endomorphisms of projective space $\mathbb{P}^k$, which turns out to be a very instructive guide for the intuition. Let f be a holomorphic self map of degree d on $\mathbb{P}^k$. There exists a natural invariant positive closed current T of bidegree $(1,1)$ satisfying $f^*T = dT$ (the Green's current), whose support is the Julia set of f [FS2]. In dimension 1, the dynamics of f is generically expanding along the Julia set. In higher dimension, the situation is more subtle in that that "the number of directions" along which the iterates are not equicontinuous can vary from 1 to k. One then introduces the following filtration of the Julia set

$$J_1 = J = \mathrm{Supp}(T) \supset \cdots \supset J_q = \mathrm{Supp}(T^q) \supset \cdots \supset J_k = \mathrm{Supp}(T^k) = \mathrm{Supp}(\mu),$$

where μ is the unique measure of maximal entropy. The dynamics on J_k is "repelling in all directions" according to the work of Briend and Duval [BD]. On the other hand for $q < k$, the dynamics along $J_q \setminus J_{q+1}$ is expected to be "Fatou in codimension q." It is not completely obvious how to formalize this precisely (see [Du3] for an account). A popular way to understand this is to conjecture that $J_q \setminus J_{q+1} = \mathrm{Supp}(T^q) \setminus \mathrm{Supp}(T^{q+1})$ is filled (in a measure theoretic sense) with holomorphic disks of codimension q along which the dynamics of (f^n) is equicontinuous.

In this section we will try to develop a similar picture for parameter spaces of polynomial and rational maps, with deformation disks playing the role of Fatou disks, and Misiurewicz parameters replacing repelling periodic points.

For cubic polynomials with marked critical points, it is also possible to draw a rather complete dictionary with the dynamics of polynomial automorphisms of $\mathbb{C}^2$ (see [Du2]).

3.1 Some general results

In this section, (f_λ) is a general holomorphic family of rational maps of degree $d \geq 2$. Our purpose here is to introduce the higher bifurcation currents T_{bif}^k and study some of their properties. We will try to demonstrate that their successive supports $\mathrm{Supp}(T_{\mathrm{bif}}^k)$, $1 \leq k \leq \dim(\Lambda)$ define a dynamically meaningful filtration of the bifurcation locus.

We first observe that it is harmless to assume that all critical points are marked. Indeed, let us take a branched cover $\pi \colon \widetilde{\Lambda} \to \Lambda$ such that the new family $\widetilde{f}(\widetilde{\lambda})$ has all critical points marked. We claim that for every $1 \leq k \leq \dim(\Lambda)$, (with obvious notation)

$$\pi^{-1} \mathrm{Supp}(T_{\mathrm{bif}}^k) = \mathrm{Supp}(\widetilde{T}_{\mathrm{bif}}^k).$$

Indeed the Lyapunov exponent function in $\widetilde{\Lambda}$ is $\widetilde{\lambda} \mapsto \chi(\pi(\widetilde{\lambda}))$. In particular, in any open subset $U \subset \widetilde{\Lambda}$ where π is a biholomorphism, for every k, we have that $\widetilde{\lambda} \in \mathrm{Supp}(\widetilde{T}_{\mathrm{bif}}^k)$ iff $\lambda = \pi(\widetilde{\lambda}) \in \mathrm{Supp}(T_{\mathrm{bif}}^k)$. Let now $\widetilde{B}$ denote the branching locus of π. Since χ is continuous, T_{bif}^k (resp. $\widetilde{T}_{\mathrm{bif}}^k$) gives no mass to analytic subsets, so we infer that

$$\mathrm{Supp}(\widetilde{T}_{\mathrm{bif}}^k) = \overline{\mathrm{Supp}(\widetilde{T}_{\mathrm{bif}}^k) \setminus \widetilde{B}} \quad \left(\text{resp. } \mathrm{Supp}(T_{\mathrm{bif}}^k) = \overline{\mathrm{Supp}(T_{\mathrm{bif}}^k) \setminus \pi(\widetilde{B})} \right).$$

Thus our claim follows.

Therefore, we assume that critical points are marked as $(c_1(\lambda), \ldots, c_{2d-2}(\lambda))$ and use $T_1, \ldots, T_{2d-2}$ to denote the respective bifurcation currents. Recall that $\mathrm{Supp}(T_i)$ is the activity locus of c_i and T_{bif} equals $\sum T_i$ by Proposition 2.10. Since the T_i are (1,1) positive closed currents with local continuous potentials, it is possible to wedge them (see [De]). Here is a first observation.

PROPOSITION 3.1. *For every* $1 \leq i \leq 2d - 2$, $T_i \wedge T_i = 0$.

PROOF. Assume that the convergence theorem 2.13 holds in Λ (e.g., if (H) holds). Then $T_i = \lim_{n \to \infty} d^{-n}[\mathrm{PerCrit}_i(n, 0)]$, where, of course, $\mathrm{PerCrit}_i(n, k)$ is the subvariety of parameters such that $f_\lambda^n(c_i(\lambda)) = f_\lambda^k(c_i(\lambda))$. Since T_i has continuous potentials, we infer that $T_i \wedge T_i = \lim_{n \to \infty} d^{-n}[\mathrm{PerCrit}_i(n, 0)] \wedge T_i$. Now for every n, $[\mathrm{PerCrit}_i(n, 0)] \wedge T_i = T_i|_{\mathrm{PerCrit}_i(n,0)}$ vanishes since c_i is passive on $\mathrm{PerCrit}_i(n, 0)$, and we conclude that $T_i \wedge T_i = 0$.

Without assuming Theorem 2.13, the proof is more involved and due to Gauthier [Ga].   □

As a consequence of this proposition, we infer that for every $1 \leq k \leq \dim(\Lambda)$,

$$T_{\mathrm{bif}}^k = k! \sum_{1 \leq i_1 < \cdots < i_k \leq 2d-2} T_{i_1} \wedge \cdots \wedge T_{i_k}. \tag{3.1}$$

In particular,

$$\mathrm{Supp}(T_{\mathrm{bif}}^k) \subset \{\lambda, \ k \text{ critical points are active at } \lambda\}. \tag{3.2}$$

As we will see in a moment (Example 3.3), this inclusion is in general *not* an equality. It thus becomes an interesting problem, still open in general, to characterize

$\mathrm{Supp}(T_{\mathrm{bif}}^{k})$. We will try to develop the idea that $\mathrm{Supp}(T_{\mathrm{bif}}^{k})$ is the set of parameters where k critical points are active and "behave independently."

The next result follows from Theorem 2.18.

THEOREM 3.2. *Let $(f_\lambda)_{\lambda \in \Lambda}$ be a holomorphic family of rational maps of degree $d \geq 2$. Then for every $k \leq \dim(\Lambda)$,*

$$\mathrm{Supp}(T_{\mathrm{bif}}^{k}) \subset \overline{\{\lambda, \, f_\lambda \text{ admits } k \text{ periodic critical points}\}}.$$

PROOF. We argue by decreasing induction on k, by using the following principle: if $V \subset \Lambda$ is a smooth analytic hypersurface and T a positive closed (1,1) current with continuous potential, then $T^k \wedge [V] = (T|_V)^k$. Here, as usual, the current $T|_V$ is defined by restricting the potential of T to V and taking dd^c.

Under the assumptions of the theorem, let $\lambda_0 \in \mathrm{Supp}(T_{\mathrm{bif}}^{k})$. Since

$$T_{\mathrm{bif}} = \lim_{n \to \infty} \frac{1}{d^n} [\mathrm{Per}(n,0)]$$

and T_{bif} has continuous potential, we infer that

$$T_{\mathrm{bif}}^{k} = \lim_{n \to \infty} \frac{1}{d^n} [\mathrm{Per}(n,0)] \wedge T_{\mathrm{bif}}^{k-1}.$$

In particular, λ_0 is approximated by parameters belonging to $\left(T_{\mathrm{bif}}|_{\mathrm{Per}(n,0)}\right)^{k-1}$ (moving slightly if necessary, we can always assume that these belong to the smooth part of $\mathrm{Per}(n,0)$).

We can now put $\Lambda_1 = \mathrm{Per}(n,0)$ and repeat the argument to find a nearby parameter belonging to $\mathrm{Supp}[\mathrm{Per}(n_1,0)] \wedge T_{\mathrm{bif}}^{k-2} \subset \Lambda_1$ for some (possibly much larger) n_1, etc.   □

We see that it is important in this argument that no special assumption on Λ is needed in Theorem 2.18. For instance if one were to replace "periodic" by "strictly preperiodic" in this theorem, and try to use Theorem 2.13, one would have to check the validity of (H) in the restricted submanifolds, which needn't be satisfied (see, however, Theorem 3.8).

We can now explain how the inclusion in (3.2) can be strict.

EXAMPLE 3.3 (Douady). In the two-dimensional space of cubic polynomials with marked critical points, let $P_0(z) = z + z^2/2 + z^3$. We claim that the two critical points are active at P_0, but P_0 does not belong to $\mathrm{Supp}(T_{\mathrm{bif}}^{2})$.

Indeed, since P_0 is real and the critical points are not real, by symmetry, both critical points are attracted to the parabolic fixed point at the origin. Since the fixed point can be perturbed to become repelling (hence does not attract any critical point), Theorem 2.1(v) implies that at least one critical point must be active. Hence, by symmetry again, both critical points are active. On the other hand it can be proven (see [DF, Example 6.13] for details) that any nearby parameter admits an attracting (and not superattracting) fixed point. In particular P_0 cannot be perturbed to make both critical points periodic, therefore $P_0 \notin \mathrm{Supp}(T_{\mathrm{bif}}^{2})$.

Denote by $(P_\lambda)_{\lambda \in \Lambda \simeq \mathbb{C}^2}$ the family of cubic polynomials with marked critical points c_1 and c_2. We see that if N is a small neighborhood of the parameter 0, the values of $(P_\lambda^n(c_1(\lambda)), P_\lambda^n(c_2(\lambda)))$ for $\lambda \in N$ avoid an open set in $\mathbb{C}^2$. Indeed for $\lambda \in N$, either c_1 or c_2 must be attracted by an attracting cycle, so for large n, $P_\lambda^n(c_1(\lambda))$ and $P_\lambda^n(c_2(\lambda))$ cannot be simultaneously large. This is a manifestation of the Fatou-Bieberbach phenomenon (failure of Montel's theorem in higher dimension). □

One can also approximate $\mathrm{Supp}(T_{\mathrm{bif}}^k)$ by parameters possessing k indifferent periodic cycles. For $N_k = (n_1, \ldots n_k) \in \mathbb{N}^k$ and $\Theta_k = (\theta_1, \cdots, \theta_k) \in (\mathbb{R}/\mathbb{Z})^k$, we denote by $\mathrm{Per}_k(N_k, e^{i\Theta_k})$ the union of codimension k irreducible components of

$$\mathrm{Per}(n_1, e^{i\theta_1}) \cap \mathrm{Per}(n_2, e^{i\theta_2}) \cap \cdots \cap \mathrm{Per}(n_k, e^{i\theta_k}),$$

and if $E \subset \mathbb{R}/\mathbb{Z}$, we let

$$\mathcal{Z}_k(E) = \bigcup_{N_k \in \mathbb{N}^k, \Theta_k \in E^k} \mathrm{Per}_k(N_k, e^{i\Theta_k}),$$

which is the (codimension p part of the) set of parameters possessing k neutral cycles with respective multipliers in E.

The following result is due to Bassanelli and Berteloot [BB1].

THEOREM 3.4 (Bassanelli-Berteloot). *Let* $(f_\lambda)_{\lambda \in \Lambda}$ *be a holomorphic family of rational maps of degree* $d \geq 2$. *If* $E \subset \mathbb{R}/\mathbb{Z}$ *is any dense subset, then for every* $k \leq \dim(\Lambda)$,

$$\mathrm{Supp}(T_{\mathrm{bif}}^k) \subset \overline{\mathcal{Z}_k(E)}.$$

PROOF (SKETCH). We argue by induction on k. For $k = 1$, the result follows from Theorem 2.1. The main idea of the induction step is as follows: first, assume that $\lambda_0 \in \mathrm{Supp}(T_{\mathrm{bif}}^k)$. Then by the induction hypothesis, $\lambda \in \overline{\mathcal{Z}_{k-1}(E)}$, so there are plenty of $(k-1)$-dimensional disks near λ_0 along which f_λ possesses $(k-1)$ neutral cycles with multipliers in E. Assume that the dynamics is J-stable along these disks. Then the Lyapunov exponent function χ is pluriharmonic there, and it follows from a general pluripotential theoretic lemma (see [FS2, Lemma 6.10]) that $(dd^c\chi)^k = 0$ in the neighborhood of λ_0, a contradiction. So the dynamics is not J-stable along the disks of $\mathcal{Z}_{k-1}(E)$; hence, Theorem 2.1 produces one more neutral cycle, thereby proving the result. □

We see that in Example 3.3, the two critical points are related in a rather subtle way. On the other hand, under a certain transversality assumption, a clever argument of similarity between parameter and dynamical spaces, due to Buff and Epstein, shows that certain parameters belong to $\mathrm{Supp}(T_{\mathrm{bif}}^k)$ [BE].

Let $\lambda_0 \in \Lambda$ be a parameter where $c_1(\lambda_0), \ldots, c_k(\lambda_0)$ fall onto repelling cycles. More precisely, we assume there are repelling periodic points $p_1(\lambda_0), \ldots, p_k(\lambda_0)$ and integers $n_1, \ldots, n_k$ such that for $1 \leq j \leq k$, $f_{\lambda_0}^{n_j}(c_j(\lambda_0)) = p_j(\lambda_0)$. The repelling orbits $p_j(\lambda_0)$ can be uniquely continued to repelling periodic orbits $p_j(\lambda)$ for λ in some neighborhood of λ_0. Fix, for each j, a coordinate chart on $\mathbb{P}^1$ containing $p_j(\lambda_0)$, so that for nearby λ, the function $\chi_j \colon \lambda \mapsto f_\lambda^{n_j}(c_j(\lambda)) - p_j(\lambda)$ is well defined. We say that the critical points $c_j(\lambda_0)$ *fall transversely* onto the respective repelling points $p_j(\lambda_0)$ if the mapping $\chi \colon \Lambda \to \mathbb{C}^k$ defined in the neighborhood of λ_0 by $\chi = (\chi_1, \ldots, \chi_k)$ has rank k at λ_0. Of course, this notion does not depend on the choice of coordinate charts.

THEOREM 3.5 (Buff-Epstein). *Let $(f_\lambda)_{\lambda \in \Lambda}$ be a holomorphic family of rational maps of degree $d \geq 2$ with marked critical points $c_1, \ldots, c_k$, and associated bifurcation currents $T_1, \ldots, T_k$. Let λ_0 be a parameter at which $c_1(\lambda_0), \ldots, c_k(\lambda_0)$ fall transversely onto repelling cycles. Then $\lambda_0 \in \mathrm{Supp}(T_1 \wedge \cdots \wedge T_k)$.*

Notice that this theorem does not appear in this form in [BE]. Our presentation borrows from [Bt2]. The validity of the transversality assumption will be discussed in various situations in Sections 3.2 and 3.3.

PROOF (SKETCH). First, a slicing argument shows that it is enough to prove that for a generic k-dimensional subspace $\Lambda' \ni \lambda_0$ (relative to some coordinate chart in Λ), the result holds by restricting to Λ'. Thus we can assume that $\dim(\Lambda) = k$ and that χ is a local biholomorphism on Λ. To simplify notation, we will assume that the f_j are polynomials, $k = 2$, $n_j = 1$ and that the p_j are fixed points.

Taking adapted coordinates (λ_1, λ_2) in Λ (in which the initial parameter λ_0 is 0) we can assume that $\chi_1(\lambda) = \lambda_1 + h.o.t.$ and $\chi_2(\lambda) = \lambda_2 + h.o.t.$ The proof, based on a renormalization argument, consists in estimating the mass, relative to the measure $T_1 \wedge T_2$, of small bidisks about 0 of carefully chosen size. Specifically, we will show that

$$\liminf d^n (T_1 \wedge T_2) \left(D\left(0, \frac{\delta}{m_1^n}\right) \times D\left(0, \frac{\delta}{m_2^n}\right) \right) > 0,$$

where the m_j are the respective multipliers of the p_j. Let δ_n be the scaling map defined by

$$\delta_n(\lambda) = \left(\frac{\lambda_1}{m_1^n}, \frac{\lambda_2}{m_2^n} \right).$$

An easy computation based on transversality and the fact that f_λ is linearizable near $p_j(\lambda)$ shows that for $j = 1, 2$, $f_{\delta_n(\lambda)}^{n+1}(c_j(\delta_n(\lambda)))$ converges as $n \to \infty$ to a non-constant map ψ_j *depending only on λ_j*, on some disk $D(0, \delta)$, with $\psi_j(0) = p_j$. Hence, since G_λ depends continuously on λ we get that $G_{\delta_n(\lambda)}(f_{\delta_n(\lambda)}^{n+1}(c_j(\delta_n(\lambda))))$ converges to $G_0 \circ \psi_j(\lambda_j)$. Using the invariance relation for the Green's function, we conclude that

$$d^{n+1} G_{\delta_n(\lambda)}(c_j(\delta_n(\lambda))) = G_{\delta_n(\lambda)}\left(f_{\delta_n(\lambda)}^{n+1}(c_j(\delta_n(\lambda)))\right) \xrightarrow[n \to \infty]{} G_0(\psi_j(\lambda_j)); \qquad (3.3)$$

hence,

$$(T_1 \wedge T_2)(\delta_n(D(0, \delta)^2))$$
$$\simeq d^{-2(n+1)} \int_{D(0,\delta)^2} dd^c \left(G_0 \circ \psi_1(\lambda_1) \right) \wedge dd^c \left(G_0 \circ \psi_2(\lambda_2) \right)$$
$$= d^{-2(n+1)} \left(\int_{D(0,\delta)} dd^c \left(G_0 \circ \psi_1(\lambda_1) \right) \right) \left(\int_{D(0,\delta)} dd^c \left(G_0 \circ \psi_2(\lambda_2) \right) \right)$$

by Fubini's theorem. Note that the first line of this equation is justified by the local uniform convergence in (3.3). Finally, the integrals on the second line are positive since G_0 is not harmonic near p_j, so $G_0 \circ \psi_j$ is not harmonic near the origin. □

Building on similar ideas, Gauthier [Ga] relaxed the transversality assumption in Theorem 3.5 as follows. Assume as before that $\lambda_0 \in \Lambda$ is a parameter where $c_1(\lambda_0), \ldots, c_k(\lambda_0)$ fall onto respective repelling periodic points $p_1(\lambda_0), \ldots p_k(\lambda_0)$. Define $\chi \colon \Lambda \to \mathbb{C}^k$ as before Theorem 3.5. We say that the critical points $c_j(\lambda_0)$ fall *properly* onto the respective repelling points $p_j(\lambda_0)$ if $\chi^{-1}(0)$ has codimension k at λ_0. To say it differently, we are requesting that in $\Lambda \times (\mathbb{P}^1)^k$ the graphs of the two mappings $\lambda \mapsto (p_1(\lambda), \ldots p_k(\lambda))$ and $\lambda \mapsto (f_\lambda^{m_1}(c_1(\lambda)), \ldots, f_\lambda^{n_k}(c_k(\lambda)))$ intersect properly[5] at $(\lambda_0, p_1(\lambda_0), \ldots p_k(\lambda_0))$.

THEOREM 3.6 (Gauthier). *Let $(f_\lambda)_{\lambda \in \Lambda}$ be a holomorphic family of rational maps of degree $d \geq 2$ with marked critical points $c_1, \ldots, c_k$, and associated bifurcation currents $T_1, \ldots, T_k$. Let λ_0 be a parameter at which $c_1(\lambda_0), \ldots, c_k(\lambda_0)$ fall properly onto repelling cycles. Then $\lambda_0 \in \mathrm{Supp}(T_1 \wedge \cdots \wedge T_k)$.*

Actually, it is enough that $c_1(\lambda_0), \ldots, c_k(\lambda_0)$ fall properly into an arbitrary hyperbolic set. We refer to [Ga] for details.

The results in this section show that for $1 \leq k \leq \dim(\Lambda)$, $\mathrm{Supp}(T_{\mathrm{bif}}^k)$ is a reasonable candidate for the locus of "bifurcations of order k." We will see in the next sections that when Λ is the space of all polynomials or rational maps and k is maximal, $\mathrm{Supp}(T_{\mathrm{bif}}^k)$ can be characterized precisely. The picture is not yet complete in intermediate codimensions. In this respect let us state a few open questions.

QUESTION 3.7. *Let $(f_\lambda)_{\lambda \in \Lambda}$ be a holomorphic family of rational maps on $\mathbb{P}^1$ of degree $d \geq 2$, with marked critical points $c_1, \ldots, c_k$.*

1. *Is it true that*

$$\mathrm{Supp}(T_1 \wedge \cdots \wedge T_k)$$
$$= \overline{\{\lambda_0, \ c_1(\lambda_0), \cdots, c_k(\lambda_0) \text{ fall transversely onto repelling cycles}\}} \ ?$$

   (By Theorem 3.5 only the inclusion $\subset$ needs to be established.)

2. *More generally, do the codimension k subvarieties*

$$\mathrm{PerCrit}_1(n, k(n)) \cap \cdots \cap \mathrm{PerCrit}_k(n, k(n))$$

   equidistribute toward $T_1 \wedge \cdots \wedge T_k$? Arithmetic methods could help here, especially when Λ is the space of polynomials and k is maximal (see the next paragraph).

3. *Is the following characterization of $\mathrm{Supp}(T_1 \wedge \cdots \wedge T_k)$ true:*
   $\lambda_0 \in \mathrm{Supp}(T_1 \wedge \cdots \wedge T_k)$ if and only if for every neighborhood U of λ_0, there exists a pluripolar subset $\mathcal{E} \subset (\mathbb{P}^1)^k$ such that the values of $f_\lambda^n(c_j(\lambda))$ for $n \in \mathbb{N}$ and $\lambda \in U$ cover $(\mathbb{P}^1)^k \setminus \mathcal{E}$?

Work in progress of the author indicates that the answer to Question 3.7(1) should be yes.

---

[5]This is the usual terminology in intersection theory; see [Ful, Ch].

3.2 The space of polynomials

In this section we specialize the discussion to the case where Λ is the space of polynomials of degree d and will mostly concentrate on the maximal exterior power of the bifurcation current (the *bifurcation measure*). Our purpose is to show that it is in many respects the right analogue in higher degree of the harmonic measure of the Mandelbrot set. All results except Theorem 3.11 come from [DF].

The space $\mathcal{P}_d$ of polynomials of degree d with marked critical points is a singular affine algebraic variety. To work on this space, in practice we use an "orbifold parameterization" (not injective) $\pi \colon \mathbf{C}^{d-1} \to \mathcal{P}_d$, defined as follows: π maps $(c_1, \cdots, c_{d-2}, a) \in \mathbf{C}^{d-1}$ to the primitive of $z \prod_1^{d-2}(z - c_i)$ whose value at 0 is a^d. In coordinates, denoting $c = (c_1, \ldots, c_{d-2})$, we get

$$\pi(c, a) = P_{c,a}(z) = \frac{1}{d} z^d + \sum_{j=2}^{d-1} (-1)^{d-j} \sigma_{d-j}(c) \frac{z^j}{j} + a^d , \qquad (3.4)$$

where $\sigma_i(c)$ is the elementary symmetric polynomial in the $\{c_j\}_1^{d-2}$ of degree i. The critical points of $P_{c,a}$ are $\{0, c_1, \cdots, c_{d-2}\}$. We put $c_0 = 0$.

The choice of this parameterization (inspired from that used by Branner and Hubbard in [BH1]) is motivated by the fact that the bifurcation currents T_i associated to the c_i have the same projective mass. Furthermore, it is well suited to understanding the behavior at infinity of certain parameter space subsets (this will be used to check condition (H) of Theorem 2.13).

We let $\mathcal{C}$ be the connectedness locus, which is compact in $\mathbf{C}^{d-1}$ by [BH1]. For $1 \leq i \leq d - 2$ we also define the closed subsets $\mathcal{C}_i$ by

$$\mathcal{C}_i = \{(c, a), \ c_i \text{ has bounded orbit}\} .$$

It is clear that $\mathcal{C} = \bigcap_i \mathcal{C}_i$ and that $\partial \mathcal{C}_i$ is the activity locus of c_i.

We also let $g_{c,a}$ be the Green's function of the polynomial $P_{c,a}$. Then $T_i = dd^c g_i$, where $g_i = g_{c,a}(c_i)$. Recall that the Manning-Przytycki formula (2.2) asserts that $\chi = \log d + \sum_{i=0}^{d-2} g_i$; hence, $T_{\mathrm{bif}} = \sum_{i=0}^{d-2} T_i$.

In this specific situation we are able to solve the problem raised after Theorem 3.2.

THEOREM 3.8.   *In $\mathcal{P}_d$, for every $1 \leq k \leq d - 1$,*

$$\mathrm{Supp}(T_{\mathrm{bif}}^k) \subset \overline{\{\lambda, \ f_\lambda \text{ admits } k \text{ strictly prepériodic critical points}\}}.$$

More precisely, for any collection of integers $i_1, \cdots, i_k \in \{0, \cdots, d - 2\}$, the analytic subset $W_{n_1,\ldots,n_k} = \bigcap_{j=1}^k \mathrm{PreperCrit}_{i_j}(n_j, n_j - 1)$ is of pure codimension k and

$$\lim_{n_k \to \infty} \cdots \lim_{n_1 \to \infty} \frac{1}{d^{n_k + \cdots + n_1}} [W_{n_1,\ldots,n_k}] = T_{i_1} \wedge \cdots \wedge T_{i_k} . \qquad (3.5)$$

Of course in (3.5) we can replace $\mathrm{PreperCrit}(n, n - 1)$ by $\mathrm{PerCrit}(n, k(n))$ for any sequence $k(n)$. As already observed, to prove this theorem one cannot simply take wedge products in Corollary 2.15 (resp. Theorem 2.13). The proof goes by induction on $\ell \leq k$, by successively applying this corollary to the parameter space $\Lambda = W_{n_1, \cdots, n_\ell}$. It is not obvious to check that assumption (H) is satisfied, and at this point the particular choice of the parameterization is useful (see also [Bt2] for neat computations). It is unknown whether in (3.5) one can take $n_1 = \cdots = n_k = n$; see Question 3.7(2).

Let us now focus on the maximal codimension case $k = d - 1$. We set

$$\mu_{\text{bif}} = T_0 \wedge \cdots \wedge T_{d-2} = \frac{1}{(d-1)!} T_{\text{bif}}^{d-1},$$

which is a probability measure supported on the boundary of the connectedness locus.

PROPOSITION 3.9. *The bifurcation measure is the pluripotential equilibrium measure of the connectedness locus. In particular,* $\text{Supp}(\mu_{\text{bif}})$ *is the Shiloff boundary of* $\mathcal{C}$.

It is important to understand that when $d \geq 3$, $\text{Supp}(\mu_{\text{bif}})$ is a proper subset of $\partial \mathcal{C}$. To get a (crude but instructive!) mental picture of the situation, think about the boundary of a polydisk in $\mathbb{C}^{d-1}$. This boundary can be decomposed into foliated pieces of varying dimension between 1 and $d - 2$, together with the unit torus, the unit torus $\mathbb{T}^{d-1}$ being the Shiloff boundary. The structure of $\partial \mathcal{C}$ should be somehow similar to this, with foliated pieces of dimension j corresponding to parts of $\partial \mathcal{C}$, where j critical points are passive. The precise picture is far from being understood, except for cubic polynomials (see Section 3.4).

It is a well-known open question whether in higher dimension the connectedness locus is the closure of its interior. Theorem 3.2 provides a partial answer to this question: if $\lambda_0 \in \partial \mathcal{C} \cap \text{Supp}(T_{\text{bif}}^j)$ and j critical points are active at λ_0, then $\lambda_0 \in \overline{\text{Int}(\mathcal{C})}$. Indeed, there exists a neighborhood $U \ni \lambda_0$ where $d - 1 - j$ critical points are passive, hence persistently do not escape, and by Theorem 3.2 there is a sequence of parameters $\lambda_n \to \lambda_0$ for which the j remaining critical points are periodic. Hence $\lambda_n \in \text{Int}(\mathcal{C})$ and we are done.

We see that the answer to the problem lies in the parameters in $\partial \mathcal{C} \cap \text{Supp}(T_{\text{bif}}^j)$ with more than j active critical points (like in Example 3.3). So far there does not seem to be any reasonable (even conjectural) understanding of the structure of this set of parameters.

We can give a satisfactory dynamical characterization of $\text{Supp}(\mu_{\text{bif}})$. A polynomial is said to be *Misiurewicz* if all critical points fall onto repelling cycles.

THEOREM 3.10. $\text{Supp}(\mu_{\text{bif}})$ *is the closure of the set of Misiurewicz parameters.*

The fact that $\text{Supp}(\mu_{\text{bif}})$ is contained in the closure of Misiurewicz polynomials follows from Theorem 3.8, since a strictly postcritically finite polynomial is automatically Misiurewicz. So the point here is to prove the converse. There are actually several proofs of this. The original one in [DF] uses landing of external rays (discussed following Theorem 3.11). Another proof goes by observing that the properness assumption of Theorem 3.6 is satisfied at every Misiurewicz parameter. Indeed, as already observed in Theorem 3.8, for every $(n_j)_{j=1,\ldots,d-1}$ and $(n_j)_{j=1,\ldots,d-1}$ with $k_j < n_j$, $\bigcap_{j=1}^{d-1} \text{PreperCrit}_j(n_j, k_j)$ is of dimension 0. Indeed, otherwise, this intersection would contain an analytic set contained in the connectedness locus, contradicting its compactness. Therefore, at a Misiurewicz point, critical points fall properly on the corresponding repelling points, and Theorem 3.6 applies.

Notice that the work of Buff and Epstein [BE] implies that these intersections are actually transverse.

Theorem 3.6 in its general form [Ga] shows that one can alternately characterize the support of μ_{bif} as being the closure of *generalized Misiurewicz polynomials,* where generalized here means that critical points fall into a hyperbolic set disjoint from the critical set. These considerations lead to a generalization in higher degree the well-known theorem of Shishikura on the Hausdorff dimension of the boundary of the Mandelbrot set [Sh]. Notice that Tan Lei [T] extended Shishikura's theorem by showing that the bifurcation locus has maximal Hausdorff dimension in any family of rational maps.

THEOREM 3.11 (Gauthier). *If U is any open set such that $U \cap \mathrm{Supp}(\mu_{\mathrm{bif}}) \neq \emptyset$, then $\mathrm{Supp}(\mu_{\mathrm{bif}}) \cap U$ has maximal Hausdorff dimension $2(d-1)$.*

We now discuss external rays, following [DF]. Let $P \in \mathcal{P}_d$ be a polynomial for which the Green's function takes the same value $r > 0$ at all critical points ($J(P)$ is then a Cantor set). The set Θ of external arguments of external rays landing at the critical points enables to describe in a natural fashion the combinatorics of P. This set of angles is known as the *critical portrait* of P. Now we can deform P by keeping Θ constant and let r vary in $\mathbb{R}_+^*$ (this is the "stretching" operation of [BH1]). This defines a ray in parameter space corresponding to the critical portrait Θ.

The set Cb of combinatorics/critical portraits is endowed with a natural Lebesgue measure μ_{Cb} coming from $\mathbb{R}/\mathbb{Z}$. It is easy to show that μ_{Cb}-a.e. ray lands as $r \to 0$ (this follows from Fatou's theorem on the existence of radial limits of bounded holomorphic functions). We thus obtain a measurable landing map $e \colon \mathrm{Cb} \to \mathcal{C}$. The measures μ_{Cb} and μ_{bif} are related as follows:

THEOREM 3.12. *μ_{bif} is the landing measure, i.e., $e_* \mu_{\mathrm{Cb}} = \mu_{\mathrm{bif}}$.*

The proof of Theorem 3.10 given in [DF] relies on a more precise landing theorem for "Misiurewicz combinatorics." A critical portrait $\Theta \in \mathrm{Cb}$ is said to be of Misiurewicz type if the external angles it contains are strictly preperiodic under multiplication by d. A combination of results due to Bielefeld, Fisher, and Hubbard [BFH] and Kiwi [Ki1] asserts that the landing map e is continuous at Misiurewicz combinatorics and that the landing point is a Misiurewicz point. It then follows from Theorem 3.12 that Misiurewicz points belong to $\mathrm{Supp}(\mu_{\mathrm{bif}})$.

The description of μ_{bif} in terms of external rays allows to generalize to higher dimensions a result of Graczyk-Świątek [GŚ] and Smirnov [Sm].

THEOREM 3.13. *The topological Collet-Eckmann property holds for μ_{bif}-almost every polynomial P.*

In particular for a μ_{bif}-a.e. P, we have that

- *all cycles are repelling;*

- *the orbit of each critical point is dense in the Julia set;*

- *$K_P = J_P$ is locally connected and its Hausdorff dimension is smaller than 2.*

Notice that Gauthier [Ga] shows that for a topologically-generic polynomial $P \in \mathrm{Supp}(\mu_{\mathrm{bif}})$, $\mathrm{HD}(J_P) = 2$. So the topologically- and metrically-generic pictures differ.

The connectedness of Cb naturally suggests the following generalization of the connectedness of the boundary of the Mandelbrot set.

QUESTION 3.14. *Is* $\mathrm{Supp}(\mu_{\mathrm{bif}})$ *connected?*

3.3 The space of rational maps

The space Rat_d of rational maps of degree d is a smooth complex manifold of dimension $2d + 1$, actually a Zariski open set of $\mathbb{P}^{2d+1}$. For $d = 2$ it is isomorphic to $\mathbb{C}^2$ [Mi2], and it was recently proven to be rational for all d by Levy [Lvy]. Automorphisms of $\mathbb{P}^1$ act by conjugation of rational maps, and the moduli space $\mathcal{M}_d$ is the quotient $\mathrm{Rat}_d / \mathrm{PSL}(2, \mathbb{C})$. Even if the action is not free, it can be shown that $\mathcal{M}_d$ is a normal quasiprojective variety of dimension $2(d - 1)$ [Sil]. Exactly as we did for polynomials, in order to work on this space it is usually convenient to work on a smooth family Λ which is transverse to the fibers of the projection $\mathrm{Rat}_d \to \mathcal{M}_d$. Through every point of Rat_d there exists such a family. Working in such a family, we can consider elements of $\mathcal{M}_d$ as rational maps rather than conjugacy classes.

In this section we give some properties of the bifurcation measure $\mu_{\mathrm{bif}} = T_{\mathrm{bif}}^{2(d-1)}$ on $\mathcal{M}_d$. The first basic result was obtained in [BB1].

PROPOSITION 3.15. *The bifurcation measure has positive and finite mass on* $\mathcal{M}_d$.

These authors also give a nice argument showing that all isolated Lattès examples belong to $\mathrm{Supp}(\mu_{\mathrm{bif}})$: it is known that a rational map is a Lattès example if and only if its Lyapunov exponent is minimal, that is, equal to $\log d/2$ [Led, Zd]. On the other hand, if u is a continuous psh function on a complex manifold of dimension k with a local minimum at x_0, then $x_0 \in \mathrm{Supp}((dd^c u)^k)$ (this follows from the so-called comparison principle [BT]). The result follows.

The precise characterization of $\mathrm{Supp}(\mu_{\mathrm{bif}})$ is due to Buff and Epstein [BE] and Buff and Gauthier [BG]. Let

- SPCF_d be the set of (conjugacy classes of) strictly post-critically finite rational maps;

- $\mathcal{Z}_d$ be the set of rational maps possessing $2d - 2$ indifferent cycles, without counting multiplicities (this is the maximal possible number).

THEOREM 3.16 (Buff-Epstein, Buff-Gauthier). *In* $\mathcal{M}_d$, $\mathrm{Supp}(\mu_{\mathrm{bif}}) = \overline{\mathrm{SPCF}_d} = \overline{\mathcal{Z}_d}$.

IDEA OF PROOF. It is no loss of generality to assume that all critical points are marked as $\{c_1, \ldots, c_{2d-2}\}$. Let SPCF_d^* be the set of (conjugacy classes of) strictly post-critically finite rational maps, which are not flexible Lattès examples.[6] We already know from Theorem 3.4 that $\mathrm{Supp}(\mu_{\mathrm{bif}}) \subset \overline{\mathcal{Z}_d}$. It should be possible to prove that $\mathrm{Supp}(\mu_{\mathrm{bif}}) \subset \overline{\mathrm{SPCF}_d}$ in the spirit of Theorem 3.8, but again it is not easy to check assumption (H). Instead Buff and Epstein prove directly that $\overline{\mathrm{SPCF}_d^*} = \overline{\mathcal{Z}_d}$ using the Mañé-Sad-Sullivan theorem.

Let us show that $\mathrm{SPCF}_d^* \subset \mathrm{Supp}(\mu_{\mathrm{bif}})$. If $f \in \mathrm{SPCF}_d^*$, it satisfies a relation of the form $f^{n_j + p_j}(c_j) = f^{n_j}(c_j)$, $j = 1, \ldots, 2d - 2$. Consider the subvariety of $\mathcal{M}_d$ defined by these algebraic equations. We claim that this subvariety is of dimension 0 at f. Indeed a theorem of McMullen [McM2] asserts that any stable algebraic family of rational maps is a family of flexible Lattès maps, hence the result. It then follows from Theorem 3.6 that $f \in \mathrm{Supp}(\mu_{\mathrm{bif}})$.

---

[6] A *flexible Lattès map* is a rational mapping descending from an integer multiplication on elliptic curve. These can be deformed with the elliptic curve, hence the terminology.

The original argument of [BE] was (under some mild restrictions on f) to check the transversality assumption of Theorem 3.5, using Teichmüller-theoretic techniques. Another proof of this fact was given by van Strien [vS].

Finally, by using an explicit deformation, Buff and Gauthier [BG] recently proved that every flexible Lattès map can be approximated by strictly post-critically finite rational maps which are not Lattès examples. Therefore, we have that $\overline{\mathrm{SPCF}_d^*} = \overline{\mathrm{SPCF}_d}$. $\hfill\square$

As in the polynomial case, in the previous result one may relax the critical finiteness assumption by requiring only that the critical points map to a hyperbolic set disjoint from the critical set. Similarly to Theorem 3.11, one thus gets the following [Ga].

THEOREM 3.17 (Gauthier). *The Hausdorff dimension of* $\mathrm{Supp}(\mu_{\mathrm{bif}})$ *equals* $2(2d - 2)$.

A famous theorem of Rees [Re] asserts that the set of rational maps that are ergodic with respect to Lebesgue measure is of positive measure in parameter space. It is then natural to ask: are these parameters inside $\mathrm{Supp}(\mu_{\mathrm{bif}})$?

3.4   Laminarity

A positive current T of bidegree (q, q) in a complex manifold M of dimension k is said to be *locally uniformly laminar* if in the neighborhood of every point of $\mathrm{Supp}(T)$ there exists a lamination by q dimensional disks embedded in M such that T is an average of integration currents over the leaves. More precisely, the restriction of T to a flow box of this lamination is of the form $\int_\tau [\Delta_t] dm(t)$, where τ is a local transversal to the lamination, m is a positive measure on τ, and Δ_t is the plaque through t.

A current T of bidegree (q, q) in M is said to be *laminar* if there exists a sequence of open subsets $\Omega_i \subset M$ and a sequence of currents T_i, respectively locally uniformly laminar in Ω_i, such that T_i increases to T as $i \to \infty$. Equivalently, T is laminar iff there exists a measured family $((\Delta_a)_{a \in \mathcal{A}}, m)$ of compatible holomorphic disks of dimension q in M such that $T = \int_{\mathcal{A}} [\Delta_a] dm(a)$. Here compatible means that the intersection of two disks in the family is relatively open (possibly empty) in each of the disks (i.e. the disks are analytic continuations of each other). It is important to note that for laminar currents there is no control on the geometry of the disks (even locally). These geometric currents appear rather frequently in holomorphic dynamics. The reader is referred to [BLS1] for a general account on this notion (see also [DG, Ca2]).

Why should we wonder about the laminarity of the bifurcation currents? We have been emphasizing the fact that $\mathrm{Supp}(T_{\mathrm{bif}}^k)$ is in a sense the locus of "bifurcations of order k." If this were true, it would mean that on $\mathrm{Supp}(T_{\mathrm{bif}}^k) \setminus \mathrm{Supp}(T_{\mathrm{bif}}^{k+1})$ we should see "stability in codimension k," that is, we should expect that it would be filled with submanifolds of codimension k where the dynamics is stable. There is a natural stratification of parameter space according to the dimension of the space of deformations, and our purpose is to compare this stratification with that of the supports of the successive bifurcation currents. Laminarity is the precise way to formulate this problem.

Let us be more specific. Throughout this section Λ is either the space of polynomials or the space of rational maps of degree d, with marked critical points if needed. We let $D = \dim(\Lambda)$. We say that two rational maps are deformations of each other if there is a J-stable family connecting them.[7]

In [McMS], McMullen and Sullivan ask for a general description of the way the deformation space of a given rational map embeds in parameter space. Our thesis is that these submanifolds tend to be organized into laminar currents.[8]

We start with a few general facts. The first easy proposition asserts that if T_{bif}^k is laminar on some set of positive measure outside $\mathrm{Supp}(T_{\mathrm{bif}}^{k+1})$, then the corresponding disks are indeed disks of deformations.

PROPOSITION 3.18. *Assume that S is a laminar current of bidegree (k,k) such that $S \leq T_{\mathrm{bif}}^k$ and $\mathrm{Supp}(S) \cap \mathrm{Supp}(T_{\mathrm{bif}}^{k+1}) = \emptyset$. Then the disks subordinate to S are disks of deformations.*

PROOF. By definition, a holomorphic disk of codimension k is said to be subordinate to S if there exists a non-zero locally uniformly laminar current $U \leq S$ such that Δ is contained in a leaf of U. Observe that since T_{bif}^{k+1} gives no mass to analytic sets, there are no isolated leaves in $\mathrm{Supp}(U)$. Since $U \leq T_{\mathrm{bif}}^k$ and $\mathrm{Supp}(U)$ is disjoint from $\mathrm{Supp}(T_{\mathrm{bif}}^{k+1})$, we see that $U \wedge T_{\mathrm{bif}} = 0$. Now if χ was not pluriharmonic along Δ, then by continuity, it wouldn't be harmonic on the nearby leaves of U, implying that $U \wedge T_{\mathrm{bif}}$ would be non-zero, a contradiction. $\square$

PROPOSITION 3.19. *For almost every parameter relative to the trace measure of T_{bif}^k, the dimension of the deformation space is at most $D - k$.*

In particular, a μ_{bif} generic parameter is rigid, that is, it admits no deformations.

We should expect the codimension to be almost everywhere equal to k on $\mathrm{Supp}(T_{\mathrm{bif}}^k) \setminus \mathrm{Supp}(T_{\mathrm{bif}}^{k+1})$.

PROOF. Similarly to Theorem 3.5, a slicing argument shows that is enough to consider the case of maximal codimension, that is, to show that for $T_{\mathrm{bif}}^D = \mu_{\mathrm{bif}}$-a.e. parameter the deformation space is zero dimensional.

If $f_0 \in \mathrm{Supp}(\mu_{\mathrm{bif}})$ possesses a disk Δ of deformations, then $\chi|_\Delta$ is harmonic. An already-mentioned pluripotential theoretic lemma [FS2, Lemma 6.10] asserts that if E is a measurable set with the property that through every point of E there exists a holomorphic disk along which χ is harmonic, then $(dd^c\chi)^D(E) = \mu_{\mathrm{bif}}(E) = 0$. The result follows. $\square$

Now let $\lambda_0 \in \mathrm{Supp}(T_{\mathrm{bif}}^k)$. There are at least k active critical points at f_{λ_0}. Assume the number is exactly k, say $c_1, \ldots, c_k$, so the remaining $D - k$ are passive ones. One might expect each of these passive critical points to give rise to a modulus of deformations of f_{λ_0}, but there is no general construction for this.

If these $D - k$ passive critical points lie in attracting basins, the existence of $D - k$ moduli of deformation for f_{λ_0} should follow from classical quasiconformal surgery techniques.

Observe that if the hyperbolicity conjecture holds, then $c_{k+1}, \ldots, c_D$ must be attracted by cycles. Indeed, let U be an open set where these points are passive.

---

[7]Notice that this is weaker than the notion considered in [McMS] (stability over the whole Riemann sphere), which introduces some distinctions which are not relevant from our point of view, like distinguishing the center from the other parameters in a hyperbolic component.

[8]We do not address the problem of the global holonomy of these laminations, which gives rise to interesting phenomena [Bra].

By Theorem 3.2, there exists $\lambda \in U$ such that $c_1(\lambda), \ldots, c_k(\lambda)$ are periodic. Hence, there is an open set $U' \subset U$ where all critical points are passive. Assuming the hyperbolicity conjecture, in U' all critical points lie in attracting basins. Thus this property persists for $c_{k+1}, \ldots, c_D$ throughout U.

When $D = 2$ and $k = 1$, one can indeed construct these deformations and relate them to the geometry of T_{bif}. The following is a combination of results of Bassanelli-Berteloot [BB2] and the author [Du2].

THEOREM 3.20. *If $\Lambda = \mathcal{P}_3$ or $\mathcal{M}_2$ and if U is an open set where one critical point is attracted by a cycle, then T_{bif} is locally uniformly laminar in U.*

In the general case one is led to the following picture.

CONJECTURE 3.21. *If $U \subset \Lambda$ is an open set where $D - k$ critical points (counted with multiplicity) lie in attracting basins, then T_{bif}^k is a laminar current in U, locally uniformly laminar outside a closed analytic set.*

The necessity of an analytic subset where the uniform laminar structure might have singularities is due to the possibility of (exceptional) critical orbit relations.

To address the question of laminarity of T_{bif}^k outside $\text{Supp}(T_{\text{bif}}^{k+1})$, we also need to analyze the structure of T_{bif}^k in the neighborhood of the parameters lying outside $\text{Supp}(T_{\text{bif}}^{k+1})$ but having more than k active critical points. As Example 3.3 shows, these parameters can admit deformations. There does not seem to be any reasonable understanding of the bifurcation current near these parameters, even for cubic polynomials. Another interesting situation is that of cubic polynomials with a Siegel disk Δ, such that a critical point falls in Δ after iteration. Of course in $\mathcal{P}_3$ both critical points are active. These parameters can be deformed by moving the critical point in the Siegel disk [Za]. We do not know whether these parameters belong to $\text{Supp}(\mu_{\text{bif}})$.

We also need to consider the possibility of "queer" passive critical points (whose existence contradicts the hyperbolicity conjecture, as seen above). In this case we have a positive result [Du2].

THEOREM 3.22. *If $\Lambda = \mathcal{P}_3$ and U is an open set where one critical point is passive, then T_{bif} is laminar in U.*

Perhaps unexpectedly, this result does not follow from the construction of some explicit deformation for a cubic polynomial with one active and one (queer) passive critical point. Instead, we use a general laminarity criterion due to De Thélin [DeT]: if a closed positive current T in $U \subset \mathbb{C}^2$ is the limit in the sense of currents of a sequence of integration currents $T = \lim d_n^{-1}[C_n]$ with $\text{genus}(C_n) = O(d_n)$, then T is laminar. We refer to [Du2] for the construction of the curves C_n.

Thus in the space of cubic polynomials, Theorems 3.20 and 3.22 show that T_{bif} is laminar outside the locus where two critical points are active (which is slightly larger than $\text{Supp}(\mu_{\text{bif}})$). In the escape locus $\mathbb{C}^2 \setminus \mathcal{C}$, where T_{bif} is uniformly laminar by Theorem 3.20, we can actually give a rather precise description of T_{bif}, which nicely complements the topological description given by Branner and Hubbard [BH1, BH2].

We are also able to show that this laminar structure really degenerates when approaching $\mathrm{Supp}(\mu_{\mathrm{bif}})$, in the sense that there cannot exist a set of positive transverse measure of deformation disks "passing through" $\mathrm{Supp}(\mu_{\mathrm{bif}})$.[9]

A consequence of this is that the genera of the curves $\mathrm{PerCrit}(n,k)$ must be asymptotically larger than 3^n near $\mathrm{Supp}(\mu_{\mathrm{bif}})$. We refer to [Du2] for details. The geometry of these curves was studied by Bonifant, Kiwi, and Milnor [Mi5, BKM] in a series of papers (see also the figures in [Mi1, §2] for some visual evidence of the complexity of the $\mathrm{PerCrit}(n,k)$ curves).

The results in this section suggest the following alternative characterization of the support of the bifurcation measure.

QUESTION 3.23. *If Λ is the moduli space of polynomials or rational maps of degree $d \geq 2$, is $\mathrm{Supp}(\mu_{\mathrm{bif}})$ equal to the closure of the set of rigid parameters?*

4   BIFURCATION CURRENTS FOR FAMILIES OF MÖBIUS SUBGROUPS

The famous Sullivan dictionary provides a deep and fruitful analogy between the dynamics of rational maps on $\mathbb{P}^1$ and that of Kleinian groups [Su1]. In this section we explain that bifurcation currents also make sense on the Kleinian group side, leading to interesting new results. Unless otherwise stated, all results are due to Deroin and the author [DD1, DD2].

4.1   Holomorphic families of subgroups of $\mathrm{PSL}(2, \mathbb{C})$

Here we gather some preliminary material and present the basic bifurcation theory of Möbius subgroups. We refer to the monographs of Beardon [Bea] and Kapovich [Kap] for basics on the theory of Kleinian groups. Throughout this section, G is a finitely generated group and Λ a connected complex manifold. We consider a holomorphic family of representations of G into $\mathrm{PSL}(2, \mathbb{C})$, that is, a mapping $\rho \colon \Lambda \times G \to \mathrm{PSL}(2, \mathbb{C})$, such that for fixed $\lambda \in \Lambda$, $\rho(\lambda, \cdot)$ is a group homomorphism, and for fixed $g \in G$ $\rho(\cdot, g)$ is holomorphic. The family will generally be denoted by $(\rho_\lambda)_{\lambda \in \Lambda}$.

We make three standing assumptions:

(R1) the family is non-trivial in the sense that there exists λ_1, λ_2 such that the representations ρ_{λ_i}, $i = 1, 2$ are not conjugate in $\mathrm{PSL}(2, \mathbb{C})$;

(R2) there exists $\lambda_0 \in \Lambda$ such that ρ_{λ_0} is faithful;

(R3) for every $\lambda \in \Lambda$, ρ_λ is non-elementary.

Assumptions (R1) and (R2) do not really restrict our scope: this is obvious for (R1), and for (R2) is suffices to take a quotient of G. Notice that under (R2), the representations ρ_λ are *generally faithful*, that is, the set of of non-faithful representations is a union of Zariski closed sets.

Recall that a representation is said to be *elementary* when it admits a finite orbit on $\mathbb{H}^3 \cup \mathbb{P}^1(\mathbb{C})$ ($\mathbb{H}^3$ is the 3-dimensional hyperbolic space). Then, either Γ fixes

---

[9]Milnor discusses in [Mi1, §3] the possibility of so-called product configurations in the connectedness locus of real cubic polynomials. Our result actually asserts that in the complex setting such configurations cannot be of positive μ_{bif} measure.

a point in $\mathbb{H}^3$ and is conjugate to a subgroup of $\mathrm{PSU}(2)$,[10] and, in particular, it contains only elliptic elements, or it has a finite orbit (with one or two elements) on $\mathbb{P}^1$. It can easily be proved that the subset of elementary representations of a given family (ρ_λ) is a real analytic subvariety E of Λ. Hence (R3) will be satisfied upon restriction to $\Lambda \setminus E$.[11]

We identify $\mathrm{PSL}(2,\mathbb{C})$ with the group of transformations which are of the form $\gamma(z) = \frac{az+b}{cz+d}$, with $\left(\begin{smallmatrix} a & b \\ c & d \end{smallmatrix}\right) \in \mathrm{SL}(2,\mathbb{C})$, and let

$$\|\gamma\| = \sigma(A^* A)^{1/2} \text{ and } \mathrm{tr}^2\, \gamma = (\mathrm{tr}\, A)^2,$$

where $A = \left(\begin{smallmatrix} a & b \\ c & d \end{smallmatrix}\right)$ is a lift of γ to $\mathrm{SL}(2,\mathbb{C})$, and $\sigma(\cdot)$ is the spectral radius. Of course these quantities do not depend on the choice of the lift.

As it is well-known, Möbius transformations are classified into three types according to the value of their trace:

- *parabolic* if $\mathrm{tr}^2(\gamma) = 4$ and $\gamma \neq \mathrm{id}$; it is then conjugate to $z \mapsto z + 1$;

- *elliptic* if $\mathrm{tr}^2(\gamma) \in [0,4)$; it is then conjugate to $z \mapsto e^{i\theta} z$ for some real number θ, and $\mathrm{tr}^2\, \gamma = 2 + 2\cos(\theta)$;

- *loxodromic* if $\mathrm{tr}^2(\gamma) \notin [0,4]$; it is then conjugate to $z \mapsto kz$, with $|k| \neq 1$.

There is a well-established notion of bifurcation for a family of Möbius subgroups, which is the translation in the Sullivan dictionary of Theorem 2.1. It has its roots in the work of Bers, Kra, Marden, Maskit, Thurston, and others on deformations of Kleinian groups.

THEOREM 4.1 (Sullivan [Su2] (see also Bers [Ber])). *Let $(\rho_\lambda)_{\lambda \in \Lambda}$ be a holomorphic family of representations of G into $\mathrm{PSL}(2,\mathbb{C})$ satisfying (R1–3), and let $\Omega \subset \Lambda$ be a connected open set. Then the following assertions are equivalent:*

i. *for every $\lambda \in \Omega$, $\rho_\lambda(G)$ is discrete;*

ii. *for every $\lambda \in \Omega$, ρ_λ is faithful;*

iii. *for every g in G, if for some $\lambda_0 \in \Omega$, $\rho_{\lambda_0}(g)$ is loxodromic (resp. parabolic, elliptic), then $\rho_\lambda(g)$ is loxodromic (resp. parabolic, elliptic) throughout Ω;*

iv. *for any λ_0, λ_1 in Ω the representations ρ_{λ_0} and ρ_{λ_1} are quasiconformally conjugate on $\mathbb{P}^1$.*

If one of these conditions is satisfied, we say that the family is *stable* in Ω. We define Stab to be the maximal such open set, and Bif to be its complement, so that $\Lambda = \mathrm{Stab} \cup \mathrm{Bif}$.

Theorem 4.1 shows that $\mathrm{Stab} = \mathrm{Int}(\mathrm{DF})$ is the interior of the set of discrete and faithful representations. One main difference with rational dynamics is that,

---

[10] Or $\mathrm{SO}(3,\mathbb{R})$ if we view $\mathbb{H}^3$ in its ball model.

[11] In order to study the space of all representations of G to $\mathrm{PSL}(2,\mathbb{C})$, it is nevertheless interesting to understand which results remain true when allowing a proper subset of elementary representations. This issue is considered in [DD1], but here for simplicity we only work with non-elementary representations.

as a consequence of the celebrated Jørgensen-Kazhdan-Margulis-Zassenhauss theorem, DF is a closed subset of parameter space. This is also referred to as Chuckrow's theorem; see [Kap, p. 170]. This implies that, whenever non-empty, Bif has non-empty interior, which is in contrast with Theorem 2.2.

The following corollary is immediate:

COROLLARY 4.2. *For every* $t \in [0, 4]$, *the set of such parameters* λ_0 *at which there exists* $g \in G$ *such that* $\mathrm{tr}^2 \rho_{\lambda_0}(g) = t$ *and* $\lambda \mapsto \mathrm{tr}^2 \rho_\lambda(g)$ *is not locally constant, is dense in* Bif.

Again, a basic motivation for the introduction of bifurcation currents is the study of the asymptotic distribution of such parameters. The most emblematic value of t is $t = 4$. In this case one either gets "accidental" new relations or new parabolic elements. Notice that when $t = 4\cos^2(\theta)$ with $\theta \in \pi\mathbb{Q}$ (e.g., $t = 0$), then if $\mathrm{tr}^2 \rho_{\lambda_0}(g) = t$, g is of finite order, so these parameters also correspond to accidental new relations in $\rho_\lambda(G)$.

A famous result in this area of research is a theorem by McMullen [McM3] which asserts that accidental parabolics are dense *in the boundary* of certain components of stability. One might also wonder what happens of Corollary 4.2 when additional assumptions are imposed on g. Here is a question (certainly well known to the experts) which was communicated to us by McMullen: if $G = \pi_1(S, *)$ is the fundamental group of a surface of finite type, does Corollary 4.2 remain true when restricting to the elements $g \in G$ corresponding to simple closed curves on S?

Another important feature of the space of all representations of G into $\mathrm{PSL}(2, \mathbb{C})$ (resp. modulo conjugacy) is that it admits a natural action of the automorphism group of $\mathrm{Aut}(G)$ (resp. the outer automorphism group $\mathrm{Out}(G)$). Despite recent advances, the dynamics of this action is not well understood (see the expository papers of Goldman [Go] and Lubotzky [Lu] for an account on this topic). There is a promising interplay between these issues and holomorphic dynamics, which was recently illustrated by the work of Cantat [Ca1].

4.2 Products of random matrices

To define a bifurcation current, we use a notion of Lyapunov exponent of a representation, arising from a random walk on G. The properties of this Lyapunov exponent will be studied using the theory of random walks on groups and random products of matrices (good references on these topics are [BL, Frm]).

Let us fix a probability measure μ on G, satisfying the following assumptions:

(A1) $\mathrm{Supp}(\mu)$ generates G as a semi-group;

(A2) there exists $s > 0$ such that $\int_G \exp(s \ \mathrm{length}(g)) d\mu(g) < \infty$.

The length in (A2) is relative to the choice of any finite system of generators of G; of course the validity of (A2) does not depend on this choice. A typical case where these assumptions are satisfied is that of a finitely supported measure on a symmetric set of generators (like in the case of the simple random walk on the associated Cayley graph).

Loosely speaking, the choice of such a measure a measure on G is somehow similar to the choice of a time parameterization for a flow, or more generally of a Riemannian metric along the leaves of a foliation.

We denote by μ^n the nth convolution power of μ, that is the image of $\mu^{\otimes n}$ under $(g_1, \ldots, g_n) \mapsto g_1 \cdots g_n$. This is the law of the nth step of the (left- or right-) random walk on G with transition probabilites given by μ.

If $\mathbf{g} = (g_n)_{n \geq 1} \in G^{\mathbb{N}}$, we let $l_n(\mathbf{g}) = g_n \cdots g_1$ be the product on the left of the g_k (resp. $r_n(\mathbf{g}) = g_1 \cdots g_n$ the product on the right). We denote by $\mu^{\mathbb{N}}$ the product measure on $G^{\mathbb{N}}$ so that $\mu^n = (l_n)_* \mu^{\mathbb{N}} = (r_n)_* \mu^{\mathbb{N}}$.

The *Lyapunov exponent* of a representation $\rho : G \to \mathrm{PSL}(2, \mathbb{C})$ is defined by the formula

$$\chi(\rho) := \lim_{n \to \infty} \frac{1}{n} \int_G \log \|\rho(g)\| \, d\mu^n(g)$$
$$= \lim_{n \to \infty} \frac{1}{n} \int \log \|\rho(g_1 \cdots g_n)\| \, d\mu(g_1) \cdots d\mu(g_n), \tag{4.1}$$

in which the limit exists by sub-additivity. It immediately follows from Kingman's sub-additive ergodic theorem that

$$\text{for } \mu^{\mathbb{N}}\text{-a.e. } \mathbf{g}, \ \lim_{n \to \infty} \frac{1}{n} \log \|\rho(l_n(\mathbf{g}))\| = \chi(\rho) \tag{4.2}$$

(this was before Kingman a theorem due to Furstenberg and Kesten [FK]).

The following fundamental theorem is due to Furstenberg [Fur1]:

THEOREM 4.3 (Furstenberg). *Let G be a finitely generated group and μ a probability measure on G satisfying (A1–2). Let $\rho : G \to \mathrm{PSL}(2, \mathbb{C})$ be a non-elementary representation. Then the Lyapunov exponent $\chi(\rho)$ is positive and depends continuously on ρ.*

The next result we need is due to Guivarc'h [Gui]:

THEOREM 4.4 (Guivarc'h). *Let G be a finitely generated group and μ a probability measure on G satisfying (A1–2). Let $\rho : G \to \mathrm{PSL}(2, \mathbb{C})$ be a non-elementary representation. Then for $\mu^{\mathbb{N}}$-a.e $\mathbf{g}$, we have that*

$$\frac{1}{n} \log |\mathrm{tr}(\rho(l_n(\mathbf{g})))| = \frac{1}{n} \log |\mathrm{tr}(\rho(g_n \cdots g_1))| \xrightarrow[n \to \infty]{} \chi(\rho). \tag{4.3}$$

A trivial remark which turns out to be a source of technical difficulties is that, in contrast with (4.2), one cannot in general integrate with respect to $\mu^{\mathbb{N}}$ in (4.3). The reason, of course, is that some words can have zero or very small trace. Conversely, if h is a function on $\mathrm{PSL}(2, \mathbb{C})$, which is bounded below and equivalent to $\log |\mathrm{tr}(\cdot)|$ as the trace tends to infinity, then one can integrate with respect to $\mu^{\mathbb{N}}$.

An example of such a function is given by the spectral radius, and under the assumptions of Theorem 4.4 we obtain that

$$\chi(\rho) = \lim_{n \to \infty} \frac{1}{n} \int \log |\sigma(\rho(g))| \, d\mu^n(g). \tag{4.4}$$

4.3 The bifurcation current

We are now ready for the introduction of the bifurcation current, following [DD1]. Let G be a finitely generated group, and $(\rho_\lambda)_{\lambda \in \Lambda}$ be a holomorphic family of representations satisfying (R1–3). Fix a probability measure μ on G satisfying (A1–2). It follows immediately from (4.1) that $\chi : \lambda \mapsto \chi(\rho_\lambda)$ is a psh function on Λ. Motivated by the analogy with rational dynamics, it is natural to suggest the following definition.

DEFINITION 4.5. *Let (G, ρ, μ) be as just described. The bifurcation current associated to (G, μ, ρ) is $T_{\text{bif}} = dd^c \chi$.*

At this point it is still unclear whether this gives rise to a meaningful concept. This will be justified by the following results.

First, it is easy to see that $\text{Supp}(T_{\text{bif}})$ is contained in the bifurcation locus. Indeed, if (ρ_λ) is stable on Ω, then the Möbius transformations $\rho_\lambda(g)$, $g \in G$, do not change type throughout Ω. In particular for every $g \in G$, $\Omega \ni \lambda \mapsto \log |\sigma(\rho_\lambda(g))|$ is plurisubharmonic. We thus infer from (4.4) that χ is plurisubharmonic on Ω.

It is a remarkable fact that, conversely, plurisubharmonicity of χ characterizes stability.

THEOREM 4.6. *Let (G, ρ, μ) be a holomorphic family of representations of G, satisfying (R1–3), endowed with a measure μ satisfying (A1–2). Then $\text{Supp}(T_{\text{bif}})$ is equal to the bifurcation locus.*

Here is a sketch of the proof, which consists of several steps and involves already encountered arguments. Without loss of generality, we may assume that $\dim(\Lambda) = 1$. We need to show that if Ω is an open set disjoint from $\text{Supp}(T_{\text{bif}})$, then (ρ_λ) is stable in Ω. The main idea is to look for a geometric interpretation of T_{bif}, in the spirit of what we did in §2.3.

For this we need a substitute for the equilibrium measure of a rational map: this will be the unique stationary measure under the action of $\rho_\lambda(G)$ on $\mathbb{P}^1$. Let us be more specific. For every representation, (G, μ) acts by convolution on the set of probability measures on $\mathbb{P}^1$ by the assignment

$$\nu \longmapsto \int \rho(g)_* \nu \, d\mu(g).$$

Any fixed point of this action is called a *stationary measure*. The following theorem is intimately related to Theorem 4.3. It is in a sense the analogue of the Brolin-Lyubich theorem in this context.

THEOREM 4.7. *Let G be a finitely generated group and μ a probability measure on G satisfying (A1–2). Let $\rho\colon G \to \text{PSL}(2, \mathbb{C})$ be a non-elementary representation. Then there exists a unique stationary probability measure ν on $\mathbb{P}^1$. Furthermore, for any $z_0 \in \mathbb{P}^1$, the sequence $\int (\rho_{\lambda_0}(g))_* \delta_{z_0} d\mu^n(g)$ converges to ν.*

In analogy with §2.3, let us now work in $\Lambda \times \mathbb{P}^1$, and consider the fibered action $\widehat{g}$ on $\Lambda \times \mathbb{P}^1$ defined by $\widehat{g}\colon (\lambda, z) \mapsto (\lambda, \rho_\lambda(g)(z))$. We seek for a current $\widehat{T}$ on $\Lambda \times \mathbb{P}^1$ "interpolating" the stationary measures. Given $z_0 \in \mathbb{P}^1$, we then introduce the sequence of positive closed currents $\widehat{T}_n$ defined by

$$\widehat{T}_n = \frac{1}{n} \int [\widehat{g}(\Lambda \times \{z_0\})] \, d\mu^n(g). \tag{4.5}$$

To understand why it is natural to have a linear normalization in (4.5), think of a polynomial family of representations. Precisely, assume that $\Lambda = \mathbb{C}$ and that for a set of generators $g^1, \ldots, g^k$ of G, the matrices $\rho_\lambda(g^j)$ are polynomial in λ. Then if w is a word in n letters in the generators, $\rho_\lambda(w)$ has degree $O(n)$ in λ; hence, in $\mathbb{C} \times \mathbb{P}^1 \subset \mathbb{P}^1 \times \mathbb{P}^1$, the degree of the graph $\widehat{w}(\Lambda \times \{z_0\})$ is $O(n)$.

It turns out that this sequence does *not* converge to a current interpolating the stationary measures (since, e.g., it vanishes above the stability locus) nevertheless it gives another crucial piece of information.

PROPOSITION 4.8. *The sequence of currents $\widehat{T}_n$ converges to $\pi_1^* T_{\text{bif}}$, where π_1 is the natural projection $\pi_1 \colon \Lambda \times \mathbb{P}^1 \to \Lambda$.*

Figure 4.9 is a visual interpretation of this result.

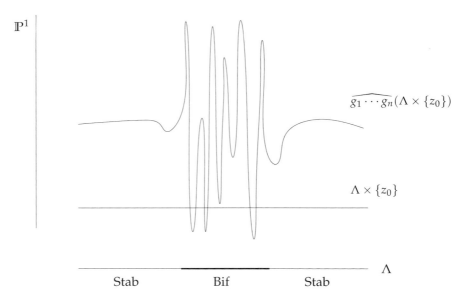

$$\widehat{g_1 \cdots g_n}(\Lambda \times \{z_0\})$$

$$\Lambda \times \{z_0\}$$

FIGURE 4.9: Fibered action of $g_1 \cdots g_n$

This proposition implies that if Ω is disjoint from $\text{Supp}(T_{\text{bif}})$, the average growth of the area of the sequence of graphs $\left(\widehat{g_1 \cdots g_n}\right)(\Lambda \times \{z_0\})$ over Ω is sublinear in n. The attentive reader will have noticed that the situation is similar to that of Section 1, except that the sequence of graphs over Λ is replaced by an average of graphs. The next result is in the spirit of Proposition 1.10.

PROPOSITION 4.10. *If Ω is an open subset of Λ such that $\overline{\Omega}$ is compact and disjoint from $\text{Supp}(T_{\text{bif}})$, then the mass of $\widehat{T}_n$ in Ω is $O\left(\frac{1}{n}\right)$. In other words:*

$$\int \text{Area}\left(\widehat{g}\left(\Lambda \times \{z_0\}\right) \cap \pi_1^{-1}(\Omega)\right) d\mu^n(g) = O(1).$$

To prove this estimate we give an analytic expression of this area:

$$\int \text{Area}(\widehat{g}(\Lambda \times \{z_0\}) \cap \pi_1^{-1}(\Omega)) d\mu^n(g)$$

$$= \int_{\pi_1^{-1}(\Omega)} \left[\widehat{g}(\Lambda \times \{z_0\})\right] \wedge (\pi_1^* \omega_\Lambda + \pi_2^* \omega_{\mathbb{P}^1}) d\mu^n(g)$$

$$= \text{Area}(\Omega) + \int_\Omega (\pi_1)_* \left(\pi_2^* \omega_{\mathbb{P}^1}|_{\widehat{g}(\Lambda \times \{z_0\})}\right) d\mu^n(g)$$

$$= \text{Area}(\Omega) + n \int_\Omega dd^c \chi_n, \text{ where } \chi_n = \frac{1}{n} \int \log \frac{\|\rho_\lambda(g)(Z_0)\|}{\|Z_0\|} d\mu^n(g).$$

Here Z_0 is a lift of z_0 to $\mathbb{C}^2$, $\|\cdot\|$ is the Hermitian norm, and the dd^c is taken w.r.t. λ. Therefore, exactly as in Theorem 2.8, we are left to prove that $\chi_n - \chi = O\left(\frac{1}{n}\right)$. This estimate in turn follows from ergodic theoretic properties of random matrix products — specifically, from the "exponential convergence of the transition operator," a result due to Le Page [LeP].

The next step is to combine this estimate on the average area of the graphs over Ω with another result of Furstenberg, which asserts that for each fixed λ, for $\mu^{\mathbb{N}}$-a.e. sequence $\mathbf{g} = (g_n)$, $\rho_\lambda(g_1 \cdots g_n)(z_0)$ converges to a point $z_{\mathbf{g}}$ (a consequence of the martingale convergence theorem). From this, as in Lemma 1.2, we infer that for a.e. $\mathbf{g} = (g_n)$, the sequence of graphs $\left(\widehat{g_1 \cdots g_n}\right)(\Lambda \times \{z_0\})$ converges to a limiting graph $\Gamma_{\mathbf{g}}$, outside a finite set of vertical bubbles. We then obtain a measurable and G-equivariant family of limiting graphs over Ω. Formally, this family is parameterized by the *Poisson boundary* $P(G, \mu)$ of (G, μ).

To show that the family of representations is stable over Ω, we need to upgrade this family of graphs into a holomorphic motion of the Poisson boundary, that is, we need to show that two distinct graphs are disjoint. For this, we use the fact that the number of intersection points between two graphs over some domain $D \subset \Omega$ is a function on $P(G, \mu) \times P(G, \mu)$ satisfying certain invariance properties. By an ergodic theorem due to Kaimanovich [Kai], this function is a.e. equal[12] to a constant ι_D on $P(G, \mu) \times P(G, \mu)$, depending only on D. The assignement $D \mapsto \iota_D$ being integer valued, we infer that there exists a locally finite set of points F such that if $D \cap F = \emptyset$, then $\iota_D = 0$. On such a D, it follows that the family of graphs is a holomorphic motion, and ultimately, that the family is stable. Finally, to show that the exceptional set F is empty and conclude that the family is stable on all D, we use the fact that the set of discrete faithful representations is closed, which implies that Bif cannot have isolated points.   □

4.4 Equidistribution of representations with an element of a given trace

Another important aspect of the bifurcation currents is that they enable to obtain equidistribution results in Corollary 4.2. This is the analogue in our context of the results of §2.5. For $t \in \mathbb{C}$ denote by $Z(g, t)$ the analytic subset of Λ defined by $Z(g, t) = \{\lambda,\ \mathrm{tr}^2(\rho_\lambda(g)) = t\}$. We study the asymptotic properties of the integration currents $[Z(g, t)]$. Note that if $\mathrm{tr}^2(\rho_\lambda(g)) \equiv t$, then $Z(g, t) = \Lambda$, and by convention, $[\Lambda] = 0$.

The first equidistribution result is the following.

THEOREM 4.11 (Equidistribution for random sequences). *Let (G, ρ, μ) be a holomorphic family of representations of G satisfying (R1–3) and endowed with a measure μ satisfying (A1–2). Fix $t \in \mathbb{C}$. Then for $\mu^{\mathbb{N}}$ a.e. sequence (g_n), we have that*

$$\frac{1}{2n}\left[Z(g_n \cdots g_1, t)\right] \xrightarrow[n\to\infty]{} T_{\mathrm{bif}}.$$

Notice that, if instead of considering a random sequence in the group, we take a word obtained by applying to $g \in G$ an iterated element of $\mathrm{Aut}(G)$, then similar equidistribution results were obtained by Cantat in [Ca1].

---

[12] As stated here, the result is true only when μ is invariant under $g \mapsto g^{-1}$. The general case needs some adaptations. Notice also that the Kaimanovich theorem can be viewed as a far-reaching generalization of the ergodicity of the geodesic flow for manifolds of constant negative curvature.

The following "deterministic" corollary makes Corollary 4.2 more precise. It seems difficult to prove it without using probabilistic methods.

COROLLARY 4.12. *Under the assumptions of Theorem 4.11, let $\varepsilon > 0$ and $\Lambda' \Subset \Lambda$. Then there exists $g \in G$ such that $\lambda \mapsto \operatorname{tr}^2(\rho_\lambda(g))$ is non-constant and $\{\lambda, \operatorname{tr}^2(\rho_\lambda(g)) = t\}$ is ε-dense in $\operatorname{Bif} \cap \Lambda'$.*

For the value $t = 4$, it is unclear which of accidental relations or parabolics prevail in $[Z(g_n \cdots g_1, 4)]$.

Theorem 4.11 is actually a consequence of a more general theorem [DD1, Theorem 4.1], which gives rise to several other random equidistribution results.

As opposed to the case of rational maps, we are able to estimate the speed of convergence, up to some averaging on g and a global assumption on Λ.

THEOREM 4.13 (Speed in the equidistribution theorem). *Let (G, ρ, μ) be a holomorphic family of representations of G, satisfying (R1–3), endowed with a measure μ satisfying (A1–2). Fix $t \in \mathbb{C}$. Assume further that one of the following hypotheses is satisfied.*

  i. *Λ is an algebraic family of representations defined over the algebraic closure $\overline{\mathbb{Q}}$ of $\mathbb{Q}$.*

  ii. *There exists a geometrically finite representation in Λ.*

Then there exists a constant C such that for every test form ϕ,

$$\left\langle \frac{1}{2n} \int [Z(g, t)] \, d\mu^n(g) - T_{\mathrm{bif}}, \phi \right\rangle \leq C \frac{\log n}{n} \|\phi\|_{C^2}.$$

The meaning of the notion of an algebraic family of representations is the following: the space $\operatorname{Hom}(G, \operatorname{PSL}(2, \mathbb{C}))$ of representations of G in $\operatorname{PSL}(2, \mathbb{C})$ admits a natural structure of an affine algebraic variety over $\mathbb{Q}$, simply by describing it as a set of matrices satisfying certain polynomial relations.[13] Changing this set of generators amounts to performing algebraic changes of coordinates, so that this structure of algebraic variety is well defined. We say that an arbitrary family of representations, viewed as a holomorphic mapping $\rho \colon \Lambda \to \operatorname{Hom}(G, \operatorname{PSL}(2, \mathbb{C}))$ is algebraic (resp. algebraic over K) if $\rho|_\Lambda$ is a dominating map to some algebraic subvariety (resp. over K) of $\operatorname{Hom}(G, \operatorname{PSL}(2, \mathbb{C}))$. To say it differently, there exists an open subset $\Omega \subset \Lambda$ such that ρ_Ω is an open subset of an algebraic subvariety of $\operatorname{Hom}(G, \operatorname{PSL}(2, \mathbb{C}))$.

These results parallel those of §2.5, and exactly as in Theorem 2.18, they become much easier after some averaging with respect to the multiplier. Let us illustrate this by proving the following result:

PROPOSITION 4.14. *Let m be the normalized Lebesgue measure on $[0, 4]$. Then under the assumptions of Theorems 4.11, we have that*

$$\frac{1}{2n} \int [Z(g, t)] \, d\mu^n(g) \, dm(t) \xrightarrow[n \to \infty]{} T_{\mathrm{bif}}.$$

---

[13]To view $\operatorname{PSL}(2, \mathbb{C})$ as a set of matrices, observe that $\operatorname{PSL}(2, \mathbb{C})$ is isomorphic to $\operatorname{SO}(3, \mathbb{C})$ by the adjoint representation.

PROOF. We prove the L^1_{loc} convergence of the potentials. Let $u(g, \cdot)$ be the psh function on Λ defined by

$$u(g, \lambda) = \int \log \left| \text{tr}^2(\rho_\lambda(g)) - t \right| dm(t) = v(\text{tr}^2(\rho_\lambda(g))),$$

where v is the logarithmic potential of m in $\mathbb{C}$. The function u is bounded below and $u(g, \lambda) \sim \log \left| \text{tr}^2(\rho_\lambda(g)) \right|$ as $\text{tr}^2(\rho_\lambda(g))$ tends to infinity.

If λ is fixed, then by Theorem 4.4, for $\mu^{\mathbb{N}}$ a.e. sequence (g_n), we have that $\frac{1}{2n} u(g_1 \cdots g_n, \lambda) \to \chi(\lambda)$. Since u is bounded below, we can apply the dominated convergence theorem and integrate with respect to $g_1, \ldots, g_n$. We conclude that

$$\frac{1}{2n} \int u(g, \lambda) d\mu^n(g) \xrightarrow[n \to \infty]{} \chi(\lambda)$$

for all $\lambda \in \Lambda$, which, by taking the dd^c in λ, implies the desired statement. □

SKETCH OF PROOF OF THEOREM 4.11. We need to show that for almost every sequence (g_n), the sequence of psh functions $\frac{1}{2n} \log \left| \text{tr}^2(\rho_\lambda(g_1 \cdots g_n)) - t \right|$ converges to χ. The point is to find a choice of random sequence (g_n) which does not depend on λ. For this, a kind of sub-additive ergodic theorem with values in the space of psh functions shows that for a.e. (g_n), $\frac{1}{n} \log \| \rho_\lambda(g_1 \cdots g_n) \|$ converges in L^1_{loc} to χ. By Theorem 4.4, it is possible to choose the sequence (g_n) so that for λ belonging to a countable dense sequence of parameters, $\frac{1}{2n} \log \left| \text{tr}^2(\rho_\lambda(g_1 \cdots g_n)) - t \right|$ converges to $\chi(\lambda)$. On the other hand,

$$\frac{1}{2n} \log \left| \text{tr}^2(\rho_\lambda(g_1 \cdots g_n)) - t \right| \leq \frac{1}{n} \log \| \rho_\lambda(g_1 \cdots g_n) \| + o(1).$$

Using the continuity of χ and the Hartogs lemma, we conclude that

$$\frac{1}{2n} \log \left| \text{tr}^2(\rho_\lambda(g_1 \cdots g_n)) - t \right| \longrightarrow \chi \quad \text{in} \quad L^1_{\text{loc}}.$$

□

For Theorem 4.13, the main difficulty is that for a given parameter we cannot, in general, integrate with respect to $g_1, \cdots, g_n$ in the almost sure convergence

$$\frac{1}{2n} \log \left| \text{tr}^2(\rho_\lambda(g_1 \cdots g_n)) - t \right| \to \chi(\lambda),$$

due to the possibility of elements with trace very close to t. This is exactly similar to the difficulty encountered in Theorem 2.18(iii). We estimate the size of the set of parameters where this exceptional phenomenon happens by using volume estimates for sub-level sets of psh functions and the global assumption (i) or (ii). In both cases this global assumption is used to show the existence of a parameter at which $\left| \text{tr}^2(\rho_\lambda(g_1 \cdots g_n)) - t \right|$ is not super-exponentially small in n. Under (i), this follows from a nice number-theoretic lemma (a generalization of the so-called Liouville inequality), which was communicated to us by P. Philippon. Another key ingredient is a large deviations estimate in Theorem 4.4, which was obtained independently by Aoun [Ao].

4.5 Canonical bifurcation currents

One might object that our definition of bifurcation currents in spaces of representations lacks of naturality, for it depends on the choice of a measure μ on G — recall however from Theorem 4.6 that the support of the bifurcation current is independent of μ. In this paragraph, following [DD2], we briefly explain how a canonical bifurcation current can be constructed under natural assumptions.

Let X be a compact Riemann surface of genus $g \geq 2$ and $G = \pi_1(X, *)$ be its fundamental group. Let (ρ_λ) be a holomorphic family of representations of G into $\mathrm{PSL}(2, \mathbb{C})$ satisfying (R1–3). We claim that there is a Lyapunov exponent function on Λ which is canonically associated to the Riemann surface structure of X (up to a multiplicative constant).

For this, let $\widetilde{X}$ be the universal cover of X (i.e., the unit disk). G embeds naturally as a subgroup of $\mathrm{Aut}(\widetilde{X})$. For any representation $\rho \in \Lambda$, consider its suspension X_ρ, that is, the quotient of $\widetilde{X} \times \mathbb{P}^1$ by the diagonal action of G. The suspension is a fiber bundle over X, with $\mathbb{P}^1$ fibers, and admits a holomorphic foliation transverse to the fibers whose holonomy is given by ρ. If γ is any path on X, we denote by h_γ its holonomy $\mathbb{P}^1_{\gamma(0)} \to \mathbb{P}^1_{\gamma(1)}$.

The Poincaré metric endows X with a natural Riemannian structure, so we can consider the Brownian motion on X. It follows from the sub-additive ergodic theorem that for a.e. Brownian path Ω (relative to the Wiener measure), the limit

$$\chi(\omega) = \lim_{t \to \infty} \frac{1}{t} \log \left\| h_{\omega(0), \omega(t)} \right\|$$

(where $\|\cdot\|$ is any smoothly varying spherical metric on the fibers) exists, and does not depend on Ω.

We define $\chi_{\mathrm{Brownian}}(\rho)$ to be this number and introduce a natural bifurcation current on Λ by putting $T_{\mathrm{bif}} = dd^c \chi_{\mathrm{Brownian}}$. We have the following theorem.

THEOREM 4.15. *Let (ρ_λ) be a holomorphic family of non-elementary representations of the fundamental group of a compact Riemann surface, satisfying (R1–3).*

Then the function χ_{Brownian} is psh on Λ and the support of $T_{\mathrm{bif}} = dd^c \chi_{\mathrm{Brownian}}$ is the bifurcation locus.

To prove this theorem, it is enough to exhibit a measure μ on G satisfying (A1–2) and such that for every ρ, $\chi_\mu(\rho) = \chi_{\mathrm{Brownian}}(\rho)$ (up to a multiplicative constant). Such a measure actually exists and was constructed using a discretization procedure by Furstenberg [Fur2]. It is non-trivial to check that μ satisfies the exponential moment condition (A2) (for instance, this measure can never be of finite support).

There is another natural family of paths on X: the geodesic trajectories. An argument similar to the previous one shows that if $(x, v) \in S^1(X)$ (the unit tangent bundle) is generic relative to the Liouville measure, and if $\gamma_{(x,v)}$ denotes the unit speed geodesic stemming from (x, v), then the limit $\lim_{t \to \infty} \frac{1}{t} \log \left\| h_{\gamma(0), \gamma(t)} \right\|$ exists and does not depend on (generic) (x, v). We use $\chi_{\mathrm{geodesic}}(\rho)$ to denote this number. It follows from the elementary properties of the Brownian motion on the hyperbolic disk that there exists a constant v depending only on X such that $\chi_{\mathrm{Brownian}} = v \chi_{\mathrm{geodesic}}$. Therefore, the associated bifurcation current is the same.

Here is a situation where these ideas naturally apply: consider the set $\mathcal{P}(X)$ of complex projective structures over a Riemann surface X, compatible with its complex structure (see [Dum] for a nice introductory text on projective structures). This is a complex affine space of dimension $3g - 3$, admitting a distinguished point, the "standard Fuchsian structure," namely, the projective structure obtained by viewing X as a quotient of the unit disk. A projective structure induces a *holonomy representation* (which is always non-elementary and defined only up to conjugacy) so the above discussion applies. We conclude that *the space of projective structures on X admits a natural bifurcation current.*

From the standard Fuchsian structure, one classically constructs an embedding of the Teichmüller space of X as a bounded open subset of $\mathcal{P}(X)$, known as the *Bers embedding* (or *Bers slice*). This open set can be defined for instance as the component of the distinguished point in the stability locus.

In [McM4], McMullen suggests the Bers slice as the analogue of the Mandelbrot set through the Sullivan dictionary. From this perspective, an interesting result in [DD2] is that the canonical Lyapunov exponent function χ_{Brownian} is *constant* on the Bers embedding, so the analogy also holds at the level of Lyapunov exponents.

4.6   Open problems

There are many interesting open questions in this area, some of them stated in [DD1, §5.2]. In the spirit of this survey, let us state only two problems related to the exterior powers of T_{bif}.

As opposite to the case of rational maps, we believe that the supports of T_{bif}^k for $k \geq 2$ do not give rise to "higher bifurcation loci."

CONJECTURE 4.16.   *Let (G, ρ, μ) be a family of representations satisfying (R1–3) and (A1–2), and assume further that two representations in Λ are never conjugate in $\mathrm{PSL}(2, \mathbb{C})$ (that is, Λ is a subset of the character variety). Then for every $k \leq \dim(\Lambda)$, we have* $\mathrm{Supp}(T_{\text{bif}}^k) = \text{Bif}.$

Here is some evidence for this conjecture: let $\theta \in \mathbb{R} \setminus \pi\mathbb{Q}$, $t = 4\cos^2(\theta)$ and consider the varieties $Z(g, t)$. Since for $\lambda \in Z(g, t)$, ρ_λ is not discrete, the bifurcation locus of $\{\rho_\lambda, \lambda \in Z(g, t)\}$ is equal to $Z(g, t)$. Hence $\mathrm{Supp}(T_{\text{bif}} \wedge [Z(g, t)]) = Z(g, t)$, which by equidistribution of $Z(g, t)$ makes the equality $\mathrm{Supp}(T_{\text{bif}}^2) = \mathrm{Supp}(T_{\text{bif}})$ reasonable.

It is also natural to look for equidistribution in higher codimension. Here is a specific question:

QUESTION 4.17.   *Let (G, ρ, μ) be a family of representations satisfying (R1–3) and (A1–2). Assume that $\dim(\Lambda) \geq 3$. Given a generic element $h \in \mathrm{PSL}(2, \mathbb{C})$, and a $\mu^{\mathbb{N}}$ generic sequence (g_n), do the solutions of the equation $g_1 \cdots g_n = h$ equidistribute (after convenient normalization) towards T_{bif}^3?*

5   FURTHER SETTINGS, FINAL REMARKS

In this section we gather some speculations about possible extensions of the results presented in the paper.

5.1   Holomorphic dynamics in higher dimension

It is likely that a substantial part of the theory of bifurcation currents for rational maps on $\mathbb{P}^1$ should remain true in higher dimension; nevertheless, little has been done so far.

Let us first discuss the case of polynomial automorphisms of $\mathbb{C}^2$. A polynomial automorphism f of degree d of $\mathbb{C}^2$ with non-trivial dynamics admits a unique measure of maximal entropy, which has two (complex) Lyapunov exponents of opposite sign $\chi^+(f) > 0 > \chi^-(f)$ and describes the asymptotic distribution of saddle periodic orbits [BLS1, BLS2]. See [Ca2] in this volume for a presentation of these results for automorphisms of compact complex surfaces. Notice that a polynomial automorphism has constant jacobian, so $\chi^+(f) + \chi^-(f) = \log |\mathrm{Jac}(f)|$ is a pluriharmonic function on parameter space. It is not difficult to see that the function $f \mapsto \chi^+(f)$ is psh (in particular, upper semi-continuous), and it was shown in [Du1] that is actually continuous (even for families degenerating to a one-dimensional map).

Since the Lyapunov exponents are well approximated by the multipliers of saddle orbits [BLS2], it follows that near any point in parameter space where $f \mapsto \chi^+(f)$ is not pluriharmonic, complicated bifurcations of saddle points occur. In the dissipative case they must become attracting. The main idea of Theorem 2.18 seems robust enough to enable some generalization to this setting.

On the other hand, a basic understanding of the phenomena responsible for the bifurcations of a family of polynomial automorphisms of $\mathbb{C}^2$ — e.g., the role of homoclinic tangencies — is still missing (see [BS] for some results in a particular case). In particular no reasonable analogue of Theorem 2.1 is available for the moment. Therefore, it seems a bit premature to hope for a characterization of the support of the bifurcation current $dd^c \chi^+$, let alone $(dd^c \chi^+)^p$.

The situation is analogous in the case of families of holomorphic endomorphisms of $\mathbb{P}^k$ (and, more generally, for families of polynomial-like mappings in higher dimension). The regularity properties of the Lyapunov exponent(s) function(s) are rather well understood, due to the work of Dinh-Sibony [DS1] and Pham [Ph] (a good account on this is in [DS3, §2.5]). In particular, it is known that the sum $L_p(f)$ of the p largest Lyapunov exponents of the maximal entropy measure is psh for $1 \leq p \leq k$ and the sum of all Lyapunov exponents is Hölder continuous. It is also known [BDM] that $L_p(f)$ is well approximated by the corresponding quantity evaluated at repelling periodic cycles, so that any point in parameter space where L_p is not pluriharmonic is accumulated by bifurcations of periodic points. Notice that the relationship between the currents $dd^c L_p$ is unclear.

Again, one may reasonably hope for equidistribution results in the spirit of Theorem 2.18.

Another interesting point is a formula, given in [BB1], for the sum L_k of Lyapunov exponents of endomorphisms of $\mathbb{P}^k$ which generalizes Przytycki's formula (2.2). From this formula one may expect to reach some understanding on the role of the critical locus towards bifurcations.

5.2 Cocycles

Yoccoz suggests in [Y] to study the geography of the (finite-dimensional) space of locally constant $\mathrm{SL}(2, \mathbb{R})$ cocycles over a transitive subshift of finite type in the same way as spaces of one-dimensional holomorphic dynamical systems, with some emphasis on the description of hyperbolic components and their boundaries. For $\mathrm{SL}(2, \mathbb{C})$ cocycles (and more generally for any cocycle with values in a complex Lie subgroup of $\mathrm{GL}(n, \mathbb{C})$) we have an explicit connection with holomorphic dynamics given by the bifurcation currents. Indeed, locally constant cocycles over a subshift are generalizations of random products of matrices, which correspond to cocycles over the full shift. In this situation we can define a Lyapunov exponent function relative to a fixed measure on the base dynamical system (the Parry measure is a natural candidate), and construct a bifurcation current by taking the dd^c.

Notice that the subharmonicity properties of Lyapunov exponents are frequently used in this area of research (an early example is [He]).

For a general holomorphic family of (say, locally constant) $SL(2, \mathbb{C})$ cocycles over a fixed subshift of finite type, one may ask the same questions as in Section 4: characterize the support of the bifurcation current, prove equidistribution theorems. Another interest of considering this setting is that it is somehow a simplified model of the tangent dynamics of 2-dimensional diffeomorphisms, so it might provide some insight on the bifurcation theory of those. In particular there is an analogue of heteroclinic tangencies in this setting ("heteroclinic connexions"), and it might be interesting to study the distribution of the corresponding parameters.

5.3   Random walks on other groups

Another obvious possible generalization of Section 4 is the study of bifurcation currents associated to holomorphic families of finitely generated subgroups of $SL(n, \mathbb{C})$. Again, if (ρ_λ) is a holomorphic family of strongly irreducible representations (see [Frm] for the definition) of a finitely generated group G endowed with a probability measure satisfying (A1–2), then Definition 4.5 makes sense, with χ being the top Lyapunov exponent. It is likely that equidistribution theorems for representations possessing an element of given trace should follow as in §4.4. More generally, one may investigate the distribution of representations with an element belonging to a given hypersurface of $SL(n, \mathbb{C})$, in the spirit of Section 1.

On the other hand, for the same reasons as in §5.1, the characterization of the support of the bifurcation current is certainly a more challenging problem.

5.4   Non-archimedian dynamics

It is a standard fact in algebraic geometry that studying families $(X_\lambda)_{\lambda \in \Lambda}$ of complex algebraic varieties often amounts to studying varieties over a field extension of $\mathbb{C}$, that is, a function field in the variable λ. The same idea applies in the dynamical context and was explored by several authors. This fact was used notably by Culler, Morgan and Shalen [CS, MS] to construct compactifications of spaces of representations into $PSL(2, \mathbb{C})$ and obtain new results on the geometry of 3-manifolds. In rational dynamics Kiwi [Ki2, Ki3] used a similar construction to study the behavior at infinity of families of cubic polynomials or quadratic rational maps (see [DeMc] for a different approach to this problem).

It would be natural to explore the interaction of bifurcation currents with these compactifications, as well as the general bifurcation theory of non-archimedian rational dynamical systems.

6   Bibliography

[Ao]   Aoun, Richard. *Random subgroups of linear groups are free.* Duke Math. J. 160 (2011), 117–173.

[BaDeM] Baker, Matthew; DeMarco, Laura. *Preperiodic points and unlikely intersections*. Duke Math. J. 159 (2011), no. 1, 1–29.

[BaH] Baker, Matthew; Hsia, Liang-Chung. *Canonical heights, transfinite diameters, and polynomial dynamics*. J. Reine Angew. Math. 585 (2005), 61–92.

[BB1] Bassanelli, Giovanni; Berteloot, François. *Bifurcation currents in holomorphic dynamics on* $\mathbf{P}^k$. J. Reine Angew. Math. 608 (2007), 201–235.

[BB2] Bassanelli, Giovanni; Berteloot, François. *Bifurcation currents and holomorphic motions in bifurcation loci*. Math. Ann. 345 (2009), 1–23.

[BB3] Bassanelli, Giovanni; Berteloot, François. *Distribution of polynomials with cycles of a given multiplier*. Nagoya Math. J. 201 (2011), 23–43.

[Bea] Beardon, Alan F. *The geometry of discrete groups*. Graduate Texts in Mathematics, 91. Springer-Verlag, New York, 1983.

[BLS1] Bedford, Eric; Lyubich, Mikhail; Smillie, John. *Polynomial diffeomorphisms of* $\mathbb{C}^2$. *IV. The measure of maximal entropy and laminar currents*. Invent. Math. 112 (1993), 77-125.

[BLS2] Bedford, Eric; Lyubich, Mikhail; Smillie, John. *Distribution of periodic points of polynomial diffeomorphisms of* $\mathbb{C}^2$. Invent. Math. 114 (1993), 277-288.

[BS] Bedford, Eric; Smillie, John. *Real polynomial diffeomorphisms with maximal entropy: Tangencies*. Ann. of Math. (2) 160 (2004), no. 1, 1–26.

[BT] Bedford, Eric; Taylor, B. Alan. *A new capacity for plurisubharmonic functions*. Acta Math. 149 (1982), no. 1-2, 1–40.

[Ber] Bers, Lipman. *Holomorphic families of isomorphisms of Möbius groups*. J. Math. Kyoto Univ. 26 (1986), no. 1, 73–76.

[Bt1] Berteloot, François. *Lyapunov exponent of a rational map and multipliers of repelling cycles*. Riv. Math. Univ. Parma, 1(2) (2010), 263–269.

[Bt2] Berteloot, François. *Bifurcation currents in one-dimensional holomorphic dynamics*. CIME Lecture notes, 2011.

[BDM] Berteloot, François; Dupont, Christophe; Molino, Laura. *Normalization of bundle holomorphic contractions and applications to dynamics*. Ann. Inst. Fourier (Grenoble) 58 (2008), no. 6, 2137–2168.

[BFH] Bielefeld, Ben; Fisher, Yuval; Hubbard, John H. *The classification of critically preperiodic polynomials as dynamical systems*. J. Amer. Math. Soc. 5 (1992), no. 4, 721–762.

[BKM] Bonifant, Araceli; Kiwi, Jan; Milnor, John. *Cubic polynomial maps with periodic critical orbit. II. Escape regions*. Conform. Geom. Dyn. 14 (2010), 68–112.

[BL] Bougerol, Philippe; Lacroix, Jean. *Products of random matrices with applications to Schrödinger operators*. Progress in Probability and Statistics, 8. Birkhäuser Boston, Inc., Boston, MA, 1985.

[Bra]   Branner, Bodil. *Cubic polynomials: Turning around the connectedness locus.* In *Topological methods in modern mathematics (Stony Brook, NY, 1991)*, 391–427, Publish or Perish, Houston, TX, 1993.

[BH1]   Branner, Bodil; Hubbard, John H. *The iteration of cubic polynomials. I. The global topology of parameter space.* Acta Math. 160 (1988), no. 3–4, 143–206.

[BH2]   Branner, Bobil; Hubbard, John H. *The iteration of cubic polynomials. II. Patterns and parapatterns.* Acta Math. 169 (1992), no. 3–4, 229–325.

[BD]   Briend, Jean-Yves; Duval, Julien. *Exposants de Liapounoff et distribution des points périodiques d'un endomorphisme de* $\mathbb{CP}^k$. Acta Math. 182 (1999), 143–157.

[Bro]   Brolin, Hans. *Invariant sets under iteration of rational functions.* Ark. Mat. 6 (1965), 103–144.

[BE]   Buff, Xavier; Epstein, Adam. *Bifurcation measure and postcritically finite rational maps.* Complex dynamics, 491–512, A K Peters, Wellesley, MA, 2009.

[BG]   Buff, Xavier; Gauthier, Thomas. *Perturbations of flexible Lattès maps.*, to appear Bull. Soc. Math. France (2013). arXiv:1111.5451.

[Ca1]   Cantat, Serge. *Bers and Hénon, Painlevé and Schrödinger.* Duke Math. J. 149 (2009), 411–460.

[Ca2]   Cantat, Serge. *Dynamics of automorphisms of compact complex surfaces.* In this volume, 463–514.

[Ch]   Chirka, Evgueny M. *Complex analytic sets.* Mathematics and its Applications (Soviet Series), 46. Kluwer Academic Publishers Group, Dordrecht, 1989.

[CS]   Culler, Marc; Shalen, Peter B. *Varieties of group representations and splittings of 3-manifolds.* Ann. of Math. (2) 117 (1983), no. 1, 109–146.

[De]   Demailly, Jean-Pierre. *Complex analytic and differential geometry, Chap. III.* Book available online at http://www-fourier.ujf-grenoble.fr/~demailly/manuscripts/agbook.pdf.

[DeM1]   DeMarco, Laura. *Dynamics of rational maps: a current on the bifurcation locus.* Math. Res. Lett. 8 (2001), no. 1–2, 57–66.

[DeM2]   DeMarco, Laura. *Dynamics of rational maps: Lyapunov exponents, bifurcations, and capacity.* Math. Ann. 326 (2003), no. 1, 43–73.

[DeMc]   DeMarco, Laura; McMullen, Curtis. *Trees and the dynamics of polynomials.* Ann. Sci. Éc. Norm. Supér. (4) 41 (2008), no. 3, 337–382.

[DD1]   Deroin, Bertrand; Dujardin, Romain. *Random walks, Kleinian groups, and bifurcation currents.* Invent. Math. 190 (2012) 57–118.

[DD2]   Deroin, Bertrand; Dujardin, Romain. *Lyapunov exponents for surface group representations.* Preprint. arXiv:1305.0049

[DeT]   De Thélin, Henry. *Sur la laminarité de certains courants.* Ann. Sci. École Norm. Sup. (4) 37 (2004), 304–311.

[Dil]   Diller, Jeffrey. *Dynamics of birational maps of* $\mathbb{P}^2$. Indiana Univ. Math. J. 45 (1996), 721–772.

[DS1]   Dinh, Tien Cuong; Sibony, Nessim. *Dynamique des applications d'allure polynomiale.* J. Math. Pures Appl. (9) 82 (2003), 367–423.

[DS2]   Dinh, Tien Cuong; Sibony, Nessim. *Distribution des valeurs de transformations méromorphes et applications.* Comment. Math. Helv. 81 (2006), no. 1, 221–258.

[DS3]   Dinh, Tien Cuong; Sibony, Nessim. *Super-potentials for currents on compact Kähler manifolds and dynamics of automorphisms.* J. Algebraic Geom. 19 (2010), no. 3, 473–529.

[DH]   Douady, Adrien; Hubbard, John H. *Étude dynamique des polynÃ´mes complexes.* Publications Mathématiques d'Orsay. Université de Paris-Sud, Orsay, 1985.

[Du1]   Dujardin, Romain. *Continuity of Lyapunov exponents for polynomial automorphisms of* $\mathbb{C}^2$. Ergodic Theory and Dynamical Systems 27 (2007), 1111–1133.

[Du2]   Dujardin, Romain. *Cubic polynomials: a measurable view on parameter space.* In *Complex Dynamics: Families and Friends*, 451–489, A. K. Peters, Wellesley, MA, 2009.

[Du3]   Dujardin, Romain. *Fatou directions along the Julia set for endomorphisms of* $\mathbb{CP}^k$. J. Math. Pures Appl. 98 (2012), 591–615.

[DF]   Dujardin, Romain; Favre, Charles. *Distribution of rational maps with a preperiodic critical point.* Amer. J. Math. 130 (2008), 979-1032.

[DG]   Dujardin, Romain; Guedj, Vincent. *Geometric properties of maximal psh functions.* In *Complex Monge-Ampère equations and geodesics in the space of Kähler metrics*, Lecture Notes in Mathematics 2038, 33–52, Springer, Heidelberg, 2012.

[Dum]   Dumas, David. *Complex projective structures.* Handbook of Teichmüller theory. Vol. II, 455–508, Eur. Math. Soc., Zürich, 2009.

[FRL]   Favre, Charles; Rivera-Letelier, Juan. *Équidistribution quantitative des points de petite hauteur sur la droite projective.* Math. Ann. 335 (2006), 311–361.

[FS2]   Fornæss, John Erik; Sibony, Nessim. *Complex dynamics in higher dimension. II*, Modern methods in complex analysis, Ann. of Math. Studies 135–182. Princeton Univ. Press.

[FLM]   Freire, Alexandre; Lopes, Artur; Mañé, Ricardo. *An invariant measure for rational maps.* Bol. Soc. Brasil. Mat. 14 (1983), 45–62.

[Ful]   Fulton, William. *Intersection theory.* Second edition. Ergebnisse der Mathematik und ihrer Grenzgebiete. 3. Folge. A Series of Modern Surveys in Mathematics. Springer-Verlag, Berlin, 1998.

[Frm]   Furman, Alex. *Random walks on groups and random transformations.* Handbook of dynamical systems, Vol. 1A, 931–1014, North-Holland, Amsterdam, 2002.

[Fur1]   Furstenberg, Hillel. *Noncommuting random products.* Trans. Amer. Math. Soc. 108 (1963) 377–428.

[Fur2]   Furstenberg, Hillel. *Random walks and discrete subgroups of Lie groups.* Advances in Probability and Related Topics, Vol. 1, 1–63 Dekker, New York, 1971.

[FK]   Furstenberg, Hillel; Kesten, Harry. *Products of random matrices.* Ann. Math. Statist. 31 (1960), 457–469.

[Ga]   Gauthier, Thomas. *Strong-bifurcation loci of full Hausdorff dimension.* Ann. Sci. Ecole Norm. Sup. 45 (2012), 947–984. arXiv:1103.2656.

[Go]   Goldman, William M. *Mapping class group dynamics on surface group representations.* Problems on mapping class groups and related topics, 189–214, Proc. Sympos. Pure Math., 74, Amer. Math. Soc., Providence, RI, 2006

[GŚ]   Graczyk, Jacek; Świątek, Gregorz. *Harmonic measure and expansion on the boundary of the connectedness locus.* Invent. Math. 142 (2000), no. 3, 605–629.

[Gue]   Guedj, Vincent. *Dynamics of polynomial mappings of $\mathbb{C}^2$.* Amer. J. Math. 124 (2002), 75–106.

[Gui]   Guivarc'h, Yves. *Produits de matrices aléatoires et applications aux propriétés géométriques des sous-groupes du groupe linéaire.* Ergodic Theory Dynam. Systems 10 (1990), no. 3, 483–512.

[He]   Herman, Michael. *Une méthode pour minorer les exposants de Lyapounov et quelques exemples montrant le caractère local d'un théorème d'Arnol'd et de Moser sur le tore de dimension* 2 Comment. Math. Helv. 58 (1983), no. 3, 453–502.

[Hö]   Hörmander, Lars. *Notions of convexity.* Progress in Math 127. Birkhäuser, Boston, MA, 1994.

[IN]   Ivashkovich, Sergey; Neji, Fethi. *Weak normality of families of meromorphic mappings and bubbling in higher dimensions.* Preprint (2011), arXiv:1104.3973.

[Kai]   Kaimanovich, Vadim A. *Double ergodicity of the Poisson boundary and applications to bounded cohomology.* Geom. Funct. Anal. 13 (2003), no. 4, 852–861.

[Kap]   Kapovich, Michael. *Hyperbolic manifolds and discrete groups.* Progress in Mathematics, 183. Birkhäuser Boston, Inc., Boston, MA, 2001.

[Ki1]   Kiwi, Jan. *Combinatorial continuity in complex polynomial dynamics.* Proc. London Math. Soc. (3) 91 (2005), no. 1, 215–248.

[Ki2]   Kiwi, Jan. *Puiseux series, polynomial dynamics, and iteration of complex cubic polynomials* Ann. Inst. Fourier (Grenoble) 56 (2006), no. 5, 1337–1404.

[Ki3]   Kiwi, Jan. *Puiseux series dynamics of quadratic rational maps.* Preprint (2011) arXiv:1106.0059.

[Led]   Ledrappier, François. *Quelques propriétés ergodiques des applications rationnelles.* C. R. Acad. Sci. Paris Sér. I Math. 299 (1) (1984), 37–40.

[LeP]   Le Page, Émile. *Théorèmes limites pour les produits de matrices aléatoires.* In *Probability measures on groups,* ed. H. Heyer. Lecture Notes in Math. no. 928 (1982), 258–303.

[Lev1]   Levin, Guennadi. *Irregular values of the parameter of a family of polynomial mappings.* Russian Math. Surveys 36:6 (1981), 189–190.

[Lev2]   Levin, Gennadi M. *On the theory of iterations of polynomial families in the complex plane.* J. Soviet Math. 52 (1990), no. 6, 3512–3522.

[Lvy]   Levy, Alon. *The space of morphisms on projective space.* Acta Arith. 146 (2011), no. 1, 13–31. arXiv:0903.1318

[Lu]   Lubotzky, Alex. *Dynamics of* $\mathrm{Aut}(F_n)$ *actions on group presentations and representations.* In Geometry, Rigidity, and Group Actions: A Festschrift in honor of Robert Zimmer's 60th birthday. ed. B. Farb and D. Fisher. Univ. of Chicago Press, 2011.

[Ly1]   Lyubich, Mikhail *Entropy properties of rational endomorphisms of the Riemann sphere.* Ergodic Theory Dynam. Systems 3 (1983), 351–385.

[Ly2]   Lyubich, Mikhail Yu. *Some typical properties of the dynamics of rational mappings.* Russian Math. Surveys 38:5 (1983), 154–155.

[Mñ]   Mañé, Ricardo. *The Hausdorff dimension of invariant probabilities of rational maps.* Dynamical systems, Valparaiso 1986, 86–117, Lecture Notes in Math., 1331, Springer, Berlin, 1988.

[MSS]   Mañé, Ricardo; Sad, Paulo; Sullivan, Dennis. *On the dynamics of rational maps.* Ann. Sci. École Norm. Sup. (4) 16 (1983), 193–217.

[Man]   Manning, Anthony. *The dimension of the maximal measure for a polynomial map.* Ann. of Math. (2) 119 (1984), 425–430.

[McM1]   McMullen, Curtis T. *The motion of the maximal measure of a polynomial.* Preliminary notes, 1985. http://nrs.harvard.edu/urn-3:HUL.InstRepos:9925393

[McM2]   McMullen, Curtis T. *Families of rational maps and iterative root-finding algorithms.* Ann. of Math. 125 (1987), 467–493.

[McM3]   McMullen, Curtis T. *Cusps are dense.* Ann. of Math. (2) 133 (1991), 217–247.

[McM4]   McMullen, Curtis T. *Renormalization and 3-manifolds which fiber over the circle.* Annals of Mathematics Studies, 142. Princeton University Press, Princeton, NJ, 1996.

[McMS]   McMullen, Curtis T.; Sullivan, Dennis P. *Quasiconformal homeomorphisms and dynamics. III. The Teichmüller space of a holomorphic dynamical system.* Adv. Math. 135 (1998), 351–395.

[Mi1]   Milnor, John. *Remarks on iterated cubic maps.* Experiment. Math. 1 (1992), no. 1, 5–24.

[Mi2]   Milnor, John. *Geometry and dynamics of quadratic rational maps.* With an appendix by the author and Lei Tan. Experiment. Math. 2 (1993), no. 1, 37–83.

[Mi3]   Milnor, John. *Hyperbolic components in spaces of polynomial maps.* With an appendix by Alfredo Poirier. In *Conformal Dynamics and Hyperbolic Dynamics*, ed. F. Bonahan, R. Devaney, F. Gardiner, and D. Šaric. Contemporary Mathematics 53, 183–232. Amer. Math. Soc. Providence RI, 2012. arXiv:1205.2668

[Mi4]   Milnor, John. *On rational maps with two critical points.* Experiment. Math. 9 (2000), no. 4, 481–522.

[Mi5]   Milnor, John. *Cubic polynomial maps with periodic critical orbit. I.* Complex dynamics, 333–411, A K Peters, Wellesley, MA, 2009.

[MS]   Morgan, John W.; Shalen, Peter B. *Valuations, trees, and degenerations of hyperbolic structures. I.* Ann. of Math. (2) 120 (1984), no. 3, 401–476.

[Ph]   Pham, Ngoc-Mai. *Lyapunov exponents and bifurcation current for polynomial-like maps.* Preprint (2005). arXiv:math:0512557

[Pr]   Przytycki, Feliks. *Hausdorff dimension of harmonic measure on the boundary of an attractive basin for a holomorphic map.* Invent. Math. 80 (1985), 161–179.

[Re]   Rees, Mary. *Positive measure sets of ergodic rational maps.* Ann. Sci. École Norm. Sup. (4) 19 (1986), 383–407.

[RSh]   Russakovskii, Alexander; Shiffman, Bernard. *Value distribution for sequences of rational mappings and complex dynamics.* Indiana Univ. Math. J. 46 (1997), 897–932.

[RSo]   Russakovskii, Alexander; Sodin, Mikhail. *Equidistribution for sequences of polynomial mappings.* Indiana Univ. Math. J. 44 (1995), no. 3, 851–882.

[Sh]   Shishikura, Mitsuhiro. *The Hausdorff dimension of the boundary of the Mandelbrot set and Julia sets.* Ann. of Math. (2) 147 (1998), no. 2, 225–267.

[Sib]   Sibony, Nessim. *Iteration of polynomials.* Lecture notes (unpublished), UCLA, 1984.

[SW]   Sibony, Nessim; Wong, Pit Mann. *Some results on global analytic sets.* Séminaire Lelong-Skoda (Analyse). Années 1978/79, pp. 221–237, Lecture Notes in Math., 822, Springer, Berlin, 1980.

[Sil]   Silverman, Joseph H. *The arithmetic of dynamical systems.* Graduate Texts in Mathematics, 241. Springer, New York, 2007.

[Sm]   Smirnov, Stanislav. *Symbolic dynamics and Collet-Eckmann conditions.* Internat. Math. Res. Notices 2000, no. 7, 333–351.

[vS]   van Strien, Sebastian. *Misiurewicz maps unfold generically (even if they are critically non-finite).* Fund. Math. 163 (2000), no. 1, 39–54.

[Su1]   Sullivan, Dennis. *Quasiconformal homeomorphisms and dynamics. I. Solution of the Fatou-Julia problem on wandering domains.* Ann. of Math. (2) 122 (1985), no. 3, 401–418.

[Su2]   Sullivan, Dennis. *Quasiconformal homeomorphisms and dynamics. II. Structural stability implies hyperbolicity for Kleinian groups.* Acta Math. 155 (1985), no. 3-4, 243–260.

[T]   Tan, Lei. *Hausdorff dimension of subsets of the parameter space for families of rational maps. (A generalization of Shishikura's result).* Nonlinearity 11 (1998), no. 2, 233–246.

[U]   Ueda, Tetsuo. *Fatou sets in complex dynamics on projective spaces.* J. Math. Soc. Japan 46 (1994), 545–555.

[Y]   Yoccoz, Jean-Christophe. *Some questions and remarks about* $SL(2, \mathbb{R})$ *cocycles.* Modern dynamical systems and applications, 447–458, Cambridge Univ. Press, Cambridge, 2004.

[Za]   Zakeri, Saeed. *Dynamics of cubic Siegel polynomials.* Comm. Math. Phys. 206 (1999), 185–233.

[Zd]   Zdunik, Anna. *Parabolic orbifolds and the dimension of the maximal measure for rational maps.* Invent. Math. 99 (3) (1990), 627–649.

Part IV

Laminations and Foliations

Entropy for hyperbolic Riemann surface laminations I

Tien-Cuong Dinh, Viet-Anh Nguyên, and Nessim Sibony

ABSTRACT. We develop a notion of entropy, using hyperbolic time, for laminations by hyperbolic Riemann surfaces. When the lamination is compact and transversally smooth, we show that the entropy is finite and at least equal to 2, and the Poincaré metric on leaves is transversally Hölder continuous. A notion of metric entropy is also introduced for harmonic measures.

Notation. Throughout the paper, $\mathbb{D}$ denotes the unit disc in $\mathbb{C}$, $r\mathbb{D}$ denotes the disc of center 0 and of radius r, and $\mathbb{D}_R \subset \mathbb{D}$ is the disc of center 0 and of radius R with respect to the Poincaré metric on $\mathbb{D}$, i.e., $\mathbb{D}_R = r\mathbb{D}$ with $R := \log[(1+r)/(1-r)]$. The Poincaré metric on a Riemann surface, in particular on $\mathbb{D}$ and on the leaves of a lamination, is given by a positive $(1,1)$-form that we denote by ω_P. The associated distance and diameter are denoted by dist_P and diam_P. A leaf through a point x of a lamination is often denoted by L_x, and $\phi_x \colon \mathbb{D} \to L_x$ denotes a universal covering map of L_x such that $\phi_x(0) = x$.

1 INTRODUCTION

The main goal of this paper is to introduce a notion of entropy for possibly singular hyperbolic laminations by Riemann surfaces. We also study the transverse regularity of the Poincaré metric and the finiteness of the entropy. In order to simplify the presentation, we will mostly focus, in this first part, on compact laminations which are transversally smooth. We will study the case of singular foliations in the second part of this paper.

The question of hyperbolicity of leaves for generic foliations in $\mathbb{P}^k$ has been adressed by many authors. We just mention here the case of a polynomial vector field in $\mathbb{C}^k$. It induces a foliation by Riemann surfaces in the complex projective space $\mathbb{P}^k$. We can consider that this foliation is the image of the foliation in $\mathbb{C}^{k+1}$ given by a holomorphic vector field

$$F(z) := \sum_{j=0}^{k} F_j(z) \frac{\partial}{\partial z_j}$$

with F_j homogeneous polynomials of degree $d \geq 2$ without common factor.

The singular set corresponds to the union of the fixed points of $f = [F_0 : \cdots : F_k]$ and the indeterminacy points of f in $\mathbb{P}^k$. The nature of the leaves as abstract Riemann surfaces has received much attention. Glutsyuk [18] and Lins Neto [28] have shown that on a generic foliation $\mathcal{F}$ of degree d, the leaves are covered by the unit disc in $\mathbb{C}$. We then say that the foliation is *hyperbolic*. More precisely, Lins Neto has shown that this is the case when all singular points have non-degenerate linear part. In [7], Candel and Gomez-Mont have shown that if all the singularities are hyperbolic, the Poincaré metric on leaves is transversally continuous. We will consider this situation in the second part of this work.

Let $(X, \mathcal{L})$ be a (transversally) smooth compact lamination by hyperbolic Riemann surfaces. We show in Section 2 that the Poincaré metric on leaves is transversally Hölder continuous. The exponent of Hölder continuity can be estimated in geometric terms. The continuity was studied by Candel [4], Ghys [19], and Verjovsky [30]. The survey [15] establishes the result as a consequence of Royden's lemma. Indeed, with his lemma, Royden proved the upper-semicontinuity of the infinitesimal Kobayashi metric in a Kobayashi hyperbolic manifold (see [26, p. 91 and p. 153]). The main tool of the proof is to use Beltrami's equation in order to compare universal covering maps of any leaf L_y near a given leaf L_x. More precisely, we first construct a non-holomorphic parametrization ψ from $\mathbb{D}_R$ to L_y which is close to a universal covering map $\phi_x : \mathbb{D} \to L_x$. Precise geometric estimates on ψ allow us to modify it, using Beltrami's equation. We then obtain a holomorphic map that we can explicitly compare with a universal covering map $\phi_y : \mathbb{D} \to L_y$.

Our second concern is to define the entropy of hyperbolic lamination possibly with singularities. A notion of geometric entropy for regular Riemannian foliations was introduced by Ghys-Langevin-Walczak [20]; see also Candel-Conlon [5], [6] and Walczak [32]. It is related to the entropy of the holonomy pseudogroup, which depends on the chosen generators. The basic idea is to quantify how much leaves get far apart transversally. The transverse regularity of the metric on leaves and the lack of singularities play a role in the finiteness of the entropy.

Ghys-Langevin-Walczak show, in particular, that when their geometric entropy vanishes, the foliation admits a transverse measure. The survey by Hurder [23] gives an account on many important results in foliation theory and contains a large bibliography.

Our notion of entropy contains a large number of classical situations. An interesting fact is that this entropy is related to an increasing family of distances as in Bowen's point of view [1]. This allows us, for example, to introduce other dynamical notions like metric entropy, local entropies, or Lyapounov exponents.

In Section 3, we first introduce a general notion of entropy on a metric space (X, d). To a given family of distances $(\mathrm{dist}_t)_{t \geq 0}$, we associate an entropy which measures the growth rate (when t tends to infinity) of the number of balls of small radius ϵ, in the metric dist_t, needed in order to cover the space X.

For hyperbolic Riemann surface laminations we define

$$\mathrm{dist}_t(x, y) := \inf_{\theta \in \mathbb{R}} \sup_{\xi \in \mathbb{D}_t} \mathrm{dist}_X(\phi_x(e^{i\theta}\xi), \phi_y(\xi)).$$

Recall that ϕ_x and ϕ_y are universal covering maps for the leaves through x and y, respectively, with $\phi_x(0) = x$ and $\phi_y(0) = y$. These maps are unique up to a rotation on $\mathbb{D}$. The metric dist_t measures how far two leaves get apart before the hyperbolic time t. It takes into account the time parametrization like in the classical case where one measures the distance of two orbits before time n, by measuring the distance at each time $i < n$. So, we are not just concerned with geometric proximity.

We will show that our entropy is finite for compact hyperbolic laminations which are transversally smooth. The notion of entropy can be extended to Riemannian foliations and a priori it is bigger than or equal to the geometric entropy introduced by Ghys, Langevin, and Walczak.

As for the tranverse regularity of the Poincaré metric, the main tool is to estimate the distance between leaves using the Beltrami equation in order to go from geometric estimates to the analytic ones needed in our definition. The advantage here is that the hyperbolic time we choose is canonical. So, the value of the entropy is unchanged under homeomorphisms between laminations which are holomorphic along leaves.

The proof that the entropy is finite for singular foliations is quite delicate and requires a careful analysis of the dynamics around the singularities. We will consider this problem in the second part of the paper [10]. We will discuss in Section 4 a notion of metric entropy for harmonic probability measures and give there some open questions.

2   POINCARÉ METRIC ON LAMINATIONS

In this section, we give some basic properties of laminations by hyperbolic Riemann surfaces. For background, see [9], [15], [19], and [29]. We will show that the Poincaré metric on leaves of a smooth compact hyperbolic lamination is transversally Hölder continuous.

Let X be a locally compact space. A *lamination* or *Riemannian lamination* $\mathcal{L}$ on X is the data of an atlas with charts

$$\Phi_i \colon \mathbb{U}_i \to \mathbb{B}_i \times \mathbb{T}_i.$$

Here, $\mathbb{T}_i$ is a locally compact metric space, $\mathbb{B}_i$ is a domain in $\mathbb{R}^n$, Φ_i is a homeomorphism defined on an open subset $\mathbb{U}_i$ of X, and all the changes of coordinates $\Phi_i \circ \Phi_j^{-1}$ are of the form

$$(x, t) \mapsto (x', t'), \quad x' = \Psi(x, t), \quad t' = \Lambda(t),$$

where Ψ, Λ are continuous maps and Ψ is smooth with respect to x, i.e., the function $t \mapsto \Psi(\cdot, t)$ is continuous for the smooth topology.

The open set $\mathbb{U}_i$ is called a *flow box* and the manifold $\Phi_i^{-1}\{t = c\}$ in $\mathbb{U}_i$ with $c \in \mathbb{T}_i$ is a *plaque*. The property of the preceding coordinate changes insures that the plaques in different flow boxes are compatible in the intersection of the boxes. Two plaques are *adjacent* if they have non-empty intersection. In what follows, we always reduce slightly flow boxes in order to avoid a bad geometry near their boundaries. For simplicity we consider only $\mathbb{B}_i$ which are homeomorphic to a ball.

A *leaf* L is a minimal connected subset of X such that if L intersects a plaque, it contains that plaque. So, a leaf L is a connected real manifold of dimension n immersed in X which is a union of plaques. It is not difficult to see that $\overline{L}$ is also a lamination. A *chain of plaques* is a sequence $P_0, \ldots, P_m$ of plaques such that P_i is adjacent to P_{i+1} for $i = 0, \ldots, m - 1$. These plaques belong necessarily to the same leaf.

A *transversal* in a flow box is a closed set of the box which intersects every plaque in one point. In particular, $\Phi_i^{-1}(\{x\} \times \mathbb{T}_i)$ is a transversal in $\mathbb{U}_i$ for any $x \in \mathbb{B}_i$. In order to simplify the notation, we often identify $\mathbb{T}_i$ with $\Phi_i^{-1}(\{x\} \times \mathbb{T}_i)$ for some $x \in \mathbb{B}_i$ or even identify $\mathbb{U}_i$ with $\mathbb{B}_i \times \mathbb{T}_i$ via the map Φ_i.

We are mostly interested in the case where the $\mathbb{T}_i$ are closed subsets of smooth real manifolds and the functions Ψ, Λ are smooth in all variables. In this case, we say that the lamination is *smooth* or *transversally smooth*. If, moreover, X is compact, we can embed it in an $\mathbb{R}^N$ in order to use the distance induced by a Riemannian metric on $\mathbb{R}^N$. When X is a Riemannian manifold and the leaves of $\mathcal{L}$ are manifolds immersed in X, we say that $(X, \mathcal{L})$ is a *foliation* and we often assume that the foliation is *transversally smooth*, i.e., the maps Φ_i are smooth.

In the definition of laminations, if all the $\mathbb{B}_i$ are domains in $\mathbb{C}$ and Ψ is holomorphic with respect to x, we say that $(X, \mathcal{L})$ is a *Riemann surface lamination*. Recall that a *Riemann surface lamination with singularities* is the data $(X, \mathcal{L}, E)$, where X is a locally compact space, E a closed subset of X and $(X \setminus E, \mathcal{L})$ is a Riemann surface lamination. The set E is *the singularity set* of the lamination and we assume that $\overline{X \setminus E} = X$, see, e.g., [9] and [14] for more details.

Consider now a smooth Riemann surface lamination $(X, \mathcal{L})$. When we do not assume that X is compact, our discussion can be applied to singular laminations by considering their regular parts. Assume that the leaves of X are all (Kobayashi) hyperbolic. Let $\phi_x \colon \mathbb{D} \to L_x$ be a universal covering map of the leaf through x with $\phi_x(0) = x$. Then, the Poincaré metric ω_P on $\mathbb{D}$, defined by

$$\omega_P(\zeta) := \frac{2}{(1 - |\zeta|^2)^2} id\zeta \wedge d\bar{\zeta}, \qquad \zeta \in \mathbb{D},$$

induces, via ϕ_x, the Poincaré metric on L_x which depends only on the leaf. The latter metric is given by a positive $(1,1)$-form on L_x that we also denote by ω_P. The associated distance is denoted by dist_P.

Let ω be a Hermitian metric on the leaves which is transversally smooth. We can construct such a metric on flow boxes and glue them using a partition of unity. We have

$$\omega = \eta^2 \omega_P, \quad \text{where} \quad \eta(x) := \|D\phi_x(0)\|.$$

Here, for the norm of the differential $D\phi_x$ we use the Poincaré metric on $\mathbb{D}$ and the Hermitian metric ω on L_x.

The extremal property of the Poincaré metric implies that

$$\eta(x) = \sup \left\{ \|D\phi(0)\|, \quad \phi \colon \mathbb{D} \to L \text{ holomorphic such that } \phi(0) = x \right\}.$$

Using a map sending $\mathbb{D}$ to a plaque, we see that the function η is locally bounded from below by a strictly positive constant. When X is compact and the leaves are hyperbolic, the classical Brody lemma (see [26, p. 100]) implies that η is also bounded from above.

Now fix a distance dist_X on X such that on flow boxes $\mathbb{U} = \mathbb{B} \times \mathbb{T}$ as above, it is locally equivalent to the distance induced by a Riemannian metric. Here is the main theorem of this section.

THEOREM 2.1. *Let $(X, \mathcal{L})$ be a smooth compact lamination by hyperbolic Riemann surfaces. Then the Poincaré metric on the leaves is Hölder continuous, that is, the function η defined earlier is Hölder continuous on X.*

The proof occupies the rest of this section. The result is also valid for laminations which are transversally of class $\mathcal{C}^{2+\alpha}$ with $\alpha > 0$. In order to simplify the notation, we embed X in an $\mathbb{R}^N$ and use the distance dist_X induced by the Euclidean metric on $\mathbb{R}^N$. Multiplying ω by a constant, we can assume that $\omega \leq \omega_P$

on leaves, i.e., $\eta \leq 1$. We also have $\omega_P \leq A\omega$, i.e., $\eta \geq 1/A$, for some fixed constant $A \geq 1$. Fix also a fine-enough atlas of X. We will consider only finite atlases which are finer than this one. For simplicity, all the plaques we consider are small and simply connected. We also use a coordinate change on $\mathbb{R}^N$ and choose A large enough such that $\text{dist}_X \leq \text{dist}_P \leq A \text{dist}_X$ on plaques. The second inequality does not hold when we deal with singular foliations.

Let ϕ and ϕ' be two maps from a space Σ to X. If K is a subset of Σ, define

$$\text{dist}_K(\phi, \phi') := \sup_{a \in K} \text{dist}_X(\phi(a), \phi'(a)).$$

Consider constants $R \gg 1$ and $0 < \delta \ll 1$ such that $e^{2R}\delta \leq 1$. We say that two points x and y in X are *conformally (R, δ)-close* if the following property is satisfied and if it also holds when we exchange x and y.

Let $\phi_x \colon \mathbb{D} \to L_x$ and $\phi_y \colon \mathbb{D} \to L_y$ be universal covering maps with $\phi_x(0) = x$ and $\phi_y(0) = y$. There is a smooth map $\psi \colon \overline{\mathbb{D}}_R \to L_y$ without critical point such that $\psi(0) = y$, $\text{dist}_{\overline{\mathbb{D}}_R}(\phi_x, \psi) \leq \delta$ and ψ is δ-conformal in the following sense. Since $\mathbb{D}_R$ is simply connected, there is a unique smooth map $\tau \colon \overline{\mathbb{D}}_R \to \mathbb{D}$ such that $\psi = \phi_y \circ \tau$ and $\tau(0) = 0$. We assume that $\|D\tau\|_\infty \leq 2A$ and the *Beltrami coefficient* μ_τ of τ satisfies $\|\mu_\tau\|_{\mathcal{C}^1} \leq \delta$. Here, we consider the norm of the differential $D\tau$ with respect to the Poincaré metric on $\mathbb{D}$ and the norm of μ_τ with respect to the Euclidean metric. Recall also that μ_τ is defined by

$$\frac{\partial \tau}{\partial \overline{\xi}} = \mu_\tau \frac{\partial \tau}{\partial \xi}.$$

Note that the preceding notion is independent of the choice of ϕ_x and ϕ_y since these maps are defined uniquely up to a rotation on $\mathbb{D}$. We have the following important estimate.

PROPOSITION 2.2. *Let x and y be conformally (R, δ)-close as before (in particular, with $\delta \leq e^{-2R}$). There is a real number θ such that if $\phi'_y(\xi) := \phi_y(e^{i\theta}\xi)$, then*

$$|\eta(x) - \eta(y)| \leq A'e^{-R} \quad and \quad \text{dist}_{\mathbb{D}_{R/3}}(\phi_x, \phi'_y) \leq A'e^{-R/3},$$

where $A' > 0$ is a constant independent of R, δ, x and y.

In what follows, we use $\lesssim$ to denote an inequality up to a multiplicative constant independent of R, δ, x and y. We will need the following quantified version of Schwarz's lemma.

LEMMA 2.3. *Let $\widetilde{\tau} \colon \mathbb{D}_R \to \mathbb{D}$ be a holomorphic map such that $\widetilde{\tau}(0) = 0$. Write $D\widetilde{\tau}(0) = \lambda e^{i\theta}$, with $\lambda > 0$ and $\theta \in \mathbb{R}$. Assume that $1 - \lambda \lesssim e^{-R}$. Then, we have*

$$\text{dist}_P(\widetilde{\tau}(\xi), e^{i\theta}\xi) \lesssim e^{-R/3} \quad for \quad \xi \in \mathbb{D}_{R/3}.$$

PROOF. We can assume that $\theta = 0$. Since R is large, we can compose $\widetilde{\tau}$ with a slight contraction in order to assume that $\widetilde{\tau}$ is defined from $\mathbb{D}$ to $\mathbb{D}$. The computation is essentially the same. So, we still have $1 - \lambda \lesssim e^{-R}$ and $1 - \lambda > 0$, by Schwarz's lemma.

Consider the holomorphic function $u \colon \mathbb{D} \to \mathbb{D}$ defined by

$$u(\xi) := \xi^{-1}\widetilde{\tau}(\xi) \quad and \quad u(0) := \lambda.$$

Observe that $1 - |\xi| \gtrsim e^{-R/3}$ for $\xi \in \mathbb{D}_{R/3}$ and $|1 - \overline{u(\xi)}|\xi|^2| \gtrsim 1 - |\xi|$. Therefore,

$$\text{dist}_P(\widetilde{\tau}(\xi), \xi) = 2\tanh^{-1} \frac{|\xi||1 - u(\xi)|}{|1 - \overline{u(\xi)}|\xi|^2|} \lesssim e^{R/3}|1 - u(\xi)|.$$

It is enough to show that $|1 - u| \lesssim e^{-2R/3}$ on $\mathbb{D}_{R/3}$.

Since u is holomorphic, it contracts the Poincaré metric on $\mathbb{D}$. So, it sends $\mathbb{D}_{R/3}$ to the disc of radius $R/3$ centered at $u(0) = \lambda$. We obtain the desired inequality using that $\text{dist}_P(0, \lambda) \geq R + o(R)$ because $0 < 1 - \lambda \lesssim e^{-R}$. □

PROOF OF PROPOSITION 2.2. We first construct a homeomorphism $\sigma \colon \mathbb{D}_R \to \mathbb{D}_R$, close to the identity, such that $\widetilde{\tau} := \tau \circ \sigma^{-1}$ is holomorphic. For this purpose, it is enough to construct σ satisfying the following Beltrami equation

$$\frac{\partial \sigma}{\partial \overline{\xi}} = \mu_\tau \frac{\partial \sigma}{\partial \xi}.$$

Indeed, it is enough to compute the derivatives of $\tau = \widetilde{\tau} \circ \sigma$ and to use the preceding equation together with the property that $\|\mu_\tau\|_\infty < 1$ in order to obtain that $\overline{\partial}\widetilde{\tau} = 0$.

It is well known from the Ahlfors-Bers theory (see, e.g., [13, p. 181]) that there is a solution such that

$$\|\sigma - \text{id}\|_{\mathcal{C}^1} \lesssim \|\mu_\tau\|_{\mathcal{C}^1} \lesssim \delta,$$

where we use the Euclidean metric on $\mathbb{D}_R$. We deduce that

$$\|\sigma^{-1} - \text{id}\|_{\mathcal{C}^1} \lesssim \delta \lesssim e^{-2R}.$$

Moreover, we can also compose σ with an automorphism of $\mathbb{D}_R$ in order to get that $\sigma(0) = 0$. Now, it is not difficult to see that

$$\text{dist}_P(\sigma^{-1}(\xi), \xi) = 2\tanh^{-1} \frac{|\sigma^{-1}(\xi) - \xi|}{|1 - \overline{\sigma^{-1}(\xi)}\xi|} \lesssim e^R \delta \quad \text{for} \quad \xi \in \mathbb{D}_R.$$

Define $\widetilde{\phi}_y := \phi_y \circ \widetilde{\tau}$. This is a holomorphic map. Recall that $\psi = \phi_y \circ \tau$ and $\|D\tau\|_\infty \leq 2A$. Therefore, since ϕ_y is isometric with respect to the Poincaré metric, we obtain from the previous estimates that

$$\text{dist}_{\mathbb{D}_R}(\widetilde{\phi}_y, \psi) \lesssim e^R \delta$$

which, coupled with the hypothesis $\text{dist}_{\mathbb{D}_R}(\phi_x, \psi) \leq \delta$, implies that

$$\text{dist}_{\mathbb{D}_R}(\phi_x, \widetilde{\phi}_y) \lesssim e^R \delta.$$

For a constant $R_0 > 0$ small enough, ϕ_x and $\widetilde{\phi}_y$ send $\mathbb{D}_{R_0}$ to the same flow box where Cauchy's formula implies that

$$\|D\phi_x(0)\| - \|D\widetilde{\phi}_y(0)\| \lesssim e^R \delta.$$

The extremal property of the Poincaré metric yields

$$\|D\widetilde{\phi}_y(0)\| \leq \frac{1}{r}\|D\phi_y(0)\| = (1 + O(e^{-R}))\|D\phi_y(0)\|,$$

where r is given by $R = \log[(1+r)/(1-r)]$, that is, $\mathbb{D}_R = r\mathbb{D}$. Recall that $\|D\phi_x(0)\| = \eta(x)$ and $\|D\phi_y(0)\| = \eta(y)$. We deduce that

$$\|D\phi_y(0)\| \geq \|D\widetilde{\phi}_y(0)\| + O(e^{-R}\eta(y)).$$

Therefore,

$$\eta(x) - \eta(y) \leq \|D\phi_x(0)\| - \|D\widetilde{\phi}_y(0)\| + O(e^{-R}\eta(y)) \leq e^R\delta + O(e^{-R}\eta(y)).$$

By symmetry we get

$$|\eta(x) - \eta(y)| \lesssim e^R\delta + e^{-R}(\eta(x) + \eta(y)).$$

This, combined with the hypothesis that $e^{2R}\delta \leq 1$ and $\frac{1}{A} \leq \eta \leq 1$, implies the first estimate in the proposition.

We deduce from the preceding inequalities that

$$\|D\phi_y(0)\| - \|D\widetilde{\phi}_y(0)\| \lesssim e^R\delta + e^{-R}(\eta(x) + \eta(y)).$$

Write $D\widetilde{\tau}(0) = \lambda e^{i\theta}$ with $\lambda \geq 0$ and $\theta \in \mathbb{R}$. Since $e^{2R}\delta \leq 1$ and $\frac{1}{A} \leq \eta \leq 1$, and ϕ_y is isometric with respect to the Poincaré metric, we obtain that $1 - \lambda \lesssim e^{-R}$. By Lemma 2.3, we have

$$\text{dist}_P(\widetilde{\tau}(\xi), e^{i\theta}\xi) \lesssim e^{-R/3} \quad \text{for} \quad \xi \in \mathbb{D}_{R/3}.$$

Define $\phi'_y(\xi) := \phi_y(e^{i\theta}\xi)$. Since ϕ_y is isometric with respect to the Poincaré metric, we obtain that

$$\text{dist}_{\mathbb{D}_{R/3}}(\phi'_y, \widetilde{\phi}_y) \lesssim e^{-R/3}.$$

This, combined with the earlier estimate on the distance between ϕ_x and $\widetilde{\phi}_y$, implies the result. $\qquad \square$

We continue the proof of Theorem 2.1. Fix a finite atlas $\mathcal{U}^l$ fine enough, another finer atlas $\mathcal{U}^n$, and a third atlas $\mathcal{U}^s$ which is finer than $\mathcal{U}^n$. The flow boxes and plaques of $\mathcal{U}^s$, $\mathcal{U}^n$, and $\mathcal{U}^l$ are said to be *small*, *normal*, and *large* respectively. Moreover, we can construct these atlases so that the following property is true for a fixed constant $0 < d \ll A^{-2}$ and for the distance dist_X on plaques:

(A1) Any disc of diameter d in a plaque is contained in a small plaque; small (resp. normal) plaques are of diameter less than $2d$ (resp. $10^4 dA$); the intersection of any large plaque with any flow box is contained in a plaque of this box.

Moreover, we can construct these atlases so that the following properties are satisfied. To each small flow box $\mathbb{U}^s$, we can associate a normal flow box $\mathbb{U}^n$ and a large flow box $\mathbb{U}^l$ such that $\mathbb{U}^s \Subset \mathbb{U}^n \Subset \mathbb{U}^l$ and for all plaques P^s, P^n, P^l in $\mathbb{U}^s$, $\mathbb{U}^n$ and $\mathbb{U}^l$ respectively, the following hold.

(A2) If P^n and P^l are adjacent, then $P^n \subset P^l$ and $\text{dist}_X(\partial P^l, P^n) \geq 10^6 dA^2$.

(A3) If P^s and P^n are adjacent, then $P^s \subset P^n$ and $\text{dist}_X(\partial P^n, P^s) \geq 10^2 dA$.

(A4) The projection Φ from P^n to P^l is well defined and smooth, and its image is compact in P^l; the projection of P^s in P^l is compact in the projection of P^n.

Here, we use dist_X in order to define the projection. Fix a constant $\kappa > 1$ large enough such that the following holds:

(A5) If x is a point in P^n, then for the $\mathcal{C}^2$-norm on P^n

$$\|\Phi - \text{id}\|_{\mathcal{C}^2} \leq e^\kappa \, \text{dist}_X(x, \Phi(x)).$$

We should note that for laminations which are transversally of class $\mathcal{C}^{2+\alpha}$ with $\alpha > 0$, property (A5) is rephrased as follows

$$\|\Phi - \text{id}\|_{\mathcal{C}^2} \leq e^\kappa \big[\text{dist}_X(x, \Phi(x)) \big]^\alpha.$$

Note that in what follows, to each small flow box $\mathbb{U}^s$ we fix a choice of the associated boxes $\mathbb{U}^n$ and $\mathbb{U}^l$ and we will only consider projections from plaques to plaques as described before. Moreover, a small, normal or large plaque is associated to a unique small, normal or large flow box. We have the following property for some fixed constant $\epsilon_0 > 0$ small enough:

(A6) Two points at distance less than ϵ_0 belong to the same small flow box and $\text{dist}_X(\partial \mathbb{U}^n, \mathbb{U}^s) > \epsilon_0$ for $\mathbb{U}^s, \mathbb{U}^n$ as defined earlier.

Consider now two points x, y such that $\text{dist}_X(x, y) \leq e^{-10\kappa d^{-1} AR}$ for $R > 0$ large enough. We will show that $|\eta(x) - \eta(y)| \lesssim e^{-R}$. This implies that η is Hölder continuous with Hölder exponent $(10\kappa d^{-1} A)^{-1}$.

Since x and y are close, by (A6), they belong to a small flow box $\mathbb{U}^s$. Consider the projection x' of x to the normal plaque containing y. We have

$$\text{dist}_X(x, x') \leq e^{-10\kappa d^{-1} AR} \quad \text{and} \quad \text{dist}_X(x', y) \leq 2e^{-10\kappa d^{-1} AR}.$$

By Proposition 2.2, it is enough to check that x and y are conformally (R, δ)-close with $\delta := e^{-2R}$. So, we have to construct the map ψ satisfying the definition of conformally (R, δ)-close points as above. We claim that it is enough to consider the case where $y = x'$.

Indeed, if we can construct a map ψ for x, x', there is a point $a \in \mathbb{D}$ such that $\text{dist}_P(0, a) \simeq \text{dist}_P(x', y) \lesssim e^{-10\kappa d^{-1} AR}$ and $\psi(a) = y$. There is an automorphism $u \colon \mathbb{D}_R \to \mathbb{D}_R$ very close to the identity such that $u(0) = a$. Therefore, if we are able to construct the map ψ for x and x', we obtain such a map for x, y by replacing ψ by $\psi \circ u$. Since u is very close to the identity, the estimates does not change too much. So, we can assume that $y = x'$.

The map ψ will be obtained by composing ϕ_x with local projections from the leaf L_x to the leaf L_y. The main problem is to show that the map is well defined. Let P_1^s be a small plaque containing x and let $\mathbb{U}_1^s$ be the associated small flow box. It is clear that y belongs to the associated normal flow box $\mathbb{U}_1^n$. Denote by Q_1^n the plaque of $\mathbb{U}_1^n$ containing y. The projection Φ_1 from P_1 to Q_1^n is well defined, as described earlier.

Consider a chain $\mathcal{P} = \{P_1^s, \dots, P_m^s\}$ of m small plaques with $m \leq 3d^{-1} AR$. Denote by $\mathbb{U}_i^s$ the small flow box associated to P_i^s and let $\mathbb{U}_i^n$ and $\mathbb{U}_i^l$ be the normal and large flow boxes associated to $\mathbb{U}_i^s$. We have the following lemma.

LEMMA 2.4. *There is a unique chain* $\mathcal{Q} = \{Q_1^n, \dots, Q_m^n\}$ *such that* Q_i^n *is a plaque of* $\mathbb{U}_i^n$. *Moreover, the projection* Φ_i *from* P_i^s *to* Q_i^n *satisfies* $\Phi_i = \Phi_{i+j}$ *on* $P_i^s \cap P_{i+j}^s$ *for* $0 \leq j \leq 10A$. *We also have* $\text{dist}_X(P_i^s, Q_i^n) \leq e^{-4\kappa R}$ *for* $i = 1, \dots, m$.

PROOF. Note that since $\mathrm{dist}_X(x, y) \leq e^{-10\kappa d^{-1}AR}$ and $m \leq 3d^{-1}AR$, the last assertion of the lemma is a consequence of the previous ones and the property (A5) applied to points in the intersections $P_i^s \cap P_{i+1}^s$. Indeed, by induction on i, we obtain

$$\mathrm{dist}_X(P_i^s, Q_i^n) \leq e^{-10\kappa d^{-1}AR} e^{i\kappa} \leq e^{-4\kappa R}.$$

We prove the other assertions by induction on m. Assume these properties for $m-1$, i.e., we already have the existence and the uniqueness of Q_i^n for $i < m$. We have to construct Q_m^n and to prove its uniqueness.

Let Q_{m-1}^l be the large plaque associated to Q_{m-1}^n. If Q_m^n exists, since it intersects Q_{m-1}^n, by (A1) and (A2), it is contained in Q_{m-1}^l and then it is the intersection of Q_{m-1}^l with $\mathbb{U}_m^n$. The uniqueness of Q_m^n follows.

Fix a point z in $P_{m-1}^s \cap P_m^s$. Since $\Phi_{m-1}(z)$ is close to z, by (A6), $\Phi_{m-1}(z)$ belongs to $\mathbb{U}_m^n$. Define Q_m^n as the plaque of $\mathbb{U}_m^n$ containing this point. So, Q_m^n intersects Q_{m-1}^n and $Q_0^n, \ldots, Q_m^n$ is a chain. By (A4), $\Phi_{m-1}(z)$ is also the projection of z to Q_{m-1}^l. Since Q_m^n is contained in Q_{m-1}^l, necessarily, the projection $\Phi_m(z)$ of z to Q_m^n coincides with $\Phi_{m-1}(z)$.

Arguing in the same way, we obtain that P_{i+j}^s is contained in P_i^n, Q_{i+j}^n is contained in Q_i^l when $j \leq 10A$ and then we obtain that $\Phi_i = \Phi_{i+j}$ on $P_i^s \cap P_{i+j}^s$. $\square$

End of the proof of Theorem 2.1. We have to show that x and y are conformally (R, δ)-close for a map ψ that we are going to construct. We call also small plaque any open set in $\mathbb{D}$ which is sent bijectively by ϕ_x to a small plaque in L_x. Let γ be a radius of $\mathbb{D}_R$. We divide γ into equal intervals γ_i of Poincaré length $\simeq d/2$. By (A1), since $\mathrm{dist}_X \leq \mathrm{dist}_P \leq A \mathrm{dist}_X$ on plaques, we can find a small plaque P_i containing γ_i such that $\mathrm{dist}_P(\gamma_i, \partial P_i) \geq d/(4A)$. So, we have a chain of $m \simeq 2d^{-1}R$ plaques which covers γ. Define $P_i^s := \phi_x(P_i)$, Q_i^n as in Lemma 2.4 and $\psi := \Phi_i \circ \phi_x$ on P_i. We will check later that ψ is well defined on $\mathbb{D}_R$.

It follows from the last assertion in Lemma 2.4 that $\mathrm{dist}_{\mathbb{D}_R}(\phi_x, \psi) \leq e^{-4\kappa R} \leq \delta$. We also deduce from (A5) that ψ has no critical point and that its Beltrami coefficient satisfies $\|\mu_\tau\|_{\mathcal{C}^1} \lesssim e^{-4\kappa R}$ for the Poincaré metric on $\mathbb{D}$ and consequently $\|\mu_\tau\|_{\mathcal{C}^1} \leq e^{-2\kappa R} \leq \delta$ for the Euclidean metric on $\mathbb{D}_R$. Here, we use that ϕ_x and ϕ_y are isometries with respect to the Poincaré metric.

The property $\|D\tau\|_\infty \leq 2A$ is also clear since we have $\|D\Phi_i\|_\infty \leq 2$, locally $\tau = \phi_y^{-1} \circ \psi = \phi_y^{-1} \circ \Phi_i \circ \phi_x$ and $\mathrm{dist}_X \leq \mathrm{dist}_P \leq A \mathrm{dist}_X$ on plaques. This implies that x, y are conformally (R, δ)-close. It remains to check that ψ is well defined.

If $P_i \cap P_{i+j} \neq \varnothing$, then by (A1), we have $\mathrm{diam}_P(P_i \cup P_{i+j}) \leq 4dA$ because $\mathrm{dist}_P \leq A \mathrm{dist}_X$. Hence, $j \leq 10A$. By Lemma 2.4, ψ is well defined on $P_i \cup P_{i+j}$. So, ψ is well defined on the union W of P_i and this union contains the radius γ. We will show later that ψ extends to the union W' of all small plaques which intersect γ. Of course, we only use projections from plaques to plaques in order to define the extension of ψ. So, the extension is unique. Let γ' be another radius of $\mathbb{D}_R$ such that the angle between γ and γ' is small enough, e.g., less than $d/(10^2 A e^{2R})$. Then, γ' is contained in W'. Observe that if we repeat the same construction of ψ for γ', the plaques P_i' used to cover γ' intersect γ because $\mathrm{dist}_P(\partial P_i', \gamma_i') \geq d/(4A)$ and $\mathrm{dist}_P(\gamma, \xi) \leq d/(4A)$ for $\xi \in \gamma'$. Therefore, the obtained values of ψ on γ' coincide with the extension to W' described earlier. A simple compactness argument implies the existence of a well-defined map ψ on $\mathbb{D}_R$.

We check now that ψ can be extended from W to W'. Consider a plaque P (in $\mathbb{D}$) which intersects γ_i and a plaque $\widetilde{P}$ which intersects γ_{i+j}. Assume that $P \cap \widetilde{P} \neq \varnothing$. It suffices to check that ψ can be extended to $W \cup P \cup \widetilde{P}$. As before, we obtain that $j \leq 10A$. This allows us by the previous arguments to see that $P^s := \phi_x(P)$, $\widetilde{P}^s := \phi_x(\widetilde{P})$ and $P_i^s, \ldots P_{i+j}^s$ belong to the normal plaque P_i^n. The projection from P_i^n to Q_i^l gives us the unique extension of ψ. The proof of Theorem 2.1 is now complete.   □

3   HYPERBOLIC ENTROPY FOR FOLIATIONS

In this section, we introduce a general notion of entropy, which permits to describe some natural situations in dynamics and in foliation theory. We will show that the entropy of any compact smooth lamination by hyperbolic Riemann surfaces is finite.

Let X be a metric space endowed with a distance dist_X. Consider a family $\mathcal{D} = \{\mathrm{dist}_t\}$ of distances on X indexed by $t \in \mathbb{R}^+$. We can also replace $\mathbb{R}^+$ by $\mathbb{N}$, and in practice we often have that $\mathrm{dist}_0 = \mathrm{dist}_X$ and that dist_t is increasing with respect to $t \geq 0$. In several interesting situations the metrics dist_t are continuous with respect to dist_X.

Let Y be a non-empty subset of X. Denote by $N(Y, t, \epsilon)$ the minimal number of balls of radius ϵ with respect to the distance dist_t needed to cover Y. Define the *entropy* of Y with respect to $\mathcal{D}$ by

$$h_{\mathcal{D}}(Y) := \sup_{\epsilon > 0} \limsup_{t \to \infty} \frac{1}{t} \log N(Y, t, \epsilon).$$

When $Y = X$ we will denote by $h_{\mathcal{D}}$ this entropy. When X is not compact, we can also consider the supremum of the entropies on compact subsets of X. Note that if Y and Y' are two subsets of X, then $h_{\mathcal{D}}(Y \cup Y') = \max(h_{\mathcal{D}}(Y), h_{\mathcal{D}}(Y'))$.

Observe that when dist_t is increasing, $N(Y, t, \epsilon)$ is increasing with respect to $t \geq 0$. Moreover,

$$\limsup_{t \to \infty} \frac{1}{t} \log N(Y, t, \epsilon)$$

is increasing when ϵ decreases. So, in the preceding definition, we can replace $\sup_{\epsilon > 0}$ by $\lim_{\epsilon \to 0^+}$. If $\mathcal{D} = \{\mathrm{dist}_t\}$ and $\mathcal{D}' = \{\mathrm{dist}_t'\}$ are two families of distances on X such that $\mathrm{dist}_t' \geq A \, \mathrm{dist}_t$ for all t with a fixed constant $A > 0$, then $h_{\mathcal{D}'} \geq h_{\mathcal{D}}$.

A subset $F \subset Y$ is said to be (t, ϵ)-*dense in* Y if the balls of radius ϵ with respect to dist_t, centered at a point in F, cover Y. Let $N'(Y, t, \epsilon)$ denote the minimal number of points in a (t, ϵ)-dense subset of Y.

Two points x and y in X are said to be (t, ϵ)-*close* if $\mathrm{dist}_t(x, y) \leq \epsilon$. A subset $F \subset X$ is said to be (t, ϵ)-*separated* if for all distinct points x, y in F we have $\mathrm{dist}_t(x, y) > \epsilon$. Let $M(Y, t, \epsilon)$ denote the maximal number of points in a (t, ϵ)-separated family $F \subset Y$. The proof of the following proposition is immediate.

PROPOSITION 3.1. *We have*

$$N(Y, t, \epsilon) \leq N'(Y, t, \epsilon) \leq M(Y, t, \epsilon) \leq N\left(Y, t, \frac{\epsilon}{2}\right).$$

In particular,

$$h_{\mathcal{D}}(Y) = \sup_{\epsilon>0} \limsup_{t\to\infty} \frac{1}{t} \log N'(Y,t,\epsilon) = \sup_{\epsilon>0} \limsup_{t\to\infty} \frac{1}{t} \log M(Y,t,\epsilon).$$

The following proposition gives a simple criterion for the finiteness of entropy. We will see that this criterion applies for smooth laminations by Riemann surfaces.

PROPOSITION 3.2. *Assume that there are positive constants A and m such that for every $\epsilon > 0$ small enough X admits a covering by fewer than $A\epsilon^{-m}$ balls of radius ϵ for the distance dist_X. Assume also that*

$$\mathrm{dist}_t \le e^{ct+d}\,\mathrm{dist}_X + \varphi(t)$$

for some constants $c,d \ge 0$ and a function φ with $\varphi(t) \to 0$ as $t \to \infty$. Then, the entropy $h_{\mathcal{D}}$ is at most equal to mc.

PROOF. Fix a constant ϵ small enough and consider only t large enough so that $\varphi(t) \le \epsilon/2$. If x and y are ϵ-separated for dist_t, then they are $\frac{1}{2}e^{-ct-d}\epsilon$-separated for dist_X. In particular, they cannot belong to a same ball of radius $\frac{1}{4}e^{-ct-d}\epsilon$ with respect to dist_X. Therefore, it follows from the hypothesis that

$$M(X,t,\epsilon) \le A4^m e^{mct+md}\epsilon^{-m}.$$

We easily deduce that $h_{\mathcal{D}} \le mc$. □

Consider now a general dynamical situation. We use the name *time space* to refer to the tuple $(\Sigma, \mathrm{dist}_\Sigma, 0_\Sigma, G)$, where $(\Sigma, \mathrm{dist}_\Sigma)$ is a metric space, 0_Σ is a point of Σ that we call *time zero*, and G is a group of isometries of Σ with 0_Σ as a common fixed point. The elements of G are called *time reparametrizations*.

In practice, the metric dist_Σ is complete, G is either $\{\mathrm{id}\}$ or the group of all the isometries fixing 0_Σ and preserving the orientation of Σ. The space Σ can be

(G1) one of the sets $\mathbb{N}, \mathbb{Z}, \mathbb{R}^+, \mathbb{R}, \mathbb{C}, \mathbb{R}^p, \mathbb{C}^p$ endowed with the usual distance;

(G2) or a group with a finite system of generators stable under inversion;

(G3) or the unit disc $\mathbb{D}$ in $\mathbb{C}$ with the Poincaré metric.

For laminations by Riemann surfaces, we will consider essentially the last case where G the group of rotations around $0 \in \mathbb{D}$. Note that the case where Σ is another symmetric domain may be also of interest.

Let (X, dist_X) be a metric space as discussed earlier. Define, for a subset $K \subset \Sigma$ and two maps ϕ, ϕ' from Σ to X,

$$\mathrm{dist}_K^G(\phi, \phi') := \inf_{\sigma,\sigma'\in G} \sup_{s\in K} \mathrm{dist}_X\left(\phi\circ\sigma(s), \phi'\circ\sigma'(s)\right).$$

When $G = \{\mathrm{id}\}$, we have

$$\mathrm{dist}_K^G(\phi, \phi') = \mathrm{dist}_K(\phi, \phi') := \sup_{s\in K} \mathrm{dist}_X\left(\phi(s), \phi'(s)\right).$$

For $t > 0$ let

$$\Sigma_t := \left\{s \in \Sigma,\ \mathrm{dist}_\Sigma(0_\Sigma, s) < t\right\}.$$

This set is invariant under the action of G. We define

$$\text{dist}_t(\phi, \phi') := \text{dist}_{\Sigma_t}^G(\phi, \phi') = \inf_{\sigma \in G} \sup_{s \in \Sigma_t} \text{dist}_X(\phi \circ \sigma(s), \phi'(s)).$$

Now consider a family $\mathcal{M}$ of maps from Σ to X satisfying the following properties:

(M1) for every $x \in X$ there is a map $\phi \in \mathcal{M}$ such that $\phi(0_\Sigma) = x$;

(M2) if ϕ, ϕ' are two maps in $\mathcal{M}$ such that $\phi(0_\Sigma) = \phi'(0_\Sigma)$, then $\phi = \phi' \circ \tau$ for some $\tau \in G$.

So, X is "laminated" by images of $\phi \in \mathcal{M}$: for every $x \in X$ there is a unique (up to time reparametrizations) map $\phi \in \mathcal{M}$ which sends the time zero 0_Σ to x. We get then a natural family $\{\text{dist}_t\}_{t \geq 0}$ on X.

Define for x and y in X

$$\text{dist}_t(x, y) := \text{dist}_t(\phi_x, \phi_y),$$

where ϕ_x, ϕ_y are in $\mathcal{M}$ such that $\phi_x(0_\Sigma) = x$ and $\phi_y(0_\Sigma) = y$. The definition is independent of the choice of ϕ_x and ϕ_y. It is clear that $\text{dist}_0 = \text{dist}_X$ and that the family $\mathcal{D} := \{\text{dist}_t\}_{t \geq 0}$ is increasing when t increases. For $Y \subset X$, denote by $h(Y)$ the associated entropy, where we drop the index $\mathcal{D}$ for simplicity.

Observe that the entropy depends on the metrics on Σ and on X. Nevertheless, the entropy does not change if we modify dist_X on a compact set keeping the same topology or if we replace dist_X by another distance dist_X' such that $A^{-1} \text{dist}_X \leq \text{dist}_X' \leq A \text{dist}_X$ for some constant $A > 0$.

We review some classical situations where we assume that $G = \{\text{id}\}$.

EXAMPLE 3.3. Consider a continuous map $f \colon X \to X$ and $f^n := f \circ \cdots \circ f$ (n times) the iterate of order n of f. For $x \in X$ define a map $\phi_x \colon \mathbb{N} \to X$ by $\phi_x(n) := f^n(x)$. For $\Sigma = \mathbb{N}$ and $\mathcal{M}$ the family of these maps ϕ_x, we obtain the *topological entropy* of f. More precisely, two points x and y in X are (n, ϵ)-separated if

$$\text{dist}_n(x, y) := \max_{0 \leq i \leq n-1} \text{dist}_X(f^i(x), f^i(y)) > \epsilon.$$

If f is K-Lipschitz continuous, then $\text{dist}_n \leq K^n \text{dist}_X$.

A subset F of X is (n, ϵ)-separated if its points are mutually (n, ϵ)-separated and the topological entropy of f is given by the formula

$$h(f) := \sup_{\epsilon > 0} \limsup_{n \to \infty} \frac{1}{n} \log \sup \{\#F, \ F \subset X \ \ (n, \epsilon)\text{-separated}\}.$$

This notion was introduced by Adler, Konheim, and McAndrew. The preceding formulation when f is uniformly continuous was introduced by Bowen [1]; see also Walters [33] or Katok-Hasselblatt [25]. When f is only continuous, Bowen considers the entropy of compact sets and then takes the supremum over compacta.

When f is a meromorphic map on a compact Kähler manifold M with indeterminacy set I, we can define $X := M \setminus \bigcup_n (f^{-1})^n(I)$. Then f is, in general, not uniformly continuous on X. However, it is shown in [12] that the entropy of $f|_X$ as defined earlier is finite. Indeed, it is dominated by the logarithm of the maximum of dynamical degrees; see also Gromov [21] and Yomdin [31] when f is holomorphic or smooth.

We can define the entropy of a map in a more general context. Suppose that f is defined only in $U \subset X$. We define $\text{dist}_n(x, y)$ only when $f^j(x)$ and $f^j(y)$ are well defined for $j < n$. Let $U_\infty := \cap f^{-n}(U)$. All the $\text{dist}_n(\cdot, \cdot)$ are well defined on $U_\infty \times U_\infty$ and we can consider the entropy of f on U_∞. This situation occurs naturally in holomorphic dynamics. See, e.g., the case of polynomial-like maps and horizontal-like maps [11, 8]. Note that the case where Σ is a tree is also interesting because it allows to consider the dynamics of correspondences.

EXAMPLE 3.4. Consider a flow $(\Phi_t)_{t \in \mathbb{R}}$ on a compact Riemannian manifold X. Define for $x \in X$ a map $\phi_x : \mathbb{R}^+ \to X$ by $\phi_x(t) := \Phi_t(x)$. If $\Sigma = \mathbb{R}^+$ and $\mathcal{M}$ is the family $\{\phi_x\}$, then the distance dist_t with $t \geq 0$ is given by

$$\text{dist}_t(x, y) := \sup_{0 \leq s < t} \text{dist}(\Phi_s(x), \Phi_s(y)).$$

We obtain the classical entropy of the flow $(\Phi_t)_{t \in \mathbb{R}}$, which is also equal to the topological entropy $h(\Phi_1)$ of Φ_1; see, e.g., Katok-Hasselblatt [25, Chapter 3, p. 112]. nnnn This notion can be extended without difficulty to complex flows.

EXAMPLE 3.5. (See Candel-Conlon [5], [6] and Walczak [32].) Let Γ be a group with a finite system of generators A. We assume that if $g \in A$, then $g^{-1} \in A$. The distance dist_Γ between g and g' in Γ is the minimal number n such that we can write $g^{-1}g'$ as a composition of n elements in A. The neutral element 1_Γ is considered as the origin. Consider an action of Γ on the left of a metric space X, that is, a representation of Γ in the group of bijections from X to X. Define for $x \in X$ the map $\phi_x : \Gamma \to X$ by $\phi_x(g) := gx$. For $\Sigma := \Gamma$ and $\mathcal{M}$ the family $\{\phi_x\}$, we obtain the entropy of the action of Γ on X. More precisely, let Γ_n be the ball of center 1_Γ and radius n in Γ with respect to the metric introduced above. Then

$$\text{dist}_n(x, y) := \sup_{g \in \Gamma_n} \text{dist}_X(gx, gy).$$

The entropy depends on the metric on Γ, i.e. on the choice of the system of generators A. We will denote it by h_A. If A' is another system of generators, there is a constant $c \geq 1$ independent of the action of Γ on X such that

$$c^{-1} h_{A'} \leq h_A \leq c h_{A'}.$$

The function describing the growth of Γ is

$$\text{lov}_A(\Gamma) := \limsup_{n \to \infty} \frac{\log \# \Gamma_n}{n}.$$

It also depends on the choice of generators. If the map $x \mapsto gx$ is uniformly Lipschitz for each generator g, we can compare $\text{lov}_A(\Gamma)$ and h_A. We get that if X has a finite box measure then $h_A \leq c \cdot \text{lov}_A(\Gamma)$ for some positive constant c.

When Γ is a hyperbolic group in the sense of Gromov [22], its Cayley graph can be compactified and the action of Γ extends to the boundary X of the Cayley graph. This allows us to define a natural notion of entropy for Γ which depends on the choice of generators.

The notion of entropy can be extended to any semi-group endowed with an invariant distance and then covers Examples 3.3 and 3.4.

We now consider the case of laminations, where the group G of time re-para-metrization is not trivial.

EXAMPLE 3.6. Let $(X, \mathcal{L})$ be a compact Riemannian laminations without singularities in a Riemannian manifold M. Assume that the lamination is transversally smooth. Ghys, Langevin, and Walczak [20] introduced and studied a notion of geometric entropy h_{GLW}. It can be summarized as follows. Define that x and y are (R, ϵ)-separated if $\delta_R(x, y) > \epsilon$ where δ_R will be defined shortly.

Denote by $\exp_x \colon \mathbb{R}^n \to L_x$ the exponential map for L_x such that $\exp_x(0) = x$. Here, we identify the tangent space of L_x at x with $\mathbb{R}^n$. So, $\exp_x$ is defined uniquely up to an element of the group $\text{SO}(n)$. Define

$$\delta'_R(x, y) := \inf_h \sup_{\xi \in B_R} \text{dist}_X(\exp_x(\xi), h(\xi)) \quad \text{and} \quad \delta_R(x, y) := \delta'_R(x, y) + \delta'_R(y, x).$$

Here, B_R denotes the ball of radius R in $\mathbb{R}^n$ and $h \colon B_R \to L_y$ is a continuous map with $h(0) = y$. This function δ_R measures the spreading of leaves. It seems that in general δ_R does not satisfy the triangle inequality. We have

$$h_{\text{GLW}} := \sup_{\epsilon > 0} \limsup_{R \to \infty} \frac{1}{R} \log \{\#F, \quad F \subset X \quad (R, \epsilon)\text{-separated as before}\}.$$

In our approach, choose $\Sigma = \mathbb{R}^n$, $G = \text{SO}(n)$ and $\mathcal{M}$ the family of all the exponential maps considered earlier. This allows us to define an entropy $h(\mathcal{L})$. Indeed, we define

$$d_R(x, y) := \inf_{g \in \text{SO}(n)} \sup_{\xi \in B_R} \text{dist}_X\big(\exp_x(g(\xi)), \exp_y(\xi)\big).$$

It is not difficult to see that

$$h_{\text{GLW}}(\mathcal{L}) \leq h(\mathcal{L}).$$

In the rest of this section, we consider a Riemann surface lamination $(X, \mathcal{L})$. We assume that all its leaves are hyperbolic. Choose $(\Sigma, 0_\Sigma) = (\mathbb{D}, 0)$ endowed with the Poincaré metric. The group G is the family of all rotations around 0. Define $\mathcal{M}$ as the family of all the universal covering maps $\phi \colon \mathbb{D} \to L$ associated to a leaf L. We obtain from the abstract formalism an entropy that we denote by $h(\mathcal{L})$. We call it the *hyperbolic entropy* of the lamination.

EXAMPLE 3.7. Consider the case where X is the Poincaré disc or a compact hyperbolic Riemann surface endowed with the Poincaré metric. Fix a constant $\epsilon > 0$ small enough. Lemma 3.8 shows that the property that x, y are (R, ϵ)-separated is almost equivalent to the property that $\text{dist}_X(x, y) \succeq e^{-R}$. It follows that the entropy of a compact subset of X is equal to its box dimension.

When X is a compact smooth lamination, we can choose a metric on X which is equivalent to the Poincaré metric on leaves. We see that moving along the leaves contributes 2 to the entropy of the lamination. This property is new in comparison with the theory of iteration of maps, but it is not verified for general non-compact laminations.

LEMMA 3.8. *Let $0 < \epsilon < 1$ be a fixed constant small enough. Then, there exist a constant $A \geq 1$ satisfying the following properties for all points a and b in $\mathbb{D}$:*

(i) *If $\operatorname{dist}_P(a, b) \leq A^{-1}e^{-R}$, then there are two automorphisms τ_a, τ_b of $\mathbb{D}$ such that $\tau_a(0) = a$, $\tau_b(0) = b$ and $\operatorname{dist}_P(\tau_a(\xi), \tau_b(\xi)) \leq \epsilon$ for every $\xi \in \mathbb{D}_R$.*

(ii) *If $Ae^{-R} \leq \operatorname{dist}_P(a, b) \leq 1$, then for all automorphisms τ_a, τ_b of $\mathbb{D}$ such that $\tau_a(0) = a$ and $\tau_b(0) = b$, we have $\operatorname{dist}_P(\tau_a(\xi), \tau_b(\xi)) > \epsilon$ for some $\xi \in \mathbb{D}_R$.*

PROOF. (i) Since we use here invariant metrics, it is enough to consider the case where $b = 0$ and $0 \leq a < 1$. We can also assume that $|a| \leq A^{-1}e^{-R}$, where $A \geq 1$ is a fixed large constant depending on ϵ. Consider the automorphisms $\tau_b := \operatorname{id}$ and $\tau_a := \tau$ with

$$\tau(z) := \frac{z + a}{1 + az}.$$

We compare them on $\mathbb{D}_R = r\mathbb{D}$, where $e^R = (1 + r)/(1 - r)$.

If r is not close to 1, the Poincaré metric is comparable with the Euclidean metric on $\mathbb{D}_R$ and on $\tau(\mathbb{D}_R)$ and it is not difficult to see that τ is close to id on $\mathbb{D}_R$. So, we can assume that r is close to 1.

We have $|a| \ll 1 - r$ and for $|z| = |x + iy| \leq r$,

$$\operatorname{dist}_P(z, \tau(z)) = 2\tanh^{-1} \frac{|\tau(z) - z|}{|1 - \bar{z}\tau(z)|} \simeq 2\tanh^{-1} \frac{a|1 - z^2|}{\sqrt{(1 - |z|^2)^2 + 4a^2y^2}}.$$

Since $|ay| \leq |a| \lesssim 1 - |z|$ and $1 - |z|^2 \simeq 1 - |z|$, the last expression is of order

$$\tanh^{-1} \frac{a|1 - z^2|}{1 - |z|} \ll 1 \quad \text{when} \quad |z| \leq r.$$

This gives the first assertion of the lemma.

(ii) As before, we can assume that $b = 0$. We can replace τ_a by $\tau_a \circ \tau_b^{-1}$ in order to assume that $\tau_b = \operatorname{id}$. Fix a constant $A > 0$ large enough depending on ϵ and assume that $Ae^{-R} \leq \operatorname{dist}_P(0, a) \leq 1$. So, R is necessarily large and r is close to 1. We first consider the case where $\tau_a = \tau$. For $z = ir$, the preceding computation gives

$$\operatorname{dist}_P(\tau(z), z) \simeq 2\tanh^{-1} \frac{2a}{\sqrt{4(1 - r^2)^2 + 4a^2}}.$$

This implies that $\operatorname{dist}_P(\tau(z), z) > 4\epsilon$ if A is large enough.

Consider now the general case where τ_a differs from τ by a rotation. There is a constant $-\pi \leq \theta \leq \pi$ such that $\tau_a(z) = \tau(e^{i\theta}z)$. Without loss of generality, we can assume that $-\pi \leq \theta \leq 0$. It is enough to show that

$$\operatorname{dist}_P(\tau_a(w), w) = \operatorname{dist}_P(\tau(z), w) > \epsilon$$

for $z = ir$ and $w = e^{-i\theta}z$.

Observe that since τ is conformal and fixes ± 1, it preserves every circle arc through $-1, 1$ and a point in $i\mathbb{R}$. Moreover, we check easily that the real part of $\tau(z)$ is positive. Since $\tau(z)$ belongs to the circle arc through $-1, 1$ and z, it follows that $\tau(z)$ is outside $\mathbb{D}_R$ and is on the right of $i\mathbb{R}$. Therefore, the geodesic segment joining $z_1 := \tau(z)$ and its projection z_2 (with respect to the Poincaré metric) to $i\mathbb{R}$

intersects $\partial\mathbb{D}_R$ at a point z_3. If $\mathrm{dist}_P(z_1, z_2) > \epsilon$, since w is on the left side of $i\mathbb{R}$, we have, necessarily, $\mathrm{dist}_P(z_1, w) > \epsilon$. Otherwise, we have $\mathrm{dist}_P(z_1, z_3) \leq \epsilon$ and since $\mathrm{dist}_P(z, z_1) > 4\epsilon$, we deduce that $\mathrm{dist}_P(z, z_3) > 3\epsilon$. Now, since w, z, z_3 are in $\partial\mathbb{D}_R$, we see, using an automorphism which sends z_3 to 0, that w is further than z from z_3, i.e., $\mathrm{dist}_P(w, z_3) > \mathrm{dist}_P(z, z_3)$. It follows that $\mathrm{dist}_P(w, z_3) > 3\epsilon$, and hence $\mathrm{dist}_P(w, z_1) > \epsilon$. This completes the proof of the lemma.   □

EXAMPLE 3.9. Let S be a hyperbolic compact Riemann surface. Let Γ denote the group of deck transformations of S, i.e., $\Gamma :\simeq \pi_1(S)$. Assume also that Γ acts on a compact metric space N as a group of homeomorphisms, that is, we have a representation of Γ into $\mathrm{Homeo}(N)$. For example, we can take $N = \partial\mathbb{D}$ or $\mathbb{P}^1$. Consider now the suspension which gives us a lamination by Riemann surfaces. More precisely, let $\widetilde{S} \simeq \mathbb{D}$ be the universal covering of S. The group Γ acts on $\widetilde{S} \times N$ by homeomorphisms

$$(\tilde{s}, x) \mapsto (\gamma\tilde{s}, \gamma x) \quad \text{with} \quad \gamma \in \Gamma.$$

This action is proper and discontinuous. The quotient $X := \Gamma\backslash(\widetilde{S} \times N)$ is compact and has a natural structure of a lamination by Riemann surfaces. Its leaves are the images of $\widetilde{S} \times \{x\}$ under the canonical projection $\pi\colon \widetilde{S} \times N \to X$.

Observe that the entropy of this lamination depends only on the representation of Γ in $\mathrm{Homeo}(N)$. So, we call it the entropy of the representation. It would be interesting to study these quantities as functions on moduli spaces of representations.

The entropy of the group Γ with respect to a system of generators is comparable with the entropy of the lamination. In particular, we can have laminations with positive entropy and with a transverse measure. More precisely, let f be a homeomorphism with positive entropy on a compact manifold N. It induces an action of $\mathbb{Z}$ on N. We then obtain an action of Γ using a group morphism $\Gamma \to \mathbb{Z}$. Indeed, if g is the genus of S, the group $\pi_1(S)$ is generated by $4g$ elements denoted by $a_i, b_i, a_i^{-1}, b_i^{-1}, 1 \leq i \leq g$, with the relation $a_1 b_1 a_1^{-1} b_1^{-1} \ldots a_g b_g a_g^{-1} b_g^{-1} = 1$. So, we can send a_1 to the homeomorphism f and the others a_i, b_i to the identity. Notice that in this case all positive $\partial\bar{\partial}$-closed currents directed by the lamination are closed; see, e.g., [16].

THEOREM 3.10. *Let $(X, \mathcal{L})$ be a smooth compact lamination by hyperbolic Riemann surfaces. Then, the entropy $h(\mathcal{L})$ is finite.*

PROOF. We will use the notations as in the proof of Theorem 2.1. Consider R large enough such that $e^{-R/3} \ll \epsilon$. We have seen in the proof of Theorem 2.1 that if $\mathrm{dist}_X(x, y) \leq e^{-10\kappa d^{-1}AR}$, then x and y are $(R/3, \epsilon)$-close. So, the maximal number of mutually $(R/3, \epsilon)$-separated points is smaller than a constant times $e^{10\kappa d^{-1}ANR}$ if the lamination is embedded in $\mathbb{R}^N$. So, the entropy $h(\mathcal{L})$ is at most equal to $30\kappa d^{-1}AN$.

Note that we can also apply Proposition 3.2 here with $\varphi = e^{-t}$. Indeed, using the arguments as before, we can show that $\mathrm{dist}_{R/3} \lesssim e^{10\kappa d^{-1}AR} \mathrm{dist}_X + \varphi(R/3)$.   □

Assume now that X is a Riemannian manifold of dimension k and $\mathcal{L}$ is transversally smooth. We can introduce various functionals in order to describe the dynamics. For example, we can introduce dimensional entropies for a foliation as is done in Buzzi [3] for maps. For an interger $1 \leq l \leq k$, consider the family $\mathcal{D}_l$ of manifolds of dimension l in X which are smooth up to the boundary. Define

$$h_l(\mathcal{L}) := \sup \{h(D) : D \in \mathcal{D}_l\},$$

where $h(D)$ is the entropy restricted to the set D as in the abstract setting. Clearly, this sequence of entropies is increasing with l.

We can also define

$$\widetilde{\chi}_l(x) := \sup_{\epsilon>0} \sup_{D} \limsup_{R\to\infty} -\frac{1}{R} \log \operatorname{vol}_l(D \cap B_R(x,\epsilon)).$$

Here, the supremum is taken over $D \in \mathcal{D}_l$ with $x \in D$. The Bowen ball $B_R(x,\epsilon)$ is associated to the lamination and is defined as in the abstract setting. The volume vol_l denotes the Hausdorff measure of dimension l. The function $\widetilde{\chi}_l$ is the analog of the sum of l largest Lyapounov exponents for dynamics of maps on manifolds. It measures how quickly the leaves get apart. We can consider this function relatively to a harmonic measure and show that it is constant when the measure is extremal. The definitions of h_l and $\widetilde{\chi}_l$ can be extended to the case of Riemannian foliations.

REMARK 3.11. Assume that the lamination admits non-hyperbolic leaves and that their union Y is a closed subset. We can consider the entropy outside Y, but this quantity can be infinite. We can in this case modify the distance outside Y, e.g., to consider

$$\operatorname{dist}'_X(x,y) = \min\Big\{ \operatorname{dist}_X(x,y), \inf_{x',y'\in Y} \operatorname{dist}_X(x,x') + \operatorname{dist}_X(y,y') \Big\}.$$

This means that we travel in Y with zero cost. The notion is natural because the Poincaré pseudo-distance vanishes on non-hyperbolic Riemann surfaces.

For foliations on $\mathbb{P}^k$, we can also consider their pull-back using generically finite holomorphic maps from a projective manifold to $\mathbb{P}^k$ in order to get hyperbolic foliations.

4   ENTROPY OF HARMONIC MEASURES

We are going to discuss a notion of entropy for harmonic measures associated to laminations. We first consider the abstract setting as in the beginning of Section 3 for a family of distances $\{\operatorname{dist}_t\}_{t\geq 0}$ on a metric space $(X, \operatorname{dist}_X)$. Let m be a probability measure on X. Fix positive constants ϵ, δ and t. Let $N_m(t,\epsilon,\delta)$ be the minimal number of balls of radius ϵ relative to the metric dist_t whose union has at least m-measure $1 - \delta$. The *entropy* of m is defined by the following formula

$$h_{\mathcal{D}}(m) := \lim_{\delta\to 0} \lim_{\epsilon\to 0} \limsup_{t\to\infty} \frac{1}{t} \log N_m(t,\epsilon,\delta).$$

We have the following general property.

PROPOSITION 4.1. *Let $(X, \operatorname{dist}_X)$ and $\mathcal{D}$ be as before. Then for any probability measure m on X, we have*

$$h_{\mathcal{D}}(m) \leq h_{\mathcal{D}}(\operatorname{supp}(m)).$$

PROOF. Define $Y := \operatorname{supp}(m)$. Choose a maximal family of (t, ϵ)-separated points x_i in Y. The family of $B_t(x_i, 2\epsilon)$ covers Y. So, for every $\delta > 0$

$$N_m(t, 2\epsilon, \delta) \leq M(Y, t, \epsilon).$$

It follows from Proposition 3.1 that $h_{\mathcal{D}}(m) \leq h_{\mathcal{D}}(Y)$.   □

As in Brin-Katok's theorem [2], we can introduce the *local entropies* of m at $x \in X$ by

$$h_{\mathcal{D}}^+(m, x, \epsilon) := \limsup_{t \to \infty} -\frac{1}{t} \log m(B_t(x, \epsilon)), \qquad h_{\mathcal{D}}^+(m, x) := \sup_{\epsilon > 0} h_{\mathcal{D}}^+(m, x, \epsilon),$$

and

$$h_{\mathcal{D}}^-(m, x, \epsilon) := \liminf_{t \to \infty} -\frac{1}{t} \log m(B_t(x, \epsilon)), \qquad h_{\mathcal{D}}^-(m, x) := \sup_{\epsilon > 0} h_{\mathcal{D}}^-(m, x, \epsilon),$$

where $B_t(x, \epsilon)$ denotes the ball centered at x of radius ϵ with respect to the distance dist_t.

Note that in the case of ergodic invariant measure associated with a continuous map on a metric compact space, the above notions of entropies coincide with the classical entropy of m; see Brin-Katok [2].

Let $(X, \mathcal{L})$ be a Riemann surface lamination such that its leaves are hyperbolic. Since we do not assume that X is compact, the following discussion can be applied to the regular part of a singular lamination.

The Poincaré metric ω_P provides a Laplacian Δ_P along the leaves. Recall that a probability measure m is *harmonic* if it is orthogonal to continuous functions ϕ which can be written $\phi = \Delta_P \psi$, where ψ is a continuous function, smooth along the leaves and having compact support in X. In a flow box $\mathbb{U} = \mathbb{B} \times \mathbb{T}$ with $\mathbb{B}$ open set in $\mathbb{C}$, we can write

$$m = \int m_s d\mu(s),$$

where μ is a positive measure on $\mathbb{T}$, $m_s = h_s \omega_P$ is a measure on $\mathbb{B} \times \{s\}$, and h_s is a positive harmonic function on this plaque. We refer to [6], [9], [14], [17], [19], and [29] for more details and for the relation with the notion of $\partial\bar{\partial}$-closed current.

Recall that in Section 3 we have associated to $(X, \mathcal{L})$ a family of distances $\{\operatorname{dist}_t\}_{t \geq 0}$. Therefore, we can associate to m a metric entropy and local entropies defined as before in the abstract setting. Recall that a harmonic probability measure m is called *extremal* if all harmonic probability measures m_1 and m_2 satisfying $m_1 + m_2 = 2m$ are equal to m. We have the following result.

THEOREM 4.2. *Let $(X, \mathcal{L})$ be a compact smooth lamination by hyperbolic Riemann surfaces. Let m be a harmonic probability measure. Then, the local entropies $h^{\pm}$ of m are constant on leaves. In particular, if m is extremal, then $h^{\pm}$ are constant m-almost everywhere.*

In fact, the result holds for compact laminations which are not necessarily smooth, provided that $A^{-1} \operatorname{dist}_X \leq \operatorname{dist}_P \leq A \operatorname{dist}_X$ with the same constant $A > 0$ for all plaques of a suitable atlas. We have seen that these inequalities hold when the lamination is smooth.

Fix a covering of X by a finite number of flow boxes $\mathbb{U} = \mathbb{B} \times \mathbb{T}$, where $\mathbb{B}$ is the disc of center 0 and of radius 2 in $\mathbb{C}$ and $\mathbb{T}$ is a ball of center s_0 and of radius 2 in a complete metric space. We assume that the boxes $\mathbb{U}' = \mathbb{D} \times \mathbb{T}$ cover X. For simplicity, in what follows, we identify the distance dist_X on $\mathbb{U}$ with the one induced by the distance on $\mathbb{T}$ and the Euclidean distance on $\mathbb{B}$. Denote by $\mathbb{T}_r$ the ball of center s_0 and of radius r in $\mathbb{T}$. Fix also a constant $\delta > 0$ such that if ϕ is a covering map of a leaf, then the image by ϕ of any subset of Poincaré diameter 2δ is contained in a flow box.

The following lemma gives us a description of the intersection of Bowen balls with plaques.

LEMMA 4.3. *Let $\epsilon > 0$ be a sufficiently small fixed constant. Then, there is a constant $A > 0$ satisfying the following properties. Let y and y' be two points in $\mathbb{D} \times \{s\}$ with $s \in \mathbb{T}_1$ and $R > 0$ be a constant. If $\mathrm{dist}_X(y, y') \leq A^{-1} e^{-R}$, then y and y' are (R, ϵ)-close. If $\mathrm{dist}_X(y, y') \geq Ae^{-R}$, then y and y' are (R, ϵ)-separated.*

PROOF. Fix a constant $A > 0$ large enough depending on ϵ. We prove the first assertion. Assume that $\mathrm{dist}_X(y, y') \leq A^{-1} e^{-R}$. Let L denote the leaf containing $\mathbb{B} \times \{s\}$. Let ϕ' be a covering map of L such that $\phi'(0) = y'$. So, there is a point $a \in \mathbb{D}$ such that $\phi'(a) = y$ and $\mathrm{dist}_P(0, a) \ll e^{-R}$. By Lemma 3.8, there is an automorphism τ of $\mathbb{D}$, close to the identity on $\mathbb{D}_R$, such that $\tau(0) = a$. Define $\phi := \phi' \circ \tau$. This is also a covering map of L. It is clear that $\mathrm{dist}_{\mathbb{D}_R}(\phi, \phi') \leq \epsilon$. Therefore, y and y' are (R, ϵ)-close.

For the second assertion, assume that $\mathrm{dist}_X(y, y') \geq Ae^{-R}$ but y and y' are (R, ϵ)-close. By definition of Bowen ball, we can find covering maps $\phi, \phi' : \mathbb{D} \to L$ such that $\phi(0) = y$, $\phi'(0) = y'$ and $\mathrm{dist}_{\mathbb{D}_R}(\phi, \phi') \leq \epsilon$. In particular, we have $\mathrm{dist}_X(y, y') \leq \epsilon$. Consequently, we can find a point $a \in \mathbb{D}$ such that $\phi'(a) = y$ and $\mathrm{dist}_P(0, a) = \mathrm{dist}_P(y, y')$. Since ϵ is small and A is large, we have

$$e^{-R} \ll \mathrm{dist}_X(y, y') \lesssim \mathrm{dist}_P(0, a) \leq \delta.$$

There is also an automorphism τ of $\mathbb{D}$ such that $\tau(0) = a$ and $\phi = \phi' \circ \tau$. The last assertion in Lemma 3.8 implies by continuity that we can find a point $z \in \mathbb{D}_R$ satisfying $\epsilon \ll \mathrm{dist}_P(z, \tau(z)) < \delta$. Finally, the property of δ implies that $\mathrm{dist}_X(\phi(z), \phi'(z)) > \epsilon$. This is a contradiction.   □

We now introduce a notion of transversal entropy which can be extended to a general lamination. In what follows, if V is a subset of $\mathbb{U}$, we denote by $\widetilde{V}$ its projection on $\mathbb{T}$. The measure m can be written in a unique way on $\mathbb{U}$ as

$$m = \int m_s d\mu(s),$$

where $m_s = h_s \omega_P$ is as before with the extra condition $h_s(0) = 1$; see [9], [15], [19], and [29].

By Harnack's principle, the family of positive harmonic functions h_s is locally uniformly bounded from above and from below by strictly positive constants. This implies that the following notions of transversal entropy do not depend on the choice of flow box. Define

$$\widetilde{h}^+(x) := \sup_{\epsilon > 0} \limsup_{R \to \infty} -\frac{1}{R} \log \mu(\widetilde{B}_R(x, \epsilon))$$

and

$$\widetilde{h}^-(x) := \sup_{\epsilon>0} \liminf_{R\to\infty} -\frac{1}{R} \log \mu(\widetilde{B}_R(x,\epsilon)).$$

Note that we can also use $\widetilde{B}_R(x,\epsilon)$ in order to define a notion of topological entropy on $\mathbb{T}$.

LEMMA 4.4. *We have* $h^\pm = \widetilde{h}^\pm + 2$.

PROOF. We can assume that x belongs to $\mathbb{D} \times \mathbb{T}_1$. By Lemma 4.3, the intersection of $B_R(x,\epsilon)$ with a plaque is of diameter at most equal to $2Ae^{-R}$. Since h_s is bounded from above uniformly on s, we deduce that

$$m(B_R(x,\epsilon)) \lesssim e^{-2R} \mu(\widetilde{B}_R(x,\epsilon)).$$

It follows that $h^\pm \geq \widetilde{h}^\pm + 2$.

We apply the first assertion in Lemma 4.3 to $\epsilon/2$ instead of ϵ. We deduce that if a plaque $\mathbb{D} \times \{s\}$ intersects $B_R(x,\epsilon/2)$, then its intersection with $B_R(x,\epsilon)$ contains a disc of radius $A^{-1}e^{-R}$. It follows that

$$m(B_R(x,\epsilon)) \gtrsim e^{-2R} \mu(\widetilde{B}_R(x,\epsilon/2)).$$

This implies that $h^\pm \leq \widetilde{h}^\pm + 2$ and completes the proof of the lemma. □

End of the proof of Theorem 4.2. Let x and y be in the same leaf L. We want to prove that $h^\pm(x) = h^\pm(y)$. It is enough to consider the case where x and y are close enough and to show that $h^\pm(x) \leq h^\pm(y)$. So, using the same notation as before, we can assume that x and y belong to $\mathbb{D} \times \{s_0\}$. We show that $\widetilde{h}^\pm(x) \leq \widetilde{h}^\pm(y)$. Fix a constant $\epsilon > 0$ small enough and a constant $\gamma > 0$ large enough. It suffices to show for large R that

$$\widetilde{B}_R(y,\epsilon) \subset \widetilde{B}_{R-\gamma}(x,\epsilon).$$

Let y' be a point in the intersection of $B_R(y,\epsilon)$ with a plaque $\mathbb{B} \times \{s_0'\}$. We have to show that $B_{R-\gamma}(x,\epsilon)$ also intersects $\mathbb{B} \times \{s_0'\}$. Let L and L' denote the leaves containing $\mathbb{B} \times \{s_0\}$ and $\mathbb{B} \times \{s_0'\}$, respectively. Consider universal maps $\phi: \mathbb{D} \to L$ and $\phi': \mathbb{D} \to L'$ such that $\phi(0) = y$, $\phi'(0) = y'$ and $\mathrm{dist}_{\mathbb{D}_R}(\phi,\phi') \leq \epsilon$. Let $a \in \mathbb{D}$ such that $\phi(a) = x$. Since x is close to y, we can find a close to 0. Let τ be an automorphism of $\mathbb{D}$ such that $\tau(0) = a$. Since a is close to 0 and R is large, the image of $\mathbb{D}_{R-\gamma}$ by τ, i.e., the disc of center a and of radius $R - \gamma$, is contained in $\mathbb{D}_R$. We deduce that $\mathrm{dist}_{\mathbb{D}_{R-\gamma}}(\phi \circ \tau, \phi' \circ \tau) \leq \epsilon$. In particular, $\phi'(a)$ is a point in $B_{R-\gamma}(x,\epsilon)$. This implies the result. □

Let $(X, \mathcal{L})$ be as in Theorem 4.2. Let m be an extremal harmonic probability measure. For simplicity, we will denote by $h^\pm(m)$ the constants associated with the local entropy functions $h^\pm$. We have the following result.

PROPOSITION 4.5. *With the preceding notation, we have*

$$h^-(m) \leq h(m) \leq h^+(m) \leq h(\mathcal{L}).$$

In particular, $h(\mathcal{L})$ is always larger or equal to 2.

PROOF. We will use the notations $h^{\pm}(x, \epsilon)$, $B_t(x, \epsilon)$, $N_m(t, \epsilon, \delta)$, and $N(X, t, \epsilon)$ as in the abstract setting. First, we will prove that $h(m) \leq h^+(m)$.

Fix constants $\alpha > 0$ and $0 < \delta < 1/4$. Given a constant $\epsilon > 0$ small enough, we can find a subset $X' \subset X$ with $m(X') \geq 1 - \delta$ such that for t large enough and for $x \in X'$, we have

$$\frac{1}{t} \log \frac{1}{m(B_t(x, \epsilon))} \leq h^+(x, \epsilon) + \frac{\alpha}{2} \leq h^+(m) + \alpha.$$

So, for such ϵ, x and t, we have

$$m(B_t(x, \epsilon)) \geq e^{-t(h^+(m)+\alpha)}.$$

Consider a maximal family of disjoint balls $B_t(x_i, \epsilon)$ with center $x_i \in X'$. The union of $B_t(x_i, 2\epsilon)$ covers X' which is of measure at least $1 - \delta$. Therefore, we have

$$1 \geq m\left(\bigcup B_t(x_i, \epsilon)\right) \geq N_m(t, 2\epsilon, \delta)e^{-t(h^+(m)+\alpha)}.$$

It follows that $N_m(t, 2\epsilon, \delta) \leq e^{t(h^+(m)+\alpha)}$ for t large enough. Since this inequality holds for every $\alpha > 0$, we deduce that $h(m) \leq h^+(m)$.

We now prove that $h^-(m) \leq h(m)$. As before, given $\epsilon > 0$, we can find a subset X'' with $m(X'') \geq 3/4$ such that for t large enough and for $x \in X''$, we have

$$m(B_t(x, 6\epsilon)) \leq e^{-t(h^-(m)-\alpha)}.$$

Consider a minimal family of balls $B_t(x_i, \epsilon)$ which covers a set of measure at least $3/4$. By removing the balls which do not intersect X'', we still have a family which covers a set of measure at least $1/2$. So, each ball $B_t(x_i, \epsilon)$ is contained in a ball $B_t(x_i', 2\epsilon)$ centered at a point $x_i' \in X''$. Vitali's covering lemma implies the existence of a finite sub-family of disjoint balls $B_t(y_j, 2\epsilon)$ such that $\bigcup B_t(y_j, 6\epsilon)$ covers $\bigcup B_t(x_i', 2\epsilon)$. Hence,

$$\frac{1}{2} \leq m\left(\bigcup B_t(y_j, 6\epsilon)\right) \leq N_m\left(t, \epsilon, \frac{3}{4}\right)e^{-t(h^-(m)-\alpha)}.$$

It follows that $2N_m(t, \epsilon, 3/4) \geq e^{t(h^-(m)-\alpha)}$ and, therefore, $h(m) \geq h^-(m)$.

It remains to show that $h^+(m) \leq h(\mathcal{L})$. Suppose in order to get a contradiction that $h(\mathcal{L}) \leq h^+(m) - 3\delta$ for some $\delta > 0$. For any $\epsilon > 0$, there exists t_0 large enough such that for all $t \geq t_0$,

$$\frac{1}{t} \log N(X, t, \epsilon) \leq h(\mathcal{L}) + \frac{\delta}{2}.$$

In particular, we have

$$N(X, t, \epsilon) \leq \frac{1}{4t^2}e^{(h(\mathcal{L})+\delta)t}.$$

Now fix an $\epsilon > 0$ small enough and then fix a t_0 large enough. Since we know $h^+(m) - \delta \geq h(\mathcal{L}) + 2\delta$, we have $m(\Lambda) > 1/2$, where

$$\Lambda := \left\{x \in X : \sup_{t \geq t_0} -\frac{1}{t} \log m(B_t(x, 2\epsilon)) \geq h(\mathcal{L}) + 2\delta\right\}.$$

Define

$$\Lambda_t := \left\{ x \in X : \ -\frac{1}{t} \log m(B_t(x, 2\epsilon)) \geq h(\mathcal{L}) + \delta \right\}$$

and

$$\Lambda'_t := \left\{ x \in X : \ -\frac{1}{t} \log m(B_t(x, 2\epsilon)) \geq h(\mathcal{L}) + 2\delta \right\}$$

Since t is large, we have $\Lambda'_t \subset \Lambda_{t+\alpha}$ for $0 \leq \alpha \leq 1$. Consider integer numbers n larger than t_0. We have

$$\Lambda = \bigcup_{t \geq t_0} \Lambda'_t \subset \bigcup_{n \geq t_0} \Lambda_n.$$

So, we can find $n \geq t_0$ such that $m(\Lambda_n) > 1/(4n^2)$. Hence, by definition of Λ_t, we get

$$N(\Lambda_n, n, 2\epsilon) > \frac{1}{4n^2} e^{(h(\mathcal{L}) + \delta)n}.$$

Therefore,

$$N(X, n, \epsilon) > \frac{1}{4n^2} e^{(h(\mathcal{L}) + \delta)n}.$$

This is a contradiction.   □

Here are some fundamental problems concerning metric entropies for Riemann surface laminations. Assume here that $(X, \mathcal{L})$ is a compact smooth lamination by hyperbolic Riemann surfaces, but the problems can be stated in a more general setting.

PROBLEM 5. *Consider extremal harmonic probability measures m. Is the following* variational principle *always true:*

$$h(\mathcal{L}) = \sup_m h(m)?$$

Even when this principle does not hold, it is interesting to consider the invariant

$$h(\mathcal{L}) - \sup_m h(m)$$

and to clarify the role of the hyperbolic time in this number.

PROBLEM 6. *If m is as described earlier, is the identity $h^+(m) = h^-(m)$ always true?*

We believe that the answer is affirmative and gives an analog of the Brin-Katok theorem.

Notice that there is a notion of entropy for harmonic measures introduced by Kaimanovich [24]. Consider a metric ω of bounded geometry on the leaves of the lamination. Then, we can consider the heat kernel $p(t, \cdot, \cdot)$ associated to the Laplacian determined by this metric. If m is a harmonic probability measure on X, Kaimanovich defines the entropy of m as

$$h_K(m) := \int dm(x) \left(\lim_{t \to \infty} -\frac{1}{t} \int p(t, x, y) \log p(t, x, y) \omega(y) \right).$$

He shows that the limit exists and is constant m-almost everywhere when m is extremal.

This notion of entropy has been extensively studied for the universal covering of a compact Riemannian manifold; see, e.g., Ledrappier [27]. It does not seem that it was studied for compact foliations with singularities. So, it would be of interest to find relations with our notions of entropy defined above. In Kaimanovich's entropy, the transverse spreading is present through the variation of the heat kernel from leaf to leaf. It would be also interesting to make this dependence more explicit.

5 Bibliography

[1]   R. Bowen, Topological entropy for noncompact sets, *Trans. Amer. Math. Soc.*, **184** (1973), 125-136.

[2]   M. Brin and A. Katok, On local entropy. Geometric dynamics (Rio de Janeiro, 1981), 30–38, *Lecture Notes in Math.*, **1007**, Springer, Berlin, 1983.

[3]   J. Buzzi, Dimensional entropies and semi-uniform hyperbolicity, *Proc. Intern. Cong. Math. Phys.*, Rio, 2006. arXiv:1102.0612

[4]   A. Candel, Uniformization of surface laminations, *Ann. Sci. École Norm. Sup.* (4), **26** (1993), no. 4, 489–516.

[5]   A. Candel and L. Conlon, *Foliations. I.* Graduate Studies in Mathematics, **23**, American Mathematical Society, Providence, RI, 2000.

[6]   A. Candel and L. Conlon, *Foliations. II.* Graduate Studies in Mathematics, **60**, American Mathematical Society, Providence, RI, 2003.

[7]   A. Candel and X. Gómez-Mont, Uniformization of the leaves of a rational vector field, *Ann. Inst. Fourier (Grenoble)*, **45** (1995), no. 4, 1123-1133.

[8]   T.-C. Dinh, V.-A. Nguyên, and N. Sibony, Dynamics of horizontal-like maps in higher dimension, *Adv. Math.*, **219** (2008), no. 5, 1689–1721.

[9]   T.-C. Dinh, V.-A. Nguyên, and N. Sibony, Heat equation and ergodic theorems for Riemann surface laminations, *Math. Ann.* **354** (2012), no. 1, 331–376. arXiv:1004.3931

[10] T.-C. Dinh, V.-A. Nguyên, and N. Sibony, Entropy for hyperbolic Riemann surface laminations II. *In this volume*, 593–621.

[11] T.-C. Dinh and N. Sibony, Dynamique des applications d'allure polynomiale, *J. Math. Pures Appl. (9)*, **82** (2003), no. 4, 367–423.

[12] T.-C. Dinh and N. Sibony, Upper bound for the topological entropy of a meromorphic correspondence, *Israel J. Math.*, **163** (2008), 29–44.

[13] C. J. Earle and A. Schatz, Teichmüller theory for surfaces with boundary, *J. Differential Geometry*, **4** (1970), 169–185.

[14] J.-E. Fornæss and N. Sibony, Harmonic currents of finite energy and laminations, *Geom. Funct. Anal.*, **15** (2005), no. 5, 962–1003.

[15] J.-E. Fornæss and N. Sibony, Riemann surface laminations with singularities, *J. Geom. Anal.*, **18** (2008), no. 2, 400–442.

[16] J.-E. Fornæss, N. Sibony, and E. Fornæss-Wold, Examples of Minimal Laminations and Associated Currents, *Math. Z.*, **269** (2011), no. 1–2, 495–520.

[17] L. Garnett, Foliations, the ergodic theorem and Brownian motion, *J. Funct. Analysis*, **51** (1983), 285–311.

[18] A. A. Glutsyuk, Hyperbolicity of the leaves of a generic one-dimensional holomorphic foliation on a nonsingular projective algebraic variety. (Russian) *Tr. Mat. Inst. Steklova,* **213** (1997), Differ. Uravn. s Veshchestv. i Kompleks. Vrem., 90–111; translation in *Proc. Steklov Inst. Math.* 1996, no. 2, **213**, 83–103.

[19] É. Ghys, Laminations par surfaces de Riemann. (French) [Laminations by Riemann surfaces] *Dynamique et géométrie complexes (Lyon, 1997),* ix, xi, 49–95, Panor. Synthèses, **8**, *Soc. Math. France, Paris,* 1999.

[20] É. Ghys, R. Langevin, and P. Walczak, Entropie géométrique des feuilletages, *Acta Math.,* **160** (1988), no. 1–2, 105–142.

[21] M. Gromov, On the entropy of holomorphic maps, *Enseignement Math.,* **49** (2003), 217–235. *Manuscript* (1977).

[22] M. Gromov, *Hyperbolic groups.* Essays in group theory, 75–263, *Math. Sci. Res. Inst. Publ.,* **8**, Springer, New York, 1987.

[23] S. Hurder, Classifying foliations. *Foliations, geometry, and topology,* 1–65, *Contemp. Math.,* **498**, Amer. Math. Soc., Providence, RI, 2009.

[24] V. Kaimanovich, Brownian motions on foliations: entropy, invariant measures, mixing, *Funct. Anal. Appl.,* **22** (1989), 326–328.

[25] A. Katok and B. Hasselblatt, *Introduction to the modern theory of dynamical systems.* Encyclopedia of Mathematics and its Applications, **54**, Cambridge University Press, Cambridge, 1995.

[26] S. Kobayashi, *Hyperbolic complex spaces.* Grundlehren der Mathematischen Wissenschaften [Fundamental Principles of Mathematical Sciences], **318**. Springer-Verlag, Berlin, 1998.

[27] F. Ledrappier, Profil d'entropie dans le cas continu. Hommage à P. A. Meyer et J. Neveu, *Astérisque,* **236** (1996), 189–198.

[28] A. Lins Neto, Uniformization and the Poincaré metric on the leaves of a foliation by curves, *Bol. Soc. Brasil. Mat. (N.S.),* **31** (2000), no. 3, 351–366.

[29] D. Sullivan, Cycles for the dynamical study of foliated manifolds and complex manifolds, *Invent. Math.,* **36** (1976), 225–255.

[30] A. Verjovsky, A uniformization theorem for holomorphic foliations. *The Lefschetz centennial conference, Part III (Mexico City, 1984),* 233–253, Contemp. Math., **58**, III, Amer. Math. Soc., Providence, RI, 1987.

[31] Y. Yomdin, Volume growth and entropy, *Israel J. Math.,* **57** (1987), 285–300.

[32] P. Walczak, *Dynamics of foliations, groups and pseudogroups.* Mathematics Institute of the Polish Academy of Sciences. Mathematical Monographs (New Series), **64**, Birkhäuser Verlag, Basel, 2004.

[33] P. Walters, *An introduction to ergodic theory,* Graduate Texts in Mathematics, **79**, Springer-Verlag, New York-Berlin, 1982.

Entropy for hyperbolic Riemann surface laminations II

Tien-Cuong Dinh, Viet-Anh Nguyên, and Nessim Sibony

ABSTRACT. Let $(X, \mathcal{L}, E)$ be a Brody hyperbolic foliation by Riemann surfaces with linearizable isolated singularities on a compact complex surface. We show that its hyperbolic entropy is finite. We also estimate the modulus of continuity of the Poincaré metric on leaves. The estimate holds for foliations on manifolds of higher dimension.

1 INTRODUCTION

In this second part, we study Riemann surface foliations with tame singular points. We say that a holomorphic vector field F in $\mathbb{C}^k$ is *generic linear* if it can be written as

$$F(z) = \sum_{j=1}^{k} \lambda_j z_j \frac{\partial}{\partial z_j}$$

where the λ_j are non-zero complex numbers. The integral curves of F define a Riemann surface foliation in $\mathbb{C}^k$. The condition $\lambda_j \neq 0$ for every j implies that the foliation has an isolated singularity at 0.

Consider a Riemann surface foliation with singularities $(X, \mathcal{L}, E)$ in a complex manifold X. We assume that the singular set E is discrete. By *foliation*, we mean that $\mathcal{L}$ is transversally holomorphic. We say that a singular point $e \in E$ is *linearizable* if there are local holomorphic coordinates on a neighborhood $\mathbb{U}$ of e in which the foliation is given by a generic linear vector field.

The purpose of the present paper is to study the notion of Brody hyperbolicity for compact Riemann surface foliations that will be given in Definition 3.1. Here is our main result.

THEOREM 1.1. *Let $(X, \mathcal{L}, E)$ be a singular foliation by Riemann surfaces on a compact complex surface X. Assume that the singularities are linearizable and that the foliation is Brody hyperbolic. Then, its hyperbolic entropy $h(\mathcal{L})$ is finite.*

The above theorem and a result by Glutsyuk [3] and Lins Neto [4] give us the following corollary. It can be applied to foliations of degree at least 2 with hyperbolic singularities.

COROLLARY 1.2. *Let $(\mathbb{P}^2, \mathcal{L}, E)$ be a singular foliation by Riemann surfaces on the complex projective plane $\mathbb{P}^2$. Assume that the singularities are linearizable. Then, the hyperbolic entropy $h(\mathcal{L})$ of $(\mathbb{P}^2, \mathcal{L}, E)$ is finite.*

In Section 2, we prove the finiteness of the entropy in the local setting near a singular point in any dimension. The main result in this section does not imply Theorem 1.1. However, it clarifies a difficulty due to singular points. We use here a division of a neighborhood of a singular point into adapted cells. The construction of these cells is crucial in the proof of Theorem 1.1.

In Section 3, we will estimate the modulus of continuity for the Poincaré metric along the leaves of the foliation. We will use the notion of conformally (R, δ)-close maps, as in the case without singularities, but there are several technical problems. Indeed, we have to control the phenomenon that leaves may go in and out singular flow boxes without any obvious rule.

The estimates we get in Section 3 hold uniformly on X. They are still far from being sufficient in order to get Theorem 1.1, i.e., the finiteness of entropy in the global setting. The proof of that result is more delicate. It is developed in the last three sections. We are able to handle the interaction of the two difficulties mentioned earlier only for foliations on complex surfaces and we conjecture that our main result is true in any dimension. A basic idea is the use of conformally (R, δ)-close maps from leaves to leaves with small Beltrami coefficient as in the case without singularities. In particular, we will glue together local orthogonal projections from leaves to leaves. However, we have to face the problem that the Poincaré metric is not bounded from above by a smooth Hermitian metric on X, and hence the Beltrami coefficient of orthogonal projections from leaves to leaves is not small near the singularities. We solve this difficulty by replacing these projections near the singularities by adapted holomorphic maps which exist thanks to the nature of the singular points.

We can introduce, for any extremal harmonic measure m, the metric entropy $h(m)$, the local entropies $h^\pm$, and the transverse local entropies $\widetilde{h}^\pm$. As in the case without singularities, we can show that $\widetilde{h}^\pm$ are constant m-almost everywhere and $h^\pm \leq \widetilde{h}^\pm + 2$. We believe that $h^\pm = \widetilde{h}^\pm + 2$ and then $h^\pm$ are constant m-almost everywhere but the question is still open. Other open problems stated for laminations without singularities in [1] can be also considered for singular foliations.

The rest of the paper is quite technical. In order to simplify the presentation, we will not try to get sharp constants in our estimates, and we prefer to give simple statements. In particular, for the following results, we assume that the hyperbolic time R is large enough and we heavily use the fact that R is larger than any fixed constants. Our notation is carefully chosen, and we invite the reader to keep in mind the following remarks and conventions.

Main notation. We use the same notation as in Part 1 of this paper [1], e.g., $\mathbb{D}$, $r\mathbb{D}$, $\mathbb{D}_R$, ω_P, dist_P, diam_P, η, L_x, $\phi_x \colon \mathbb{D} \to L_x$, dist_R, $\text{dist}_{\mathbb{D}_R}$, and μ_τ. Let $\mathbb{D}(\xi, R)$ denote the disc of center ξ and of radius R in the Poincaré disc $\mathbb{D}$ and set $\log^\star(\cdot) := 1 + |\log(\cdot)|$, a log-type function. Conformally (R, δ)-close points are defined as in [1] with an adapted constant $A := 3c_1^2$ and for $\delta \leq e^{-2R}$. If D is a disc, a polydisc, or a ball of center a and ρ is a positive number, then ρD is the image of D by the homothety $x \mapsto \rho x$, where x is an affine coordinate system centered at a. In particular, $\rho \mathbb{U}_i$ and $\rho \mathbb{U}_e$ correspond, respectively, to a regular flow box and to a flow box at a singular point $e \in E$.

Flow boxes and metric. We consider only flow boxes which are biholomorphic to $\mathbb{D}^k$. For regular flow boxes, i.e., flow boxes outside the singularities, the plaques are identified with the discs parallel to the first coordinate axis. Singular flow boxes are identified with their models described in Section 2. In particular, the leaves in a singular flow box are parametrized in a canonical way using holomorphic maps $\varphi_x \colon \Pi_x \to L_x$, where Π_x is a convex polygon in $\mathbb{C}$.

For each singular point $e \in E$, we fix a singular flow box $\mathbb{U}_e$ such that for $e \neq e'$, we have $3\mathbb{U}_e \cap 3\mathbb{U}_{e'} = \varnothing$. We also cover $X \setminus \cup \frac{1}{2}\mathbb{U}_e$ by regular flow boxes $\frac{1}{2}\mathbb{U}_i$ which are fine enough. In particular, each $\mathbb{U}_i$ is contained in a larger regular flow box $2\mathbb{U}_i$, with $2\mathbb{U}_i \cap \frac{1}{4}\mathbb{U}_e = \varnothing$ such that if $2\mathbb{U}_i$ intersects $\mathbb{U}_e$, it is contained in $2\mathbb{U}_e$. More consequences of the small size of $\mathbb{U}_i$ will be given when needed. We identify $\{0\} \times \mathbb{D}^{k-1}$ with a transversal $\mathbb{T}_i$ of $\mathbb{U}_i$ and call it the *distinguished transversal*. For $x \in \mathbb{U}_i$, denote by $\Delta(x, \rho)$ the disc of center x and of radius ρ contained in the plaque of x.

Fix a Hermitian metric ω on X which coincides with the standard Euclidean metric on each singular flow box $2\mathbb{U}_e \simeq 2\mathbb{D}^k$.

Other notation. Denote by $L_x[\epsilon]$ the intersection of L_x with the ball of center x and of radius ϵ with respect to the metric induced on L_x by the Hermitian metric on X. It should be distinguished from $L_x(\epsilon) := \phi_x(\mathbb{D}_\epsilon)$. We only use ϵ small enough so that $L_x[\epsilon]$ is connected and simply connected. Fix a constant $\epsilon_0 > 0$ small enough so that if x is outside the singular flow boxes $\frac{1}{2}\mathbb{U}_e$, then $L_x[\epsilon_0]$ is contained in a plaque of a regular flow box $\frac{1}{2}\mathbb{U}_i$. If $L_x(\epsilon_0) := \phi_x(\mathbb{D}_{\epsilon_0})$ is not contained in a singular flow box $\frac{1}{4}\mathbb{U}_e$, then ϕ_x is injective on $\mathbb{D}_{\epsilon_0}$.

The constant γ is given in Lemma 2.10. The constant λ in Section 2 can be equal to λ_* but we will later use the case with a large constant λ. Define $\alpha_1 := e^{-e^{7\lambda R}}$ and $\alpha_2 := e^{-e^{23\lambda R}}$. The set Σ and the maps $J_l, \Phi_{x,y}, \Psi_{x,y}, \widetilde{\Psi}_{x,y}$ are introduced in Section 2. The constants A and c_i are introduced in Section 3 with $A := 3c_1^2$. The constants $m_0, m_1, \hbar, p$ are fixed just before Lemma 3.9, the constant t just after that lemma. They satisfy $c_1 \ll m_0, c_1 m_0 \ll m_1, m_1 \hbar \ll \epsilon_0$ and $p := m_1^4$.

Orthogonal projections. For $x = (x_1, \ldots, x_k) \in \mathbb{C}^k$, define the norm $\|x\|_1$ as $\max |x_j|$. We choose $\epsilon_0 > 0$ small enough so that if x and y are two points outside the singular flow boxes $\frac{1}{4}\mathbb{U}_e$ such that $\mathrm{dist}(x, y) \leq \epsilon_0$, then for the Euclidean metric, the local orthogonal projection Φ from $L_x[\epsilon_0]$ to $L_y[10\epsilon_0]$ is well defined and its image is contained in $L_y[3\epsilon_0]$. In a singular flow box $\mathbb{U}_e \simeq \mathbb{D}^k$, since the metric and the foliation are invariant under homotheties, when $\mathrm{dist}(x, y) \leq \epsilon_0\|x\|_1$, the local orthogonal projection is well defined from $L_x[\epsilon_0\|x\|_1]$ to $L_y[10\epsilon_0\|x\|_1]$ with image in $L_y[3\epsilon_0\|x\|_1]$. If, moreover, x, y are very close to each other and are outside the coordinate hyperplanes, a global orthogonal projection $\Phi_{x,y}$ from L_x to L_y is constructed in Lemma 2.11. All projections described here are called the *basic projections* associated to x and y. They are not holomorphic in general.

In order to construct a map ψ satisfying the definition of conformally (R, δ)-close points, we have to glue together basic projections. In the case without singularities [1], we have carefully shown that the gluing is possible, i.e., there is no monodromy problem. The same arguments work in the case with singularities. We sometimes skip the details on this point in order to simplify the presentation.

2   LOCAL MODELS FOR SINGULAR POINTS

In this section, we give a description of the local model for linearizable singularities. We also prove the finiteness of entropy in this setting. The construction of cells and other auxiliary results given at the end of the section will be used in the proof of Theorem 1.1.

Consider the foliation $(\mathbb{D}^k, \mathcal{L}, \{0\})$, which is the restriction to $\mathbb{D}^k$ of the foliation associated to the vector field

$$F(z) = \sum_{j=1}^{k} \lambda_j z_j \frac{\partial}{\partial z_j}$$

with $\lambda_j \in \mathbb{C}^*$. The foliation is singular at the origin. We use here the Euclidean metric on $\mathbb{D}^k$. The following notations, $\widehat{L}_x$, $\widehat{\phi}_x$, $\widehat{\eta}$, dist_R, and $h(\cdot)$, are defined as in the case of general foliations. Here, we use a hat for some notation in order to avoid the confusion with the analogous notation that we will use later in the global setting.

Define

$$\lambda_* := \frac{\max\{|\lambda_1|, \ldots, |\lambda_k|\}}{\min\{|\lambda_1|, \ldots, |\lambda_k|\}}.$$

THEOREM 2.1. *For every compact subset K of $\mathbb{D}^k$, the hyperbolic entropy $h(K)$ of K is bounded from above by $70k\lambda_*$.*

Observe that the entropy of K is bounded independently of K. For the proof of this result, we will construct a division of $\mathbb{D}^k$ into cells whose shape changes according to their position with respect to the singular point and to the coordinate hyperplanes. These cells are shown to be contained in Bowen (R, e^{-R})-balls and we obtain an upper bound of $h(K)$ using an estimate on the number of such cells needed to cover K. We start with a description of the leaves of the foliation.

For simplicity, we multiply F with a constant in order to assume that

$$\min\{|\lambda_1|, \ldots, |\lambda_k|\} = 1 \quad \text{and} \quad \max\{|\lambda_1|, \ldots, |\lambda_k|\} = \lambda_*.$$

This does not change the foliation. Write $\lambda_j = s_j + it_j$ with $s_j, t_j \in \mathbb{R}$. For any $x = (x_1, \ldots, x_k) \in \mathbb{D}^k \setminus \{0\}$, define the holomorphic map $\varphi_x : \mathbb{C} \to \mathbb{C}^k \setminus \{0\}$ by

$$\varphi_x(\zeta) := \left(x_1 e^{\lambda_1 \zeta}, \ldots, x_k e^{\lambda_k \zeta}\right) \quad \text{for} \quad \zeta \in \mathbb{C}.$$

It is easy to see that $\varphi_x(\mathbb{C})$ is the integral curve of F which contains $\varphi_x(0) = x$.

Write $\zeta = u + iv$ with $u, v \in \mathbb{R}$. The domain $\Pi_x := \varphi_x^{-1}(\mathbb{D}^k)$ in $\mathbb{C}$ is defined by the inequalities

$$s_j u - t_j v < -\log|x_j| \quad \text{for} \quad j = 1, \ldots, k.$$

So, Π_x is a convex polygon which is not necessarily bounded. It contains 0 since $\varphi_x(0) = x$. Moreover, we have

$$\mathrm{dist}(0, \partial\Pi_x) = \min\left\{-\frac{\log|x_1|}{|\lambda_1|}, \ldots, -\frac{\log|x_k|}{|\lambda_k|}\right\}.$$

Thus, we obtain the following useful estimates:

$$-\lambda_*^{-1}\log\|x\|_1 \leq \mathrm{dist}(0, \partial\Pi_x) \leq -\log\|x\|_1.$$

The leaf of $\mathcal{L}$ through x is given by $\widehat{L}_x := \varphi_x(\Pi_x)$. Observe that when the ratio λ_i/λ_j are not all rational and all the coordinates of x do not vanish, $\varphi_x : \Pi_x \to \widehat{L}_x$ is bijective and, hence, $\widehat{L}_x$ is simply connected. Otherwise, when the ratios λ_i/λ_j are rational, all the leaves are closed submanifolds of $\mathbb{D}^k \setminus \{0\}$ and are biholomorphic to annuli.

Let $\tau_x \colon \mathbb{D} \to \Pi_x$ be a biholomorphic map such that $\tau_x(0) = 0$. Then $\widehat{\phi}_x := \varphi_x \circ \tau_x$ is a map from $\mathbb{D}$ to $\widehat{L}_x$ and is a universal covering map of $\widehat{L}_x$ such that $\widehat{\phi}_x(0) = x$. If ω_P and ω_0 denote the Hermitian forms associated to the Poincaré metric and the Euclidean metric on Π_x, define the function ϑ_x by

$$\omega_0 = \vartheta_x^2 \omega_P.$$

The following lemma describes the Poincaré metric on Π_x. The first assertion is probably known and is still valid if we replace Π_x by an arbitrary convex domain in $\mathbb{C}$. Recall that $\widehat{\eta}$ is given by $\omega = \widehat{\eta}^2 \omega_P$ on $\widehat{L}_x$.

LEMMA 2.2. *We have, for any $a \in \Pi_x$,*

$$\frac{1}{2}\operatorname{dist}(a, \partial \Pi_x) \leq \vartheta_x(a) \leq \operatorname{dist}(a, \partial \Pi_x).$$

In particular, we have

$$-\frac{1}{2}\lambda_*^{-1}\|x\|_1 \log \|x\|_1 \leq \widehat{\eta}(x) \leq -k\lambda_* \|x\|_1 \log \|x\|_1.$$

PROOF. For each $a \in \Pi_x$, consider the family of holomorphic maps $\sigma \colon \mathbb{D} \to \Pi_x$ such that $\sigma(0) = a$. Denote by $\|D\sigma(0)\|$ the norm of the differential of σ at 0 with respect to the Euclidean metrics on $\mathbb{D}$ and on Π_x. The extremal property of ω_P implies that

$$\vartheta_x(a) = \frac{1}{2}\max_\sigma \|D\sigma(0)\|.$$

Now, since the disc with center a and radius $\operatorname{dist}(a, \partial \Pi_x)$ is contained in Π_x, the first estimate in the lemma follows. Let l be a side of Π_x which is tangent to this disc. Consider the halfplane $\mathbb{H}$ containing Π_x such that l is contained in the boundary of $\mathbb{H}$. By Schwarz's lemma, the Poincaré metric on Π_x is larger than the one on $\mathbb{H}$. The Poincaré metric on $\mathbb{H}$ is associated to the Hermitian form $\operatorname{dist}(\cdot, \partial \mathbb{H})^{-2}\omega_0$. This implies the second estimate.

For the second assertion in the lemma, we have

$$\widehat{\eta}(x) = \vartheta_x(0)\|D\varphi_x(0)\|.$$

This, the first assertion, and the estimates on $\operatorname{dist}(a, \partial \Pi_x)$ imply the result because we get from the definition of φ_x that $\|x\|_1 \leq \|D\varphi_x(0)\| \leq k\lambda_* \|x\|_1$. $\square$

Fix a constant λ such that $\lambda \geq \lambda_*$. For the main results in this section, it is enough to take $\lambda = \lambda_*$, but we will later use the case with a large constant λ. Fix also a constant $0 < \rho < 1$ such that K is strictly contained in $\rho \mathbb{D}^k$. Denote by $\Omega_x \subset \Pi_x$ the set of points $\zeta := u + iv$ such that

$$\begin{cases} s_j u - t_j v < -\log|x_j| - e^{-20\lambda R} & \text{for} \quad j = 1, \ldots, k, \\ |\zeta| \leq e^{20\lambda R}. \end{cases}$$

Recall from the introduction that $\alpha_1 := e^{-e^{7\lambda R}}$ and $\alpha_2 := e^{-e^{23\lambda R}}$.

LEMMA 2.3. *Assume that $\alpha_1 \leq \|x\|_1 \leq \rho$. Then, $\tau_x(\mathbb{D}_{7R})$ is contained in Ω_x. In particular, $\widehat{\phi}_x(\mathbb{D}_{7R})$ is contained in $\rho'\mathbb{D}^k$ with $\rho' := e^{-e^{-21\lambda R}} \simeq 1 - e^{-21\lambda R}$.*

PROOF. It is not difficult to see that $\varphi_x(\Omega_x)$ is contained in $\rho'\mathbb{D}^k$. So, the second assertion in the lemma is a direct consequence of the first one. We now prove the first assertion. Consider a point $\zeta \in \tau_x(\mathbb{D}_{7R})$. By the second inequality in Lemma 2.2, the Poincaré distance between 0 and ζ in Π_x is at least equal to

$$\int_0^{|\zeta|} \frac{dt}{t + \operatorname{dist}(0, \partial\Pi_x)} \geq \int_0^{|\zeta|} \frac{dt}{t - \log\|x\|_1} = \log\left(1 - \frac{|\zeta|}{\log\|x\|_1}\right).$$

Since ζ is a point in $\tau_x(\mathbb{D}_{7R})$, this distance is at most equal to $7R$. Therefore, using that $\|x\|_1 \geq \alpha_1$, we obtain

$$|\zeta| \leq -\log\|x\|_1 e^{7R} \leq e^{20\lambda R}.$$

So, if the lemma were false, there would be a ζ with $\operatorname{dist}_P(0, \zeta) \leq 7R$ such that

$$s_j u - t_j v = -\log|x_j| - e^{-20\lambda R}$$

for some j. It follows that $\operatorname{dist}(\zeta, \partial\Pi_x) \lesssim e^{-20\lambda R}$. Hence, using Lemma 2.2 and the estimate

$$\operatorname{dist}(0, \partial\Pi_x) \geq -\lambda^{-1}\log\|x\|_1 \geq -\lambda^{-1}\log\rho,$$

we obtain

$$\operatorname{dist}_P(0, \zeta) \geq \left|\int_{\operatorname{dist}(\zeta, \partial\Pi_x)}^{\operatorname{dist}(0, \partial\Pi_x)} \frac{dt}{t}\right| = \left|\log\operatorname{dist}(0, \partial\Pi_x) - \log\operatorname{dist}(\zeta, \partial\Pi_x)\right| > 7R.$$

This is a contradiction. $\qquad\square$

In order to prove Theorem 2.1, we need to study carefully the points which are close to a coordinate plane in $\mathbb{C}^k$. We have the following lemma.

LEMMA 2.4. *Let x be a point in $\rho\mathbb{D}^k$ such that $\|x\|_1 \geq \alpha_1$. Let m be an integer such that $1 \leq m \leq k$ and $|x_j| \leq 2\alpha_2$ for $j = m+1, \ldots, k$. Then, $\widehat{\phi}_x(\mathbb{D}_{7R})$ is contained in $\mathbb{D}^m \times e^{-3R}\mathbb{D}^{k-m}$.*

PROOF. Let ξ be a point in $\mathbb{D}_{7R}$. Define $\zeta := \tau_x(\xi)$ and $x' := \widehat{\phi}_x(\xi) = \varphi_x(\zeta)$. We have to show that $|x'_j| < e^{-3R}$ for $j \geq m+1$. By definition of φ_x, we have

$$|x'_j| = |x_j e^{\lambda_j \zeta}| \leq 2\alpha_2 e^{\lambda|\zeta|}.$$

This, combined with the first assertion of Lemma 2.3, implies the result. $\qquad\square$

We consider now the situation near the singular point.

LEMMA 2.5. *If $\|x\|_1 \leq 2\alpha_1$ and $\|y\|_1 \leq 2\alpha_1$, then x and y are (R, e^{-R})-close.*

PROOF. It is enough to show that $\widehat{\phi}_x(\mathbb{D}_R) \subset e^{-2R}\mathbb{D}^k$. This and the similar property for y imply the lemma. Consider ξ, $\zeta = u + iv$ and x' as before with $\xi \in \mathbb{D}_R$. A computation as in the end of Lemma 2.3 implies that

$$R \geq \operatorname{dist}_P(0, \zeta) \geq \log\operatorname{dist}(0, \partial\Pi_x) - \log\operatorname{dist}(\zeta, \partial\Pi_x).$$

Since $\|x\|_1 \leq 2\alpha_1$, we have $\log \text{dist}(0, \partial\Pi_x) > 6\lambda R$ and then $\text{dist}(\zeta, \partial\Pi_x) \geq 3R$. It follows that

$$|x'_j| = e^{\log|x_j| + s_j u - t_j v} \leq e^{-|\lambda_j| \text{dist}(\zeta, \partial\Pi_x)} \leq e^{-2R}$$

for every j. The result follows. □

LEMMA 2.6. *Let x be a point in $\rho\mathbb{D}^k$ and $1 \leq m \leq k$ be an integer such that $\|x\|_1 > \alpha_1$ and $|x_j| \leq 2\alpha_2$ for $j = m+1, \ldots, k$. If $x' := (x_1, \ldots, x_m, 0, \ldots, 0)$, then x and x' are (R, e^{-2R})-close.*

PROOF. Fix a point $\xi \in \mathbb{D}_R$. We have to show that $\text{dist}(\widehat{\phi}_x(\xi), \widehat{\phi}_{x'}(\xi)) \leq e^{-2R}$. Observe that by hypotheses $\Omega_{x'} = \Omega_x \subset \Pi_x$. So, by Lemma 2.3, $\tau_{x'}$ defines a holomorphic map from $\mathbb{D}_{7R}$ to Π_x. Observe that the Euclidean radius of $\mathbb{D}_{7R}$ is larger than $1 - 2e^{-7R}$. Hence, using the extremal property of the Poincaré metric, we deduce that

$$\vartheta_{x'}(0) \leq (1 + 3e^{-7R})\vartheta_x(0).$$

Consider the map $\tau := \tau_{x'}^{-1} \circ \tau_x$ from $\mathbb{D}_{7R}$ to $\mathbb{D}$. We have $\|D\tau(0)\| \geq 1 - 3e^{-7R}$. Composing τ_x with a suitable rotation allows us to assume that $D\tau(0)$ is a positive real number. By Lemma 2.3 in [1] applied to τ, there is a point ξ' such that $\text{dist}_P(\xi, \xi') \ll e^{-2R}$ and $\tau_{x'}(\xi') = \tau_x(\xi)$. Observe that the first m coordinates of $\widehat{\phi}_x(\xi)$ are equal to the ones of $\widehat{\phi}_{x'}(\xi')$. Therefore, by Lemma 2.4 applied to x and to x', the distance between $\widehat{\phi}_x(\xi)$ and $\widehat{\phi}_{x'}(\xi')$ is less than ke^{-3R}. On the other hand,

$$\text{dist}(\widehat{\phi}_{x'}(\xi), \widehat{\phi}_{x'}(\xi')) \lesssim \text{dist}_P(\widehat{\phi}_{x'}(\xi), \widehat{\phi}_{x'}(\xi')) = \text{dist}_P(\xi, \xi') \ll e^{-2R}.$$

The lemma follows. □

Consider now two points x and y in $\rho\mathbb{D}^k$ such that for each $1 \leq j \leq k$, one of the following properties holds:

(S1) $|x_j| < \alpha_2$ and $|y_j| < \alpha_2$;

(S2) $x_j, y_j \neq 0$ and $\left|\frac{x_j}{y_j} - 1\right| < e^{-22\lambda R}$ and $\left|\frac{y_j}{x_j} - 1\right| < e^{-22\lambda R}$.

We have the following proposition.

PROPOSITION 2.7. *Under the preceding conditions, x and y are (R, e^{-R})-close.*

PROOF. If $\|x\|_1 \leq 2\alpha_1$ and $\|y\|_1 \leq 2\alpha_1$, then Lemma 2.5 implies the result. Assume this is not the case. By condition (S2), we have $\|x\|_1 \geq \alpha_1$ and $\|y\|_1 \geq \alpha_1$. Moreover, up to a permutation of coordinates, we can find $1 \leq m \leq k$ such that $|x_j| \geq \alpha_2$, $|y_j| \geq \alpha_2$ for $j \leq m$ and $|x_j| \leq 2\alpha_2$, $|y_j| \leq 2\alpha_2$ for $j \geq m+1$. By Lemma 2.6, we can assume that $x_j = y_j = 0$ for $j \geq m+1$. Now, in order to simplify the notation, we can assume without loss of generality that $m = k$. So, we have $|x_j| \geq \alpha_2$ and $|y_j| \geq \alpha_2$ for every j.

Define the linear holomorphic map $\Psi_{x,y}: \mathbb{C}^k \to \mathbb{C}^k$ by

$$\Psi_{x,y}(z) := \left(\frac{y_1}{x_1}z_1, \ldots, \frac{y_k}{x_k}z_k\right), \qquad z = (z_1, \ldots, z_k) \in \mathbb{C}^k.$$

This map preserves the foliation and sends x to y. Moreover, the property (S2) implies that for such x, y, $\|\Psi_{x,y} - \text{id}\| \ll e^{-21\lambda R}$ on $\mathbb{D}^k$.

Define also $\widetilde{\phi}_y := \Psi_{x,y} \circ \widehat{\phi}_x$. It follows from the last assertion in Lemma 2.3 that this map is well defined on $\mathbb{D}_{7R}$ with image in $\widehat{L}_y$ and we have $\widetilde{\phi}_y(0) = y$. We also deduce from (S2) that $\mathrm{dist}(\widehat{\phi}_x(\xi), \widetilde{\phi}_y(\xi)) \ll e^{-R}$. It remains to compare $\widehat{\phi}_y$ and $\widetilde{\phi}_y$ on $\mathbb{D}_R$.

Using the extremal property of the Poincaré metric, we obtain

$$\|D\widehat{\phi}_y(0)\| \geq (1 - e^{-6R})\|D\widetilde{\phi}_y(0)\| \geq (1 - e^{-5R})\|D\widehat{\phi}_x(0)\|.$$

By symmetry, we deduce that $\|D\widehat{\phi}_x(0)\|$, $\|D\widehat{\phi}_y(0)\|$ and $\|D\widetilde{\phi}_y(0)\|$ are close, i.e., their ratios are bounded by $1 + e^{-4R}$.

Since $\widehat{\phi}_y$ is a universal covering map, there is a unique holomorphic map $\tau \colon \mathbb{D}_{7R} \to \mathbb{D}$ such that $\tau(0) = 0$ and $\widehat{\phi}_y \circ \tau = \widetilde{\phi}_y$. Composing $\widehat{\phi}_y$ with a suitable rotation allows us to assume that $D\tau(0)$ is a positive real number. It follows from the preceding discussion that $|1 - \tau'(0)| \leq e^{-4R}$. Lemma 2.3 in [1] implies that $\mathrm{dist}_P(\xi, \tau(\xi)) \ll e^{-R}$ on $\mathbb{D}_R$. Hence, $\mathrm{dist}_P(\widehat{\phi}_y(\xi), \widetilde{\phi}_y(\xi)) \ll e^{-R}$ for ξ in $\mathbb{D}_R$. The proposition follows.   □

End of the proof of Theorem 2.1. The idea is to divide $\mathbb{D}^k$ into cells which are contained in Bowen (R, e^{-R})-balls. The sizes of these cells are very different and this is one of the main difficulties in the proof of Theorem 1.1.

We first divide $\mathbb{D}$ into rings using the circle of center 0 and of radius $\alpha_2 e^{ne^{-23\lambda R}}$ for $n = 1, \ldots, e^{46\lambda R}$. In fact, we have to take the integer part of the last number but we will not write it in order to simplify the notation. Then, we divide these rings into cells using $e^{23\lambda R}$ half-lines starting at 0 which are equidistributed in $\mathbb{C}$. We obtain fewer than $e^{70\lambda R}$ cells and we denote by Σ the set of the vertices, i.e., the intersection of circles and half-lines. Except those at 0, if a cell contains a point a, it looks like a rectangle whose sides are approximatively $|a|e^{-23\lambda R}$. Consider the product of k copies of $\mathbb{D}$ together with the preceding division, we obtain a division of $\mathbb{D}^k$ into fewer than $e^{70\lambda kR}$ cells.

Consider two points x, y in $\rho\mathbb{D}^k$ which belong to the same cell. They satisfy the conditions (S1) and (S2). So, by Proposition 2.7, they are (R, e^{-R})-close. It follows that if a cell is contained in $\rho\mathbb{D}^k$, it is contained in a Bowen (R, e^{-R})-ball. We deduce that the entropy of K is bounded by $70\lambda k$. The estimate holds for $\lambda = \lambda_*$. □

PROPOSITION 2.8. *The function $\widehat{\eta}$ is locally Hölder continuous outside the coordinate hyperplanes $\{x_j = 0\}$, $1 \leq j \leq k$, with Hölder exponent $(6\lambda_*)^{-1}$.*

PROOF. We consider x in a fixed compact outside the coordinate hyperplanes. So, as in the case without singularities [1], we can show that $\widehat{\eta}(x)$ is bounded from above and from below by strictly positive constants. Using the comparison between $D\widehat{\phi}_x(0)$ and $D\widehat{\phi}_y(0)$ in the proof of Proposition 2.7, we deduce that if $\mathrm{dist}(x, y) \leq e^{-23\lambda R}$, then $|\widehat{\eta}(x) - \widehat{\eta}(y)| \leq e^{-4R}$. The result follows.   □

Note that the forementioned division of $\mathbb{D}^k$ respects the invariance of the foliation under the homotheties. However, it is important to observe that the Poincaré metric on leaves is not invariant under the homotheties, see Lemma 2.2. As a consequence, the plaques in the cells are very small in the sense that their Poincaré diameters tend to 0 when R tends to infinity. We will heavily use properties of $(\mathbb{D}^k, \mathcal{L}, \{0\})$, in particular, this specific division into cells, as a model of singular flow boxes in our study of global foliations. We now give a construction that will be used later.

Recall that Σ is the intersection of the circles and the half-lines used in the end of the proof of Theorem 2.1. Note that this set depends on R. We now introduce p maps J_l, $0 \leq l \leq p - 1$, from $\Sigma^k \cap \frac{3}{4}\mathbb{D}^k$ to Σ^k, where p is a large integer that will be fixed just before Lemma 3.9. These maps describe, roughly, the displacement of points when we travel along leaves following p given directions after a certain time, which is independent of R.

For $x \in \Sigma^k$ and $0 \leq l \leq p - 1$, define

$$z^l := \varphi_x\Big(-t \log \|x\|_1 e^{\frac{2i\pi l}{p}} \Big),$$

where the constant $0 < t \ll \lambda_*^{-1}$ will be fixed just after Lemma 3.9 below. For the moment, we will need that the real part of $te^{2i\pi l/p}\lambda_j$ is not equal to -1 for all j and l. This property simplifies the proof of Lemma 2.9. We choose $J_l(x)$ a vertex of a cell of $\mathbb{D}^k$ which contains z^l. Note that the choice is not unique, but this is not important for our problem.

The point x is displaced to the points z^l when we travel following p directions after a certain time. The points $J_l(x)$ give us an approximation of z^l and allow us to understand the displacement of points near x using the lattice Σ^k. We will use these maps J_l in order to make an induction on hyperbolic time R which is an important step in the proof of Theorem 1.1; see also Lemmas 4.5 and 4.6.

LEMMA 2.9. *Let $y = (y_1, \ldots, y_k)$ be a point in Σ^k and $0 \leq l \leq p - 1$ be an integer. There is a constant $M > 0$ independent of R such that if $|y_j| > \alpha_1$ for every j, then $J_l^{-1}(y)$ contains at most M points.*

PROOF. Let $x = (x_1, \ldots, x_k)$ be such that $J_l(x) = y$. For simplicity, write z for the point $z^l = \varphi_x\big(-t \log \|x\|_1 e^{2i\pi l/p} \big)$. Since z and y belong to the same cell, z_j/y_j is very close to 1. We deduce that $|z_j| \geq \frac{1}{2}\alpha_1$ for every j. Using the definition of φ_x, we obtain

$$\frac{1}{2}\alpha_1 \leq |z_j| = |x_j|\big|e^{-t \log \|x\|_1 e^{2i\pi l/p}}\big| \leq |x_j|\|x\|_1^{-t} \leq |x_j|^{1-t}.$$

Hence, since t is small, we infer $|x_j| \geq \alpha_2$ for every j.

We will consider only the case where $\|x\|_1 = |x_1|$. The other cases are treated in the same way. In order to simplify the notation, assume also that $l = 0$. Consider another point x' satisfying similar properties. It is enough to show that the number of such points x' is bounded. Define $z' := \varphi_{x'}\big(-t \log |x_1'| \big)$.

We have

$$z = (x_1|x_1|^{s\lambda_1}, x_2|x_1|^{s\lambda_2}, \ldots, x_k|x_1|^{s\lambda_k})$$

and

$$z' = (x_1'|x_1'|^{s\lambda_1}, x_2'|x_1'|^{s\lambda_2}, \ldots, x_k'|x_1'|^{s\lambda_k}).$$

Since z, y belong to the same cell and z', y satisfy the same property, we have $|z_j'/z_j - 1| \lesssim e^{-23\lambda R}$ and $|z_j/z_j' - 1| \lesssim e^{-23\lambda R}$. Recall that $|x_j| \geq \alpha_2$ and $|x_j'| \geq \alpha_2$ and x_j, x_j' are in Σ. Using that the real part of $t\lambda_j$ is not equal to -1, we easily see that for each j there is a bounded number of x_j' satisfying the preceding properties. For this purpose, we can also use a homothety in order to reduce the problem to the case where $|x_j| \simeq 1$. $\qquad\square$

In general, the point $J_l(x)$ does not belong to a plaque containing z^l but it is, however, very close to such a plaque. Write $J_l(x) = w = (w_1, \ldots, w_k)$. We have the following lemma.

LEMMA 2.10. *Assume that $|z_j^l| > \alpha_2$ for every j. Then, the plaque $L_{z^l}[\|z^l\|_1 \epsilon_0]$ intersects $\{w_1\} \times \mathbb{D}^{k-1}$ at a unique point w^l. Moreover, there is a constant $\gamma \geq 1$ independent of λ and R such that $|w_j/w_j^l - 1| \leq \gamma e^{-23\lambda R}$ and $|z_j^l/w_j^l - 1| \leq \gamma e^{-23\lambda R}$ for every j.*

PROOF. We show that the lemma is true for a general point z with $|z_j| > \alpha_2$ for every j and for any vertex w of a cell containing z. Observe that $w_j \neq 0$ for every j. We first consider the case where $|z_j| \simeq 1/2$ for every j. In this case, $L_z[\epsilon_0/2]$ is the graph of a map with bounded derivatives over a domain on the first coordinate axis. The cell containing z looks like a cube of size $\simeq e^{-23\lambda R}$. So, the lemma is clear in this case.

Consider now the general case. Observe that the foliation and the set Σ^k are invariant when we multiply a coordinate by a power of $e^{e^{-23\lambda R}}$. Therefore, multiplying the coordinates by a same constant allows us to assume that $\|z\|_1 = |z_j| \simeq 1/2$ for some index j. Then, $L_z[\epsilon_0/2]$ is a graph of a map with bounded derivatives over a domain D in the jth axis. We claim that if we multiply each coordinate z_l with $l \neq j$ with an appropriate power of $e^{e^{-23\lambda R}}$, the problem is reduced to the first case. Indeed, the image of $L_z[\epsilon_0/2]$ is still the graph over D of a map with bounded derivatives because this graph is contained in a plaque. Therefore, the size of $L_z[\epsilon_0/2]$ changes with a factor bounded independently of R and λ. Since ϵ_0 is small enough, the same arguments as in the first case give the result. □

Observe that in the proof of Theorem 2.1, the holomorphic maps $\Psi_{x,y}$ are used in order to control the distance between the leaves. Recall that for global foliations without singularities, we have to use some orthogonal projections in order to construct a parametrization ψ as in the definition of conformally (R, δ)-close points. These maps are not holomorphic, but we can correct them using Beltrami's equation. In order to prove Theorem 1.1 for the case with singularities, we will use both kinds of maps depending if we are near or far from singular points. We shall describe some relations between these two kinds of maps and show that one can glue them near the boundaries of singular flow boxes.

Recall that for a constant ϵ_0 small enough, if x, y are two points in $\frac{3}{4}\mathbb{D}^k \setminus \frac{1}{4}\mathbb{D}^k$ such that $\text{dist}(x, y) \leq \epsilon_0$, then the basic projection Φ from $L_x[\epsilon_0]$ to $L_y[10\epsilon_0]$ is well defined with image in $L_y[3\epsilon_0]$. Fix a constant $\kappa > 1$ large enough such that the following estimate holds on $L_x[\epsilon_0]$:

$$\|\Phi - \text{id}\|_{\mathcal{C}^2} \leq \kappa \, \text{dist}(\Phi(x), x),$$

where we compute the norm using the Euclidean metric on $\widehat{L}_x$ and $\widehat{L}_y$

The situation is more delicate near the singularities. We use the fact that the foliation in $\mathbb{D}^k$ is invariant under homotheties $z \mapsto tz$ with $|t| \leq 1$. Such a homothety multiplies the distance with $|t|$. For $x \in \frac{3}{4}\mathbb{D}^k$ and $t := \frac{4}{3}\|x\|_1$, some point in the boundary of $\frac{3}{4}\mathbb{D}^k$ is sent to x. Thus, if $\text{dist}(x, y) \leq \|x\|_1 \epsilon_0$, we can define the orthogonal projection Φ from $L_x[\|x\|_1 \epsilon_0]$ to $L_y[10\|x\|_1 \epsilon_0]$ with image in $L_y[3\|x\|_1 \epsilon_0]$. Moreover, we have

$$\|\Phi - \text{id}\|_{\mathcal{C}^0} \leq \kappa \, \text{dist}(\Phi(x), x) \quad \text{and} \quad \|\Phi - \text{id}\|_{\mathcal{C}^2} \leq \kappa \, \text{dist}(\Phi(x), x) \|x\|_1^{-2}.$$

Note that $L_x[\|x\|_1\epsilon_0]$ and $L_y[3\|x\|_1\epsilon_0]$ are connected and simply connected.

We see that when we are near the singularity, the Beltrami coefficient of Φ may have a large $\mathcal{C}^1$-norm and cannot satisfy the estimates in the definition of conformally (R, δ)-close points. This is the reason why we have to use the holomorphic maps $\Psi_{x,y}$ defined before. Nevertheless, the control of the $\mathcal{C}^0$-norm $\|\Phi - \text{id}\|_{\mathcal{C}^0}$ is still good near the singular point. Therefore, in the proof of Theorem 1.1, it is convenient to use first the orthogonal projections and then correct the parametrization using the following two lemmas.

LEMMA 2.11. *Let x, y be two points in $\frac{1}{2}\mathbb{D}^k$ such that $y_j \neq 0$ and $|x_j/y_j - 1| \leq e^{-10R}$ for every j. Then, there is an orthogonal projection $\Phi_{x,y}$ from $\widehat{L}_x \cap \frac{3}{4}\mathbb{D}^k$ to $\widehat{L}_y$ which coincides on $L_z[\|z\|_1\epsilon_0]$ with the basic projection associated to z and to $w := \Psi_{x,y}(z)$ for every $z \in \widehat{L}_x \cap \frac{3}{4}\mathbb{D}^k$. Moreover, we have $\|\Phi_{x,y} - \text{id}\|_{\mathcal{C}^0} \leq e^{-9R}$.*

PROOF. Observe that
$$|z_j/w_j - 1| = |x_j/y_j - 1| \leq e^{-10R}.$$

Therefore, there is a basic projection from $L_z[\|z\|_1\epsilon_0]$ to $L_w[10\|z\|_1\epsilon_0]$ with image in $L_z[3\|z\|_1\epsilon_0]$ as described earlier. It satisfies the estimate in the lemma. It is not difficult to see that such projections coincide on the intersection of their domains of definition. Indeed, the property is clear near the boundary of $\frac{3}{4}\mathbb{D}^k$. The general case can be reduced to that case using a homothety. The lemma follows. □

LEMMA 2.12. *Let x, y be as in Lemma 2.11. Then, there is a map $\widetilde{\Psi}_{x,y}$ from $\widehat{L}_x \cap \frac{3}{4}\mathbb{D}^k$ to $\widehat{L}_y$ such that $\widetilde{\Psi}_{x,y} = \Psi_{x,y}$ on $\widehat{L}_x \cap \frac{1}{4}\mathbb{D}^k$ and $\widetilde{\Psi}_{x,y} = \Phi_{x,y}$ on $\widehat{L}_x \cap \frac{3}{4}\mathbb{D}^k \setminus \frac{1}{2}\mathbb{D}^k$. In particular, $\widetilde{\Psi}_{x,y}$ is holomorphic in $\frac{1}{4}\mathbb{D}^k$. Moreover, we have $\|\widetilde{\Psi}_{x,y} - \text{id}\|_{\mathcal{C}^0} \leq e^{-8R}$ on $\widehat{L}_x \cap \frac{3}{4}\mathbb{D}^k$ and $\|\widetilde{\Psi}_{x,y} - \text{id}\|_{\mathcal{C}^2} \leq e^{-8R}$ on $\widehat{L}_x \cap \frac{3}{4}\mathbb{D}^k \setminus \frac{1}{4}\mathbb{D}^k$.*

PROOF. Define $z' := \Phi_{x,y}(z)$ and $w := \Psi_{x,y}(z)$. So, z' belongs to a small plaque containing w. Using a homothety and the definition of φ_w, we can check that $|z'_i/w_i - 1| \lesssim e^{-10R}$. Therefore, the function $\log(z'_i/w_i)$ is well defined.

Let $0 \leq \chi \leq 1$ be a smooth function equal to 0 on $\frac{1}{4}\mathbb{D}^k$ and equal to 1 outside $\frac{1}{2}\mathbb{D}^k$. Define $\chi_i(z) := \chi(z)\log(z'_i/w_i)$ and
$$\widetilde{\Psi}_{x,y}(z) := \left(e^{\chi_1(z)}w_1, \ldots, e^{\chi_k(z)}w_k\right).$$

Clearly, $\widetilde{\Psi}_{x,y} = \Psi_{x,y}$ on $\widehat{L}_x \cap \frac{1}{4}\mathbb{D}^k$ and $\widetilde{\Psi}_{x,y} = \Phi_{x,y}$ on $\widehat{L}_x \cap \frac{3}{4}\mathbb{D}^k \setminus \frac{1}{2}\mathbb{D}^k$.

Observe that a point $w' = (w'_1, \ldots, w'_k)$ belongs to a small plaque containing w if and only if
$$\lambda_j \log(w'_i/w_i) = \lambda_i \log(w'_j/w_j).$$

This holds in particular for $w' = z'$ and, hence, $\widetilde{\Psi}_{x,y}(z)$ also satisfies this criterion. It follows that $\widetilde{\Psi}_{x,y}$ has values in $\widehat{L}_y$.

We deduce from the definition of $\Psi_{x,y}$ that $\|\Psi_{x,y} - \text{id}\|_{\mathcal{C}^0} \leq e^{-9R}$ on $\widehat{L}_x \cap \frac{3}{4}\mathbb{D}^k$ and $\|\Psi_{x,y} - \text{id}\|_{\mathcal{C}^2} \leq e^{-9R}$ on $\widehat{L}_x \cap \frac{3}{4}\mathbb{D}^k \setminus \frac{1}{4}\mathbb{D}^k$. Recall that $\Phi_{x,y}$ satisfies similar estimates. So, $\widetilde{\Psi}_{x,y}$ satisfies the estimates in the lemma. □

3 POINCARÉ METRIC ON LEAVES

In this section, we will meet another important difficulty for our study: a leaf of the foliation may visit singular flow boxes without any obvious rule. However, we analyze the behavior and get an explicit estimate on the modulus of continuity of the Poincaré metric on leaves. We are concerned with the following class of foliations.

DEFINITION 3.1. *A Riemann surface foliation with singularities* $(X, \mathcal{L}, E)$ *on a Hermitian compact complex manifold* X *is said to be* Brody hyperbolic *if there is a constant* $c_0 > 0$ *such that*

$$\|D\phi(0)\| \leq c_0$$

for all holomorphic maps ϕ *from* $\mathbb{D}$ *into a leaf.*

It is clear that if the foliation is Brody hyperbolic then its leaves are hyperbolic in the sense of Kobayashi. Conversely, the Brody hyperbolicity is a consequence of the non-existence of holomorphic non-constant maps $\mathbb{C} \to X$ such that out of E the image of $\mathbb{C}$ is locally contained in leaves, see [2, Theorem 15].

On the other hand, Lins Neto proved in [4] that for every holomorphic foliation of degree larger than 1 in $\mathbb{P}^k$, with non-degenerate singularities, there is a smooth metric with negative curvature on its tangent bundle; see also Glutsyuk [3]. Hence, these foliations are Brody hyperbolic. Consequently, holomorphic foliations in $\mathbb{P}^k$ are generically Brody hyperbolic; see also [5].

From now on, we assume that $(X, \mathcal{L}, E)$ is Brody hyperbolic. We have the following result with the notation introduced earlier.

THEOREM 3.2. *Let* $(X, \mathcal{L}, E)$ *be a Brody hyperbolic foliation by Riemann surfaces on a Hermitian compact complex manifold* X. *Assume that the singular set* E *is finite and that all points of* E *are linearizable. Then, there are constants* $c > 0$ *and* $0 < \alpha < 1$ *such that*

$$|\eta(x) - \eta(y)| \leq c \left(\frac{\max\{\log^\star \operatorname{dist}(x, E), \log^\star \operatorname{dist}(y, E)\}}{\log^\star \operatorname{dist}(x, y)} \right)^\alpha$$

for all x, y *in* $X \setminus E$.

The following estimates are crucial in our study.

PROPOSITION 3.3. *Under the hypotheses of Theorem 3.2, there exists a constant* $c_1 > 1$ *such that* $\eta \leq c_1$ *on* X, $\eta \geq c_1^{-1}$ *outside the singular flow boxes* $\frac{1}{4}\mathbb{U}_e$ *and*

$$c_1^{-1} s \log^\star s \leq \eta(x) \leq c_1 s \log^\star s$$

for $x \in X \setminus E$ *and* $s := \operatorname{dist}(x, E)$.

PROOF. It is enough to prove the last assertion. Without loss of generality, we only need to show it for all x in a singular flow box $\frac{1}{2}\mathbb{U}_e$. We identify $\mathbb{U}_e$ with the model in Section 2 and use the notation introduced there. Recall that for simplicity, we use here a metric on X whose restriction to $\mathbb{U}_e \simeq \mathbb{D}^k$ coincides with the Euclidean metric.

Define the map $\tau \colon \mathbb{D} \to L_x$ by

$$\tau(\xi) := \varphi_x(-c^{-1} \log \|x\|_1 \xi)$$

for some constant $c > 1$ large enough. Since $\|D\varphi_x(0)\|$ is bounded from below by $\|x\|_1$, we deduce that

$$\eta(x) \geq \|D\tau(0)\| \gtrsim -\log \|x\|_1 \|x\|_1 \gtrsim s \log^\star s.$$

This gives us the first inequality in the last assertion of the proposition.

Next, since the foliation is Brody hyperbolic, the function η is bounded from above. It follows that there is a constant $R_0 > 0$ independent of $x \in \frac{1}{2}\mathbb{U}_e$ such that the disc of center x in L_x with radius R_0 with respect to the Poincaré metric is contained in $\mathbb{U}_e$. Let $r_0 \in (0, 1)$ such that $\mathbb{D}_{R_0} = r_0\mathbb{D}$. We have $\phi_x(r_0\mathbb{D}) \subset \mathbb{D}^k$. We then deduce from the extremal property of the Poincaré metric that $\eta(x) \leq r_0^{-1}\widehat{\eta}(x)$. Lemma 2.2 implies the result. $\qquad\square$

The following lemma gives us a speed estimate when we travel in a singular flow box along a geodesic with respect to the Poincaré metric on leaves. We denote by $[0, \xi]$ the segment joining 0 and ξ in $\mathbb{D}$.

LEMMA 3.4. *There is a constant $c_2 > 0$ independent of x in $\frac{1}{2}\mathbb{U}_e \simeq \frac{1}{2}\mathbb{D}^k$ with the following property. If $\xi \in \mathbb{D}_R$ such that $\phi_x([0, \xi]) \subset \frac{1}{2}\mathbb{D}^k$ and if $y := \phi_x(\xi)$, then*

$$\|x\|_1^{e^{c_2 R}} \leq \|y\|_1 \leq \|x\|_1^{e^{-c_2 R}}.$$

PROOF. We have to prove only the first inequality. The second one is obtained by exchanging x and y. So, we need to consider only the case where $\|y\|_1 \leq \|x\|_1$. By Proposition 3.3, we have

$$R \geq \text{dist}_P(x, y) \gtrsim \int_{\|y\|_1}^{\|x\|_1} \frac{dt}{t|\log t|} = \log \frac{\log \|y\|_1}{\log \|x\|_1}.$$

The lemma follows. $\qquad\square$

The following lemma shows us how deep a leaf can go into a singular flow box before the hyperbolic time R.

LEMMA 3.5. *There is a constant $c_3 > 0$ such that for every $x \in X \setminus E$, we have*

$$\text{dist}(\phi_x(\mathbb{D}_R), E) \geq e^{-\log^\star \text{dist}(x, E)e^{c_3 R}}.$$

PROOF. We have only to estimate $\text{dist}(\phi_x(\xi), E)$ for a point $\xi \in \mathbb{D}_R$ such that $\phi_x(\xi)$ is close to a singular point e. So, we can replace x by a suitable point in $\phi_x([0, \xi])$ in order to assume that $\phi_x([0, \xi]) \subset \frac{1}{2}\mathbb{U}_e$. Consequently, Lemma 3.4 implies the result. $\qquad\square$

Recall that we define the notion of conformally (R, δ)-close as in [1] using the constant $A := 3c_1^2$ and a number $0 < \delta \leq e^{-2R}$. In order to prove Theorem 3.2, we follow the approach of that paper. We will need the following result for a large enough constant c_4.

PROPOSITION 3.6. *Let $c_4 > 1$ be a fixed constant. Let x and y be conformally (R, δ)-close such that $c_4^{-1}\eta(y) \leq \eta(x) \leq c_4\eta(y)$ and $e^{2R}\delta \leq \eta(y)$. Then, there is a real number θ such that if $\phi'_y(\xi) := \phi_y(e^{i\theta}\xi)$, we have*

$$|\eta(x) - \eta(y)| \leq e^{-R/4} \quad \text{and} \quad \text{dist}_{\mathbb{D}_{R/3}}(\phi_x, \phi'_y) \leq e^{-R/4}.$$

PROOF. We argue as in the proof of Proposition 2.2 in [1] with the same notation. The difference with the non-singular case is that we have a condition on the relative size of $\eta(x)$ and $\eta(y)$. We get that

$$\|D\phi_y(0)\| - \|D\widetilde{\phi}_y(0)\| \lesssim e^R \delta + e^{-R}(\eta(x) + \eta(y)).$$

Recall that $\eta(y) = \|D\phi_y(0)\|$. We deduce from the hypotheses that the constant λ used in the proof of Proposition 2.2 in [1] satisfies

$$1 - \lambda \lesssim \frac{e^R \delta + e^{-R}(\eta(x) + \eta(y))}{\eta(y)} \lesssim e^{-R}.$$

Now, it is enough to follow the proof of the preceding proposition. We use here that $e^{-R/3} \ll e^{-R/4}$ since we consider only R large enough. $\square$

In order to apply the last proposition, we have to show that if x and y are close enough, they are conformally (R, δ)-close. So, we need to construct a map ψ as in the definition of conformally (R, δ)-close points. As in the case without singularities, ψ will be obtained by composing ϕ_x with basic projections from L_x to L_y. There are two main steps. The first one is to show that up to time R, the leaves L_x and L_y are still close enough in order to define basic projections. The second one is that we can glue these projections together in order to get a well-defined map on $\mathbb{D}_R$. The second step can be treated as in the case without singularities. So, for simplicity, in what follows, we consider only the first problem.

Recall that if x, y are two points outside the singular flow boxes $\frac{1}{2}U_e$ such that $\mathrm{dist}(x, y) \leq \epsilon_0$, then the basic projection Φ from $L_x[\epsilon_0]$ to $L_y[3\epsilon_0]$ is well defined. Moreover, we have the following estimate on $L_x[\epsilon_0]$ for a fixed constant $\kappa > 1$:

$$\|\Phi - \mathrm{id}\|_{\mathcal{C}^2} \leq \kappa \, \mathrm{dist}(\Phi(x), x).$$

Inside the singular flow boxes, the foliation and the metric are invariant under homotheties. Therefore, since ϵ_0 is small enough, we deduce that for all $x, y \in X$ with $\mathrm{dist}(x, y) \leq \mathrm{dist}(x, E)\epsilon_0$, there is a basic projection Φ from $L_x[\mathrm{dist}(x, E)\epsilon_0]$ to $L_y[3 \, \mathrm{dist}(x, E)\epsilon_0]$ which satisfies

$$\|\Phi - \mathrm{id}\|_{\mathcal{C}^2} \leq \kappa \frac{\mathrm{dist}(\Phi(x), x)}{\mathrm{dist}(x, E)^2}.$$

It is not difficult to obtain the following useful estimates, where we change the constant κ if necessary:

$$\frac{\mathrm{dist}(\Phi(z), z)}{\mathrm{dist}(z, E)^2} \leq \kappa \frac{\mathrm{dist}(\Phi(x), x)}{\mathrm{dist}(x, E)^2} \quad \text{and} \quad \frac{\mathrm{dist}(\Phi(z), z)}{\mathrm{dist}(z, E)^6} \leq \kappa \frac{\mathrm{dist}(\Phi(x), x)}{\mathrm{dist}(x, E)^6}$$

for every z in $L_x[\mathrm{dist}(x, E)\epsilon_0]$.

PROPOSITION 3.7. *There exists a constant $c_5 > 1$ with the following property. Let R be sufficiently large, $x, y \in X \setminus E$, and let $\delta := \mathrm{dist}(x, y)e^{\log^\star \mathrm{dist}(x, E)e^{c_5 R}}$ be such that $\delta \leq e^{-2R}$. Then, x and y are conformally (R, δ)-close.*

Taking this proposition for granted, we now complete the proof of Theorem 3.2.

End of the proof of Theorem 3.2. We can assume that x, y are close and

$$|\log \mathrm{dist}(x, y)| \gg \log^\star \mathrm{dist}(x, E) \quad \text{and} \quad |\log \mathrm{dist}(x, y)| \gg \log^\star \mathrm{dist}(y, E).$$

In particular, we have

$$\mathrm{dist}(x,y) \ll \mathrm{dist}(x,E) \quad \text{and} \quad \mathrm{dist}(x,y) \ll \mathrm{dist}(y,E).$$

We will apply Propositions 3.6 and 3.7. Choose $R > 1$ such that

$$|\log \mathrm{dist}(x,y)| = \log^\star \mathrm{dist}(x,E) e^{2c_5 R}.$$

So, R is a large number. By Proposition 3.7, x and y are conformally (R, δ)-close with

$$\delta := \mathrm{dist}(x,y) e^{\log^\star \mathrm{dist}(x,E) e^{c_5 R}} = e^{\log^\star \mathrm{dist}(x,E)(e^{c_5 R} - e^{2c_5 R})}.$$

It is clear that $\delta \leq e^{-2R}$. The preceding identities, together with Proposition 3.3, also imply that $e^{2R}\delta \leq \eta(y)$. Therefore, x, y, R and δ satisfy the hypotheses of Proposition 3.6. Consequently,

$$|\eta(x) - \eta(y)| \leq e^{-R/4} = \left(\frac{\log^\star \mathrm{dist}(x,E)}{|\log \mathrm{dist}(x,y)|} \right)^{1/(8c_5)}.$$

The result follows. $\qquad\qquad\qquad\qquad\qquad\qquad\qquad\qquad\qquad\qquad\qquad\qquad\qquad$ $\square$

We prove now Proposition 3.7. Consider x, y, R and δ as in this proposition. So, x, y are close and satisfy

$$|\log \mathrm{dist}(x,y)| \geq \log^\star \mathrm{dist}(x,E) e^{c_5 R} \quad \text{and} \quad |\log \mathrm{dist}(x,y)| \geq \log^\star \mathrm{dist}(y,E) e^{c_5 R}.$$

Now we have to construct a map $\psi \colon \overline{\mathbb{D}}_R \to L_y$ as in the definition of conformally (R, δ)-close points. We first consider the case where y is the orthogonal projection of x to $L_y[3\,\mathrm{dist}(x,E)\epsilon_0]$.

Fix a point $\xi \in \overline{\mathbb{D}}_R$. We will construct the map ψ on a neighborhood of $[0, \xi]$ using basic projections from L_x to L_y. This allows us to prove that ψ is well defined on $\overline{\mathbb{D}}_R$ as in the case without singularities. The details for this last point are the same as in the non-singular case.

We divide $[0, \xi]$ into a finite number of segments $[\xi^j, \xi^{j+1}]$ with $0 \leq j \leq n-1$ and define $x^j := \phi_x(\xi^j)$. The points ξ^j are chosen by induction: $\xi^0 := 0$ and ξ^{j+1} is the closest point to ξ^j satisfying

$$\mathrm{dist}(x^{j+1}, x^j) = \mathrm{dist}(x^j, E)\epsilon_1$$

with a fixed constant $\epsilon_1 \ll \epsilon_0$. The integer n satisfies $\xi \in [\xi^{n-1}, \xi^n]$.

Also define by induction the points y^j in L_y with $y^0 := y$ and y^{j+1} is the image of x^{j+1} by the basic projection $\Phi_j \colon L_{x^j}[\mathrm{dist}(x^j, E)\epsilon_0] \to L_{y^j}[3\,\mathrm{dist}(x^j, E)\epsilon_0]$. Note that since $\epsilon_1 \ll \epsilon_0$, the point y^{j+1} is also the image of x^{j+1} by Φ_{j+1}. We deduce from the properties of basic projections that

$$\frac{\mathrm{dist}(y^{j+1}, x^{j+1})}{\mathrm{dist}(x^{j+1}, E)^6} \leq \kappa \frac{\mathrm{dist}(y^j, x^j)}{\mathrm{dist}(x^j, E)^6}.$$

The following lemma guarantees, by induction on j, the existence of the projections Φ_j.

LEMMA 3.8. *We have*

$$\mathrm{dist}(y^j, x^j) \leq e^{-2R}\delta\, \mathrm{dist}(x^j, E)^6$$

for $j = 0, \ldots, n$.

PROOF. We deduce from the above discussion that

$$\mathrm{dist}(y^j, x^j) \leq \kappa^n \,\mathrm{dist}(x, y)\, \mathrm{dist}(x^j, E)^6\, \mathrm{dist}(x, E)^{-6}.$$

Since c_5 is large, it is enough to show that $n \leq \log^\star \mathrm{dist}(x, E)Re^{cR}$ for some constant $c > 0$. For this purpose, we have to check only that

$$\mathrm{dist}_P(\xi^j, \xi^{j+1}) \geq \frac{e^{-cR}}{\log^\star \mathrm{dist}(x, E)}.$$

By definition of ξ^j, we have

$$\epsilon_1 \,\mathrm{dist}(x^j, E) = \mathrm{dist}(x^j, x^{j+1}) \leq \mathrm{dist}_P(\xi^j, \xi^{j+1}) \max_{t \in [\xi^j, \xi^{j+1}]} \eta(\phi_x(t)).$$

Proposition 3.3 and Lemma 3.5 imply that

$$\eta(\phi_x(t)) \simeq \log^\star \mathrm{dist}(x^j, E)\, \mathrm{dist}(x^j, E) \lesssim \log^\star \mathrm{dist}(x, E)e^{c_3 R}\, \mathrm{dist}(x^j, E).$$

The lemma follows.   □

So, the map ψ is well defined on $[0, \xi]$. We obtain as in [1] that it is well defined on $\overline{\mathbb{D}}_R$. It is not difficult to see, using the last lemma, that $\mathrm{dist}_{\overline{\mathbb{D}}_R}(\phi_x, \psi) \lesssim e^{-2R}\delta$. Since ϕ_y is a universal covering map, there is a unique map $\tau \colon \overline{\mathbb{D}}_R \to \mathbb{D}$ such that $\psi = \phi_y \circ \tau$ and $\tau(0) = 0$. In order to show that ψ satisfies the definition of conformally (R, δ)-close points, it remains to check that $\|D\tau\|_\infty \leq A$ and $\|\mu_\tau\|_{\mathcal{C}^1} \leq \delta$. It is enough to prove these properties at the point ξ considered earlier.

In a neighborhood of $[\xi^{n-1}, \xi^n]$, the maps ψ and τ are given by

$$\psi = \Phi_n \circ \phi_x \quad \text{and} \quad \Phi_n \circ \phi_x = \phi_y \circ \tau.$$

So, we can write locally

$$\tau = \phi_y^{-1} \circ \Phi_n \circ \phi_x.$$

Define $z := \phi_x(\xi)$ and $w := \Phi_n(z)$. Then, the norm $\|D\tau(\xi)\|$ with respect to the Poincaré metric on $\mathbb{D}$ is bounded by

$$\eta(z)\|D\Phi_n(z)\|\eta(w)^{-1}.$$

Since z and w are very close to x^n, by Proposition 3.3, $\eta(z)\eta(w)^{-1}$ is between $1/(2c_1^2)$ and $2c_1^2$. Moreover, we have

$$\|\Phi_n - \mathrm{id}\|_{\mathcal{C}^2} \lesssim \frac{\mathrm{dist}(\Phi(x^n), x^n)}{\mathrm{dist}(x^n, E)^2}.$$

By Lemma 3.8, the last quantity is very small. Consequently, we deduce that $\|D\tau(\xi)\| \ll 3c_1^2 = A$. We also deduce the following useful estimate

$$\|\overline{\partial}\Phi_n\|_{\mathcal{C}^1} \lesssim \frac{\mathrm{dist}(\Phi(x^n), x^n)}{\mathrm{dist}(x^n, E)^2}.$$

It remains to bound $\|\mu_\tau\|_{\mathscr{C}^1}$, where μ_τ is the Beltrami coefficient defined as in [1]. Recall that this norm is computed with the Euclidean metric on $\mathbb{D}$. We first rescale the maps ϕ_x and ϕ_y using the function η. This takes into account our distance to the singular points. Define $\zeta := \tau(\xi)$ and

$$\widetilde{\phi}_x(t) := \phi_x(\xi + \eta(z)^{-1}t) \quad \text{and} \quad \widetilde{\phi}_y(t) := \phi_y(\zeta + \eta(w)^{-1}t).$$

Define also

$$\widetilde{\tau} := \widetilde{\phi}_y^{-1} \circ \Phi_n \circ \widetilde{\phi}_x.$$

Observe that $\eta(z)$ and $\eta(w)$ are comparable with $\log^\star \operatorname{dist}(x^n, E)\operatorname{dist}(x^n, E)$, which is bounded from below by $\operatorname{dist}(x^n, E)$. Fix a constant $0 < \epsilon_2 \ll \epsilon_0$ small enough. Then, the image of $D_1 := \epsilon_2 \operatorname{dist}(x^n, E)\mathbb{D}$ by ϕ_x is small and its distance to E is comparable with $\operatorname{dist}(x^n, E)$. Therefore, $\Phi_n(\widetilde{\phi}_x(D_1))$ is close to x^n; hence, it is contained in the image by $\widetilde{\phi}_y$ of a fixed small disc D_2 centered at 0.

The images of D_1 and D_2 by $\widetilde{\phi}_x$ and $\widetilde{\phi}_y$ are small and contained in some chart of X. By Cauchy's formula, we obtain that

$$\|\widetilde{\phi}_x\|_{\mathscr{C}^2} \lesssim \operatorname{dist}(x^n, E)^{-2} \quad \text{and} \quad \|\widetilde{\phi}_y\|_{\mathscr{C}^2} \lesssim \operatorname{dist}(x^n, E)^{-2}.$$

Since $\|D\widetilde{\phi}_y(0)\| = 1$ and $\widetilde{\phi}_y(0) = w$, we deduce that

$$\|D\widetilde{\phi}_y^{-1}(w)\| = 1 \quad \text{and} \quad \|D^2\widetilde{\phi}_y^{-1}(w)\| \lesssim \operatorname{dist}(x^n, E)^{-2}.$$

All these estimates together with the ones on $\|\Phi_n - \operatorname{id}\|_{\mathscr{C}^2}$ and $\|\overline{\partial}\Phi_n\|_{\mathscr{C}^1}$ imply that the Beltrami coefficient of $\widetilde{\tau}$ satisfies the following estimate at 0

$$\|\mu_{\widetilde{\tau}}\|_{\mathscr{C}^1} \lesssim \frac{\operatorname{dist}(\Phi(x^n), x^n)}{\operatorname{dist}(x^n, E)^4}.$$

Recall that $\eta(z)$ and $\eta(w)$ are comparable with $\log^\star \operatorname{dist}(x^n, E)\operatorname{dist}(x^n, E)$. Hence, we can deduce from the definitions of τ and $\widetilde{\tau}$ the following estimate at the point ξ:

$$\|\mu_\tau\|_{\mathscr{C}^1} \lesssim \frac{\operatorname{dist}(\Phi(x^n), x^n)}{\operatorname{dist}(x^n, E)^6}.$$

By Lemma 3.8, the last quantity is smaller than $e^{-R}\delta$. So, the proof of Proposition 3.7 is complete for the case where y is the orthogonal projection of x to $L_y[3\operatorname{dist}(x, E)\epsilon_0]$.

End of the proof of Proposition 3.7. We consider the general case where y is not necessarily equal to the orthogonal projection x' of x to $L_y[3\operatorname{dist}(x, E)\epsilon_0]$. By hypotheses, x, y are close and satisfy

$$|\log\operatorname{dist}(x, y)| \geq \log^\star \operatorname{dist}(x, E)e^{c_5 R} \quad \text{and} \quad |\log\operatorname{dist}(x, y)| \geq \log^\star \operatorname{dist}(y, E)e^{c_5 R}.$$

We have $\operatorname{dist}(x, x') \leq \operatorname{dist}(x, y)$ and hence $\operatorname{dist}(x', y) \leq 2\operatorname{dist}(x, y)$.

Applying the earlier construction to x and x', we obtain a map $\psi': \mathbb{D}_R \to L_y$ with $\psi'(0) = x'$ and $\operatorname{dist}_{\mathbb{D}_R}(\phi_x, \psi') \ll \delta$, and such that the associated map τ' satisfies

$$\|D\tau'\|_\infty \ll A \quad \text{and} \quad \|\mu_{\tau'}\|_{\mathscr{C}^1} \ll \delta.$$

We have to construct a good map ψ associated to x and y.

Observe that $\mathrm{dist}(x',y) \ll \delta \, \mathrm{dist}(x',E) e^{-e^R}$ since c_5 is large. Moreover, ψ' is locally the composition of ϕ_x with basic projections from L_x to L_y. By Proposition 3.3, there is a point $a \in \mathbb{D}$ with $|a| \leq \delta e^{-e^R}$ such that $\psi'(a) = y$. So, we can find an automorphism $u \colon \mathbb{D}_R \to \mathbb{D}_R$ such that $u(0) = a$ and $\|u - \mathrm{id}\|_{e^2} \leq \delta e^{-2R}$. Define $\psi := \psi' \circ u$. We have $\mathrm{dist}_{\mathbb{D}_R}(\phi_x \circ u, \psi) = \mathrm{dist}_{\mathbb{D}_R}(\phi_x, \psi') \ll \delta$. Since η is bounded from above, we also have $\mathrm{dist}_{\mathbb{D}_R}(\phi_x, \phi_x \circ u) \ll \delta$. Therefore, $\mathrm{dist}_{\mathbb{D}_R}(\phi_x, \psi) \ll \delta$.

The map τ associated to ψ is given by $\tau := \tau' \circ u$. It is not difficult to see that $\psi(0) = y$ and

$$\|D\tau\|_\infty \leq A \quad \text{and} \quad \|\mu_\tau\|_{e^1} \leq \delta.$$

So, ψ satisfies the definition of conformally (R, δ)-close points.   □

We end this section with some technical results that will be used later. Let m_0 and m_1 be two integers large enough and $\hbar$ be a constant small enough such that $c_1 \ll m_0$, $c_1 m_0 \ll m_1$ and $m_1 \hbar \ll \epsilon_0$. Define $p := m_1^4$.

We divide the annulus $\mathbb{D}_{m_1 \hbar} \setminus \mathbb{D}_{3\hbar}$ into $40 m_1$ equal sectors S_j ($1 \leq j \leq 40 m_1$), using $40 m_1$ half-lines starting at 0 which are equidistributed in $\mathbb{C}$. For x in a regular flow box $\mathbb{U}_i$, denote by $S_l(x)$ the sector of the points ξ in $\Delta(x, 3m_0 \hbar) \setminus \Delta(x, m_0 \hbar)$ such that $2\pi l / p \leq \arg(\xi) \leq 2\pi(l+1)/p$ for $0 \leq l \leq p - 1$. Here, $\arg(\xi)$ is defined using the natural coordinate on $\Delta(x, 3m_0 \hbar)$ centered at x.

LEMMA 3.9. *Let x be a point in a regular flow box $\mathbb{U}_i$. Then, for every j, $\phi_x(S_j)$ contains a sector $S_l(x)$. In particular, $\phi_x(S_j)$ intersects the restriction of $m_1^{-5} \hbar(\mathbb{Z} + i\mathbb{Z}) \times \mathbb{D}^{k-1}$ to a regular flow box $\mathbb{U}_{i'} \simeq \mathbb{D}^k$.*

PROOF. This is a simple consequence of the first assertion in Proposition 3.3 and that $c_1 \ll m_0$ and $c_1 m_0 \ll m_1$. Recall that ϕ_x is injective on the disc $\mathbb{D}_{\epsilon_0}$ which contains the sectors S_j since $m_1 \hbar \ll \epsilon_0$.   □

Consider now a point x in a singular flow box $\mathbb{U}_e$. We deduce from basic properties of the universal covering that there is a unique map $u \colon \Pi_x \to \mathbb{D}$ such that $\varphi_x = \phi_x \circ u$ and $u(0) = 0$. Fix a constant t satisfying the condition stated just before Lemma 2.9 such that $m_0 \hbar < t < 2m_0 \hbar$. Define $\zeta_l := -t \log \|x\|_1 e^{2i\pi l / p}$ and $\xi_l := u(\zeta_l)$. We use the construction in Section 2 for a fixed constant λ large enough. The following two lemmas are related to Lemmas 2.9 and 2.10.

LEMMA 3.10. *Let x be a point in $\frac{3}{4} \mathbb{U}_e$ satisfying the hypotheses of Lemma 2.10. Let w^l be as in that lemma. Then, the image Γ of the boundary of the disc $D := -t \log \|x\|_1 \mathbb{D}$ by u satisfies $\Gamma \cap \mathbb{D}_{4\hbar} = \varnothing$ and $\Gamma \Subset \mathbb{D}_{m_1 \hbar}$. The points ξ_l divide Γ into p arcs of length smaller than $m_1^{-2} \hbar$. Moreover, for each l, there is a point ξ_l' such that $\mathrm{dist}_P(\xi_l', \xi_l) \leq m_1^{-2} \hbar$ and $\phi_x(\xi_l') = w^l$.*

PROOF. Since $m_0 \ll m_1$ and $t \leq 2m_0 \hbar \ll \epsilon_0$, by Lemma 2.2, D is contained in the disc of center 0 and of radius $m_1 \hbar$ with respect to the Poincaré metric on Π_x. Since holomorphic maps contract the Poincaré metric, $u(D)$ is contained in $\mathbb{D}_{m_1 \hbar}$.

The points ζ_l divide the boundary of D into p arcs. We deduce from Lemma 2.2 that the length of each arc with respect to the Poincaré metric is smaller than $4\lambda_*(-\log \|x\|_1)^{-1}$ times its length with respect to the Euclidean metric. In particular, it is smaller than $m_1^{-3} \hbar$ since $p = m_1^4$. It follows that the p arcs of Γ have length smaller than $m_1^{-2} \hbar$.

Let ξ be a point in the boundary of $\mathbb{D}_{4\hbar}$. The length of $\phi_x([0,\xi])$ with respect to the Poincaré metric on L_x is $4\hbar$. By Proposition 3.3 and Lemma 2.2, we have $\eta \lesssim c_1\hat{\eta}$ on $\frac{7}{8}\mathbb{U}_e$. Note that $\frac{7}{8}\mathbb{U}_e$ contains $\phi_x([0,\xi])$ since $\hbar$ is small. Therefore, the length of $\phi_x([0,\xi])$ with respect to the Poincaré metric on $\widehat{L}_x = \varphi_x(\Pi_x)$ is bounded by a constant times $c_1\hbar$. So, the lift of $\phi_x([0,\xi])$ to a curve starting at 0 in Π_x is contained in D because $m_0 \gg c_1$. This completes the proof of the first assertion.

Again using Lemma 2.2 and the last assertion in Lemma 2.10, we obtain that the distance between z^l and w^l with respect to the Poincaré metric on $\widehat{L}_x = \varphi_x(\Pi_x)$ is bounded by a constant times $e^{-23\lambda R}$. By Proposition 3.3, this still holds for the Poincaré metric on L_x. Therefore, there is a point ξ'_l such that $\mathrm{dist}_P(\xi'_l, \xi_l) \leq m_1^{-2}\hbar$ and $\phi_x(\xi'_l) = w^l$. $\qquad\square$

LEMMA 3.11. *Let x be as in Lemma 3.10. Assume that x is outside the singular flow boxes $\frac{1}{2}\mathbb{U}_e$. Then, there is a point ξ''_l such that $\mathrm{dist}_P(\xi_l, \xi''_l) \leq m_1^{-4}\hbar$ and $\phi_x(\xi''_l)$ is contained in the restriction of $m_1^{-5}\hbar(\mathbb{Z} + i\mathbb{Z}) \times \mathbb{D}^{k-1}$ to a regular flow box $\mathbb{U}_i \simeq \mathbb{D}^k$.*

PROOF. By hypotheses, x belongs to a regular flow box $\frac{1}{2}\mathbb{U}_i$. Hence, we obtain the result using the same arguments as in the previous lemmas. $\qquad\square$

4   FINITENESS OF ENTROPY: THE STRATEGY

In this section, we will present the strategy for the proof of Theorem 1.1. We have seen a simpler situation in Section 2, where we approximate Bowen balls by cells. The sizes of these cells, with respect to the Hermitian metric and also with respect to the Poincaré metric along the leaves, are very different. In the global setting, we have considered, in Section 3, the difficulty that a leaf may visit singular flow boxes several times without any obvious rule.

For the proof of Theorem 1.1, we can imagine that the following two phenomena may happen while going along the leaves, up to hyperbolic time R:

- some cells of the singular flow boxes are transported into the regular part of the foliation;

- some cells are transported back to the singular flow boxes, but their shape changes in different directions.

The interaction of these phenomena increases when R goes to infinity, and it is quite hard to get a rough image of the Bowen balls for large time R. We are not able to solve this problem, but we can obtain in dimension 2 a good estimate for the number of Bowen balls in the definition of entropy.

We start with a criterion for the finiteness of entropy which allows us to reduce the problem to the bound of the entropy of a finite number of well-chosen transversals.

PROPOSITION 4.1. *Let $\mathbb{T}$ denote the union of the distinguished transversals of the regular flow boxes $\frac{1}{2}\mathbb{U}_i$. Assume that the entropy of $\mathbb{T}$ is finite. Then, the entropy of X is also finite.*

We first prove the following weaker property.

LEMMA 4.2. *Let Y denote the complement of the union of the singular flow boxes $\frac{1}{2}\mathbb{U}_e$. Then, under the hypothesis of Proposition 4.1, the entropy of Y is finite.*

PROOF. Fix a constant $\epsilon > 0$. For each flow box $\mathbb{U}_i$, consider a $(2R, \epsilon/2)$-dense subset F_i of $\mathbb{T}_i$ with minimal cardinal. Define $F := \cup F_i$. Choose a set $\Lambda \subset \mathbb{D}_R$ of fewer than e^{10R} points which is e^{-4R}-dense in $\mathbb{D}_R$, i.e., the discs with centers in Λ and with Poincaré radius e^{-4R} cover $\mathbb{D}_R$. Let G denote the union of the sets $\phi_x(\Lambda)$ with $x \in F$. Since G contains at most $e^{10R}\#F$ points, in order to show that the entropy of Y is finite, it suffices to check that any point z in a regular box $\frac{1}{2}\mathbb{U}_i$ is (R, ϵ)-close to a point of G.

Consider the plaque P_z of $\frac{1}{2}\mathbb{U}_i$ which contains z. Denote by y the intersection of P_z with the transversal $\mathbb{T}_i$. There is a point $x \in F_i$ such that $\text{dist}_{2R}(x, y) \le \epsilon/2$. Up to a re-parametrization of the leaves, we can assume without loss of generality that $\text{dist}_{2R}(\phi_x, \phi_y) \le \epsilon/2$. Since R is large, $\phi_y(\mathbb{D}_R)$ contains P_z. So, there is a point $\xi \in \mathbb{D}_R$ such that $\phi_y(\xi) = z$. It is clear that z and $w := \phi_x(\xi)$ are $(R, \epsilon/2)$-close. Choose a point $\xi' \in \Lambda$ such that $\text{dist}_P(\xi, \xi') \le e^{-4R}$. It is enough to show that w and $w' := \phi_x(\xi')$ are $(R, \epsilon/2)$-close because $w' \in G$.

Observe that there is an automorphism τ of $\mathbb{D}$ such that $\tau(\xi) = \xi'$ and that $\text{dist}_P(\tau(a), a) \le e^{-R}$ on $\mathbb{D}(\xi, R)$. If u is an automorphism such that $u(0) = \xi$, then $\phi_w := \phi_x \circ u$ is a covering map of L_w which sends 0 to w and $\phi_{w'} := \phi_x \circ \tau \circ u$ is a covering map of $L_{w'}$ which sends 0 to w'. Since the Poincaré metric is invariant, we have $\text{dist}_P(\phi_w(a), \phi_{w'}(a)) \le e^{-R} \ll \epsilon$ on $\mathbb{D}_R$. Now, the fact that η is bounded from above implies that w and w' are $(R, \epsilon/2)$-close. Hence, the entropy of Y is finite. $\quad\square$

We will also use the following lemma.

LEMMA 4.3. *Let x be a point in $X \setminus E$. Assume that $\phi_x(\mathbb{D}_{2R})$ is contained in a singular flow box $\frac{1}{2}\mathbb{U}_e$. Let $\epsilon > 0$ be a fixed number. If R is large enough, then $\phi_x(\mathbb{D}_R)$ is contained in $\epsilon \mathbb{U}_e$.*

PROOF. As before, we identify $\mathbb{U}_e$ with $\mathbb{D}^k$. Assume that $\phi_x(\mathbb{D}_R)$ contains a point y such that $\|y\|_1 \ge \epsilon$. Without loss of generality, assume that the first coordinate y_1 of y is a positive number and $\|y\|_1 = y_1 \ge \epsilon$. By hypothesis, $\phi_y(\mathbb{D}_R) \subset \frac{1}{2}\mathbb{D}^k$. The real curve l defined by

$$y^t := \left(t, y_2(t/y_1)^{\lambda_2/\lambda_1}, \ldots, y_k(t/y_1)^{\lambda_k/\lambda_1}\right) \quad \text{with} \quad t \in [y_1, 1/2]$$

is contained in L_y but not in $\phi_y(\mathbb{D}_R)$. Therefore, its Poincaré length is at least equal to R. On the other hand, since $\|y^t\|_1 \ge t \ge \epsilon$, we deduce from Proposition 3.3 that this length is bounded by a constant depending on ϵ. This is a contradiction since R is large. $\quad\square$

End of the proof of Proposition 4.1. Fix a constant $\epsilon > 0$ and consider $R > 0$ large enough. Let $F \subset Y$ be a $(3R, \epsilon/2)$-dense family in Y. Choose a set $W \subset \mathbb{D}_{2R}$ of cardinal e^{10R} which is e^{-4R}-dense in $\mathbb{D}_{2R}$. Consider the union F' of the sets $\phi_x(W)$ with $x \in F$. By Lemma 4.2, it is enough to show that $F' \cup E$ is (R, ϵ)-dense in X, i.e., the Bowen (R, ϵ)-balls centered at the points of F' cover X.

Consider a point $z \in X$. If $\phi_z(\mathbb{D}_{2R})$ is contained in a singular flow box $\frac{1}{2}\mathbb{U}_e$, by Lemma 4.3, we have $\text{dist}_R(e, z) < \epsilon$. So, for R large enough, z belongs to the Bowen (R, ϵ)-ball centered at e. It remains to consider the case where $\phi_z(\mathbb{D}_{2R})$ contains a point $y \in Y$.

Let x be a point in F such that $\text{dist}_{3R}(x, y) \le \epsilon/2$. So, we can find covering maps ϕ_x and ϕ_y such that $\text{dist}_{3R}(\phi_x, \phi_y) \le \epsilon/2$. Since $y \in \phi_z(\mathbb{D}_{2R})$, there is a point

$\xi \in \mathbb{D}_{2R}$ such that $\phi_y(\xi) = z$. It is clear that z and $w := \phi_x(\xi)$ are $(R, \epsilon/2)$-close. Let ξ' be a point in W such that $\mathrm{dist}_P(\xi, \xi') \leq e^{-4R}$. The point $w' := \phi_x(\xi')$ belongs to F'. We show as in Lemma 4.2 that w and w' are $(R, \epsilon/2)$-close. Therefore, z belongs to the Bowen (R, ϵ)-ball of center w'.   □

From now on, assume that $\dim X = 2$ and, hence, $\dim \mathbb{T} = 1$. Our strategy is to construct an adapted covering of $\mathbb{T}$ by open discs. Consider a hyperbolic time R large enough. For simplicity, assume that $R = N\hbar$ with N integer. We will construct a set $\widetilde{\mathbb{T}}$ which contains $\mathbb{T}$ and other transversals ouside and inside the singular flow boxes. Then, we will construct by induction on m, with $2m_1 \leq m \leq N$, a covering $\mathcal{V}_m$ of $\widetilde{\mathbb{T}}$ by discs such that if two points x, y belong to the same disc, the associated leaves L_x and L_y are close until time $m\hbar$. Of course, we have to control the cardinality of $\mathcal{V}_m$ in order to deduce the finiteness of entropy of $\mathbb{T}$ by taking $m = N$.

The covering $\mathcal{V}_m$ is obtained by refining a finite number of other coverings of $\widetilde{\mathbb{T}}$. We will use the following technical lemma in order to estimate the cardinal of $\mathcal{V}_m$ and to show that the cardinal of $\mathcal{V}_N$ grows at most exponentially when N tends to infinity.

LEMMA 4.4. *Let K be a finite family of sets such that each of them is contained in a complex plane, i.e., a copy of $\mathbb{C}$. Let $\mathcal{V}^i$ with $1 \leq i \leq n$ be n coverings of K by fewer than M discs. Then, we can cover K with a family $\mathcal{V}$ of fewer than $200^n M$ discs such that $\mathcal{V} \prec \mathcal{V}^i$ in the sense that every disc $D \in \mathcal{V}$ satisfies $2D \subset 2D_1 \cap \cdots \cap 2D_n$ for some $D_i \in \mathcal{V}^i$.*

PROOF. By induction, it is enough to consider the case $n = 2$ and to find a covering $\mathcal{V}$ with fewer than $200M$ discs. The covering $\mathcal{V}$ contains two kinds of discs that we construct next.

An element D_2 of $\mathcal{V}^2$ belongs to $\mathcal{V}$ iff there is $D_1 \in \mathcal{V}^1$ such that $D_1 \cap D_2 \neq \emptyset$ and $\mathrm{radius}(D_1) > 2\,\mathrm{radius}(D_2)$. Clearly, these discs satisfy the last condition in the lemma.

Consider now $D_j \in \mathcal{V}^j$ such that $D_1 \cap D_2 \neq \emptyset$ and $\mathrm{radius}(D_1) \leq 2\,\mathrm{radius}(D_2)$. Denote by $4\rho_1$ the radius of D_1. A disc D of the second kind is a disc of radius ρ_1 centered at a point in $\rho_1(\mathbb{Z} + i\mathbb{Z})$ which intersects $D_1 \cap D_2$. It is clear that $2D \subset 2D_1 \cap 2D_2$. Note that such a disc D can be associated to several discs D_2.

Observe that each disc D_1 is associated to fewer than 100 discs of the second kind. Therefore, $\mathcal{V}$ contains fewer than $200M$ discs. This family covers K since it covers $D_1 \cap D_2$ for all $D_j \in \mathcal{V}^j$. This completes the proof.   □

As we mentioned before, the covering $\mathcal{V}_m$ will be obtained by induction on m. We will construct p coverings $\mathcal{V}_m(l)$ of $\widetilde{\mathbb{T}}$ by discs which are obtained using the images of discs in $\mathcal{V}_{m-1}$ by some holonomy maps (the integer $p = m_1^4$ was fixed previously). Near a leaf, such a holonomy map looks like a displacement following a given direction with a small hyperbolic time. The covering $\mathcal{V}_m$ will be obtained by refining the $p + 1$ coverings $\mathcal{V}_{m-1}$ and $\mathcal{V}_m(l)$, with $0 \leq l \leq p - 1$, thanks to Lemma 4.4. The details will be given in Section 5.

A crucial property of $\mathcal{V}_m$ is that when two points x, y belong to the same element of $\mathcal{V}_m$, the maps ϕ_x and ϕ_y are close on $\mathbb{D}_{m\hbar}$. This property will be obtained by induction, i.e., from a similar property of $\mathcal{V}_{m-1}$. For this purpose, we need to cover $\mathbb{D}_{m\hbar}$ and $\phi_y(\mathbb{D}_{m\hbar})$ by discs of radius $(m-1)\hbar$ in order to apply the induction argument. The following lemma shows that we need only a fixed number of such discs.

LEMMA 4.5. *Let ξ_j be a point in S_j with $1 \leq j \leq 40m_1$. Then, for $m \geq 2m_1$, the disc $\mathbb{D}(0, (m-1)\hbar)$ and the $40m_1$ discs $\mathbb{D}(\xi_j, (m-2)\hbar)$ cover the disc $\mathbb{D}_{m\hbar}$.*

PROOF. Let ξ be a real number such that $3\hbar \leq \xi < 2m_1\hbar$. We claim that it is sufficient to show that $\mathbb{D}(\xi, (m-2)\hbar)$ contains all points $\zeta \in \mathbb{D}_{m\hbar} \setminus \mathbb{D}_{(m-1)\hbar}$ such that $|\arg(\zeta)| \leq \pi/(20m_1)$. Indeed, this property implies that $\mathbb{D}(\xi_j, (m-2)\hbar)$ contains the sector $|\arg(\zeta) - \arg(\xi_j)| \leq \pi/(5m_1)$ in $\mathbb{D}_{m\hbar} \setminus \mathbb{D}_{(m-1)\hbar}$. The union of these sectors covers $\mathbb{D}_{m\hbar} \setminus \mathbb{D}_{(m-1)\hbar}$.

Denote by ζ' the intersection of the half-line through ζ started at 0 with the circle of center 0 through ξ. Since $2m_1\hbar$ is small, it is easy to see that $\mathrm{dist}_P(\xi, \zeta') < \hbar$. Moreover, we have

$$\mathrm{dist}_P(\zeta, \zeta') = \mathrm{dist}_P(0, \zeta) - \mathrm{dist}_P(0, \xi) \leq (m-3)\hbar.$$

Therefore,

$$\mathrm{dist}_P(\xi, \zeta) \leq \mathrm{dist}_P(\xi, \zeta') + \mathrm{dist}_P(\zeta, \zeta') < (m-2)\hbar.$$

The lemma follows. □

We deduce the following result from the last lemma, which is more adapted to our problem when we are near a singular point.

LEMMA 4.6. *Let Γ be a closed curve in $\mathbb{D}_{m_1\hbar} \setminus \mathbb{D}_{4\hbar}$ such that 0 does not belong to the unbounded component of $\mathbb{C} \setminus \Gamma$. We divide it into p arcs $\widehat{\xi_j\xi_{j+1}}$ with $0 \leq j \leq p$ and $\xi_p = \xi_0$. Assume that the length of $\widehat{\xi_j\xi_{j+1}}$ with respect to the Poincaré metric on $\mathbb{D}$ is smaller than $m_1^{-2}\hbar$. Let ξ_j' be a point in $\mathbb{D}$ such that $\mathrm{dist}_P(\xi_j, \xi_j') \leq m_1^{-2}\hbar$. Then, the disc $\mathbb{D}(0, (m-1)\hbar)$ and the p discs $\mathbb{D}(\xi_j', (m-1)\hbar)$ cover the disc $\mathbb{D}_{m\hbar}$.*

PROOF. Observe that $\Gamma \cap S_j$ admits a point ξ such that the disc $\mathbb{D}(\xi, 3m_1^{-2}\hbar)$ is contained in S_j. If ξ belongs to $\widehat{\xi_i\xi_{i+1}}$, then this arc and also the point ξ_i' are contained in S_j. Lemma 4.5 implies the result. □

5 ADAPTED TRANSVERSALS AND THEIR COVERINGS

In this section, assume that X is a compact complex surface. We will construct, for every integer N large enough, the family of transversals $\widetilde{\mathbb{T}} = \mathbb{T}^{\mathrm{reg}} \cup \mathbb{T}^{\mathrm{sing}}$ for $R = N\hbar$ and its coverings $\mathcal{V}_m = \mathcal{V}_m^{\mathrm{reg}} \cup \mathcal{V}_m^{\mathrm{sing}}$ with $2m_1 \leq m \leq N$. We will use the constants introduced in the list given in the introduction.

Recall that each regular flow box $\mathbb{U}_i$ is identified to $\mathbb{D}^2$ and is associated with the distinguished transversal $\{0\} \times \mathbb{D}$. Define $\mathbb{T}_i^a := \{a\} \times \mathbb{D}$ with a in the lattice $m_1^{-5}\hbar(\mathbb{Z} + i\mathbb{Z}) \cap \mathbb{D}$. Denote by $\mathbb{T}^{\mathrm{reg}}$ the family of these transversals. Note that this family does not depend on R. We will, however, consider each element $\mathbb{T}_i^a$ of $\mathbb{T}^{\mathrm{reg}}$ with multiplicity $e^{46\lambda R}$, and we denote them by $\mathbb{T}_i^a(1), \ldots, \mathbb{T}_i^a(e^{46\lambda R})$. These transversals are considered as distinct. Later, we will construct the covering $\mathcal{V}_m$ of $\widetilde{\mathbb{T}}$ by induction on m using Lemma 4.4 applied to $\mathcal{V}_{m-1}$ and p other coverings of $\widetilde{\mathbb{T}}$. The discs used to cover $\mathbb{T}_i^a(j)$ depend on the index j. The multiplicities allow us to get a good bound for the number of discs in $\mathcal{V}_m$; see Proposition 5.1.

Consider now a singular flow box $\mathbb{U}_e \simeq \mathbb{D}^2$ and define $\mathbb{T}_e^a := \{a\} \times \frac{3}{4}\mathbb{D}$ with $a \in \Sigma \cap \frac{3}{4}\mathbb{D}$. Recall that Σ was constructed in Section 2, and we use here a large fixed constant λ. Denote by $\mathbb{T}^{\text{sing}}$ the family of these transversals $\mathbb{T}_e^a$, where each element is counted only one time. Note that $\mathbb{T}^{\text{sing}}$ depends on R. Define $\widetilde{\mathbb{T}}$ the union of $\mathbb{T}^{\text{reg}}$ and $\mathbb{T}^{\text{sing}}$.

We now construct the covering $\mathcal{V}_{2m_1}$ of $\widetilde{\mathbb{T}}$. Choose for $\mathbb{T}^{\text{reg}}$ a covering $\mathcal{V}_{2m_1}^{\text{reg}}$ by fewer than $e^{70\lambda R}$ discs of radius e^{-10R}, where we count the multiplicities of transversals. Consider the family $\mathcal{V}_{2m_1}^{\text{sing}}$ of the discs in $\mathbb{T}_e^a$ centered at $(a, b) \in \Sigma^2$ with radius $100e^{-23\lambda R}|b|$ if $b \neq 0$ and of radius α_1 if $b = 0$. It is not difficult to check that this family covers $\mathbb{T}^{\text{sing}}$. Define $\mathcal{V}_{2m_1}$ as the union of $\mathcal{V}_{2m_1}^{\text{reg}}$ and $\mathcal{V}_{2m_1}^{\text{sing}}$. The total number of discs used here is bounded by $e^{200\lambda R}$. Recall that we consider only a large R. So, we have $m_1 \ll R$.

For each point $x \in \mathbb{T}_i^a$ and ρ small enough, denote by $D(x, \rho)$ the disc of center x and of radius ρ in $2\mathbb{T}_i^a$ and by $\Delta(x, \rho)$ the disc of center x and of radius ρ in the plaque of $2\mathbb{U}_i$ containing x. The notation $D(x, \rho)$ can be used for x in a transversal $\mathbb{T}_e^a$. Since $m_1 \hbar \ll \epsilon_0$, we have the following useful properties for large R, where we use that the discs are of size less than e^{-10R}:

(H1) If x, y are in $2D$ for some D in $\mathcal{V}_{2m_1}^{\text{reg}}$, then the basic projection Φ associated to x and y exists and sends $\Delta(x, 2m_1\hbar)$ to $\Delta(y, 6m_1\hbar)$. Moreover, we have $\|\Phi - \text{id}\|_{\mathcal{C}^2} \leq e^{-9R}$ on $\Delta(x, 2m_1\hbar)$.

(H1)′ If x, y are in $2D$ for some D in $\mathcal{V}_{2m_1}^{\text{sing}}$, then the basic projection Φ associated to x and y exists as in Lemma 2.11. It satisfies $\|\Phi - \text{id}\|_{\mathcal{C}^0} \leq e^{-9R}$ on $L_x \cap \frac{3}{4}\mathbb{D}^2$.

(H2) Consider two transversals $\mathbb{T}_i^a$ and $\mathbb{T}_j^b$. If x is in $\mathbb{T}_i^a$ such that $\Delta(x, 2m_1\hbar)$ intersects $\mathbb{T}_j^b$ at a point y, then the holonomy map π from $2\mathbb{T}_j^b$ to $2\mathbb{T}_i^a$ is well defined on $D(y, 4\hbar)$ with image in $\frac{3}{2}\mathbb{T}_i^a$.

(H3) If D is a disc contained in $D(y, \hbar)$, then $\pi(D)$ is *quasi-round*, i.e., there is a disc $D' \subset \frac{3}{2}\mathbb{T}_i^a$ such that $D' \subset \pi(D) \subset \frac{11}{10}D'$ and $2D' \subset \pi(2D)$.

For the property (H3), we use the fact that the holonomy map is holomorphic with no critical point. So, on small discs, it is close to homotheties.

The construction of $\mathcal{V}_m = \mathcal{V}_m^{\text{reg}} \cup \mathcal{V}_m^{\text{sing}}$ will be obtained by induction on m. It contains only small discs of diameter less than e^{-10R}. Assume that the construction is done for $m - 1$. In order to obtain $\mathcal{V}_m$, we will apply Lemma 4.4 to $K := \widetilde{\mathbb{T}}$, to the covering $\mathcal{V}_{m-1}$ and p other coverings $\mathcal{V}_m(l) = \mathcal{V}_m^{\text{reg}}(l) \cup \mathcal{V}_m^{\text{sing}}(l)$ with $0 \leq l \leq p - 1$. Lemma 4.4 allows us to obtain a covering $\mathcal{V}_m$ such that $\mathcal{V}_m \prec \mathcal{V}_{m-1}$ and $\mathcal{V}_m \prec \mathcal{V}_m(l)$. Roughly speaking, we will cover each disc in $\mathcal{V}_{m-1}$ by smaller discs, so that after traveling a fixed time in some direction starting from such a small disc, we arrive at a disc of $\mathcal{V}_{m-1}$ in another transversal.

We explain now the construction of $\mathcal{V}_m^{\text{reg}}(l)$. There are two cases to consider. Recall that $S_l(x)$ is defined just before Lemma 3.9.

Case 1a. Assume that there is a point $x_0 \in \mathbb{T}_i^a$ such that $S_l(x_0)$ intersects a singular flow box $\frac{1}{2}\mathbb{U}_e$. Consider an arbitrary point $x \in \mathbb{T}_i^a$. Since we are still far from singular points, $S_l(x)$ contains and is contained in small discs of size independent of R. Thus, there are about a constant times $e^{46\lambda R}$ transversals $\mathbb{T}_e^b$ which intersect $S_l(x)$. These transversals $\mathbb{T}_e^b$ are still far from the singularities. So, the following properties hold because the regular flow boxes are of small size and $\hbar$ is small:

(H2)′ There is a well-defined holonomy map from $2\mathbb{T}_i^a$ to $\mathbb{T}_e^b$. Denote by π its inverse.

(H3)′ If D is a disc in $\mathbb{T}_e^b$ of radius less than $\hbar$ which intersects $\pi^{-1}(\mathbb{T}_i^a)$, then π is defined on $2D$ and $\pi(D)$ is quasi-round, i.e., there is a disc $D' \subset 2\mathbb{T}_i^a$ such that $D' \subset \pi(D) \subset \frac{11}{10}D'$ and $2D' \subset \pi(2D)$.

The property (H3)′ allows us to cover $\pi(D)$ by D' and its 100 *satellites*, which are 100 discs D_n', $0 \le n \le 99$, 10 times smaller than D', and such that $D_n' \cap D' \ne \varnothing$. Notice that $2D_n' \subset \pi(2D)$ for all n. We will use this important property later. So, we can cover each $\mathbb{T}_i^a(s)$ with the discs D' obtained earlier for $D \in \mathcal{V}_{m-1}$, together with their satellites D_n'. The choice of $\mathbb{T}_e^b$ depends on s. Thanks to the multiplicities of the $\mathbb{T}_i^a$, we can use each $\mathbb{T}_e^b$ only a bounded number of times. The reason for introducing those multiplicities is that in the intersection of regular and singular flow boxes the $\mathbb{T}_e^b$ are more dense than the $\mathbb{T}_i^a$.

Case 1b. Assume that $S_l(x)$ does not intersect any singular flow box $\frac{1}{2}\mathbb{U}_e$ for every $x \in \mathbb{T}_i^a$. For all holonomy maps π and discs D in $\mathcal{V}_{m-1}$ satisfying (H2) and (H3) for some x and y such that $y \in S_l(x)$, we choose a disc D' as in (H3) and 100 satellites of D' in order to cover $\pi(D)$. Note that for any $x \in \mathbb{T}_i^a$, there exists a choice of y, π, D such that $x \in \pi(D)$, see also Lemma 3.9. It follows that the construction gives us a covering of $\mathbb{T}_i^a(s)$ using the elements of $\mathcal{V}_{m-1}$ which cover the transversals $\mathbb{T}_i^b(s)$. We make sure to use here the same index s, in particular, each disc in $\mathcal{V}_{m-1}$ is used a bounded number of times. This ends the construction of the covering $\mathcal{V}_m^{\mathrm{reg}}(l)$ of $\mathbb{T}^{\mathrm{reg}}$.

The construction of $\mathcal{V}_m^{\mathrm{sing}}(l)$ is more delicate. First, we always add to $\mathcal{V}_m^{\mathrm{sing}}(l)$ the disc of center $(a,0)$ and of radius α_1 in $\mathbb{T}_e^a$ and call it *an exceptional disc*. If $|a| \le \alpha_1$, we just choose $\mathcal{V}_m^{\mathrm{sing}}(l)$ equal to $\mathcal{V}_{m-1}$ on $\mathbb{T}_e^a$. In this case, we also say that $\mathbb{T}_e^a$ is an *an exceptional transversal*. Consider now a transversal $\mathbb{T}_e^a$ with $\alpha_1 < |a| \le 3/4$. Consider a point $x = (a,d) \in \mathbb{T}_e^a \cap \Sigma^2$ with $|d| \ge \alpha_1$. Recall that the map J_l is defined just before Lemma 2.9. By Lemma 3.5, the point $w := J_l(x)$ satisfies the hypotheses of Lemmas 2.10 and 3.10 since λ is a large constant. We also distinguish two cases.

Case 2a. Assume that w belongs to $\frac{5}{8}\mathbb{U}_e$ (note that $\frac{5}{8}\mathbb{U}_e \subset \frac{3}{4}\mathbb{U}_e$). Consider the transversal $\mathbb{T}_e^b$ which contains w and write $w = (b,v)$. If $D \subset \mathbb{T}_e^b$ is an element of $\mathcal{V}_{m-1}^{\mathrm{sing}}$ such that $\mathrm{dist}(w, D) \le 100\gamma|v|e^{-23\lambda R}$, denote by D' the disc on $\mathbb{T}_e^a$ which is the image of D by $\Psi_{w,x}$ since this map preserves Σ^2. By Lemma 2.10, the obtained discs D' cover the disc of center x and of radius $100|d|e^{-23\lambda R}$ in $\mathbb{T}_e^a$. They are elements of $\mathcal{V}_m^{\mathrm{sing}}(l)$. Note that we use here the property that $\Psi_{w,x}$ is conformal. This is the only point where the hypothesis $\dim X = 2$ is essential.

Case 2b. Assume that w is not in $\frac{5}{8}\mathbb{U}_e$. Since $\hbar$ is small, by Lemma 3.10, x, w are outside $\frac{1}{2}\mathbb{U}_e$ and $\varphi_x(-t\log\|x\|_1\mathbb{D})$ is contained in a plaque of a regular flow box.

By Lemma 3.11, we can find a transversal $\mathbb{T}_i^b$ which intersects $\varphi_x(-t \log \|x\|_1 \mathbb{D})$ at a point y near $\varphi_x(-t \log \|x\|_1 e^{2i\pi l/p})$, i.e., the distance between these two points is less than $m_1^{-3}\hbar$. We have the following properties:

(H2)″ There is a well-defined holonomy map π from $2\mathbb{T}_i^b$ to $\mathbb{T}_e^a$.

(H3)″ If D is a disc in $2\mathbb{T}_i^b$ of radius less than $\hbar$ which intersects $\mathbb{T}_i^b$, then π is defined on $2D$ and $\pi(D)$ is quasi-round in the sense that there is a disc $D' \subset \mathbb{T}_e^a$ such that $D' \subset \pi(D) \subset \frac{11}{10}D'$ and $2D' \subset \pi(2D)$.

We cover a neighborhood of x with the discs D' and their satellites as before with $D \in \mathcal{V}_{m-1}$. We make sure that for each $\mathbb{T}_e^a$ we only use discs from $\mathbb{T}_i^b(s)$ for a fixed s. It is important to observe that the last construction concerns a constant times $e^{46\lambda R}$ transversals $\mathbb{T}_e^a$. Therefore, we can choose the index s so that each disc in $\mathcal{V}_{m-1}^{\text{reg}}$ is used a bounded number of times. We also fix a choice which does not depend on m. This ends the construction of the covering $\mathcal{V}_m^{\text{sing}}(l)$.

We have the following crucial proposition.

PROPOSITION 5.1. *There is a constant $c > 1$ independent of R such that the cardinal of $\mathcal{V}_m$ is smaller than c^R for $2m_1 \leq m \leq N$.*

PROOF. By Lemma 4.4 applied to $n := p+1$, it is enough to show that in the preceding construction of $\mathcal{V}_m(l)$, each disc in $\mathcal{V}_{m-1}$ is used fewer than c' times where c' is a constant. This can be checked step by step in our construction. For one of these steps, we use Lemma 2.9. We obtain by induction that

$$\#\mathcal{V}_m \leq (200^{p+1}c')^{m-2m_1}\#\mathcal{V}_{2m_1} \leq (200^{p+1}c')^{m-2m_1}e^{200\lambda R}.$$

This implies the proposition.   □

The last proposition shows that the cardinality of $\mathcal{V}_N$ is smaller than $c^R = c^{N\hbar}$.

Consider a disc D in $\mathcal{V}_N$. It is constructed by induction using the holonomy maps π as in (H2), (H2)′, (H2)″ or the map $\Psi_{w,x}$. This corresponds to the four cases described previously. By Proposition 4.1, in order to obtain Theorem 1.1, we will show in Proposition 6.1 that two points in the same disc D are $(R/3, e^{-R/4})$-close. This will be done using the notion of conformally (R, δ)-close points.

To such a disc D, we will associate a tree F_D which partially encodes the construction of $\mathcal{V}_N$. Its combinatorial and metric properties (see Lemma 5.2) will allow us to construct conformally (R, δ)-close maps from leaves to leaves following the tree. Points in D are associated to some isomorphic trees and the isomorphisms are coherent with the dynamics of the foliation.

The set of vertices of the tree F_D will consist of the union

$$F_D(0) \cup F_D(1) \cup \cdots \cup F_D(N - 2m_1),$$

where $F_D(0) = \{D\}$ and $F_D(m) \subset \mathcal{V}_{N-m}$. Moreover, each point in $F_D(m)$ is joined to a unique point in $F_D(m-1)$ and to at most p points in $F_D(m+1)$. We now give the construction of F_D by induction.

If D belongs to an exceptional disc or an exceptional transversal, we just take $F_D(1) = \cdots = F_D(N - 2m_1) = \varnothing$. Otherwise, D is obtained using p holonomy maps π_i as specified in (H2), (H2)′, (H2)″, or the map $\Psi_{w,x}$, as before. By construction, there are D_i in $\mathcal{V}_{N-1}$ such that $2D$ is contained in $\pi_i(2D_i)$. We choose $F_D(1) := \{D_1, \ldots, D_p\}$. Each D_i is joined to D. We then obtain a part of the tree.

In order to obtain $F_D(2)$, we will repeat the preceding construction, but for each $D_i \in F_D(1)$ instead of D. If D_i belongs to an exceptional disc or an exceptional transversal, then it is not joined to any element in $F_D(2)$. Otherwise, it is joined to p elements in $F_D(2) \subset \mathcal{V}_{N-2}$. We then continue the same construction in order to obtain $F_D(3), \ldots, F_D(N - 2m_1)$. Note that F_D is not uniquely determined by D but we fix here a choice for each disc D.

Now, for each point $x \in 2D$, we construct a tree $F_x \subset \mathbb{D}$ which is canonically isomorphic to F_D. The set of vertices of F_x will be $F_x(0) \cup F_x(1) \cup \cdots \cup F_x(N - 2m_1)$ such that $F_x(0) = \{0\}$ and so that ϕ_x sends each point in $F_x(m)$ to a disc $2D'$ with $D' \in F_D(m)$ and defines a bijection between $F_x(m)$ and $F_D(m)$. Moreover, a point in $F_x(m + 1)$ and a point in $F_x(m)$ are joined by an edge if and only if the associated vertices in F_D are also joined by an edge. We have to give here some details because ϕ_x is not injective in general.

We obtain $F_x(1)$ as follows. If D belongs to an exceptional disc or an exceptional transversal, then we take $F_x(1) = \varnothing$. Otherwise, we distinguish two cases. When D is outside the singular flow boxes $\frac{1}{4}\mathbb{U}_e$, then ϕ_x is injective on $\mathbb{D}_{\epsilon_0}$ and the image of this disc intersects $2D_i$ at a unique point for any $D_i \in F_D(1)$. Therefore, it is enough to define $F_x(1)$ as the pull-back by ϕ_x of these intersection points in $\mathbb{D}_{\epsilon_0}$. Consider now the case where D intersects a singular flow box $\frac{1}{4}\mathbb{U}_e$. By construction, there is a set G of p points very close to the points $-t \log \|x\|_1 e^{2i\pi l/p}$ in Π_x which are sent by φ_x to the discs in $F_D(1)$. If $\tau : \Pi_x \to \mathbb{D}$ is the unique holomorphic map such that $\tau(0) = 0$ and $\varphi_x = \phi_x \circ \tau$, define $F_x(1) := \tau(G)$. Clearly, ϕ_x defines a bijection between $F_x(1)$ and $F_D(1)$. Each point in $F_x(1)$ is joined to $F_x(0) = \{0\}$.

In order to obtain $F_x(2)$, it is enough to repeat the same construction to each point $\phi_x(a)$ with $a \in F_x(1)$. We use that ϕ_x is injective on $\mathbb{D}(a, \epsilon_0)$ if $\phi_x(a)$ is outside the singular flow boxes $\frac{1}{4}\mathbb{U}_e$, and otherwise there is a unique holomorphic map $\tau : \Pi_x \to \mathbb{D}$ such that $\tau(0) = a$ and $\varphi_a = \phi_x \circ \tau$. By induction, we obtain $F_x(3), \ldots, F_x(N - 2m_1)$ satisfying the stated properties.

The following lemma is essential for the proof of Theorem 1.1.

LEMMA 5.2. *The set of vertices $F_x(m)$ is contained in $\mathbb{D}_{m_1 m \hbar}$ for every m. If a is a point in F_x which is joined to p points $a_1, \ldots, a_p$, then the union of $\mathbb{D}(a, (m-1)\hbar)$ and $\mathbb{D}(a_i, (m-1)\hbar)$ contains $\mathbb{D}(a, m\hbar)$ for every $m \geq 2m_1$.*

PROOF. We prove the first assertion. It is enough to check that if a point ζ in $F_x(n-1)$ is joined to a point ξ in $F_x(n)$, then $\mathrm{dist}_P(\xi, \zeta) \leq m_1 \hbar$. For simplicity, we consider the case where $n = 1$, the general case is obtained in the same way. So, we have that $\zeta = 0$ and ξ belongs to $F_x(1)$. Define $y := \phi_x(\xi)$. This is the preimage of x by a holonomy map π as in (H2), (H2)′, (H2)″, or by a map $\Psi_{w,x}$, as before.

Now, if y is given by π as in (H2), the distance between x and y is smaller than $3m_0 \hbar$. By Proposition 3.3, the Poincaré distance between 0 and ξ is smaller than $m_1 \hbar$ since $m_1 \gg c_1 m_0$.

For the other cases, y is very close to a point $y' := \varphi_x(-t \log \|x\|_1 e^{2i\pi l/p})$ with respect to the Poincaré metric. Recall that $m_0 \hbar < t < 2m_0 \hbar \ll 1$. The Poincaré distance between x and y' is bounded by the distance between 0 and $-t \log \|x\|_1 e^{2i\pi l/p}$ with respect to the Poincaré metric on Π_x. Therefore, by Lemma

2.2, we deduce that $\text{dist}_P(x, y')$ is bounded by a constant times $m_0 \hbar$. It follows easily that $\text{dist}_P(0, \xi)$, which is equal to $\text{dist}_P(x, y)$, is smaller than $m_1 \hbar$. This completes the proof of the first assertion.

We now prove the second assertion. By Lemmas 4.5 and 4.6, we have to check only that $\{a, a_1, \ldots, a_p\}$ contains a subset satisfying the hypotheses of those lemmas. If a belongs to $\mathbb{T}^{\text{reg}}$, this is a consequence of Lemma 3.9. If a belongs to $\mathbb{T}^{\text{sing}}$, Lemmas 3.10 and 3.11 imply the result.   □

6   FINITENESS OF ENTROPY: END OF THE PROOF

In this section, we complete the proof of Theorem 1.1. By Proposition 4.1, it is enough to show that the entropy of $\mathbb{T}^{\text{reg}}$ is finite. By Proposition 5.1, we have to check only that each disc in $\mathcal{V}_N^{\text{reg}}$ is contained in a Bowen $(R/3, e^{-R/4})$-ball. So, Theorem 1.1 is a consequence of the following proposition.

PROPOSITION 6.1. *Let x, y be two points of a disc $2D \subset 2\mathbb{T}_i^a$ with D in $\mathcal{V}_N$. Then, they are $(R/3, e^{-R/4})$-close.*

The proof of this result uses Proposition 3.6 and occupies the rest of this section. Recall that by Proposition 3.3, we have $c_1^{-1} \leq |\eta| \leq c_1$ on $\mathbb{T}^{\text{reg}}$. So, it is enough to check that x, y are conformally (R, e^{-3R})-close, and we have to construct a map ψ from $\mathbb{D}_R$ to L_y which is close to ϕ_x as in the definition of conformally (R, e^{-3R})-close points. Proposition 6.1 is a direct consequence of Lemmas 6.3 and 6.4, which correspond, respectively, to the case where the leaves are far from the separatrices and to the case where the leaves are close to some separatrix.

We have the following lemma, which holds for x, y in $2D$ with an arbitrary disc D in $\mathcal{V}_N = \mathcal{V}_N^{\text{reg}} \cup \mathcal{V}_N^{\text{sing}}$. Recall that the trees F_x and F_y are isomorphic to F_D. Therefore, there is an isomorphism σ from F_x to F_y. Denote by x' the image of x by the basic projection associated to x and y. We say that the tree F_D is *complete* if each element in $F_D(m-1)$ is joined to p elements in $F_D(m)$ for any $1 \leq m \leq N - 2m_1$.

LEMMA 6.2. *Assume that the tree F_D is complete. Then there exists a unique map $\psi' : \mathbb{D}_{N\hbar} \to L_y$ which is locally the composition of ϕ_x with basic projections from leaves to leaves and is equal, in a neighborhood of any point a in $F_x \cap \mathbb{D}_{N\hbar}$, to the composition of ϕ_x with the basic projection associated to $\phi_x(a)$ and $\phi_y(\sigma(a))$.*

PROOF. Observe that by continuity, if such a map ψ' exists, it must be unique. We show by induction on m that such a projection exists on $\mathbb{D}(a, m\hbar)$ for all a in $F_x(0) \cup \cdots \cup F_x(N-m)$ with $2m_1 \leq m \leq N$. Define $\widetilde{x} := \phi_x(a)$ and $\widetilde{y} := \phi_y(\sigma(a))$. Also denote by $\widetilde{D}$ the element of F_D associated to $\widetilde{x}$ and $\widetilde{y}$. This is an element of $F_D(0) \cup \ldots \cup F_D(N-m)$ such that $\widetilde{x}$ and $\widetilde{y}$ belong to $2\widetilde{D}$.

If $\widetilde{x}$ or $\widetilde{y}$ is outside the singular flow boxes $\frac{1}{2}\mathbb{U}_e$, since $2m_1\hbar$ is small, ϕ_x sends $\mathbb{D}(a, 2m_1\hbar)$ bijectively to a disc in $L_{\widetilde{x}}[\epsilon_0]$. So, it is not difficult to see that the desired property holds in this case for $m = 2m_1$. When $\widetilde{x}$ and $\widetilde{y}$ belong to a singular flow box $\frac{1}{2}\mathbb{U}_e$, the property for $m = 2m_1$ is a consequence of Lemma 2.11. Assuming now the property for $m - 1$, we have to show it for m.

Recall that $2\widetilde{D}$ is contained in $2D' \cap \pi_0(2D_0) \cap \cdots \cap \pi_{p-1}(2D_{p-1})$. Here, D' and D_i are elements of $\mathcal{V}_{m-1} \cup \cdots \cup \mathcal{V}_{N-1}$ and the π_i are holonomy maps from some transversals in $\mathbb{T}$ to the one containing $\widetilde{D}$ or a map $\Psi_{w,x}$ as in Section 5. Define $x^i := \pi_i^{-1}(\widetilde{x})$ and $y^i := \pi_i^{-1}(\widetilde{y})$. Denote by ξ^i the points in F_x such that $\phi_x(\xi^i) = x^i$.

The induction hypothesis implies that there is a map ψ_i' from $\mathbb{D}(\xi^i, (m-1)\hbar)$ to L_y which is the composition of ϕ_x with basic projections from leaves to leaves. Moreover, it is equal, in a neighborhood of any point b in $F_x \cap \mathbb{D}(\xi^i, (m-1)\hbar)$, to the composition of ϕ_x with the basic projection associated to $\phi_x(b)$ and $\phi_y(\sigma(b))$. Also by the induction hypothesis, an analogous map is defined on $\mathbb{D}(a, (m-1)\hbar)$. From the uniqueness of these maps, we can glue them together. Using the second assertion of Lemma 5.2, we obtain a map ψ' defined on $\mathbb{D}(a, m\hbar)$. $\qquad \square$

LEMMA 6.3. *Let* x, y, D *be as in Proposition 6.1. Assume that the tree* F_D *is complete. Then,* x *and* y *are conformally* (R, e^{-3R})*-close.*

PROOF. We have to construct a map ψ satisfying the definition of conformally (R, e^{-3R})-close points. Let ψ' be the map constructed in Lemma 6.2. We deduce from the construction using (H1), (H1)′, and Lemma 2.11 that $\|\psi' - \phi_x\|_{\mathcal{C}^0} \leq e^{-9R}$ on $\mathbb{D}_R$ and that $\|\psi' - \phi_x\|_{\mathcal{C}^2} \leq e^{-6R}$ on $\mathbb{D}_R \setminus \phi_x^{-1}(\cup \frac{1}{4}\mathbb{U}_e)$. Here, we use that $\|\phi_x\|_{\mathcal{C}^2} \lesssim e^{2R}$ on $\mathbb{D}_R$ when we consider the Euclidean metric on $\mathbb{D}_R$ and the Hermitian metric on L_x. So, a priori, on $\phi_x^{-1}(\cup \frac{1}{4}\mathbb{U}_e)$, the Beltrami coefficient associated to this map does not satisfy the condition required for conformally close (R, e^{-3R})-points.

Using the maps $\widetilde{\Psi}_{x,y}$ as in Lemma 2.12, we can correct ψ' in each connected component of $\phi_x^{-1}(\frac{3}{4}\mathbb{U}_e)$ in order to obtain a map ψ'' such that $\|\psi'' - \phi_x\|_{\mathcal{C}^0} \leq e^{-8R}$ on $\mathbb{D}_R$ and $\|\psi'' - \phi_x\|_{\mathcal{C}^2} \leq e^{-5R}$ on $\mathbb{D}_R \setminus \phi_x^{-1}(\cup \frac{1}{4}\mathbb{U}_e)$. Moreover, ψ'' is holomorphic on $\phi_x^{-1}(\cup \frac{1}{2}\mathbb{U}_e)$. Therefore, its Beltrami coefficient vanishes on $\phi_x^{-1}(\cup \frac{1}{2}\mathbb{U}_e)$ and hence satisfies the required property. It remains to modify ψ'' in order to obtain a map ψ with $\psi(0) = y$. But this can be done by composition with an automorphism of $\mathbb{D}$, close to the identity, as in the end of the proof of Proposition 3.7 or in the proof of Theorem 2.1 in [1]. $\qquad \square$

The following lemma, together with Lemma 6.3, completes the proof of Proposition 6.1.

LEMMA 6.4. *Let* x, y, D *be as in Proposition 6.1. Assume that the tree* F_D *is not complete. Then,* x *and* y *are* (R, e^{-R})*-close.*

PROOF. Since F_D is not complete, there is a path $(\xi^0, \ldots, \xi^m)$ of the graph F_x joining $\xi^0 = 0$ to a vertex ξ^m such that $\phi_x(\xi^m)$ belongs to an exceptional transversal or to $2D'$ with D' an exceptional disc. Define $x^i := \phi_x(\xi^i)$.

The image of $(\xi^0, \ldots, \xi^m)$ by σ is a path $(\zeta^0, \ldots, \zeta^m)$ of F_y. We have seen in the proof of Lemma 5.2 that $\mathrm{dist}_P(\xi^i, \xi^{i+1})$ and $\mathrm{dist}_P(\zeta^i, \zeta^{i+1})$ are smaller than $m_1 \hbar \ll \epsilon_0$. Define $y^i := \phi_y(\zeta^i)$. There is a disc D_i in $\mathcal{V}_{N-i}$ such that x^i and y^i belong to $2D_i$. So, there is a basic projection associated to x^i and y^i. Denote by z^i the image of x^i by this projection.

Observe that $\mathrm{dist}(x^m, z^m) \leq 10\alpha_1$. Recall that $\alpha_1 := e^{-e^{7\lambda R}}$. Therefore, using Proposition 3.7 applied to λ large enough, we obtain that x^m and z^m are conformally $(2Nm_1\hbar, e^{-4R})$-close. Moreover, the map ψ associated to these conformally close points is obtained using basic projections as in (H1), (H1)′ and Lemma 2.11. So, we can follow the path $(\xi^0, \ldots, \xi^m)$ and see that ξ^i is sent by ψ to z^i. So, 0 is sent by ψ to $z := z^0$. By Lemma 5.2, the Poincaré distance between 0 and ξ^m is at most equal to $mm_1\hbar$. Therefore, x and z are conformally $(Nm_1\hbar, e^{-4R})$-close. It follows that x and z are (R, e^{-2R})-close. Moreover, z belongs to a small plaque containing

y and $\mathrm{dist}(z,y) \le 2\,\mathrm{dist}(x,y) \le e^{-3R}$. We deduce that z and y are (R, e^{-2R})-close. This implies that x and y are (R, e^{-R})-close.   □

7 Bibliography

[1] T.-C. Dinh, V.-A. Nguyên, and N. Sibony, Entropy for hyperbolic Riemann surface laminations I. *In this volume*, 569–592.

[2] J.-E. Fornæss and N. Sibony, Riemann surface laminations with singularities, *J. Geom. Anal.*, **18** (2008), no. 2, 400–442.

[3] A. A. Glutsyuk, Hyperbolicity of the leaves of a generic one-dimensional holomorphic foliation on a nonsingular projective algebraic variety. (Russian) *Tr. Mat. Inst. Steklova*, **213** (1997), Differ. Uravn. s Veshchestv. i Kompleks. Vrem., 90–111; translation in *Proc. Steklov Inst. Math.* **213** (1996), no. 2, 83–103.

[4] A. Lins Neto, Uniformization and the Poincaré metric on the leaves of a foliation by curves, *Bol. Soc. Brasil. Mat. (N.S.)*, **31** (2000), no. 3, 351–366.

[5] A. Lins Neto and M. G. Soares, Algebraic solutions of one-dimensional foliations, *J. Differential Geom.*, **43** (1996), no. 3, 652–673.

Intersection theory for ergodic solenoids

Vicente Muñoz and Ricardo Pérez-Marco

ABSTRACT. We develop the intersection theory associated to immersed, oriented and measured solenoids, which were introduced in [3].

1 INTRODUCTION

In [3], the authors define the concept of k-solenoid as an abstract laminated space and prove that a solenoid with a transversal measure immersed in a smooth manifold defines a generalized Ruelle-Sullivan current, as considered in [8]. Considering as object the abstract space with the inmersion provides a more versatile concept than previously considered by other authors (see, for example, [1], [9], or [10]) that is suitable for results like the representation theorem in [5]. The purpose of the current paper is to study the intersection theory of such objects.

If M is a smooth manifold, any closed oriented submanifold $N \subset M$ of dimension k determines a homology class in $H_k(M, \mathbb{Z})$. This homology class in $H_k(M, \mathbb{R})$, as dual of de Rham cohomology, is explicitly given by integration of the restriction to N of differential k-forms on M. For representing real homology classes, we need to consider more general objects. In [3], we define a k-solenoid to be a Hausdorff compact space foliated by k-dimensional leaves with finite dimensional transverse structure (see the precise definition in Section 2). For these oriented solenoids we can consider k-forms that we can integrate provided that we are given a transverse measure invariant by the holonomy group. We define an immersion of a solenoid S into M to be a regular map $f \colon S \to M$ that is an immersion on each leaf. If the solenoid S is endowed with a transverse measure μ, then any smooth k-form in M can be pulled back to S by f and integrated. This defines a closed current that we denote by (f, S_μ) and call a generalized current. The associated homology class is denoted as $[f, S_\mu] \in H_k(M, \mathbb{R})$. This construction generalizes the currents introduced by Ruelle and Sullivan in [8].

In [5], we prove that every real homology class in $H_k(M, \mathbb{R})$ can be realized by a generalized current (f, S_μ), where S_μ is an oriented, minimal, uniquely ergodic solenoid. It is a known result that not all real homology classes can be represented by *embedded* solenoids [2], hence the importance of using immersions $f \colon S \to M$ to represent real homology classes. Uniquely ergodic solenoids are defined in [3]. These are minimal solenoids which possess a unique transverse measure. The importance of such solenoids stems from the fact that their topology, or more precisely, the recurrence of any leaf, determines its solenoidal structure. This is formulated in a precise way through the Schwartzman measures in [4]. Finally, in [6] we prove that the generalized currents associated to oriented, minimal, uniquely ergodic, immersed solenoids are dense in the space of closed currents.

It is relevant to note here that there is some motivation to study these immersed solenoids in relation with the Hodge conjecture. This is given by the *Solenoidal*

Partially supported through Spanish MEC grant MTM2007-63582.

Hodge Conjecture, stated in [5]. This conjecture asserts that for a compact Kähler manifold M, any homology class in $H_{p,p}(M) \subset H_{2p}(M, \mathbb{R})$ is represented by a complex immersed solenoid of dimension $2p$.

Let us review the contents of the paper. In Section 2, we recall the definition of solenoids and of the generalized current associated to an immersed, oriented and measured solenoid. In Section 3 we prove that the generalized current is invariant by perturbation (homotopy) of a solenoid. Section 4 is devoted to defining the intersection of two solenoids S_{1,μ_1}, S_{2,μ_2} which intersect transversally (i.e., when the leaves intersect transversally). In this case, the intersection is a solenoid and it has a natural transverse measure associated to the measures of the two given solenoids. The generalized current of the intersection solenoid is the product of the generalized currents of the given solenoids. We also prove that we may homotope a solenoid to make it intersect transversally a submanifold whenever the transverse structure is Cantor.

In general, it is not possible to perturb solenoids immersed in a differentiable manifold to make them intersect transversally. This is also clearly impossible in the case of foliations. It is also impossible to create such transverse intersections via smooth perturbations in the case of solenoides with transverse Cantor structure, as the persistence of homoclinic tangencies for stable and unstable foliations shows (see [7]). So the concept of almost everywhere transversality introduced in Section 5 is very useful. In the case of complementary dimensions, two solenoids are said to intersect almost everywhere transversally if $(\mu_1 \times \mu_2)$-almost all leaves intersect transversally and the other leaves intersect in isolated points. We define a measure for the intersection and the integral of the measure (taking also into account the intersection index) equals the product of the generalized currents. In the case of non-complementary dimensions, we shall require that the models are conjugate to analytic solenoids. This allows us to define an intersection current supported on the intersection of the solenoids. This is worked out in Section 6.

Acknowledgements. We would like to thank the referee for interesting comments on the paper.

2 MEASURED SOLENOIDS AND GENERALIZED CURRENTS

Let us recall the definition of a k-solenoid from [3].

DEFINITION 2.1. *Let $0 \leq s, r \leq \omega$, $r \geq s$, and let $k, l \geq 0$ be two integers. A foliated manifold (of dimension $k + l$, with k-dimensional leaves, of regularity $C^{r,s}$) is a smooth manifold W of dimension $k + l$ endowed with an atlas $\mathcal{A} = \{(U_i, \varphi_i)\}$, $\varphi_i: U_i \to \mathbb{R}^k \times \mathbb{R}^l$, whose changes of charts*

$$\varphi_{ij} = \varphi_i \circ \varphi_j^{-1} \colon \varphi_j(U_i \cap U_j) \to \varphi_i(U_i \cap U_j),$$

are of the form $\varphi_{ij}(x, y) = (X_{ij}(x, y), Y_{ij}(y))$, where $Y_{ij}(y)$ is of class C^s and $X_{ij}(x, y)$ is of class $C^{r,s}$ (i.e., X_{ij} admits r (continuous) partial derivatives with respect to x-variables and s partial derivatives with respect to y-variables).

A flow box for W is a pair (U, φ) consisting of an open subset $U \subset W$ and a map $\varphi: U \to \mathbb{R}^k \times \mathbb{R}^l$ such that $\mathcal{A} \cup \{(U, \varphi)\}$ is still an atlas for W.

(Here C^ω is the space of analytic functions.)

Given two foliated manifolds W_1, W_2 of dimension $k + l$, with k-dimensional leaves, and of regularity $C^{r,s}$, a regular map $f \colon W_1 \to W_2$ is a continuous map which is locally, in flow boxes, of the form $f(x, y) = (X(x, y), Y(y))$, where Y is of class C^s and X is of class $C^{r,s}$. A diffeomorphism $\phi \colon W_1 \to W_2$ is a homeomorphism such that ϕ and ϕ^{-1} are both regular maps.

DEFINITION 2.2 (k-solenoid). *Let $0 \leq r \leq s \leq \omega$, and let $k, l \geq 0$ be two integers. A pre-solenoid of dimension k, of class $C^{r,s}$, and transverse dimension l is a pair (S, W), where W is a foliated manifold and $S \subset W$ is a compact subspace which is a collection of leaves.*

Two pre-solenoids (S, W_1) and (S, W_2) are called equivalent *if there are open subsets $U_1 \subset W_1$ and $U_2 \subset W_2$ with $S \subset U_1$ and $S \subset U_2$, and a diffeomorphism $f \colon U_1 \to U_2$ preserving the foliations such that f is the identity on S.*

A k-solenoid of class $C^{r,s}$ and transverse dimension l (or just a k-solenoid, or a solenoid) is an equivalence class of pre-solenoids.

We usually denote a solenoid by S, without making explicit mention of W. We shall say that W defines the solenoid structure of S.

DEFINITION 2.3 (flow box). *Let S be a solenoid. A* flow box *for S is a pair (U, φ) formed by an open subset $U \subset S$ and a homeomorphism*

$$\varphi \colon U \to D^k \times K(U),$$

where D^k is the k-dimensional open ball and $K(U) \subset \mathbb{R}^l$, such that there exists a foliated manifold W defining the solenoid structure of S, $S \subset W$, and a flow box $\hat{\varphi} \colon \hat{U} \to \mathbb{R}^k \times \mathbb{R}^l$ for W, with $U = \hat{U} \cap S$, $\hat{\varphi}(U) = D^k \times K(U) \subset \mathbb{R}^k \times \mathbb{R}^l$ and $\varphi = \hat{\varphi}_{|U}$.

The set $K(U)$ is the transverse space *of the flow box. The dimension l is the* transverse dimension.

As S is locally compact, any point of S is contained in a flow box U whose closure $\overline{U}$ is contained in a bigger flow box. For such flow boxes, $\overline{U} \cong \overline{D}^k \times \overline{K}(U)$, where $\overline{D}^k$ is the closed unit ball, $\overline{K}(U)$ is some compact subspace of $\mathbb{R}^l$, and the flow box $U = D^k \times K(U) \subset \overline{D}^k \times \overline{K}(U)$. All flow boxes that we shall use are of this type, without making further explicit mention.

DEFINITION 2.4 (Leaf). *A* leaf *of a k-solenoid S is a leaf l of any foliated manifold W inducing the solenoid structure of S, such that $l \subset S$. Note that this notion is independent of W.*

DEFINITION 2.5 (Oriented solenoid). *An* oriented solenoid *is a solenoid S such that there is a foliated manifold $W \supset S$ inducing the solenoid structure of S, where W has oriented leaves (in a transversally continuous way).*

A solenoid is minimal if it does not contain a proper sub-solenoid. It is a classical result that there are always minimal solenoids inside any solenoid.

DEFINITION 2.6 (Transversal). *Let S be a k-solenoid. A* local transversal *at a point $p \in S$ is a subset T of S with $p \in T$, such that there is a flow box (U, φ) of S with U a neighborhood of p containing T and such that*

$$\varphi(T) = \{0\} \times K(U).$$

A transversal *T of S is a compact subset of S such that for each $p \in T$, there is an open neighborhood V of p such that $V \cap T$ is a local transversal at p.*

If S is a k-solenoid of class $C^{r,s}$, then any transversal T inherits an l-dimensional C^s-Whitney structure.

DEFINITION 2.7 (Global transversal). *A transversal T of S is a* global transversal *if all leaves intersect T.*

DEFINITION 2.8 (Holonomy). *Given two points p_1 and p_2 in the same leaf, two local transversals T_1 and T_2, at p_1 and p_2, respectively, and a path $\gamma\colon [0,1] \to S$, contained in the leaf with endpoints $\gamma(0) = p_1$ and $\gamma(1) = p_2$, we define a germ of a map (the* holonomy map*) as*

$$h_\gamma\colon (T_1, p_1) \to (T_2, p_2),$$

by lifting γ to nearby leaves.

We denote by $\mathrm{Hol}_S(T_1, T_2)$ the set of germs of holonomy maps from T_1 to T_2. These form the holonomy pseudo-group.

DEFINITION 2.9 (Transverse measure). *Let S be a k-solenoid. A* transverse measure *$\mu = (\mu_T)$ for S associates to any local transversal T a locally finite measure μ_T supported on T, which are invariant by the holonomy pseudogroup. More precisely, if T_1 and T_2 are two transversals and $h\colon V \subset T_1 \to T_2$ is a holonomy map, then*

$$h_*(\mu_{T_1}|_V) = \mu_{T_2}|_{h(V)}.$$

We assume that a transverse measure μ is non-trivial, i.e., for some T, μ_T is non-zero.

We denote by S_μ a k-solenoid S endowed with a transverse measure $\mu = (\mu_T)$. We refer to S_μ as a *measured solenoid*.

We fix now a C^∞ manifold M of dimension n.

DEFINITION 2.10 (Immersion and embedding of solenoids). *Let S be a k-solenoid of class $C^{r,s}$ with $r \geq 1$. An* immersion

$$f\colon S \to M$$

is a regular map (that is, it has an extension $\hat{f}\colon W \to M$ of class $C^{r,s}$, where W is a foliated manifold which defines the solenoid structure of S), such that the differential restricted to the tangent spaces of leaves has rank k at every point of S. We say that $f\colon S \to M$ is an immersed solenoid.

Let $r, s \geq 1$. A transversally immersed solenoid *$f\colon S \to M$ is a regular map $f\colon S \to M$ such that it admits an extension $\hat{f}\colon W \to M$ which is an immersion (of a $(k + l)$-dimensional manifold into an n-dimensional one) of class $C^{r,s}$ such that the images of the leaves intersect transversally in M.*

An embedded solenoid *$f\colon S \to M$ is a transversally immersed solenoid of class $C^{r,s}$, with $r, s \geq 1$, with injective f, that is, the leaves do not intersect or self-intersect.*

Note that under a transversal immersion (resp. an embedding) $f\colon S \to M$, the images of the leaves are immersed (resp. injectively immersed) submanifolds.

Let M be a smooth manifold. We shall denote the space of compactly supported currents of dimension k by

$$\mathcal{C}_k(M).$$

These currents are functionals $T\colon \Omega^k(M) \to \mathbb{R}$. A current $T \in \mathcal{C}_k(M)$ is closed if $T(d\alpha) = 0$ for any $\alpha \in \Omega^{k-1}(M)$. Therefore, by restricting to the closed forms, a closed current T defines a linear map

$$[T]\colon H^k(M, \mathbb{R}) \longrightarrow \mathbb{R}.$$

By duality, T defines a real homology class $[T] \in H_k(M, \mathbb{R})$.

DEFINITION 2.11 (Generalized currents). *Let S be an oriented k-solenoid of class $C^{r,s}$, $r \geq 1$, endowed with a transverse measure $\mu = (\mu_T)$. An immersion*

$$f : S \to M$$

defines a current $(f, S_\mu) \in \mathcal{C}_k(M)$, called a generalized Ruelle-Sullivan current *(or just a* generalized current*), as follows. Let ω be an k-differential form in M. The pullback $f^*\omega$ defines a k-differential form on the leaves of S.*

Let $S = \bigcup_i U_i$ be an open cover of the solenoid. Take a partition of unity $\{\rho_i\}$ subordinate to the covering $\{U_i\}$. We define

$$\langle (f, S_\mu), \omega \rangle = \sum_i \int_{K(U_i)} \left(\int_{L_y} \rho_i f^* \omega \right) d\mu_{K(U_i)}(y),$$

where L_y denotes the horizontal disk of the flow box.

The current (f, S_μ) is closed; hence, it defines a real homology class

$$[f, S_\mu] \in H_k(M, \mathbb{R}),$$

called the Ruelle-Sullivan homology class.

From now on, we shall consider a C^∞ compact and oriented manifold M of dimension n. Let $f : S_\mu \to M$ be an oriented measured k-solenoid immersed in M. We shall denote

$$[f, S_\mu]^* \in H^{n-k}(M, \mathbb{R}),$$

the dual of $[f, S_\mu]$ under the Poincaré duality isomorphism between $H_k(M, \mathbb{R})$ and $H^{n-k}(M, \mathbb{R})$.

3   HOMOTOPY OF SOLENOIDS

Let us see that the Ruelle-Sullivan homology class defined by an immersed oriented measured k-solenoid does not change by perturbations.

DEFINITION 3.1 (Solenoid with boundary). *Let $0 \leq s \leq r \leq \omega$, and let $k, l \geq 0$ be two integers. A* foliated manifold with boundary *(of dimension $k + l$, with k-dimensional leaves, of class $C^{r,s}$) is a smooth manifold W with boundary, of dimension $k + l$, endowed with an atlas $\{(U_i, \varphi_i)\}$ of charts*

$$\varphi_i : U_i \to \varphi_i(U_i) \subset \mathbb{R}^{k+l}_+ = \{(x_1, \ldots, x_k, y_1, \ldots, y_l \, ; \, x_1 \geq 0)\},$$

whose changes of charts are of the form $\varphi_i \circ \varphi_j^{-1}(x, y) = (X_{ij}(x, y), Y_{ij}(y))$, where $Y_{ij}(y)$ is of class C^s and $X_{ij}(x, y)$ is of class $C^{r,s}$.

A pre-solenoid with boundary *is a pair (S, W), where W is a foliated manifold with boundary and $S \subset W$ is a compact subspace which is a collection of leaves.*

Two pre-solenoids with boundary (S, W_1) and (S, W_2) are equivalent *if there are open subsets $U_1 \subset W_1$, $U_2 \subset W_2$ with $S \subset U_1$ and $S \subset U_2$, and a diffeomorphism $f : U_1 \to U_2$ (preserving leaves, of class $C^{r,s}$) which is the identity on S.*

A k-solenoid with boundary *is an equivalence class of pre-solenoids with boundary.*

Note that any manifold with boundary is a solenoid with boundary.

The boundary of a k-solenoid with boundary S is the $(k-1)$-solenoid (without boundary) ∂S defined by the foliated manifold ∂W, where W is a foliated manifold with boundary defining the solenoid structure of S.

A k-solenoid with boundary S has two types of flow boxes. If $p \in S - \partial S$ is an interior point, then there is a flow box (U, φ) with $p \in U$, of the form $\varphi \colon U \to D^k \times K(U)$. If $p \in \partial S$ is a boundary point, then there is a flow box (U, φ) with $p \in U$ such that φ is a homeomorphism

$$\varphi \colon U \to D_+^k \times K(U),$$

where $D_+^k = \{(x_1, \ldots, x_k) \in D_k \,;\, x_1 \geq 0\}$, and $K(U) \subset \mathbb{R}^l$, $\varphi(p) = (0, \ldots, 0, y_0)$ for some $y_0 \in K$. Note that writing

$$U' = \partial S \cap U = \varphi^{-1}(D^{k-1} \times K(U)),$$

where $D^{k-1} = \{(0, x_2, \ldots, x_k) \in D_k\} \subset D_k^+$, we have that $(U', \varphi_{|U'})$ is a flow box for ∂S. Therefore, if T is a transversal for ∂S, then it is also a transversal for S.

For a solenoid with boundary S, there is also a well-defined notion of holonomy pseudo-group. If T is a local transversal for ∂S and $h \colon T \to T$ is a holonomy map for ∂S defined by a path in ∂S, then h lies in the holonomy pseudo-group of S. So

$$\mathrm{Hol}_{\partial S}(T) \subset \mathrm{Hol}_S(T),$$

but they are not equal in general. In particular, if S is connected and minimal with non-empty boundary, then

$$\mathcal{M}_T(S) \subset \mathcal{M}_T(\partial S).$$

That is, if $\mu = (\mu_T)$ is a transverse measure for S, then it yields a transverse measure for ∂S, by considering only those transversals T which are transversals for ∂S. We denote this transverse measure by μ again.

If S comes equipped with an orientation, then ∂S has a natural induced orientation. Note that any leaf $l \subset S$ is a manifold with boundary and each connected component of ∂l is a leaf of ∂S.

THEOREM 3.2 (Stokes theorem). *Let $f \colon S_\mu \to M$ be an oriented $(k+1)$-solenoid with boundary, endowed with a transverse measure, and immersed into a smooth manifold M. Let ω be a k-form on M. Then*

$$\langle [f, S_\mu], d\omega \rangle = \langle [f_{|\partial S}, \partial S_\mu], \omega \rangle.$$

PROOF. Let $\{U_i\}$ be a covering of S by flow boxes, and let $\{\rho_i\}$ be a partition of unity subordinated to it. Adding up the equalities

$$\int_{K(U_i)} \left(\int_{L_y} d\rho_i \wedge f^*\omega \right) d\mu_{K(U_i)}(y) + \int_{K(U_i)} \left(\int_{L_y} \rho_i f^* d\omega \right) d\mu_{K(U_i)}(y)$$

$$= \int_{K(U_i)} \left(\int_{L_y} d(\rho_i f^*\omega) \right) d\mu_{K(U_i)}(y)$$

$$= \int_{K(U_i)} \left(\int_{\partial L_y} \rho_i f^*\omega \right) d\mu_{K(U_i)}(y),$$

for all i, and using that $\sum d\rho_i \equiv 0$, we get

$$\langle [f, S_\mu], d\omega \rangle = \sum_i \int_{K(U_i)} \left(\int_{L_y} \rho_i f^* d\omega \right) d\mu_{K(U_i)}(y)$$

$$= \sum_i \int_{K(U_i)} \left(\int_{\partial L_y} \rho_i f^* \omega \right) d\mu_{K(U_i)}(y) = \langle [f_{|\partial S}, \partial S_\mu], \omega \rangle . \qquad \square$$

Let S be a k-solenoid of class $C^{r,s}$. We endow $S \times I = S \times [0,1]$ with the structure of a natural $(k+1)$-solenoid with boundary of the same class by taking a foliated manifold $W \supset S$ defining the solenoid structure of S, and foliating $W \times I$ with the leaves $l \times I$, $l \subset W$ being a leaf of W. Then $S \times I \subset W \times I$ is a $(k+1)$-solenoid with boundary. The boundary of $S \times I$ is

$$(S \times \{0\}) \sqcup (S \times \{1\}).$$

If S is oriented then $S \times I$ is naturally oriented and its boundary consists of $S \times \{0\} \cong S$ with orientation reversed, and $S \times \{1\} \cong S$ with orientation preserved.

Moreover if T is a transversal for S, then $T' = T \times \{0\}$ is a transversal for $S' = S \times I$. The following is immediate.

LEMMA 3.3. *There is an identification of the holonomies of S and $S \times I$. More precisely, under the identification $T \cong T' = T \times \{0\}$,*

$$\text{Hol}_S(T) = \text{Hol}_{S \times I}(T') .$$

In particular,

$$\mathcal{M}_T(S) = \mathcal{M}_T(S \times I).$$

DEFINITION 3.4 (Equivalence of immersions). *Two solenoid immersions $f_0 \colon S_0 \to M$ and $f_1 \colon S_1 \to M$ of class $C^{r,s}$ in M are* immersed equivalent *if there is a $C^{r,s}$-diffeomorphism $h \colon S_0 \to S_1$ such that*

$$f_0 = f_1 \circ h .$$

Two measured solenoid immersions are immersed equivalent *if h can be chosen to preserve the transverse measures.*

DEFINITION 3.5 (Homotopy of immersions). *Let S be a k-solenoid of class $C^{r,s}$ with $r \geq 1$. A* homotopy *between immersions $f_0 \colon S \to M$ and $f_1 \colon S \to M$ is an immersion of solenoids $f \colon S \times I \to M$ such that $f_0(x) = f(x,0)$ and $f_1(x) = f(x,1)$.*

DEFINITION 3.6 (Cobordism of solenoids). *Let S_0 and S_1 be two $C^{r,s}$-solenoids. A* cobordism of solenoids *is a $(k+1)$-solenoid S with boundary $\partial S = S_0 \sqcup S_1$.*

If S_0 and S_1 are oriented, then an oriented cobordism *is a cobordism S which is an oriented solenoid such that it induces the given orientation on S_1 and the reversed orientation on S_0.*

If S_0 and S_1 have transverse measures μ_0 and μ_1, respectively, then a measured cobordism *is a cobordism S endowed with a transverse measure μ inducing the measures μ_0 and μ_1 on S_0 and S_1, respectively.*

DEFINITION 3.7 (Homology equivalence). *Let $f_0: S \to M$ and $f_1: S \to M$ be two immersed solenoids in M. We say that they are* homology equivalent *if there exists a cobordism of solenoids S between S_0 and S_1 and a solenoid immersion $f: S \to M$ with $f_{|S_0} = f_0$, $f_{|S_1} = f_1$. We call $f: S \to M$ a* homology *between $f_0: S \to M$ and $f_1: S \to M$.*

Let $f_0: S_{0,\mu_0} \to M$ and $f_1: S_{1,\mu_1} \to M$ be two immersed oriented measured solenoids. They are homology equivalent *if there exists an immersed oriented measured solenoid $f: S_\mu \to M$ such that $f: S \to M$ is a homology between $f_0: S \to M$ and $f_1: S \to M$ and S_μ is a measured oriented cobordism from S_0 to S_1.*

Clearly, two homotopic immersions of a solenoid give homology equivalent immersions.

THEOREM 3.8. *Suppose that two oriented measured solenoids $f_0: S_{0,\mu_0} \to M$ and $f_1: S_{1,\mu_1} \to M$ immersed in M are homology equivalent. Then the generalized currents coincide*

$$[f_0, S_{0,\mu_0}] = [f_1, S_{1,\mu_1}] .$$

The same happens if they are immersed equivalent.

PROOF. In the first case, let ω be a closed k-form on M, then Stokes' theorem gives

$$\langle [f_1, S_{1,\mu_1}], \omega \rangle - \langle [f_0, S_{0,\mu_0}], \omega \rangle = \langle [f_{|\partial S}, \partial S_\mu], \omega \rangle = \langle [f, S_\mu], d\omega \rangle = 0 .$$

In the second case, $f_0 = f_1 \circ h$ implies that the actions of the generalized currents over a closed form on M coincide, since the pull-back of the form to the solenoids agree through the diffeomorphism h, and the integrals over the transverse measure gives the same numbers, since the measures correspond by h. □

REMARK 3.9. In both Definitions 3.5 and 3.6, we do not need to require that f be an immersion. Actually, the generalized current $[f, S_\mu]$ makes sense for any measured solenoid S_μ and any regular map $f: S \to M$, of class $C^{r,s}$ with $r \geq 1$. Theorem 3.8 still holds with these extended notions.

4   INTERSECTION THEORY OF SOLENOIDS

Let M be a smooth C^∞ oriented manifold.

DEFINITION 4.1 (Transverse intersection). *Let $f_1: S_1 \to M$, $f_2: S_2 \to M$ be two immersed solenoids in M. We say that they* intersect transversally *if, for every $p_1 \in S_1$, $p_2 \in S_2$ such that $f_1(p_1) = f_2(p_2)$, the images of the leaves through p_1 and p_2 intersect transversally.*

If two immersed solenoids $f_1: S_1 \to M$ and $f_2: S_2 \to M$, of dimensions k_1 and k_2, respectively, intersect transversally, we define the intersection solenoid $f: S \to M$ as follows. The solenoid S is

$$S = \{(p_1, p_2) \in S_1 \times S_2 ; f_1(p_1) = f_2(p_2)\} , \tag{4.1}$$

and the map $f: S \to M$ is given by

$$f(p_1, p_2) = f_1(p_1) = f_2(p_2), \quad (p_1, p_2) \in S . \tag{4.2}$$

We will see that S, the intersection solenoid, is indeed a solenoid. Also the intersection $f: S \to M$ of the two immersed solenoids $f_1: S_1 \to M$, $f_2: S_2 \to M$ is an immersed solenoid. In order to prove this, we consider the intersection of the product solenoid $F = f_1 \times f_2: S_1 \times S_2 \to M \times M$ with the diagonal $\Delta \subset M \times M$. So we have to analyze first the case of the intersection of an immersed solenoid with a submanifold. The notion of transverse intersection given in Definition 4.1 applies to this case (a submanifold is an embedded solenoid).

LEMMA 4.2. *Let $f: S \to M$ be an immersed k-solenoid in M intersecting an embedded closed submanifold $N \subset M$ of codimension q transversally. If $S' = f^{-1}(N) \subset S$ is non-empty, then $f_{|S'}: S' \to M$ is an immersed $(k - q)$-solenoid in N.*

If S and N are oriented, so is S'.

If S has a transverse measure μ, then S' inherits a natural transverse measure, also denoted by μ.

PROOF. First of all, note that S' is a compact and Hausdorff space.

Let W be a foliated manifold defining the solenoid structure of S such that there is a smooth map $\hat{f}: W \to M$ of class $C^{r,s}$, extending $f: S \to M$, which is an immersion on leaves. By definition, for any leaf $l \subset S$, $f(l)$ is transverse to N. Thus, reducing W if necessary, the same transversality property occurs for any leaf of W. The transversality of the leaves implies that the map $\hat{f}: W \to M$ is transverse to the submanifold $N \subset M$, meaning that for any $p \in W$ such that $\hat{f}(p) \in N$,

$$d\hat{f}(p)(T_p W) + T_{\hat{f}(p)} N = T_{\hat{f}(p)} M.$$

This implies that $W' = \hat{f}^{-1}(N)$ is a submanifold of W of codimension q (in particular, $k - q \geq 0$). Moreover, it is foliated by the connected components l' of $l \cap \hat{f}^{-1}(N) = (\hat{f}_{|l})^{-1}(N)$, where l are the leaves of W. By transversality of $\hat{f}$ along the leaves, l' is a $(k - q)$-dimensional submanifold of l. So W' is a foliated manifold with leaves of dimension $k - q$. This gives the required solenoid structure to $S' = S \cap \hat{f}^{-1}(N) = f^{-1}(N)$.

Clearly, $f_{|S'}: S' \to N$ is an immersion (of class $C^{r,s}$) since $\hat{f}_{|W'}: W' \to N$ is a smooth map which is an immersion on leaves.

If S and N are oriented, then each intersection $l' = l \cap \hat{f}^{-1}(N)$ is also oriented (using that M is oriented as well). Therefore, the leaves of S' are oriented, and hence S' is an oriented solenoid.

Let $p \in S'$ and let $U \cong D^k \times K(U)$ be a flow box for S around p. We can take U small enough so that $f(U)$ is contained in a chart of M in which N is defined by equations $x_1 = \ldots = x_q = 0$. By the transversality property, the differentials $dx_1, \ldots, dx_q$ are linearly independent on each leaf $f(D^k \times \{y\})$, $y \in K(U)$. Therefore, $x_1, \ldots, x_q$ can be completed to a set of functions $x_1, \ldots, x_k$ such that $dx_1, \ldots, dx_k$ are a basis of the cotangent space for each leaf (reducing U if necessary). Thus the pull-back of $x = (x_1, \ldots, x_k)$ to U give coordinates functions so that, using the coordinate $y \in K(U)$ for the transverse direction, (x, y) are coordinates for U, and $f^{-1}(N)$ is defined as $x_1 = \ldots = x_q = 0$. This means that

$$S' \cap U \cong \{(0, \ldots, 0, x_{q+1}, \ldots, x_k, y) \in D^k \times K(U)\} \cong D^{k-q} \times K(U).$$

Therefore, any local transversal T for S' is a local transversal for S, and any holonomy map for S' is a holonomy map for S. So

$$\mathrm{Hol}_{S'}(T) \subset \mathrm{Hol}_S(T).$$

Hence, a transverse measure for S gives a transverse measure for S'. □

Now we can address the general case.

PROPOSITION 4.3. *Suppose that* $f_1 \colon S_1 \to M$, $f_2 \colon S_2 \to M$ *are two immersed solenoids in M intersecting transversally, and let S be its intersection solenoid defined in (4.1) and let f be the map (4.2). If $S \neq \emptyset$, then $f \colon S \to M$ is an immersed solenoid of dimension $k = k_1 + k_2 - n$ (in particular, k is a non-negative number).*

If S_1 and S_2 are both oriented, then S is also oriented.

If S_1 and S_2 are endowed with transverse measures μ_1 and μ_2 respectively, then S has an induced measure μ.

PROOF. The product $S_1 \times S_2$ is a $(k_1 + k_2)$-solenoid and

$$F = f_1 \times f_2 \colon S_1 \times S_2 \to M \times M$$

is an immersion. Let $\Delta \subset M \times M$ be the diagonal. There is an identification (as sets)

$$S = (S_1 \times S_2) \cap F^{-1}(\Delta).$$

The condition that $f_1 \colon S_1 \to M$, $f_2 \colon S_2 \to M$ intersect transversally can be translated into that $F \colon S_1 \times S_2 \to M \times M$ and Δ intersect transversally in $M \times M$.

Therefore applying Lemma 4.2, we know $(S, F_{|S})$ is an immersed k-solenoid, where $F_{|S} \colon S \to \Delta$ is defined as $F(x_1, x_2) = f_1(x_1)$. Using the diffeomorphism $M \cong \Delta$, $x \mapsto (x, x)$, $F_{|S}$ corresponds to $f \colon S \to M$. So $f \colon S \to M$ is an immersed k-solenoid.

If S_1 and S_2 are both oriented, then $S_1 \times S_2$ is also oriented. By Lemma 4.2, S inherits an orientation.

If S_1 and S_2 are endowed with transverse measures μ_1 and μ_2, then $S_1 \times S_2$ has a product transverse measure μ. For any local transversals T_1 and T_2 to S_1 and S_2, respectively, $T = T_1 \times T_2$ is a local transversal to $S_1 \times S_2$ (and conversely). We define

$$\mu_T = \mu_{1,T_1} \times \mu_{2,T_2}. \tag{4.3}$$

Now Lemma 4.2 applies to give the transverse measure for S. Note that the local transversals to S are of the form $T_1 \times T_2$, for some local transversals T_1 and T_2 to S_1 and S_2. □

REMARK 4.4. If $k_1 + k_2 = n$ then S is a 0-solenoid. For a 0-solenoid S, an orientation is a continuous assignment $\epsilon \colon S \to \{\pm 1\}$ of sign to each point of S.

Note also that for a 0-solenoid S, $T = S$ is a transversal and a transverse measure is a Borel measure on S.

Let $f_1 \colon S_1 \to M$, $f_2 \colon S_2 \to M$ be two immersed solenoids in M intersecting transversally, with $f \colon S \to M$ its intersection solenoid. Let $p = (p_1, p_2) \in S$.

Then we can choose flow boxes $U_1 = D^{k_1} \times K(U_1)$ for S_1 around p_1 with coordinates $(x_1, \ldots, x_{k_1}, y)$, and $U_2 = D^{k_2} \times K(U_2)$ for S_2 around p_2 with coordinates $(x_1, \ldots, x_{k_2}, z)$, and coordinates for M around $f(p)$, such that

$$f_1(x, y) = (x_1, \ldots, x_{k_1+k_2-n}, x_{k_1+k_2-n+1}, \ldots, x_{k_1}, B_1(x, y), \ldots, B_{n-k_1}(x, y)),$$
$$f_2(x, z) = (x_1, \ldots, x_{k_1+k_2-n}, C_1(x, z), \ldots, C_{n-k_2}(x, z), x_{k_1+k_2-n+1}, \ldots, x_{k_2}).$$

Then S is defined locally as $D^{k_1+k_2-n} \times K(U_1) \times K(U_2)$. The coordinates are given by $(x, y, z) = (x_1, \ldots, x_{k_1+k_2-n}, y, z)$ and

$$f(x_1, \ldots, x_{k_1+k_2-n}, y, z)$$
$$= (x_1, \ldots, x_{k_1+k_2-n}, C_1(x, z), \ldots, C_{n-k_2}(x, z), B_1(x, y), \ldots, B_{n-k_1}(x, y)).$$

THEOREM 4.5. *Let $f \colon S_\mu \to M$ be an oriented measured k-solenoid immersed in M intersecting transversally a closed submanifold $i \colon N \hookrightarrow M$ of codimension q, such that $S' = f^{-1}(N) \subset S$ is non-empty. Consider the oriented measured $(k - q)$-solenoid immersed in N, $f' \colon S' \to M$, where $f' = f_{|S'}$. Then, under the restriction map*

$$i^* \colon H^{n-k}(M) \to H^{(n-q)-(k-q)}(N), \tag{4.4}$$

the dual of the Ruelle-Sullivan homology class $[f, S_\mu]^*$ maps to $[f', S'_\mu]^*$.

PROOF. Let $U \subset M$ be a tubular neighborhood of N with projection $\pi \colon U \to N$. Note that U is diffeomorphic to the unit disc bundle of the normal bundle of N in M. Let τ be a Thom form for $N \subset M$, that is a closed form $\tau \in \Omega^q(M)$ supported in U, whose integral in any normal space $\pi^{-1}(n)$, $n \in N$, is one. The dual of the map (4.4) under Poincaré duality is the map

$$H^{k-q}(N) \to H^k(M),$$

which sends $[\beta] \in H^{k-q}(N)$ to $[\tilde{\beta}]$, where $\tilde{\beta} = \pi^*\beta \wedge \tau$ (this form is extended from U to the whole of M by zero). So we need only to see that

$$\langle [f, S_\mu], \tilde{\beta} \rangle = \langle [f', S'_\mu], \beta \rangle.$$

Take a covering of S by flow boxes $U_i \cong D^k \times K(U_i) \cong D^q \times D^{k-q} \times K(U_i)$ so that $U'_i = U_i \cap S'$ is given by $x_1 = \ldots = x_q = 0$. Making the tubular neighborhood $U \supset N$ smaller if necessary, we can arrange that $f^{-1}(U) \cap U_i$ is contained in $D_r^q \times D^{k-q} \times K(U_i)$, for some $r < 1$. Now construct a map $\tilde{\pi} \colon f^{-1}(U) \to f^{-1}(N)$ which consists of projecting in the normal directions along the leaves. Then $f \circ \tilde{\pi}$ and $\pi \circ f$ are homotopic.

Let S'_i be a measurable partition of S' with $S'_i \subset U'_i$. We may assume that $S_i = \tilde{\pi}^{-1}(S'_i)$ is contained in U_i. The sets S_i form a measurable partition containing $f^{-1}(U)$, the support of $f^*\tilde{\beta} = f^*(\pi^*\beta \wedge \tau)$. Then

$$\langle [f, S_\mu], \tilde{\beta} \rangle = \sum_i \int_{K(U_i)} \left(\int_{S_i \cap (D^k \times \{y\})} f^*(\pi^*\beta \wedge \tau) \right) d\mu_{K(U_i)}(y)$$

$$= \sum_i \int_{K(U_i)} \left(\int_{S_i \cap (D^k \times \{y\})} \tilde{\pi}^* f^*\beta \wedge f^*\tau \right) d\mu_{K(U_i)}(y)$$

$$= \sum_i \int_{K(U_i)} \left(\int_{S_i' \cap (D^{k-q} \times \{y\})} f^*\beta \right) \left(\int_{f(D^q)} \tau \right) d\mu_{K(U_i)}(y)$$

$$= \sum_i \int_{K(U_i')} \left(\int_{S_i' \cap (D^{k-q} \times \{y\})} f^*\beta \right) d\mu_{K(U_i')}(y) = \langle [f', S_\mu'], \beta \rangle . \qquad \square$$

THEOREM 4.6. *Suppose that* $f_1 \colon S_{1,\mu_1} \to M$ *and* $f_2 \colon S_{2,\mu_2} \to M$ *are two oriented measured immersed solenoids in* M *intersecting transversally, and let* $f \colon S_\mu \to M$ *be the intersection solenoid. Then the duals of the Ruelle-Sullivan homology classes satisfy*

$$[f, S_\mu]^* = [f_1, S_{1,\mu_1}]^* \cup [f_2, S_{2,\mu_2}]^* .$$

PROOF. Note that $[f_1, S_{1,\mu_1}]^* \in H^{n-k_1}(M)$ and $[f_2, S_{2,\mu_2}]^* \in H^{n-k_2}(M)$, so $[f, S]^*$ and $[f_1, S_{1,\mu_1}]^* \cup [f_2, S_{2,\mu_2}]^*$ both live in

$$H^{n-k_1+n-k_2}(M) = H^{n-k}(M) .$$

Consider the immersed solenoid $(F, S_1 \times S_2)$, where $S_1 \times S_2$ has the transverse measure μ given by (4.3), and $F = f_1 \times f_2 \colon S_1 \times S_2 \to M \times M$. Let us see that the following equality, involving the respective generalized currents,

$$[F, (S_1 \times S_2)_\mu] = [f_1, S_{1,\mu_1}] \otimes [f_2, S_{2,\mu_2}] \in H_{k_1+k_2}(M \times M)$$

holds. We prove this by applying both sides to $(k_1 + k_2)$-cohomology classes in $M \times M$. Using the Künneth decomposition, it is enough to evaluate on a form $\beta = p_1^*\beta_1 \wedge p_2^*\beta_2$, where $\beta_1, \beta_2 \in H^*(M)$ are closed forms and $p_1, p_2 \colon M \times M \to M$ are the two projections. Let $\{U_i\}, \{V_j\}$ be open covers of S_1 and S_2, respectively, by flow boxes, and let $\{\rho_{1,i}\}, \{\rho_{2,j}\}$ be partitions of unity subordinated to such covers. Then

$$\langle [F, (S_1 \times S_2)_\mu], \beta \rangle$$

$$= \sum_{i,j} \int_{K(U_i) \times K(V_j)} \left(\int_{L_y \times L_z} (p_1^*\rho_{1,i})(p_2^*\rho_{2,j}) F^*(p_1^*\beta_1 \wedge p_2^*\beta_2) \right) d\mu_{K(U_i) \times K(V_j)}(y, z)$$

$$= \sum_{i,j} \int_{K(U_i) \times K(V_j)} \left(\int_{L_y \times L_z} p_1^*(\rho_{1,i} f_1^*\beta_1) \wedge p_2^*(\rho_{2,j} f_2^*\beta_2) \right) d\mu_{1,K(U_i)}(y) \, d\mu_{2,K(V_j)}(z)$$

$$= \left(\sum_i \int_{K(U_i)} \left(\int_{L_y} \rho_{1,i} f_1^*\beta_1 \right) d\mu_{1,K(U_i)}(y) \right) \left(\sum_j \int_{K(V_j)} \left(\int_{L_z} \rho_{1,j} f_2^*\beta_2 \right) d\mu_{2,K(V_j)}(y) \right)$$

$$= \langle [f_1, S_{1,\mu_1}], \beta_1 \rangle \langle [f_2, S_{2,\mu_2}], \beta_2 \rangle ,$$

as required.

Now we are ready to prove the statement of the theorem. Let $\varphi\colon M \to \Delta$ be the natural diffeomorphism of M with the diagonal $\Delta \subset M \times M$, and let $i\colon \Delta \hookrightarrow M \times M$ be the inclusion. Then, using Theorem 4.6,

$$
\begin{aligned}
[f, S_\mu]^* &= [\varphi \circ f, S_\mu]^* \\
&= i^*([F, (S_1 \times S_2)_\mu]^*) \\
&= i^*([f_1, S_{1,\mu_1}]^* \otimes [f_2, S_{2,\mu_2}]^*) \\
&= [f_1, S_{1,\mu_1}]^* \cup [f_2, S_{2,\mu_2}]^* . \qquad \square
\end{aligned}
$$

Let us look more closely at the case where $k_1 + k_2 = n$. Here, we assume that $f_1\colon S_{1,\mu_1} \to M$ and $f_2\colon S_{2,\mu_2} \to M$ are two oriented immersed measured solenoids of dimensions k_1 and k_2, respectively, which intersect transversally. Let $f\colon S_\mu \to M$ be the intersection 0-solenoid of $f_1\colon S_{1,\mu_1} \to M$ and $f_2\colon S_{2,\mu_2} \to M$.

DEFINITION 4.7 (Intersection index). *At each point $x = (x_1, x_2) \in S$, the intersection index $\epsilon(x_1, x_2) \in \{\pm 1\}$ is the sign of the intersection of the leaf of S_1 through x_1 with the leaf of S_2 through x_2. The continuous function $\epsilon\colon S \to \{\pm 1\}$ gives the orientation of S.*

Recall that the 0-solenoid $f\colon S_\mu \to M$ comes equipped with a natural measure μ (for a 0-solenoid the notions of measure and transverse measure coincide). If $x = (x_1, x_2) \in S$, then locally around x, S is homeomorphic to $T = T_1 \times T_2$, where T_1 and T_2 are small local transversals of S_1 and S_2 at x_1 and x_2, respectively. The measure μ_T is the product measure $\mu_{1,T_1} \times \mu_{2,T_2}$.

DEFINITION 4.8 (Intersection measure). *The is the transverse measure μ of the intersection solenoid $f\colon S_\mu \to M$, induced by those of $f_1\colon S_{1,\mu_1} \to M$ and $f_2\colon S_{2,\mu_2} \to M$.*

DEFINITION 4.9 (Intersection pairing). *We define the intersection pairing as the real number*

$$
(f_1, S_{1,\mu_1}) \cdot (f_2, S_{2,\mu_2}) = \int_S \epsilon \, d\mu .
$$

THEOREM 4.10. *If $f_1\colon S_{1,\mu_1} \to M$ and $f_2\colon S_{2,\mu_2} \to M$ are two oriented immersed measured solenoids of dimensions k_1 and k_2, respectively, which intersect transversally such that $k_1 + k_2 = n$. Then*

$$
(f_1, S_{1,\mu_1}) \cdot (f_2, S_{2,\mu_2}) = [f_1, S_{1,\mu_1}]^* \cdot [f_2, S_{2,\mu_2}]^* .
$$

PROOF. By Theorem 4.6,

$$
[f_1, S_{1,\mu_1}]^* \cup [f_2, S_{2,\mu_2}]^* = [f, S_\mu]^* \in H_c^n(M, \mathbb{R}) .
$$

The intersection product $[f_1, S_{1,\mu_1}]^* \cdot [f_2, S_{2,\mu_2}]^*$ is obtained by evaluating this cup product on the element $1 \in H^0(M, \mathbb{R})$, i.e.,

$$
[f_1, S_{1,\mu_1}]^* \cdot [f_2, S_{2,\mu_2}]^* = \langle [f, S_\mu], 1 \rangle = \int_S f^*(1) d\mu(x) = \int_S \epsilon \, d\mu ,
$$

since the pull-back of a function gets multiplied by the orientation of S, which is the function ϵ. $\qquad \square$

When the solenoids are uniquely ergodic, we can sometimes recover this intersection index by a natural limiting procedure. Recall that we say that a Riemannian solenoid S is of *controlled growth* (see definition 3.3 in [4]) if there is a leaf $l \subset S$ and a point $p \in l$ such that the Riemannian balls $l_n \subset l$, of some radius $R_n \to \infty$, satisfy the property that for each flow box U in a finite covering of S, the number of incomplete horizontal discs in $U \cap l_n$ is negligible with respect to the number of complete horizontal discs in $U \cap l_n$. Then, if μ_n is the normalized measure corresponding to l_n, the limit $\mu = \lim_{n \to \infty} \mu_n$ is the unique Schwartzman measure (corollary 3.7 in [4]).

THEOREM 4.11. *Let $f_1 \colon S_{1,\mu_1} \to M$ and $f_2 \colon S_{2,\mu_2} \to M$ be two immersed, oriented, uniquely ergodic solenoids with controlled growth transversally intersecting. Let $l_1 \subset S_1$ and $l_2 \subset S_2$ be two arbitrary leaves. Choose two base points $x_1 \in l_1$ and $x_2 \in l_2$, and fix Riemannian exhaustions $(l_{1,n})$ and $(l_{2,n})$. Define*

$$(f_1, l_{1,n}) \cdot (f_2, l_{2,n}) = \frac{1}{M_n} \sum_{\substack{p=(p_1,p_2)\in l_{1,n} \times l_{2,n} \\ f_1(p_1)=f_2(p_2)}} \epsilon(p), \quad \text{where } M_n = \mathrm{Vol}_{k_1}(l_{1,n}) \cdot \mathrm{Vol}_{k_2}(l_{2,n}).$$

Then

$$\lim_{n \to +\infty} (f_1, l_{1,n}) \cdot (f_2, l_{2,n}) = (f_1, S_{1\mu_1}) \cdot (f_2, S_{2,\mu_2}).$$

In particular, the limit exists and is independent of the choices of l_1, l_2, x_1, x_2 and the radius of the Riemannian exhaustions.

PROOF. The key observation is that because of the unique ergodicity, the atomic transverse measures associated to the normalized k-volume of the Riemannian exhaustions (name them $\mu_{1,n}$ and $\mu_{2,n}$) are converging to μ_1 and μ_2, respectively. In particular, in each local flow box we have

$$\mu_{1,n} \times \mu_{2,n} \to \mu_1 \times \mu_2 = \mu.$$

Therefore, the average defining $(f_1, l_{1,n}) \cdot (f_2, l_{2,n})$ converges to the integral defining $(f_1, S_{1\mu_1}) \cdot (f_2, S_{2,\mu_2})$ since ϵ is a continuous and integrable function (indeed bounded by 1). $\qquad\square$

REMARK 4.12. The previous theorem and proof work in the same form for ergodic solenoids, provided that we know that the Schwartzman limit measure for almost all leaves is the given ergodic measure. This is simple to prove for ergodic solenoids with trapping regions mapping to a contractible ball in M (compare Theorem 7.12 in [4]).

We end this section with a perturbation result. We want to prove that we can achieve transversality for a large class of solenoids by a suitable homotopy. The solenoids that we have in mind are those whose transversal is a Cantor set.

THEOREM 4.13. *Let $f \colon S \to M$ be a solenoid whose transversals are Cantor sets, and let $N \subset M$ be a smooth closed submanifold. Then we can homotope $f \colon S \to M$ so that N and S intersect transversely.*

PROOF. Let $f: S \to M$ be an immersion of the solenoid into a manifold M. Recall that this means that the differential of f along leaves is injective.

Let p be a point in the solenoid. We want to perturb f in a neighborhood of p. Consider a flow box $U \subset S$ of the form $D_{1+r}^k \times T$, where T is a Cantor set, and D_{1+r}^k is a k-disc of radius $1 + r$, for some small real number $r > 0$. Consider also a coordinate chart $(V(x_1, \ldots, x_n))$ for M so that N is given by $x_1 = \ldots = x_q = 0$, and $p = (0, y_0)$. Define the composition

$$f: D_{1+r}^k \times T \to V \to \mathbb{R}^p .$$

The transversality of the leaf $L_y = D_1^k \times \{y\}$ to N is equivalent to the transversality of the map $f_y := f(\,\cdot\,, y)$ to zero.

For any $\epsilon > 0$ small enough, there is a vector $v \in \mathbb{R}^p$ so that f_{y_0} is transverse to $-v$. Therefore, $f_{y_0} + v$ is transverse to zero (that is, if $f_{y_0}(x) + v = 0$, then $df_{y_0}(x)$ is surjective). Moreover, there is an open neighborhood of y_0, $T_0 \subset T$, where this transversality still holds.

Now take a bump function $\rho(x)$ which is one on D_1^k and is zero near the boundary of D_{1+r}^k. The map

$$\hat{f}(x, y) = \begin{cases} f(x, y) + \rho(x)v, & x \in D_{1+r}^k, y \in T_0 \\ f(x, y), & \text{otherwise} \end{cases}$$

is smooth (here is where we use that T is a Cantor set) and transverse to N along $D_1^k \times T_0$.

Repeating this process, we can find a finite cover $T = \sqcup T_j$, and define the perturbations independently on $D_{1+r}^k \times T_j$.

Finally, we have managed to achieve transversality on $V = D_1^k \times T$, perturbing on $D_{1+r}^k \times T$. What we do now is to use a finite cover of S with subsets as V and perturb successively. At each step we take a perturbation of norm small enough so that this does not destroy the perturbation over the set where it was previously achieved. $\qquad \square$

REMARK 4.14.   We can construct an example where it is not possible to perturb two solenoids (at least in a differentiable way) with transverse Cantor sets so that they intersect transversally.

Consider $M = \mathbb{R}^2$, and let $K_1, K_2 \subset [0, 1]$ be two Cantor sets. Let S_1 be given by the leaves (x, y) with $x \in \mathbb{R}$, $y \in K_1$. Let S_2 be given by the leaves $(x, x^2 + z)$ with $x \in \mathbb{R}$, $z \in K_2$. These two solenoids intersect non-transversally at the points determined by $x = 0$, $y = z \in K_1 \cap K_2$.

Suppose that we have small perturbations S_1', S_2' of S_1, S_2, respectively. Then S_1' is defined by leaves of the form $(x, y + f_1(x, y))$, $x \in \mathbb{R}$, $y \in K_1$, and S_2' by leaves of the form $(x, x^2 + z + f_2(x, z))$, $x \in \mathbb{R}$, $z \in K_2$, where f_j is a smooth function on $\mathbb{R} \times K_j$, having small norm, $j = 1, 2$ (recall that a smooth function on $\mathbb{R} \times K_j$ extends as a smooth function on some neighborhood of it). Composing with a suitable diffeomorphism of $\mathbb{R}^2$, we can suppose that $f_1 = 0$. So we are looking for non-transverse intersections of $S_1' = \{(x, y) \mid x \in \mathbb{R}, y \in K_1\}$ and

$S'_2 = \{(x, x^2 + z + g(x,z)) \mid x \in \mathbb{R}, z \in K_2\}$, g some small smooth function on $\mathbb{R} \times K_2$. These are obtained by solving

$$2x + g_x(x,z) = 0, \qquad y = x^2 + z + g(x,z).$$

The equation $2x + g_x(x,z) = 0$ can be solved as $x = \phi(z)$, for some (small) smooth function ϕ. Write $r(z) = \phi(z)^2 + z + g(\phi(z), z)$, which is a smooth function on K_2 close to $r_0(z) = z$. This defines a smooth isotopy of K_2. Let $K'_2 = r(K_2)$. The points of non-transverse intersections of S'_1 and S'_2 are given by solving $y = r(z)$, so they correspond to the points in $K_1 \cap K'_2$.

To guarantee that $K_1 \cap K'_2 \neq \emptyset$, just choose K_1, K_2 to be two Cantor sets with positive Lebesgue measure such that $\mu(K_1) + \mu(K_2) > 1$. As K'_2 is a small smooth perturbation of K_2, the measure of K'_2 is close to that of K_2. So K_1 and K'_2 must intersect.

5   ALMOST EVERYWHERE TRANSVERSALITY

The intersection theory developed in Section 4 is not fully satisfactory since there are examples of solenoids (e.g., foliations) which do not intersect transversally and cannot be perturbed to do so. Even in the case of solenoids whose transverse structure is Cantor, sometimes it is not possible to smoothly perturb the solenoid to make them intersect transversally, as Remark 4.14 shows. Another example is given by the persistence of homoclinic tangencies for stable and unstable foliations shows (see [7]).

However, a weaker notion is enough to develop intersection theory for solenoids. Indeed, the intersection pairing can also be defined for oriented, measured solenoids $f_1 \colon S_{1,\mu_1} \to M$ and $f_2 \colon S_{2,\mu_2} \to M$, immersed in an oriented n-manifold M, with $k_1 + k_2 = n$, $k_1 = \dim S_1$, $k_2 = \dim S_2$, which intersect transversally almost everywhere in the following sense.

DEFINITION 5.1 (Almost everywhere transversality). *Suppose* $f_1 \colon S_{1,\mu_1} \to M$ *and* $f_2 \colon S_{2,\mu_2} \to M$ *are two measured immersed oriented solenoids. Then they intersect almost everywhere transversally if the set*

$$F = \Big\{(p_1, p_2) \in S_1 \times S_2 ;$$
$$f_1(p_1) = f_2(p_2), df_1(p_1)(T_{p_1}S_1) + df_2(p_2)(T_{p_2}S_2) \neq T_{f_1(p_1)}M \Big\} \subset S_1 \times S_2$$

of non-transverse intersection points satisfies the following.

(1) *Every point* $p \in F$ *is an isolated point of*

$$S = \{(p_1, p_2) \in S_1 \times S_2 ; f_1(p_1) = f_2(p_2)\} \subset S_1 \times S_2$$

in the leaf of $S_1 \times S_2$ *through* p.

(2) *F is null-transverse in* $S_1 \times S_2$ *(with the natural product transverse measure μ), i.e., if the set of leaves of* $S_1 \times S_2$ *intersecting F has zero μ-measure.*

It is useful to translate the meaning of almost everywhere transversality to $S_1 \times S_2$.

DEFINITION 5.2 (Almost everywhere transversality). *Let $f: S_\mu \to M$ be a measured immersed oriented k-solenoid and $N \subset M$ a closed submanifold of codimension k. They intersect almost everywhere transversally if the set*

$$F = \left\{ p \in S ;\, f(p) \in N,\, df(p)(T_pS) + T_{f(p)}N \neq T_{f(p)}M \right\} \subset S$$

of non-transverse intersection points satisfies the following.

(1) *Every point $p \in F$ is isolated as a point of S' in the leaf of S through p.*

(2) *$F \subset S_\mu$ is null-transverse, i.e., for any flow box $U = D^k \times K(U)$, the projection by $\pi: U = D^k \times K(U) \to K(U)$ of the intersection $F \cap U$, that is, $\pi(F \cap U) \subset K(U)$, is of zero $\mu_{K(U)}$-measure in $K(U)$.*

Note that a set $F \subset S_\mu$ in a measured solenoid is null-transverse if for any local transversal T, the set of leaves passing through F intersects T in a set of zero μ_T-measure.

Every point of $S' - F$ is automatically isolated as a point of S' in the leaf of S through it. Therefore, condition (1) is equivalent to saying that every point of S' is isolated in the corresponding leaf.

Then we have the following straightforward lemma.

LEMMA 5.3. *The solenoids $f_1: S_{1,\mu_1} \to M$ and $f_2: S_{2,\mu_2} \to M$ are almost everywhere transversal if and only if $f_1 \times f_2: (S_1 \times S_2)_\mu \to M \times M$ and the diagonal $\Delta \subset M \times M$ intersect almost everywhere transversally.*

Let $f: S_\mu \to M$ be an immersed solenoid which is almost everywhere transversal to a closed submanifold $N \subset M$. Write $S' = f^{-1}(N)$ and let $F \subset S'$ be the subset of non-transverse points. Note that $S'_{reg} = S' - F$ is open in S' and F is closed. Moreover, S'_{reg} consists of the transverse intersections, so the intersection index $\epsilon: S'_{reg} \to \{\pm 1\}$ is well defined and continuous. We define the *intersection number* as

$$\int_{S'-F} \epsilon(x)d\mu(x) .$$

THEOREM 5.4. *Suppose that an immersed measured oriented k-solenoid $f: S_\mu \to M$ and a submanifold $N \subset M$ of codimension k intersect almost everywhere transversally. Then*

$$[f, S_\mu]^* \cdot [N] = \int_{S'-F} \epsilon \, d\mu .$$

PROOF. Fix an accessory Riemannian metric on M. By pull-back, this gives a metric on S.

Let $p \in F$. Then, by assumption, there is some $\eta > 0$ such that

$$B_{\frac{3}{2}\eta}(p) \cap F = B_{\frac{3}{2}\eta}(p) \cap S' = \{p\},$$

where $B_r(p)$ is the Riemannian ball in the leaf centered at p and of radius $r > 0$. It is easy to construct a flow box $U = D^k \times K(U)$ with $p = (0, y_0) \in U$ so that

(i) $D^k \times \{y_0\} = B_{\frac{3}{2}\eta}(p)$,

(ii) $D^k_{3/4} \times \{y_0\} = B_\eta(p)$, ($D^k_r$ denotes the open disc of radius $r > 0$),

(iii) the open annulus $A = (D^k - \bar{D}^k_{1/2})$ satisfies that $(A \times K(U)) \cap S' = \emptyset$,

(iv) the intersection number $[f(D^k_{3/4} \times \{y\})] \cdot [N]$ is constant for $y \in K(U)$.

To achieve this, take $K(U)$ small enough. Note that the intersection number in (iv) is well defined since $f(\partial(D^k_{3/4}) \times \{y\})$ does not touch N; and it is locally constant by continuity. We fix a finite covering $\{U_i\}$ of F with such flow boxes.

Let $\pi_i \colon U_i = D^k \times K(U_i) \to K(U_i)$ be the projection onto the second factor. By hypothesis, $\pi_i(F \cap U_i)$ is of zero measure. We may take a nested sequence $(V_{i,n})$ of open neighborhoods of $\pi_i(F \cap U_i)$ in $K(U_i)$ such that $\bigcap_{n \geq 1} V_{i,n} = \pi_i(F \cap U_i)$. Let

$$U_{i,n} = D^k_{3/4} \times V_{i,n}$$

and

$$U_n = \bigcup_i U_{i,n}.$$

Then (U_n) is a nested sequence of open neighborhoods of F in S. It may happen that $\bigcap_{n \geq 1} U_n$ contains points of $S' - F$, but this is a set of μ-measure zero. So

$$\int_{S'-U_n} \epsilon \, d\mu \longrightarrow \int_{S'-\bigcap_{n\geq1} U_n} \epsilon \, d\mu = \int_{S'-F} \epsilon \, d\mu.$$

As $S' - U_n$ is compact, the angle of intersection in $S' - U_n$ between $f(S)$ and N is bounded below, so there is a small $\rho > 0$ (depending on n) such that if U_ρ is the ρ-tubular neighborhood of N in M, then for each intersection point $x \in S' - U_n$, there is a (topological) disc D_x contained in a local leaf through x, which is exactly the path component of $f^{-1}(U_\rho)$ through x. Making ρ smaller we can assume that D_x is as small as we want. Note that (iii) guarantees that D_x does not touch $D^k_{3/4} \times V_{i,n} = U_{i,n}$ for any i. So, $D_x \subset S - U_n$.

Let τ_ρ be a Thom form for $N \subset M$, that is, a closed k-form supported in U_ρ, whose integral in the normal space to N is one. Then $\int_{D_x} \tau_\rho = 1$ for any $x \in S' - U_n$. So

$$\int_{S'-U_n} \epsilon \, d\mu = \int_{S-U_n} f^* \tau_\rho.$$

On the other hand,

$$\int_{U_{i,n}} f^* \tau_\rho = \int_{V_{i,n}} \left(\int_{D^k_{3/4} \times \{y\}} f^* \tau_\rho \right) d\mu_{K(U_i)} \leq C \, \mu_{K(U_i)}(V_{i,n}) \to 0,$$

where C is a bound for all the intersection numbers in (iv) for all U_i simultaneously. Then

$$\int_{U_n} f^* \tau_\rho \to 0,$$

when $n \to \infty$.

Putting everything together,

$$[f, S_\mu]^* \cdot [N] = \langle [f, S_\mu], [\tau_\rho] \rangle = \int_{S_\mu} f^* \tau_\rho$$

$$= \int_{S - U_n} f^* \tau_\rho + \int_{U_n} f^* \tau_\rho$$

$$= \int_{S' - U_n} \epsilon \, d\mu + \int_{U_n} f^* \tau_\rho \to \int_{S' - F} \epsilon \, d\mu. \qquad \square$$

REMARK 5.5. It is not true that, without further restrictions, the measure μ is finite on $S'_{reg} = S' - F$. For instance, it may happen that around a point $p \in F$, there are leaves of S (leaves which do not go through F) with arbitrary large number of positive and negative intersections with N (near p). Obviously, the difference between positive and negative intersections is bounded. Therefore, there is no current associated to (S'_{reg}, μ).

Consider now two immersed measured oriented solenoids $f_1 \colon S_{1,\mu_1} \to M$, $f_2 \colon S_{2,\mu_2} \to M$ intersecting almost everywhere transversally. Let $F \subset S_1 \times S_2$ be the subspace of non-transversal intersection points, which has null-transversal measure in $S_1 \times S_2$. Set $S = (S_1 \times S_2) \cap \tilde{f}^{-1}(\Delta)$, where $\tilde{f} = f_1 \times f_2$. Then there is an intersection index $\epsilon(x)$ for each $x \in S - F$ and an intersection measure μ on $S - F$. We define the *intersection product* as

$$\int_{S - F} \epsilon \, d\mu.$$

Then Theorem 5.4 implies the following theorem.

THEOREM 5.6. *In the situation just described, we have that*

$$[f_1, S_{1,\mu_1}]^* \cdot [f_2, S_{2,\mu_2}]^* = \int_{S - F} \epsilon(x) \, d\mu(x).$$

6   INTERSECTION OF ANALYTIC SOLENOIDS

It is now our intention to translate the theory of solenoids intersecting almost-everywhere transversally to the case where the dimensions are not complementary, that is, when $k_1 + k_2 > n$.

DEFINITION 6.1 (Almost everywhere transversality). *Suppose $f_1 \colon S_{1,\mu_1} \to M$ and $f_2 \colon S_{2,\mu_2} \to M$ are two measured immersed oriented solenoids in an oriented n-manifold M, with $k_1 + k_2 \geq n$, $k_1 = \dim S_1$, $k_2 = \dim S_2$. They intersect almost everywhere transversally if the set*

$$F = \Big\{ (p_1, p_2) \in S_1 \times S_2;$$

$$f_1(p_1) = f_2(p_2), \, df_1(p_1)(T_{p_1} S_1) + df_2(p_2)(T_{p_2} S_2) \neq T_{f_1(p_1)} M \Big\} \subset S_1 \times S_2$$

of non-transversal intersection points satisfies

(1) *every point $p \in F$ is an isolated point of F in the leaf of $S_1 \times S_2$ through p;*

(2) *the set*
$$S = \{(p_1, p_2) \in S_1 \times S_2 \,;\, f_1(p_1) = f_2(p_2)\} \subset S_1 \times S_2$$
is $C^{1,0}$-conjugate, locally near any $p \in F$, to a leafwise (real) analytic set (i.e., it is of class $C^{\omega,0}$);

(3) *F is null-transverse in $S_1 \times S_2$ (with the natural product transverse measure μ), i.e., if the set of leaves of $S_1 \times S_2$ intersecting F has zero μ-measure.*

Then it is useful to translate the meaning of almost everywhere transversality to $S_1 \times S_2$.

DEFINITION 6.2 (Almost everywhere transversality). *Let $f \colon S_\mu \to M$ be a measured immersed oriented k-solenoid and $N \subset M$ be a closed submanifold of codimension q. They intersect almost everywhere transversally if the set*
$$F = \{p \in S \,;\, f(p) \in N,\, df(p)(T_p S) + T_{f(p)} N \neq T_{f(p)} M\} \subset S$$
of non-transversal intersection points satisfies the following:

(1) *every point $p \in F$ is isolated as a point of F in the leaf of S through p;*

(2) *the set*
$$S' = \{p \in S \,;\, f(p) \in N\} \subset S$$
is $C^{1,0}$-conjugate, locally near any $p \in F$, to a leafwise (real) analytic set (i.e., it is of class $C^{\omega,0}$);

(3) *$F \subset S_\mu$ is null-transverse, i.e., for any flow box $U = D^k \times K(U)$, the projection by $\pi \colon U = D^k \times K(U) \to K(U)$ of the intersection $F \cap U$, that is, $\pi(F \cap U) \subset K(U)$ is of zero $\mu_{K(U)}$-measure in $K(U)$.*

We have the following straightforward lemma.

LEMMA 6.3. *The solenoids $f_1 \colon S_{1,\mu_1} \to M$ and $f_2 \colon S_{2,\mu_2} \to M$ are almost everywhere transverse if and only if $\tilde{f} = f_1 \times f_2 \colon (S_1 \times S_2)_\mu \to M \times M$ and the diagonal $\Delta \subset M \times M$ intersect almost everywhere transversally.*

Let $f \colon S_\mu \to M$ be an immersed oriented k-solenoid intersecting transversally almost everywhere a closed oriented submanifold $N \subset M$ of codimension q. Write $S' = f^{-1}(N)$ and let $F \subset S'$ be the subset of non-transverse points. Note that $S'_{reg} = S' - F$ is open in S' and F is closed. Let $p \in S'_{reg} = S' - F$. Then the transversal intersection property implies that there exists a flow box $U = D^k \times K(U)$ for S around p, with coordinates $(x_1, \ldots, x_k, y_1, \ldots, y_l)$ such that N is defined by $x_1 = \ldots = x_q = 0$. Thus S'_{reg} locally has the structure of $(k-q)$-dimensional oriented solenoid, with a transverse measure induced by μ (which we also denote by μ) and invariant by holonomy. Note that S'_{reg} is not a solenoid because it is not compact.

THEOREM 6.4. *Suppose that a k-solenoid $f \colon S_\mu \to M$ and a submanifold $N \subset M$ of codimension q intersect almost everywhere transversally. Let $\omega \in \Omega^{k-q}(M)$ be a closed form. Then*
$$\langle [f, S_\mu]^* \cup [\omega], [N] \rangle = \int_{S'-F} \omega .$$

PROOF. Let us take a Thom form τ_ρ for N, and consider the current

$$(f, S_\mu) \wedge \tau_\rho \tag{6.1}$$

defined as the wedge of the generalized current with the smooth form τ_ρ. Let us see that there is a limit for (6.1) when $\rho \to 0$.

Let us define the current of integration (S_{reg}). This is obviously well defined off F. Now suppose that ω is a $(k-q)$-form supported in a small ball around a point $p \in F$. Let $U = D^k \times K(U)$ be a flow box around p, where ω is defined. Then, after taking a $C^{1,0}$-diffeomorphism, we can suppose that S' is defined as $f(x, y) = 0$, for some $f \colon D^k \times K(U) \to \mathbb{R}^q$ of class $C^{\omega,0}$. Then $S_y = S'_{reg} \cap L_y = f_y^{-1}(0)$ is an analytic subset. Hence the integral of ω on S_y is bounded:

$$\int_{S_y} \omega \leq C \|\omega\|,$$

where C is a constant that we can suppose valid for all $y \in K(U)$ by continuity.

Moreover, making the radius of the ball smaller, we have that $C \to 0$. Hence,

$$\langle (S_{reg}), \omega \rangle := \int_{K(U)} \left(\int_{S_y} \omega \right) d\mu_{K(U)}(y)$$

is well defined.

Now, to see that

$$(f, S_\mu) \wedge \tau_\rho \to (S_{reg}),$$

we apply both sides to a $(k-q)$-form ω. For ω supported in a flow box off F, we have that

$$\langle (f, S_\mu) \wedge \tau_\rho, \omega \rangle = \int_{K(U)} \int_{L_y} \tau_\rho \wedge \omega \, d\mu_{K(U)}(y) \to \int_{K(U)} \int_{S_y} \omega \, d\mu_{K(U)}(y) = \langle (S_{reg}), \omega \rangle$$

Now let ω be supported in an ϵ-ball around $p \in F$. Then $\langle (S_{reg}), \omega \rangle$ is as small as we want, and

$$\lim_{\rho \to 0} \langle (f, S_\mu) \wedge \tau_\rho, \omega \rangle = \lim_{\rho \to 0} \int_{K(U)} \int_{L_y} \tau_\rho \wedge \omega$$

is small. Since $\mu_{K(U)}(K(U))$ is small, it remains only to see that

$$\lim_{\rho \to 0} \int_{L_y} \tau_\rho \wedge \omega = \int_{S_y} \omega$$

is bounded for y off the bad locus. This is bounded by the area of S_y times the norm of ω, and both these quantities are bounded (the first one is bounded due to the transverse continuity). $\qquad\square$

Now consider two immersed measured oriented solenoids $f_1 \colon S_{1,\mu_1} \to M$, $f_2 \colon S_{2,\mu_2} \to M$ intersecting almost everywhere transversally. Let $F \subset S_1 \times S_2$ be the subspace of non-transverse intersection points, which has null-transversal measure in $S_1 \times S_2$. Set $S = (S_1 \times S_2) \cap \tilde{f}^{-1}(\Delta)$, where $\tilde{f} = f_1 \times f_2 \colon S_1 \times S_2 \to M \times M$. Then Theorem 6.4 implies that

$$\langle [f_1, S_{1,\mu_1}]^* \cup [f_2, S_{2,\mu_2}]^*, [\omega] \rangle = \int_{S-F} \omega$$

for any closed form ω of degree $k_1 + k_2 - n$.

COROLLARY 6.5. *Let M be an analytic manifold, and let $f_1 \colon S_{1,\mu_1} \to M$, $f_2 \colon S_{2,\mu_2} \to M$ be two immersed measured oriented solenoids of class $C^{\omega,0}$ (that is, with analytic leaves). Let $S = \{(p_1, p_2) \in S_1 \times S_2 \,;\, f_1(p_1) = f_2(p_2)\}$, and let $F \subset S$ consist of points (p_1, p_2) such that the leaves of S_1 and S_2 at p_1 and p_2 do not intersect transversally. Suppose that*

(1) *every point $p \in F$ is an isolated point of F in the leaf of $S_1 \times S_2$ through p;*

(2) *F is null-transverse in $S_1 \times S_2$.*

Then

$$\langle [f_1, S_{1,\mu_1}]^* \cup [f_2, S_{2,\mu_2}]^*, [\omega] \rangle = \int_{S-F} \omega$$

for any $[\omega] \in H^{k_1+k_2-n}(M)$.

PROOF. We need to note only that condition (2) in Definition 6.1 is automatic. $\square$

7 Bibliography

[1] Ghys, E. *Laminations par surfaces de Riemann.* Dynamique et géométrie complexes (Lyon, 1997), ix, xi, Panor. Synthèses, 8, Soc. Math. France, Paris (1999), 49–95,

[2] Hurder, S.; Mitsumatsu, Y. *The intersection product of transverse invariant measures.* Indiana Univ. Math. J. 40, no. 4, (1991), 1169–1183.

[3] Muñoz, V.; Pérez-Marco, R. *Ergodic solenoids and generalized currents.* Revista Matemática Complutense, 24 (2011), 493–525.

[4] Muñoz, V.; Pérez-Marco, R. *Schwartzman cycles and ergodic solenoids.* In Essays in Mathematics and its Applications. Dedicated to Stephen Smale (eds. P. Pardalos and Th. M. Rassias), Springer, 2012, 295–333.

[5] Muñoz, V.; Pérez-Marco, R. *Ergodic solenoidal homology: Realization theorem.* Comm. Math. Physics, 302 (2011), 737–753.

[6] Muñoz, V.; Pérez-Marco, R. *Ergodic solenoidal homology II: Density of ergodic solenoids.* Australian J. Math. Anal. and Appl. 6, no. 1, Article 11 (2009), 1–8..

[7] Palis, J.; Takens, F. Hyperbolic and Sensitive Chaotic Dynamics at Homoclinic Bifurcations, Cambridge Studies in Advanced Mathematics, Vol. 35, Cambridge University Press, 1993.

[8] Ruelle, D.; Sullivan, D. *Currents, flows and diffeomorphisms.* Topology 14 (1975), 319–327.

[9] Schwartzman, S. *Asymptotic cycles.* Ann. of Math. (2) 66 (1957), 270–284.

[10] Sullivan, D. *Linking the universalities of Milnor-Thurston, Feigenbaum and Ahlfors-Bers.* In Topological methods in modern mathematics, Stony Brook 1991. Publish or Perish, Houston, 1993, 543–564.

Invariants of four-manifolds with flows via cohomological field theory

Hugo García-Compeán, Roberto Santos-Silva, and Alberto Verjovsky

ABSTRACT. The Jones-Witten invariants can be generalized for smooth, non-singular vector fields with invariant probability measure on three-manifolds, giving rise to new invariants of dynamical systems [22]. After a short survey of cohomological field theory for Yang-Mills fields, Donaldson-Witten invariants are generalized to four-dimensional manifolds with non-singular smooth flows generated by homologically non-trivial p-vector fields. These invariants have the information of the flows and they are interpreted as the intersection number of these flow orbits and constitute invariants of smooth four-manifolds admitting global flows. We study the case of Kähler manifolds by using the Witten's consideration of the strong coupling dynamics of $\mathcal{N} = 1$ supersymmetric Yang-Mills theories. The whole construction is performed by implementing the notion of higher-dimensional asymptotic cycles à la Schwartzman [18]. In the process Seiberg-Witten invariants are also described within this context. Finally, we give an interpretation of our asymptotic observables of four-manifolds in the context of string theory with flows.

1 INTRODUCTION

Quantum field theory is not only a framework to describe the physics of elementary particles and condensed matter systems, but it has been useful to describe mathematical structures and their subtle interrelations. One of the most famous examples is perhaps the description of knot and link invariants through the correlation functions of products of Wilson line operators in the Chern-Simons gauge theory [1]. These invariants are the Jones-Witten invariants or Vassiliev invariants, depending on whether the coupling constant is weak or strong, respectively. Very recently some aspects of gauge and string theories found a strong relation with Khovanov homology [2].

In four dimensions the Donaldson invariants are invariants of the smooth structure on a closed four-manifold. This is in the sense that if two homeomorphic differentiable manifolds have different Donaldson invariants, then they are not diffeomorphic [3, 4]. These invariants were reinterpreted by Witten in terms of the correlation functions of suitable observables of a cohomological Yang-Mills field theory in four dimensions [5]. Such a theory can be obtained from an appropriate topological twist on the global symmetries of the $\mathcal{N} = 2$ supersymmetric Yang-Mills theory in Minkowski space with global R-symmetry $SU(2)$ that rotates the supercharges. A gravitational analog of the Donaldson theory is given by the topological gravity in four and two dimensions [6]. The computation of Donaldson invariants for Kähler manifolds has been done from the mathematical point of view in [7] and [8]. These Donaldson invariants were later reproduced in [9] by using the strong coupling dynamics of $\mathcal{N} = 1$ supersymmetric gauge theories in four dimensions. Precisely, a deeper understanding of the dynamics of strong coupling $\mathcal{N} = 2$ supersymmetric Yang-Mills theories in four dimensions [10], including the notion of S-duality, allowed an alternative approach to Donaldson theory in terms of the low-energy effective abelian gauge theory coupled to magnetic monopoles [11]. For a recent account of all these developments, see [12].

Moreover, the topological twist was applied to other theories such as string theory, resulting in the so-called topological sigma models [13]. The two possible twists of the global symmetries of the world-sheet theory leads to the so-called A- and B-models, whose correlation functions give rise to a description of the moduli space in terms of only the Kähler cone or only the moduli of complex structures of a target space Calabi-Yau manifold. A-models give rise to Gromov-Witten invariants. Mirror symmetry is realized through the interchanging of A- and B-models of two Calabi-Yau manifolds related by the interchanging of Betti numbers [14]. For a recent survey of all these topological field theories and their interrelations, see [15], for instance.

On the other hand, it is well known that topology and symplectic geometry play a very important role in the theory of dynamical systems [16]. Some years ago, Schwartzman introduced homology 1-cycles associated to a foliation known as *asymptotic cycles* [17]. These 1-cycles are genuine homology cycles, and they represent an important tool to study some properties of dynamical systems. Moreover, the generalization to p-cycles, with $p > 1$, was done in [18]. Such generalization was achieved by using some concepts of dynamical systems such as flow boxes and geometric currents [19], [20]. The definition of asymptotic cycles for non-compact spaces was discussed in [21]. In particular, the results [17] were used to define the Jones-Witten polynomial for a dynamical system [22]. More recently, the ideas from [18], [19], and [20] were used to find new suitable higher-dimensional generalizations of the asymptotic linking number starting from a topological BF theory (see [23] and references therein).

In the present paper we also use the notion of asymptotic p-cycles to extend the Donaldson-Witten and Seiberg-Witten invariants when smooth p-vector fields are incorporated globally on the underlying four-manifold. The asymptotic p-cycles associated to p-vectors on the manifolds define real homology p-cycles on these manifolds. They will constitute refined topological invariants of dynamical systems which distinguish the triplet $(M, \mathcal{F}, \mu)$, where M is a four-manifold, $\mathcal{F}$ is the foliation (possibly singular) associated to a p-vector and μ is a transverse measure of $\mathcal{F}$ which is invariant under holonomy. Two triplets $(M_1, \mathcal{F}_1, \mu_1)$ and $(M_2, \mathcal{F}_2, \mu_2)$ are *differentiably equivalent* if there is a diffeomorphism from M_1 to M_2, which sends the leaves of $\mathcal{F}_1$ to the leaves of $\mathcal{M}_2$ and the push-forward of μ_1 is μ_2. Moreover, these invariants will constitute a generalization of the Donaldson-Witten invariants for such triplets. For instance, one of the main results here is that our invariants will distinguish triplets: if two triplets $(M_i, \mathcal{F}_i, \mu_i)$ ($i = 1, 2$) have the property that the four-dimensional Donaldson-Witten invariants of M_1 and M_2 are equal but our invariants are different, then the corresponding systems of flows on them are not differentiably equivalent.

On the other hand, it is well known that Donaldson-Witten invariants can be interpreted in terms of the scattering amplitude (at zero momentum) of an axion with a NS5-brane in the heterotic string theory [24]. This paper would suggest a possible physical interpretation of our invariants involving flows in terms of an averaged propagation of a closed string in a target space described in terms of the moduli space of positions of a NS5-brane. That means a "continuous" flux of closed strings (propagating in the transverse space to the worldvolume of the NS5 brane) giving rise to an asymptotic 2-cycle. The diffuseness of the asymptotic cycle is determined by a flow (or set of flows) in the target space given by some field in the target space, for instance, the NS B-field whose associated 2-vector field gives the 2-foliation on the target.

The organization of the present paper is as follows: Section 2 is devoted to a brief review of asymptotic p-cycles with $p > 1$. In Section 3, we give an overview of cohomological field theory for Donaldson-Witten theory. In Section 4 we define the Donaldson-Witten invariant for four-dimensional manifolds in the presence of a smooth and nowhere vanishing p-vector field over the underlying spacetime manifold. It is also verified that this invariant is well defined as a limiting average of the standard definition. Section 5 is devoted to describe the procedure for Kähler four-manifolds. This is done by using a physical procedure through the incorporation of a mass term which breaks the supersymmetry to $\mathcal{N} = 1$ theories allowing the existence of a mass gap. In Section 6 we survey the Seiberg-Witten invariants. We focus mainly on the case of abelian magnetic monopoles. Non-abelian monopoles are also briefly described. In Section 7 we derive the Seiberg-Witten invariants in the presence of flows. Section 8 is devoted to explain how the Donaldson-Witten invariants for flows can be derived from a suitable system of strings in non-trivial flows on the spacetime target space. Finally, in Section 9, our final remarks and conclusions close the paper.

2   ASYMPTOTIC CYCLES AND CURRENTS

In this section we give a brief overview of asymptotic p-cycles with $p \geq 1$. Our aim is not to provide an extensive review of this material but to introduce the notations and conventions of the relevant structures, which will be needed in the subsequent sections. For a more complete treatment, see [17], [18], [19], [20], and [25].

In order to study the main aim of the paper, which is a generalization of invariants of four-manifolds in the presence of a non-singular flows over a closed four-dimensional manifold M, it is necessary to consider asymptotic homology p-cycles of the flow on M with values of p greater than one. Here we will have two possibilities. The first one corresponds to a flow generated by a p-vector field which is not localized in the homology p-cycles of M. The second possibility is when the p-vector field is defined only on the tangent space of the p-cycles of M. Of course, we could have a mixed situation. We also consider a set of flow invariant probability measures supported on the whole underlying manifold M. In this case the cycles constitute some "diffuse" cycles depending on the flow and the measure. The invariants constructed from these cycles detecting the differentiable structure of the four-manifolds with flows will be the asymptotic polynomial invariants of M. These invariants will coincide with the standard Donaldson-Witten invariants when the measure set is supported on the homology p-cycles γ_p of M. For simply connected closed four-manifolds we will be interested in cycles of dimension $p = 0, 2, 4$. From physical reasons $p = 4$ is not an interesting case, since it gives a topological term that can be added to the classical Lagrangian, while that for $p = 0$ it is a trivial cycle. Thus, the only relevant cycle will be for $p = 2$. In this section we define and interpret the observables as currents in terms of the winding number of asymptotic cycles.

The case $p = 1$ was discussed in detail by Schwartzman in [17]. In [22] asymptotic cycles were applied to the Jones-Witten theory in order to find refined invariants of dynamical systems. Recently, these ideas were generalized to higher dimensions with foliations of dimension greater than one using the BF theory without a cosmological constant [23].

A current on a compact manifold M of dimension n is a linear and continuous functional in the de Rham complex $\Omega^*(M)$, i.e., satisfying

$$C[a_1\omega_1 + a_2\omega_2] = a_1 C[\omega_1] + a_2 C[\omega_2] \tag{2.1}$$

for all ω_1 and ω_2 differential forms and a_1 and a_2 scalars. As an example, we define the following current $\gamma_p[\omega] = \int_{\gamma_p} \omega$, where γ_p is a p-cycle of M and ω is a p-form on M. Moreover, a closed $(n - p)$-form α also defines a current in the following way:

$$\alpha[\omega] = \int_M \alpha \wedge \omega. \tag{2.2}$$

Another example is the contraction of a p-vector field v_p and a p-form ω. Let $v_p = v^{i_1 \dots i_p} \partial_{i_1} \wedge \cdots \wedge \partial_{i_p}$ be a p-vector field and $\omega = \omega_{i_1 \dots i_p} dx^{i_1} \wedge \cdots \wedge dx^{i_p}$, then we have

$$v_p[\omega] = \omega_{i_1 \dots i_p} v^{i_1 \dots i_p}. \tag{2.3}$$

A current restricted to the space of smooth m-forms is called an m-current. Let $\mathcal{D}_m$ denote the topological vector space (with the weak* topology) of m-currents. Then in [25], a series of boundary operators $\partial_m : \mathcal{D}_m \to \mathcal{D}_{m-1}$, defined on arbitrary m-currents, were constrcted; these define a chain complex and thus a homology theory which is dual to the de Rham cohomology.

2.1 p-Cycles and Geometrical Currents

The definition of asymptotic cycles for higher-dimensional foliations (under suitable hypothesis) starts by considering a closed subset S of a n-dimensional manifold M, a family of submanifolds L_α of dimension p, such that $S = \cup_\alpha L_\alpha$ defines a *partial foliation* $\mathcal{F}_p$ (or *lamination* [26, chapter 10]) of dimension p. Now we cover all M (including the interior) with a collection of closed disks $\mathbf{D}^p \times \mathbf{D}^{n-p}$ (horizontal and vertical disks, respectively); these collections are called *flow boxes* and they are defined in such a way that they intersect each L_α in a set of horizontal disks $\{\mathbf{D}^p \times \{y\}\}$. The disks are smoothly embedded such that the tangent planes vary continuously on the flow boxes.

An $(n - p)$-submanifold T of M is called a *transversal* if it is transverse to each L_α of the foliation $\mathcal{F}_p$. A *transversal Borel measure* of the foliation $\mathcal{F}_p$ assigns to each small[1] transverse submanifold T a measure $\mu_{p,T}$. We assume that the measures are holonomy invariant and that they are finite on compact subsets of the transversals [27].

Thus a *geometrical current* is the triple (L_α, μ_T, ν), with the entries being the objects defined earlier and ν is the orientation of L_α, which is assigned to every point.

---

[1]A submanifold is said to be *small* if it is contained in a single flow box.

Assume that M is covered by a system of flow boxes endowed with partitions of unity. Then every p-form ω can be decomposed into a finite sum $\omega = \sum_i \omega_i$, where each ω_i has his own support in the ith flow box. We proceed to integrate out every ω_i over each horizontal disk $(\mathbf{D}^p \times \{y\})_i$. Thus we obtain, using the transverse measure, a continuous function f_i over $(\mathbf{D}^{n-p})_i$. In this way we define a geometric current as

$$\langle (L_\alpha, \mu_T, \nu), \omega \rangle = \sum_i \int_{(\mathbf{D}^{n-p})_i} \mu_T(y) \left(\int_{(\mathbf{D}^p \times \{y\})_i} \omega_i \right). \tag{2.4}$$

This current is closed in the sense of de Rham [25], i.e., if $\omega = d\phi$, where ϕ has compact support, then $\langle (L_\alpha, \mu_T, \nu), d\phi \rangle = 0$, since we can write $\phi = \sum_i \phi_i$. Ruelle and Sullivan [19] have shown that this current determines precisely an element of the pth cohomology group of M. It does not depend of the choice of flow boxes. Recall that any $(n - p)$-form ρ on M, determines a p-dimensional current by Poincaré duality $\langle \rho, \omega \rangle = \int_M \rho \wedge \omega$. Thus (L, μ, ν) determines an element in $\mathrm{Hom}(H^k(M, \mathbb{R}), \mathbb{R})$ which is isomorphic to $H_k(M, \mathbb{R})$ and, therefore, gives the asymptotic cycle.

Now consider an example of a geometrical current. Let μ_p be an invariant (under $\mathbf{X}_p$) volume n-form and $\mathbf{X}_p$ is a p-vector field nowhere vanishing on M. This defines a current in the de Rham sense via the $(n - p)$-form $\eta = i_{\mathbf{X}_p}(\mu_p)$. The current is given by

$$W_{\mu, \mathbf{X}_p}(\beta) = \int_M i_{\mathbf{X}_p}(\mu_p) \wedge \beta. \tag{2.5}$$

This current is not closed in general, but it will be closed, for instance, if the p-vector $\mathbf{X}_p$ consists of vector fields corresponding to one-parameter subgroups of an action of a Lie group which preserves the volume form μ_p. More precisely, one can obtain asymptotic cycles for values of $p > 1$ [18], as follows. Consider the action of a connected Lie group G on a smooth compact oriented manifold M, whose orbits are of the same dimension p. A *quantifier* is a continuous field of p-vectors on M everywhere tangent to the orbits and invariant under the action of G via the differential.

A quantifier is said to be *positive* if it is nowhere vanishing and determines the orientation of the tangent space. A *preferred action* is an oriented action of a connected Lie group G such that for any $x \in M$ the isotropy group I_x of x is a normal subgroup of G and G/I_x is unimodular.

In [18] it was proved that a preferred action possesses a positive quantifier, and given a positive quantifier, we can define a one-to-one correspondence between finite invariant measures and transversal invariant measures. An important result which will be used in the next sections is the following theorem (Schwartzman [18]) that states: Let $\mathbf{X}_p$ be a positive definite quantifier (i.e., a p-vector field) and μ_p an invariant measure given by a the volume n-form, then $i_{\mathbf{X}_p}(\mu_p)$ is a closed $(n - p)$-form, and the asymptotic cycle $W_{\mu, \mathbf{X}_p}$ will be obtained by Poincaré duality of an element of $H^{n-p}(M, \mathbb{R})$ determined by $i_{\mathbf{X}_p}(\mu_p)$.

If $W_{\mu, \mathbf{X}_p}$ in $H_p(M, \mathbb{R})$ is an asymptotic cycle, the theorem gives an explicit way to construct asymptotic cycles and interpret currents as winding cycles if the preceding conditions are satisfied. In [20], Sullivan gave another way of specifying a foliation, using structures of p-cones and operators acting over vectors on these cones.

One concrete example of this is the following: let G be a connected abelian Lie group (for instance $\mathbb{R}^n$ or a compact torus $\mathbb{T}^n$) acting differentiably and locally freely (i.e., the isotropy group of every point is a discrete subgroup of G) on the smooth closed manifold M. Then, the orbits of the action determines a foliation with leaves of the same dimension as G. Since the group is abelian it has an invariant volume form, and we obtain a natural foliated cycle.

3 OVERVIEW OF COHOMOLOGICAL QUANTUM FIELD THEORY: DONALDSON-WITTEN INVARIANTS

In this section we give a brief overview of the Donaldson-Witten invariants for a closed, oriented, and Riemannian four-manifold M [3], [4] representing our spacetime. We will focus on the Witten description [5] in terms of correlation functions (expectation values of some BRST-invariant operators). A cohomological field theory is a field theory with a BRST-like operator $\mathcal{Q}$ transforming as a scalar with respect to the spacetime symmetries. This operator represents a symmetry of the theory and it is constructed such that $\mathcal{Q}^2 = 0$. The Lagrangians of these theories can be written as a BRST commutator (BRST-exact) $L = \{\mathcal{Q}, V\}$ for some functional V. Given the properties of $\mathcal{Q}$, it implies that the Lagrangian is invariant under the $\mathcal{Q}$ symmetry $\{\mathcal{Q}, L\} = 0$, i.e., the Lagrangian is $\mathcal{Q}$-closed. In general, all observables $\mathcal{O}$ in the theory are BRST invariant, and they define cohomology classes given by $\mathcal{O} \sim \mathcal{O} + \{\mathcal{Q}, \lambda\}$ for some λ. Here $\mathcal{O}$ are the observables of the theory, which are invariant polynomials of the fields under the symmetry generator $\mathcal{Q}$. The observables are given by local field operators; thus, they depend on the point $x \in M$. Sometimes, in order to simplify the notation, we will explicitly omit this dependence.

Usually the relevant topological (twisted) Lagrangian can be derived from a physical Lagrangian, which may depend on the Riemannian metric $g_{\mu\nu}$ of the underlying spacetime manifold M; consequently, there exists an energy-momentum tensor $T_{\mu\nu}$ which is also BRST-exact, i.e., $T_{\mu\nu} = \{\mathcal{Q}, \lambda_{\mu\nu}\}$ for some $\lambda_{\mu\nu}$. It was proved for any BRST-exact operator $\mathcal{O}$ that the correlation function $\langle\{\mathcal{Q}, \mathcal{O}\}\rangle$ vanishes and the partition function is also independent of the metric and the physical parameters encoded in the Lagrangian. Thus the correlation functions are topological invariants.

There are several examples of these kind of theories. In particular, the theories that we are interested in are the Donaldson-Witten and the Seiberg-Witten ones. From the physical point of view these theories come from a suitable twist to the Lorentz group of the $\mathcal{N} = 2$ supersymmetric Yang-Mills theories with a compact Lie group (for definiteness we will use $SU(2)$, though its generalization to higher-dimensional groups is not difficult). The supercharges also are affected by the twist which gives rise to our $\mathcal{Q}$ transforming as a scalar in the new assignation of the representations of spacetime global symmetries.

In the path integral formalism the Donaldson-Witten polynomials are given by correlation functions in the Euclidean signature

$$\langle \mathcal{O} \rangle = \int \mathcal{DX} \exp\left(-L_{DW}/e^2\right)\mathcal{O}, \tag{3.1}$$

where L_{DW} is the Donaldson-Witten Lagrangian, e is the coupling constant, the $\mathcal{D}\mathcal{X}$ represent the measure of the fields in the theory, which includes a non-abelian gauge $A^a_\mu(x)$, scalar $\phi(x)$, fermionic $\psi(x)$, and ghost (anti-ghost) fields, all of them taking values in the adjoint representation of the gauge group. Fields $(A^a_\mu(x)$, $\psi(x), \phi(x))$ (with associated ghost numbers $U = (0, 1, 2)$) constitute a fermionic BRST multiplet. We note that there is a nice mathematical interpretation of the mentioned ingredients of the theory. For instance, the fields will represent differential forms on the moduli space of the theory, the ghost number of the fields corresponds with the degree of these differential forms, and the BRST charge $\mathcal{Q}$, which changes the ghost number in a unit, can be regarded as the exterior derivative.

It is possible to see that the change of the correlation functions with respect to the coupling constant e is $\mathcal{Q}$-exact. Due to the property mentioned earlier, $\langle \{\mathcal{Q}, V\} \rangle = 0$ for some V, the correlation functions are independent of e. Consequently, they can be computed in the semi-classical regime when e is small, and they can be evaluated by the stationary phase method. The path integration in the space $\mathcal{C} = \mathcal{B}/\mathcal{G}$ (of all gauge connections $\mathcal{B}$ modulo gauge transformations $\mathcal{G}$) localizes precisely in the space of gauge fields satisfying the instanton equation (anti-self-dual Yang-Mills equations): $\widetilde{F}_{\mu\nu} = -F_{\mu\nu}$, i.e., the instanton moduli space $\mathcal{M}_D$ of dimension $d(\mathcal{M}_D) = 8p_1(E) - \frac{3}{2}(\chi + \sigma))$ [28]. Here χ and σ are the Euler characteristic and the signature of M, respectively.

It is worth mentioning that in general $\mathcal{M}_D$ has singularities which are associated with the reducible connections or the zero size instantons (small instantons). If one considers four-manifolds with $b_2^+(M) > 1$, the moduli space $\mathcal{M}_D$ behaves as an smooth, orientable, and compact manifold [3], [4]. But for general $b_2^+(M)$, it is more involved. Thus, in general, they are usually neglected by assuming that the only zero modes come from the gauge connection $A^a_\mu(x)$ and its BRST partner $\psi^a_\mu(x)$. The scalar field $\phi^a(x)$ has zero modes in the singularities, and this would lead to a modification of the observables. The Donaldson map $H_p(M) \to H^{4-p}(\mathcal{M}_D)$ $(p = 0, \ldots, 4)$ is given by $\gamma_p \mapsto \int_{\gamma_p} c_2(\mathcal{P})$, where $\mathcal{P}$ is the universal bundle over $M \times \mathcal{M}_D \subset M \times \mathcal{B}/\mathcal{G}$ and c_2 is the second Chern class of $\mathcal{P}$.

For the gauge group $SU(2)$, the observables are

$$\mathcal{O}^{\gamma_p} \equiv \int_{\gamma_p} W_{\gamma_p}, \tag{3.2}$$

where γ_p is a p-homology cycle of M and W_{γ_p} is a p-form over M given by

$$W_{\gamma_0}(x) = \frac{1}{8\pi^2} \mathrm{Tr}\phi^2, \quad W_{\gamma_1} = \frac{1}{4\pi^2} \mathrm{Tr}(\phi \wedge \psi), \tag{3.3}$$

$$W_{\gamma_2} = \frac{1}{4\pi^2} \mathrm{Tr}(-i\psi \wedge \psi + \phi F), \quad W_{\gamma_3} = \frac{1}{4\pi^2} \mathrm{Tr}(\psi \wedge F), \quad W_{\gamma_4} = \frac{1}{8\pi^2} \mathrm{Tr}(F \wedge F),$$

where $W_{\gamma_0}(x)$ is, by construction, a Lorentz and $\mathcal{Q}$ invariant operator. These observables have ghost number $U = (4, 3, 2, 1, 0)$, respectively, and they are constructed as descendants which can be obtained from the relation

$$dW_{\gamma_p} = \{\mathcal{Q}, W_{\gamma_{p+1}}\}. \tag{3.4}$$

This construction establishes an isomorphism between the BRST cohomology $H^*_{BRST}(\mathcal{Q})$ and the de Rham cohomology $H^*_{dR}(M)$. One can check that $\mathcal{O}^{\gamma_p}$ is BRST-invariant (BRST-closed):

$$\{\mathcal{Q}, \mathcal{O}^{\gamma_p}\} = \int_{\gamma_p} \{\mathcal{Q}, W_{\gamma_p}\} = \int_{\gamma_p} dW_{\gamma_{p-1}} = 0. \tag{3.5}$$

For that reason the BRST commutator of $\mathcal{O}^{\gamma_p}$ depends only on the homology class of γ_p. Indeed, suppose that $\gamma_p = \partial \beta_{p+1}$; then we get (BRST-exact)

$$\mathcal{O}^{\gamma_p} = \int_{\gamma_p} W_{\gamma_p} = \int_{\beta_{p+1}} dW_{\gamma_p} = \int_{\beta_{p+1}} \{\mathcal{Q}, W_{\gamma_{p+1}}\} = \{\mathcal{Q}, \int_{\beta_{p+1}} W_{\gamma_{p+1}}\}. \tag{3.6}$$

Then the correlation functions are written as

$$\langle \mathcal{O}^{\gamma_{p_1}} \cdots \mathcal{O}^{\gamma_{p_r}} \rangle = \left\langle \prod_{j=1}^{r} \int_{\gamma_{p_j}} W_{\gamma_{p_j}} \right\rangle$$

$$= \int \mathcal{D}\mathcal{X} \exp(-L_{DW}/e^2) \prod_{j=1}^{r} \int_{\gamma_{p_j}} W_{\gamma_{p_j}}. \tag{3.7}$$

These are the Donaldson-Witten invariants in the path integral representation. They are invariants of the smooth structure of M. For simply connected manifolds $\pi_1(M) = 0$, the relevant cycles are of the dimensions $p = 0, 2$, and 4.

Consider a simply connected four-manifold M. The correlation functions of r observables $\mathcal{O}_{\Sigma_1}, \ldots, \mathcal{O}_{\Sigma_r}$ are given by

$$\langle \mathcal{O}_{\Sigma_1}(x_1) \cdots \mathcal{O}_{\Sigma_r}(x_r) \rangle = \left\langle \prod_{j=1}^{r} \mathcal{O}_{\Sigma_j}(x_j) \right\rangle$$

$$= \int \mathcal{D}\mathcal{X} \exp(-L_{DW}/e^2) \prod_{j=1}^{r} \mathcal{O}_{\Sigma_j}(x_j), \tag{3.8}$$

where we have considered only r arbitrary 2-cycles, i.e., $\gamma_{2_j} = \Sigma_j$ with $j = 1, \ldots, r$. Since $\mathcal{O}_{\Sigma_j}(x_j)$ has ghost number $U = 2$, the preceding correlation function has $U = 2r$. In terms of the zero modes, one can write each $\mathcal{O}_{\Sigma_j}(x_j) = \Phi_{i_1 i_2}(a_i)\psi^{i_1}\psi^{i_2}$, which absorbs two zero modes; consequently, in the weak coupling limit we have

$$\langle \mathcal{O}_{\Sigma_1}(x_1) \cdots \mathcal{O}_{\Sigma_r}(x_r) \rangle = \int_{\mathcal{M}_D} \Phi_{\Sigma_1} \wedge \cdots \wedge \Phi_{\Sigma_r}. \tag{3.9}$$

Thus we have $\Sigma \in H_2(M) \to \Phi_\Sigma \in H^2(\mathcal{M}_D)$ and Eq. (3.9) becomes

$$\langle \mathcal{O}_{\Sigma_1}(x_1) \cdots \mathcal{O}_{\Sigma_r}(x_r) \rangle = \#(H_{\Sigma_1} \cap \cdots \cap H_{\Sigma_r}), \tag{3.10}$$

where H_{Σ_j} is the Poincaré dual to Φ_{Σ_j} and represents a $(d(\mathcal{M}_D) - 2)$-homology cycle of the instanton moduli space $\mathcal{M}_D$. Equation (3.10) is interpreted as the intersection number of these homology cycles in the moduli space.

The topological invariance is not evident from Eq. (3.7). However, the preceding construction has a natural interpretation in terms of equivariant cohomology

[29]. Moreover, Atiyah and Jeffrey [30] showed that this expression can be reinterpreted in terms of the Euler class of a suitable infinite dimensional vector bundle in the Mathai-Quillen formalism [31]. This construction requires a real vector bundle $\mathcal{E}$ over the quotient space $\mathcal{C}$ of the space of all connections $\mathcal{B}$ modulo gauge transformations $\mathcal{G}$. This bundle is such that the fibers are the space of sections $\Gamma(\Lambda^{2,+} \otimes \mathrm{ad}(P))$. Here P is the SU(2)-principal bundle over M with gauge connection $A_\mu^a(x)$.

Moreover, a section s of $\mathcal{E}$ is given by $s = -F^+$, i.e., the locus $s^{-1}(0)$ is precisely the anti-self-dual moduli space $\mathcal{M}_D \subset \mathcal{C}$. The Euler class $e(\mathcal{E})$ is the pullback $s^*\Phi(\mathcal{E})$ of the Thom class $\Phi(\mathcal{E})$ of $\mathcal{E}$, under the section s. If $d(\mathcal{C})$ is the dimension of $\mathcal{C}$, the work of Mathai-Quillen [31] allows to gave a gaussian representative for the associated Thom class given by $e_{s,\nabla}(\mathcal{E}) = \exp[-\frac{1}{e^2}|s|^2 + \cdots]$ (which is given by a $2m$ differential form on $\mathcal{C}$), such that the Euler class is given by

$$\int_{\mathcal{C}} e_{s,\nabla}(\mathcal{E}) \wedge \alpha, \tag{3.11}$$

where α is an appropriate form of co-dimension $2m$, i.e., $\alpha \in H^{d(\mathcal{C})-2m}(\mathcal{C})$. This Euler class is, of course, independent of the connection ∇ used in this construction. The Euler class (3.11), for an appropriate α, represents the Donaldson-Witten invariants (3.7).

4   DONALDSON-WITTEN INVARIANTS FOR FLOWS

In this section we study the Donaldson-Witten invariants when there exist flows associated to a p-vector field over the spacetime manifold M equipped with an invariant probability measure μ (normalized such that $\int_M \mu = 1$) and a non-singular p-vector field $\mathbf{Y}_p = Y_1 \wedge \cdots \wedge Y_p$, where $Y_i = Y_i^\mu \partial_\mu$ with $\mu = 1, \ldots, 4$ and $i = 1, \ldots, p$. We require that the probability measure $\mu_{T,p}$ be invariant under $\mathbf{Y}_p$ for every p. Thus, the global information is encoded in the set of triplets $\{(M, \mathcal{F}_p, \mu_{T,p}), \ p = 0, \ldots, 4\}$, where $\mathcal{F}_p$ is the foliation generated by the p-vector field $\mathbf{Y}_p$. Each triplet $(M, \mathcal{F}_p, \mu_{T,p})$ determines an asymptotic p-cycle that we denote as $\widetilde{\gamma}_p$.

Now we define the generalized Lie derivative $\mathcal{L}_{\mathbf{Y}_p}$ for p-vectors, which is a graded operator defined as follows:

$$\mathcal{L}_{\mathbf{Y}_p}\omega = [d, i_{\mathbf{Y}_p}]\omega = d(i_{\mathbf{Y}_p}\omega) + (-1)^{\deg d \cdot \deg i_Y} i_{\mathbf{Y}_p}(d\omega), \tag{4.1}$$

where $[d, i_{\mathbf{Y}_p}]$ is a graded commutator and ω is a p-form. For further details on the formalism of multi-vector field see [32] and [33]. Here there are two possibilities:

- The homology groups associated to the orbits of the p-vector fields $\mathbf{Y}_p$ are trivial; therefore, they do not give relevant information of the four-manifold. However, we can use these trivial asymptotic homology cycles to describe some particular interesting configurations of flows. This case corresponds to the situations found in [22] and [23]. In particular, if ω is the volume form invariant under $\mathbf{Y}_p$ for every p, the last term of the previous equation is zero; then in the Lie derivative $\mathcal{L}_{\mathbf{Y}_p}\mu_p = d(i_{\mathbf{Y}_p}\mu_p)$, the term $i_{\mathbf{Y}_p}\mu_p$ looks like the expression from the Schwartzman theorem at the end of Section 2. In order for $i_{\mathbf{Y}_p}\mu_p$ to be a cohomology class, it needs to be closed. This requirement is established by the following equation:

$$\mathcal{L}_{\mathbf{Y}_p}\mu_p = 0. \tag{4.2}$$

If this condition implies $d(i_{\mathbf{Y}_p}\mu_p) = 0$, i.e., $i_{\mathbf{Y}_p}\mu_p$ is closed, then this element defines an element of the $(n-p)$-cohomology group.

- The other possibility corresponds to the case when the $\widetilde{\gamma}_p$ are orbits of the flow generated by the p-vector fields $\mathbf{Y}_p$ for each value of p. These cycles are non-trivial. In this case it gives rise to a generalization of the four-manifold invariants as the measure μ is supported in the whole manifold M.

In the present paper we will focus mainly on this second possibility.

4.1   The Definition of Observables

Now we will introduce flows over M and promote its homology cycles to asymptotic cycles. We define the asymptotic observable for a p-vector field ($p = 1,\ldots,4$) according to the expression

$$\widetilde{\mathcal{O}}_{\mathbf{Y}_p}(\mu_p) = \int_{\widetilde{\gamma}_p} W_{\gamma_p} := \int_M i_{\mathbf{Y}_p}(W_{\gamma_p})\mu_p(x), \tag{4.3}$$

where μ_p is the volume form of M invariant under $\mathbf{Y}_p$, $i_{\mathbf{Y}_p}(W_{\gamma_p})$ denotes the contraction, and Tr is the trace of the adjoint representation of the gauge group.

We can think of the observable as an average winding number of the asymptotic cycle. The observables are related to the asymptotic cycles, so then they carry information about the flow whether it is trivial or not.

Let $\mathbf{Y}_1,\ldots,\mathbf{Y}_4$ be p-vector fields ($p = 1,\ldots,4$) and, together with the expressions (3.2) and (3.3), we define the asymptotic observables as

$$\widetilde{\mathcal{O}}_{\mathbf{Y}_0}(\mu_0) \equiv \mathcal{O}^{\gamma_0}(x) = \frac{1}{8\pi^2}\mathrm{Tr}\phi^2, \tag{4.4}$$

$$\widetilde{\mathcal{O}}_{\mathbf{Y}_1}(\mu_1) = \int_M \mathrm{Tr}\frac{1}{4\pi^2}i_{\mathbf{Y}_1}(\phi\psi)\,\mu_1, \tag{4.5}$$

$$\widetilde{\mathcal{O}}_{\mathbf{Y}_2}(\mu_2) = \int_M \mathrm{Tr}\frac{1}{4\pi^2}i_{\mathbf{Y}_2}(-i\psi\wedge\psi+\phi F)\,\mu_2, \tag{4.6}$$

$$\widetilde{\mathcal{O}}_{\mathbf{Y}_3}(\mu_3) = \int_M \mathrm{Tr}\frac{1}{4\pi^2}i_{\mathbf{Y}_3}(\psi\wedge F)\,\mu_3, \tag{4.7}$$

$$\widetilde{\mathcal{O}}_{\mathbf{Y}_4}(\mu_4) = \int_M \mathrm{Tr}\frac{1}{8\pi^2}i_{\mathbf{Y}_4}(F\wedge F)\,\mu_4. \tag{4.8}$$

It is an easy matter to check that these asymptotic observables $\widetilde{\mathcal{O}}_{\mathbf{Y}_p}(\mu_p)$ are $\mathcal{Q}$-invariant:

$$\begin{aligned}
\{\mathcal{Q},\widetilde{\mathcal{O}}_{\mathbf{Y}_{p+1}}(\mu_{p+1})\} &= \{\mathcal{Q},\int_M i_{\mathbf{Y}_{p+1}}(W_{\gamma_{p+1}})\mu_{p+1}\} \\
&= \int_M \{\mathcal{Q},i_{\mathbf{Y}_{p+1}}(W_{\gamma_{p+1}})\}\mu_{p+1} \\
&= \int_M i_{\mathbf{Y}_{p+1}}(dW_{\gamma_p})\mu_{p+1} = 0.
\end{aligned} \tag{4.9}$$

Here we have used the fact that the measure is invariant under the flow, i.e., these observables are closed in the de Rham sense (see [18, Theorem 2A]) and the fact that the BRST charge $\mathcal{Q}$ commutes with the contraction operation $i_{\mathbf{Y}_p}$. Then these asymptotic observables are BRST invariant; therefore, they will give rise to topological invariants of dynamical system through a generalization of the Donaldson-Witten invariants.

Moreover, we observe that the problem arising in the Jones-Witten case [22], which distinguishes strongly the abelian and non-abelian cases is absent here and for the present case, the non-abelian case can be treated exhaustively. Even if the theory is non-abelian, our observables are Lie algebra–valued p-forms and the group and space-time information decouples. For the gauge group $SU(N)$ with Lie algebra $su(N)$, we take, for instance,

$$
\begin{aligned}
\widetilde{\mathcal{O}}_{\mathbf{Y}_2}(\mu_2) &= \int_M \frac{1}{4\pi^2} \mathrm{Tr} i_{\mathbf{Y}_2}(-i\psi \wedge \psi + \phi F)\, \mu_2 \\
&= \int_M \frac{1}{4\pi^2} \mathrm{Tr}\{i_{\mathbf{Y}_2}(-i\psi^a \wedge \psi^b + \phi^a F^b)t_a t_b\}\, \mu_2,
\end{aligned}
\tag{4.10}
$$

where if t_a and t_b are generators of $su(2)$; then they satisfy the normalization condition: $\mathrm{Tr}(t_a t_b) = \frac{1}{2}\delta_{ab}$. Then, the last expression takes the form

$$
\widetilde{\mathcal{O}}_{\mathbf{Y}_2}(\mu_2) = \int_M \frac{1}{8\pi^2} i_{\mathbf{Y}_2}(-i\psi^a \wedge \psi^a + \phi^a F^a)\, \mu_2.
\tag{4.11}
$$

We will use the following notation for r components of p-cycles of different dimension. The observables will be denoted by $\widetilde{\mathcal{O}}_{\mathbf{Y}_{p_j}}(\mu_{p_j})$, where p_j take values $1, \ldots, 4$, $j = 1, \ldots, r$, such that they satisfy $\sum_{p,j} p_j = d(\mathcal{M}_D)$, which is the dimension of the moduli space of instantons.

4.2 Donaldson-Witten Invariants of Four-Manifolds for Flows

For an oriented manifold M with p_j-vectors fields $\mathbf{Y}_{p_j}$, with probability invariant measure μ_{p_j}, the r-point correlation functions (Donaldson-Witten invariants) for flows $\mathbf{Y}_{p_j}$ is given by

$$
\left\langle \widetilde{\mathcal{O}}_{\mathbf{Y}_{p_1}}(\mu_{p_1}) \cdots \widetilde{\mathcal{O}}_{\mathbf{Y}_{p_r}}(\mu_{p_r}) \right\rangle = \int \mathcal{D}\mathcal{X} \exp(-L_{DW}/e^2) \prod_{j=1}^{r} \int_M i_{\mathbf{Y}_{p_j}}(W_{\gamma_{p_j}})\mu_{p_j}(x).
\tag{4.12}
$$

This expression reduces to the ordinary Donaldson-Witten invariants (3.7) in the case when the measure is supported on the cycles. One can think of this set of measures $\{\mu_{p_j}\}$ as Dirac measures on the set of $(p-j)$-cycles $\{\widetilde{\gamma}_{p_j}\}$ if we consider the invariant probability measure $\mu_p = \sum_j^r \mu_{p_j}$, where each μ_{p_j} is supported on $\widetilde{\gamma}_{p_j}$ and they are uniformly distributed with respect the coordinates of $\{\widetilde{\gamma}_{p_j}\}$. In other words, μ_{p_j} is supported on γ_{p_j} and it coincides with the normalized area form of the surface γ_{p_j}. We need to normalize in order to have μ be a probability measure.

We want to remark that the underlying p_j-fields $\mathbf{Y}_{p_j}$ will be considered here just as spectator fields. That is, they are background fields that are not of dynamical nature and don't represent additional degrees of freedom of the underlying theory. Thus, they don't contribute to the measure $\mathcal{D}\mathcal{X}$, to the Lagrangian L_{DW}, nor to the counting of zero modes; consequently, they do not lead to a modification of the dimension of the moduli space of instantons.[2] There will be an influence of these p_j-vector fields on our systems modifying mainly the structure of the observables of the theory. The structure of the vacuum also remains unchanged, i.e., the mass gap and the chiral symmetry breaking are still playing an important role in the definition of invariants.

For the moment we will consider an arbitrary operator $\widetilde{\mathcal{O}}_{\mathbf{Y}_{p_j}}(\mu_{p_j})$ with $p_j \geq 1$ (because in the case $p_j = 0$ there is not a flow). Now we proceed to perform the integral over the non-zero modes, as in the case without flows.

We assume that the only zero modes correspond to the gauge field A_μ and those associated to ψ_μ. Denote this observable by

$$\widetilde{\mathcal{O}}_{\mathbf{Y}_{p_j}}(\mu_{p_j}) = \widetilde{\Phi}_{i_1 \cdots i_n}(a_i, \mathbf{Y}_{p_j}) \psi^{i_1} \cdots \psi^{i_n},$$

where the a_i denote the zero modes of the gauge field and the ψs are the zero modes of the fermionic field; $\widetilde{\Phi}(a_i, \mathbf{Y}_{p_j})$ is a function that depends only on the zero modes of the gauge field and contains the information about the flow. As in the standard case the partition function is zero; the integrals which are non-zero are of the form (4.12), where $\widetilde{\mathcal{O}}$ absorbs the zero modes.

Performing the functional integration over the non-zero modes in the weak coupling limit, we get $\widetilde{\Phi}_{i_1 \cdots i_n}(a_i, \mathbf{Y}_{p_j})$ is a skew-symmetric tensor; then $\widetilde{\mathcal{O}}$ can be regarded as a $n = d(\mathcal{M}_D)$-form in $\mathcal{M}_D$. Consequently, the correlation functions of one observable $\widetilde{\mathcal{O}}$ reads

$$\left\langle \widetilde{\mathcal{O}}_{\mathbf{Y}_{p_j}}(\mu_{p_j}) \right\rangle = \int_{\mathcal{M}_D} da_1 \dots da_n d\psi^1 \dots d\psi^n \widetilde{\Phi}_{i_1 \cdots i_n}(a_i, \mathbf{Y}_{p_j}) \psi^{i_1} \dots \psi^{i_n}$$

$$= \int_{\mathcal{M}_D} \widetilde{\Phi}_{\mathbf{Y}_{d(\mathcal{M}_D)}}, \tag{4.13}$$

where we integrate out the a_i's and obtain a n-form $\widetilde{\Phi}$ defined in the moduli space $\mathcal{M}_D$. If one considers a product of observables $\widetilde{\mathcal{O}} = \widetilde{\mathcal{O}}_{\mathbf{Y}_{p_1}}(\mu_{p_1}) \cdots \widetilde{\mathcal{O}}_{\mathbf{Y}_{p_r}}(\mu_{p_r})$ with $\sum_{p,j} p_j = n = d(\mathcal{M}_D)$ and p_j being the number of zero modes of $\widetilde{\mathcal{O}}_{\mathbf{Y}_{p_j}}(\mu_{p_j})$, then, in analogy to [23] one obtains

$$\left\langle \widetilde{\mathcal{O}}_{\mathbf{Y}_{p_1}}(\mu_{p_1}) \cdots \widetilde{\mathcal{O}}_{\mathbf{Y}_{p_r}}(\mu_{p_r}) \right\rangle = \int_{\mathcal{M}_D} \widetilde{\Phi}_{\mathbf{Y}_{p_1}} \wedge \cdots \wedge \widetilde{\Phi}_{\mathbf{Y}_{p_r}}. \tag{4.14}$$

These correlation functions are the asymptotic Donaldson-Witten invariants, and they are invariants of the triplet $(M, \mathcal{F}, \mu)$. In order to compute the observables we integrate out the zero modes. This is completely analogous to the case without

---

[2]Thus we compute the effect of these spectator vector fields on the invariants. At this stage it is not possible to compute the back reaction of all dynamical fields to $\mathbf{Y}_{p_j}$.

flows because the measure of the path integral does not include the **Y**s

$$\widetilde{\Phi}_{\mathbf{Y}_0} = \frac{1}{8\pi^2} \mathrm{Tr}\langle\phi\rangle^2, \tag{4.15}$$

$$\widetilde{\Phi}_{\mathbf{Y}_1} = \int_M \mathrm{Tr}\frac{1}{4\pi^2} i_{\mathbf{Y}_1}(\langle\phi\rangle\psi)\mu_1, \tag{4.16}$$

$$\widetilde{\Phi}_{\mathbf{Y}_2} = \int_M \mathrm{Tr}\frac{1}{4\pi^2} i_{\mathbf{Y}_2}(-i\psi\wedge\psi + \langle\phi\rangle F)\mu_2, \tag{4.17}$$

$$\widetilde{\Phi}_{\mathbf{Y}_3} = \int_M \mathrm{Tr}\frac{1}{4\pi^2} i_{\mathbf{Y}_3}(\psi\wedge F)\mu_3, \tag{4.18}$$

$$\widetilde{\Phi}_{\mathbf{Y}_4} = \int_M \mathrm{Tr}\frac{1}{8\pi^2} i_{\mathbf{Y}_4}(F\wedge F)\mu_4. \tag{4.19}$$

Now we define the intersection number in a way analogous to the case without flows.

For the simply connected case ($\pi_1(M) = 0$), we have that the important observables are those associated with cycles of dimension 0, 2 and 4[3]. In general, a p_j-cycle has associated an operator (form) with ghost number $U = 4 - p_j$, corresponding to the Donaldson map $\mu_D \colon H_p(M) \to H^{4-p}(\mathcal{M}_D)$. In [5] Witten constructed this map, interpreted as intersection number of cycles in the four manifold M,

$$\left\langle \widetilde{I}_{\mathbf{Y}_{2_1}}(\mu_{2_1})(x_1) \cdots \widetilde{I}_{\mathbf{Y}_{2_r}}(\mu_{2_r})(x_r) \right\rangle = \int_{\mathcal{M}_D} \nu_{\mathbf{Y}_{2_1}} \wedge \cdots \wedge \nu_{\mathbf{Y}_{2_r}}$$

$$= \#(H_{\widetilde{\Sigma}_{\mathbf{Y}_{2_1}}} \cap \cdots \cap H_{\widetilde{\Sigma}_{\mathbf{Y}_{2_r}}}), \tag{4.20}$$

where $H_{\widetilde{\Sigma}_{\mathbf{Y}_{2_r}}}$ is the Poincaré dual of codimension 2. If the observables $\widetilde{I}_{\mathbf{Y}_{2_j}}$ are denoted by $\widetilde{I}_{\mathbf{Y}_j}$ and μ_{2_j} is denoted by μ_j, then Eq. (4.20) represents the asymptotic intersection linking numbers of the 2-flows in the moduli space $\mathcal{M}_D$ determined by the integration of differential two-forms ν_j on $\mathcal{M}_D$, depending on the set of 2-vector fields $\{\mathbf{Y}_j\}_{j=1,\ldots,r}$ with $r = d/2$. In terms of the asymptotic cycles Eq. (4.20) represents the asymptotic intersection number of r asymptotic homology 2-cycles $\widetilde{\Sigma}_{\mathbf{Y}_j}$ in M.

Donaldson-Witten invariants (3.7) are defined for $b_2^+(M) > 1$. It is very interesting to know the analogous condition for defining the existence of the corresponding asymptotic invariants for foliations. The analog of the wall-crossing that does exist in the Donaldson case for $b_2^+(M) = 1$ will be also of interest in the context of foliations. We leave this question for future work.

4.3 Asymptotic Intersection Numbers

In this subsection we use the dynamics of strongly coupled supersymmetric gauge theories. In particular, we use some features such as the existence of a mass gap, the cluster decomposition, and a structure of the vacua degeneracy consisting of a finite number of discrete states obtained after a chiral symmetry breaking due to gaugino condensation. We proceed to find an interpretation of the intersection

---

[3]0- and 4-cycles are related by Hodge duality, so we will consider one of them, say, 0-cycles.

number for asymptotic cycles described in Eq. (4.20) with the aid of the mentioned features. In order to do that, we are going to compute the two-point correlation function of a pair of the observables $\tilde{I}_{\mathbf{Y}_1}(\mu)$ at different points

$$\left\langle \tilde{I}_{\mathbf{Y}_1}(\mu_1)(x_1)\tilde{I}_{\mathbf{Y}_2}(\mu_2)(x_2) \right\rangle = \#(H_{\widetilde{\Sigma}_{\mathbf{Y}_1}} \cap H_{\widetilde{\Sigma}_{\mathbf{Y}_2}}), \tag{4.21}$$

where the x_i's are points on M and $\#(H_{\widetilde{\Sigma}_{\mathbf{Y}_1}} \cap H_{\widetilde{\Sigma}_{\mathbf{Y}_2}})$ represents the asymptotic intersection number of the asymptotic cycles $\widetilde{\Sigma}_{\mathbf{Y}_1}$ and $\widetilde{\Sigma}_{\mathbf{Y}_2}$.

The one-point correlation function $\langle \tilde{I}_{\mathbf{Y}}(\mu)(x) \rangle$ is also zero, as in the standard case without flows and for the same reason. In $M = \mathbb{R}^4$ it vanishes by Lorentz invariance with the measure invariant under the flow; this yields

$$\langle \tilde{I}_{\mathbf{Y}}(\mu)(x) \rangle = \int_{\widetilde{\Sigma}} d\sigma^{mn} \langle Z_{mn} \rangle$$
$$= \int_M \langle i_{\mathbf{Y}}(Z) \rangle \mu = \int_M \langle Y^{mn} Z_{mn} \rangle \mu. \tag{4.22}$$

As the $\mathbf{Y}$s are not dynamical fields, their expectation values is given by $Y^{mn}\langle Z_{mn} \rangle$ and as $\langle Z_{mn} \rangle$ vanishes by Lorentz invariance, then consequently $\langle \tilde{I}_{\mathbf{Y}}(\mu)(x) \rangle$ also vanishes in flat spacetime. However for a general four-manifold in a theory with a mass gap (it is known that $\mathcal{N} = 2$ gauge field theories in four dimensions don't have a mass gap; however, we assume, following Witten [9], that is indeed the case);[4] the expectation value of the operator $Z_{mn}Y^{mn}$ is expanded, as in [9] in terms of local invariants of the Riemannian geometry of M:

$$\langle Z_{mn}(x)Y^{mn}(x) \rangle = D_m R D_n D_s D^s R \cdot Y^{mn} \pm \dots. \tag{4.23}$$

Under the metric scaling $g \to tg$ with t positive, the volume form μ scales as t^4; then $\langle Z_{mn}(x)Y^{mn}(x) \rangle$ should scale faster than $1/t^4$. This is precisely achieved by the mass in the case we have for flows. Thus, in general, $\langle \int_M i_{\mathbf{Y}}(Z)\,\mu \rangle$ vanishes as $t \to \infty$.

Now we want to compute

$$\left\langle \tilde{I}_{\mathbf{Y}_1}(\mu_1)(x_1)\tilde{I}_{\mathbf{Y}_2}(\mu_2)(x_2) \right\rangle = \int_{M_1 \times M_2} G_{\mathbf{Y}_1,\mathbf{Y}_2}(x_1,x_2)\,\mu_1(x_1)\,\mu_2(x_2), \tag{4.24}$$

where

$$G_{\mathbf{Y}_1,\mathbf{Y}_2}(x_1,x_2) = \langle i_{\mathbf{Y}_1}(Z)(x_1) \cdot i_{\mathbf{Y}_2}(Z)(x_2) \rangle. \tag{4.25}$$

Considering the properties of $i_{X_1 \wedge \cdots \wedge X_p}$, one can see that the next formula holds:

$$i_{X_1 \wedge \cdots \wedge X_p} B_p \wedge \mu_n - (-1)^{\frac{p}{2}(3+p)} B_p \wedge i_{X_p \wedge \cdots \wedge X_1} \mu_n = 0, \tag{4.26}$$

---

[4]In the process of obtaining invariants of smooth manifolds from physical theories, the dynamics of these theories is an important guide. However the topological invariants are independent on the metric and the coupling constant; one can compute these invariants in the limit where the theory is under control. The assumption of a mass gap for $\mathcal{N} = 2$ theories is justified as it allows us to compute invariants for some four-manifolds. However, it is observed that the own theory tells us that one has to consider the full dynamics (including the supersymmetry breaking) in order to find the right invariants.

where B_p is any p-form. After some work it is easy to see that using the previous equation, we have

$$\left\langle \widetilde{I}_{\mathbf{Y}_1}(\mu_1)(x_1)\widetilde{I}_{\mathbf{Y}_2}(\mu_2)(x_2) \right\rangle = \int_{M_1 \times M_2} (\Theta_{\mathbf{Y}_1} \wedge Z)(x_1) \wedge (\Theta_{\mathbf{Y}_2} \wedge Z)(x_2) \cdot \delta(x_1 - x_2)$$

(4.27)

where $\Theta_{\mathbf{Y}_1} = i_{\mathbf{Y}_1}(\mu_1)$ and $\Theta_{\mathbf{Y}_2} = i_{\mathbf{Y}_2}(\mu_2)$ are the Poincaré dual to $\widetilde{\Sigma}_1$ and $\widetilde{\Sigma}_2$, respectively. This is a double form [25]; consequently, $G_{\mathbf{Y}_1,\mathbf{Y}_2}(x_1, x_2)$ is proportional to $\delta(x_1 - x_2)$ and the only non-vanishing contributions come from the points $x_1 = x_2$.

Equivalently, one can follow a dimensional analysis with $g \to tg$, with $t \to \infty$. For $x_1 \neq x_2$, $\langle i_{\mathbf{Y}_1}(Z(x_1))\, i_{\mathbf{Y}_2}(Z(x_2)) \rangle$ vanishes faster than $1/t^8$. The only possible non-vanishing contribution is localized around $x_1 = x_2$ as $t \to \infty$. These are precisely the intersection points of the asymptotic cycles. This reduces to the transverse intersection of the flows at finitely many points. Thus, we have

$$\left\langle \widetilde{I}_{\mathbf{Y}_1}(\mu_1)(x_1)\widetilde{I}_{\mathbf{Y}_2}(\mu_2)(x_2) \right\rangle = \eta \cdot \#(\widetilde{\Sigma}_{\mathbf{Y}_1} \cap \widetilde{\Sigma}_{\mathbf{Y}_2}) \cdot \langle 1 \rangle,$$

(4.28)

where η is a constant, $\langle 1 \rangle = \exp(a\chi(M) + b\sigma(M))$ with a,b being constants, and $\chi(M)$ and $\sigma(M)$ the Euler characteristic and signature of M, respectively. In analogy with the definition of asymptotic linking number, we define the asymptotic intersection number of two 2-flows generated by the 2-vector fields $\mathbf{Y}_1$ and $\mathbf{Y}_2$. Thus, $\langle \widetilde{I}_{\mathbf{Y}_1}(\mu_1)(x_1)\widetilde{I}_{\mathbf{Y}_2}(\mu_2)(x_2) \rangle$ can be interpreted as the average intersection number.

With the aid of cluster decomposition property for a vacua consisting of only one state, Eq. (4.28) can be used to write the generating functional of the correlation functions of observables associated to r arbitrary two-dimensional asymptotic cycles $\widetilde{\Sigma}_{\mathbf{Y}_1}, \ldots, \widetilde{\Sigma}_{\mathbf{Y}_r}$

$$\left\langle \exp\left(\sum_a \alpha_a \widetilde{I}_{\mathbf{Y}_a}(\mu_a) \right) \right\rangle = \exp\left(\frac{\eta}{2} \sum_{a,b} \alpha_a \alpha_b \#(\widetilde{\Sigma}_{\mathbf{Y}_a} \cap \widetilde{\Sigma}_{\mathbf{Y}_b}) \right) \cdot \langle 1 \rangle.$$

(4.29)

This is given in terms of the pairwise intersection between the corresponding asymptotic cycles. If one incorporates the operators $\widetilde{\mathcal{O}}$ and takes into account that the vacua consist of a finite set $\mathcal{S}$ of discrete states, it can be modified as follows

$$\left\langle \exp\left(\sum_a \alpha_a \widetilde{I}_{\mathbf{Y}_a}(\mu_a) + \lambda\widetilde{\mathcal{O}} \right) \right\rangle = \sum_{\rho \in \mathcal{S}} C_\rho \exp\left(\frac{\eta_\rho}{2} \sum_{a,b} \alpha_a \alpha_b \#(\widetilde{\Sigma}_{\mathbf{Y}_a} \cap \widetilde{\Sigma}_{\mathbf{Y}_b}) + \lambda\langle \mathcal{O} \rangle_\rho \right),$$

(4.30)

where C_ρ is a constant including the gravitational contribution of the curvature invariants $\chi(M)$ and $\sigma(M)$ coming from $< 1 >$. When the gauge group is SU(2) the chiral symmetry breaking tells that the set $\mathcal{S}$ is precisely $\mathbb{Z}_2$. Thus the preceding formula consists of two terms.

5 DONALDSON-WITTEN INVARIANTS FOR KÄHLER MANIFOLDS WITH FLOWS

In this section we discuss the Donaldson invariants on a Kähler four-manifold M. We follow closely [9], from which we take the notation and conventions. We use the dynamics of strong coupling $\mathcal{N} = 1$ supersymmetric gauge theories in four dimensions in the infrared. In particular, the perturbation of the $\mathcal{N} = 2$ theory by adding a mass term[5] breaks supersymmetry, leaving a theory with a remnant $\mathcal{N} = 1$ supersymmetry. As we mentioned before, the properties of strong coupled gauge theories (confinement, mass gap and chiral symmetry breaking) are an important subject in order to compute the invariants of Kähler manifolds admitting a non-trivial canonical divisor ($H^{(2,0)}(M) \neq 0$) defining the choice of the corresponding class of a global holomorphic $(2,0)$ form ω.[6] To be more precise, we consider an $\mathcal{N} = 1$ gauge theory with the introduction of a mass term defined by the choice of a canonical divisor. We focus on its strong coupling dynamics and take, in addition, a series of non-singular smooth flows generated by 2-vector fields $\mathbf{X}$ and $\mathbf{Y}$ over M. We want to describe how Donaldson-Witten invariants of Kähler manifolds will be modified in the presence of these flows.[7]

The theory on $\mathbb{R}^4$ in Euclidean coordinates $(y^1, \ldots, y^4)$, with $z_1 = y^1 + iy^2$ and $z_2 = y^3 + iy^4$, suggests that the theory written in terms of $\mathcal{N} = 1$ multiplets implies that the observables Z are given by

$$Z^{(2,0)} = \psi\psi + \overline{\omega}\overline{B}B, \quad Z^{(1,1)} = \lambda\psi + BF, \quad Z^{(0,2)} = \lambda\lambda, \tag{5.1}$$

where $\overline{\omega}$ is an anti-holomorphic 2-form. These observables are BRST-invariant with respect to the remnant supercharge Q_1 (after the supersymmetry breaking). This structure of observables comes from decomposition of a $\mathcal{N} = 2$ vector multiplet in terms of $\mathcal{N} = 1$ gauge multiplet (A_m, λ) and a complex matter multiplet given by the scalar superfield $\Phi = (B, \psi)$. Here A_m is a gauge field, B is a complex scalar field, and λ and ψ are spinor fields, with all of them in the adjoint representation of the gauge group. The observables $Z^{(1,1)}$ and $Z^{(0,2)}$ come from the mentioned decomposition. However the presence of the term $\overline{\omega}\overline{B}B$ in $Z^{(2,0)}$ is a direct manifestation of the mass term that is added to the $\mathcal{N} = 2$ Lagrangian in order to break supersymmetry. The introduction of a mass term in $\mathbb{R}^4$ reads

$$\Delta L = -m \int d^4x d^2\theta \mathrm{Tr}\Phi^2 - \text{h.c.}, \tag{5.2}$$

where the volume form is $d^4x d^2\theta = d^2z d^2\overline{z} d^2\theta$. This term preserves only $\mathcal{N} = 1$ supersymmetry. It was proved in [9] that this perturbation ΔL to the Lagrangian is of the form: $\sum_a \alpha_a I(\Sigma_a) + \{Q_1, \cdot\}$. The canonical divisor $C \subset M$ is defined as the zero locus of ω. Thus, in general on a curved Kähler manifold with $H^{(2,0)}(M) \neq 0$, the perturbed Lagrangian ΔL can be rewritten in terms of ω being a non-zero holomorphic form in $H^{(2,0)}(M)$ such that it vanishes on C. Consequently, the mass term vanishes precisely in the zeros of ω over the divisor (global cosmic strings).[8]

Assume that $C = \bigcup_y C_y$, where C_y is a Riemann surface for each y such that ω has at most simple zeroes on C_y. One would estimate the contribution of the di-

---

[5]The mass term consist of a quadratic term of a scalar superfield in the adjoint representation of the gauge group.

[6]The condition of the existence of a canonical divisor is related to $b_2^+(M)$ by $b_2^+(M) = 2\dim H^{(2,0)}(M) + 1$. Thus, the condition $H^{(2,0)}(M) \neq 0$ is equivalent to $b_2^+(M) > 1$.

[7]Remember that the 2-vector fields $\mathbf{X}$ and $\mathbf{Y}$ are not dynamical, and they do not contribute to the Feynman integral to compute correlation functions. The analysis of zero modes is also unchanged and the dimension of the moduli space of instantons remains the same. The only change will be reflected in the definition of the observables.

[8]This cosmic string indeed captures chiral fermion zero modes of the field ψ which propagates on the canonical divisor C.

visor (cosmic string) to the Donaldson-Witten invariants by considering the intersections $\Sigma \cap C_y \neq \emptyset$. In the intersection (which is assumed to be transverse) points P, one can insert operators $V_y(P)$. Thus, if $\#(\Sigma \cap C_y)$ is the intersection number of Σ and C_y is given by

$$\#(\Sigma \cap C_y) = \int_M \theta_\Sigma \wedge \theta_{C_y}, \tag{5.3}$$

where θ_Σ is the Poincaré dual of Σ and θ_{C_y} is the Poincaré dual of C_y. Then the operators $I(\Sigma)$ must be replaced by $\sum_y \#(\Sigma \cap C_y)V_y$. Here $V_y = V_y(P)$ is a local operator inserted in the intersection points P's between Σ and C_y.

For the theory on the worldsheet (cosmic string) C it is assumed that it has a mass gap and a chiral symmetry breaking with vacuum degeneracy determined by $\mathbb{Z}_2$. Fermionic zero modes on the divisor lead also to a non-vanishing anomaly inside the theory on C_y which should cancel by other trapped fields along the string. Thus these chiral fermions contribute to the path integral measure by a factor $t_y = (-1)^{d\varepsilon_y}$ where $\varepsilon_y = 0, 1$ and d is the dimension of the gauge group. Gathering everything together in [9], it was found that the Donaldson-Witten invariant is of the form

$$\left\langle \exp\left(\sum_a \alpha_a I(\Sigma) + \lambda \mathcal{O} \right) \right\rangle \tag{5.4}$$

$$= 2^{\frac{1}{4}(7\chi + 11\sigma)} \exp\left(\frac{1}{2} \sum_{a,b} \alpha_a \alpha_b \#(\Sigma_a \cap \Sigma_b) + 2\lambda \right) \cdot \prod_y \left(e^{\phi_y} + t_y e^{-\phi_y} \right)$$

$$+ i^\Delta 2^{1 + \frac{1}{4}(7\chi + 11\sigma)} \exp\left(-\frac{1}{2} \sum_{a,b} \alpha_a \alpha_b \#(\Sigma_a \cap \Sigma_b) - 2\lambda \right) \cdot \prod_y \left(e^{-\phi_y} + t_y e^{\phi_y} \right),$$

where

$$\phi_y = \sum_a \alpha_a \#(\Sigma_a \cap C_y) \tag{5.5}$$

and $\Delta = \frac{1}{2}d(\mathcal{M}_D)$.

The invariants (5.4) can be further generalized in the case the divisor components C_y have singularities. Also, the consideration of higher-dimensional gauge groups leads to interesting generalizations. Both extensions were discussed in [9].

In summary, in the process of obtaining (5.4), a series of physical considerations were made. We assumed cluster decomposition with a set of vacuum states, mass gap, chiral symmetry breaking, and the smooth breaking of supersymmetry by the introduction of a mass term in the matter multiplet. All these assumptions are reasonable except the mass gap. It is well known that supersymmetric $\mathcal{N} = 2$ Yang-Mills theory doesn't have a mass gap. However, this assumption makes sense as one adds terms in the Lagrangian of the original $\mathcal{N} = 2$ theory, which leaves only one unbroken supersymmetry. The theory is $\mathcal{N} = 1$; the mass gap is allowed, and it gives precisely the necessary ingredient to interpret (5.4) as the Donaldson-Witten invariants of Kähler manifolds. This is the subject of the following subsection.

5.1 Asymptotic Observables in Kähler Manifolds

Before we proceed with the case of Kähler manifolds, we make some considerations of general character about the asymptotic intersection of two asymptotic cycles. When a complex structure is defined in M, every form and vector field can be, in general, decomposed into holomorphic, mixed, and anti-holomorphic parts. Let $(z^n, z^{\bar{n}})$ be complex coordinates on M, let $\{dz^n, dz^{\bar{n}}\}$ be a basis for the cotangent space $T_x^* M$, and let $\{\partial_n, \partial_{\bar{n}}\}$ be a basis for the tangent space $T_x M$ at the point x. Then we can decompose our observable as [34]:

$$Z = Z^{(0,2)} + Z^{(1,1)} + Z^{(2,0)}, \tag{5.6}$$

where in complex coordinates it looks like

$$Z^{(0,2)} = Z_{mn}dz^m \wedge dz^n, \quad Z^{(1,1)} = Z_{m\bar{n}}dz^n \wedge dz^{\bar{n}} \quad \text{and} \quad Z^{(2,0)} = Z_{\overline{mn}}dz^{\overline{m}} \wedge dz^{\bar{n}}.$$

In general, every element corresponds to the decomposition of $\Omega^p(M)$ into the direct sum $\oplus_{p=r+s}\Omega^{(r,s)}(M)$, where r and s stand for the degrees of the corresponding holomorphic and anti-holomorphic components. One has a direct sum decomposition of p-vector fields $\mathcal{H}^p(M) = \oplus_{p=r+s}\mathcal{H}^{(r,s)}(M)$. Thus, for $p = 2$:

$$\mathbf{Y} = \mathbf{Y}^{(0,2)} + \mathbf{Y}^{(1,1)} + \mathbf{Y}^{(2,0)}, \tag{5.7}$$

where $\mathbf{Y}^{(0,2)} = Y^{mn}\partial_m \wedge \partial_n$, $\mathbf{Y}^{(1,1)} = Y^{m\bar{n}}\partial_n \wedge \partial_{\bar{n}}$ and $\mathbf{Y}^{(2,0)} = Y^{\overline{mn}}\partial_{\overline{m}} \wedge \partial_{\bar{n}}$. Now a 2-vector field can be constructed from vector fields as $\mathbf{Y} = Y_1 \wedge Y_2$, where each Y_i (for $i = 1, 2$) is a vector field. Each Y_i can be decomposed as the sum of a holomorphic and a anti-holomorphic part as follows: $Y_i = Y_i^m\partial_m + Y_i^{\overline{m}}\partial_{\overline{m}}$; then, the 2-vector field $\mathbf{Y}$ takes the form

$$\mathbf{Y} = Y_1^m Y_2^n \partial_m \wedge \partial_n + (Y_1^m Y_2^{\bar{n}} - Y_1^{\bar{n}} Y_2^m)\partial_m \wedge \partial_{\bar{n}} + Y_1^{\overline{m}} Y_2^{\bar{n}} \partial_{\overline{m}} \wedge \partial_{\bar{n}}. \tag{5.8}$$

Thus, we can calculate the contraction of the 2-vector field $\mathbf{Y}$ and the observable Z. By the orthogonality relations between the basis, the asymptotic observable takes the following form:

$$\widetilde{I}_{\mathbf{Y}}(\mu) = \int_M \left[i_{\mathbf{Y}^{(0,2)}}(Z^{(0,2)}) + i_{\mathbf{Y}^{(1,1)}}(Z^{(1,1)}) + i_{\mathbf{Y}^{(2,0)}}(Z^{(2,0)}) \right] \mu$$

$$= \int_M \left[Z_{mn} \cdot Y^{mn} + Z_{m\bar{n}} \cdot Y^{m\bar{n}} + Z_{\overline{mn}} \cdot Y^{\overline{mn}} \right] \mu. \tag{5.9}$$

If the 2-vector fields are coming from the product of vector fields, we have

$$\widetilde{I}_{\mathbf{Y}}(\mu) = \int_M \left[Z_{mn} \cdot Y_1^m Y_2^n + Z_{m\bar{n}} \cdot (Y_1^m Y_2^{\bar{n}} - Y_1^{\bar{n}} Y_2^m) + Z_{\overline{mn}} \cdot Y_1^{\overline{m}} Y_2^{\bar{n}} \right] \mu. \tag{5.10}$$

Moreover, we can decompose the asymptotic observables into three different parts, one associated to a completely holomorphic part, one to a completely anti-holomorphic part, and one to the mixed component.

For physical reasons [9], there are only three types of relevant observables. These are given by the following:

- The usual observable $\widetilde{I}(\Sigma)$ that contributes to the asymptotic intersection number (4.24) is given by $\int_M i_{\mathbf{Y}^{(2,0)}}(Z^{(2,0)})\mu + \text{h.c.}$, with $Z^{(2,0)}$ given in (5.1).

- The observable $I(\omega) = \int_\Sigma \omega$ arises from a non-vanishing mass term which breaks supersymmetry (from $\mathcal{N} = 2$ to $\mathcal{N} = 1$). The asymptotic version is written as $\int_M i_{\mathbf{Y}^{(2,0)}}(\omega^{(2,0)})\mu + $ h.c. and it contributes to the asymptotic intersection number (5.3).

- Near intersection points of Σ and C_y, an operator $V(P)$ is assumed to be inserted, yielding a term of the form $I(\theta) = \int_M \theta \wedge Z$. The natural form that couples to the asymptotic canonical divisor $\widetilde{C}_y$ is given by the first Chern class $J = c_1(\widetilde{C}_y)$; it is a 2-form of type $(1,1)$, and can be associated to a vector field $\mathbf{X}^{(1,1)}$. Thus, we have that the observable is given by $\int_M i_{\mathbf{X}^{(1,1)}}(J)\mu$.

5.2   Invariants for Kähler Manifolds

Let $\widetilde{C}$ be the disjoint union of a finite number of $\widetilde{C}_y$, where $\widetilde{C}_y$ is an asymptotic Riemann surface for each y with simple zeroes. This asymptotic cycle can be defined in terms of a divergence-free 2-vector field $\mathbf{X}$ and is given by

$$\widetilde{C}_y(\mathbf{X}) = \int_M i_{\mathbf{X}}(J)\mu, \qquad (5.11)$$

where $J = c_1(C_y)$ and $\mathbf{X} \in \mathcal{H}^{(1,1)}$.

One can estimate the contribution of the divisor (cosmic string) to the Donaldson-Witten invariants of M with flows. The calculation of the contribution of the intersections $\widetilde{\Sigma}_\mathbf{Y} \cap \widetilde{C}_y \neq \varnothing$ can be done as follows. The operators $\widetilde{I}_\mathbf{Y}(\mu)(x)$ inserted on the canonical divisor contribute only to the intersections of the cycles $\widetilde{\Sigma}$, with the canonical divisor $\widetilde{C}_y$ multiplied by a local operator $V(P)$. Thus, one should make the replacement

$$\widetilde{I}_\mathbf{Y}(\mu)(x) \rightarrow \sum_y \#(\widetilde{\Sigma}_\mathbf{Y} \cap \widetilde{C}_y)V_y + \text{terms involving intersections of } \Sigma\text{s}. \qquad (5.12)$$

The expectation values in the vacua of operators $V(P)$ are fixed by normalization. Thus the main contribution comes from $\#(\widetilde{\Sigma}_\mathbf{Y} \cap \widetilde{C}_y)$, which is given by the intersection number of two flows, one of them associated to the $\widetilde{\Sigma}$ and the other one to the asymptotic canonical divisor $\widetilde{C}_y$

$$\#(\widetilde{\Sigma}_\mathbf{Y} \cap \widetilde{C}_y) = \int_{M_1 \times M_2} (Z \wedge \Theta_\mathbf{Y})(x_1) \wedge (J \wedge \Theta_\mathbf{X})(x_2) \cdot \delta(x_1 - x_2), \qquad (5.13)$$

where $\Theta_\mathbf{Y} = i_\mathbf{Y}(\mu)$ and $\Theta_\mathbf{X} = i_\mathbf{X}(\mu)$ are the Poincaré duals of $\widetilde{\Sigma}_\mathbf{Y}$ and $\widetilde{C}_y$, respectively. Here the measure μ is also invariant under the flow $\mathbf{X}$ generating the canonical divisor.

Gathering together all the previous considerations, we have a form for the

Donaldson-Witten invariants for Kähler manifolds with canonical divisor:

$$\left\langle \exp\left(\sum_a \alpha_a \tilde{I}_{\mathbf{Y}}(\mu) + \lambda \mathcal{O} \right) \right\rangle \tag{5.14}$$

$$= 2^{\frac{1}{4}(7\chi + 11\sigma)} \exp\left(\frac{1}{2} \sum_{a,b} \alpha_a \alpha_b \#(\tilde{\Sigma}_{\mathbf{Y}_a} \cap \tilde{\Sigma}_{\mathbf{Y}_b}) + 2\lambda \right) \cdot \prod_y \left(e^{\tilde{\phi}_y} + t_y e^{-\tilde{\phi}_y} \right)$$

$$+ i^{\Delta} 2^{1 + \frac{1}{4}(7\chi + 11\sigma)} \exp\left(-\frac{1}{2} \sum_{a,b} \alpha_a \alpha_b \#(\tilde{\Sigma}_{\mathbf{Y}_a} \cap \tilde{\Sigma}_{\mathbf{Y}_b}) - 2\lambda \right) \cdot \prod_y \left(e^{-\tilde{\phi}_y} + t_y e^{\tilde{\phi}_y} \right).$$

where $\tilde{\phi}_y$ is given by Eq. (5.13) and $\Delta = \frac{1}{2} d(\mathcal{M}_D)$.

Finally, it is worth mentioning that one can generalize these expressions for more general Kähler manifolds with canonical divisors that don't have simple zeroes. For generalizations to non–simply connected manifolds with $\pi_1(M) \neq 0$ and for higher-dimensional gauge groups, the reader can see, for instance, [35].

5.3   Examples

1.   Flows on hyper-Kähler manifolds with $H^{(2,0)}(M) \neq 0$. We start with r divergence-free 2-vector fields Y_a with $a = 1, \ldots, r$ on a 4-torus $\mathbf{T}^4$. Then the asymptotic invariants associated to every 2-vector field can be computed through the correlation functions of a product of operators

$$\tilde{I}(\mathbf{Y}_a) = \int_M i_{\mathbf{Y}_a}(Z^{(2,0)})\mu_a + \text{h.c} = \int_M i_{\mathbf{Y}_a}(\lambda\psi + \overline{\omega}\overline{B}B)\mu_a + \text{h.c.}$$

Thus, to this configuration of flows we get the following invariant by using the Hodge structure of the torus: $h^{1,0} = 2$, $h^{2,0} = 1$. The invariant is given by

$$\left\langle \exp\left(\sum_a \alpha_a \tilde{I}(Y_a) + \lambda \mathcal{O} \right) \right\rangle = \exp\left(\frac{1}{2} \sum_{a,b} \alpha_a \alpha_b \#(\tilde{\Sigma}_{\mathbf{Y}_a} \cap \tilde{\Sigma}_{\mathbf{Y}_b}) + 2\lambda \right)$$

$$+ \exp\left(-\frac{1}{2} \sum_{a,b} \alpha_a \alpha_b \#(\tilde{\Sigma}_{\mathbf{Y}_a} \cap \tilde{\Sigma}_{\mathbf{Y}_b}) - 2\lambda \right) \tag{5.15}$$

where $\#(\tilde{\Sigma}_{\mathbf{Y}_a} \cap \tilde{\Sigma}_{\mathbf{Y}_b})$ is the asymptotic intersection number. Thus, we find that $\#(\tilde{\Sigma}_{\mathbf{Y}_a} \cap \tilde{\Sigma}_{\mathbf{Y}_b})$ is given by Eq. (4.28). In the specific case of $\mathbf{T}^4$, we have 2 cycles of dimension two, one holomorphic and the other anti-holomorphic. The vector fields $\mathbf{Y}_a$ are wrapped on these homology cycles. There are one-dimensional homology cycles, and one can introduce vector fields whose orbits coincide with these cycles. One can construct asymptotic invariants associated with them and compute their contribution to the correlation functions. But we are not interested in this addition in the present paper. However it is interesting to remark that 2-vector fields can be constructed from the wedge product of two of these vector fields and we have constructed observables by using (5.10). Thus, we find that our invariant can be computed by using Eq. (4.24).

For the case of K3, where $h^{1,0} = 0$, $h^{2,0} = 1$, we don't have nonsingular vector fields (since the Euler characteristic is 24). However we have only intrinsic 2-vector fields. Thus, there will not be 2-vector fields constructed from 1-vector fields, as in the previous example. In this case the invariant is given by

$$\left\langle \exp\left(\sum_a \alpha_a \tilde{I}(Y_a) + \lambda \mathcal{O}\right)\right\rangle = C\left[\exp\left(\frac{1}{2}\sum_{a,b}\alpha_a\alpha_b \#(\tilde{\Sigma}_{Y_a}\cap\tilde{\Sigma}_{Y_b}) + 2\lambda\right)\right.$$

$$\left. - \exp\left(-\frac{1}{2}\sum_{a,b}\alpha_a\alpha_b \#(\tilde{\Sigma}_{Y_a}\cap\tilde{\Sigma}_{Y_b}) - 2\lambda\right)\right] \quad (5.16)$$

where $C = 1/4$. The asymptotic intersection number is also given by Eq. (4.28), and the sum involves also normal and asymptotic self-intersection numbers.

2. Hilbert Modular surfaces [36]. Let $\mathbb{H} \subset \mathbb{C}$ denote the upper half-plane with the Poincaré metric. Let $K := \mathbb{Q}(\sqrt{2})$ be the totally real quadratic number field obtained by adding $\sqrt{2}$ to $\mathbb{Q}$. The ring of integers $\mathbb{Z}(\sqrt{2})$ is the ring of real numbers of the form $m + n\sqrt{2}$, $m, n \in \mathbb{Z}$. Let σ be the nontrivial Galois automorphism of K given explicitly by $\sigma(a + b\sqrt{2}) = a - b\sqrt{2}$, $a, b \in \mathbb{Q}$. Let us consider the group $\Gamma := PSL(2, \mathbb{Z}(\sqrt{2}))$. Let $\bar{\sigma}\colon \Gamma \to \Gamma$ be the induced automorphism on Γ.

Γ acts properly and discontinuously on $\mathbb{H} \times \mathbb{H}$ as $\gamma(z, w) = (\gamma(z), \sigma(\gamma)(w))$. The quotient is an orbifold of dimension four which is not compact, but it has finite volume (with respect to the induced metric coming from the product metric on $\mathbb{H} \times \mathbb{H}$). The ends (cusps) of Hilbert modular surfaces of real quadratic number fields are manifolds which are of form $M^3 \times [0, \infty)$, where M^3 is a compact solvable three-manifold which fibers over the circle with fibre a torus $\mathbb{T}^2$. The number of such ends is equal to the class number of the field [36]. For $\mathbb{Q}(\sqrt{2})$ there is only one cusp, and M^3 is the mapping torus of the automorphism of the 2-torus induced by the matrix

$$A = \begin{pmatrix} 2 & 1 \\ 1 & 1 \end{pmatrix}.$$

Thus, M^3 fibers over the circle. In fact, the sugbroup $\Lambda \subset \Gamma$ consisting of affine transformations of the form $z \mapsto (1 + \sqrt{2})^r z + m + n\sqrt{2}$, $r, m, n \in \mathbb{Z}$ is the semidirect product $\mathbb{Z} \ltimes_A (\mathbb{Z} \times \mathbb{Z})$, which is the solvable fundamental group of M^3. The universal cover of M^3 is a solvable simply connected 3-dimensional Lie group whose Lie algebra is generated by three left-invariant vector fields $\mathbf{X}$, $\mathbf{Y}$, and $\mathbf{Z}$ whose Lie brackets satisfy $[\mathbf{X}, \mathbf{Y}] = a\mathbf{Y}$, $[\mathbf{X}, \mathbf{Z}] = -a\mathbf{Z}$ and $[\mathbf{Y}, \mathbf{Z}] = 0$, for some constant $a > 0$. The commuting vector fields $\mathbf{Y}$ and $\mathbf{Z}$ descend to vector fields in M^3 which are tangent to the torus fibers of the fibration of M^3 over the circle. The flow generated by $\mathbf{X}$ is an Anosov flow. The vector fields $\mathbf{Y}$ and $\mathbf{Z}$ generate two flows tangent to 1-dimensional foliations L_1 and L_2. These flows are homologous to zero, and the Arnold self-linking number of both is zero.

By a theorem of Selberg [37], Γ contains a finite index subgroup $\tilde{\Gamma}$ which acts freely on $\mathbb{H} \times \mathbb{H}$. The quotient manifold $M^4(\tilde{\Gamma}) := \tilde{\Gamma}/\mathbb{H} \times \mathbb{H}$ is a non-compact manifold of finite volume with a finite number of cusps depending upon the Selberg subgroup. The action of $PSL(2, \mathbb{Z}(\sqrt{2}))$ preserves the natural foliations of $\mathbb{H} \times \mathbb{H}$ whose leaves are, respectively, of the form $\mathbb{H} \times \{w\}$ and $\{z\} \times \mathbb{H}$. These foliations descend to $M^4(\tilde{\Gamma})$ to a pair of 2-dimensional foliations $\mathcal{F}_{horizontal}$ and $\mathcal{F}_{vertical}$ which are mutually transverse, and each has dense leaves. Furthermore, since the action of Γ is by isometries, the foliations $\mathcal{F}_{horizontal}$ and $\mathcal{F}_{vertical}$ are transversally Riemannian and thus both have natural transverse measures.

Now we can do two things to obtain examples of four-manifolds with (possibly singular) foliations.

- We can compactify $M^4(\tilde{\Gamma})$ à la Hirzebruch [38] by adding one point at infinity for each cusp. The resulting space is an algebraic surface with singularities at the cusps and the link of each singularity is the corresponding solvmanifold. After desingularizing one obtains a smooth algebraic surface which is, therefore, a Kähler surface. The foliations $\mathcal{F}_{horizontal}$ and $\mathcal{F}_{vertical}$ lift to the desingularized manifold to foliations with singularities at the cusps. There are important relations of these constructions with $K3$ surfaces [39].

- One can "cut" the manifold at each cusp to obtain a compact manifold with boundary, and each component of the boundary is a solvmanifold as described before. In other words, we remove a conic open neighborhood of each cusp whose boundary is the corresponding solvmanifold at the cusp. Now we can take the double to obtain a compact closed manifold with a pair of transversally Riemannian foliations with dense leaves (since in the double the foliations can be glued differentiably). Both foliations meet the solvmanifolds transversally and determine two flows in them. For the case $K = \mathbb{Q}(\sqrt{2})$ we obtain a compact four-manifold with two transversally Riemannian foliations which meet the solvmanifold in the foliations L_1 and L_2. Therefore *each of the foliations $\mathcal{F}_{horizontal}$ and $\mathcal{F}_{vertical}$ has self-intersection zero* (since L_1 and L_2 have self intersection zero).

Of course, one can construct examples like these using any totally real quadratic field and the group $PSL(2, \mathfrak{O}_K)$, where $\mathfrak{O}_K$ is the ring of integers of K.

3. Elliptic K3 surfaces and elliptic surfaces. Let S be an elliptic surface with Kodaira fibration $\pi\colon S \to \Sigma_g$, where Σ_g is an algebraic curve of genus g. The fibers are elliptic curves except for a finite number of singular fibers, which are rational curves. The fibration provides us with a singular foliation as mentioned in the preceding remark. There is a canonical choice for a transverse measure μ which is obtained from the Poincaré metric via the uniformization theorem applied to Σ_g: if τ is a 2-disk which is transverse to the regular part of the foliation, its measure is the hyperbolic area of $\pi(\tau)$. Then we can apply our results to the triple $(S, \mathcal{F}, \mu)$. One modification of elliptic surfaces can be obtained by the so-called *logarithmic transformation*. Using the logarithmic transformation one can change the Kodaira dimension and turn an algebraic surface into a non-algebraic surface.

Particular cases of elliptic surfaces are the $K3$ surfaces, Enriques surfaces, and the Dolgachev surfaces. We recall that Dolgachev X_p surfaces depending on an integer p were used by Donaldson to obtain the first examples X_2 and X_3 of manifolds which are homeomorphic but not diffeomorphic [39], [4]. From this, the following two questions arise:

- How do our invariants change after performing a logarithmic transformation on an elliptic surface?

- Can we detect exotic differentiable structures by our invariants?

4. Symplectic four-manifolds and Lefschetz fibrations and pencils. By a result of Donaldson [40], every symplectic four-manifold admits a Lefschetz fibration and these fibrations are an essential tool for the study of symplectic four-manifolds. As in the previous example, one has a triple $(M^4, \mathcal{F}, \mu)$, where $\mathcal{F}$ is the (possibly singular) foliation determined by the Lefschetz fibration and μ is a transverse measure coming from a choice of an area form from a Riemannian metric on the base surface. The question is how to compute our invariants and how they can be used to study symplectic manifolds.

6   SURVEY ON SEIBERG-WITTEN INVARIANTS

Another example of the theories which can be constructed through the Mathai-Quillen formalism is the cohomological field theory describing Seiberg-Witten monopoles [11]. The geometric data consists of the square root of the (determinant) line bundle $L^{1/2}$ over a four-manifold M with an abelian gauge connection A with curvature $F_A = dA$. We also have the tensor product $S^{\pm} \otimes L^{1/2}$ of $L^{1/2}$ with the spin bundle $S^{\pm}$, which exist whenever M is a spin manifold, i.e., $w_2(M) = 0$ (for more details on the spin structure, see [41] and [42]). This tensor product is even well defined if M is not a spin manifold. In addition we have a section $\psi_\alpha \in \Gamma(S^+ \otimes L^{1/2})$. The Seiberg-Witten equations are

$$F_{\alpha\beta}^+ = -\frac{i}{2}\overline{\psi}_{(\alpha}\psi_{\beta)}, \qquad D_{\alpha\dot{\alpha}}\psi^\alpha = 0, \tag{6.1}$$

where F^+ is the self-dual part of the curvature F_A. Here α and β are spinorial indices instead of vector ones μ and they are related by $A_\mu = \sigma_\mu^{\alpha\dot{\alpha}}A_{\alpha\dot{\alpha}}$.

The moduli space $\mathcal{M}_{SW}$ of solutions to the Seiberg-Witten equations will be denoted as $\mathcal{M}_{SW} \subset \mathcal{A} \times \Gamma(S^+ \otimes L^{1/2})/\mathcal{G}$, where $\mathcal{A}$ is the space of abelian connections on $L^{1/2}$ and $\mathcal{G}$ is the gauge group of the $U(1)$-bundle, i.e., $\mathcal{G} = \text{Map}(M, U(1))$. This moduli problem can be described in terms of the Mathai-Quillen construction. In this case the vector bundle is also trivial $\mathcal{V} = \mathcal{M} \times \mathcal{F}$, where $\mathcal{F}$ is the fiber. For the monopole case $\mathcal{F} = \Lambda^{2,+}(M) \otimes \Gamma(S^- \otimes L^{1/2})$. The section s is given by

$$s(A, \psi) = \left(\frac{1}{\sqrt{2}}(F_{\alpha\beta}^+ + \frac{i}{2}\overline{\psi}_{(\alpha}\psi_{\beta)}), D_{\alpha\dot{\alpha}}\psi^\alpha\right), \tag{6.2}$$

where $D_{\alpha\dot{\alpha}}\psi_\beta = \sigma_{\alpha\dot{\alpha}}^\mu(\partial_\mu + iA_\mu)\psi_\beta$. The zero section determines precisely the Seiberg-Witten equations.

The dimension $d(\mathcal{M}_{SW})$ of the moduli space $\mathcal{M}_{SW}$ can be obtained from an index theorem

$$d(\mathcal{M}_{SW}) = \lambda^2 - \frac{2\chi + 3\sigma}{4}, \tag{6.3}$$

where $\lambda = \frac{1}{2}c_1(L)$ (being $c_1(L)$ the first Chern class), χ and σ are the Euler characteristic and the signature of M, respectively. The Mathai-Quillen construction [31] provides a set of fields A_μ, ψ_μ, ϕ, $\chi_{\mu\nu}$, $H_{\mu\nu}$, η, ψ_α, μ_α, $v_{\dot{\alpha}}$, and $h_{\dot{\alpha}}$ of different ghost number. This set of fields will be denoted for short as $\mathcal{X}$. The Lagrangian can be read off from the exponential of the Thom class and is given by [12] and [43]:

$$
L_{SW} = \int_M e\left(g^{\mu\nu} D_\mu \overline{\psi}^\alpha D_\nu \psi_\alpha + \frac{1}{4} R\overline{\psi}^\alpha \psi_\alpha + \frac{1}{2} F^{+\alpha\beta} F^+_{\alpha\beta} - \frac{1}{8}\overline{\psi}^{(\alpha}\psi^{\beta)}\overline{\psi}_{(\alpha}\psi_{\beta)} \right)
$$

$$
+ i \int_M \left(\lambda \wedge *d^* d\phi - \frac{1}{\sqrt{2}}\chi \wedge *\rho^+ d\psi \right)
$$

$$
+ \int_M \left[i\phi\lambda\overline{\psi}^\alpha \psi_\alpha + \frac{1}{2\sqrt{2}}\chi^{\alpha\beta}(\overline{\psi}_{(\alpha}\mu_{\beta)} + \overline{\mu}_{(\alpha}\psi_{\beta)}) - \frac{i}{2}(v^{\dot\alpha} D_{\alpha\dot\alpha}\mu^\alpha - \mu^\alpha D_{\alpha\dot\alpha}v^{\dot\alpha}) \right.
$$

$$
\left. - \frac{i}{2}[\overline{\psi}^\alpha \psi_{\alpha\dot\alpha} v^{\dot\alpha}] + \frac{1}{2}\eta(\overline{\mu}^\alpha \psi_\alpha) - \overline{\psi}^\alpha \mu_\alpha) + \frac{i}{4}\phi\overline{v}^{\dot\alpha} v_{\dot\alpha} - \lambda\overline{\mu}^\alpha \mu_\alpha \right].
$$

$$(6.4)$$

The observables are products of BRST invariant operators which are cohomologically non-trivial:

$$
d\Theta^n_p = \{\mathcal{Q}, \Theta^n_{p+1}\}. \tag{6.5}
$$

The $\mathcal{Q}$-invariant operators are [43]

$$
\mathcal{O}^{\gamma_0}_n = \Theta^n_0(x),
$$

$$
\mathcal{O}^{\gamma_1}_n = \int_{\gamma_1} \Theta^n_1, \qquad \mathcal{O}^{\gamma_2}_n = \int_{\gamma_2} \Theta^n_2,
$$

$$
\mathcal{O}^{\gamma_3}_n = \int_{\gamma_3} \Theta^n_3, \qquad \mathcal{O}^{\gamma_4}_n = \int_M \Theta^n_4, \tag{6.6}
$$

where

$$
\Theta^n_0 = \binom{n}{0} \phi^n,
$$

$$
\Theta^n_1 = \binom{n}{1} \phi^{n-1}\psi,
$$

$$
\Theta^n_2 = \binom{n}{2} \phi^{n-2}\psi \wedge \psi + \binom{n}{1} \phi^{n-1}\psi \wedge F, \tag{6.7}
$$

$$
\Theta^n_3 = \binom{n}{3} \phi^{n-3}\psi \wedge \psi \wedge \psi + 2\binom{n}{2} \phi^{n-2}\psi \wedge F,
$$

$$
\Theta^n_4 = \binom{n}{4} \phi^{n-4}\psi \wedge \psi \wedge \psi \wedge \psi + 3\binom{n}{3} \phi^{n-3}\psi \wedge \psi \wedge F + \binom{n}{2} \phi^{n-2}F \wedge F.
$$

Here $\Theta^n_0(x)$ is constructed with a gauge and $\mathcal{Q}$ invariant field ϕ. All other observables are descendants obtained from it [43]. As in the Donaldson-Witten case, the construction establishes an isomorphism between the BRST cohomology $H^*_{BRST}(\mathcal{Q})$ and the de Rham cohomology $H^*_{dR}(M)$. To be more precise, the analogue to the Donaldson map is $\delta_{SW}: H_p(M) \to H^{2-p}(\mathcal{M}_{SW})$, given in terms of the first Chern class of $\mathcal{V}$. The observables (6.6) are BRST invariant (BRST closed) and the BRST commutator depends only on the homology class. This can be shown by following a procedure similar to that in the Donaldson case (see Eqs. (3.5) and (3.6)).

The correlation functions of r operators are written as

$$
\langle \mathcal{O}_n^{\gamma_{p_1}} \cdots \mathcal{O}_n^{\gamma_{p_r}} \rangle = \left\langle \prod_{j=1}^{r} \int_{\gamma_{p_j}} \Theta_{p_j}^n \right\rangle
$$

$$
= \int \mathcal{D}\mathcal{X} \exp(-L_{SW}/e^2) \prod_{j=1}^{r} \int_{\gamma_{p_j}} \Theta_{p_j}^n. \tag{6.8}
$$

These are the Seiberg-Witten invariants in the path integral representation [12], [43]. They are topological invariants and also invariants of the smooth structure of M. After integration over the non-zero modes one has

$$
\langle \mathcal{O}_1^{\gamma_{p_1}} \cdots \mathcal{O}_r^{\gamma_{p_r}} \rangle = \left\langle \prod_{j=1}^{r} \int_{\gamma_{p_j}} \Theta_{p_j}^n \right\rangle
$$

$$
= \int_{\mathcal{M}_{SW}} \nu_{p_1} \wedge \cdots \wedge \nu_{p_r}, \tag{6.9}
$$

where $\nu_{p_j} = \delta_{SW}(\gamma_{p_j})$. The possible values of p_j are $0, 1, 2$. Thus, for $p_j = 1, 2$, we can rewrite the previous equation as

$$
\langle \mathcal{O}_1^{\gamma_1} \cdots \mathcal{O}_r^{\gamma_1} \cdot \mathcal{O}_1^{\gamma_2} \cdots \mathcal{O}_{d/2}^{\gamma_2} \rangle = \int_{\mathcal{M}_{SW}} \nu_{1_1} \wedge \cdots \wedge \nu_{1_r} \wedge \phi_\Sigma^{d/2}, \tag{6.10}
$$

where ϕ_Σ are 2-forms on the SW moduli space. For simply connected manifolds $\pi_1(M) = 0$, the relevant cycles are of dimension $p = 2$. The Seiberg-Witten invariants can be also written in terms of differential forms in the moduli space $\mathcal{M}_{SW}$ in the form [12], [43]:

$$
\langle \mathcal{O}_1^{\gamma_2} \cdots \mathcal{O}_{d/2}^{\gamma_2} \rangle = \int_{\mathcal{M}_{SW}} \phi_\Sigma^{d/2}. \tag{6.11}
$$

7 SEIBERG-WITTEN INVARIANTS FOR FLOWS

In order to incorporate flows in the Seiberg-Witten theory, we define

$$
\widetilde{\mathcal{O}}_{\mathbf{Y}_p}(\mu) = \int_M i_{\mathbf{Y}_p}(\Theta_p^n)\mu_p(x), \tag{7.1}
$$

where μ_p is the invariant volume form. We can interpret the integral as the averaged asymptotic cycles on M by the Schwartzman theorem [18], $i_{\mathbf{Y}_p}(\mu)$ is a closed $(4-p)$-form, from which we will obtain a asymptotic p-cycle $\widetilde{\gamma}_p$ by Poincaré duality (an element of the $H_p(M, \mathbb{R})$).

Let $\mathbf{Y}_p$ be p-vector fields with $p = 0, 1, 2$, then the expression (7.1) defines the asymptotic observables as

$$
\widetilde{\mathcal{O}}_{\mathbf{Y}_1}^n(\mu_1) = \int_M i_{\mathbf{Y}_1}(\Theta_1^n)\,\mu_1, \qquad \widetilde{\mathcal{O}}_{\mathbf{Y}_2}^n(\mu_2) = \int_M i_{\mathbf{Y}_2}(\Theta_2^n)\,\mu_2, \tag{7.2}
$$

where Θ_p^n are given by (6.7).

Following the procedure we did in Eq. (4.9), it is an easy matter to check that these asymptotic observables $\widetilde{\mathcal{O}}_{\mathbf{Y}_p}(\mu_p)$ are $\mathcal{Q}$-invariant (BRST).

For an oriented manifold M with p_j-vectors fields $\mathbf{Y}_{p_j}$, with $\sum_{j=1}^{r} p_j = d(\mathcal{M}_{SW})$ and probability invariant measure, the r-point correlation functions for the flow generated $\mathbf{Y}_{p_j}$ and μ_{p_j} are given by

$$\left\langle \widetilde{\mathcal{O}}^n_{\mathbf{Y}_{p_1}}(\mu_{p_1}) \cdots \widetilde{\mathcal{O}}^n_{\mathbf{Y}_{p_r}}(\mu_{p_r}) \right\rangle = \int (\mathcal{D}\mathcal{X}) \exp(-L_{SW}/e^2) \prod_{j=1}^{r} \int_M i_{\mathbf{Y}_{p_j}}(\Theta^n_{p_j}) \mu_{p_j}.$$

(7.3)

This expression is reduced to the ordinary Seiberg-Witten invariants (6.9), when the measure is supported on the cycles. This means $\mu_p = \sum_{j=1}^{r} \mu_{p_j}$, where each μ_p is distributed uniformly over the cycles of M.

As in the Donaldson case, let us assume that the only zero modes correspond to the abelian gauge field A_μ and its BRST-like companion ψ_μ. Following a similar procedure, as in the Donaldson case, to compute the partition function, we get

$$\begin{aligned}
\left\langle \widetilde{\mathcal{O}}_{\mathbf{Y}_{p_j}}(\mu_{p_j}) \right\rangle &= \int_{\mathcal{M}_{SW}} da_1 \ldots da_n d\psi^1 \ldots d\psi^n \widetilde{\Phi}_{i_1 \cdots i_n}(a_i, \mathbf{Y}_{p_j}) \psi^{i_1} \ldots \psi^{i_n} \\
&= \int_{\mathcal{M}_{SW}} \widetilde{\Phi}_{\mathbf{Y}_{d(M)'}}
\end{aligned}$$

(7.4)

where $\widetilde{\Phi}_{\mathbf{Y}_{p_j}}(\mu_{p_j}) = \widetilde{\Phi}_{i_1 \cdots i_n}(a_i, \mathbf{Y}_{p_j}) \psi^{i_1} \ldots \psi^{i_n}$, the a_i are the zero modes of the gauge field and ψs are the zero modes of the fermionic field, and $\widetilde{\Phi}(a, \mathbf{Y}_{p_j})$ is a function that depends only on the zero modes of the gauge field and contains the information of the flow.

We integrate out a_i's and obtain a n-form $\widetilde{\Phi}$ defined in the moduli space. Now suppose that $\widetilde{\mathcal{O}} = \widetilde{\mathcal{O}}_{\mathbf{Y}_{p_1}}(\mu_{p_1}) \cdots \widetilde{\mathcal{O}}_{\mathbf{Y}_{p_r}}(\mu_{p_r})$ with $\sum_{p,j} p_j = n = d(\mathcal{M}_{SW})$ and p_j is the number of zero modes of $\widetilde{\mathcal{O}}_{\mathbf{Y}_{p_j}}(\mu_{p_j})$. These functions define forms in the moduli space in the following way:

$$\begin{aligned}
\left\langle \widetilde{\mathcal{O}}_{\mathbf{Y}_{p_1}}(\mu_{p_1}) \cdots \widetilde{\mathcal{O}}_{\mathbf{Y}_{p_r}}(\mu_{p_r}) \right\rangle &= \int_{\mathcal{M}_{SW}} \widetilde{\Phi}_{\mathbf{Y}_{p_1}} \wedge \cdots \wedge \widetilde{\Phi}_{\mathbf{Y}_{p_r}} \\
&= \int_{\mathcal{M}_{SW}} \widetilde{\nu}_{\mathbf{Y}_{p_1}} \wedge \cdots \wedge \widetilde{\nu}_{\mathbf{Y}_{p_r}}.
\end{aligned}$$

(7.5)

In the simply connected case ($\pi_1(M) = 0$), the important observables are those associated with cycles of zero dimension γ_0 and of dimension two γ_2. In general, a k_γ-cycle has associated an operator (form) with ghost number $U = 2 - k_\gamma$; this is the analog of the Donaldson map $H_k(M) \to H^{2-k}(\mathcal{M}_{SW})$. Finally, it is easy to see that for $k = 2$ the product of r operators yields

$$\begin{aligned}
\left\langle \widetilde{\mathcal{O}}_{\mathbf{Y}_1}(\mu_1)(x_1) \cdots \widetilde{\mathcal{O}}_{\mathbf{Y}_{d/2}}(\mu_{d/2}) \right\rangle &= \int_{\mathcal{M}_{SW}} \widetilde{\nu}_{\mathbf{Y}_1} \wedge \cdots \wedge \widetilde{\nu}_{\mathbf{Y}_{d/2}} \\
&= \# \left(H_{\widetilde{\Sigma}_{\mathbf{Y}_1}} \cap \ldots \cap H_{\widetilde{\Sigma}_{\mathbf{Y}_{d/2}}} \right),
\end{aligned}$$

(7.6)

where we have assumed the notation $\mathbf{Y}_{2_j} = \mathbf{Y}_j$ and $\mu_{2_j} = \mu_j$ for asymptotic 2-cycles.

These ideas can be generalized by considering a twisted version of the Yang-Mills theory with non-abelian gauge group, and this leads to non-abelian monopole equations [44], [45]. We consider the general case, when M is not Spin but a Spin$_c$ manifold. For one hypermultiplet the equations for a non-abelian connection A_μ coupled to a spinor $M_\alpha \in \Gamma(S^+ \otimes L^{1/2} \otimes E)$, where S^+ is the spin bundle, $L^{1/2}$ is the determinant line bundle of the Spin$_c$ structure and E is the vector bundle associated to a principal G-bundle via some representation of the gauge group. Then the equations for the moduli space are given by

$$F^{+a}_{\dot\alpha\dot\beta} + 4i\overline{M}_{(\dot\alpha}(T^a)M_{\dot\beta)} = 0, \quad (\nabla_E^{\alpha\dot\alpha}\overline{M}_{\dot\alpha}) = 0, \tag{7.7}$$

where T^a are the generators of the Lie algebra, and $\nabla_E^{\alpha\dot\alpha}$ is the Dirac operator constructed with the covariant derivative with respect to the gauge connection A_μ. Thus, it is possible to extend the set of observables (7.2) for the non-abelian case. Similar computations can be done to obtain the Seiberg-Witten invariants associated to non-abelian monopoles [44] for the case of compact Kähler manifolds, following [45] and using a procedure similar to the one described in Section 5.2.

7.1   Relation to Donaldson Invariants

The computation of Donaldson-Witten and Seiberg-Witten invariants for Kähler manifolds can be obtained by physical methods. The difference lies in the underlying physics of the dynamics of $\mathcal{N} = 2$ supersymmetric gauge theories. For the Donaldson-Witten case, it was necessary to add a mass term to soft breaking supersymmetry. This introduces a non-trivial canonical divisor defined as the zero locus of the mass term. However, for the Seiberg-Witten case, this is not necessary. The underlying dynamics describing the strong coupling limit of the gauge theory was elucidated in [10] through the implementation of S-duality. The dual theory was used later by Witten in [11] to find new invariants of four-manifolds, the Seiberg-Witten invariants. In that paper a relation between both invariants was found. This subsection contains the description of this relation when there are non-singular global flows on the manifold.

The relevant ingredients are the operators $\widetilde{I}_{\mathbf{Y}_a}$ and $\mathcal{O}$ inserted in M. Then the Donaldson invariants take the following form:

$$\left\langle \exp\left(\sum_a \alpha_a \widetilde{I}_{\mathbf{Y}_a}(\mu_a) + \lambda\mathcal{O} \right) \right\rangle = 2^{1 + \frac{1}{4}(7\chi + 11\sigma)} \left[\exp\left(\frac{1}{2}\widetilde{v}^2 + 2\lambda \right) \sum_{\widetilde{x}} \widetilde{SW}(\widetilde{x})e^{\widetilde{v}\cdot\widetilde{x}} \right.$$
$$\left. + i^\Delta \exp\left(-\frac{1}{2}\widetilde{v}^2 - 2\lambda \right) \cdot \sum_{\widetilde{x}} \widetilde{SW}(\widetilde{x})e^{-i\widetilde{v}\cdot\widetilde{x}} \right] \tag{7.8}$$

where α_a and λ are complex numbers,

$$\widetilde{v}^2 = \sum_{a,b} \alpha_a\alpha_b \#(\widetilde{\Sigma}_{\mathbf{Y}_a} \cap \widetilde{\Sigma}_{\mathbf{Y}_b}), \tag{7.9}$$

$$\widetilde{v} \cdot \widetilde{x} = \sum_a \alpha_a \#(\widetilde{\Sigma}_{\mathbf{Y}_a} \cap \widetilde{x}) \tag{7.10}$$

and $\widetilde{SW}$ is the asymptotic version of the Seiberg-Witten invariant (7.6).

Here $\#(\widetilde{\Sigma}_{\mathbf{Y}_a} \cap \widetilde{x})$ is the asymptotic intersection number between $\widetilde{\Sigma}_{\mathbf{Y}_a}$ and $\widetilde{x}$; it is given by

$$\#(\widetilde{\Sigma}_{\mathbf{Y}_a} \cap \widetilde{x}) = \int_{M_1 \times M_2} (\Theta_2 \wedge \eta_{\mathbf{Y}_a})(x_1) \wedge (x \wedge \eta_{\mathbf{X}})(x_2) \cdot \delta(x_1 - x_2), \qquad (7.11)$$

where $\eta_{\mathbf{Y}_a}$ and $\eta_{\mathbf{X}}$ are the Poicaré dual of $\widetilde{\Sigma}_a$ and $\widetilde{x}$.

In the previous equations, we used the definitions $x = -2c_1(L \otimes L) \in H^2(M, \mathbb{Z})$ and $\widetilde{x}_{\mathbf{X}} = \int_M i_{\mathbf{X}}(x)\mu$ with $\mathbf{X}$ being the 2-vector field wrapping $\widetilde{x}$.

8   A PHYSICAL INTERPRETATION

In this paper we have assumed the existence of "diffused" cycles in a suitable four-manifold. Due their relevance in dynamical systems on simply-connected four-manifolds, we have considered asymptotic 2-cycles $\widetilde{\Sigma}_{\mathbf{Y}_1}, \ldots, \widetilde{\Sigma}_{\mathbf{Y}_r}$ together with a set of invariant probability measures $\mu = \mu_1 + \cdots + \mu_r$, with μ_i supported in $\widetilde{\Sigma}_{\mathbf{Y}_i}$ on M. The present section is devoted to interpret this system in terms of string theory. Thus one would wonder if these 2-cycles can be interpreted as closed string probes propagating on the underlying four-manifold (which would be compact or non-compact). For this issue there is a nice response through the computation of the scattering amplitudes of an axion at zero momentum with a NS5-brane in the heterotic string theory [24]. In the present paper we follow this direction and we argue that Eq. (4.20) is a consequence of these considerations.

Let us consider a spacetime manifold $M^{1,9}$ provided with a set of Borel probability measures invariant under a non-singular smooth 2-flow generated by a 2-vector field Y. Moreover, we take the following splitting $M^{1,9} = M^{1,5} \times M$, where $M^{1,5}$ is a flat Minkowski space and M is the transverse space. We consider a NS fivebrane (NS5) as a solitonic object filling the space $M^{1,5}$. Thus, the transverse space M consists of a four-manifold parameterizing the positions of the NS5-brane. Our flow 2-orbit can be regarded as a closed string propagating on $M^{1,9}$ without necessarily being localized in a homology cycle of $M^{1,9}$. For the purposes of this paper we focus on a special situation by limiting the 2-flow to be defined only in the transverse space M and supported in the whole M. Consequently the Borel measures will be defined only on M. In this case we can think of the 2-flow as an asymptotic cycle $\widetilde{\Sigma}_{\mathbf{Y}}$) representing a "diffuse" closed string propagating in M and viewing the NS5 as a scattering center. If the probability measures are totally on M, then the NS5-brane sector will remain unchanged and the moduli space of instantons will remain the same. The effective action described in [24] is a non-linear sigma model on the worldvolume of the NS5 brane W and with target space the space $\mathcal{M}_N$. This latter space represents the space of static NS5 brane solutions, and it is equal to the moduli space of N Yang-Mills instantons $\mathcal{M}_N(M)$ over M. Thus, the ground states correspond to cohomology classes on $\mathcal{M}_N(M)$. If we identify the "diffuse" heterotic closed string with the asymptotic 2-cycles on M i.e. $\widetilde{\Sigma}_{\mathbf{Y}}$, then the action is given by

$$S = \int_M i_{\mathbf{Y}}(B)\mu. \qquad (8.1)$$

Then from the interacting terms coupling the B-field with the gauginos of the heterotic supergravity action we have

$$\widetilde{O}_{ij,\mathbf{Y}} = \int_M i_\mathbf{Y}(Z_{ij})\mu, \tag{8.2}$$

where $Z = \mathrm{tr}(\delta_i A \wedge \delta_j A - \phi_{ij}F)$.

Now we would like to consider multiple axion scattering with zero momentum. Then the transitions among the quantum ground states of the worldvolume W, for instance, from $|0\rangle$ to $|m\rangle$, induced by the scattering of r axions with the NS5-branes, is described by the scattering amplitude. The r axions represent r closed heterotic strings wrapping the homology cycles $\widetilde{\Sigma}_{\mathbf{Y}_1}, \dots, \widetilde{\Sigma}_{\mathbf{Y}_r}$, associated with the 2-vector fields $\mathbf{Y}_1, \dots, \mathbf{Y}_r$. Thus, the scattering amplitude is given by

$$\begin{aligned} \mathcal{A}(\mathbf{Y}_1, \cdots, \mathbf{Y}_r) &= \langle m|\widetilde{O}_{\mathbf{Y}_1} \cdots \widetilde{O}_{\mathbf{Y}_r}|0\rangle \\ &= \int_{\mathcal{M}_N} \widetilde{O}_{\mathbf{Y}_1} \wedge \cdots \wedge \widetilde{O}_{\mathbf{Y}_r}. \end{aligned} \tag{8.3}$$

Thus, we have deduced Eq. (4.20) from string theory.

Previously, we assumed that the set of Borel measures is distributed along the space M. This is consistent with the fact we have in the transverse directions (along $M^{1,5}$) filled with the NS5-brane, which is a solitonic object and consequently is very heavy in the perturbative regime; consequently, they are very difficult to excite.

S-duality between the heterotic and type I string interchanges NS5 by a D5-brane and axions by D1-branes [46]. The self-dual gauge field on M leads to the ADHM construction of instantons [47]. The interactions are now given by the Type I string action. It would be interesting to make a description of the asymptotic cycles within this context.

9 FINAL REMARKS

In the present paper we look for the implementation of the procedure followed in [22] for Jones-Witten invariants to compute invariants for flows in higher-dimensional manifolds. In this situation the relevant invariants of interest were the smooth invariants of four-manifolds, i.e., the Donaldson-Witten and the Seiberg-Witten invariants. We were able to obtain these invariants when some flows generated by non-singular and non-divergence-free smooth p-vector fields are globally defined on the four-manifold. We assumed that the homology cycles of M are described by the asymptotic cycles. We focus our work on simply-connected four-manifolds; thus, the only relevant flows are the 2-flows, though the invariants can be defined for any other p-flows. This is the situation that leads to a generalization of invariants of four-manifolds with flows.

In order to implement these considerations, we use the Witten cohomological field theory, whose observables are cohomology classes of M. In the presence of flows these observables were constructed as geometric currents underlying asymptotic cycles and foliations introduced by Ruelle and Sullivan [19] and Schwartzman [18]. Thus, the asymptotic observables and their correlation functions give rise to new smooth invariants for four-manifolds with a dynamical system with an invariant probability measure. That is, they represent smooth invariants for foliations, i.e., triples $(M, \mathcal{F}, \mu)$. This was done for Donaldson-Witten invariants (4.20) as well as for Seiberg-Witten invariants (7.6). Donaldson-Witten invariants are also obtained for the case of Kähler manifolds with flows, and some examples were described in Section 5.3. Finally, we attempt to give a physical interpretation in terms of string theory. We used the procedure outlined in [24] to obtain the invariants (4.20) as scattering amplitudes of r axions at the zero momentum with N coincident NS5-branes in the heterotic string theory. These axions are of special character, and they represent r 2-flows wrapping r homology 2-cycles of M. We gave only general remarks on this subject, and a further detailed analysis must be performed. This includes the incorporation of S-duality between the heterotic and type I string theory [46] and the uses of this structure [47] to construct an ADHM construction of instantons with flows. It would be interesting to include flows in terms of proper dynamical fields such as is the case in string theory such as the NS B-field and the RR fields and make similar consideration as the present paper, but this time in terms of specific interactions of the flow degrees of freedom. In this case it is possible to compute the back reaction of the fields of the theory to the flow.

There are several possible further generalizations of our work. One of them is the extension of asymptotic invariants to quantum cohomology [13], [14], [48] by considering asymptotic cycles in the target space of a topological non-linear sigma model of types A or B [14]. Of special interest is the possibility to define an asymptotic version of the Rozansky-Witten invariants [49]. This is due to the fact that their construction involves a topological sigma model on a three-manifold and target space being an hyper-Kähler manifold. The theory leads to link-invariants on the three-manifold with underlying structure group labeled by the hyper-Kähler structure. We would like to establish a relation with the results obtained in [22].

Another possibility is the consideration of flows generated by p-vector fields on supermanifolds. The analysis involves the computation of correlation functions with even and odd operators. This will constitute a supersymmetric extension of the work considered in the present paper. One more possible direction constitutes the implementation of the procedure to the computation of correlation functions of observables in the eight-dimensional generalization of the cohomological field theory [50], [51]. Some of these issues are already under current investigation and will be reported elsewhere.

Acknowledgements. We would like to thank the referee for carefully reading our manuscript and for giving us very important suggestions. The work of H. G.-C. is supported in part by the CONACyT grant 128761. The work of R. S. was supported by a CONACyT graduate fellowship. The work of A. V. was partially supported by CONACyT grant number 129280 and DGAPA-PAPIIT, Universidad Nacional Autónoma de México.

10　Bibliography

[1]　E. Witten, Commun. Math. Phys. **121**, 351 (1989).

[2]　E. Witten, "Fivebranes and Knots," arXiv:1101.3216 [hep-th].

[3]   S. K. Donaldson, Topology **29** 257 (1990).

[4]   S. K. Donaldson and P. B. Kronheimer, *The Geometry of Four-manifolds*, Oxford University Press, New York (1990).

[5]   E. Witten, Commun. Math. Phys. **117**, 353–386 (1988).

[6]   E. Witten, Phys. Lett. B **206**, 601 (1988); J. M. F. Labastida, M. Pernici and E. Witten, Nucl. Phys. B **310**, 611 (1988); E. Witten, Nucl. Phys. B **340**, 281 (1990).

[7]   K. G. O'Grady, J. Diff. Geom. **35**, 415–427 (1992).

[8]   P. B. Kronheimer and T. S. Mrowka, Bull. Am. Math. Soc. **30**, 215 (1994); J. Diff. Geom. **41**, 573 (1995).

[9]   E. Witten, J. Math. Phys. **35** (10), 5101–5135 (1994).

[10] N. Seiberg and E. Witten, Nucl. Phys. B **426**, 19 (1994); erratum, ibid. B **430**, 485 (1994); Nucl. Phys. B **431**, 484 (1994).

[11] E. Witten, Math. Res. Lett. **1**, 769 (1994).

[12] J.M. Labastida and M. Mariño, *Topological Quantum Field Theory and Four Manifolds*, Springer-Verlag (2005).

[13] E. Witten, Commun. Math. Phys. **118**, 411 (1988).

[14] E. Witten, "Mirror manifolds and topological field theory,"   arXiv:hep-th/9112056.

[15] M. Mariño, *Chern-Simons Theory, Matrix Models, and Topological Strings*, Oxford, Clarendon, Oxford, UK (2005).

[16] V. I. Arnold and B. A. Khesin, *Topological Methods in Hydrodynamics*, Springer-Verlag (1998).

[17] S. Schwartzman, Ann. Math. **66**(2), 270–284 (1957).

[18] S. Schwartzman, Canad. J. Math. **55**(3), 636–648 (2003).

[19] D. Ruelle and D. Sullivan, Topology **14**, 319–327 (1975).

[20] D. Sullivan, Invent. Math. **36**, 225–255 (1976).

[21] S. Schwartzman, Bull. London Math. Soc. **29**, 350–352 (1997).

[22] A. Verjovsky, R. Vila Freyer, Commun. Math. Phys. **163**, 73–88 (1994).

[23] H. García-Compeán and R. Santos-Silva, J. Math. Phys. **51**, 063506 (2010).

[24] J. A. Harvey and A. Strominger, Commun. Math. Phys. **151**, 221 (1993).

[25] G. de Rham, *Differentiable Manifolds: Forms, Currents, Harmonic Forms*, Springer-Verlag, Berlin, Heidelberg (1984).

[26] A. Candel, L. Conlon, *Foliations I*, Graduate Studies in Mathematics, **23**. American Mathematical Society, Providence, RI (2000).

[27] J. Plante, *Foliations With Measure Preserving Holonomy*, Annals of Math. Second Series, **102**(2), 327–361 (Sept., 1975).

[28] M. F. Atiyah, N. J. Hitchin and I. M. Singer, Proc. R. Soc. Lond. A **362**, 425 (1978).

[29] H. Kanno, Z. Phys. C **43**, 477 (1989).

[30] M. F. Atiyah, L. Jeffrey, J. Geom. Phys. **7**(1), 120 (1990).

[31] V. Mathai and D. Quillen, Topology **25**, 85 (1986).

[32] I. Vaisman, *Lectures on the Geometry of Poisson Manifolds*, Birkhauser, Boston, MA (1994).

[33] M. Holm, "New insights in brane and Kaluza-Klein theory through almost product structures," arXiv:hep-th/9812168v1.

[34] J. S. Park, Commun. Math. Phys. **163**, 113 (1994).

[35] P. C. Argyres and A. E. Faraggi, Phys. Rev. Lett. **74**, 3931 (1995).

[36] G. van der Geer, *Hilbert modular surfaces*. Ergebnisse der Mathematik und ihrer Grenzgebiete. **3**, 16. Springer-Verlag, Berlin (1988).

[37] A. Selberg, *On discontinuous groups in higher-dimensional symmetric spaces*. Contributions to function theory (internat. Colloq. Function Theory, Bombay, 1960) pp. 147–164. Tata Institute of Fundamental Research, Bombay.

[38] F. Hirzebruch, Enseignement Math. **2** 19, 183–281 (1973).

[39] R. Friedman, J. Morgan, *Smooth four-manifolds and complex surfaces*, Ergebnisse der Mathematik und ihrer Grenzgebiete **3**, 27. Springer-Verlag, Berlin (1994).

[40] S. K. Donaldson, J. Diff. Geom. **53**, no. 2, 205–236, (1999).

[41] H. B. Lawson and M.-L. Michelsohn, *Spin Geometry*, Princeton Mathematical Series **38**. Princeton University Press, Princeton, NJ (1989),

[42] D. Salamon, *Spin Geometry and Seiberg-Witten Invariants*, University of Warwick (1995).

[43] J. M. F. Labastida and M. Marino, Phys. Lett. B **351**, 146 (1995).

[44] J. M. F. Labastida and M. Marino, Nucl. Phys. B **448**, 373 (1995).

[45] J. M. F. Labastida and M. Marino, Nucl. Phys. B **456**, 633 (1995).

[46] J. Polchinski and E. Witten, Nucl. Phys. B **460**, 525 (1996).

[47] E. Witten, Nucl. Phys. B **460**, 541 (1996).

[48] M. Kontsevich and Yu. Manin, Commun. Math. Phys. **164**, 525 (1994).

[49] L. Rozansky and E. Witten, Selecta Math. **3**, 401 (1997).

[50] L. Baulieu, H. Kanno, and I. M. Singer, Commun. Math. Phys. **194**, 149 (1998).

[51] B. S. Acharya, J. M. Figueroa-O'Farrill, B. J. Spence and M. O'Loughlin, Nucl. Phys. B **514**, 583 (1998).

Color Plates

PLATE 1: Jack Milnor at Lake Louise, February 2011.

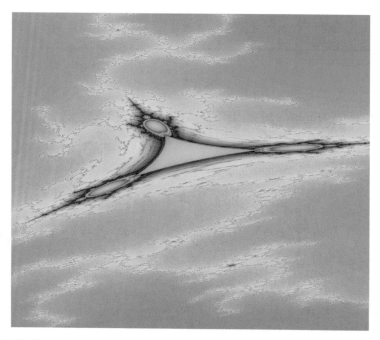

PLATE 2: A "little tricorn" within the tricorn $\mathcal{M}_2^*$ illustrating that the "umbilical cord" converges to the little tricorn without landing at it. *(See page 73.)*

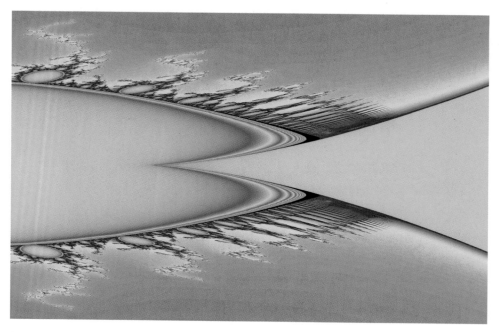

PLATE 3: Decorations at a period 2 component of the tricorn that accumulate along arcs on the boundary of the period 1 component. *(See page 100.)*

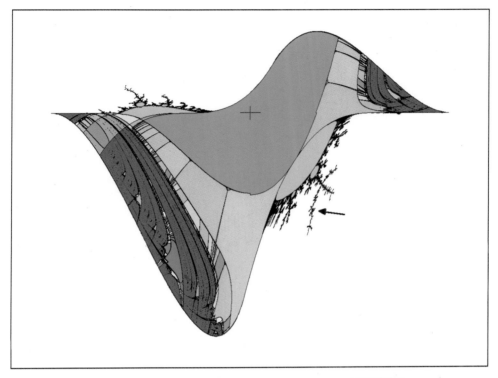

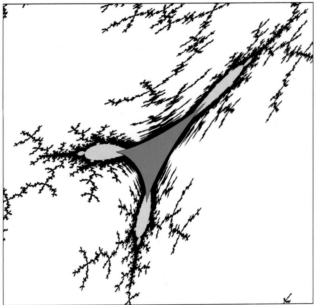

PLATE 4: The connectedness locus of real cubic polynomials and a detail from the southeast quadrant, showing a tricorn-like structure. *(See page 75.)*
Figures courtesy J. Milnor.

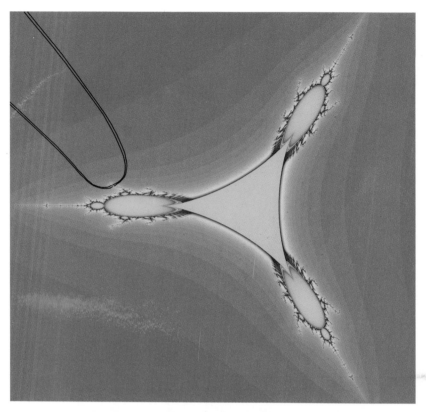

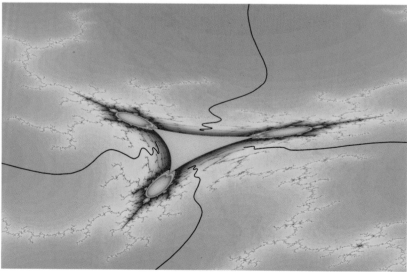

PLATE 5: The tricorn and a blow-up showing a "small tricorn" of period 5. Shown in both pictures are the four parameter rays accumulating at the boundary of the period 5 hyperbolic component (at angles 371/1023, 12/33, 13/33, and 1004/1023). The wiggly features of these non-landing rays are clearly visible in the blow-up. *(See page 76.)*

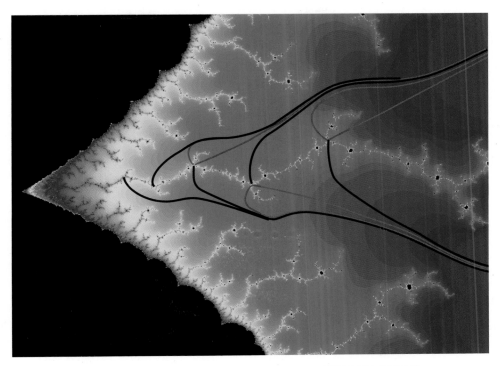

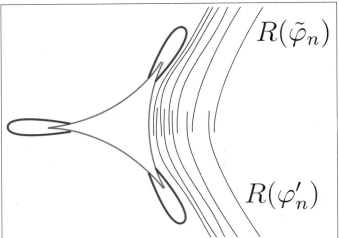

PLATE 6: Loss of pathwise connectivity in the tricorn because of approximating overlapping rays. Top: approximating preperiodic dynamic rays in the dynamic plane with a parabolic orbit. Only the rays drawn by heavy lines are used in the argument; other rays landing at the same points are drawn in grey. Below: symbolic sketch of the situation in the parameter space. *(See page 96.)*

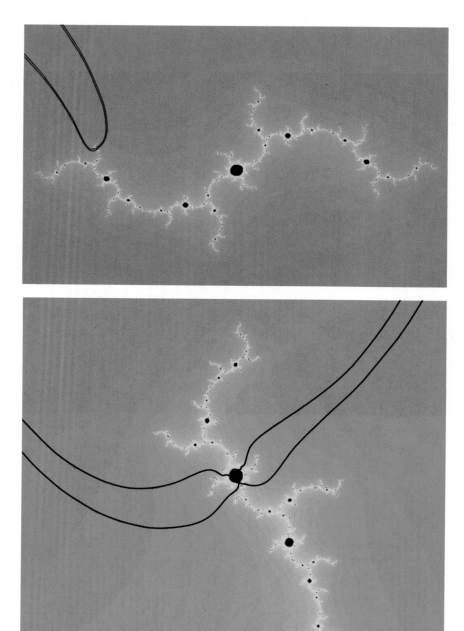

PLATE 7: An antiholomorphic map $p_c(z) = \bar{z}^2 + c$ of degree $d = 2$ with an attracting cycle of period 5. The Fatou component containing the critical value has $d + 1 = 3$ boundary points that are fixed under $p_c^{\circ 5}$, and together these are the landing points of $d + 2 = 4$ dynamic rays: the dynamic root is the landing point of 2 rays (here, at angles $371/1023$ and $404/1023$ of period 10, and the two dynamic 2-roots are the landing points of one ray each (at angles $12/33$ and $13/33$ of period 5). Above is the entire Julia set with the four rays indicated; the lower figure is a blow-up of a neighborhood of the critical value Fatou component where the four rays can be distinguished. *(See page 90.)*

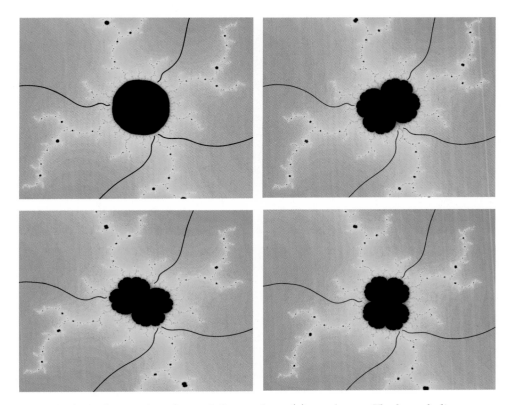

PLATE 8: As in the previous figure, Julia sets for $p_c(z)$ are shown. The hyperbolic component containing the parameter c is bounded by $d + 1$ parabolic arcs (see Plate 5): one arc
contains the accumulation set of the parameter rays at angles $12/33$ and $13/33$ (the root
arc), and the other two arcs contain the accumulation sets of one parameter ray each (at
angles $12/33$ and $13/33$ respectively). The four pictures above show blow-ups near the
critical value for parameters at the center (top left), from the parabolic root arc (top right)
and from the two parabolic co-root arcs (bottom row). *(See page 90.)*

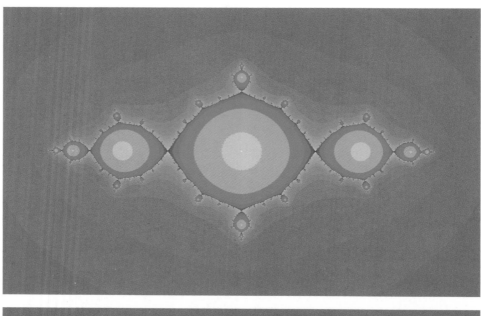

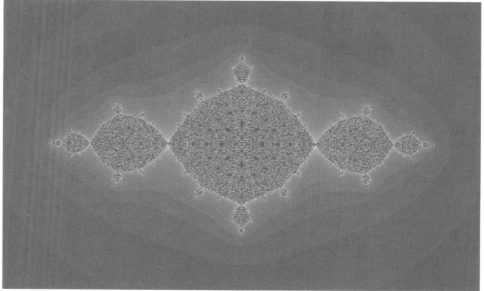

PLATE 9: The Julia sets for $z^2 - 1 + \lambda/z^2$ where $\lambda = 0$ and $\lambda = -0.00001$. *(See page 131.)*

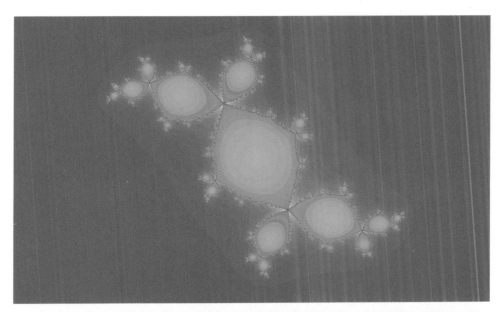

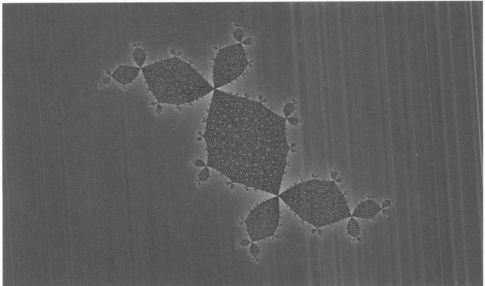

PLATE 10: The Julia sets for $z^2 - 0.122 + 0.745i + \lambda/z^2$ where $\lambda = 0$ and $\lambda = -0.000001$. *(See page 132.)*

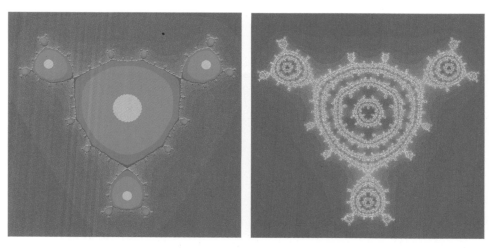

PLATE 11: The Julia set for the unperturbed map $z^3 - i$. and for $z^3 - i + 0.0001/z^3$. *(See page 133.)*

PLATE 12: A magnification of the Julia set for $z^3 - i + 0.0001/z^3$. *(See page 133.)*

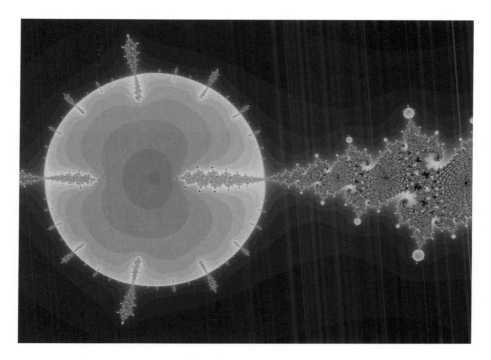

PLATE 13: The Julia set of $g_a(z) = z^3(z-a)/(1-az)$ for a value a near 5. *(See page 145.)*

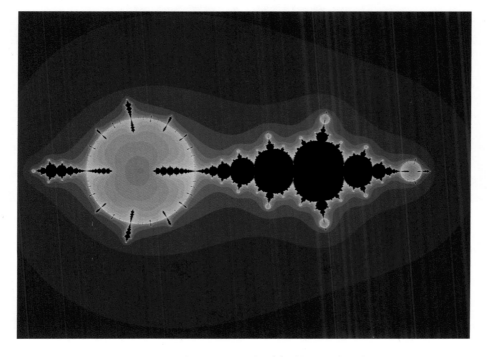

PLATE 14: The Julia set of $g_5(z)$. *(See page 145.)*

PLATE 15: The Julia set of $f_t(z) = e^{2i\pi t}z^2(z-4)/(1-4z)$, where t approximates a map with a non-Brjuno rotation number. One can imagine the hairy circle. *(See page 151.)* Image courtesy H. Inou.

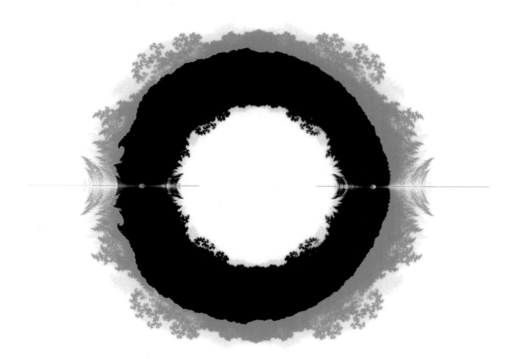

PLATE 16: Pictured here is the connectedness locus of a class of affine fractals, drawn from a sample size of about 10^9; all Galois conjugates of $\exp(h(f))$ for a postcritically finite quadratic map f are in the set. The limits of roots of integer polynomials with coefficients ± 1 are indicated in black or green, and the limits of roots of integer polynomials with coefficients $1, 0, -1$ are shown in black, green, or yellow. *(See page 340.)*

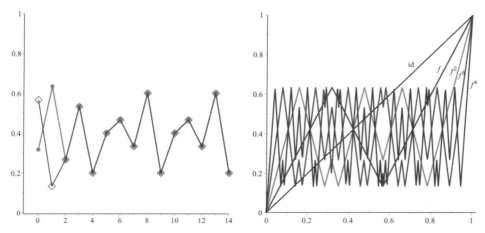

PLATE 17: The diagram at left shows the postcritical orbits for the Pisot construction where $\lambda = 2, d = 3$, with the critical points chosen as $19/60$ and $17/30$. On the right is the plot of the first four iterates of this piecewise linear function of entropy $\log(2)$. *(See page 348.)*

$(\lambda, \lambda') = (1.61803, -.61803)$

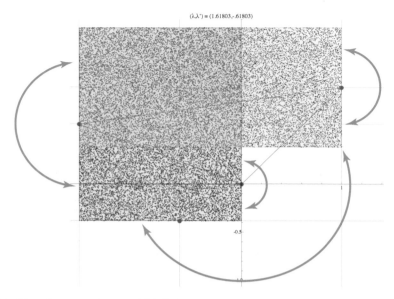

PLATE 18: This diagram shows $L(f^{1+\sigma})$ where $\lambda = 1.61803\ldots$, the golden ratio, is a Pisot number, and $\sigma(\lambda) = -1/\lambda$. It was drawn by taking a random point near the origin, first iterating it 15,000 times and then plotting the next 30,000 images. The dynamics is hyperbolic and ergodic, so almost any point would give a very similar picture. The critical point is periodic of period 3, and the limit set is a finite union of rectangles with a critical point on each vertical side. The dynamics reflects the light blue rectangle on the upper right in a horizontal line, and arranges it as the pink rectangle on the lower left. The big rectangle formed by the two stacked rectangles at left is reflected in a vertical line, stretched horizontally and squeezed vertically, and arranged as the two side-by-side rectangles on the top. Although the dynamics is discontinuous, the sides of the figure can be identified, as indicated (partially) by the green arrows to form a tetrahedron to make it continuous. *(See page 361.)*

PLATE 19: This is the diagram from the figure above taped together into a tetrahedron, where the dynamics acts continuously (but reverses orientation). The map $f^{1+\sigma}$ acts on the tetrahedron as an Anosov diffeomorphism. There are coordinates in which it becomes the Fibonacci recursion $(s, t) \mapsto (t, s + t)$, modulo a $(2, 2, 2, 2)$ symmetry group generated by $180°$ rotations about lattice points.

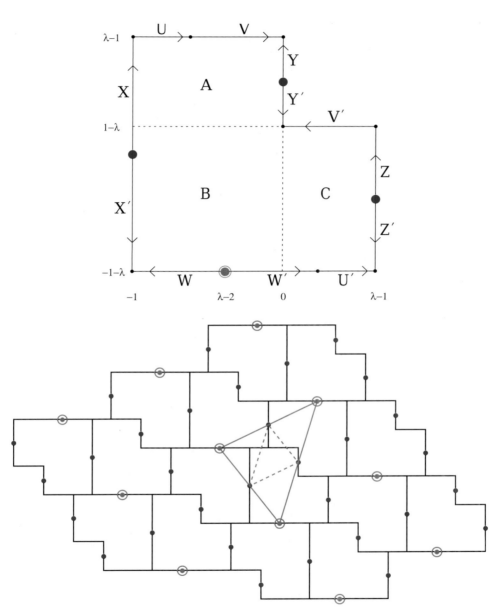

PLATE 20: Here is an alternative way of describing Thurston's construction as shown in Plates 18 and 19. The fundamental domain (top) is a reflected version of Thurston's original, interchanging top and bottom. The period three orbit is marked in blue, and the fixed point in red.

Now a tiling of the plane can be formed by applying the group of Euclidean motions generated by 180° rotation about the four marked periodic points. After translating the fixed point to the origin, the extended map is given simply by $(x, y) \mapsto (-\lambda x, (\lambda - 1)y)$. The red triangle forms an alternative fundamental domain for this action, and folding along the dotted lines gives the tetrahedron model of plate 19. *(See page 381.)*

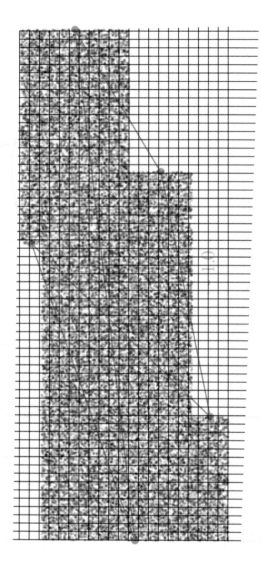

PLATE 21: This diagram shows $L(f^{1+\sigma})$ where $\lambda = 1.7220838\ldots$ is a Salem number of degree 4 satisfying $\lambda^4 - \lambda^3 - \lambda^2 + \lambda - 1 = 0$, and $\sigma(\lambda) = 1/\lambda$. The critical point is periodic of period 5, and the limit set is a finite union of rectangles with a critical point on each vertical side. The dynamics multiplies the left half (to the left of the vertical line through the uppermost red dot) by the diagonal matrix with entries $(-\lambda, -1/\lambda)$, then translates until the lowermost red dot goes to the uppermost red dot. The right half of the figure is multiplied by the diagonal matrix $(\lambda, 1/\lambda)$, and translated to fit in the lower left. As with the example in the previous figure, it can be folded up, starting by folding the vertical sides at the red dots, to form (topologically) an S^2 on which the dynamics acts continuously, a pseudo-Anosov map of the $(2, 2, 2, 2, 2)$-orbifold. This phenomenon has been explored, in greater generality, by André de Carvalho and Toby Hall, and it is also related to the concept of the dual of a hyperbolic groupoid developed by Nekrashevych. *(See page 362.)*

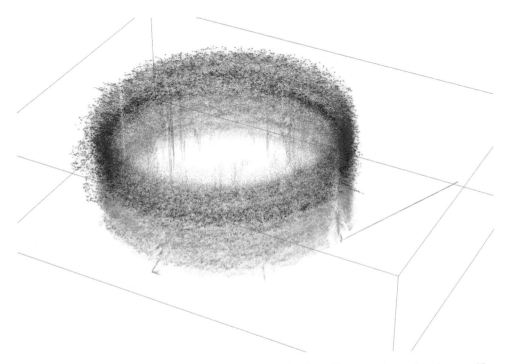

PLATE 22: This picture shows the parameter values for friendly extensions of tent maps (the space $F(1,2)$). The vertical direction is the λ-axis. The horizontal plane is $\mathbf{C}$, and the points shown are roots of minimal polynomials for postcritically finite examples. The points are color-coded by value of λ (height). With a finer resolution and expanded vertical scale, the friendly extensions would appear as a network of very frizzy hairs (usually of Hausdorff dimension > 1), sometimes joining and splitting, but always transverse to the horizontal planes. The diagonal line on the right is λ itself. *(See page 364.)*

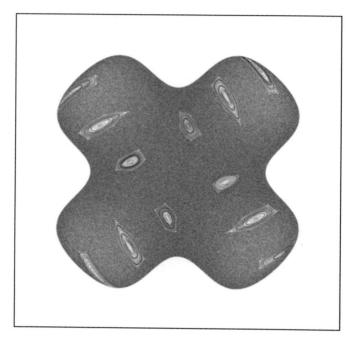

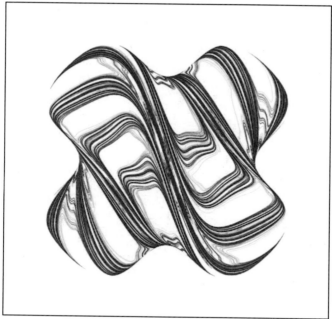

PLATE 23: A surface X defined by a polynomial with real coefficients. The automorphism $f = s_1 \circ s_2 \circ s_3$ preserves the real part $X(\mathbb{R})$. On the top, several orbits are plotted, while below, an approximation of a stable manifold of a saddle fixed point is drawn. *(See page 465.)* Picture realized by V. Pit, based on a program by C. T. McMullen.

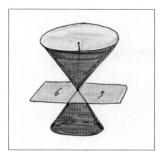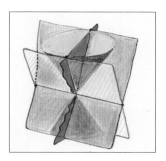

PLATE 24: Three types of automorphisms (from left to right): elliptic, parabolic, and loxo-dromic. Elliptic isometries preserve a point in $\mathbb{H}_m$ and act as a rotation on the orthogonal complement. Parabolic isometries fix an isotropic vector v; all positive and negative orbits in $\mathbb{H}_m$ converge toward the line $\mathbb{R}v$; the orthogonal complement of $\mathbb{R}v$ contains it and is tangent to the isotropic cone. Loxodromic isometries dilate an isotropic line, contract an-other one, and act as a rotation on the intersection of the planes tangents to the isotropic cone along those lines. *(See page 472.)*

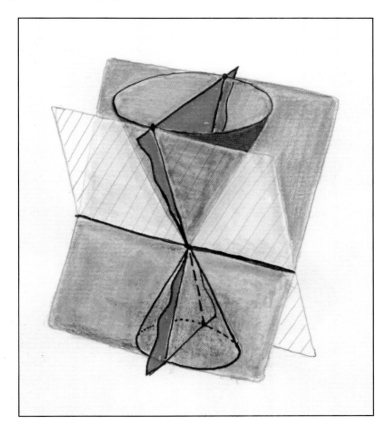

PLATE 25: Action of a loxodromic automorphism on $H^{1,1}(X, \mathbb{R})$. The two invariant isotropic lines $\mathbb{R}v_f^+$ and $\mathbb{R}v_f^-$ generate an orange plane; the vector spaces $(\mathbb{R}v_f^+)^\perp$ and $(\mathbb{R}v_f^-)^\perp$ are tangent to the isotropic cone, and their intersection N_f is the orthogonal com-plement to the orange plane. *(See page 473.)*

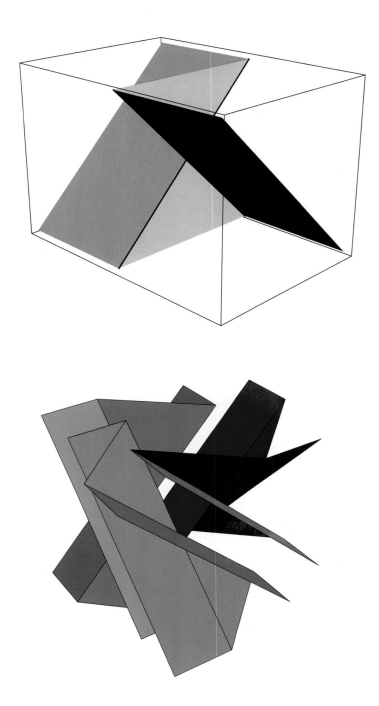

PLATE 26: A crooked plane, and a family of three pairwise disjoint crooked planes. *(See page 693.)*

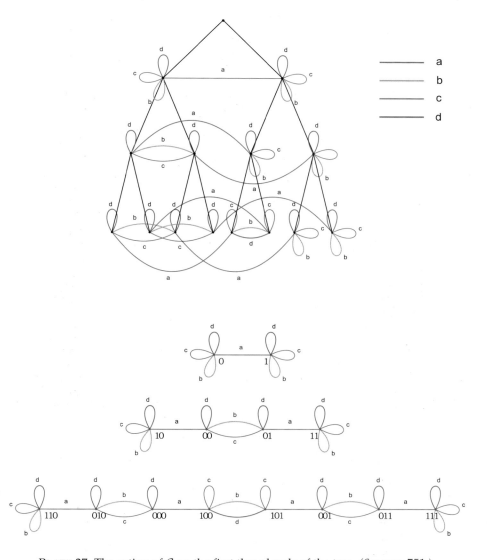

PLATE 27: The action of $\mathcal{G}$ on the first three levels of the tree. *(See page 751.)*

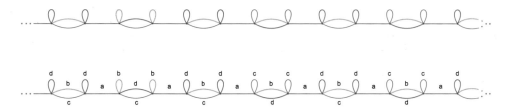

PLATE 28: Typical Schreier graph of the boundary action. *(See page 751.)*

PLATE 29: Attendees at the "Frontiers in Complex Dynamics" conference held in Banff, February 2011. Pictured are (roughly left to right):
Adam Epstein, Nikita Selinger, Sarah Koch, Hexi Ye, Bill Goldman, Misha Yampolsky, Jan Kiwi, Igors Gorbovickis Jasmin Raissy, John Hubbard, Mitsu Shishikura, Tommin Shishikura, Turgay Bayraktar, Tanya Firsova, André de Carvalho, Jarek Kwapisz, Tan Lei, Jan-Li Lin, Liz Vivas, Rodrigo Robles Montero, Pascale Roesch, Slava Grigorchuk, Stas Smirnov, Serge Cantat, Genadi Levin, Ilia Binder, Charles Favre, Alberto Verjovsky, David Ni, Vladlen Timorin, Daniel Smania, Matthieu Arfeux, Qian Yin, Dzmitry Dudko, Paul Blanchard, Joshua Bowman, Volodymyr Nekrashevych, Paul Reschke, Linda Keen, Ross Ptacek, Sebastian van Strien, Nishant Chandgotia, Grzegorz Świrszcz, Lucille Meci, Lex Oversteegen, Neil Dobbs, Sylvain Bonnot, Jack Milnor, Gerri Sciulli, Roman Dujardin, Krzysztof Barański, Magnus Aspenberg, Victor Kleptsyn, Dierk Schleicher.

PLATE 30: Attendees the "Frontiers in Complex Dynamics" conference (continued): Michael Benedicks, Nikolay Dimitrov, Anna Benini, Tom Milnor, Luka Bok Thaler, Goran Drazic, Polina Vytnova, Stergios Antonakoudis, Misha Lyubich, Jeremy Kahn, Hiroyuki Inou, Han Peters, Shaun Bullett, Monica Moreno Rocha, Artem Dudko, Daniel Meyer, Rodrigo Perez, Mary Rees, Trevor Clark, Todd Woodard, Anna Zdunik, Eva Uhre, Bill Thurston, Ángeles Sandoval-Romero, Clinton Curry, Ricardo Pérez-Marco, Remus Radu, Logan Hoehn, Bob Devaney, Arnaud Chéritat, Raluca Tanase, Hannah Lyubich, John Mayer, Vaibhav Gadre, Mark Comerford, Nessim Sibony, Clara Sullivan, Dusa McDuff, Alexander Blokh, Curt McMullin, Eric Bedford, Thomas Gauthier, Luna Lomanaco, Xavier Buff, Dennis Sullivan, Bruce Kitchens, Elizabeth Russell, Giulio Tiozzo, Carsten Petersen, Roland Roeder, Dan Cuzzocreo, Shrihari Sridharan, Elizabeth Fitzgibbon, Phil Mummert, Araceli Bonifant, Lilya Lyubich, Scott Sutherland, Yunping Jiang, Hrant Hakobyan.

Frontiers in Complex Dynamics

Celebrating John Milnor's 80th Birthday

Feb. 21-25, 2011
Banff, Canada

For further details, please see the conference webpage at

http://www.math.sunysb.edu/jackfest

Support has been kindly provided by the Simons Foundation, the National Science Foundation, the University of Rhode Island, the Banff International Research Station, Stony Brook University, IBM, and the Clay Mathematics Institute.

Part V

Geometry and Algebra

Two papers which changed my life: Milnor's seminal work on flat manifolds and bundles

William M. Goldman

ABSTRACT. We survey developments arising from Milnor's 1958 paper, "On the existence of a connection with curvature zero" and his 1977 paper, "On fundamental groups of complete affinely flat manifolds."

1 INTRODUCTION

For a young student studying topology at Princeton in the mid-1970s, John Milnor was a inspiring presence. The excitement of hearing him lecture at the Institute for Advanced Study and reading his books and unpublished lecture notes available in Fine Library made a deep impact on me. One heard rumors of exciting breakthroughs in the Milnor-Thurston collaborations on invariants of 3-manifolds and the theory of kneading in 1-dimensional dynamics. The topological significance of volume in hyperbolic 3-space and Gromov's proof of Mostow rigidity using simplicial volume were in the air at the time (later to be written up in Thurston's notes [69]). When I began studying geometric structures on manifolds, my mentors Bill Thurston and Dennis Sullivan directed me to two of Milnor's papers [63, 61]. Like many mathematicians of my generation, his papers and books, such as *Morse theory, Characteristic Classes,* and *Singular Points of Complex Hypersurfaces,* were very influential for my training. Furthermore, his lucid writing style made his papers and books role models for exposition.

I first met Jack in person when I was a graduate student in Berkeley and he was visiting his son. Several years later I was extremely flattered when I received a letter from him, where he, very politely, pointed out a technical error in my Bulletin Announcement [41]. It is a great pleasure and honor for me to express my gratitude to Jack Milnor for his inspiration and insight, in celebration of his eightieth birthday.

2 GAUSS-BONNET BEGINNINGS

2.1 Connections and characteristic classes

The basic topological invariant of a closed orientable surface M^2 is its Euler characteristic $\chi(M)$. If M has genus g, then $\chi(M) = 2 - 2g$. Give M a Riemannian metric — then the Gauss-Bonnet theorem identifies $\chi(M^2)$ as $1/2\pi$ of the total Gaussian curvature of M, a geometric invariant. This provides a fundamental topological restriction of what kind of geometry M may support.

This paper was presented at the workshop "Frontiers in Complex Dynamics" at the Banff International Research Station, in Banff, Alberta, Canada. I gratefully acknowledge partial support from National Science Foundation grant DMS070781.

For example, if the metric is flat — that is, locally Euclidean — then its Gaussian curvature vanishes and, therefore, $\chi(M) = 0$. Since M is orientable, M must be homeomorphic to a torus.

In 1944 Chern [26] proved his intrinsic Gauss-Bonnet theorem, which expresses the topological invariant $\chi(M^n)$ as the integral of a differential form constructed from the curvature of a Riemannian metric. More generally, if ξ is an oriented n-plane bundle over M^n, then its *Euler number*,

$$e(\xi) \in H^n(M, \mathbb{Z}) \cong \mathbb{Z},$$

can be computed as an integral of an expression derived from an *orthogonal connection* ∇ on ξ. (Compare Milnor-Stasheff [64].) For example, take to M to be a (pseudo-) Riemannian manifold, with tangent bundle $\xi = TM$ and ∇ the Levi-Civita connection. If ∇ has curvature zero, then, according to Chern, $\chi(M) = 0$.

This paradigm generalizes. When M is a complex manifold, its tangent bundle is a holomorphic vector bundle and the Chern classes can be computed from the curvature of a holomorphic connection. In particular, $\chi(M)$ is a Chernnumber. Therefore, if TM has a flat holomorphic connection, then $\chi(M) = 0$. If M has a flat pseudo-Riemannian metric, then a similar Gauss-Bonnet theorem holds (Chern [27]) and $\chi(M) = 0$. Therefore, it is natural to ask whether a compact manifold whose tangent bundle admits a flat *linear connection* has Euler characteristic zero.

2.2 Smillie's examples

In 1976, John Smillie [65] constructed, in every even dimension $n > 2$, a compact n-manifold such that TM admits a flat connection ∇. However, the torsion of ∇ is (presumably) nonzero. (None of Smillie's examples are *aspherical*; it would be interesting to construct a closed aspherical manifold with flat tangent bundle; compare Bucher-Gelander [14].)

Requiring the torsion of ∇ to vanish is a natural condition. When both curvature and torsion vanish, the connection arises from an *affine structure* on M, that is, the structure defined by a coordinate atlas of coordinate charts into an affine space E such that the coordinate changes on overlapping coordinate patches are locally affine. A manifold together with such a geometric structure is called an *affine manifold.* The coordinate charts globalize into a *developing map*

$$\tilde{M} \xrightarrow{\text{dev}} \mathsf{E},$$

where $\tilde{M} \longrightarrow M$ is a universal covering space. The developing map dev is a local diffeomorphism (although generally not a covering space onto its image), which defines the affine structure. Furthermore dev is equivariant with respect to a homomorphism

$$\pi_1(M) \xrightarrow{\rho} \mathsf{Aff}(\mathbb{R}^2)$$

(the *affine holonomy representation*) where the fundamental group $\pi_1(M)$ acts by deck transformations of $\tilde{M}$. Just as dev globalizes the coordinate charts, ρ globalizes the *coordinate changes.* The flat connection on TM arises from the representation ρ in the standard way: TM identifies with the fiber product

$$\xi_\rho := \tilde{M} \times_\rho \mathbb{R}^n = (\tilde{M} \times \mathbb{R}^n)/(\pi_1(M)),$$

where $\pi_1(M)$ acts *diagonally* — by deck transformations on the $\tilde{M}$ factor and via ρ on the $\mathbb{R}^n$-factor. The differential of dev defines an isomorphism of TM with ξ_ρ.

We may interpret Smillie's examples in this description as follows. In each even dimension $n > 2$, Smillie constructs an n-manifold M^n and a representation $\pi_1(M) \xrightarrow{\rho} \mathrm{Aff}(\mathbb{R}^n)$ such that ξ_ρ is isomorphic to the tangent bundle TM. Sections ξ_ρ correspond to *singular developing maps* which may be smooth, but *not* necessarily local diffeomorphisms.

Despite the many partial results, we know no example of a closed affine manifold with *nonzero Euler characteristic*.

2.3   Benzécri's theorem on flat surfaces

In dimension two, a complete answer is known, due to the work of Benzécri [10]. This work was part of his 1955 thesis at Princeton, and Milnor served on his thesis committee.

THEOREM (Benzécri). *A closed surface M admits an affine structure if and only if* $\chi(M) = 0$.

(Since every connected orientable open surface can be immersed in $\mathbb{R}^2$, pulling back the affine structure from $\mathbb{R}^2$ by this immersion gives an affine structure. With a small modification of this technique, every connected nonorientable open surface can also be given an affine structure.)

Benzécri's proof is geometric and starts with a fundamental polygon Δ for $\pi_1(M)$ acting on $\tilde{M}$. The boundary $\partial\Delta$ consists of various edges, which are paired by homeomorphisms, reconstructing M as the quotient space by these identifications. One standard setup for a surface of genus g uses a $4g$-gon for Δ, where the sides are alternately paired to give the presentation

$$\pi_1(M) = \langle A_1, B_1, \ldots, A_g, B_g \mid A_1 B_1 A_1^{-1} B_1^{-1} \ldots A_g B_g A_g^{-1} B_g^{-1} = 1 \rangle. \qquad (2.1)$$

The developing map dev immerses Δ into $\mathbb{R}^2$, and the identifications between the edges of $\partial\Delta$ are realized by orientation-preserving affine transformations.

Immersions of S^1 into $\mathbb{R}^2$ are classified up to regular homotopy by their *turning number* (the Whitney-Graustein theorem [77]), which measures the total angle the tangent vector (the velocity) turns as the curve is traversed. Since the restriction $\mathrm{dev}|_{\partial\Delta}$ extends to an immersion of the disc Δ, its turning number (after choosing compatible orientations) is

$$\tau(\mathrm{dev}|_{\partial\Delta}) = 2\pi. \qquad (2.2)$$

However, Benzécri shows that for any smooth immersion $[0,1] \xrightarrow{f} \mathbb{R}^2$ and orientation-preserving affine transformation γ,

$$|\tau(f) - \tau(\gamma \circ f)| < \pi. \qquad (2.3)$$

Using the fact that $\partial\Delta$ consists of $2g$ pairs of edges which are paired by $2g$ orientation-preserving affine transformations, combining (2.2) and (2.3) implies $g = 1$.

Milnor realized the algebraic-topological ideas underlying Benzécri's proof, thereby initiating the theory of characteristic classes of flat bundles.

3 THE MILNOR-WOOD INEQUALITY

3.1 "On the existence of a connection with curvature zero"

In his 1958 paper [63], Milnor shows that a closed 2-manifold M has a flat tangent bundle if and only if $\chi(M) = 0$. This immediately implies Benzécri's theorem, although it doesn't use the fact that the developing map is nonsingular (or, equivalently, the associated flat connection is torsion-free). In this investigation, Milnor discovered, remarkably, flat, oriented $\mathbb{R}^2$-bundles over surfaces M with *nonzero Euler class*. In particular the Euler class *cannot* be computed from the curvature of a linear connection.

Oriented $\mathbb{R}^2$-bundles ξ over M are classified up to isomorphism by their Euler class

$$e(\xi) \in H^2(M; \mathbb{Z}),$$

and if M is an orientable surface, an orientation on M identifies $H^2(M; \mathbb{Z})$ with $\mathbb{Z}$. (See Milnor-Stasheff [64] for details.) An oriented $\mathbb{R}^2$-bundle ξ admits a flat structure if and only if it arises from a representation

$$\pi_1(M) \overset{\rho}{\to} \mathrm{GL}^+(2, \mathbb{R})$$

(where $\mathrm{GL}^+(2, \mathbb{R})$ denotes the group of orientation-preserving linear automorphisms of $\mathbb{R}^2$). Milnor shows that ξ admits a flat structure if and only if its Euler number satisfies

$$|e(\xi)| < g. \tag{3.1}$$

Since $e(TM) = \chi(M) = 2 - 2g$, Milnor's inequality (3.1) implies that $g = 1$.

The classification of S^1-bundles is basically equivalent to the classification of rank 2 vector bundles but is somewhat more general. To any vector bundle ξ with fiber $\mathbb{R}^n$ is associated an S^{n-1}-bundle: the fiber of the S^{n-1}-bundle over a point x consists of all directions in the fiber $\xi_x \approx \mathbb{R}^n$. In particular, two $\mathbb{R}^2$-bundles are isomorphic if and only if their associated S^1-bundles are isomorphic. Therefore, we henceforth work with S^1-bundles, slightly abusing notation by writing ξ for the S^1-bundle associated to ξ.

3.2 Wood's extension and foliations

In 1971, John W. Wood [78] extended Milnor's classification of flat 2-plane bundles to flat S^1-bundles. Circle bundles with structure group $\mathrm{GL}^+(2, \mathbb{R})$ have an important special property. The *antipodal map* associates a direction in a vector space its opposite direction. Since all linear transformations commute with it, the antipodal map defines an involution on any vector bundle or associated sphere bundle ξ. The quotient is the associated $\mathbb{RP}^1$-bundle $\hat{\xi}$, and the quotient map $\xi \longrightarrow \hat{\xi}$ is a double covering. This is also an oriented S^1-bundle, with Euler class

$$e(\hat{\xi}) = 2e(\xi).$$

Wood [78] determines the flat oriented S^1-bundles for an arbitrary homomorphism

$$\pi_1(\Sigma) \overset{\rho}{\to} \mathrm{Homeo}^+(S^1).$$

He proves the Euler number satisfies the following inequality:

$$|e(\rho)| \leq -\chi(\Sigma) \tag{3.2}$$

(now known as the *Milnor-Wood inequality*). Furthermore, every integer in the interval $[\chi(\Sigma), -\chi(\Sigma)]$ occurs as $e(\rho)$ for some homomorphism ρ.

Milnor's proof interprets the Euler class as the obstruction for lifting the holonomy representation ρ from the group $GL^+(2, \mathbb{R})$ of linear transformations of $\mathbb{R}^2$ with positive determinant to its universal covering group $\widetilde{GL^+(2, \mathbb{R})}$. Suppose G is a Lie group with universal covering $\tilde{G} \longrightarrow G$. If S_g is a closed oriented surface of genus $g > 1$, then its fundamental group admits a presentation (2.1) Let

$$\pi_1(S_g) \overset{\rho}{\to} G$$

be a representation; then the *obstruction* $o_2(\rho)$ for lifting ρ to $\tilde{G}$ is obtained as follows. Choose lifts $\widetilde{\rho(A_i)}$, $\widetilde{\rho(B_i)}$ of $\rho(A_i)$ and $\rho(B_i)$ to $\tilde{G}$, respectively. Then

$$o_2(\rho) := [\widetilde{\rho(A_1)}, \widetilde{\rho(B_1)}] \ldots [\widetilde{\rho(A_g)}, \widetilde{\rho(B_g)}] \qquad (3.3)$$

is independent of the chosen lifts and lies in

$$\pi_1(G) = \ker(\tilde{G} \longrightarrow G).$$

It vanishes precisely when ρ lifts to $\tilde{G}$. When $G = GL^+(2, \mathbb{R})$, the obstruction class $o_2(\rho)$ is just the Euler class $e(\rho)$.

To identify the element of $\pi_1(GL^+(2, \mathbb{R}))$ corresponding to $e(\rho)$, Milnor and Wood estimate the *translation number* of the lifts of generators to $\tilde{G}$, which is based on the rotation number of orientation-preserving circle homeomorphisms. Milnor uses a retraction $GL^+(2, \mathbb{R}) \overset{r}{\to} SO(2)$ (say, the one arising from the Iwasawa decomposition), which lifts to a retraction

$$\tilde{G} \overset{\theta}{\to} \widetilde{SO(2)} \cong \mathbb{R}$$

and proves the estimate

$$|\theta(\gamma_1\gamma_2) - \theta(\gamma_1) - \theta(\gamma_2)| < \frac{\pi}{2}. \qquad (3.4)$$

Wood considers a more general retraction θ defined on $\tilde{G} = \widetilde{Homeo^+(S^1)}$ and shows a similar estimate, sharpened by a factor of two. Applying this to (3.3), he shows that if an m-fold product of commutators in $\tilde{G}$ is translation by a, then

$$|a| < 2m - 1. \qquad (3.5)$$

The estimate (3.5) extends Benzécri's original estimate (2.3) in a stronger and more abstract context. This — the boundedness of the Euler class of flat bundles — may be regarded as one of the roots of the theory of bounded cohomology. The fundamental role of the Euler class as a *bounded cohomology class* was discovered by Ghys [39]. In particular, he showed that the *bounded Euler class* characterizes orientation-preserving actions of surface groups on the circle up to quasi-conjugacy.

For other generalizations of the Milnor-Wood inequality, compare Dupont [36, 37], Sullivan [67], Domic-Toledo [31], and Smillie [66]. For more information, see Burger-Iozzi-Wienhard [17] and the second chapter of Calegari [18]. The question of when a foliation on the total space of a circle bundle over a surface is isotopic to a flat bundle is the subject of Thurston's thesis [68].

4   MAXIMAL REPRESENTATIONS

4.1   A converse to the Milnor-Wood inequality

Equality in (3.2) has special and deep significance. Let M be a closed oriented surface. Then, just as described earlier for affine structures, every hyperbolic structure on M determines a *developing pair* (dev, ρ), where

$$\tilde{M} \xrightarrow{\mathrm{dev}} \mathsf{H}^2,$$

$$\pi_1(M) \xrightarrow{\rho} \mathsf{Isom}^+(\mathsf{H}^2),$$

by globalizing the coordinate charts and coordinate charts in an atlas defining the hyperbolic structure. The flat $(\mathsf{Isom}^+(\mathsf{H}^2), \mathsf{H}^2)$-bundle $E_M \longrightarrow M$ corresponding to ρ has a section δ_M corresponding to dev, which is transverse to the flat structure on E_M. Consequently the normal bundle of $\delta_M \subset E_M$ (by the tubular neighborhood theorem) is isomorphic to the tangent bundle TM, and, therefore,

$$e(\rho) = e(TM) = \chi(M),$$

proving sharpness in the Milnor-Wood inequality. By conjugating ρ with an orientation-reversing isometry of H^2, one obtains a representation ρ with

$$e(\rho) = -\chi(M).$$

The converse statement was proved in my doctoral dissertation [40]. Say that a representation is *maximal* if $e(\rho) = \pm\chi(M)$.

THEOREM. *Let* $\rho \in \mathsf{Hom}(\pi_1(M), \mathsf{PSL}(2, \mathbb{R}))$. *Then the following are equivalent:*

- *ρ is the holonomy of a hyperbolic structure on M;*

- *ρ is an embedding onto a discrete subgroup of $\mathsf{Isom}^+(\mathsf{H}^2) \cong \mathsf{PSL}(2, \mathbb{R})$;*

- *for every $\gamma \in \pi_1(M)$ with $\gamma \neq 1$, the holonomy $\rho(\gamma)$ is a hyperbolic element of $\mathsf{PSL}(2, \mathbb{R})$.*

4.2   Kneser's theorem on surface maps

A special case follows from the classical theorem of Kneser [54]:

THEOREM. *Let M, N be closed oriented surfaces, N having genus > 1. Suppose that $M \xrightarrow{f} N$ is a continuous map of degree d. Then*

$$d|\chi(N)| \leq |\chi(M)|.$$

Furthermore, $d|\chi(N)| = |\chi(M)|$ if and only if f is homotopic to a covering space.

The theorem follows by giving N a hyperbolic structure, with holonomy representation ρ. Then the composition

$$\pi_1(M) \xrightarrow{f_*} \pi_1(N) \xrightarrow{\rho} \mathsf{PSL}(2, \mathbb{R})$$

has Euler number $d\chi(N)$. Now apply Milnor-Wood and its converse statement to the composition.

4.3 Components of the representation variety

Since the space of hyperbolic structures on M is connected, the Euler class defines a continuous map

$$\mathrm{Hom}\big(\pi_1(M), \mathrm{PSL}(2, \mathbb{R})\big) \longrightarrow H^2(M; \mathbb{Z}) \cong \mathbb{Z}.$$

Reversing the orientation on M reverses the sign of the Euler number. Therefore, the maximal representations constitute *two* connected components of $\mathrm{Hom}\big(\pi_1(M), \mathrm{PSL}(2, \mathbb{R})\big)$. In general the connected components are the fibers of this map (Goldman [41, 43], Hitchin [53]). In particular the space of representations has $4g - 3$ connected components. Each component has dimension $6g - 6$. Only the Euler class 0 component is not a smooth manifold. Furthermore Hitchin relates the component corresponding to Euler class $2 - 2g + k$ to the kth symmetric power of M. See Hitchin [53], Bradlow-Garcìa-Prada- Gothen [12, 13], and my expository article [45].

4.4 Rigidity and flexibility

This characterization of maximal representations is a kind of rigidity for surface group representations. Dupont [36, 37], Turaev [75, 76], and Toledo [72] defined obstruction classes o_2 for Lie groups G of automorphisms of Hermitian symmetric spaces. In particular, Toledo proved the following rigidity theorem.

THEOREM (Toledo [72]). *Suppose that $\pi_1(M) \xrightarrow{\rho} \mathrm{PU}(n, 1)$ is a representation. Equality is attained in the generalized Milnor-Wood inequality*

$$|o_2(\rho)| \leq \frac{|\chi(M)|}{2}.$$

Then ρ embeds $\pi_1(M)$ as a discrete subgroup of the stabilizer (conjugate to

$$\mathrm{U}(1, 1) \times \mathrm{U}(n - 1)$$

of a holomorphic totally geodesic curve C in $\mathbf{H}_{\mathbb{C}}^2$. In particular $C / \mathrm{Image}(\rho)$ is a hyperbolic surface diffeomorphic to M.

Recently these results have been extended to higher rank in the work of Burger-Iozzi-Wienhard [15, 16, 17].

As maximality of the Euler class in the Milnor-Wood inequality implies rigidity, various values of the Euler class imply various kinds of flexibility [42]. If $\pi = \pi_1(M)$ is the fundamental group of a compact Kähler manifold M and G is a reductive algebraic Lie group, then Goldman-Millson [51] gives a complete description of the the analytic germ of the space of representations $\mathrm{Hom}(\pi, G)$ at a reductive representation ρ. Specifically, ρ has an open neighborhood in $\mathrm{Hom}(\pi, G)$ analytically equivalent to the quadratic cone defined by the symmetric bilinear form

$$Z^1(\pi, \mathfrak{g}_{\mathrm{Ad}\rho}) \times Z^1(\pi, \mathfrak{g}_{\mathrm{Ad}\rho}) \longrightarrow H^2(\pi, \mathfrak{g}_{\mathrm{Ad}\rho})$$

obtained by combining cup product on π with Lie bracket,

$$\mathfrak{g}_{\mathrm{Ad}\rho} \times \mathfrak{g}_{\mathrm{Ad}\rho} \xrightarrow{[,]} \mathfrak{g}_{\mathrm{Ad}\rho},$$

as coefficient pairing.

Consider the special case when M is a closed hyperbolic surface and a representation $\pi \xrightarrow{\rho_0} SU(1,1) \cong SL(2, \mathbb{R})$. We assume that ρ_0 has Zariski-dense image, which in this case simply means that its image is non-solvable. In turn, this means the corresponding action on H^2 fixes no point in $H^2 \cup \partial H^2$. In that case ρ_0 is reductive and defines a smooth point of the $\mathbb{R}$-algebraic set $Hom(\pi, SU(1,1))$. Extend the action to an isometric action on the complex hyperbolic plane $H^3_{\mathbb{C}}$ via the composition ρ defined by

$$\pi \xrightarrow{\rho_0} SU(1,1) \hookrightarrow PU(2,1). \tag{4.1}$$

For $A \in U(1,1)$, taking the equivalence class of the direct sum

$$A \oplus 1 := \begin{bmatrix} A & 0 \\ 0 & 1 \end{bmatrix} \in U(2,1)$$

in $PU(2,1)$ defines an embedding

$$SU(1,1) \hookrightarrow PU(2,1).$$

This representation stabilizes a *complex hyperbolic line* $H^3_{\mathbb{C}} \subset H^3_{\mathbb{C}}$ inside the complex hyperbolic plane. What are the local deformations of ρ in $Hom(\pi, PU(2,1))$?

The representation ρ is maximal if and only if ρ_0 is maximal, which occurs when $e(\rho) = \pm\chi(M)$. In that case any representation $\pi \longrightarrow PU(2,1)$ near ρ stabilizes $H^3_{\mathbb{C}}$, that is, it lies in the subgroup $U(1,1) \subset PU(2,1)$.

In general, representations $\pi \longrightarrow U(1,1)$ can be easily understood in terms of their composition with the projectivization homomorphism $U(1,1) \longrightarrow PU(1,1)$. The corresponding map on representation varieties,

$$Hom(\pi, U(1,1)) \longrightarrow Hom(\pi, PU(1,1)),$$

is a torus fibration, where the points of the fiber correspond to different actions in the normal directions to $H^3_{\mathbb{C}} \subset H^3_{\mathbb{C}}$ (which are described by characters $\pi \longrightarrow U(1)$. In particular ρ defines a smooth point of $Hom(\pi, U(1,1))$ with tangent space

$$Z^1(\pi, \mathfrak{su}(1,1)_{Ad\rho_0}) \oplus Z^1(\pi, \mathbb{R})$$

since $\mathfrak{u}(1)_{Ad\rho_0})$ equals the ordinary coefficient system $\mathbb{R}$ (where π acts by the identity).

From the general deformation result, the analytic germ of $Hom(\pi, PU(2,1))$ near ρ looks like the Cartesian product of the (smooth) analytic germ of

$$Hom(\pi, U(1,1)) \times PU(2,1)/U(1,1)$$

with the quadratic cone $\mathcal{Q}_\rho$ in $Z^1(\pi, \mathfrak{u}(1,1)_{Ad\rho}) \cong \mathbb{C}^{2g}$ defined by the cup-product pairing

$$Z^1(\pi, \mathfrak{u}(1,1)_{Ad\rho}) \times Z^1(\pi, \mathfrak{u}(1,1)_{Ad\rho}) \longrightarrow H^2(\pi, \mathbb{R}) \cong \mathbb{R},$$

where $\mathfrak{u}(1,1)_\rho$ denotes the π-module defined by the standard 2-dimensional complex representation of $U(1,1)$. The coefficient pairing (which is derived from the Lie bracket on $\mathfrak{su}(2,1)$) is just the imaginary part of the indefinite Hermitian form on $\mathfrak{u}(1,1)$, and is skew-symmetric. In particular the space of *coboundaries*,

$$B^1(\pi, \mathfrak{u}(1,1)_{Ad\rho}) \subset Z^1(\pi, \mathfrak{u}(1,1)_{Ad\rho}),$$

is isotropic. Thus, we reduce to the symmetric bilinear form obtained from the cohomology pairing

$$H^1(\pi, \mathfrak{u}(1,1)_{\mathsf{Ad}\rho}) \times H^1(\pi, \mathfrak{u}(1,1)_{\mathsf{Ad}\rho}) \longrightarrow H^2(\pi, \mathbb{R}) \cong \mathbb{R}. \qquad (4.2)$$

The real dimension of $H^1(\pi, \mathfrak{u}(1,1)_\rho)$ equals

$$-2\chi(M) = 8(g-1).$$

By the signature theorem of Meyer [60], the quadratic form corresponding to (4.2) has signature $8e(\rho_0)$. Thus, near ρ, the $\mathbb{R}$-algebraic set $\mathsf{Hom}(\pi, \mathsf{PU}(2,1))$ is analytically equivalent to the Cartesian product of a manifold with a cone on $\mathbb{R}^{8(g-1)}$ defined by a quadratic form of signature $8e(\rho)$.

Meyer's theorem immediately gives a proof of Milnor's inequality (3.1), since the signature of a quadratic form is bounded by the dimension of the ambient vector space. Furthermore, ρ is maximal if and only if the quadratic form is definite, in which the quadratic cone has no real points, and any small deformation of ρ must stabilize a complex geodesic.

5   COMPLETE AFFINE MANIFOLDS

We return to the subject of flat affine manifolds and the second [63] of Milnor's papers on this subject.

5.1   The Auslander-Milnor question

An affine manifold M is *complete* if some (and hence every) developing map is bijective. In that case $\tilde{M}$ identifies with E, and M arises as the quotient $\Gamma \backslash E$ by a discrete subgroup $\Gamma \subset \mathsf{Aff}(E)$ acting properly and freely on E. The affine holonomy representation

$$\pi_1(M) \overset{\rho}{\hookrightarrow} \mathsf{Aff}(E)$$

embeds $\pi_1(M)$ onto Γ.

Equivalently, an affine manifold is complete if and only if the corresponding affine connection is *geodesically complete,* that is, every geodesic extends infinitely in both directions.

A simple example of an *incomplete* affine structure on a closed manifold is a *Hopf manifold M,* obtained as the quotient of $\mathbb{R}^n \setminus \{0\}$ by a cyclic group $\langle A \rangle$. Here the generator A must be a *a linear expansion,* that is, an element $A \in \mathsf{GL}(n, \mathbb{R})$ such that every eigenvalue has modulus > 1. Such a quotient is diffeomorphic to $S^{n-1} \times S^1$. A geodesic aimed at the origin winds seemingly faster and faster around the S^1-factor, although it's travelling with zero acceleration with respect to the flat affine connection. In finite time, it "runs off the edge" of the manifold.

If $M = \Gamma \backslash E$ is a complete affine manifold, then $\Gamma \subset \mathsf{Aff}(E)$ is a discrete subgroup acting properly and freely on E. However, in the previous example $\langle A \rangle$ is a discrete subgroup which doesn't act properly. A proper action of a discrete group is the usual notion of a *properly discontinuous action.* If the action is also free (that is, no fixed points), then the quotient is a (Hausdorff) smooth manifold, and the quotient map $E \longrightarrow \Gamma \backslash E$ is a covering space. A properly discontinuous action whose quotient is compact as well as Hausdorff is said to be *crystallographic,* in analogy with the classical notion of a *crystallographic group*: a *Euclidean crystallographic group* is a discrete cocompact group of Euclidean isometries. Its quotient space is a Euclidean orbifold. Since such groups act isometrically on metric spaces, discreteness here does imply properness; this dramatically fails for more general discrete groups of *affine transformations*.

L. Auslander [6] claimed to prove that the Euler characteristic vanishes for a compact complete affine manifold, but his proof was flawed. It rested upon the following question, which in [38] was demoted to a "conjecture" and is now known as the "Auslander conjecture."

CONJECTURE 5.1. *Let M be a compact complete affine manifold. Then $\pi_1(M)$ is virtually polycyclic.*

In that case the affine holonomy group $\Gamma \cong \pi_1(M)$ embeds in a closed Lie subgroup $G \subset \mathsf{Aff}(\mathsf{E})$ satisfying

- G has finitely many connected components;

- the identity component G^0 acts simply transitively on E.

Then $M = \Gamma \backslash \mathsf{E}$ admits a finite covering space $M^0 := \Gamma^0 \backslash \mathsf{E}$ where

$$\Gamma^0 := \Gamma \cap G^0.$$

The simply transitive action of G^0 defines a complete *left-invariant affine structure* on G^0. (The developing map is just the evaluation map of this action.) Necessarily, G^0 is a 1-connected solvable Lie group and M^0 is affinely isomorphic to the *complete affine solvmanifold* $\Gamma^0 \backslash G^0$. In particular, $\chi(M^0) = 0$ and thus $\chi(M) = 0$.

This theorem is the natural extension of Bieberbach's theorems describing the structure of flat Riemannian (or Euclidean) manifolds; see Milnor [62] for an exposition of this theory and its historical importance. Every flat Riemannian manifold is finitely covered by a *flat torus*, the quotient of E by a lattice of translations. In the more general case, G^0 plays the role of the group of translations of an affine space and the solvmanifold M^0 plays the role of the flat torus. The importance of Conjecture 5.1 is that it would provide a detailed and computable structure theory for compact complete affine manifolds.

Conjecture 5.1 was established in dimension 3 in Fried-Goldman [38]. The proof involves classifying the possible Zariski closures $A(\mathsf{L}(\Gamma))$ of the linear holonomy group inside $\mathsf{GL}(\mathsf{E})$. Goldman-Kamishima [47] prove Conjecture 5.1 for flat Lorentz manifolds. Grunewald-Margulis [52] establish Conjecture 5.1 when the Levi component of $\mathsf{L}(\Gamma)$ lies in a real rank-one subgroup of $\mathsf{GL}(\mathsf{E})$. See Tomanov [73, 74] and Abels-Margulis-Soifer [2, 3, 4] for further results. The conjecture is now known in all dimensions ≤ 6 (Abels-Margulis-Soifer [5]).

5.2 The Kostant-Sullivan theorem

Although Conjecture 5.1 remains unknown in general, the question which motivated it was proved by Kostant and Sullivan [55].

THEOREM (Kostant-Sullivan). *Let M be a compact complete affine manifold. Then*

$$\chi(M) = 0.$$

Their ingenious proof uses an elementary fact about free affine actions and Chern-Weil theory. The first step is that if $\Gamma \subset \mathsf{Aff}(E)$ is a group of affine transformations acting freely on E, then the Zariski closure $A(\Gamma)$ of Γ in $\mathsf{Aff}(E)$ has the property that every element $g \in A(\Gamma)$ has 1 as an eigenvalue. To this end, suppose that $\Gamma \subset \mathsf{Aff}(E)$ acts freely. Then solving for a fixed point

$$\gamma(x) = \mathsf{L}(\gamma)(x) + u(\gamma) = x$$

implies that $\mathsf{L}(\gamma)$ has 1 as an eigenvalue for every $\gamma \in \Gamma$. Thus, every element $\gamma \in \Gamma$ satisfies the polynomial condition

$$\det(\mathsf{L}(\gamma) - I) = 0,$$

which extends to the Zariski closure $A(\Gamma)$ of Γ in $\mathsf{Aff}(E)$.

Next one finds a Riemannian metric (or more accurately, an orthogonal connection) to which Chern-Weil applies. Passing to a finite covering, using the finiteness of $\pi_0(A(\Gamma))$, we may assume the holonomy group lies in the identity component $A(\Gamma)^0$, which is a connected Lie group. Since every connected Lie group deformation retracts to a maximal compact subgroup, the structure group of TM reduces from Γ to a maximal compact subgroup $K \subset A(\Gamma)^0$. This reduction of structure group gives an orthogonal connection ∇ taking values in the Lie algebra $\mathfrak{k}$ of K. Since every compact group of affine transformations fixes a point, we may assume that $K \subset \mathsf{GL}(E)$. Since every element of $A(\Gamma)$ (and hence K) has 1 as an eigenvalue, every element of $\mathfrak{k}$ has determinant zero. Thus the Pfaffian polynomial (the square root of the determinant) vanishes on $\mathfrak{k}$. Since the curvature of ∇ takes values in $\mathfrak{k}$ and the Euler form is the Pfaffian of the curvature tensor, the Euler form is zero. Now apply the Chern-Gauss-Bonnet theorem [26]. Integrating over M gives $\chi(M) = 0$, as claimed.

5.3 "On fundamental groups of complete affinely flat manifolds"

In his 1977 paper [63], Milnor set the record straight caused by the confusion surrounding Auslander's flawed proof of Conjecture 5.1. Influenced by the work of Tits [71] on free subgroups of linear groups and amenability, Milnor observed that for an affine space E of given dimension, the following conditions are all equivalent:

- Every discrete subgroup of $\mathsf{Aff}(E)$ which acts properly on E is amenable.

- Every discrete subgroup of $\mathsf{Aff}(E)$ which acts properly on E is virtually solvable.

- Every discrete subgroup of $\mathsf{Aff}(E)$ which acts properly on E is virtually polycyclic.

- A nonabelian free subgroup of $\mathsf{Aff}(E)$ cannot act properly on E.

- The Euler characteristic $\chi(\Gamma \backslash E)$ (when defined) of a complete affine manifold $\Gamma \backslash E$ must vanish (unless $\Gamma = \{1\}$, of course).

- A complete affine manifold $\Gamma \backslash E$ has finitely generated fundamental group Γ.

(If these conditions were met, one would have a satisfying structure theory similar to, but somewhat more involved than, the Bieberbach structure theory for flat Riemannian manifolds.)

In [63], Milnor provides abundant "evidence" for this "conjecture." For example, the *infinitesimal version:* namely, let $G \subset \mathrm{Aff}(\mathsf{E})$ be a connected Lie group which acts properly on E. Then G must be an amenable Lie group, which simply means that it is a compact extension of a solvable Lie group. (Equivalently, its Levi subgroup is compact.) Furthermore, he provides a *converse:* Milnor shows that every virtually polycyclic group admits a proper affine action. (However, Milnor's actions do *not* have compact quotient. Benoist [9] found finitely generated nilpotent groups which admit no affine crystallographic action. Benoist's examples are 11-dimensional.)

However convincing as his "evidence" is, Milnor still proposes a possible way of constructing counterexamples:

> Start with a free discrete subgroup of $O(2,1)$ and add translation components to obtain a group of affine transformations which acts freely. However it seems difficult to decide whether the resulting group action is properly discontinuous.

This is clearly a geometric problem: As Schottky showed in 1907, free groups act properly by isometries on hyperbolic 3-space and, hence, by diffeomorphisms of E^3. These actions are *not* affine.

One might try to construct a proper affine action of a free group by a construction like Schottky's. Recall that a *Schottky group of rank g* is defined by a system of g open half-spaces $H_1, \ldots, H_g$ and isometries $A_1, \ldots, A_g$ such that the $2g$ half-spaces

$$H_1, \ldots, H_g, A_1(H_1^c), \ldots, A_g(H_g^c)$$

are all disjoint (where H^c denotes the *complement* of the closure $\bar{H}$ of H). The *slab*

$$\mathrm{Slab}_i := H_i^c \cap A_i(H_i)$$

is a fundamental domain for the action of the cyclic group $\langle A_i \rangle$. The *ping-pong lemma* then asserts that the intersection of all the slabs

$$\Delta := \mathrm{Slab}_1 \cap \cdots \cap \mathrm{Slab}_g$$

is a fundamental domain for the group $\Gamma := \langle A_1, \ldots, A_g \rangle$. Furthermore, Γ is freely generated by $A_1, \ldots, A_g$. The basic idea is the following. Let $B_i^+ := A_i(H_i^c)$ (respectively, $B_i^- := H_i$) denote the *attracting basin* for A_i (respectively, A_i^{-1}). That is, A_i maps all of H_i^c to B_i^+ and A_i^{-1} maps all of $A_i(H_i)$ to B_i^-. Let $w(a_1, \ldots, a_g)$ be a reduced word in abstract generators $a_1, \ldots, a_g$, with initial letter $a_i^\pm$. Then

$$w(A_1, \ldots, A_g)(\Delta) \subset B_i^\pm.$$

Since all the basins $B_i^\pm$ are disjoint, $w(A_1, \ldots, A_g)$ maps Δ off itself. Therefore, $w(A_1, \ldots, A_g) \neq 1$.

Freely acting discrete cyclic groups of affine transformations have fundamental domains which are *parallel slabs*, that is, regions bounded by two parallel affine hyperplanes. One might try to combine such slabs to form "affine Schottky groups," but immediately one sees this idea is doomed, if one uses parallel slabs for Schottky's construction: parallel slabs have disjoint complements only if they are parallel to each other, in which case the group is necessarily cyclic anyway. From this viewpoint, a discrete group of affine transformations seems very unlikely to act properly.

6 MARGULIS SPACETIMES

In the early 1980's Margulis, while trying to prove that a nonabelian free group can't act properly by affine transformations, discovered that discrete free groups of affine transformations can indeed act properly!

Around the same time, David Fried and I were also working on these questions and reduced Milnor's question in dimension 3 to what seemed at the time to be one annoying case which we could not handle. Namely, we showed the following: Let E be a three-dimensional affine space and $\Gamma \subset \mathsf{Aff}(\mathsf{E})$. Suppose that Γ acts properly on E. Then, either Γ is polycyclic or the restriction of the linear holonomy homomorphism

$$\Gamma \xrightarrow{\mathsf{L}} \mathsf{GL}(\mathsf{E})$$

discretely embeds Γ onto a subgroup of $\mathsf{GL}(\mathsf{E})$ conjugate to the orthogonal group $\mathsf{O}(2,1)$.

In particular, the complete affine manifold $M^3 = \Gamma \backslash \mathsf{E}$ is a *complete flat Lorentz 3-manifold* after one passes to a finite-index torsionfree subgroup of Γ to ensure that Γ acts freely. In particular the restriction $\mathsf{L}|_\Gamma$ defines a free properly discrete isometric action of Γ on the hyperbolic plane H^2, and the quotient $\Sigma^2 := \mathsf{H}^2/\mathsf{L}(\Gamma)$ is a complete hyperbolic surface with a homotopy equivalence

$$M^3 := \Gamma \backslash \mathsf{E} \simeq \mathsf{H}^2/\mathsf{L}(\Gamma) =: \Sigma^2.$$

Already this excludes the case when M^3 is compact, since Γ is the fundamental group of a closed aspherical 3-manifold (and has cohomological dimension 3) and the fundamental group of a hyperbolic surface (and has cohomological dimension ≤ 2). This is a crucial step in the proof of Conjecture 5.1 in dimension 3.

That the hyperbolic surface Σ^2 is *noncompact* is a much deeper result due to Geoffrey Mess [59]. Later proofs and a generalization have been found by Goldman-Margulis [50] and Labourie [56]. (Compare the discussion in §6.3.) Since the fundamental group of a noncompact surface is free, Γ is a free group. Furthermore $\mathsf{L}|_\Gamma$ embeds Γ as a free discrete group of isometries of hyperbolic space. Thus Milnor's suggestion is the *only* way to construct nonsolvable examples *in dimension 3*.

6.1 Affine boosts and crooked planes

Since L embeds Γ_0 as the fundamental group of a hyperbolic surface, $\mathsf{L}(\gamma)$ is elliptic only if $\gamma = 1$. Thus, if $\gamma \neq 1$, then $\mathsf{L}(\gamma)$ is either hyperbolic or parabolic. Furthermore $\mathsf{L}(\gamma)$ is hyperbolic for most $\gamma \in \Gamma_0$.

When $\mathsf{L}(\gamma)$ is hyperbolic, γ is an *affine boost*, that is, it has the form

$$\gamma = \begin{bmatrix} e^{\ell(\gamma)} & 0 & 0 \\ 0 & 1 & 0 \\ 0 & 0 & e^{-\ell(\gamma)} \end{bmatrix} \begin{bmatrix} 0 \\ \alpha(\gamma) \\ 0 \end{bmatrix} \tag{6.1}$$

in a suitable coordinate system. (Here the 3×3 matrix represents the linear part, and the column 3-vector represents the translational part.) γ leaves invariant a unique (spacelike) line C_γ (the second coordinate line in (6.1). Its image in $\mathbb{E}_1^3/\Gamma$ is a *closed geodesic* $C_\gamma/\langle\gamma\rangle$. Just as for hyperbolic surfaces, most loops in M^3 are freely homotopic to such closed geodesics. (For a more direct relationship between the dynamics of the geodesic flows on Σ^2 and M^3, compare Goldman-Labourie [48]).

Margulis observed that C_γ inherits a natural orientation and metric, arising from an orientation on E, as follows. Choose repelling and attracting eigenvectors $L(\gamma)^\pm$ for $L(\gamma)$, respectively; choose them so they lie in the same component of the nullcone. Then the orientation and metric on C_γ is determined by a choice of nonzero vector $L(\gamma)^0$ spanning $Fix(L(\gamma))$. This vector is uniquely specified by requiring that

- $L(\gamma)^0 \cdot L(\gamma)^0 = 1$;

- $(L(\gamma)^0, L(\gamma)^-, L(\gamma)^+)$ is a positively oriented basis.

The restriction of γ to C_γ is a translation by displacement $\alpha(\gamma)$ with respect to this natural orientation and metric.

Compare this to the more familiar *geodesic length function* $\ell(\gamma)$ associated to a class γ of closed curves on the hyperbolic surface Σ. The linear part $L(\gamma)$ acts by *transvection* along a geodesic $c_{L(\gamma)} \subset H^2$. The quantity $\ell(\gamma) > 0$ measures how far $L(\gamma)$ moves points of $c_{L(\gamma)}$.

This pair of quantities

$$(\ell(\gamma), \alpha(\gamma)) \in \mathbb{R}_+ \times \mathbb{R}$$

is a complete invariant of the isometry type of the *flat Lorentz cylinder* $E/\langle\gamma\rangle$. The absolute value $|\alpha(\gamma)|$ is the length of the unique primitive closed geodesic in $E/\langle\gamma\rangle$.

A fundamental domain is the parallel slab

$$(\Pi_{C_\gamma})^{-1} \left(p_0 + [0, \alpha(\gamma)] \, \gamma^0 \right),$$

where

$$E \xrightarrow{\Pi_{C_\gamma}} C_\gamma$$

denotes orthogonal projection onto

$$C_\gamma = p_0 + \mathbb{R}\gamma^0.$$

As noted before, however, parallel slabs can't be combined to form fundamental domains for Schottky groups, since their complementary half-spaces are rarely disjoint.

In retrospect this is believable, since these fundamental domains are fashioned from the dynamics of the translational part (using the projection Π_{C_γ}). While the effect of the translational part is properness, the dynamical behavior affecting most points is influenced by the *linear part:* while points on C_γ are displaced by γ at a polynomial rate, all other points move at an exponential rate.

Furthermore, parallel slabs are less robust than slabs in H^2: while small perturbations of one boundary component extend to fundamental domains, this is no longer true for parallel slabs. Thus, one might look for other types of fundamental domains better adapted to the exponential growth dynamics given by the linear holonomy $L(\gamma)$.

Todd Drumm, in his 1990 Maryland thesis [32], defined more flexible polyhedral surfaces, which can be combined to form fundamental domains for *Schottky groups* of 3-dimensional affine transformations. A *crooked plane* is a PL surface in E, separating E into two *crooked half-spaces*. The complement of two disjoint crooked half-space is a *crooked slab,* which forms a fundamental domain for a cyclic group generated by an affine boost. Drumm proved the remarkable theorem that if $S_1, \dots, S_g$ are crooked slabs whose complements have disjoint interiors, then given any collection of affine boosts γ_i with S_i as fundamental domain, then the intersection $S_1 \cap \cdots \cap S_g$ is a fundamental domain for $\langle \gamma_1, \dots, \gamma_g \rangle$ acting on *all* of E.

Modeling a crooked fundamental domain for Γ acting on E on a fundamental polygon for Γ_0 acting on H^2, Drumm proved the following sharp result:

THEOREM (Drumm [32, 33]). *Every* noncocompact *torsionfree Fuchsian group* Γ_0 *admits a proper affine deformation* Γ *whose quotient is a solid handlebody.*

(Compare also [34, 25].)

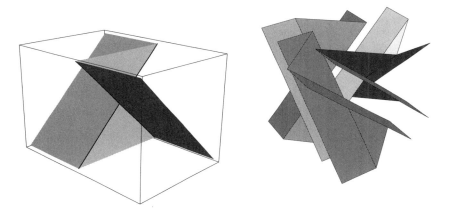

FIGURE 6.1: A crooked plane, and a family of three pairwise disjoint crooked planes. (See also Plate 26.)

6.2 Marked length spectra

We now combine the geodesic length function $\ell(\gamma)$ describing the geometry of the hyperbolic surface Σ with the Margulis invariant $\alpha(\gamma)$ describing the Lorentzian geometry of the flat affine 3-manifold M.

As noted by Margulis, $\alpha(\gamma) = \alpha(\gamma^{-1})$ and, more generally,

$$\alpha(\gamma^n) = |n|\alpha(\gamma).$$

The invariant ℓ satisfies the same homogeneity condition, and, therefore,

$$\frac{\alpha(\gamma^n)}{\ell(\gamma^n)} = \frac{\alpha(\gamma)}{\ell(\gamma)}$$

is constant along hyperbolic cyclic subgroups. Hyperbolic cyclic subgroups correspond to periodic orbits of the geodesic flow ϕ on the unit tangent bundle $U\Sigma$.

Periodic orbits, in turn, define ϕ-invariant probability measures on $U\Sigma$. Goldman-Labourie-Margulis [49] prove that, for any affine deformation, this function extends to a continuous function Y_Γ on the space $\mathcal{C}(\Sigma)$ of ϕ-invariant probability measures on $U\Sigma$. Furthermore, when Γ_0 is convex cocompact (that is, contains no parabolic elements), then the affine deformation Γ acts properly if and only if Y_Γ never vanishes. Since $\mathcal{C}(\Sigma)$ is connected, nonvanishing implies either all $Y_\Gamma(\mu) > 0$ or all $Y_\Gamma(\mu) < 0$. From this follows Margulis's *Opposite Sign Lemma*, first proved in [57, 58] and extended to groups with parabolics by Charette and Drumm [20]:

THEOREM (Margulis). *If Γ acts properly, then all the numbers $\alpha(\gamma)$ have the same sign.*

For an excellent treatment of the original proof of this fact, see the survey article of Abels [1].

6.3 Deformations of hyperbolic surfaces

The Margulis invariant may be interpreted in terms of deformations of hyperbolic structures as follows [50, 44]).

Suppose Γ_0 is a Fuchsian group with quotient hyperbolic surface $\Sigma_0 = \Gamma_0 \backslash \mathsf{H}^2$. Let $\mathfrak{g}_{\mathsf{Ad}}$ be the Γ_0-module defined by the adjoint representation applied to the embedding $\Gamma_0 \hookrightarrow \mathsf{O}(2,1)$. The coefficient module $\mathfrak{g}_{\mathsf{Ad}}$ corresponds to the Lie algebra of *right-invariant* vector fields on $\mathsf{O}(2,1)$ with the action of $\mathsf{O}(2,1)$ by left-multiplication. Geometrically these vector fields correspond to the infinitesimal isometries of H^2.

A family of hyperbolic surfaces Σ_t smoothly varying with respect to a parameter t determines an *infinitesimal deformation*, which is a cohomology class

$$[u] \in H^1(\Gamma_0, \mathfrak{g}_{\mathsf{Ad}}).$$

The cohomology group $H^1(\Gamma_0, \mathfrak{g}_{\mathsf{Ad}})$ corresponds to *infinitesimal deformations* of the hyperbolic surface Σ_0. In particular, the tangent vector to the path Σ_t of marked hyperbolic structures smoothly varying with respect to a parameter t defines a cohomology class $[u] \in H^1(\Gamma_0, \mathfrak{g}_{\mathsf{Ad}})$.

The same cohomology group parametrizes affine deformations. The translational part u of a linear representations of Γ_0 is a cocycle of the group Γ_0 taking values in the corresponding Γ_0-module V. Moreover, two cocycles define affine deformations which are conjugate by a translation if and only if their translational parts are cohomologous cocycles. Therefore, translational conjugacy classes of affine deformations form the cohomology group $H^1(\Gamma_0, \mathsf{V})$. Inside $H^1(\Gamma_0, \mathsf{V})$ is the subset Proper corresponding to *proper affine deformations.*

The adjoint representation Ad of $\mathsf{O}(2,1)$ identifies with the orthogonal representation of $\mathsf{O}(2,1)$ on V. Therefore, the cohomology group $H^1(\Gamma_0, \mathsf{V})$ consisting of translational conjugacy classes of affine deformations of Γ_0 can be identified with the cohomology group $H^1(\Gamma_0, \mathfrak{g}_{\mathsf{Ad}})$ corresponding to infinitesimal deformations of Σ_0.

THEOREM. *Suppose $u \in Z^1(\Gamma_0, \mathfrak{g}_{\mathsf{Ad}})$ defines an* infinitesimal deformation *tangent to a smooth deformation Σ_t of Σ.*

- *The marked length spectrum ℓ_t of Σ_t varies smoothly with t.*

- *Margulis's invariant $\alpha_u(\gamma)$ represents the derivative*

$$\left.\frac{d}{dt}\right|_{t=0} \ell_t(\gamma)\,.$$

- *(Opposite sign lemma) If $[u] \in$ Proper, then all closed geodesics lengthen (or shorten) under the deformation Σ_t.*

Since closed hyperbolic surfaces do not support deformations in which *every* closed geodesic shortens, such deformations exist only when Σ_0 is noncompact. This leads to a new proof [50] of Mess's theorem that Σ_0 is not compact. (For another, somewhat similar proof, which generalizes to higher dimensions, see Labourie [56].)

The tangent bundle TG of any Lie group G has a natural structure as a Lie group, where the fibration $TG \overset{\Pi}{\to} G$ is a homomorphism of Lie groups, and the tangent spaces

$$T_x G = \Pi^{-1}(x) \subset TG$$

are vector groups. The deformations of a representation $\Gamma_0 \overset{\rho_0}{\to} G$ correspond to representations $\Gamma_0 \overset{\rho}{\to} TG$ such that $\Pi \circ \rho = \rho_0$. In our case, affine deformations of $\Gamma_0 \hookrightarrow O(2,1)$ correspond to representations in the tangent bundle $TO(2,1)$. When G is the group $G(\mathbb{R})$ of $\mathbb{R}$-points of an algebraic group G defined over $\mathbb{R}$, then

$$TG \cong G(\mathbb{R}[\epsilon])\,,$$

where ϵ is an indeterminate with $\epsilon^2 = 0$. (Compare [44].) This is reminiscent of the classical theory of quasi-Fuchsian deformations of Fuchsian groups, where one deforms a Fuchsian subgroup of $SL(2,\mathbb{R})$ in

$$SL(2,\mathbb{C}) = SL(2,\mathbb{R}[i])\,,$$

where $i^2 = -1$.

6.4 Classification

In light of Drumm's theorem, classifying Margulis spacetimes M^3 begins with the classification of hyperbolic structures Σ^2. Thus, the deformation space of Margulis spacetimes maps to the Fricke space $\mathfrak{F}(\Sigma)$ of marked hyperbolic structures on the underlying topology of Σ.

The main result of [49] is that the positivity (or negativity) of Y_Γ on on $\mathcal{C}(\Sigma)$ is necessary and sufficient for properness of Γ. (For simplicity we restrict ourselves to the case when $L(\Gamma)$ contains no parabolics — that is, when Γ_0 is convex cocompact.) Thus the proper affine deformation space Proper identifies with the open convex cone in $H^1(\Gamma_0, V)$ defined by the linear functionals Y_μ, for μ in the compact space $\mathcal{C}(\Sigma)$. These give uncountably many linear conditions on $H^1(\Gamma_0, V)$, one for each $\mu \in \mathcal{C}(\Sigma)$. Since the invariant probability measures arising from periodic orbits are dense in $\mathcal{C}(\Sigma)$, the cone Proper is the interior of half-spaces defined by the countable set of functional Y_γ, where $\gamma \in \Gamma_0$.

The zero level sets $Y_\gamma^{-1}(0)$ correspond to affine deformations where γ does not act freely. Therefore, Proper defines a component of the subset of $H^1(\Gamma_0, V)$ corresponding to affine deformations which are *free* actions.

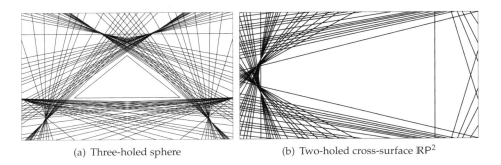

(a) Three-holed sphere   (b) Two-holed cross-surface $\mathbb{RP}^2$

FIGURE 6.2: Finite-sided deformation spaces for surfaces with $\chi(\Sigma)$

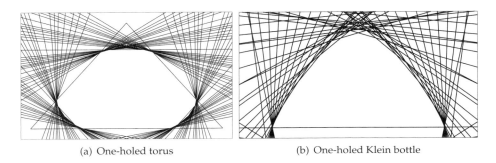

(a) One-holed torus   (b) One-holed Klein bottle

FIGURE 6.3: Infinite-sided deformation spaces for surfaces with $\chi(\Sigma)$.

Actually, one may go further. An argument inspired by Thurston [70], reduces the consideration to only those measures arising from *multicurves*, that is, unions of disjoint *simple* closed curves. These measures (after scaling) are dense in the *Thurston cone* $\mathcal{ML}(\Sigma)$ of *measured geodesic laminations* on Σ. One sees the combinatorial structure of the Thurston cone replicated on the boundary of

$$\mathsf{Proper} \subset H^1(\Gamma_0, \mathsf{V}) \,.$$

(Compare Figures 6.2 and 6.3.)

Two particular cases are notable. When Σ is a 3-holed sphere or a 2-holed cross-surface (real projective plane), then the Thurston cone degenerates to a finite-sided polyhedral cone. In particular properness is characterized by 3 Margulis functionals for the 3-holed sphere and 4 for the 2-holed cross-surface. Thus, the deformation space of equivalence classes of proper affine deformations is either a cone on a triangle or a convex quadrilateral, respectively.

When Σ is a 3-holed sphere, these functionals correspond to the three components of $\partial\Sigma$. The half-spaces defined by the corresponding three Margulis functionals cut off the deformation space (which is a polyhedral cone with 3 faces). The Margulis functionals for the other curves define half-spaces which strictly contain this cone.

When Σ is a 2-holed cross-surface, these functionals correspond to the two components of $\partial\Sigma$ and the two orientation-reversing simple closed curves in the interior of Σ. The four Margulis functionals describe a polyhedral cone with 4 faces. All other closed curves on Σ define a half-space strictly containing this cone.

In both cases, an ideal triangulation for Σ models a crooked fundamental domain for M, and Γ is an affine Schottky group, and M is an open solid handlebody of genus 2 (Charette-Drumm-Goldman [21],[22],[23]). Figure 6.2 depicts these finite-sided deformation spaces.

For the other surfaces where $\pi_1(\Sigma)$ is free of rank two (or, equivalently, where $\chi(\Sigma) = -1$), infinitely many functionals Y_μ are needed to define the deformation space, which necessarily has infinitely many sides. In these cases M^3 admits crooked fundamental domains corresponding to ideal triangulations of Σ, although unlike the preceding cases there is no single ideal triangulation which works for all proper affine deformations. Once again M^3 is a genus two handlebody. Figure 6.3 depicts these infinite-sided deformation spaces.

Since this paper was originally written, Danciger-Guéritaud-Kassel [29] and Choi-Goldman [28] independently proved that every Margulis spacetime is homeomorphic to an open solid handlebody, provided the associated hyperbolic surface is convex cocompact. More recently Danciger-Guéritaud-Kassel [30] have announced a construction for crooked fundamental polyhedra for such manifolds.

6.5 An arithmetic example

These examples are everywhere. As often happens in mathematics, finding the first example of generic behavior can be quite difficult. However, once the basic phenomena are recognized, examples of this generic behavior abound. The following example, taken from [22], shows how a proper affine deformation sits inside the symplectic group $\mathsf{Sp}(4, \mathbb{Z})$.

Begin with a 2-dimensional vector space L_0 over $\mathbb{R}$ with the group of linear automorphisms $\mathsf{GL}(L_0)$. Let V denote the vector space of *symmetric bilinear maps* $L_0 \times L_0 \xrightarrow{b} \mathbb{R}$ with the induced action of $\mathsf{GL}(L)$. Identifying V with symmetric 2×2 real matrices, the negative of the determinant defines an invariant Lorentzian inner product on V. In particular this defines a local embedding $\mathsf{GL}(L_0) \longrightarrow \mathsf{O}(2, 1)$.

Let $L_\infty := L^*$ denote the vector space dual to L_0 and $\mathsf{W} := L_0 \oplus L_\infty$ the direct sum. Then W admits a unique symplectic structure ω such that L_0 and L_∞ are Lagrangian subspaces and the restriction of ω to $L_0 \times L_\infty$ is the duality pairing. Let $\mathsf{Sp}(4, \mathbb{R})$ denote the group of linear symplectomorphisms of (W, ω). It acts naturally on the homogeneous space $\mathcal{L}(\mathsf{W}, \omega)$ of Lagrangian 2-planes L in (W, ω).

The Minkowski space E associated to V consists of Lagrangians $L \in \mathcal{L}(\mathsf{W}, \omega)$ which are transverse to L_∞. This is a torsor for the Lorentzian vector space V as follows. V consists of symmetric bilinear forms on L_0, and these can be identified with *self-adjoint* linear maps

$$L_0 \xrightarrow{f} L_\infty \cong (\mathsf{L}_0)^*.$$

A 2-dimensional linear subspace of V which is transverse to L_∞ is the graph

$$L := \mathsf{graph}(f)$$

of a linear map $L_0 \xrightarrow{f} L_\infty$. Moreover, L is Lagrangian if and only if f is self-adjoint. Furthermore, since V is a vector space, it acts simply transitively on the space E

FIGURE 6.4: A proper affine deformation of level 2 congruence subgroup of $SL(2, \mathbb{Z})$.

of such graphs by addition. In terms of 4×4 symplectic matrices (2×2 block matrices using the decomposition $W = L_0 \oplus L_\infty$), these translations correspond to *shears*:

$$\begin{bmatrix} I_2 & f_2 \\ 0 & I_2 \end{bmatrix}, \tag{6.2}$$

where the corresponding symmetric 2×2 matrix corresponding to f is denoted f_2. The corresponding subgroup of $Sp(4, \mathbb{R})$ consists of linear symplectomorphisms of (W, ω) which preserve L_∞, and act identically both on L_∞ and on its quotient W/L_∞.

As the translations of E are represented by shears in block upper-triangular form (6.2), the linear isometries are represented by the block diagonal matrices arising from $SL(L_0)$. More generally, the *Lorentz similarities* of E correspond to $GL(L_0)$ as follows. A linear automorphism $L_0 \xrightarrow{g} L_0$ induces a linear symplectomorphism $g \oplus (g^\dagger)^{-1}$ of $W = L_0 \oplus L_\infty$:

$$g \oplus (g^\dagger)^{-1} = \begin{bmatrix} g & 0 \\ 0 & (g^\dagger)^{-1} \end{bmatrix}.$$

These linear symplectomorphisms can be characterized as those which preserve the Lagrangian 2-planes L_0 and L_∞. Furthermore g induces an *isometry* of E with the flat Lorentzian structure if and only if $\text{Det}(g) = \pm 1$.

Here is our example. The level two congruence subgroup Γ_0 is the subgroup of $GL(2, \mathbb{Z})$ generated by

$$\begin{bmatrix} -1 & -2 \\ 0 & -1 \end{bmatrix}, \begin{bmatrix} -1 & 0 \\ 2 & -1 \end{bmatrix}$$

and the corresponding hyperbolic surface is a triply punctured sphere. For i in $\{1, 2, 3\}$, choose three positive integers μ_1, μ_2, μ_3 (the coordinates of the translational parts). Then the subgroup Γ of $\mathrm{Sp}(4, \mathbb{Z})$ generated by

$$\begin{bmatrix} -1 & -2 & \mu_1 + \mu_2 - \mu_3 & 0 \\ 0 & -1 & 2\mu_1 & -\mu_1 \\ 0 & 0 & -1 & 0 \\ 0 & 0 & 2 & -1 \end{bmatrix}, \begin{bmatrix} -1 & 0 & -\mu_2 & -2\mu_2 \\ 2 & -1 & 0 & 0 \\ 0 & 0 & -1 & -2 \\ 0 & 0 & 0 & -1 \end{bmatrix}$$

defines a affine deformation of a Γ_0.

By the main result of Charette-Drumm-Goldman [21], this affine deformation is proper with a fundamental polyhedron bounded by crooked planes. The quotient 3-manifold $M^3 = \Gamma \backslash E$ is homeomorphic to a genus two handlebody. Figure 6.4 depicts the intersections of crooked fundamental domains for this group (when $\mu_1 = \mu_2 = \mu_3 = 1$) with a spacelike plane. Note the parallel line segments cutting off fundamental domains for the cusps of Σ; the parallelism results from the parabolicity of the holonomy around the cusps.

7 Bibliography

[1] Abels, H., *Properly discontinuous groups of affine transformations: a survey,* Geom. Ded. **87** (2001), no. 1–3, 309–333.

[2] Abels, H., Margulis, G., and Soifer, G., *The Auslander conjecture for groups leaving a form of signature* $(n - 2, 2)$ *invariant,* Probability in mathematics. Israel J. Math. **148** (2005), 11–21.

[3] Abels, H., Margulis, G., and Soifer, G., *On the Zariski closure of the linear part of a properly discontinuous group of affine transformations,* J. Diff. Geo. **60** (2002), no. 2, 315–344.

[4] Abels, H., Margulis, G., and Soifer, G., *The linear part of an affine group acting properly discontinuously and leaving a quadratic form invariant,* Geom. Ded. **153** (2011), 1–46.

[5] Abels, H., Margulis, G., and Soifer, G., *The Auslander conjecture for dimension less than 7,* preprint (2011). arXiv:1211.2525

[6] Auslander, L., *The structure of complete locally affine manifolds,* Topology **3** (1964), suppl. 1, 131–139

[7] Auslander, L., *Simply transitive groups of affine motions,* Amer. J. Math. **99** (1977), no. 4, 809–826.

[8] Barbot, T., Charette, V., Drumm, T., Goldman, W., and Melnick, K., *A primer on the Einstein (2+1)-universe,* in "Recent Developments in Pseudo-Riemannian Geometry," D. Alekseevsky and H. Baum (eds.), Erwin Schrödinger Institute Lectures in Mathematics and Physics, Eur. Math. Soc. (2008), 179–221, arXiv:0706.3055 .

[9] Benoist, Y., *Une nilvariété non affine.* J. Diff. Geo. **41** (1995), no. 1, 21–52.

[10] Benzécri, J. P., *Variétés localement affines,* Sem. Topologie et Géom. Diff., Ch. Ehresmann (1958–60), no. 7 (Mai 1959).

[11] Bradlow, S., García-Prada, O., and Gothen, P., *Surface group representations and* $\mathrm{U}(p, q)$-*Higgs bundles,* J. Diff. Geo. **64** (2003), no. 1, 111–170.

[12] Bradlow, S., García-Prada, O., and Gothen, P., *Maximal surface group representations in isometry groups of classical Hermitian symmetric spaces*, Geom. Ded. **122** (2006), 185–213.

[13] Bradlow, S., García-Prada, O., and Gothen, P., *What is a Higgs bundle?* Notices Amer. Math. Soc. **54** (2007), no. 8 980–981.

[14] Bucher, M., and Gelander, T., *Milnor-Wood inequalities for manifolds locally isometric to a product of hyperbolic planes*, C. R. Acad. Sci. Paris **346** (2008), no. 11–12, 661–666.

[15] Burger, M., Iozzi, A., Labourie, F., and Wienhard, A., *Maximal representations of surface groups: Symplectic Anosov structures*, Pure and Applied Mathematics Quarterly, Special Issue: In Memory of Armand Borel Part 2 of 3,**1** (2005), no. 3, 555–601.

[16] Burger, M., Iozzi, A., and Wienhard, A., *Surface group representations with maximal Toledo invariant*, C. R. Acad. Sci. Paris , Sér. I **336** (2003), 387–390; *Representations of surface groups with maximal Toledo invariant*, Ann. Math. (2) **172** (2010), no. 1, 517–566,

[17] Burger, M., Iozzi, A., and Wienhard, A., *Higher Teichmüller spaces from* SL$(2, \mathbb{R})$ *to other Lie groups, handbook of Teichmüller theory, vol. III* (A. Papadopoulos, ed.), IRMA Lectures in Mathematics and Theoretical Physics European Mathematical Society (to appear).

[18] Calegari, D., "Foliations and the geometry of 3-manifolds," Oxford Mathematical Monographs. Oxford University Press, Oxford, 2007.

[19] Charette, V., *Proper actions of discrete groups on* $(2 + 1)$-*spacetime*, Doctoral dissertation, University of Maryland (2000).

[20] Charette, V., and Drumm, T., *The Margulis invariant for parabolic transformations*, Proc. Amer. Math. Soc. **133** (2005), no. 8, 2439–2447.

[21] Charette, V., Drumm, T., and Goldman, W., *Affine of deformations of the three-holed sphere*, Geometry & Topology **14** (2010), 1355–1382. arXiv:0907.0690

[22] Charette, V., Drumm, T., and Goldman, W., *Finite-sided deformation spaces for complete affine 3-manifolds* J. Topology, to appear. arXiv:1107.2862

[23] Charette, V., Drumm, T., and Goldman, W. *Proper affine deformations of a two-generator Fuchsian group* (in preparation).

[24] Charette, V., Drumm, T., Goldman, W., and Morrill, M., *Complete flat affine and Lorentzian manifolds*, Geom. Ded. **97** (2003), 187–198.

[25] Charette, V., and Goldman, W., *Affine Schottky groups and crooked tilings*, Contemp. Math. **262**, (2000), 69–98.

[26] Chern, S., *A simple intrinsic proof of the Gauss-Bonnet formula for closed Riemannian manifolds*. Ann. of Math. (2) **45** (1944), 747–752.

[27] Chern, S., *Pseudo-Riemannian geometry and the Gauss-Bonnet formula*. An. Acad. Brasil. Ci. **35** (1963), 17–26.

[28] Choi, S., and Goldman, W., *Topological tameness of Margulis spacetimes.* arXiv:1204.5308

[29] Danciger, J., Guéritaud, F., and Kassel, F., *Geometry and topology of complete Lorentz spacetimes of constant curvature.* arXiv:1306.2240

[30] Danciger, J., Guéritaud, F., and Kassel, F., *Margulis spacetimes via the arc complex* (in preparation).

[31] Domic, A., and Toledo, D., *The Gromov norm of the Kähler class of symmetric domains.* Math. Ann. **276** (1987), no. 3, 425–432.

[32] Drumm, T., *Fundamental polyhedra for Margulis space-times,* Doctoral dissertation, University of Maryland (1990); Topology **31** (1992), no. 4, 677–683.

[33] Drumm, T., *Linear holonomy of Margulis space-times,* J. Diff. Geo. **38** (1993), 679–691.

[34] Drumm, T., and Goldman, W., *Complete flat Lorentz 3-manifolds with free fundamental group.* Internat. J. Math. **1** (1990), no. 2, 149–161.

[35] Drumm, T., and Goldman, W., *The geometry of crooked planes,* Topology **38** (1999), no.2, 323–351.

[36] Dupont, J., *Bounds for characteristic numbers of flat bundles,* in "Algebraic topology, Aarhus 1978 (Proc. Sympos., Univ. Aarhus, Aarhus, 1978)," Lecture Notes in Math., **763**, Springer, Berlin, (1979), 109–119.

[37] Dupont, J., "Curvature and characteristic classes." Lecture Notes in Mathematics **640**. Springer-Verlag, Berlin-New York, 1978.

[38] Fried, D., and Goldman, W., *Three-dimensional affine crystallographic groups,* Adv. Math. **47** (1983), 1–49.

[39] Ghys, E., *Groupes d'homéomorphismes du cercle et cohomologie bornée,* in "The Lefschetz centennial conference, Part III, (Mexico City 1984)," Contemp. Math. **58**, Amer. Math. Soc., Providence RI (1987), 81–106.

[40] Goldman, W., *Discontinuous groups and the Euler class,* Doctoral dissertation, University of California, Berkeley (1980).

[41] Goldman, W., *Characteristic classes and representations of discrete subgroups of Lie groups.* Bull. Amer. Math. Soc. (N.S.) **6** (1982), no. 1, 91–94.

[42] Goldman, W., *Representations of fundamental groups of surfaces,* in "Geometry and Topology, Proceedings, University of Maryland 1983-1984," J. Alexander and J. Harer (eds.), Lecture Notes in Mathematics **1167** (1985), 95–117, Springer-Verlag, New York.

[43] Goldman, W., *Topological components of spaces of representations.* Invent. Math. **93** (1988), no. 3, 557–607.

[44] Goldman, W., *The Margulis Invariant of Isometric Actions on Minkowski (2+1)-Space,* in "Ergodic Theory, Geometric Rigidity and Number Theory," Springer-Verlag (2002), 149–164.

[45] Goldman, W., *Higgs bundles and geometric structures on surfaces,* in "The Many Facets of Geometry: A tribute to Nigel Hitchin," O. García-Prada, J. P. Bourgignon, and S. Salamon (eds.), Oxford University Press (2010), 129–163, arXiv:0805.1793.

[46] Goldman, W., *Locally homogeneous geometric manifolds,* Proceedings of the 2010 International Congress of Mathematicians, Hyderabad, India, Hindustan Book Agency, New Delhi, India (2010), 717–744.

[47] Goldman, W., and Kamishima, Y., *The fundamental group of a compact flat Lorentz space form is virtually polycyclic.* J. Diff. Geo. **19** (1984), no. 1, 233–240.

[48] Goldman, W., and Labourie, F., *Geodesics in Margulis spacetimes,* Ergodic Theory and Dynamical Systems, Special memorial issue for Daniel Rudolph **32** (2012), 643–651. arXiv:1102.0431.

[49] Goldman, W., Labourie, F., and Margulis, G., *Proper affine actions and geodesic flows of hyperbolic surfaces,* Ann. Math. **170** (2009), 1051–1083, arXiv:math.DG/0406247.

[50] Goldman, W., and Margulis, G., *Flat Lorentz 3-manifolds and cocompact Fuchsian groups,* in "Crystallographic Groups and their Generalizations," Contemp. Math. **262** (2000), 135–146, Amer. Math. Soc.

[51] Goldman, W., and Millson, J., *The deformation theory of representations of fundamental groups of compact Kähler manifolds.* Publ. Math. I.H.E.S. *67* (1988), 43–96.

[52] Grunewald, F., and Margulis, G., *Transitive and quasitransitive actions of affine groups preserving a generalized Lorentz-structure.* J. Geom. Phys. **5** (1988), no. 4, 493–531.

[53] Hitchin, N., *The self-duality equations on Riemann surfaces,* Proc. Lond. Math. Soc. **55** (1987), 59–126.

[54] Kneser, H., *Die kleinste Bedeckungszahl innerhalb einer Klasse von Flächenabbildungen,* Math. Ann. **103** (1930), no. 1, 347–358.

[55] Kostant, B., and Sullivan, D., *The Euler characteristic of a compact affine space form is zero,* Bull. Amer. Math. Soc. **81** (1975).

[56] Labourie, F., *Fuchsian affine actions of surface groups,* J. Diff. Geo. **59** (2001), no. 1, 15–31.

[57] Margulis, G., *Free properly discontinuous groups of affine transformations,* Dokl. Akad. Nauk SSSR **272** (1983), 937–940.

[58] Margulis, G., *Complete affine locally flat manifolds with a free fundamental group,* J. Soviet Math. **134** (1987), 129–134.

[59] Mess, G., *Lorentz spacetimes of constant curvature,* Geom. Ded. **126** (2007), 3–45.

[60] Meyer, W., *Die Signatur von lokalen Koeffizientensystemen und Faserbündeln.* Bonn. Math. Schr. **53** (1972), viii–59.

[61] Milnor, J., *On the existence of a connection with curvature zero,* Comm. Math. Helv. **32** (1958), 215–223.

[62] Milnor, J., *Hilbert's problem 18: on crystallographic groups, fundamental domains, and on sphere packing. Mathematical developments arising from Hilbert problems,* Proc. Sympos. Pure Math., Vol. XXVIII, Amer. Math. Soc., Providence, RI (1976), 491–506.

[63] Milnor, J., *On fundamental groups of complete affinely flat manifolds,* Adv. Math., **25** (1977), 178–187.

[64] Milnor, J., and Stasheff, J., "Characteristic classes." Annals of Mathematics Studies **76**, Princeton University Press, Princeton, NJ, 1974.

[65] Smillie, J., *Flat manifolds with nonzero Euler characteristic,* Comm. Math. Helv. **52** (1977), 453–456.

[66] Smillie, J., *The Euler characteristic of flat bundles* (preprint).

[67] Sullivan, D., *A generalization of Milnor's inequality concerning affine foliations and affine manifolds.* Comm. Math. Helv. **51** (1976), no. 2, 183–189.

[68] Thurston, W., *Foliations of three-manifolds which are circle bundles,* Doctoral dissertation, University of California, Berkeley, 1972.

[69] Thurston, W., *The geometry and topology of 3-manifolds,* mimeographed notes, Princeton University, 1979.

[70] Thurston, W., *Minimal stretch maps between hyperbolic surfaces,* arXiv:math.GT/9801039.

[71] Tits, J., *Free subgroups in linear groups,* J. Algebra **20** (1972), 250–270.

[72] Toledo, D., *Representations of surface groups in complex hyperbolic space,* J. Diff. Geo. **29** (1989), no. 1, 125–133.

[73] Tomanov, G., *The virtual solvability of the fundamental group of a generalized Lorentz space form.* J. Diff. Geo. **32** (1990), no. 2, 539–547.

[74] Tomanov, G., *On a conjecture of L. Auslander.* C. R. Acad. Bulgare Sci. **43** (1990), no. 2, 9–12.

[75] Turaev, V. G., *A cocycle of the symplectic first Chern class and Maslov indices,* Funktsional. Anal. i Prilozhen. **18** (1984), no. 1, 43–48.

[76] Turaev, V. G., *The first symplectic Chern class and Maslov indices,* Studies in topology, V. Zap. Nauchn. Sem. Leningrad. Otdel. Mat. Inst. Steklov. (LOMI) **143** (1985), 110D129, 178.

[77] Whitney, H., *On regular closed curves in the plane.* Compositio Mathematica **4** (1937), 276–284.

[78] Wood, J. W., *Bundles with totally disconnected structure group.* Comm. Math. Helv. **46** (1971), 257–273.

Milnor's problem on the growth of groups and its consequences

Rostislav Grigorchuk

ABSTRACT. We present a survey of results related to Milnor's problem on group growth. We discuss the cases of polynomial growth and exponential but not uniformly exponential growth; the main part of the article is devoted to the intermediate (between polynomial and exponential) growth case. A number of related topics (growth of manifolds, amenability, asymptotic behavior of random walks) are considered, and a number of open problems are suggested.

1 INTRODUCTION

The notion of the growth of a finitely generated group was introduced by A. S. Schwarz (also spelled Schvarts and Švarc) [Š55] and independently by Milnor [Mil68b, Mil68a]. Particular studies of group growth and their use in various situations have appeared in the works of Krause [Kra53], Adelson-Velskii and Shreider [AVŠ57], Dixmier [Dix60], Dye [Dye59], [Dye63], Arnold and Krylov [AK63], Kirillov [Kir67], Avez [Ave70], Guivarc'h [Gui70], [Gui71], [Gui73], Hartley, Margulis, Tempelman, and other researchers. The note of Schwarz did not attract a lot of attention in the mathematical community and was essentially unknown to mathematicians both in the USSR and the West (the same happened with papers of Adelson-Velskii, Dixmier, and of some other mathematicians). By contrast, the note of Milnor [Mil68a], and especially the problem raised by him in [Mil68b], initiated a lot of activity and opened new directions in group theory and areas of its applications.

The motivation for the studies of Schwarz and Milnor on the growth of groups was of geometric character. For instance, it was observed by Schwarz that the rate of volume growth of the universal cover $\tilde{M}$ of a compact Riemannian manifold M coincides with the rate of growth of the fundamental group $\pi_1(M)$ [Š55]. At the same time, Milnor and Wolf demonstrated that growth type of the fundamental group gives some important information about the curvature of the manifold. A relation between the growth of a group and its amenability was discovered by Adel'son-Vel'skii and Shreider (spelled also as Šreǐder) [AVŠ57]; specifically, subexponential growth implies the amenability of the group, i.e., the existence of an invariant mean).

The problem posed by Milnor focuses on two main questions:

1. Are there groups of intermediate growth between polynomial and exponential?

2. What are the groups with polynomial growth?

The author is supported by NSF grant DMS-1207688, ERS grant GA 257110 "RaWG", and by the Simons Foundation

Moreover, Milnor formulated a remarkable conjecture about the coincidence of the class of groups with polynomial growth and the class of groups containing a nilpotent subgroup of finite index (i.e., with the class of virtually nilpotent groups) which was later proved by Gromov [Gro81a].

The first part of Milnor's question was formulated originally by him in the following form: "Is it true that the growth function of every finitely generated group is necessarily equivalent to a polynomial or to the function 2^n?" This was answered in the negative by the author in 1983 [Gri83], [Gri84b]. Despite the negative character of the answer, the existence of groups of intermediate growth made group theory and the areas of its applications much richer. Eventually it led to the appearance of new directions in group theory: self-similar groups [GNS00], [Nek05], branch groups [Gri00a], [Gri00b], [BGŠ03], and iterated monodromy groups [BGN03], [Nek05]. Moreover, completely new methods, used in the study of groups of intermediate growth, stimulated intensive studies of groups generated by finite automata — a direction on the border between computer science and algebra initiated by V. M. Glushkov in the beginning of the 1960s [Glu61], [KAP85], [GNS00], [BGK$^+$08]. Groups of intermediate growth were used and continue to be used in many different situations: in the study of amenable groups [Gri84b], [Gri98], in topology [FT95], in theory of random walks on groups [Ers10], [Kai05], [Ers04a], in theory of operator algebras [Gri05], [GN07], [Gri], in dynamical systems [BN08], [BP06], in percolation [MP01b], in the study of cellular automata [MM93], in the theory of Riemannian surfaces [Gri89b], etc.

As was already mentioned, the study of the growth of groups was introduced by Schwarz and Milnor. However, some preliminary cases were considered by H. U. Krause in his Ph.D thesis, defended in Zurich in 1953 [Kra53], and geometric growth was considered around 1953 by V. A. Efremovich [Efr53]. Later, the question of group growth was considered in certain situations by Adelson-Velskii and Shreider, Dixmier, Dye, Arnold and Krylov, Guivarc'h, Kirillov, Margulis, Tempelman, and other researchers; however, as has been already mentioned, the systematic study of group growth started only with Milnor's work in 1968.

The first period of studies concerning group growth (in the 1960s and 1970s) was dedicated to the study of growth of nilpotent and solvable groups. During a short period of time, Milnor, Wolf, Hartley, Guivarc'h and Bass discovered that nilpotent groups have polynomial growth of integer degree, given by a number (6.2) expressed in terms of the lower central series. The converse fact, namely, that polynomial growth implies the virtual nilpotence of the group, was proven by M. Gromov in his remarkable paper [Gro81a], which stimulated a lot of activity in different areas of mathematics. Gromov's proof uses the idea of the limit of a sequence of metric spaces, as well as Montgomery and Zippin's solution of Hilbert's 5th problem [MZ74]. Van den Dries and Wilkie [vdDW84] used methods of nonstandard analysis to explore Gromov's idea in order to slightly improve his result (in particular, to outline a broader approach to the notion of a cone of a group), while B. Kleiner, using the ideas of Colding-Minicozzi [CM97] and applying the techniques of harmonic functions, gave a proof of Gromov's theorem which doesn't rely on the techniques in the Montgomery-Zipin solution to Hilbert's 5th problem. This approach was explored by Y. Shalom and T. Tao to get effective and quantitative results about the polynomial growth case [ST10].

The author's paper [Gri89a], followed by the ICM Kyoto paper [Gri91], raised a very interesting question, which we discuss in detail in this article (section 10) and call the *gap conjecture*. The conjecture states that if the growth degree of a finitely generated group is strictly smaller than $e^{\sqrt{n}}$, then it is polynomial and the group is virtually nilpotent. If proven, this conjecture would give a far-reaching generalization of Gromov's polynomial growth theorem. There are several results, some presented here, supporting the conjecture.

It is interesting that the *p*-adic analogue of Hilbert's 5th Problem, solved by M. Lazard [Laz65], was used by the author [Gri89a, Gri91] to obtain results related to the gap conjecture; these are discussed in section 10. See also the work of A. Lubotzky, A. Mann, and D. Segal [LM91, LMS93] for a description of the groups with polynomial subgroup growth.

The notion of growth can be defined for many algebraic and combinatorial objects, in particular for semigroups, associative and Lie algebras, graphs, and discrete metric spaces as discussed, for instance, in [dlHGCS99]. There is much more freedom in the asymptotic behavior of growth in each of these classes of algebraic objects, and there is no hope to obtain a general result characterizing polynomial growth similar to Gromov's theorem. But in [Gri88], the author was able to extend the theorem of Gromov to the case of cancellative semigroups (using the notion of nilpotent semigroup introduced by A. I. Malcev [Mal53]).

Let us say a few words about the exponential growth case. Perhaps typically, a finitely generated group has exponential growth. The fact that any solvable group which is not virtually nilpotent has exponential growth was established by Milnor and Wolf in 1968 [Mil68c], [Wol68]. A direct consequence of the Tits alternative [Tit72] is that a finitely generated linear group has exponential growth if it is not virtually nilpotent.

The exponential growth of a group immediately follows from the existence of a free subgroup on two generators, or even a free sub-semigroup on two generators. There are many classes of groups (i.e., solvable non-virtually nilpotent groups, non-elementary Gromov hyperbolic groups, linear groups without free subgroup on two generators, etc.) which are known to contain such objects. But many infinite torsion groups also have exponential growth, as was first proved by S. I. Adian in the case of free Burnside groups of large odd exponent [Adi79].

The second period of studies of group growth begins in the 1980s and can be divided into three directions: the study of analytic properties of growth series $\Gamma(z) = \sum_{n=0}^{\infty} \gamma(n)z^n$, the study of groups of intermediate growth, and the study around Gromov's problem on the existence of groups of exponential but not uniformly exponential growth [Gro81b]. This paper describes the main developments of the second and third directions in detail, while the direction of study of growth series is only briefly mentioned toward the end of Section 3.

It was discovered that for many groups and even for classes of groups, the function represented by the growth series $\Gamma(z)$ is rational. For instance, this holds for virtually abelian groups [Kla81a], [Kla81b], [Ben83], for Gromov hyperbolic groups (with any finite system of generators) [Gro87], [CDP90a], and for Coxeter groups with the canonical system of generators [Par91]. But the growth series can be irrational even in the case of nilpotent groups of nilpotency degree 2 (for instance, for some higher rank Heisenberg groups), and it can be rational for one system of generators and irrational for another [Sto96]. A very interesting approach to the study of growth functions, together with applications, is suggested by Kyoji Saito [Sai10], [Sai11].

All known examples of groups of intermediate growth are groups of branch type and are self-similar groups, or have some self-similarity features (for example, groups from the class $\mathcal{G}_\omega, \omega \in \Omega$ discussed in section 9), or arise from constructions based on the use of branch groups of self-similar type. These groups act on spherically homogeneous rooted trees and can be studied using techniques which have been developed during the last three decades.

Until recently, there were no examples of groups of intermediate growth with a precise estimate on the growth rate (there were only upper and lower bounds on the growth as discussed in detail in section 11). Results about oscillatory behavior of the growth function were presented in [Gri84b] and [Gri85b] in connection with the existence of groups with incomparable growth. Recent articles of L. Bartholdi and A. Erschler [BE10], [BE11] provide examples of groups with explicitly computed intermediate behavior of the type $\exp n^\alpha, \alpha < 1$ and many other types of growth, as given in Theorem 13.5.

During the last three decades, many remarkable properties of groups of intermediate growth were found, and some of these properties will be listed shortly. But, unfortunately, we do not know yet if there are finitely presented groups of intermediate growth, and this is the main open problem in the field.

The paper is organized as follows. After providing some background about group growth in Section 3, and formulation of Milnor's problem in Section 4, we discuss geometric motivations for the group growth and the relation of growth with amenability (Sections 5 and 7). Section 6 describes results obtained before 1981 and Section 8 contains an account of results about the polynomial growth case.

We describe the main construction of groups of intermediate growth, methods of obtaining upper and lower bounds, and the Bartholdi-Erschler construction in Sections 9, 11, and 13). The gap conjecture is the subject of Section 10. In Section 12, we discuss the relation between the growth of groups and asymptotic characteristics of random walks on them. The final Section 14 contains a short discussion concerning Gromov's problem on groups of uniformly exponential growth, on the oscillation phenomenon that holds in the intermediate growth case, and on the role of just-infinite groups in the study of group growth.

A number of open problems are included in the text. In our opinion they are among the most important problems in the field of group growth. We hope that these problems will stimulate further studies of group growth and related topics.

During the last two decades, a number of expositions of various topics related to the growth of groups and their application have been published. These include the following sources [Wag93], [Nav11], [GdlH97], [CSMS01], [GP08]. We especially recommend the book of P. de la Harpe [dlH00], which is a comprehensive source of information about finitely generated groups in general and, in particular, about their growth, as well as the recent books of T. Ceccherini-Silberstein and M. Coornaert [CSC10] and A. Mann [Man12], which contains additional material.

The theory of growth of groups is a part of a bigger area of mathematics that studies coarse asymptotic properties of various algebraic and geometric objects. Some views in this direction can be found in the following sources: [Lub94], [BH99], [Roe03], [BdlHV08], and [NY12].

The forementioned sources are recommended to the reader for an introduction to the subject, to learn more details about some of the topics considered in this article, or to get information about some other directions of research involving the growth of groups.

2 ACKNOWLEDGMENTS

The author would like to express his thanks to L. Bartholdi, A. Bonifant, T. Chec-cherini-Silberstein, T. Delzant, A. Erschler, V. Kaimanovich, A. Mann, T. Nag-nibeda, P. de la Harpe, I. Pak, M. Sapir, and J. S. Wilson. My special thanks to S. Sutherland and Z. Sunic for tremendous help with preparation of this paper and for valuable remarks, comments, and suggestions, and to M. Lyubich for encouraging me to write this survey. Parts of this work were completed during a visit to the Institut Mittag-Leffler (Djursholm, Sweden) during the program Geometric and Analytic Aspects of Group Theory.

3 PRELIMINARY FACTS

Let G be a finitely generated group with a system of generators $A = \{a_1, a_2, \ldots, a_m\}$ (throughout the paper we consider only infinite finitely generated groups and only finite systems of generators). The *length* $|g| = |g|_A$ of the element $g \in G$ with respect to A is the length n of the shortest presentation of g in the form

$$g = a_{i_1}^{\pm 1} a_{i_2}^{\pm 1} \ldots a_{i_n}^{\pm 1},$$

where a_{i_j} are elements in A. This depends on the set of generators, but for any two systems of generators A and B there is a constant $C \in \mathbb{N}$ such that the inequalities

$$|g|_A \leq C|g|_B, \qquad |g|_B \leq C|g|_A \tag{3.1}$$

hold. To justify this, it is enough to express each a-generator as a word in b-generators, and vice versa: $a_i = A_i(b_\mu), b_j = B_j(a_\nu)$. Then $C = max_{i,j}\{|A|_i, |B|_j\}$, where $|W|$ denotes the length of the word W. In addition to the length, one can also introduce a *word metric* $d_A(g, h) = |g^{-1}h|, g, h \in G$ on a group which is left invariant. More general types of left-invariant metrics and length functions on groups can be considered and studied as well. For instance, one can assign different positive weights to generators and define the length and metric according to these weights.

The *growth function* of a group G with respect to the generating set A is the function

$$\gamma_G^A(n) = |\{g \in G : |g|_A \leq n\}|,$$

where $|E|$ denotes the cardinality of the set E, and n is a natural number.

If $\Gamma = \Gamma(G, A)$ is the Cayley graph of a group G with respect to a generating set A, then $|g|$ is the combinatorial distance from vertex g to vertex e (represented by identity element $e \in G$), and $\gamma_G^A(n)$ counts the number of vertices at combinatorial distance $\leq n$ from e (i.e., it counts the number of elements in the ball of radius n with center at the identity element).

It follows from (3.1) that the growth functions $\gamma_G^A(n), \gamma_G^B(n)$ satisfy the inequalities

$$\gamma_G^A(n) \leq \gamma_G^B(Cn), \qquad \gamma_G^B(n) \leq \gamma_G^A(Cn). \tag{3.2}$$

The dependence of the growth function on the generating set is an inconvenience, and it is customary to avoid it by using the following trick. Following

J. Milnor, two functions on the naturals $\gamma_1(n)$ and $\gamma_2(n)$ are called *equivalent* (written $\gamma_1(n) \sim \gamma_2(n)$) if there is a constant $C \in \mathbb{N}$ such that $\gamma_1(n) \leq \gamma_2(Cn)$, $\gamma_2(n) \leq \gamma_1(Cn)$ for all $n \geq 1$. Then according to (3.2), the growth functions constructed with respect to two different systems of generators are equivalent. The class of equivalence $[\gamma_G^A(n)]$ of the growth function is called the *degree of growth* (the *growth degree*), or the *rate of growth* of a group G. It is an invariant of a group not only up to isomorphism but also up to a weaker equivalence relation *quasi-isometry*.

Recall that two finitely generated groups G and H with generating sets A and B, respectively, are quasi-isometric if the metric spaces (G, d_A) and (H, d_B) are quasi-isometric. Here d_A and d_B are the *word metrics* on G and H defined as $d_A(f, g) = |f^{-1}g|_A, d_B(h, l) = |h^{-1}l|_B$, with $f, g \in G$ and $h, l \in H$. Two metric spaces $(X, d_1), (Y, d_2)$ are quasi-isometric if there is a map $\phi: X \to Y$ and constants $C \geq 1, D \geq 0$ such that

$$\frac{1}{C}d_1(x_1, x_2) - D \leq d_2(\phi(x_1), \phi(x_2)) \leq Cd_1(x_1, x_2) + D$$

for all $x_1, x_2 \in X$; a further requirement is that there is a constant $L > 0$ so that for any point $y \in Y$, there is a point $x \in X$ with

$$d_2(y, \phi(x)) \leq L.$$

This concept is due to M. Gromov [Gro87] and is one of the most important notions in geometric group theory, allowing the study of groups from coarse point of view.

It is easy to see that the growth of a group coincides with the growth of a subgroup of finite index and that the growth of a group is not smaller than the growth of a finitely generated subgroup or a factor group. We will say that a group is *virtually nilpotent* (resp. virtually solvable) if it contains a nilpotent (solvable) subgroup of finite index.

We will also consider a preorder $\preceq$ on the set of growth functions:

$$\gamma_1(n) \preceq \gamma_2(n) \tag{3.3}$$

if there is an integer $C > 1$ such that $\gamma_1(n) \leq \gamma_2(Cn)$ for all $n \geq 1$. This makes a set $\mathcal{W}$ of growth degrees of finitely generated groups a partially ordered set. The notation $\prec$ will be used in this article to indicate a strict inequality.

Observe that Schwarz in his note [Š55] used formally weaker equivalence relation $\sim_1$ given by inequalities

$$\gamma_G^A(n) \leq C\gamma_G^B(Cn), \qquad \gamma_G^B(n) \leq C\gamma_G^A(Cn). \tag{3.4}$$

However, both equivalence relations coincide when restricted to the set of growth functions of infinite finitely generated groups, as was observed in [Gri84b, Proposition 3.1]. Therefore, we will use either of them, depending on the situation.

Because of the independence of the growth rate on a generating set, we will usually omit subscripts in the notation.

Let us list the main examples of growth rates that will be used later:

- The power functions n^α belong to different equivalence classes for different $\alpha \geq 0$.

- The polynomial function $P_d(n) = c_d n^d + \cdots + c_1 n + c_0$, where $c_d \neq 0$ is equivalent to the power function n^d.

- All exponential functions $\lambda^n, \lambda > 1$ are equivalent and belong to the class $[2^n]$ (or to $[e^n]$.)

- All functions of *intermediate type* $e^{n^\alpha}, 0 < \alpha < 1$, belong to different equivalence classes.

Observe that this is not a complete list of rates of growth that a group may have.

A free group F_m of rank m has $2m(2m-1)^{n-1}$ elements of length n with respect to any free system of generators A and

$$\gamma^A_{F_m}(n) = 1 + 2m + 2m(2m-1) + \cdots + 2m(2m-1)^{n-1} \sim 2^n.$$

Since a group with m generators can be presented as a quotient group of a free group of rank m, the growth of a finitely generated group cannot be faster than exponential (i.e., it can not be superexponential). Therefore, we can split the growth types into three classes:

- *Polynomial* growth. A group G has a *polynomial* growth is there are constants $C > 0$ and $d > 0$ such that $\gamma(n) < Cn^d$ for all $n \geq 1$. This is equivalent to

$$\overline{lim}_{n \to \infty} \frac{\log \gamma(n)}{\log n} < \infty. \tag{3.5}$$

A group G has *weakly polynomial* growth if

$$\underline{lim}_{n \to \infty} \frac{\log \gamma(n)}{\log n} < \infty. \tag{3.6}$$

(as we will see later (3.5) is equivalent to (3.6)).

- *Intermediate* growth. A group G has *intermediate* growth if $\gamma(n)$ grows faster than any polynomial but slower than any exponent function $\lambda^n, \lambda > 1$ (i.e., $\gamma(n) \prec e^n$).

- *Exponential growth.* A group G has *exponential* growth if $\gamma(n)$ is equivalent to 2^n.

The case of exponential growth can be redefined in the following way. Because of the obvious semi-multiplicativity

$$\gamma(m+n) \leq \gamma(m)\gamma(n),$$

the limit

$$\lim_{n \to \infty} \sqrt[n]{\gamma(n)} = \kappa \tag{3.7}$$

exists. If $\kappa > 1$, then the growth is exponential. If $\kappa = 1$, then the growth is subexponential and therefore is either polynomial or intermediate.

REMARK 3.1. In some situations it is reasonable to extend the domain of growth function to all nonnegative real numbers. This can be done, for instance, by setting $\gamma(x) = \gamma([x])$, where $[x]$ is the integer part of x.

Milnor's equivalence relation $\sim$ on the set of growth functions is a coarse approach to the study of growth. Sometimes, in order to study the growth, weaker equivalence relations on the set of monotone functions are used, for instance, when the factor C appears not only in the argument of the growth function, but also in front of it (as in the case of the Schwarz equivalence relation). Additionally, in some situations, an additive term appears in the form $C\gamma(Cn) + Cn$ (for instance, in the study of Dehn functions of finitely presented groups, as suggested by Gromov [Gro87]). Instead of the constant C one can use the equivalence relation with $C = C(n)$, where $C(n)$ is a slowly growing function, for instance, a polynomial. In contrast with the suggested ways of weakening the equivalence relation leading to a loss of some information about the growth, the more precise evaluation of the rate of growth requires tools from analysis based on classical asymptotic methods.

There is a standard way to associate a *growth series* to a growth function, defined by

$$\Gamma(z) = \sum_{n=0}^{\infty} \gamma(n)z^n.$$

The radius of converges of the series is $R = 1/\kappa \geq 1$. If the analytic function represented by this series is a rational function, or more generally, an algebraic function, then the coefficients $\gamma(n)$ grow either polynomially or exponentially and, therefore, intermediate growth is impossible in this case. For some classes of groups (e.g., abelian groups or Gromov hyperbolic groups), the growth series is always a rational function [Can80, Gro87]. There are groups or classes of groups for which there is a system of generators with rational or algebraic growth series; for instance, Coxeter groups have this property with respect to the canonical system of generators [Par91]. There are examples of groups for which the growth series is a rational function for one system of generators but is not rational for another system of generators [Sto96]. Typically, the growth series of a group is a transcendental function.

There is also interest in the study of the *complete* growth function

$$\Gamma_*(z) = \sum_{g \in G} g z^{|g|} \in \mathbb{Z}[G][[z]]$$

viewed as an element of the ring of formal power series with coefficients in a group ring $\mathbb{Z}[G]$ of a group, or an *operator* growth function

$$\Gamma_\pi(z) = \sum_{g \in G} \pi(g) z^{|g|},$$

where $\pi(g)$ is a representation of the group in a Hilbert space (or, more generally, a Banach space) [Lia96, GN97]. An important case is when π is a left regular representation of the group in $l^2(G)$.

REMARK 3.2. The growth can be defined for many algebraic, combinatorial, geometric, probabilistic and dynamical objects. For instance, one can speak about the growth of a connected locally finite graph Γ, by which we mean the growth of the function $\gamma_{\Gamma,v}(n)$, which counts the number of vertices at combinatorial distance not more than n from a base vertex v. Growth of such graphs can be superexponential, but if the graph has uniformly bounded degree (for instance is a regular graph), the growth is at most exponential. The growth does not depend on the choice of v.

The growth of a Riemannian manifold M is, by definition, the growth as $r \to \infty$ of the function $\gamma(r) = Vol(B_x(r))$ expressing the volume of a ball of radius r with center at fixed point $x \in M$. The rate of growth is independent of the choice of x.

The definition of growth of a semigroup is similar to the group case. Also, one can define growth of finitely generated associative algebras, graded associative algebras, Lie algebras, etc. For instance, if $\mathcal{A} = \bigoplus_{n=0}^{\infty} \mathcal{A}_n$ is a finitely generated associative graded algebra defined over a field $\mathbb{F}$, then the growth of dimensions $d_n = \dim_{\mathbb{F}} \mathcal{A}_n$ determines the growth of $\mathcal{A}_n$ and the corresponding growth series

$$\mathcal{H}(z) = \sum_{n=0}^{\infty} d_n z^n \tag{3.8}$$

usually is called a *Hilbert-Poincaré* series.

Given a countable group G and a probabilistic measure μ on it whose support generates G, one can consider a right random walk on G which begins at the identity element $e \in G$ and such that the transitions $g \to gh$ happen with probability $\mu(h)$. One of the main characteristics of such random process is the probability $P_{e,e}^{(n)}$ of return after n steps. This probability may decay exponentially (as $r^n, r < 1$) or subexponentially. The value

$$r = \overline{lim}_{n \to \infty} \sqrt[n]{P_{e,e}^{(n)}} \tag{3.9}$$

is called the *spectral radius* of the random walk (observe that $\lim_{n \to \infty} \sqrt[2n]{P_{e,e}^{(2n)}}$ exists). It was introduced by H. Kesten in [Kes59b]. In the case of a symmetric measure (i.e., when for any $g \in G$, the equality $\mu(g) = \mu(g^{-1})$ holds), the spectral radius coincides with the norm of the Markov operator of the random walk, and the subexponential decay of $P_{e,e}^{(n)}$ (i.e., the equality $r = 1$) holds if and only if the group is amenable in the von Neumann sense (i.e., when G has a left invariant mean) [vN29, Kes59a]. We will discuss this topic in more detail later in this paper.

4   THE PROBLEM AND THE CONJECTURE OF MILNOR

In his note published in the *American Mathematical Monthly* [Mil68b], Milnor formulated a remarkable problem concerning the growth of groups, as well as an ingenious conjecture concerning the polynomial growth case. We reproduce them here, but divide the problem into two parts (which was not done in [Mil68b]). As before, $\gamma(n)$ denotes the growth function of a finitely generated group (we shall keep this notation through the paper).

MILNOR'S PROBLEM.

(I) *Is the function $\gamma(n)$ necessarily equivalent either to a power of n or to exponential function 2^n?*

(II) *In particular, is the growth exponent*

$$d = \lim_{n \to \infty} \frac{\log \gamma(n)}{\log n} \qquad (4.1)$$

always either a well-defined integer or infinity? For which groups is $d < \infty$?

MILNOR'S CONJECTURE. *A possible conjecture would be that $d < \infty$ if and only if G contains a nilpotent subgroup of finite index.*

The first part of Milnor's problem is a question on the existence of groups of intermediate growth. The second part and the conjecture are oriented toward the study of the polynomial growth case and the regularity properties of growth functions.

Clearly the motivation for suggesting such a problem and a conjecture was based on Milnor's background in the area of group growth as of 1968. Later, we will provide more information on what was known about group growth around 1968.

In short, the history of solutions of Milnor's problem and his conjecture is the following. The conjecture was confirmed by M. Gromov [Gro81a]. This, together with the results of Guivarc'h [Gui70], [Gui71], B. Hartley (unpublished, but see "Added in Proof" in [Wol68]), and H. Bass [Bas72], showed that the upper limit (3.5) is a non-negative integer and that the finiteness of the limit (4.1) implies that the group is virtually nilpotent. The existence of the limit (4.1) also follows from results of the mathematicians cited here, giving the complete solution of the second part of Milnor's problem. In the case of nilpotent groups, the existence of the limit

$$\lim_{n \to \infty} \frac{\gamma(n)}{n^d} \qquad (4.2)$$

(where d is the degree of polynomial growth) was proved by P. Pansu [Pan83] and is an additional bonus in the study of the polynomial growth case.

The first part of Milnor's problem was solved in the negative by the author in 1983 [Gri83], [Gri84b], [Gri85a]. It was shown that there are groups of intermediate growth and, moreover, that there are uncountably many of them. Also, it was shown that there are pairs of groups with incomparable growth in the sense of the order $\preceq$. In addition, many other results about growth and algebraic properties of groups of intermediate growth were obtained around 1984 and later. Despite a negative solution of the first part of Milnor's problem, the fact that there are groups of intermediate growth made group theory richer and substantially extended the area of its applications.

All known examples of groups of intermediate growth are infinitely presented groups.

PROBLEM 1. *Is it true that the growth function of a finitely presented group is equivalent either to a polynomial or to the exponential function 2^n?*

In [GP08], it is conjectured that there are no finitely presented groups of intermediate growth. At the same time, the author suggests even a stronger conjecture.

CONJECTURE 4.1. *A finitely presented group either contains a free subsemigroup on two generators or is virtually nilpotent.*

Problem 1 is the main remaining open problem concerning group growth. As we will see later, there are many recursively presented groups of intermediate growth, in particular, the groups $\mathcal{G} = \mathcal{G}_\xi$ and $\mathcal{G}_\eta$ described in Section 9 are recursively presented; we note that the group $\mathcal{G}$ will serve in this text as the main illustrating example. By Higman's embedding theorem [LS77] such groups embed into finitely presented groups. It is worth mentioning that for the group $\mathcal{G}_\xi$, there is a very precise and nice embedding based on the use of Lysionok presentation (9.1), as was observed in [Gri98] (a similar claim holds for $\mathcal{G}_\eta$). In fact, a similar embedding exists for all groups with finite L-presentation that are defined and studied in [BGŠ03, Bar03a]. The corresponding finitely presented group (which we denote here by $\tilde{\mathcal{G}}_\xi$) is an ascending HNN-extension of $\mathcal{G}_\xi$. It has a normal subgroup N which is ascending union of conjugates of a subgroup isomorphic to $\tilde{\mathcal{G}}_\xi$ with the quotient $\tilde{\mathcal{G}}_\xi / N$ isomorphic to a infinite cyclic group. Unfortunately (or fortunately), the group $\tilde{\mathcal{G}}_\xi$ has exponential growth but shares the property of amenability with $\mathcal{G}_\xi$. The quotients of this group are described in [SW02]. The idea of finding a finitely presented group containing $\mathcal{G}_\xi$ as a normal subgroup (with the hope of thus obtaining a finitely presented group of intermediate growth) fails, as was observed by M. Sapir, because $\tilde{\mathcal{G}}_\xi$ cannot serve as a normal subgroup of any finitely presented groups (see the argument in `http://mathoverflow.net/questions/73076/higman-embedding-theorem/`).

Moreover, it was observed by P. de la Harpe and the author that any finitely presented group $\hat{\mathcal{G}}$ that can be homomorphically mapped onto $\mathcal{G}$ contains a free subgroup on two generators and hence is of exponential growth. In view of these facts, it would be interesting to find a finitely presented group with a normal subgroup of intermediate growth and to find a finitely presented group without a free subgroup on two generators (or perhaps even a finitely presented amenable group) that can be mapped onto a group of intermediate growth. This is discussed in detail in [BGH13].

For the group $\mathcal{G}_\eta$ and many other groups possessing a presentation of the type (9.1) (i.e., a presentation involving a finite set of relators and their iterations by one or more, but finitely many, substitutions), embeddings into finitely presented groups similar to the one for $\mathcal{G}$ also exist [BGŠ03, Bar03a].

5 RELATIONS BETWEEN GROUP GROWTH AND RIEMANNIAN GEOMETRY

One of the first results showing the usefulness of the notion of group growth was the result of A. S. Schwarz, who proved the following theorem in 1957.

THEOREM 5.1. *Let $\tilde{M}$ be the universal cover of a compact Riemannian manifold M. Then the rate of growth of $\tilde{M}$ is equal to the rate of growth of the fundamental group $\pi_1(M)$.*

In addition to the obvious examples of groups with polynomial or exponential growth (free abelian groups and free noncommutative groups, respectively), in his note Schwarz produced examples of solvable (and even metabelian) groups of exponential growth. These are defined as a semidirect product of $\mathbb{Z}^d, d \geq 2$ and a cyclic group generated by an automorphism $\varphi \in SL_d(\mathbb{Z})$ given by a matrix with at least one eigenvalue off the unit circle.

In Theorem 5.1, the comparison of the growth rates of a manifold and a group is considered with respect to the equivalence relation

$$\gamma_1(n) \sim_1 \gamma_2(n) \quad \Leftrightarrow \quad \exists C > 0 \, \forall n \; \gamma_1(n) \geq C\gamma_2(Cn) \, \& \, \gamma_2(n) \geq C\gamma_1(Cn); \quad (5.1)$$

the domain of $\gamma_{\pi_1(M)}(n)$ is extended in the natural way to $\mathbb{R}_+$. Of course, the equivalence relations $\sim$ and $\sim_1$ defined on the set of monotone functions of natural argument are different. But, as was already mentioned in the previous section, they coincide on the set of growth functions of finitely generated infinite groups.

The theorem of Schwarz relates the growth of groups with the volume growth of the universal cover of a compact Riemannian manifold. In fact a more general statement which deals with non-universal and even non-regular coverings holds; we shall formulate this shortly.

Milnor's investigation into the relation between growth and curvature led him to the following two results [Mil68a].

THEOREM 5.2. *If M is a complete d-dimensional Riemannian manifold whose mean curvature tensor R_{ij} is everywhere positive semi-definite, then the growth function $\gamma(n)$ associated with any finitely generated subgroup of the fundamental group $\pi_1(M)$ must satisfy $\gamma(n) < Cn^d$ for some positive constant C.*

THEOREM 5.3. *If M is a compact Riemannian manifold with all sectional curvatures less than zero, then the growth function of the fundamental group $\pi_1(M)$ is exponential: $\gamma(n) > a^n$ for some constant $a > 1$.*

The next example considered by Milnor was the first step in the direction toward understanding the growth of nilpotent groups. Let G be the nilpotent Lie group consisting of all 3×3 triangular real matrices with 1s on the diagonal, and let $\mathcal{H}_3$ be the subgroup consisting of all integer matrices of the same form. Then the coset space $G/\mathcal{H}_3$ is a compact 3-dimensional manifold with fundamental group $\mathcal{H}_3$.

LEMMA 5.4. *The growth function of $\mathcal{H}_3$ is quartic:*

$$C_1 n^4 < \gamma_{\mathcal{H}_3}(n) < C_2 n^4$$

with $0 < C_1 < C_2$.

COROLLARY 5.5. *No Riemannian metric on $G/\mathcal{H}_3$ can satisfy either the hypothesis of Theorem 5.2 or the hypothesis of Theorem 5.3.*

The ideas of Milnor were used by A. Aves [Ave70] to get a partial answer to a conjecture of E. Hopf [Hop48]: Let M be a compact, connected Riemannian manifold without focal points. Then either the fundamental group $\pi_1(M)$ has exponential growth or M is flat.

Now let us go back to the theorem of Schwarz and present a more general statement. But before that, we need to recall some notions from geometric group theory. Let G be a finitely generated group with a system of generators A and let $H < G$ be a subgroup. Let $\Gamma = \Gamma(G, H, A)$ be the Schreier graph determined by the triple (G, H, A). The vertices of Γ are in bijection with cosets $gH, g \in G$, and two vertices gH and hH are joined by oriented edge (labeled by $a \in A$) if $hH = agH$. This notion is a generalization of the notion of a Cayley graph of a group (Cayley graphs correspond to the case when $H = \{e\}$). $\Gamma(G, H, A)$ is a

2m-regular graph, where m is the cardinality of the generating set. As was already defined, the growth function $\gamma_\Gamma(n)$ of a graph counts the number of vertices at combinatorial distance $\leq n$ from the base vertex v (we remind the reader that the choice of v does not play a role; it is natural for a Schreier graph to choose $v = 1H$).

If a group G with a generating set A acts transitively on a set X, then $\Gamma(G, H, A)$ is isomorphic to the graph of the action, i.e., the graph Γ_* with set of vertices X and set of edges consisting of pairs $(x, a(x)), x \in X, a \in A$. In this case the subgroup H coincides with the stabilizer $st_G(x)$.

THEOREM 5.6. *Let M be a compact Riemannian manifold, let H be a subgroup of the fundamental group $\pi_1(M)$, and let $\tilde{M}$ be a cover of M corresponding to H supplied by a Riemannian metric lifted from M. Then the growth function $\gamma_{\tilde{M}}(r)$ is $\sim_1$ equivalent to the growth function $\gamma_{\Gamma(\pi_1(M), H, A)}(r)$ (with domain naturally extended to $\mathbb{R}_+$) of the Schreier graph $\Gamma(\pi_1(M), H, A)$ where A is a finite system of generators of $\pi_1(M)$.*

PROOF. Triangulate M, lift the triangulation to $\tilde{M}$, and make the comparison of the volume growth of $\tilde{M}$ with the growth function $\gamma_{\Gamma(\pi_1(M), H, A)}(r)$ using the graph of the action of $\pi_1(M)$ on the preimage $p^{-1}(x)$, where $p \colon \tilde{M} \to M$ is the canonical projection, and then apply the arguments from [Š55]. □

A particular case of this theorem mentioned in [Gri89b] is the case of a regular (i.e., Galois) cover, i.e., when H is a normal subgroup of $\pi_1(M)$. In this case, the growth of the covering manifold $\tilde{M}$ coincides with the growth of the quotient group $\pi_1(M)/H$, and the latter is isomorphic to the group of deck transformations of the cover. This fact together with the results about groups of intermediate growth obtained in [Gri84b] allowed the author to construct uncountably many Riemannian surfaces supplied with groups of isometries acting on them cocompactly, having a topological type of oriented surface of infinite genus with one end, and which are not pairwise quasi-isometric [Gri89b].

6   RESULTS ABOUT GROUP GROWTH OBTAINED BEFORE 1981

Among the first publications to use of the notion of group growth were [Š55] and [AVŠ57]. The result of the article of Adelson-Velskii and Shreider relates growth with amenability and will be discussed in the next section. Then after more than a decade of sporadic appearances of group growth in various articles (partly listed in the introduction), the papers of Milnor [Mil68a], [Mil68c] and his note [Mil68b] appeared.

These publications, followed by Wolf's article [Wol68], attracted attention of many researchers and a flurry of activity occurred during a short period. The fact that nilpotent groups have polynomial growth was already observed in [Mil68a] (the example of Heisenberg group) and in full generality was studied by Wolf [Wol68] and B. Hartly (see "Added in Proof" in [Wol68]), Guivarc'h [Gui70, Gui71], and by H. Bass [Bas72], who showed that for nilpotent groups, the growth function satisfies the inequalities

$$C_1 n^d \leq \gamma(n) \leq C_2 n^d. \tag{6.1}$$

Here

$$d = \sum_i i \cdot rank_{\mathbb{Q}}(\gamma_i(G)/\gamma_{i+1}(G)), \tag{6.2}$$

where $\gamma_i(G)$ is ith member of the lower central series of the group, C_1 and C_2 are positive constants, and $\mathrm{rank}_\mathbb{Q}(A)$ is the torsion-free rank of the abelian group A.

In [Wol68], Wolf proved that a polycyclic group either contains a nilpotent subgroup of finite index and has polynomial growth or the growth is exponential. At the same time, Milnor observed that a solvable but not polycyclic group has exponential growth. This, in combination with Wolf's result, led to the fact that the growth of a solvable group is exponential except for the case when the group is virtually nilpotent.

A self-contained proof of the main result about growth of solvable groups was given by Tits in the appendix to Gromov's paper [Gro81a]. In [Rose74] J. Rosenblatt showed that groups of subexponential growth are *supramenable* (the class of supramenable groups is a subclass of the class of amenable groups; see the next section for the definition), and indicated that a solvable group of exponential growth contains a *free subsemigroup* on two generators (perhaps the latter was known before, but the author has no corresponding reference).

The remarkable theorem of Tits (usually called the *Tits alternative*) implies that a finitely generated subgroup of a linear group (i.e., a subgroup of $GL_n(\mathbb{F}), n \geq 1$, $\mathbb{F}$ a field), either contains a free subgroup on two generators or is virtually solvable [Tit72]. This, together with the results about the growth of solvable groups, implies that the growth of a finitely generated linear group is either polynomial or exponential.

The first part of Milnor's problem on growth was included by S. I. Adian in his monograph [Adi79], dedicated to the one of the most famous problems in algebra — the *Burnside problem on periodic groups*. Adian showed that the free Burnside group

$$B(m,n) = \langle a_1, a_2, \ldots, a_m | X^n = 1 \rangle, \tag{6.3}$$

of exponent $n \geq 665$ (n odd) with $m \geq 2$ has exponential growth. This is a stronger result than the result of P. S. Novikov and S. I. Adian [NA68b] about the infiniteness of $B(m,n)$ in the case $m \geq 2$ and odd $n \geq 4381$. A number of results about growth of semigroups was obtained by V. Trofimov [Tro80] (for more recent developments about growth of semigroups, see [BRS06], [Shn04], and [Shn05]). However, we are not going to get much into the details of growth in the semigroup case.

A useful fact about groups of intermediate growth is due to S. Rosset [Ros76].

THEOREM 6.1. *If G is a finitely generated group which does not grow exponentially and H is a normal subgroup such that G/H is solvable, then H is finitely generated.*

The basic tool for the proof of this theorem is *Milnor's lemma*.

LEMMA 6.2. *Let G be a finitely generated group with subexponential growth. If $x, y \in G$, then the group generated by the set of conjugates $y, xyx^{-1}, x^2yx^{-2}, \ldots$ is finitely generated.*

The theorem of Rosset was stated in 1976 for groups of subexponential growth, but its real application is to the case of groups of intermediate growth, because by a theorem of Gromov the groups of polynomial growth are virtually nilpotent and all subgroups in such groups are finitely generated.

The theorem of Rosset can be generalized. In the next statement we use the notion of elementary amenable group, which is defined in the next section. Observe that solvable groups constitute a subclass of the class of elementary amenable groups.

THEOREM 6.3. *Let G be a finitely generated group with no free subsemigroups on two generators and let the quotient G/N be an elementary amenable group. Then the kernel N is a finitely generated group.*

The proof of this fact is based on the use of a version of Milnor's lemma [LMR95, Lemma 1] and transfinite induction on the "complexity" of elementary amenable groups, defined shortly. Observe that this induction was used by Chou to prove the absence of groups of intermediate growth, infinite finitely generated torsion groups, and infinite finitely generated simple groups in the class of elementary amenable groups [Cho80]. Interesting results about algebraic properties of "generalized" elementary groups are obtained by D. Osin [Osi02].

In the introduction to his paper published in 1984 [Gri84b], the author wrote: "In the past decade, in group theory there appeared a direction that could be called 'Asymptotic Group Theory'" (perhaps this was the first time when the name "asymptotic group theory" was used in the mathematical literature). Milnor is one of the pioneers of this direction and his ideas and results contributed a lot to its formation. Asymptotic group theory studies various asymptotic invariants of groups, first of all asymptotic characteristics of groups (like growth), many of which were defined and studied during the last three decades. For instance, the notion of *cogrowth* (or *relative growth*) was introduced by author in [Gri80a] and later was used by him, Olshanskii, and Adian to answer some questions related to the *von Neumann Conjecture* on non-amenable groups [Gri79], [Ol'80], [Adi82]. The notion of *subgroup growth* was introduced by Grunewald, Segal, and Smith in [GSS88] and studied by many mathematicians (see [LS03] and the literature cited there). Dehn functions and their growth were introduced by Gromov and also happen to be a popular subject for investigation (see [BH99] and citations therein). Many asymptotic invariants of groups were introduced and studied by Gromov in [Gro93], and many other asymptotic invariants have been introduced since then. The asymptotic methods in the case of algebras (including the topics of self-similarity) are discussed in survey of E. Zelmanov [Zel05], [Zel07] and in [Gro08]. The direction of asymptotic group theory is flourishing at present time, and the author has no doubt that the situation will not change in the next few decades.

7 GROWTH AND AMENABILITY

DEFINITION 7.1 (John Von Neumann (1929)). *A group G is called* amenable *if there is a finitely additive measure μ defined on the algebra of all subsets of G and such that*

- $\mu(G) = 1, 0 \leq \mu(E) \leq 1, \forall E \subset G;$

- μ *is left invariant, i.e.,* $\forall E \subset G$:

$$\mu(E) = \mu(gE), \qquad \forall g \in G, \forall E \subset G.$$

The measure μ determines an invariant, positive, normalized functional m on Banach space $l^\infty(G)$ defined by

$$m(f) = \int_G f \, d\mu$$

and is called a *left invariant mean* (*LIM*). And, vice versa, any invariant, positive, normalized functional m determines a measure μ by restriction of its values on the characteristic functions of subsets. In a similar way, von Neumann defined amenability of action of a group on a set [vN29].

The simplest example of a non-amenable group is the free group F_2 on two generators. The simplest examples of amenable groups are finite groups and commutative groups.

In the preceding definition the group G is assumed to be a group with discrete topology. There is a version of this definition due to N. N. Bogolyubov [Bog], M. Day, and others for general topological groups [HR79] but it depends on the choice of the space of functions (bounded continuous or bounded uniformly continuous functions) as was observed by P. de la Harpe [dlH73]. In the case of locally compact groups one can use any of the mentioned spaces to define amenability [HR79], [Gre69]. Growth of groups also can be defined for locally compact, compactly generated groups, but we will not consider the case of topological groups here. We recommend to the reader the survey [dlHGCS99] on amenability and literature there.

Following M. Day [Day57], we use AG to denote the class of amenable groups. This class is extremely important for various topics in mathematics. In this section we discuss this notion briefly, because of its relation with growth. By the theorem of Adelson-Velskii and Schreider [AVŠ57], each finitely generated group of subexponential growth belongs to the class AG. This class contains finite groups and commutative groups and is closed under the following operations:

1. taking a *subgroup*;

2. taking a *quotient group*;

3. *extensions* (i.e., an extension of amenable group by amenable group is amenable);

4. *directed unions*:
    $G_\alpha \in AG, G_\alpha \subset G_\beta$ if $\alpha < \beta \Rightarrow \cup_\alpha G_\alpha \in AG$ (here we assume that the family of groups $\{G_\alpha\}$ is a directed family, i.e., for any α and β there is γ with $G_\alpha \leq G_\gamma$ and $G_\beta \leq G_\gamma$).

Let EG be the class of *elementary* amenable groups. This is the smallest class of groups which contains finite groups and commutative groups, and that is closed with respect to the operations (1)–(4). For such groups a kind of complexity can be defined in the following way, as was suggested by C. Chou [Cho80]. For each ordinal α, define a subclass EG_α of EG. EG_0 consists of finite groups and commutative groups. If α is a limit ordinal, then

$$EG_\alpha = \bigcup_{\beta \preceq \alpha} EG_\beta.$$

Further, $EG_{\alpha+1}$ is defined as the set of groups which are extensions of groups from the set EG_α by groups from the same set. The *elementary complexity* of a group $G \in EG$ is the smallest α such that $G \in EG_\alpha$.

One more class of subexponentially amenable groups, denoted here SG, was defined in [Gri98] as the smallest class which contains all groups of subexponential growth and is closed with respect to the operations (1)–(4). It is contained in the class of "good" groups, in the terminology of M. Freedman and P. Teichner [FT95]. Observe that the inclusions

$EG \subset SG \subset AG$ hold. A general concept of elementary classes is developed by D. Osin [Osi02], along with an indication of their relation to the Kurosh-Chernikov classes. We will mention the class SG again in Section 12.

The question of Day [Day57] concerning coincidence of the classes AG and EG was answered in the negative by the author as a result of the solution of the first part of Milnor's problem [Gri84b]. In fact, each group of intermediate growth belongs to the complement $AG \setminus EG$ as was showed in [Cho80]. And as there are uncountably many 2-generated groups of intermediate growth, the cardinality of the set $AG \setminus EG$ is the cardinality of the continuum.

The second question of Day about coincidence of classes AG and NF (the latter is the class of groups that do not contain a free subgroup on two generators) was answered in the negative by A. Olshanskii [Ol'80] and S. Adian [Adi82]. Sometimes this problem is formulated in the form of a conjecture, with attribution to von Neumann. A counterexample to a stronger version of the von Neumann conjecture was constructed in [Gri79], where a subgroup H of a free group F_2 with the property that the action of G on G/H is nonamenable and some nonzero power of each element in F_2 belongs to some conjugate of H. All three articles [Gri79], [Ol'80], [Adi82] use the cogrowth criterion of amenability from [Gri80a].

It is important to mention that both problems of Day had negative solutions, not only for the class of finitely generated groups but also for the class of finitely presented groups, as was established by the author and Olshanskii and Sapir respectively [Gri98, OS02]. In other words, there are examples of finitely presented amenable but non-elementary amenable groups, and there are examples of finitely presented non-amenable groups without a free subgroup on two generators. Therefore the situation with amenable/non-amenable groups is in a sense better than the situation with groups of intermediate growth where the existence of finitely presented groups is unknown (see problem 1).

The variety of different (but equivalent) definitions of the notion of amenable group is tremendous. Perhaps there is no other notion in mathematics which may compete with the notion of amenable group in the number of different definitions. The criteria of amenability of Tarski, Fölner, Kesten, Reiter, and many others were found during the eight decades of studies of amenability. Kesten's criterion expresses amenability in terms of spectral radius r of random walk on a group in the following way: a group G is amenable if and only if $r = 1$ (here we should assume that the measure defining the random walk is symmetric and its support generates a group; more on this in Section 12).

In the final part of his article [Mil68a], Milnor raised the following question.

> *Consider a random walk on the fundamental group of compact manifold of negative curvature. Is the spectral radius r necessarily less than 1?*

In view of Kesten's criterion, this is equivalent to the question about non-amenability of the fundamental group of a compact manifold of negative curvature. This problem was positively solved by P. Eberlain [Ebe73]. Related results are obtained by Yau [Yau75] (see also Proposition 3 on page 98 of the book by Gromov, Pansu, and Lafontaine [Gro81b]) and Chen [Che78]. As observed by T. Delzant, the solution of Milnor's problem can also be deduced from the results and techniques in the paper of A. Avez [Ave70]. A theorem, due to M. Anderson, considers the case of non-positive curvature (i.e., the curvature can be 0) [And87]. It is proved that amenable subgroups of such a group must be virtually abelian and must be the fundamental group of a flat totally geodesic submanifold. In fact, after the work of Gromov [Gro87] on hyperbolic groups, followed by the books [GH90] and [CDP90a], it became clear that fundamental groups of compact manifolds of negative curvature are non-elementary hyperbolic. Such groups contain a free subgroup on two generators and hence are nonamenable.

In 1974, J. Rosenblatt introduced the interesting notion of a *supramenable* group and proved that a group of subexponential growth is supramenable [Rose74]. A group G is called supramenable if for any G-set X and any nonempty subset $E \subset X$ there is a G-invariant finitely additive measure on the algebra of all subsets of X taking values in $[0, +\infty)$ and normalized by condition $\mu(E) = 1$. This property implies there is an invariant Radon measure for any cocompact continuous action on a locally compact Hausdorff space [KMR13]. It is worth mentioning that amenability of a group can be characterized by the property of a group having an invariant probability measure for any continuous action on a compact space, because of the theorem of Bogolyubov [Bog] (which generalizes a theorem of Bogolyubov-Krylov [Gre69]). Supramenability is discussed in the book [Wag93]. In the next section we will mention it again.

8 POLYNOMIAL GROWTH

Recall that a group G has polynomial growth if there are constants $C > 0$ and $d > 0$ such that $\gamma(n) < Cn^d$ for all $n \geq 1$, i.e., (3.5) holds. As was discussed in Section 6, a nilpotent group has polynomial growth and, therefore, a virtually nilpotent group also has polynomial growth. In his remarkable paper [Gro81a], Gromov established the converse: polynomial growth implies the virtual nilpotence of the group.

THEOREM 8.1 (Gromov 1981). *If a finitely generated group G has polynomial growth, then G contains a nilpotent subgroup of finite index.*

This theorem, together with known information about growth of nilpotent groups, gives a wonderful characterization of groups of polynomial growth in algebraic terms: a finitely generated group has polynomial growth if and only if it is virtually nilpotent.

The proof of Gromov's theorem is very geometric by its nature, but it also uses some fundamental facts from algebra and analysis. Roughly speaking, the idea of Gromov consists of considering the group as a metric space and at looking at this space via a "macroscope." The implementation of this idea goes through the development of the technique of limits of sequences of metric spaces. Given a group of polynomial growth G, this method associates with it a Lie group $\mathcal{L}_G$ which has finitely many connected components and a nontrivial homomorphism $\phi \colon G \to \mathcal{L}_G$. The rest of the proof is a matter of mathematical culture modulo known facts from algebra and analysis, including the Jordan theorem, Tits alternative, and the Milnor-Wolf theorem on growth of solvable groups. The skeleton of Gromov's proof has been adapted to many other proofs. In particular, it initiated the study of *asymptotic cones* of groups [vdDW84, Gro93].

Combining his theorem with the results of Shub and Franks [Shu70] on properties of fundamental group of a compact manifold admitting an expanding self-covering, Gromov deduced the following statement.

COROLLARY 8.2. *An expanding self-map of a compact manifold is topologically conjugate to an ultra-nil-endomorphism.*

In fact, Gromov obtained a stronger result about polynomial growth.

THEOREM 8.3 (Gromov's effective polynomial growth theorem). *For any positive integers d and k, there exist positive integers R, N, and q with the following property. If a group G with a fixed system of generators satisfies the inequality $\gamma(n) \leq kn^d$ for $n = 1, 2, \ldots, R$, then G contains a nilpotent subgroup H of index at most q and whose degree of nilpotence is at most N.*

In [Man07], A. Mann made the first step in the direction of getting a concrete effective bound on nilpotency class in Gromov's Theorem 8.1 by showing that if G is a finitely generated group of polynomial growth of degree d, then G contains a finite index nilpotent subgroup of nilpotence class at most $\sqrt{2d}$. This is a corollary of the Bass-Guivarch formula (6.2). Mann also showed that G contains a finite-by-nilpotent subgroup of index at most $g(d)$, the latter function being the maximal order of a finite subgroup of $GL_d(\mathbb{Z})$, which is known to be $2^n n!$ in most cases (e.g., for $d = 1, 3, 5$ and $d \geq 11$) [Man12, Theorem 9.8].

Another fact obtained by Gromov in [Gro81a] is the following: "If an infinitely generated group G has no torsion and each finitely generated subgroup $G_1 < G$ has polynomial growth of degree at most d, then G contains a nilpotent subgroup of finite index." On the other hand, A. Mann showed in [Man07] that the property of an infinitely generated group G being virtually nilpotent and of finite rank is equivalent to the property that there are some (fixed) constants C and d such that $\gamma_H^A(n) \leq Cn^d, n = 1, 2, \ldots$ for every finitely generated subgroup $H < G$ with finite system of generators A.

Gromov in [Gro81a] raised the following question.

GROMOV'S PROBLEM ON GROWTH (I). *What is the dependence of the numbers R, N and q on d and k? In particular, does there exist an effective estimate of these numbers in terms of d and k?*

The second part of this question was answered by Y. Shalom and T. Tao in their "quantitative Gromov theorem" [ST10, Theorem 1.8]; we shall say more on results of Shalom and Tao shortly.

The theorem of Gromov was improved by van den Dries and Wilkie [vdDW84]. They showed that polynomial growth takes place under a weaker assumption: that the inequality $\gamma(n) \leq Cn^d$ holds for an infinite number of values of the argument n (i.e., that the group has weakly polynomial growth in our terminology, see (3.6)). Van den Dries and Wilkie applied techniques of nonstandard analysis and generalized the notion of asymptotic cone using ultrafilters.

The original proof of Gromov (as well as its modification by van den Dries and Wilkie; see also the paper of Tits [Tit81]) is based on the use of the Gleason, Montgomery, Zippin, Yamabe structural theory of locally compact groups [MZ74]. For a long time there was a hope of finding a proof of Gromov's theorem which would avoid the use of this remarkable (but technically very complicated) machinery which describes a relation between locally compact groups and Lie groups.

This goal was achieved by B. Kleiner [Kle10], who proved Gromov's theorem using a completely different approach that involves harmonic functions. The core of Kleiner's arguments is in the new proof of (a slight modification of) the theorem of Colding-Minicozzi [CM97].

THEOREM 8.4 (Kleiner 2010). *Let G be a group of weakly polynomial growth and suppose $l \in (0, \infty)$. Then the space of harmonic functions on Γ with polynomial growth at most l is finite dimensional.*

Recall that a function f on a group G is called μ-harmonic if it satisfies $Mf = f$, where M is the Markov operator of the random walk on G determined by the measure μ. A function f has at most polynomial growth of degree l if for some positive constant C the inequality $|f(g)| \leq C|g|^l$ holds, for all $g \in G$.

Gromov's effective polynomial growth theorem was improved by Y. Shalom and T. Tao in different directions [ST10]. They showed that for some absolute (and explicit) constant C, the following holds for every finitely generated group G, and all $d > 0$: If there is some $R_0 > \exp(\exp(Cd^C))$ for which $\gamma_G(R_0) < R_0^d$, then G has a finite index subgroup which is nilpotent of nilpotency degree less than C^d. In addition, an effective upper bound on the index is provided in [ST10] if "nilpotent" is replaced by "polycyclic."

In [ST10], a pair (G, A) (a group G with a finite generating set A) is called an (R_0, d)-growth group if

$$\gamma_G^A(R_0) \leq R_0^d. \tag{8.1}$$

THEOREM 8.5 (Fully quantitative weak Gromov theorem). *Let $d > 0$ and $R_0 > 0$, and assume that $R_0 > \exp(\exp(cd^c))$ for some sufficiently large absolute constant c. Then every (R_0, d)-growth group has a normal subgroup of index at most $\exp(R_0 \exp(\exp(d^c)))$ which is polycyclic.*

COROLLARY 8.6 (Slightly super-polynomial growth implies virtual nilpotency). *Let (G, A) be a finitely generated group such that $\gamma_G^A(n) < n^{c(\log \log n)^c}$ for some $n > 1/c$, where $c > 0$ is a sufficiently small absolute constant. Then G is virtually nilpotent.*

This result shows that if the growth of a group is less than that of $n^{(\log \log n)^c}$, then it is polynomial.

Gromov's original proof implies the existence of a function $v(n)$ growing faster than any polynomial and such that if $\gamma_G(n) \prec v(n)$, then the growth of G is polynomial. Therefore, there is a *gap* in the scale of growth degrees of finitely generated groups. In fact, as there are groups with incomparable growth functions, it may happen that there can be many gaps between polynomial and intermediate growth.

The function $n^{(\log \log n)^c}$ is the first concrete example of the superpolynomial bound separating the polynomial growth case from the intermediate one. Finding the border(s) between polynomial and intermediate growth is one of the main open problems about the growth of groups. We call this the *gap problem*, and there is the associated *gap conjecture*, which we discuss in more detail in Section 11.

One more result in the direction of study of polynomial growth is the characterization of finitely generated cancellative semigroups of polynomial growth given in [Gri88]. Recall that a semigroup S is cancellative if the left and right cancellative laws hold: $\forall a, b, c \in S, ab = ac \Rightarrow b = c, ba = ca \Rightarrow b = c$. Cancellation is a necessary condition for embedding of a semigroup into a group. In 1957, A. I. Malcev introduced the notion of a nilpotent semigroup [Mal53]. A semigroup is nilpotent if for some $n \geq 1$, it satisfies the identity $X_n = Y_n$, where $X_n, Y_n, n = 0, 1, \ldots$ are words over the alphabet $x, y, z_1, z_2, \ldots, z_n, \ldots$ defined inductively: $X_0 = x, Y_0 = y$ and

$$X_{n+1} = X_n z_n Y_n, \qquad Y_{n+1} = Y_n z_n X_n.$$

A semigroup is of nilpotency degree n if it satisfies the identity $X_n = Y_n$ but not the identity $X_{n-1} = Y_{n-1}$. Malcev proved that a group G is nilpotent of degree n if and only if it is nilpotent of degree n as a semigroup.

We say that a subsemigroup $L \leq S$ has a finite (left) index in S if there is a finite subset $E = \{e_1, \ldots, e_k\} \subset S$ such that

$$S = \bigcup_{i=1}^{k} e_i L.$$

THEOREM 8.7 ([Gri88]). *Let S be a cancellative semigroup of polynomial growth of degree d, i.e.,*

$$d = \overline{\lim}_{n \to \infty} \frac{\log \gamma_S(n)}{\log n} < \infty. \tag{8.2}$$

Then the following hold:

(i) *S has a group $G = S^{-1}S$ of left quotients which also has polynomial growth of degree d and, therefore, is virtually nilpotent. In particular, d is a nonnegative integer.*

(ii) *The semigroup S is virtually nilpotent in Malcev sense. More precisely, G contains a nilpotent subgroup H of finite index such that $S_1 = H \cap S$ is a nilpotent semigroup of the same degree of nilpotency as H, and S_1 is of finite (left or right) index in S.*

The proof of this theorem relies on the fact that a cancellative semigroup S of subexponential growth satisfies the Ore condition and hence has a group of left quotients G (this also holds for cancellative right amenable semigroups). The next step consists in getting of an upper bound of polynomial type for the growth of G, and an application of Gromov's theorem finishes the argument.

Observe that much less is known about growth of nilpotent cancellative semigroups than nilpotent groups. For instance, it is unknown if there is a constant $c > 0$ such that

$$cn^d \leq \gamma_S(n), n = 1, 2, \ldots$$

in the case when S is a cancellative semigroup of polynomial growth of degree d. In the case of a semigroup without cancellation, the growth can be equivalent to the growth of an arbitrary function $\gamma(n)$ satisfying reasonable restrictions, as was shown by V. Trofimov in [Tro80]. Also, a semigroup of polynomial growth need not be virtually nilpotent in Malcev sense.

We conclude this section with the following questions.

PROBLEM 2. *Let S be a cancellative semigroup of polynomial growth of degree d. Is it true that the limit*

$$d = \lim_{n \to \infty} \frac{\gamma_S(n)}{n^d} \tag{8.3}$$

exists?

PROBLEM 3. *Does there exist a finitely generated cancellative semigroup S of subexponential growth such that the group $G = S^{-1}S$ of left quotients has exponential growth?*

A positive answer to the previous question would provide an example of a supramenable group of exponential growth and an answer to the J. Rosenblatt's question from [Rose74]. Also observe that in the case $\gamma_S(n) \prec e^{\sqrt{n}}$, the group of quotients $G = S^{-1}S$ has subexponential growth which justifies the following question.

PROBLEM 4.

(a) *Does there exist a finitely generated cancellative semigroup S of superpolynomial growth which is strictly less that the growth of $e^{\sqrt{n}}$?*

(b) *Does there exist a finitely generated cancellative semigroup S of growth equivalent to the growth of $e^{\sqrt{n}}$?*

This problem is related to Problem 10 and the gap conjecture discussed in Section 10.

9 INTERMEDIATE GROWTH: THE CONSTRUCTION

The first part of Milnor's problem was solved by the author in 1984.

THEOREM 9.1 ([Gri84b]).

- *There are finitely generated groups of intermediate growth.*

- *The partially ordered set $\mathcal{W}$ of growth degrees of 2-generated groups contains a chain of the cardinality of the continuum, and contains an antichain of the cardinality of the continuum.*

- *The previous statement holds for the class of 2-generated p-groups for any prime p.*

COROLLARY 9.2. *There are uncountably many 2-generated groups, up to quasi-isometry.*

The main example of a group of intermediate growth — which will be denoted through the whole paper by $\mathcal{G}$ — is the group generated by four involutions a, b, c, d, which are defined by the next figure, the meaning of which will be explained soon.

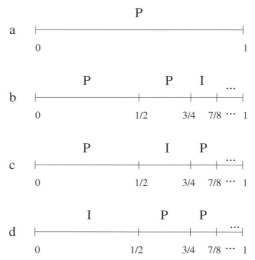

This group was constructed in 1980 in the author's note [Gri80b] as a simple example of a finitely generated infinite torsion group. Recall that the question on the existence of such groups was the subject of the *general Burnside problem* raised by Burnside in 1904. The problem was solved by E. Golod in 1964 on the basis of the Golod-Shafarevich theorem [GŠ64], [Gol64]. The bounded version of the Burnside problem (when orders of all elements are assumed to be uniformly bounded) was solved by S. Novikov and S. Adian [NA68a], [NA68c], [NA68d], [Adi79]). A third version of the Burnside problem, known as the *restricted Burnside problem*, was solved by E. Zelmanov [Zel91a], [Zel90], [Zel91b].

The note [Gri80b] contains two examples of Burnside groups, and the second group B presented there was, in fact, historically constructed before group G. Therefore, it can be considered as the first example of a nonelementary self-similar group (self-similar groups are discussed in the next section). The group B shares with G the property of being an infinite torsion group, but it is unknown whether it has intermediate growth. Also it is unknown if the Gupta-Sidki p-groups [GS83], whose construction and properties are similar to B in many aspects, have intermediate growth.

The group G acts on the unit interval $[0,1]$ with the dyadic rational points removed, i.e., on the set $\Delta = [0,1] \setminus \{m/2^n, \ 0 \le m \le 2^n, n = 1,2,\dots\}$ (if needed, the action of G can be extended to the entire interval $[0,1)$). The generator a is a permutation of the two halves of the interval, and b,c,d are also interval exchange type transformations (but involve a partition of Δ into infinitely many subintervals). They preserve the partition of the interval indicated in the figure, and their action is described by the labeling of the atoms of partition by symbols $\{I,P\}$. The letter I written over an interval means the action is the identity transformation there, and the letter P indicates the two halves of the interval are permuted as indicated in the following figure.

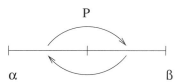

Now we are going to describe a general construction of uncountably many groups of intermediate growth. These groups are of the form

$$G_\omega = \langle a, b_\omega, c_\omega, d_\omega \rangle,$$

where the generator a is defined as before, and $b_\omega, c_\omega, d_\omega$ are transformations of the set Δ defined using an *oracle* $\omega \in \Omega = \{0,1,2\}^{\mathbb{N}}$ in the same manner as b,c,d. To describe the construction of groups G_ω completely, let us start with the space $\Omega = \{0,1,2\}^{\mathbb{N}}$ of infinite sequences over alphabet $\{0,1,2\}$. The bijection

$$0 \longleftrightarrow \begin{pmatrix} P \\ P \\ I \end{pmatrix} \qquad 1 \longleftrightarrow \begin{pmatrix} P \\ I \\ P \end{pmatrix} \qquad 2 \longleftrightarrow \begin{pmatrix} I \\ P \\ P \end{pmatrix}$$

between symbols of the alphabet $\{0,1,2\}$ and corresponding columns is used in the construction. Namely, given a sequence $\omega = \omega_1 \omega_2 \dots \in \Omega$, replace each ω_i,

$i = 1, 2, \ldots$ by the corresponding column to get a vector

$$\begin{pmatrix} U_\omega \\ V_\omega \\ W_\omega \end{pmatrix}$$

consisting of the three infinite words U_ω, V_ω, and W_ω over the alphabet $\{I, P\}$. For instance, the triple of words

$$U_\xi = PPI\,PPI\ldots,$$
$$V_\xi = PIP\,PIP\ldots,$$
$$W_\xi = IPP\,IPP\ldots$$

corresponds to the sequence $\xi = 012\,012\ \ldots$, while the triple

$$U_\eta = PPP\,PPP\ldots,$$
$$V_\eta = PIP\,IPI\ldots,$$
$$W_\eta = IPI\,PIP\ldots$$

corresponds to the sequence $\eta = 01\,01\ \ldots$.

Using the words U_ω, V_ω, and W_ω, define transformations b_ω, c_ω, and d_ω of the set Δ as in the case of the sequence ξ; the sequence ξ corresponds to the group $\mathcal{G}$. Then $\mathcal{G}_\omega = \langle a, b_\omega, c_\omega, d_\omega \rangle$. The most interesting groups from this family correspond to sequences ξ and η, which are respectively the groups $\mathcal{G}$ and $\mathcal{E} = \mathcal{G}_{(01)^\infty}$ studied by A. Erschler [Ers04a].

The generators of $\mathcal{G}_\omega$ satisfy the relations

$$a^2 = b_\omega^2 = c_\omega^2 = d_\omega^2 = 1,$$
$$b_\omega c_\omega = c_\omega b_\omega = d_\omega,$$
$$b_\omega d_\omega = d_\omega b_\omega = c_\omega,$$
$$c_\omega d_\omega = d_\omega c_\omega = b_\omega,$$

(this is only a partial list of the relations), from which it follows that the groups $\mathcal{G}_\omega$ are 3-generated (and even 2-generated in some degenerate cases, for example, the case of the sequence $0\,0\ldots0\ldots$). Nevertheless, we prefer to consider them as 4-generated groups, with the system $\{a, b_\omega, c_\omega, d_\omega\}$ viewed as a canonical system of generators.

While the groups $\mathcal{G}_\omega$ are 3-generated, a simple trick used in [Gri84b] produces a family $\mathcal{H}_\omega$, $\omega \in \Omega$ of 2-generated groups with the property that for $\omega \in \Omega_1$ (the set Ω_1 will be defined shortly), the growth of $\mathcal{H}_\omega$ is the same as growth of $\mathcal{G}_\omega$ raised to the fourth power.

Alternatively, the groups $\mathcal{G}_\omega$ can be defined as groups acting by automorphism on an infinite binary rooted tree T depicted (in part) in the following figure.

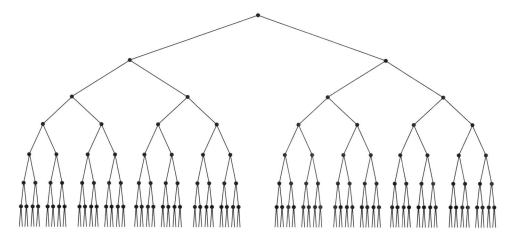

For instance, let us explain how generators of $\mathcal{G}$ act on T. The generator a acts by permuting the two rooted subtrees T_0 and T_1 which have roots at the first level. The generator b acts on the left subtree T_0 as a acts on T (here we use the self-similarity property of a binary tree), and acts on the right subtree T_1 as generator c (whose action on the whole tree is shifted to the right subtree). Similarly, c acts on T_0 as a on T and on T_1 as d acts on T. And finally, d acts on T_0 as the identity automorphism and on T_1 as b on T. This gives us a recursive definition of automorphisms b, c, d as in the figure below, where dotted arrows show the action of a generator on the corresponding vertex.

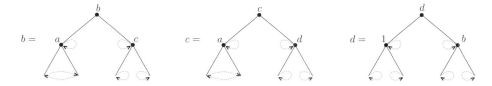

Making the identification of vertices of T with finite binary sequences, identifying the boundary ∂T of the tree with the set of infinite sequences from $\{0, 1\}^{\mathbb{N}}$, and representing dyadic irrational points by the corresponding binary expansions, we obtain an isomorphism of the dynamical system $(\mathcal{G}, [0, 1], m)$ (where m is Lebesgue measure) with the system $(\mathcal{G}, \partial T, \nu)$, where ν is a uniform measure on the boundary ∂T (i.e., a $\{\frac{1}{2}, \frac{1}{2}\}$ Bernoulli measure if the boundary is identified with the space of binary sequences). A similar description of the action on a binary tree can be given for all groups $\mathcal{G}_\omega$, although self-similarity should be used in a more general sense.

There are many other ways to define the groups $\mathcal{G}_\omega$. For instance, a very nice approach was suggested by Z. Sunic which allows one to define the groups $\mathcal{G}$ and $\mathcal{E}$ by use of irreducible (in the ring $\mathbb{F}_2[x]$) polynomials $x^2 + x + 1$ and $x^2 + 1$, respectively. This approach allows one to associate to any irreducible polynomial a self-similar group [BGŠ03].

One more definition of the group $\mathcal{G}$ can be given by the following presentation found by I. Lysenok [Lys85]:

$$\mathcal{G} = \langle a, b, c, d \mid a^2, b^2, c^2, d^2, bcd, \sigma^k((ad)^4), \sigma^k((adacac)^4), k \geq 0 \rangle, \qquad (9.1)$$

where

$$\sigma : \begin{cases} a \to aca, \\ b \to d, \\ c \to b, \\ d \to c \end{cases}$$

is a substitution. A similar presentation (called an L-presentation) can be found for $\mathcal{E}$ and many other self-similar groups.

THEOREM 9.3 ([Gri80b, Gri84b]).

1. $\mathcal{G}$ is a residually finite torsion 2-group, which is not finitely presentable.

2. $\mathcal{G}$ is of intermediate growth with bounds

$$e^{n^{1/2}} \preceq \gamma_{\mathcal{G}}(n) \preceq e^{n^\beta},$$

   where $\beta = \log_{32} 31 < 1$.

3. (A. Erschler [Ers04a]) For any $\epsilon > 0$, the following upper and lower bounds hold:

$$\exp \frac{n}{\log^{2+\epsilon} n} \prec \gamma_{\mathcal{E}}(n) \prec \exp \frac{n}{\log^{1-\epsilon} n}.$$

4. ([Gri85a]) There is a torsion-free extension $\hat{\mathcal{G}}$ of abelian group of infinite rank by $\mathcal{G}$ which has intermediate growth.

The first part of this theorem shows that there are examples of groups of Burnside type already in the class of groups of intermediate growth. The last part shows that there are torsion-free groups of intermediate growth.

The question of finding the precise (in Milnor-Schwarz sense) growth degree of any group of intermediate growth was open until recently, when in their excellent paper [BE10], Bartholdi and Erschler constructed groups with growth of the type $\exp n^\alpha$ for infinitely many $\alpha \in (\alpha_0, 1)$ (more on their result in Section 13). Moreover, they were able to find the growth degree of the torsion-free group $\hat{\mathcal{G}}$ by showing that

$$\gamma_{\hat{\mathcal{G}}}(n) \sim n^{n^{\alpha_0}},$$

with $\alpha_0 = \log(2)/\log(2/\rho) \approx 0.7674$, where ρ is the real root of the polynomial $x^3 + x^2 + x - 2$. In addition to the properties of $\hat{\mathcal{G}}$ listed earlier, we have the following fact, which was established in the paper of A. Machi and the author [GM93].

THEOREM 9.4. The group $\hat{\mathcal{G}}$ is left orderable and hence acts by orientation preserving homeomorphisms of the line $\mathbb{R}$ (or, equivalently, of the interval $(0,1)$).

It is unknown if there are orderable groups (i.e., groups possessing a two-sided invariant order) of intermediate growth.

PROBLEM 5. Does there exist a finitely generated, orderable group of intermediate growth?

A possible approach to this problem could be the following. By a theorem of M. I. Zaiceva [KK74], if a factor group F/A of a free group F has an infrainvariant system of subgroups (for the definition, see [KK74]) with torsion-free abelian quotients, then the group G/A is left orderable. Moreover, by D. M. Smirnov's theorem [Smi64], [KK74] the group $F/[A, A]$ is totally orderable in this case. It is known that the group $\hat{\mathcal{G}}$ has an infrainvariant system of subgroups with torsion-free abelian quotients. Representing the group $\hat{\mathcal{G}}$ in the form F/A, we obtain an interesting totally orderable group $F/[A, A]$. The question is whether it is of intermediate growth.

The last theorem shows that $\mathcal{G}$ embeds into the group $Homeo^+([0, 1])$ of homeomorphisms of the interval. The notion of growth is useful for study of codimension one foliations, and group growth as well as growth of Schreier graphs play an important role. The question concerning the smoothness of the action of a group on a manifold arises in many situations. Many topics related to the subject of actions on the interval and on the circle can be found in [Ghy01], [Nav11], [Bek04], and [Bek08]. The following result of A. Navas [Nav08] gives important information about what one can expect in the one-dimensional case.

THEOREM 9.5. *The group $\mathcal{G}$ embeds into $Diff^1([0, 1])$. However, for every $\epsilon > 0$, every subgroup of $Diff^{1+\epsilon}([0, 1])$ without a free subsemigroup on two generators is virtually nilpotent.*

Until 2004, all known examples of groups of intermediate growth were residually finite groups. The question about existence of non-residually finite groups of intermediate growth was open for a while. In [Ers04b], Erschler constructed an uncountable family of non-residually finite groups of intermediate growth. These groups are extensions of an elementary 2-group of infinite rank by $\mathcal{G}$.

Let $\Omega_0 \subset \Omega$ be the subset consisting of sequences in which all three symbols $0, 1, 2$ occur infinitely often, let $\Omega_1 \subset \Omega$ be the subset consisting of sequences in which at least two of the symbols $0, 1, 2$ occur infinitely often, and let $\Theta \subset \Omega$ be the subset consisting of sequences ω with the property that there is a natural number $C = C(\omega)$ such that for every n, each of symbols $0, 1, 2$ appears among any C consecutive symbols $\omega_n \omega_{n+1} \cdots \omega_{n+C-1}$. Observe that the inclusions $\Theta \subset \Omega_0 \subset \Omega_1$ hold.

Recall that a group is called *just-infinite* if it is infinite but every proper quotient is finite. We postpone the definition of *branch* groups (as well as the notions of *self-similar* group and *contracting* group) until Section 9.

THEOREM 9.6 ([Gri84b]).

- *The word problem for the group $\mathcal{G}_\omega$ is solvable if and only if the sequence ω is recursive.*

- *For each $\omega \in \Omega_0$, the group $\mathcal{G}_\omega$ is a just-infinite torsion 2-group.*

- *For each $\omega \in \Omega_1$, the group $\mathcal{G}_\omega$ is infinitely presented and branch.*

- *For each $\omega \in \Omega_1$, the group $\mathcal{G}_\omega$ has intermediate growth and the lower bound*

$$e^{n^{1/2}} \preceq \gamma_{G_\omega}(n)$$

holds.

- *For each $\theta \in \Theta$ there is a $\beta = \beta(\theta) < 1$ such that*

$$\gamma_{G_\omega}(n) \preceq e^{n^\beta}$$

holds.

The first statement of this theorem shows that there are many groups of intermediate growth for which the word problem is decidable, and, therefore, there are recursively presented groups of intermediate growth. At the same time there are recursively presented residually finite groups of intermediate growth for which the word problem is unsolvable, as is shown in [Gri85b].

Formally speaking, [Gri84b] does not contain the proof of the last part of the last theorem. But it can be easily obtained following the same line as the proof of theorem 3.2 from [Gri84b].

The lower bound by the function $e^{\sqrt{n}}$ for the growth of $\mathcal{G}$ was improved by L. Barthodi [Bar01] and Y. Leonov [Leo01], who showed that $\exp n^{0.5157} \preceq \gamma_{\mathcal{G}}(n)$.

The upper bound for $\mathcal{G}$ given by Theorem 9.3 was improved by L. Bartholdi [Bar98] by showing that

$$\gamma_{\mathcal{G}}(n) \preceq e^{n^{\alpha_0}}, \tag{9.2}$$

with $\alpha_0 = \log 2 / \log(2/\rho) \approx 0.7674$. The method used consists of consideration of a more general type of length function, arising from prescribing positive weights to generators, and counting the length using the weighted contribution of each generator. The method relies on the idea of giving the optimal weights to generators that lead to the best upper bound. It happened to be also useful for the construction of groups of intermediate growth with explicitly computed growth, as recently was demonstrated by Bartholdi and Erschler in [BE10] (more on this in Section 13).

Another technical tool was explored by R. Muchnik and I. Pak [MP01a] to get an upper bound on growth for the whole family of groups $\{\mathcal{G}_\omega\}$. Surprisingly, in the case of $\mathcal{G}$ their approach gives the same upper bound as (9.2), so the question of improving it is quite interesting (see Problem 7).

Unfortunately, even after three decades of study of the group $\mathcal{G}$ and other groups $\mathcal{G}_\omega$, we still do not know the precise growth rate of any group of intermediate growth from the family $\{\mathcal{G}_\omega\}$.

PROBLEM 6.

1. *Does there exist α such that $\gamma_{\mathcal{G}}(n) \sim e^{n^\alpha}$?*

2. *If the answer to the previous question is* yes, *what is the value of α?*

The question of whether the upper bound obtained in [BE10] and [MP01a] is optimal (i.e., it coincides with the growth rate of the group $\mathcal{G}$) is currently of significant interest, and we formulate it as the next problem. The point is that all known groups whose growth is explicitly computed up to equivalence by $\sim$ (i.e., groups considered in [BE10, BE11]) have growth not smaller than $e^{n^{\alpha_0}}$. And it looks that if the growth of $\mathcal{G}$ is less than $e^{n^{\alpha_1}}$ with $\alpha_1 < \alpha_0$, then, using the results of the cited papers and of the paper by Kassabov and Pak [KP11], one can extend the range of possible growth rates from the "interval" $[e^{n^{\alpha_0}}, e^n]$ to the "interval" $[e^{n^{\alpha_1}}, e^n]$. But so far $e^{n^{\alpha_0}}$ is a kind of a "mountain" which "closes the sky" for people working in the area of group growth.

In view of recent results from [BE10] and [Bri11], even obtaining a weaker result that would answer the next question seems to be interesting in its own right.

PROBLEM 7.

1. *Is the upper bound (9.2) the best possible for $\mathcal{G}$?*

2. *Does there exist a group of intermediate growth whose growth is less than $e^{n^{\alpha_1}}$, where $\alpha_1 < \alpha_0$ and $\alpha_0 = \log 2 / \log(2/\rho)$ is the constant defined earlier?*

   In Section 11, we will give a brief account of methods that can be used to obtain upper and lower bounds for intermediate growth.

   The groups $\mathcal{G}$ and $\mathcal{E}$ belong to the class of self-similar groups, that is, groups generated by automata of Mealy type, which are discussed a bit in the next section. An important quantitative characteristic of such groups is the pair (m, n) which is a rough indication of the *complexity* of the group, where m is the cardinality of the alphabet and n is the cardinality of the set of the states of automaton. From this point of view, $\mathcal{G}_\xi$ and $\mathcal{G}_\eta$ are $(2,5)$-groups (the Moore diagrams of corresponding automata are presented in the next section). And there is even a group of intermediate growth of complexity $(2,4)$ (namely, the iterated monodromy group $IMG(z^2 + i)$ in the sense of Nekrashevych [Nek05] of quadratic polynomial $z^2 + i$), as was showed by K. Bux and R. Peres [BP06].

PROBLEM 8.

   (i) *Find the growth degree of each of the groups $\mathcal{G}_\xi$, $\mathcal{G}_\eta$, and $IMG(z^2 + i)$.*

   (ii) *Are there groups of intermediate growth of complexity $(2,3)$?*

   (iii) *Determine all automata of complexity $(2,3), (2,4)$, and $(2,5)$ which generate groups of intermediate growth.*

   (iv) *Determine all possible types of growth of self-similar groups generated by finite automata.*

   It is a kind of a miracle that an automaton with a small number of states can generate a group with very complicated algebraic structure and asymptotic behavior. Therefore, it is not surprising that some of the automata groups studied prior to 1983 (when the the first examples of groups of intermediate growth were found) are also of intermediate growth. Specifically, the 2-group of Aleshin [Ale72] (generated by two automata with 3 and 7 states) and the p-groups of V. Sushchanskii (generated by automata with a number of states growing as a quadratic function of p [Sus79]) were shown to have intermediate growth by the author in [Gri84b, page 280] and [Gri85a, page 197], respectively (see also [BS07], where the Sushchanskii group is treated in detail). It is worth mentioning that the papers of Aleshin and Sushchanskii deal exclusively with the question of construction of finitely generated infinite torsion groups (contributing to the *general Burnside problem*), and Milnor's problem is not considered in these articles at all.

   As all the groups of intermediate growth from Theorem 9.6 have only finite quotients (and, consequently, at most a countable set of quotients), in 1983 it was reasonable to ask if there are groups of intermediate growth with uncountably many homomorphic images, one of the properties that a finitely generated virtually nilpotent group does not have. This was affirmatively answered in [Gri84a]. The next theorem gives a hint to the main result of [Gri84a] and its proof.

Let $\Lambda \subset \Omega$ be the subset consisting of sequences that are products of blocks $012, 120, 201$, and let $\mathcal{G}_\lambda$ be presented as the quotient $F_4/\mathcal{N}_\lambda$ of the free group F_4 of rank 4 by a normal subgroup $\mathcal{N}_\lambda$. Call the group $\mathcal{U}_\Lambda = F_4/\bigcap_{\lambda \in \Lambda} \mathcal{N}_\lambda$ Λ-universal.

THEOREM 9.7. *The Λ-universal group $\mathcal{U}_\Lambda$ has intermediate growth and has uncountably many quotients which are pairwise non-isomorphic.*

The first part of the theorem follows from [Gri84b, Theorem 3.2].

As each of the groups $\mathcal{G}_\lambda$ is a homomorphic image of $\mathcal{U}_\Lambda$, the second part of the theorem is obvious modulo the fact that the classes of isomorphisms of groups from the family $\mathcal{G}_\omega$, $\omega \in \Omega$ are at most countable. It was shown by Nekrashevych in [Nek07] that $\mathcal{G}_\omega \simeq \mathcal{G}_\zeta$ if and only if the sequence ζ can be obtained from ω by the diagonal action at all coordinates of an element from the symmetric group $Sym(3)$ acting on the set $\{0, 1, 2\}$. (In fact, it is enough to quote [Gri84b, Theorem 5.1], which states that for each ω there are at most countably many groups $\mathcal{G}_\eta$ isomorphic to $\mathcal{G}_\omega$).

In contrast to the Λ-universal group $\mathcal{U}_\Lambda$, the growth of the Ω-universal group $\mathcal{U}_\Omega = F_4/\bigcap_{\omega \in \Omega} \mathcal{N}_\omega$ is exponential [Muc05]. It is known that $\mathcal{U}_\Omega$ does not contain a free subgroup on two generators, is self-similar (of complexity $(6, 5)$ [Gri05]), weakly branch, and contracting. However, the following question is still open (unfortunately the article [Muc05], which contains the claim about amenability of $\mathcal{U}_\Omega$, has a mistake).

PROBLEM 9. *Is $\mathcal{U}_\Omega$ amenable or not?*

In [Gri84b] the author proved that for any monotone function $\rho(n)$ growing slower than exponential functions, there is a group with growth not slower than $\rho(n)$ (so either $\rho(n) \preceq \gamma_G(n)$ or $\rho(n)$ and $\gamma_G(n)$ are incomparable with respect to the preorder $\preceq$). This result was improved by Erschler .

THEOREM 9.8 (Erschler [Èrs05b]). *For any increasing function $\rho(n)$ growing slower than an exponential function, there is a finitely generated group G of intermediate growth with $\rho(n) \preceq \gamma_G(n)$.*

The last result shows that there is no upper bound for intermediate growth, in contrast with the lower bound given by the Shalom-Tao function $n^{c(\log \log n)^c}$ (c a constant), as discussed in Section 8.

10   THE GAP CONJECTURE

The history of the gap conjecture is as follows. While reading Gromov's paper on polynomial growth in 1982 (soon after its publication), the author realized that the effective version of Gromov's polynomial growth theorem (Theorem 8.1) implies the existence of a function $v(n)$ growing faster than any polynomial such that if $\gamma_G(n) \prec v(n)$, then the growth of G is polynomial. Indeed, taking a sequence $\{k_i, d_i\}_{i=1}^{\infty}$ with $k_i \to \infty$ and $d_i \to \infty$ as $i \to \infty$ along with the corresponding sequence $\{R_i\}_{i=1}^{\infty}$ (whose existence follows from Theorem 8.3), one can build a function $v(n)$ which coincides with the polynomial $k_i n^{d_i}$ on the interval $[R_{i-1} + 1, R_i]$. The constructed function $v(n)$ grows faster than any polynomial and separates polynomial growth from intermediate growth. In fact, as was already mentioned in Section 8 and at the end of previous section, the function $n^{c(\log\log n)^c}$ (c a constant) separates polynomial growth from exponential [ST10]. As, according to Theorem 9.1, the set of growth degrees is not linearly ordered, it may happen that there is more than one "gap" between polynomial growth and intermediate growth. But in any case, it would be nice to obtain the best possible estimate of the asymptotics of a function which "uniformly" separates the polynomial and intermediate growth.

Approximately at the same time, while reading Gromov's paper (thus around 1982), the author was establishing his results on groups of intermediate growth discussed in the previous section and in 1983–1985 published [Gri83], [Gri84b], and [Gri85a]. The lower bound of the type $e^{\sqrt{n}}$ for all groups $\mathcal{G}_\omega$ of intermediate growth established in those papers and in his habilitation [Gri85b] allowed to author to guess that the equivalence class of the function $e^{\sqrt{n}}$ could be a good candidate for a "border" between polynomial and exponential growth. This guess became stronger in 1988 when the author obtained the results published in [Gri89a] (see Theorem 10.2). In the ICM Kyoto paper [Gri91], the author raised a question of whether the function $e^{\sqrt{n}}$ gives a universal lower bound for all groups of intermediate growth. Moreover, at approximately the same time, he conjectured that indeed this is the case and stated this later at numerous talks.

CONJECTURE 10.1 (Gap Conjecture). *If the growth function $\gamma_G(n)$ of a finitely generated group G is strictly bounded from above by $e^{\sqrt{n}}$ (i.e., if $\gamma_G(n) \prec e^{\sqrt{n}}$), then the growth of G is polynomial.*

We are also interested to know whether there is a group, or more generally a cancellative semigroup, with growth equivalent to $e^{\sqrt{n}}$.

In this section we formulate several results in the direction of confirmation of the gap conjecture and suggest slightly different versions of it. Later in Section 12 we will formulate analogous conjectures about some other asymptotic characteristics of groups.

The next few results, together with the results about lower bounds on growth discussed in the next section, are the main source of support of the gap conjecture. Recall that a group G is said to be a residually finite-p group (sometimes also called residually finite p-group) if it is approximated by finite p-groups, i.e., for any $g \in G$, there is a finite p-group H and a homomorphism $\phi: G \to H$ with $\phi(g) \neq 1$. This class is, of course, smaller than the class of residually finite groups, but it is pretty large. For instance, it contains Golod-Shafarevich groups, the p-groups $\mathcal{G}_\omega$ from [Gri84b], [Gri85a], and many other groups.

THEOREM 10.2 ([Gri89a]). *Let G be a finitely generated residually finite-p group. If $\gamma_G(n) \prec e^{\sqrt{n}}$, then G has polynomial growth.*

As was established by the author in a discussion with A. Lubotzky and A. Mann during the conference on profinite groups in Oberwolfach in 1990, the same arguments as given in [Gri84b] combined with Lemma 1.7 from [LM91] allow one to prove a stronger version of the preceding theorem (see also the remark after Theorem 1.8 in [LM91], but be aware that capital O has to be replaced by lower-case o).

THEOREM 10.3. *Let G be a residually nilpotent finitely generated group. If $\gamma_G(n) \prec e^{\sqrt{n}}$, then G has polynomial growth.*

The main goal of the paper [LM91], followed by the article [LMS93], was to give a complete description of finitely generated groups with polynomial sub-group growth (the growth of the function which counts the number of subgroups of given finite index). The remarkable result achieved in [LMS93] shows that such groups are precisely the *solvable groups of finite rank*.

As was already mentioned in the introduction, the original proof of Gromov's polynomial growth theorem is based on the use of the solution of Hilbert's 5th problem by Montgomery and Zippin, concerning the isometric actions of locally compact groups and their relation to Lie groups. Surprisingly, in the proofs of the results stated in Theorems 10.2 and 10.3, as well as in the results from [LM91, LMS93] about polynomial subgroup growth, M. Lazard's solution of the p-adic analog of Hilbert's 5th problem [Laz65] plays an important role. The result of Lazard gives a characterization of analytic pro-p-groups. After a long period of search, a proof of Gromov's Theorem which avoids the use of the 5th Hilbert problem was found by B. Kleiner [Kle10]. Now we will formulate a theorem (Theorem 10.4), which generalizes Theorems 10.2 and 10.3 and whose proof is based on the techniques of J. S. Wilson from [Wil05, Wil11] and some other results. Wilson's arguments how to handle with growth of residually finite groups are quite original (the techniques of ultraproducts is used at some point), but eventually they reduce the arguments to the case of residually nilpotent groups (i.e., to the previous theorem). It would be interesting to find a proof of Theorem 10.2 which avoids the use of the p-adic version of Hilbert's 5th problem.

Recall that a group is called *supersolvable* if it has a finite normal descending chain of subgroups with cyclic quotients. Every finitely generated nilpotent group is supersolvable [Rob96]; therefore, the next theorem improves Theorem 10.3.

THEOREM 10.4. *The gap conjecture holds for residually supersolvable groups.*

The proof of this theorem is based on the techniques of J. S. Wilson developed in [Wil05, Wil11] for studies around gap conjecture. Currently, it is not known if the main conjecture holds for residually polycyclic and, more generally, residually solvable groups. However, it is quite plausible that it does.

CONJECTURE 10.5 (Gap Conjecture with Parameter β). *There exists β, $0 < \beta < 1$, such that if the growth function $\gamma_G(n)$ of a finitely generated group G is strictly bounded from above by e^{n^β} (i.e., if $\gamma(n) \prec e^{n^\beta}$), then the growth of G is polynomial.*

Thus the gap conjecture with parameter $1/2$ is just the gap conjecture (Conjecture 10.1). If $\beta < 1/2$, then the gap conjecture with parameter β is weaker than the gap conjecture, and if $\beta > 1/2$ then it is stronger than the gap conjecture.

CONJECTURE 10.6 (Weak Gap Conjecture). *There is a $\beta < 1$, such that if $\gamma_G(n) \prec e^{n^\beta}$, then the gap conjecture with parameter β holds.*

As was already mentioned, there are some results of J. S. Wilson in the direction of confirmation of the gap conjecture. He showed that if G is a residually solvable group whose growth is strictly less than $e^{n^{1/6}}$, then it has polynomial growth [Wil05], [Wil11]. Therefore the gap conjecture with parameter $1/6$ holds for residually solvable groups. The proof of Wilson's result is based on the estimate of the rank of chief factors of finite solvable quotients of G. The methods of [Wil05] and [Wil11], combined with the theorem of Morris [Mor06] and theorem of Rosset [Ros76], can be used to prove the following statement.

THEOREM 10.7.

(i) *The gap conjecture with parameter* $1/6$ *holds for left orderable groups.*

(ii) *The gap conjecture holds for left orderable groups if it holds for residually polycyclic groups.*

Let us also formulate an open problem which is related to the above discussion.

PROBLEM 10.

(i) *Does there exist* α, $0 < \alpha < 1$ *such that if the growth of a finitely generated group is strictly less than the growth of* e^{n^α}, *then it is polynomial?*

(ii) *If such* α *exists, what is its maximal value? Is it* $< 1/2, = 1/2, or > 1/2$?

(iii) *In the case* α *exists (and is chosen to be maximal), is there a group (or a cancellative semigroup) with growth equivalent to* e^{n^α}?

(iv) *Is there a finitely generated group approximated by nilpotent groups with growth equivalent to* $e^{\sqrt{n}}$? *Is there a residually finite-p group with growth equivalent to* $e^{\sqrt{n}}$ *(p a fixed prime)?*

There is some evidence based on considerations presented in the last section of this article and some additional arguments that these conjectures and problem (parts (i), (ii), (iii)) can be reduced to consideration of three classes of groups: *simple* groups, *branch* groups, and *hereditary just-infinite* groups. These three types of groups appear in a natural division of the class of just-infinite groups into three subclasses described in Theorem 14.3. Branch groups are defined in Section 11, and a hereditary just-infinite group is a residually finite group with the property that every proper quotient of every subgroup of finite index (including the group itself) is just-infinite. In any case, the following theorem holds (observe that branch groups and hereditary just-infinite groups are residually finite groups).

THEOREM 10.8. *If the gap conjecture holds for the classes of residually finite groups and simple groups, then it holds for the class of all groups.*

This theorem is a corollary of the main result of [BM07]. A different proof is suggested in [Gri12]. It is adapted to the needs of the proof of Theorem 14.6, which is discussed at the end of the article.

As it is quite plausible that the gap conjecture could be proved for residually finite groups (the classification of finite simple groups may help), we suspect that the validity of the gap conjecture depends on its validity for the class of simple groups. We will return to just-infinite groups at the end of the article and state one more reduction of the gap conjecture.

It is unknown if there are simple groups of intermediate growth (Problem 14), but a recent article of K. Medynets and the author [GM11] shows that there are infinite finitely generated simple groups which belongs to the class *LEF* (locally embeddable into finite groups); the authors conjectured that these groups are amenable. K. Juschenko and N. Monod [JM12] recently confirmed their amenability.

This gives some hope that groups of intermediate growth may exist within the subgroups of the groups considered in [GM11] (namely, among subgroups of *full topological groups* $[[T]]$ associated with minimal homeomorphisms T of a Cantor set). Algebraic properties of $[[T]]$ were studied by H. Matui [Mat06], who showed that their commutator subgroup $[[T]]'$ is simple and is finitely generated in the case when the homeomorphism T is a minimal subshift over a finite alphabet. Observe that $[[T]]'$ always has exponential growth, as was recently shown in [Mat11]. Therefore, the only hope is that groups of intermediate growth may exist among subgroups of $[[T]]$.

There are several other gap conjectures related to various asymptotic characteristics of groups. We list some of them in Section 12 and discuss briefly their relation to the problems and conjectures considered in this section.

11   INTERMEDIATE GROWTH: THE UPPER AND LOWER BOUNDS

Here we give an overview of the main methods of getting upper and lower bounds of growth in the intermediate growth case. We begin with upper bounds. For establishing if a group is of intermediate growth, it is more important to have tools to obtain upper bounds because as soon as it is known that a group is not virtually nilpotent (and usually this is not difficult to check), its growth is known to be superpolynomial (by Gromov's theorem). Therefore, if a finitely generated infinite group possesses any property such as being simple, torsion, not residually finite, nonhopfian etc., one immediately knows that the growth is superpolynomial.

Finding a lower bound for the growth is of interest not only because of its connection with the gap problem (Problem 10) discussed earlier, but also because of the connection with various topics in the theory of random walks on groups and spectral theory of the discrete Laplace operator.

The method for getting upper bounds for the growth in the intermediate case that we are going to describe was used in [Gri84b, Gri85a]. Roughly, the idea consists of encoding each element g of a group G by a set of d elements $g_1, \ldots, g_d$, $g_i \in G$ ($d \geq 2$ and fixed) in such a way that for some fixed constants C and λ, $0 < \lambda < 1$ (independent of g), the inequality

$$\sum_{i=1}^{d} |g_i| \leq \lambda |g| + C \tag{11.1}$$

holds. The meaning of this inequality is that an element of (large) length n is coded by a set of d elements of the total length strictly less than n (with coefficient of the reduction λ). In dynamics, this situation corresponds to the case when the entropy of the system is zero, while in the context of growth it corresponds to the case when the constant κ given by (3.7) is equal to 1 (i.e., when the growth is subexponential).

There are some variations of condition (11.1). For instance, in some cases the coefficient λ can be taken to be equal to 1, but some additional conditions on the group have to be satisfied in order to claim that the growth is subexponential. For instance this happens in the case of the group $\mathcal{E}$ and, more generally, in the case of the groups $\mathcal{G}_\omega$, $\omega \in \Omega_1$.

At the moment, this idea is realized only in the case of certain groups acting on *regular rooted trees* that are *self-similar* or have certain self-similarity features. The simplest property that leads to intermediate growth is the *strong contracting property* that we are going to describe.

The set of vertices of a d-regular rooted tree $T = T_d$ is in a natural bijection with the set of finite words over an alphabet X of cardinality d (usually one of the alphabets $\{0, 1, \ldots, d-1\}$ or $\{1, 2, \ldots, d\}$ is used). The set of vertices of the tree is graded by levels $n = 0, 1, 2, \ldots$, and vertices of the nth level are in natural bijection with the words of length n (i.e., with the elements of the set X^n) listed in lexicographical order. Let $V = X^* = \bigcup_{n=1}^{\infty} X^n$ be the set of all vertices.

For the full group $Aut(T)$ of automorphisms of a d-regular rooted tree T, a natural decomposition into a semidirect product

$$(Aut(T) \times \cdots \times Aut(T)) \rtimes Sym(d) \qquad (11.2)$$

(with d factors) holds, as well as a corresponding decomposition of any element $g \in Aut(T)$

$$g = (g_1, g_2, \ldots, g_d)\sigma. \qquad (11.3)$$

Here the element σ of the symmetric group $Sym(d)$ shows how g acts on the vertices of the first level, while the *projections* g_i show how g acts on the subtree T_i which has its root at vertex i of the first level (we identify the subtree T_i with T using the canonical self-similarity of the regular rooted tree). In a similar way, the projection g_v of an element $g \in Aut(T)$ can be defined for an arbitrary vertex $v \in V$.

DEFINITION 11.1. *The action of a group G on a regular rooted tree T is called* self-similar *if for any vertex v, the projection g_v is again an element of G modulo the canonical identification of the subtree T_v with the original tree T. A group is called self-similar if it has a faithful self-similar action.*

For instance, the groups $\mathcal{G}$ and $\mathcal{E}$ are self-similar. For generators a, b, c, d of $\mathcal{G}$ the following relations of the type (11.3) hold

$$a = (1,1)\sigma, \quad b = (a,c)e, \quad c = (a,d)e, \quad d = (1,b)e, \qquad (11.4)$$

where e is the identity element and σ is a permutation (both are elements of $Sym(2)$ acting on the alphabet $\{0, 1\}$). Observe that the equalities (11.4) are used in the simplified form

$$a = \sigma, \quad b = (a,c), \quad c = (a,d), \quad d = (1,b). \qquad (11.5)$$

An equivalent definition of a self-similar group is that it is a group generated by states of non-initial Mealy type automaton with the operation of composition of automata. We are not going to explain here what the Mealy automaton is, nor the group defined by it. For this approach to self-similarity, see [GNS00, Nek05]). For instance, the groups $\mathcal{G}$ and $\mathcal{E}$ are groups generated by the following automata.

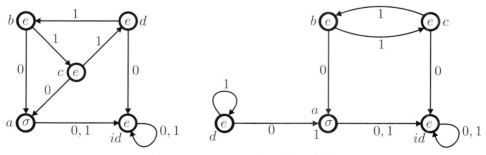

Automata generating $\mathcal{G}$ and $\mathcal{E}$.

A challenging problem is to understand what the class of self-similar groups is, and especially what constitutes the subclass consisting of groups that are generated by finite automata. There is a description of groups generated by 2-state automata over the alphabet on two letters (there are 6 such groups), see [GNS00]. There are not more than 115 groups generated by 3-state automata over an alphabet on two letters (see [BGK$^+$08] where an *Atlas* of self-similar groups is started).

Another important notion useful when studying growth is the notion of a *branch* group, introduced in [Gri00a], [Gri00b]. With each sequence $\bar{m} = \{m_n\}_{n=1}^\infty$ consisting of integers $m_n \geq 2$, one can associate a spherically homogeneous rooted tree $T_{\bar{m}}$ as done in [BOERT96], [Gri00b], [BGŠ03]. Given a group G acting faithfully by automorphisms on such a tree $T = T_{\bar{m}}$ and a vertex $v \in V(T)$, the rigid stabilizer $rist_G(v)$ is defined as the subgroup in G consisting of elements which act trivially outside the subtree T_v with its root at v. The rigid stabilizer $rist_G(n)$ of level n is the subgroup generated by the rigid stabilizers of vertices of nth level (it is their internal direct product).

DEFINITION 11.2. *A group G is called branch if it has a faithful, level-transitive action on some spherically homogeneous rooted tree $T_{\bar{m}}$ with the property that*

$$[G : rist_G(n)] < \infty$$

for each $n = 1, 2, \ldots$.

This is a geometric definition of branch groups. There is also an algebraic definition discussed in [Gri00b, BGŠ03] which gives a slightly larger class of groups.

As was already mentioned in the introduction, branch groups of self-similar type are basically the only source for the constructions of groups of intermediate growth and, as is stated in Theorem 9.6, the groups $\mathcal{G}_\omega$, $\omega \in \Omega_1$, are branch. Branch groups constitute one of three classes in which the class of just-infinite groups naturally splits. The role of this class in the theory of growth of groups will be emphasized once more in the last section of the paper.

DEFINITION 11.3. *A finitely generated self-similar group is called* contracting *with parameters $\lambda < 1$ and C if the inequality*

$$|g_i| < \lambda|g| + C \tag{11.6}$$

holds for every element $g \in G$ and any of its section g_i on the first level.

It follows from (11.6) that
$$|g_i| < |g|$$
if $|g| > C/(1 - \lambda)$. Therefore, for elements of sufficiently large length the projections on the vertices of the first level are shorter.

The contracting property and self-similarity allow one to study algebraical properties of the group. This was first used in [Gri80b] to show that the groups $\mathcal{G}$ and $\mathcal{B}$ are torsion, and later many other interesting and sometimes unusual properties of the group $\mathcal{G}$ and similar type groups (for instance Gupta-Sidki p-groups [GS83]) were established.

To be more precise, the groups $\mathcal{G}_\omega$ are not self-similar for a typical ω. But in fact, the whole family $\{\mathcal{G}_\omega\}_{\omega \in \Omega}$ of groups is self-similar, because the projections of any element $g \in \mathcal{G}_\omega$ on vertices of arbitrary level n belong to the group $\mathcal{G}_{\tau^n(\omega)}$

where τ is the shift in the space of sequences. Moreover the contracting property holds for this family with respect to the canonical generating sets $\{a, b_\omega, c_\omega, d_\omega\}$. This allows one to get various results (including the estimates of the growth) using a simultaneous induction on the length of elements for infinite shift invariant families of groups from the class $\{\mathcal{G}_\omega\}_{\omega \in \Omega}$.

DEFINITION 11.4. *A finitely generated self-similar group acting on a d-regular rooted tree T is called* strictly contracting *(or* sum contracting*) with parameters $k \geq 1$, $\lambda < 1$ and C if there is a level k of the tree such that the inequality*

$$\sum_{v \in V_k} |g_v| \leq \lambda |g| + C \tag{11.7}$$

holds for an arbitrary element $g \in G$ (V_k denotes the set of vertices of level k).

The idea of encoding of the elements of length n by tuples of elements of total length less than n discussed earlier has a clear implementation for strictly contracting groups. Namely, every element $g \in G$ can be encoded by the set $\{\bar{g}, g_1, g_2, \ldots, g_{d^k}\}$, where $\bar{g}$ is a shortest representative of the coset $g st_G(k)$ of the stabilizer $st_G(k)$ of kth level, and $g_1, g_2, \ldots, g_{d^k}$ are projections on the vertices of k-th level (observe that the length of $\bar{g}$ is uniformly bounded by some constant independent of the choice of the element).

PROPOSITION 11.5 ([Gri84b]). *Let G be a strictly contracting group with parameters k, λ and C. Then there is a constant $\beta = \beta(\lambda, k) < 1$ such that*

$$\gamma_G(n) \preceq e^{n^\beta}.$$

The strict contracting property holds, for instance, for the groups $\mathcal{G}_\omega, \omega \in \Theta$. For the groups $\mathcal{G}_\omega, \omega \in \Omega_1 \setminus \Omega_0$, condition (11.7) does not hold in general. Regardless, a modification of the arguments allows one to show that the growth is intermediate in this case as well and even to get an upper bound of the type shown in the third part of the Theorem 9.3.

Now let us turn to lower bounds. There is a much wider variety of methods used to obtain lower bounds than for upper bounds. We are going to mention the *branch commensurability* property, and give a short account (partly here, partly in the next section) of other methods: the *anti-contracting property* (which is a kind of opposite to the contracting property), the *Lie method*, the *boundary of random walks* method [Ers04a], and the *Munchhausen trick* method [BV05], [Kai05], [GN07].

The first observation is that the group $\mathcal{G}$ and many other self-similar groups of branch type — for instance, the so-called *regularly branch* groups [BGŠ03] — have the property of being *abstractly commensurable* to some power $d \geq 2$ of itself (i.e., to a direct product of d copies of the group, in the case of $\mathcal{G}$, the power $d = 2$). Recall that two groups are said to be abstractly commensurable if they have isomorphic subgroups of finite index. Observe that there are even finitely generated groups G which are isomorphic to their proper powers G^d, $d \geq 2$ [Jon74, BM88]. Such groups are non-hopfian (and hence not residually finite), in contrast to regularly branch groups, but among them, groups of intermediate growth have not yet been detected.

PROBLEM 11. *Does there exist a finitely generated group G of intermediate growth isomorphic to some power G^d, $d \geq 2$?*

For a group G commensurable with G^d, there is a lower bound on growth given by the inequality

$$e^{n^\alpha} \preceq \gamma_G(n) \tag{11.8}$$

for some positive α. Verifying this is an easy exercise if one keeps in mind that the growth of a group coincides with that of a subgroup of finite index and that

$$\gamma_{G^d}(n) \sim \gamma_G^d(n).$$

Observe that, in general, we do not have control over α in this type of argument.

Another condition that allows us to get a lower bound of type (11.7) is the following. Let G be a self-similar group acting on a d-regular tree.

DEFINITION 11.6. *A group G satisfies the* anti-contracting *property with parameters k, μ, and C if for arbitrary $g \in G$, the inequality*

$$|g| \leq \mu \sum_{v \in V_k} |g_v| + C \tag{11.9}$$

holds, where V_k is the set of vertices of level k and g_v is the projection of g on v.

Inequality (11.9) is a kind of opposite to inequality (11.7). Using the methods of [Gri84b] one can prove the following fact.

THEOREM 11.7. *Let G be a group satisfying the anti-contracting property with parameters μ and C. Then there is $\alpha = \alpha(k, \mu) > 0$ such that inequality (11.8) holds.*

The lower bound for α in terms of k and μ (C is not important) can be explicitly written. In the case of the group $\mathcal{G}$, the *anti-contracting* property holds with parameters $k = 1, \mu = 2, C = 1$ and $\alpha = 1/2$. In fact, a lower bound of type $e^{\sqrt{n}}$ holds for all groups from the family $\mathcal{G}_\omega, \omega \in \Omega_1$.

Another approach to obtain lower bounds is via an idea of W. Magnus: using Lie algebras and associative algebras associated with a group. Given a descending central series $\{G_n\}_{n=0}^\infty$ of a group G, one can construct the graded Lie ring $\mathcal{L} = \oplus_{n=0}^\infty G_n / G_{n+1}$ and the graded associative algebra $\mathcal{A} = \oplus_{n=0}^\infty A_n$, where $A_n = \Delta^n / \Delta^{n+1}$, and Δ is the fundamental ideal of the group algebra $\mathbb{F}[G]$ ($\mathbb{F}$ a field). The Lie operation in $\mathcal{L}$ is induced by the operation of taking commutators of pairs of elements in the group (defined first on the abelian quotients G_n / G_{n+1} and then extended by linearity to the whole ring). The important cases are given by the lower central series and the Jennings-Lazard-Zassenhaus lower p-central series, which in the case of a simple field of characteristic p can be defined as $G_n = \{g \in G : 1 - g \in \Delta^n\}$.

By Quillen's theorem, the algebra $\mathcal{A}$ is the universal enveloping algebra of $\mathcal{L}$ (or p-universal in the case *char* $\mathbb{F} = p$). There is a close relationship between the growth of the algebras $\mathcal{A}$ and $\mathcal{L}$, i.e., the growth of the dimensions of the homogeneous components of these algebras. Namely, $\mathcal{L}$ has exponential growth if and only if $\mathcal{A}$ has exponential growth, and if $\mathcal{L}$ has polynomial growth of degree d then $\mathcal{A}$ has intermediate growth of type e^{n^α} with $\alpha = \frac{d+1}{d+2}$ (for details see, e.g., [Gri89a], [BG00b]). More information about growth of algebras can be found in [KL00]. Observe that finitely generated Lie algebras may be of fractional (and even irrational) power growth, as is shown in [PSZ10].

The following fact shows that the growth of $\mathcal{A}$ gives a universal lower bound for the growth of a group independently of the system of generators.

PROPOSITION 11.8 ([Gri89a]). *Let G be a finitely generated group with a finite system S of semigroup generators (i.e., each element of G can be expressed as a product of elements from S). Let $\gamma_G^S(n)$ be the growth function of G with respect to S and $a_n = \dim_{\mathbb{F}} A_n$. Then, for any $n \in \mathbb{N}$,*

$$\gamma_G^S(n) \geq a_n. \tag{11.10}$$

If the algebra $\mathcal{L}$ is infinite dimensional, then the growth of $\mathcal{A}$ is at least $e^{\sqrt{n}}$, and therefore the growth of the group is at least $e^{\sqrt{n}}$. The Lie algebras approach was used to show that gap conjecture holds for residually nilpotent groups, as was already mentioned.

Observe that in order to have an example of a residually-p finite group (i.e., a group approximated by finite p-groups) whose growth is exactly $e^{\sqrt{n}}$, the ranks of the consecutive quotients G_n/G_{n+1} must be uniformly bounded (i.e., the group G has to have *bounded width*). But this condition is not enough. For instance, the group $\mathcal{G}$ has finite width, as it was proved in [BG00b], but its growth is bounded from below by $e^{n^{0.51}}$. The growth of Lie algebra $\mathcal{L}$ associated with the Gupta-Sidki 3-group $\mathcal{S}$ is linear, which implies that the group $\mathcal{S}$ has growth at least $e^{n^{2/3}}$ (it is not known yet if Gupta-Sidki p-groups have intermediate growth or not). Inequality (11.10) also gives a way to prove that a group has uniformly exponential growth [Gri89a], [BG00b].

Recall that a group G is said to have *uniformly exponential* growth if

$$\kappa_* = \inf_A \kappa_A > 1, \tag{11.11}$$

where κ_A denotes the base of exponential growth with respect to the system of generators A (κ is defined by relation (3.7)), and the infimum is taken over all finite systems of generators. An immediate corollary of Proposition 11.8 is the following.

COROLLARY 11.9. *Assume that the Lie algebra $\mathcal{L}$ associated to the group G has exponential growth. Then G has uniformly exponential growth.*

It follows that Golod-Shafarevich groups [Gol64], [GŠ64] have uniformly exponential growth.

12 ASYMPTOTIC INVARIANTS OF PROBABILISTIC AND ANALYTIC NATURE AND CORRESPONDING GAP-TYPE CONJECTURES

In this section we discuss the relation between group growth and asymptotic behavior of random walks on a group. At the end of it we formulate gap-type conjectures related to the asymptotic characteristics of random walks and discuss their relation with the growth gap conjecture from Section 10.

The ICM paper of A. Erschler [Ers10], which we recommend to the reader, contains important material related to the subject of random walks and growth. Also, we recommend the book of W. Woess [Woe00], the paper of Kaimanovich and Vershik [KV83], the article [BPS09], and the unfinished manuscript of C. Pittet and L. Saloff-Coste [PSC].

Let G be a finitely generated group and μ be a probability measure on G whose support A generates the group. Consider a random walk (G, μ) on G which starts at the identity element e and the transitions $g \to ga$ take place with probability $\mu(a)$. Let $P(n) = P_{e,e}^{(n)}$ be the probability of return after n steps. Observe that

$P(n) = \mu^{*n}(e)$, where μ^{*n} denotes the nth convolution of μ. In the case of a symmetric measure (i.e., when $\mu(a) = \mu(a^{-1})$, for every $a \in A$) the inequality

$$\frac{1}{P(2n)} \leq \gamma_G^A(2n) \tag{12.1}$$

holds, since the maximal mass of μ^{*2n} is concentrated at the identity element e [Woe00]. In (12.1) the probability $P(n)$ is evaluated only for even values of n because for odd values it can be zero (this happens when the identity element of the group can not be expressed as a product of an odd number of generators). From now on, when we discuss the rate of decay of the probabilities $P(n)$ as $n \to \infty$, we assume that the argument n takes only even values. We will use the comparison $\preceq$ of the rate of decay of $P(n)$, or the rate of growth when $n \to \infty$ of some other functions that will be introduced later, in the sense of the definition given in Section 3.

The rate of decay of the probabilities $P(n)$ can obey a power law (of the type $n^\alpha, \alpha < 0$), be exponential (of type $\lambda^n, 0 < \lambda < 1$), or can be intermediate between the two. A power law holds if and only if the group has polynomial growth. This follows from a combination of Gromov's theorem and results of Varopoulos on random walks ([Var91] and [VSCC92, Theorem VI.5.1 on p. 84]).

An important characteristic of random walks, introduced by Kesten [Kes59b], is the spectral radius defined by relation (3.9), which in the case of symmetric measures coincides with the norm of the Markov operator

$$Mf(x) = \sum_{g \in G} \mu(g) f(xg).$$

Observe that this is also the operator given by right convolution with a measure μ acting on $l^2(G)$. By Kesten's criterion a group is amenable if and only if for some ("some" can be replaced by "every") symmetric measure μ whose support generates a group, the spectral radius takes its maximal possible value $r = 1$ [Kes59a]. Therefore, amenable groups have subexponential rate of decay of return probabilities, and the rate of decay is exponential in the case of nonamenable groups.

For groups of exponential growth, the rate of decay is not slower than $e^{-\sqrt[3]{n}}$. In other words, the upper bound

$$P(n) \preceq e^{-\sqrt[3]{n}} \tag{12.2}$$

holds [Var91]. This result cannot be improved, as there are groups of exponential growth for which the upper bound (12.2) is sharp (for instance, the lamplighter group $\mathcal{L} = \mathbb{Z}_2 \wr \mathbb{Z}$ or the Baumslag-Solitar solvable groups $BS(1, n), n \geq 2$).

Because of inequality (12.1), if a bound of the type

$$P(n) \preceq e^{-n^\alpha} \tag{12.3}$$

holds, then

$$\gamma_G(n) \succeq e^{n^\alpha}. \tag{12.4}$$

On the other hand if (12.4) holds then

$$P(n) \preceq e^{-n^{\alpha/(\alpha+2)}} \tag{12.5}$$

[Var91], [Woe00], [BPS09].

This leads to the following natural questions. What are the slowest and fastest rates of decay of probabilities $P(n)$ for groups of intermediate growth? Is it of type e^{-n^α} (resp. e^{-n^β}) for some positive α and β? What are the values of α and β? The values $1/2$, $1/3$ and $1/5$ are the first candidates for these numbers. Is there a group of intermediate growth with rate of decay $P(n) \succ e^{-n^{1/3}}$? If the rate of decay of $P(n)$ for a group of intermediate growth cannot be slower than $e^{-n^{1/3}}$, then the weak version of gap conjecture 10.1 holds with parameter $1/3$. Later we will formulate a conjecture related to the preceding discussion.

A new approach for obtaining lower bounds for growth based on techniques of random walks is developed by A. Erschler in [Ers04a] and [Ers05a]. Without getting into details, let us briefly outline some features of her approach. For a random walk given by pair (G, μ), the *entropy* $h = h(G, \mu)$ and the *drift* (or the *rate of escape*) $l = l(G, \mu)$ are defined as

$$h = \lim_{n \to \infty} \frac{H(n)}{n}, \qquad (12.6)$$

where $H(n) = H(\mu^{*n})$ and $H(\mu) = -\sum_{g \in G} \mu(g) \log \mu(g)$ is Shannon entropy, and

$$l = \lim_{n \to \infty} \frac{L(n)}{n}, \qquad (12.7)$$

where $L(n) = \sum_{g \in G} |g| \mu^{*n}(g)$ is the expectation of the length $|g|$ of a random element at the nth moment of the random walk (the length $|g|$ is considered with respect to the system of generators given by the support of μ). By the Guivarc'h inequality [Gui80], in the case of symmetric measure with finite support (or more generally with finite first moment $\sum_{g \in G_\omega} |g| \mu(g)$), the numbers h, l, and κ (the base of exponential growth defined by (3.7)) are related as

$$h \le l\kappa. \qquad (12.8)$$

Therefore, the equality $l = 0$ or $k = 0$ implies $h = 0$.

An important notion due to Furstenberg is the Poisson boundary (which we will call the Poisson-Furstenberg boundary). It is a triple $(G, \mathcal{B}, \nu)$ consisting of a G-space $(\mathcal{B}, \nu)$ with a μ-stationary probability measure ν (i.e., $\mu * \nu = \nu$). This boundary describes space of *bounded μ-harmonic functions*:

$$f(g) = \int_\mathcal{B} \phi(gx) d\nu(x) \qquad (12.9)$$

(the Poisson integral). The left-hand side of the last equality takes values in the space of bounded μ-harmonic functions while ϕ belongs to the space $L^\infty(\mathcal{B})$ (see [KV83] for details). The Liouville property of a group (more precisely, of a pair (G, μ)) is that every bounded μ-harmonic function is constant; this property is equivalent to the triviality of the Poisson-Furstenberg boundary.

The entropy criterion due to Avez-Derriennic-Kaimanovich-Vershik [KV83] states the following. Let G be a countable group and let μ be a probability measure on G with finite entropy $H(\mu)$. Under this assumption the Poisson-Furstenberg boundary is trivial if and only if the entropy h of the random walk is equal to zero.

If μ is symmetric and has finite first moment with respect to some (and hence with respect to every) word metric on G, then the entropy h is positive if and only if the rate of escape l of the random walk determined by (G, μ) is positive. In one direction this follows from the Guivarc'h inequality (12.8). The converse was proved by Varopoulos [Var91] for finitely supported measures and then extended by Karlsson and Ledrappier to the case of a measure with finite first moment [KL07]. It is known that for a group of intermediate growth and a measure with finite first moment, the entropy is zero; therefore, the drift is also zero and the Poisson-Furstenberg boundary is trivial. The vanishing of the entropy, and hence the triviality of the boundary, can easily be deduced for instance from inequality (18) in [KV83]. In the case of a nonsymmetric measure, the drift l can be nonzero, even on a group of polynomial growth (for instance, for a $(p, 1-p)$ random walk on $\mathbb{Z}$ with $p < 1/2$). But, in the case of a symmetric measure, it is zero if and only if $h = 0$. Therefore, in the case of groups of subexponential growth and symmetric measures with finite first moment $h = l = 0$ and the functions $H(n)$ and $L(n)$ grow sublinearly. The Poisson-Furstenberg boundary is also trivial for each group of polynomial growth and any measure μ; this follows from Gromov's result on groups of polynomial growth and the theorem of Dynkin and Malyutov concerning Martin boundaries of nilpotent groups (from which the triviality of the Poisson-Furstenberg boundary follows in this case [DM61]; see also the work of G. Margulis [Mar66]).

To obtain a lower bound for growth for some groups from the family G_ω and to obtain new results about the Poisson-Furstenberg boundary of random walks on groups of intermediate growth, in [Ers04a] Erschler introduced the so called "strong condition" $(*)$ for some type actions on the interval $(0, 1]$. She proved that if a group G satisfies condition $(*)$ and the group $germ(G)$ of germs of G (also defined in [Ers04a]) satisfies some extra condition, then G admits a symmetric probabilistic measure μ with finite entropy $H(\mu)$ and nontrivial Poisson-Furstenberg boundary. These conditions are satisfied for all groups $G_\omega, \omega \in \Omega_4$, where the set $\Omega_4 = \Omega_1 \setminus \Omega_0$ consists of the sequences containing only two symbols from $\{0, 1, 2\}$, with each of them occurring infinitely many times in the sequence. Recall that by Theorem 9.6 all such groups have intermediate growth. The groups $G_\omega, \omega \in \Omega_4$ are first examples of groups of intermediate growth possessing a symmetric measure with nontrivial Poisson-Furstenberg boundary. Also for all these groups and any $\epsilon > 0$, a lower bound on growth of type

$$\exp \frac{n}{\log^{2+\epsilon} n} \preceq \gamma_{G_\omega}(n) \tag{12.10}$$

holds [Ers04a]. The proof of this result uses the existence of a special element $g \in G$ of infinite order. It is based on the combination of facts of existence of measure with non-trivial Poisson boundary and the analogue of inequality (12.8) for measures with infinite first moment.

Interesting results concerning growth and triviality of the Poisson-Furstenberg boundary were obtained by Karlsson, Ledrappier and Erschler [KL07, ErsK10]. These results led to upper bounds on growth of $H(n)$ and $L(n)$ for the groups G_ω and show that, under certain conditions, non-vanishing of the drift implies that the group is indicable (i.e., existence of a surjective homomorphism onto $\mathbb{Z}$).

Now, let us mention a new method of studying of asymptotic properties of self-similar groups discovered by L. Barthlodi and B. Virag in [BV05]. It received further development in the paper of V. Kaimanovich [Kai05], where it was called the *Munchhausen trick*, and in the papers of Bartholdi, Kaimanovich, and Nekrashevych [BKN10] and of Amir, Angel, and Virag [AAV09]. In [Kai05] the entropy arguments were used to prove amenability, the notion of a self-similar measure was introduced, and the map ψ in the space of probabilistic measures on a self-similar group was defined. It allows one to describe self-similar measures as fixed points of this map. The relation between ψ and the classical tool of linear algebra known as the Schur complement was established in [GN07].

The Munchhausen trick has been used to prove amenability of certain self-similar groups of exponential growth [BV05], [BKN10], [AAV09]. For the first time, this method was applied to prove the amenability of the group B, named "Basilica," which was introduced in the paper of A. Żuk and the author [GŻ02], and can be defined as the *iterated monodromy group* of the polynomial $z^2 - 1$, or alternatively as the group generated by the automaton $\mathcal{A}_{852}$ from the *atlas* of self-similar complexity $(2,3)$ groups [BGK$^+$08]. Observe that B has exponential growth. The amenability of B allows us to separate the class AG from the class SG of subexponentially amenable groups that was mentioned in Section 7. It was originally defined in [Gri98] (where the question about the possible coincidence of classes AG and SG was raised). It is currently unknown whether the cardinality of the set $AG \setminus SG$ is the cardinality of the continuum or not.

Unfortunately, the Munchausen trick has not been used so far to obtain new information on growth of groups. But it was used in [BV05], [Kai05], and [BKN10] to obtain interesting results about the rate of growth of the functions $H(n)$, $L(n)$, and the rate of decay of $P(n)$. For instance, in the case of the Basilica group, $P(n) \succeq e^{-n^{2/3}}$ [BV05], while for the group of intermediate growth $\mathcal{G}$, the lower bound is $P(n) \succeq e^{-n^{1/2-\epsilon}}$ for any positive number ϵ (this follows for instance from the results of the paper by Bartholdi, Kaimanovich, and Nekrasevych [BKN10]). Using inequalities (12.2) and (12.5) we obtain the following estimates:

$$e^{-n^{2/3}} \preceq P_B(n) \preceq e^{-n^{1/3}},$$

$$e^{-n^{1/2-\epsilon}} \preceq P_{\mathcal{G}}(n) \preceq e^{-n^{1/5}}.$$

It would be interesting to find the asymptotics of the rate of decay of $P(n)$ for each of the groups B and $\mathcal{G}$ (as well as the rate of growth of the functions $H(n)$ and $L(n)$).

Behind the idea of the Munchausen trick is the conversion of the self-similarity of the group into self-similarity of the random walk on the group. Let G be a self-similar group acting level transitively on a d-regular rooted tree T_d, and let μ be a probability measure on G. Denote by $H = st_G(x)$ the stabilizer of a vertex x on the first level. Then $[G : H] = d$. Let $p_x : H \to G$ be the projection homomorphism of H on the subtree T_x with root x ($p_x(g) = g_x$, where g_x is the section of g at vertex x), and let μ_H be the probability distribution on H given by the probability of the first hit of H by a random walk on G determined by μ. Denote by $\mu_x = (p_x)_*(\mu_H)$ the image of the measure μ_H under the projection p_x.

DEFINITION 12.1. *A measure μ is called* self-similar *if for some vertex x of the first level*

$$\mu_x = (1 - \lambda)\delta_e + \lambda\mu$$

for some $\lambda, 0 < \lambda < 1$.

THEOREM 12.2 ([BV05], [Kai05]). *If a self-similar group G possesses a self-similar symmetric probability measure μ with finite entropy and contracting coefficient $\lambda, 0 < \lambda < 1$, then the entropy h of the corresponding random walk on G is zero and, therefore, the group G is amenable.*

For instance, for $\mathcal{G}$ the measure $\mu = \frac{4}{7}a + \frac{1}{7}(b + c + d)$ is self-similar with contracting coefficient $\lambda = 1/2$ [Kai05].

Let G be an amenable group with a finite generating system A. Then one can associate with (G, A) a function

$$F_G(n) = \min\{n : \text{ there is a finite subset } F \subset G \text{ s.t. } \frac{|F \triangle aF|}{|F|} < \frac{1}{n}, \forall a \in A\},$$

which is called the Følner function of G with respect to A (because of Følner's criterion of amenability [Føl57]). This function was introduced by Vershik in the appendix to the Russian edition of Greenleaf's book on amenability [Gre69]. The growth type of this function does not depend on the generating set. By a result of Coulhon and Saloff-Coste [CSC93], the growth function and the Følner function are related by the inequality

$$\gamma(n) \preceq F(n). \tag{12.11}$$

On the other hand, for groups with $\gamma(n) \preceq e^{n^\alpha}$, the Følner function can be estimated as in [Ers06, Lemma 3.1]

$$F(n) \preceq e^{n^{\frac{\alpha}{1-\alpha}}}. \tag{12.12}$$

It is proved by Erschler that for an arbitrary function $f \colon \mathbb{N} \to \mathbb{N}$, there is a group of intermediate growth with Følner function $F(n) \succeq f(n)$. The method of construction of such groups is based on the "oscillation"-type techniques that we discuss briefly in the last section. Very interesting results about the asymptotics of random walks on Schreier graphs associated with finitely generated groups, and in particular with $\mathcal{G}$, are obtained by Erschler in [Ers05a]. We shall reformulate the following question soon in the form of a conjecture.

PROBLEM 12. *Can a Følner function grow strictly slower than the exponential function but faster than any polynomial?*

In view of the inequality (12.12), the gap conjecture holds if the answer to the last problem is negative.

The next conjecture (consisting of three subconjectures) was formulated by P. Pansu and the author in 2000 in an unpublished note. Let G be an amenable finitely generated group, μ a symmetric probability measure with finite support that generates G, M a Markov operator of the associated random walk on G given by μ, and $P(n)$ the probability of return after n steps for this random walk. Let $F(n)$ be the Følner function, and $\mathcal{N}(\lambda)$ be the spectral density defined by the relation

$$\mathcal{N}(\lambda) = tr_{vN}\Big(\chi_{(-\infty,\lambda]}(\Delta)\Big),$$

where tr_{vN} is the von Neumann trace defined on the von Neumann algebra $\mathcal{N}(G)$ of G, generated by the right regular representation, $\Delta = I - M \in \mathcal{N}(G)$ is the discrete Laplace operator on G, and $\chi_{(-\infty,\lambda]}(\Delta)$ is the projection obtained by application of the characteristic function $\chi_{(-\infty,\lambda]}$ to Δ. We are interested in the asymptotic behavior of $P(n)$ and $F(n)$ when $n \to \infty$ and of $\mathcal{N}(\lambda)$ when $\lambda \to 0+$. Their asymptotic behavior does not depend on the choice of the measure μ [BPS09].

CONJECTURE 12.3.

(i) Gap Conjecture for the Heat Kernel: *The function $P(n)$ is either of power rate of decay or satisfies*

$$P(n) \preceq e^{-\sqrt[3]{n}}.$$

(ii) Gap Conjecture for the Følner Function: *The Følner function $F(n)$ has either polynomial growth or the growth is at least exponential.*

(iii) Gap Conjecture for Spectral Density: *The spectral density $\mathcal{N}(\lambda)$ either has power decay of type $\lambda^{d/2}$ for some $d \in \mathbb{N}$ when $\lambda \to 0$ or*

$$\mathcal{N}(\lambda) \preceq e^{-1/\sqrt[4]{\lambda}}.$$

Let us also formulate a modified version of the previous conjecture. We guess that for each of three conjectures stated next there is a number $\beta > 0$ for which it holds.

CONJECTURE 12.4 (Gap Conjectures with Parameter β, $\beta > 0$).

(i) *The function $P(n)$ is either of power rate of decay or satisfies*

$$P(n) \preceq e^{-n^{\beta}}.$$

(ii) *The Følner function $F(n)$ has either polynomial growth or the growth is not less than $e^{n^{\beta}}$.*

(iii) *The spectral density $\mathcal{N}(\lambda)$ either has power decay of type $\lambda^{d/2}$ for some $d \in \mathbb{N}$ or*

$$\mathcal{N}(\lambda) \preceq e^{-\lambda^{-\beta}}.$$

Each of the preceding alternatives separates the case of polynomial growth or power rate of decay from that of intermediate growth (or decay).

Perhaps gap-type conjectures can also be formulated in a reasonable way for some other asymptotic characteristics of groups such as the isoperimetric profile, L^2-isoperimetric profile, entropy function $H(n)$, drift function $L(n)$, etc.

There are relations between all the conjectures stated here to the growth gap conjecture from Section 10 (which we will call here the *growth gap conjecture*), and to its generalization, Conjecture 10.5. For instance, using Inequality (12.12) we conclude that the gap conjecture for the Følner function implies the growth gap conjecture. On the other hand, inequality (12.11) shows that the growth gap conjecture with parameter β implies the gap conjecture with parameter β for the Følner function. Therefore, the weak gap conjecture for growth is equivalent to the weak gap conjecture for the Følner function (the latter is formulated similarly to Conjecture 10.6).

The growth gap conjecture with parameter α implies the gap conjecture with parameter $\frac{\alpha}{\alpha+2}$ for the return probabilities $P(n)$ (because of Inequality (12.5)), etc. It would be interesting to find the relation between all stated conjectures.

The results of [Gri89a], [LM91], [Wil11] and some statements from this article provide the first classes of groups for which the gap conjectures of the given type hold. For instance, the gap conjecture with parameter $\frac{1}{5}$ for return probabilities $P(n)$ and with parameter $\frac{1}{2}$ for the Følner function $F(n)$ hold for residually supersolvable groups.

13   INVERSE ORBIT GROWTH AND EXAMPLES WITH EXPLICIT GROWTH

Until recently there was no exact computation of the intermediate growth in the sense of the Schwarz-Milnor equivalence. The first such examples were produced recently by L. Bartholdi and A. Erschler [BE10]. The idea is very nice, and we shall explain it briefly.

Observe that the notion of growth can be defined for transitive group actions. Namely, if a finitely generated group G with a system of generators A acts transitively on a set X and a base point $x \in X$ is selected, then the growth function $\gamma_{X,x}^A(n)$ counts the number of points in X that can be reached from x by consecutive applications of at most n elements from set $A \cup A^{-1}$. The growth type of this function (in the sense of the equivalence $\sim$) does not depend on the choice of x. The triple (G, X, x) can be encoded by the *Schreier graph* (or the *graph of the action*) Γ, with set of vertices X and a set of oriented edges (labeled by the elements of A) consisting of pairs $(x, a(x)), x \in X, a \in A$. The growth function of the action is the same as the growth function of the graph Γ, which is a $2d$-regular graph (viewed as a non-oriented graph).

It is easy to construct actions with intermediate growth between polynomial and exponential. An interesting topic is the study of growth of Schreier graphs associated with actions of self-similar groups on corresponding rooted trees and their boundaries. The graphs we allude to here are of the form $\Gamma = \Gamma(G, H, A)$, where G is a finitely generated self-similar group with generating set A and $H = St_G(\xi)$, where ξ is a point on the boundary of the tree. Such graphs are isomorphic to the corresponding graphs of the action of the group on the orbit of the base point (namely, ξ). For contracting groups they have polynomial growth, which can be of fractional and even irrational degree [BG00a], [Bon07]. There are examples of actions with quite exotic intermediate behavior, like $n^{\log^m n}$ for some $m > 0$ [BH05], [GŠ06], [Bon11], [BCSDN11]. At the same time there are examples with quite regular polynomial-type orbit growth. For instance, for the group $\mathcal{G}$ the orbit growth is linear. Schreier graphs of the action of $\mathcal{G}$ on the first three levels of the binary tree are shown in Figure 13.1, and the infinite graph of the action on the orbit of a typical point of the boundary is given in Figure 13.2 (drawn in two versions: with labels and without).

In [BE10],[BE11] the notion of *inverted orbit growth function* has been introduced. Let us briefly explain the idea and formulate some of the results of these articles. Let G be a group acting on the right on a set X, let S be a generating set for G (viewed as a monoid), and let $x \in X$ be a base point. Denote by S^* the set of (finite) words over the alphabet S. For a word $w = w_1...w_l \in S^*$, its inverted orbit

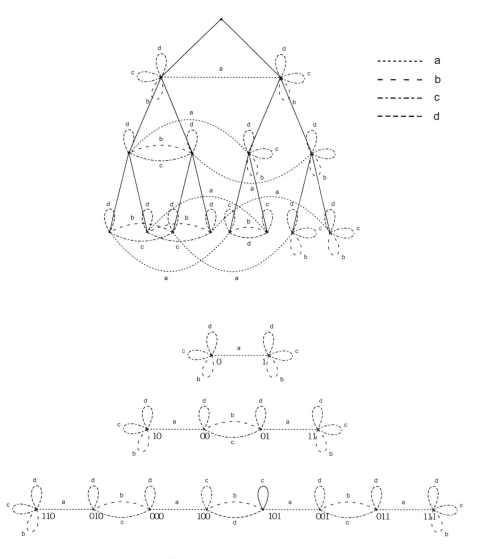

FIGURE 13.1: The action of $\mathcal{G}$ on the first three levels of the tree. (See also Plate 27.)

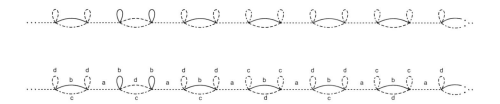

FIGURE 13.2: Typical Schreier graph of the boundary action. (See also Plate 28.)

is

$$\mathcal{O}(w) = \{x, xw_l, xw_{l-1}w_l, xw_1 \ldots w_{l-1}w_l\},$$

and the inverted orbit growth of w is $\delta(w) = |\mathcal{O}(w)|$. The inverted orbit growth function of G is the function

$$\Delta(n) = \Delta_{(G,X,x)}(n) = \max_w \{\delta(w) | |w| = n\}.$$

Clearly $\Delta_{(G,X,x)}(n) \preceq \gamma_{(G,X,x)}(n)$.

The notions of *wreath product* and *permutational wreath product* are standard in group theory. Consider groups A, G and a G-set X, such that G acts on X on the right. The wreath product $W = A \wr_X G$ is a semidirect product of $\sum_X A$ (a direct sum of copies of A indexed by X) with G acting on $\sum_X A$ by automorphisms induced by the corresponding permutations of X. In other words, view elements of $\sum_X A$ as finitely supported functions $X \to A$. A left action of G on $\sum_X A$ by automorphisms is then defined by $(gf)(x) = f(xg)$. There are two versions of (permutational) wreath products, restricted and unrestricted, and here we use the *restricted* version (because of the assumption of finiteness of the support of the elements from the base group $\sum_X A$).

In the next two theorems, proved in [BE10], X is a $\mathcal{G}$-orbit of an arbitrary point of the boundary ∂T of the binary rooted tree T that does not belong to the orbit of the point 1^∞. Observe that the orbit of 1^∞ consists of the sequences cofinal to 1^∞, where two sequences are cofinal if they coincide starting with some coordinate, and, more generally, two points of the boundary of T are in the same orbit if they are cofinal [BG00a], [BG02a], [Gri05]. The graph of the action of $\mathcal{G}$ (after deletion of the labels) looks similar to the one shown in the last figure. Recall that the number $\alpha_0 \approx 0.7674$ was defined earlier and is included in (9.2).

THEOREM 13.3. *Consider the sequence of groups which is given by* $K_0 = \mathbb{Z}/2\mathbb{Z}$ *and* $K_{k+1} = K_k \wr_X \mathcal{G}$. *Then every* K_k *is a finitely generated infinite torsion group, with growth function*

$$\gamma_{K_k}(n) \sim exp(n^{[1-(1-\alpha_0)^k)]}).$$

Recall that the torsion-free group of intermediate growth $\hat{\mathcal{G}}$, constructed in [Gri85a], was mentioned in Section 9 (Theorem 9.3).

THEOREM 13.4. *Consider the following sequence of groups:* $H_0 = \mathbb{Z}$, $H_1 = \hat{\mathcal{G}}$ *and* $H_{k+1} = H_k \wr_X \mathcal{G}, k \geq 1$. *Their growth functions satisfy*

$$\gamma_{H_k}(n) \sim \exp(log(n)n^{1-(1-\alpha_0)^k}).$$

It is remarkable that not only two infinite series of groups of intermediate growth with precisely computed growth have been constructed, but also that the precise growth of the torsion-free group $\hat{\mathcal{G}}$, constructed more than 25 years ago, has finally been evaluated.

The proof of these two theorems is based on the calculation of the inverse orbit growth of the action of $\mathcal{G}$ on X, which is n^{α_0} in this case. In the realization of the first step of this program, the technique used previously by Bartholdi [Bar98] for improving the upper bound in the case of the group $\mathcal{G}$ from the value $\log_{32} 31$ used in [Gri84b] to the value α_0 established in [Bar98], [MP01a] is explored again.

The technique is based on assigning positive weights to the canonical generators a, b, c, d and finding the values that give the best possible (for this approach) upper bound for growth.

The preceding ideas and some other tools (in particular, the dynamics of partially continuous self-maps of simplices) have been used by Bartholdi and Erschler in [BE11] to construct a family of groups with growth of the type e^{n^α}, with α belonging to the interval $(\alpha_0, 1)$. Moreover, they presented the following impressive result. Let $\eta_+ \approx 2.4675$ be the positive root of $x^3 - x^2 - 2x - 4$.

THEOREM 13.5. *Let* $f: \mathbb{R} \to \mathbb{R}$ *be a function satisfying*

$$f(2x) \le f(x)^2 \le f(\eta_+ x)$$

for all x large enough. Then there exists a finitely generated group with growth equivalent to the growth of f.

This theorem provides a large class of growth functions of finitely generated groups that "fill" the "interval" $[e^{n^{\alpha_0}}, e^n]$.

14 MISCELLANEOUS

Recall that a group has uniformly exponential growth if it has exponential growth and moreover the number κ_* (the base of exponential growth) defined by (11.11) is > 1.

In [Gro81b] Gromov raised the following question.

GROMOV'S PROBLEM ON GROWTH (II). *Are there groups of exponential but not uniformly exponential growth?*

Some preliminary results concerning uniformly exponential growth were obtained by P. de la Harpe [dlH02]. The problem of Gromov (II) was solved by J. S. Wilson [Wil04b], [Wil04a] by providing an example of such a group. A shorter solution was found later by Bartholdi in [Bar03b]. Tree-like constructions and techniques of self-similar groups naturally leading to such examples were explored by V. Nekrashevych [Nek10]. Similar to the intermediate growth case, all known examples of groups of exponential but not uniformly exponential growth are based on the use of self-similar groups of branch type.

There are results that show that groups within certain classes of groups of exponential growth are of uniformly exponential growth. This holds, for instance, for hyperbolic groups (M. Koubi [Kou98]), one-relator groups, solvable groups, linear groups over fields of characteristic 0 (A. Eskin, S. Mozes and H. Oh [EMO05]), subgroups of the mapping class group which have exponential growth (J. Mangahas [Mang10]), and some other groups and classes of groups.

The fact that solvable groups of exponential growth have uniformly exponential growth was proved by D. Osin [Osi03] and he generalized this result to elementary amenable groups [Osi04]. E. Breuillard gave a "ping pong"-type proof of Osin's result [Bre07]. A connection between the so-called "slow growth" and the Lehmer conjecture is another result in Breuillard's paper. The fact that one relator group of exponential growth has uniformly exponential growth is proved by P. de la Harpe and the author [GdlH01], and that a one relator group has either polynomial growth or exponential growth is shown in the paper of T. Ceccherini-Silberstein and the author [CSG97].

A second topic of this section is the *oscillation* phenomenon that exists in the world of growth of finitely generated groups. The meaning of this is that there are groups whose growth in a certain sense may oscillate between two types of growth. This was first discovered by the author in [Gri84b] and was further developed in his habilitation [Gri85b]. The goal achieved in that work was the construction of a chain and an anti-chain of cardinality of the continuum in the space of growth degrees of finitely generated groups. The "trick" used for this purpose can be described briefly as follows.

The groups $\mathcal{G}_\omega, \omega \in \Omega \setminus \Omega_1$ (recall that $\Omega \setminus \Omega_1$ consists of sequences that are constant at infinity) are virtually abelian, while the rest of the groups $\mathcal{G}_\omega, \omega \in \Omega_1$ have intermediate growth and the set $\{\mathcal{G}_\omega, \omega \in \Omega_1\}$ has the property that if two sequences $\lambda, \mu \in \omega \in \Omega_1$ have the same prefix of length n then the subgraphs with vertices in the balls $B_{\mathcal{G}_\lambda}(2^{n-1})$ and $B_{\mathcal{G}_\nu}(2^{n-1})$ of radius 2^{n-1} with centers at the identity elements in the Cayley graphs of these groups are isomorphic. It was suggested in [Gri84b] to replace the groups from the set $\{\mathcal{G}_\omega, \omega \in \Omega \setminus \Omega_1\}$ by the set of accumulation points of $\{\mathcal{G}_\omega, \omega \in \Omega_1\}$ in the space of the Cayley graphs of 4-generated groups, supplied with a natural topology introduced in the article. Then each of deleted groups $\{\mathcal{G}_\omega, \omega \in \Omega \setminus \Omega_1\}$ is replaced by some virtually metabelian group (for which we will keep the same notation) of exponential growth and the modified set of groups $\{\mathcal{G}_\omega, \omega \in \Omega\}$ becomes a closed subset in the space of groups homeomorphic to a Cantor set.

Consider a sequence θ of the type

$$\theta = (012)^{m_1} 0^{k_1} (012)^{m_2} 0^{k_2} \ldots \tag{14.1}$$

with a very fast growing sequence $m_1, k_1, m, k_2, \ldots, m_i, k_i, \ldots$ of parameters. In view of the property that the balls of radius 2^{n-1} coincide for the considered groups when they are determined by sequences with the same prefix of length n and the fact that the group $\mathcal{G}_\omega, \omega \in \Omega_1$ is (abstractly) commensurable to the group $\mathcal{G}_{\tau^n(\omega)}^{2^n}$ (where τ is the shift in space of sequences), we conclude that for initial values of n when $n \in [1, R(m_1)]$ ($R(m_1)$ determined by m_1) the group $\mathcal{G}_\theta$ behaves as a group of intermediate growth (similar to $\mathcal{G} = \mathcal{G}_{(012)^\infty}$), but then for larger values of $n, n \in [R(m_1) + 1, R(m_1, k_1)]$ (with $R(m_1, k_1)$ determined by m_1 and k_1), it starts to grow as the group $\mathcal{G}_{0^\infty}^{2^{3m_1-1}}$ (i.e., exponentially), but then again the growth slows down and behaves in intermediate fashion, etc. Taking sequences of the type (14.1) but determined by various sequences of parameters, one can construct a chain and an anti-chain of cardinality of the continuum in the space of growth degrees of 3-generated groups; this was done in [Gri84b] and [Gri85b].

The possibility of application of the oscillation technique is based on the use of a topology in the space of Cayley graphs (or what is the same, in the space of marked groups) introduced in [Gri84b] (this topology is a relative to Chabauty topology known in the theory of locally compact groups and geometric topology [dlH00]). The space of marked groups is a compact totally disconnected space, and one of the major problems is to find its Cantor-Bendixson rank (for more on this and related problems see [Gri05]). Although the fact that this space has a nontrivial perfect core follows from the result established by B. Neumann in 1937 (a construction of uncountably many 2-generated groups, up to isomorphism), there is a considerable interest in finding Cantor subsets in the space of marked groups consisting of interesting families of groups. The first such subset was identified in

[Gri84b], which, with exception of a countable set, consists of groups of intermediate growth $\mathcal{G}_\omega, \omega \in \Omega_1$ (its construction was described in Section 9).

Oscillation techniques received further development in the paper of Erschler
[Ers06] (which was already mentioned at the end of Section 12), where the notion
of piecewise automatic group was introduced. Using this notion, Erschler constructed groups of intermediate growth with arbitrary fast growth of the Fölner
function. Moreover, the asymptotic entropy of a random walk on Erschler groups
can be arbitrarily close to a linear function, while at the same time, the Poisson boundary can be trivial. Oscillation techniques, in combination with ideas
from [BE10], were also used by J. Brieussel [Bri11]. The most recent result of "oscillation character" is due to M. Kassabov and I. Pak [KP11]. They demonstrated
a very unusual oscillation phenomenon for groups of intermediate growth, and
their result is based on a new idea which we are going to explain briefly.

The groups G_ω, as well as groups of branch type, act on spherically homoge-
neous rooted trees, that is, trees $T_{\bar{k}}$ defined by sequences of integers $\bar{k} = k_1, k_2, \ldots$,
with $k_i \geq 2$; these sequences $\bar{k}$ are called the *branch index* (k_i is the branching number for the ith level of the tree). Kassabov and Pak suggested modifying this approach by considering actions on *decorated* trees by attaching to some levels of $T_{\bar{k}}$
finite subtrees with actions of suitably chosen finite groups (each vertex of the
corresponding level is decorated by the same structure). The sequence $\{F_i\}$ of attached groups has to satisfy certain properties (in particular the groups need to be
generated by four involutions, three of which commute as in case of $\mathcal{G}_\omega$), but the
main property is that the groups F_i must behave as expanders, in a certain sense.
Namely, their Cayley graphs have to have diameters d_i growing as $i \to \infty$ as the
logarithm of the size of the group, and for values of n in the range $1 \leq n \leq C d_i$
($C, 0 < C < 1$ some constant independent of i), the growth functions $\gamma_i(n)$ have to
behave as the exponential function.

As already mentioned in the introduction, there are various types of asymptotic characteristics that can be associated with algebraic objects. In addition to
the group growth, and other characteristics considered in Section 12, the *subgroup*
growth, the *conjugacy* growth, the *geodesic* growth, and many other types of growth
have been studied. The subgroup growth was already discussed a little bit, and we
refer the reader to the comprehensive book on this and other subjects by Lubotzky
and Segal [LS03] and the literature cited therein.

The conjugacy growth counts the number of conjugacy classes of length at
most n, $n = 1, 2, \ldots$, where the length of a conjugacy class is the length of the
shortest representative of this class. Interesting results on this subject are obtained
by I. Babenko [Bab88] (who was perhaps the first who introduced this notion),
M. Coornaert and G. Knieper, who studied the hyperbolic groups case [CK02],
Breuillard and Coornaert [BdC10] (the case of solvable groups), and M. Hull and
D. Osin [HO11] who showed that basically any monotone function growing not
faster than an exponential function is equivalent to the conjugacy growth function. This is a far from complete list of papers and results on this subject (for a
more complete list see the literature in the cited papers).

The geodesic growth of a pair (G, A) (a group and a finite system of generators) is a growth of the language of geodesic words over alphabet $A \cup A^{-1}$ (i.e.,
words over an alphabet of generators that represent geodesics in the Cayley graph

$\Gamma(G, A)$ with the origin at the identity element). It can be polynomial with respect to some system of generators but exponential with respect to other systems of generators, and it is unclear if it can be intermediate between polynomial and exponential. This notion was studied in [Can84], [NS95], [GN97], [BBES11]. The following question was discussed by M. Shapiro and the author around 1993.

PROBLEM 13. *Are there pairs (G, A) consisting of a group of intermediate growth G and a finite system of generators A with intermediate geodesic growth?*

All known groups of intermediate growth have exponential geodesic growth. For instance, for $\mathcal{G}$ this follows from the fact that the Schreier graph presented in Figure 13.2 has exponential geodesic growth, which is obvious.

The study of geodesic growth is a particular case of study of growth of formal languages. Such questions originated in the work of Schützenberger in the 1950s. The literature on this subject related to group theory can be found in [BrG02b], [CSW02], [CSW03], [Gil05]. Observe that for regular and for context-free languages, growth can be only polynomial or exponential (R. Incitti [Inc01], M. Bridson and R. Gilman [BrG02b]), while the intermediate type behavior is possible for indexed languages (the next in the language hierarchy type of languages after the context-free languages), as is shown in the note of A. Machi and the author [GM99].

There are interesting studies about growth of regular graphs. One of the first publications on this subject is the article of V. Trofimov [Tro84] where, under certain conditions on the group of automorphisms of the graph, the case of polynomial growth is studied. In the last decade, the study of Schreier graphs of finitely generated groups, and, in particular, of their growth and amenability properties, has been intensified. Some results were already mentioned in previous sections. We mention here the paper of Bartholdi and the author [BG00a], where it is observed that Schreier graphs of self-similar contracting groups are of polynomial growth, and that in this case the degree of the polynomial (more precisely of *power*) growth can be non-integer and can even be a transcendental number. Interesting results on the growth of Schreier graphs are obtained by I. Bondarenko [Bon07] (see also [BH05], [BCSDN11]). The amenability of Schreier graphs associated with actions of almost finitary groups on the boundary of rooted tree is proven in [GN05].

Finally, let us return to the discussion on the role of just-infinite groups in the study of growth. Recall that a group G is called just-infinite if it is infinite but every proper quotient is finite. Such groups are on the border between finite groups and infinite groups, and surely they should play an important role in investigations around various gap type conjectures considered in this article. The following statement is an easy application of Zorn's lemma.

PROPOSITION 14.1. *Let G be a finitely generated infinite group. Then G has a just-infinite quotient.*

COROLLARY 14.2. *Let $\mathcal{P}$ be a group theoretical property preserved under taking quotients. If there is a finitely generated group satisfying the property $\mathcal{P}$, then there is a just-infinite group satisfying this property.*

Although the property of a group having intermediate growth is not preserved when passing to a quotient group (the image may have polynomial growth), by theorems of Gromov [Gro81a] and Rosset [Ros76], if the quotient G/H of a group G of intermediate growth is a virtually nilpotent group, then H is a finitely generated group of intermediate growth and one may look for a just-infinite quotient of H and iterate this process in order to represent G as a consecutive extension of a chain of groups that are virtually nilpotent or just-infinite groups. This observation is the base of the arguments for statements given by theorems 10.7, 10.4, 10.8 and 14.6.

The next theorem was derived by the author from a result of J. S. Wilson [Wil71].

THEOREM 14.3 ([Gri00b]). *The class of just-infinite groups naturally splits into three subclasses:*

  (B) *algebraically branch just-infinite groups,*

  (H) *hereditary just-infinite groups, and*

  (S) *near-simple just-infinite groups.*

Recall that branch groups were already defined in Section 9. The definition of algebraically branch groups can be found in [Gri00a, BGŠ03]. Every geometrically branch group is algebraically branch, but not vice versa. The difference between the two versions of the definitions is not large, but there is still no complete understanding of how much the two classes differ. Not every branch group is just-infinite, but every proper quotient of a branch group is virtually abelian. Therefore, branch groups are "almost just-infinite," and most of the known finitely generated branch groups are just-infinite.

DEFINITION 14.4. *A group G is* hereditary just-infinite *if it is infinite, it is residually finite, and every subgroup $H < G$ of finite index is just-infinite.*

For instance, $\mathbb{Z}$, D_∞, and $PSL(n, \mathbb{Z})$, $n \geq 3$ (by a result of G. Margulis [Mar91]) are hereditary just-infinite groups.

DEFINITION 14.5. *We call a group G* near-simple *if it contains a subgroup of finite index H which is a direct product*

$$H = P \times P \times \cdots \times P$$

of finitely many copies of a simple *group P.*

We already know that there are finitely generated branch groups of intermediate growth (for instance, groups $G_\omega, \omega \in \Omega_1$). The question on the existence of non-elementary amenable hereditary just-infinite groups is still open (observe that the only elementary amenable hereditary just-infinite groups are $\mathbb{Z}$ and D_∞).

PROBLEM 14. *Are there finitely generated hereditary just-infinite groups of intermediate growth?*

PROBLEM 15. *Are there finitely generated simple groups of intermediate growth?*

As it was already mentioned in Section 10, we believe that there is a reduction of the Gap conjecture to the class of just-infinite groups, that is, to the classes of (just-infinite) branch groups, hereditary just-infinite groups and simple groups. The corresponding result would hold if the gap conjecture holds for residually solvable groups. Using the results of Wilson [Wil11], one can prove the following result.

THEOREM 14.6 ([Gri12]).

(i) *If the gap conjecture with parameter 1/6 holds for just-infinite groups, then it holds for all groups.*

(ii) *If the gap conjecture holds for residually polycyclic groups and for just-infinite groups, then it holds for all groups.*

Therefore, to obtain a complete reduction of the gap conjecture to just-infinite groups, it is enough to prove it for residually polycyclic groups, which is quite plausible. Similar reductions hold for some other gap type conjectures stated in this article.

As was already mentioned in Section 10, uncountably many finitely generated simple groups that belong to the class *LEF* were recently constructed by K. Medynets and the author [GM11]. It may happen that among the subgroups considered in [GM11] (they are commutator subgroups of *topological full groups* of *subshifts of finite type*), there are groups of intermediate growth, but this has to be checked. On the other hand, it may happen that there are no simple groups of intermediate growth and that there are no hereditary just-infinite groups of intermediate growth at all. In this case the gap conjecture would be reduced to the case of branch groups. In [Gri84b, Gri85a] the author proved that growth functions of all p-groups of intermediate growth $\mathcal{G}_\omega$ discussed in Section 9 satisfy the lower bound $\gamma_{\mathcal{G}_\omega}(n) \succeq e^{\sqrt{n}}$, and this was proved by direct computations based on the anti-contracting property given by definition 11.7. This gives some hope that the gap conjecture can be proved for the class of branch groups by a similar method.

15 Bibliography

[AVŠ57]   G. M. Adel'son-Vel'skiĭ and Yu. A. Šreĭder, *The Banach mean on groups*, Uspehi Mat. Nauk (N.S.) **12** (1957), no. 6(78), 131–136. MR 0094726 (20 #1238).

[Adi79]   S. I. Adian, *The Burnside problem and identities in groups*, Ergebnisse der Mathematik und ihrer Grenzgebiete [Results in Mathematics and Related Areas], vol. 95, Springer-Verlag, Berlin, 1979. MR 80d:20035.

[Adi82]   S. I. Adian, *Random walks on free periodic groups*, Izv. Akad. Nauk SSSR Ser. Mat. **46** (1982), no. 6, 1139–1149, 1343. MR 84m:43001.

[Ale72]   S. V. Alešin, *Finite automata and the Burnside problem for periodic groups*, Mat. Zametki **11** (1972), 319–328. MR 0301107 (46 #265).

[AAV09]   A. Amir, O. Angel, and B. Virag, *Amenability of linear-activity automaton groups*, 2009. arXiv:0905.2007

[And87]   Michael T. Anderson, *On the fundamental group of nonpositively curved manifolds*, Math. Ann. **276** (1987), no. 2, 269–278. MR 870965 (88b:53046).

[AK63]   V. I. Arnol'd and A. L. Krylov, *Uniform distribution of points on a sphere and certain ergodic properties of solutions of linear ordinary differential equations in a complex domain*, Dokl. Akad. Nauk SSSR **148** (1963), 9–12. MR 0150374 (27 #375).

[Ave70]   André Avez, *Variétés Riemanniennes sans points focaux*, C. R. Acad. Sci. Paris Sér. A-B **270** (1970), A188–A191. MR 0256305 (41 #961).

[Bab88]   I. K. Babenko, *Closed geodesics, asymptotic volume and the characteristics of growth of groups*, Izv. Akad. Nauk SSSR Ser. Mat. **52** (1988), no. 4, 675–711, 895. MR 966980 (90b:58219).

[BM07]   B. Bajorska and O. Macedońska, *A note on groups of intermediate growth*, Comm. Algebra **35** (2007), no. 12, 4112–4115. MR 2372323 (2008j:20082).

[Bar98]   Laurent Bartholdi, *The growth of Grigorchuk's torsion group*, Internat. Math. Res. Notices (1998), no. 20, 1049–1054. MR 1656258 (99i:20049).

[Bar01]   Laurent Bartholdi, *Lower bounds on the growth of a group acting on the binary rooted tree*, Internat. J. Algebra Comput. **11** (2001), no. 1, 73–88. MR 1818662 (2001m:20044).

[Bar03a]   Laurent Bartholdi, *Endomorphic presentations of branch groups*, J. Algebra **268** (2003), no. 2, 419–443. MR 2009317 (2004h:20044).

[Bar03b]   Laurent Bartholdi, *A Wilson group of non-uniformly exponential growth*, C. R. Math. Acad. Sci. Paris **336** (2003), no. 7, 549–554. MR 1981466 (2004c:20051).

[BE10]   Laurent Bartholdi and Anna Erschler, *Growth of permutational extensions*, Invent. Math. **189** (2010), no. 2, 431–455. MR 2947548.

[BE11]   Laurent Bartholdi and Anna Erschler, *Groups of given intermediate word growth*, 2011. arXiv:1110.3650

[BG00a]   Laurent Bartholdi and Rostislav I. Grigorchuk, *On the spectrum of Hecke type operators related to some fractal groups*, Tr. Mat. Inst. Steklova **231** (2000), no. Din. Sist., Avtom. i Beskon. Gruppy, 5–45. MR 1841750 (2002d:37017).

[BG00b]   Laurent Bartholdi and Rostislav I. Grigorchuk, *Lie methods in growth of groups and groups of finite width*, Computational and geometric aspects of modern algebra (Edinburgh, 1998), London Math. Soc. Lecture Note Ser., vol. 275, Cambridge Univ. Press, Cambridge, 2000, pp. 1–27. MR 1776763 (2001h:20046).

[BG02a]   Laurent Bartholdi and Rostislav I. Grigorchuk, *On parabolic subgroups and Hecke algebras of some fractal groups*, Serdica Math. J. **28** (2002), no. 1, 47–90. MR 1899368 (2003c:20027).

[BGN03]   Laurent Bartholdi, Rostislav I. Grigorchuk, and Volodymyr Nekrashevych, *From fractal groups to fractal sets*, Fractals in Graz 2001, Trends Math., Birkhäuser, Basel, 2003, pp. 25–118. MR 2091700.

[BGŠ03]   Laurent Bartholdi, Rostislav I. Grigorchuk, and Zoran Šuniḱ, *Branch groups*, Handbook of algebra, Vol. 3, North-Holland, Amsterdam, 2003, pp. 989–1112. MR 2035113.

[BKN10]   Laurent Bartholdi, Vadim A. Kaimanovich, and Volodymyr V. Nekrashevych, *On amenability of automata groups*, Duke Math. J. **154** (2010), no. 3, 575–598. MR 2730578.

[BN08]   Laurent Bartholdi and Volodymyr V. Nekrashevych, *Iterated mon-odromy groups of quadratic polynomials. I*, Groups Geom. Dyn. **2** (2008), no. 3, 309–336.

[BRS06]   L. Bartholdi, I. I. Reznykov, and V. I. Sushchansky, *The smallest Mealy automaton of intermediate growth*, J. Algebra **295** (2006), no. 2, 387–414. MR 2194959 (2006i:68060).

[BV05]   Laurent Bartholdi and Bálint Virág, *Amenability via random walks*, Duke Math. J. **130** (2005), no. 1, 39–56. MR 2176547 (2006h:43001).

[Bas72]   H. Bass, *The degree of polynomial growth of finitely generated nilpotent groups*, Proc. London Math. Soc. (3) **25** (1972), 603–614. MR 0379672 (52 #577).

[BOERT96]   Hyman Bass, Maria Victoria Otero-Espinar, Daniel Rockmore, and Charles Tresser, *Cyclic renormalization and automorphism groups of rooted trees*, Lecture Notes in Mathematics, vol. 1621, Springer-Verlag, Berlin, 1996. MR 1392694 (97k:58058).

[BM88]   Gilbert Baumslag and Charles F. Miller, III, *Some odd finitely presented groups*, Bull. London Math. Soc. **20** (1988), no. 3, 239–244. MR 931184 (89e:20059).

[BdlHV08]   Bachir Bekka, Pierre de la Harpe, and Alain Valette, *Kazhdan's property (T)*, New Mathematical Monographs, vol. 11, Cambridge University Press, Cambridge, 2008. MR 2415834 (2009i:22001).

[Bek04]   L. A. Beklaryan, *Groups of homeomorphisms of the line and the circle. Topological characteristics and metric invariants*, Uspekhi Mat. Nauk **59** (2004), no. 4(358), 3–68. MR 2106645 (2005i:37044).

[Bek08]   Levon A. Beklaryan, *Groups homeomorphisms: Topological characteristics, invariant measures and classifications*, Quasigroups Related Systems **16** (2008), no. 2, 155–174. MR 2494874 (2010d:54054).

[BH05]   Itai Benjamini and Christopher Hoffman, ω-*periodic graphs*, Electron. J. Combin. **12** (2005), Research Paper 46, 12 pp. (electronic). MR 2176522 (2006f:05151).

[BPS09]   A. Bendikov, C. Pittet, and R. Sauer, *Spectral distribution and l^2-isoperimetric profile of Laplace operators on groups*, Math. Ann. **354** (2012), no. 1, 43–72. arXiv:0901.0271

[BGH13]   M. Benli, R. Grigorchuk, and P. de la Harpe, *Amenable groups without finitely presented covers*. Bull. Math. Sci. **3** (2013), no. 1, 73–131.

[Ben83]   M. Benson, *Growth series of finite extensions of $\mathbb{Z}^n$ are rational*, Invent. Math. **73** (1983), no. 2, 251–269. MR 714092 (85e:20026).

[Bog]   N. N. Bogolyubov, *On some ergodic properties of continious groups of transformations*, Nauk. Zap. Kiïv Derzh. Univ. im. T. G. Shevchenka, v. IV, (1939) no. 5, 45–52.

[Bon07]   I. Bondarenko, *Groups generated by bounded automata and their Schreier graphs*, (Dissertation), 2007.

[Bon11]   I. Bondarenko, *Growth of Schreier graphs of automaton groups*, 2011. arXiv:1101.3200

[BCSDN11]   I. Bondarenko, T. Ceccherini-Silberstein, A. Donno, and V. Nekrashevych, *On a family of Schreier graphs of intermediate growth associated with a self-similar group*, 2011. arXiv:1106.3979

[BGK+08]   Ievgen Bondarenko, Rostislav Grigorchuk, Rostyslav Kravchenko, Yevgen Muntyan, Volodymyr Nekrashevych, Dmytro Savchuk, and Zoran Šunić, *Classification of groups generated by 3-state automata over 2-letter alphabet*, Algebra Discrete Math. (2008), no. 1, 1–163. arXiv:0803.3555

[BBES11]   M. Bridson, J. Burillo, M. Elder, and Z. Šunić, *On groups whose geodesic growth is polynomial*, Internat. J. Algebra Comput. **22** (2012), no. 5, 1250048, 13 pp.

[BrG02b]   Martin R. Bridson and Robert H. Gilman, *Context-free languages of sub-exponential growth*, J. Comput. System Sci. **64** (2002), no. 2, 308–310. MR 1906807 (2003d:68132).

[BH99]   Martin R. Bridson and André Haefliger, *Metric spaces of non-positive curvature*, Grundlehren der Mathematischen Wissenschaften [Fundamental Principles of Mathematical Sciences], vol. 319, Springer-Verlag, Berlin, 1999. MR 1744486 (2000k:53038).

[Bre07]   Emmanuel Breuillard, *On uniform exponential growth for solvable groups*, Pure Appl. Math. Q. **3** (2007), no. 4, part 1, 949–967. MR 2402591 (2009d:20073).

[BdC10]   Emmanuel Breuillard and Yves de Cornulier, *On conjugacy growth for solvable groups*, Illinois J. Math. **54** (2010), no. 1, 389–395. MR 2777001.

[Bri11]   J. Brieussel, *Growth behaviors in the range e^{r^α}*, 2011. arXiv:1107.1632

[BS07]   Ievgen V. Bondarenko and Dmytro M. Savchuk, *On Sushchansky p-groups*, Algebra Discrete Math. (2007), no. 2, 22–42. MR 2364061 (2008i:20049).

[BP06]   Kai-Uwe Bux and Rodrigo Pérez, *On the growth of iterated monodromy groups*, Topological and asymptotic aspects of group theory, Contemp. Math., vol. 394, Amer. Math. Soc., Providence, RI, 2006, pp. 61–76. MR 2216706 (2006m:20062). arXiv:math.GR/0405456

[Can80]   J. Cannon, *The growth of the closed surface groups and compact hyperbolic Coxeter groups* (preprint), 1980.

[Can84]   James W. Cannon, *The combinatorial structure of cocompact discrete hyperbolic groups*, Geom. Dedicata **16** (1984), no. 2, 123–148. MR 758901 (86j:20032).

[CSC10]   Tullio Ceccherini-Silberstein and Michel Coornaert, *Cellular automata and groups*, Springer Monographs in Mathematics, Springer-Verlag, Berlin, 2010. MR 2683112.

[CSG97]   Tullio G. Ceccherini-Silberstein and Rostislav I. Grigorchuk, *Amenability and growth of one-relator groups*, Enseign. Math. (2) **43** (1997), no. 3–4, 337–354. MR 1489891 (99b:20057).

[CSMS01]   Tullio Ceccherini-Silberstein, Antonio Machì, and Fabio Scarabotti, *The Grigorchuk group of intermediate growth*, Rend. Circ. Mat. Palermo (2) **50** (2001), no. 1, 67–102. MR 1825671 (2002a:20044).

[CSW02]   Tullio Ceccherini-Silberstein and Wolfgang Woess, *Growth and ergodicity of context-free languages*, Trans. Amer. Math. Soc. **354** (2002), no. 11, 4597–4625. MR 1926891 (2003g:68067).

[CSW03]   Tullio Ceccherini-Silberstein and Wolfgang Woess, *Growth-sensitivity of context-free languages*, Theoret. Comput. Sci. **307** (2003), no. 1, 103–116, Words. MR 2022843 (2005h:68066).

[Che78]   Su Shing Chen, *On the fundamental group of a compact negatively curved manifold*, Proc. Amer. Math. Soc. **71** (1978), no. 1, 119–122. MR 0514740 (58 #24117).

[Cho80]   Ching Chou, *Elementary amenable groups*, Illinois J. Math. **24** (1980), no. 3, 396–407. MR 573475 (81h:43004).

[CM97]   Tobias H. Colding and William P. Minicozzi, II, *Harmonic functions on manifolds*, Ann. of Math. (2) **146** (1997), no. 3, 725–747. MR 1491451 (98m:53052).

[CDP90a]   M. Coornaert, T. Delzant, and A. Papadopoulos, *Géométrie et théorie des groupes*, Lecture Notes in Mathematics, vol. 1441, Springer-Verlag, Berlin, 1990, Les groupes hyperboliques de Gromov. [Gromov hyperbolic groups], With an English summary. MR 1075994 (92f:57003).

[CK02]   M. Coornaert and G. Knieper, *Growth of conjugacy classes in Gromov hyperbolic groups*, Geom. Funct. Anal. **12** (2002), no. 3, 464–478. MR 1924369 (2003f:20071).

[CSC93]   Thierry Coulhon and Laurent Saloff-Coste, *Isopérimétrie pour les groupes et les variétés*, Rev. Mat. Iberoamericana **9** (1993), no. 2, 293–314. MR 1232845 (94g:58263).

[Day57]   Mahlon M. Day, *Amenable semigroups*, Illinois J. Math. **1** (1957), 509–544. MR 19,1067c.

[dlH73]   Pierre de la Harpe, *Moyennabilité de quelques groupes topologiques de dimension infinie*, C. R. Acad. Sci. Paris Sér. A-B **277** (1973), A1037–A1040. MR 0333060 (48 #11385).

[dlH00]   Pierre de la Harpe, *Topics in geometric group theory*, Chicago Lectures in Mathematics, University of Chicago Press, Chicago, 2000. MR 1786869 (2001i:20081).

[dlH02]   Pierre de la Harpe, *Uniform growth in groups of exponential growth*, Proceedings of the Conference on Geometric and Combinatorial Group Theory, Part II (Haifa, 2000), vol. 95, 2002, pp. 1–17. MR 1950882 (2003k:20031).

[dlHGCS99] P. de la Harpe, R. I. Grigorchuk, and T. Chekerini-Sil'berstaïn, *Amenability and paradoxical decompositions for pseudogroups and discrete metric spaces*, Tr. Mat. Inst. Steklova **224** (1999), no. Algebra. Topol. Differ. Uravn. i ikh Prilozh., 68–111. MR 1721355 (2001h:43001).

[Dix60]   Jacques Dixmier, *Opérateurs de rang fini dans les représentations uni-taires*, Inst. Hautes Études Sci. Publ. Math. (1960), no. 6, 13–25. MR 0136684 (25 #149).

[Dye59]   H. A. Dye, *On groups of measure preserving transformation. I*, Amer. J. Math. **81** (1959), 119–159. MR 0131516 (24 #A1366).

[Dye63]   H. A. Dye, *On groups of measure preserving transformations. II*, Amer. J. Math. **85** (1963), 551–576. MR 0158048 (28 #1275).

[DM61]   E. B. Dynkin and M. B. Maljutov, *Random walk on groups with a finite number of generators*, Dokl. Akad. Nauk SSSR **137** (1961), 1042–1045. MR 0131904 (24 #A1751).

[Ebe73]   Patrick Eberlein, *Some properties of the fundamental group of a Fuchsian manifold*, Invent. Math. **19** (1973), 5–13. MR 0400250 (53 #4085).

[Efr53]   V. A. Efremovič, *The proximity geometry of Riemannien manifolds.*, Uspekhi Math. Nauk. **8** (1953), 189.

[Ers04a]   Anna Erschler, *Boundary behavior for groups of subexponential growth*, Annals of Math. **160** (2004), no. 3, 1183–1210.

[Ers04b]   Anna Erschler, *Not residually finite groups of intermediate growth, commensurability and non-geometricity*, J. Algebra **272** (2004), no. 1, 154–172. MR 2029029 (2004j:20066).

[Ers05a]   Anna Erschler, *Critical constants for recurrence of random walks on G-spaces*, Ann. Inst. Fourier (Grenoble) **55** (2005), no. 2, 493–509. MR 2147898 (2006c:20085).

[Èrs05b]   A. G. Èrshler, *On the degrees of growth of finitely generated groups*, Funktsional. Anal. i Prilozhen. **39** (2005), no. 4, 86–89. MR 2197519 (2006k:20056).

[Ers06]   Anna Erschler, *Piecewise automatic groups*, Duke Math. J. **134** (2006), no. 3, 591–613. MR 2254627 (2007k:20086).

[Ers10]   Anna Erschler, *Poisson-Furstenberg boundaries, large-scale geometry and growth of groups*, Proc. ICM, Vol III. Hindustan Book Agency, New Dehli, 2010, 681–704. MR 2827814 (2012h:60016).

[ErsK10]   Anna Erschler and Anders Karlsson, *Homomorphisms to* $\mathbb{R}$ *constructed from random walks*, Ann. Inst. Fourier (Grenoble) **60** (2010), no. 6, 2095–2113. MR 2791651.

[EMO05]   Alex Eskin, Shahar Mozes, and Hee Oh, *On uniform exponential growth for linear groups*, Invent. Math. **160** (2005), no. 1, 1–30. MR 2129706 (2006a:20081).

[Føl57]   Erling Følner, *Note on groups with and without full Banach mean value*, Math. Scand. **5** (1957), 5–11. MR 0094725 (20 #1237).

[FT95]   Michael H. Freedman and Peter Teichner, *4-manifold topology. I. Subexponential groups*, Invent. Math. **122** (1995), no. 3, 509–529. MR 1359602 (96k:57015).

[Ghy01]   Étienne Ghys, *Groups acting on the circle*, Enseign. Math. (2) **47** (2001), no. 3-4, 329–407. MR 1876932 (2003a:37032).

[GH90]   Étienne Ghys and André Haefliger, *Groupes de torsion*, Sur les groupes hyperboliques d'après Mikhael Gromov (Bern, 1988), Progr. Math., vol. 83, Birkhäuser Boston, Boston, MA, 1990, pp. 215–226. MR 1086660.

[Gil05]   Robert H. Gilman, *Formal languages and their application to combinatorial group theory*, Groups, languages, algorithms, Contemp. Math., vol. 378, Amer. Math. Soc., Providence, RI, 2005, pp. 1–36. MR 2159313 (2006g:68142).

[Glu61]   V. M. Gluškov, *Abstract theory of automata*, Uspehi Mat. Nauk **16** (1961), no. 5 (101), 3–62. MR 0138529 (25 #1976).

[Gol64]   E. S. Golod, *On nil-algebras and finitely approximable p-groups*, Izv. Akad. Nauk SSSR Ser. Mat. **28** (1964), 273–276. MR 0161878 (28 #5082).

[GŠ64]   E. S. Golod and I. R. Šafarevič, *On the class field tower*, Izv. Akad. Nauk SSSR Ser. Mat. **28** (1964), 261–272. MR 0161852 (28 #5056).

[Gre69]   Frederick P. Greenleaf, *Invariant means on topological groups and their applications*, Van Nostrand Mathematical Studies, No. 16, Van Nostrand Reinhold Co., New York, 1969.

[Gri]   Rostislav Grigorchuk, *Some Topics in the Dynamics of Group Actions on Rooted Trees*, Proceedings of the Steklov Institute of Mathematics.

[Gri79]   R. I. Grigorčuk, *Invariant measures on homogeneous spaces*, Ukrain. Mat. Zh. **31** (1979), no. 5, 490–497, 618. MR 552478 (81k:60073).

[Gri80a] R. I. Grigorchuk, *Symmetrical random walks on discrete groups*, Multicomponent random systems, Adv. Probab. Related Topics, vol. 6, Dekker, New York, 1980, pp. 285–325. MR 599539 (83k:60016).

[Gri80b] R. I. Grigorčuk, *On Burnside's problem on periodic groups*, Funktsional. Anal. i Prilozhen. **14** (1980), no. 1, 53–54. MR 565099 (81m:20045).

[Gri83]   R. I. Grigorchuk, *On the Milnor problem of group growth*, Dokl. Akad. Nauk SSSR **271** (1983), no. 1, 30–33. MR 712546 (85g:20042).

[Gri84a] R. I. Grigorchuk, *Construction of p-groups of intermediate growth that have a continuum of factor-groups*, Algebra i Logika **23** (1984), no. 4, 383–394, 478. MR 781246 (86h:20058).

[Gri84b] R. I. Grigorchuk, *Degrees of growth of finitely generated groups and the theory of invariant means*, Izv. Akad. Nauk SSSR Ser. Mat. **48** (1984), no. 5, 939–985. MR 764305 (86h:20041).

[Gri85a] R. I. Grigorchuk, *Degrees of growth of p-groups and torsion-free groups*, Mat. Sb. (N.S.) **126(168)** (1985), no. 2, 194–214, 286. MR 784354 (86m:20046).

[Gri85b] R. I. Grigorchuk, *Groups with intermediate growth function and their applications*, Habilitation, Steklov Institute of Mathematics, 1985.

[Gri88]   R. I. Grigorchuk, *Semigroups with cancellations of polynomial growth*, Mat. Zametki **43** (1988), no. 3, 305–319, 428. MR 941053 (89f:20065).

[Gri89a]   R. I. Grigorchuk, *On the Hilbert-Poincaré series of graded algebras that are associated with groups*, Mat. Sb. **180** (1989), no. 2, 207–225, 304. MR 993455 (90j:20063).

[Gri89b]   R. I. Grigorchuk, *Topological and metric types of surfaces that regularly cover a closed surface*, Izv. Akad. Nauk SSSR Ser. Mat. **53** (1989), no. 3, 498–536, 671. MR 1013710 (90j:57002).

[Gri91]   Rostislav I. Grigorchuk, *On growth in group theory*, Proceedings of the International Congress of Mathematicians, Vol. I, II (Kyoto, 1990) (Tokyo), Math. Soc. Japan, 1991, pp. 325–338. MR 1159221 (93e:20001).

[Gri98]   R. I. Grigorchuk, *An example of a finitely presented amenable group that does not belong to the class EG*, Mat. Sb. **189** (1998), no. 1, 79–100. MR 1616436 (99b:20055).

[Gri00a]   R. I. Grigorchuk, *Branch groups*, Mat. Zametki **67** (2000), no. 6, 852–858. MR 1820639 (2001i:20057).

[Gri00b]   R. I. Grigorchuk, *Just infinite branch groups*, New horizons in pro-*p* groups, Progr. Math., vol. 184, Birkhäuser Boston, Boston, MA, 2000, pp. 121–179. MR 1765119 (2002f:20044).

[Gri05]   Rostislav Grigorchuk, *Solved and unsolved problems around one group*, Infinite groups: geometric, combinatorial and dynamical aspects, Progr. Math., vol. 248, Birkhäuser, Basel, 2005, pp. 117–218. MR 2195454.

[Gri12]   Rostislav Grigorchuk, *On the gap conjecture concerning group growth*, Bull. Math. Sciences, online first, October 2012. doi:10.1007/s13373-012-0029-4. arXiv:1202.6044

[GdlH97]   R. Grigorchuk and P. de la Harpe, *On problems related to growth, entropy, and spectrum in group theory*, J. Dynam. Control Systems **3** (1997), no. 1, 51–89. MR 1436550 (98d:20039).

[GdlH01]   R. I. Grigorchuk and P. de la Harpe, *One-relator groups of exponential growth have uniformly exponential growth*, Mat. Zametki **69** (2001), no. 4, 628–630. MR 1846003 (2002b:20041).

[GM93]   R. I. Grigorchuk and A. Machì, *On a group of intermediate growth that acts on a line by homeomorphisms*, Mat. Zametki **53** (1993), no. 2, 46–63. MR 1220809 (94c:20008).

[GM99]   R. I. Grigorchuk and A. Machì, *An example of an indexed language of intermediate growth*, Theoret. Comput. Sci. **215** (1999), no. 1–2, 325–327. MR 1678812 (99k:68092).

[GM11]   R. Gigorchuk and K. Medynets, *Topological full groups are locally embeddable into finite groups*, 2011. arXiv:1105.0719

[GN97]   Rostislav Grigorchuk and Tatiana Nagnibeda, *Complete growth functions of hyperbolic groups*, Invent. Math. **130** (1997), no. 1, 159–188. MR 1471889 (98i:20038).

[GN05]   R. Grigorchuk and V. Nekrashevych, *Amenable actions of nonamenable groups*, Zap. Nauchn. Sem. S.-Peterburg. Otdel. Mat. Inst. Steklov. (POMI) **326** (2005), no. Teor. Predst. Din. Sist. Komb. i Algoritm. Metody. 13, 85–96, 281. MR 2183217 (2006j:43005).

[GN07]   Rostislav Grigorchuk and Volodymyr Nekrashevych, *Self-similar groups, operator algebras and Schur complement*, J. Modern Dyn. **1** (2007), no. 3, 323–370.

[GNS00]   R. I. Grigorchuk, V. V. Nekrashevich, and V. I. Sushchanskiĭ, *Automata, dynamical systems, and infinite groups*, Tr. Mat. Inst. Steklova **231** (2000), no. Din. Sist., Avtom. i Beskon. Gruppy, 134–214. MR 1841755 (2002m:37016).

[GP08]   Rostislav Grigorchuk and Igor Pak, *Groups of intermediate growth: an introduction*, Enseign. Math. (2) **54** (2008), no. 3-4, 251–272. MR 2478087 (2009k:20101).

[GŻ02]   Rostislav I. Grigorchuk and Andrzej Żuk, *On a torsion-free weakly branch group defined by a three state automaton*, Internat. J. Algebra Comput. **12** (2002), no. 1–2, 223–246. MR 1902367 (2003c:20048).

[GŠ06]   Rostislav Grigorchuk and Zoran Šunik, *Asymptotic aspects of Schreier graphs and Hanoi Towers groups*, C. R. Math. Acad. Sci. Paris **342** (2006), no. 8, 545–550. MR 2217913.

[Gro81a]   Mikhael Gromov, *Groups of polynomial growth and expanding maps*, Inst. Hautes Études Sci. Publ. Math. (1981), no. 53, 53–73. MR 623534 (83b:53041).

[Gro81b]   Mikhael Gromov, *Structures métriques pour les variétés riemanniennes*, Textes Mathématiques [Mathematical Texts], vol. 1, CEDIC, Paris, 1981, Edited by J. Lafontaine and P. Pansu. MR 682063 (85e:53051).

[Gro87]   M. Gromov, *Hyperbolic groups*, Essays in group theory, Math. Sci. Res. Inst. Publ., vol. 8, Springer, New York, 1987, pp. 75–263. MR 919829 (89e:20070).

[Gro93]   M. Gromov, *Asymptotic invariants of infinite groups*, Geometric group theory, Vol. 2 (Sussex, 1991), London Math. Soc. Lecture Note Ser., vol. 182, Cambridge Univ. Press, Cambridge, 1993, pp. 1–295. MR 1253544 (95m:20041).

[Gro08]   Misha Gromov, *Entropy and isoperimetry for linear and non-linear group actions*, Groups Geom. Dyn. **2** (2008), no. 4, 499–593. MR 2442946 (2010h:37011).

[GSS88]   F. J. Grunewald, D. Segal, and G. C. Smith, *Subgroups of finite index in nilpotent groups*, Invent. Math. **93** (1988), no. 1, 185–223. MR 943928 (89m:11084).

[Gui70]   Yves Guivarc'h, *Groupes de Lie à croissance polynomiale*, C. R. Acad. Sci. Paris Sér. A-B **271** (1970), A237–A239. MR 0272943 (42 #7824).

[Gui71]   Yves Guivarc'h, *Groupes de Lie à croissance polynomiale*, C. R. Acad. Sci. Paris Sér. A-B **272** (1971), A1695–A1696. MR 0302819 (46 #1962).

[Gui73]   Yves Guivarc'h, *Croissance polynomiale et périodes des fonctions harmoniques*, Bull. Soc. Math. France **101** (1973), 333–379. MR 0369608 (51 #5841).

[Gui80]   Y. Guivarc'h, *Sur la loi des grands nombres et le rayon spectral d'une marche aléatoire*, Conference on Random Walks (Kleebach, 1979) (French), Astérisque, vol. 74, Soc. Math. France, Paris, 1980, pp. 47–98, 3. MR 588157 (82g:60016).

[GS83]   Narain Gupta and Saïd Sidki, *On the Burnside problem for periodic groups*, Math. Z. **182** (1983), no. 3, 385–388. MR 696534 (84g:20075).

[HR79]   Edwin Hewitt and Kenneth A. Ross, *Abstract harmonic analysis. Vol. I*, second ed., Grundlehren der Mathematischen Wissenschaften [Fundamental Principles of Mathematical Sciences], vol. 115, Springer-Verlag, Berlin, 1979, Structure of topological groups, integration theory, group representations. MR 551496 (81k:43001).

[Hop48]   Eberhard Hopf, *Closed surfaces without conjugate points*, Proc. Nat. Acad. Sci. U.S.A. **34** (1948), 47–51. MR 0023591 (9,378d).

[HO11]   M. Hull and D. Osin, *Conjugacy growth of finitely generated groups*, 2011. arXiv:1107.1826

[Inc01]   Roberto Incitti, *The growth function of context-free languages*, Theoret. Comput. Sci. **255** (2001), no. 1–2, 601–605. MR 1819093 (2001m:68098).

[Jon74]   J. M. Tyrer Jones, *Direct products and the Hopf property*, J. Austral. Math. Soc. **17** (1974), 174–196, Collection of articles dedicated to the memory of Hanna Neumann, VI. MR 0349855 (50 #2348).

[JM12]   Kate Juschenko and Nicolas Monod, *Cantor systems, piecewise translations and simple amenable groups*, 2012. arXiv:1204.2132

[Kai05]   Vadim A. Kaimanovich, *"Münchhausen trick" and amenability of self-similar groups*, Internat. J. Algebra Comput. **15** (2005), no. 5–6, 907–937. MR 2197814.

[KV83]   V. A. Kaĭmanovich and A. M. Vershik, *Random walks on discrete groups: boundary and entropy*, Ann. Probab. **11** (1983), no. 3, 457–490. MR 704539 (85d:60024).

[KL07]   Anders Karlsson and François Ledrappier, *Linear drift and Poisson boundary for random walks*, Pure Appl. Math. Q. **3** (2007), no. 4, Special Issue: In honor of Grigory Margulis. Part 1, 1027–1036. MR 2402595 (2009d:60133).

[KP11]   M. Kassabov and I. Pak, *Groups of oscillating intermediate growth*, Annals Math, **177** (2013), no. 3, 1113–1145. arXiv:1108.0268

[KMR13]   J. Kellerhls, N. Monod, and M. Rørdam, *Non-supramenable groups acting on locally compact spaces*, 2013. arXiv:1305.5375

[Kes59a]   Harry Kesten, *Full Banach mean values on countable groups*, Math. Scand. **7** (1959), 146–156. MR 0112053 (22 #2911).

[Kes59b]   Harry Kesten, *Symmetric random walks on groups*, Trans. Amer. Math. Soc. **92** (1959), 336–354. MR 0109367 (22 #253).

[Kir67]   A. A. Kirillov, *Dynamical systems, factors and group representations*, Uspehi Mat. Nauk **22** (1967), no. 5 (137), 67–80. MR 0217256 (36 #347).

[Kla81a]   D. A. Klarner, *Mathematical crystal growth. I*, Discrete Appl. Math. **3** (1981), no. 1, 47–52. MR 604265 (82e:05016).

[Kla81b]   D. A. Klarner, *Mathematical crystal growth. II*, Discrete Appl. Math. **3** (1981), no. 2, 113–117. MR 607910 (83a:05018).

[Kle10]   Bruce Kleiner, *A new proof of Gromov's theorem on groups of polynomial growth*, J. Amer. Math. Soc. **23** (2010), no. 3, 815–829. MR 2629989.

[KK74]   Ali Ivanovič Kokorin and Valerii Matveevič Kopytov, *Fully ordered groups*, Halsted Press [John Wiley & Sons], New York–Toronto, Ont., 1974, Translated from the Russian by D. Louvish. MR 0364051 (51 #306).

[Kou98]   Malik Koubi, *Croissance uniforme dans les groupes hyperboliques*, Ann. Inst. Fourier (Grenoble) **48** (1998), no. 5, 1441–1453. MR 1662255 (99m:20080).

[KL00]   Günter R. Krause and Thomas H. Lenagan, *Growth of algebras and Gelfand-Kirillov dimension*, revised ed., Graduate Studies in Mathematics, vol. 22, American Mathematical Society, Providence, RI, 2000. MR 1721834 (2000j:16035).

[Kra53]   Hans Ulrich Krause, *Gruppenstruktur und Gruppenbild*, Thesis, Eidgenössische Technische Hochschule, Zürich, 1953. MR 0056599 (15,99b).

[KAP85]   V. B. Kudryavtsev, S. V. Aleshin, and A. S. Podkolzin, *Vvedenie v teoriyu avtomatov*, "Nauka," Moscow, 1985. MR 837853 (87k:68112).

[Laz65]   Michel Lazard, *Groupes analytiques p-adiques*, Inst. Hautes Études Sci. Publ. Math. (1965), no. 26, 389–603. MR 0209286 (35 #188).

[Leo01]   Yu. G. Leonov, *On a lower bound for the growth of a 3-generator 2-group*, Mat. Sb. **192** (2001), no. 11, 77–92. MR 1886371 (2003a:20050).

[Lia96]   F. Liardet, *Croissance dans les groupes virtuellement abéliens*, Thèse, Université de Genève, 1996.

[LMR95]   P. Longobardi, M. Maj, and A. H. Rhemtulla, *Groups with no free subsemigroups*, Trans. Amer. Math. Soc. **347** (1995), no. 4, 1419–1427. MR 1277124 (95g:20043).

[LM91]   Alexander Lubotzky and Avinoam Mann, *On groups of polynomial subgroup growth*, Invent. Math. **104** (1991), no. 3, 521–533. MR 1106747 (92d:20038).

[LMS93]   Alexander Lubotzky, Avinoam Mann, and Dan Segal, *Finitely generated groups of polynomial subgroup growth*, Israel J. Math. **82** (1993), no. 1–3, 363–371. MR 1239055 (95b:20051).

[Lub94]   Alexander Lubotzky, *Discrete groups, expanding graphs and invariant measures*, Progress in Mathematics, vol. 125, Birkhäuser Verlag, Basel, 1994, With an appendix by Jonathan D. Rogawski. MR 1308046 (96g:22018).

[LS03]   Alexander Lubotzky and Dan Segal, *Subgroup growth*, Progress in Mathematics, vol. 212, Birkhäuser Verlag, Basel, 2003. MR 1978431 (2004k:20055).

[LS77]   Roger C. Lyndon and Paul E. Schupp, *Combinatorial group theory*, Springer-Verlag, Berlin, 1977, Ergebnisse der Mathematik und ihrer Grenzgebiete, Band 89. MR 0577064 (58 #28182).

[Lys85] I. G. Lysënok, *A set of defining relations for the Grigorchuk group*, Mat. Zametki **38** (1985), no. 4, 503–516, 634. MR 819415 (87g:20062).

[MM93] Antonio Machì and Filippo Mignosi, *Garden of Eden configurations for cellular automata on Cayley graphs of groups*, SIAM J. Discrete Math. **6** (1993), no. 1, 44–56. MR 1201989 (95a:68084).

[Mal53] A. I. Mal'cev, *Nilpotent semigroups*, Ivanov. Gos. Ped. Inst. Uč. Zap. Fiz.-Mat. Nauki **4** (1953), 107–111. MR 0075959 (17,825d).

[Man07] Avinoam Mann, *Growth conditions in infinitely generated groups*, Groups Geom. Dyn. **1** (2007), no. 4, 613–622. MR 2357485 (2008k:20101).

[Mang10] Johanna Mangahas, *Uniform exponential growth of subgroups of the mapping class group*, Geom. Funct. Anal. **19** (2010), no. 5, 1468–1480. MR 2585580 (2011d:57002).

[Man12] Avinoam Mann, *How groups grow*, Cambridge Univ. Press, 2012, London Math. Soc. Lecture Notes, v. 395.

[Mar66] G. A. Margulis, *Positive harmonic functions on nilpotent groups*, Soviet Math. Dokl. **7** (1966), 241–244. MR 0222217 (36 #5269).

[Mar91] G. A. Margulis, *Discrete subgroups of semisimple Lie groups*, Ergebnisse der Mathematik und ihrer Grenzgebiete (3) [Results in Mathematics and Related Areas (3)], vol. 17, Springer-Verlag, Berlin, 1991. MR 1090825 (92h:22021).

[Mat06] Hiroki Matui, *Some remarks on topological full groups of Cantor minimal systems*, Internat. J. Math. **17** (2006), no. 2, 231–251. MR 2205435 (2007f:37011).

[Mat11] Hiroki Matui, *Some remarks on topological full groups of Cantor minimal systems II*, 2011. arXiv:1111.3134

[Mil68a] J. Milnor, *A note on curvature and fundamental group*, J. Differential Geometry **2** (1968), 1–7. MR 0232311 (38 #636).

[Mil68b] J. Milnor, *Problem 5603*, Amer. Math. Monthly **75** (1968), 685–686.

[Mil68c] John Milnor, *Growth of finitely generated solvable groups*, J. Differential Geometry **2** (1968), 447–449. MR 0244899 (39 #6212).

[Mor06] Dave Witte Morris, *Amenable groups that act on the line*, Algebr. Geom. Topol. **6** (2006), 2509–2518. MR 2286034 (2008c:20078).

[Muc05] Roman Muchnik, *Amenability of universal 2-Grigorchuk group*, 2005. arXiv:0802.2837

[MP01a] Roman Muchnik and Igor Pak, *On growth of Grigorchuk groups*, Internat. J. Algebra Comput. **11** (2001), no. 1, 1–17. MR 1818659 (2002e:20066).

[MP01b] Roman Muchnik and Igor Pak, *Percolation on Grigorchuk groups*, Comm. Algebra **29** (2001), no. 2, 661–671. MR 1841989 (2002e:82033).

[MZ74] Deane Montgomery and Leo Zippin, *Topological transformation groups*, Robert E. Krieger Publishing Co., Huntington, NY, 1974, Reprint of the 1955 original. MR 0379739 (52 #644).

[Nav08]   Andrés Navas, *Growth of groups and diffeomorphisms of the interval*, Geom. Funct. Anal. **18** (2008), no. 3, 988–1028. MR 2439001 (2010c:37086).

[Nav11]   Andrés Navas, *Groups of circle diffeomorphisms*, spanish ed., Chicago Lectures in Mathematics, University of Chicago Press, Chicago, 2011. MR 2809110.

[Nek05]   Volodymyr Nekrashevych, *Self-similar groups*, Mathematical Surveys and Monographs, vol. 117, American Mathematical Society, Providence, RI, 2005. MR 2162164.

[Nek07]   Volodymyr Nekrashevych, *A minimal Cantor set in the space of 3-generated groups*, Geom. Dedicata **124** (2007), 153–190. MR 2318543 (2008d:20075).

[Nek10]   Volodymyr Nekrashevych, *A group of non-uniform exponential growth locally isomorphic to* $IMG(z^2 + i)$, Trans. Amer. Math. Soc. **362** (2010), no. 1, 389–398. MR 2550156 (2010m:20067).

[NS95]   Walter D. Neumann and Michael Shapiro, *Automatic structures, rational growth, and geometrically finite hyperbolic groups*, Invent. Math. **120** (1995), no. 2, 259–287. MR 1329042 (96c:20066).

[NA68a]   P. S. Novikov and S. I. Adjan, *Infinite periodic groups. I*, Izv. Akad. Nauk SSSR Ser. Mat. **32** (1968), 212–244. MR 0240178 (39 #1532a).

[NA68b]   P. S. Novikov and S. I. Adjan, *Infinite periodic groups. II*, Izv. Akad. Nauk SSSR Ser. Mat. **32** (1968), 251–524. MR 0240179 (39 #1532b).

[NA68c]   P. S. Novikov and S. I. Adjan, *Infinite periodic groups. II*, Izv. Akad. Nauk SSSR Ser. Mat. **32** (1968), 251–524. MR 0240179 (39 #1532b).

[NA68d]   P. S. Novikov and S. I. Adjan, *Infinite periodic groups. III*, Izv. Akad. Nauk SSSR Ser. Mat. **32** (1968), 709–731. MR 0240180 (39 #1532c).

[NY12]   Piotor W. Nowak and Guoliang Yu, *Large scale geometry*, EMS Textbooks in Mathematics, European Mathematical Scociety, Zürich, 2012. MR 2986138.

[Ol'80]   A. Ju. Ol'šanskiĭ, *On the question of the existence of an invariant mean on a group*, Uspekhi Mat. Nauk **35** (1980), no. 4(214), 199–200. MR 586204 (82b:43002).

[OS02]   Alexander Yu. Ol'shanskii and Mark V. Sapir, *Non-amenable finitely presented torsion-by-cyclic groups*, Publ. Math. Inst. Hautes Études Sci. (2002), no. 96, 43–169 (2003). MR 1985031 (2004f:20061).

[Osi02]   D. V. Osin, *Elementary classes of groups*, Mat. Zametki **72** (2002), no. 1, 84–93. MR 1942584 (2003h:20062).

[Osi03]   D. V. Osin, *The entropy of solvable groups*, Ergodic Theory Dynam. Systems **23** (2003), no. 3, 907–918. MR 1992670 (2004f:20065).

[Osi04]   D. V. Osin, *Algebraic entropy of elementary amenable groups*, Geom. Dedicata **107** (2004), 133–151. MR 2110759 (2006b:37010).

[Pan83]   Pierre Pansu, *Croissance des boules et des géodésiques fermées dans les nilvariétés*, Ergodic Theory Dynam. Systems **3** (1983), no. 3, 415–445. MR 741395 (85m:53040).

[Par91]   Luis Paris, *Growth series of Coxeter groups*, Group theory from a geo-metrical viewpoint (Trieste, 1990), World Sci. Publ., River Edge, NJ, 1991, pp. 302–310. MR 1170370 (93g:20081).

[PSZ10]   Victor M. Petrogradsky, Ivan P. Shestakov, and Efim Zelmanov, *Nil graded self-similar algebras*, Groups Geom. Dyn. **4** (2010), no. 4, 873–900. MR 2727670 (2011i:17028).

[PSC]   Christophe Pittet and Laurent Saloff-Coste, *A survey on the relation-ships between volume growth, isoperimetry, and the behavior of simple ran-dom walk on Cayley graphs, with examples*, manuscript. http://www.math.cornell.edu/~lsc/articles.html.

[Rob96]   Derek J. S. Robinson, *A course in the theory of groups*, second ed., Grad-uate Texts in Mathematics, vol. 80, Springer-Verlag, New York, 1996. MR 1357169 (96f:20001).

[Roe03]   John Roe, *Lectures on coarse geometry*, University Lecture Series, vol. 31, American Mathematical Society, Providence, RI, 2003. MR 2007488 (2004g:53050).

[Rose74]   Joseph Max Rosenblatt, *Invariant measures and growth conditions*, Trans. Amer. Math. Soc. **193** (1974), 33–53. MR 0342955 (49 #7699).

[Ros76]   Shmuel Rosset, *A property of groups of non-exponential growth*, Proc. Amer. Math. Soc. **54** (1976), 24–26. MR 0387420 (52 #8263).

[Sai10]   Kyoji Saito, *Limit elements in the configuration algebra for a cancella-tive monoid*, Publ. Res. Inst. Math. Sci. **46** (2010), no. 1, 37–113. MR 2662615 (2011i:05095).

[Sai11]   Kyoji Saito, *Opposite power series*, European J. Combin. **33** (2012), no. 7, 1653–1671. MR 2923475.

[SW02]   Mark Sapir and Daniel T. Wise, *Ascending HNN extensions of resid-ually finite groups can be non-Hopfian and can have very few finite quo-tients*, J. Pure Appl. Algebra **166** (2002), no. 1–2, 191–202. MR 1868545 (2002i:20038).

[ST10]   Yehuda Shalom and Terence Tao, *A finitary version of Gromov's polyno-mial growth theorem*, Geom. Funct. Anal. **20** (2010), no. 6, 1502–1547. MR 2739001.

[Shn04]   L. M. Shneerson, *On semigroups of intermediate growth*, Comm. Alge-bra **32** (2004), no. 5, 1793–1803. MR 2099701 (2005h:20131).

[Shn05]   L. M. Shneerson, *Types of growth and identities of semigroups*, Inter-nat. J. Algebra Comput. **15** (2005), no. 5-6, 1189–1204. MR 2197827 (2006j:20086).

[Shu70]   Michael Shub, *Expanding maps*, Global Analysis (Proc. Sympos. Pure Math., Vol. XIV, Berkeley, CA, 1968), Amer. Math. Soc., Providence, RI, 1970, pp. 273–276. MR 0266251 (42 #1158).

[Smi64]   D. M. Smirnov, *Generalized soluble groups and their group rings*, Dokl. Akad. Nauk SSSR **155** (1964), 535–537. MR 0163961 (29 #1260).

[Sto96]   Michael Stoll, *Rational and transcendental growth series for the higher Heisenberg groups*, Invent. Math. **126** (1996), no. 1, 85–109. MR 1408557 (98d:20033).

[Sus79]   V. I. Sushchansky, *Periodic permutation p-groups and the unrestricted Burnside problem*, DAN SSSR. **247** (1979), no. 3, 557–562, (in Russian).

[Š55]   A. S. Švarc, *A volume invariant of covering*, Dokl. Akad. Nauj SSSR (N.S.), **105** (1955), 32–34. MR 0075634 (17,781d).

[Tit72]   J. Tits, *Free subgroups in linear groups*, J. Algebra **20** (1972), 250–270. MR 0286898 (44 #4105).

[Tit81]   Jacques Tits, *Groupes à croissance polynomiale (d'après M. Gromov et al.)*, Bourbaki Seminar, Vol. 1980/81, Lecture Notes in Math., vol. 901, Springer, Berlin, 1981, pp. 176–188. MR 647496 (83i:53065).

[Tro80]   V. I. Trofimov, *The growth functions of finitely generated semigroups*, Semigroup Forum **21** (1980), no. 4, 351–360. MR 597500 (82g:20094).

[Tro84]   V. I. Trofimov, *Graphs with polynomial growth*, Mat. Sb. (N.S.) **123(165)** (1984), no. 3, 407–421. MR 735714 (85m:05041).

[Var91]   Nicholas Th. Varopoulos, *Analysis and geometry on groups*, Proceedings of the International Congress of Mathematicians, Vols. I, II (Kyoto, 1990), Math. Soc. Japan, Tokyo, 1991, pp. 951–957. MR 1159280 (93k:22006).

[VSCC92]   N. Th. Varopoulos, L. Saloff-Coste, and T. Coulhon, *Analysis and geometry on groups*, Cambridge Tracts in Mathematics, vol. 100, Cambridge University Press, Cambridge, 1992. MR 1218884 (95f:43008).

[vdDW84]   L. van den Dries and A. J. Wilkie, *Gromov's theorem on groups of polynomial growth and elementary logic*, J. Algebra **89** (1984), no. 2, 349–374. MR 751150 (85k:20101).

[vN29]   John von Neumann, *Zurr allgemeinen theorie des masses*, Fund. Math. **13** (1929), 73–116.

[Wag93]   Stan Wagon, *The Banach-Tarski paradox*, Cambridge University Press, Cambridge, 1993, With a foreword by Jan Mycielski, Corrected reprint of the 1985 original. MR 1251963 (94g:04005).

[Wil71]   J. S. Wilson, *Groups with every proper quotient finite*, Proc. Cambridge Philos. Soc. **69** (1971), 373–391. MR 0274575 (43 #338).

[Wil04a]   John S. Wilson, *Further groups that do not have uniformly exponential growth*, J. Algebra **279** (2004), no. 1, 292–301. MR 2078400 (2005e:20066).

[Wil04b]   John S. Wilson, *On exponential growth and uniformly exponential growth for groups*, Invent. Math. **155** (2004), no. 2, 287–303. MR 2 031 429.

[Wil05]   John S. Wilson, *On the growth of residually soluble groups*, J. London Math. Soc. (2) **71** (2005), no. 1, 121–132. MR 2108251 (2006a:20069).

[Wil11]   J. Wilson, *The gap in the growth of residually soluble groups*, Bull. Lond. Math. Soc. **43** (2011), no. 3, 576–582. MR 2820146 (2012f:20105).

[Woe00]   Wolfgang Woess, *Random walks on infinite graphs and groups*, Cambridge Tracts in Mathematics, vol. 138, Cambridge University Press, Cambridge, 2000. MR 1743100 (2001k:60006).

[Wol68]   Joseph A. Wolf, *Growth of finitely generated solvable groups and curvature of Riemanniann manifolds*, J. Differential Geometry **2** (1968), 421–446. MR 0248688 (40 #1939).

[Yau75]   Shing Tung Yau, *Harmonic functions on complete Riemannian manifolds*, Comm. Pure Appl. Math. **28** (1975), 201–228. MR 0431040 (55 #4042).

[Zel90]   E. I. Zel'manov, *Solution of the restricted Burnside problem for groups of odd exponent*, Izv. Akad. Nauk SSSR Ser. Mat. **54** (1990), no. 1, 42–59, 221. MR 1044047 (91i:20037).

[Zel91a]   E. I. Zel'manov, *Solution of the restricted Burnside problem for 2-groups*, Mat. Sb. **182** (1991), no. 4, 568–592. MR 1119009 (93a:20063).

[Zel91b]   Efim I. Zelmanov, *On the restricted Burnside problem*, Proceedings of the International Congress of Mathematicians, Vols. I, II (Kyoto, 1990), Math. Soc. Japan, Tokyo, 1991, pp. 395–402. MR 1159227 (93d:20076).

[Zel05]   Efim Zelmanov, *Infinite algebras and pro-p groups*, Infinite groups: geometric, combinatorial and dynamical aspects, Progr. Math., vol. 248, Birkhäuser, Basel, 2005, pp. 403–413. MR 2195460 (2006k:20053).

[Zel07]   Efim Zelmanov, *Some open problems in the theory of infinite dimensional algebras*, J. Korean Math. Soc. **44** (2007), no. 5, 1185–1195. MR 2348741 (2008g:16053).

Contributors

MARCO ABATE
Dipartimento di Matematica, Università di Pisa, Largo Pontecorvo 5, 56127 Pisa, Italy
abate@dm.unipi.it

MARCO ARIZZI
marco.arizzi@gmail.com

ALEXANDER BLOKH
Department of Mathematics, University of Alabama at Birmingham, Birmingham, AL 35294-1170, USA
ablokh@math.uab.edu

THIERRY BOUSCH
Laboratoire de Mathématique (UMR 8628 du CNRS), bât. 425/430, Université de Paris-Sud, 91405 Orsay Cedex, France
Thierry.Bousch@math.u-psud.fr

XAVIER BUFF
Université Paul Sabatier, Institut de Mathématiques de Toulouse, 118 route de Narbonne, 31062 Toulouse Cedex 9, France
xavier.buff@math.univ-toulouse.fr

SERGE CANTAT
Institut de Recherche Mathématique de Rennes (UMR 6625 du CNRS), Université de Rennes 1, Campus de Beaulieu, bâtiment 22-23, 35042 Rennes cedex, France
serge.cantat@univ-rennes1.fr

TAO CHEN
Department of Mathematics, Engineering and Computer Science, Laguardia Community College, CUNY, 31-10 Thomson Ave., Long Island City, NY 11101, USA
chentaofdh@gmail.com

ROBERT DEVANEY
Mathematics Department, 111 Cummington Street, Boston University, Boston MA 02215, USA
bob@math.bu.edu

ALEXANDRE DEZOTTI
Department of Mathematical Sciences, The University of Liverpool, Mathematical Sciences Building, Liverpool, L69 7ZL, UK
dezotti.alexandre@gmail.com

TIEN-CUONG DINH
Université Pierre et Marie Curie (Université Paris 6), UMR 7586, Institut de Mathématiques de Jussieu, 4 place Jussieu, F-75005 Paris, France
dinh@math.jussieu.fr

ROMAIN DUJARDIN
Laboratoire d'Analyse et de Mathématiques Appliquées, Université de Marne-la-
Vallée, 5 boulevard Descartes, Cité Descartes – Champs-sur-Marne, 77454 Marne-
la-Vallée cedex 2, France
romain.dujardin@univ-mlv.fr

HUGO GARCÍA-COMPEÁN
Departamento de Física, Centro de Investigación y de Estudios Avanzados del
IPN, P.O. Box 14-740, 07000 México D.F., México
compean@fis.cinvestav.mx

WILLIAM GOLDMAN
3106 Math Building, University of Maryland, College Park, MD 20742 USA
wmg@math.umd.edu

ROSTISLAV GRIGORCHUK
Department of Mathematics, Mailstop 3368, Texas A&M University, College Station, TX 77843-3368, USA
grigorch@math.tamu.edu

JOHN HUBBARD
Department of Mathematics, Malott Hall, Cornell University, Ithaca, NY 14853-
4201, USA and CMI, 39 Rue Joliet Curie, 13453 Marseille Cedex 13, France.
jhh8@cornell.edu

YUNPING JIANG
Department of Mathematics, Queens College of CUNY, Flushing, NY 11367, USA
and Mathematics Dept., CUNY Graduate Center, New York, NY 10016, USA
yunping.jiang@qc.cuny.edu

LINDA KEEN
Department of Mathematics, Graduate Center and Lehman College of CUNY, New
York, NY 10016, USA
LKeen@gc.cuny.edu

JAN KIWI
Facultad de Matemáticas, Pontificia Universidad Católica de Chile, Casilla 306,
Correo 22, Santiago, Chile
jkiwi@puc.cl

GENADI LEVIN
Institute of Mathematics, the Hebrew University of Jerusalem, Israel
levin@math.huji.ac.il

DANIEL MEYER
Jacobs University, School of Engineering and Science, Campus Ring 1, 28759 Bremen, Germany
dmeyermail@gmail.com

JOHN MILNOR
Institute for Mathematical Sciences, Stony Brook University, Stony Brook NY, 11794-3661, USA
jack@math.sunysb.edu

CARLOS MOREIRA
Instituto Nacional de Matemática Pura e Aplicada (IMPA), Estrada Dona Castorina 110, Jardim Botânico, Rio de Janeiro-RJ, CEP 22460-320 Brazil
gugu@impa.br

VICENTE MUÑOZ
Facultad de Matemáticas, Universidad Complutense de Madrid, Plaza de Ciencias 3, 28040 Madrid, Spain
vicente.munoz@mat.ucm.es

VIET-ANH NGUYÊN
Mathématique-Bâtiment 425, UMR 8628, Université Paris-Sud, 91405 Orsay, France
VietAnh.Nguyen@math.u-psud.fr

LEX OVERSTEEGEN
Department of Mathematics, University of Alabama at Birmingham, Birmingham, AL 35294-1170
overstee@math.uab.edu

RICARDO PÉREZ-MARCO
CNRS, LAGA UMR 7539, Université Paris XIII, 99 Avenue J.-B. Clément, 93430-Villetaneuse, France
ricardo@math.univ-paris13.fr

ROSS PTACEK
Faculty of Mathematics, Laboratory of Algebraic Geometry and its Applications, Higher School of Economics, Vavilova St. 7, 112312 Moscow, Russia
rptacek@uab.edu

JASMIN RAISSY
Université Paul Sabatier, Institut de Mathématiques de Toulouse, 118 route de Narbonne, 31062 Toulouse Cedex 9, France
jraissy@math.univ-toulouse.fr

PASCALE ROESCH
Centre de Mathématiques et Informatique (CMI), Technopôle de Château-Gombert 39, rue F. Joliot Curie, 13453 Marseille Cedex 13, France
pascale.roesch@cmi.univ-mrs.fr

ROBERTO SANTOS-SILVA
Departamento de Física, Centro de Investigación y de Estudios Avanzados del IPN, P.O. Box 14-740, 07000 México D.F., México
rsantos@fis.cinvestav.mx

DIERK SCHLEICHER
Jacobs University, Research I, Postfach 750 561, 28725 Bremen, Germany
d.schleicher@jacobs-university.de

NESSIM SIBONY
Mathématique-Bâtiment 425, UMR 8628, Université Paris-Sud, 91405 Orsay, France
Nessim.Sibony@math.u-psud.fr

DANIEL SMANIA
Departamento de Matemática, Instituto de Ciências Matemáticas e de Computação USP Campus de São Carlos, Caixa Postal 668, CEP 13560-970, São Carlos SP, Brazil
smania@icmc.usp.br

SEBASTIAN VAN STRIEN
Mathematics Department, Imperial College, London SW7 2AZ, UK
svanstrien@googlemail.com

TAN LEI
Université d'Angers, Faculté des sciences, LAREMA (UMR CNRS 6093), 2 Boulevard Lavoisier, 49045 Angers cedex 01, France
tanlei@math.univ-angers.fr

WILLIAM THURSTON

VLADLEN TIMORIN
Faculty of Mathematics, Laboratory of Algebraic Geometry and its Applications, Higher School of Economics, Vavilova St. 7, 112312 Moscow, Russia
vtimorin@hse.ru

ALBERTO VERJOVSKY
Instituto de Matemáticas, Universidad Nacional Autónoma de México, Unidad Cuernavaca, Av. Universidad s/n, Col. Lomas de Chamilpa, c.p. 62210, Cuernavaca Morelos, México
alberto@matcuer.unam.mx

Index